SUSTAINABLE NANOTECHNOLOGY FOR ENVIRONMENTAL REMEDIATION

SUSTAINABLE NANOTECHNOLOGY FOR ENVIRONMENTAL REMEDIATION

Edited by

RAMA RAO KARRI

Petroleum and Chemical Engineering, Faculty of Engineering, Universiti Teknologi Brunei, Brunei Darussalam

JANARDHAN REDDY KODURU

Department of Environmental Engineering, Kwangwoon University, Nowon-gu, Seoul, Korea

NABISAB MUJAWAR MUBARAK

Petroleum and Chemical Engineering, Faculty of Engineering, Universiti Teknologi Brunei, Brunei Darussalam

ERICK R. BANDALA

Division of Hydrologic Sciences, Desert Research Institute, Las Vegas, Nevada, USA

ELSEVIER

Elsevier
Radarweg 29, PO Box 211, 1000 AE Amsterdam, Netherlands
The Boulevard, Langford Lane, Kidlington, Oxford OX5 1GB, United Kingdom
50 Hampshire Street, 5th Floor, Cambridge, MA 02139, United States

British Library Cataloguing-in-Publication Data
A catalogue record for this book is available from the British Library

Library of Congress Cataloging-in-Publication Data
A catalog record for this book is available from the Library of Congress

ISBN: 978-0-12-824547-7

For Information on all Elsevier publications visit our website at
https://www.elsevier.com/books-and-journals

Publisher: Matthew Deans
Acquisitions Editor: Simon Holt
Editorial Project Manager: Gabriela D. Capille
Production Project Manager: Sojan P. Pazhayattil
Cover Designer: Victoria Pearson

Typeset by Aptara, New Delhi, India

Dedication

I dedicate this book to my mother Karri Kannathalli who protected, guided, and supported me in all these years. She was my inspiration for driving me to achieve the best. She was a superwoman and constant inspiration.

I also thank my lovely wife, Soni, my lovely children Yajna & Jay, My In-laws L.V. Rao & Prameela without their support and understanding, this book as well as my research achievements were not possible.

Dr. Rama Rao Karri

I dedicate this book to my parents Late Smt. Narayanamma Koduru and Sri. Narayana Reddy Koduru who protected, guided, and supported me in all these years. They were my inspiration for driving me to achieve the best.

I also thank my lovely wife, Dr. Lakshmi Prasanna Lingamdinne, without her support, this book as well as my academic and research achievements are not possible.

Dr. Janardhan Reddy Koduru

In the name of Allah, the Most Gracious and the Most Merciful. First of all, I would like to raise my heartfelt gratitude and appreciation to Allah SWT for the permission, guidance, wisdom, and blessing for all these years till now, where I have reached to this important destination of my journey in life to accomplish my goal. Finally, I would like to present my most heartfelt and warmest appreciation to the great parents and parents in law (may ALLAH SWT bless and reward them), brothers, and sisters who always encouraged and supported me during the completion of the book. Special and heartiest gratitude to my dearest wife, Muna Tasnim Mukhtaruddin, and kids, Muhammad Fayyad, Muhammad Fawwaz, and Mulaika Faleeha, for their invariable encouragement, endless sacrifices, patience, understanding, ideas, and inspirations from time to time in finishing the book smoothly and timely.

Dr. Nabisab Mujawar Mubarak

To my wife Annie, with love.

Dr. Erick Bandala

CONTENTS

Contributors

Rashid Abro
Department of Chemical Engineering, Dawood University of Engineering and Technology, Karachi, Pakistan

Adeyemi Adesina
Civil and Environmental Engineering, University of Windsor, Windsor, Canada

Akil Ahmad
Centre of Lipids Engineering and Applied Research, Universiti Teknologi Malaysia, Johor Bahru, Malaysia

Shoaib Ahmed
Department of Chemical Engineering, Dawood University of Engineering and Technology, Karachi, Pakistan

Zahoor Ahamd
Department of Chemistry, Mirpur University of Science and Technology, Mirpur AJK, Pakistan

Faheem Akhter
Department of Chemical Engineering, Quaid-e-Awam University of Engineering, Science & Technology, Nawabshah, Pakistan; Department of Chemical Engineering, Mehran University of Engineering and Technology, Jamshoro, Sindh, Pakistan

Esfandyar Ali
Department of Chemical Engineering, Dawood University of Engineering and Technology, Karachi, Pakistan

Imran Ali
Department of Environmental Science and Engineering, College of Chemistry and Environmental Engineering, Shenzhen University, Shenzhen, China

Muhammad Anees
Department of Microbiology, Kohat University of Science & Technology (KUST), Kohat, Khyber Pakhtunkhwa, Pakistan

Viswadevarayalu Annavaram
Department of Chemistry , Annamacharya Institute of Science & Technology, Rajampet, Andhra Pradesh , India

Paul O. Awoyera
Department of Civil Engineering, Covenant University, Ota, Nigeria

Muhammad Ashar Ayub
Institute of Soil and Environmental Sciences, Faculty of Agriculture, University of Agriculture Faisalabad, Pakistan; Indian River Research and Education Center, Institute of Food and Agricultural Science, University of Florida, USA

Alexander V. Babkin
Tambov State Technical University, Tambov, Russian Federation

Humair Ahmed Baloch
School of Engineering, RMIT University, Melbourne, Australia

Erick R. Bandala
Division of Hydrologic Sciences, Desert Research Institute, Las Vegas, Nevada, USA

Sushmita Banerjee
School of Basic Science and Research, Department of Environmental Sciences, Sharda University, Greater Noida, Uttar Pradesh, India

Pranta Barua
Department of Electronic Materials Engineering, Kwangwoon University, Seoul-01891, South Korea

Jigna R. Bhamore
Department of Chemistry, Sardar Vallabhbai National Institute of Technology, Surat, Gujarat–395007, India; Research Institute of Chem-Bio Diagnostic Technology, Department of Chemistry, Chung-Ang University, 84 Heukseok-ro, Dongjak-gu, Seoul 06974, Republic of Korea

Alexander E. Burakov
Tambov State Technical University, Tambov, Russian Federation

Irina V. Burakova
Tambov State Technical University, Tambov, Russian Federation

Heena Chauhan
Department of Environmental Sciences, J.C. Bose University of Science and Technology, YMCA, Faridabad, Haryana, India

Guilherme C.F. Cruz
Chemical Biology Laboratory, Department of Organic Chemistry, Institute of Chemistry, University of Campinas (UNICAMP), Campinas, SP, Brazil

Ana Belen Cueva-Sola
Convergence Research Center for Development of Mineral Resources (DMR), Korea Institute of Geoscience and Mineral Resources (KIGAM), Daejeon, Republic of Korea; Resources Recycling Major, Korea University of Science and Technology (UST), Daejeon, Republic of Korea

Sumistha Das
Amity Institute of Biotechnology, Amity University Haryana, Gurugram, Haryana, India

Nitai Debnath
Amity Institute of Biotechnology, Amity University Haryana, Gurugram, Haryana, India

Shikha Dhiman
Amity Institute of Biotechnology, Amity University Haryana, Gurugram, Haryana, India

Zia Ur Rahman Farooqi
Institute of Soil and Environmental Sciences, Faculty of Agriculture, University of Agriculture Faisalabad, Pakistan

Hina Fatima
School of Applied Biosciences, Kyungpook National University, Daegu, South Korea

Vinod Kumar Garg
Department of Environmental Science and Technology, Central University of Punjab, Bathinda, Punjab, India

Pavan Kumar Gautam
Department of Applied Sciences, Indian Institute of Information Technology Allahabad, Allahabad, Uttar Pradesh, India

Renuka Gupta
Department of Environmental Sciences, J.C. Bose University of Science and Technology, YMCA, Faridabad, Haryana, India

Wasim M.K. Helal
Department of Mechanical Engineering, Faculty of Engineering, Kafrelsheikh University, Kafrelsheikh, Egypt

Nazia Hossain
School of Engineering, RMIT University, Melbourne VIC, Australia

Mohamad Nasir Mohd Ibrahim
School of Chemical Sciences, Universiti Sains Malaysia, Penang, Malaysia

Eberechukwu Laura Ikechukwu
Department of Biochemistry, College of Natural Sciences, Michael Okpara University of Agriculture Umudike, Umuahia, Nigeria

Tariqul Islam
Department of Agricultural Construction and Environmental Engineering, Sylhet Agricultural University, Sylhet, Bangladesh

Abdul Sattar Jatoi
Department of Chemical Engineering, Dawood University of Engineering and Technology, Karachi, Pakistan

Jong-Ryul Jeong
Department of Materials Science and Engineering, Graduate School of Energy Science and Technology, Chungnam National University, Daejeon, South Korea

Juliana John
Department of Civil Engineering, National Institute of Technology Tiruchirapalli, Tamil Nadu, India

Sumalatha Jorepalli
Nirmala College of Pharmacy, Ukkayapalli, Kadapa, Andhra Pradesh, India

Rajesh Kumar Jyothi
Convergence Research Center for Development of Mineral Resources (DMR), Korea Institute of Geoscience and Mineral Resources (KIGAM), Daejeon, Republic of Korea; Resources Recycling Major, Korea University of Science and Technology (UST), Daejeon, Republic of Korea

Suresh Kumar Kailasa
Department of Chemistry, Sardar Vallabhbai National Institute of Technology, Surat – 395007, Gujarat, India

Rama Rao Karri
Petroleum and Chemical Engineering, Faculty of Engineering, Universiti Teknologi Brunei, Brunei Darussalam

Navish Kataria
Department of Environmental Sciences, J.C. Bose University of Science and Technology, YMCA, Faridabad, Haryana, India

Md Shahidullah Kayshar
Department of Food Engineering and Technology, Sylhet Agricultural University, Sylhet, Bangladesh

Dongsoo Kim
Convergence Research Center for Development of Mineral Resources, Korea Institute of Geoscience and Mineral Resources, Daejeon, South Korea; Powder & Ceramics Division, Korea Institute of Materials Science, Changwon, Gyeongnam, South Korea

Mehmet Serkan Kırgız
İstanbul Üniversitesi-Cerrahpaşa, Avcılar, İstanbul, Turkey

Janardhan Reddy Koduru
Department of Environmental Engineering, Kwangwoon University, Nowon-gu, Seoul, Korea

Rambabu Kuchi
Convergence Research Center for Development of Mineral Resources, Korea Institute of Geoscience and Mineral Resources, Daejeon, South Korea; Powder & Ceramics Division, Korea Institute of Materials Science, Changwon, Gyeongnam, South Korea

Jai Kumar
State Key Laboratory of Organic-Inorganic Composites, Key Laboratory of Electrochemical Process and Technology for Materials, Beijing University of Chemical Technology, Beijing, China

Kashif Latif
Department of Microbiology, Quaid-i-Azam University, Islamabad, Pakistan

Jin-Young Lee
Convergence Research Center for Development of Mineral Resources (DMR), Korea Institute of Geoscience and Mineral Resources (KIGAM), Daejeon, Republic of Korea; Resources Recycling Major, Korea University of Science and Technology (UST), Daejeon, Republic of Korea

Lucas G. Martins
Chemical Biology Laboratory, Department of Organic Chemistry, Institute of Chemistry, University of Campinas (UNICAMP), Campinas, SP, Brazil

Shaukat Ali Mazari
Department of Chemical Engineering, Dawood University of Engineering and Technology, Karachi, Pakistan

Daniel A. Medina-Orendain
Departamento de Química y Bioquímica, Tecnológico Nacional de México-Instituto Tecnológico de Tepic, Av. Tecnológico # 2595, Lagos del Country, 63175, Tepic, Nayarit, México

Alexander V. Melezhik
Tambov State Technical University, Tambov, Russian Federation

Abdul Qayoom Memon
Department of Chemical Engineering, Dawood University of Engineering and Technology, Karachi, Pakistan

Nabisab Mujawar Mubarak
Petroleum, and Chemical Engineering, Faculty of Engineering, Universiti Teknologi Brunei, Bandar Seri Begawan, Brunei Darussalam

Atta Muhammad
Department of Chemical Engineering, Dawood University of Engineering and Technology, Karachi, Pakistan

Riaz Muhammad
Government College Peshawar, Khyber Pakhtunkhwa, Pakistan

Muhammad Nadeem
Institute of Soil and Environmental Sciences, Faculty of Agriculture, University of Agriculture Faisalabad, Pakistan

Asif Naeem
Institute of Plant Nutrition and Soil Science, Christian-Albrechts-Universität zu Kiel, Kiel, Germany

Iffat Naz
Department of Biology, Deanship of Educational Services, Qassim University, Buraidah, Kingdom of Saudi Arabia (KSA)

Elena A. Neskoromnaya
Tambov State Technical University, Tambov, Russian Federation

Sabzoi Nizamuddin
School of Engineering, RMIT University, Melbourne VIC, Australia

Chidozie Charles Nnaji
Department of Civil Engineering, University of Nigeria, Nsukka, Nigeria; Faculty of Engineering and Built Environment, University of Johannesburg, South Africa

Amina Othmani
Department of Chemistry, Faculty of Sciences, University of Monastir, Monastir, Tunisia

V.C. Padmanaban
Department of Biotechnology, Kamaraj College of Engineering & Technology, Tamil Nadu, India

Krishna Kumar Pandey
Department of Physics, School of Basic Sciences and Research, Sharda University, Greater Noida, Uttar Pradesh, India

Pankaj Kumar Parhi
Department of Chemistry, Fakir Mohan University, Balasore, Odisha, India

Tae-Jung Park
Research Institute of Chem-Bio Diagnostic Technology, Department of Chemistry, Chung-Ang University, 84 Heukseok-ro, Dongjak-gu, Seoul 06974, Republic of Korea

Irwing Ramirez
School of Engineering and Innovation, The Open University, Milton Keynes, MK76AA UK

Mohd Rashid
School of Chemical Sciences, Universiti Sains Malaysia, Penang, Malaysia

Shalu Rawat
Department of Environmental Science, Babasaheb Bhimrao Ambedkar University, Lucknow, Uttar Pradesh, India

Abdul Rehman
Department of Microbiology, Kohat University of Science & Technology (KUST), Kohat, Khyber Pakhtunkhwa, Pakistan

Muhammad Zia ur Rehman
Institute of Soil and Environmental Sciences, Faculty of Agriculture, University of Agriculture Faisalabad, Pakistan

Muhammad Rizwan
Department of Environmental Sciences & Engineering, Government College University Faisalabad, Faisalabad, Pakistan

Oscar M. Rodríguez-Narvaez
Dirección de investigación y soluciones tecnológicas, Centro de Innovación Aplicada en Tecnologías Competitivas (CIATEC), Omega 201, Col. Industrial Delta, León, Guanajuato, C.P. 37545, México

Nizamuddin Sabzoi
School of Engineering, RMIT University, Melbourne, Australia

Sintu Kumar Samanta
Department of Applied Sciences, Indian Institute of Information Technology Allahabad, Allahabad, Uttar Pradesh, India

Raluca Savu
Centre for Semiconductor Components and Nanotechnology (CCS-Nano), University of Campinas (UNICAMP), Campinas, SP, Brazil

Shama Sehar
College of Science, University of Bahrain, Sakhir, Kingdom of Bahrain

Muhammad Shabaan
Institute of Soil and Environmental Sciences, Faculty of Agriculture, University of Agriculture Faisalabad, Pakistan

Bhanu Shrestha
Department of Electronics Engineering, Kwangwoon University, Seoul, Korea

MTH Sidddiqui
School of Engineering, RMIT University, Melbourne VIC, Australia

Geoffrey S. Simate
School of Chemical and Metallurgical Engineering, University of the Witwatersrand, Johannesburg, South Africa

Bharti Singh
Department of Electronics Engineering, Ramswaroop Memorial Group of Professional Colleges, Lucknow, Uttar Pradesh, India

Jiwan Singh
Department of Environmental Science, Babasaheb Bhimrao Ambedkar University, Lucknow, Uttar Pradesh, India

Megha Singh
Amity Institute of Biotechnology, Amity University Haryana, Gurugram, Haryana, India

Ved Vati Singh
Department of Chemistry, Indian Institute of Technology Delhi, New Delhi, India

Adinarayana Reddy Somala
Department of Materials Science and Nanotechnology, Yogi Vemana University, Kadapa, Andhra Pradesh, India

Danijela Stanisic
Chemical Biology Laboratory, Department of Organic Chemistry, Institute of Chemistry, University of Campinas (UNICAMP), Campinas, SP, Brazil

Ljubica Tasic
Chemical Biology Laboratory, Department of Organic Chemistry, Institute of Chemistry, University of Campinas (UNICAMP), Campinas, SP, Brazil

Alexey G. Tkachev
Tambov State Technical University, Tambov, Russian Federation

Manoj Tripathi
Department of Physics and Materials Science and Engineering, Jaypee Institute of Information Technology, Noida, Uttar Pradesh, India

Emmanuel Ikechukwu Ugwu
Department of Civil Engineering, College of Engineering and Engineering Technology, Michael Okpara University of Agriculture Umudike, Nigeria

Muhammad Umair
Institute of Soil and Environmental Sciences, Faculty of Agriculture, University of Agriculture Faisalabad, Pakistan

Wajid Umar
School of Environmental Science, Hungarian University of Agriculture and Life Sciences, Gödöllő, Hungary

Deborah L. Villaseñor-Basulto
Departamento de Química. División de Ciencias Naturales y Exactas, Campus Guanajuato, Universidad de Guanajuato, Guanajuato, Gto. México

Asim Ali Yaqoob
School of Chemical Sciences, Universiti Sains Malaysia, Penang, Malaysia

Adnan Younis
Department of Physics, College of Science, University of Bahrain, Sakhir, Kingdom of Bahrain

Husnain Zia
Institute of Soil and Environmental Sciences, Faculty of Agriculture, University of Agriculture Faisalabad, Pakistan

About the editors

Dr. Rama Rao Karri is a Professor (Sr. Asst) in Faculty of Engineering, Universiti Teknologi Brunei, Brunei Darussalam. He has PhD from Indian Institute of Technology (IIT) Delhi, Masters from IIT Kanpur in Chemical Engineering. He has worked as Post-Doctoral research fellow at NUS, Singapore for about six years and has over 18 years of working experience in Academics, Industry, and Research. He has experience of working in multidisciplinary fields and has expertise in various evolutionary optimization techniques and process modeling. He has published 110+ research articles in reputed journals, book chapters and conference proceedings with a combined Impact factor of 264.74 and has an h-index of 20 (Scopus and Google Scholar). He is in editorial board member in 7 renowned journals and peer-review member in more than 81 reputed journals. He also has the distinction of being listed in the top 2% of the world's most influential scientists in the area of environmental sciences and chemical for Year 2021. The List of the Top 2% Scientists in the World compiled and published by Stanford University is based on their international scientific publications, number of scientific citations for research, and participation in the review and editing of scientific research. He held a position as Editor-in-Chief (2019–2021) in International Journal of Chemoinformatics and Chemical Engineering, IGI Global, USA. He is also Associate editor in Scientific Reports, Springer Nature & International Journal of Energy and Water Resources (IJEWR), Springer Inc. He is also a Managing Guest editor for Spl. Issues: 1) "Magnetic nano composites and emerging applications", in Journal of Environmental Chemical Engineering (IF: 5.909), 2) "Novel CoronaVirus (COVID-19) in Environmental Engineering Perspective", in Journal of Environmental Science and Pollution Research (IF: 4.223), Springer. 3) "Nanocomposites for the Sustainable Environment", in Applied Sciences Journal (IF: 2.679), MDPI. He along with his mentor, Prof. Venkateswarlu is authoring an Elsevier book, "Optimal state estimation for process monitoring, diagnosis and control". He is also co-editor and managing editor for 8 Elsevier, 1 Springer and 1 CRC edited books. Elsevier: 1) Sustainable Nanotechnology for Environmental Remediation, 2) Soft computing techniques in solid waste and wastewater management, 3) Green technologies for the defluoridation of water, 4) Environmental and health management of novel coronavirus disease (COVID-19), 5) Pesticides remediation technologies from water and wastewater: Health effects and environmental remediation, 6) Hybrid Nanomaterials for Sustainable

Applications, 7) Sustainable materials for sensing and remediation of noxious pollutants. Springer: 1) Industrial wastewater treatment using emerging technologies for sustainability. CRC: 1) Recent Trends in Advanced Oxidation Processes (AOPs) for micropollutant removal.

Address: Faculty of Engineering, Universiti Teknologi Brunei, Gadong, BE1410, Brunei.
E-mail: kramarao.iitd@gmail.com; karri.rao@utb.edu.bn
Research Interests: Process modeling and simulation, process optimization, machine learning, wastewater treatment, nanotechnology.

Prof. Janardhan Reddy Koduru is a Professor of Environmental Engineering at Kwangwoon University, Seoul, South Korea. He has over 19 years of experience in academics and research. Dr. Koduru expertizes in the development of sustainable nanocomposites for environmental and energy applications. He has published over 100 papers including 7 book chapters and one book and holds three Korean patents. He has received over 14 awards for his excellent research and academic achievements. He is an editorial board member in three renewed journals, and a peer-reviewed member for more than 100 scientific journals. His works (75 articles/papers) have been published in various international conference proceedings. He received the top 1% Publons Peer Reviewer Awards for multidisciplinary and environment and ecology continuously during 2016–2019. Dr. Koduru is an editorial board member for *Current Analytical Chemistry*, *Indian Journal of Advances in Chemical Science*, *IOSR Journal of Applied Chemistry*, and *Current Research in Materials Chemistry*. Also, he is a guest editor for various scientific journals including *Journal Environmental Management* (Special Issue: Environment & Energy), *Applied Sciences Journal*, MDPI (Special Issue: Nanocomposites for the Sustainable Environment). He is a guest editor for ongoing Special Issue: "Magnetic nano composites and emerging applications," in *Journal of Environmental Chemical Engineering*, Elsevier. He is also a coeditor for one ongoing Elsevier edited books.

Address: Department of Environmental Engineering, Kwangwoon University, Seoul, 01897, Republic of Korea.
E-mail: reddyjchem@gmail.com; reddyjchem@kw.ac.kr
Research Interests: Sustainable nanomaterials, wastewater treatment, sensors, energy, separation & purification techniques, enviro-analytical techniques.

Dr Nabisab Mujawar Mubarak is an Associate Professor in Faculty of Engineering, Universiti Teknologi Brunei, Brunei Darussalam. He serves as a scientific reviewer in numerous journals in the area of chemical engineering and nano technology. In research, Dr. Mubarak has published more than 200 journal papers, 30 conference proceedings, and authored 30 book chapters and the H-index is 40. His area of interest is carbon nanomaterials synthesis, magnetic biochar production using microwave, and wastewater treatment using advanced materials. He is a recipient of the Curtin Malaysia Most Productive Research award, outstanding faculty of Chemical Engineering award, Best Scientific Research Award London, and outstanding scientist in publication and citation by i- Proclaim, Malaysia. He also has the distinction of being listed in the top 2% of the world's most influential scientists in the area of chemical and energy. The *List of the Top 2% Scientists in the World* compiled and published by Stanford University is based on their international scientific publications, number of scientific citations for research, and participation in the review and editing of scientific research. Dr. Mubarak is a Fellow Member of the Institution of Engineers Australia, Chartered Professional Engineer (CPEng) of The Institution of Engineers Australia, and also a Chartered Chemical Engineer of the Institute of Chemical Engineering (IChemE) United Kingdom. He is a coeditor for four ongoing Elsevier edited books: (1) Sustainable Nanotechnology for Environmental Remediation, (2) Nanomaterials for Carbon Capture and Conversion Technique, (3) Advanced nanomaterials and nanocomposites for Bio electrochemical Systems, and (4) Green Mediated Synthesis-based Nanomaterials for Photocatalysis.

Address: Petroleum, and Chemical Engineering, Faculty of Engineering, Universiti Teknologi Brunei, Bandar Seri Begawan, Brunei Darussalam
E-mail: mubarak.yaseen@gmail.com
Research Interests: Advanced carbon nanomaterials synthesis via microwave technology, graphene/CNT buckypaper for strain sensor application, biofuels, magnetic buckypaper, immobilization of enzymes, protein purification, magnetic biochar production using microwave, and wastewater treatment using advanced materials.

Dr. Erick R. Bandala is an Assistant Research Professor for Advanced Water Technologies at Desert Research Institute. Dr. Bandala holds a Ph.D. in environmental engineering, M.Sc. in organic chemistry, and B.Eng. in chemical engineering. His research interests in environmental engineering include (1) The Water-Energy-Food NEXUS; (2) Water Security; (3) International Water, Sanitation, and Hygiene (IWASH); (4) Advanced oxidation processes (AOPs) for environmental restoration; (5) Synthesis, characterization, and application of nanomaterials for environmental restoration; (6) Development of Climate Change adaptation methodologies for water security. Dr. Bandala has authored or coauthored 108 peer-reviewed papers in international journals (average impact factor 3.5 >3540 citations, H-index 30); 5 books, 28 book chapters, and 65 works published in proceedings of international conferences.

Address: 755 E. Flamingo Road, Las Vegas, Nevada, 89119, USA
E-mail: erick.bandala@dri.edu
Research Interests: Water treatment, water quality, water reuse, climate change adaptation.

Foreword

Nanotechnology could be successfully used for various applications that support environmental and climate protection. The book entitled "*Sustainable Nanotechnology for Environmental Remediation*" edited by Dr. Rama Rao Karri, Dr. Janardhan Reddy Koduru, Dr. Mubarak Nabisab Mujawar and Dr. Erick Bandala, very clearly demonstrates the use of nanotechnology to achieve sustainable environmental goals. The book is brilliantly presented in three sections: Section 1: Insights, Synthesis and Properties; Section 2: Environmental Remediation Applications and Future Prospective; Section 3: Miscellaneous Applications. The chapters under each section are contributed by world leaders who are experts in their fields. The book is very much useful for chemists, physicists, biologists, environmental scientists, and chemical engineers who are using nanotechnology platforms for environmental remediation. As we all know, nanostructured materials are commonly used for environmental remediation due to their high surface to volume ratio, and enormous flexibility towards both in situ and ex-situ applications in aqueous systems. The book has an interdisciplinary and multidisciplinary approach in understanding and solving the problems. The book discusses syntheses, modification, characterization, and fabrication of nanomaterials and their nanocomposites for environmental remediation. From this book, one can really understand the relationship between synthesis, morphology, properties, and applications. Very specifically, the book deals with the latest advancements in the use of nanostructured materials for the removal of toxic pollutants from our ecosystem and it very clearly depicts gaps, challenges, and future opportunities. I strongly recommend the book for students, researchers, industrial experts and university professors who are active in the domain of nanotechnology for environmental remediation. I congratulate all the editors for their excellent contribution.

Prof. (Dr.) Sabu Thomas, D.Sc., Ph.D., FRSC, FEurASc
Vice-Chancellor;
Director, School of Energy Materials;
Professor, School of Chemical Sciences & International and Inter-University
Centre for Nanoscience and Nanotechnology,
Mahatma Gandhi University,
Kerala, India.

Preface

With the continuous rise in the global human population, a sufficient supply of resources has become meager. The development of pollution-free processes and tools for environmental remediation and safe and clean energy supplies for the sustainable growth of human society is the need of the hour. Nanotechnology can offer considerable impact on introducing "greener" and "cleaner" processes and tools with extensive health and environmental advantages. The applications of nanotechnology are being studied for their potential to offer solutions to control, alleviate, and clean up water, air, and land contamination; moreover, enhance the performance of traditional processes and tools employed in environmental clean-up. Green nanotechnology is the branch of nanotechnology that envisions sustainability via different applications.

This book *Sustainable Nanotechnology for Environmental Remediation* presents exciting new novel materials and efficient techniques to achieve sustainable environmental remediation. The organization of the book contents provides a one-hand solution to researchers working/affiliated in environmental, wastewater management, biological and composite nanomaterials applications. It also addresses the potential environmental risks and uncertainties surrounding the use of nanomaterials for environmental remediation, giving an understanding of their impact on ecological receptors in addition to their potential benefits. This book presents the latest developments that can be used for the precise characterization of the materials, analyzing their applications for the removal of toxic pollutants, identifying current knowledge gaps, and providing a framework for future fueling a new direction of research.

This book thus provides an exhaustive application of the state-of-the-art processes currently available to synthesize advanced green nanocomposite materials and biogenic nanomaterials. It also discusses a wide range of promising approaches for green nanotechnologies and nanocomposites preparations. The case study chapters included in this book connect the materials engineering and technology to the social context for a sustainable environment. The applications and different case studies provide solutions to the challenges faced by the industries; thus, minimize the negative social impacts and contribute in achieving a clean and sustainable environment. The book provides chapters from authors who have embarked on research programs in the exciting sectors of cleaning the environmental pollution through employing green nanomaterial and composites. Most consideration has been given to compile the necessary resources to design this book comprehensively informative for the readers.

For better readability, the book is segregated into three sections: Section 1: Insights, Synthesis, and Properties; Section 2: Environmental Remediation Applications and

Future Prospective; and Section 3: Miscellaneous Applications. Utmost care is taken to compile the required resources to make this book fully informative and collates key techniques to synthesize advanced materials that can be used in multidisciplinary applications.

The introduction, Chapter 1, provides an up-to-date and comprehensive understanding of sustainable nanotechnology for environmental remediation. Chapter 2 discusses the scope of green nanotechnology in environmental remediation. The use of nanotechnology to resolve environmental concerns such as recycle products and eco-friendly nanomaterials. Chapter 3 reveals the understandings of green and biosynthesis of nanoparticles. Chapter 4 summarizes various traditional approaches for the production of nanomaterials, for instance, microwave irradiation, chemical vapor deposition, and so forth. Chapter 5 reveals the fabrication approaches of metal nanoparticles using the green approach, particularly for environmental remediation. Chapter 6 discusses different sustainable nanocomposites for use in engineering materials. Chapter 7 reveals the approaches for the production of biogenic magnetic nanoparticles and their remediation applications, specifically in water and wastewater treatment. Chapter 8 provides the characterization techniques of green nanoparticles and their incorporation with different materials. Chapter 9 shows the features, physical and chemical, of nanoparticles and hybrid materials.

Chapter 10 reveals the applications, merits, and demerits of nanotechnology as an effective tool in water and wastewater treatment. Chapter 11 provides the potential of green biocomposites materials for remediation applications, such as biopolymer-clays composites. Chapter 12 provides the advancement in the production of green nanocomposites materials, particularly for wastewater treatment applications. Chapter 13 discusses the use of green nanocomposites for the elimination of various aqueous pollutants that occurred in different water sources, for example, heavy metal ions, dyes, and PPCs. Chapter 14 presents different surface modification approaches that have been designed for green carbon-based material for applications related to the environment. Chapter 15 describes an in-depth analysis of the feasibility and potential of biogenic nanomaterials. The photocatalytic activity of biogenic nanomaterials for environmental applications has gained substantial attention. Chapter 16 presents the different fabrication approaches and photocatalytic applications of CuO/ZnO for environmental remediation. Chapter 17 discusses the management of water contamination through phytogenic nanoparticles. Chapter 18 describes the use of magnetic nanoparticles for environmental applications. Chapter 19 defines the current barrier and prospect of green nanocomposites, particularly for environmental remediation applications.

Chapter 20 discusses primarily the involvement of nanotechnology in biosensors applications; Chapter 21 presents mainly the sensing and bioimaging application using ultra-thin fluorescent nanomaterials such as metal nanoclusters, carbon nanomaterials, and quantum dots. Chapter 22 describes the fabrication of advance carbon-based nanocomposites, primarily for biomedical applications. Chapter 23 deals with the different fabrication techniques of metal oxide–based nanocomposites for different applications, mainly energy storage. Chapter 24 discusses the engineered applications of nanomaterials for green cementitious materials. Chapter 25 presents different carbon nanomaterial with extensive specific surface areas for applications related to liquid adsorption. Chapter 26 describes the prospect of magnetic nanoparticles and nanocomposites for the elimination of electromagnetic interference contamination. Chapter 27 reveals the part of nanotechnology in improving the production and quality of crops. The eco-friendly techniques to reuse the used selective catalytic reduction catalyst are described in Chapter 28.

We are fortunate to have gathered contributions from reputable authors and sincerely thank all of them. The enthusiasm and efforts of the contributors in the field of environmental remediation with nanotechnology have culminated in this book, which we believe will be vital reading to the growing number of researchers in this field. Our heartfelt thanks to all the elite reviewers for their time and consideration in reviewing the manuscript and giving their suggestions and feedback. Their support has supported the book to meet international standards. Our special thanks to the editorial team of Elsevier (Holt Simon and Capille, Gabriela), who handhold us in every stage of initiation, operation, and completion of the book, including the production process.

Editors
Dr. Rama Rao Karri
Dr. Janardhan Reddy Koduru
Dr. Nabisab Mujawar Mubarak
Dr. Erick Bandala

Acknowledgments

I thank Prof. Zohrah, Vice Chancellor, Universiti. Teknologi Brunei and higher management for the support. I also thank my coeditors; without their support and co-operation, this book is not possible. I also thank all the authors who contributed chapters consisting of their valuable research.

Dr. Rama Rao Karri

I would like thank to the higher management of Kwangwoon University and Department of Environmental Engineering, Seoul, Korea for continuous support. I also thank my coeditors; without their support and co-operation this book has not got its shape. I am grateful to all the authors and reviewers for their valuable contribution and support.

Dr. Janardhan Reddy Koduru

I would like to take this opportunity to express my sincere gratitude to higher management of Curtin University Malaysia and Department of Chemical Engineering, where I previously worked. I also thank my present affliation, Universiti Teknologi Brunei for the encouragement. My special thanks go to all my coeditors as well as authors for their valuable contribution.

Dr. Nabisab Mujawar Mubarak

I am grateful to contributors, reviewers, and coeditors for their effort. Thank you all!

Dr. Erick Bandala

SECTION 1

Insights, Synthesis and Properties

CHAPTER 1

Appraisal of nanotechnology for sustainable environmental remediation

Tariqul Islam[a], Imran Ali[b], Iffat Naz[c] and Md Shahidullah Kayshar[d]

[a]Department of Agricultural Construction and Environmental Engineering, Sylhet Agricultural University, Sylhet, Bangladesh
[b]Department of Environmental Science and Engineering, College of Chemistry and Environmental Engineering, Shenzhen University, Shenzhen, China
[c]Department of Biology, Deanship of Educational Services, Qassim University, Buraidah, Kingdom of Saudi Arabia (KSA)
[d]Department of Food Engineering and Technology, Sylhet Agricultural University, Sylhet, Bangladesh

1.1 Introduction

Environmental pollution is one of the most critical problems of the modern world due to rapid industrialization, urbanization, and deforestation. All major components of the environment such as water, soil, and air are being affected with different kinds of pollutants such as heavy metals, toxic dyes, pesticides, fertilizers, industrial effluents, inorganic and organic compounds, etc. (Guerra et al., 2018; Jyothi et al., 2007; Fahad Saleem Ahmed Khan, Mubarak, Khalid, et al., 2020; Fahad Saleem Ahmed Khan, Mubarak, Tan, et al., 2020; F.S.A. Khan et al., 2020). For remediation of these environmental pollutants, different kinds of treatment technologies (adsorption, filtration, aerobic/anaerobic digestion, coagulation, biological filters, activated sludge, chemical precipitation, photocatalysis, advanced oxidation process, etc.) with various different approaches are being explored (Cheng et al., 2020; Moriwaki et al., 2020). However, nanotechnology-based techniques have higher advantages over the other technologies for environmental remediation (Cheng et al., 2020; Dehghani et al., 2020; Karri et al., 2019). The nanotechnologists/environmental engineers are trying to design novel materials with desired morphology or characteristics by manipulating physical and biological chemistry, that should be able to remove various kinds of environmental hazardous pollutants by achieving superior performance using minimum costs and energy (J. Chen et al., 2020; Siyal et al., 2019; Tian et al., 2019). Presently, various kinds of materials such as nanoparticles (NPs), nanomaterial, ceramics, graphene oxide (GO), metal-organic frameworks, carbon nanostructures, zeolites, biochar, synthetic natural polymers membrane, etc. are obtained from different sources for remediation of toxic environmental pollutants (I. Ali et al., 2017; Ali et al., 2019a; Sharifi et al., 2019; Yang et al., 2019). These materials are attracting much research attention due to their exceptional qualities such as their unique physical and chemical characteristics, lower mass,

Sustainable Nanotechnology for Environmental Remediation
DOI: https://doi.org/10.1016/B978-0-12-824547-7.00010-2

higher surface area, chemical reactivity, regeneration efficiency, and cost-effectiveness (Ali et al., 2019b; Ali et al., 2019c; K. Zhang et al., 2020). Further, they have vast opportunities to be grafted with different functional groups from both organic and inorganic sources (plants, bacteria, fungi, polymers, etc.) (Ali et al., 2019a; Ali et al., 2019a). However, the performance efficiencies of the nanomaterials being affected directly by their tunable physical properties such as size, porosity, morphology, and other chemical properties (M. Y. Zhang et al., 2010). Thus, the functionalization ability of nanomaterials, by targeting specific pollutants, increases their capability and efficiency in environmental remediation (Guerra et al., 2017). The nanomaterials are applied for the removal of environmental contaminants by different mechanisms namely, adsorption, absorption, photocatalysis, filtration, chemical reaction, etc. (Dehghani et al., 2020; Guerra et al., 2018; Karri et al., 2019; Fahad Saleem Ahmed Khan, Mubarak, Khalid, et al., 2020; Santhosh et al., 2016). In the environmental context, it is important that secondary or intermediate pollutants should not be created by the employed materials used for cleaning the environment. Therefore, the nanomaterials from environmentally friendly (biodegradable) sources are getting more popularity in the field of nanoscience for environmental remediation (Guerra et al., 2018). Additionally, the use of the waste materials such as biodegradable industrial and agricultural waste (e.g., rice husk, fallen plant leaves, etc.) is getting much interest to be employed for the synthesis of nanomaterials to reduce the cost of their fabrication (T. Islam et al., 2020; T. Islam, Liu, et al., 2018; Tariqul & Changsheng, 2019). To develop the new nanomaterials for environmental remediation, their green synthesis process, zero toxicity, recyclability, regeneration capability, biodegradability, and most importantly the sustainability are the main key challenges that need to be addressed (Guerra et al., 2018). To find the sustainable nanotechnology-based processes for environmental application, it is required to understand the different types of nanomaterials, which are being used for water, soil, or air reclamations, and their manufacturing process according to the sources of the baseline materials (plant, microbes, and other inorganic substances) (I. Ali et al., 2017). Thus, this chapter is designed to describe the different types of nanomaterials employed for environmental remediation, their fabrication techniques, sustainability, and usefulness in the future. The nanomaterials employed for environmental remediation are classified into different groups (Fig. 1.1), although there is no specific classification. Fig. 1.2 shows the schematic mechanisms for sustainable environmental remediation through nanotechnologies (Guerra et al., 2018).

1.2 Application of nanotechnology for remediation of different environmental components

1.2.1 Nanotechnology-mediated water/wastewater remediation from pollutants

Water is one of the most noteworthy components of the globe and significantly related with the existence of living beings. Though a major portion of the globe is full

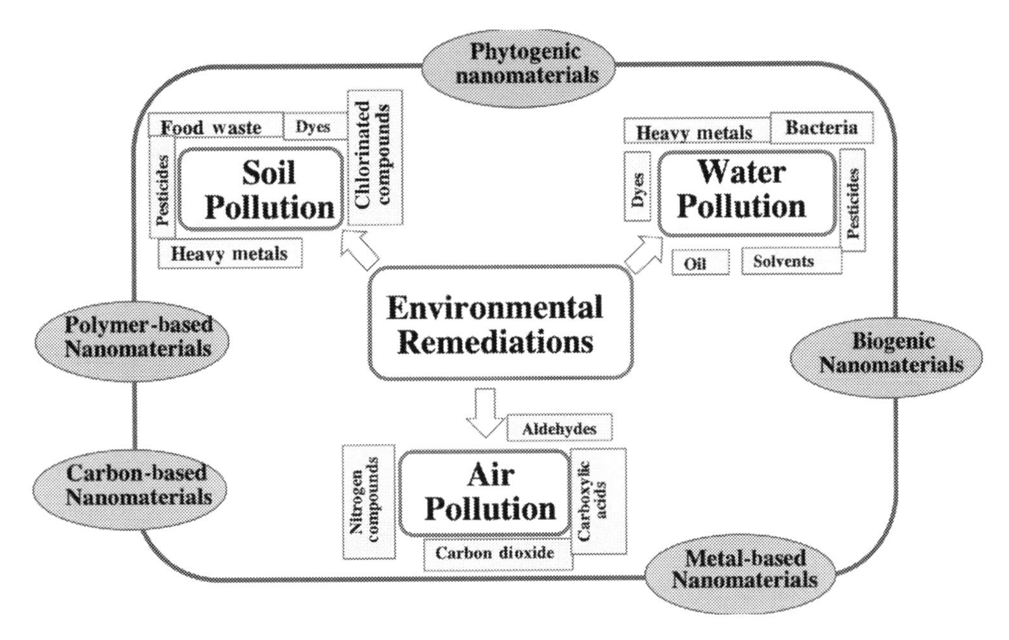

Figure 1.1 Environmental remediation by different type of nanomaterials.

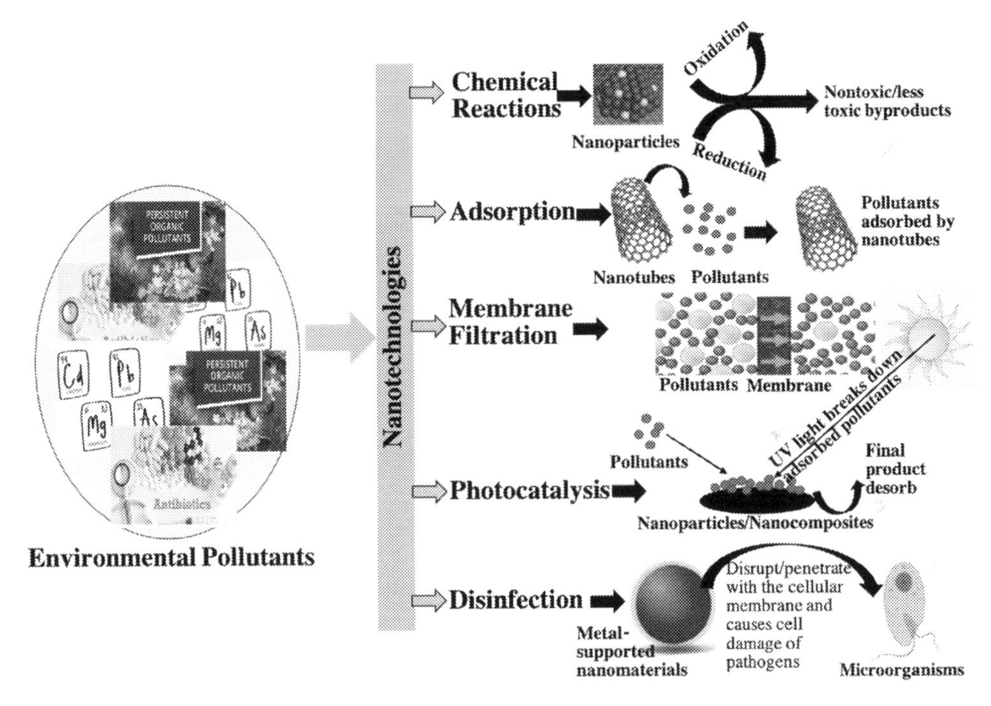

Figure 1.2 Schematic mechanisms for sustainable environment remediation through nanotechnologies. From Guerra, F., Attia, M., Whitehead, D., & Alexis, F. (c. 2018). Nanotechnology for environmental remediation: Materials and applications. Molecules, *23*(7). https://doi.org/10.3390/molecules23071760.

of water but only 2.5% is freshwater (Simonovic & Fahmy, 1999). Around 70% of the freshwater is frozen in polar areas and the other major volume of water exists either as moisture or in deep aquifers, which is out of the reach from human beings (T. Islam, Liu, et al., 2018). while, just 1% of the total freshwater is available as drinking water, which shows the significance of the water resource for the existence of the globe (T. Islam, Ye, et al., 2018). Few decades ago water was one of the free natural resources, though nowadays the world is facing severe crisis of water due to unplanned withdrawal, misuse of water, and the over- pollution of water resources (Zimmerman et al., 2008). To control this water environment pollution, the remediation of water from different toxic elements such as heavy metals, dyes, pharmaceuticals is urgently needed (T. Islam, Liu, et al., 2018; T. Islam, Peng, et al., 2018). Recently the different materials such as activated carbon, biochar from different sources, and different low-cost adsorbents are used conventionally by many researchers (T. Islam et al., 2020; Karri et al., 2020; Karri & Sahu, 2018). Due to unique nature and higher performances, nanostructured materials get immense attention in water/wastewater reclamations (Gautam et al., 2019). Nanomaterials have been reported by different researchers as highly capable to degrade and/or adsorb different pollutants such as dyes, heavy metals, pharmaceutical and other toxic pollutants from water/wastewater/soil/air (Ali et al., 2019a). Different nanomaterials, such as nanoadsorbents, nanocatalyst, and nanomembrane, are reported as successful candidate for environmental remediation (Gautam et al., 2019; Jun, Karri, Mubarak, et al., 2020).

1.2.1.1 Nanoadsorbents

NPs are used as potential adsorbents due to their higher surface area with comparatively lower mass which increase the chemical activity and binding sites of the surface (Kalfa et al., 2009). Further, nanoadsorbents are applied in different forms such as carbon nanotubes (CNTs), graphene, activated carbon, manganese oxide, and titanium and ferric oxides (Gupta et al., 2015). Innate surface and diversified external functionalization makes the nanomaterials dynamic in characteristics (Ali et al., 2019a; F.S.A. Khan et al., 2020; Lingamdinne et al., 2019). The extrinsic structure, intrinsic composition, and apparent size make the nanomaterials ideal adsorbents. The nanomaterials used to remove heavy metals or dyes or other pollutants from the aqueous environment might be affected by different parameters. It should be considered that the nanomaterials should be nontoxic, adsorption capacity should be higher, efficient even in low concentrations of pollutants, and also pollutants could be easily separable from the surface of the nanomaterials after treatments (Botes and Cloete, 2010). There are diversified types of NPs depending on their adsorption process including metallic NPs, magnetic NPs, and nanostructured mixed oxides. Additionally, the carbonaceous nanomaterials including carbon NPs, CNTs, and carbon nanosheets are the recently used and developed by the researchers as adsorbents. As well as the silicon nanomaterials including silicon NPs, silicon nanotubes, and silicon nanosheets have also been reported to be employed as the adsorbents

(Anjum et al., 2016). The other nanoadsorbents used as adsorbents for the adsorption of dye and heavy metals from aqueous solutions are nanofibers, nanoclays, aerogels, etc. (Anjum et al., 2016). Size, shape, dimensions, agglomeration state, solubility, crystal structure, and chemical compositions of the NPs are the main parameters that control the properties of the nanoadsorbents. Fine sizes and the significant chemical activities are making the nanomaterials prominent as adsorbents compared with the others (Kalfa et al., 2009). Further, the nanomaterials can be modified with different reagents to change their properties and to enhance their efficiencies (Kalfa et al., 2009).

1.2.1.2 *Nanocatalysts*

Nanomaterials, especially the inorganic materials such as metal oxides and semiconductors are also gaining interest as a catalyst in wastewater treatment (Ali et al., 2019a). Different types of nanocatalysts, such as electrocatalysts, photocatalysts and Fenton-based catalysts for chemical oxidation and antimicrobial actions, have been reported for wastewater reclamation (Ma et al., 2015). The NPs having a photocatalytic reaction in the presence of light are classified as nanocatalysts and have a broad range of application for wastewater for diminution of various contaminants (Akhavan, 2009). These photocatalysts consist of a semiconductor metal to degrade the pollutants such as dyes, heavy metals, pesticides, detergent, etc. from wastewater (Adeleye et al., 2016).

1.2.2 Nanomembranes

Membrane filtration processes associated with nanomaterials are recognized as one of the most effective methods for wastewater treatments (F. Zhang et al., 2013). The concept of nanotechnology in the membrane filtration makes this technology more functional with high permeability, fouling resistance, catalytic reactivity, etc. (Pendergast & Hoek, 2011). The nanomembranes have been adopted for wastewater treatment due to the effective disinfection, quality of water and suitable size of the filtration plant, high efficiency, economical and simple in design (Guo et al., 2016). The nanomembrane filtration has been reported with effective removal efficiency for heavy metals, dyes, and other organic or inorganic contaminants (Jie et al., 2015; Jun, Karri, Mubarak, et al., 2020; Jun, Karri, Yon, et al., 2020). These are basically composed of one-dimensional organic and/or inorganic nanomaterials such as nanofibers, nanotubes and nanoribbons (Liu et al., 2014).

1.2.3 Nanotechnology-mediated soil remediation from pollutants

Soil is another important part of the environment to support existence of life on the globe and is the only medium for plant growth or food production (Abhilash et al., 2012). For the safeguard of agricultural production and maintaining the environmental balance it is crucial to manage the soil health (Rattan, 2015). Presently soil pollution and degradation is the serious environmental problem due to different reasons such as urbanization, soil erosion, deforestation, overdoses of fertilizer, etc. (D. & W., 2014). Additionally, different chemical contaminants such as heavy metals, pesticides, and other industrial pollutants are also responsible for the soil pollution (Ghormade et al., 2011). Due to different manmade reasons about 30% of land has been either polluted or degraded all over the world and the pollution of soil/land is rising day by day (Abhilash et al., 2012). Thus, it is instantly desired to work for remediation of pollutants from land. Nanotechnology has great scope to be employed for remediation of soil just like water/wastewater reclamation. Various research investigators have reported different kinds of nanomaterials (nZVI, TiO_2, CNTs, etc.) and have applied them successfully for soil remediation (Cai et al., 2019). Nanotechnology-based processes are much more applicable for soil remediation due to their tunable nature and special characteristics for specific pollutants (Tang et al., 2014). As shown in Table 1.1, different kinds of heavy metals such as Arsenic (As), Lead (Pb), Chromium (Cr), Zinc (Zn), Nickel (Ni), etc. have been reported to be present as a contaminant in soil from different anthropogenic and industrial sources (Alloway, 2012). Many techniques have been revealed to eliminate these heavy metals from the soil to rescue the agricultural land from the toxicity and to increase the fertility. Nanotechnology-based nanomaterials proved as an ideal candidate for the soil remediation due their extreme mobilizing capability and the higher adsorption capabilities. The higher adsorption capacity of the nanomaterials makes them ideal adsorbents for the heavy metal elimination from the soil (H. Ali et al., 2013). The nZVI has been found as a successful nanocandidate to reduce the As, Cr, Pb, Cd, and Zn from the multimetal-contaminated soil (Gil-Dıaz et al., 2016). It was also found that different stabilizers such as starch, carboxymethyl cellulose have been investigated for soil remediation (Virkutyte et al., 2014). Various different nanocomposites are reported to remove up to 94%–98% arsenic from the polluted soil (B. An & Zhao, 2012). Biochar with the support of nZVI is another successful agent for up to 100% Cr removal from the contaminated soil (H. Su et al., 2016). NPs with biochar improve the soil fertility by improving pH condition and finally positively influence the plant growth and prevent the excessive iron release to the environment (Zhu et al., 2016). The employment of the water treatment residual plant NPs for the soil remediation has great impact on the Hg and Cr release from the soil (M. Y. Zhang et al., 2010). Further, Na-zeolite nanostuff was also found as a successful agent for sorption of Cd from soil (Ghrair et al., 2010). NPs also significantly remove different toxic chemicals from soil, released from different types of pesticides. Photocatalysis has been occured for the degradation of the pesticides in the soil. In this process the NPs undergo a process where chemical pollutants go through the reaction to convert the harmful pesticides' chemical into harmless pollutants. Titanium dioxide and zinc oxide are considered as good photocatalysts for the soil remediation (Fujishima et al., 2008). Presently the nanomaterials-based sensor or nanosensor is employed for improvement of the soil quality for higher crop production. Basically, the nanosensors are applied to monitor and detect the toxic materials, pesticides, etc. in the field or in production unit

Table 1.1 Nanomaterials employed for soil remediation from various different pollutants.

Nanomaterial used	Targeted pollutants	Type of pollutants	References
nZVI	As, Cr, Pb, Cd, Zn	Heavy metals	Gil-Díaz et al. 2016
Multi-walled carbon nanotubes (MWCNT)/ Polyaniline (PANI)/Acetylcholinesterase (AChE)	Carbaryl and methomyl	Pesticide residue	Cesarino et al., 2012
Nanoscale zero valent iron	DDT	Pesticides	El-Temsah et al., 2016
Biochar-supported nZVI	Cr (VI)	Heavy metals	Su et al. 2016
Electrochemically reduced graphene oxide (Er-GRO)-nafion nanocomposite	Dichlorvos	Pesticide residue	Wu et al., 2013
Fe (II) phosphate NPs	Cr (VI)	Heavy metals	Liu and Zhao, 2007
Na-zeolite nanostuff	Cd	Heavy metals	Ghrair et al. 2010
Anatase TiO_2	Phenanthrene (PAHs)	Pesticides	Gu et al., 2012
Water treatment residual NPs	Hg, Cr	Heavy metals	Moharem et al., 2019
Rhenium (Re^{+3})-doped nano-TiO_2	Carbofuran	Pesticides	Rui et al., 2010
Au NP–coated Si nanowires	Dichlorvos	Pesticide residue	Su et al., 2008
Al_2O_3, SiO_2, TiO_2 NPs	Zn, Cd, Ni	Heavy metals	Peikam and Jalali, 2019
AChE/RGO (reduced graphene oxide)-Au nanocomposite	Organophosphate and carbamate	Pesticide residue	Liu et al., 2011
CMC (carboxymethyl cellulose)- stabilized nano-Pd/Fe	Pentachlorophenol	Pesticides	Yuan et al., 2012
AgNP–coated Si nanowires	Carbaryl	Pesticide residue	Wang et al., 2010
nZVI and Pd/Fe bimetallic NPs	Polychlorinated biphenyls (PCBs)	Pesticides	Chen et al., 2014
FeSNPs	Lindane	Pesticide	Paknikar et al., 2005
MWCNT-chitosan nanocomposites	Methyl parathion	Pesticide residue	Dong et al., 2013
Organophosphate hydrolase–conjugated AuNPs	Paraoxon	Pesticide residue	Simonian et al., 2005

(Baruah & Dutta, 2009). Nanosensors can significantly contribute to improve the wide range of soil quality along with the pesticide residual detection. The residual impact of pesticides is significantly needed to be addressed to prevent the foodborne and other long-term illness (Ghormade et al., 2011).

1.2.4 Nanotechnology-mediated air remediation from pollutants

Air is another precious gift of nature and an inevitable part of the environment, but due to different manmade reasons air pollution with dust particles, gases, heavy metals, etc. is a common matter of the entire globe. Control of air pollution and remediation is very much crucial and considered as a significant necessity of the day. Thus, nanotechnology-based techniques are playing vital role to treat, detect, or prevent the air pollution (Yadav et al., 2017; Yunus et al., 2012). These processes can treat the pollution of air via adsorption with nanoadsorbents, can carry degradation of contaminants with nanocatalyst, and can do separation of pollutants by nanomembranes (Table 1.2). Various different dimensional nanoadsorbents have been reported for treatment of air pollution such as fullerene, CNT, graphene, graphite, etc. (F., 2017). Greenhouse gases have dangerous effect on the environment, which are continuously released to the environment in different ways. However, nanotechnology-based processes can be an easy solution for remediation of air from these dangerous greenhouse gases (Bergmann & Machado, 2015). Nanoadsorbents have high interaction with organic compound with electrostatic force, pie pie bond, hydrogen bond, van der Waals forces, and hydrophobic interactions (Ren et al., 2011). Further, CNTs have higher adsorption capacity than the activated carbon due to their cylindrical pores (Wei et al., 2014), whereas the nanocatalyst can be used for indoor air pollutant remedies with the effect of the frequency of light (Özkar, 2009). Titanium oxide NPs are being reported to have a self-cleaning coating for depletion of atmospheric contaminants such as nitrogen oxides and other gaseous pollutants (Shen et al., 2015). Nanofiber-coated filter media has been reported for filtration of air of industrial plants and inlet air of turbines (Muralikrishnan et al., 2014). For the prevention of air pollution, the detection of various different impurities in the air is a very much important step. The conventional methods (physical, chemical, and biological) are very laborious, time consuming, and costly compared to the nanosensors. Further, these nanosensors have been reported to detect several toxic compounds at the ppm and ppb level (V. et al., 2016; Zhou et al., 2015).

1.3 Different sources of nanomaterials

Various different materials from different sources have been used for fabrication of nanomaterials. These nanomaterials can be classified on the bases of their source materials such as plant-based nanomaterials (phytogenic nanomaterials), microbe-based

Table 1.2 Nanotechnology used for air remediation.

Nanomaterials used	Targeted pollutants	Type of nanomaterials	References
SWNTs and MWNTs	Mixture of NO, and NO_2	Carbon nanotube	Zhang et al., 2013
Modified CNTs using 3-aminopropyltriethoxysilane	CO_2	Carbon nanotube	Su et al., 2009
SWNTs/NaClO	Isopropyl vapor	Carbon nanotube	Hsu and Lu, 2007
CNTs deposited on quartz filters	VOCs	Carbon nanotube	Amade et al., 2014
Si- and Boron-doped SWCNTs	CO, and CH_3OH gases	Carbon nanotube	Azama et al., 2017
Fullerene B40	CO_2	Fullerene	Dong et al., 2015
Fullerene-like boron nitride nanocage	N_2O	Fullerene	Esrafili, 2017
Graphene oxide (GO)/nanocomposites	CO_2, NH_3, SO_2, H_2S, and N_2	Fullerene	Seredych and Bandosz, 2012
TiO_2	NO_2, and VOC	Nanofilm	Su et al., 2009
Silver (Ag)	Antibacterial biocide	Nanofilter	Su et al., 2009

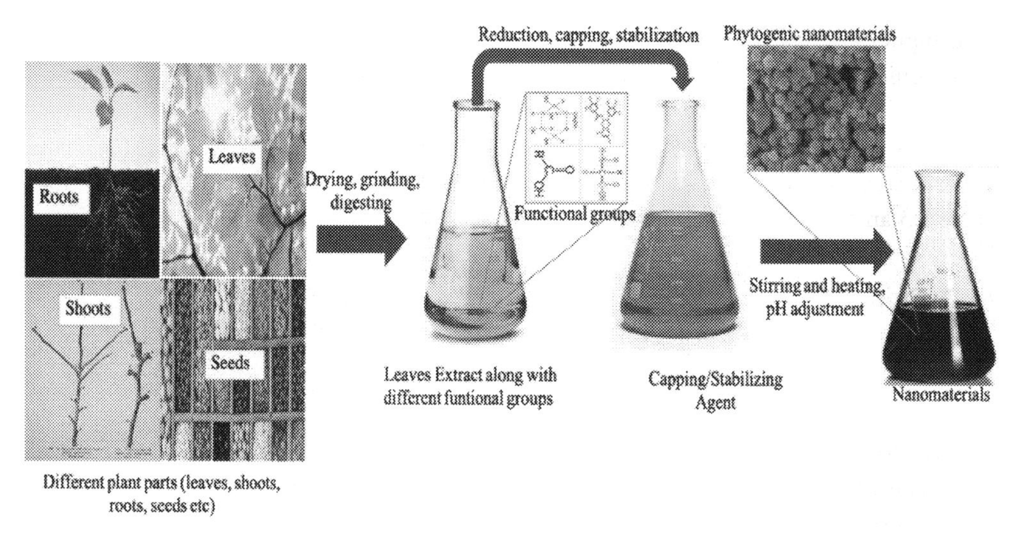

Figure 1.3 Synthesis process of phytogenic nanomaterials.

nanomaterials (biogenic nanomaterials), carbon-based nanomaterials, and polymer-based nanomaterials.

1.3.1 Phytogenic nanomaterials

Phytogenic nanomaterials are basically synthesized from different plant sources such as shoots, roots, leaves, or other parts of the plants. These nanomaterials are getting more popularity for their feasible performances and green manufacturing methods that do not required external energy (high pressure and temperature) (Mittal et al., 2013). The synthesis process of phytogenic nanomaterials has been shown in Fig. 1.3. Plant extract has been used as stabilizing or capping agents due the presence of different functional groups along with the chemical compounds such as proteins, polysaccharides, enzymes, etc.(Rajeshkumar & Bharath, 2017). The morphology of the phytogenic nanomaterials depend on the chemical compound present in the respective plants (Gautam et al., 2019). These nanomaterials take less time to prepare in comparison with the biogenic nanomaterials on the basis of the functional groups. Thus, it is very much important step to find out suitable plant species for synthesis of nanomaterials for the specific targeted pollutants (Parsons et al., 2007). The application of phytogenic nanomaterials for environmental remediation such as water/wastewater treatment is rising sharply due to its simple fabrication process, nontoxic sources/ingredient substances, cost effectiveness, fast removal efficiency, and eco-friendliness (I. Ali et al., 2017). Different types of tree ferns are available all over the world with cheap cost, while various functional groups present in plants extract such as polyphenols, hydroxyl, amino, aromatic, carbonyl, carboxyl, etc. make them ideal nanomaterials, and provide them greater functionality with higher adsorption capacity. Fig. 1.3 illustrates the general manufacturing process of phytogenic nanomaterials, (Islam et al., 2020). The green phytogenic adsorbents basically eradicate the pollutants via the polar ion exchange and are also involved in environmental remediation by a photocatalyst (Ali et al., 2017). Phytogenic nanomaterials reported by numerous researchers for remediation of various different targeted pollutants are listed in Table 1.3 and Fig. 1.3.

1.3.2 Biogenic nanomaterials

Bioaccumulation, bioleaching, and biomineralization are the bioprocess performed by bacterial cells to solubilize the metal ions through oxidation and/or reduction (Klaus-Joerger et al., 2001; J. N. Sahu et al., 2019). The biotransformation nature of bacteria has attracted the attention of researchers to use the bacterial cells as nanofactories for manufacturing the biogenic nanomaterials (Prakash et al., 2009). The natural protective strategies of different bacteria from the naturally soluble metal ions make them the bio-nanocomposites through biomineralization (Karri et al., 2019; Ramanathan et al., 2013). The manipulation of the different microbes is comparatively easy, which makes it a more suitable candidate for extracellular and/or intracellular biosynthesis of bionanomaterials; and the nanomaterials of extracellular synthesis of bacteria are easier to handle and separate (Prasad et al., 2016). Though the synthesis of nanomaterials through bacteria is time consuming than the physical or chemical methods, time can be reduced by coupling with microwave and ultrasound. The different microbes reported by various researchers in the literature are shown in Table 1.4. Additionally, the synthesis process of biogenic nanomaterials is demonstrated in Fig. 1.4 (Cotica et al., 2018; Islam et al., 2020).

1.3.3 Carbon-based nanomaterials

There are different types of carbon-based nanomaterials on the basis of their dimensions employed for the remediation of environmental pollutants. Thus the nanomaterials are termed as zero dimensional (0D), when its all of the three dimension are less than 100 nm such as quantum dots, fullerene, etc. (Han et al., 2016), while they are termed as one-dimensional (1D) nanomaterials, when its only one dimension is greater than 100 nm but the other two are less than 100 nm such as carbon and titanium nanotube (Lee et al., 2014). Further, the nanomaterials are termed as two-dimensional (2D), when its two dimensions are greater than 100 nm like graphene. Finally, the materials are termed as three dimensional (3D) which have all of the three dimensions larger than 100 nm such as graphite nanocomposite (P., 2011). Fullerenes counted as an ideal material for their high electron affinity, hydrophobic character, and high surface volume ratio. They have numerous different applications such as solar cells, semiconductors, surface coating, etc. (Gokhale & Somani, 2015). Its employment for water pollution remediation by killing the pathogenic microorganisms was reported previously (Baby et al., 2019).

Table 1.3 Phytogenic nanomaterials for environmental remediation.

Source plants	Surface area (m^2/g)	Capping agent	Adsorption capacity (mg/g)	References
Green tea (*Camellia sinensis*)	27.5	Polyphenols, caffeine	-	Hoag et al., 2009
Orange peel	65.19	Carboxyl group	68.1	Gupta and Nayak, 2012
Soya bean sprouts	-	Protein, amino acid, carbohydrates, polysaccharides	37.1	Cai et al., 2010
Aloe vera	-	-	71.678	Phumying et al., 2013
Plantain peel	11.31	Polyphenols, protein, lipid, fibers	15.8	Venkateswarlu et al., 2014
Syzygium cumini	3.517	Carbohydrates, polyphenols, protein, lipid, fibers	13.6	Venkateswarlu et al., 2014
Punica granatum	10.88	Carbohydrates, polyphenols, protein, lipid, fibers	22.7	Venkateswarlu et al., 2014
Citrullus lanatus	9.58	Carbohydrates and polyphenols	24.7	Venkateswarlu et al., 2015
Vitis vinifera	-	Carbohydrates and polyphenols	15.74	Venkateswarlu et al., 2015
Ananas comosus	11.25	Sugars, esters, polyphenols, flavonoid	16.9	Venkateswarlu et al., 2015
Lagerstroemia speciosa	52.791	Hydroxyl group, carboxyl group, polysaccharides	-	Srivastava et al., 2017
Allium cepa	-	Carbohydrates and polyphenols	52.6	Venkateswarlu et al., 2014
Pisum sativum peels	17.6	-	64.2	Prasad et al., 2017
Lonicera japonica	23.5	Aromatic and aliphatic groups	36.4	Lingamdinne et al., 2016
Ridge gourd peels	26.21	-	17.3	Cheera et al., 2016
Cnidium monnieri	122.54	Phenylpropanoids, aromatic olens, hydroxyl	54.6	Lingamdinne et al., 2017

Table 1.4 Biogenic nanomaterials for remediation of different environmental pollutants.

Source microbes	Type/class	Nanomaterials	Size (nm)	Shape	Removed pollutant	References
Trichoderma viride, Hypocrea lixii	Fungus	Au	20–30	Spherical	4-nitrophenol, 4-aminophenol	Hazra et al., 2013
Escherichia coli K12	Bacteria	Au	50	Spherical	4-nitrophenol	Su et al., 2014
Shewanella oneidensis MR-1	Bacteria	Na_2PdCl_4	28	-	Trichloroethylene	Hazra et al., 2013
Itajahia sp.	Fungus	Fe	˜200	Rod like	Lindane	Hazra et al., 2013
Pseudomonas putida MnB6	Bacteria	$MnCl_2$	-	-	Diclofenac, 2-anilinoph-enylacetate	Meerburg et al., 2012
Bacillus marisflavi	Bacteria	Au	8-30	Spherical	Congo red, methylene blue	Ahluwalia et al., 2016
Microbial granules (ABCSBR)	-	Na_2PdCl_4	4.78	Short rods	Cr(VI)	Suja et al., 2014
LAR-2 and *Geobacter* sp. EB1	Bacteria	Ferrihydrite	<30	-	Co^{2+}, Mn^{2+}, Ni^{2+}, Zn^{2+}	Iwahori et al., 2014
Cladosporium oxysporum AJP03	Fungus	Au	72.32 ± 21.80	Quasi-spherical	Rhodamine B	Hazra et al., 2013

(continued on next page)

Table 1.4 Biogenic nanomaterials for remediation of different environmental pollutants—cont'd

Source microbes	Type/class	Nanomaterials	Size (nm)	Shape	Removed pollutant	References
Desulfovibrio vulgaris	Bacteria	Pt (IV)	-	=-	Ciprofloxacin	Martins et al., 2017
Chlorella pyrenoidosa	Algae	Ag	2–15	Spherical, oval	Methylene blue	Iwahori et al., 2014
Trichosporon montevideense WIN	Fungus	$HAuCl_4$	53–12	Spherical	2-nitrophenol (2-NP),	Shen et al., 2016
Streptomyces griseobrunneus FSHH12	Bacteria	Se	50–500	Spherical	Bromothymol blue	Hazra et al., 2013
Desmodesmus sp. WR1	Algae	Mn	-	Irregular geometry	Bisphenol	Hazra et al., 2013
Shewanella oneidensis MR-1	Bacteria	Pd/Pt	4.5–59.9	Spherical, flower like	4-nitrophenol	Hazra et al., 2013
Pseudomonas aeruginosa BS01	Bacteria	$ZnCl_2$	10–15	Spherical	Direct brown MR dye	Hazra et al., 2013
Fermentative bacteria consortium	Bacteria	Pd	5–30	Irregular	Congo Red, Evans Blue, and Orange II	Pei et al., 2013
Marinobacter sp. MnI7-9	Bacteria	$MnCl_2$	-	-	Ag	Pei et al., 2013
Aspergillum sp.	Fungus	Au	10–100	Hexagonal, irregular	Nitroaromatic pollutants and dyes	Qu et al., 2017
Aspergillus sp.	Fungus	Au	25–35	Spherical	4-nitrophenol	Qu et al., 2017

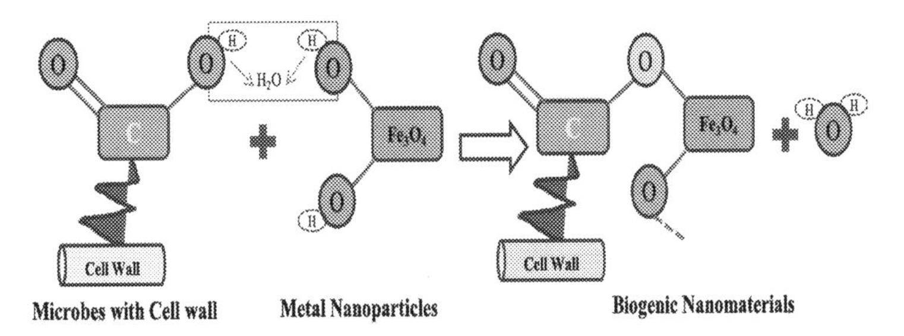

Figure 1.4 Synthesis process of biogenic nanomaterials Cotica, L., & Prasad, R. (Eds.). (2018). Nanobiocomposites: synthesis and environmental applications. In *Fungal Nanobionics: Principles and Applications* (pp. 1–19). Springer Nature Singapore Pte Ltd. https://doi.org/10.1007/978-981-10-8666-3_1.

Different dimensional CNTs are widely used for pollution control due to their higher surface area, porous structure, easier surface functionalization, and tunable nanosize. These characteristics make them promising for the water treatment and other environmental remedies (Alijani & Shariatinia, 2018). Further, CNTs have also shown promising results in the environmental application due to their higher surface area because of multiple layers of graphene (Baby et al., 2019). Graphene or its modified forms have been investigated for environmental remediation of various targeted pollutants for example pristine graphene has been reported as effective adsorbents to remove fluoride from water (Wu et al., 2011). Higher adsorption capacity (35.59 mg/g) of graphene for the fluoride removal makes them one of promising carbon-based materials for fluoride adsorption from aqueous solutions. Modified graphene such as pristine graphene and others have been considered as favorable practices for the remediation of different other compounds (X. An & Yu, 2011). Further, by modifying the surface of the graphene with a layer of pristine can make them more efficient with the decrease of aggregation (X. An & Yu, 2011), while GO is another material that has been reported by researchers as a promising material for environmental remediations by adsorption of different gaseous (sulfur dioxide, hydrogen sulfide, volatile organic compounds, etc.) and water (pharmaceuticals' pollutants, heavy metals, pesticides, etc.) compounds (Koduru et al., 2019; Lingamdinne et al., 2019; Wang et al., 2013). Diverse functional groups, such as epoxides, carboxylic acids, hydroxyl, etc., present on the surface of the GO make them prominent adsorbents for diverse environment pollutants (Seredych & Bandosz, 2007). Moreover, GO has been proved to have higher capability for cationic metals but with the modification of organic or metal oxides it can be implied for the anionic metal adsorption (X. An & Yu, 2011). Titanium oxide and graphene nanocomposites have also proved as a potential adsorbent for the benzene remediations from environment (Y. Zhang et al., 2010). However the photolytic activity of the material is significantly related with the ratio of the graphene and titanium oxide. After a certain ratio of graphene, the composite reduce their efficiencies (Liu et al., 2011). Further, ZnO-graphene and CdS-graphene are also reported as the efficient adsorbents for the removal of heavy metals from water

(F. Zhang et al., 2013). The modification of the graphene with different metal oxide makes them more stable to degrade the pollutants from aqueous solutions or any other substances. However, CNTs are another type of carbon-based nanotechnology widely used nowadays for the environmental remediations. Mostly the open-end CNTs have been developed to enhance the adsorption capacity (Lithoxoos et al., 2010). Generally, CNTs are arranged in a hexagonal arrangement by bonding bundle of heterogeneous and porous tubes. The adsorption on the surface of the nanotubes reaches equilibrium earlier then the internal surface of the tube (Ren et al., 2011). Closed-end CNTs can be more capable as adsorbent due their more binding sites. Carbon-based nanomaterials have great scope with the desired modifications for the environmental remediations.

1.3.4 Polymer-based nanomaterials

Polymer-based nanomaterials are also employed for environmental remediation to enhance the stability of the materials as well as to increase its performances. The approach of the polymer-based nanomaterials is an alternative to boost the stability of the materials to employ the host materials for specific purposes (Zhao et al., 2011). Different polymer-based nanomaterials have been reported in the remediation of contaminants such as amphiphilic polyurethane NPs, Polyamidoamine (PAMAM) dendrimers, polyamine-modified cellulose volatile organic compounds (VOCs), polymer nanocomposites (metal ions, microorganisms), etc. (Guerra et al., 2018). Amphiphilic polyurethane NPs promote the mobility of the soil and reported to remove phenanthrene up to 80% (Tungittiplakorn et al., 2004), while PAMAM dendrimers were reported for the treatment of wastewater treatment due to its functional groups such as amine, hydroxamates, carboxylates, etc., which have facilitated the encapsulation of a broad range of cations (Diallo et al., 2005). Polymer-based nanomaterials are also reported to reduce the gaseous pollutants along with different types of heavy metals both from soil and water (C. Su et al., 2012). Though these types of nanomaterials have been applied for environmental remediation but it has major disadvantages as they can create other types of pollutants due to their nonbiodegradable nature (Guerra et al., 2018).

1.3.5 Metal-based nanomaterials

Metal and metal-based nanomaterials have been described as efficient elements for environmental remediation (Santhosh et al., 2016). The majority of the metal-based nanomaterials have been proved as competent adsorbents for the removal of heavy metals and other organic pollutants from aqueous solutions (Das et al., 2015). Nanomaterials used in environmental remediations are very much popular due to their both *in-situ* and *ex-situ* applications in aqueous system (Das et al., 2015). Table 1.5 shows the different metal-based nanomaterials used for water remediations targeting various different pollutants. For the synthesis of metal oxide–based nanomaterials coprecipitation, thermal decomposition,

Table 1.5 Metal-based nanomaterials applied for remediation of environmental pollution.

Nanomaterials	Targeted pollutants/ contaminants	Type of environmental components	References
Bimetallic nanoparticles	Chlorinated and brominated contaminates	Soil and water	Chen et al., 2008
Titanium oxide nanoparticles	*E. coli*, aromatic hydrocarbons, biological nitrogen, phenanthrene	Soil and water disinfectant	Cho et al., 2005
Iron-based NMs	Heavy metals, chlorinated organic solvents	Water	Hooshyar et al., 2013
Ag nanoparticles/ Ag ions	*E. coli*	Water	Cho et al., 2005
Metal-doped titanium oxide	Rhodamine B, *E. coli*, *Staphylococcus aureus*, endotoxin	Water contaminants	Park and Lee, 2014
Binary mixed oxide	Methylene blue dye	Water	Rasalingam et al., 2014
Titanate nanotube	Nitric oxide	Gaseous	Chen et al., 2013

reduction, and hydrothermal synthetic protocols were so much popular during last decades (Cushing et al., 2004). These methods are easily scalable with the higher yields of shape controlled, stable, and monodisperse metal oxide (Willis et al., 2005). Further, bimetallic nanomaterials are widely used to overcome the limitations of monometallic NPs such as low stability and propensity toward aggregation. Usually different stabilizing agents have been used to increase the stability of the metallic oxide, but the addition of the second metal in the formulation can make them more stable (Khin et al., 2012). Increasing the stability of the metallic NPs could increase efficiencies of the manufactured materials (Tao, 2012). Different metal such as nickel (Ni), lead (Pb), or copper (Cu) are reported as second metal to increase the stability of the zero-valent iron (nZVI) (Dehghani et al., 2020; N. Sahu et al., 2019; Zheng et al., 2009). Titanium oxide NPs are also reported for the environmental remediation, especially to remove the pollutants from the water (Ding et al., 2014). For the catalytic reduction of NO with ammonia, titanium nanotubes have been reported to be manufactured by thermal method (X. Chen et al., 2013). Mn-doped titanate nanosheets, titanate nanotubes, and titanate nanorods have been investigated in different neutral, basic, and acidic media. Iron NPs are the nZVI comprising the Fe (II) and Fe (III) oxides with the formation of shells (Kharisov et al., 2012). Through the electron donation from the nZVI core, different heavy metals (HMs) and chlorinated compound can be reduced. Additionally, silver NPs (AgNPs) are also reported as promising nanomaterials (NMs) for their significant antifungal, antibacterial, and antiviral activity as a water disinfectant (Chou et al., 2005). Triangular and large particle sizes make them affordable for the antibacterial and antiviral activities (Pal et al., 2007). Moreover, binary mixed magnetic nanomaterials are also reported as highly capable to reduce different dyes and organic materials (Ding et al., 2014).

1.3.6 Silica nanomaterials

Silica nanomaterials are mesoporous and they have higher surface area, tunable pore size, and large pore volume. This versatile nature makes them unique in environmental applications as adsorbents and catalysts (Tsai et al., 2016). These silica nanomaterials have been reported for the remediation of contaminant in the gaseous phase (Son et al., 2008). However, amine-modified xerogels, aluminosilicates, and porous silica are also reported for the gaseous pollutant (CO_2, H_2S, aldehydes, ketones) removal (Guerra et al., 2018). Further, carboxylic acid–, amino-, and thiol-functionalized mesoporous silica have been reported for dyes and heavy metal contaminants removal from water (Guerra et al., 2018). The silica materials can be easily modified to change their surface phenomenon due to the hydroxyl groups present on their surface. To design new material such as adsorbent and catalyst grafting functional group onto the pore walls is a technically sound approach (Huang et al., 2003). Amine surface–modified silica xerogels and the mesoporous silica have been reported as an efficient material for the removal of up to 80% CO_2 within 30 min, whereas 80% H_2S adsorption has been achieved within 35 min (Huang et al., 2003). Amine-modified aluminosilicates have been reported for the capture of the CO_2 and other carbonyl compounds through reversible adsorption (Nomura & Jones, 2013). This material is so much efficient and could be an easy alternate with the other expensive silica-supported materials (Qi et al., 2011). Moreover, mesoporous silica functionalized with carboxyl groups can be dynamic adsorbents for dyes and heavy metals from water (Tsai et al., 2016).

1.4 Future aspects

Following points should be explored in future to improve status for the use of nanomaterials in the environmental remediation:

1. The appoaches for degrading nanoparticles following their use in environmental rehabilitation should be defined, or else they may generate a new environmental danger.
2. Environmental remediation with the biogenic and phytogenic nanomaterials should be studied more to meet up the challenges on their size control, aggregation, and stability.
3. Still there have enormous species (plants, bacteria, fungus, algae, and other microorganisms) to be explored for manufacturing environmental friendly nanomaterials.

4. Most of the nanotechnology-based processes are costly, so consuming waste materials to manufacture environmental nanomaterials will be the unique approach to minimize their cost.

1.5 Conclusion

Various types of nanomaterials such as phytogenic, biogenic, carbonaceous, and polymeric nanomaterials are widely used for the environmental remediations. Specific nanomaterials are applied either for the remediation of water, air, or soil pollution. Selection of the nanomaterials for the successful implication on the environment (water/soil/air) depends on the type of the contaminants, accessibility of the contaminated sites, reusability efficiencies, availability of the raw materials for its fabrication, environmental friendliness of manufactured materials. Each material used in different parts of environmental remediation has its own advantages. In this study, we analyzed the materials on the basis of their manufacturing sources, and their applicability on three major parts of the environment such as water, soil, and air. Nanotechnology-based environmental remediation is gaining much attention these days. Previously, researchers have reported carbonaceous and polymeric nanomaterials for environmental remediation. Though these processes have shown higher removal efficiencies but have presented major disvantages. Most important disadvantage including their manufacturing processes may create further toxicity to the environment and they are not biodegradable. Consequently, other environmental approaches such as plant- and the microbe-based nanomaterials got much consideration over the world. Different types of the plant species or the microbes have been used to synthesize environmental friendly nanomaterials. These nanomaterials from the organic sources also have disadvantages, as they are not much stable like the polymeric substances, so may disintegrate during application. Thus, still there is a great scope to fill this gap to improve these techniques for its successful application for the environmental remediation.

References

Abhilash, P.C., Powell, J.R., Singh, H.B., Singh, B.K., 2012. Plant-microbe interactions: novel applications for exploitation in multipurpose remediation technologies. Trends Biotechnol. 30 (8), 416–420. https://doi.org/10.1016/j.tibtech.2012.04.004.

Adeleye, A.S., Conway, J.R., Garner, K., Huang, Y., Su, Y., Keller, A.A., 2016. Engineered nanomaterials for water treatment and remediation: costs, benefits, and applicability. Chem. Eng. J. 286, 640–662. https://doi.org/10.1016/j.cej.2015.10.105.

Ahluwalia, S., Prakash, N.T., Prakash, R., Pal, B., 2016. Improved degradation of methyl orange dye using bio-co-catalyst Se nanoparticles impregnated ZnS photocatalyst nder UV irradiation. Chem. Eng. J. 306, 1041–1048. doi:https://doi.org/10.1016/j.cej.2016.08.028.

Akhavan, O., 2009. Lasting antibacterial activities of Ag-TiO_2/Ag/a-TiO_2 nanocomposite thin film photocatalysts under solar light irradiation. J. Colloid Interf. Sci. 336 (1), 117–124. https://doi.org/10.1016/j.jcis.2009.03.018.

Ali, H., Khan, E., Sajad, M.A., 2013. Phytoremediation of heavy metals-Concepts and applications. Chemosphere 91 (7), 869–881. https://doi.org/10.1016/j.chemosphere.2013.01.075.

Ali, I., Peng, C., Khan, Z.M., Naz, I., Sultan, M., Ali, M., Abbasi, I.A., Islam, T., Ye, T., 2019a. Overview of microbes based fabricated biogenic nanoparticles for water and wastewater treatment. J. Environ. Manage. 230, 128–150. https://doi.org/10.1016/j.jenvman.2018.09.073.

Ali, I., Peng, C., Lin, D., Saroj, D.P., Naz, I., Khan, Z.M., Sultan, M., Ali, M., 2019b. Encapsulated green magnetic nanoparticles for the removal of toxic Pb2+ and Cd2+ from water: development, characterization and application. J. Environ. Manage. 234, 273–289. https://doi.org/10.1016/j.jenvman.2018.12.112.

Ali, I., Peng, C., Naz, I., Khan, Z.M., Sultan, M., Islam, T., Abbasi, I.A., 2017. Phytogenic magnetic nanoparticles for wastewater treatment: a review. RSC Adv., 7 (64), 40158–40178. https://doi.org/10.1039/c7ra04738j.

Ali, I., Peng, C., Naz, I., Lin, D., Saroj, D.P., Ali, M., 2019c. Development and application of novel bio-magnetic membrane capsules for the removal of the cationic dye malachite green in wastewater treatment. RSC Adv. 9 (7), 3625–3646. https://doi.org/10.1039/c8ra09275c.

Alijani, H., Shariatinia, Z., 2018. Synthesis of high growth rate SWCNTs and their magnetite cobalt sulfide nanohybrid as super-adsorbent for mercury removal. Chem. Eng. Res. Des. 129, 132–149. https://doi.org/10.1016/j.cherd.2017.11.014.

Alloway, B.J., 2012. Heavy Metals in Soils: Trace Metals and Metalloids in Soils and Their Bio- availability, Vol. 22. Elsevier, Amsterdam.

An, B., Zhao, D., 2012. Immobilization of As(III) in soil and groundwater using a new class of polysaccharide stabilized Fe-Mn oxide nanoparticles. J. Hazard. Mater. 211–212, 332–341. https://doi.org/10.1016/j.jhazmat.2011.10.062.

Amade, R., Hussain, S., Ocana, I.R., Bertran, E., 2014. Growth and functionalization of carbon nanotubes on quartz filter for environmental applications. J. Environ. Engg. Ecol. Sci. 3 (2), 1–7. https://doi.org/10.7243/2050-1323-3-2.

An, X., Yu, J.C., 2011. Graphene-based photocatalytic composites. RSC Adv. 1 (8), 1426–1434. https://doi.org/10.1039/c1ra00382h.

Anjum, M., Miandad, R., Waqas, M., Gehany, F., Barakat, M.A., 2016. Remediation of wastewater using various nano-materials. Arab. J. Chem. 12 (8), 30175–30177. doi:10.1016/j.arabjc.2016.10.004.

Azama, M.A., Aliasa, F.M., Tacka, L.W., Amalina, R.N., Mohamad, R.S, Taibb, F.M., 2017. Electronic properties and gas adsorption behaviour of pristine, silicon and boron-doped (8, 0) single-walled carbon nanotube: A first principles study. J. Mol. Graphics Modell. 75, 85–93. https://doi.org/10.1016/j.jmgm.2017.05.003.

Baby, R., Saifullah, B., Hussein, M.Z., 2019. Carbon Nanomaterials for the treatment of heavy metal-contaminated water and environmental remediation. Nanoscale Res. Lett. 14 (1), 341. https://doi.org/10.1186/s11671-019-3167-8.

Baruah, S., Dutta, J., 2009. Nanotechnology applications in pollution sensing and degradation in agriculture. Environ. Chem. Lett. 7 (3), 191–204. https://doi.org/10.1007/s10311-009-0228-8.

Bergmann, C.P., Machado, F., 2015. Carbon Nanostructures (Paulo Araujo, Tuscaloosa, AL, USA), Library of Congress, Springer Cham Heidelberg, New York, Dordrecht, London. In: Bergmann, C.P., Machado, F. (Eds.), Carbon Nanomaterials as Adsorbents for Environmental and Biological Applications, pp. 1–122.

Botes, M., Cloete, T.E., 2010. The Potential of Nanofibers and Nanobiocides in Water Purification. Nanotechnology in Water Treatment Applications, In: Cloete, T.E., de Kwaadsteniet, M., Botes, M., López-Romero, J.M. (Eds.), Nanotechnology in Water Treatment Applications, 13th ed, 196. Horizon Scientific Press, New York, USA, p. 9781904455660.

Cai, Y., Shen, Y., Xie, A., Li, S., Wang, X., 2010. Green synthesis of soya bean sprouts-mediated superparamagnetic Fe3O4 nanoparticles. J. Magn. Magn. Mater. 322, 2938–2943. https://doi.org/10.1016/j.jmmm.2010.05.009.

Cai, C., Zhao, M., Yu, Z., Rong, H., Zhang, C., 2019. Utilization of nanomaterials for in-situ remediation of heavy metal(loid) contaminated sediments: a review. Sci. Total Environ. 662, 205–217. https://doi.org/10.1016/j.scitotenv.2019.01.180.

Chen, J., Fan, T., Xie, Z., Zeng, Q., Xue, P., Zheng, T., Chen, Y., Luo, X., Zhang, H., 2020. Advances in nanomaterials for photodynamic therapy applications: status and challenges. Biomaterials 237, 119827. https://doi.org/10.1016/j.biomaterials.2020.119827.

Cesarino, I., Moraes, F.C., Lanza, M.R., Machado, S.A., 2012. Electrochemical detection of carbamate pesticides in fruit and vegetables with a biosensor based on acetylcholinesterase immobilised on a composite of polyaniline–carbon nanotubes. Food Chemistry 135 (3), 873–879. https://doi.org/10.1016/j.foodchem.2012.04.147.

Cheera, P., Karlapudi, S., Sellola, G., Ponneri, V., 2016. A facile green synthesis of spherical Fe_3O_4 magnetic nanoparticles and their effect on degradation of methylene blue in aqueous solution. J. Mol. Liq. 221, 993–998. https://doi.org/10.1016/j.molliq.2016.06.006.

Chen, X., Cen, C., Tang, Z., Zeng, W., Chen, D., Fang, P., Chen, Z., 2013. The key role of pH value in the synthesis of titanate nanotubes-loaded manganese oxides as a superior catalyst for the selective catalytic reduction of NO with NH_3. J. Nanomater. 2013, 871528.

Chen, C., Wang, X., Chang, Y., Liu, H., 2008. Dechlorination of disinfection by-product monochloroacetic acid in drinking water by nanoscale palladized iron bimetallic particle. J. Environ. Sci. 20, 945–951. doi:10.1016/s1001-0742(08)62191-9.

Chen, X., Yao, X., Yu, C., Su, X., Shen, C., Chen, C., 2014. Hydrodechlorination of polychlorinated biphenyls in contaminated soil from an e-waste recycling area, using nanoscale zerovalent iron and Pd/Fe bimetallic nanoparticles. Environ. Sci. Pollut. Res. 21 (7), 5201–5210. doi:10.1007/s11356-013-2089-8.

Cheng, H.H., Narindri, B., Chu, H., Whang, L.M., 2020. Recent advancement on biological technologies and strategies for resource recovery from swine wastewater. Bioresour. Technol. 303. https://doi.org/10.1016/j.biortech.2020.122861.

Cho, M., Chung, H., Choi, W., Yoon, J., 2005. Different inactivation behaviors of MS-2 phage and *Escherichia coli* in TiO2 photocatalytic disinfection. Appl. Environ. Microbiol. 71, 270–275. https://doi.org/10.1128/AEM.71.1.270-275.2005.

Chou, K.S., Lu, Y.C., Lee, H.H., 2005. Effect of alkaline ion on the mechanism and kinetics of chemical reduction of silver. Mater. Chem. Phys. 94 (2–3), 429–433. https://doi.org/10.1016/j.matchemphys.2005.05.029.

Cotica, L.F., Garcia, A., Polli, A.D., Bini, R.D., de Chaves, T., de Oliveira Junior, V.A., Pamphile, J.A..

Cushing, B.L., Kolesnichenko, V.L., O'Connor, C.J, 2004. Recent advances in the liquid-phase syntheses of inorganic nanoparticles. Chem. Rev. 104 (9), 3893–3946. https://doi.org/10.1021/cr030027b.

D., L.J., W., O.N, 2014. Role of litter turnover in soil quality in tropical degraded lands of Colombia. Sci. World J. 1–11. https://doi.org/10.1155/2014/693981.

Das, S., Sen, B., & Debnath, N. (2015). Recent trends in nanomaterials applications in environmental monitoring and remediation. Environ. Sci. Pollut. Res., 22(23), 18333–18344. https://doi.org/10.1007/s11356-015-5491-6

Dehghani, M.H., Karri, R.R., Alimohammadi, M., Nazmara, S., Zarei, A., Saeedi, Z., 2020. Insights into endocrine-disrupting Bisphenol-A adsorption from pharmaceutical effluent by chitosan immobilized nanoscale zero-valent iron nanoparticles. J. Mol. Liquids 311,, 113317. https://doi.org/10.1016/j.molliq.2020.113317.

Diallo, M.S., Christie, S., Swaminathan, P., Johnson, J.H., Goddard, W.A., 2005. Dendrimer enhanced ultrafiltration. 1. Recovery of Cu(II) from aqueous solutions using PAMAM dendrimers with ethylene diamine core and terminal NH_2 groups. Environ.Sci. Technol. 39 (5), 1366–1377. https://doi.org/10.1021/es048961r.

Ding, S., Zhao, L., Qi, Y., Lv, Q.Q., 2014. Preparation and characterization of lecithin-nano Ni/Fe for effective removal of PCB77. J. Nanomater. 2014. https://doi.org/10.1155/2014/678489.

Dong, J., Fan, X., Qiao, F., Ai, S., Xin, H., 2013. A novel protocol for ultra-trace detection of pesticides: Combined electrochemical reduction of Ellman's reagent with acetylcholinesterase inhibition. Analytica Chimica Acta 761, 78–83 doi:10.1016/j.aca.2012.11.042.

Dong, H., Lin, B., Gilmore, K., Hou, T., Lee, S.T., Li, Y., 2015. B40 fullerene: An efficient material for CO2 capture, storage and separation. Curr. Appl. Phys. 1084–1089. https://doi.org/10.1016/j.cap.2015.06.008.

El-Temsah, Y.S., Sevcu, A., Bobcikova, K., Cernik, M., Joner, E.J., 2016. DDT degradation efficiency and ecotoxicological effects of two types of nano-sized zero-valent iron (nZVI) in water and soil Chemosphere 144, 2221–2228. https://doi.org/10.1016/j.chemosphere.2015.10.122.

Esrafili, M.D., 2017. N2O reduction over a fullerene-like boron nitride nanocage: A DFT study. Phys. Lett. A 381 (25–26), 2085–2091. https://doi.org/10.1016/j.physleta.2017.04.009.

F., M.E, 2017. Nanotechnology: future of environmental air pollution control.. In: Environ. Manage. Sustain. Dev., 6. pp. 429–454. https://doi.org/10.5296/emsd.v6i2.12047.

Fujishima, A., Zhang, X., Tryk, D.A., 2008. TiO_2 photocatalysis and related surface phenomena. Surf. Sci. Rep. 63 (12), 515–582. https://doi.org/10.1016/j.surfrep.2008.10.001.

Gautam, P.K., Singh, A., Misra, K., Sahoo, A.K., Samanta, S.K., 2019. Synthesis and applications of biogenic nanomaterials in drinking and wastewater treatment. J. Environ. Manage. 231, 734–748. https://doi.org/10.1016/j.jenvman.2018.10.104.

Ghormade, V., Deshpande, M.V., Paknikar, K.M., 2011. Perspectives for nano-biotechnology enabled protection and nutrition of plants. Biotechnol. Adv. 29 (6), 792–803. https://doi.org/10.1016/j.biotechadv.2011.06.007.

Ghrair, A.M., Ingwersen, J., Streck, T., 2010. Immobilization of heavy metals in soils amended by nanoparticulate zeolitic tuff: sorption-desorption of cadmium. J. Plant Nutr. Soil Sci. 173 (6), 852–860. https://doi.org/10.1002/jpln.200900053.

Gil-Dıaz, M., Pinilla, P., Alonso, J., Lobo, M.C., 2016. Viability of a nanoremediation process in single or multi-metal (loid) contaminated soils. J. Hazard. Mater. 321, 812–819.

Gokhale, M. M., & Somani, R. R. (2015). Fullerenes: chemistry and its applications. Mini-Rev. Org. Chem., 12(4), 355–366. https://doi.org/10.2174/1570193x12666150930224428

Gu, J., Dong, D., Kong, L., Zheng, Y., Li, X., 2012. Photocatalytic degradation of phenanthrene on soil surfaces in the presence of nanometer anatase TiO2 under UV-light. J. Environ. Sci. 24 (12), 2122–2126. doi:10.1016/s1001-0742(11)61063-2.

Guerra, F.D., Attia, M.F., Whitehead, D.C., Alexis, F., 2018. Nanotechnology for environmental remediation: materials and applications. Molecules 23 (7). https://doi.org/10.3390/molecules23071760.

Guerra, F.D., Campbell, M.L., Whitehead, D.C., Alexis, F., 2017. Tunable properties of functional nanoparticles for efficient capture of VOCs. ChemistrySelect 2 (31), 9889–9894. https://doi.org/10.1002/slct.201701736.

Guo, J., Zhang, Q., Cai, Z., Zhao, K., 2016. Preparation and dye filtration property of electrospun polyhydroxybutyrate–calcium alginate/carbon nanotubes composite nanofibrous filtration membrane. Sep. Purif. Technol. 161, 69–79. https://doi.org/10.1016/j.seppur.2016.01.036.

Gupta, V.K., Nayak, A., 2012. Cadmium removal and recovery from aqueous solutions by novel adsorbents prepared from orange peel and Fe_2O_3 nanoparticles. Chem. Eng. J. 180, 81–90. https://doi.org/10.1016/j.cej.2011.11.006.

Gupta, V.K., Tyagi, I., Sadegh, H., Shahryari-Ghoshekand, R., Makhlouf, A.S.H., Maazinejad, B., 2015. Nanoparticles as adsorbent; a positive approach for removal of noxious metal ions: a review. Sci. Technol. Dev. 34 (3), 195–214.

Han, X., Li, S., Peng, Z., Al-Yuobi, A.O., Bashammakh, A.S.O., El-Shahawi, M.S., Leblanc, R.M., 2016. Interactions between carbon nanomaterials and biomolecules. J. Oleo Sci. 65 (1), 1–7. https://doi.org/10.5650/jos.ess15248.

Hazra, C., Kundu, D., Chaudhari, A., Jana, T., 2013. Biogenic synthesis, characterization, toxicity and photocatalysis of zinc sulfide nanoparticles using rhamnolipids from Pseudomonas aeruginosa BS01 as capping and stabilizing agent. J. Chem. Technol. Biotechnol. 88 (6), 1039–1048. https://doi.org/10.1002/jctb.3934.

Hoag, G.E., Collins, J.B., Holcomb, J.L., Hoag, J.R., Nadagouda, M.N., Varma, R.S., 2009. Degradation of bromothymol blue by 'greener' nano-scale zero-valent iron synthesized using tea polyphenols. Mater. Chem. 19, 8671–8677. doi:10.1039/B909148C.

Hooshyar, Z., Rezanejade, B.G., Ghayeb, Y., 2013. Sonication enhanced removal of nickel and cobalt ions from polluted water using an iron based sorbent. J. Chem. 786954, 1–5. https://doi.org/10.1155/2013/786954.

Hsu, S., Lu, C., 2007. Modification of single-walled carbon nanotubes for enhancing isopropyl alcohol vapor adsorption from water streams. Sep. Sci. Technol. 42, 2751–2766. https://doi.org/10.1080/01496390701515060.

Huang, H.Y., Yang, R.T., Chinn, D., Munson, C.L., 2003. Amine-grafted MCM-48 and silica xerogel as superior sorbents for acidic gas removal from natural gas. Ind. Eng. Chem. Res. 42 (12), 2427–2433. https://doi.org/10.1021/ie020440u.

Islam, T., Ali, I., Naz, I., Peng, C., Zahid, M.K., Amjad, M.A., 2020.). Aquananotechnology: Applications of Nanomaterials for Water Purification.

Islam, T., Liu, J., Shen, G., Ye, T., Peng, C., 2018. Synthesis of chemically modified carbon embedded silica and zeolite from rice husk to adsorb crystal violet dye from aqueous solution. Appl. Ecol. Environ. Res. 16 (4), 3955–3967. https://doi.org/10.15666/aeer/1604_39553967.

Islam, T., Peng, C., Ali, I., Abbasi, I.A., 2018. Comparative study on anionic and cationic dyes removal from aqueous solutions using different plant mediated magnetic nano particles. Indian J. Geo-Marine Sci. 47 (3), 598–603. http://nopr.niscair.res.in/bitstream/123456789/44124/1/IJMS%2047(3)%20598-603.pdf.

Islam, T., Peng, C., Ali, I., Li, J., Khan, Z.M., Sultan, M., Naz, I., 2020. Synthesis of rice husk-derived magnetic biochar through liquefaction to adsorb anionic and cationic dyes from aqueous solutions. Arab. J. Sci. Eng 46, 233–246. https://doi.org/10.1007/s13369-020-04537-z.

Islam, T., Ye, T., Peng, C., 2018. Adsorption of eriochrome black T dye using alginate crosslinked *Prunus avium* leaf mediated nanoparticles: characterization, optimization and equilibrium studies. Desalin. Water Treat. 131, 305–316. https://doi.org/10.5004/dwt.2018.22944.

Iwahori, K., Watanabe, J.I., Tani, Y., Seyama, H., Miyata, N., 2014. Removal of heavy metal cations by biogenic magnetite nanoparticles produced in Fe (III)-reducingm microbial enrichment cultures. J. Biosci. Bioeng. 117 (3), 333–335. doi:10.1016/j.jbiosc.2013.08.013.

Jie, G., Kongyin, Z., Xinxin, Z., Zhijiang, C., Min, C., Tian, C., Junfu, W., 2015. Preparation and characterization of carboxyl multi-walled carbon nanotubes/calcium alginate composite hydrogel nano-filtration membrane. Mater. Lett. 157, 112–115. https://doi.org/10.1016/j.matlet.2015.05.080.

Jun, L.Y., Karri, R.R., Mubarak, N.M., Yon, L.S., Bing, C.H., Khalid, M., Jagadish, P., Abdullah, E.C., 2020. Modelling of methylene blue adsorption using peroxidase immobilized functionalized Buckypaper/polyvinyl alcohol membrane via ant colony optimization. Environ. Pollut. 259, 113940. https://doi.org/10.1016/j.envpol.2020.113940.

Jun, L.Y., Karri, R.R., Yon, L.S., Mubarak, N.M., Bing, C.H., Mohammad, K., Jagadish, P., Abdullah, E.C., 2020. Modeling and optimization by particle swarm embedded neural network for adsorption of methylene blue by jicama peroxidase immobilized on buckypaper/polyvinyl alcohol membrane. Environ. Res. 183,, 109158. https://doi.org/10.1016/j.envres.2020.109158.

Jyothi, N., Vaclavikova, M., Vaseashta, Gallios, G., Roy, P., Pummakarnchana, O., Venkateswarlu, S., Kumar, B., N., Prathima, B., Anitha, K, 2007. A novel green synthesis of Fe_3O_4-Ag core shell recyclable nanoparticles using *Vitis vinifera* stem extract and its enhanced antibacterial performance. Sci. Technol. Adv. Mater 8, 30–35.

Kalfa, O.M., Yalçinkaya, O., Türker, A.R., 2009. Synthesis of nano B_2O_3/TiO_2 composite material as a new solid phase extractor and its application to preconcentration and separation of cadmium. J. Hazard. Mater. 166 (1), 455–461. https://doi.org/10.1016/j.jhazmat.2008.11.112.

Karri, R.R., Sahu, J.N., 2018. Process optimization and adsorption modeling using activated carbon derived from palm oil kernel shell for Zn (II) disposal from the aqueous environment using differential evolution embedded neural network. J. Mol. Liq. 265, 592–602. https://doi.org/10.1016/j.molliq.2018.06.040.

Karri, R.R., Sahu, J.N., Meikap, B.C., 2020. Improving efficacy of Cr (VI) adsorption process on sustainable adsorbent derived from waste biomass (sugarcane bagasse) with help of ant colony optimization. Ind. Crop. Prod. 143,, 111927. https://doi.org/10.1016/j.indcrop.2019.111927.

Karri, R.R., Shams, S., Sahu, J.NAhsan, A., Ismail, A.F. (Eds.), 2019. Overview of Potential Applications of Nano-Biotechnology in Wastewater and Effluent Treatment. Nanotechnology in Water and Wastewater Treatment 87–100.

Khan, Fahad Saleem Ahmed, Mubarak, N.M., Khalid, M., Walvekar, R., Abdullah, E.C., Mazari, S.A., Nizamuddin, S., Karri, R.R, 2020. Magnetic nanoadsorbents' potential route for heavy metals removal—a review. Environ. Sci. Pollut. Res. 27 (19), 24342–24356. https://doi.org/10.1007/s11356-020-08711-6.

Khan, Fahad Saleem Ahmed, Mubarak, N.M., Tan, Y.H., Karri, R.R., Khalid, M., Walvekar, R., Abdullah, E.C., Mazari, S.A., Nizamuddin, S, 2020. Magnetic nanoparticles incorporation into different substrates for dyes and heavy metals removal—a review. Environ. Sci. Pollut. Res. 27 (35), 43526–43541. https://doi.org/10.1007/s11356-020-10482-z.

Khan, F.S.A., Mubarak, N.M., Khalid, M., Walvekar, R., Abdullah, E.C., Mazari, S.A., Nizamuddin, S., Karri, R.R., 2020. Magnetic nanoadsorbents' potential route for heavy metals removal—a review. Environ. Sci. Pollut. Res. 27 (19), 24342–24356. https://doi.org/10.1007/s11356-020-08711-6.

Kharisov, B.I., Rasika Dias, H.V., Kharissova, O.V., Manuel Jiménez-Pérez, V., Olvera Pérez, B., Muñoz Flores, B., 2012. Iron-containing nanomaterials: synthesis, properties, and environmental applications. RSC Adv. 2 (25), 9325–9358. https://doi.org/10.1039/c2ra20812a.

Khin, M.M., Nair, A.S., Babu, V.J., Murugan, R., Ramakrishna, S., 2012. A review on nanomaterials for environmental remediation. Energy Environ. Sci. 5 (8), 8075–8109. https://doi.org/10.1039/c2ee21818f.

Klaus-Joerger, T., Joerger, R., Olsson, E., Granqvist, C.G, 2001. Bacteria as workers in the living factory: metal-accumulating bacteria and their potential for materials science. Trends Biotechnol. 19 (1), 15–20. https://doi.org/10.1016/S0167-7799(00)01514-6.

Koduru, J.R., Karri, R.R., Mubarak, N.M., 2019. Smart materials, magnetic graphene oxide-based nanocomposites for sustainable water purification. In: Inamuddin, M., Thomas, S., Mishra, R.K., Asiri, A.M. (Eds.), Sustainable Polymer Composites and Nanocomposites. Springer International Publishing, New York, pp. 759–781. https://doi.org/10.1007/978-3-030-05399-4_26.

Lee, K., Mazare, A., Schmuki, P., 2014. One-dimensional titanium dioxide nanomaterials: nanotubes. Chem. Rev. 114 (19), 9385–9454. https://doi.org/10.1021/cr500061m.

Lingamdinne, L.P., Chang, Y.Y., Yang, J.K., Singh, J., Choi, E.H., Shiratani, M., Attri, P., 2017. Biogenic reductive preparation of magnetic inverse spinel iron oxide nanoparticles for the adsorption removal of heavy metals. Chem. Eng. J. 307, 74–84. https://doi.org/10.1016/j.cej.2016.08.067.

Lingamdinne, L.P., Koduru, J.R., Karri, R.R., 2019. A comprehensive review of applications of magnetic graphene oxide based nanocomposites for sustainable water purification. J. Environ. Manage. 231, 622–634. https://doi.org/10.1016/j.jenvman.2018.10.063.

Lingamdinne, L.P., Yang, J.K., Chang, Y.Y., Koduru, J.R., 2016. Studies on removal of Pb (II) and Cr (III) using graphene oxide based inverse spinel nickel ferrite nano-composite as sorbent. Hydrometallurgy 165, 81–89. https://doi.org/10.1016/j.hydromet.2015.11.005.

Lithoxoos, G.P., Labropoulos, A., Peristeras, L.D., Kanellopoulos, N., Samios, J., Economou, I.G., 2010. Adsorption of N_2, CH_4, CO and CO_2 gases in single walled carbon nanotubes: a combined experimental and Monte Carlo molecular simulation study. J. Supercrit. Fluids 55 (2), 510–523. https://doi.org/10.1016/j.supflu.2010.09.017.

Liu, T., Li, B., Hao, Y., Yao, Z., 2014. MoO_3-nanowire membrane and $Bi_2Mo_3O_{12}/MoO_3$ nano-heterostructural photocatalyst for wastewater treatment. Chem. Eng. J. 244, 382–390. https://doi.org/10.1016/j.cej.2014.01.070.

Liu, T., Su, H., Qu, X., Ju, P., Cui, L., Ai, S., 2011. Acetylcholinesterase biosensor based on 3- carboxyphenylboronic acid/reduced graphene oxide–gold nanocomposites modified electrode for amperometric detection of organophosphorus and carbamate pesticides. Sens. Actuat. B: Chem. 160 (1), 1255–1261. doi:10.1016/J.SNB.2011.09.059.

Liu, R., Zhao, D., 2007. In situ immobilization of Cu (II) in soils using a new class of iron phosphate nanoparticles. Chemosphere 68 (10), 1867–1876. doi:10.1016/j.chemosphere.2007.03.010.

Ma, H., Wang, H., Na, C., 2015. Microwave-assisted optimization of platinum-nickel nanoalloys for catalytic water treatment. Appl. Catal. B Environ. 163, 198–204. https://doi.org/10.1016/j.apcatb.2014.07.062.

Martins, M., Mourato, C., Sanches, S., Noronha, J.P., Crespo, M.B., Pereira, I.A., 2017. Biogenic platinum and palladium nanoparticles as new catalysts for the removal of pharmaceutical compounds. Water Res. 108, 160–168. https://doi.org/10.1016/j.watres.2016.10.071.

Meerburg, F., Hennebel, T., Vanhaecke, L., Verstraete, W., Boon, N., 2012. Diclofenac and 2-anilinophenylacetate degradation by combined activity of biogenic manganese oxides and silver. Microb. Biotechnol. 5 (3), 388–395. doi:10.1111/j.1751-7915.2011.00323.x.

Mittal, A.K., Chisti, Y., Banerjee, U.C., 2013. Synthesis of metallic nanoparticles using plant extracts. Biotechnol. Adv. 31 (2), 346–356. https://doi.org/10.1016/j.biotechadv.2013.01.003.

Moharem, M., Elkhatib, E., Mesalem, M., 2019. Remediation of chromium and mercury polluted calcareous soils using nanoparticles: Sorption–desorption kinetics, speciation and fractionation. Environmental Research 170, 366–373. https://doi.org/10.1016/j.envres.2018.12.054.

Moriwaki, H., Fujii, S., Oshima, M., 2020. Simple method for removal of pollutants from water using freeze-thaw treatment. Sep. Purif. Technol. 237, 116382. https://doi.org/10.1016/j.seppur.2019.116382.

Muralikrishnan, R., Swarnalakshmi, M., Nakkeeran, E., 2014. Nanoparticle-membrane filtration of vehicular exhaust to reduce air pollution—a review. Int. J. Environ. Sci. 3 (4), 82–86.

Nomura, A., Jones, C.W., 2013. Amine-functionalized porous silicas as adsorbents for aldehyde abatement. ACS Appl. Mater. Interf. 5 (12), 5569–5577. https://doi.org/10.1021/am400810s.

Özkar, S., 2009. Enhancement of catalytic activity by increasing surface area in heterogeneous catalysis. Appl. Surf. Sci. 256 (5), 1272–1277. https://doi.org/10.1016/j.apsusc.2009.10.036.

P., S.R., 2011. Prospects of nanobiomaterials for biosensing. Int. J. Electrochem. 2011, 125487. https://doi.org/10.4061/2011/125487.

Paknikar, K.M., Nagpal, V., Pethkar, A.V., Rajwade, J.M., 2005. Degradation of lindane from aqueous solutions using iron sulfide nanoparticles stabilized by biopolymers. Sci. Technol. Adv. Mat. 6 (3–4), 370–374. https://doi.org/10.1016/j.stam.2005.02.016.

Pal, S., Tak, Y.K., Song, J.M., 2007. Does the antibacterial activity of silver nanoparticles depend on the shape of the nanoparticle? A study of the gram-negative bacterium *Escherichia coli*. Appl. Environ. Microbiol. 73 (6), 1712–1720. https://doi.org/10.1128/AEM.02218-06.

Park, J.Y., Lee, I.H., 2014. Photocatalytic degradation of 2-chlorophenol using Ag-doped TiO2 nanofibers and a near-UV light-emitting diode system J. Nanomater. 250803, 1–6. https://doi.org/10.1155/2014/250803.

Parsons, J.G., Peralta-Videa, J.R., Gardea-Torresdey, J.L., 2007. Use of plants in biotechnology: synthesis of metal nanoparticles by inactivated plant tissues, plant extracts, and living plants. In: Sarkar, D., Datta, R., Hannigan, R. (Eds.). In: Developments in Environmental Science, 5. Elsevier, Amsterdam, pp. 463–485. https://doi.org/10.1016/S1474-8177(07)05021-8.

Pei, Y., Chen, X., Xiong, D., Liao, S., Wang, G., 2013. Removal and recovery of toxic silver ion using deep-sea bacterial generated biogenic manganese oxides. PLoS One 8 (12), e81627. https://doi.org/10.1371/journal.pone.0081627.

Peikam, E.N., Jalali, M., 2019. Application of three nanoparticles (Al2O3, SiO2 and TiO2) for metal-contaminated soil remediation (measuring and modeling). Intl. J. Enviro. Sci. Tech. 16 (16), 1–14. doi:10.1007/s13762-018-2134-8.

Pendergast, M.M., Hoek, E.M.V., 2011. A review of water treatment membrane nanotechnologies. Energy Environ. Sci. 4 (6), 1946–1971. https://doi.org/10.1039/c0ee00541j.

Phumying, S., Labuayai, S., Thomas, C., Amornkitbamrung, V., Swatsitang, E., Maensiri, S., 2013. Aloe vera plant-extracted solution hydrothermal synthesis and magnetic properties of magnetite (Fe3O4) nanoparticles. Appl. Phys. A 111, 1187–1193. https://doi.org/10.1007/s00339-012-7340-5.

Prakash, N.T., Sharma, N., Prakash, R., Raina, K.K., Fellowes, J., Pearce, C.I., Lloyd, J.R., Pattrick, R.A.D., 2009. Aerobic microbial manufacture of nanoscale selenium: exploiting nature's bio-nanomineralization potential. Biotechnol. Lett. 31 (12), 1857–1862. https://doi.org/10.1007/s10529-009-0096-0.

Prasad, R., Pandey, R., & Barman, I. (2016). Engineering tailored nanoparticles with microbes: quo vadis? WIREs Nanomed. Nanobiotechnol., 8(2), 316–330. https://doi.org/10.1002/wnan.1363

Prasad, C., Yuvaraja, G., Venkateswarlu, P., 2017. Biogenic synthesis of Fe3O4 magnetic nanoparticles using Pisum sativum peels extract and its effect on magnetic and Methyl orange dye degradation studies. J. Magn. Magn. Mater. 424, 376–381. doi:10.1016/J.JMMM.2016.10.084.

Qi, G., Wang, Y., Estevez, L., Duan, X., Anako, N., Park, A.H.A., Li, W., Jones, C.W., Giannelis, E.P., 2011. High efficiency nanocomposite sorbents for CO_2 capture based on amine-functionalized mesoporous capsules. Energy Environ. Sci. 4 (2), 444–452. https://doi.org/10.1039/c0ee00213e.

Qu, Y., Pei, X., Shen, W., Zhang, X., Wang, J., Zhang, Z., Li, S., You, S., Ma, F., Zhou, J., 2017. Biosynthesis of gold nanoparticles by Aspergillum sp. WL-Au for degradation of aromatic pollutants. Phys. E Low-dimens. Syst. Nanostruct. 88, 133–141. https://doi.org/10.1016/j.physe.2017.01.010.

Rajeshkumar, S., Bharath, L.V., 2017. Mechanism of plant-mediated synthesis of silver nanoparticles—a review on biomolecules involved, characterisation and antibacterial activity. Chem-Biol. Interact. 273, 219–227. https://doi.org/10.1016/j.cbi.2017.06.019.

Ramanathan, R., Field, M.R., O'Mullane, A.P., Smooker, P.M., Bhargava, S.K., Bansal, V., 2013. Aqueous phase synthesis of copper nanoparticles: a link between heavy metal resistance and nanoparticle synthesis ability in bacterial systems. Nanoscale 5 (6), 2300–2306. https://doi.org/10.1039/c2nr32887a.

Rasalingam, S., Peng, R., Koodali, R.T., 2014. Removal of hazardous pollutants from wastewaters: Applications of TiO2-SiO2 mixed oxide materials. J. Nanomater. 617405, 1–42. https://doi.org/10.1155/2014/617405.

Rattan, L., 2015. Restoring soil quality to mitigate soil degradation. Sustainability 5875–5895. https://doi.org/10.3390/su7055875.

Ren, X., Chen, C., Nagatsu, M., Wang, X., 2011. Carbon nanotubes as adsorbents in environmental pollution management: a review. Chem. Eng. J. 170 (2–3), 395–410. https://doi.org/10.1016/j.cej.2010.08.045.

Rui, Z., Jingguo, W., Jianyu, C., Lin, H., Kangguo, M., 2010. Photocatalytic degradation of pesticide residues with RE^{3+}-doped nano-TiO_2. J. Rare Earths 28, 353–356. doi:10.1016/S1002-0721(10)60329-8.

Sahu, J.N., Zabed, H., Karri, R.R., Shams, S., Qi, X., 2019. Applications of nano-biotechnology for sustainable water purification. In: Thomas, S., Grohens, Y., Pottathara, Y.B. (Eds.), Industrial Applications of Nanomaterials. Elsevier, Amsterdam, pp. 313–340. https://doi.org/10.1016/B978-0-12-815749-7.00011-6.

Sahu, N., Rawat, S., Singh, J., Karri, R.R., Lee, S., Choi, J.S., Koduru, J.R., 2019. Process optimization and modeling of methylene blue adsorption using zero-valent iron nanoparticles synthesized from sweet lime pulp. Appl. Sci. (Switzerland) 9 (23), 5112. https://doi.org/10.3390/app9235112.

Santhosh, C., Velmurugan, V., Jacob, G., Jeong, S.K., Grace, A.N., Bhatnagar, A., 2016. Role of nanomaterials in water treatment applications: a review. Chem. Eng. J. 306, 1116–1137. https://doi.org/10.1016/j.cej.2016.08.053.

Seredych, M., Bandosz, T.J., 2007. Removal of ammonia by graphite oxide via its intercalation and reactive adsorption. Carbon 45 (10), 2130–2132. https://doi.org/10.1016/j.carbon.2007.06.007.

Seredych, M., Bandosz, T.J., 2012. Manganese oxide and graphite oxide/MnO2 composites as reactive adsorbents of ammonia at ambient conditions. Micropor. Mesopor. Mat. 150, 55–63. https://doi.org/10.1016/j.micromeso.2011.09.010.

Sharifi, S., Vahed, S.Z., Ahmadian, E., Dizaj, S.M., Eftekhari, A., Khalilov, R., Ahmadi, M., Hamidi-Asl, E., Labib, M., 2019. Detection of pathogenic bacteria via nanomaterials-modified aptasensors. Biosens. Bioelectron., 150, 111933.

Shen, W., Qu, Y., Pei, X., Zhang, X., Ma, Q., Zhang, Z., Li, S., Zhou, J., 2016. Green synthesis of gold nanoparticles by a newly isolated strain Trichosporon montevideense for catalytic hydrogenation of nitroaromatics. Biotechnol. Lett. 38 (9), 1503–1508. doi:10.1007/s10529-016-2120-5.

Shen, W., Zhang, C., Li, Q., Zhang, W., Cao, L., Ye, J., 2015. Preparation of titanium dioxide nano particle modified photocatalytic self-cleaning concrete. J. Clean. Prod. 87 (1), 762–765. https://doi.org/10.1016/j.jclepro.2014.09.014.

Simonian, A.L., Good, T.A., Wang, S.S., Wild, J.R., 2005. Nanoparticle-based optical biosensors for the direct detection of organophosphate chemical warfare agents and pesticides. Analytica Chimica Acta 534 (1), 69–77. doi:10.1016/j.aca.2004.06.056.

Simonovic, S.P., Fahmy, H., 1999. A new modeling approach for water resources policy analysis. Water Resour. Res. 35 (1), 295–304. https://doi.org/10.1029/1998WR900023.

Siyal, M.I., Lee, C.K., Park, C., Khan, A.A., Kim, J.O., 2019. A review of membrane development in membrane distillation for emulsified industrial or shale gas wastewater treatments with feed containing hybrid impurities. J. Environ. Manage. 243, 45–66. https://doi.org/10.1016/j.jenvman.2019.04.105.

Son, W.J., Choi, J.S., Ahn, W.S., 2008. Adsorptive removal of carbon dioxide using polyethyleneimine-loaded mesoporous silica materials. Micropor. Mesopor. Mater. 113 (1–3), 31–40. https://doi.org/10.1016/j.micromeso.2007.10.049.

Su, S., He, Y., Zhang, M., Yang, K., Song, S., Zhang, X., 2008. High-sensitivity pesticide detection via silicon nanowires-supported acetylcholinesterase-based electrochemical sensors. Appl. Phys. Lett. 93 (2), 023113. doi:10.1063/1.2959827.

Su, F., Lu, C., Chen, W., Bai, H., Hwang, J.F., 2009. Capture of CO2 from flue gas via multiwalled carbon nanotubes. Sci. Tot. Environ. 407 (8), 3017–3023. https://doi.org/10.1016/j.scitotenv.2009.01.007.

Su, C., Puls, R.W., Krug, T.A., Watling, M.T., O'Hara, S.K., Quinn, J.W., Ruiz, N.E., 2012. A two and half-year-performance evaluation of a field test on treatment of source zone tetrachloroethene and its chlorinated daughter products using emulsified zero valent iron nanoparticles. Water Res. 46 (16), 5071–5084. https://doi.org/10.1016/j.watres.2012.06.051.

Srivastava, S., Agrawal, S.B., Mondal, M.K., 2017. Synthesis, characterization and application of Lagerstroemia speciosa embedded magnetic nanoparticle for Cr (VI) adsorption from aqueous solution. Environ. Sci. 55, 283–293. doi:10.1016/j.jes.2016.08.012.

Su, J., Deng, L., Huang, L., Guo, S., Liu, F., He, J., 2014. Catalytic oxidation of manganese II) by multicopper oxidase CueO and characterization of the biogenic Mn oxide. Water Res. 56, 304–313. doi:0.1016/j.watres.2014.03.013.

Su, H., Fang, Z., Tsang, P.E., Zheng, L., Cheng, W., Fang, J., Zhao, D., 2016. Remediation of hexavalent chromium contaminated soil by biochar-supported zero-valent iron nanoparticles. J. Hazard. Mater. 318, 533–540. https://doi.org/10.1016/j.jhazmat.2016.07.039.

Suja, E., Nancharaiah, Y.V., Venugopalan, V.P., 2014. Biogenic nanopalladium production by self-immobilized granular biomass: application for contaminant remediation. Water Res. 65, 395–401. doi:10.1016/j.watres.2014.08.005.

Tang, W.W., Zeng, G.M., Gong, J.L., Liang, J., Xu, P., Zhang, C., Huang, B.B., 2014. Impact of humic/fulvic acid on the removal of heavy metals from aqueous solutions using nanomaterials: a review. Sci. Total Environ. 468–469, 1014–1027. https://doi.org/10.1016/j.scitotenv.2013.09.044.

Tao, F., 2012. Synthesis, catalysis, surface chemistry and structure of bimetallic nanocatalysts. Chem. Soc. Rev. 41 (24), 7977–7979. https://doi.org/10.1039/c2cs90093a.

Tariqul, I., Changsheng, P., 2019. Synthesis of carbon embedded silica and zeolite from rice husk to remove trace element from aqueous solutions: characterization, optimization and equilibrium studies. Sep. Sci. Technol. 55 (16), 2890–2903. https://doi.org/10.1080/01496395.2019.1658781.

Tian, X., Song, Y., Shen, Z., Zhou, Y., Wang, K., Jin, X., Han, Z., & Liu, T. (2019). A comprehensive review on toxic petrochemical wastewater pretreatment and advanced treatment. J. Clean. Prod., 245, 118692

Tsai, C.H., Chang, W.C., Saikia, D., Wu, C.E., Kao, H.M., 2016. Functionalization of cubic mesoporous silica SBA-16 with carboxylic acid via one-pot synthesis route for effective removal of cationic dyes. J. Hazard. Mater. 309, 236–248. https://doi.org/10.1016/j.jhazmat.2015.08.051.

Tungittiplakorn, W., Lion, L.W., Cohen, C., Kim, J.Y., 2004. Engineered polymeric nanoparticles for soil remediation. Environ. Sci. Technol. 38 (5), 1605–1610. https://doi.org/10.1021/es0348997.

V., Z.I., P., B.N., N., P.Y., V., K.L, 2016. Carbon nanotubes: sensor properties. A review. Mod. Electron. Mater. 95–105. https://doi.org/10.1016/j.moem.2017.02.002.

Venkateswarlu, S., Kumar, B.N., Prasad, C.H., Venkateswarlu, P., Jyothi, N.V.V., 2014. Bio-inspired green synthesis of Fe_3O_4 spherical magnetic nanoparticles using Syzygium cumini seed extract. Phys. B: Condensed Matter 449, 67–71. https://doi.org/10.1016/j.physb.2014.04.031.

Venkateswarlu, S., Kumar, B.N., Prathima, B., Anitha, K., Jyothi, N.V.V., 2015. A novel green synthesis of Fe3O4-Ag core shell recyclable nanoparticles using Vitis vinifera stem extract and its enhanced antibacterial performance. Phys. B 457, 30–35. https://doi.org/10.1016/j.physb.2014.09.007.

Virkutyte, J., Al-Abed, S.R., Choi, H., Bennett-Stamper, C., 2014. Distinct structural behavior and transport of TiO_2 nano- and nanostructured particles in sand. Colloids Surf. A Physicochem. Eng. Asp. 443, 188–194. https://doi.org/10.1016/j.colsurfa.2013.11.004.

Wang, X.T., Shi, W.S., She, G.W., Mu, L.X., Lee, S.T., 2010. High-performance surface-enhanced Raman scattering sensors based on Ag nanoparticles-coated Si nanowire arrays for quantitative detection of pesticides. Appl. Phys. Lett. 96 (5), 053104. https://doi.org/10.1063/1.3300837.

Wang, S., Sun, H., Ang, H.M., Tadé, M.O., 2013. Adsorptive remediation of environmental pollutants using novel graphene-based nanomaterials. Chem. Eng. J. 226, 336–347. https://doi.org/10.1016/j.cej.2013.04.070.

Wei, W., Susan, L., Ming, L., Qian, Z., Yuhe, Z., 2014. Polymer composites reinforced by nanotubes as scaffolds for tissue engineering. Int. J. Polym. Sci. 1–14. https://doi.org/10.1155/2014/805634.

Willis, A.L., Turro, N.J., O'Brien, S, 2005. Spectroscopic characterization of the surface of iron oxide nanocrystals. Chem. Mater. 17 (24), 5970–5975. https://doi.org/10.1021/cm051370v.

Wu, S., Huang, F., Lan, X., Wang, X., Wang, J., Meng, C., 2013. Electrochemically reduced graphene oxide and Nafion nanocomposite for ultralow potential detection of organophosphate pesticide. Sensors and Actuators B: Chemical 177, 724–729. https://doi.org/10.1016/j.snb.2012.11.069.

Wu, D., Zhang, P., Du, Q., Peng, X., Liu, T., Wang, Z., Xia, Y., Zhang, W., Wang, K., Zhu, H., 2011. Adsorption of fluoride from aqueous solution by graphene. J. Colloid Interf. Sci. 363 (1), 348–354. https://doi.org/10.1016/j.jcis.2011.07.032.

Yadav, K.K., Singh, J.K., Gupta, N., Kumar, V., 2017. A review of nanobioremediation technologies for environmental cleanup: a novel biological approach. J. Mater. Environ. Sci. 8 (2), 740–757. http://www.jmaterenvironsci.com/Document/vol8/vol8_N2/78-JMES-2831-Yadav.pdf.

Yang, B., Wei, Y., Liu, Q., Luo, Y., Qiu, S., Shi, Z., 2019. Polyvinylpyrrolidone functionalized magnetic graphene-based composites for highly efficient removal of lead from wastewater. Colloids Surf. A Physicochem. Eng. Asp. 582, 123927. https://doi.org/10.1016/j.colsurfa.2019.123927.

Yuan, S., Long, H., Xie, W., Liao, P., Tong, M., 2012. Electrokinetic transport of CMC-stabilized Pd/Fe nano- particles for the remediation of PCP-contaminated soil. Geoderma 185, 18–25. https://doi.org/10.1016/j.geoderma.2012.03.028.

Yunus, I.S., Harwin Kurniawan, A., Adityawarman, D., Indarto, A, 2012. Nanotechnologies in water and air pollution treatment. Environ. Technol. Rev. 1 (1), 136–148. https://doi.org/10.1080/21622515.2012.733966.

Zhang, F., Ge, Z., Grimaud, J., Hurst, J., He, Z., 2013. Long-term performance of liter-scale microbial fuel cells treating primary effluent installed in a municipal wastewater treatment facility. Environ. Sci. Technol. 47 (9), 4941–4948. https://doi.org/10.1021/es400631r.

Zhang, K., Khan, A., Sun, P., Zhang, Y., Taraqqi-A-Kamal, A., Zhang, Y., 2020. Simultaneous reduction of Cr (VI) and oxidization of organic pollutants by rice husk derived biochar and the interactive influences of coexisting Cr (VI). Sci. Total Environ. 706, 135763.

Zhang, M.Y., Wang, Y., Zhao, D.Y., Pan, G., 2010. Immobilization of arsenic in soils by stabilized nanoscale zero-valent iron, iron sulfide (FeS), and magnetite (Fe_3O_4) particles. Chin. Sci. Bull. 55 (4), 365–372. https://doi.org/10.1007/s11434-009-0703-4.

Zhang, Y., Tang, Z.R., Fu, X., Xu, Y.J., 2010. TiO_2-graphene nanocomposites for gas-phase photocatalytic degradation of volatile aromatic pollutant: is TiO_2-graphene truly different from other TiO_2-carbon composite materials? ACS Nano 4 (12), 7303–7314. https://doi.org/10.1021/nn1024219.

Zhang, N., Yang, M.Q., Tang, Z.R., Xu, Y.J., 2013. CdS graphene nanocomposites as visible light photocatalyst for redox reactions in water: A green route for selective transformation and environmental remediation. J. Catal. 303, 60–69. https://doi.org/10.1016/j.jcat.2013.02.026.

Zhao, X., Lv, L., Pan, B., Zhang, W., Zhang, S., Zhang, Q., 2011. Polymer-supported nanocomposites for environmental application: a review. Chem. Eng. J. 170 (2–3), 381–394. https://doi.org/10.1016/j.cej.2011.02.071.

Zheng, Z., Yuan, S., Liu, Y., Lu, X., Wan, J., Wu, X., Chen, J., 2009. Reductive dechlorination of hexachlorobenzene by Cu/Fe bimetal in the presence of nonionic surfactant. J. Hazard. Mater. 170 (2–3), 895–901. https://doi.org/10.1016/j.jhazmat.2009.05.052.

Zhou, R., Hu, G., Yu, R., Pan, C., Wang, Z.L., 2015. Piezotronic effect enhanced detection of flammable/toxic gases by ZnO micro/nanowire sensors. Nano Energy 12, 588–596. https://doi.org/10.1016/j.nanoen.2015.01.036.

Zhu, F., Li, L., Ma, S., Shang, Z., 2016. Effect factors, kinetics and thermodynamics of remediation in the chromium contaminated soils by nanoscale zero valent Fe/Cu bimetallic particles. Chem. Engg. J. 302, 663–669. doi:10.1016/j.cej.2016.05.072.

Zimmerman, J.B., Mihelcic, J.R., Smith, J., 2008. Global stressors on water quality and quantity. Environ. Sci. Technol. 42 (12), 4247–4254. https://doi.org/10.1021/es0871457.

CHAPTER 2

Green nanotechnology for environmental remediation

Ved Vati Singh
Department of Chemistry, Indian Institute of Technology Delhi, New Delhi, India

2.1 Introduction

Environment pollution is undoubtedly a universal menace, and it is increasing continuously due to change in lifestyles of the people, urbanization, and heavy industrialization. Thus, getting clean drinking water and fresh air has become a great challenge before humanity (Schwarzenbach et al., 2010). So, the development of useful techniques that can curve/remediate the environment pollution is strictly needed. Due to the highly interdisciplinary nature of nanotechnology, it has a wide range of potential applications such as in engineering (chemical and electrical), physics (solid-state physics, biophysics, materials science, etc.), biochemistry, and biotechnology. Due to enhancement in surface to volume ratio and unique functionalities from bulk to nanoscale, nanotechnology can offer immense opportunities as practical tools for environmental remediation. The use of nanomaterials to remediate environment is another area where nanotechnology can play a crucial role because properties of nanoparticles (NPs) (such as electronic, optical, magnetic, and catalytic properties) can be engineered by the engineering of size and shape of the NPs (Dehghani et al., 2020; Khan et al., 2020). Nanotechnology renders an opportunity to manipulate shape, size, and physical and chemical properties such as surface to volume ratio, structure, dispersibility, reactivity, etc. Nanomaterials have the capacity to increase reactivity and leverage surface chemistry compared to traditional approaches. Modification of nanomaterials is possible by functionalizing or grafting with different functional groups, which can be very useful to target specific pollutants. Thus, the remediation process can be very efficient. Tunable parameters/properties in case of nanomaterials such as size, morphology, porosity make them advantageous over conventional methods. Hence, techniques based on nanomaterials are beneficial for the treatment of natural/industrial/domestic wastewater, soils, sediments, mine tailings, and polluted atmosphere (Peralta-Videa et al., 2011).

Due to large surface to volume ratio and high reactivity, NPs have also shown excellent properties such as catalysts, absorbers, and sensors. Development of new technologies, new products, replacement of existing equipment, reformulation of new materials and chemicals for better performance in terms of less consumption of energy and materials

Sustainable Nanotechnology for Environmental Remediation
DOI: https://doi.org/10.1016/B978-0-12-824547-7.00017-5

becomes possible with nanotechnology. The synthesis of nanomaterials using a green approach reduces the environment pollution during the process of NP synthesis. Thus, green nanotechnology has attracted the attention of the scientific community to design new methods for environmental remediation. The focus of the present chapter is on the applications of nanomaterials in environmental remediation (Lowry et al., 2012). Green nanotechnology involves designing of nanomaterials to minimize pollution without using hazardous and toxic chemicals that are harmful to the environment (Guo et al., 2008, Daniel and Astruc, 2004, Hutchison, 2008) and offers a green and safe alternative to remove pollutants in the environment. Green technology has shown the potential to develop cost-effective, high thermal and chemical stable, less toxic and high degradable, active nanomaterials desirable for environmental remedy. The utmost goals of green technology are the development of eco-friendly designs, products, and processes (Nath and Banerjee, 2013) to reduce environmental pollution and its human health implications. Green nanotechnology can control many processes due to control over particle size at a small scale. Green nanotechnology can develop new processes or products for renewable energy generation, nanoscale sensors and pollution monitoring, photocatalysis environment treatment, and efficient membrane for treatment of wastewater with the help of multifunctional nanomaterials (Guo, 2012).

In this chapter, we will discuss some of the green technology techniques being used for environmental remediation. The main principles of green technology are based on green chemistry and green engineering. Green chemistry gives the insight about the reduction of unwanted hazards (Guo, 2012). A comparison between green and conventional technologies shows that green technique is far ahead than the other technologies in terms of design and development of new processes and products to eliminate or minimize the pollution. In this technique, the elimination of pollution starts from synthesis of nontoxic nanomaterials to their utilization for the solution of the existing environmental problems. The life-cycle assessment shows that environmental impacts during product chain in this technique are also relatively low (Christian et al., 2008). One of the most critical segments of green technology is the development of low-cost, least toxic, multifunctional, efficient product using ecofriendly process. Another part is to enhance its performance over the conventional technologies already being used in environmental remediation. Thus, utmost goals of green nanotechnologies are to develop the eco-friendly process and end product for the reduction of environmental contamination and health implications. The chapter is divided into two parts: (1) the synthesis of NPs and (2) their application in environmental remediation. The main objectives of green nanotechnology are to create eco-friendly designs at nanoscales to minimize environmental pollution and consecutively the health hazards. The state-of-the-art solutions are being offered for the current and future challenges related to sustainability and susceptibility faced in the area of biomedical sciences, food sciences, agricultural sciences, medicine, and energy sector. There are 12 fundamental principles of green chemistry for the researchers, scientists, chemical technologists, and chemists around the world to develop less harming chemical products and byproducts that is reduction of derivatives, use catalysts, efficient production, designing for safer chemicals and products, environment friendly, safer reaction conditions and chemicals used, renewable feedstock, increase energy efficiency, prevention of pollution and waste, design for degradation, safer chemicals synthesis and maximize atom economy. The main benefits of green nanotechnology are increased energy efficiency, reduced waste and greenhouse gas emissions, and minimized consumption of nonrenewable raw materials.

2.1.1 Environmental remediation

All remediation methods being used to overcome environmental risks to the human being due to environmental pollution are known as environmental remediation.

2.1.2 Types of NPs

On the basis of size, morphology, physical and chemical properties, NPs can be classified into different types such as carbon-based, ceramic-based, metal-based, semiconductor-based, polymeric-based, lipid-based, dendrimers, and composites (Pan and Xing, 2012, Bour et al., 2015). Of the aforementioned type, we will discuss the following four types of NPs.

- Carbon-based NPs
- Metal-based NPs
- Nanodendrimers
- Nanocomposites

2.1.2.1 Carbon-based NPs

Carbon-based NPs are of two types: (1) Carbon nanotubes (CNTs) and 2) fullerenes. They are made of carbon mainly, having the shapes of hollow spheres, ellipsoids, and tubes/hollow cylinders. Of which the hollow cylinders or tubes are known as CNTs. The CNTs are of two types: (1) single-walled CNTs (SWCNTs) and (2) multiwalled CNTs (MWCNTs) (Fig. 2.1).

The carbon NP in the shapes of sphere and ellipsoid is known as fullerenes (Fig. 2.2). Hollow cages with greater than or equal to 60 carbon atoms fall in the category of fullerenes or Buckminsterfullerene, which appears like a hollow football. The shapes of fullerenes are usually pentagonal and hexagonal arrangement. The carbon-based NPs have enormous potential applications in stronger and lighter materials, coating materials, etc. (Khan et al., 2021b, 2021a). Carbon-based nanomaterials show good electrical conductivity, high electron affinity, good strength, and therefore have outstanding applications in electronics (Hu et al., 2019, Xing et al., 2019, Xing et al., 2016) (Fig. 2.3).

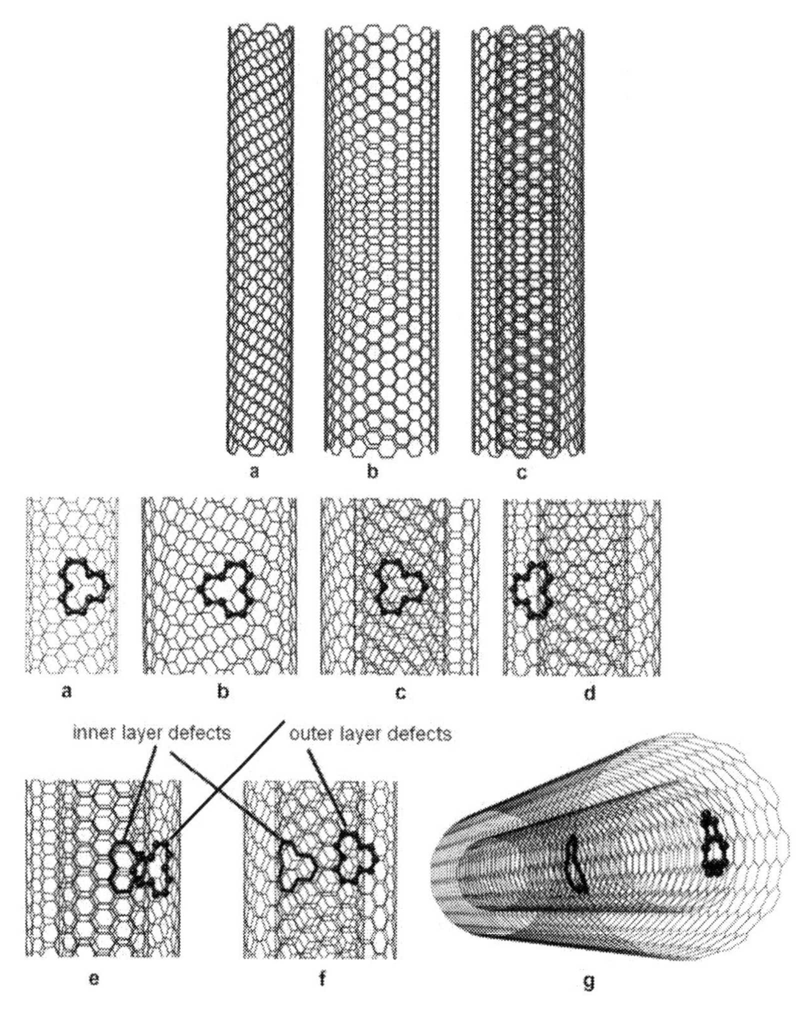

Figure 2.1 Configurations of perfect (7, 7), (12, 12) SWCNTs and ((7, 7), (12, 12)) double-walled carbon nanotubes (DWCNT), and Models of the defects in defective (7, 7), (12, 12) SWCNTs and ((7, 7), (12, 12)) DWCNTs. From Xin, H., Han, Q., & Yao, X. (2008). Buckling of defective single-walled and double-walled carbon nanotubes under axial compression by molecular dynamics simulation. Compos. Sci. Technol., 1809–1814. https://doi.org/10.1016/j.compscitech.2008.01.013.

2.1.2.2 Metal-based NPs

Metal-based NPs can be synthesized by electrochemical methods, photochemical methods, and chemical methods by using metal precursors. In the case of the chemical approach, NPs (Fan et al., 2017, Wu et al., 2019) are synthesized by reduction of metal ion precursors with the help of chemical reducing agents. The surface energy of metal NPs is relatively high. These particles can absorb small molecules. Both quantum dots

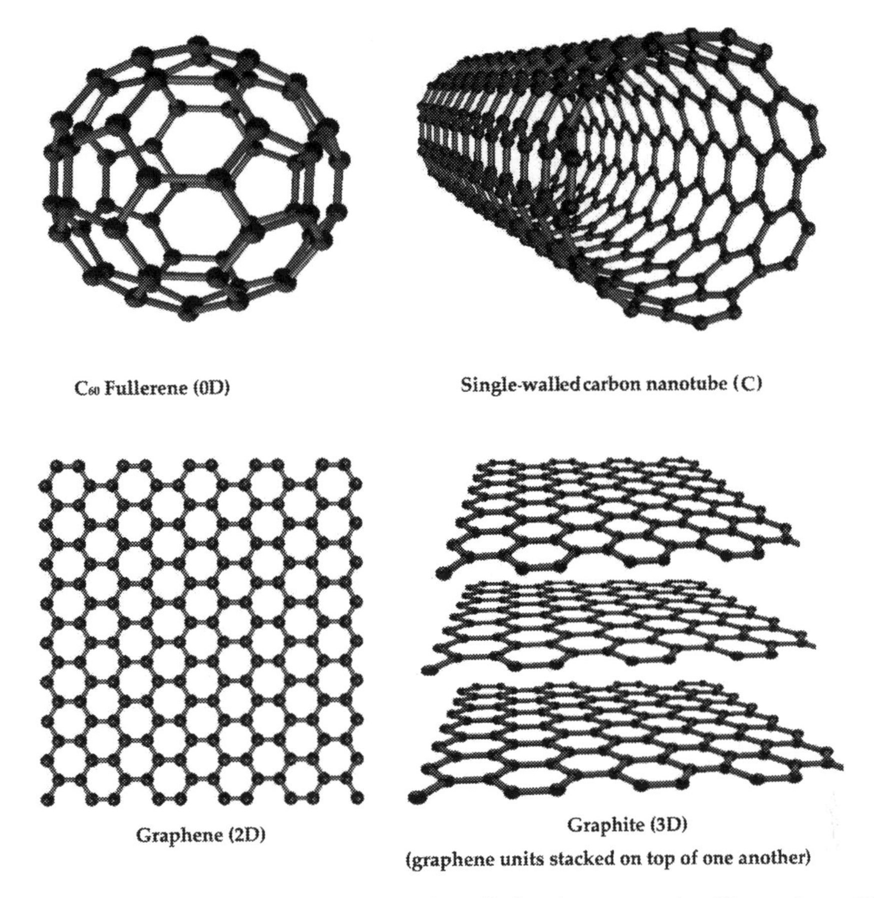

Figure 2.2 Carbon materials: fullerene 0D, single-walled carbon nanotube 1D, graphene 2D, and graphite 3D. From Xin, H., Han, Q., & Yao, X. (2008). Buckling of defective single-walled and double-walled carbon nanotubes under axial compression by molecular dynamics simulation. Compos. Sci. Technol., 1809–1814. https://doi.org/10.1016/j.compscitech.2008.01.013.

and metal oxides belong to metal nanomaterials. Quantum dots are defined as a closely packed semiconductor crystal with particle size from few nanometers to few hundred nanometers. Optical properties of metal NPs depend on the size of quantum dots and changes with the change in the size of quantum dots. Metal NPs are being used in the detection and biomolecule imaging. These particles also have applications in environmental and bioanalytical sciences (Figs. 2.4 and 2.5).

2.1.2.3 Dendrimers-shaped nanomaterials

Dendrimers are nano-sized ploymers, radially symmetric molecules with well-defined, homogeneous, and monodisperse structure consisting of branched units. The dendrimers (Khin et al., 2012, Ghasemzadeh et al., 2014) contain 3D interior cavities where other

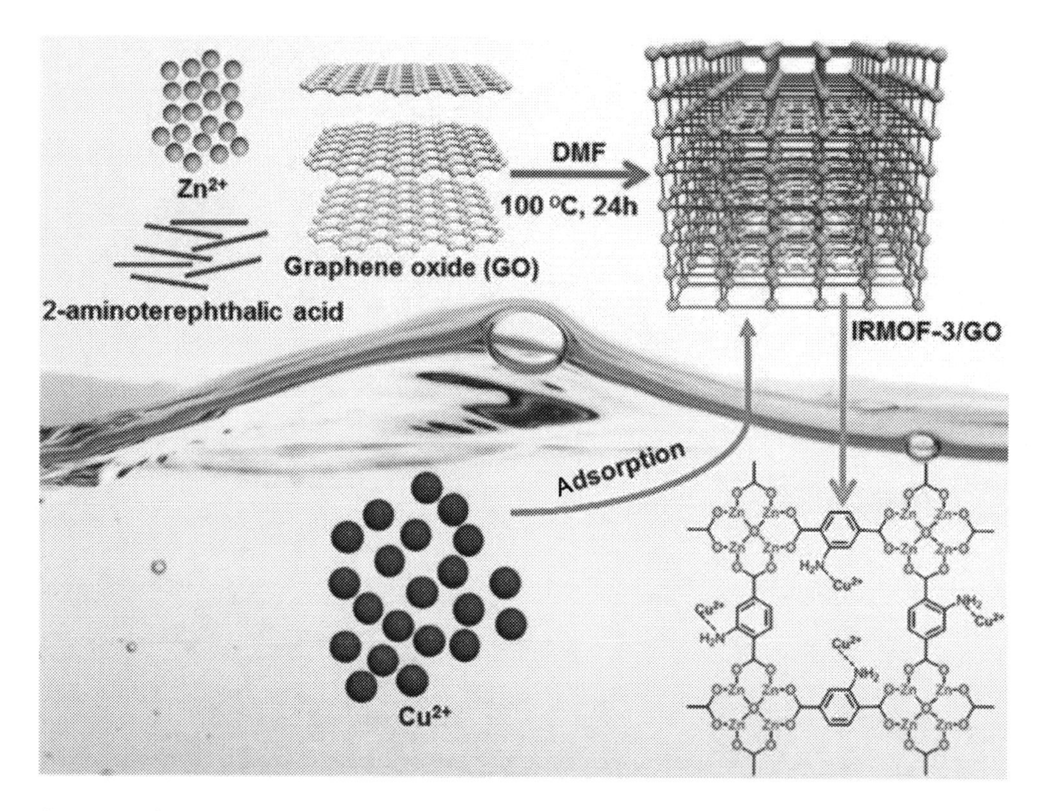

Figure 2.3 The preparation of amino-functionalized metal-organic framework/graphene oxide composite (IRMOF-3/GO). From Lina, Z., Lingjun, M., Jiaqi, S., Jinhai, L., Xuesheng, Z., & Mingbao, F. (2019). Metal-organic frameworks/carbon-based materials for environmental remediation: a state-of-the-art mini-review. J. Environ. Manage., 964–977. https://doi.org/10.1016/j.jenvman.2018.12.004.

molecules could be attached, which offer the right choice for targeted drug delivery (Fig. 2.6). The surface of a dendrimer contains various chain ends that can be manipulated as a specific requirement of the chemical activity and as a result, are suitable for catalysis (Fig. 2.7).

2.1.2.4 Nanocomposites

Nanocomposites are a combination of two or more than two types of NPs or NPs with bulk materials. A combination of different types of NPs provides unique properties to the nanocomposites such as mechanical, electrical, magnetic, thermal, catalytic properties for essential applications in commercial, environmental, medical, and military sectors (Lingamdinne et al., 2019; Mehmood et al., 2021). For example, nano-sized clays are being used for making nanocomposites for the automobile industry and packaging industry to enhance thermal and mechanical properties of the material and to enhance barrier and

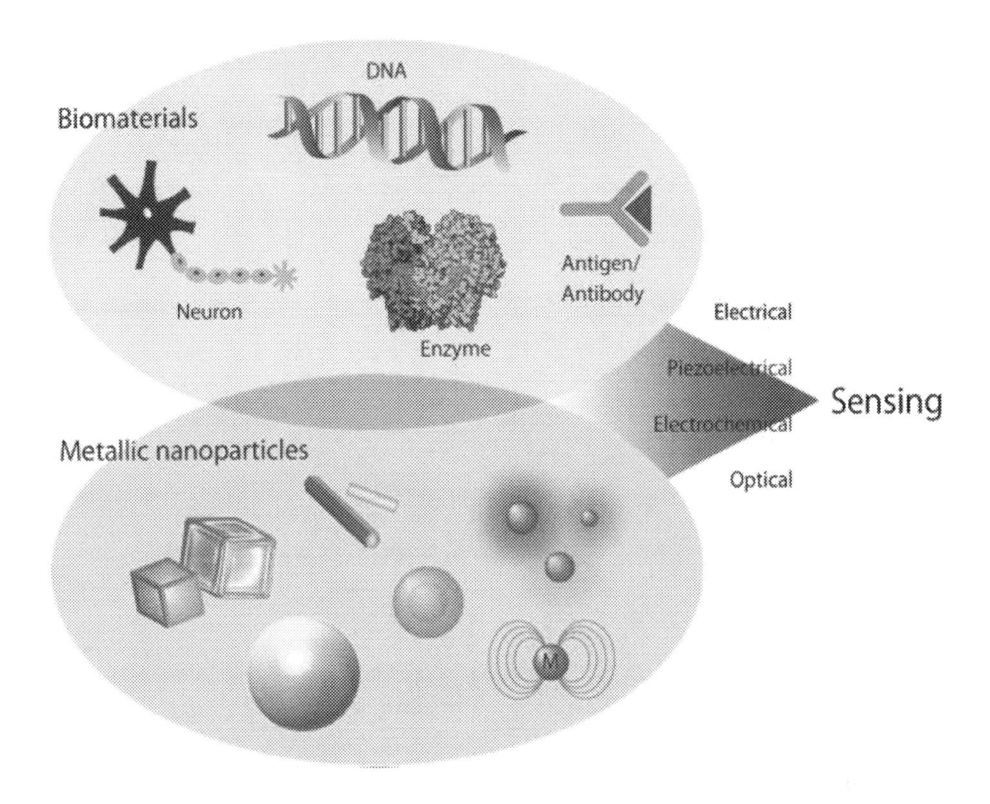

Figure 2.4 Biomaterials with metallic nanoparticles for biosensor. From Shiho, T., Yojiro, Y., Hiroshi, S., & Tsutomu, N. (2012). Synthesis and bioanalytical applications of specific-shaped metallic nanostructures: a review. Anal. Chim. Acta, 76–91. https://doi.org/10.1016/j.aca.2011.12.025.

flame-retardant properties (Wu et al., 2020, Maleki, 2016, Tesh and Scott, 2014). A more complex nanostructured system can be explored by these materials for new applications. The research toward the development of new nanocomposites with identified unique properties for specific applications is continuously growing.

2.2 Classification of synthesis approaches of nanomaterials

On the basis of size conversion, approaches for the synthesis of nanomaterials are broadly classified into two approaches (Hulkoti and Taranath, 2014, Akbarzadeh et al., 2012) namely: (1) Top-down approach and (2) bottom-up approach. In the case of top-down approaches, NPs are generated by size reduction from bulk materials, whereas in the bottom-up approach, NPs are synthesized by a combination of small entities such as

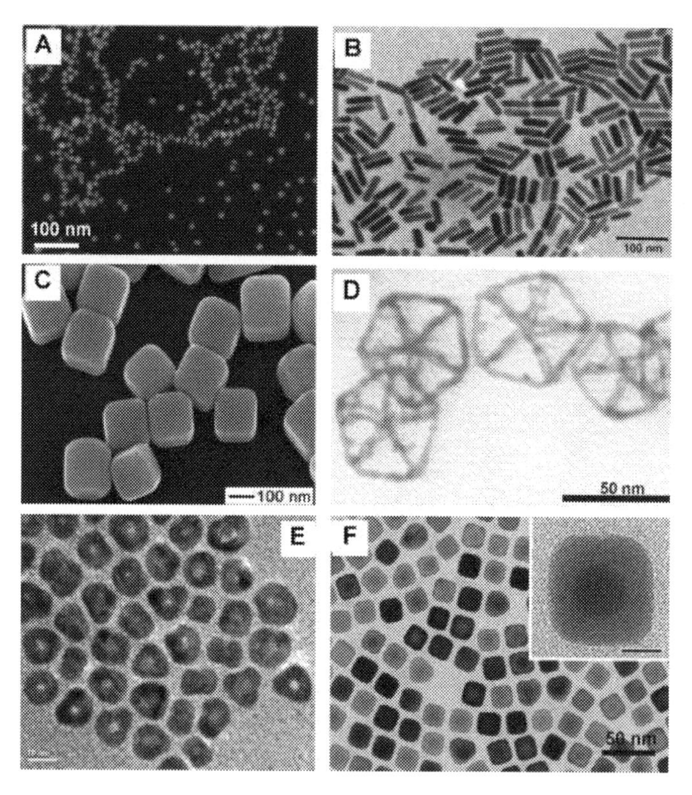

Figure 2.5 Transmission electron microscopy images of (A) gold nanoparticles, (B) gold nanorods, (C) silver nanocubes, (D) gold nanoframes, (E) AgAu core-shell nanoparticles, and (F) Aug core-shell nanocubes. From Shiho, T., Yojiro, Y., Hiroshi, S., & Tsutomu, N. (2012). Synthesis and bioanalytical applications of specific-shaped metallic nanostructures: a review. Anal. Chim. Acta, 76–91. https://doi.org/10.1016/j.aca.2011.12.025.

atoms and molecules. Both of these approaches are equally important in nanotechnology. All methods of synthesis at nanoscale follow either of these two approaches (Meyers et al., 2006, Thakkar et al., 2010).

2.2.1 Methods of NPs generation or fabrication

There are three basic methods, physical, chemical, and biological, being used for the synthesis of NPs (Zhou et al., 2005, Liu et al., 2005, Faraday, 1857, Bönnemann and Richards, 2001, Toshima and Yonezawa, 1998, Patzke et al., 2011, Cushing et al., 2004, Chen et al., 2002).

2.2.1.1 NPs by physical methods

In physical methods, a top-down approach is usually followed, where larger (macroscopic)/bulk materials are reducing to the nanoscale. This process is an extremely

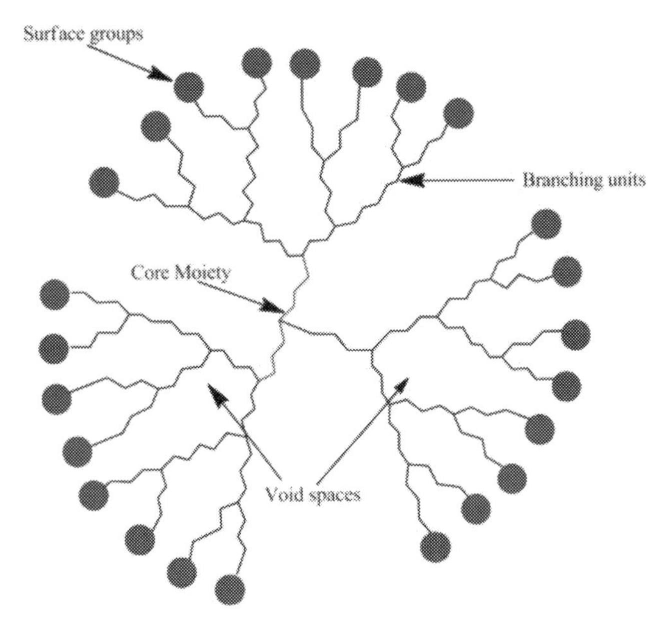

Figure 2.6 Structure of dendrimer. From Vunain, E., Mishra, A. K., & Mamba, B. B. (2016). Dendrimers, mesoporous silicas and chitosan-based nanosorbents for the removal of heavy-metal ions: a review. Int. J. Biol. Macromol., 86, 570–586. https://doi.org/10.1016/j.ijbiomac.2016.02.005.

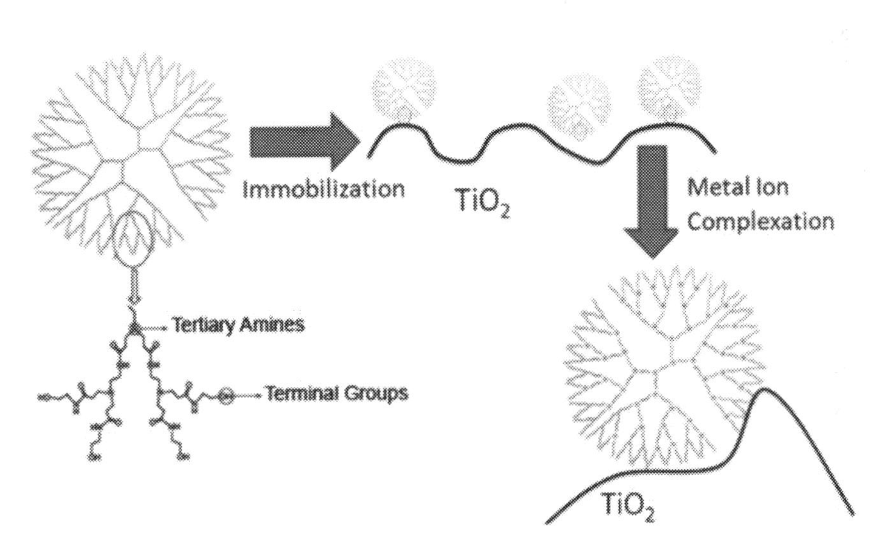

Figure 2.7 The synthesis method for immobilization of G4-OH dendrimers on titania for metal ion separations. From Vunain, E., Mishra, A. K., & Mamba, B. B. (2016). Dendrimers, mesoporous silicas and chitosan-based nanosorbents for the removal of heavy-metal ions: a review. Int. J. Biol. Macromol., 86, 570–586. https://doi.org/10.1016/j.ijbiomac.2016.02.005.

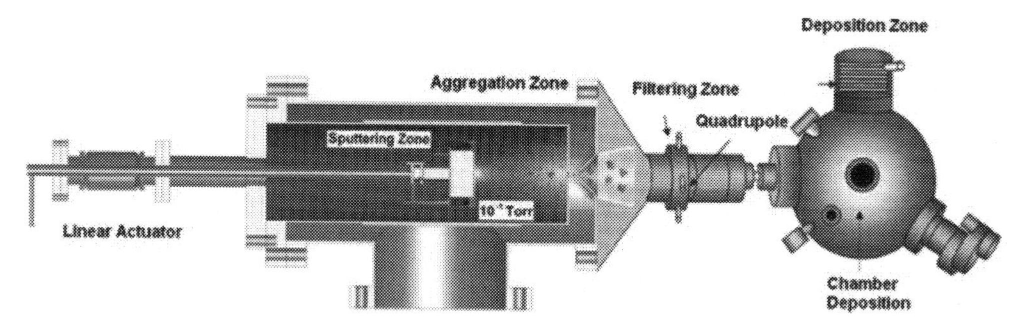

Figure 2.8 Schematic diagram of the nanocluster source From M., G.-P., E., M., Silva, V. G., & E., P.-T. (2010). Deposition of size-selected Cu nanoparticles by inert gas condensation. Nanoscale Res. Lett., 180–188. https://doi.org/10.1007/s11671-009-9462-z.

controlled process. The most commonly used physical methods for the synthesis of NPs (Scaramuzza et al., 2015, Kuang et al., 2015, Tsuji et al., 2002) are as follows:

- Inert gas condensation
- Electric arc discharge
- Radiofrequency (RF) plasma method
- Pulsed laser method, sputtering
- Laser ablation
- The ball milling method
- Molecular beam epitaxy
- Chemical vapor deposition.

We are giving only a brief description of all these physical techniques. Inert gas condensation method involves only two steps. In the first step, evaporation of the materials takes place, while in the second step a rapidly controlled condensation is used to produce the NPs of the required size (Figs. 2.8 and 2.9).

In electrical arc discharge method, an electric arc is used to vaporize the bulk material using a direct laser pulse (Fig. 2.10). Only 10%–20% of the clusters formed are ions, which are used to separate cluster ionization stage. The essential parameters in this method, gas pressure and the arc current are the crucial parameters that need to be controlled during evaporation to obtain the desired particle size of nanomaterials.

RF heating coils are used in RF plasma methods. In this approach, high-voltage RF is used to heat the metal above its evaporation point (Fig. 2.11). Helium gas is used to form plasma at high temperature in the region heating coils. The nucleation of metal vapor takes place on helium atoms, which diffuses to a cold collector rod where NPs are collected and thereafter passivated with oxygen gas.

In a pulsed laser method is capable of high rate of production of 3 gm/min. This method is used mainly to synthesize the silver NPs (AgNPs). In this method, a laser pulse

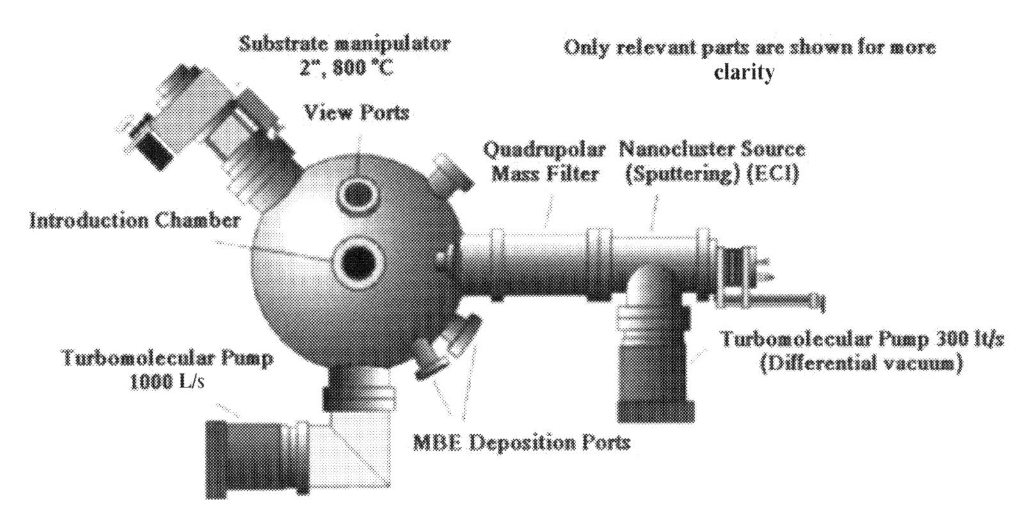

Figure 2.9 Integral system to synthesize Cu nanoparticles. From M., G.-P., E., M., Silva, V. G., & E., P.-T. (2010). Deposition of size-selected Cu nanoparticles by inert gas condensation. Nanoscale Res. Lett., 180–188. https://doi.org/10.1007/s11671-009-9462-z.

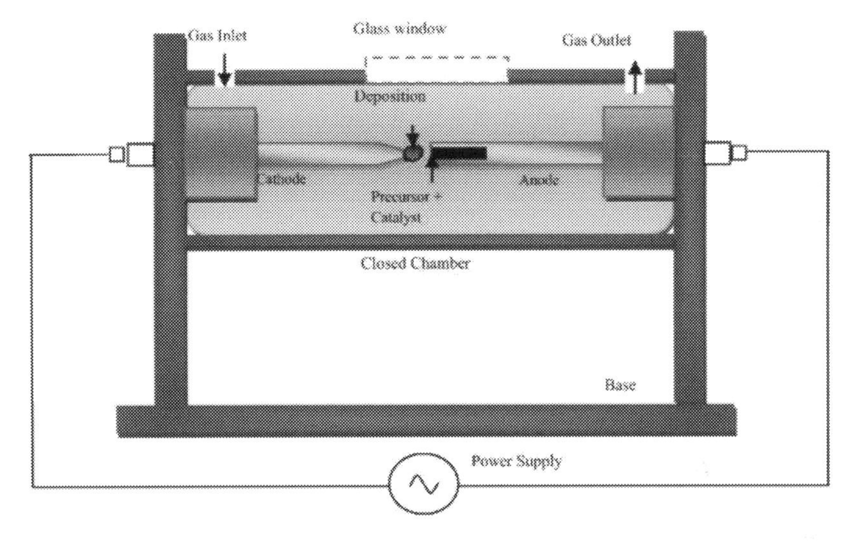

Figure 2.10 Schematic of an arc discharge setup From Neha, A., & N.N., S. (2014). Arc discharge synthesis of carbon nanotubes: comprehensive review. Diam. Relat. Mater., 135–150. https://doi.org/10.1016/j.diamond.2014.10.001.

is shined on a rotating solid disc with a solution that creates a hot spot on the surface of the disc. At the hot spot, silver nitrate reacts with a reducing agent and as a result an AgNP is formed, which is separated using a centrifuge. Both energies of the laser and angular velocity of rotating disc control the size of NPs (Fig. 2.12).

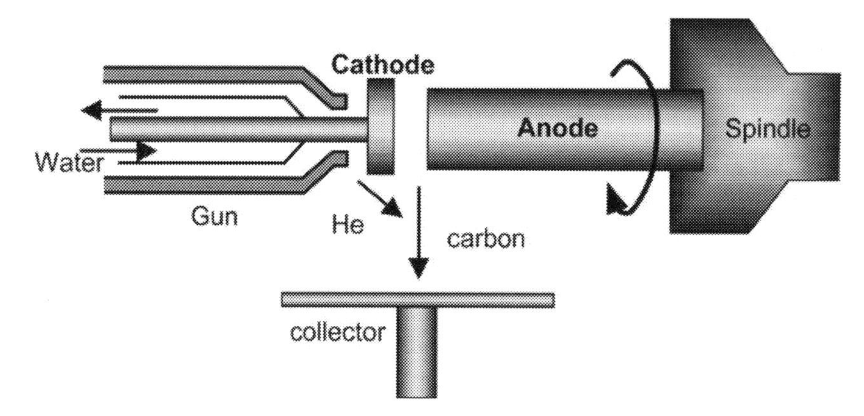

Figure 2.11 Schematic diagram of the plasma rotating electrode process system From Seung, J. L., Hong, K. B., Jae-eun, Y., & Jong, H. H. (2002). Large scale synthesis of carbon nanotubes by plasma rotating arc discharge technique. Diam. Relat. Mater., 914–917. https://doi.org/10.1016/s0925-9635(01)00639-2.

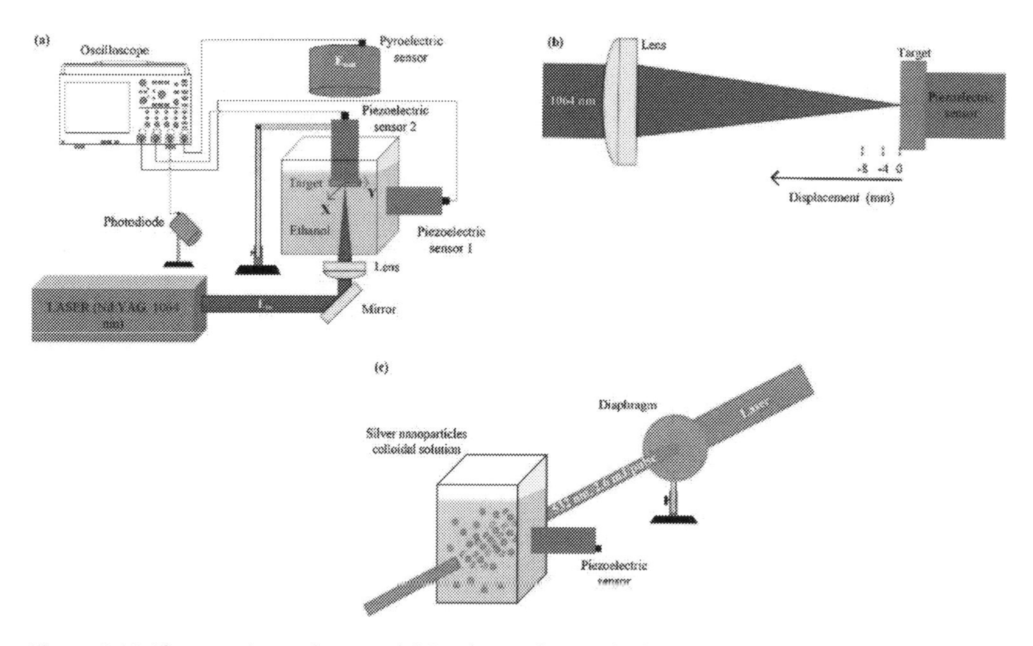

Figure 2.12 The experimental setups: (A) Synthesis of AgNPs by laser ablation in ethanol, (B) the effect of the laser spot area and laser pulse energy on the pulsed photoacoustic (PA) signal during the ablation of the silver target in the air, and (C) the effect the silver colloids concentration on PA signal at low laser fluence. From Valverde-Alva, M. A., García-Fernández, T., Villagrán-Muniz, M., Sánchez-Aké, C., Castañeda-Guzmán, R., Esparza-Alegría, E., Sánchez-Valdés, C. F., Llamazares, J. L. S., & Herrera, C. E. M. (2015). Synthesis of silver nanoparticles by laser ablation in ethanol: a pulsed photoacoustic study. Appl. Surf. Sci., 355, 341–349. https://doi.org/10.1016/j.apsusc.2015.07.133.

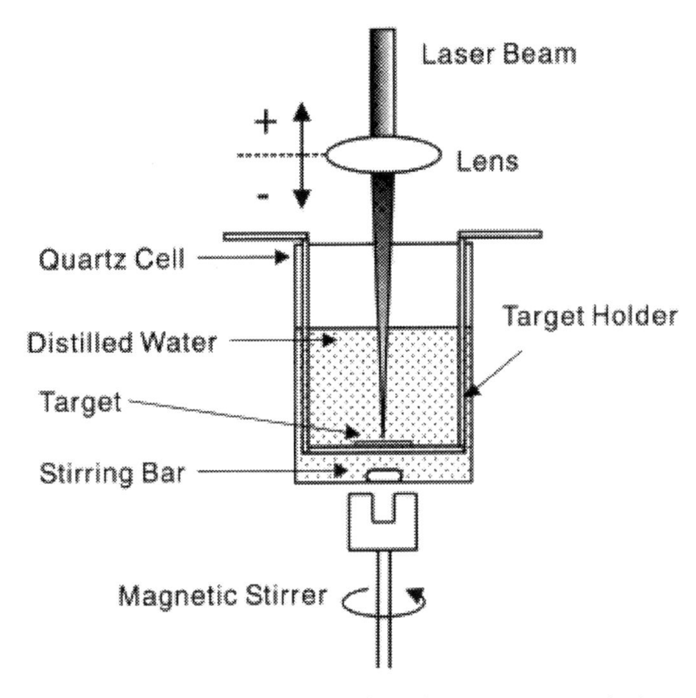

Figure 2.13 Experimental setup for laser ablation of metals in water. From Takeshi, T., Kenzo, I., Yukio, N., & Masaharu, T. (2001). Preparation of metal colloids by a laser ablation technique in solution: influence of laser wavelength on the ablation efficiency (II). J. Photochem. Photobiol. A Chem., 201–207. https://doi.org/10.1016/s1010-6030(01)00583-4.

Sputtering is another essential physical technique, which is based on the bombardment of a target with high-velocity inert gas ions, which causes the ejection of atoms and clusters and as a result vaporize the material from a solid surface. High vacuum inside the sputtering chamber is needed to avoid the effect of pressure on transportation of the sputtered materials. In the process of evaporation during the sputtering process, hot positive, negative and neutral clusters are produced, which cool down during the evaporation flight.

In laser ablation technique, a laser beam is focused on a target surface where ablation from the material surface from the irradiated zone takes place. This technique has been used for the production of NPs, deposition of thin films of metallic and dielectric materials, fabrication of superconducting materials, welding, metal parts binding, and micromachining to develop micro-electromechanical systems (MEMS) structures (Fig. 2.13).

The ball milling method is usually used to reduce the size of NPs. In this method, milling of small iron, hardened steel, silicon carbide, or tungsten carbide balls is carried out inside a drum till powder of nanoscale particles is formed. The powder is then taken inside a container of stainless steel.

Molecular beam epitaxy is another essential physical method, which is used where highly controlled evaporation of a variety of source under ultrahigh vacuum conditions is required. For example, highly controlled evaporation of a variety of source for a single crystal growth is carried out using a molecular beam epitaxy technique. This method is also used to synthesize high-quality thin films.

In chemical vapor deposition technique, one or more volatile precursors are transported in the vapor phase to the reaction chamber where they decompose is a chemical process. This technique is used to produce high-quality, high-performance NPs.

2.2.1.2 NPs by chemical method

In this approach, a process of self-assembly of the physical forces operating at the nanoscale is used to join the basic units to generate bigger stable nanostructures. The most commonly used bottom-up approach or chemical methods include solvothermal synthesis, sol-gel technique, an aerosol-based process, atomic or molecular condensation, hydrothermal synthesis, spray drying, cryochemical synthesis.

The solvothermal synthesis method is used for the synthesis of nanoscale materials such as semiconductors, polymers, ceramics, and metals. In this method, solvent interaction takes place under moderate-to-high pressure and temperature. If the solvent used in the solvothermal synthesis method is water, then this method is known as a hydrothermal synthesis method. NPs synthesized by this method are thermodynamically stable or thermodynamically metastable that is difficult to synthesize by other synthesis routes.

Sol-gel technique is another chemical technique used to form colloidal NP from the liquid phase in the industrial process. This method is basically used for the production of advanced coatings. Oxide NPs and their composites are being synthesized by this approach. Processing in this method is carried out at low temperature under versatility and flexible rheology conditions for the purpose of easy shaping and embedding (Fig. 2.14). This also provides an opportunity for the use of organic and inorganic materials. Alkoxides precursors are mostly used for the synthesis of oxides, because of their abundant availability and high liability of M-OR bond that allow *in-situ* facile tailorings during the process.

An aerosol-based process is a standard chemical method for the large-scale production of NPs. If solid or liquid particles with a particle size up to 100 μm are suspended in air or another gaseous environment, then it is known as an aerosol. Pigments of carbon black particle and titania particle are examples of aerosols. These aerosol particles are used as reinforcements for car tires. These aerosol particles are also being used for paints and plastics production. The aerosol particles such as silica and titania are synthesized from respective tetrachlorides by flame pyrolysis and optical fibers, respectively.

Atomic or molecular condensation is another chemical technique mainly used for the synthesis of NPs that contain metal. A stream of vaporized matter is produced by the

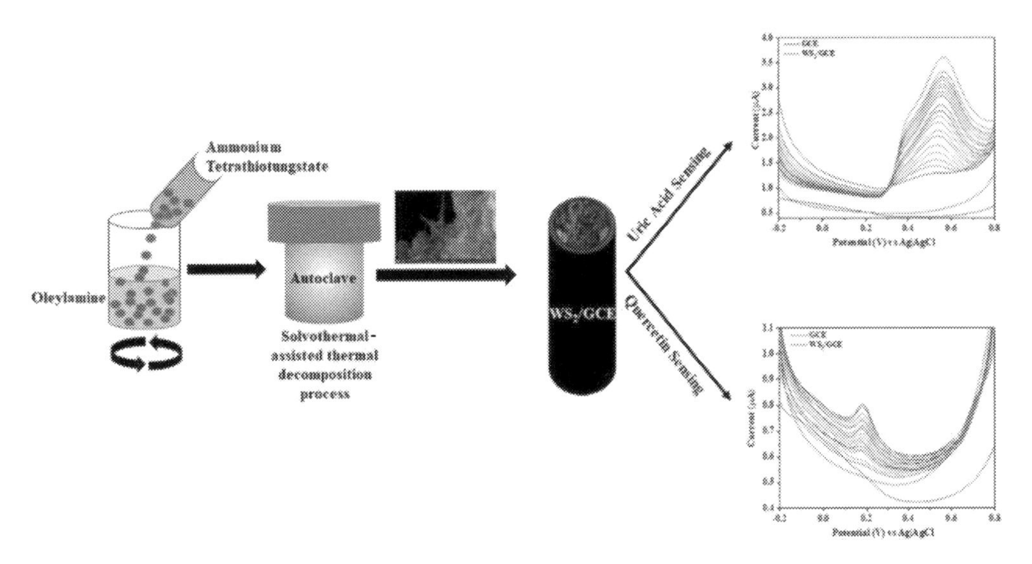

Figure 2.14 The mechanism for sensing of WS2-based nanoflake-nanorod hybrid nanomaterial modified glassy carbon electrode (GCE) for uric acid (UA) and quercetin. From Lignesh, D., Yi, K. C., & Sushmee, B. (2020). One-step solvothermal synthesis of nanoflake-nanorod WS2 hybrid for non-enzymatic detection of uric acid and quercetin in blood serum. Mater. Sci. Eng. C, 110217. https://doi.org/10.1016/j.msec.2019.110217.

heating of bulk material in a vacuum, and then directed to an inert or reactive gas-filled chamber. Metal atoms transfer their thermal energies to the gas molecules during the collisions with gas molecules, which leads to rapid cooling and as a result condensation and formation of NPs. If a reactive gas like oxygen is filled in the chamber, then metal oxide NPs can be produced.

In hydrothermal chemical synthesis, an aqueous solution of different chemical compounds is used in a closed system at more than 100 °C temperature and 1 atm pressure to produce different chemical compounds and materials. The ability of water and aqueous solutions to dilute substances at high temperature (500 °C) and pressure (10–80 MPa) for those substances, which are insoluble under normal conditions is utilized in this method. Some oxides, silicates, and sulfides are synthesized by this method.

The spray drying method is used for removing the solvent from solutions and suspensions. In Spray drying method, liquid droplets are injected into stream of carrier gas (usually air) at 100–300 °C temperatures, which results in separation of solid particles. The main difference between this method and the method of pyrolysis of aerosols is that (1) this method works at a lower temperature of carrier gas, which does not allow complete thermal decomposition of the saline components of the solution, and (2) that nozzles are used instead of aerosol generators; as a result, larger size droplets are produced with much better performance. Food and pharmaceutical industries are using spray drying methods to produce dry products and drugs.

Cryochemical synthesis is a set of methods used to synthesize nanomaterials. This is based on low-temperature chemical processes. Aqueous solutions are used in cryochemical synthesis techniques. In this method, chemical coprecipitation of initial solution components containing cations of the synthesized nanoscale materials in a stoichiometric proportion are frozen rapidly and dried in a vacuum after which thermal decomposition takes place. The end products in this method are generally oxide powders with crystallite sizes 40–3000 nm.

Lots of time and a large amount of energy is involved in the synthesis of NPs using both physical and chemical methods, which can raise the environmental temperature. Therefore, both these methods are not environment friendly and appropriate for the synthesis of NPs. The main disadvantage of both methods is that they involve a lot of toxic chemicals, which creates a lot of environmental issues. Due to the aforesaid reasons, the preferred methods for the synthesis of NPs are biological methods.

2.2.1.3 NPs by biological methods

In this section, we will discuss the synthesis of NPs using a biological route. Biological approach is ecofriendly, and NPs synthesized are less toxic. Size and morphology of NPs synthesized by this method are controllable. Several natural sources such as plants, bacteria, fungi, etc. are being used for NP synthesis. For example, the synthesis of intracellular and extracellular inorganic NPs has been carried out using unicellular and multicultural organisms. Thus, the development of eco-friendly biological methods for the synthesis of NPs has always been an interesting area for the scientific community. This is an alternate technique for NP synthesis to the comparatively more toxic physical and chemical methods. In the biological method, synthesis of NPs is carried out using plants (such as leaves, roots, flowers) and fruits and microorganisms (such as bacteria, algae, and fungus). As this method is eco-friendly, therefore, it is beneficial for environmental regulation, control, clean up, and remediation process (Sahu et al., 2021). It prevents toxic waste, reduces pollution, and produces renewable feedstock. Metallic NPs synthesized using biological materials such as bacteria, fungi, algae, and plant extracts are called as biogenic metal NPs. Biological precursors used in the process of green synthesis depend on several reaction parameters such as pH, pressure, temperature, and solvent. Due to availability of various useful phytochemicals such as ketones, carboxylic acids aldehydes, terpenoids, flavones, phenols, amides, ascorbic acids, in different plant extracts, biodiverse plants have been considered for metal NPs synthesis. Metal salts are reduced to metal NPs with the help of these phytochemical components. These biogenic nanomaterials have significant applications in antimicrobial activities, molecular sensing, biomedical diagnostics, biological system labeling, optical sensing, and environmental remediation (Fig. 2.15).

In this chapter, our focus will be on the systematic description of mechanisms for green synthesis, their components, and their applications for environmental remediation. Requirements of chemical and physical synthesis approaches are high intensity and

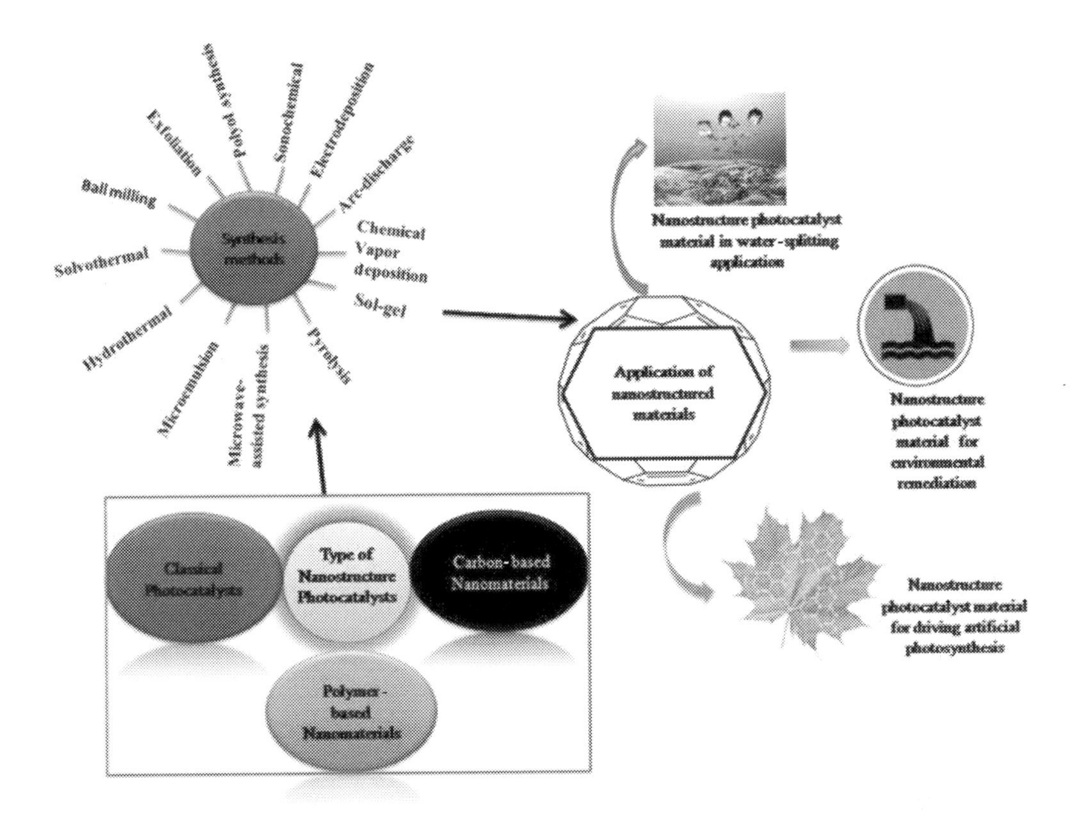

Figure 2.15 Different process for the synthesis of nanostructured materials for energy harnessing and environmental remediation From Ankita, R., Rajesh, R., Uttkarshni, S., Priya, M., Priyanka, M., Aneek, K., Ching, S. L., & Pichiah, S. (2018). A review on the progress of nanostructure materials for energy harnessing and environmental remediation. J. Nanostruct. Chem., 255–291. https://doi.org/10.1007/s40097-018-0278-1.

stabilizing agents that are highly toxic, which have a severe adverse effect on humans and aquatic life. On the other hand, the biological synthesis approach is eco-friendly, cost-effective, and one pot. The reduction process needs relatively low energy to initiate the reaction. In the past, the emergence of biotechnology as an elementary division has attracted the global attention of material scientists. The details of various biological routes for the synthesis of NPs are given next.

2.2.1.3.1 NPs synthesis by plant extracts

The synthesis of NPs using plants is a single-step biosynthesis process. The protocols involving plant sources for synthesis of NPs are nontoxic as plants supply natural capping agents rather than artificial toxic chemicals. In the synthesis process, the leaf extract is mixed with the solution of the metal precursor at diverse operating conditions. The

Figure 2.16 Synthesis of metal nanoparticles with various types of plants. From Mittal, A. K., Chisti, Y., & Banerjee, U. C. (2013). Synthesis of metallic nanoparticles using plant extracts. Biotechnol. Adv., 31(2), 346–356. https://doi.org/10.1016/j.biotechadv.2013.01.003.

plant extract parameters such as phytochemical type and concentration, pH, metal salt concentration, and reaction temperature determine the rate, yield, and stability of NP synthesis. The phytochemicals in the plant leaf extracts have supernatural potential to reduce metal ions into metal NPs in a short time compared to fungi and bacteria (which required longer incubation time). The rate of reaction is fast, therefore, it is considered as an excellent source for the synthesis of metal and metal oxide NP. Phytochemicals of leaf extract act as important reducing and stabilizing agents. Different leaf extracts can have different concentrations of phytochemicals. Phytochemicals such as flavones, ketones, carboxylic acids, amides, aldehydes, sugars, and terpenoids are mainly present in the plant extract. Several studies for the synthesis of NPs using plant extract have been carried out (Fig. 2.16).

Among these techniques is a single-step green synthesis process using plant extract biomolecules, which has many advantages: (1) Biogenic reduction rate from metal ion to base metal is relatively high; (2) reaction works at room temperature and pressure; (3) easy scale-up of reaction; (4) environmentally benign; (e) reducing agents involved water-soluble plant metabolites such as alkaloids, phenolic compounds, terpenoids, and coenzymes.

The synthesis of silver (Ag), gold (Au), platinum (Pt), and palladium (Pd) NPs has been reported using this technique (Mittal et al., 2013, Akhtar et al., 2013). Different shapes and sizes of the NPs of silver (Ag), gold (Au), platinum, and palladium have been explored (Fig. 2.17).

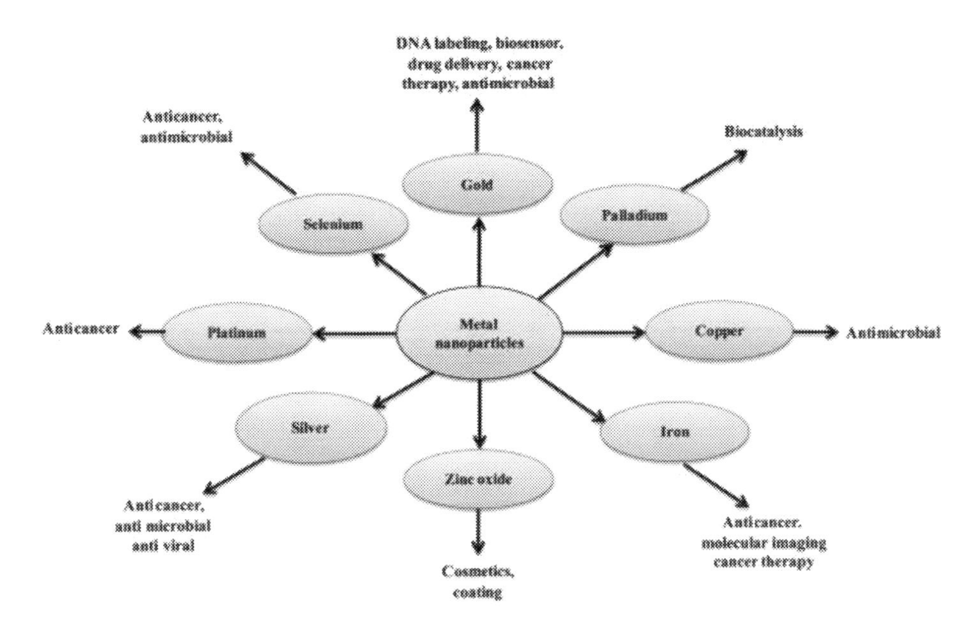

Figure 2.17 Types of metal nanoparticles and their applications in biotechnology, From Mittal, A. K., Chisti, Y., & Banerjee, U. C. (2013). Synthesis of metallic nanoparticles using plant extracts. Biotechnol. Adv., 31(2), 346–356. https://doi.org/10.1016/j.biotechadv.2013.01.003.

Four different plant extracts obtained from *Lippia citriodora, Punica granatum, Salvia officinalis*, and *Pelargonium graveolens* have been explored as reducing and stabilizing agents for the synthesis of gold NPs (GNPs) (Elia et al., 2014). Scientific efforts for the development of new technologies for the diagnosis and treatment of cancer are underway. Due to simple, one-pot, fast, safer, energy-efficient, economical, and biocompatibility of the bio-nanotechnology, this approach has attracted the attention of researcher over the conventional synthesis methods. Medicinally active plants are the best reservoirs of diverse phytochemicals for the synthesis of biogenic AgNPs. Recently biogenic AgNPs have been developed as cancer theranostic. It is predicted that the proposed action mechanism of AgNPs biogenic can be a potential cancer theranostic agent in the future (Ovais et al., 2016). Biogenic AgNPs have antibacterial properties. People have developed an environment-friendly route for the synthesis of AgNPs using plant extracts, biodegradable polymers, enzymes, bacteria. Microwave irradiation is being used as alternative energy whose photon energy is relatively low (pose least human risk), which makes the process efficient with reduced energy consumption and uniform particle size due to prevention of agglomeration of ensuing NPs (Hebbalalu et al., 2013). Leaf extracts of pine, persimmon, ginkgo, magnolia, and platanus have been used for AgNPs synthesis.

For example, an aqueous solution of $AgNO_3$ when treated with plant leaf extracts as a reducing agent gives stable AgNPs. Among the above-mentioned plant extracts, the magnolia leaf is the best reducing agent as far as synthesis rate (reaction time 11

Figure 2.18 Various plants for the synthesis of ZnO-NPs; *Cassia auriculata* (A), *Parthenium hysterophorous* (B), Aloe vera (C), *Acalypha indica* (D), *Calotropis gigantean* (E), and *Abrus precatorius* (F). From Hamed, M., & Majid, D. (2017). Zinc oxide nanoparticles: Biological synthesis and biomedical applications. Ceram. Int., 907–914. https://doi.org/10.1016/j.ceramint.2016.10.051.

min) and conversion (90%) to AgNPs are concerned. A reaction temperature of 95°C is required. The average particle size was 15– 500 nm. Particle size depends on the reaction temperature and concentrations of leaf broth and $AgNO_3$. Biogenic AgNPs also have potential applications in cosmetics, food industry, and medical (Song and Kim, 2009). Another example is the green synthesis of zero-valent iron NPs and their application for Cr(VI) removal from aqueous solutions. In this synthesis process extracts of *Thymus vulgaris* (TV), *Urtica dioica* (UD), and *Rosa damascene* (RD) are used in the presence of reducing agents such as polyphenols, proteins, and organic acids, which also serve as stabilizing agents to reduce the possibility of aggregation of NPs. These biogenic iron NPs show very high adsorption capacity. These NPs have yielded > 90% removal efficiency with a 0.2 g/L NP dose for 10 min contact time. The respective removal efficiency was obtained 94.87%, 83.48%, and 86.8% for RD-Fe, UD-Fe, and TV-Fe using 1 min contact time. It is reported that this efficiency increases to 100% for TV-Fe and UD-Fe with an increase in removal time to 25 min. The contact time required for 100% removal of Cr(VI) by RD-RF is 30 min (Fazlzadeh et al., 2017).

The biogenic synthesis process has also been extended for the synthesis of gold, copper oxide, and Zinc oxide NPs (Fig. 2.18). The biogenic NPs are being used for environmental remediation as removal of pollutants heavy metal ions and dyes, catalytic activity, antimicrobial activity, etc. (Singh et al., 2018).

The biogenic synthesis of iron NPs has several advantages over conventional methods due to (1) eco-friendliness and simple synthesis process, (2) cost-effectiveness, and (3) its nontoxic properties. Iron NPs are good for antioxidant and antibacterial activity and also suitable for environmental remediation to the magnetization of sediments. The synthesis of iron NPs has been reported using extracts of plants such as green tea, *Amaranthus dubius, Eichhornia crassipes, Cynometra ramiflora, Eucalyptus tereticornis, Melaleuca nesophila,* and *Rosemarinus officinalis*. These Iron NPs have been used (Thacker et al., 2019) for degradation of dyes, removal of nitrate, removal of hexavalent chromium, removal of arsenate, removal of arsenite, and used for chemical oxygen demand. These novel applications of biogenic iron NPs compel the scientific community to develop new and efficient methods for their environmental applications. Thus, extensive study of the principles of green chemistry, waste prevention, energy efficiency, harmless precursor, and solvent become significant. Synthesis of iron (Fe) NPs using microorganisms (fungi, bacteria, actinomycetes, and viruses, algae, plant and their extracts) has been rigorously investigated and applied for environmental remediation to reduce toxicological implications (Bolade et al., 2020).

Biogenic tin (IV) oxide NPs (SnO_2NPs) have shown outstanding results in environmental remediation. These NPs also have various other applications such as in sensing, catalysis, and in lithium-ion batteries. Green biosynthesis approach for the synthesis of SnO_2NPs has been reported using various plants (Matussin et al., 2020).

2.2.1.3.2 NP synthesis by bacteria

The synthesis of various metal NPs has been carried out using bacterial species. The NPs synthesized using bacterial species have been extensively tested for various biotechnological applications such as genetic engineering, bioleaching, and bioremediation, (Gericke and Pinches, 2006). Due to its reducing ability, bacteria are the momentous candidate for NPs synthesis (Iravani, 2014). Researchers have investigated a different variety of bacterial species for the synthesis of NPs. For example, synthesizing metal/metal oxide NPs using prokaryotic and actinomycetes bacteria has been investigated due to the relative ease of manipulating the bacteria (Thakkar et al., 2010). Bacterial strains that have been used for the synthesis of bio-reduced AgNPs with distinct size/shape morphologies are, *Bacillus cecembensis, Enterobacter cloacae, Corynebacterium* sp. SH09, *Shewanella oneidensis, Bacillus amyloliquefaciens, Bacillus indicus, Geobacter* spp., *Arthrobacter gangotriensis, Phaeocystis antarctica, Pseudomonas proteolytic, Lactobacillus casei, Bacillus cereus, Escherichia coli, Aeromonas* sp. SH10,

Bacterial species that has been used for the synthesis of GNPs are *Bacillus subtilis* 168, *Shewanella alga, Desulfovibrio desulfuricans, Bacillus megatherium* D01, *E. coli* DH5a, *Plectonema boryanum* UTEX 485, and *Rhodopseudomonas capsulate*. Biosynthesis of copper NPs (CuNPs) in aqueous phase using bacterium *Morganella morganii* is reported. The mechanical aspect of copper ion reduction using bacterial has been carried out with the help of electrochemical analysis of bacterial cells that are exposed to copper ions, which indicates a strong link between silver and copper resistance machinery of bacteria with respect to the metal ion reduction (Ramanathan et al., 2013). The significant advantages of NPs synthesized using bacteria are cost-effectiveness, eco-friendliness and nontoxicity.

2.2.1.3.3 NP synthesizes by fungi

Another method, green synthesis of NPs, is based on fungi. The following properties reception toward toxicity, higher bioaccumulation, comparatively economical, easy synthesis and simple downstream processing and biomass handling make this approach very attractive. Based on different fungi, the biosynthesis of monodispersed metal and metal oxide NPs of Ag, Au, TiO_2, and ZnO has been reported (Chen et al., 2009, Mohanpuria et al., 2008). Due to the presence of enzymes/reducing agents/proteins in the cells of *Aspergillus niger, Fusarium solani*, and *Aspergillus oryzae*, these fungi have been studied for NP synthesis (Narayanan and Sakthivel, 2011).

2.2.1.3.4 NPs synthesis by yeast

Eukaryotic cells have a single-celled microorganism called yeasts (Sun et al., 2019). Researchers have identified more than 1500 yeast species for NPs synthesis. *Saccharomyces cerevisiae* and *Yarrowia lipolytica* have been studied for Ag and AuNPs. A large number of other yeasts have been used for different NP synthesis (Marchiol, 2012).

2.2.1.4 NPs synthesis using biological particles

Bioparticles such as virus, protein, peptides, carbohydrates, enzyme, and coenzyme have been extensively investigated for the synthesis of different metal NPs to reduce metal salt into metal NPs. Plants such as coriander (*Coriandrum sativum*), mustard (*Brassica juncea*), alfalfa (*Medicago sativa*), neem (*Azadirachta indica*), lemon (*Citrus limon*), tulsi (*Ocimum sanctum*), Oat (*Avena sativa*), lemongrass (*Cymbopogon flexuosus)*, and aloe vera (*Aloe barbadensis Miller*) have been studied for this purpose. The NPs of zinc, nickel, cobalt, copper, and zinc oxide have been also synthesized by living plants, plant leaf extracts such as extracts of plants such as sunflower (*Helianthus annuus)*, mustard (*B. juncea*), alfalfa (*M. sativa*), (Maensiri et al., 2008), green tea (*Camellia sinensis)* (Gnanasangeetha and Saralathambavani, 2014), copper leaf (*Acalypha indica*) (Devi and Singh, 2014), China rose (*Hibiscus rosa-sinensis)* (Devi and Singh, 2014), aloe leaf broth extract (*A. barbadensis Miller*) (Gunalan et al., 2012), coriander (*C. sativum*) (Anastas and Warner, 1998), and crown flower (*Calotropis gigantean*) (Vidya et al., 2013) have been used for the synthesis of Zn, Ni, Co, Cu, and ZnO.

2.3 Role of NPs in environmental remediation

2.3.1 Antimicrobial activities of NPs

The antimicrobial effectiveness of the metal NPs depends on their particle size and materials involved in the process of synthesis. Penicillin-resistant, vancomycin-resistant sulfonamide-resistant, and Methicillin-resistant properties are found in antimicrobial drug-resistant bacteria (Fair and Tor, 2014). The challenges such as combatting multidrug-resistant mutants and biofilms are faced by the antibiotics. Due to the drug-resistant nature of microbes, the effectiveness of antibiotic dilutes rapidly. A bacterial disease can persist in living beings even if a large dose of antibiotic is given. Biofilms are another critical way, which can provide multidrug resistance against antibiotics. Thus, antibiotic drug resistance in case of bacteria is a big challenge. NP-based drugs offer a promising solution to the drug-resistant issue with a bacterial disease. Various NP-based mechanisms have been developed to overcome the problem of multidrug-resistance and biofilm formation. Several types of metal NPs (such as NO-releasing NPs and chitosan-based NPs) have been investigated to develop multiple mechanisms concurrently to fight microbes. NPs used multiple mechanisms to overcome the drug resistance problem. This mechanism is successful because microbes require multiple gene mutations in their cell to counter this mechanism, which is rare. AgNPs have shown remarkable result as efficient antiviral, antifungal, anti-inflammatory, and antimicrobial agents (Zinjarde, 2012). The AuNPs are also highly useful as effective antibacterial agents due to their nontoxic nature, strange functionalization ability, and polyvalent effects (Lima et al., 2013- (Zhou et al., 2012). The antimicrobial potential of ZnO, CuO, and Fe_2O_3 NPs toward *E. coli, Pseudomonas aeruginosa* (gram-negative bacteria), *Staphylococcus aureus* and *B. subtilis* (gram-positive bacteria) has been reported (Cui et al., 2012). Antimicrobial activities of CuONPs on several bacterial species such as *Shigella Salmonella* paratyphoid, *P. aeruginosa*, and *Klebsiella pneumonia* have been tested (Mahapatra et al., 2008) and claimed suitable antibacterial activity of CuONPs against these bacteria. They assumed that crucial enzymes of bacteria were damaged by NPs by crossing the bacterial cell membrane, which induces death of the bacteria cell. It has been reported that antimicrobial activity of biogenic NPs is far better compared to chemically synthesized or commercial NPs.

2.3.2 Catalytic activities of NPs

Herbicides, insecticides, and synthetic dyestuffs manufactured using 4-nitrophenol and its derivatives have an adverse effect on the ecosystem. They are toxic and inhibitory in nature and also common organic pollutants of wastewater. The 4-nitrophenol reduction product has been used as a crucial intermediary for the synthesis of paracetamol, sulfur dyes, rubber antioxidants. It has also been used for the synthesis of black and white film developers, precursors in antipyretic, analgesic drugs, and corrosion inhibitors (Panigrahi et al., 2007, Woo et al., 2012). $NaBH_4$ is used as the simplest and most effective reductant to reduce 4-nitrophenol. It has also been used as a metal catalyst (NPs) of Au (Lim et al., 2016), Ag (Rostami-Vartooni et al., 2016), CuO (Sharma et al., 2015), and Pd (Gopalakrishnan et al., 2017). The NPs of metals can promote the rate of reaction by increasing the adsorption of reactants on their surface, thereby diminishing activation energy barriers (Gangula et al., 2011, Singh et al., 2017).

2.3.3 Removal of pollutant dyes using NPs

Anionic and cationic dyes are being rigorously used in plastic, pharmaceuticals, food, leather, paper mills, printing, and textiles industries (Habibi and Rezvani, 2015). These dyes are basically organic pollutants. Pigmentation for many fabrics (Chatterjee et al., 2020) in the textiles industry used about 60% of dyes. Fifteen percent of dyes are discharged into the hydrosphere, which causes pollution due to their recalcitrance nature (Padhi, 2012). Manufacturing industries are mostly responsible for ecological pollution. The undesirable turbidity of water developed due to industrial waste obstruct sunlight penetration and as a result resists photochemical synthesis and are responsible for biological attacks on aquatic and marine life (Dutta et al., 2014, Gonawala and Mehta, 2014, Jyoti and Singh, 2016). Thus, management of dye effluents is a tremendous challenge for environmental chemist (Wesenberg et al., 2003). Metal, metal oxides, and semiconductor nanomaterials are the potential candidates for oxidizing toxic pollutants (Fowsiya et al., 2016, Nakkala et al., 2015, Varadavenkatesan et al., 2016). Further, nanoscale semiconductors show superior photocatalytic activity relative to the bulk materials. Metal oxide NPs (such as CuO, WO_3, SnO_2, TiO_2, and ZnO) have shown good photocatalytic activity to oxidize synthetic dyes (Bhuyan et al., 2015, Stan et al., 2015, Thandapani et al., 2017). The merits of nano-photocatalysts such as zinc oxide and titanium oxide NPs show better adsorption of organic pollutants due to their large surface area to mass ratio. The rate of contaminant removal at low concentrations in case of nanomaterials is much higher than that of bulk materials due to a large number of useful reactive sites on the surfaces of NPs. Consequently, a small amount of nanocatalyst is needed to treat polluted water than the bulk material (Ajithaa et al., 2015, Dror et al., 2005, Pradeep, 2009, Tsuda and Konduru, 2016).

2.3.4 Role of NPs for heavy metal ion removal

Metals such as Pb, Cr, Zn, Hg, Ni, Cr, Co, Cd, Cu, Fe, and Mn are known as heavy metals. These metals are well-known pollutants in water, soil, and even in the air. Mining waste, automobile emissions, natural gas burning, dye, coal, plastic, and paper industries (Zhang et al., 2012) are the known sources of heavy metal pollution. Pb, Cu, Ca, and Hg are extremely toxic. Even a trace of these toxic metals in the biological and aquatic environment can be very dangerous and requires immediate and proper remediation (Aragay et al., 2011, Nolan and Lippard, 2008, Ray, 2010). Researchers prefer metallic NPs for the detection of heavy metal ions in polluted water systems due to tunable size and distance-dependent optical properties (Annadhasan et al., 2014, Maiti et al., 2016). Thus, metal NPs are being used as colorimetric sensors for heavy metal ions in the environmental. Their synthesis process is simple, cheaper, and highly sensitive at sub-ppm levels. AgNPs (Karthiga and Anthony, 2013) synthesized using plant extracts are excellent colorimetric sensors for heavy metal ions such as Cd, Cr, Hg, Ca, and Zn.

2.3.5 Removal of biological contaminants using NPs

Shape and morphology such as tubes, wires, fibers, etc. of nanomaterials play an essential role as adsorbents and catalysts. Toxic gases such as SO_2, CO, NO_x can be detected and removed by NP composites with polymers. Toxic chemicals such as arsenic, iron, manganese, nitrate, heavy metals, etc. can also be removed by polymer nanocomposites. Polymer nanocomposites are also capable of removing the organic pollutants such as aliphatic and aromatic hydrocarbons and biological substances such as viruses, bacteria, parasites, and antibiotics. Nano-remediation incorporates nanometals NPs, nanomembranes, and nanopowders for detection as well as for the removal of toxic metals and organic compounds from the environment. The removal of toxic metals and organic compounds is carried out by adsorption, transformation, photocatalysis, and catalytic reduction using several inorganic and organic nanoscale materials. Heavy metals and organic pollutants have been detected by nanobiosensors (Das et al., 2018).

2.4 Conclusion and future prospects

The synthesis of different types of NPs using a green approach is summarized in this chapter. Various recent developments in this direction have also been incorporated. Comparison based on merits and demerits with conventional physical and chemical methods has been discussed. Various kinds of natural extracts such as bacteria, yeast, fungi, and plant extracts have been discussed as an efficient source for the synthesis of eco-friendly biogenic nanoscale synthesis. This chapter was organized to encompass the "state-of-the-art" research on the green synthesis of various NPs and their applications in environmental remediation. Future research in the perspective of green NP synthesis should be directed from extending laboratory-based work to an industrial scale, keeping present issues such as health and environment in mind. The author believes that green NP synthesized using a green approach based on biocomponent-derived materials will be extensively used in the field of environmental remediation and other essential areas such as food, cosmetic and pharmaceutical industries.

References

Ajithaa, B., Reddy, Y.A.K., Reddy, P.S., 2015. Biosynthesis of silver nanoparticles using *Momordica charantia* leaf broth: evaluation of their innate antimicrobial and catalytic activities. J. Photochem. Photobiol. B Biol. 146, 1–9.

Akbarzadeh, A., Samiei, M., Davaran, S., 2012. Magnetic nanoparticles: preparation, physical properties, and applications in biomedicine. Nanoscale Res. Lett. 7, 144.

Akhtar, S., Panwar, J., Yun, Y-S, 2013. Biogenic synthesis of metallic nanoparticles by plant extracts. ACS Sustain. Chem. Eng. 6 (1), 591–602.

Anastas, P.T., Warner, J.C., 1998. 12 Principles of Green Chemistry. Green Chemistry: Theory and Practice. Oxford University Press, Oxford.

Annadhasan, M., Muthukumarasamyvel, T., Sankar Babu, V.R., Rajendiran, N., 2014. Green synthesized silver and gold nanoparticles for colorimetric detection f Hg^{2+}, Pb^{2+}, and Mn^{2+} in aqueous medium. ACS Sustain. Chem. Eng 2, 887–896.

Aragay, G., Pons, J., Merkoçi, A., 2011. Recent trends in macro-, micro-, and nanomaterial-based tools and strategies for heavy-metal detection. Chem. Rev 111, 3433–3458.

Bönnemann, H., Richards, R.M., 2001. Nanoscopic metal particles—synthetic methods and potential applications. Eur. J. Inorg. Chem. 2001 (10), 2455–2480.

Bhuyan, T., Mishra, K., Khanuja, M., et al., 2015. Biosynthesis of zinc oxide nanoparticles from *Azadirachta indica* for antibacterial and photocatalytic applications. Mater. Sci. Semicond. Process 32, 55–61.

Bolade, O.P., Williams, A.B., Benson, N.U., 2020. Green synthesis of iron based nanomaterials for environmental remediation: A Review. Environ. Nanotechnol. Monit. Manage., 13.

Bour, A., Mouchet, F., Silvestre, J., Gauthier, L., Pinelli, E., 2015. Environmentally relevant approaches to assess nanoparticles eco-toxicity: a review. J. Hazard. Mat. 283, 764–777.

Chatterjee, Arka, Kar, Prasenjit, Wulferding, Dirk, Lemmens, Peter, 2020. Samir Kumar Pal, Flower-like BiOI microspheres decorated with plasmonic gold nanoparticles for dual detoxification of organic and inorganic water pollutants. ACS Appl. Nano Mater. 3 (3), 2733–2744.

Chen, W.X., Lee, J.Y., Liu, Z., 2002. Microwave-assisted synthesis of carbon-supported Pt nanoparticles for fuel cell applications. Chem. Comm. 8 (21), 2588–2589.

Chen, Y.-L., Tuan, H.-Y., Tien, C.-W., et al., 2009. Augmented biosynthesis of cadmium sulfide nanoparticles by genetically engineered *Escherichia coli*. Biotechnol. Prog 25, 1260–1266.

Christian, P., der., Kammer F.Von, Baalousha, M., Th., Hofmann, 2008. Nanoparticles: structure, properties, preparation and behaviour in environmental media. Ecotoxicology 17, 326–343.

Cui, Y., Zhao, Y., Tian, Y., et al., 2012. The molecular mechanism of action of bactericidal gold nanoparticles on *Escherichia coli*. Biomaterials 33, 2327–2333.

Cushing, B.L., Kolesnichenko, V.L., O'Connor, C.J., 2004. Recent advances in the liquid-phase syntheses of inorganic nanoparticles. Chem. Rev. 104 (9), 3893–3946.

Daniel, M.C., Astruc, D., 2004. Gold nanoparticles: assembly, supramolecular chemistry, quantum-size-related properties, and applications towards biology, catalysis, and nanotechnology. Chem. Rev. 104, 293–346.

Das, S., Chakraborty, J., Chatterjee, S., Kumar, H., 2018. Prospects of biosynthesized nanomaterials for the remediation of organic and inorganic environmental contaminants. Environ. Sci. Nano 5, 2784–2808.

Dehghani, M.H., Karri, R.R., Alimohammadi, M., Nazmara, S., Zarei, A., Saeedi, Z., 2020. Insights into endocrine-disrupting Bisphenol-A adsorption from pharmaceutical effluent by chitosan immobilized nanoscale zero-valent iron nanoparticles. J. Mol. Liq. 311, 113317.

Devi, H.S., Singh, T.D., 2014. Synthesis of copper oxide nanoparticles by a novel method and its application in the degradation of methyl orange. Adv. Electron. Electr. Eng 4, 83–88.

Dror, I., Baram, D., Berkowitz, B., 2005. Use of nanosized catalysts for transformation of chloro-organic pollutants. Environ. Sci. Technol 39, 1283–1290.

Dutta, A.K., Maji, S.K., Adhikary, B., 2014. γ-Fe_2O_3 nanoparticles: an easily recoverable effective photocatalyst for the degradation of rose bengal and methylene blue dyes in the waste-water treatment plant. Mater. Res. Bull 49, 28–34.

Elia, P., Zach, R., Hazan, S., Kolusheva, S., Porat, Z., Zeiri, Y., 2014. Green synthesis of gold nanoparticles using plant extracts as reducing agents. Int. J. Nanomed 9, 4007–4021.

Fair, R.J., Tor, Y., 2014,. Antibiotics and bacterial resistance in the 21st century. Perspect. Med. Chem 6, 25–64.

Fan, D., Lan, Y., Tratnyek, P.G., Johnson, R.L., Filip, J., O'Carroll, D.M., Garcia, A.N., Agrawal, A., 2017. Sulfidation of iron-based materials: a review of processes and implications for water treatment and remediation. Environ. Sci. Technol. 51 (22), 13070–13085.

Faraday, M., 1857. The Bakerian lecture: experimental relations of gold (and other metals) to light. Philos. Trans. R. Soc. 147, 145–181.

Fazlzadeh, M., Rahmani, K., Zarei, A., Abdoallahzadeh, H., Nasiri, F., Khosravi, R., 2017. A novel green synthesis of zero valent iron nanoparticles (NZVI) using 5 three plant extracts and their efficient application for removal of 6 Cr(VI) from aqueous solutions. Adv. Pow. Tech. 28 (1), 122–130.

Fowsiya, J., Madhumitha, G., Al-Dhabi, N.A., Arasu, M.V., 2016. Photocatalytic degradation of Congo red using *Carissa edulis* extract capped zinc oxide nanoparticles. J. Photochem. Photobiol. B Biol 162, 395–401.

Gangula, A., Podila, R., Rao, A.M., et al., 2011. Catalytic reduction of 4-nitrophenol using biogenic gold and silver nanoparticles derived from *Breynia rhamnoides*. Langmuir 27, 15268–15274.

Gericke, M., Pinches, A., 2006. Microbial production of gold nanoparticles. Gold Bull 39, 22–28.

Ghasemzadeh, G., Momenpour, M., Omidi, F., Hosseini, M.R., Ahani, M., Barzegari, A., 2014. Applications of nanomaterials in water treatment and environmental remediation. Front. Environ. Sci. Eng. 8, 471–482.

Gnanasangeetha, D., Saralathambavani, D., 2014. Biogenic production of zinc oxide nanoparticles using *Acalypha indica*. J. Chem. Biol. Phys. Sci 4, 238–246.

Gonawala, K.H., Mehta, M.J., 2014. Removal of colour from different dye wastewater by using ferric oxide as an adsorbent. Int. J. Eng. Res. Appl 4, 102–109.

Gopalakrishnan, R., Loganathan, B., Dinesh, S., Raghu, K., 2017. Strategic green synthesis, characterization and catalytic application to 4-nitrophenol reduction of palladium nanoparticles. J. Clust. Sci 28, 2123–2131.

Gunalan, S., Sivaraj, R., Rajendran, V., 2012. Green synthesized ZnO nanoparticles against bacterial and fungal pathogens. Prog. Nat. Sci. Mater. Int 22, 693–700.

Guo, J.Z., Cui, H., Zhou, W., Wang, W., 2008. Ag nanoparticle-catalyzed chemiluminescent reaction between luminal and hydrogen peroxide. J. Photochem. Photobiol. A 193, 89–96.

Guo, K.W., 2012. Green nanotechnology of trends in future energy: a review. Int. J. Eng. Res. 1 (36), 1–17.

Habibi, M.H., Rezvani, Z., 2015. Photocatalytic degradation of an azo textile dye (C.I. Reactive Red 195 (3BF)) in aqueous solution over copper cobaltite nanocomposite coated on glass by doctor blade method. Spectrochim. Acta. A Mol. Biomol. Spectrosc 147, 173–177.

Hebbalalu, D., Lalley, J., Nadagouda, M.N., Varma, R.S., 2013. Greener techniques for the synthesis of silver nanoparticles using plant extracts, enzymes, bacteria, biodegradable polymers and microwaves. ACS Sustain. Chem. Eng. 7 (1), 703–712.

Hu, C., Lin, Y., Connell, J.W., Cheng, H.-M., Gogotsi, Y., Titirici, M.-M., Dai, L., 2019. Carbon-based metal-free catalysts for energy storage and environmental remediation. Adv. Mat. 31 (13), 1806128.

Hulkoti, N.I., Taranath, T.C., 2014. Biosynthesis of nanoparticles using microbes—a review. Colloids Surf. B Biointerf. 121, 474–483.

Hutchison, J.E., 2008. Greener nanoscience: a proactive approach to advancing applications and reducing implications of nanotechnology. ACS Nano 2, 395–402.

Iravani, S., 2014. Bacteria in nanoparticle synthesis: current status and future prospects. Int. Sch. Res. Not 2014, 1–18.

Jyoti, K, Singh, A., 2016. Green synthesis of nanostructured silver particles and their catalytic application in dye degradation. J. Genet. Eng. Biotechnol 14, 311–317.

Karthiga, D., Anthony, S.P., 2013. Selective colorimetric sensing of toxic metal cations by green synthesized silver nanoparticles over a wide pH range. RSC Adv 3, 16765–16774.

Khan, F.S.A., Mubarak, N.M., Khalid, M., Khan, M.M., Tan, Y.H., Walvekar, R., Abdullah, E.C., Karri, R.R., Rahman, M.E., 2021. Comprehensive review on carbon nanotubes embedded in different metal and polymer matrix: fabrications and applications. Critical Rev. Solid State Mater. Sci. 1–28.

Khan, F.S.A., Mubarak, N.M., Tan, Y.H., Karri, R.R., Khalid, M., Walvekar, R., Abdullah, E.C., Mazari, S.A., Nizamuddin, S., 2020. Magnetic nanoparticles incorporation into different substrates for dyes and heavy metals removal—A review. Env. Sci. Poll. Res. 1–16.

Khan, F.S.A., Mubarak, N.M., Tan, Y.H., Khalid, M., Karri, R.R., Walvekar, R., Abdullah, E.C., Nizamuddin, S., Mazari, S.A., 2021. A comprehensive review on magnetic carbon nanotubes and carbon nanotube-based buckypaper-heavy metal and dyes removal. J. Hazard. Mater. 125375.

Khin, M.M., Nair, A.S., Babu, V.J., Murugan, R., Ramakrishna, S., 2012. A review on nanomaterials for environmental remediation. Energy Environ. Sci. 5, 8075–8109.

Kuang, L., Mitchell, B.S., Fink, M.J., 2015. Silicon nanoparticles synthesised through reactive high- energy ball milling: enhancement of optical properties from the removal of iron impurities. J. Exp. Nanosci. 10 (16), 1214–1222.

Lim, S.H., Ahn, E.-Y., Park, Y., 2016. Green synthesis and catalytic activity of gold nanoparticles synthesized by *Artemisia capillaris* water extract. Nanoscale Res. Lett 11, 474.

Lima, E., Guerra, R., Lara, V., Guzmán, A., 2013,. Gold nanoparticles as efficient antimicrobial agents for *Escherichia coli* and *Salmonella typhi*. Chem. Cent. J 7, 11.

Lingamdinne, L.P., Koduru, J.R., Karri, R.R., 2019. A comprehensive review of applications of magnetic graphene oxide based nanocomposites for sustainable water purification. J. Env. Manag. 231, 622–634.

Liu, Z., Gan, L.M., Hong, L., Chen, W., Lee, J.Y., 2005. Carbon supported Pt nanoparticles as catalysts for proton exchange membrane fuel cells. J. Power Sour. 139 (1-2), 73–78.

Lowry, G.V., Gregory, K.B., Apte, S.C., Lead, J.R, 2012. Transformations of nanomaterials in the environment. Environ. Sci. Technol. 46, 6893–6899.

Maensiri, S., Laokul, P., Klinkaewnarong, J., et al., 2008. Indium oxide (In_2O_3) nanoparticles using aloe vera plant extract: synthesis and optical properties. J. Optoelectron. Adv. Mater 10, 161–165.

Mahapatra, O., Bhagat, M., Gopalakrishnan, C., Arunachalam, K.D., 2008. Ultrafine dispersed CuO nanoparticles and their antibacterial activity. J. Exp. Nanosci 3, 185–193.

Maiti, S., Gadadhar, B., Laha, J.K., 2016. Detection of heavy metals (Cu^{+2}, Hg^{+2}) by biosynthesized silver nanoparticles. Appl. Nanosci 6, 529–538.

Maleki, H., 2016. Recent advances in aerogels for environmental remediation applications: a review. Chem. Eng. J. 98–118.

Marchiol, L., 2012. Synthesis of metal nanoparticles in living plants. Ital. J. Agron 7, 274–282.

Matussin, S., Harunsani, M.H., Tan, A.L., Khan, M.M., 2020. Plant extract mediated SnO_2 nanoparticles synthesis and application. ACS Sustain. Chem. Eng. 8 (8), 3040–3054.

Mehmood, A., Khan, F.S.A., Mubarak, N.M., Tan, Y.H., Karri, R.R., Khalid, M., Walvekar, R., Abdullah, E.C., Nizamuddin, S., Mazari, S.A., 2021. Magnetic nanocomposites for sustainable water purification—a comprehensive review. Env. Sci. Poll. Res. 1–26.

Meyers, M.A., Mishra, A., Benson, D.J., 2006. Mechanical properties of nanocrystalline materials. Prog. Mater. Sci. 51 (4), 427–556.

Mittal, A.K., Chisti, Y., Banerjee, U.C., 2013. Synthesis of metallic nanoparticles using plant extracts. Biotech. Adv. 13 (2), 346–356.

Mohanpuria, P., Rana, N.K., Yadav, S.K., 2008. Biosynthesis of nanoparticles: technological concepts and future applications. J. Nanoparticle Res 10, 507–517.

Nakkala, J.R., Bhagat, E., Suchiang, K., Sadras, SR., 2015. Comparative study of antioxidant and catalytic activity of silver and gold nanoparticles synthesized from *Costus pictus* leaf extract. J. Mater. Sci. Technol 31, 986–994.

Narayanan, K.B., Sakthivel, N., 2011. Synthesis and characterization of nanogold composite using *Cylindrocladium floridanum* and its heterogeneous catalysis in the degradation of 4-nitrophenol. J. Hazard. Mater 189, 519–525.

Nath, D., Banerjee, P., 2013. Green nanotechnology – a new hope for medical biology. Environ. Toxicol. Pharmacol. 36 (3), 997–1014.

Nolan, E.M., Lippard, S.J., 2008. Tools and tactics for the optical detection of mercuric ion. Chem. Rev 108, 3443–3480.

Ovais, M., Khalil, A.T., Raza, A., Khan, M.A., Ahmad, I., Islam, N.U., Saravanan, Muthupandian, Ubaid, Muhammad Furqan, Ali, Muhammad, Khan Shinwari, Zabta, 2016. Green synthesis of silver nanoparticles via plant extract: beginning a new era in cancer theranostics. Nanomedicine 11 (23), 3157–3177.

Padhi, B.S.Ratna, 2012. Pollution due to synthetic dyes toxicity and carcinogenicity studies and remediation. Int. J. Environ. Sci 3, 940–955.

Pan, B., Xing, B., 2012. Applications and implications of manufactured nanoparticles in soils: a review. Euro. J. Soil Sci. 63 (4), 437–456.

Panigrahi, S., Basu, S., Praharaj, S., et al., 2007. Synthesis and size-selective catalysis by supported gold nanoparticles: study on heterogeneous and homogeneous catalytic process. J. Phys. Chem. C 111, 4596–4605.

Patzke, G.R., Zhou, Y., Kontic, R., Conrad, F., 2011. Oxide nanomaterials: synthetic developments, mechanistic studies, and technological innovations. Angew. Chem. Int. Ed. 50 (4), 826–859.

Peralta-Videa, J.R., Zhao, L., Lopez-Moreno, M.L., la, Rosa G.de, Jie, Hong, Gardea-Torresdey, J.L., 2011. Nanomaterials and the environment: a review for the biennium 2008–2010. J. Hazard. Mater. 186, 1–15.

Pradeep, T., 2009. Anshup: noble metal nanoparticles for water purification: a critical review. Thin Solid Films 517, 6441–6478.

Ramanathan, R., Field, M.R., O'Mullane, Anthony P., Smooker, P.M., Bhargava, S.K., Bansal, V., 2013. Aqueous phase synthesis of copper nanoparticles: a link between heavy metal resistance and nanoparticles synthesis ability in bacterial systems. Nanoscale (5) 2300–2306.

Ray, P.C., 2010. Size and shape dependent second order nonlinear optical properties of nanomaterials and their application in biological and chemical sensing. Chem. Rev 110, 5332–5365.

Rostami-Vartooni, A., Nasrollahzadeh, M., Alizadeh, M., 2016. Green synthesis of perlite supported silver nanoparticles using *Hamamelis virginiana* leaf extract and investigation of its catalytic activity for the reduction of 4-nitrophenol and Congo red. J. Alloys Compd 680, 309–314.

Sahu, J.N., Karri, R.R., Zabed, H.M., Shams, S., Qi, X., 2021. Current perspectives and future prospects of nano-biotechnology in wastewater treatment. Separation & Purification Rev. 50 (2), 139–158.

Scaramuzza, S., Agnoli, S., Amendola, V., 2015. Metastable alloy nanoparticles, metal-oxide nanocrescents and nanoshells generated by laser ablation in liquid solution: influence of the chemical environment on structure and composition. Phys. Chem. Chem. Phys. 17 (42), 28076–28087.

Schwarzenbach, R.P., Egli, T., Hofstetter, T.B., Gunten, U.V., Wehrli, B., 2010. Global water pollution and human health. Annu. Rev. Environ. Resour. 35, 109–136.

Sharma, J.K., Akhtar, M.S., Ameen, S., et al., 2015. Green synthesis of CuO nanoparticles with leaf extract of *Calotropis gigantea* and its dye-sensitized solar cells applications. J. Alloys Compd 632, 321–325.

Singh, J., Kukkar, P., Sammi, H., et al., 2017. Enhanced catalytic reduction of 4-nitrophenol and congo red dye by silver nanoparticles prepared from *Azadirachta indica* leaf extract under direct sunlight exposure. Part. Sci. Technol 37, 434–443.

Singh, J., Dutta, T., Kim, K-H, Rawat, M., Samddar, P., Kumar, P., 2018. Green synthesis of metals and their oxide nanoparticles: applications for environmental remediation. J. Nanobiotechnol 16, 84.

Song, J.Y., Kim, B.S., 2009. Rapid biological synthesis of silver nanoparticles using plant leaf extracts. Bioprocess Biosyst Eng 32, 79–84.

Stan, M., Popa, A., Toloman, D., et al., 2015. Enhanced photocatalytic degradation properties of zinc oxide nanoparticles synthesized by using plant extracts. Mater. Sci. Semicond. Process 39, 23–29.

Sun, G.L., Reynolds, Erin.E., Belcher, A.M., 2019. Designing yeast as plant-like hyperaccumulators for heavy metals. Nat. Comm 10, 5080.

Tesh, S.J., Scott, T.B., 2014. Nano-composites for water remediation: a review. Adv. Mat. 26 (35), 6056–6068.

Thacker, H., Ram, V., Dave, P.N., 2019. Plant mediated synthesis of iron nanoparticles and their applications: A review. Prog. Chem. Biochem. Res. 2 (3), 84–91.

Thakkar, K.N., Mhatre, S.S, Parikh, R.Y., 2010. Biological synthesis of metallic nanoparticles. Nanomed. Nanotechnol. Biol. Med 6, 257–262.

Thakkar, K.N., Mhatre, S.S., Parikh, R.Y., 2010. Biological synthesis of metallic nanoparticles. Nanomed. Nanotechnol. Biol. Med. 6 (2), 257–262.

Thandapani, K., Kathiravan, M., Namasivayam, E., et al., 2017. Enhanced larvicidal, antibacterial, and photocatalytic efficacy of TiO_2 nanohybrids green synthesized using the aqueous leaf extract of *Parthenium hysterophorus*. Environ. Sci. Pollut. Res 25, 1–12.

Toshima, N., Yonezawa, T., 1998. Bimetallic nanoparticles—novel materials for chemical and physical applications. New J. Chem. 22 (11), 1179–1201.

Tsuda, A., Konduru, N.V., 2016. The role of natural processes and surface energy of inhaled engineered nanoparticles on aggregation and corona formation. NanoImpact 2, 38–44.

Tsuji, T., Iryo, K., Watanabe, N., Tsuji, M., 2002. Preparation of silver nanoparticles by laser ablation in solution: influence of laser wavelength on particle size. Appl. Surf. Sci. 202 (1-2), 80–85.

Varadavenkatesan, T., Selvaraj, R., Vinayagam, R., 2016. Phytosynthesis of silver nanoparticles from *Mussaenda erythrophylla* leaf extract and their application in catalytic degradation of methyl orange dye. J. Mol. Liquids 221, 1063–1070.

Vidya, C., Hiremath, S., Chandraprabha, M.N., et al., 2013. Green synthesis of ZnO nanoparticles by *Calotropis gigantea*. Int. J. Curr. Eng. Technol (1) 118–120.

Wesenberg, D., Kyriakides, I., Agathos, S.N., 2003. White-rot fungi and their enzymes for the treatment of industrial dye effluents. Biotechnol. Adv 22, 161–187.

Woo, Y., Lai, D.Y., Bingham, E., Cohrssen, B., Powell, C.H., 2012. Aromatic amino and nitro–amino compounds and their halogenated derivatives. Patty's Toxicology. Wiley.
Wu, T., Liu, X., Liu, Y., Cheng, M., Liu, Z., Zeng, G., Shao, B., Liang, Q., Zhang, W., He, Q., Zhang, W., 2020. Application of QD-MOF composites for photocatalysis: energy production and environmental remediation. Coord. Chem. Rev. 403, 213097.
Wu, Y., Pang, H., Liu, Y., Wang, X., Yu, S., Fu, D., Chen, J., Wang, X., 2019. Environmental remediation of heavy metal ions by novel-nanomaterials: a review. Environ. Pollut. 246, 608–620.
Xing, B., Zhu, L., Meng, L., Shi, J., Li, J., Zhang, X., Feng, M., 2019. Metal-organic frameworks/carbon-based materials for environmental remediation: a state-of-the-art mini-review. J. Environ. Manag. 232, 964–977.
Xing, B., Zou, Y., Wang, X., Khan, A., Wang, P., Liu, Y., Alsaedi, A., Hayat, T., Wang, X., 2016. Environmental remediation and application of nanoscale zero-valent iron and its composites for the removal of heavy metal ions: a review. Environ. Sci. Technol. 50 (14), 7290–7304.
Zhang, M., Liu, Y.-Q., Ye, B.-C., 2012. Colorimetric assay for parallel detection of Cd^{2+}, Ni^{2+} and Co^{2+} using peptide-modified gold nanoparticles. Analyst 137, 601–607.
Zhou, B., Balee, R., Groenendaal, R., 2005. Nanoparticle and nanostructure catalysts: technologies and markets. Nanotech. Law Busi. 2 (3), 222–322.
Zhou, Y., Kong, Y., Kundu, S., et al., 2012. Antibacterial activities of gold and silver nanoparticles against *Escherichia coli* and bacillus Calmette-Guérin. J. Nanobiotechnol 1 (1).
Zinjarde, S., 2012. Bio-inspired nanomaterials and their applications as antimicrobial agents. Chron. Young Sci 3 (1), 1–74.

CHAPTER 3

Insights of green and biosynthesis of nanoparticles

Ljubica Tasic[a], Danijela Stanisic[a], Lucas G. Martins[a], Guilherme C.F. Cruz[a] and Raluca Savu[b]
[a]Chemical Biology Laboratory, Department of Organic Chemistry, Institute of Chemistry, University of Campinas (UNICAMP), Campinas, SP, Brazil
[b]Centre for Semiconductor Components and Nanotechnology (CCS-Nano), University of Campinas (UNICAMP), Campinas, SP, Brazil

3.1 Introduction

Nanoparticles (NPs) are materials that at least in 1D show nanometric size, from 1 nm to 100 nm, with characteristic large surface to volume ratios (Barros & Eoin, 2020; Bandala et al., 2020; Tripathi & Chung, 2019a). Among NPs are found biogenic NPs (BNPs) that are synthesized using biological routes and show some very specific properties caused by their stabilizing agents, which are called capping or corona biomolecules (Ballotin et al., 2016) that make them more diverse when compared to other NPs. BNPs show many forms (Gericke & Pinches, 2006a) such as sphere, cube, rod, triangular, flower, and many sizes, which range from just few nanometers in all directions (3D) to nanofibers that are thin but long such as spiders net, Fig. 3.1. BNPs are exclusively obtained through bottom-up protocols, intra- or extracellularly, or using biomimetic processes. Bottom-up, or self-assembly, is approach to nanofabrication that uses chemical or physical forces operating at the nanoscale to assemble basic units into larger structures. BNPs are intensively studied because of interesting properties and applications (Barabadi, 2017; Barabadi et al., 2020; Bandala et al., 2020), and the most studied are metallic NPs (Durán et al., 2011), metallic oxide NPs (Durán & Seabra, 2012), metallic sulfide NPs, and bimetallic NPs (Durán & Seabra, 2018). Biogenic syntheses (El-Shabasy et al., 2019; Durán et al., 2005, Durán et al., 2011) of metallic NPs based on the reduction of metal ions, or hydroxides can be used, by plant extracts, bacterial and fungal extracellular and intracellular syntheses, to self-assemble Me^0 valence into a suitable form and size (Durán & Seabra, 2018). The size and form of MNPs can variate and, generally, there is certain heterogeneity that rises because of stabilization of the NPs during Me^0 growth. Consequently, different biogenic stabilization agents, mostly proteins, lead to different theories on biogenic metallic NPs (BMNPs) growth mechanisms (Durán & Seabra, 2012).

Biogenic nanomaterials (BNMs) have at least one dimension smaller than 100 nm and different forms such as spherical, rod- like or flowers. Capping biomolecules make spherical NPs look like classical core/shell NPs, but their external layers are called capping

Sustainable Nanotechnology for Environmental Remediation
DOI: https://doi.org/10.1016/B978-0-12-824547-7.00020-5

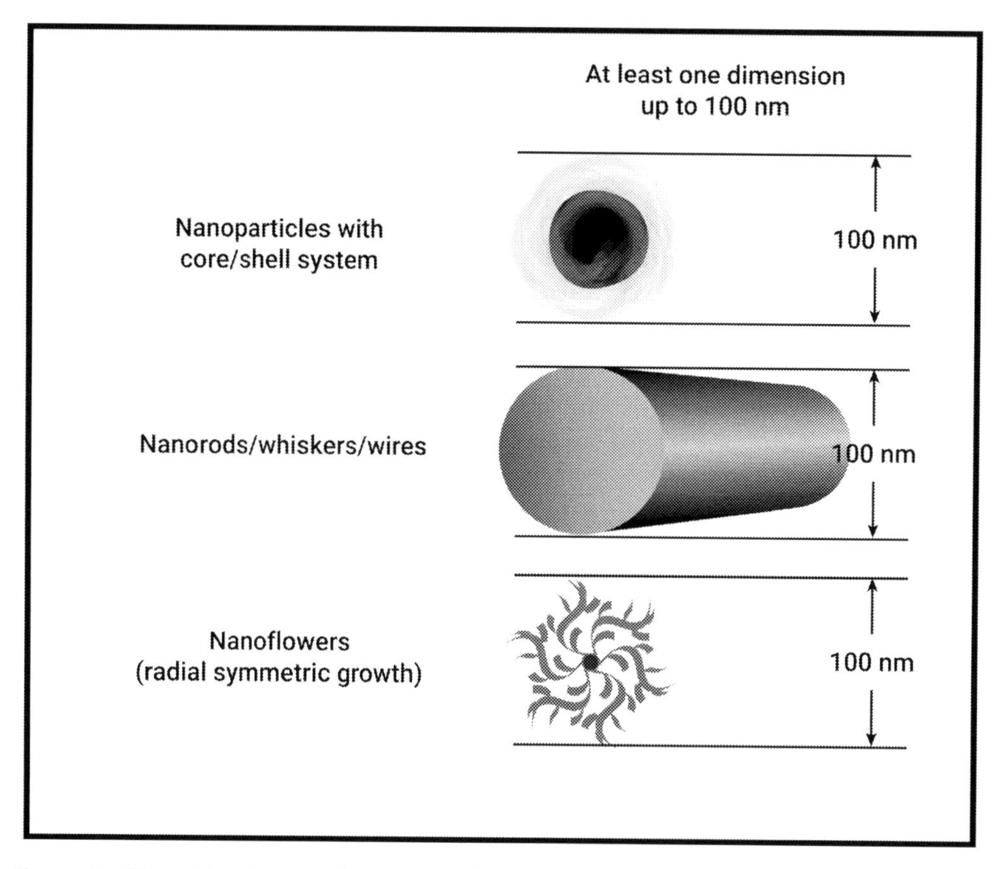

Figure 3.1 Illustration of biogenic nanoparticles (BNP) forms. BNMs show at least 1D smaller than 100 nm and different forms, such as spherical, rod like or flowers. Capping biomolecules make spherical nanoparticles (NPs) look like classical core/shell NPs, but their external layers are called capping or corona layers because of their complexity and distribution, which are different from classical shell systems.

or corona layers because of their complexity and distribution, which are different from classical shell systems.

Gold (Gericke & Pinches, 2006; Lee et al., 2019; Lingamdinne et al., 2020), palladium (Gioria et al., 2020), and silver (Lee & Jun, 2019) NPs show many fascinating properties that are explored for biomedical, nanotechnical, and environment applications (trace metal detection).

There are numerous ways to synthesize BNMs, but most of the BNMs account for silver, gold, and copper NPs, while other Me^0NPs are less explored. Table 3.1 presents just some literature data on bacterial and fungal syntheses that are discussed in following sections. AgNPs and AuNPs are the most studied BNMs and can be prepared using different bacterial and fungal strains.

Table 3.1 Biogenic synthesis of metallic nanoparticles.

Source	References	BNMs
Bacteria		
Bacillus cereus	Ganesh & Gunasekaran, 2009	AgNPs
Morganella morganii strain RP42	Ramanathan et al., 2013	CuNPs
Escherichia coli	Shivashankarappa & Sanjay, 2020	CdSNPs
Aeromonas sp. THG-FG1.2	Singh et al., 2017	AgNPs
Lactobacillus strains	Sintubin et al. 2009	AgNPs
Corynebacterium	Zhang et al., 2005	AgNPs
Staphylococcus aureus	Nanda and Saravanan, 2009	AgNPs
Pseudomonas putida	Avendaño et al., 2016	SeNPs
Pseudomonas aeruginosa	Kumari et al., 2017	AgNPs
Fungi and yeasts		
Aspergillus flavus	Vigneshwaran et al., 2017	AgNPs
Verticillium sp.	Mukherjee et al., 2001	AuNPs
Fusarium oxysporum	Durán et al., 2005	AgNPs
Fusarium oxysporum	Mirzadeh et al., 2013	ZnS-NPs
Thermophilic filamentous fungi	Molnár et al., 2018	AuNPs
Trichoderma viride	Thakkar et al., 2010	Metallic NPs
Neurospora crassa	Castro-Longoria et al., 2011	AgNPs, AuNPs, Ag/AuNPs
Fusarium acuminatum	Ingle et al., 2008	AgNPs
Botrytis cinerea	Castro et al., 2014	AuNPs
Rhizopus oryzae	Vala, 2014	AuNPs
Extremophilic yeast	Mourato et al., 2011	AgNPs, AuNPs
Phoma macrostoma	Sheikhloo & Salouti, 2012	AuNPs
Aspergillus tubingensis	Rodrigues et al., 2012	AgNPs
Bionectria ochroleuca	Rodrigues et al., 2012	AgNPs
Trichothecium sp.	Absar et al., 2005	AuNPs

Note: There are also many plants, algal- and viruses-based synthetic strategies as summarized in Table 3.2.

The most biosynthetic methods (Table 3.2) use reduction agents to promote metallic ions transformation into the MNPs and tailor MNP shapes, while the use of stabilizing agents ensures that MNPs do not aggregate and controls their sizes (Alamri et al., 2018; Ahmad et al., 2019; Fang et al., 2019; Garibo et al., 2020; Ghosh & Paria, 2012; Mehmood et al., 2021). Biological synthesis often counts on mild conditions and all-in-one reduction/oxidation and stabilization (Hulkoti & Taranath, 2014; Luo et al., 2015; Maxwell et al., 2017; Mittal et al., 2013; Yugav et al., 2020; Sharma et al., 2020), but some explore two or more steps for MNP production.

Intracellular synthesis uses fungi, bacteria, or algae for MNP reduction, while extracellular synthesis uses reduction, biosorption, enzyme, or latex-mediated processes and is less laborious because of no need to recover and purify MNPs (Mukherjee et al., 2001; Narayann & Sakthivel, 2010; Patil & Chandrasekaran, 2020; Singh et al., 2015).

Table 3.2 BNM syntheses with plants, algae, and viruses

Source	References	BNMs
Algal		
Fucus vesiculosus	Mata et al., 2009	AuNPs
Tetraselmis kochinensis	Senapati et al., 2012	AuNPs
Gracilaria edulis—macro alga	Priyadharshini et al., 2014	AgNPs, ZnONPs
Brown seaweed *Sargassum wightii*	Govindaraju et al., 2009	AgNPs
Microalga *Chlorella vulgaris*	Luangpipat et al., 2011	AuNPs
Plants based		
Citrus sinensis	Barros et al., 2018a Barros et al., 2018b	AgNPs
Citrus sinensis	Adama et al., 2020	TiO_2NPs
Plants based*	Chung et al., 2016 and references therein	AgNPs
Plants based*	Durán et al., 2011 and references therein	AgNPs
Plants based*	Durán et al., 2018 and references therein	Au/AgNPs, bimetallic
Extracts from plants	El-Seedi et al., 2019 and references therein	AgNPs. AuNPs, PdNPs, Au-AgNPs, ZnO
Clove buds extract	Sharma et al., 2020	Au/AgNPs, Bimetallic
Viruses		
Tobacco mosaic virus (TMV)	Shenton et al., 1999	SiO_2, CdS, PbS, Fe_2O_3
M13 bacteriophage	Mao et al., 2003	ZnS, CdS
Viruses with icosahedron geometry	Narayanan and Han, 2017	MNPs
Plant virus *Squash leaf* curl *China virus* (SLCCNV)	Thangavelu et al., 2020	AuNPs and AgNPs

* Many different extracts from plants such as from flowers, fruits, leaves, etc. can be used.

Bio-based methods mimic biological synthesis by using isolated compounds from plants, bacteria, and fungi (Barros et al., 2018a; Barros et al., 2018b). Other NPs can also be prepared to apply biosynthetic processes, such as metal-oxide particles (Nitin et al., 2021; Sackey at al., 2021), which gained attention because of interesting applications, such as magnetic NPs (Lima et al., 2020; Jung et al., 2012).

In regard of feasibility and costs for production of BNMs, extracellular synthesis based on extracts from waste biomass, plants, algae, fungal (Rajput et al., 2016; Ballotin et al.,) or bacteria might figure as best solutions. Green synthesis is used for successful synthesis of numerous BNPs. From the standpoint of process scalability and easiness to perform, it stands out as bio-based methods, using food and agro residues (Barros et al., 2018a; Barros et al., 2018b). Biomolecules isolated from waste commonly lead to the fast, low-cost, and green approaches. Reused, recycled waste is easily available and does not require rigorous processing methods. These can be directly used in the synthesis of BNPs and also pave the way for the management of the waste. However, the preparation of algae culture requires time and the costs are high, which is a limitation of maintaining the cultures over a period of time. The toxicity factors of some of the species of the algae are also to be considered before using them for the synthesis of NPs as certain compounds present could induce toxicity to the NPs and thus limit their use for biomedical applications. On the other side, fungal- and bacterial-based synthesis must take in consideration the process costs, environmental issues, and contamination of the cultures with other microorganisms, which can be the bottlenecks for the successful BNM production.

Nevertheless, the most of the cited processes explore nonhazardous reagents and a great majority is environmentally friendly, clean, and reproducible in small-scale production, but large-scale production of BNM is still a challenge, and there are no commercially available BNMs. Therefore, current trends need to tackle the challenges of green biosynthesis of nanomaterial (NMs) to their wide-scale fabrication.

There are many fields for applications of biosynthesized nanomaterials, and one of those is biomedical area of research, followed with catharsis and analytics, such as photocatalysis (Letsholathebe, et al., 2020), sensors, etc. The properties of the BNMs exhibit broad new spectra of potential applications (Prasad et al., 2011; Pramila & Kumar, 2018; Maheshwaran et al., 2020; Letsholathebe, et al., 2020; Gudikandula et al., 2017; Chen et al., 2014; Barros et al., 2018a; Ballotin et al., 2017).

3.2 Biosynthesis of nanomaterials

3.2.1 Green methods

Many NPs can be synthesized through eco-friendly, sustainable, low-cost, and/or quick methods (Hung et al., 2020; Shukla & Iravani, 2017; Rolim et al., 2019) that are gaining in popularity and in near future may replace chemical methods for NP synthesis that mostly employ routes and chemicals that can be toxic or generate hazardous materials to the environment. Applied biological processes for BNP synthesis include use of cells, whole microorganisms or even viruses, or their derivatives for the syntheses of metallic NPs (MNPs), metallic oxide NPs, metallic sulfide NPs, and others. Almost all biogenic methods are considered green because they explore renewable resources for the BNM syntheses and minimize use of potentially hazardous materials. Details on the most applied biogenic methods for BNP syntheses follow.

3.2.2 BNM bacterial biosynthesis

Many bacteria and bacterial subproducts are successfully used as mini bioreactors for production of BNMs. The most explored in the literature are *Bacillus, Escherichia,*

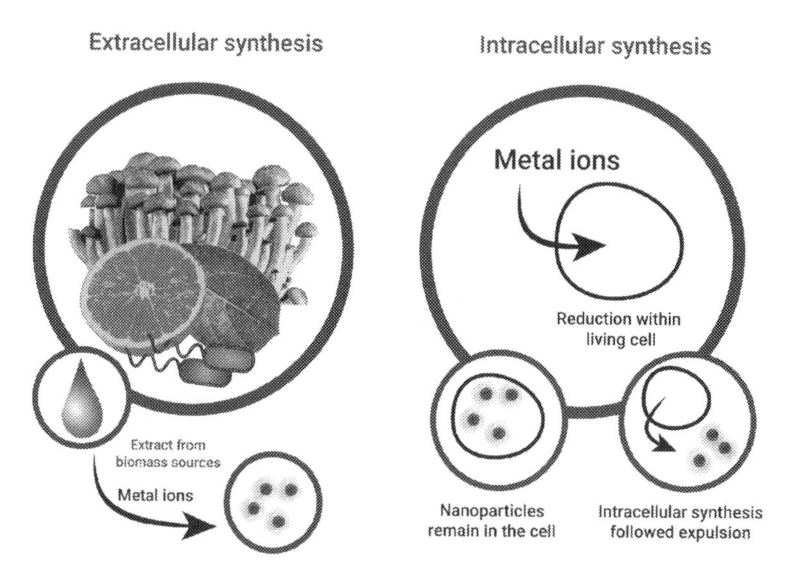

Figure 3.2 BNPs can be synthesized using different strategies. There are two types of BNP synthesis: extracellular and intracellular, within which or cells' extracts are used (left panel) or BNPs are expelled after synthesis.

Pseudomonas, Serratia, Shewanella, Rhodopseudomonas, Lactobacillus, and *Actinobacter*, but other bacteria can also be used for NPs production (Luo et al., 2015; Shukla & Iravani, 2017; Dhondiia & Chakraborty, 2012).

There are two methods for bacterial synthesis named bacterial in- and outgrowth of BNMs as illustrated in Fig. 3.2 Intracellular synthesis occurs inside the bacterial cell while the extracellular process happens outside the bacterial cell (Fig. 3.2). Also, some bio-based or biomimetic processes may include the use of bacterial extracts (such as total proteins) or isolated target extracted components, for example, bovine serum albumin (Murawala et al., 2009), or other compounds (Capeness et al., 2019), for BNM synthesis.

Intracellular biosynthesis occurs inside the bacterial cells and, in many cases, it is necessary to release the produced BNMs from the bacterial cell biomass, then to purify and wash the BNMs before any characterization or further use. In the intracellular synthesis, the most important is that the employed bacterial strain shows certain resistance to metal ion presence (Barros et al., 2018). Although the exact mechanism for the BNM synthesis is not known, bacterial proteins' groups such as amino (-NH_2), hydroxy (-OH), thiol (-SH), and carboxyl groups (-COOH) may take part in the reduction of metal ions and posterior stabilization of the formed metal NPs (MNPs).

But before that can even happen, metal ions and bacteria must collide, then metal ion must be imported into bacterial cell and then bacteria may conduct different pathways and synthesize metal or metal oxide of metal sulfide NPs. There are many bacterial compartments that can host BNM synthesis, which is the reason for nonuniform BNMs produced in regard of their sizes and types. For instance, copper ion might be used to synthesize CuNPs as observed for *Morganella morganii* strain RP42, which is Ag(I)-resistant strain, guided biosynthesis (Ramanathan et al., 2013), where just CuNPs were obtained, synthesized intracellularly but expelled using protein transporters. The capping CuNP proteins might have protected them from oxidation to CuO-NPs. The proteins' capping of BNMs occurs by sulfur-metal and nitrogen-metal interactions (Ballottin et al., 2016a), and by additional proteins' layer due to protein–protein interactions (Ramanathan et al., 2013). Other studies reported silver oxide-NPs synthesis using bacterial strains and different silver oxide types such as Ag_2O to Ag_3O_4 (Dhoondia & Chakraborty, 2012).

In intracellular synthesis, metal ions are absorbed by bacterial cells and expelled in nanoforms after reduction and stabilization. To be reduced, metal ions must receive electrons coming from the donor groups, which usually involve proteins and their thiol groups, peptides with thiol amino acid (Cys) in their sequence, but also, cytochromes, DNA, RNA, NAD(P)H, and enzymes from the family of oxidoreductases can take part in such synthesis. Anyway, it is necessary to additionally clean BNMs after bioreduction and stabilization (Shukla & Iravani, 2017; Singh et al., 2015).

Different bacterial cells' supernatants can be also used for this kind of synthesis. But if intercellular bacterial process is discussed, then metal ions can undergo reduction and stabilization outside, inside, or in bacteria periplasmic space. Besides, there are cytoplasm components such as NADH/NAD(P)H, vitamins, and organic acids that might be electron donors. Capping proteins usually stabilize the newly formed NPs and prevent their oxidation or agglomeration and nanomaterials' growth to bigger sizes (Fang et al., 2019). By avoiding the chemical reduction, an environmentally friendly technology was introduced by Singh and colleagues (Singh et al., 2017; Singh et al., 2015) and *Aeromonas sp.* THG-FG1.2 was employed to AgNP production. The Ag^0 biosynthesized NPs were spherical with average size of 8–15 nm.

Pseudomonas aeruginosa is a well-known pathogenic strain, therefore its extracted secondary metabolite, pyoverdine, effectively reduced silver(I) nitrate to spherical and pseudo-spherical AgNPs (Kumari et al., 2017). *Pseudomonas putida* was for the first time reported as a soil strain that has the ability to reduce selenite to the nano-Se (Avendaño et al., 2016). The process occurs aerobically and the NPs, in range size from 100 to 500 nm, were bounded to cell membrane and dispersed in surrounding medium. According to authors, reduction of selenium oxyanions involves two steps: selenate is reduced to selenite, and then selenite is reduced to elemental selenium. Nitrate and/or nitrite reductase take part in the mechanism, reduction of selenate to selenite, $SeO_4^{-2} \rightarrow SeO_3^{-2}$ (Avendaño et al., 2016).

3.2.3 Plant BNM synthesis

Plant extracts (Fig. 3.2) are very common choice when BNMs are being prepared and act as reducing agents and stabilizers (Adama et al., 2020; Barros et al., 2018; de Barros et al., 2018; Chung et al., 2016; Akintelu et al., 2020; Vigneshwaran et al., 2017). The hypothesized mechanism of reduction of metallic salt solution starts with the extracts' components complexation and interaction with the metal ion and then reduction to formation of the nanometal particles or nanometal oxides (Akintelu et al., 2020). The plant extracts have many polyphenols capable to reduce metal ions and stabilize MNP during nucleation and growth. For example, *Mimosa hamata* flower extract is used for the biosynthesis of CuONPs. The reducing agent suggested by authors, present in the *Mimosa* plant are 4-ethyl gallic acid and 3,4,5-trihydroxybenzoic acid. The reduction of cupric acetate to cubic CuO-NPs, and an average size ~20 nm was also performed successfully (Sackey et al., 2021).

Tetragonal TiO_2NPs were synthesized using green method with natural extract of orange peel (*Citrus sinensis*). Titanium (IV) bis (ammonium lactate) dihydroxide was reduced by water extract from peels, and tetragonal NPs were consequently dried at high temperature (700–900 °C). Water extract of orange peel is rich in polyphenols and citric acid that demonstrated excellent reducing abilities (Adama et al., 2020). Zinc oxide (ZnO) NPs were produced using *Moringa Oleifera* leaf extract by hydrothermal method, starting from zinc nitrate hexahydrate ($Zn_2N_2O_{12}H_{12}$). Those ZnONPs have successful application in photocatalytic degradation of Titan yellow dye, which is widely used in paper, textile, wood and leather dyeing (Letsholathebe et al., 2020; Nitin et al., 2021).

Greener approach for nano-Ag_2O production was applied by reducing silver nitrate promoted with *Zephyranthes Rosea* flower extract. The possible reducing agent in this type of synthesis is probably by lycorine and other compounds present in flower extract, as suggested by authors. *Zephyranthes* consists of flavonoids, lectins, gibberellins, alkaloids, sterols, and others, among those alkaloids that have different skeletons such as lycorine, crinane, pancratistatin, and hemanthamine (Maheshwaran et al., 2020).

Single-step synthesis of bimetallic Au/AgNPs approach emploies clove buds extract and, from gold (III) trichloride trihydrate ($HAuCl_4 \cdot 3H_2O$) and silver(I) nitrate ($AgNO_3$), the final product, bimetallic NPs, presents average sizes around 18 nm. Au(III) and Ag(I) reductions were completed by eugenol, the main polyphenol present in clove buds (Sharma et al., 2020).

The biomass residues are a new area of interest for development of economical strategies in nanotechnology. Water hyacinth (*Eichhoria crassipes*) extract was used for reduction of $FeCl_3 \cdot 6H_2O$ and $Co(NO_3)_2 \cdot 6H_2O$ and synthesis of magnetic hybrid material for removal of organic (ibuprofen) and inorganic (Zn, Cu, and Ni) pollutants from aqueous media in different pH ranges. In the United States of America, the hyacinth is listed as an invasive harmful weed, while in Brazil, there are already methods to reduce such species. Therefore, authors suggested exploitation of the exotic invasive plant species for production of MNPs and their application as metal adsorbents and for the removal of pharmaceutical, agricultural, and industrial effluents (Lima et al., 2020).

Another example is β-D-glucose, soluble starch, and solution of H_2PdCl_4, which were tested in batch and continuous synthesis of PdNPs. The average size of PdNPs was from 4 up to 20 nm produced by green protocols and microfluidic devices (Gioria et al., 2020).

When dealing with the BNM synthesis using plants or plant biomass extracts, the major drawback is reproducibility as to maintain similar chemical composition of extracts and, consequently, standard ratios among reducing and stabilizing agents. This is difficult because of a significant variation in chemical compositions of plant extract even if extracts were obtained using the standardized procedure and the same plant species that show differences when grown in different parts of the world. Therefore, it would be recommended to elucidate mechanism of BNM synthesis, identify the biomolecules responsible for reduction/oxidation and BNM stabilization as to overcome the problem of reproducibility.

Another frequently encountered problem is that capping agents vary in number, type, and the way that interact or bound to metallic surfaces. It is not known how many layers protect BNMs from oxidation and further growth into micro-sized materials. Another important feature is BNMs interaction with other biomolecules (Murawala et al., 2009; Paula et al., 2014)

3.2.4 BNM fungal synthesis

As already discussed, in MNPs synthesis, metal ions must be reduced to their zero oxidation state and this reduction is mediated by reducing agents. In cases where the reaction occurs in a biomimetic form, especially using fungi as mediators, it will be the biomolecules produced by the fungi that will act as reducing agents in the NP synthesis process (Hulkoti & Taranath, 2014; Gade et al., 2008).

The exact mechanism of MNPs formation through fungi is still not completely elucidated; however it can occur intra- or extracellularly. For the synthesis of NP to take place via the intracellular route (Fig. 3.2), it is necessary that there is a special transport of ions into the fungi cell. In this case, the cell wall of the fungus is of paramount importance for the synthesis to occur. This is because the negative charge of the cell wall interacts electrostatically with the positive charge of the metal ion, allowing the enzymes present in the cell wall to reduce the ion to NPs and then they diffuse through the cell wall. The metal is then trapped, bioreduced, and capped (Mukherjee et al., 2001). NPs produced within fungal cells may be smaller compared to the size of the NPs produced using the fungal filtrate (Fig. 3.3). Nucleation within cells can be more limited, which reduces

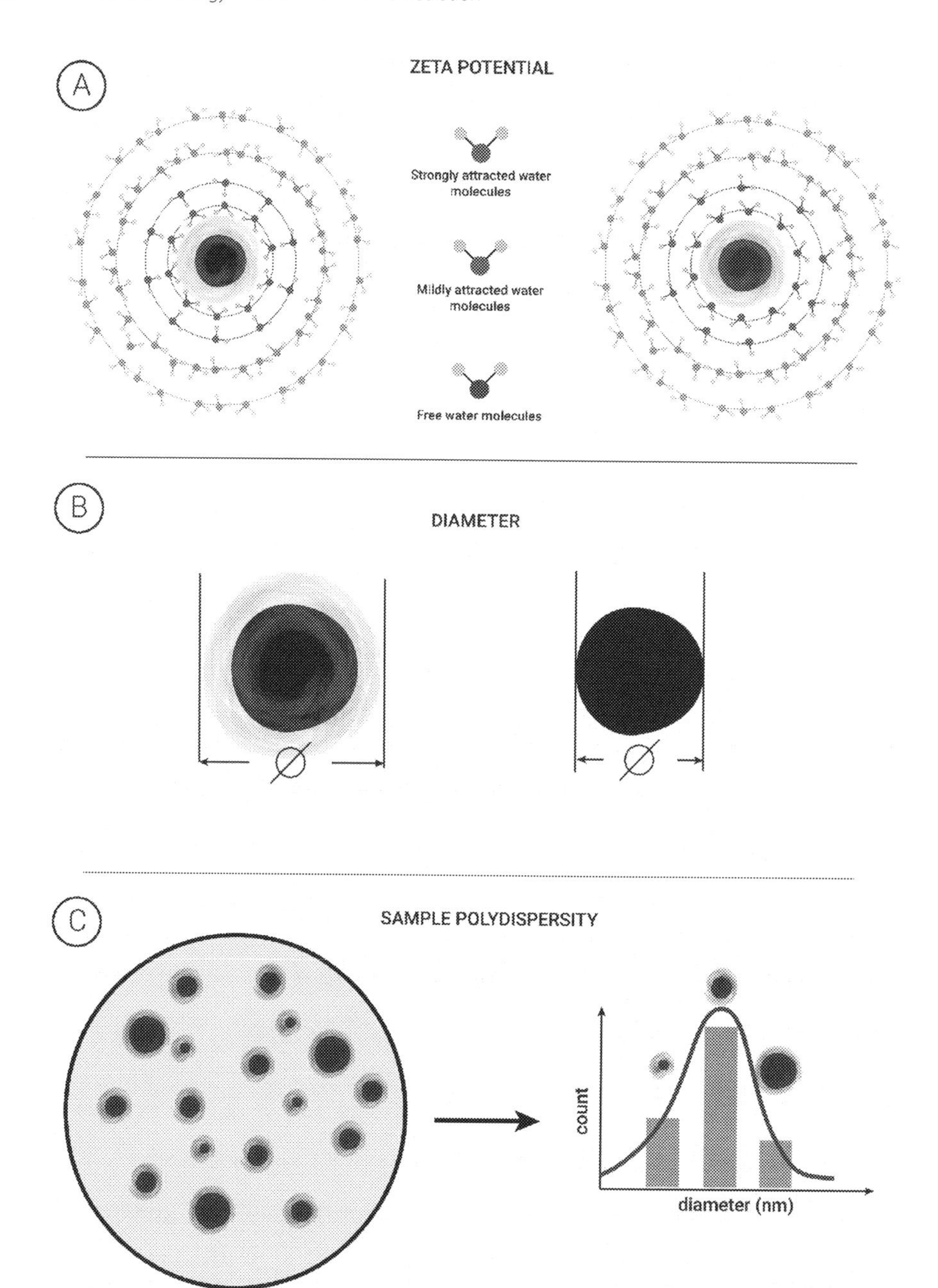

Figure 3.3 Physical properties of BNM illustrated in the case of spherical NP. NP properties define their action and use. (A) NPs show mostly negative zeta potentials and are considered stable if their zeta potential are lower than -30 mV. But some NPs show positive zeta potentials. And as illustrated, in suspensions of NPs in the first few water layers are organized in a way to contribute to NP stability, see that water molecule orientation depends on NP zeta potential. (B) Stabilizing agents make capping that can be measured and observed by TEM, therefore NP core is smaller that NP diameter. (C) BNMs are synthesized in a manner that their sizes are different from smaller to greater nanomaterials and show sample polydispersity (PDI).

the final size of the NP (Narayanan & Sakthivel, 2010). There are several reports of intracellular synthesis of gold NPs, produced by *V. luteoalbum* (Gericke & Pinches, 2006a, 2006b), *Trichothecium* sp. (Ahmad et al., 2019), *Verticillium* sp. (Ahmad, et al., 2001), *Phoma macrostoma* (Sheikhloo & Salouti, 2012), and *Rhizopus oryzae* (Vala, 2014).

There is also evidence of obtaining silver NPs (AgNPs) intracellularly by *Verticillium* sp. (Ahmad et al., 2001) *Aspergillus flavus* (Vigneshwaran et al., 2017), by the yeast *Meyerozyma guilliermondii* (Alamri et al., 2018) and even bimetallic gold and AgNPs obtained by *Neurospora crassa* (Castro-Longoria et al., 2011), all showing a size range from around 9 nm to 32 nm. Although the NPs synthesized through intracellular mechanisms of fungal cells have good size distribution, this method of obtaining NPs has been used less and less due to the difficulty and cost of purifying them, as it is necessary to lyse the cells (Rajput et al., 2016) and separate the NPs from all biomolecules present (Molnár et al., 2018).

NP properties define their action and use. In Fig. 3.3A NPs show mostly negative zeta potentials and are considered stable if their zeta potential are lower than -30 mV. But some NPs show positive zeta potentials. As illustrated, for suspensions of NPs, the first few water layers are organized in a way to contribute to NP stability; see that water molecule orientation depends on NP zeta potential. In Fig. 3.3B, stabilizing agents make capping that can be measured and observed by TEM, therefore NP core is smaller than NP diameter. In Fig. 3.3C, BNMs are synthesized in a manner that their sizes are different from smaller to greater nanomaterials and show sample polydispersity.

The mechanism of extracellular synthesis of NPs mediated by fungi is very convenient, as the purification of the products is facilitated because all the unsecreted biomolecules remain inside the cells, which are easily separated from the medium. This mechanism basically consists of catalysis mediated by nitrate reductase, that is, the enzyme nitrate reductase secreted by fungi, which helps in the bioreduction of metal ions and the synthesis of NPs. Several tests were carried out to prove the presence of nitrate reductase in the fungal filtrate, which was later used for the synthesis of NPs (Absar et al., 2005; Durán et al., 2011a, 2011b; Gade et al., 2008; Ingle et al., 2008). The size of the NPs obtained depends on the synthesis conditions, such as the temperature of the synthesis, the pH, the culture medium and, mainly, the species of the fungus, as was possible to see for AgNPs (Pramila & Kumar, 2018; Thakkar et al., 2010).

There are many biomolecules that can interact with metal ions and assist in the formation of NPs. As the synthesis of NPs is an oxidation–reduction reaction, all molecules that participate in electron transfer processes, such as conversion from NAD(P)H and NADH to $NAD(P)^+$ and NAD^+, can play an important role in the mechanism of NP synthesis (Durán et al., 2005; Gudikandula et al., 2017). Therefore, nicotinamide adenine dinucleotide (NADH) and NADH-dependent nitrate reductase enzymes are considered the most important in the biogenic synthesis of MNPs mediated by fungi and realized through extracellular mechanism (Durán et al., 2005; Gudikandula et al., 2017). However, it was reported that an NP formation reaction occurred through the action of NAD(P)H in the absence of any enzyme (Hietzschold et al., 2019), which indicates that nitrate reductases are not always responsible for the extracellular formation of fungal-mediated NPs.

Although the majority of NPs produced through fungi are AgNPs, there are a multitude of other NPs that are also produced with the aid of fungi. It is possible to mention gold NPs produced by filamentous fungi (Kumar et al., 2007; Molnár et al., 2018) or *Botrytis cinerea* (Castro et al., 2014); cadmium sulfide NPs produced by *Phanerochaete chrysosporium* (Chen et al., 2014); zinc sulfide NPs (Mirzadeh et al., 2013) and copper sulfide (Hosseini et al., 2012) synthesized by *Fusarium oxysporum* and others.

BNM synthesis using fungi as bioreactors, either by intracellular or extracellular processes, has advantages in handling and downstream processing. There is vast literature on syntheses of metals, some metal sulfides, and few metal oxides. Biogenic methods are safe and relatively inexpensive with a wide range of benefits of fungal-mediated synthesis of metal NPs, still there are a number of limitations and challenges to overcome before it can be used practically. For example, there are no standardized methods for size, shape, dispersity, and stability control of the BNM. Thus, BNM synthetic methodology must improve through development of new strategies and make production processes commercially feasible.

3.2.5 BNM algal and viral-based biosynthesis

Algal-biosynthesis is not very common biological synthesis of noble metal NPs. Nevertheless, some green, blue-green, and brown algae were explored for AgNPs and AuNPs syntheses. For example, intracellular synthesis of biomaterials was reported with *Chlorella vulgaris*, which produced algal-bound gold NPs. AuNPs were formed in the inner and outer parts of cell surfaces with tetrahedral, decahedral, and icosahedral structures (Luangpipat et al., 2011). The extracellular synthesis of *Spirulina platensis* was used to synthesize gold, silver, and Au/Ag bi-MNPs in extracellular synthesis (Govindaraju et al., 2009). Metal bio-NPs can be also prepared with *Sargassum wightii, Kappaphycus alvarezii, Tetraselmis kochinensis* in extracellular synthesis (Senapati et al., 2012). Biomass of the brown alga *Fucus vesiculosus* (Mata et al., 2009) was also used for preparing AuNPs.

On the other side, viruses (Narayanan & Han, 2017; Thanagvelu et al., 2020), biomolecules, and their assemblies can be used for inorganic nanomaterials synthesis as an alternative for bio-based synthetic processes. Viruses are not often used for BNM synthesis, and there are some highly cited articles, such as by Shenton et al. (1999) and Mao et al. (2003). For instance, Tobacco mosaic virus (TMV) was used as a "core" or template (Shenton et al., 1999) for synthesizing NMs, such as CdS, PbS, SiO_2, Fe_2O_3, and Fe_3O_4 materials, which were deposited onto TMV. Thus, nanotubes with different nanoshell materials were reported. Specific bottom-up nucleation was achieved and

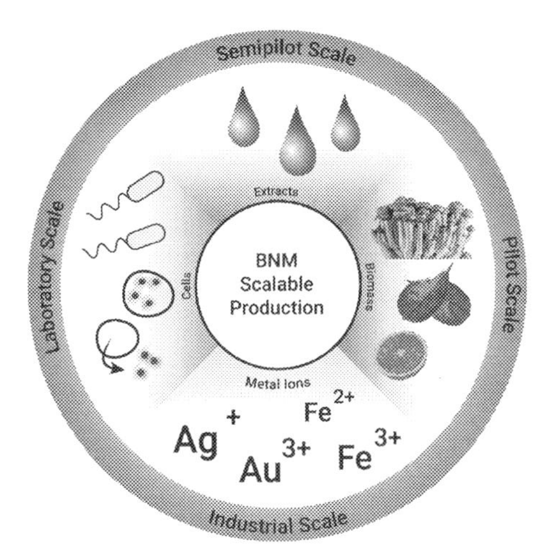

Figure 3.4 Scale-up of biogenic synthesis of nanomaterials. Laboratory scale synthesis cannot be easily turned into industrial scale production of BNMs. There are some steps to be tested such as semi-pilot and pilot production to evaluate proportion of metal ion salts or hydroxides to biomass that is going to be used, also, temperature, contact, time, and best process, such as intra-, extracellular or biomimetic. Estimative of costs, residues, and steps that must be optimised are calculated and tested previous any experimental test.

different sized viral-particle-encapsulated inorganic nanowires were synthesized. Virus M13 bacteriophage was explored by Mao et al. (2003), as it is as well as modified, enriched with recombinant proteins, for the syntheses of ZnS and CdS nanoshells that were very thin (3–5 nm). One more time, the virus was used as a template for the bottom-up growth of nanomaterials. The virus core stabilized the inorganic nanomaterials that interacted with its surface proteins, and amino acids. There are also reports on synthesis of gold and AgNPs using plant virus (Thangavelu et al., 2020). So, this strategy is opposite from the discussed for other biosynthetic processes, because the bio-organic core is a template that is covered with the appropriate thin inorganic shell that may vary in thickness from 3 to 7 nm.

3.2.6 Scale up

As explained in previous sections, BNMs can be obtained by bottom-up approaches using intracellular, extracellular syntheses and biomimetic processes, but all reported data principally discuss BNMs' laboratory-scale production (Fig. 3.4). Although the BNM large-scale synthesis is of great interest, and one of the new green nanotechnology targets, there are rare reports on such processes, because of the difficulty to control them. For example, the availability and costs of large-scale raw materials, environmental friendliness, difficulties in reproducing yields, composition, physical-chemical properties, and/or bioactivity of BNMs (Jung et al., 2012; Lee & Jun, 2019).

Before discussing some literature examples, it is important to define scale ups in BNMs production terms (Fig. 3.4). Scales of BNM production are bench scale (laboratory), semipilot, pilot, and industrial scale. A semipilot scale is smaller than pilot scale, but with the production capacity greater than the bench scale, commonly tested in the laboratory as a concept test design. A pilot scale is a small-scale preliminary study conducted to evaluate feasibility, time, cost, scaling factors, unpredicted results, further development a process, which must be evaluated to properly design and implement a full-scale commercial process. Not rare, this scale integrates process operations and equipment that might be available in a larger scale for a commercial plant. As size increases, system properties that depend on quantity of matter may change. For example, the surface area to liquid ratio that on a small chemical scale, in a flask, is represented with relatively large surface area to liquid ratio. However, if the reaction is scaled up to fit in a 100-L reactor, the surface area to liquid ratio decreases. Therefore, thermodynamics, the reaction kinetics and the process change in a nonlinear fashion. This is why a reaction in an Erlenmeyer flask can behave differently from the same reaction in a large-scale production process. When discussing BNM production, there are many steps to be optimized, starting from substances that are going to be used for the synthesis—cells, extracts, biomass, than ions and salts that are going to be used, proportion among those, before even discussing the scale up.

A study using microfluidics showed promising results for potential scale-up. Most of the references use up to 2 L of extract such as green tea extract (Lee, 2019), or up to 7 L (El-Moslamy, 2018) for production of MgO-NPs with up to 320 g/L. Maybe, the easiest to scale up would be biomimetic processes that are just based on use of one or two components isolated from the extracts, with at least one reductive and stabilizing agent, such as bioflavonoids. For example, zinc oxide nanoflowers were produced using maltose as a reducing and stabilizing agent in a green method process (Din et al., 2020) that used single biomolecule–based synthesis.

When dealing with the prospection of the process for the scaling-up, basal medium is first evaluated and optimized to medium for the scaling-up production, and the biogenic synthesis reaction must be optimized and tested gradually, for example, in ten-folds of biomass: from 2.5 g/L to 20 g/L, the last one to 120 g/L, and so on. It is also important to evaluate the ratio of metal ion to biomass, and reaction conditions, such as stirring, oxygen supply, temperature, pH, ionic staring, and time. There are many factors to be evaluated theoretically, predicted in a multiparameter evaluation.

Therefore, if wished to use the green synthesis and production for bionanomaterials (BNMs), it is advisable to develop method that would be cost inexpensive, easy scalable for mass production, and that can provide highly stable and uniform BNMs. There are rich and almost limitless bioresources in the kingdoms of microbes and plants, yet, for their use, many bottlenecks must be overcome. For example, if wished to effectively utilize microbes, their handling in a scale-up process, cultivation and processing, mutation, contamination with other microbes, among other problems can be a challenge for successful application (Patil & Chandrasekaran, 2020). On the other side, plants and their extracts are rich biosources of many active compounds, and for BNM synthesis, the most useful are flavonoids (Levchenko et al., 2011), phenols, some alkaloids, terpenoids, saponins, tannins, and polysaccharides, which can be used as oxidative, reductive agents (Kumar et al., 2007; Hietzschold et al., 2019; Hosseini et al., 2012; Gade et al., 2008), and stabilizers. Moreover, using plants as a resource for synthesis offers advantages such as plant material availability, costs—inexpensive, scalable for mass production. Only proper and optimized use of biological entities for the synthesis of BNPs will produce well-characterized and highly stable NPs.

Laboratory-scale synthesis cannot be easily turned into industrial-scale production of BNMs. There are some steps to be tested, such as semipilot and pilot production, to evaluate proportion of metal ion salts or hydroxides to biomass that is going to be use, temperature, contact, time, intra-, extra-, or bio-mimetic process. Estimation of costs, residues, and steps that must be optimized are calculated and tested previous to any experimental test.

BNMs' production needs to be adopted for large scale, which is industrial-scale (Mohammadinejad & Mansoori, 2020) production through: (1) optimization of raw materials that must be abundant, and available in large scale; (2) raw materials, products, byproducts, and solid, semisolid, and water wastes must be environmentally friendly; (3) process must be well defined, as to be controlled automatically; and (4) process must be economically viable to justify investment.

3.3 Physical methods for BNM characterization

NPs became a very important and studied material shape as the world of nanotechnology grew and advanced in technological applications. MNPs are one of the most common particles (Huang et al., 2020; Mittal et al., 2013; Rolim et al., 2019; Yugav et al., 2020; Vigneshwaran et al., 2017; El-Seedi et al., 2019; Biswal & Misra, 2020; Tripathi & Chung, 2019; Vijayaraghavan & Ashokkumar, 2017; Sharma et al., 2019), synthesized both by physical and chemical routes, but semiconductor oxide and organic NPs (Tripathi & Chung, 2019; Vijayaraghavan & Ashokkumar, 2017; Sharma et al., 2019; Chaudhuri & Paria, 2012; Durán et al., 2016; Durán & Seabra, 2012) attracted the interest of last decades' research in materials science and engineering. As the demand for greener synthesis has grown, the biogenic routes increased drastically (Barros et al., 2018a; Barros et al., 2018b; de Barros et al., 2018; Barros & Eoin, 2020). We can see BNPs, especially metallic, being synthesized using plants, fruits, and trees extracts (Ahmad et al., 2019) as well as bacterial and fungal biomasses (Sharma et al., 2019; Durán & Seabra, 2012; Barros et al., 2018; Prasad et al., 2011) as already described.

All biological synthesized NPs show core that is inorganic and stabilizing corona or capping layer on their superficies. Regardless of the biological agent used for MNPs formation (Chen et al., 2014; El-Seedi et al., 2019; Maxwell et al., 2017), the most used physical methods for material characterization are electronic microscopies (scanning electron microscopy (SEM) and transmission electron microscopy (TEM), X-ray diffraction (XRD), energy-dispersive X-ray spectroscopy (EDX or EDS), and X-ray photoelectron spectroscopy (XPS).

The microscopical techniques are mostly used for size and shape observations, as well to evaluate surface defects, and suspensions or colloids dispersivity. Together with the morphological characterization, the XRD graphs help to see the crystallinity of the materials, and to confirm crystallite sizes. Most of the time coupled to the electron microscopy equipment, EDS helps in quantifying the material composition and, also, to elemental mapping of the samples.

A newer tool for a finer understanding of surface composition is the XPS, giving rich information about element bonding. It must be pointed out here that all physical methods presented are noninvasive and nondestructive techniques performed in vacuum.

3.3.1 Scanning electron microscopy

SEM is the first analytical method used for a rapid investigation of nanomaterial morphology with adequate resolution. The technique consists in focusing an electron beam onto the probe and subsequently raster scanned over a certain small rectangular area. The interaction of the beam with the matter (sample) creates various signals, all of which can be appropriately detected using the right technology. Secondary electron and backscattered electron (BSE) images as well as elemental X-ray maps are the main forms of presenting the beam/matter interaction by an SEM. Secondary and BSEs are conventionally separated according to their energies and produced by different mechanisms (Brundle et al., 1992). Secondary electrons or secondaries are produced when emitted electron with energy lower than about 50 eV, resulting from inelastic scattering of the beam with atomic electrons, exceeds the work function of the material and exits the solid. Most of the emitted secondaries are produced within the first few nanometer of the surface. BSE are electrons having energy greater than 50 eV, most of them comparable to the energy of the primary beam, which helps to identify the material according to its atomic number as backscattering is most likely to occur as atomic number of the material (Brundle et al., 1992). Thus, if the composition of the material changes, this change can be identified by BSE images. Normally, this technique does not require difficult sample preparation especially for conductive samples. Maximum, a very thin metallic coating ($\sim$5 nm) must be applied to insulating samples for promoting surface discharge.

The SEM technique became a very popular tool for nanomaterials characterization as it can give fruitful information, especially about material morphology and composition.

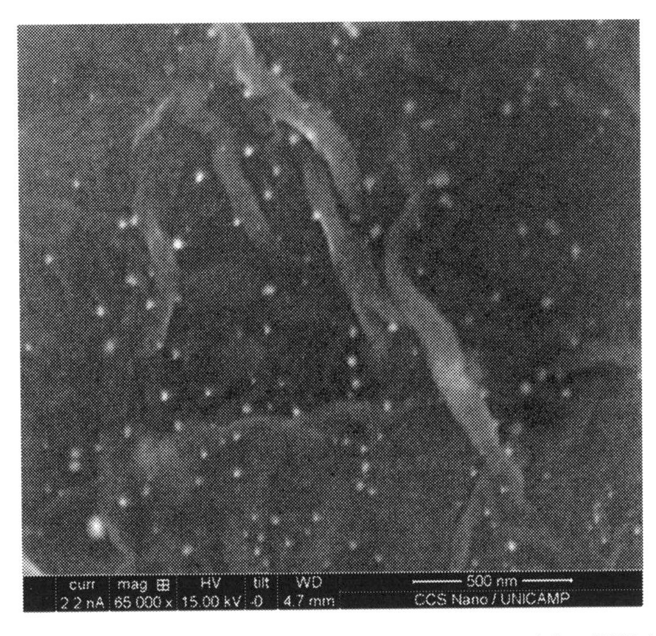

Figure 3.5 Example of material decorated with biogenic nanoparticles (BNPs). HR-SEM images of silver nanoparticles (Ag NPs) that decorate graphene oxide (GO); the Ag NPs were obtained by a biogenic synthesis using hesperetin (HST) as reducing and capping agent, and then mixed with GO suspension by ultrasound (*data not published yet*).

All publications enclose now SEM images for size distribution and shape characterization of BNPs (Huang et al., 2020; El-Seeidi et al., 2019; Tripathi & Chung, 2019; Shukla & Iravani, 2017; Vijayaraghavan & Ashokkumar, 2017; Sharma et al., 2019; Chaudhuri & Paria, 2012; Durán et al., 2016; Durán & Seabra, 2012; Prasad et al., 2011). It is important to point out that to differentiate from the core-shell nanostructures, mostly seen on inorganic materials, the biogenic materials are called capped nanostructures as the biological agent is recognized as a capping agent.

In the biochemical synthesis, the reducing agent and the capping agent are the same, meaning that the biomass reduces the metallic ions and stabilize the particles after their formation, avoiding agglomeration. Depending on the metal core and on the reducing agent used, the shapes of the NPs can be swept from round to triangular and stared ones. SEM characterization is manly used for observing AgNP shapes and sizes (Garibo et al., 2020; Ghosh Chaudhuri & Paria, 2012; Prasad et al., 2011), carrying out NPs size distribution, and observing nanosuspensions dispersivity and BNM surface characteristics (Garibo et al., 2020). Regarding the AgNP shape, it was observed that the main shape is spherical, with sizes ranging between 5 and 100 nm and an excellent dispersivity due to the existence of capping agents (Chung et al., 2016; Shukla & Iravani, 2017). Fig. 3.5 shows an example of AgNP application that is in the decoration of graphene oxide (GO) sheets for obtaining virucidal and sterile filters with controlled porosity. For obtaining high-resolution SEM images of AgNPs that decorate GO, the AgNPs were obtained by a biogenic synthesis using hesperitin (HST) as a reducing and capping agent and then mixed with GO suspension by ultrasound.

3.3.2 Transmission electron microscopy (TEM)

TEM differs a bit from the SEM, but is used with the same objective to deliver information about material morphology. Mainly, this technique confirms SEM and delivers higher resolution images that contain more information, especially about surface defects and crystalline nature. As for the SEM, TEM is based on electron beam interaction with a solid specimen. A series of electromagnetic lenses magnifies the transmitted electron signal and projects on a screen for image formation. These images are formed due to scattering mechanisms associated with interactions between electrons and the atomic constituents of the sample (Tekade, 2019). Compared with SEM, TEM resolution reaches atomic level permitting to obtain information on materials atomic structure, crystallinity, and degree of defects.

Selected area electron diffraction is used to determine the atomic structure of the nanomaterial. TEM specimen preparation is not so time demanding, especially for observing individual NPs that were suspended in liquid media, but a bit more complex than SEM. Metallic grids, normally Cu ones, are used for NPs suspension dripping for specimen preparation, being subsequently dried and submitted to microscope investigation. For thicker samples, the process is rather more complex as is necessary sample thinning by cutting and polishing to a nanometer scale. Even though it requires a somewhat more detailed preparation, TEM brings a lot of rich information about the analyzed samples, as presented in several recent articles focusing on biogenic synthesis of MNPs, especially AgNPs.

TEM results show that biogenic routes lead to fabrication of narrow size distribution particles, with elevated degree of crystallinity and few defects (Yugav et al., 2020; Huang et al., 2020; Gol et al., 2020; Tripathi & Chung, 2019b; Vijayaraghavan & Ashokkumar, 2017; Sharma et al., 2019; Chaudhuri & Paria, 2012; Durán et al., 2016; Durán & Seabra, 2018; Garibo et al., 2020). More, the capping agent, difficult to observe by SEM microscopy, can be detected in the TEM images (Fig. 3.6), indicating its double reducing-coating role. TEM images of AgNPs were obtained by a biogenic synthesis using HST as a reducing and capping agent.

3.3.3 Energy-dispersive X-ray spectroscopy (EDX or EDS)

EDX or EDS is an analytical technique used to identify and characterize the elemental composition of the sample material. It registers X-rays produced by an atom ionized

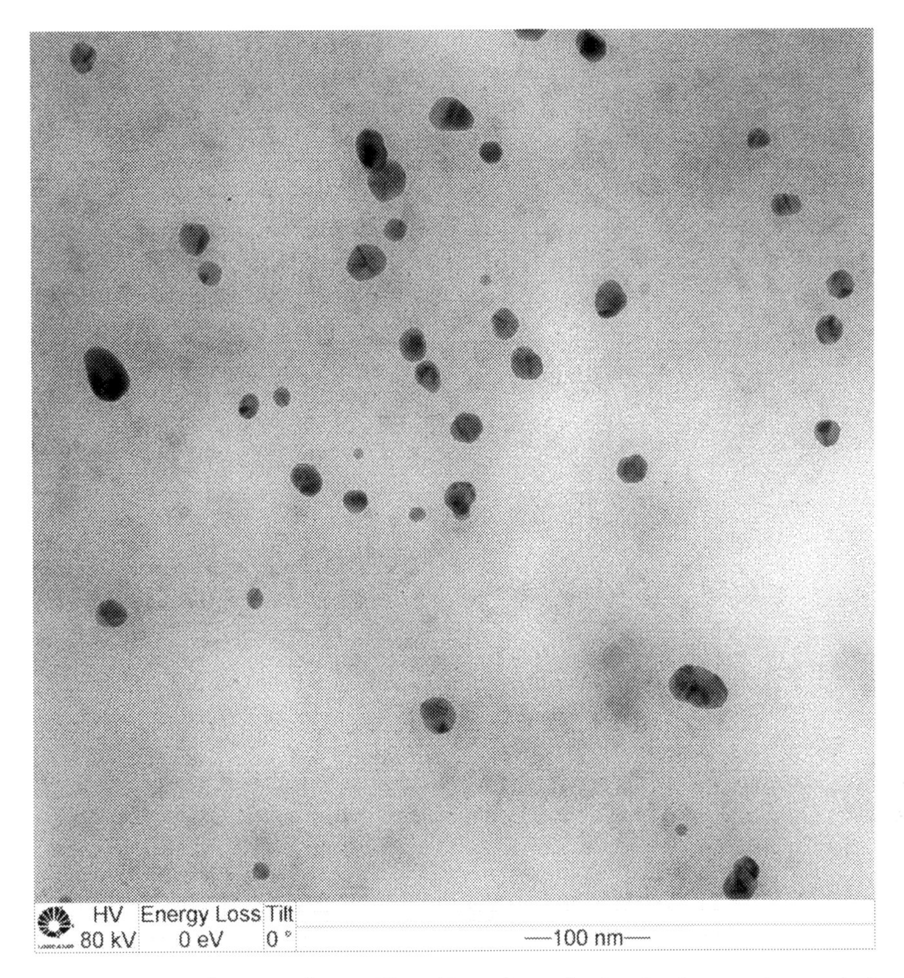

Figure 3.6 TEM image of Ag NPs obtained by a biogenic synthesis using hesperetin as reducing and capping agent (*data not published yet*).

by an electron beam, where an inner shell electron is removed. In order for the atom to return to its ground state, the vacancy left by the ejected electron is filled by an electron from a higher energy outer shell. When this occurs, the energy equal to the potential energy difference between the two shells is emitted either as X-ray photon or is self-absorbed and emitted as an Auger electron. The X-ray emitted is specific of each transition, thus, can be used to identify the elements in the sample (Brundle et al., 1992; Tekade, 2019). Both electron microscope techniques (SEM and TEM) can be equipped with an EDX detector for better sample characterization. It is important to point out that none of these techniques are invasive or destructive, allowing a complete morphological and compositional sample characterization. Some typical examples of EDX image and

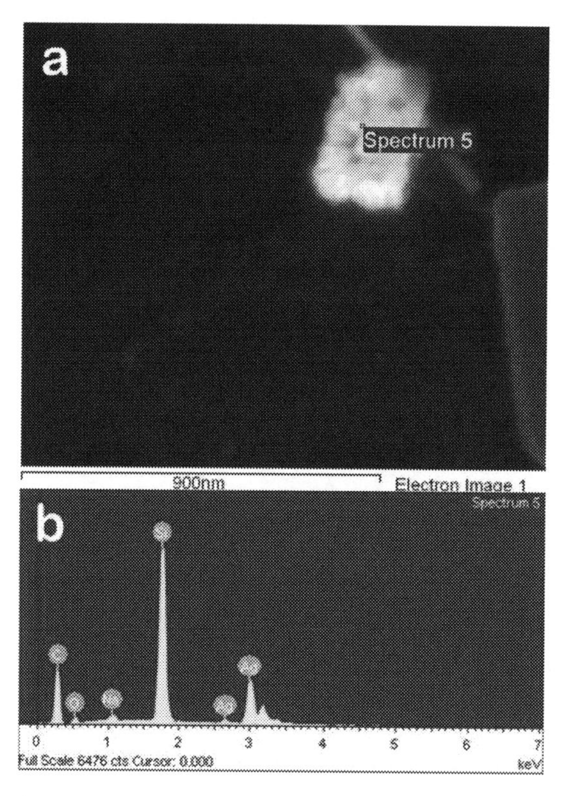

Figure 3.7 Example of typical data obtained in Ag NP characterisation. EDS spectra of AgNPs obtained by a biogenic extracellular synthesis using hesperetin as a capping agent (*data not published yet*).

spectra are presented in Fig. 3.7A that presents the SEM image indicating the spot used for spectral registration (Fig. 3.7) indicating sample compositions. In this particular example, we can see signal from the silicon substrate, carbon from the decorated expanded graphite, and silver from the cluster indicated in Fig. 3.7B. Thus, this technique helps to identify sample composition, detect impurities if so, and contributes to a complete BNPs characterization (Huang et al., 2020; Durán & Seabra, 2018). EDS spectra of AgNPs were obtained by a biogenic extracellular synthesis using hesperitin as a capping agent.

3.3.4 X-ray diffraction

One of the oldest material characterization techniques is the XRD technique. Mostly used for identifying sample crystallinity and preferred crystal orientation, this method is based on constructive interference of a monochromatic beam of X-rays scattered at specific angles by a crystal lattice. The technique is based on Bragg´s law (Eq. 3.1):

$$n\lambda = 2d \sin \theta \tag{3.1}$$

where

n—order of diffraction;

λ—incident X-Ray wavelength;

d—interplanar spacing;

θ—angle of incidence.

The Bragg family, father and son, earned the 1915 Nobel Prize in Physics for their contribution to the characterization of the crystalline structure employing X-ray.

Normally, the XRD equipment presents a spectrum of 2θ versus peaks intensity, and, using the databases already well established for this technique, one can identify material crystalline nature and preferential orientation behavior. Most of the manuscripts based on biogenic metallic NPs fabrication present the X-ray pattern as a strong indication of material crystallinity and crystallite size. The last one is possible to evaluate using Scherrer's equation (Eq. 3.2):

$$\tau = \frac{K\lambda}{\beta \cos \theta} \tag{3.2}$$

where

τ—mean size of crystalline domains (grains or particles);

K—dimensionless shape factor, with typical value of about 0.9.

λ—incident X-Ray wavelength;

β—peak´s full width to half maximum;

θ—angle of incidence;

It must be pointed out that this equation, determined for grain distribution sizes, allows a coarse evaluation of particle sizes and should always be used in correlation with the electron microscopy images obtained for the sample. Usually, what is observed for powders formed of NPs, this includes biogenic AgNPs, are smaller intensity broader peaks, indicating a small crystallite size (Garibo et al., 2020; Barros et al., 2018; Durán & Seabra, 2012; Yugav et al., 2020).

Most of the manuscripts on BMNPs fabrication present the X-ray pattern as a strong indication of material crystallinity and crystallite size (Absar et al., 2005; Adama et al., 2020; Akintelu et al., 2020; Avendaño et al., 2016; Ballottin et al., 2016a, 2016b, 2017; Barabadi et al., 2020; Barros, Stanisic, et al., 2018; Capeness et al., 2019; Castro et al., 2014; Chen et al., 2014; Chung et al., 2016; Dhoondia & Chakraborty, 2012). The last one is possible to evaluate using Schemer´s equation together with the electron microscopy images obtained for the sample. Usually, what is observed for powders formed from NPs, also for the biogenic AgNPs, is the presence of low intensity broad peaks, indicating a small crystallite size.

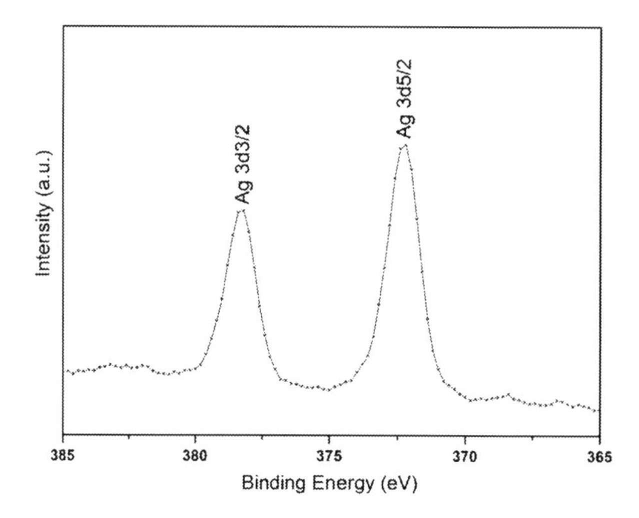

Figure 3.8 Metal core BNM analysis involves XPS showed for biogenic Ag NP characterisation. A typical Ag 3d high-resolution XPS spectra for AgNPs obtained by a biogenic extracellular synthesis using bioflavonoid hesperetin as a capping agent (*data not published yet*).

3.3.5 X-ray photoelectron spectroscopy (XPS)

One of the most used physical characterization techniques for qualitative material composition is XPS. This technique, also denominated electron spectroscopy for chemical analysis, is based on the photoelectric effect. Known monoenergetic X rays are focused onto the material surface cause electrons to be ejected. Using the photoelectric effect equation:

$$KE = h\nu - BE \tag{3.3}$$

where

KE—kinetic energy of the photoelectron;

hν—photon energy;

BE—binding energy of the electron to the atom.

elemental identification of the samples is made by measuring the kinetic energy of the ejected photoelectron and thus, determining its binding energy. XPS is a very fine analytical technique for surface characterization as it penetrates just up to 20 atoms layer (Brundle et al., 1992). Almost all the manuscripts focusing on nanomaterials fabrication present spectra of chemical state surface analyses (Garibo et al., 2020; Brundle et al., 1992). In particular, biogenic silver nanomaterials are investigated especially for carbon (capping agent), oxygen (capping agent and metal oxidation state), and silver (Garibo et al., 2020; Gol et al., 2020; Huang et al., 2020; Rolim et al., 2019), leading to high-resolution XPS spectra as one presented in Fig. 3.8 for Ag 3d. A typical Ag 3d high-resolution XPS spectrum for AgNPs was obtained by a biogenic extracellular synthesis using bioflavonoid hesperitin as a capping agent.

3.4 Conclusions and perspectives for BNM synthesis

Herein, green and biogenic processes for nanomaterial synthesis were reviewed and the main highlights follow. The most explored are extracellular routes for synthesis of BMNPs, with extracts from different sources as leaders in green methods. Plants extracts are prevailing, followed with fungal-based synthesis, and bacterial production. Somewhat less explored are algal- and viruses-based syntheses. All BNMs, except those produced using viruses and biological molecules assemblies, show specific structures of nanomaterial core, metallic mostly, metal sulfides, oxides, or other compounds and capped with biogenic molecules, proteins, peptides, sugars, flavones. Thus, a vast variety of BNMs can be found. Besides, BNMs can be synthesized using nanoscale assemblies of biomolecules, viruses, or modified viruses, which serve as templates for core synthesis, makeing different shaped bionanomaterials with an inverse structure; biomolecules still stabilize such BNMs, but make their core structure.

There are some knowledge gaps on mechanisms that take part in the synthesis processes for BNPs and in understanding how and which agents from the extracts used for the synthesis take part in capping of the newly formed BNPs. Also, the quality control was identified as still to be improved to permit comparisons among BNPs.

At the moment, BNP characteristics and properties are as different as protocols used for their synthesis. The most uniformed results were reported for one-flask synthesis that is based on just one reducing and capping agent as in bioflavonoids promoted synthesis of AgNPs. Nevertheless many challenges must be overcome as green and biological synthesis are expected to lead to scale up and industrial production of BNMs as the next-generation alternative to industrialization.

Some of the great advantages in synthesis of these BNMs are for sure ecologically friendly processes, which use renewable materials, recycle waste material and take advantage of mild reaction conditions. For such, plant, bacterial and fungal extracts, extracellular or intracellular processes can be recognized as the most used methods for BNP synthesis in a bottom-up approach. NMs produced in such ways are stabilized with capping biomolecules, which provide very interesting surface properties and enhance their activity and applications. The diversity of BNM shapes, such as sphere, cube, rod, and triangular, and different ranges of size depend on the final application of the BNPs, as much as it is influenced by biological synthesis methods. BNM application diversity can be found in the area of pharmacy, biomedicine, photocatalysis, agricultural, and electronic fields.

Therewith, eco-friendly and sustainable, low-cost, and quick methods are developed in recent years, and are expected to replace other nanomaterials syntheses that employ hazardous materials.

Acknowledgments

Authors acknowledge the INCTBio, FAPESP (Fundação de Amparo à Pesquisa do Estado de São Paulo) and CNPq.

References

Absar, A., Satyajyoti, S., Islam, K.M., Rajiv, K., Murali, S., 2005. Extra-intracellular biosynthesis of gold nanoparticles by an alkalotolerant fungus, *Trichothecium* sp. J. Biomed. Nanotechnol. 1 (1), 47–53. https://doi.org/10.1166/jbn.2005.012.

Adama, F., Ibrahima, N., Moussa, B., Fatou, S.N., Hamza, E.A.M., et al., 2020. Biosynthesis of TiO_2 nanoparticles by using natural extract of *Citrus sinensis*. 36. pp. 349–356. https://doi.org/10.1016/j.matpr.2020.04.131.

Ahmad, S., Munir, S., Zeb, N., Ullah, A., Khan, B., Ali, J., Bilal, M., Omer, M., Alamzeb, M., Salman, S.M., Ali, S., 2019. Green nanotechnology: a review on green synthesis of silver nanoparticles—an ecofriendly approach. Int. J. Nanomed. 14, 5087–5107. https://doi.org/10.2147/IJN.S200254.

Akintelu, S.A., Folorunso, A.S., Folorunso, F.A., Oyebamiji, A.K., 2020. Green synthesis of copper oxide nanoparticles for biomedical application and environmental remediation. In: Heliyon, 6. p. e04508. https://doi.org/10.1016/j.heliyon.2020.e04508.

Alamri, S.A.M., Hashem, M., Nafady, N.A., Sayed, M.A., Alshehri, A.M., Alshaboury, G., 2018. Controllable biogenic synthesis of intracellular silver/silver chloride nanoparticles by *Meyerozyma guilliermondii* KX008616. J. Microbiol. Biotechnol. 28, 917–930. https://doi.org/10.4014/jmb.1802.02010.

Avendaño, R., Chaves, N., Fuentes, P., Sánchez, E., Jiménez, J.I., Chavarría, M., 2016. Production of selenium nanoparticles in *Pseudomonas putida* KT2440. Sci. Rep. 6,, 37155. https://doi.org/10.1038/srep37155.

Bandala., E., Stanisic, D., Tasic, L, 2020. Biogenic nanomaterials for photocatalytic degradation and water disinfection: a review. Environ. Sci. Water Res. Technol 6, 3195–3213. https://doi.org/10.1039/d0ew00705f.

Ballottin, D., Fulaz, S., Cabrini, F., Tsukamoto, J., Durán, N., Alves, O.L., Tasic, L., 2017. Antimicrobial textiles: biogenic silver nanoparticles against Candida and XANTHOMONAS. Mater. Sci. Eng. C 75, 582–589. https://doi.org/10.1016/j.msec.2017.02.110.

Ballottin, D., Fulaz, S., Souza, M.L., Corio, P., Rodrigues, A.G., Souza, A.O., Gaspari, P.M., Gomes, A.F., Gozzo, F., Tasic, L., 2016a. Elucidating protein involvement in the stabilization of the biogenic silver nanoparticles. Nanoscale Res. Lett. 11. https://doi.org/10.1186/s11671-016-1538-y.

Barabadi, H., 2017. Nanobiotechnology: a promising scope of gold biotechnology. Cell. Mol. Biol. 63, 3–4. https://doi.org/10.14715/cmb/2017.63.12.2.

Barabadi, H., Webster, T.J., Vahidi, H., Sabori, H., Kamalid, D.K., Shoushtarie, F.J., Mahjoub, M.A., Rashedig, M., Mostafavi, E., Cruzb, D.M., Hosseinih, O., Saravanan, M., 2020. Green nanotechnology-based gold nanomaterials for hepatic cancer therapeutics: a systematic review. Iran. J. Pharma. Res. https://doi.org/10.22037/IJPR.2020.113820.14504.

Barros, C, Eoin, C., 2020. A review of nanomaterials and technologies for enhancing the antibiofilm activity of natural products and phytochemicals. ACS Appl. Nanomater. 8537–8556. https://doi.org/10.1021/acsanm.0c01586.

Barros, C.H.N., Cruz, G.C.F., Mayrink, W., Tasic, L., 2018a. Bio-based synthesis of silver nanoparticles from orange waste: effects of distinct biomolecule coatings on size, morphology, and antimicrobial activity. Nanotechnol. Sci. Appl. 11, 1–14. https://doi.org/10.2147/NSA.S156115.

Barros, C.H.N., Fulaz, S., Stanisic, D., Tasic, L., 2018b. Biogenic nanosilver against multidrug-resistant bacteria (MDRB). In: Antibiotics, 7. p. 69. https://doi.org/10.3390/antibiotics7030069.

Barros, C.H.N., Stanisic, D., Morais, B.F., Tasic, L., 2018. Soda lignin from *Citrus sinensis* bagasse: extraction, NMR characterization and application in bio-based synthesis of silver nanoparticles. Energy Ecol. Environ. 3, 87–94. https://doi.org/10.1007/s40974-017-0078-3.

Biswal, A.K., Misra, P.K., 2020. Biosynthesis and characterization of silver nanoparticles for prospective application in food packaging and biomedical fields. Mat. Chem. Phys. 250, 123014. 10.1016/j.matchemphys.2020.123014.

Brundle, C.R., Evans Jr., C.A., Wihon, S, 1992. Encyclopedia of Materials Characterization. Butterworth-Heinemann, Oxford, UK ISBN CL7506-9168-9.

Capeness, M.J., Echavarri-Bravo, V., Horsfall, L.E., 2019. Production of biogenic nanoparticles for the reduction of 4-nitrophenol and oxidative laccase-like reactions https://doi.org/10.3389/fmicb.2019.00997.

Castro, M.E., Cottet, L., Castillo, A., 2014. Biosynthesis of gold nanoparticles by extracellular molecules produced by the phytopathogenic fungus *Botrytis cinerea*. Mater. Lett. 115, 42–44. https://doi.org/10.1016/j.matlet.2013.10.020.

Castro-Longoria, E., Vilchis-Nestor, A.R., Avalos-Borja, M., 2011. Biosynthesis of silver, gold and bimetallic nanoparticles using the filamentous fungus *Neurospora crassa*. Colloids Surf. B Biointerf. 83, 42–48. https://doi.org/10.1016/j.colsurfb.2010.10.035.

Chen, G., Yi, B., Zeng, G., Niu, Q., Yan, M., Chen, A., Du, J., Huang, J., Zhang, Q., 2014. Facile green extracellular biosynthesis of CdS quantum dots by white rot fungus *Phanerochaete chrysosporium*. Colloids Surf. B Biointerf. 117, 199–205. https://doi.org/10.1016/j.colsurfb.2014.02.027.

Chaudhuri, R.G., Paria, S., 2012. Core/shell nanoparticles: classes, properties, synthesis mechanisms, characterization, and applications. Chem. Rev. 112, 2373–2433. 10.1021/cr100449n.

Chung, I.M., Park, I., Seung-Hyun, K., Thiruvengadam, M., Rajakumar, G., 2016. Plant-mediated synthesis of silver nanoparticles: their characteristic properties and therapeutic applications. Nanoscale Res. Lett. 11, 1–14. https://doi.org/10.1186/s11671-016-1257-4.

Dhoondia, Z.H., Chakraborty, H., 2012. Lactobacillus mediated synthesis of silver oxide nanoparticles. Nanomater. Nanotechnol. 2. https://doi.org/10.5772/55741.

Din, M.I., Najeeb, J., Hussain, Z., Khalid, R, Ahmad, G, 2020. Biogenic scale up synthesis of ZnO nano-flowers with superior nano-photocatalytic performance. Inorgan. Nanomet. Chem. 50, 613–619.

Durán, N., Nakazato, G., Seabra, A.B., 2016. Antimicrobial activity of biogenic silver nanoparticles, and silver chloride nanoparticles: an overview and comments. Appl. Microbiol. Biotechnol. 100, 6555–6570. 10.1007/s00253-016-7657-7.

Durán, N., Marcato, P.D., Alves, O.L., De Souza, G.I.H., Esposito, E., 2005. Mechanistic aspects of biosynthesis of silver nanoparticles by several *Fusarium oxysporum* strains. J. Nanobiotechnol. 3, 8. https://doi.org/10.1186/1477-3155-3-8.

Durán, N., Marcato, P.D., Durán, M., Yadav, A., Gade, A., Rai, M., 2011. Mechanistic aspects in the biogenic synthesis of extracellular metal nanoparticles by peptides, bacteria, fungi, and plants. Appl. Microbiol. Biotechnol. 90, 1609–1624. https://doi.org/10.1007/s00253-011-3249-8.

Durán, N., Seabra, A.B., 2012. Metallic oxide nanoparticles: state of the art in biogenic syntheses and their mechanisms. Appl. Microbiol. Biotechnol. 95, 275–288. https://doi.org/10.1007/s00253-012-4118-9.

Durán, N., Seabra, A.B., 2018. Biogenic synthesized Ag/Au nanoparticles: production, characterization, and applications. Curr. Nanosci 14, 82–94. https://doi.org/10.2174/1573413714666171207160637.

EL-Moslamy, S.H., 2018. Bioprocessing strategies for cost-effective large-scale biogenic synthesis of nano-MgO from endophytic *Streptomyces coelicolor* strain E72 as an anti-multidrug-resistant pathogens agent. Sci. Rep. 8, 3820. https://doi.org/10.1038/s41598-018-22134-x.

El-Seedi, H.R., El-Shabasy, R.M., Khalifa, S.A.M., Saeed, A., Shah, A., Shah, R., Iftikhar, F.J., Abdel-Daim, M.M., Omri, A., Hajrahand, N.H., Sabir, J.S.M., Zou, X., Halabi, M.F., Sarhan, W., Guo, W., 2019. Metal nanoparticles fabricated by green chemistry using natural extracts: biosynthesis, mechanisms, and applications. RSC. Adv. 9, 24539–24559. https://doi.org/10.1039/c9ra02225b.

Fang, X., Wang, Y., Wang, Z., Jiang, Z., Dong, M., 2019. Microorganism assisted synthesized nanoparticles for catalytic applications. Energies 12. https://doi.org/10.3390/en12010190.

Gade, A.K., Bonde, P., Ingle, A.P., Marcato, P.D., Durán, N., Rai, M.K., 2008. Exploitation of *Aspergillus niger* for synthesis of silver nanoparticles. J. Biobased Mater. Bioenergy 2, 243–247. https://doi.org/10.1166/jbmb.2008.401.

Ganesh, B., Gunasekaran, P., 2009. Production and structural characterization of crystalline silver nanoparticles from *Bacillus cereus* isolate. Colloids Surf. B Biointerf. 74, 191–195.

Garibo, D., Borbón-Nuñez, H.A., de León, J.N.D., García Mendoza, E., Estrada, I., Toledano-Magaña, Y., Tiznado, H., Ovalle-Marroquin, M., Soto-Ramos, A.G., Blanco, A., Rodríguez, J.A., Romo, O.A., Chávez-Almazán, L.A., Susarrey-Arce, A., 2020. Green synthesis of silver nanoparticles using *Lysiloma acapulcensis* exhibit high-antimicrobial activity. Sci. Rep 10. https://doi.org/10.1038/s41598-020-69606-7.

Gericke, M., Pinches, A., 2006a. Biological synthesis of metal nanoparticles. Hydrometallurgy 83, 132–140. https://doi.org/10.1016/j.hydromet.2006.03.019.

Gericke, M., Pinches, A., 2006b. Microbial production of gold nanoparticles. Gold Bull. 39, 22–28. https://doi.org/10.1007/BF03215529.

Ghosh Chaudhuri, R., Paria, S., 2012. Core/shell nanoparticles: classes, properties, synthesis mechanisms, characterization, and applications. Chem. Rev. 112, 2373–2433. https://doi.org/10.1021/cr100449n.

Gioria, E., Signorini, C., Wisniewski, F., Gutierrez, L., 2020. Green synthesis of time-stable palladium nanoparticles using microfluidic devices. J. Environ. Chem. Eng. 8. https://doi.org/10.1016/j.jece.2020.104096.

Göl, F., Aygün, A., Seyrankaya, A., Gür, T., Yenikaya, C., & Şen, F. (2020) Green synthesis and characterization of *Camellia sinensis* mediated silver nanoparticles for antibacterial ceramic applications, Mater. Chem. Phys., 250, 23037. doi: 10.1016/j.matchemphys.2020.123037.

Govindaraju, K., Kiruthiga, V., Kumar, V.G., Singaravelu, G., 2009. Extracellular synthesis of silver nanoparticles by a marine alga, *Sargassum Wightii Grevilli* and their antibacterial effects. J. Nanosci. Nanotechnol. 9, 5497—5501. https://doi.org/10.1166/jnn.2009.1199.

Gudikandula, K., Vadapally, P., Singara Charya, M.A., 2017. Biogenic synthesis of silver nanoparticles from white rot fungi: their characterization and antibacterial studies. OpenNano 2, 64–78. https://doi.org/10.1016/j.onano.2017.07.002.

Hietzschold, S., Walter, A., Davis, C., Taylor, A.A., Sepunaru, L., 2019. Does nitrate reductase play a role in silver nanoparticle synthesis? Evidence for NADPH as the sole reducing agent. ACS Sustain. Chem. Eng. 7, 8070–8076. https://doi.org/10.1021/acssuschemeng.9b00506.

Hosseini, M.R., Schaffie, M., Pazouki, M., Darezereshki, E., Ranjbar, M., 2012. Biologically synthesized copper sulfide nanoparticles: production and characterization. Mater. Sci. Semicondut. Process. 15, 222–225. https://doi.org/10.1016/j.mssp.2012.03.012.

Huang, X., Chang, L., Lu, Y., Li, Z., Kang, Z., Zhang, X., Liu, M., Yang, D.P., 2020. Plant-mediated synthesis of dual-functional eggshell/Ag nanocomposites towards catalysis and antibacterial applications. Mater. Sci. Eng. C 113. https://doi.org/10.1016/j.msec.2020.111015.

Hulkoti, N.I., Taranath, T.C., 2014. Biosynthesis of nanoparticles using microbes—a review. Colloids. Surf. B. Biointerf. 121, 474–483. https://doi.org/10.1016/j.colsurfb.2014.05.027.

Ingle, A., Gade, A., Pierrat, S., Sönnichsen, C., Rai, M., 2008. Mycosynthesis of silver nanoparticles using the fungus *Fusarium acuminatum* and its activity against some human pathogenic bacteria. Curr. Nanosci. 4, 141–144. https://doi.org/10.2174/157341308784340804.

Jung, J.H., Park, T.J., Lee, S.Y., Seo, T.S., 2012. Homogeneous biogenic paramagnetic nanoparticle synthesis based on a microfluidic droplet generator. Angew. Chem. Int. Ed. 51, 5634–5637. https://doi.org/10.1002/anie.201108977.

Kumar, S.A., Abyaneh, M.K., Gosavi, S.W., Kulkarni, S.K., Pasricha, R., Ahmad, A., Khan, M.I., 2007. Nitrate reductase-mediated synthesis of silver nanoparticles from $AgNO_3$. Biotechnol. Lett. 29, 439–445. https://doi.org/10.1007/s10529-006-9256-7.

Kumari, R., Barsainya, M., Singh, D.P., 2017. Biogenic synthesis of silver nanoparticle by using secondary metabolites from *Pseudomonas aeruginosa* DM1 and its anti-algal effect on *Chlorella vulgaris* and *Chlorella pyrenoidosa*. Environ. Sci. Pollut. Res. 24, 4645–4654. https://doi.org/10.1007/s11356-016-8170-3.

Lee, Y.J., Ahn, E., Park, Y., 2019. Shape-dependent cytotoxicity and cellular uptake of gold nanoparticles synthesized using green tea extract. Nanoscale Res. Lett 14, 129. https://doi.org/10.1186/s11671-019-2967-1.

Lee, S.H., Jun, B.H., 2019. Silver nanoparticles: synthesis and application for nanomedicine. Int. J. Mol. Sci. 20 (4), 865. https://doi.org/10.3390/ijms20040865.

Letsholathebe, D., Thema, F., Mphale, K., Maabong, K., Maria Magdalane, C., 2020. Green synthesis of ZnO doped *Moringa oleifera* leaf extract using Titon yellow dye for photocatalytic applications, 36, pp. 475–479.

Levchenko, L., Golovanova, S.A., Lariontseva, N.V., Sadkov, A., Voilov, D.N., Shul'ga, Y.M., Nikitenko, N.G., Shestakov, A.F., 2011. Synthesis and study of gold nanoparticles stabilized by bioflavonoids. Russ. Chem. Bull. 60, 426–433. https://doi.org/10.1007/s11172-011-0067-1.

Lima, J.R.A., De Farias, D.L., Menezes, T.H.S., Oliveira, R.V.M., Silva, I.A.A., Da Costa Cunha, G., Romão, L.P.C., 2020. Potential of a magnetic hybrid material produced using water hyacinth (*Eichhornia crassipes)* for removal of inorganic and organic pollutants from aqueous media. J. Environ. Chem. Eng. 8,, 104100. https://doi.org/10.1016/j.jece.2020.104100.

Lingamdinne, L.P., Choi, J.S., Choi, Y.L., Chang, Y.Y., Yang, J.K., Karri, R.R., Koduru, J.R., 2020. Process modeling and optimization of an iron oxide immobilized graphene oxide gadolinium nanocomposite for arsenic adsorption. J. Mol. Liq. 299, 112261.

Luangpipat, T., Beattie, I.R., Chisti, Y., et al., 2011. Gold nanoparticles produced in a microalga. J. Nanopart. Res. 13, 6439–6445. https://doi.org/10.1007/s11051-011-0397-9.

Luo, C.H., Shanmugam, V., Yeh, C.S., 2015. Nanoparticle biosynthesis using unicellular and subcellular supports. NPG Asia Mater. 7, e209. https://doi.org/10.1038/am.2015.90.

Maheshwaran, G., Nivedhitha Bharathi, A., Malai Selvi, M., Krishna Kumar, M., Mohan Kumar, R., Sudhahar, S, 2020. Green synthesis of silver oxide nanoparticles using *Zephyranthes Rosea* flower extract and evaluation of biological activities. J. Environ. Chem. Eng. 8, 104137. https://doi.org/10.1016/j.jece.2020.104137.

Mata, Y.N., Blázquez, M.L., Ballester, A., González, F., Muñoz, J.A., 2009. Gold biosorption and bioreduction with brown alga *Fucus vesiculosus*. J. Hazard. Mater. 166, 612–618. https://doi.org/10.1016/j.jhazmat.2008.11.064.

Mao, C., Flynn, C.E., Hayhurst, A., Sweeney, R., Qi, J., Georgiou, G., Iverson, B., Belcher, A.M., 2003. Viral assembly of oriented quantum dot nanowires. Proc. Natl. Acad. Sci. USA. 100, 6946–6951. https://doi.org/10.1073/pnas.0832310100.

Maxwell, B., Huang, F., Rogelj, S., 2017. Rapid one-step synthesis of gold nanoparticles using the ubiquitous coenzyme NADH. Matters. https://doi.org/10.19185/matters.201705000007.

Mehmood, A., Khan, F.S.A., Mubarak, N.M., Tan, Y.H., Karri, R.R., Khalid, M., Walvekar, R., Abdullah, E.C., Nizamuddin, S., Mazari, S.A., 2021. Magnetic nanocomposites for sustainable water purification—a comprehensive review. Environ. Sci. Pollut. Res., 1–26.

Mirgane, Nitin A., Shivankar, Vitthal S., Kotwal, Sandip B., Wadhawa, Gurumeet C., Sonawale, Maryappa C..

Mirzadeh, S., Darezereshki, E., Bakhtiari, F., Fazaelipoor, M.H., Hosseini, M.R., 2013. Characterization of zinc sulfide (ZnS) nanoparticles biosynthesized by *Fusarium oxysporum*. Mater. Sci. Semiconduct. Process. 16, 374–378. https://doi.org/10.1016/j.mssp.2012.09.008.

Mittal, A.K., Chisti, Y., Banerjee, U.C., 2013. Synthesis of metallic nanoparticles using plant extracts. Biotechnol. Adv. 31, 346–356. https://doi.org/10.1016/j.biotechadv.2013.01.003.

Mohammadinejad, R., Mansoori, G.A., 2020. Large-scale production/biosynthesis of biogenic nanoparticles. In: Ghorbanpour, M., Bhargava, P., Varma, A., Choudhary, D. (Eds.), Biogenic Nano-particles and Their Use in Agro-ecosystems. Springer, Singapore https://doi.org/10.1007/978-981-15-2985-6_5.

Molnár, Z., Bódai, V., Szakacs, G., Erdélyi, B., Fogarassy, Z., Sáfrán, G., Varga, T., Kónya, Z., Tóth-Szeles, E., Szucs, R., Lagzi, I., 2018. Green synthesis of gold nanoparticles by thermophilic filamentous fungi. Sci. Rep. 8, 3943. https://doi.org/10.1038/s41598-018-22112-3.

Mourato, A, Gadanho, M, Lino, AR, Tenreiro, R, 2011. Biosynthesis of crystalline silver and gold nanoparticles by extremophilic yeasts. Bioinorg. Chem. Appl., 546074. https://doi.org/546074.10.1155/2011/546074.

Mukherjee, P., Ahmad, A., Mandal, D., Senapati, S., Sainkar, S.R., Khan, M.I., Parishcha, R., Ajaykumar, P.V., Alam, M., Kumar, R., Sastry, M., 2001. Fungus-mediated synthesis of silver nanoparticles and their immobilization in the mycelial matrix: a novel biological approach to nanoparticle synthesis. Nano Lett. 1, 515–519. https://doi.org/10.1021/nl0155274.

Mukherjee, P., Ahmad, A., Mandal, D., Senapati, S., Sainkar, S.R., Khan, M.I., Ramani, R., Parischa, R., Ajayakumar, P.V., Alam, M., Sastry, M., Kumar, R., 2001. Bioreduction of $AuCl_4$ - ions by the fungus, *Verticillium* sp. and surface trapping of the gold nanoparticles formed. Angew. Chem. Int. Ed. 40, 3585–3588. https://doi.org/10.1002/1521-3773(20011001)40:19<3585::AID-ANIE3585>3.0.CO;2-K.

Murawala, P., Phadnis, S.M., Bhonde, R.R., Prasad, B.L.V., 2009. In situ synthesis of water dispersible bovine serum albumin capped gold and silver nanoparticles and their cytocompatibility studies. Colloid. Surf. B. Biointerf 73, 224–228. https://doi.org/10.1016/j.colsurfb.2009.05.029.

Nanda, A., Saravanan, M., 2009. Biosynthesis of silver nanoparticles from *Staphylococcus aureus* and its antimicrobial activity against MRSA and MRSE. Nanomedicine 5, 452–456. https://doi.org/10.1016/j.nano.2009.01.012.

Narayanan, K.B., Han, S.S., 2017. Icosahedral plant viral nanoparticles-bioinspired synthesis of nanomaterials/nanostructures. Adv. Colloid Interf. Sci 248, 1–19. 10.1016/j.cis.2017.08.005.

Narayanan, K.B., Sakthivel, N., 2010. Biological synthesis of metal nanoparticles by microbes. Adv. Colloid Interf. Sci. 156, 1–13. https://doi.org/10.1016/j.cis.2010.02.001.

Patil, S., Chandrasekaran, R., 2020. Biogenic nanoparticles: a comprehensive perspective in synthesis, characterization, application and its challenges. J. Genet. Eng. Biotechnol. 18, 67. https://doi.org/10.1186/s43141-020-00081-3.

Paula, A.J., Silveira, C.P., Martinez, D.S.T., Souza Filho, A.G., Romero, F.V., Fonseca, L.C., Tasic, L., Alves, O.L., Durán, N., 2014. Topography-driven bionano-interactions on colloidal silica nanoparticles. ACS Appl. Mater. Interf. 6, 3437–3447. https://doi.org/10.1021/am405594q.

Pramila, K., Kumar, S.S., 2018. Mycogenic nanoparticles and their bio-prospective applications: current status and future challenges. J. Nanostruct. Chem. 369–391. https://doi.org/10.1007/s40097-018-0285-2.

Prasad, T., Kambala, V., Naidu, R., 2011. A critical review on biogenic silver nanoparticles and their antimicrobial activity. Curr. Nanosci. 7, 531–544. https://doi.org/10.2174/157341311796196736.

Priyadharshini, R.I., Prasannaraj, G., Geetha, N., Venkatachalam, P., 2014. Microwave-mediated extracellular synthesis of metallic silver and zinc oxide nanoparticles using macro-algae (*Gracilaria edulis*) extracts and its anticancer activity against human PC3 cell lines. Appl. Biochem. Biotechnol. 174, 2777–2790. http://doi.org/10.1007/s12010-014-1225-3.

Rajput, S., Werezuk, R., Lange, R.M., Mcdermott, M.T., 2016. Fungal isolate optimized for biogenesis of silver nanoparticles with enhanced colloidal stability. Langmuir 32, 8688–8697. https://doi.org/10.1021/acs.langmuir.6b01813.

Ramanathan, R., Field, M.R., O'Mullane, A.P., Smooker, P.M., Bhargava, S.K., Bansal, V., 2013. Aqueous phase synthesis of copper nanoparticles: a link between heavy metal resistance and nanoparticle synthesis ability in bacterial systems. Nanoscale 5, 2300–2306. https://doi.org/10.1039/c2nr32887a.

Rodrigues, A., Ping, L., Marcato, P., Alves, O., Silva, M., Ruiz, R., Melo, I., Tasic, L., de Souza, A.O., 2012. Biogenic antimicrobial silver nanoparticles produced by fungi. Appl. Microbiol. Biotechnol. 97, 775–782. https://10.1007/s00253-012-4209-7.

Rolim, W.R., Pelegrino, M.T., de Araújo Lima, B., Ferraz, L.S., Costa, F.N., Bernardes, J.S., Rodigues, T., Brocchi, M., Seabra, A.B., 2019. Green tea extract mediated biogenic synthesis of silver nanoparticles: characterization, cytotoxicity evaluation and antibacterial activity. Appl. Surf. Sci. 463, 66–74. https://doi.org/10.1016/j.apsusc.2018.08.203.

Sackey, J., Razanamahandry, L., Ntwampe, S., Mlungisi, N., Fall, A., Kaonga, C., Nuru, Z., 2021. Biosynthesis of CuO nanoparticles using *Mimosa hamata* extracts, 36, pp. 540–548.

Senapati, S., Syed, A., Moeez, S., Kumar, A., Ahmad, A., 2012. Intracellular synthesis of gold nanoparticles using alga *Tetraselmis kochinensis*. Mater. Lett. 79, 116–118. https://doi.org/10.1016/j.matlet.2012.04.009.

Sharma, C., Ansari, S., Ansari, M.S., Satsangee, S.P., Srivastava, M.M., 2020. Single-step green route synthesis of Au/Ag bimetallic nanoparticles using clove buds extract: enhancement in antioxidant bio-efficacy and catalytic activity. Mater. Sci. Eng. C 116. https://doi.org/10.1016/j.msec.2020.111153.

Sharma, D., Kanchi, S., Bisetty, K., 2019. Biogenic synthesis of nanoparticles: a review. Arab. J. Chem. 12, 3576–3600. 10.1016/j.arabjc.2015.11.002.

Sheikhloo, Z., Salouti, M., 2012. Intracellular biosynthesis of gold nanoparticles by fungus *Phoma macrostoma*. Syn. React. Inorgan. MetaOrg. Nanometal Chem. 42, 65–67. https://doi.org/10.1080/15533174.2011.609230.

Shenton, W., Douglas, T., Young, M., Stubbs, G., Mann, S., 1999. Inorganic-organic nanotube composites from template mineralization of tobacco mosaic virus. Adv. Mater 11, 253–256. https://doi.org/10.1002/(SICI)1521-4095(199903)11:3<253::AID-ADMA253>3.0.CO;2-7.

Shukla, A.K., Iravani, S., 2017. Metallic nanoparticles: green synthesis and spectroscopic characterization. Environ. Chem. Lett. 15, 223–231. https://doi.org/10.1007/s10311-017-0618-2.

Singh, H., Du, J., Yi, T.H., 2017. Biosynthesis of silver nanoparticles using *Aeromonas* sp. THG-FG1.2 and its antibacterial activity against pathogenic microbes. artificial cells. Nanomed. Biotechnol. 45, 584–590. https://doi.org/10.3109/21691401.2016.1163715.

Singh, R., Shedbalkar, U.U., Wadhwani, S.A., Chopade, B.A., 2015. Bacteriagenic silver nanoparticles: synthesis, mechanism, and applications. Appl. Microbiol. Biotechnol. 99 (11), 4579–4593. https://doi.org/10.1007/s00253-015-6622-1.

Sintubin, L., De Windt, W., Dick, J., Mast, J., van der Ha, D., et al., 2009. Lactic acid bacteria as reducing and capping agent for the fast and efficient production of silver nanoparticles. Appl. Microbiol. Biotechnol 84, 741–749. https://doi.org/10.1007/s00253-009-2032-6.

Shivashankarappa, A., Sanjay, K.R., 2020. *Escherichia coli*-based synthesis of cadmium sulfide nanoparticles, characterization, antimicrobial and cytotoxicity studies. Braz. J. Microbiol. 51, 939–948. https://doi.org/10.1007/s42770-020-00238-9.

Shukla, A.K., Iravani, S., 2017. Metallic nanoparticles: green synthesis and spectroscopic characterization. Environ. Chem. Lett. 15, 223–231. 10.1007/s10311-017-0618-2.

Tekade, R.K., 2019. Biomaterials and Bionanotechnology: A Volume in Advances in Pharmaceutical Product Development and Research. Academic Press, MA https://doi.org/10.1016/B978-0-12-814427-5.00015-9.

Thakkar, K.N., Mhatre, S.S., Parikh, R.Y., 2010. Biological synthesis of metallic nanoparticles. Nanomed. Nanotechnol. Biol. Med. 6 (2), 257–262. https://doi.org/10.1016/j.nano.2009.07.002.

Thangavelu, R.M., Ganapathy, R., Ramasamy, P., Krishnan, K., 2020. Fabrication of virus metal hybrid nanomaterials: an ideal reference for bio semiconductors. Arab. J. Chem. 13, 2750–2765. http://doi.org/10.1016/j.arabjc.2018.07.006.

Tripathi, R.M., Chung, S.J., 2019. Biogenic nanomaterials: synthesis, characterization, growth mechanism, and biomedical applications. J. Microbiol. Methods 157, 65–80. https://doi.org/10.1016/j.mimet.2018.12.008.

Vala, A.K., 2014. Intra-and extracellular biosynthesis of gold nanoparticles by a marine-derived fungus *Rhizopus oryzae*. Synthesis and reactivity in inorganic. Metal-Org. Nano-Met. Chem. 44 (9), 1243–1246. https://doi.org/10.1080/15533174.2013.799492.

Vigneshwaran, N., Ashtaputre, N.M., Varadarajan, P.V., Nachane, R.P., Paralikar, K.M., Balasubramanya, R.H., 2017. Biological synthesis of silver nanoparticles using the fungus *Aspergillus flavus*. Mater. Lett. 61, 1413–1418. https://doi.org/10.1016/j.matlet.2006.07.042.

Vijayaraghavan, K., Ashokkumar, T., 2017. Plant-mediated biosynthesis of metallic nanoparticles: A review of literature, factors affecting synthesis, characterization techniques and applications. J. Env. Chem. Eng. 5, 4866–4883. 10.1016/j.jece.2017.09.026.

Yugay, Y., Usoltseva, R., Silant'ev, V., Egorova, A., Karabtsov, A., Kumeiko, V., Ermakova, S., Bulgakov, V., Shkryl, Y., 2020. Synthesis of bioactive silver nanoparticles using alginate, fucoidan and laminaran from brown algae as a reducing and stabilizing agent. Carbohyd. Polym. 254, 116547. https://doi.org/10.1016/j.carbpol.2020.116547.

Zhang, H., Li, Q., Lu, Y., Sun, D., Lin, X., et al., 2005. Biosorption and bioreduction of diamine silver complex by Corynebacterium. J. Chem. Technol. Biotechnol. 80, 285–290. https://doi.org/10.1002/jctb.1191.

CHAPTER 4

Conventional techniques for nanomaterials preparation

Abdul Sattar Jatoi[a], Faheem Akhter[b,e], Nabisab Mujawar Mubarak[c], Shaukat Ali Mazari[a], Shoaib Ahmed[a], Nizamuddin Sabzoi[d], Abdul Qayoom Memon[a], Humair Ahmed Baloch[d], Rashid Abro[a] and Atta Muhammad[a]

[a]Department of Chemical Engineering, Dawood University of engineering and technology, Karachi, Pakistan
[b]Department of Chemical Engineering, Quaid-e-Awam University of Engineering, Science & Technology, Nawabshah, Pakistan
[c]Petroleum, and Chemical Engineering, Faculty of Engineering, Universiti Teknologi Brunei, Bandar Seri Begawan, Brunei Darussalam
[d]School of Engineering, RMIT University, Melbourne, Australia
[e]Department of Chemical Engineering, Mehran University of Engineering and Technology, Jamshoro, Sindh, Pakistan

4.1 Introduction

Today, nanotechnology has become one of the most advanced and fastest growing technologies, and has made significant contributions to food technology (Lowry et al., 2019), agriculture (Fortunati et al., 2019), and drug delivery (Saeedi et al., 2019). Nanoparticles can vary in size from a thousand to one thousand nanometers and have varying physiochemical and electronic characteristics, such as higher quantum effects, greater surface to volume ratio, and higher surface reactivity (Mathew et al., 2019; Rajitha et al., 2019; Zhang et al., 2019). Briefly, the nanoparticles can be well explained as the particles reduced to 100 nm in their 3D structure. However, at the nanoscale, these particles can have internal as well as external structure. Therefore, nanomaterials can be divided into nanolayers (with one dimension on the nanoscale), nanowires, nanofibers, carbon nanotubes (CNTs) (with two dimensions on the nanoscale) (Khan et al., 2021), and nanoparticles (all three dimensions on the nanoscale) (Deng et al., 2019; Fiyadh et al., 2019). Nanoparticles are considered to be the bridge between atoms and the molecular structure of the bulk material. Currently, various physical, chemical, and biological technologies have been used to synthesize nanomaterials. On the other hand, biotechnology based on the action of microorganisms and plant or chemical extracts used to synthesize nanomaterials has attracted researchers from all over the world.

The main factor in preparing nanoparticles is to choose nontoxic, nonantigen, biocompatible, and biodegradable polymers. Examples of some natural polymers used to synthesize nanoparticles are albumin, gelatin, sodium alginate, and chitosan to name a few (Rajput, 2015). However, the most widely used synthetic polymers are polyanhydride, polyvinyl alcohol, polylactide, polyglycolide, polyacrylic acid, polyethylene glycol,

Sustainable Nanotechnology for Environmental Remediation
DOI: https://doi.org/10.1016/B978-0-12-824547-7.00001-1

poly-N-vinylpyrrolidone, polycyanoacrylate, polyorthoester, Poly (malic acid), polyglutamic acid, polycaprolactone, and polymethylmethacrylate. Several methods have been investigated to synthesis nanoparticles. Among those methods, emulsification, nanoprecipitation, dialysis, salting-out solvent evaporation, and supercritical fluid technology are most widely considered. Apart from these technologies, various physiochemical techniques have also been used to synthesize nanoparticles. It takes two steps to prepare the nanoparticles; precipitation and emulsification. However, laser ablation and evaporation condensation are the major physical methods to synthesize nanoparticles. Sometimes, the physical method for nanoparticle preparation may involve studies of long-term inhalation toxicity and calibration of devices (Gang et al., 2011; Jonoobi et al., 2015). Furthermore, laser ablation of metal-rich materials in a homogenous solution phase also produces nanoparticles. This method can be used to synthesize silver nanospheres, colloidal particles, and CNTs without the generation of any agglomerates. On the other hand, chemical methods include reduction reactions using organic and inorganic reagents. Numerous reducing substrates are presently being utilized for the synthesis of metal nanoparticles such as Tollens reagent, N, N-dimethylformamide, and elemental hydrogen (Jelinkova et al., 2019; Pejjai et al., 2017).

During the chemical reduction process, diverse polymer components including polyvinyl alcohol polyvinylpyrrolidone and polymethacrylic acid are effectively used as defensive agents to ensure the stability of the prepared nanoparticles (Jain et al., 2020). Spherical silver nanoparticles with moderate dispersibility synthesized according to the improved injection precursor method were observed. Owing to chemical reaction between the nanoparticles, some colloidal microemulsion also known as nanodispersion is formed (Soudagar et al., 2018). Brownian motion of the formed droplets proposes that the particles form molten dimer after collision. The nanoparticles made of metals can also be prepared through electrochemical and radiation techniques. Several monodisperse silver nanospheres have also been developed using active precipitation of the zeolite-altered electrodes along with the electrochemical reduction. Others include the synthesis of metal nanoparticles based on laser irradiation and short-term photosensitivity at different intervals.

4.2 Overview of nanomaterials

4.2.1 Carbon-based nanomaterials

The carbon is an essential component of this type of nanomaterial, fullerenes and CNTs are the common examples. In these nanomaterials, CNTs are combined with graphene sheets and then rolled up in a tube to produce CNTs. CNTs hold the characteristics of high tensile strength and are considered stronger than steel. The CNTs are present in both single-walled CNTs (SWCNTs) and multiwalled CNTs (MWCNTs)(Khan et al., 2021a, 2021c, 2021b). Fullerenes, being the allotropes of carbon, are the particles with hollow cage structure and may comprise of 60 or more carbon atoms. Its structure contains carbon units with hexagonal and pentagonal regular patterns. They are conductive in nature and have good electron affinity and high resistance (Maiti et al., 2019; Xie et al., 2019).

4.2.2 Metal-based nanomaterials

Divalent and trivalent metal ions comprise the starting metal nanomaterials. Numerous chemical or photochemical methods have been proposed to synthesize metal nanoparticles. For example, reducing agents can be used to reduce metal ions into metal nanoparticles (Dehghani et al., 2020). These particles possess good small molecule adsorption capacity and high surface area. Their applications have been found in various research fields such as environmental and biological research. Not only a single nanoparticle can be realized, but also a mixture of two or more kinds of nanoparticles with size control can be realized. The doping can change the main characteristics of the element, even the rare-earth metals. By doping different elements with different ingredients, their characteristics will also change (Jin & Maduraiveeran 2019; Makvandi et al., 2020; Mao et al., 2019; Xia & Guo 2019).

4.2.3 Nanomaterials-based semiconductors

Semiconductor nanomaterials possess metallic and nonmetallic properties. By modifying them, they have a wider band gap and have several characteristics. These nanomaterials are widely accepted for electronic and photocatalytic equipment. For instance, ZnS (Xiao et al., 2019), ZnO (Wang et al., 2019b), CdS (Di et al., 2019), CdSe (Wang et al., 2019a), CdTe (Pan et al., 2019) are related to group II–VI semiconductor materials. GaN (Gang et al., 2011), GaP (Gürsoy, 2020), InP (Hao et al., 2003), InAs (Hao et al., 2003) are from group III–V. In recent times, graphene can be used to enhance the chemical and physical properties of the semiconductor. The graphene composites can be used for improving gas-sensing sensitivity and piezoelectric properties.

4.2.4 Nanocomposites

Nanocomposites are solid multiphase materials, one of the phases has one, two, or three dimensions less than 100 nm. Nanocomposites are different than the typical composite materials as, comparatively, they show high surface area to volume ratio. Depending on size and shape, its physical and chemical properties may vary, as shown below (Baker, 1989; Jelinkova et al., 2019). Various kinds of nanocomposite materials have been developed such as polymer matrix nanocomposite, ceramics matrix nanocomposite, and metal matrix composite. Recently, graphene-based polymer matrix composite materials have been developed and investigated on a larger scale. Graphene is composed of carbon fragments along with single layer of carbon atoms arranged in a hexagonal matrix (Malaki et al., 2019; Vatansever et al., 2019). It holds the characteristics of a zero-band gap and electrons are like massless particles, composed of the correct two-dimensional dielectric

(He et al., 2019; Magalhães et al., 2020). Graphene oxide (GO) is the precursor of graphene but it has a low electronic conductivity. Hence, the reduced GO can be more useful as it shows a better conductivity (Ngidi et al., 2019). Various methods can be used to convert GO to rGO such as chemical vapor deposition (CVD) method, exfoliation method, thermal reduction, chemical reduction, and multi-step reduction. Graphene family nanocomposites include metal chalcogenide and metal-oxide graphene nanocomposites. Metal oxides (TiO_2, Fe_2O_3, ZnO, MnO_2, In_2O_3) have found the multifaceted applications, mainly the ones involving matter interaction with the surroundings, such as drug delivery, photovoltaic, photocatalytic, and cytotoxicity.

4.2.5 Conventional technique for preparation of nanomaterials

4.2.5.1 Chemical vapor deposition

In terms of product purity and mass production, CVD technology is one of the efficient techniques to synthesize carbon nanomaterials (CNM). The synthesis method under low temperature and ambient pressure is simple and economical. CVD allows experimenters to bypass the process of separating nanotubes/nanofibers/nanobeads, etc. The experimental parameters strongly affect the CVD reaction such as pressure, reactor temperature, composition of precursor composition, and concentration (Karim et al., 2019; Shi et al., 2019).

Owing to the influence of temperature and the volume ratio limitation between the sample and the source of carbon, gas-phase technologies (such as CVD) have been substituted with laser ablation methods. H, the arc discharge CVD is a controllable process for preparing the nanotubes with pre-forming processes (Christy, 2019; Thamaraiselvan et al., 2019). The CVD method is the result of tremendous efforts to synthesize nanotubes in a controlled manner by breaking down gas. However, despite of higher output level achieved by arc discharge method, it contains unpurified CNTs (Hong et al., 2020).

The CVD is performed in two distinct steps: Catalyst deposition on substrate followed by nucleation via chemical etching or thermal annealing

A chamber contains a carbon source in the form of gas phase such as carbon monoxide, acetylene, or methane. In this chamber, first, the conversion of carbon molecules into atoms takes place through plasma process or hot coil followed by diffusion of carbon onto the substrate coated with a catalyst, hence resulting in growth of nanotubes over the metal catalyst. Numerous CVD methods (Fig. 4.1) are available for nanomaterial formation, that is, plasma-enhanced CVD, thermal CVD, hot/cold wall CVD, etc. (Ma et al., 2019; Tian et al., 2019). The advantages and disadvantages of CVD, plasama vapor deposition (PVD), and plasma-enhanced CVD (PECVD) are given in Table 4.1.

4.3 CVD enhanced by plasma

The advance development in science and technology is one of the greatest milestones, which allows us to investigate the matter and change the structure of atoms. Over the

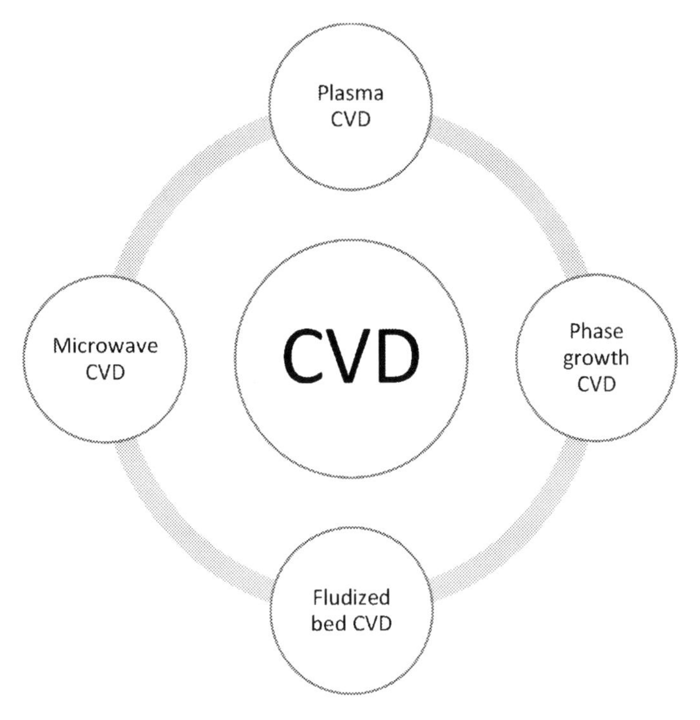

Figure 4.1 Chemical vapor deposition based assisted technique.

past decades, different nanoscale materials and their surfaces have been altered. The synthesis of carbon intangibles (CNM) and their surface changes offers the opportunity to strengthen scientific efforts to build a more ingenious international society capable of meeting challenges. The functional CNM have a numerous application in an extensive field. In CNM, CNT and graphene show superior characteristics such as excellent thermal and mechanical properties, high electrical conductivity, high aspect ratio, low weight, and extraordinary optical applications (Wu et al., 2020; Xu et al., 2020).

The functionalization of the surface of CNMs has promoted the CNTs application in numerous fields, and the importance of modifying the surface of nanomaterials has been fully demonstrated in scientific research in the last decade (Li et al., 2019; Verma et al., 2019). The physicochemical properties of these functionalized nanomaterials have been found in various applications in the fields of energy, biotechnology, medicine, and aerospace. As an example, the medical-related applications include cancer treatment, drug transportation, and antiviral drug development. Furthermore, theoretically, extensive efforts have been taken for the optimization of functionalized nanomaterials. Different techniques are utilized for the production of CNMs through which analytical and experimental work could be performed by utilizing modified methods. In addition, the

Table.4.1 Advantages and disadvantages of CVD, PVD, and PECVD.

Methods	CVD	PVD	PECVD
Advantages	Avoids line of sight High deposition rate Production of thick coating layers Co-deposition of material at the same time	Atomic level control of chemical deposition Not requiring the usage of special precursor Safer than CVD due to the absence of toxic precursor or byproducts	Avoids the line of sight issue to certain extent High deposition rate Low temperatures Both organic and inorganic materials are precursors Unique chemical properties of deposition films Thermal and chemical stability High solvent and corrosion resistance No limitation on substances complicated geometries and composition
Disadvantages	Requirements of high temperature Possibility of toxicity of precursors Mostly inorganic material have been used	Line of sight deposition Low deposition rate Production of thin coating layers Requirement of annealing time	Instability against humidity and aging Existence of compressive and residual stresses in the films Time consuming specially for super lattice structure Existence of toxic, explosives gases in plasma stream High cost of equipment

modified technique such as plasma-enhanced CVD can be used to recover the nanomaterials. Nanomaterials such as CNTs and nanofibers can be synthesized and produced using the plasma technique. Another name for plasma-enhanced CVD technique is glow discharge CVD. This technique follows the activation method using electron energy (plasma) to initiate the deposition at a reasonable rate and low temperature. At reduced pressures, a high voltage from an electrical power source is supplied to a gas.

The plasma-assisted CVD of CNT and carbon nanofiller (CNF) is a very complex process with multiple coupling phenomena: plasma chemistry, neutral and ionic reaction, catalyst growth, aggregation/separation of catalyst particles/heating migration, plasma, electric field effect, ion bombardment; the parameters that determine the growth characteristics include the nature of the raw material, the diluent (such as H_2, Ar, N_2, etc.), the composition, and flow rate of the raw gas. The nature of the plasma energy source, the input power of the plasma, the pressure, the temperature of the substrate, the nature of the catalyst and its mode of application, the lower metallic layer or diffusion barrier between the catalyst and the substrate (if any) and the nature of the catalyst pretreatment

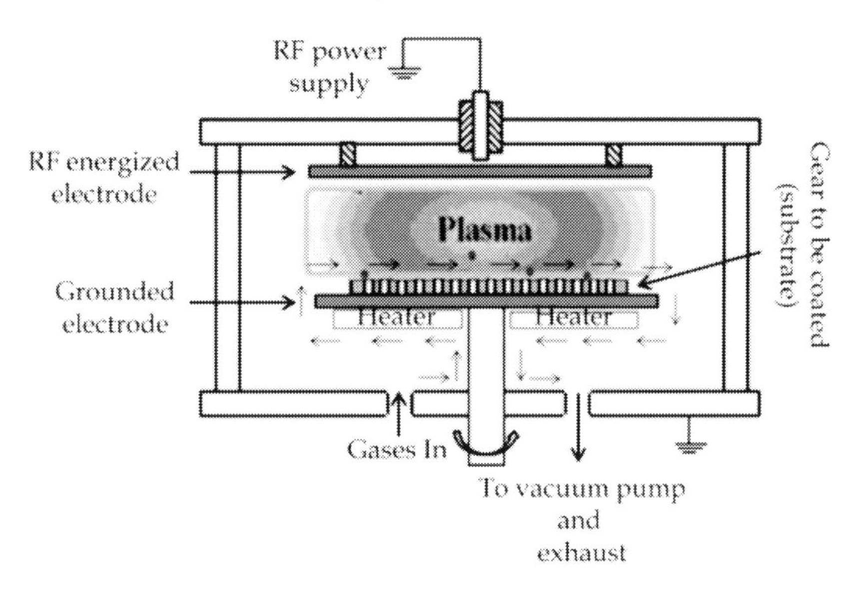

Figure 4.2 Plasma enhanced CVD.

(if any). At the start of adoption of this method in the CNT community in 2003, the enhanced CVD status of plasma during CNT growth was examined (Lai et al., 2019). The plasma CVD reactor is mainly classified on the basis of a plasma source that generates the loaded gas discharge (Gürsoy, 2020) and the small dots consist of a 7-nm nickel film that is deposited simultaneously on the whole sample (Fig. 4.2). The height difference of CNTs is more than an order of magnitude, only due to the different sizes of CNTs catalyst model. Therefore, if all the CNTs on the substrate have the same height, the lines and areas of the catalyst can also be drawn as individual dots, as shown by the lines.

In this case, the single CNT is surrounded by rectangular-shaped pattern dots of e-beam. Fig. 4.2E–G shows the pattern of catalyst dots before the CNT growth, which ensures the need of parameter adjustment for individual CNTs. The final high temperature of plasma growth did not indicate any growth on the part of small catalyst dots; however, the reproducible growth was seen at low temperature on both catalyst layers and arrays of individual dots. This shows the sensitivity of small particles to passivation at high temperatures despite possessing high surface to volume ratio. Commonly used methods to produce nanotubes include laser vaporization, arc discharge, laser vaporization, and CVD.

4.3.1 Microwave-enhanced CVD

Microwave-enhanced CVD is a well-known method used to synthesize CNTs (Wang et al., 2004) and carbon nanowalls (Wu et al., 2002). Similarly, foil-like carbon nanostructure can be manufactured using another method known as PECVD (Hirao et al., 2001). The potential of this method lies in its synthesis ability for low-temperature and

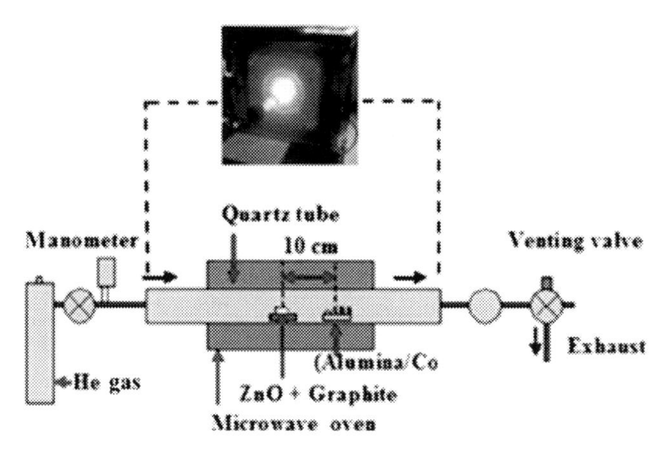

Figure 4.3 Microwave enhanced CVD.

highly graphitized vertically aligned CNTs (Amama et al., 2006; Meyyappan et al., 2003). The growth characteristics are affected by the process itself and the subsequent transport to the growth surface (Yen et al., 2005). One of the key parameters of this technique that are shown to be advantageous over CVD are growth at low temperature, chirality, diameter control, and CNT alignment. Various microwave-enhanced CVD studies show that there is no loss of energy as a result of plasma density and resonance field capable of concentrating the plasma. In various studies for microwave-enhanced CVD, the source of plasma used is microwave energy, owing to its characteristics of a high plasma density and a resonance field able to concentrate the plasma, thereby ensuring the reduced electron loss in the surrounding area. In various microwave-enhanced CVD studies, a plasma source is utilized as an energy source owing to its characteristics of high plasma density and the ability of resonance field to concentrate the plasma, thereby confirming loss of reduced electron in adjacent areas (Fig. 4.3). The parameters that control the intensity of the plasma and the ionic flux to the substrate are microwave power and the applied DC bias voltage. These parameters can change the CNTs properties as they work independently. This method carries various advantages over the traditional methods, such as it is more ecological, clean, fast, and noninvasive in comparison. Usually, the microwave is known to accelerate and facilitate the reactions, thereby improving the relative yields. In the CNTs functionalized by microwave-assisted techniques, the reaction time is shown to reduce and products with a higher degree of functionalization are produced. Interestingly, the competitive effect of microwave radiation has been proposed, which both promotes functionalization and removes certain functional groups that are initially present. On the other hand, Vazquez and his colleagues showed the combination of solvent-free technology and microwave radiation can generate functionalized nanotubes in just 1 h, paving the way for mass production of functionalized (Nüchter et al., 2004). Although the way the material is heated by microwave is different, the maximum temperature is mainly determined by the dielectric properties of the receiver. With local temperatures above 1100 °C, these carbon-based materials will generate hotspots in the form of small sparks or arcs. Microwave radiation sensitizer is used in organic synthesis, environmental regeneration, catalyst preparation, and carbon nanostructure fields (Kharissova, 2004).

In this regard, Méndez et al. (2003) used microwave heating (approx. 1200 °C for 60 min) by applying microwave power of 800 W for the synthesis of CNTs from graphite. The study showed no need for catalyst during the conversion process of graphite other than microwave heating. Along with graphite, the nanotubes, nanofibers, and microparticles for boric acid and sucrose graphite along with graphite were also observed. Hong et al. (2003) successfully synthesized the CNTs using microwave irradiation (2.45 GHz, 800 W). They achieved this by loading catalysts on numerous supports such as silica powder organic polymer, and carbon black. Similarly, Kharissova et al. (2004) developed a highly effective single-step method to achieve aligned and long CNTs with or without iron filling. Catalysts used during the microwave irradiation were ferrocene, Fe, and silica fused. On the other hand, to better understand the influence of chemical structures and processing conditions on CNTs properties, Lee Benito et al. (2009) conducted the study involving synthesis of CNTs on various supports including organic polymer substrates. Following this, Zajíčková et al. (2009) evaluated the effect of microwaves-hydrothermal treatment on physicochemical properties of the Zl, Zn, and Co catalyst obtained via calcination of double-layered hydroxide. The treatment was shown to distribute the cations within the layers of precursors, the same effect being observed in the catalysts as well. According to a kinetic analysis of carbon growth, this action improves the activity and stability of the Co, Zn, Zl catalyst. However, there were no remarkable differences between different aged catalysts, which is in agreement with results of other studies. On the other hand, results of transmission electron microscopy show that the properties of carbon products also depend upon the duration of microwave-hydrothermal treatment. The duration effect brings about change in the type of nanofilament and the quantity of amorphous carbon produced. Furthermore, Lee (2007) successfully applied microwave operating in the atmospheric pressure to fast decompose MWCNTs on the substrate without heating or vacuum equipment. The study was able to generate dense straight standing nanotubes on silicon substrates with or without a SiO_x barrier. Also the study produced CNTs directly on conductive silicon, hence use them as electron-emitting electrodes of the gas pressure sensor. Moreover, Reddy et al. (2020) studied morphology of CNTs and microwave-assisted methane pyrolysis on the growth without using any catalyst. The experiment was conducted by placing a vertical quartz tube filled with carbon/carbon absorber material without any mineral content (catalyst) in a domestic microwave oven. The reaction involved mixed gas flow CH_4/N_2 (1:4) for 60 min followed by analyzing the substance deposited on the quartz as a result. The results indicated the microwave-assisted pyrolysis to be a novel and promising approach for carbon nanostructure synthesis. Furthermore, the structure of the CNTs was affected by the experimental conditions such as methane to nitrogen ratio, temperature, and flow rate of the gas. Vivas-Castro et al. (2011) reported the microwave irradiation technique to be cost effective and simple. The study obtained nanostructural material from graphite/iron acetate powder using numerous preparation conditions and a commercial microwave oven as a source of energy. The microwave adsorption results in decomposition of iron acetate, which provides the metallic iron nanoparticles that act as a catalyst during synthesis of carbon nanostructures. Altering the preparation conditions can give different types of carbon nanostructures, such as direct irradiation of quartz ampoules and attenuated irradiation. The reactions during this technique occur very fast and can be explosive; hence the maximum radiation intensity ought to be determined beforehand. The growth of MWCNT is well-oriented and aligned when exposed to static temperature and microwave gradients. Following this, Bekarevich et al. (2012) used microwave-excited surface plasma technique to successfully grow MWCNTs and few-layered graphene sheets on Si and polyimide substrates at relatively low temperatures. Using graphite-encapsulated Ni nanoparticles, the study showed that the CNM structure is highly influenced by nature of the substrate and bias voltage. Similarly, Konno et al. (2013) indicated the direct preparation of CNTs using microwave plasma decomposition of methane over Fe/Si activated by biased hydrogen plasma. The researchers decomposed methane into H_2 and CNTs using microwave, whereas Fe/Si was used as a catalyst. It was accomplished inside a reactor exposed to 500 W for 30 min reaction time and 600 °C with both gases (methane/hydrogen) flowing. The minimum and maximum CNT diameters observed at 20 min and 30 min were 22.5 nm and 9.8 nm, respectively. However, increasing the plasma treatment time also increased the maximum height of MWCNTs, reaching 30.7 nm at 10 min.

4.3.1.1 CVD by fluidized bed

CNTs are tubular carbon structures that possess excellent chemical, mechanical, electrical, and optical properties, making these materials potentially valuable for several applications. At present, the large-scale production of these materials is hindered owing to lack of appropriate and cost-efficient technique, thereby making this an active ongoing research. CNTs can be synthesized via three methods: (1) CVD (2) arc discharge, and (3) laser ablation. Among all these techniques, CVD has shown to be economically promising for large-scale production as shown by the literature. The example of this is the fluidized CVD, which is an inherently similar technique and has explicitly shown to produce high-quality CNTs in large amounts and with low cost. Fluidized bed technology is a well-recognized technique used widely to perform gas-solid reaction on an industrial scale. When coupled with CVD, this technique carries a great potential to modify or create new materials. In this case, Fig. 4.4 shows the schematic representation of fluidized bed CVD. The nano-sized powders have been investigated to enhance the overall reaction

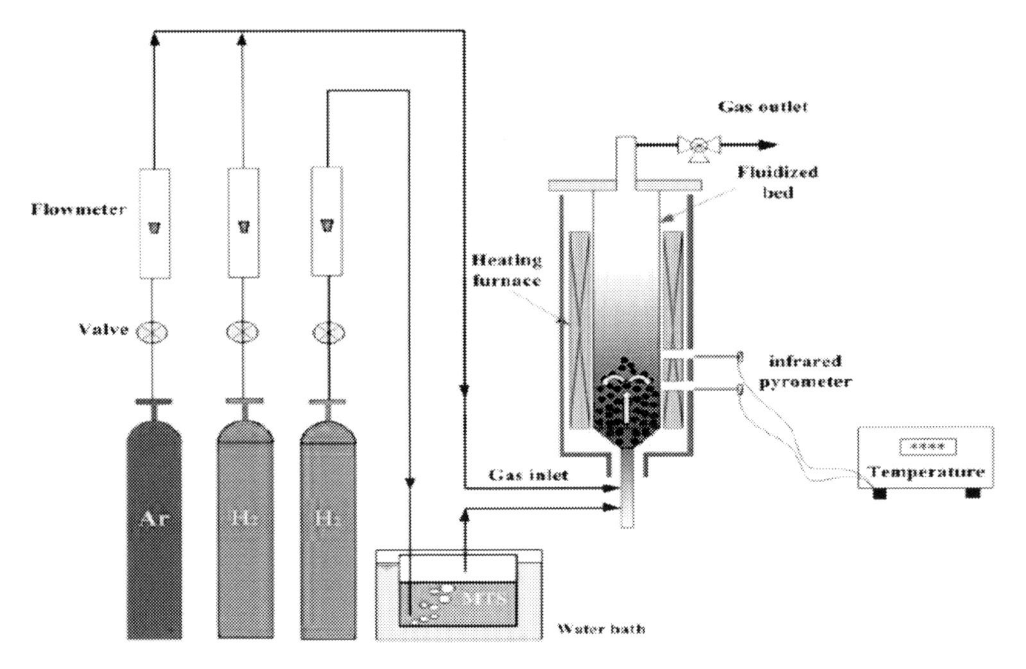

Figure 4.4 Fluidized bed CVD.

efficiency in various powder-based applications such as drug delivery, electronics, and semiconductors. Nevertheless, the fluidization of larger bulky counterparts (100 μm) mass is more complicated than its ultrafine particles (1 μm) as the former particles exhibit different characteristics. The nanoparticles tend to form agglomerates as a result of increased cohesive forces caused by high surface area and mobility. Frequently, fluidization of nanoparticles is achieved by agglomeration of these nanoparticles. These agglomerates follow a complex formation mechanism and are recognized by their fractal structures. It was observed that the agglomerate having the size between 230 and 231 μm was formed by fluidizations of silica powder containing particle size of 7–16 nm. Fluidization of such agglomerates occurred in the absence of bubbles and involved a large bed expansion. These agglomerates are considered a complex and different fluidized agglomerate because the mechanism of fluidization is complicated than the discrete particles. Moreover, these agglomerates are broken down by applying some external force to contribute in fluidization. The formation of these agglomerates has been reported in various studies. Yao et al. (2002) described the agglomerate formation and called them as "loose" agglomerates, whereas Hao et al. (2003) observed that the size of these agglomerates increased with time. However, the results reported are matched with observations by Yao et al. (2002). Thus, the formation of agglomerates during manufacturing of CNT through fluidization process is a fundamental in fluidization bed.

4.3.1.2 Vapor phase growth chemical vapor deposition

The improvement in CNT production can efficiently be achieved by understanding the growth mechanism comprehensively. Several theories have been put forward that try to explain this growth. As per those theories, two mechanisms are involved in CNTs by means of CVD method: base growth and tip growth. The catalyst stays on the surface of the substrate in the complete growth process when mechanism involved follow the base growth. The existence of strong van der Waals forces between nanotubes and substrates is one of the essential reasons behind the limited growth of nanotubes. This limitation occurs when the nanotubes reach a certain length. As a result, force responsible for the movement of whole nanotubes becomes limited therefore, growth stops eventually. This problem does not occur in case of tip-growth mechanisms as catalysts stay on the top of the substrate. Another reason for the length difference between the tip and base growth mechanisms may be the diffusion of feeding gas to the catalyst surface. The surface of the substrate receives the feeding gas at a much slower flow rate compared to above the surface. Sinnott et al. (1999) proposed a mechanism where they developed CNTs in the presence of metal catalyst. During the process, diffusion of carbon into the catalytic particles having size in nanometers and when solubility limits reached within the metal, the carbon precipitates out with graphitic structure. Other studies (Baker 1989; Oberlin et al., 1976) witnessed a direct relationship between the deposited/diffused carbon over particle surface and the length of the tube. Describing the growth of CNTs in vapor phase using metal catalysts, Zhang et al. (2002) showed that initially catalyst particles rise with the CNTs growth; however, it becomes fixed ahead due to friction between tube walls and particles. Furthermore, Deck & Vecchio (2005) investigated the vapor-phase CVD growth mechanism by a spray pyrolysis and found that the condensation of particles is responsible for CNTs growth.

4.3.2 Arc discharge methods

The arc discharge technique for synthesis of CNTs is one of the most widely used and easiest methods in which current is passed between two graphite electrodes to ultimately vaporize the graphite. During this condensation, few amounts of the graphite condenses on the wall while some on the cathode (Fig. 4.5). That which is deposited on the wall contains CNTs. The addition of metals such as Co and Ni onto the anode results in SWCNTs. Various studies have shown that it is possible to synthesize CNTs using gas. At first, the gaseous molecule is catalyzed into carbon by the catalyst followed by tube growth at the tip. In 1991, the arc discharge method was used to prepare CNTs with the needle structure. Using the direct current, the needle CNTs were synthesized by evaporating the carbon in the vessel; however, the CNTs produced were of poor quality compared to the ones produced via arc evaporation method. Great improvements have been witnessed in recent years pertaining to CNTs. For example, Choi et al. (2006) developed an

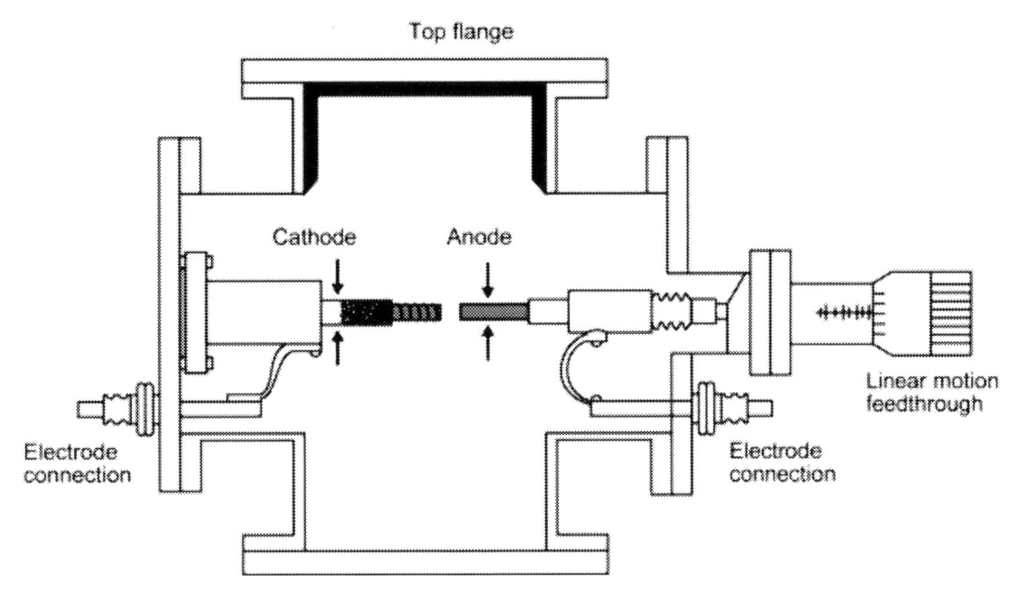

Figure 4.5 Arc discharge method.

economical technique to synthesize a high purity SWCNTs by using FeS as a catalyst and an argon DC arc discharge from charcoal. The third method that can give high yield of SWCNTs is vaporizing the metal graphite through powerful laser.

4.3.3 Laser ablation

The first SWCNT was synthesized in 1996 by Smalley's research group at Rice University at a larger scale. The process involved evaporation of carbon atoms from graphite followed by their deposition on cooled substrate (Fig. 4.6). To vaporize the composite target in a high temperature reactor, a continuous or pulse laser was used (Ravi-Kumar et al., 2019). This resulted in the formation of vaporized graphite feather which subsequently catalyzed the SWCNTs growth. As the vaporized temperature of the species cools down, the fullerenes formed are the result of carbon atoms/molecules condensing together. During the process of condensation, the catalysts are combined in clusters to avoid the cage structure. The short-range electrostatic forces are responsible for the gathering of synthesized SWCNTs. The prepared nanotubes are gathered from the composite target downstream via condensation. Both laser ablation and arc discharge methods utilize graphite therefore both are comparable. Similarly, MWCNTs can also be manufactured either by arc-discharge or laser ablation methods. Among these two methods, arc discharge method is quite efficient for MWCNTs because it gives longer tubes in comparison to laser ablation method. In addition, the laser ablation method requires high-energy consumption, therefore it is expensive and unfavorable for bulk

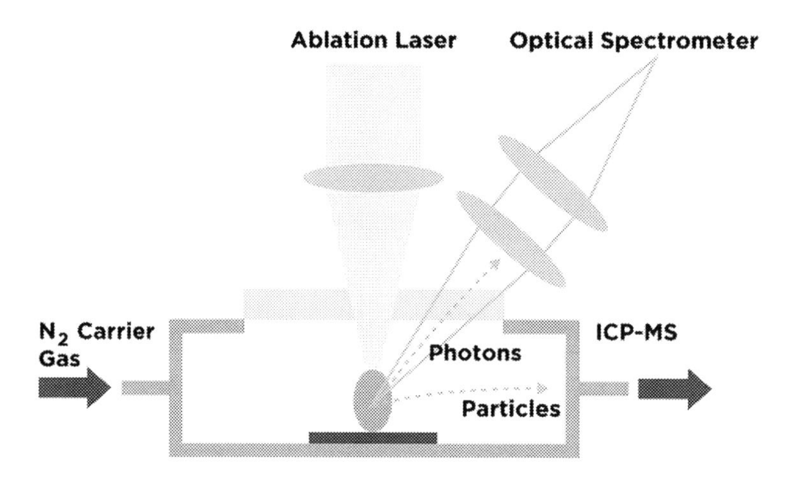

Figure 4.6 Laser ablation.

production. However, the yield capability of this method has enabled its use to the present day (Menazea, 2020).

4.4 Discussion and perspectives

The incentive to synthesize nanomaterials with distinguished features and unusual properties attracts great attention. Owing to their remarkable applications, researchers are focusing on their utilization in numerous fields, such as optical application, drug delivery, energy storage, capturing carbon dioxide, hydrogen storage, biodegradation, composite design and nanoelectronics, catalytic activities, and environmental remediations. Several techniques are available for nanomaterials preparation, among them few that possess great potentials. Available techniques for the preparation of nanomaterials are arc discharges, vapor phase CVD, laser ablation, and fluidized bed CVD. Microwave technology can be used for blending of proprietary carbon nanostructures, including MWCNT, SWCNT, graphene, flare, and other controlled forms in numerous production methods. Follow up with more practical, profitable, and fast kinetics at first glance, the most widely accepted method are microwaves to define the performance in synthesis phase. This step is integral part of the synthesis process; heat transfer rate is another integral part in the thermochemical process. Plus, they also open new perspectives on the development of new CNMs with high specific features that can be used in various new fields. However, for successful implementations, several trials and complications still need to be addressed. One of the questions frequently asked is about the material that must be environmentally friendly, flexible, and naturally available. Answer to this question is to laminate the hazardous gases from the environment to produce a sustainable green product. One challenge may be to manage and provide catalyst use.Expensive much catalyst recovery in search of more economical and more efficient catalysts for improving the economy as a whole performance. Till now, nanomaterials made of carbons have been investigated in lab scale for various applications therefore, there is a dire need to utilize these materials at larger scale for industrial applications. Moreover, owing to the certain complications such as cost, and smaller particle size, it always restricted for commercial application therefore, the study on the cost effectiveness also needs to be performed to design a sustainable method. sustainable as well as econmical aspects, manufacturers may facee technical and administrative problems that promote these new types of products to replace traditional one. Besides that, by performing a significance modification, these nanomaterials can also be utilized for adsorption process. Another problem with nanoscale products is their safety and the right handling, and the level of awareness on this particular issue is quite high. Smaller and larger efforts may be required in this area. More precisely, the effects of CNMs should be eliminated from the environment point of view. In addition, a proper standard test system is missing. More in-depth work is needed to highlight the different nanometric carbons in the laboratory. conventional technques produce unique CNMs, their stability, shape, control structure, and multi-features can play and prove an important role be an innovation for different sectors

4.5 Conclusions

Nanomaterials with wide application play an important role in various fields as researched by different researchers. Nanomaterials having extremely exceptional properties in thermal stability, catalytic, adsorption, and mechanical strength depend mainly on size, diameter and morphology. This chapter studied the use of various techniques for nanomaterials synthesis, and their application along with their future perspectives. Studied techniques in this chapter include CVD enhanced with microwave, plasma, fluidized bed, phase growth approach, laser ablation, disc charge, etc. From this chapter it is concluded that microwave technique appears a great approach in form cost effectiveness and easy to synthesis nanomaterials.

References

Amama, P.B., Ogebule, O., Maschmann, M.R., Sands, T.D., Fisher, T.S., 2006. Dendrimer-assisted low-temperature growth of carbon nanotubes by plasma-enhanced chemical vapor deposition. Chem. Commun. 2899–2901.

Baker, R., 1989. Catalytic growth of carbon filaments. Carbon 27, 315–323.

Bekarevich, R., Miura, S., Ogino, A., Rogachev, A.V., Nagatsu, M., 2012. Low temperature growth of carbon nanomaterials on the polymer substrate using ion assisted microwave plasma CVD. J. Photopolym. Sci. Technol. 25, 545–549.

Benito, P., Herrero, M., Labajos, F., Rives, V., Royo, C., Latorre, N., Monzon, A., 2009. Production of carbon nanotubes from methane: use of Co-Zn-Al catalysts prepared by microwave-assisted synthesis. Chem. Eng. J. 149, 455–462.

Choi, T.-Y., Poulikakos, D., Tharian, J., Sennhauser, U., 2006. Measurement of the thermal conductivity of individual carbon nanotubes by the four-point three-ω method. Nano Lett. 6, 1589–1593.

Christy, L.A., 2019. Carbon Nanotube Film Structure and Method for Making. Google Patents inventor; GENERAL NANO LLC, assignee, United States patent application US 16/323,001..

Deck, C.P., Vecchio, K., 2005. Growth mechanism of vapor phase CVD-grown multi-walled carbon nanotubes. Carbon 43, 2608–2617.

Dehghani, M.H., Karri, R.R., Alimohammadi, M., Nazmara, S., Zarei, A., Saeedi, Z., 2020. Insights into endocrine-disrupting Bisphenol-A adsorption from pharmaceutical effluent by chitosan immobilized nanoscale zero-valent iron nanoparticles. J. Mol. Liq. 311, 113317.

Deng, Y., Ok, Y.S., Mohan, D., Pittman Jr, C.U., Dou, X., 2019. Carbamazepine removal from water by carbon dot-modified magnetic carbon nanotubes. Environ. Res. 169, 434–444.

Di, T., Cheng, B., Ho, W., Yu, J., Tang, H., 2019. Hierarchically CdS–Ag_2S nanocomposites for efficient photocatalytic H_2 production. Appl. Surf. Sci. 470, 196–204.

Fiyadh, S.S., AlSaadi, M.A., Jaafar, W.Z., AlOmar, M.K., Fayaed, S.S., Mohd, N.S., Hin, L.S., El-Shafie, A., 2019. Review on heavy metal adsorption processes by carbon nanotubes. J. Clean. Prod. 230, 783–793.

Fortunati, E., Mazzaglia, A., Balestra, G.M., 2019. Sustainable control strategies for plant protection and food packaging sectors by natural substances and novel nanotechnological approaches. J. Sci. Food Agric. 99, 986–1000.

Gang, L., Xiaohong, L., Zhijun, Z., 2011. Preparation methods of copper nanomaterials. Prog. Chem. 23, 1644.

Gürsoy, M., 2020. Vapor deposition polymerization of synthetic rubber thin film in a plasma enhanced chemical vapor deposition reactor. J. Appl. Polym. Sci. 49722.

Hao, Y., Qunfeng, Z., Fei, W., Weizhong, Q., Guohua, L., 2003. Agglomerated CNTs synthesized in a fluidized bed reactor: agglomerate structure and formation mechanism. Carbon 41, 2855–2863.

He, J., Li, D., Ying, Y., Feng, C., He, J., Zhong, C., Zhou, H., Zhou, P., Zhang, G., 2019. Orbitally driven giant thermal conductance associated with abnormal strain dependence in hydrogenated graphene-like borophene. NPJ Comput. Mater. 5, 1–8.

Hirao, T., Ito, K., Furuta, H., Yap, Y.K., Ikuno, T., Honda, S.-i., Mori, Y., Sasaki, T., Oura, K., 2001. Formation of vertically aligned carbon nanotubes by dual-RF-plasma chemical vapor deposition. Jpn J. Appl. Phys. 40, L631.

Hong, E.H., Lee, K.H., Oh, S.H., Park, C.G., 2003. Synthesis of carbon nanotubes using microwave radiation. Adv. Funct. Mater. 13, 961–966.

Hong, Y.-L., Liu, Z., Wang, L., Zhou, T., Ma, W., Xu, C., Feng, S., Chen, L., Chen, M.-L., Sun, D.-M., 2020. Chemical vapor deposition of layered two-dimensional $MoSi_2N_4$ materials. Science 369, 670–674.

Jain, P., Patidar, B., Bhawsar, J., 2020. Potential of nanoparticles as a corrosion inhibitor: a review. . J. Bio-Tribo-Corros. 6, 43.

Jelinkova, P., Mazumdar, A., Sur, V.P., Kociova, S., Dolezelikova, K., Jimenez, A.M.J., Koudelkova, Z., Mishra, P.K., Smerkova, K., Heger, Z., Vaculovicova, M., Moulick, A., Adam, V., 2019. Nanoparticle-drug conjugates treating bacterial infections. J. Control. Release 307, 166–185.

Jin, W., Maduraiveeran, G., 2019. Recent advances of porous transition metal-based nanomaterials for electrochemical energy conversion and storage applications. Mater. Today Energy 13, 64–84.

Jonoobi, M., Oladi, R., Davoudpour, Y., Oksman, K., Dufresne, A., Hamzeh, Y., Davoodi, R., 2015. Different preparation methods and properties of nanostructured cellulose from various natural resources and residues: a review. Cellulose 22, 935–969.

Karim, M.R., Jayatunga, B.H.D., Feng, Z., Kash, K., Zhao, H., 2019. Metal–organic chemical vapor deposition growth of $ZnGeN_2$ films on sapphire. Cryst. Growth Des. 19, 4661–4666.

Khan, F.S.A., Mubarak, N.M., Khalid, M., Khan, M.M., Tan, Y.H., Walvekar, R., Abdullah, E.C., Karri, R.R., Rahman, M.E., 2021. Comprehensive review on carbon nanotubes embedded in different metal and polymer matrix: fabrications and applications. Critical Reviews in Solid State and Materials Sciences 1–28. doi:10.1080/10408436.2021.1935713.

Khan, F.S.A., Mubarak, N.M., Khalid, M., Walvekar, R., Abdullah, E.C., Ahmad, A., Karri, R.R., Pakalapati, H., 2021. Functionalized multi-walled carbon nanotubes and hydroxyapatite nanorods reinforced with polypropylene for biomedical application. Scientific Reports 11 (1), 1–10. doi:10.1038/s41598-020-80767-3.

Khan, F.S.A., Mubarak, N.M., Tan, Y.H., Khalid, M., Karri, R.R., Walvekar, R., Abdullah, E.C., Nizamuddin, S., Mazari, S.A., 2021. A comprehensive review on magnetic carbon nanotubes and carbon nanotube-based buckypaper-heavy metal and dyes removal. J. Hazard. Mater. 125375. doi:10.1016/j.jhazmat.2021.125375.

Kharissova, O.V., 2004. Vertically aligned carbon nanotubes fabricated by microwaves. Rev. Adv. Mater. Sci 7, 50–54.

Konno, K., Onoe, K., Takiguchi, Y., Yamaguchi, T., 2013. Direct preparation of hydrogen and carbon nanotubes by microwave plasma decomposition of methane over Fe/Si activated by biased hydrogen plasma. Green Sustain. Chem., 3, pp. 19–25.

Lai, G.S., Lau, W.J., Goh, P.S., Karaman, M., Gürsoy, M., Ismail, A.F., 2019. Development of thin film nanocomposite membrane incorporated with plasma enhanced chemical vapor deposition-modified hydrous manganese oxide for nanofiltration process. Compos. B Eng. 176, 107328.

Lee, K.-H., 2007. Low Temperature Synthesis of Carbon Nanotubes by Direct Microwave Irradiation. Department of Chemical Engineering, Pohang University of Science and Technology (Korea South).

Li, Z., Wang, L., Li, Y., Feng, Y., Feng, W., 2019. Carbon-based functional nanomaterials: preparation, properties and applications. Compos. Sci. Technol. 179, 10–40.

Lowry, G.V., Avellan, A., Gilbertson, L.M., 2019. Opportunities and challenges for nanotechnology in the agri-tech revolution. Nat. Nanotechnol. 14, 517–522.

Ma, Z., Chai, S., Feng, Q., Li, L., Li, X., Huang, L., Liu, D., Sun, J., Jiang, R., Zhai, T., 2019. Chemical vapor deposition growth of high crystallinity Sb_2Se_3 nanowire with strong anisotropy for near-infrared photodetectors. Small 15, 1805307.

Magalhães, G.C., Alves, V.S., Marino, E.C., Nascimento, L.O., 2020. Pseudo quantum electrodynamics and Chern-Simons theory coupled to two-dimensional electrons. Phys. Rev. D 101, 116005.

Maiti, D., Tong, X., Mou, X., Yang, K., 2019. Carbon-based nanomaterials for biomedical applications: a recent study. Front. Pharmacol. 9, 1401.

Makvandi, P., Wang, Cy, Zare, E.N., Borzacchiello, A., Niu, Ln, Tay, F.R., 2020. Metal-based nanomaterials in biomedical applications: antimicrobial activity and cytotoxicity aspects. Adv. Funct. Mater. 30 (22), 1910021.

Malaki, M., Xu, W., Kasar, A.K., Menezes, P.L., Dieringa, H., Varma, R.S., Gupta, M., 2019. Advanced metal matrix nanocomposites. Metals 9, 330.

Mao, J., Li, J., Pei, J., Liu, Y., Wang, D., Li, Y., 2019. Structure regulation of noble-metal-based nanomaterials at an atomic level. Nano Today 26, 164–175.

Mathew, J., Joy, J., George, S.C., 2019. Potential applications of nanotechnology in transportation: a review. J. King Saud Univ. Sci. 31, 586–594.

Menazea, A.A., 2020. Femtosecond laser ablation-assisted synthesis of silver nanoparticles in organic and inorganic liquids medium and their antibacterial efficiency. Radiat. Phys. Chem. 168, 108616.

Méndez, U.O., Kharissova, O.V., Rodríguez, M., 2003. Synthesis and morphology of nanostructures via microwave heating. Rev. Adv. Mater. Sci 5, 398.

Meyyappan, M., Delzeit, L., Cassell, A., Hash, D., 2003. Carbon nanotube growth by PECVD: a review. Plasma Sour. Sci. Technol. 12, 205.

Ngidi, N.P., Ollengo, M.A., VO, N.y.a.m.o.r.i., 2019. Effect of doping temperatures and nitrogen precursors on the physicochemical, optical, and electrical conductivity properties of nitrogen-doped reduced graphene oxide. Materials 12, 3376.

Nüchter, M., Ondruschka, B., Bonrath, W., Gum, A., 2004. Microwave assisted synthesis–a critical technology overview. Green Chem. 6, 128–141.

Oberlin, A., Endo, M., Koyama, T., 1976. Filamentous growth of carbon through benzene decomposition. J. Cryst. Growth 32, 335–349.

Pan, D., Chen, K., Zhou, Q., Zhao, J., Xue, H., Zhang, Y., Shen, Y., 2019. Engineering of CdTe/SiO_2 nanocomposites: Enhanced signal amplification and biocompatibility for electrochemiluminescent immunoassay of alpha-fetoprotein. Biosensors Bioelectron. 131, 178–184.

Pejjai, B., Minnam Reddy, V.R., Gedi, S., Park, C., 2017. Status review on earth-abundant and environmentally green Sn-X (X = Se, S) nanoparticle synthesis by solution methods for photovoltaic applications. Int. J. Hydrogen Energy 42, 2790–2831.

Rajitha, B., Malla, R.R., Vadde, R., Kasa, P., Prasad, G.L.V., Farran, B., Kumari, S., Pavitra, E., Kamal, M.A., Raju, G.S.R., 2019. Horizons of Nanotechnology Applications in Female Specific Cancers, Seminars in Cancer Biology. Elsevier,, Amsterdam.

Rajput, N., 2015. Methods of preparation of nanoparticles-a review. Int.J. Adv. Eng. Technol. 7, 1806.

Ravi-Kumar, S., Lies, B., Zhang, X., Lyu, H., Qin, H., 2019. Laser ablation of polymers: a review. Polym. Int. 68, 1391–1401.

Reddy, B.R., Ashok, I., Vinu, R., 2020. Preparation of carbon nanostructures from medium and high ash Indian coals via microwave-assisted pyrolysis. Adv. Powder Technol. 31, 1229–1240.

Saeedi, M., Eslamifar, M., Khezri, K., Dizaj, S.M., 2019. Applications of nanotechnology in drug delivery to the central nervous system. Biomed. Pharmacother. 111, 666–675.

Shi, J., Huan, Y., Hong, M., Xu, R., Yang, P., Zhang, Z., Zou, X., Zhang, Y., 2019. Chemical vapor deposition grown large-scale atomically thin platinum diselenide with semimetal–semiconductor transition. ACS Nano 13, 8442–8451.

Sinnott, S., Andrews, R., Qian, D., Rao, A.M., Mao, Z., Dickey, E., Derbyshire, F., 1999. Model of carbon nanotube growth through chemical vapor deposition. Chem. Phys. Lett. 315, 25–30.

Soudagar, M.E.M., Nik-Ghazali, N.-N., Abul Kalam, M., Badruddin, I.A., Banapurmath, N.R., Akram, N., 2018. The effect of nano-additives in diesel-biodiesel fuel blends: a comprehensive review on stability, engine performance and emission characteristics. Energy Convers. Manage. 178, 146–177.

Thamaraiselvan, C., Wang, J., James, D.K., Narkhede, P., Singh, S.P., Jassby, D., Tour, J.M., Arnusch, C.J., 2019. Laser-induced graphene and carbon nanotubes as conductive carbon-based materials in environmental technology. Mater. Today, 34, 155 -131.

Tian, C., Wang, F., Wang, Y., Yang, Z., Chen, X., Mei, J., Liu, H., Zhao, D., 2019. Chemical vapor deposition method grown all-inorganic perovskite microcrystals for self-powered photodetectors. ACS Appl. Mater. Interf. 11, 15804–15812.

Vatansever, E., Arslan, D., Nofar, M., 2019. Polylactide cellulose-based nanocomposites. Int. J. Biol. Macromol. 137, 912–938.

Verma, S.K., Das, A.K., Gantait, S., Kumar, V., Gurel, E., 2019. Applications of carbon nanomaterials in the plant system: a perspective view on the pros and cons. Sci. Total Environ. 667, 485–499.

Vivas-Castro, J., Rueda-Morales, G., Ortega-Cervantez, G., Moreno-Ruiz, L., Ortega-Aviles, M., Ortiz-Lopez, J.Yellampalli, S. (Ed.), 2011. Synthesis of carbon nanostructures by microwave irradiation. Carbon Nanotubes: Synthesis Characterization, Applications 47–60.

Wang, H., Gao, Y., Liu, J., Li, X., Ji, M., Zhang, E., Cheng, X., Xu, M., Liu, J., Rong, H., 2019a. Efficient plasmonic Au/CdSe nanodumbbell for photoelectrochemical hydrogen generation beyond visible region. Adv. Energy Mater. 9, 1803889.

Wang, L., Li, X., Li, Q., Yu, X., Zhao, Y., Zhang, J., Wang, M., Che, R., 2019b. Oriented polarization tuning broadband absorption from flexible hierarchical ZnO arrays vertically supported on carbon cloth. Small 15, 1900900.

Wang, Y., Gupta, S., Nemanich, R., 2004. Role of thin Fe catalyst in the synthesis of double-and single-wall carbon nanotubes via microwave chemical vapor deposition. Appl. Phys. Lett. 85, 2601–2603.

Wu, C., Guo, D., Zhang, L., Li, P., Zhang, F., Tan, C., Wang, S., Liu, A., Wu, F., Tang, W., 2020. Systematic investigation of the growth kinetics of β-Ga_2O_3 epilayer by plasma enhanced chemical vapor deposition. Appl. Phys. Lett. 116, 072102.

Wu, Y., Qiao, P., Chong, T., Shen, Z., 2002. Carbon nanowalls grown by microwave plasma enhanced chemical vapor deposition. Adv. Mater. 14, 64–67.

Xia, Z., Guo, S., 2019. Strain engineering of metal-based nanomaterials for energy electrocatalysis. Chem. Soc. Rev. 48, 3265–3278.

Xiao, S., Dai, W., Liu, X., Pan, D., Zou, H., Li, G., Zhang, G., Su, C., Zhang, D., Chen, W., 2019. Microwave-induced metal dissolution synthesis of core–shell copper nanowires/ZnS for visible light photocatalytic H_2 evolution. Adv. Energy Mater. 9, 1900775.

Xie, F., Yang, M., Jiang, M., Huang, X.-J., Liu, W.-Q., Xie, P.-H., 2019. Carbon-based nanomaterials–a promising electrochemical sensor toward persistent toxic substance. TrAC Trends Anal. Chem. 119, 115624.

Xu, K., Liu, H., Shi, Y.-C., You, J.-Y., Ma, X.-Y., Cui, H.-J., Yan, Q.-B., Chen, G.-C., Su, G., 2020. Preparation of T-carbon by plasma enhanced chemical vapor deposition. Carbon 157, 270–276.

Yao, W., Guangsheng, G., Fei, W., Jun, W., 2002. Fluidization and agglomerate structure of SiO_2 nanoparticles. Powder Technol. 124, 152–159.

Yen, J., Leu, I.-C., Lin, C., Hon, M.-H., 2005. Synthesis of well-aligned carbon nanotubes by inductively coupled plasma chemical vapor deposition. Appl. Phys. A 80, 415–421.

Zajíčková, L., Jašek, O., Synek, P., Eliáš, M., Kudrle, V., Kadlečíková, M., Breza, J., Hanzlíková, R., 2009. Synthesis of Carbon Nanotubes in MW Plasma Torch with Different Methods of Catalyst Layer Preparation and Their Applications. Proc. NANOCON, Czech Republic..

Zhang, K., Gao, H., Deng, R., Li, J., 2019. Emerging applications of nanotechnology for controlling cell-surface receptor clustering. Angew. Chem. Int. Ed. 58, 4790–4799.

Zhang, X., Cao, A., Wei, B., Li, Y., Wei, J., Xu, C., Wu, D., 2002. Rapid growth of well-aligned carbon nanotube arrays. Chem. Phys. Lett. 362, 285–290.

CHAPTER 5

Green synthesis of metal nanoparticles for environmental remediation

Sumalatha Jorepalli[a], Adinarayana Reddy Somala[b], Viswadevarayalu Annavaram[c] and Janardhan Reddy Koduru[d]
[a]Nirmala College of Pharmacy, Ukkayapalli, Kadapa, Andhra Pradesh, India
[b]Department of Materials Science and Nanotechnology, Yogi Vemana University, Kadapa, Andhra Pradesh, India
[c]Department of Chemistry, Annamacharya Institute of Science & Technology, Rajampet, Andhra Pradesh, India
[d]Department of Environmental Engineering, Kwangwoon University, Nowon-gu, Seoul, Korea

5.1 Introduction

From the 1980s onwards, there has been a fast expansion in the discipline currently identified as nanoscience. The innovative study zone combines physical science, natural science, chemical science, engineered, medical materials science, and effects on various needs too. Nanoscience addresses innumerable issues with an enormous segment of these having the potential for novel particular applications. Precisely the consideration changes from fundamental science regarding the applications, the word nanotechnology is much more regularly used (Ashraf et al., 2016; Gade et al., 2010; Han et al., 2017; Lanone & Boczkowski, 2006; Samadi et al., 2018; Santhosh et al., 2016). Nanoscience is commonly characterized as the assessment of miracles of the science, which in the size is ranged 1–100 nm, despite how it is generally helpful to expand the show up slightly at both ends (Lingamdinne et al., 2018; Ruthiraan et al., 2019). Nanotechnology has contained a limit to prepare, control, and create matter on this scale to make novel materials that have explicit properties (Keshri & Agarwal, 2012). The historical backdrop of nanomaterials is very long. Taking everything into account, critical enhancements inside nanoscience have happened in the midst of the latest twentieth years. The idea of nanotechnology was at first featured by Noble laureate Richard Feynman, in his famous talk at the California Institute of Technology, on 29th December 1959 "There is plenty of room at the bottom" observed the idea of nanomaterials (Feynman, 2018). He pointed out that on the off chance that a bit of information required only 100 particles, and afterward all of the books ever created could be taken care of in a robust shape with sides 0.02 inch long. In 1970, Norio Taniguchi first described the term "nanotechnology" (Vincent, 2008). According to him, "nanotechnology chiefly comprises of the handling, division, distortion, and combination of material by one atom or by one molecule." Also, in 1980, another

Sustainable Nanotechnology for Environmental Remediation
DOI: https://doi.org/10.1016/B978-0-12-824547-7.00011-4

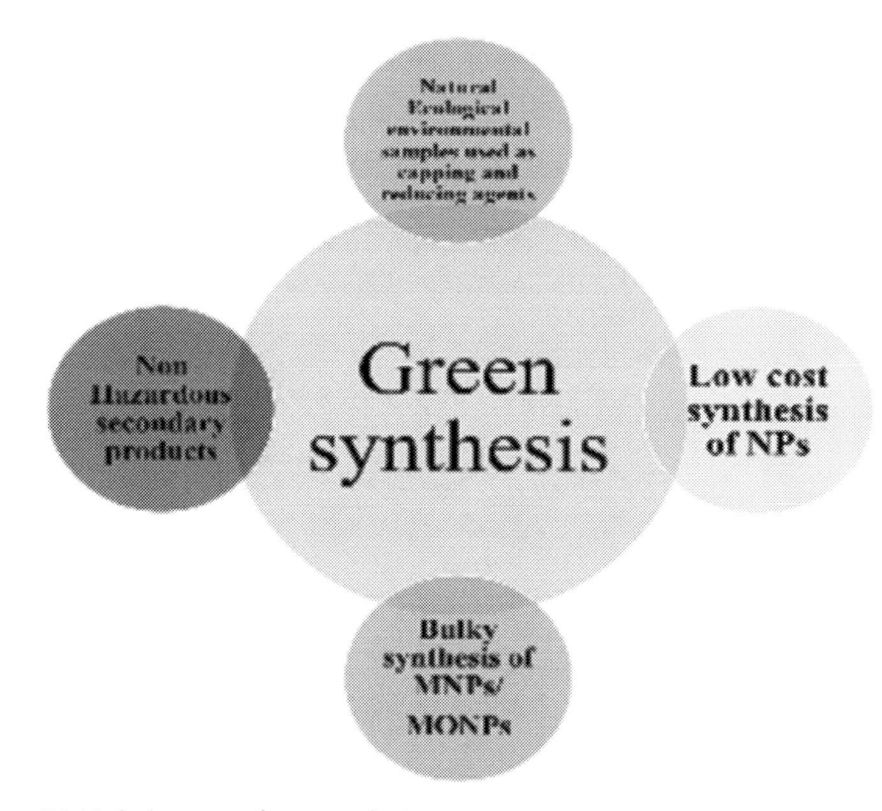

Figure 5.1 Vital advantages of green synthesis.

technologist, K. Eric Drexler tended to innovative impact in nanoscale. The basic rule to find nanoparticles (NPs) is that the NPs' properties are extremely not equivalent to bulk-scale properties. NPs are used as a section of a particular stream together with genetic, fabric, and chemical science fit as a fiddle and the dimension of colloidal MNPs/MONPs expects a significant role in various applications, and also in preparation of electronic contraptions, wound patching, magnetic materials, the carcinogenicity of gene expression, microorganisms in the amalgamation of biocomposites (Leydecker, 2008; Moon et al., 2012). Based on MNPs/MONPs have optical and reactant electromagnetic properties which are dependent on the size and shape of the NPs. Nanoscience and nanotechnology have the ability to follow or to make designer materials, contraptions, and frameworks in the nanometer range (Fig. 5.1) (Wang et al., 2018). The thoughts and considerations from chemical science, physical science, engineered and bioscience are consolidated combinedly to design an innovative material with favored properties. The functions and properties of nanomaterials might be particular as their size varies from the nano-dimension to the micro-sized organization ultimately to bulky size. Nanotechnology is known to be having unacceptable development through current materials, in every way that usefulness is everything like material and system made by man which have various features fit for different innovative applications that are different for auxiliary and practical *in-vivo* biomedical applications (Sahu et al., 2019a, 2019b). The auxiliary course of action of particles and the dimension of the materials basically are two factors, when manufactured legally at nano size, that induce variation in the characteristics of the material as opposed to its bulk structure (Cerrillo et al., 2017; Zaib & Iqbal, 2019). Fig. 5.1 illustrates the general approach to the benefits of green synthetic materials. The subject of nanotechnology is one of the most vibrant areas of research in today's material science. NPs are spotted as new or improved features considering specific features such as dimensions, assignment, and morphology (Madkour, 2019). These materials have significant advances in the stream of nanomaterials and nanotechnology in modern times, with different approaches designed to turn NPs into distinct morphology and size requirements. The new consumption of NPs is rapidly expanding (McCray, 2005).

To identify practical gadgets using their kind of property with microscopic materials, NPs must coordinate with different sizes, shapes, and associations. Unexpectedly, major accomplishments have been made in this method and NPs with bulky size and dimensions can be orchestrated today over a wide range of arrangements. It seems assorted shapes are coordinated at the various sizes of NPs by bottom-up and top-down systems. In the top-down method, the NPs are mixed together by actually cutting or scraping the mass material until the ideal dimension is accomplished. This method is polished by Neanderthals before 400,000 years during the Old Stone Age period when mankind first discovered how to make tools (Mellars, 1996). Throughout the development, the individual engages in this work by having the opportunity to understand the shape of the subatomic scale utilizing various refining methods, for example, laser-prompted material scratching, ball processing, etc. (Landis, 2013; Mellars, 1996). Bottom-up procedures for the manufacture of nanostructures and NPs include essentially manufactured and simple frameworks. These strategies involve the controlled growth of soluble particles confined to the center of the synthetic reaction. The limit, size, and shape of the development or progression prompt a cell plan (Gao et al., 2014). However, the combination of microscopic materials with uniform size and shape can be inconvenient, not as small as the compounding of the molecules with the desired structure. Therefore, the combination of a large scope of nanomaterials with a specific alignment and uniformly sized shape can actually be a debilitating task (Moon et al., 2012). In an attempt to design small gadgets, it has been understood that a decrease in the size of the material prompts changes in properties, for example, electrical conductivity, color, mechanical stiffness, magnetic performance, and liquefying point, against which bulk counter (Wang et al., 2018) can be considered significant in nature. The excitement was back. Gold envelopes were used since ancient times to give reddish color to glass—the famous Lycurgus container dating back to the fourth century AD. Although described in Roman times, gold-ruby glass production was not rediscovered until the seventeenth century. Nowadays the application of micronutrients reaches huge zones, for example, catalyst (Zhu et al., 2019), biosensing (Cardenas-Benitez et al., 2018), drug delivery (Jacob et al., 2018), diagnostics (Loh et al., 2018), solar cell (Jacob et al., 2018), optoelectronics (Gong et al., 2018), cell labeling and image illustration (Xing et al., 2014), photonic bandgap materials (Vasudevan et al., 2019), single-electron optical equipment to list some of the linear (Zhang & Wang, 2017) and surface-enhanced Raman spectroscopy (Annavaram et al., 2019), etc. Extending data about the novel properties of NPs has reenergized them for possible applications. AuNPs have been utilized from antiquated occasions to offer reddish glass—the acclaimed fourth-century Lycurgus compartment. Although portrayed in the Roman period, gold-ruby glass creation was not reinvented until the seventeenth century. Considerable red china was made by including colloidal gold and stain hydroxide "Purple of Cassius" to the base glass. These days the utilization of minuscule materials arrives at immense zones, for instance, impetus, optical gadgets to show a portion of the nonlinear and so forth. An interesting thing about nanomaterials is the number of various factors that may affect their observable properties just as to turn them into different parts of our everyday life. The transformation in discernible properties of nanomaterials, for example, optical, color, and electronic conduct, and the magnetic reaction is because of the way that as the size methodologies nuclear measurements, vitality level groups are gradually changed into quantized discrete vitality levels. Subsequently, the modifications in the electronic arrangement occur in the nanometer region, it gives an understanding regarding how the properties develop from the subatomic or nuclear level to the mass. Also, the decrease in dimension of NPs would keep the electronic development, which will impact the physical and substance properties of the material. The transformation in physical properties of nanomaterials is a result of their estimations being for all intents and purposes indistinguishable from the De Broglie frequency of the charge carriers and their high surface to volume ratio (Beydoun et al., 1999).

5.2 Metal NPs

Well-known scientists have begun to focus on optical properties of MNPs, while the mid-20th century with Gustav Mie's works (Mie 1908). Nevertheless, the usage of their astounding properties is considerably more prepared and returns to a couple of hundreds of years prior (Horvath, 2009; Wriedt, 2008). Metallic NPs have dazzled scientists for longer than a century and are right now enthusiastically utilized in engineering and biomedical sciences. A portion of the antiquated MNPs/MONPs on glass coatings (Schaming & Remita, 2015; Reddy et al., 2013) was arranged in Table 5.1. They are having a focal point of enthusiasm because of their colossal potential in nanotechnology. Nowadays, these materials can be coordinated and changed with various invention pragmatic social occasions to be formed with antibodies, ligands, and medications (Brewer et al., 2007) of intrigue and thusly opening a broad assortment of possible applications in biotechnology (Alatrash et al., 2013; Zhang et al., 2020), alluring separation and

Table 5.1 Metallic and metal oxide nanoparticles utilized in antiquated days in various glass covering materials.

Metallic NPs	Applications	Era	Reference
Cu and Ag	Italian Renaissance pottery	14th–16th century	Bontempi et al., 2006
Au	Andreas Cassius glass	15th Century	Borkow & Gabbay, 2009; Hunt, 1981; Schaming & Remita, 2015
Pt	Lüdersdorff ' lustre	Roman period	Ricciardi et al., 2009 Silvestri et al., 2014
Ag and Au	Glass cup	Roman period	Raman Martín Dombrowski, A. 2017; Gartia et al., 2013
Cu	Reddish glass	19th Century	Nakai et al., 1999; Ruivo et al., 2008
Metal oxide nanoparticles			
SnO_2	Cassiterite	9th	Sciau et al., 2016
ZrO_2	Zirconia blue	20th	Sciau et al., 2016
TiO_2	Titanium white	19th	Sciau et al., 2016
ZnO	Light blue	15th	Sciau et al., 2016

pregathering of target analytes zeroed in taking drugs movement, and vehicles for quality and medication transport and even more fundamentally suggestive imaging. Additionally, unique imaging modalities have been made over the time period, for instance, MRI, CT, PET (Positron Emission Tomography), ultrasound, SERS (Surface Enhanced Raman Spectroscopy), and optical imaging as a manual for picture distinctive ailment states. These imaging modalities shift in the two techniques and instrumentation and even more fundamentally require offset masters with one of kind physiochemical properties. This provoked the advancement of various nanoparticulated contrast masters, for instance, Ag, Au, Cu, and PdNPs for their application in these imaging modalities (Sahu et al., 2019b; Jain et al., 2008). Furthermore, distinct imaging approaches in coupled more recent multifunctional nanoshells and nanocages have been developed, as well as a level of the examples of ancient materials (Hubert et al., 2005; Singh et al., 2018).

Assessments using various system exhibited that red glasses of the previous Bronze Age (1200–1000 BCE) from Frattesina di Rovigo (Italy) was concealed because of the excitation of plasmon surface techniques for various NPs (Hasan, 2015). By exposing the material to deteriorating conditions, the protohistoric wing of this region induced glass-manufacturing and may induce the solution of metallic Cu, precious gravels in the upper layer of glass. In the Roman period, metallic particles were used to hide glass surfaces. It is accurately the example of the most likely comprehended Roman Lycurgus Cup in glass dated from the fourth CE, which is located in the British Museum. The glass holder is dichroic and takes after jade with a dinky greenish-yellow color, anyway when light emanates on the glass (sent light) it changes into a clear ruby concealing. It is clearly indicated that the explosive concealing change is achieved by colloidal metal and even

Figure 5.2 Sustainable (green) synthesis of MNPs/MONPs utilizing the green contents from the leaves of different plant species prompting the creation of structures with different composites, shapes, and sizes.

more totally by nanocrystals of an Ag-Au compound dispersed all through the cleaned network (Gamucci et al., 2014). During medieval times, glass manufactured broadened and astonishingly, especially notifies the enthusiasm for colored glass. This headway was joined by a development in the sort of colloidal metal used for concealing glass. The amassing method of reddish glass was used around the world (Gamucci et al., 2014). The mainstream Satsuma glasses made in Japan during the nineteenth century era were got using a practically identical framework and its ruby color comes from the digestion properties of MNPs. Detailed strategies for the dispersion of NPs existed much before the seventeenth century era when old Hindus used pulverized AuNPs generally called "Suvarna bhasma" in natural medicine for the treatment of rheumatoid arthritis (Biswas et al., 2020). Nevertheless, the technique utilized by Hindus was awkward and utilized a "top-down methodology." From Fig. 5.2 assessments of the diverse MNPs/MONPs were combined by various green materials with various shapes and sizes. The various green syntheses of MNPs are listed in Table 5.2 (Annavaram et al., 2015).

5.3 Metal oxide NPs

Metal oxide NPs (MONPs) have numerous ecological applications, including as a catalyst for the expulsion of poisonous components. Other innovative uses of metal oxide NPs incorporate microelectronic creation, piezoelectric segments, energy units, and surface coatings. The exceptional properties of metal oxide NPs are because of their size and high surface locales. A portion of the prominent MONPs are TiO_2, CuO, ZnO, and SnO, Fe_3O_4, which are utilized in different organic applications (Lingaraju et al., 2015; Mohapatra et al., 2018). Nevertheless, the production of NPs has started a few worries because of their carcinogenic nature to humans (Albrecht et al., 2006; Handy & Shaw, 2007). The size of the NP is the primary issue for the carcinogenic impact as NPs can undoubtedly go through physiological obstructions and contact intracellular compartments, for example, the cytoplasm and core, which can prompt the decimation of the cell (Handy & Shaw, 2007). It has been accounted for the surface harmfulness that can be settled with the assistance of topping specialists or an appropriate network. In any case, the carcinogenicity is caused because of the amalgamation technique (e.g., choosing harmful solvents and reducing agents) must be settled by the cautious decision of composite groups. For the most part, the chemical procedure contains some harmful synthetic compounds that lead to natural poisonousness; and methods, for example, UV illumination, photochemical decrease, the airborne technique, the ultrasonic strategy, lithography, and laser ablation are costly and include poisonous synthetic chemicals. These poisonous synthetic compounds, for example, hydrazine hydrate, dimethylformamide, sodium borohydride, and ethylene glycol, lead to natural and organic issues. Consequently, it is basic to create measures that are "greener" and all the more ecologically neighborly because of the issues identifying with natural defilement. Henceforth, scientists are zeroing in on poisonous free solvents (e.g., water) and shut reactor green procedures. Eco-friendly synthesis of MNPs/MONPs, synthetic compounds, for example, nutrients, microorganisms, sugars, biopolymers, and plant materials is generally utilized as a topping and reducing agent. As of late, metal oxide NPs have been integrated utilizing plant tissue, plant extracts, and different parts of the plant materials (Mohapatra et al., 2018). Among these techniques, the plant-based process is more reasonable for the huge production green combination of NPs. The decreasing rate of metallic or MONPs in the presence of plant material is thought to be much faster than that of microorganisms, and it also provides particles. Some of green synthesis of metal/MONPs and their shape and size are summarized in Table 5.3. The plants contain free revolutionary rummaging particles, for example, phenols, nitrogen functional agents, terpenoids, and different metabolites that have cancer prevention agents (Annavaram et al., 2017; Baro et al., 2013; Snehal et al., 2016). The plants utilized for NP amalgamation

Table 5.2 Green synthesis of MNPs.

Plant	Metallic nanoparticles	Size (nm)	Shape	Reference
Azadirachta indica	Ag NPs	2–40	Spherical	Hasan, 2015
Ocimum	Ag NPs	8–10	Spherical	Mallikarjuna et al., 2011
Argemone mexicana	Ag NPs	10–90	Spherical	Singh et al., 2010
Coffea arabica	Ag NPs	30–40	Spherical	Dhand et al., 2016
olive	Ag NPs	90	Spherical	Khalil et al., 2014
Cycas	Ag NPs	35–55	Spherical	Jha & Prasad, 2010
Artocarpus heterophyllus Lam.	Ag NPs	21–25	Spherical and cubic	Jagtap et al., 2013
Ficus benghalensis	Ag NPs	20–30	Spherical	Saxena et al., 2012
Murraya koenigii	Ag NPs	50–90	Spherical	Philip, et al. 2011
Pinus eldarica	Ag NPs	87	Spherical	Iravani & Zolfaghari, 2013
Berberis vulgaris	Ag NPs	5–10	Quasi-spherical	Behravan et al., 2019
marigold flower	Ag NPs	36	Spherical, and Triangular	Padalia et al., 2015
Pulicaria glutinosa	Ag NPs	40	Spherical	Khan et al., 2013
Tea	Ag NPs	18	Hexagonal	Sun et al., 2014
L. sativa & M. sativa	Cu NPs	50	Dendritic, Plate-like and Irregular	Hong et al., 2015
henna leaf	Cu NPs	27	Spherical	Cheirmadurai et al., 2014
E. esula L	Cu NPs	50	Spherical	Nasrollahzadeh et al., 2014
A. marmelos	Cu NPs	<100	Spherical	Kulkarni & Kulkarni, 2014
E. Wallichii	Cu NPs	12	Spherical	Atarod et al., 2015
E. prostrata	Pd NPs	60	Spherical	Rajakumar et al., 2015
Watermelon rind	Pd NPs	97	Spherical	R. et al., 2015
Piper betle	Pd NPs	50	Spherical	K. et al., 2013
T. cacao L.	Pd NPs	35	Spherical	Nasrollahzadeh, Sajadi, Rostami-Vartooni, et al., 2015
A. annua	Pd NPs	25	Spherical	Edayadulla et al., 2015
A. blackberry	Pd NPs	55	Decahedron	Kumar et al., 2015
Banana	Pd NPs	40	Spherical	Bankar et al., 2010
Sour Cherry tree gum	Pd NPs	5	Spherical	Perumalsamy et al., 2017
E. condylocarpa M. bieb	Pd NPs	33	Spherical	Nasrollahzadeh, Mohammad Sajadi, Rostami-Vartooni, & Khalaj, 2015
H. hamnoides Linn	Pd NPs	5	Spherical	Nasrollahzadeh, Sajadi, & Maham, 2015
A. occidentale	Pd NPs	3.5	Spherical	Sheny et al., 2012
T. chebula	Pd NPs	<100	Triangular and pentagonal	Kumar et al., 2013

(*continued on next page*)

Table 5.2 Green synthesis of MNPs—cont'd

Plant	Metallic nanoparticles	Size (nm)	Shape	Reference
C. roseus	Pd NPs	39	Spherical	Kalaiselvi et al., 2015
Perilla frutescens	Pd NPs	11	Spherical	Basavegowda et al., 2015
P. longum	Pd NPs	25	Spherical	Nasrollahzadeh, Mohammad Sajadi, Maham, & Ehsani, 2015
U. davidiana	Pd NPs	1–4	Spherical	Ankamwar et al., 2020
C. esculenta	Pd NPs	20	Irregular	Arsiya et al., 2017
H. sabdariffa	Au NPs	22	Spherical	Mishra et al., 2016
B. oleracea L.	Au NPs	28	Spherical	Kuppusamy et al., 2015
Mango peel extract	Au NPs	21	Spherical	Yang et al., 2014
E. ulmoides	Au NPs	19	Spherical	Guo et al., 2015
P. roxburghii	Au NPs	22	Quasi-Spherical	Paul et al., 2016
S. nigrum	Au NPs	25	Spherical	Muthuvel et al., 2014
B. cinerea	Au NPs	50	Triangular, Hexagonal, Spherical, Decahedral and Pyramidal	Castro et al., 2014
S. swartzii	Au NPs	30	Spherical	Dash et al., 2014
Black Cardamom	Au NPs	18	Spherical	A. K. Singh & Srivastava, 2015
A. augusta Linn	Au NPs	34	Triangular, Pentagonal and Hexagonal	Das et al., 2015
M. elengi	Au NPs	12	Spherical	Majumdar et al., 2016
L. camara Linn	Au NPs	7	Spherical	Dash et al., 2015
Pomegranate Seed Oil	Au NPs	5–10	Spherical	Sadrolhosseini et al., 2014
Z. officinale	Au NPs	10–22	Spherical	K. P. Kumar et al., 2011
C. nucifera	Au NPs	19	Triangular and Spherical	Roopan & Elango, 2015
Z. mauritiana	Au NPs	30	Spherical	Babak, 2015
S. indica	Au NPs	21	Triangular, Pentagonal, Hexagonal and Spherical	Dash et al., 2014
G. sylvestre	Au NPs	24	Spherical	Castillo-López & Pal, 2014
Potato	Au NPs	18	Spherical	Reddy et al., 2015

Table 5.3 Green synthesis of MNPs/MONPs.

Plant	Metallic nanoparticles	Size (nm)	Shape	Reference
Green peas	CuO NPs	35	Spherical	Nair and Chung, 2015
A. lebbeck	CuO NPs	<100	Spherical	Jayakumarai et al., 2015
R. serpentina	CuO NPs	10–20	Sponge	Lingaraju et al., 2015
C. medica L	Cu NPs	10–60	Spherical	Shende et al., 2015
Aloe vera	CuO NPs	20–30	Spherical	Kumar et al., 2015
G. biloba L	Cu NPs	30–40	Spherical	Nasrollahzadeh & Mohammad Sajadi, 2015
A. xylopoda	Cu NPs	20	Spherical	Nasrollahzadeh, Sajadi, & Hatamifard, 2015
G. superba L	CuO NPs	5–10	Spherical	Naika et al., 2015
E. esula L	Cu NPs	<32	Spherical	Naika et al., 2015
F. religiosa	CuO NPs	577	Spherical	Sankar et al., 2014
Citrus limon	CuO NPs	5–20	Spherical	Mohan et al., 2015
C. papaya	CuO NPs	140	Rod	Sankar et al., 2014
A. indica	CuO NPs	26–30	Spherical	Sivaraj et al., 2014
Aloe Vera	Zno NPs	15	Hexagonal and Spherical	D. et al., 2017
Coptidis Rhizoma	Zno NPs	17.5	Rod and Spherical	Nemati et al., 2019
Ocimum basilicum	Zno NPs	45	Rod	Tantiwatcharothai et al., 2019
E. crassipes	Zno NPs	34	Spherical and aggregation	Agarwal et al., 2017
S. nigrum	Zno NPs	26	Quasi Spherical	Muthuvel et al., 2020
E. crassipes	Fe_2O_3, Fe_3O_4 NPs	42	Spherical	Jagathesan and Rajiv 2018; Ding et al., 2019
Avicennia marina	Fe_2O_3	20–35	Irregular spherical	Karpagavinayagam and P, Vedhi 2019
Lagenaria siceraria	Fe_2O_3	45–50	Spherical	Kanagasubbulakshmi and Kadirvelu, 2017
Daphne mezereum	Fe_2O_3	10–55	Spherical	Beheshtkhoo et al., 2018
Platanus orientalis	Fe_2O_3	33	Spherical	Devi et al., 2019
Piper nigrum	SnO_2	10–22	Spherical	Tammina et al., 2017

are accounted to have polyols and cancer prevention agents. The hydroxyl group and the carboxylic groups present in these composite are used as stabilizers and reducing agents (Annavaram et al., 2016) (Fig. 5.3). Additional benefits are: (1) plant inoculum are modest, (2) they do not need extraordinary loading environments, (3) less carcinogenic, and (4) truly constant against unfavorable environments (high temperature, pH). During plant-intervened NPs preparation, the inoculum is combined with precursor salt at different temperatures on various occasions. It has been accounted for that the phenols and

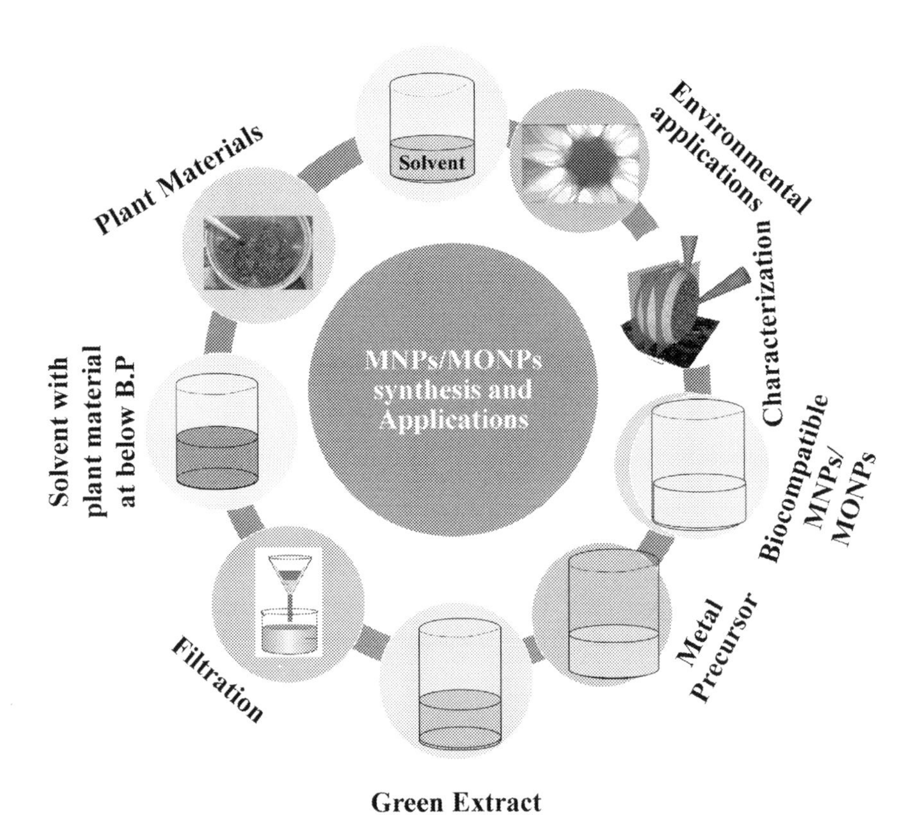

Figure 5.3 A schematic delineation of plants as a hotspot for the eco-friendly synthesis of nanoparticles and the properties and biomedical execution of nanoparticles (Singh et al., 2018).

alkaloids balance out and passivate the size of bulk material into NPs. The growth in steadiness is because of the development of a connection between the NPs and the phytochemicals present in the plant inoculum (Viswadevarayalu, Venkata Ramana, et al., 2016). The reduction of metal particles prompts the arrangement of nucleation focuses, which sequester extra metal particles and lead to the development of NPs (Singh et al., 2016).

5.4 Environmental remediation

Ecological contamination is one of the significant issues around the world. Nanomaterials, for example, Fe, TiO_2, Si, ZnO, carbon nanotubes, have particular physical and compound properties and can be utilized to treat air, water, and soil contamination. The properties of MNPs/MONPs are ideal materials for nanocatalysis, where the response yield and selectivity are dependent on the method of the catalyst surface (Fig. 5.4). Gold is, in general, liberally more affordable and fundamentally more sufficient than platinum and

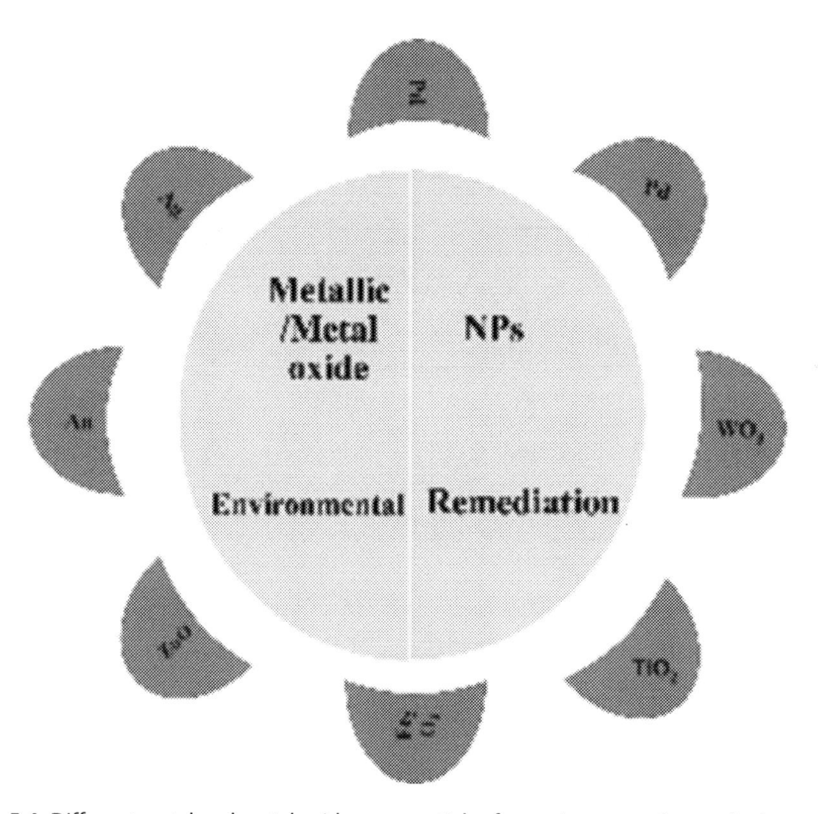

Figure 5.4 Different metal and metal oxide nanoparticles for environmental remediation applications.

palladium. These soapnut shells interceded gold NPs were displayed to have extraordinary synergist activity for the decrease of para nitroaniline (Nasrollahzadeh et al., 2015). The electro reactant use of silver for the decrease of benzyl chloride shows high synergist action contrasted and mass Ag (Jebakumar Immanuel Edison & Sethuraman, 2013).The dehalogenation of palladium NPs on chlorophenol (CP) shows great activity (Redwood et al., 2008). The removal of chromium (VI) shows great synergist movement for the reductant Fe^0 NPs (Karri et al., 2020; Vemula et al., 2012).

For marking, electron fascinating properties of the MNPs are explored to deliver contrast. The gold NPs earnestly acclimatize electrons and thus make them reasonable as a differentiating specialist in TEM (Transmission Electron Microscopy). Moreover, NPs have a similar size gap as proteins which make nanomaterials reasonable for bio-labeling (OV, 2004). In view of their small size and functionalizing properties, for example with antibodies (immunostaining), gold NPs provide for an extraordinary degree of high spatial assurance and are associated with various marking applications. The enormous surface-area-to-volume, extent property of NPs expects a basic part in the synergist reactions used to degrade pesticides. The optical properties of NPs are related to their size and surface-induced changes in electronic structure, which help in the acknowledgment of pesticides. For the destruction of pesticides, a photocatalytic oxidation methodology used with TiO_2 NPs (Tran et al., 2013).

Especially, Fe (0) NPs have been regularly utilized in treating groundwater and soil contaminates with accurate pollutants, for example, As and Cr or perchloroethylene and trichloroethylene in view of its utilization in penetrable-responsive barriers (Ali et al., 2018; Tran et al., 2013). A few researchers have prepared various NPs that associate more effectively with toxins and decay them into less carcinogenic substances (Machado et al., 2013; O'Carroll et al., 2013). As of late, iron oxide NPs settled by biopolymers indicated 92% lindane reduction effectiveness in 88 h of incubation (Paknikar et al., 2005). Likewise, iron-palladium NPs have been concentrated under the high-impact and anaerobic conditions for the corruption of lindane, in which the anaerobic method was discovered to be more effective (Joo & Zhao, 2008). Analysts led an investigation joining dechlorination measures by palladium or iron NPs and the reductive activity of S. wittichii to decontaminate a contaminant solvent of 2,3,7,8-tetrachlorodibenzo-p-dioxin. In this review, after dechlorination, the secondary product acquired (dibenzo-p-dioxin) was not manageable to remediation by NPs and was completely corrupted by microbial digestion within 10 h of exposure (Lakshminarasimhan et al., 2012). Thus, another researcher has led a coordinated remediation procedure utilizing consolidated CMC-palladium/iron nanostructures and Sphingomonas sp. for the treatment of soil polluted with lindane (Y. X. Zhang & Wang, 2017). Notwithstanding, further exploration ought to be committed to the exemplification of NPs, and bioreactor improvement in ceaseless mode activity by which NPs can be used proficiently to control ecological contamination.

5.5 Future prospects of green synthesized metal/MONPs for environmental remediation

As of late, numerous plants have been ceaselessly investigated for the green component combination of NPs and have been effectively utilized in expansive applications because of their ample bounty in nature. At present, plant extricates have been assessed for their viability to the synthesis of metallic/MONPs, which is subject to the synthesis of the perplexing and different phytoconstituent functional groups just as the plant sources. Be that as it may, the explicit biomolecules causing the reduction, stabilizing, and topping agents of the NPs in the biosynthesis component are as yet not completely understood; therefore, more investigations are expected to decide these details. Recommended techniques that could possibly decide the capable functional groups going about as reducing and stabilizing specialists techniques for pure mixtures to distinguish explicit biomolecules. And also, the arrangement of phytoconstituents in plant concentrates can be inspected through analytical instrumental strategies, for example, GC-MS (Gas Chromatography - Mass Spectrometry), HPLC (High Pressure Liquid Chromatography),

ICP-AES (Inductively Coupled Plasma - Atomic Emission Spectroscopy), and NMR (Nuclear Magnetic Resonance Spectroscopy), and a few qualitative and quantitative compound examinations can be utilized to decide the assortment of biomolecule metabolites through phenol-sulfuric acid assays, 3,5-dinitrosalicylic acid colorimetric examines, Coomassie blue tests, and so forth. The primary imperative and challenge might be in deciding the profile of functional groups responsible for going about as reducing agents of the metal particles. Despite the numerous focal points of plant extricates, there are a few different hindrances that ought to be thought of before they can be applied for all intents and purposes, for example, the very much characterized control of the size, shape, structure, crystallinity, what is more, monodispersity of the plant-integrated NPs. This template morphology is additionally identified with the phytoconstituent molecules present in a given plant separate and their starting point. Likewise, different components that impact the morphology of NPs incorporate pH, response temperature, response brooding time, plant extricate fixation, metal particle focus, and the proportion of plant concentrate to the metal substrate. Additionally, as referenced, the capacity to accomplish a high yielding of metal NPs is additionally influenced, just like the reduction force of a given plant extricate. Also, the connection of some biomolecules from the plant remove onto the outside of the biosynthesized NPs may include the irreversible restricting of other utilitarian gatherings, which could frustrate the ensuing functionalization of the combined NPs and breaking point their utilization for different explicit applications. The dependability of biosynthesized NPs is another boundary to consider. It is critical to guarantee that the plant-combined NPs can stay stable with no adjustments in morphology for a long period upon capacity. The data show that the organic limit of plant extricates incorporated NPs is significantly expanded in view of the natural kind disposition of this methodology and the extraordinary single-step response with a system including synergistic reduction, stabilization, and the topping of the NPs during the synthesis process. More orderly, thorough, and coordinated exploration is as yet needed to totally characterize the natural and human harmfulness profile of biosynthesized NPs and to set up a stable framework for the creation of NPs with critical homogeneity and all-around characterized morphology and also, size. For business purposes, it is valuable to have a consolidated test study including the instruments of biotechnology, nanotechnology, bioprocess designing, substance designing, hereditary designing, and plant physiology for a nitty-gritty examination of biosynthesized NPs.

5.6 Conclusion

Though the way that nanoscience has ascending as an unprecedented control of study, the novelty is still truly energetic and the field remains roughly described, indicating an opportunity to intertwine novel thoughts within the current structure. Utilizing green science principles in both the design and use of nanomaterials ensured that this novelty is moved closer in a trustworthy manner by limiting perils to prosperity and the earth while upgrading general profitability without compromising quality and execution. As the field is still at its outset, the conditions uphold the affirmation of greener contemplations and principles within standard scientists, as their passion is not as is normally done coupled to the replacement or retrofitting of dug-in engineered strategies. The benefits of greener nano-synthesis are inclined to intensify well past the innovative method, as green principles are seen emphatically by individuals by and large, and thusly their utilization will assist open with a review of nanotechnology. The improvement of nanoformulations and their various applications make the green synthesis of MNPs and MONPs cheerful and ideal territory for different examines. The components that affect the green combination of MNPs and MONPs from plant extricates, particularly plant leaf inoculum, distinctive kinetics, and NPs creation system are as yet not well thoroughly comprehended, in spite of their few focal points such as carcinogenic decrease capacity, and the simplicity of the synthesis method execution, which make it a broadly used hotspot for NP union with effective, incredibly mixed, nearly modest, ordinary and "generally secure" reductive agent that was utilized for a long time to get synthesized different natural particles and oxidized particles effectively. The entire green amalgamation of NPs requires the composition of other safer, manageable, and trustworthy reductive and topping agents, and finally maintaining to avoid unsafe synthetic compounds, for example, $NaBH_4$ and hydrazine. The designed green NPs are utilized in numerous applications, for example, biomedicine, science, materials science, hardware, biosensors, drug, food, and corrective ventures, and eco-remediation scope.

References

Agarwal, H., Kumar, S., Rajeshkumar, S., 2017. A review on green synthesis of zinc oxide nanoparticles–an eco-friendly approach. Resource-Efficient Technol. 3 (4), 406–413.

Alatrash, G., Jakher, H., Stafford, P.D., Mittendorf, E.A., 2013. Cancer immunotherapies, their safety and toxicity. Exp. Opin. Drug Saf. 12 (5), 631–645. https://doi.org/10.1517/14740338.2013.795944.

Albrecht, M.A., Evans, C.W., Raston, C.L., 2006. Green chemistry and the health implications of nanoparticles. Green Chem. 8 (5), 417–432. https://doi.org/10.1039/b517131h.

Ali, A., Phull, A.R., Zia, M., 2018. Elemental zinc to zinc nanoparticles: is ZnO NPs crucial for life? synthesis, toxicological, and environmental concerns. Nanotechnol. Rev. 7 (5), 413–441. https://doi.org/10.1515/ntrev-2018-0067.

Ankamwar, B., Kirtiwar, S., Shukla, A.C., 2020. Plant-mediated green synthesis of nanoparticles. In: Patra, J.K., Shukla, A.C., Das, G. (Eds.), Advances in Pharmaceutical Biotechnology: Recent Progress and Future Applications. Springer, Singapore, pp. 221–234.

Annavaram, V., Kutsanedzie, Y.H., F., Agyekum A., A., Shah, S., A., Zareef, M., Hassan, M.M., Waqas, A., Ouyang, Q., Chen, Q, 2019. NaYF 4 @Yb,Ho,Au/GO-nanohybrid materials for SERS applications—Pb (II) detection and prediction. Colloids Surf. B Biointerfaces 174, 598–606. https://doi.org/10.1016/j.colsurfb.2018.11.039.

Annavaram, V., Posa, V.R., Jorepalli, S., A.R., Somala, 2016. Biocompatible synthesis of palladium nanoparticles and their impact on fungal species. J. Nanosci. Technol 2 (3), 169–172.

Annavaram, V., Posa, V.R., D., Vijaya Lakshmi, Jorepalli, S., Somala, A.R., 2017. *Terminalia bellirica* fruit extract-mediated synthesis of gold nanoparticles (AuNPs) and studies on antimicrobial and antioxidant activity. Inorganic Nano-Metal Chem. 47 (5), 681–687. https://doi.org/10.1080/15533174.2016.1212219.

Annavaram, V., Posa, V.R., Uppara, V.G., Jorepalli, S., Somala, A.R., 2015. Facile green synthesis of silver nanoparticles using *Limonia Acidissima* leaf extract and its antibacterial activity. BioNanoScience 5 (2), 97–103. https://doi.org/10.1007/s12668-015-0168-7.

Arsiya, F., Sayadi, M.H., Sobhani, S., 2017. Green synthesis of palladium nanoparticles using *Chlorella vulgaris*. Mater. Lett. 186, 113–115. https://doi.org/10.1016/j.matlet.2016.09.101.

Ashraf, S., Pelaz, B., Del Pino, P., Carril, M., Escudero, A., Parak, W.J., Soliman, M.G., Zhang, Q., Carrillo-Carrion, C., 2016. Gold-based nanomaterials for applications in nanomedicine. In: Sortino, S. (Ed.), Light-Responsive Nanostructured Systems for Applications in Nanomedicine. Topics in Current Chemistry, 370. Springer Verlag, Cham, pp. 169–202.

Atarod, M., Nasrollahzadeh, M., Sajadi, S.M., 2015. Green synthesis of a Cu/reduced graphene oxide/Fe_3O_4 nanocomposite using Euphorbia wallichii leaf extract and its application as a recyclable and heterogeneous catalyst for the reduction of 4-nitrophenol and rhodamine B. RSC Adv., 5 (111), 91532–91543. https://doi.org/10.1039/c5ra17269a.

Babak, S., 2015. *Zizyphus mauritiana* extract-mediated green and rapid synthesis of gold nanoparticles and its antibacterial activity J. Nanostruct. Chem., 265–273.

Bankar, A., Joshi, B., Kumar, A.R., Zinjarde, S., 2010. Banana peel extract mediated novel route for the synthesis of palladium nanoparticles. Mater. Lett. 64 (18), 1951–1953. https://doi.org/10.1016/j.matlet.2010.06.021.

Baro, M., Nayak, P., Baby, T.T., Ramaprabhu, S., 2013. Green approach for the large-scale synthesis of metal/metal oxide nanoparticle decorated multiwalled carbon nanotubes. J. Mater. Chem. A 1 (3), 482–486. https://doi.org/10.1039/c2ta00483f.

Basavegowda, N., Mishra, K., Lee, Y.R., 2015. Ultrasonic-assisted green synthesis of palladium nanoparticles and their nanocatalytic application in multicomponent reaction. New J. Chem. 39 (2), 972–977. https://doi.org/10.1039/c4nj01543f.

Beheshtkhoo, N., Kouhbanani, M.A.J., Savardashtaki, A., Amani, A.M., Taghizadeh, S., 2018. Green synthesis of iron oxide nanoparticles by aqueous leaf extract of *Daphne mezereum* as a novel dye removing material. Appl. Phys. A Mater. Sci. Process. 124 (5). https://doi.org/10.1007/s00339-018-1782-3.

Behravan, M., Hossein Panahi, A., Naghizadeh, A., Ziaee, M., Mahdavi, R., Mirzapour, A., 2019. Facile green synthesis of silver nanoparticles using *Berberis vulgaris* leaf and root aqueous extract and its antibacterial activity. Int. J. Biol. Macromol. 124, 148–154. https://doi.org/10.1016/j.ijbiomac.2018.11.101.

Beydoun, D., Amal, R., Low, G., McEvoy, S., 1999. Role of nanoparticles in photocatalysis. J. Nanoparticle Res. 1 (4), 439–458. https://doi.org/10.1023/A:1010044830871.

Biswas, S., Dhumal, R., Selkar, N., Bhagat, S., Chawda, M., Thakur, K., Gudi, R., Vanage, G., Bellare, J., 2020. Physicochemical characterization of Suvarna Bhasma, its toxicity profiling in rat and behavioural assessment in zebrafish model. J. Ethnopharmacol. 249. https://doi.org/10.1016/j.jep.2019.112388.

Bontempi, E., Colombi, P., Depero, L.E., Cartechini, L., Presciutti, F., Brunetti, B.G., Sgamellotti, A., 2006. Glancing-incidence X-ray diffraction of Ag nanoparticles in gold lustre decoration of Italian Renaissance pottery. Appl. Phys. A Mater. Sci. Process. 83 (4), 543–546. https://doi.org/10.1007/s00339-006-3524-1.

Borkow, G., Gabbay, J., 2009. Copper, an ancient remedy returning to fight microbial, fungal and viral infections. Curr. Chem. Biol. 3 (3), 272–278. https://doi.org/10.2174/187231309789054887.

Brewer, M., Zhang, T., Dong, W., Rutherford, M., Tian, Z.R., 2007. Future approaches of nanomedicine in clinical science. Med. Clin. North Am. 91 (5), 963–1016. https://doi.org/10.1016/j.mcna.2007.05.006.

Cardenas-Benitez Djordjevic, I., Hosseini, S., Madou, M., SOJJotES, M.-C.

Castillo-López, D.N., Pal, U., 2014. Green synthesis of Au nanoparticles using potato extract: stability and growth mechanism. J. Nanoparticle Res. 16 (8). https://doi.org/10.1007/s11051-014-2571-3.

Cerrillo, C., Barandika, G., Igartua, A., Areitioaurtena, O., Mendoza, G., 2017. Key challenges for nanotechnology: standardization of ecotoxicity testing. J. Environ. Sci. Health 35 (2), 104–126. https://doi.org/10.1080/10590501.2017.1298361.

Cheirmadurai, K., Biswas, S., Murali, R., Thanikaivelan, P., 2014. Green synthesis of copper nanoparticles and conducting nanobiocomposites using plant and animal sources. RSC Adv. 4 (37), 19507–19511. https://doi.org/10.1039/c4ra01414f.

Das, S., Bag, B.G., Basu, R., 2015. *Abroma augusta* Linn bark extract-mediated green synthesis of gold nanoparticles and its application in catalytic reduction. Appl. Nanosci. (Switzerland) 5 (7), 867–873. https://doi.org/10.1007/s13204-014-0384-4.

Dash, S.S., Bag, B.G., Hota, P., 2015. *Lantana camara* Linn leaf extract mediated green synthesis of gold nanoparticles and study of its catalytic activity. Appl. Nanosci. (Switzerland) 5 (3), 343–350. https://doi.org/10.1007/s13204-014-0323-4.

Dash, S.S., Majumdar, R., Sikder, A.K., Bag, B.G., Patra, B.K., 2014. Saraca indica bark extract mediated green synthesis of polyshaped gold nanoparticles and its application in catalytic reduction. Appl. Nanosci. (Switzerland) 4 (4), 485–490. https://doi.org/10.1007/s13204-013-0223-z.

Devi, H.S., Boda, M.A., Shah, M.A., Parveen, S., Wani, A.H., 2019. Green synthesis of iron oxide nanoparticles using *Platanus orientalis* leaf extract for antifungal activity. Green Process. Synth. 8 (1), 38–45. https://doi.org/10.1515/gps-2017-0145.

Dhand, V., Soumya, L., Bharadwaj, S., Chakra, S., Bhatt, D., Sreedhar, B., 2016. Green synthesis of silver nanoparticles using Coffea arabica seed extract and its antibacterial activity. Mater. Sci. Eng. C 58, 36–43. https://doi.org/10.1016/j.msec.2015.08.018.

Ding, Y., Bai, X., Ye, Z., Ma, L., Liang, L., 2019. Toxicological responses of Fe_3O_4 nanoparticles on *Eichhornia crassipes* and associated plant transportation. Sci. Total Environ. 671, 558–567. https://doi.org/10.1016/j.scitotenv.2019.03.344.

Edayadulla, N., Basavegowda, N., Lee, Y.R., 2015. Green synthesis and characterization of palladium nanoparticles and their catalytic performance for the efficient synthesis of biologically interesting di (indolyl)indolin-2-ones. J. Ind. Eng. Chem. 21, 1365–1372. https://doi.org/10.1016/j.jiec.2014.06.007.

Feynman, R., 2018. There's plenty of room at the bottom. In: Anthony, H. (Ed.), Feynman and Computation. CRC Press, Boca Raton, FL, pp. 1–464.

Gade, A., Ingle, A., Whiteley, C., Rai, M., 2010. Mycogenic metal nanoparticles: progress and applications. Biotechnol. Lett. 32 (5), 593–600. https://doi.org/10.1007/s10529-009-0197-9.

Gamucci, O., Bertero, A., Gagliardi, M., Bardi, G., 2014. Biomedical nanoparticles: overview of their surface immune-compatibility. Coatings 4 (1), 139–159. https://doi.org/10.3390/coatings4010139.

Gao, W.Y., Chrzanowski, M., Ma, S., 2014. Metal-metalloporphyrin frameworks: a resurging class of functional materials. Chem. Soc. Rev. 43 (16), 5841–5866. https://doi.org/10.1039/c4cs00001c.

Gartia, M.R., Hsiao, A., Pokhriyal, A., Seo, S., Kulsharova, G., Cunningham, B.T., Bond, T.C., Liu, G.L., 2013. Colorimetric plasmon resonance imaging using nano Lycurgus cup arrays. Adv. Optical Mater. 1 (1), 68–76. https://doi.org/10.1002/adom.201200040.

Gong, C., Hu, K., Wang, X., Wangyang, P., Yan, C., Chu, J., Liao, M., Dai, L., Zhai, T., Wang, C., Li, L., Xiong, J., 2018. 2D nanomaterial arrays for electronics and optoelectronics. Adv. Functional Mater. 28 (16), 1706559. https://doi.org/10.1002/adfm.201706559.

Guo, M., Li, W., Yang, F., Liu, H., 2015. Controllable biosynthesis of gold nanoparticles from a *Eucommia ulmoides* bark aqueous extract. Spectrochim. Acta A Mol. Biomol. Spectrosc. 142, 73–79. https://doi.org/10.1016/j.saa.2015.01.109.

Han, X.X., Ji, W., Zhao, B., Ozaki, Y., 2017. Semiconductor-enhanced Raman scattering: active nanomaterials and applications. Nanoscale 9 (15), 4847–4861. https://doi.org/10.1039/c6nr08693d.

Handy, R.D., Shaw, B.J., 2007. Toxic effects of nanoparticles and nanomaterials: Implications for public health, risk assessment and the public perception of nanotechnology. Health Risk Soc. 9 (2), 125–144. https://doi.org/10.1080/13698570701306807.

Hasan, S., 2015. A review on nanoparticles: their synthesis and types. Res. J. Recent Sci 4, 2277.

Hong, J., Rico, C.M., Zhao, L., Adeleye, A.S., Keller, A.A., Peralta-Videa, J.R., Gardea-Torresdey, J.L., 2015. Toxic effects of copper-based nanoparticles or compounds to lettuce (*Lactuca sativa*) and alfalfa (*Medicago sativa*). Environ. Sci. Processes Impacts 17 (1), 177–185. https://doi.org/10.1039/c4em00551a.

Horvath, H., 2009. Gustav Mie and the scattering and absorption of light by particles: historic developments and basics. J. Quant. Spectrosc. Radiat. Transfer 110 (11), 787–799. https://doi.org/10.1016/j.jqsrt.2009.02.022.

Hubert, C., Rumyantseva, A., Lerondel, G., Grand, J., Kostcheev, S., Billot, L., Vial, A., Bachelot, R., Royer, P., Chang, S.H., Gray, S.K., Wiederrecht, G.P., Schatz, G.C., 2005. Near-field photochemical imaging of noble metal nanostructures. Nano Lett. 5 (4), 615–619. https://doi.org/10.1021/nl047956i.

Hunt, L.B., 1981. Gold based glass and enamel colours. Endeavour 5 (2), 61–67. https://doi.org/10.1016/0160-9327(81)90150-2.

Iravani, S., Zolfaghari, B., 2013. Green synthesis of silver nanoparticles using *Pinus eldarica* bark extract. BioMed Res. Int. 2013. https://doi.org/10.1155/2013/639725.

Jacob, J., Haponiuk, J.T., Thomas, S., Gopi, S., 2018. Biopolymer based nanomaterials in drug delivery systems: a review. Mater. Today Chem. 9, 43–55. https://doi.org/10.1016/j.mtchem.2018.05.002.

Jagathesan, G., Rajiv, P., 2018. Biosynthesis and characterization of iron oxide nanoparticles using *Eichhornia crassipes* leaf extract and assessing their antibacterial activity. Biocatal. Agric. Biotechnol. 13, 90–94. https://doi.org/10.1016/j.bcab.2017.11.014.

Jagtap, U.B., Bapat, V.A., 2013. Green synthesis of silver nanoparticles using *Artocarpus heterophyllus* Lam. seed extract and its antibacterial activity. Ind. Crops Prod. 46, 132–137. https://doi.org/10.1016/j.indcrop.2013.01.019.

Jain, P.K., Xiaohua, H., El-Sayed, I.H., El-Sayed, M.A., 2008. Noble metals on the nanoscale: optical and photothermal properties and some applications in imaging, sensing. Biol. Med. Acc. Chem. Res. 41 (12), 1578–1586. https://doi.org/10.1021/ar7002804.

Jayakumarai, G., Gokulpriya, C., Sudhapriya, R., Sharmila, G., Muthukumaran, C., 2015. Phytofabrication and characterization of monodisperse copper oxide nanoparticles using Albizia lebbeck leaf extract. Appl. Nanosci. (Switzerland) 5 (8), 1017–1021. https://doi.org/10.1007/s13204-015-0402-1.

Jebakumar Immanuel Edison, T.N., Sethuraman, M.G, 2013. Electrocatalytic reduction of benzyl chloride by green synthesized silver nanoparticles using pod extract of *Acacia nilotica*. ACS Sustain. Chem. Eng. 1 (10), 1326–1332. https://doi.org/10.1021/sc4001725.

Jha, A.K., Prasad, K., 2010. Green synthesis of silver nanoparticles using Cycas leaf. Int. J. Green Nanotechnol. Phys. Chem. 1 (2), P110–P117. https://doi.org/10.1080/19430871003684572.

Joo, S.H., Zhao, D., 2008. Destruction of lindane and atrazine using stabilized iron nanoparticles under aerobic and anaerobic conditions: effects of catalyst and stabilizer. Chemosphere 70 (3), 418–425. https://doi.org/10.1016/j.chemosphere.2007.06.070 .

K., M., N., J.S., B.V., S.R., G., N., B., D.P.R, 2013. Palladium nanoparticles: single-step plant-mediated green chemical procedure using Piper betle leaves broth and their anti-fungal studies. Int. J. Chem. Anal. Sci. 4 (1), 14–18. https://doi.org/10.1016/j.ijcas.2013.03.006.

Kalaiselvi, A., Roopan, S.M., Madhumitha, G., Ramalingam, C., Elango, G., 2015. Synthesis and characterization of palladium nanoparticles using *Catharanthus roseus* leaf extract and its application in the photo-catalytic degradation. Spectrochim. Acta A Mol. Biomol. Spectrosc. 135, 116–119. https://doi.org/10.1016/j.saa.2014.07.010.

Kanagasubbulakshmi, S., Kadirvelu, K, 2017. Green synthesis of iron oxide nanoparticles using *Lagenaria siceraria* and evaluation of its antimicrobial activity. Defence Life Sci. J. 2 (4), 422. https://doi.org/10.14429/dlsj.2.12277.

Karpagavinayagam, P., Vedhi, C., 2019. Green synthesis of iron oxide nanoparticles using *Avicennia marina* flower extract. Vacuum 160, 286–292. https://doi.org/10.1016/j.vacuum.2018.11.043.

Karri, R.R., Sahu, J.N., Meikap, B.C., 2020. Improving efficacy of Cr (VI) adsorption process on sustainable adsorbent derived from waste biomass (sugarcane bagasse) with help of ant colony optimization. Ind. Crops Products 143, 111927. https://doi.org/10.1016/j.indcrop.2019.111927.

Keshri, A.K., Agarwal, A., 2012. Plasma processing of nanomaterials for functional applications—a review. Nanosci. Nanotechnol. Lett. 4 (3), 228–250. https://doi.org/10.1166/nnl.2012.1324.

Khalil, M.M.H., Ismail, E.H., El-Baghdady, K.Z., Mohamed, D., 2014. Green synthesis of silver nanoparticles using olive leaf extract and its antibacterial activity. Arab.J. Chem. 7 (6), 1131–1139. https://doi.org/10.1016/j.arabjc.2013.04.007.

Khan, M., Khan, M., Adil, S.F., Tahir, M.N., Tremel, W., Alkhathlan, H.Z., Al-Warthan, A., Siddiqui, M.R.H., 2013. Green synthesis of silver nanoparticles mediated by *Pulicaria glutinosa* extract. Int. J. Nanomed. 8, 1507–1516. https://doi.org/10.2147/IJN.S43309.

Kulkarni, V., Kulkarni, P., 2014. Synthesis of copper nanoparticles with aegle marmelos leaf extract. Nano Sci Nano Technol. 8, 401–404.

Kumar, P.P.N. Vijay, Shameem, U., Kollu, Pratap, Kalyani, R.L., Pammi, S.V.N., 2015. Green synthesis of copper oxide nanoparticles using aloe vera leaf extract and its antibacterial activity against fish bacterial pathogens. BioNanoScience 5 (3), 135–139. https://doi.org/10.1007/s12668-015-0171-z.

Kumar, B., Smita, K., Cumbal, L., Debut, A., 2015. Ultrasound agitated phytofabrication of palladium nanoparticles using Andean blackberry leaf and its photocatalytic activity. J. Saudi Chem. Soc. 19 (5), 574–580. https://doi.org/10.1016/j.jscs.2015.05.008.

Kumar, K.M., Mandal, B.K., Koppala, S.K., Pamanji, S.R., Sreedhar, B., 2013. Biobased green method to synthesise palladium and iron nanoparticles using Terminalia chebula aqueous extract. Spectroscopy B 102, 128–133.

Kumar, K.P., Paul, W., Sharma, C.P., 2011. Green synthesis of gold nanoparticles with *Zingiber officinale* extract: characterization and blood compatibility. Process Biochem. 46 (10), 2007–2013. https://doi.org/10.1016/j.procbio.2011.07.011.

Kuppusamy, P., Ichwan, S.J.A., Parine, N.R., Yusoff, M.M., Maniam, G.P., Govindan, N., 2015. Intracellular biosynthesis of Au and Ag nanoparticles using ethanolic extract of *Brassica oleracea* L. and studies on their physicochemical and biological properties. J. Environ. Sci. (China) 29, 151–157. https://doi.org/10.1016/j.jes.2014.06.050.

Lakshminarasimhan, N., Bokare, A.D., Choi, W., 2012. Effect of agglomerated state in mesoporous TiO_2 on the morphology of photodeposited Pt and photocatalytic activity. J. Phys. Chem. C 116 (33), 17531–17539. https://doi.org/10.1021/jp303118q.

Landis, S. (2013). Nano Lithography, Wiley, NJ.

Lanone, S., Boczkowski, J., 2006. Biomedical applications and potential health risks of nanomaterials: molecular mechanisms. Curr. Mol. Med. 6 (6), 651–663. https://doi.org/10.2174/156652406778195026.

Leydecker, S., 2008. Nano Materials: In Architecture, Interior Architecture and Design. Birkhäuser, Berlin.

Lingamdinne, L.P., Koduru, J.R., Chang, Y.-Y., Karri, R.R., 2018. Process optimization and adsorption modeling of Pb (II) on nickel ferrite-reduced graphene oxide nano-composite. J. Mol. Liquids 250, 202–211. https://doi.org/10.1016/j.molliq.2017.11.174.

Lingaraju, K., Raja Naika, H., Manjunath, K., Nagaraju, G., Suresh, D., Nagabhushana, H., 2015. Rauvolfia serpentina-mediated green synthesis of CuO nanoparticles and its multidisciplinary studies. Acta Metall. Sin. (English Lett.) 28 (9), 1134–1140. https://doi.org/10.1007/s40195-015-0304-y.

Loh, K.P., Ho, D., Chiu, G.N.C., Leong, D.T., Pastorin, G., Chow, E.K.H., 2018. Clinical applications of carbon nanomaterials in diagnostics and therapy. Adv. Mater. 30 (47), 1802368. https://doi.org/10.1002/adma.201802368.

Machado, S., Stawiński, W., Slonina, P., Pinto, A.R., Grosso, J.P., Nouws, H.P.A., Albergaria, J.T., Delerue-Matos, C., 2013. Application of green zero-valent iron nanoparticles to the remediation of soils contaminated with ibuprofen. Sci. Total Environ. 461–462, 323–329. https://doi.org/10.1016/j.scitotenv.2013.05.016.

Madkour, L.H., 2019. Introduction to nanotechnology (NT) and nanomaterials (NMs). In: Madkour, L.H. (Ed.), Advanced Structured Materials, 116. Springer Verlag, NY, pp. 1–47.

Mahendiran, D., Subash, G., Arumai Selvan, D., Dilaveez, R., Senthil Kumar, R., Kalilur Rahiman, A., 2017. Biosynthesis of zinc oxide nanoparticles using plant extracts of aloe vera and *Hibiscus sabdariffa*: phytochemical, antibacterial, antioxidant and anti-proliferative studies. BioNanoScience 530–545. https://doi.org/10.1007/s12668-017-0418-y.

Majumdar, R., Bag, B.G., Ghosh, P., 2016. Mimusops elengi bark extract mediated green synthesis of gold nanoparticles and study of its catalytic activity. Appl. Nanosci. (Switzerland) 6 (4), 521–528. https://doi.org/10.1007/s13204-015-0454-2.

Mallikarjuna, K., Narasimha, G., Dillip, G.R., Praveen, B., Shreedhar, B., Sree Lakshmi, C., Reddy, B.V.S., Deva Prasad Raju, B., 2011. Green synthesis of silver nanoparticles using Ocimum leaf extract and their characterization. Digest J. Nanomater. Biostruct. 6 (1), 181–186. http://www.chalcogen.infim.ro/181_Mallikarjuna.pdf.

McCray, W.P., 2005. Will small be beautiful? Making policies for our nanotech future. History Technol. 21 (2), 177–203. https://doi.org/10.1080/07341510500103735.

Mellars, P., 1996. The Neanderthal legacy: an archaeological perspective from Western Europe. Princeton University Press, NJ.

Mishra, P., Ray, S., Sinha, S., Das, B., Khan, M.I., Behera, S.K., Yun, S.I., Tripathy, S.K., Mishra, A., 2016. Facile bio-synthesis of gold nanoparticles by using extract of *Hibiscus sabdariffa* and evaluation of its cytotoxicity against U87 glioblastoma cells under hyperglycemic condition. Biochem. Eng. J. 105, 264–272. https://doi.org/10.1016/j.bej.2015.09.021.

Mohan, S., Singh, Y., Verma, D.K., Hasan, S.H., 2015. Synthesis of CuO nanoparticles through green route using Citrus limon juice and its application as nanosorbent for Cr (VI) remediation: process optimization with RSM and ANN-GA based model. Process Saf. Environ. Protection 96, 156–166. https://doi.org/10.1016/j.psep.2015.05.005.

Mohapatra, S., Nguyen, T.A., Nguyen-Tri, P., 2018. Noble Metal-Metal Oxide Hybrid Nanoparticles: Fundamentals and Applications. Elsevier, Amsterdam, pp. 1–674.

Moon, J.J., Huang, B., Irvine, D.J., 2012. Engineering nano- and microparticles to tune immunity. Adv. Mater. 24 (28), 3724–3746. https://doi.org/10.1002/adma.201200446.

Muthuvel, A., Adavallan, K., Balamurugan, K., Krishnakumar, N., 2014. Biosynthesis of gold nanoparticles using *Solanum nigrum* leaf extract and screening their free radical scavenging and antibacterial properties. Biomed. Prev. Nutr. 4 (2), 325–332. https://doi.org/10.1016/j.bionut.2014.03.004.

Muthuvel, A., Jothibas, M., Manoharan, C., 2020. Effect of chemically synthesis compared to biosynthesized ZnO-NPs using *Solanum nigrum* leaf extract and their photocatalytic, antibacterial and in-vitro antioxidant activity. J. Environ. Chem. Eng. 8 (2), 103705. https://doi.org/10.1016/j.jece.2020.103705.

Naika, H., Lingaraju, K., Manjunath, K., Kumar, D., Nagaraju, G., Suresh, D., HJJoTUfS, N., 2015. Green synthesis of CuO nanoparticles using *Gloriosa superba* L. extract and their antibacterial activity. J. Taibah Univ. Sci. 9, 7–12.

Nair, P.M.G., Chung, I.M., 2015. The responses of germinating seedlings of green peas to copper oxide nanoparticles. Biol. Planta. 59 (3), 591–595. https://doi.org/10.1007/s10535-015-0494-1.

Nakai, I., Numako, C., Hosono, H., Yamasaki, K., 1999. Origin of the red color of satsuma copper-ruby glass as determined by EXAFS and optical absorption spectroscopy. J. Am. Ceram. Soc. 82 (3), 689–695. https://doi.org/10.1111/j.1151-2916.1999.tb01818.x.

Nasrollahzadeh, M., Mohammad Sajadi, S., 2015. Green synthesis of copper nanoparticles using *Ginkgo biloba* L. leaf extract and their catalytic activity for the Huisgen [3+2] cycloaddition of azides and alkynes at room temperature. J. Colloid Interf. Sci. 457, 141–147. https://doi.org/10.1016/j.jcis.2015.07.004.

Nasrollahzadeh, M., Mohammad Sajadi, S., Maham, M., Ehsani, A., 2015. Facile and surfactant-free synthesis of Pd nanoparticles by the extract of the fruits of *Piper longum* and their catalytic performance for the Sonogashira coupling reaction in water under ligand- and copper-free conditions. RSC Adv. 5 (4), 2562–2567. https://doi.org/10.1039/c4ra12875c.

Nasrollahzadeh, M., Sajadi, S.M., Hatamifard, A., 2015. Anthemis xylopoda flowers aqueous extract assisted in situ green synthesis of Cu nanoparticles supported on natural Natrolite zeolite for N-formylation of amines at room temperature under environmentally benign reaction conditions. J. Colloid Interf. Sci. 460, 146–153. https://doi.org/10.1016/j.jcis.2015.08.040.

Nasrollahzadeh, M., Sajadi, S.M., Khalaj, M., 2014. Green synthesis of copper nanoparticles using aqueous extract of the leaves of *Euphorbia esula* L and their catalytic activity for ligand-free Ullmann-coupling reaction and reduction of 4-nitrophenol. RSC Adv. 4 (88), 47313–47318. https://doi.org/10.1039/c4ra08863h.

Nasrollahzadeh, M., Sajadi, S.M., Maham, M., 2015. Green synthesis of palladium nanoparticles using *Hippophae rhamnoides* Linn leaf extract and their catalytic activity for the Suzuki-Miyaura coupling in water. J. Mol. Catal. A Chem. 396, 297–303. https://doi.org/10.1016/j.molcata.2014.10.019.

Nasrollahzadeh, M., Sajadi, S.M., Rostami-Vartooni, A., Bagherzadeh, M., 2015. Green synthesis of Pd/CuO nanoparticles by *Theobroma cacao* L. seeds extract and their catalytic performance for the reduction of 4-nitrophenol and phosphine-free Heck coupling reaction under aerobic conditions. J. Colloid Interf. Sci. 448, 106–113. https://doi.org/10.1016/j.jcis.2015.02.009.

Nemati, S., Hosseini, H.A., Hashemzadeh, A., Mohajeri, M., Sabouri, Z., Darroudi, M., Kazemi Oskuee, R., 2019. Cytotoxicity and photocatalytic applications of biosynthesized ZnO nanoparticles by Rheum turketanicum rhizome extract. Mater. Res. Exp. 6 (12), 125016. https://doi.org/10.1088/2053-1591/ab46fb.

O'Carroll, D., Sleep, B., Krol, M., Boparai, H., Kocur, C., 2013. Nanoscale zero valent iron and bimetallic particles for contaminated site remediation. Adv. Water Resour. 51, 104–122. https://doi.org/10.1016/j.advwatres.2012.02.005.

OV, S., 2004. Applications of nanoparticles in biology and medicine. J. Nanobiotechnol. 2, 3. https://doi.org/10.1186/1477-3155-2-3.

Padalia, H., Moteriya, P., Chanda, S., 2015. Green synthesis of silver nanoparticles from marigold flower and its synergistic antimicrobial potential. Arab. J. Chem. 8 (5), 732–741. https://doi.org/10.1016/j.arabjc.2014.11.015.

Paknikar, K.M., Nagpal, V., Pethkar, A.V., Rajwade, J.M., 2005. Degradation of lindane from aqueous solutions using iron sulfide nanoparticles stabilized by biopolymers. Sci. Technol. Adv. Mater. 6 (3–4)), 370–374. https://doi.org/10.1016/j.stam.2005.02.016.

Paul, B., Bhuyan, B., Purkayastha, D.D., Dhar, S.S., 2016. Photocatalytic and antibacterial activities of gold and silver nanoparticles synthesized using biomass of *Parkia roxburghii* leaf. J. Photochem. Photobiol. B Biol. 154, 1–7. https://doi.org/10.1016/j.jphotobiol.2015.11.004.

Perumalsamy, V., Singaravelu, V., Sankaralingam, M., 2017. Plant-mediated biogenic synthesis of palladium nanoparticles: recent trends and emerging opportunities. ChemBioEng Rev. 4 (1), 18–36. https://doi.org/10.1002/cben.201600017.

Philip, D., Unni, C., Aromal, S.A., Vidhu, V.K., 2011. *Murraya Koenigii* leaf-assisted rapid green synthesis of silver and gold nanoparticles. Spectrochim. Acta A Mol. Biomol. Spectrosc. 78 (2), 899–904. https://doi.org/10.1016/j.saa.2010.12.060.

Lakshmipathy, R., Palakshi Reddy, B., N. Sarada, C., Chidambaram, K., Khadeer Pasha, S.K., 2015. Watermelon rind-mediated green synthesis of noble palladium nanoparticles: catalytic application. Appl. Nanosci. 5, 223–228. https://doi.org/10.1007/s13204-014-0309-2.

Rajakumar, G., Rahuman, A.A., Chung, I.M., Kirthi, A.V., Marimuthu, S., Anbarasan, K., 2015. Antiplasmodial activity of eco-friendly synthesized palladium nanoparticles using *Eclipta prostrata* extract against *Plasmodium berghei* in Swiss albino mice. Parasitol. Res. 114 (4), 1397–1406. https://doi.org/10.1007/s00436-015-4318-1.

Raman Martín Dombrowski, A. (2017). Applications of dissipative particle dynamics on nanostructures: understanding the behaviour of multifunctional gold nanoparticles. Available at: https://hdl.handle.net/10803/458132.

Reddy, G.B., Madhusudhan, A., Ramakrishna, D., Ayodhya, D., Venkatesham, M., Veerabhadram, G., 2015. Catalytic reduction of methylene blue and Congo red dyes using green synthesized gold nanoparticles capped by salmalia malabarica gum. Int. Nano Lett. 5, 215–222. https://doi.org/10.1007/s40089-015-0158-3.

Reddy, V., Torati, R.S., Oh, S., Kim, C., 2013. Biosynthesis of gold nanoparticles assisted by *Sapindus mukorossi* Gaertn. Fruit pericarp and their catalytic application for the reduction of p-nitroaniline. Ind. Eng. Chem. Res. 52 (2), 556–564. https://doi.org/10.1021/ie302037c.

Redwood, M.D., Deplanche, K., Baxter-Plant, V.S., Macaskie, L.E., 2008. Biomass-supported palladium catalysts on *Desulfovibrio desulfuricans* and *Rhodobacter sphaeroides*. Biotechnol. Bioeng. 99 (5), 1045–1054. https://doi.org/10.1002/bit.21689.

Ricciardi, P., Colomban, P., Tournié, A., Macchiarola, M., Ayed, N., 2009. A non-invasive study of Roman Age mosaic glass tesserae by means of Raman spectroscopy. J. Archaeol. Sci. 36 (11), 2551–2559. https://doi.org/10.1016/j.jas.2009.07.008.

Roopan, S.M., Elango, G., 2015. Exploitation of *Cocos nucifera* a non-food toward the biological and nanobiotechnology field. Ind. Crops Prod. 67, 130–136. https://doi.org/10.1016/j.indcrop.2015.01.008.

Ruivo, A., Gomes, C., Lima, A., Botelho, M.L., Melo, R., Belchior, A., Pires de Matos, A., 2008. Gold nanoparticles in ancient and contemporary ruby glass. J. Cult. Herit. 9, e134–e137. https://doi.org/10.1016/j.culher.2008.08.003.

Ruthiraan, M., Mubarak, N.M., Abdullah, E.C., Khalid, M., Nizamuddin, S., Walvekar, R., Karri, R.R., 2019. An overview of magnetic material: preparation and adsorption removal of heavy metals from wastewater. In: Abd-Elsalam, K., Mohamed, M., Prasad, R. (Eds.), Magnetic Nanostructures. Nanotechnology in the Life Sciences. Springer, Cham, pp. 131–159.

Sadrolhosseini, A.R., Noor, A.S.M., Husin, M.S., Sairi, N.A., 2014. Green synthesis of gold nanoparticles in pomegranate seed oil stabilized using laser ablation. J. Inorgan. Organometall. Polym. Mater. 24 (6), 1009–1013. https://doi.org/10.1007/s10904-014-0090-4.

Sahu, J.N., Karri, R.R., Zabed, H.M., Shams, S., Qi, X., 2019a. Current perspectives and future prospects of nano-biotechnology in wastewater treatment. Sep. Purif. Rev. 50 (2), 139–158. https://doi.org/10.1080/15422119.2019.1630430.

Sahu, J.N., Zabed, H., Karri, R.R., Shams, S., Qi, X., 2019b. Applications of nano-biotechnology for sustainable water purification. In: Sabu Thomas, Yves Grohens, Yasir Beeran Pottathara (Eds.). Micro and Nano Technologies-Industrial Applications of Nanomaterials. Elsevier, Amsterdam, pp. 313–340. https://doi.org/10.1016/B978-0-12-815749-7.00011-6.

Samadi, M., Sarikhani, N., Zirak, M., Zhang, H., Zhang, H.L., Moshfegh, A.Z., 2018. Group 6 transition metal dichalcogenide nanomaterials: synthesis, applications and future perspectives. Nanoscale Horizons 3 (2), 90–204. https://doi.org/10.1039/c7nh00137a.

Sankar, R., Maheswari, R., Karthik, S., Shivashangari, K.S., Ravikumar, V., 2014. Anticancer activity of *Ficus religiosa* engineered copper oxide nanoparticles. Mater. Sci. Eng. C 44, 234–239. https://doi.org/10.1016/j.msec.2014.08.030.

Sankar, R., Manikandan, P., Malarvizhi, V., Fathima, T., Shivashangari, K.S., Ravikumar, V., 2014. Green synthesis of colloidal copper oxide nanoparticles using *Carica papaya* and its application in photocatalytic dye degradation. Spectrochim. Acta A Mol. Biomol. Spectrosc. 121, 746–750. https://doi.org/10.1016/j.saa.2013.12.020.

Santhosh, C., Velmurugan, V., Jacob, G., Jeong, S.K., Grace, A.N., Bhatnagar, A., 2016. Role of nanomaterials in water treatment applications: a review. Chem. Eng. J. 306, 1116–1137. https://doi.org/10.1016/j.cej.2016.08.053.

Saxena, A., Tripathi, R.M., Zafar, F., Singh, P., 2012. Green synthesis of silver nanoparticles using aqueous solution of *Ficus benghalensis* leaf extract and characterization of their antibacterial activity. Mater. Lett. 67 (1), 91–94. https://doi.org/10.1016/j.matlet.2011.09.038.

Schaming, D., Remita, H., 2015. Nanotechnology: from the ancient time to nowadays. Found. Chem. 17 (3), 187–205. https://doi.org/10.1007/s10698-015-9235-y.

Sciau, P., Noé, L., Colomban, P., 2016. Metal nanoparticles in contemporary potters' master pieces: lustre and red "pigeon blood" potteries as models to understand the ancient pottery. Ceram. Int. 42 (14), 15349–15357. https://doi.org/10.1016/j.ceramint.2016.06.179.

Shende, S., Ingle, A.P., Gade, A., Rai, M., 2015. Green synthesis of copper nanoparticles by *Citrus medica* Linn. (Idilimbu) juice and its antimicrobial activity. World J. Microbiol. Biotechnol. 31 (6), 865–873. https://doi.org/10.1007/s11274-015-1840-3.

Sheny, D.S., Philip, D., Mathew, J., 2012. Rapid green synthesis of palladium nanoparticles using the dried leaf of *Anacardium occidentale*. Spectrochim. Acta A Mol. Biomol. Spectrosc. 91, 35–38. https://doi.org/10.1016/j.saa.2012.01.063.

Silvestri, A., Tonietto, S., Molin, G., Guerriero, P., 2014. The palaeo-Christian glass mosaic of St. Prosdocimus (Padova, Italy): Archaeometric characterisation of tesserae with copper- or tin-based opacifiers. J. Archaeol. Sci. 42 (1), 51–67. https://doi.org/10.1016/j.jas.2013.10.018.

Singh, A.K., Srivastava, O.N., 2015. One-step green synthesis of gold nanoparticles using black cardamom and effect of pH on its synthesis. Nanoscale Res. Lett. 10 (1), 353. https://doi.org/10.1186/s11671-015-1055-4.

Singh, A., Jain, D., Upadhyay, M.K., Khandelwal, N., Verma, H.N., 2010. Green synthesis of silver nanoparticles using *Argemone Mexicana* leaf extract and evaluation of their antimicrobial activities. Digest J. Nanomater. Biostruct. 5 (2), 483–489. http://www.chalcogen.infim.ro/483_Singh.pdf.

Singh, J., Dutta, T., Kim, K.-H., Rawat, M., Samddar, P., Kumar, P.J.Jon, 2018. 'Green'synthesis of metals and their oxide nanoparticles: applications for environmental remediation. J. Nanobiotechnol. 16 (1), 84. https://doi.org/10.1186/s12951-018-0408-4.

Singh, P., Kim, Y.J., Zhang, D., Yang, D.C., 2016. Biological synthesis of nanoparticles from plants and microorganisms. Trends Biotechnol. 34 (7), 588–599. https://doi.org/10.1016/j.tibtech.2016.02.006.

Sivaraj, R., Rahman, P.K.S.M., Rajiv, P., Narendhran, S., Venckatesh, R., 2014. Biosynthesis and characterization of *Acalypha indica* mediated copper oxide nanoparticles and evaluation of its antimicrobial and anticancer activity. Spectrochim. Acta A Mol. Biomol. Spectrosc. 129, 255–258. https://doi.org/10.1016/j.saa.2014.03.027.

Snehal, Y., Chandra, M., Prakash, M., 2016. Biosynthesis of zinc oxide nanoparticles using Ixora coccinea leaf extract—a green approach. Open J. Synth. Theory Appl. 5, 1–14. https://doi.org/10.4236/ojsta.2016.51001.

Sun, Q., Cai, X., Li, J., Zheng, M., Chen, Z., Yu, C.P., 2014. Green synthesis of silver nanoparticles using tea leaf extract and evaluation of their stability and antibacterial activity. Colloid. Surf. A Physicochem. Eng. Asp. 444, 226–231. https://doi.org/10.1016/j.colsurfa.2013.12.065.

Tammina, S.K., Mandal, B.K., Ranjan, S., Dasgupta, N., 2017. Cytotoxicity study of *Piper nigrum* seed mediated synthesized SnO_2 nanoparticles towards colorectal (HCT116) and lung cancer (A549) cell lines. J. Photochem. Photobiol. B Biol. 166, 158–168. https://doi.org/10.1016/j.jphotobiol.2016.11.017.

Tantiwatcharothai, S., Prachayawarakorn, J., 2019. Characterization of an antibacterial wound dressing from basil seed (*Ocimum basilicum* L.) mucilage-ZnO nanocomposite. Int. J. Biol. Macromol. 135, 133–140. https://doi.org/10.1016/j.ijbiomac.2019.05.118.

Tran, Q.H., Nguyen, V.Q., Le, A.T., 2013. Silver nanoparticles: synthesis, properties, toxicology, applications and perspectives. Adv. Nat. Sci: Nanosci. Nanotechnol., 4 https://doi.org/10.1088/2043-6262/4/3/033001.

Vasudevan, K., Divyasree, M.C., Chandrasekharan, K., 2019. Enhanced nonlinear optical properties of ZnS nanoparticles in 1D polymer photonic crystal cavity. Optics Laser Technol. 114, 35–39. https://doi.org/10.1016/j.optlastec.2019.01.027.

Vemula, M., Reddy, A., Reddy, K., Gajulappli, M., 2012. A simple method for the determination of efficiency of stabilized Fe0 nanoparticles for detoxification of chromium (VI) in water. J. Chem. Pharma. Res. 4 (3), 1539–1545.

Vincent, D., 2008. The invisible revolution. Nature 451, 770–771. https://doi.org/10.1038/451770a.

Viswadevarayalu, A., Venkata Ramana, P., Sreenivasa Kumar, G., Rathna sylvia, L., Sumalatha, J., Adinarayana Reddy, S., 2016. Fine ultrasmall copper nanoparticle (UCuNPs) synthesis by using *Terminalia bellirica* fruit extract and its antimicrobial activity. J. Cluster Sci. 27 (1), 155–168. https://doi.org/10.1007/s10876-015-0917-3.

Wang, Q., Yu, Y., Liu, J., 2018. Preparations, characteristics and applications of the functional liquid metal materials. Adv. Eng. Mater. 20 (5), 1700781. https://doi.org/10.1002/adem.201700781.

Wriedt, T., 2008. Mie theory 1908, on the mobile phone 2008. J. Quant. Spectrosc. Radiat Transfer 109 (8), 1543–1548. https://doi.org/10.1016/j.jqsrt.2008.01.009.

Xing, Y., Zhao, J., Conti, P.S., Chen, K., 2014. Radiolabeled nanoparticles for multimodality tumor imaging. Theranostics 4 (3), 290–306. https://doi.org/10.7150/thno.7341.

Yang, N., Weihong, L., Hao, L., 2014. Biosynthesis of Au nanoparticles using agricultural waste mango peel extract and its in vitro cytotoxic effect on two normal cells. Mater. Lett. 134, 67–70. https://doi.org/10.1016/j.matlet.2014.07.025.

Zaib, S., Iqbal, J., 2019. Nanotechnology: Applications, techniques, approaches, & the advancement in toxicology and environmental impact of engineered nanomaterials. Importance & Applications of Nanotechnology 8 (1), 1–8.

Zhang, W., Lin, Z., Pang, S., Bhatt, P., Chen, S., 2020. Insights into the biodegradation of lindane (γ-hexachlorocyclohexane) using a microbial system. Front. Microbiol. 11, 522. https://doi.org/10.3389/fmicb.2020.00522.

Zhang, Y.X., Wang, Y.H., 2017. Nonlinear optical properties of metal nanoparticles: a review. RSC Adv., 7 (71), 45129–45144. https://doi.org/10.1039/c7ra07551k.

Zhu, W., Chen, Z., Pan, Y., Dai, R., Wu, Y., Zhuang, Z., Wang, D., Peng, Q., Chen, C., Li, Y., 2019. Functionalization of hollow nanomaterials for catalytic applications: nanoreactor construction. Adv. Mater. 31 (38), 1800426. https://doi.org/10.1002/adma.201800426.

CHAPTER 6

Synthesis of green nanocomposite material for engineering application

Manoj Tripathi[a] **and Bharti Singh**[b]
[a]Department of Physics and Materials Science and Engineering, Jaypee Institute of Information Technology, Noida, Uttar Pradesh, India
[b]Department of Electronics Engineering, Ramswaroop Memorial Group of Professional Colleges, Lucknow, Uttar Pradesh, India

6.1 Introduction

In the year 1959, Nobel laureate Richard P. Feynman delivered a lecture titled "There is plenty of room at the bottom." There he envisioned the role of building nano-objects atom by atom to explore and control things at nanoscales. The idea of making particles at nanoscale is not new. A gold ruby glass made with dispersion of gold nanoparticles in glass matrix was first produced by Assyrians, invented by Kunkel in Leipzig in the seventeenth century. Other examples include, the medieval era stained glasses containing silver and gold nanoparticles. Chinese were known to use gold nanoparticles as an inorganic dye to introduce red color in their porcelain in their ceramics. These unique examples of use nanoparticles in history remained of limited use for centuries. The new interest in the technology emanates with the drive to have electronic devices with reduced size and explore enhanced properties of materials shown at nanoscales (Li et al., 2016).

By the mid end of twentieth century nanomaterials started to play an active role in different fields of engineering and rightly predicted to be the materials with a wide range of applications. Thereafter, nanocomposites were synthesized by dispersing the nanofillers in a suitable matrix. The main concern of the synthesis of nanocomposites is to have adequate dispersion of the filler nanocomposites in the matrix (Fu et al., 2019). Many properties such as electrical and mechanical strength depend on the shape, size, and direction of filler material (Dey & Tripathi, 2010). Most of the natural polymer matrix shows very weak mechanical properties such as tensile strength and Young's modulus thus being of little practical relevance in most of the high-end engineering applications. Addition of very small amount of nanofiller (reinforcement) shows tremendous improvement of overall mechanical, electrical, and optical properties of nanocomposite material hence making it useful in the field of major engineering applications (Tripathi et al., 2016). Over the years, scientists have searched for the ways to enhance properties of nanocomposite materials while retaining ease of processing. Materials show properties different at nanoscales than at larger sizes. One of the reasons for such behavior is particle's large surface to volume ratio at nanoscales and their better interaction with surrounding

Sustainable Nanotechnology for Environmental Remediation
DOI: https://doi.org/10.1016/B978-0-12-824547-7.00012-6

materials. Surface to volume ratio and specific surface area are inversely proportional to size of a particle and increase remarkably as we reduce particle size. In bulk solids atoms, fillers residing at the surface are less (<1%) (Eliezer & Gileadi, 2019), whereas in nanostructures more atoms start to reside at the surface and hence give more active sites for reaction to take place. Some of these behavioral changes are quite surprising for example inert metals such as platinum and gold become catalyst and silver shows increased antimicrobial properties (Nizammudin et al., 2016; Karri et al., 2019; Sahu et al., 2019).

There is much literature available to the synthesis of material of interest such as thin film techniques for the fabrication of solid-state materials for use in electronic devices. Metal oxides are used for the fabrication of microelectronic circuits, fuel cells, and sensors. The metal oxides mainly used for these applications are ZnO, TiO_2, WO_3, etc. Metal oxide such as cupric oxide (CuO) thin films show properties such as magnetic, optical, semiconducting, and electronic and optoelectronics (Bhoopal et al., 2015). There are various methods to fabricate thin films of metal oxides for many applications and fabrication routes are decided according to the structural properties needed in a suggested application (Greene, 2017; Park et al., 2015).

The green nanocomposites are obtained from various plant materials, agricultural wastes, and byproducts of the agricultural wastes (Lingamdinne et al., 2022; Sahu et al., 2021). Organic fibers are very popular for the synthesis of green nanocomposite. Moreover fungi and bacteria are also used for the green nanocomposite synthesis (Jain et al., 2020). One of the major advantages of these materials are that these are biodegradable and renewable (Sunday et al., 2012). Moreover these materials are derived from agricultural waste so these are inexpensive as well. Thus the use of green nanocomposite not only is eco-friendly but is an inexpensive material that has the ability to be used in so many different applications such as sensors, packaging, biomedical, energy, and many more.

6.2 Green nanocomposite materials

Green nanocomposites are the nanocomposites that use biological particles as a filler or matrix(Magendran et al., 2019). The use of these green nanocomposites is increasing with each passing day. These materials not only are easy to mold but also are cost efficient. Fabrication of green nanocomposite is easy compared to other materials. These are environment friendly. All these properties promote the use of green nanocomposite materials in many applications. The green nanocomposites are obtained from various plant materials, agricultural wastes, and byproducts of the agricultural wastes. Organic fibers are very popular for the synthesis of green nanocomposite. Moreover fungi and bacteria are also used for the green nanocomposite synthesis. One of the major advantages of these materials is that these are biodegradable and renewable. Moreover these materials are derived from agricultural waste, so these are inexpensive as well. Thus the use of green nanocomposite not only eco-friendly but is an inexpensive material that has the ability to be used in so many different applications such as sensors, packaging, biomedical, energy, and many more applications. For composite material matrix to be mechanically strong as well as eco-friendly is particularly challenging. Thus a need for novel techniques to produce nanocomposite materials with superior electrical, mechanical, and optical properties and it is one of the most popular areas of research these days. Some of the prominent biodegradable polymer sources are biopolymers from natural sources, lignin, cellulose acetate, starch, polylactic acid (PLA), polyhydroxyalkanoates (PHA), polyhydroxybutyrate, etc.

Cellulose (polysaccharide, $(C_6H_{10}O_5)_n$) offers good mechanical strength and low mass density. It is easy to process cellulose from agricultural waste therefore it is most economically feasible biomaterial. Cellulose is also identified as biopolymer that can be used as a substitute for petroleum polymers. Most of the celluloses are hygroscopic in nature but do not dissolve in water (Ray and Bousmina, 2005). Alemdar et al. reported extraction of cellulose from wheat straw and soy hulls, by a chemomechanical technique (Alemdar & Sain, 2008). Cellulose can be separated from various organic compounds such as soybean stock, cotton, chitosan, etc. In some cases these cellulose fibers can have length up to 100 nm with a width of few nanometers. Nanocellulose are mainly used in paper and packaging industry; however, new horizons are discovered with a wide range of applications such as construction, automotive, furniture, electronics, pharmacy, and medical field.

Chitin $(C_8H_{13}O_5N)_n$ is the second most abundant biopolymer following cellulose. It is found in exoskeleton of crabs, shrimps, prawn, and insect and cell wall of mushrooms. Chitin is a long-chain polymer of N-acytylglucosamine and is a derivative of glucose. Chitin nanofibers can be extracted from inorganic materials by series of chemical treatment of small flakes of crab shell followed by mechanical grinding and drying (Ifuku et al., 2009).

Chitin nanofibers have high mechanical strength and high Young's modulus. Therefore, they are good candidate for reinforcing filler material for various engineering nanocomposite materials (Chang et al., 2021)(Rasoulzadeh et al., 2020). Chitin-reinforced acrylic resin optically transparent sheets show better mechanical and thermal properties. Chitin is useful for several medicinal, industrial, and biotechnological purposes.

Starch is another widely used material used for the synthesis of biopolymer-based green nanocomposites. The source of starch occurs naturally in cereal grains (corn, wheat, and rice), seeds, legumes (lentils), and potatoes. The main sources of starch are maize (82%), wheat (8%), potatoes (5%), and cassava (5%) from which tapioca starch is derived (Angellier et al., 2004). It primarily is composed of two components—amylose (straight chain) and amylopectin (branched chain). The ratio between amylose and amylopectin varies depending on the starch source (L. et al., 2011). Starch-based nanocomposites are prepared using methods same as used for the preparation of synthetic thermoplastics.

Extrusion in the presence of high shear forces with temperature and pressure is applied to melt starch.

Advantages of using starch-based modified materials are improved mechanical strength, barrier properties, and better thermal stability. Starch-based films are mainly in use to improve food shelf-life. Their applications generally remain limited due to their low mechanical strength and easy permeability. Therefore, nanocomposites using various nanofillers with starch matrix are a major area of research these days.

PHA is a naturally occurring biodegradable polysters produced by various organisms. The existence of PHA in bacteria was first reported in 1920s by French Scientist Lemoigne. PHAs are storage form of carbon and energy for bacteria that produce them. Steps involving PHA production are (1) fermentation, (2) separation of the biomass from the broth, (3) biomass drying, and (iv) PHA extraction and drying. PHA-based materials show a high mechanical strength with a good thermal stability. It is widely used in the food packaging.

PLA is another thermoplastic polymer used for the synthesis of green nanocomposite. This is biodegradable polymer (via hydrolysis, thermal, UV radiation induced degradation). PLA is extractable from agricultural resources such as corn and sugarcane. PLA is substantially used in the manufacturing of biomedical devices and packaging industry. PLA is high mechanical strength polymer far better than polypropylene. A major drawback of PLA is its brittleness. However, it can be used in packaging industry in place of polystyrene (PS) with biodegradability being major incentive.

6.3 Synthesis of green nanocomposite materials

The synthesis of nanocomposite materials adopts two approaches—top-down and bottom-up. In the top-down approach, a suitable raw material reduced in size to produce fine nanoparticles ultimately, while in the bottom-up approach we find nanoparticles through continuous built-up of growth of atoms. The major drawbacks of the top-down approach are large amount of energy requirement and crystallographic imperfections.

Due to these drawbacks the bottom-up approach is mostly followed, which can produce uniform size shape and crystalline structure of nanoshape particles thus having better control of mechanical and electrical properties of fabricated nanostructures. Next we discuss four main routes for the preparation of nanocomposite materials.

6.3.1 Physical vapor deposition

Physical vapor deposition is a method to transfer material on the surface of substrate in the form of a thin film. Thickness and morphology of deposits depend on factors such as rate of supply of atoms to the region of deposition, rate of removal of energy from saturated atoms, and rate of removal of recent nucleation. Thickness of deposits varies from 10^{-10} m to 10^{-3} m range. There are two methods of removal of growth from source: (1) Evaporation and (2) sputtering. In evaporation, atoms from source are removed by thermal process, and deposition of thin film is done at very low pressure so that the mean free path of collision of atoms and molecules in vapor remains essentially high compared to the distance between source and substrate. Another means of the removal of growth atoms from a source is by sputtering that is, by the means of high-energy ions beams. Sputtering is a phenomenon in which particles are dislodged from the surface of substrate sample by the bombardment of high-energy particles such as ions, electrons, plasma, or gas. Sputtering is used in case of high melting point source such as ceramics. To get the fine grain size (less than 100 nm) distribution another popular method called inert gas condensation is used by Siegel (1993). In this method, the evaporation of material from source is carried out in an ultra-high vacuum chamber in the presence of some inert gas (Tripathi et al., 2020). These vaporized atoms then lose energy in collision with inert gas molecules. These collisions limit mean free path. This vapor is transported to some cooler part of system that collects these particles. These newly collected nanoparticles on the surface of a collecting device (cold finger) can be scraped off the finger and collected through funnel to a compaction device. This method is well suited for it provides controlled particle size distribution. Physical vapor deposition (PVD) has its limitations due to high cost of production.

6.3.2 Chemical vapor deposition

In chemical vapor deposition process, some precursor volatile gases come in contact with heated substrate where chemical reactions take place between precursor gases with the surface of substrate. Now decomposition of gases forms solid phase deposits on the surface of substrate. Substrate temperature determines the kind of reaction to take place at the surface of substrate. Surface temperature are typically in the range of 300–1200 °C. Typical chemical vapor deposition (CVD) apparatus consists of a reaction chamber, source of energy to heat substrate, precursor gas supply to reaction chamber, exhaust system to remove unwanted byproducts, and loading and removing substrate (Chang et al., 1994; Phasinam et al., 2021).

CVD is very versatile process and it is employed to produce almost any metallic and ceramic compounds. CVD process is widely discussed due to it being principal method for fabrication of thin films for solid state devices in microelectronics. There are many and well-documented processes of synthesis available across the literature (Choy, 2003; Ser et al., 2002). Owing to its simplicity for production of fine grained homogenous coatings on substrates and low cost, it is used extensively. Nanomaterials developed by this CVD method are seen to have a uniform coating of nanoparticles and films. Limitations of this process include requirement of higher temperatures and difficulty to scale up (Sudarshan, 2004).

6.3.3 Spray pyrolysis

In spray pyrolysis method, precursor solution droplets are generated using a nebulizer containing a desired solute nanomaterial. This precursor solution is sprayed on hot surface of substrate. Decomposition of solvent droplets forms desired nanostructure on the surface of substrate. The nanostructure particle size, shape, and thickness are controlled by different controlling parameters such as nebulizer energy, vapor pressures of gases, the temperature of furnace, and distance between the spray gun and the substrate. The steps involved in spray pyrolysis technique are precursor solution composition, aerosol generation transport, and synthesis. Each of these steps is calibrated as per the final chemical and physical properties required in the end. The preferred choice for solvent in precursor solution is water and alcohol, which dissolves many inorganic salts. For organic salts organic solutions are chosen. Aerosol-generation mechanism determines droplets size distribution (Hou & Choy, 2006; Avaru et al., 2006).

The main advantages of the ultrasonic aerosol generation systems are the narrow distribution of drops and the control on the average size of particles. In the third stage, whether the material synthesized is the powder or a thin film is determined by if the reaction takes place in vapor phase or on the surface of hot substrate. This technique is cost effective and has simple equipment for fabrication; however, a drawback is difficulty in scaling up. Spray pyrolysis technique is used to obtain many fine nanostructured particles that are used in applications such as solar cells, sensors, and solid oxide fuel cells.

6.3.4 Chemical methods

In chemical methods nanoparticles are nucleate to form nanostructures of different sizes and shapes. This is called bottom-up approach. Shape and size of nanomaterials depends on chemicals used and conditions of reaction. Major preparation strategies from chemical methods are (1) Colloidal and precipitation, (2) Sol gel method, and (3) template synthesis. A brief discussion of each is given next.

6.3.4.1 Colloidal and precipitation method

Colloidal method of synthesis is a well-known process and was introduce in 1857. M. Faraday reported wet chemical method to synthesis colloidal gold particles (Avaru et al., 2006). In this method an aqueous solution of gold is prepared and this aqueous solution is then reduced. Colloidal gold particles can be produced by reduction of $HAuCl_4$ with phosphorous. Colloidal particles are synthesized through reduction reaction, which can be controlled by the type of precursor and type of reducing reagents used. Steps involved in the process are nucleation, condensation, surface reduction, and stabilization.

Strong reduction reagent favors fast reaction and forms smaller size nanoparticles. The size and shape of nanoparticles are influenced by the choice of solvent and polymeric stabilizer. Coagulation of particles should be hindered by electrostatic repulsions, and steric hindrance as nanoparticles has strong tendency to coagulate or agglomerate. Chemical capping agents are used to control size distribution of nanoparticles. A variety of capping agents such as thio-phenol and mercapto-ethanol are used.

For the synthesis of gold nanoparticles a reaction is carried out in the presence of inert gas such as nitrogen. Dilute solution of chloroauric acid is prepared in water. Then few drops of ˜1 mL of 0.5% trisodium citrate is added in boiling solution until color of solution changes.

$$HAuCl_4 + Na_3C_6H_5O_7 \rightarrow Au^+ + C_6H_5O_7 + HCl + 3NaCl \quad (6.1)$$

According to reduction method used, other methods for the synthesis of gold nanoparticles are Irradiation method, sonochemical, laser, and photochemical techniques (Hatling et al., 2007). In particular, gold nanoparticles are used in diagnosis and drug delivery. Nanoparticles typically in the size range up to 10 nm can be easily cleared by human Kidneys. Nanoparticles-enhanced drug delivery is also used in the treatment of brain diseases such as Alzheimer and Parkinson due to their capability to cross blood-brain barrier. Other metal nanoparticles such as silver, copper, palladium, or alloys can be synthesized by appropriate precursor solution under controlled pH and temperature with certain chemical capping material solutions.

6.3.4.2 Sol gel method

In a sol gel technique, a precursor solution (metal alkoxide, carbide, and nitride) is hydrolyzed and then condensed to form sol. Sol is a colloidal suspension of solid particles in a liquid, which neither gets agglomerated nor gets sediment in solutionVander Waal forces are responsible for the agglomeration of particles. It causes a reduction in surface energy. This Vander Waal force is weak and has a small range of few nanometers. Electrostatic repulsion must be established to negate this weak force so that agglomeration can be stopped. Steric hindrance in the solution can further be countered by coating nanoparticle by some surfactant molecules (organic molecules). For this, the particles are adsorbed into some organic solvent. This basically prohibits the molecules from approaching to each other. Thus the Vander Waal forces can be reduced.

Once this sol is stabilized it can be further condensed to produce networked structure of gel in the solvent. The gel takes the shape of container or mold in which it is prepared. The gel is termed as aqua-gel (or aqua-sol) when water is used as solvent. If we use alcohol in place of water then the gel is called alco-gel. To remove liquid from gel either supercritical drying or evaporative drying is used resulting in aerogel or xerogel, respectively. Fig. 6.1 gives a diagram of the synthesis using the sol-gel technique.

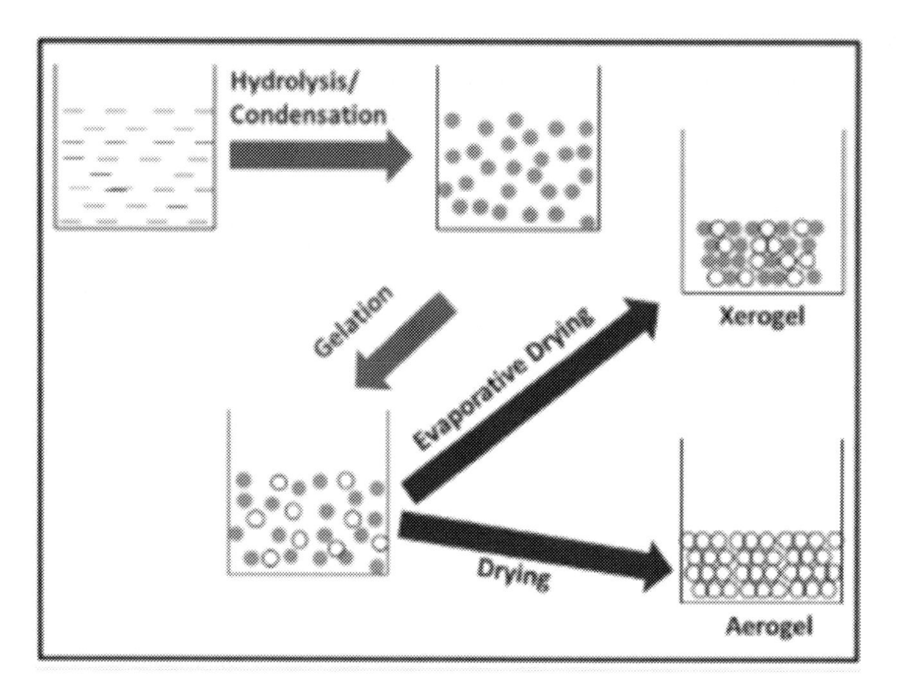

Figure 6.1 Synthesis technique using the sol gel method.

This method has found its applications due to its superiority over the conventional methods. This method can be used to make highly surface to volume ratio porous materials and nanostructure materials. This method has advantage of low cost (Balek et al., 1996). Many researchers have reported the synthesis of nanocomposites such as silica nanoparticles, alumina gel–based nanocomposite containing CaO, FeO, TiO_2, etc. using this technique successfully (Stöber et al., 1968; Hamada et al., 1996).

6.3.4.3 Template synthesis

In nano-scale synthesis more and more emphasis is laid on the requirement of well-controlled size, shape, and spatial arrangement. Template synthesis is well known for centuries where ceramic artifacts and utensils have been made by the use of wood template; it is removed after reduction of the size of article by drying or calcination. General steps involving the template synthesis include (1) Template preparation, (2) synthesis of required nanostructure, and (3) removal of template. However, some of the templates are naturally occurring such as bacterial cell surface protein, nano-sized nucleic acid compound, hollow biomolecular compartments such as viruses and others are made via synthetic processes.

Template preparation from a synthetic route involves surface coating and channel replication. Templates can take the form of 0D nanoparticles, 1D nanowires, and 2D and

3D nanostructures. Nearly all solid material in principle can be prepared by this route. However, there are some concerns to consider such as whether the depositing material wet the pore surface? Moreover, blockage of pore site and stability of template under reaction conditions always remains an issue (Hulteen & Martin, 1997).

This method is applicable mainly on electrically conducting material such as metal, alloys, semiconductors of low band gap, and electrically conducting polymers (Pérez-Page et al., 2016). In this method there are electrodes in the chemical solution. We pass current through them then there is deposition of solid material on an electrode. There are two ways to do this, one is by using negative template method. in this mathod, the direction of current is opposite to the metal electrode. Prefabricated nanopores in solids (like aluminum), which work as templates, are filled with various methods. Electrodeposition of growth species can take place on electrodes with membrane working as a host material to get free-standing nanowires. Afterwards, host material is dissolved (Rahman et al., 2012).

Many advantages of this method are highly conducting nanowires and better control of aspect ratio. Aspect ratio depends on the total amount of charge passed. By changing the deposition time, the length of nanowires or nanotubes can be changed (Seki et al., 2016). It also controls wall thickness of nanotubes (Pérez-Page et al., 2016). Other method is a positive template method to get nanowires or nanotubes growth on the outer surface of wire like nanostructure (template of DNA or carbon nanotubes, as the case may be). DNA is one of the templates may be used to fabricate further nanostructure; it is a wire with the diameter of approximately 2 nm. Diameter of nanowire is not dependent on template size. Silica (SiO_2) nanotubes can be prepared using silver nanowires; silver nanowires are coated with silica by using reagent tetra-ethoxy-silane by sol-gel method. After removing silver by ammonia solution, silica nanotubes are obtained (Ramanathan et al., 2005; Seki et al., 2016).

In other arrangement DNA strand is used as a template between two electrodes having some potential difference; shape of DNA strand guides nanostructure to grow along this strand. DNA consists of nucleic acid that may be dissolved later by chemical etching. A similar technique known as electrophoretic technique is another useful technique. In this technique, depositing material need not be electrically conductive essentially. In this method electrical charge is generated on the surface of nonconducting nanoparticles by using chemicals. Charges induced on the surface of nanoparticles by the use of chemical make them respond to externally applied electric field. Further, it is the same as in electrochemical deposition method. The size of the nanorods depends on the size of template used.

On applying a suitable electric field to a colloidal solution or sol, charge particles present in the colloidal solution start to move. These nano-sized particles in colloidal dispersions are stabilized by electrostatic or electro steric mechanisms. This motion is referred as electrophoresis (Sugunan et al., 2006).

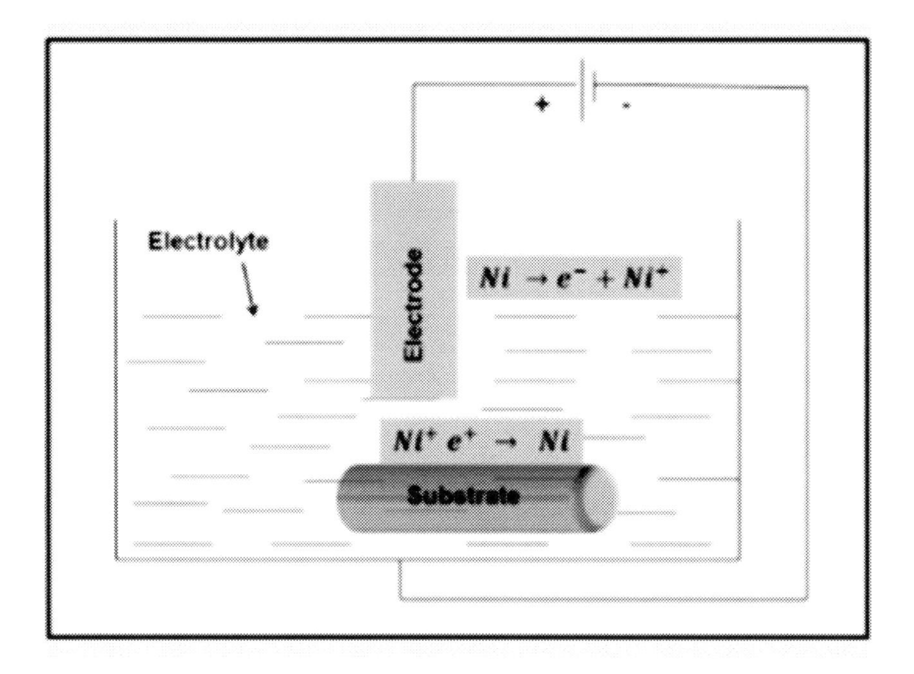

Figure 6.2 Diagram of nickel electro-deposition.

6.3.5 Electrodeposition

This method is useful in deposition of thin film on the substrate using electric current. In this method, the substrate is dipped into an electrolyte. This electrolyte contains a salt of metal that is to be deposited. The substrate is connected to the negative pole of cell and a current is passed through. The substrate works as a cathode in the electrolytic solution. One of examples of this method is electro-deposition of nickel onto the substrate of a negatively charged surface. If we take nickel chloride salt in the aqueous solution Ni dissociates itself as a positively charged cation from a negatively charged chloride anion. The negatively charged substrate attracts positively charged nickel (cation). The electrons flow from substrate to the cation. At the anode electrons are removed from the Ni metal to oxidize it to Ni cation thus dissolving nickel into the solution. Electro-deposition depends on concentration of the electrolyte, pH of the solution, and temperature. These parameters may be controlled to get a fine grained material as deposit. A schematic of the nickel electro-deposition is given in Fig. 6.2. Here an electrode is immersed in an electrolyte solution. A suitable pulse of current is applied and the chemical reactions during the electrolysis of the electrolyte allow the deposition of the nickel on the substrate. It is a very effective technique and by changing the current pulse the deposition of nickel can be somewhat controlled also.

A popular method for deposition of nanoparticles is pulse plating. In this method, a square wave–shaped current is used. The current is controlled by parameters such as peak current density, ON and OFF time of the current pulse. Generally, ON time of pulse is much shorter than OFF time of pulse. It is kept so such that metal ion concentration which reduces significantly during pulse ON time may be recovered during relatively a longer period of pulse OFF time. In this method large pulse ON/OFF time ratio and saccharine is used for refinement of grain growth of Ni deposits (Tjong & Chen, 2004; Natter et al., 1998).

Conventional powder method is one of the very old techniques for the synthesized ceramic nanocomposites. In this method, we start by selecting raw material with a high degree of purity and homogenous particle size distribution of ceramic powder. High purity of constituents is required to avoid the formation of secondary phase during sintering. On the other hand, ultra-fine, loosely dispersed powder is needed for both matrix and reinforcing phases. Steps involved in the process to synthesize are mixing, drying, and consolidation by high temperature and pressure (sintering). This method is used conventionally for being simple but major drawbacks include agglomeration, poor phase dispersion, formation of secondary phases, for these reasons this method has limited acceptance for the synthesis of high purity materials for exploiting fine attributes. This method is mainly used for the synthesis of Al_2O_3/SiC nanocomposites (Nakahira & Niihara, 1992; Stearns et al., 1992).

Liquid metal Infiltration process is the preferred process for industrial production of metal matrix composites. In this process, porous reinforcement is taken which is invaded by liquid metal matrix and finally metal matrix solidifies to take the structure of original preform. Due to low viscosity of metals, this method is widely used for rapid production of metal matrix composites. Thus there are three stages involved in the production namely preparation of self-sustaining preform, infiltration, and matrix solidification. There are three kinds of infiltration process: capillarity, transport, and solidification. Of these, capillarity is the most common phenomenon. It dictates if infiltration will be spontaneous and if it drives or impedes the process. The rate of flow of liquid metal is determined by local average pressure in liquid metal, metal density, and viscosity. Preform should be self-sustaining to withstand pressure if pressure infiltration is used. Matrix solidification in composite is same as in unreinforced metals. Solidification starts by nucleation, and advances in growth of solid phase.

Rapid solidification has become a field of interest lately because method promises prospects of new engineering materials with advance properties. This is a well-known process since 1960, introduced by Duwez et al. In this technique molten metal alloy is cooled very rapidly (10^6 K/s) to form amorphous and metallic glass alloys. Such high cooling rates have been achieved by techniques such as melt spinning droplet methods or surface melting technologies.

Melt spinning: In the melt spinning method small amount of alloy is melt inside crucible (generally heating through induction) these molten drops are fizzled through fine nozzle pressurized through some inert gas onto the fast rotating (5000–7000 rpm) copper wheel, such a high spinning rate cools down molten metal into amorphous state.

Droplet method: In this method, atomization of molten metal is achieved through gas atomization method (conversion of molten metal into aerosol particles) by forcing through nozzle by high pressure applied through some inert gas. Further they are cooled down by fall in inert tower. Later, this fine powder may be consolidated in a bulk nanostructured material (Liu et al., 2005; Surreddi et al., 2010).

Surface melting: This method involves rapid melting at the surface followed by rapid freezing sustained by rapid heat extraction into the unmelted bulk. In this process, a bulk material block is melted by some traversing heat source (such as laser or electron beam) which results in different microstructure of the surface of metal, and metal block acts as a heat sink for rapid cooling (Jones, 2001).

To achieve such a rapid solidification rates, it is required to have one dimension of the specimen of the order of 20–50 μm. Metals and alloys are mainly crystalline in nature and this crystallinity is absent in amorphous phase.

6.4 Processing techniques of polymer nanocomposites

Many natural polymer matrices show very weak mechanical properties such as low tensile strength and Young's modulus, poor thermal conductivity, and a relatively high coefficient of thermal expansion, thus being of little practical relevance in the majority of engineering applications. However, these properties are needed but even a very small amount of nanofiller reinforcement shows tremendous improvement of overall mechanical, electrical, and optical properties of nanocomposite material making it useful in the field of engineering applications. The main concern of the synthesis of nanocomposites is the maximum dispersion of the filler nanocomposites in the matrix. Nanofillers in nanocomposite may be 1D, 2D, or 3D in size. The one important feature of nanofiller is its aspect ratio. It is the ratio of longest to shortest dimension of nanofiller particles. The mechanical, thermal, and electrical properties of nanocomposite materials are strongly related to the interaction between nanofillers and the polymer matrix and also dependent on morphology of nanocomposite polymers. Various types of nanoparticles are carbon nanotubes, graphene, nanocellulose, halloysites, etc.

A major requirement for improvement in properties of nanocomposite materials is uniform distribution of reinforcement phase within a polymer matrix. Main routes for the preparation of nanocomposite polymers are (1) melt intercalation, (2) template synthesis, (3) *in-situ* polymerization, and (4) exfoliation adsorption.

6.4.1 Melt intercalation

This method is usually preferred for thermoplastic polymers. In this method, nanofillers are dispersed into melt polymer. In this method, polymer matrix is annealed at a high temperature to soften polymer matrix, then fillers additives are added and the composite is knead to achieve uniform distribution. Temperature needs to be better controlled to achieve good mixing and avoid degradation of filler material surface. Amorphous polymers are processed above their glass transition temperature, while semicrystalline polymers processing temperature is fixed above their melting temperature. Finally bulk samples can be produced by extrusion, injection, or compression molding. It is desirable to have good dispersion of filler nanomaterial in the polymer matrix and better wetting of reinforcement.

Alig et al. (2012) discussed the relation between processing conditions and morphologies obtained for carbon nano tube (CNT) nanocomposites. Pavlidou and Papaspyrides (2008) explained the thermodynamics behind the process, and the effects of multiple conditions on melt intercalation for polymer/layered silicates. This technique is quite useful and the nanocomposite can be developed smoothly using it. Researchers have used this method to prepare polymer nanocomposite such as high-impact PS/organoclay nanocomposites (Chimanowsky Junior et al., 2014), Polycaprolactone (PCL) – multi-walled carbon nanotubes (MWCNT) mixture followed by the synthesis of PCL–MWCNT nanocomposite (Maiti et al., 2014). It also has been observed that the many characteristics of material often depend upon the process used to synthesize the material. These properties mainly are the function of interface interaction between the filler and matrix. Moreover, the dispersion of the fillers and adhesion properties also are vital in predicting the properties of nanocomposite.

Annala et al. (2012) reported the synthesis of poly(methyl methacrylate) (PMMA)/MWCNT and PS/MWCNT using *in-situ* polymerized master batches that were to be used in corotating twin screw mini-extruder with the capacity of 16 cm^3 and screw length of 150 mm. They investigated and deduced that rheological properties and interfacial interaction between components affect the properties of composites significantly. Wang et al. (2002) synthesized phthalocyanine (Pc)/MWCNT nanocomposites by placing the prepared master batch in a preheated mold at 250 °C and cured at controlled elevated temperatures for 4 h. A novel approach of melt spinning layered double hydroxide/high-density polyethylene nanocomposites prepared by melt extrusion was reported by Kutlu et al. (2013). This method is benign environmentally, it has easy process and low cost. Drawbacks include use of high temperature that may damage/degrade filler material surface.

6.4.2 Template synthesis

This method is also called sol gel technique for the preparation of silicate in the aqueous solution. In this process, polymers act as a template for the nucleation and growth of silicates in a polymer matrix. It is limited by the requirement that it is useful with water-soluble polymers only. In principle, the method can be used for the one-step dispersion of silicate layers. Requirement of high temperature is one of the major limitations of this method. Another issue is related to the aggregation of the silicate layers. However, few reports are present in the literature in which hectorite-type clay mineral has been synthesized under mild conditions (Pavlidou & Papaspyrides, 2008).

6.4.3 *In-situ* polymerization

In this method, a dispersion of clay (swelling of filler) in a liquid monomer solution is done with the help of a catalyst or initiator. A low molecular weight monomer is then polymerized using heat or radiation thus forming an intercalated nanocomposite polymer. Various polymer nanocomposites produced using this method include silica-nylon 6, silica-poly 2-hydroxyethylmethacrylate, alumina- PMMA, titania-PMMA, and $CaCO_3$-PMMA. This method achieves easier dispersion of filler material. Graphene is a material that has been used for the synthesis of a wide range of nanocomposites using *in-situ* techniques.

Recently polyolefin nanocomposites have generated vast amount of interest in researchers due to their high potential as materials with novel properties. Walter Kaminsky reported a technique of synthesis of polyolefin nanocomposites by *in-situ* polymerization using soluble metallocene/methylaluminoxane (MAO) as catalysts. The synthesis was a two-step process. In the first step, metallocene/MAO was allowed to deposit and adsorb on nanofiller surface. Once the adsorption is properly done, a polyolefin film is developed using ethene or propene. This film covers the nanoparticle-layered silicates. These polyethylene and polypropylene nanocomposites were studied for their mechanical properties. The result showed that the particles synthesized by this method offer improved mechanical properties compared to the particles synthesized by mechanical blending (Kaminsky, 2018).

In-situ polymerization does not demand any rigorous thermodynamic treatment and analysis. This makes the process more suitable and effective and interesting. Abedi and Abdouss (2014) state that *in-situ* polymerization is the most suitable preparation method for polyolefin/clay nanocomposites because of its lack of rigorous thermodynamic requirement compared to other methods. Physical and chemical properties change by adding just a small fraction of nanoparticles including $CaCO_3$ (Andricic et al., 2008; Kozlov et al., 2012), clay (Park et al., 2015), talc (Bocchini & Frache, 2013), carbon nanotube (Liu et al., 2005), carbon black (Alig et al., 2012; Wang et al., 2013; Zois et al., 2012), silica (Ray and Bousmina, 2005), etc. to the polymer matrix. *In-situ* electrochemical synthesis is considered useful for the quick detection of current–voltage characteristics.

6.4.4 Exfoliation adsorption

Exfoliation adsorption technique is useful if both the nanofillers and polymer are dispersible in a common solvent. This method is mostly used when the synthesis of layered silicate is required. It involves the intercalation of polymers from solution. One thing that should be taken care of during the selection of the polymer should be solubility in solvent. Also the silicate layers should be sellable in solvent. In this method, layered silicate layers are swelled and dispersed in solvent and then mixed with polymer solution. Finally on removal of solvent by filtration or precipitation from solvent, a multilayered structure is formed. Desorption of the solvent actually increases the entropy of the solvent. This increase in entropy works as the driving force in exfoliation adsorption process. The polymers have almost no polarity on using this synthesis technique, which is an additional advantage of this technique (Pavlidou & Papaspyrides, 2008). This technique is widely used with water-soluble polymers such as Polyethylene Oxide (PEO), Polyvinyl Ether (PVE), Polyvinyl Pyrrolidone (PVP), and Polyacrylic Acid (PAA). The polymers dissolvable in common organic solvents are polycarbonate, PS, PMMA, polyvinyl alcohol, etc. The process requires a large amount of solvent, which is one of the disadvantage of the exfoliation adsorption process (Ray and Bousmina, 2005).

6.5 Applications of nanocomposite materials

6.5.1 Energy applications

As discussed earlier, hydrogen can be useful fuel for future; however, storage of hydrogen is a challenge, the best possible solution may be to store it in hydride forms. The magnesium hydride has quiet high hydrogen storage capacity MgH_2 (i.e., 6.5×1022 H atoms/cm^3 or 7.6 wt% hydrogen). CNT comes to the rescue felicitating diffusion of hydrogen in metal by CNT working as a host.

Another application is in the field of solar cells (photovoltaic cells). These solar cells are made from a PN junction devices made by materials such as gallium arsenide, gallium phosphide; however, a higher cost of production is major hurdle to make these cells easily marketable. Titania (TiO_2) films derived from sol gel method have shown excellent energy conversion efficiency with an advantage of a low cost. This large conversion efficiency is achieved due to large surface area at which photo-electrochemical process takes place.

In another application area of energy storage devices, lithium sulfur rechargeable batteries can replace currently used Li-ion batteries in near future due to low cost of sulfur. Nanostructure-based lithium sulfur battery consisting of graphene–sulfur–carbon nanocomposite multilayered structure gives excellent charge–discharge performance. Nowadays the batteries with specific capacitance as high as 500 Wh/kg are available.

Wind energy production is increasing day by day and taking a large chunk of energy production throughout the world. Nanocomposite materials have large strength to weight ratio and therefore can be utilized to make lithium-ion batteries. Longer and stronger blades increase conversion efficiency of wind energy.

6.5.2 Automobile applications

Polymer/clay nanocomposites are used in the car part industry. It is required to use light material in the manufacturing of cars for saving on fuel. Polymer/clay nanocomposites are preferred for their lightweight mechanical stiffness, thermal, and mechanical resistance. Various parts of external body parts of vehicles such as handles, gas tanks, mirrors, etc., are made from nanocomposites.

Car parts, such as handles, rear view mirror, timing belt, components of the gas tank, engine cover, bumper, etc. are made by nanocomposite materials such as nylon (polyamide). In near-future, car glasses can be replaced by transparent nanocomposite materials having good abrasion resistance property. Some car makers are using innovative nanocomposite materials for manufacturing of car structure that is light but has better strength and more scratch and dent resistant. One application of the nanomaterials in the field of automotive industry is as nanofluids. To overcome poor heat extraction efficiency from engine requires the use of fluids in radiator and brake system. The use of nanofluids as a coolant allows to use smaller size radiators thus saving cost on fuel. Many new applications in the manufacturing of various components in automotive industry are the use of nanocomposite materials in the tire rubber and use of nanoadditives in the fuel to reduce fuel consumption (Presting & König, 2003).

Research is going on in the fuel cell technology and hydrogen storage. Hydrogen is an excellent fuel and is therefore used in the launch of space shuttle as it has excellent energy to weight ratio. However, it is highly combustible when comes in contact of open air. Therefore a major hurdle is the storage of hydrogen fuel in the vehicle tank. From recent research in the field of nanotechnology it is known that CNT fuel cells have potential to store hydrogen. The use of hydrogen fuel in the combustion releases clean water vapor in the environment. Therefore, it has no bad environmental implications.

6.5.3 Packaging applications

Silver has been known for centuries for its antibacterial, antifungal, and antiviral properties. Due to its large surface to volume ratio, even a small concentration of nanoparticle provides antibacterial activity. Owing to these properties, silver nanoparticles coatings may be employed in biomedical devices, food packaging, textiles, and medical implants.

Relatively low price and convenience have made plastic suitable for packaging applications over the years; however, disposal of plastic waste is a matter of great concern and hazardous to nature. Therefore, biodegradable nanocomposite polymers showing good tensile strength, low permeability and better degradable properties produced by cellulose have been shown to replace synthetic plastic (polyvinyl chloride, PS, polypropylene), responsible to cause environmental hazard. Clay and inorganic solids have been used as additives to improve barrier properties of cellulose.

6.5.4 Biomedical applications

One important application in the field of bio-nanocomposite materials is its use in tissue engineering. At present, in severe bone injuries permanent biomedical material may be used which may corrode, wear, and cause infection. Additionally, there is bone resorption around the implant because of stress shielding by mechanically more sturdy material than original bone (Khan et al., 2021). Therefore there is a need of more tissue engineering by having tunable mechanical properties of the implant.

Therefore, polymers having biodegradable properties are in use extensively due to greater design flexibility because the polymers have ability to be tailored according to need. Degradable polymeric materials are suitable for the use in drug delivery vehicles, temporary prosthetic material, and scaffolds for tissue engineering. Nanocellulose applications are extensively researched due to their superior tensile strength (Klemm et al., 2006).

6.5.5 Other applications

Nanomaterials have a huge potential in the sensor fabrication as they exhibit a wide range of properties that can be used for the sensing applications. For example, giant magnetoresistance (GMR) sensor that can be used in computer hard disk drives. Sensors made from the nanomaterials can be used to monitor pressure, conductivity, humidity, resistance, and many other properties. Nanocomposite materials can be utilized as surface coatings to produce easy to clean surface thus saving cost on water and energy. Nanocomposites such as nylon-6 reinforced with silica nanoparticles have tremendous improvement in mechanical properties. Therefore, these composites are a promising candidate for the use as an energy-absorbing, high-impact material for clothing of defense personnel (Ou et al., 1998). Another area of defense research has been microwave absorption by the use of nanocomposite materials coatings over defense equipment. Various new materials such as carbon nanotube, carbon fibers, and metal oxide–reinforced polymer composites show promises to achieve desired stealth feature. These are only few areas where the nanomaterials are being used extensively but the use of nanomaterials is not limited to the mentioned field. The ease in molding and capability of altering the properties of nanomaterials by choosing a suitable matrix-filler combination in appropriate ratio can have diverse applications of nanomaterials. Addition of nanomaterial to the matrix improves the properties of the composites to a high extent. Addition of a suitable filler-matrix combination in adequate amount can improve the properties of the composite significantly. Table 6.1 summarizes the effect of adding a suitable filler in an appropriate matrix. It shows that the addition of the filler improves the thermal conductivity, thermal diffusivity, electrical conductivity, mechanical strength, hardness, and other properties of the composite material.

Table 6.1 Effect of fillers on different properties of polymer composite.

Matrix	Filler	Affected properties	Reference
CNT	Fe_2O_3	Enhance electrical conductivity	Jiang and gao, 2003
Cu	Al_2O_3	Enhance micro hardness	Ying and Zhang, 2000
Fe23C6	Fe_5B	Enhance hardness	Branagan and Tang, 2002
Al	SiC	Enhance hardness	Marchal et al., 1996
Al	AlN	Enhance compression resistance	Goujon and Goeuriot, 2001
Cu	Nb	Enhance hardness	Xiaochun et al., 2004
Ni	PSz	Enhance Strength and hardness	Kamigaito, 1991
Ni	YSz	Enhance catalytic activity	Aruna and Rajam, 2003
Ag	Au	Enhance catalytic activity	Chen et al., 2003
Polyamide	Boron Nitride (BN)	Enhanced thermal conductivity, thermal diffusivity	Wang et al., 2018
$NaNO_3$-KNO_3	SiO_2	Enhanced heat capacity and mechanical strength	Chieruzzi et al., 2013
$NaNO_3$-KNO_3	Al_2O_3	Enhanced heat capacity and mechanical strength	Chieruzzi et al., 2013
$NaNO_3$-KNO_3	TiO_2	Enhanced heat capacity and mechanical strength	Chieruzzi et al., 2013
$NaNO_3$-KNO_3	SO_2-Al_2O_3	Enhanced heat capacity and mechanical strength	Chieruzzi et al., 2013
Fluorinated polysiloxane	TiO_2	Enhanced hydro-affinity	Xiaofeng et al., 2011
Polydimethylsiloxane	SiO_2	Enhanced hydro-affinity	Facio and Mosquera, 2013

6.6 Summary

Nanocomposites show properties that are unmatched by conventional materials. The synthesis of desired structured material requires an understanding of mechanism of interaction between particles in the nanoscale. The synthesis technique influence the material property also so it is highly recommended to choose the synthesis technique judiciously. Among the available techniques, chemical vapor technique is the most widely used technique for the synthesis of nanocomposite materials. This technique is simpler and quick. However, the selection may depend on the constituents being used for the nanocomposites. Their properties, chemical reactivity, stability are few parameters that should be taken care of while selecting the synthesis technique. The researcher needs a good understanding of conventional as well as new routes of synthesis. Further, it is required to calibrate the route of synthesis (modeling) according to the need of fabrication design of nanomaterials. Therefore, the major routes discussed here to synthesis nanomaterials are not exhaustive rather require further investigation. It may be noted that applications discussed here provide a ready reference for the vast amount of research in different fields that can be seen as a precursor to explore further. This chapter may be seen as a first step toward rather long journey.

References

Abedi, S., Abdouss, M., 2014. A review of clay-supported Ziegler-Natta catalysts for production of polyolefin/clay nanocomposites through in situ polymerization. Appl. Catal. A 475, 386–409. https://doi.org/10.1016/j.apcata.2014.01.028.

Alemdar, A., Sain, M., 2008. Isolation and characterization of nanofibers from agricultural residues - wheat straw and soy hulls. Bioresour. Technol. 99 (6), 1664–1671. https://doi.org/10.1016/j.biortech.2007.04.029.

Alig, I., Pötschke, P., Lellinger, D., Skipa, T., Pegel, S., Kasaliwal, G.R., Villmow, T., 2012. Establishment, morphology and properties of carbon nanotube networks in polymer melts. Polymer 53 (1), 4–28. https://doi.org/10.1016/j.polymer.2011.10.063.

Andricic, B., Kovacic, T., Perinovic, S., Grigic, A., 2008. Thermal Properties of Poly(L-lactide)/Calcium Carbonate Nanocomposites. Macromol. Symp. 263, 96–101. https://doi.org/10.1002/masy.200850312.

Angellier, H., Choisnard, L., Molina-Boisseau, S., Ozil, P., Dufresne, A., 2004. Optimization of the preparation of aqueous suspensions of waxy maize starch nanocrystals using a response surface methodology. Biomacromolecules 5 (4), 1545–1551. https://doi.org/10.1021/bm049914u.

Annala, M., Lahelin, M., Seppälä, J., 2012. Utilization of poly(methyl methacrylate) - carbon nanotube and polystyrene - carbon nanotube in situ polymerized composites as masterbatches for melt mixing. Exp. Polym. Lett. 6 (10), 814–825. https://doi.org/10.3144/expresspolymlett.2012.87.

Aruna, S.T., Rajam, K.S., 2003. Synthesis, characterisation and properties of Ni/PSZ and Ni/YSZ nanocomposites. Scr. Mater. 48 (5), 504–512. https://doi.org/10.1016/S1359-6462(02)00498-0.

Avaru, Balasubrahmanyam, Patil, M.N., Gogate, P.R., Pandit, A.B., 2006. Ultrasonic atomization: effect of liquid phase properties. Ultrasonics 44 (2), 146–158. https://doi.org/10.1016/j.ultras.2005.09.003.

Balek, V., Klosova, E., Murat, M., Camargo, N.A., 1996. Emanation thermal analysis of Al2O3/SiC nanocomposite xerogel powders. Am. Ceram. Soc. Bull. 75 (9), 75.

Bhoopal, R.S., Tripathi, M., Sharma, P.K., Singh, R., 2015. Modelling of effective thermal conductivity: A comparison of artificial neural networks and theoretical models. Adv. Eng. Appl. Sci.: An Int. J. 2 (2), 21–27.

Bocchini, S., Frache, A., 2013. Comparative study of filler influence on polylactide photooxidation. Express Polym. Lett. 7 (5), 431–442. https://doi.org/10.3144/expresspolymlett.2013.40.

Branagan, D.J., Tang, Y., 2002. Developing extreme hardness (>15 GPa) in iron based nanocomposites. Compos. Part A Appl. Sci. Manuf. 33 (6), 855–859. https://doi.org/10.1016/S1359-835X(02)00028-3.

Chang, X.X., Mubarak, N.M., Mazari, S.A., Jatoi, A.S., Ahmad, A., Khalid, M., Walvekar, R., Abdullah, E.C., Karri, R.R., Siddiqui, M.T.H., Nizamuddin, S., 2021. A review on the properties and applications of chitosan, cellulose and deep eutectic solvent in green chemistry. J. Ind. Eng. Chem.

Chang, W., Skandan, G., Danforth, S.C., Kear, B.H., Hahn, H., 1994. Chemical vapor processing and applications for nanostructured ceramic powders and whiskers. Nanostruct. Mater. 4 (5), 507–520. https://doi.org/10.1016/0965-9773(94)90058-2.

Chen, W., Zhang, J., Cai, W., 2003. Sonochemical Preparation of Au, Ag, Pd/SiO2 Mesoporous Nanocomposites. Scr. Mater. 48 (8), 1061–1066. https://doi.org/10.1016/S1359-6462(02)00635-8.

Chieruzzi, M., Cerritelli, G.F., Miliozzi, A., Kenny, M., 2013. Effect of nanoparticles on heat capacity of nanofluids based on molten salts as PCM for thermal energy storage. Nanoscale Res. Lett. 8, 448. https://doi.org/10.1186/1556-276X-8-448.

Chimanowsky Junior, J.P., Soares, I.L., Luetkmeyer, L., Tavares, M.I.B, 2014. Preparation of high-impact polystyrene nanocomposites with organoclay by melt intercalation and characterization by low-field nuclear magnetic resonance. Chem. Eng. Process. 77, 66–76. https://doi.org/10.1016/j.cep.2013.11.012.

Choy, K.L., 2003. Chemical vapour deposition of coatings. Prog. Mater Sci. 48 (2), 57–170. https://doi.org/10.1016/S0079-6425(01)00009-3.

Dey, T.K., Tripathi, M., 2010. Thermal properties of silicon powder filled high-density polyethylene composites. Thermochim. Acta 502 (1), 35–42. https://doi.org/10.1016/j.tca.2010.02.002.

Eliezer, N., Gileadi, E., 2019. Physical Electrochemistry: Fundamentals, Techniques, and Applications. Multi-step Electrode Reactions. John Wiley & Sons, pp. 67–81.

Facio, D.S., Mosquera, M.J., 2013. Simple Strategy for Producing Superhydrophobic Nanocomposite Coatings In Situ on a Building Substrate. ACS Appl. Mater. Interfaces 5 (15), 7517–7536. https://doi.org/10.1021/am401826g.

Fu, S., Sun, Z., Huang, P., Li, Y., Hu, N., 2019. Some basic aspects of polymer nanocomposites: a critical review. Nano Mater. Sci. 1 (1), 2–30. https://doi.org/10.1016/j.nanoms.2019.02.006.

Goujon, C., Goeuriot, P., 2001. Solid state sintering and high temperature compression properties of Al-alloy5000/AlN nanocomposites. Mater. Sci. Eng. A 315 (1–2), 180–188. https://doi.org/10.1016/S0921-5093(01)01139-X.

Greene, J.E., 2017. Review Article: Tracing the recorded history of thin-film sputter deposition: From the 1800s to 2017. J. Vac. Sci. Technol. A 35 O5C204–1-O5C204-60. https://doi.org/10.1116/1.4998940.

Hamada, H., Kintaichi, M., Inaba, M., Tabata, T., Yoshinari, H., Tsuchida, 1996. Role of supported metals in the selective reduction of nitrogen monoxide with hydrocarbons over metal/alumina catalysts. Catal Today 29 (1–4), 53–57. https://doi.org/10.1016/0920-5861(95)00263-4.

Hatling, M.A., Glover, D.J., Whitehead, C., Zhang, K., 2007. Industrial Scale Destruction of Environmental Pollutants using a Novel Plasma Reactor. J. Phys. Chem. C. 47 (16), 5856–5860. https://doi.org/10.1021/ie8001364.

Hou, X., Choy, K.L., 2006. Processing and applications of aerosol-assisted chemical vapor deposition. Chem. Vap. Depos. 12 (10), 583–596. https://doi.org/10.1002/cvde.200600033.

Hulteen, J.C., Martin, C.R., 1997. A general template-based method for the preparation of nanomaterials. J. Mater. Chem. 7, 1075–1087. https://doi.org/10.1021/ie8001364.

Ifuku, S., Nogi, M., Abe, K., Yoshioka, M., Morimoto, M., Saimoto, H., Yano, H., 2009. Preparation of chitin nanofibers with a uniform width as α-chitin from crab shells. Biomacromolecules 10 (6), 1584–1588. https://doi.org/10.1021/bm900163d.

Jain, A., Banik, B.K., Banik, B.K., 2020. 8 - Green synthesis of nanoparticles and nanocomposites: Medicinal aspects. Adv. Green Chem.. Elsevier, Amsterdam, pp. 231–242.

Jiang, L., gao, L., 2003. Carbon Nanotubes-Magnetite Nanocomposites from Solvothermal Processes: Formation, Characterization, and Enhanced Electrical Properties. Hem. Mater. 15 (14), 2848–2853. https://doi.org/10.1021/cm030007q.

Jones, H., 2001. A perspective on the developed of rapid solidification and nonequilibrium processing and its future. Mater. Sci. Eng. A 304–306 (1–2), 11–19. https://doi.org/10.1016/S0921-5093(00)01552-5.

Kamigaito, O., 1991. What Can Be Improved by Nanometer Composites. J. Japan Soc. Powder Powder Metall. 38, 315–321. doi:10.2497/JJSPM.38.315.

Kaminsky, W., 2018. Polyolefin-nanocomposites with special properties by in-situ polymerization. Front. Chem. Sci. Eng. 12 (3), 555–563. https://doi.org/10.1007/s11705-018-1715-x.

Karri, R.R., Shams, S., Sahu, J.N., 2019. Overview of potential applications of nano-biotechnology in wastewater and effluent treatment. In: Nanotechnology in water and wastewater treatment. Elsevier, pp. 87–100.

Khan, F.S.A., Mubarak, N.M., Khalid, M., Walvekar, R., Abdullah, E.C., Ahmad, A., Karri, R.R., Pakalapati, H., 2021. Functionalized multi-walled carbon nanotubes and hydroxyapatite nanorods reinforced with polypropylene for biomedical application. Sci. Rep. 11 (1), 1–10.

Klemm, D., Schumann, D., Kramer, F., Heßler, N., Hornung, M., Schmauder, H.P., Marsch, S., 2006. Nanocelluloses as innovative polymers in research and application. Adv. Polym. Sci. 205 (1), 49–96. https://doi.org/10.1007/12_097.

Kozlov, G., Aphashagova, Z., Malamatov, A., Zaikov, G., 2012. The behaviour features of polymer nanocomposites filled with calcium carbonate. Chem. Chem. Technol. 6 (1), 113–118. https://doi.org/10.23939/chcht06.01.113.

Kutlu, B., Meinl, J., Leuteritz, A., Brünig, H., Heinrich, G., 2013. Melt-spinning of LDH/HDPE nanocomposites. Polymer 54 (21), 5712–5718. https://doi.org/10.1016/j.polymer.2013.08.015.

L., G.N., Laura, R., Alain, D., Mirta, A., Silvia, G, 2011. Effect of glycerol on the morphology of nanocomposites made from thermoplastic starch and starch nanocrystals. Carbohydr. Polym. 84 (1), 203–210. https://doi.org/10.1016/j.carbpol.2010.11.024.

Li, Y., Yang, Y., Zhu, J., Zhang, X., Jiang, S., Zhang, Z., Yao, Z., Solbrekken, G.L., 2016. Colour-generating mechanism of copper-red porcelain from Changsha Kiln (A.D. 7th–10th century), China. Ceram. Int. 42 (7), 8495–8500. https://doi.org/10.1016/j.ceramint.2016.02.072.

Lingamdinne, L.P., Choi, J.S., Angaru, G.K.R., Karri, R.R., Yang, J.K., Chang, Y.Y., Koduru, J.R., 2022. Magnetic-watermelon rinds biochar for uranium-contaminated water treatment using an electromagnetic semi-batch column with removal mechanistic investigations. Chemosphere 286, 131776.

Liu, Y., Liu, Z., Guo, S., Du, Y., Huang, B., Chen, S., et al., 2005. Amorphous and nanocrystalline Al82Ni10Y8 alloy powder prepared by gas atomization. Intermetallics 13 (3–4), 393–398. https://doi.org/10.1016/j.intermet.2004.07.026.

Magendran, S.S., Khan, F.S.A., Mubarak, N.M., Khalid, M., Walvekar, R., Abdullah, E.C., Nizamuddin, S., Karri, R.R., 2019. Synthesis of organic phase change materials by using carbon nanotubes as filler material. Nano-Structures & Nano-Objects 19, 100361.

Maiti, S., Suin, S., Shrivastava, N.K., Khatua, B.B., 2014. Low percolation threshold and high electrical conductivity in melt-blended polycarbonate/multiwall carbon nanotube nanocomposites in the presence of poly(ε-caprolactone). Polym. Eng. Sci. 54 (3), 646–659. https://doi.org/10.1002/pen.23600.

Marchal, Y., Delannay, F., Froyen, L., 1996. The essential work of fracture as a means for characterizing the influence of particle size and volume fraction on the fracture toughness of plates of Al/SiC composites. Scr. Mater. 35 (2), 193–198. https://doi.org/10.1016/1359-6462(96)00128-5.

Nakahira, A., Niihara, K., 1992. Sintering behaviors and consolidation process for Al_2O_3/SiC nanocomposites. J. Ceram. Soc. Jpn. 100 (1160), 448–453. https://doi.org/10.2109/jcersj.100.448.

Natter, H., Schmelzer, M., Hempelmann, R., 1998. Nanocrystalline nickel and nickel-copper alloys: synthesis, characterization, and thermal stability. J. Mater. Res. 13 (5), 1186–1197. https://doi.org/10.1557/JMR.1998.0169.

Nizammudin, S., Mubarak, N.M., Tripathi, M., Jaykumar, N.S., Sahu, J.N., ganesan, P.B., 2016. Chemical, dielectric and structural characterization of optimized hydrochar produced from hydrothermal carbonization of palm shell. Fuel 163, 88–97. https://doi.org/10.1016/j.fuel.2015.08.057.

Ou, Y., Yang, F., Yu, Z.Z., 1998. A new conception on the toughness of nylon 6/silica nanocomposite prepared via in situ polymerization. J. Polym. Sci. B Polym. Phys. 36 (5), 789–795. https://doi.org/10.1002/(SICI)1099-0488(19980415)36:5<789::AID-POLB6>3.0.CO;2-G.

Park, N.W., Lee, W.Y., Kim, J.A., Song, K., Lim, H., Kim, W.D., Yoon, S.G., Lee1, S.K., 2015. Effect of grain size on thermal transport in post-annealed antimony telluride thin films. Nanoscale Res. Lett 10, 0733. https://doi.org/10.1186/s11671-015-0733-6.

Pavlidou, S., Papaspyrides, C.D., 2008. A review on polymer-layered silicate nanocomposites. Progr. Polym. Sci. (Oxford) 33 (12), 1119–1198. https://doi.org/10.1016/j.progpolymsci.2008.07.008.

Pérez-Page, M., Yu, E., Li, J., Rahman, M., Dryden, D.M., Vidu, R., Stroeve, P., 2016. Template-based syntheses for shape controlled nanostructures. Adv. Colloid Interf. Sci. 234, 51–79. https://doi.org/10.1016/j.cis.2016.04.001.

Phasinam, K., Kassanuk, T., Tripathi, M., Khan, Z.A., Salahuddin, G., Nomani, M.Z.M., 2021. The Sustainable Development of Food Production in Agriculture Based on the Innovation in Nano-Science with Implication on Health and Environment. J. Qualit. Inq. 12 (3), 550–560.

Presting, H., König, U., 2003. Future nanotechnology developments for automotive applications. Mater. Sci. Eng. C 23 (6–8), 737–741. https://doi.org/10.1016/j.msec.2003.09.120.

Rahman, S.A., Song, I., Leung, M.K.H., 2012. Negative space template: a novel feature to describe activities in video. The International Joint Conference on Neural Networks (IJCNN). Proceedings, pp. 1–7. https://doi.org/10.1109/IJCNN.2012.6252666.

Ramanathan, K., Bangar, M.A., Yun, M., Chen, W., Myung, N.V., Mulchandani, A., 2005. Bioaffinity sensing using biologically functionalized conducting-polymer nanowire. J. Am. Chem. Soc. 127 (2), 496–497. https://doi.org/10.1021/ja044486l.

Rasoulzadeh, H., Dehghani, M.H., Mohammadi, A.S., Karri, R.R., Nabizadeh, R., Nazmara, S., Kim, K.H., Sahu, J.N., 2020. Parametric modelling of Pb (II) adsorption onto chitosan-coated Fe3O4 particles through RSM and DE hybrid evolutionary optimization framework. J. Mol. Liq. 297, 111893.

Ray, S.S., Bousmina, M., 2005. Biodegradable polymers and their layered silicate nanocomposites: in greening the 21st century materials world. Prog. Mater. Sci. 50 (8), 962–1079. doi:10.1016/j.pmatsci.2005.05.002.

Sahu, J.N., Karri, R.R., Jayakumar, N.S., 2021. Improvement in phenol adsorption capacity on eco-friendly biosorbent derived from waste Palm-oil shells using optimized parametric modelling of isotherms and kinetics by differential evolution. Ind. Crops Prod. 164, 113333.

Sahu, J.N., Zabed, H., Karri, R.R., Shams, S., Qi, X., 2019. Applications of nano-biotechnology for sustainable water purification. In: Industrial Applications of Nanomaterials. Elsevier, pp. 313–340.

Seki, S., Xie, Y., Kocaefe, D., Chen, C., Kocaefe, Y., 2016. Review of research on template methods in preparation of nanomaterials. J. Nanomater. 2016, 2302595. https://doi.org/10.1155/2016/2302595.

Ser, P., Kalck, P., Feurer, R., 2002. Chemical Vapor Deposition Methods for the Controlled Preparation of Supported Catalytic Materials. Chem. Rev. 102 (9), 3085–3128. https://doi.org/10.1021/cr9903508.

Siegel, R.W., 1993. Synthesis and properties of nanophase materials. Mater. Sci. Eng. A 168 (2), 189–197. https://doi.org/10.1016/0921-5093(93)90726-U.

Stearns, L.C., Zhao, J., Harmer, M.P., 1992. Processing and microstructure development in Al_2O_3-SiC "nanocomposites. J. Eur. Ceram. Soc. 10 (6), 473–477. https://doi.org/10.1016/0955-2219(92)90022-6.

Stöber, W., Fink, A., Bohn, E., 1968. Controlled growth of monodisperse silica spheres in the micron size range. J. Colloid Interf. Sci. 26 (1), 62–69. https://doi.org/10.1016/0021-9797(68)90272-5.

Sudarshan, T.S., 2004. Coated powders - New horizons and applications. In: Proc: International Conference on Advances in Surface Treatment: Research and Applications, ASTRA, 2004, pp. 412–422.

Sugunan, A., Warad, H., Boman, M., Dutta, J., 2006. Zinc Oxide Nanowires in Chemical Bath on Seeded Substrates. J. Sol-Gel Sci. Technol. 39 (1), 49–56. doi:10.1007/s10971-006-6969-y.

Sunday, A., Balogun, S., Akpan, E., 2012. Review of green polymer nanocomposites. J. Minerals Mater. Characterization Eng. 11. https://doi.org/10.4236/jmmce.2012.114028.

Surreddi, K.B., Scudino, S., Sakaliyska, M., Prashanth, K.G., Sordelet, D.J., Eckert, J., 2010. Crystallization behavior and consolidation of gas-atomized Al84Gd6Ni7Co3 glassy powder. J. Alloys Comp. 491 (1–2), 137–142. https://doi.org/10.1016/j.jallcom.2009.10.178.

Tjong, S.C., Chen, H., 2004. Nanocrystalline materials and coatings. Mater. Sci. Eng. R Rep. 45 (1–2), 1–88. https://doi.org/10.1016/j.mser.2004.07.001.

Tripathi, M., Bhatnagar, A., Mubarak, N.M., Sahu, J.N., Ganesan, Poo Balan, 2020. RSM optimization of microwave pyrolysis parameters to produce OPS char with high yield and large BET surface area. Fuel 277, 118184. https://doi.org/10.1016/j.fuel.2020.118184.

Tripathi, M., Ganesan, P., Jewaratnam, J., Sahu, J.N., 2016. Thermophysical properties of a novel copper reinforced oil palm shell (OPS) composite: effect of copper volume fraction and temperature. Mater. Chem. Phys. 182, 418–428. https://doi.org/10.1016/j.matchemphys.2016.07.050.

Wang, X., Zhao, J., Chen, M., Ma, L., Zhao, X., Dang, Z.M., Wang, Z., 2013. Improved self-healing of polyethylene/carbon black nanocomposites by their shape memory effect. J. Phys. Chem. B 117 (5), 1467–1474. https://doi.org/10.1021/jp3098796.

Wang, J., Li, Q., Liu, D., Chen, C., Chen, Z., Hao, J., Li, Y., Zhang, J., Naebe, M., Lei, W., 2018. High temperature thermally conductive nanocomposite textile by "green" electrospinning. Nanoscale 10 (35), 16868–16872. doi:10.1039/C8NR05167D.

Wang, Z., Yang, X., Wei, J., Xu, M., Tong, L., Zhao, R., Liu, X., 2012. Morphological, electrical, thermal and mechanical properties of phthalocyanine/multi-wall carbon nanotubes nanocomposites prepared by masterbatch dilution. J. Polym. Res. 19 (9), 9969. https://doi.org/10.1007/s10965-012-9969-3.

Xiaochun, L., Yang, Y., Cheng, X., 2004. Ultrasonic-assisted fabrication of metal matrix nanocomposites. J. Mater. Sci. 39, 3211–3212. https://doi.org/10.1023/B:JMSC.0000025862.23609.6f.

Xiaofeng, D., Shuxue, Z., Guangxin, G., Limin, W., 2011. A facile and large-area fabrication method of superhydrophobic self-cleaning fluorinated polysiloxane/TiO_2 nanocomposite coatings with long-term durability. J. Mater. Chem. 21 (17), 6161–6164. doi:10.1039/c0jm04546b.

Ying, D.Y., Zhang, D.L., 2000. Processing of Cu-Al_2O_3 metal matrix nanocomposite materials by using high energy ball milling. Mater. Sci. Eng.: A 282 (1), 152–156. https://doi.org/10.1016/S0921-5093(00)00627-4.

Zois, H., Kanapitsas, A., Stimoniaris, A.Z., Delides, C.G., 2012. Thermal and thermomechanical properties of epoxy resin/carbon black nanocomposites. Proc. Technical Proceedings of the 2012 NSTI Nanotechnology Conference and Expo, NSTI-Nanotech 2012, pp. 582–585.

CHAPTER 7

Sustainable approaches for synthesis of biogenic magnetic nanoparticles and their water remediation applications

Pavan Kumar Gautam[a], Sushmita Banerjee[b] and Sintu Kumar Samanta[a]
[a]Department of Applied Sciences, Indian Institute of Information Technology Allahabad, Allahabad, Uttar Pradesh, India
[b]School of Basic Science and Research, Department of Environmental Sciences, Sharda University, Greater Noida, Uttar Pradesh, India

7.1 Introduction

In the last few decades, the role of nanotechnology has become well established and skillfully documented in approximately all areas of science and technology (Khin et al., 2012; Wagner et al., 2014). Particularly, in environmental applications, these nanomaterials play a significant role over the past decades (Taghipour et al., 2019; Lingamdinne et al., 2019). The recent utilization of nanomaterials in wastewater purification has attracted much interest in the scientific community. The presence of hazardous heavy metals such as lead, arsenic, cadmium, chromium, etc. in natural aquatic bodies is a matter of great concern due to the toxicity associated with them. Due to the high aqueous solubility and ability to bioaccumulate, the heavy metals cause severe harm to the aquatic ecosystem as well as humans. Acute effects of heavy metals pollution include diarrhea, dehydration, abdominal pain, pulmonary discomfort, gastrointestinal problems, etc., While long-term exposure to heavy metals affects the functioning of vital organs such as the liver, kidney, heart, and lungs. Heavy metals also impede the activity of enzymes, which creates abnormality in the metabolic pathways of living beings (Karri et al., 2020). Naturally heavy metals are introduced into the hydrosphere by volcanic eruptions, forest fires, and natural weathering of rocks. However, the uncontrolled urbanization, industrial and modern agricultural practices, mining activities, and discharge of untreated or partially treated effluents have increased the heavy metal load in the environment.

Nanomaterials have tremendous environmental decontamination efficiency that is mainly credited to their alluring physicochemical qualities such as high dispersibility, fairly large surface area, adequate surface area to volume ratio, high surface reactivity, and porosity (Khan et al., 2019; Karri et al., 2019). Recently, studies on the synthesis of

Sustainable Nanotechnology for Environmental Remediation
DOI: https://doi.org/10.1016/B978-0-12-824547-7.00021-7

biogenic nanoparticles (NPs) from plant extract for scavenging pollutant species from waste effluents have received huge attention from scientific communities. Increasing research interest in the manufacturing of biogenic magnetic NPs (BMNPs) and their appliance in the water/wastewater remediation proved to be highly desirable and encouraging from the point of environmental perspective and cost-effectiveness (Tripathi and Chung, 2019, Sahu et al., 2019). Green synthesis of BMNPs) is a very simple, inexpensive, and hassle-free process; however, in contrast, conventional methods generally include physicochemical treatments that are very hazardous, expensive, and require complicated synthesis technique, which in turn possesses various environmental-related risks (Asmathunisha and Kathiresan, 2013; Shivalkar et al., 2021). So, the use of biogenic methods for green fabrication of magnetic NPs seems to be very perspective due to the nonhazardous, environment-friendly, fast, and economical nature. Moreover, the synthesis can be carried out at ambient temperature and pressure. The synthesis mechanism involves treating dissolved metal ions with natural metabolites such as polysaccharides, sugars, amino acids, terpenoids, alkaloids, proteins, carbonyl, carboxyl, and polyphenols that are supposed to benefit by acting as reducing and capping agents, facilitating the reduction of metal ions as a function of their reductive potential (Shamaila et al., 2016; Khodadadi et al., 2017), and also allow the particles to grow in the segregated state.

Furthermore, it is also easy to achieve the standard and desirable morphology of the BMNPs by simply modifying the fabrication process and by tailoring the NPs through the appropriate selection of organic functional groups for the sequestration of specific contaminants from contaminated water (Akhtar et al., 2013; Kumari et al., 2017). Additionally, these BMNPs exhibit superparamagnetic behavior with a strong saturation magnetization value that allows ease of separation from the final effluents by the application of magnetic field, and through chemical treatment, it can be subsequently reused in the consecutive treatment cycles. Despite the fact that the synthesis method is very sustainable and economical, this technique does not gain wide acceptance, as it experiences lots of drawbacks for large-scale applications. The fabrication process designed for lab-scale synthesis of nanoparticles is associated with some shortcomings such as controlling morphology, size, biocompatibility issues; activities of organic functional groups etc. (Hennebel et al., 2008; Hamouda et al., 2019). The integrity and stability of BMNPs is a matter of significant concern, as the degradation of organic functional groups may result in the leaching of metallic contents into the effluents, which in turn put an obvious question about the efficacy of wastewater treatment. So, it is very necessary to ensure safe, economical, and affordable water treatment.

Overall this chapter enlightens the state-of-the-art on production methodology as well as the prospective application of diverse forms of BMNPs for the decontamination of metallic species from the polluted water. Special attention has been given to the most toxic metals such as Pb(II), Cd(II), Cr(VI), and As(III) and As(V). In addition, it also throws light on its future prospects and various challenges associated with the system in wastewater treatment.

7.2 Biogenic synthesis of magnetic NP

7.2.1 Plant-based (phytogenic) synthesis

The plant-assisted production of magnetic NPs is receiving a great deal of acceptance due to synthetic viability. The cellular extract of different plant parts contains a variety of bioactive natural compounds namely amino acids, vitamins, peptides, phenolics, alkaloids, carbohydrates, and other phytochemicals that are exploited as strong reducing agents (Herlekar et al., 2014). Extraction of phytochemicals from different plant parts such as leaf, peel, flower, seed, root, and stem is less time consuming and easy to perform in the laboratory. Generally, plant biomass is chopped into small pieces and heated in an aqueous solution at 80–100 °C for 30–60 min. The extract is filtered and employed as a green agent that brings about the reduction and simultaneously capping of the synthesized Fe_3O_4, Fe_2O_3, and FeONPs (Gautam et al., 2019). The amount of extract required to fabricate the magnetic NPs with desired size, shape, and crystallinity is a crucial parameter of concern. Fatimah and coworkers have recently fabricated crystalline bioengineered Fe_3O_4 and Fe_2O_3NPs (10–80 nm in size) by using *Parkia speciosa husk* pod extract (Fatimah et al., 2020). These NPs showed excellent catalytic activity against a toxic dye bromophenol blue. In another report, the aqueous leaf extract of *Wedelia urticifolia* was utilized to form weak ferromagnetic NPs having a size of 15–20 nm (Rather and Sundarapandian, 2020). These NPs were found to be a very effective catalyst for methylene blue and more than 80% dye was degraded within a short period of time. A general mechanism of synthesis of BMNPs with cellular extract of different organisms is depicted in Fig. 7.1.

7.2.2 Algae-based (phycogenic) synthesis

The algal extract is a rich source of several biochemicals such as chlorophylls, vitamins, amino acids, and various kinds of pigments. Plenty of workers have successfully explored different kinds of algae for the synthesis of stable, crystalline, and monodispersed magnetic Fe_3O_4NPs (Dahoumane,et al., 2017). In a report, the environmentally benign route was followed to prepare magnetite NPs from *Shewanella oneidensis* (Perez-Gonzalez et al., 2010). High-resolution transmission electron microscopy (HR-TEM) exposed that the prepared NPs were found to be of 40–50 nm with spherical, rectangular, hexagonal, and rhombic shapes. Mashjoor et al. (2018) demonstrated the formation of Fe_3O_4NPs (mean diameter of 12.3 nm) with superparamagnetic behaviors from culture extract of species *Alga flexuosa*. Fourier transform infrared spectroscopy (FTIR) bands of the microalga extract certified the presence of polyphenolic groups, pectin-like carbohydrates, and glycoproteins that participated in the formation of magnetic NPs.

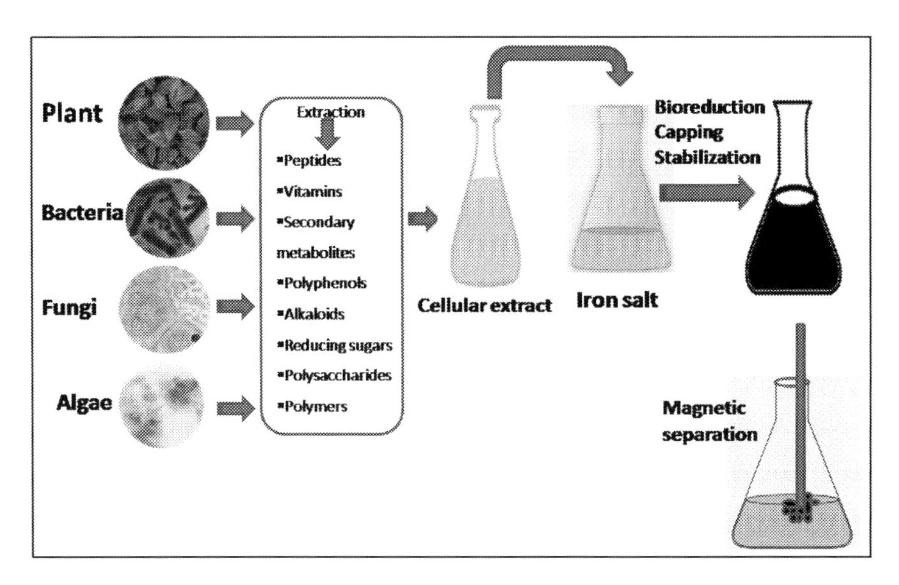

Figure 7.1 Schematic illustration for synthesis of BMNPs from different living organisms.

In another example, *Kappaphycus alvarezii*, a microalga (seaweed species) was used for the eco-friendly preparation of magnetite NPs (size in 12–20 nm) with characteristics crystalline in nature (Yew et al., 2016).

7.2.3 Fungi-based (mycogenic) fabrication

The utilization of fungal species for the production of NPs is becoming popular due to abundance and biological diversity among fungi. Nearly 70,000 species of fungi have been identified to date. Fungi are well known for their ability to produce magnetic NPs as the cellular extract contains numerous bioactive secondary metabolites. Mahanty et al. (2019) demonstrated fungi-mediated synthesis of Fe_3O_4NPs from fungal species such as *Trichoderma asperellum, Phialemoniopsis ocularis*, and *Fusarium incarnatum*. The shape of iron NP was observed to be globular with particle size ranging between 25 ± 3.94 nm for T. asperellum, 13.13 ± 4.32 nm for *P. ocularis*, and 30.56 ± 8.68 nm for *F. incarnatum*. Majumdar and Haloi (2017) exploited *Pleurotus* sp.to assess their ability to form magnetic NPs. FTIR analysis of the fungal extract identified several proteins and extracellular matrix accountable for the reduction and functionalization of the NPs. Pavani and Kumar (2013) explored *Aspergillus* sp. for the fabrication of magnetic NPs (50–200 nm in size). The chemoattractants present on the cell wall of *Aspergillus* sp. attracted Fe ions toward the fungal cell wall where they react with chelating chemicals of the cell wall to form NPs. A praiseworthy effort to synthesize magnetic NPs by using fungal species was made by Mahanty et al. (2019). His research group prepared spherical shaped, crystalline, and ferromagnetic Fe_2O_3NPs by using a fungal extract of *Aspergillus tubingensis* (STSP 25) isolated from sediments of Sundarbans. These sustainably produced NPs demonstrated a high capacity to remove several metallic pollutants from wastewater. Nearly 98.00%, 96.45%, 92.20%, and 93.99% elimination was observed for Ni^{2+}, Pb^{2+}, Zn^{2+}, and Cu^{2+}, respectively. The regeneration ability of these myco-synthesized NPs was also very impressive and their sorption efficiency remained unaffected for five repetitive cycles.

7.2.4 Bacteria-mediated fabrication

The bacterial cell utilizes several biochemical pathways such as mineralization, chelation, complexation, etc. as a defense strategy against various metallic species (Gautam et al., 2019). These biochemical reactions are triggered by various natural bioactive compounds present inside the bacterial cell. This unique characteristic of the bacterial cell intrigued the scientist to use the bacterial cells as an efficient factory for magnetic NPs formation (Singh et al., 2020). Bharde et al. (2005) demonstrated the bacterial-mediated synthesis of quasi-spherical-shaped magnetite NPs (10–40 nm in size) by using *Actinobacter* spp. FTIR study demonstrated that the hydrolysis of the iron precursor by surface-bound proteins of bacterium facilitated the reduction process. This bacteria-driven biotransformation of iron salt took place under an acidic and entirely aerobic environment. The same group performed biofabrication of superparamagnetic maghemite (Υ-Fe_2O_3) and greigite (Fe_3S_4) from the culture broth of *Actinobacter* sp. (Bharde et al., 2008). Synthesized NPs were highly crystalline in nature with 50 nm in size. The FTIR study revealed that the NPs were capped with bacterial proteins.

7.3 Factors influencing synthesis of BMNPs

7.3.1 Solution pH

Solution pH exerts a significant impact on the biological production of NPs. This may be due to the fact that the pH of the solution can regulate the size, shape, and synthesis rate of the NPs. Moreover it was also reported that high pH favors the NPs synthesis process that particularly credited to the formation of nucleation centers. These nucleation centers are rich in reaction sites that promote reduction transformation of metallic ions into metal NPs. Solution pH also governs the activity of various functional groups present in the biological extract through protonation and deprotonation process. The relationship between pH and size of NPs was well established by Herrera-Becerra et al. (2007). It was investigated that the Fe_3O_4NPs were prepared with a relatively larger size and in lower number at low pH value (pH 3, 5, and 7); but at the same time when the value of solution pH raised to 10, it favors production of high numbers and small-sized NPs. It was hypothesized that low pH discourages nucleation process that is supposed to be responsible for the formation of new and small-sized particles. On the other hand, highly protonated condition promotes aggregation tendency that leads to the creation of bigger sized Fe_3O_4NPs. However, with a slight increase in the solution pH condition, certain functional groups such as carbonyl and hydroxyl become active and become more available for interacting with the Fe ions leading to the formation of complexes. Some experiments also demonstrated that the pH also determines the magnetic property of synthesized magnetic NPs. In a report, green fabrication of magnetic Fe_3O_4NPs was demonstrated by using marine algae *Chaetomorpha antennina* (Siji et al., 2018). The pH-dependent synthesis was performed to observe the effect on size, extent of aggregation, and magnetic property. It was noticed that the NPs synthesized at pH 8 exhibited highest remanence. The study also revealed that the NPs produced at pH 6 showed low magnetic property, while higher ferromagnetic NPs were achieved at pH 8. The highest yield of NPs was also observed at pH 8. Reduction rate as a function of solution pH was also studied by Jacob et al. (2012) for the biosynthesis of Au and Ag ions. They discovered that the reduction efficiency was very poor for both Au and Ag ions at pH 3.3; but both these ions started to reduce simultaneously as the pH value of the system raises to 10. Further the research group also conclude that alkaline pH condition allows formation of AgO species and these species interacted strongly with the functional groups particularly hydroxyl moiety present in the plant extract and completely ensheaths the surface of the NPs through the interaction.

7.3.2 Temperature

Reaction temperature is also one of the vital parameters that strongly regulate the shape, volume, and rate of creation of nanomaterials. Temperature also plays important role in the nucleation process by promoting the formation of active nucleation centers, which further facilitates the rate of biosynthesis. A study reported by Hassan et al. (2018) showed a valid connection between the shape of NP and temperature. The hematite (α-Fe_2O_3) NPs were synthesized by using aqueous floral extracts of *Callistemon viminalis* at various temperatures. The TEM analysis revealed that the annealing temperature subjected for the synthesis of NPs played a major decisive aspect in regulating the size, agglomeration, and crystallinity of the NPs. The X-ray diffraction (XRD) experiment showed that the NPs synthesized at 300 °C were poorly crystalline, while the highly crystalline NPs were produced at 500 °C. NPs with smallest mean diameter (20.79 nm) were formed at 500 °C. Further, the synthesis of NPs as a function of temperature and amount of plant extract required was investigated by Sheny et al. (2011). It was observed that higher quantity of *Anacardium occidentale* plant leaf extract was vital for the reduction of Au and Ag ions at low reaction temperature (27 °C), while at high reaction temperature (100 °C), less quantity of plant extract was found sufficient for the effective biosynthesis of the ions with enhanced stability. Iravani and Zolfaghari (2013) had investigated the effect of four different reaction temperatures that is, 25, 50, 100, and 150 °C on the biosynthesis of silver NPs from *Pinus eldarica* bark extract. The outcome of their study indicated negative correlation between reaction temperature and size of silver NPs.

7.4 Application of BMNPs for metal removal

7.4.1 Lead removal

Lead is considered one of the most noxious heavy metals due to its perilous impact on human health and aquatic life. Usually, lead enters into the water by corrosion of lead-containing plumbing material. Besides this, wastewater discharged from various industries such as electroplating, battery, paint, smelting is also the major contributor of lead pollution of aquatic bodies. Recently, several attempts have been made to minimize the concentration of lead by using biologically prepared magnetic NPs. Al-Kassas et al. (2016) synthesized biogenic and highly magnetic Fe_3O_4NPs from the extract of two seaweeds: *Padina pavonica* (Linnaeus) Thivy and *Sargassum acinarium* which were employed for the adsorptive separation of Pb(II) from water. It was evident that sulfated polysaccharides were the chief phytochemicals present in the algal extracts that played a dual function that is, reduction as well as stabilization of phytogenic Fe_3O_4NPs. The removal experiment deciphered that the NPs prepared from *P. pavonica* showed better removal (91%) of Pb(II) in comparison to NPs synthesized from *S. acinarium* (78%) within a reaction time of 75 min. In another work phyto-inspired Fe_3O_4NPs were produced from *Trigonella foenum-graecum* leaf extract (Das and Rebecca, 2018). The potential of the prepared magnetite NPs for the removal of Pb(II) from water and wastewater was assessed. The NPs were globular with an average size range of 51.6–215.7 nm. Batch adsorption experiment demonstrated total 93.0% removal of Pb(II) within 60 min with 0.4 g of prepared nanoadsorbent. Lingamdinne and coworkers developed inverse spinel iron oxide NPs (superparamagnetic in nature) by using seed extract of *Cnidium monnieri (L.) Cuss* and applied them for adsorptive removal of Pb(II) (Lingamdinne et al., 2017). The FTIR analysis demonstrated that different functional groups namely OH, R-CH_2, R-CH were capped over the surface of the NPs and facilitated the removal of Pb(II). Sorption of Pb(II) fitted to pseudo-second-order kinetics and more than 90% removal was achieved. Moreover, a reusability study showed that the prepared NPs could be applied for five repetitive cycles. The same research group performed the bioreductive synthesis of highly magnetic Fe_3O_4NPs from tangerine peel extract and harnessed them for preconcentration of Pb(II) from synthetic wastewater (Lingamdinne et al., 2020). Nearly 95% of Pb(II) adsorption removal was achieved with a Fe_3O_4 adsorbent dose of 0.625 mg/L and an initial concentration of Pb(II) at 32.5 ppm. The pH of the solution exerted an important effect and the maximum removal was observed at pH 4.5 within 95 min of contact time. Adsorption data were suited better onto the Langmuir model that advocated the monolayer coverage of Pb(II) molecules on synthesized NPs. The maximum adsorption capacity was estimated to be 100 mg/g. Gautam et al. (2020) prepared superparamagnetic Fe_3O_4NPs (60–100 nm in size) from *Moringa oleifera* leaf extract which were further explored for removal of Pb(II). High magnetization saturation value (66.99 emu/g) ensured the superparamagnetic behavior of synthesized NPs. Developed NPs worked efficiently with removal of 94% Pb(II) within 60 min of adsorption time. The sorption of Pb(II) obeyed pseudo-second-order kinetics and equilibrium adsorption data fitted to the Freundlich isotherm model. Very recently, an effort was made to fabricate magnetite NPs by using a lichen, *Ramalina sinensis* for the removal of Pb(II) from water (Arjaghi et al., 2021). The size of the synthesized NPs was in the range of 20–40 nm with crystalline in nature. A total 82% of Pb(II) was removed in the pH range of 4–5. These NPs also showed good uptake capacity (77%) for Cd(II) as well. The sorption of both the metallic pollutants was explained by pseudo-second-order kinetics. According to a latest report, the tea extract was employed for the production of iron NPs (Lin et al., 2020). These nanoadsorbents were used for the preconcentration of Pb(II). The presence of polyphenols, catechols, and amino acids in tea extract was predicted by FTIR which served as reducing and capping molecules. Nearly, 100% of Pb(II) was adsorbed onto the surface of NPs within 90 min of contact time. Nearly 100.0 mg/g was found to be the maximum adsorption capacity at 313 K. In another novel attempt, *Punica granatum* rind extract capped magnetite NPs were developed and further coated with dimercaptosuccinic acid for Pb(II) removal. Synthesized magnetites were rod shaped with 40 nm diameter and 200 nm length. Developed NPs performed well and more than 90% of Pb(II) was sequestrated within 60 min with 46.18 mg/g of maximum adsorption capacity. Sorption data obeyed pseudo-second-order of kinetics. The graphical representation of the procedure for the elimination of metallic contaminants from water by using biosynthesized magnetic NPs has been described in Fig. 7.2.

7.4.2 Cadmium removal

Cadmium (Cd) is an enormously noxious heavy metal, even in a lesser quantity. Cd is a highly deleterious industrial and environmental pollutant categorized as a human carcinogen. Long-term Cd inhalation through air leads to lung cancer. It is primarily used for metal plating and coating operations. It is also applied as a coating agent to avert the process of corrosion and electroplating onto the steel. Cd sulfide and selenide are generally utilized as colorants in plastics. Cd-based materials are exploited in electric batteries, electronic devices, and nuclear reactors. It is primarily used for metal plating and coating operations. Long-term exposure to Cd causes renal damage. Cd affects the metabolism of calcium and cases of prostate and lung cancer have been evidenced in cases of high Cd exposure (Idrees et al., 2018). Gupta and Nayak (2012) prepared a novel biogenic Fe_2O_3 magnetic NP from orange peel for the scavenging of Cd(II) from wastewater. Synthesized NPs were 25–29 nm in diameter. The study indicated

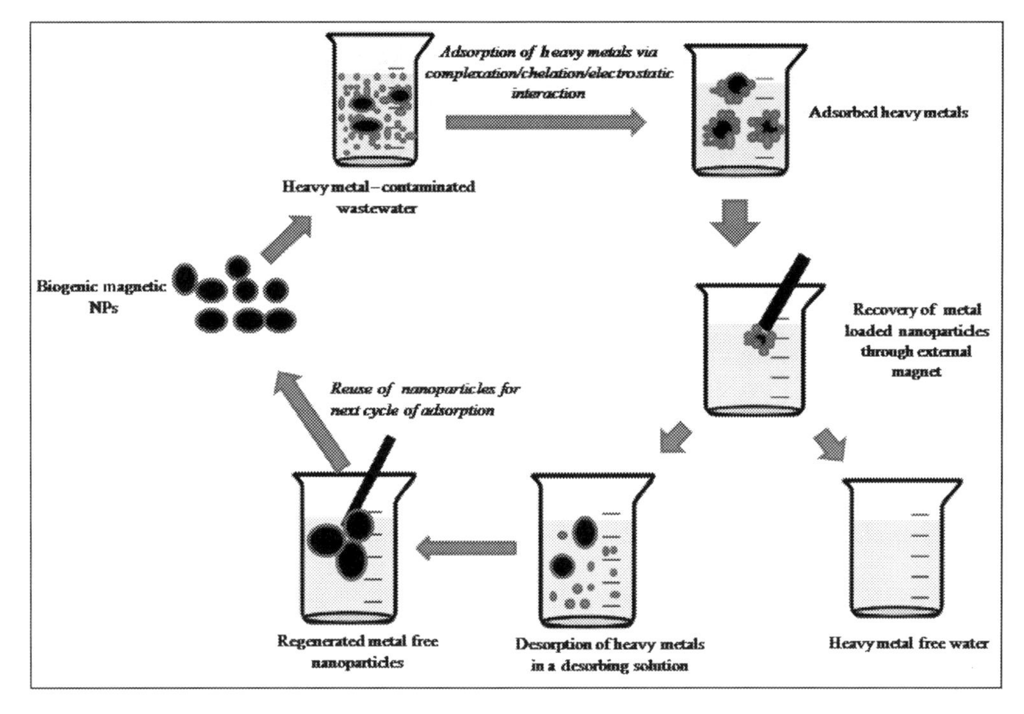

Figure 7.2 Pictorial depiction of employment of BMNPs for treatment of heavy metal-contaminated wastewater and regeneration of spent BMNPs for reuse.

that high temperature favored the extent of Cd(II) removal and the highest removal was acquired at 318 K. It will be worthwhile to mention here that the efficiency of prepared nanocomposite was also tested for electroplating industrial effluent. Surprisingly, the Fe_2O_3NPs showed 82% removal of Cd(II) regardless of the presence of other competitive metallic species such as nickel and zinc. In another work, an aqueous extract of sawdust obtained from poplar tree was used as a green reducing biocompound to prepare Fe_3O_4NPs for the separation of Cd(II) from aqueous solution (Katria et al., 2018). The FTIR analysis demonstrated that various bioactive compounds such as lignin, hemicellulose, and cellulose acted as a capping agent. To enhance the removal efficiency, the prepared NPs were further coated with ethylenediamine tetraacetic acid (EDTA). It was noticed that EDTA-coated Fe_3O_4NPs performed better than the uncoated ones. It was attributed to the strong electrostatic and chemical interaction between EDTA-coated Fe_3O_4NPs and Cd molecules. The thermodynamic study of the adsorption reaction professed that the sorption of Cd(II) was feasible, spontaneous, and endothermic in nature. In another report, an aqueous extract of Tangerine peel was utilized for the bioreductive production of magnetic NPs (Ehrampoush et al., 2015). These biologically manufactured NPs were proved to be proficient adsorbent for the remediation of Cd(II). Nearly 87% of removal of Cd(II) was ascertained within 90 min. In another significant work, a fully green protocol was adopted to fabricate highly magnetic Fe_3O_4NPs to remove Cd(II) (Lung et al., 2018). The NPs were capped with *Citrus limon* extract. It was found that the *C. limon* extract–mediated NPs showed superior removal of Cd(II) than bare NPs. Additionally, the prepared NPs showed good removal capacity with the presence of several coexisting cations. Sebastian et al. (2018) tested the removal efficacy of magnetite NPs for Cd(II) that were synthesized from coconut husk extract. Total 40% removal for Cd(II) was achieved in 120 min of equilibrium time. Sorption of Cd(II) fitted onto the Langmuir isotherm model and the maximum adsorption capacity was found to be 9.6 mg/g. In a sustainable approach, a natural biocompatible resin Shellac was used as a capping agent for Fe_3O_4NPs (Gong et al., 2012). These shellac-coated magnetite NPs (20 nm in size) were applied to minimize the quantity of Cd(II) from highly saline water. The maximum sorption capacity of the manufactured NPs was estimated to be 18.8 mg/g at pH 7.0. Another work represents the synthesis of magnetic maghemite (γ-Fe_2O_3) NPs from aqueous extract of *Tridax* (Yadav and Fulekar, 2018). These bioinspired maghemite NPs were applied for the separation of Cd(II) from fly ash. Scanning electron microscopy (SEM) images certified the creation of highly polydispersed NPs with a size distribution range of 50–120 nm. Participation of flavonoids and terpenoids in the reduction of iron salt, as well as capping of synthesized NPs, was assured by the FTIR analysis. The sorption study showed 67.8% removal of Cd(II) within 60 min of time span.

7.4.3 Arsenic removal

Arsenic is a metalloid element that is ubiquitously distributed all over the earth's crust and groundwater. Arsenic poisoning has become a global concern. The level of arsenic may vary across different geographical locations. Arsenic enters into the aquatic environment through the dissolution of rocks, minerals, and ores, from industrial effluents, including mining wastes, and via atmospheric deposition. Arsenic pollution in groundwater is one of the most crippling matters in the drinking water scenario of India. This may be due to human activities such as the use of pesticides and metal mining. Long-term consumption of arsenic-contaminated water is highly detrimental for health and it causes kidney, bladder, skin, and lung cancer as well as pigmentation disorders, hyperkeratosis, neurological problems, muscular weakness, and nausea (Roy and Saha, 2002). Biologically produced magnetic NPs found to be satisfactorily employed for the removal of different arsenic species present in the aqueous phase. Nikic and coworkers adopted an eco-friendly approach to fabricate magnetite NPs by using onion peel and corn silk extract (Nikic et al., 2019). Both magnetite NPs exhibited excellent removal efficiency in the separation of arsenic from groundwater samples. Brunauer-Emette-Teller (BET) surface area study concluded that the surface area of developed Fe_3O_4NPs was as high as 243–261 m^2/g. The FTIR analysis verified the presence of different biologically active compounds that served as a stabilizing agent. The removal study showed that most of the arsenic species present in the groundwater samples were efficiently captured by both magnetite NPs. Mukharjee and coworkers utilized an aqueous extract of aloe vera leaf to fabricate Fe_2O_3NPs that were used for minimization of arsenic from water (Mukherjee et al., 2016). The FTIR study demonstrated the occurrence of various amino acids, polysaccharides, and anthraquinones which provided stability to the synthesized Fe_2O_3NPs. These NPs effectively sequestrated arsenic with a maximum adsorption capacity of 38.47 mg/g at 20 °C. In another noteworthy attempt, the superparamagnetic iron oxide NPs was produced from tea waste extract as a green reducing medium (Lunge et al., 2014). Formed NPs showed exceptional potency in the removal of arsenic. The evaluated adsorption capacities of the NPs were as high as 188.69 and 153.8 mg^{-1} for As(III) and As(V), respectively. Very recently, the cellular extracts of fungal species such as *Fusarium oxysporum* were harnessed for the production of magnetite NPs for the successful arsenic removal from arsenic-rich groundwater samples (Balakrishnan et al., 2020). The adsorption experiment revealed that 91% of As (III) and 95.12% of As (V) were eliminated by synthesized NPs at a pH value of 5.5. Casentini et al. (2019) synthesized BMNPs inside the exopolysaccharide obtained from *Klebsiella oxytoca* strain DSM 29614. The applicability of synthesized BMNPs for removing As(III) and As(V) from arsenic-rich groundwater was tested. The size of the NPs was in the range of 9–15 nm. Nearly 95% and 65% removal was observed for As(V) and As(III), respectively. Adsorption of both arsenic species onto BMNPs followed pseudo-second-order reaction kinetics. BMNPs specifically synthesized from plants, their size along with maximum adsorption capacity for the respective heavy metals are listed in Table 7.1

7.4.4 Chromium removal

Chromium (Cr) mainly exist in the earth in two oxidation forms: Cr(III) and Cr(VI) and well known for its lethal effects on living organisms. Industries such as electroplating, leather tanning, and textile discharge huge amounts of Cr in surface water. Preconcentration of Cr in drinking water as well as natural aquatic bodies has become a major concern for scientific communities. As per the reports, biologically developed NPs have been found effective for the remediation of Cr-polluted water. Subramaniyam et al. (2015) synthesized magnetic Fe_3O_4NPs from culture extract of soil microalga, *Chlorococcum* sp. MM11 which could effectively remove hexavalent Cr from water. The FTIR analysis demonstrated that carbonyl and amine bonds from polysaccharides and glycoproteins present on algal cell wall facilitated the reduction of iron salt. Nearly 92% removal of Cr(VI) was achieved with this environmentally benign magnetite NPs. Kim and Roh (2019) utilized a metal-reducing bacteria with akaganeite (β-FeOH) as an electron acceptor and glucose as an electron donor to form magnetite NPs. They noticed that the net charge on the surface of bio-nanoadsorbents and ion affinity governed

Table 7.1 Applications of BMNPs synthesized from plant sources for heavy metal removal

Plant source	Type of BMNPs	Size (nm)	Adsorbed metal	Maximum adsorption capacity (mg/g)	Reference
M. oleifera	Fe_3O_4	80–100	Pb(II)	64.97	Gautam et al., (2020)
Cnidium monnieri (L.) Cuss	Fe_3O_4	35–45	Pb(II)	105.6	Lingamdinne et al., 2017
Tangerine peel	Fe_3O_4	50	Pb(II)	100.0	Lingamdinne et al., 2020
Punica Granatum	Fe_3O_4	40	Pb(II)	46.18	Venkates warlu et al. (2014)
Onion peel and corn silk	Fe_3O_4	26 and 28	Total arsenic [As(III and As(V)]	1.86, 2.79	Nikic et al., 2019
Orange peel powder	Fe_3O_4	32–35	Cd(II)	40.0	Gupta and Nayak, 2012
Poplar tree	Fe_3O_4	10–30	Cd(II)	63.29	Kataria and Garg, 2018
Aloe vera	α-Fe_2O_3	100	As(V)	38.47	Mukherjee et al., 2016
P. granatum	Fe_3O_4	100–200	Cr(VI)	156.3	Rao et al., 2013
Cynodon dactylon and *Muraya koenigii*	Fe_3O_4	20	Cr(VI)	34.7	Vishnu and Dhandapani, 2020
Eriobotrya japonica	Fe_3O_4	89	Cr(VI)	312.5	Önal et al., 2019
Citrullus lanatus	Fe_3O_4	5–20	Hg(II)	52.1	Venkateswarlu and Yoon, 2015
Lantana camara	Fe_3O_4	28	Ni(II)	227.20	Nithya et al., 2018

the sorption of Cr(III) and Cr(VI). These magnetite NPs also showed good removal capacity for both the arsenic species such as As(III) and As(V) over a broad pH regime. In another report, aqueous P. granatum extract was utilized to produce bioengineered magnetic NPs (Rao et al., 2013). After fabrication, the surfaces of prepared NPs were further modified by two distinct strains of a *Yarrowia lipolytica* namely, NCIM 3589 and NCIM 3590. Both NPs were applied for the sequestration of hexavalent Cr from water. Acidic environment facilitated the degree of sorption and optimum removal occurred at pH 2.0. The adsorption of Cr(VI) was well described by the Langmuir isotherm model. Nearly 125.0 and 156.3 mg/g were the maximum adsorption capacities for NCIM 3589 and NCIM 3590, respectively. In another recent report, manglicolous (mangrove) fungus *Aspergillus niger* BSC-1 was harnessed for the biosynthesis of Fe_3O_4NPs that were found very effective adsorbent for Cr(VI) (Chatterjee et al., 2020). The FTIR analysis of fungal extracts proved that the amide I and amide II bonds present in the extracellular enzymes were actively involved in the synthesis of NPs. More than 80% removal was achieved at pH 3.5. Moreover, these NPs were recyclable up to five consecutive cycles. Recently, silica-coated amino-functionalized magnetic NP (AF-MnP) was prepared from the extracts of plants such as *Cynodon dactylon* and *Murraya koenigii* (Vishnu and Dhandapani, 2020). This biologically prepared nanocomposite showed tremendous adsorption capacity for hexavalent Cr. The prepared nanocomposite was found to be nearly 20 nm in size. High magnetic saturation value (45 emu/g) ensured the superparamagnetic behavior of the synthesized nanocomposite. The maximum adsorption capacity of the developed nanocomposite was computed to be 34.7 mg/g. Interestingly, the prepared nanocomposite was nine times recyclable with a minimal loss of removal efficacy. Onal and coworkers demonstrated bioproduction of Fe_3O_4NPs using aqueous leave extract of loquat (*Eriobotrya japonica*) (Onal et al., 2019). The prepared bionanomaterials with mean particle size of 89 nm were used as an efficient adsorbent for the removal of Cr(VI). Sorption experiment showed more than 90% removal of Cr(VI) over a wide concentration range (50–500 mg/L) and the maximum adsorption capacity was computed to be 312.5 mg/g.

7.4.5 Other metals

Biofabricated magnetic NPs have also proven their effectiveness for other heavy metals as well. Nithya et al. (2017) followed a green procedure and developed superparamagnetic Fe_3O_4NPs from *Lantana camara* leaf extract. A spectroscopic analysis identified the predicted involvement of hydroxyl and carboxylate groups present in the green extract in the synthesis of the NPs. Developed NPs showed efficient removal of Ni(II). It was found that 60% of the Ni(II) was removed within 10 min only. The maximum adsorption capacity of the NPs was as high as 227.20 mg/g. In an eco-friendly attempt, Es'haghi et al. (2016) tested the potential of olive oil–coated Fe_3O_4NPs to remove Ni(II) from simulated urban and purified drinking water. Size of this magnetite nanocomposite was approximately 80–100 nm. The optimum pH was 7.0 for Ni(II) removal. The spectroscopy analysis demonstrated the presence of the carboxylic acid group in olive oil that was responsible for the high removal capacity of the prepared nanocomposite. In a report, a natural consortium of iron-reducing bacteria was applied to produce biogenic iron oxide NPs that were used (Castro et al., 2018) for the removal of Cu and Zn from wastewater. Removal investigation indicates that nearly 95.5% and 60.3% removal for Cu and Zn, respectively, was obtained. Kinetic modeling of adsorption revealed that the sorption of both metallic pollutants followed pseudo-second-order kinetics. Kandasamy et al. (2017) adopted a novel approach to synthesize iron oxide magnetic NPs by applying the culture extract of *Streptomyces thermolineatus*. These BMNPs were applied to remove Cu from wastewater discharged from the pigment industry.

Total 85% removal for Cu was observed. Recycling performance revealed that the spent NPs could be applied four times and insignificant loss was noticed in their adsorption efficacy. In a similar attempt, culture supernatant produced by *S. thermolineatus* was applied for functionalizing Fe_3O_4NPs, which was applied for the removal of Cu(II) ions (Sugnathi S. and Kandasamy, 2017). These BMNPs were found to be superparamagnetic in nature. From the FTIR study, it was predicted that carboxylic, amine, and hydroxyl groups present on the surface of BMNPs participated in binding with Cu(II), which resulted in its high removal. Very recently, microwave-mediated phytogenic production of Fe_3O_4NPs was demonstrated by applying *Ficus benghalensis* plant leaf extract (Lagashetty et al., 2020). These greenly prepared BMNPs were employed for the elimination of Hg(II) from wastewater. The FTIR analysis of leaf extract ratified the involvement of a range of reducing compounds such as amino acids, polysaccharides, phytochemicals, vitamins, and enzymes present in the cellular components of plants in the ancillary reduction of iron salt into iron oxide. A study demonstrated more than 50% reduction of Hg(II) ion from contaminated water. In another report, Venkateswarlu and Yoon (2015) demonstrated green production of Fe_3O_4NPs from watermelon (*Citrullus lanatus*) peel extract. Further, the synthesized NPs were treated with 3,4-dihydroxyphenethylcarbamodithioate and employed for the exclusion of Hg(II) from water. Removal study showed that 98% of Hg(II) got adsorbed within 60 min and the maximum sorption capacity was estimated as 52.1 mg/g. Magnetic NPs developed from various organisms other than plants and their details for the sequestration of metallic pollutants are listed in Table 7.2.

7.5 Factors affecting removal of heavy metals

The capacity for removal of toxic metallic species is governed by many factors. Surface area is regarded as one of the imperative factors that govern the performance of BMNPs as an adsorbent. NPs are blessed with a high surface-area-to-volume ratio that enhances the availability of active sites and facilitates various physical and chemical interactions

Table 7.2 Applications of BMNPs synthesized from various microorganisms for heavy metal removal.

Biological source	Type of BMNPs	Size (nm)	Adsorbed metal	Maximum adsorption capacity (mg/g)	Reference
Klebsiella oxytoca DSM 29614	Fe_3O_4	9–15	As(III) and As(V)	Not reported	Casentini et al. (2019)
Metal-reducing bacteria (Geocha-1)	Fe_3O_4	5–15	Cr(III), Cr(VI)	Not reported	Kim and Roh, 2019
Chlorococcum sp. *MM11*—	Fe_3O_4	20–50	Cr(VI)	Not reported	Subramaniyam et al., 2015
Aspergillus niger BSC-1	Fe_3O_4	20–40	Cr(VI)	15.57	Chatterjee et al., 2020
Eriobotrya japonica	Fe_3O_4	89	Cr(VI)	312.5	Önal et al., 2019

between surface and pollutant (Tang and lo, 2013; Lingamdinne et al., 2018). More active sites offer superior removal of contaminants. Therefore, smaller sized BMNPs are more proficient than the larger ones. The extent of aggregation and dispersity of the synthesized BMNPs also influence their metal uptake capacity. High surface area is available in case of monodispersed NPs, which allows greater interaction with metallic species leading to better removal efficiency. Temperature is also a crucial factor that affects the uptake of the metallic pollutants onto magnetic NPs. Generally high temperature favors the extent of removal. High temperature reduces the surface tension and viscosity of the wastewater, which allows greater mass transfer at solid-liquid interface (Gautam et al., 2015). pH of the wastewater also governs the extent of removal of metallic species by BMNPs. Some of the key factors such as solubility and speciation of metal species are greatly influenced by pH of the solution. pH also exerts major impact on surface characteristics of the NPs. Most of the metallic species such as Pb(II), Cd(II), Cu(II), etc. are positively charged and their optimum removal is achieved between pH range 4–5 due to the strong electrostatic interaction between negatively charged surface of BMNPs and cationic metal species (Arjaghi et al., 2021). Beyond pH 6, the metals get precipitated by forming their respective hydroxides. Temperature is another factor that influences the process of adsorptive removal of metals onto BMNPs. From published reports, it has been observed that the higher temperature enhances the rate of removal. High temperature positively favors the rate of mass transfer of solute molecules (metals) toward the solid-liquid interface and facilitates the greater interaction of metallic contaminants with BMNPs, which is amenable to higher removal rate. Chatterjee et al. (2020) studied the removal of Cr(VI) onto *A. niger* BSC-1 mediated Fe_3O_4NPs. Batch adsorption experiments revealed that the process of adsorption was temperature dependent. The maximum sorption capacities were found to be 9.95, 11.75, and 15.75 mg/g at 20, 30, and 40 °C, respectively. The scavenging efficiency of magnetic NPs is also affected by contact time. Pragmatically, the removal rate is high at initial stages of adsorption due to the proper availability of active sites present on the surface of NPs. As the time increases, the metal uptake capacity of NPs decreases because of the slackening of vacant sites available for the binding of metals species and finally reaches equilibrium. Equilibrium is the time at which no further sorption takes place due to the unavailability of binding sites necessary for the capturing of metals ions (Tamez et al., 2016.)

7.6 Reusability of BMNPs

Recyclability of the adsorbent for repetitive cycles in water treatment must be taken into consideration with high priority. Reuse of adsorbent is highly beneficial in terms of cost and energy. According to the published works, biologically produced NPs exhibited exceptional recyclability. There are several factors that contribute to the high regeneration competency of spent BMNPs and their reuse for decontamination. The surface of BMNPs is functionalized by biochemicals explored from living organisms. The important advantage of such functionalization is to provide stability to the developed NPs through addition of distinct functional groups. These functional groups assist the process of adsorption by the formation of double bond, hydrogen bond, covalent bond, and electrostatic interaction between NPs and metallic contaminants (Patra et al., 2017). The bare magnetic NPs are much prone to oxidation and hence the recycling capacity also becomes very poor. It is evident from the published literature that the BMNPs can be reused up to five to seven times without any substantial loss in their removal ability. Therefore, both the size and dispersity of NPs have become very important parameters that need to be taken care of during their synthesis.

7.7 Advantages and limitations of BMNPs

There are numerous advantages related to the application of BMNPs over chemically synthesized NPs for the water treatment processes:

- The synthesis of BMNPs is ecologically sound as the synthesis is carried out without any harsh and volatile chemical reactants.
- The BMNPs are synthesized in a short period of time. Usually the NPs are synthesized within 1 h of reaction time.
- Energy efficiency is another important advantage associated with the production of BMNPs. Synthesis does not require high temperature and pressure.
- BMNPs are developed at low cost therefore they can be produced at a large scale.
- Due to the magnetic nature, they can easily be recovered from wastewater by an external magnet.
- BMNPs are easy to combine with other NPs to design composites. This is usually done to enhance the treatment efficiency and posttreatment separation due to the magnetic component.

The important limitation related to BMNPs is the high extent of aggregation of particles during synthesis. Larger aggregates reduce the exposed surface area, thereby remediation efficiency of the NPs also gets affected. However, this problem can be addressed by optimizing the synthetic parameters such as temperature, pH, amount of extract dose, and reaction time (Sadhasivam et al., 2020).

7.8 Concluding remarks and future perspectives

The acceptance of biologically synthesized magnetic NPs for the remediation of toxic metallic species from wastewater and drinking water is rapidly increasing due to their environmental benignity, economic feasibility, and extraordinary performance. The application of naturally occurring bioactive compounds and secondary metabolites extracted from plants, fungi, algae, and bacteria have established their applicability as a green and environmentally safer reducing and stabilizing agent. This green and eco-friendly reaction media have provided the best alternative to manmade toxic chemicals and volatile solvents employed in the conventional fabrication of magnetic NPs. More importantly, the easy recovery of BMNPs and astounding recyclability for reuse confer an energy-efficient and inexpensive wastewater treatment strategy for the future. These BMNPs can be effectively applied for the treatment of metal-loaded effluents released from electroplating, tanning, and metallurgical industries. These NPs may also be utilized in water purification equipment, membranes, and filters used for the purification of potable water supply. BMNPs may also be explored for the design and fabrication of engineered nanobots for the removal of metallic as well as organic pollutants and oil spills.

References

Akhtar, M.S., Panwar, J., Yun, Y-Sang., 2013. Biogenic synthesis of metallic nanoparticles by plant extracts. ACS Sustain. Chem. Eng. 1, 591–602.

Arjaghi, S.K., Alasl, M.K., Sajjadi, N., Fataei, E., Rajaei, G.E., 2021. Green synthesis of iron oxide nanoparticles by RS lichen extract and its application in removing heavy metals of lead and cadmium. Biol. Trace Elem. Res. 199 (2), 763–768.

Asmathunisha, N., Kathiresan, K., 2013. A review on biosynthesis of nanoparticles by marine organisms. Colloids Surf. B 103, 283–287.

Balakrishnan, G.S., Rajendran, K., Kalirajan, J., 2020. Microbial synthesis of magnetite nanoparticles for arsenic removal. J. Appl. Biol. Biotechnol. 8 (03), 70–75.

Bharde, A., Wani, A., Shouche, Y., Joy, P.A., Prasad, B.L., Sastry, M., 2005. Bacterial aerobic synthesis of nanocrystalline magnetite. J. Am. Chem. Soc. 127 (26), 9326–9327.

Bharde, A.A., Parikh, R.Y., Baidakova, M., Jouen, S., Hannoyer, B., Enoki, T., Prasad, B.L.V., Shouche, Y.S., Ogale, S., Sastry, M., 2008. Bacteria-mediated precursor-dependent biosynthesis of superparamagnetic iron oxide and iron sulfide nanoparticles. Langmuir 24 (11), 5787–5794.

Casentini, B., Gallo, M., Baldi, F., 2019. Arsenate and arsenite removal from contaminated water by iron oxides nanoparticles formed inside a bacterial exopolysaccharide. J. Environ. Chem. Eng. 7 (1), 102908.

Castro, L., Blázquez, M.L., González, F., Muñoz, J.A., Ballester, A., 2018. Heavy metal adsorption using biogenic iron compounds. Hydrometallurgy 179, 44–51.

Chatterjee, S., Mahanty, S., Das, P., Chaudhuri, P., Das, S., 2020. Biofabrication of iron oxide nanoparticles using manglicolous fungus *Aspergillus niger* BSC-1 and removal of Cr (VI) from aqueous solution. Chem. Eng. J. 385, 123790.

Dahoumane, S.A., Mechouet, M., Wijesekera, K., Filipe, C.D., Sicard, C., Bazylinski, D.A., Jeffryes, C., 2017. Algae-mediated biosynthesis of inorganic nanomaterials as a promising route in nanobiotechnologya review. Green Chem. 19 (3), 552–587.

Das, M.P., Rebecca, L.J., 2018. Removal of lead (II) by phyto-inspired iron oxide nanoparticles. Nat. Environ. Pollut. Technol. 17 (2), 569–574.

Ehrampoush, M.H., Miria, M., Salmani, M.H., Mahvi, A.H., 2015. Cadmium removal from aqueous solution by green synthesis iron oxide nanoparticles with tangerine peel extract. J. Environ. Health Sci. Eng. 13 (1), 84.

El-Kassas, H.Y., Aly-Eldeen, M.A., Gharib, S.M., 2016. Green synthesis of iron oxide (Fe_3O_4) nanoparticles using two selected brown seaweeds: characterization and application for lead bioremediation. Acta Oceanol. Sin. 35 (8), 89–98.

Es'haghi, Z., Vafaeinezhad, F., Hooshmand, S., 2016. Green synthesis of magnetic iron nanoparticles coated by olive oil and verifying its efficiency in extraction of nickel from environmental samples via UV–vis spectrophotometry. Process Saf. Environ. Prot. 102, 403–409.

Fatimah, I., Pratiwi, E.Z., Wicaksono, W.P., 2020. Synthesis of magnetic nanoparticles using *Parkia speciosa* Hask pod extract and photocatalytic activity for Bromophenol blue degradation. Egypt. J. Aquat. Res., 46 (1), 35–40.

Gautam, P.K., Shivalkar, S., Banerjee, S., 2020. Synthesis of *M. oleifera* leaf extract capped magnetic nanoparticles for effective lead [Pb (II)] removal from solution: kinetics, isotherm and reusability study. J. Mol. Liq. 305 (69), 112811.

Gautam, P.K., Shivapriya, P.M., Banerjee, S., Sahoo, A.K., Samanta, S.K., 2020. Biogenic fabrication of iron nanoadsorbents from mixed waste biomass for aqueous phase removal of alizarin red S and tartrazine: Kinetics, isotherm, and thermodynamic investigation. Environ. Prog. Sustainable Energy 39 (2), e13326.

Gautam, P.K., Singh, A., Misra, K., Sahoo, A.K., Samanta, S.K., 2019. Synthesis and applications of biogenic nanomaterials in drinking and wastewater treatment. J. Environ. Manage. 231, 734–748.

Gautam, R.K., Gautam, P.K., Banerjee, S., Soni, S., Singh, S.K., Chattopadhyaya, M.C., 2015. Removal of Ni (II) by magnetic nanoparticles. J. Mol. Liq. 204, 60–69.

Gong, J., Chen, L., Zeng, G., Long, F., Deng, J., Niu, Q., He, X., 2012. Shellac-coated iron oxide nanoparticles for removal of cadmium (II) ions from aqueous solution. J. Environ. Sci. 24 (7), 1165–1173.

Gupta, V.K., Nayak, A., 2012. Cadmium removal and recovery from aqueous solutions by novel adsorbents prepared from orange peel and Fe_2O_3 nanoparticles. Chem. Eng. J. 180, 81–90.

Hamouda, R.A., Hussein, M.H., Abo-elmagd, R.A., Bawazir, S.S., 2019. Synthesis and biological characterization of silver nanoparticles derived from the cyanobacterium *Oscillatoria limnetica*. Sci. Rep. 9, 13071.

Hassan, D., Khalil, A.T., Saleem, J., Diallo, A., Khamlich, S., Shinwari, Z.K., Maaza, M., 2018. Biosynthesis of pure hematite phase magnetic iron oxide nanoparticles using floral extracts of *Callistemon viminalis* (bottlebrush): their physical properties and novel biological applications. Artific. Cells Nanomed. Biotechnol. 46 (sup1), 693–707.

Hennebel, T., Gusseme, B.D., Boon, Nico., Verstraete, W., 2008. Biogenic metals in advanced water treatment. Trends Biotechnol. 27, 90–98.

Herlekar, M., Barve, S., Kumar, R., 2014. Plant-mediated green synthesis of iron nanoparticles. J. Nanoparticles 2014, 140614.

Herrera-Becerra, R., Zorrilla, C., Ascencio, J.A., 2007. Production of iron oxide nanoparticles by a biosynthesis method: an environmentally friendly route. J. Phys. Chem. C 111 (44), 16147–16153.

Idrees, N., Tabassum, B., Abd_Allah, E.F., Hashem, A., Sarah, R., Hashim, M., 2018. Groundwater contamination with cadmium concentrations in some West UP Regions, India. Saudi J. Biol. Sci. 25 (7), 1365–1368.

Iravani, S., Zolfaghari, B., 2013. Green synthesis of silver nanoparticles using *Pinus eldarica* bark extract. Biomed Res. Int. 2013, 639725.

Jacob, J., Mukherjee, T., Kapoor, S., 2012. A simple approach for facile synthesis of Ag, anisotropic Au and bimetallic (Ag/Au) nanoparticles using cruciferous vegetable extracts. Mater. Sci. Eng. C 32 (7), 1827–1834.

Kandasamy, R., 2017. A novel single step synthesis and surface functionalization of iron oxide magnetic nanoparticles and thereof for the copper removal from pigment industry effluent. Sep. Purif. Technol. 188, 458–467.

Karri, R.R., Sahu, J.N., Meikap, B.C., 2020. Improving efficacy of Cr (VI) adsorption process on sustainable adsorbent derived from waste biomass (sugarcane bagasse) with help of ant colony optimization Ind. Crops Prod. 143, 111927.

Karri, R.R., Shams, S., Sahu, J.N., 2019. Overview of potential applications of nano-biotechnology in wastewater and effluent treatment. In: Ahsan, A., Ismail, A. (Eds.), Nanotechnology in Water and Wastewater Treatment,. Elsevier, Amsterdam, pp. 87–100.

Kataria, N., Garg, V.K., 2018. Green synthesis of Fe_3O_4 nanoparticles loaded sawdust carbon for cadmium (II) removal from water: regeneration and mechanism. Chemosphere 208, 818–828.

Khan, I., Saeed, K., Khan, Idrees., 2019. Nanoparticles: properties, applications and toxicities. Arab. J. Chem. 12, 908–931.

Khin, M.M., Nair, S.A., Babu, J.V., Murugan, R., Ramakrishna, S., 2012. A review on nanomaterials for environmental remediation. Energy Environ. Sci. 5, 8075–8109.

Khodadadi, B., Bordbar, M., Nasrollahzadeh, M., 2017. Green synthesis of Pd nanoparticles at Apricot kernel shell substrate using *Salvia hydrangea* extract: catalytic activity for reduction of organic dyes. J. Coll. Interf. Sci. 490, 1–10.

Kim, Y., Roh, Y., 2019. Environmental application of biogenic magnetite nanoparticles to remediate chromium (III/VI)-contaminated water. Minerals 9 (5), 260.

Kumari, M., Pandey, Shipra., Giri, V.P., Bhattacharya, A., Shukla, R., Mishra, A., Nautiyal, C.S., 2017. Tailoring shape and size of biogenic silver nanoparticles to enhance antimicrobial efficacy against MDR bacteria. Microb. Pathog. 105, 346–355.

Lagashetty, A., Ganiger, S.K., Preeti, R.K., Reddy, S., Pari, M., 2020. Microwave-assisted green synthesis, characterization and adsorption studies on metal oxide nanoparticles synthesized using *Ficus benghalensis* plant leaf extracts. New J. Chem. 44, 14095–14102.

Lin, Z., Weng, X., Owens, G., Chen, Z., 2020. Simultaneous removal of Pb (II) and rifampicin from wastewater by iron nanoparticles synthesized by a tea extract. J. Clean. Prod. 242, 118476.

Lingamdinne, L.P., Chang, Y.Y., Yang, J.K., Singh, J., Choi, E.H., Shiratani, M., Koduru, J.R., Attri, P., 2017. Biogenic reductive preparation of magnetic inverse spinel iron oxide nanoparticles for the adsorption removal of heavy metals. Chem. Eng. J. 307, 74–84.

Lingamdinne, L.P., Koduru, J.R., Chang, Y.Y., Karri, R.R., 2018. Process optimization and adsorption modeling of Pb (II) on nickel ferrite-reduced graphene oxide nano-composite. J. Mol. Liq. 250, 202–211.

Lingamdinne, L.P., Koduru, J.R., Karri, R.R., 2019. A comprehensive review of applications of magnetic graphene oxide based nanocomposites for sustainable water purification. J. Environ. Manage. 231, 622–634.

Lingamdinne, L.P., Vemula, K.R., Chang, Y.Y., Yang, J.K., Karri, R.R., Koduru, J.R., 2020. Process optimization and modeling of lead removal using iron oxide nanocomposites generated from bio-waste mass. Chemosphere 243, 125257.

Lung, I., Stan, M., Opris, O., Soran, M.L., Senila, M., Stefan, M., 2018. Removal of lead (II), cadmium (II), and arsenic (III) from aqueous solution using magnetite nanoparticles prepared by green synthesis with Box–Behnken design. Anal. Lett. 51 (16), 2519–2531.

Lunge, S., Singh, S., Sinha, A., 2014. Magnetic iron oxide (Fe_3O_4) nanoparticles from tea waste for arsenic removal. J. Magn. Magn. Mater. 356, 21–31.

Mahanty, S., Bakshi, M., Ghosh, S., Chatterjee, S., Bhattacharyya, S., Das, P., Das, S., Chaudhuri, P., 2019. Green synthesis of iron oxide nanoparticles mediated by filamentous fungi isolated from Sundarban Mangrove ecosystem, India. Bio. Nano. Sci. 9 (3), 637–651.

Mashjoor, S., Yousefzadi, M., Zolgharnain, H., Kamrani, E., Alishahi, M., 2018. Organic and inorganic nano-Fe_3O_4: alga ulva flexuosa-based synthesis, antimicrobial effects and acute toxicity to briny water rotifer *Brachionus rotundiformis*. Environ. Pollut. 237, 50–64.

Mazumdar, H., Haloi, N., 2017. A study on biosynthesis of iron nanoparticles by *Pleurotus* sp. J. Microbiol. Biotechnol. Res. 1 (3), 39–49.

Mukherjee, D., Ghosh, S., Majumdar, S., Annapurna, K., 2016. Green synthesis of α-Fe_2O_3 nanoparticles for arsenic (V) remediation with a novel aspect for sludge management. J. Environ. Chem. Eng. 4 (1), 639–650.

Nikić, J., Tubić, A., Watson, M., Maletić, S., Šolić, M., Majkić, T., Agbaba, J., 2019. Arsenic removal from water by green synthesized magnetic nanoparticles. Water 11 (12), 2520.

Nithya, K., Sathish, A., Kumar, P.S., Ramachandran, T., 2018. Fast kinetics and high adsorption capacity of green extract capped superparamagnetic iron oxide nanoparticles for the adsorption of Ni (II) ions. J. Ind. Eng. Chem. 59, 230–241.

Önal, E.S., Yatkın, T., Aslanov, T., Ergüt, M., Özer, A., 2019. Biosynthesis and characterization of iron nanoparticles for effective adsorption of Cr (VI). Int. J. Chem. Eng. 2019,, 1 13.

Patra, J.K., Baek, K.H., 2017. Green biosynthesis of magnetic iron oxide (Fe_3O_4) nanoparticles using the aqueous extracts of food processing wastes under photo-catalyzed condition and investigation of their antimicrobial and antioxidant activity. J. Photochem. Photobiol. B 173, 291–300.

Pavani, K.V., Kumar, N.S., 2013. Adsorption of iron and synthesis of iron nanoparticles by *Aspergillus* species kvp 12. Am. J. Nanomater. 1 (2), 24–26.

Perez-Gonzalez, T., Jimenez-Lopez, C., Neal, A.L., Rull-Perez, F., Rodriguez-Navarro, A., Fernandez-Vivas, A., Iañez-Pareja, E., 2010. Magnetite biomineralization induced by *Shewanellaoneidensis*. Geochim. Cosmochim. Acta 74 (3), 967–979.

Rao, A., Bankar, A., Kumar, A.R., Gosavi, S., Zinjarde, S., 2013. Removal of hexavalent chromium ions by *Yarrowiali polytica* cells modified with phyto-inspired Fe0/Fe_3O_4 nanoparticles. J. Contam. Hydrol. 146, 63–73.
Rather, M.Y., Sundarapandian, S., 2020. Magnetic iron oxide nanorod synthesis by *Wedeli aurticifolia* (Blume) DC. leaf extract for methylene blue dye degradation. Appl. Nanosci. 10 (7), 2219–2227.
Roy, P., Saha, A., 2002. Metabolism and toxicity of arsenic: a human carcinogen. Curr. Sci. 82 (1), 38–45.
Sadhasivam, S., Vinayagam, V., Balasubramaniyan, M., 2020. Recent advancement in biogenic synthesis of iron nanoparticles. J. Mol. Struct. 1217, 128372.
Sahu, J.N., Zabed, H., Karri, R.R., Shams, S., Qi, X., 2019. Applications of nano-biotechnology for sustainable water purification. In: Thomas, S., Grohens, Y., Pottathara, Y.B. (Eds.), Industrial Applications of Nanomaterials,. Elsevier, Amsterdam, pp. 313–340.
Sebastian, A., Nangia, A., Prasad, M.N.V., 2018. A green synthetic route to phenolics fabricated magnetite nanoparticles from coconut husk extract: implications to treat metal contaminated water and heavy metal stress in *Oryza sativa* L. J. Cleaner Prod. 174, 355–366.
Shamaila, S., Sajjad, A.K.L., Ryma, N., Farooqi, S.A., Jabeen, N., Majeed, S., Farooq, I., 2016. Advancements in nanoparticle fabrication by hazard free eco-friendly green routes. Appl. Mater. Today 5, 150–199.
Sheny, D.S., Mathew, J., Philip, D., 2011. Phytosynthesis of Au, Ag and Au–Ag bimetallic nanoparticles using aqueous extract and dried leaf of *Anacardium occidentale*. Spectrochim. Acta Part A 79 (1), 254–262.
Siji, S., Njana, J., Amrita, P.J., Vishnudasan, D., 2018. Biogenic synthesis of iron oxide nanoparticles from marine algae. TKM Int. J. Multidiscip. Res. 1 (1), 1–7.
Singh, A., Gautam, P.K., Verma, A., Singh, V., Shivapriya, P.M., Shivalkar, S., Sahoo, A.K., Samanta, S.K., 2020. Green synthesis of metallic nanoparticles as effective alternatives to treat antibiotics resistant bacterial infections: a review. Biotechnol. Rep. 25, e00427.
Subramaniyam, V., Subashchandrabose, S.R., Thavamani, P., Megharaj, M., Chen, Z., Naidu, R., 2015. *Chlorococcum* sp. MM11—a novel phyco-nanofactory for the synthesis of iron nanoparticles. J. Appl. Phycol. 27 (5), 1861–1869.
Taghipour, S., Hossseini, S.M, Ashtiani, B., 2019. Engineering nanomaterials for water and wastewater treatment: review of classifications, properties and applications. New J. Chem. 43, 7902–7927.
Tamez, C., Hernandez, R., Parsons, J.G., 2016. Removal of Cu (II) and Pb (II) from aqueous solution using engineered iron oxide nanoparticles. Microchem. J. 125, 97–104.
Tang, S.C., Lo, I.M., 2013. Magnetic nanoparticles: essential factors for sustainable environmental applications. Water Res. 47 (8), 2613–2632.
Tripathi, R.M., Chung, S.J., 2019. Biogenic nanomaterials: synthesis, characterization, growth mechanism, and biomedical applications. J. Micriobio. Methods 157, 65–80.
Venkateswarlu, S., Kumar, B.N., Prathima, B., SubbaRao, Y., Jyothi, N.V.V., 2014. A novel green synthesis of Fe_3O_4 magnetic nanorods using *Punica granatum* rind extract and its application for removal of Pb (II) from aqueous environment. Arab. J. Chem. 12 (4), 588–596.
Venkateswarlu, S., Yoon, M., 2015. Surfactant-free green synthesis of Fe_3O_4 nanoparticles capped with 3, 4-dihydroxyphenethylcarbamodithioate: stable recyclable magnetic nanoparticles for the rapid and efficient removal of Hg (II) ions from water. Dalton Trans. 44 (42), 18427–18437.
Vishnu, D., Dhandapani, B., 2020. Integration of Cynodon dactylon and *Muraya koenigii* plant extracts in amino-functionalised silica-coated magnetic nanoparticle as an effective sorbent for the removal of chromium (VI) metal pollutants. IET Nanobiotechnol. 14 (6), 449–456.
Wagner, S., Gondikas, A., Neubauer, E., Hofmann, T., Kammer, F., 2014. Spot the difference: engineered and natural nanoparticles in the environment—release, behavior, and fate. Angew. Chem. Int. Ed. 53, 12398–12419.
Yadav, V.K., Fulekar, M.H., 2018. Biogenic synthesis of maghemite nanoparticles (γ-Fe_2O_3) using Tridax leaf extract and its application for removal of fly ash heavy metals (Pb, Cd). Mater. Today Proc. 5 (9), 20704–20710.
Yew, Y.P., Shameli, K., Miyake, M., Kuwano, N., Khairudin, N.B.B.A., Mohamad, S.E.B., Lee, K.X., 2016. Green synthesis of magnetite (Fe_3O_4) nanoparticles using seaweed (*Kappaphycus alvarezii*) extract. Nanoscale Res. Lett. 11 (1), 1–7.

CHAPTER 8

Nanoscale texture characterization of green nanoparticles and their hybrids

Manoj Tripathi[a], Krishna Kumar Pandey[b] and Bharti Singh[c]
[a]Department of Physics and Materials Science and Engineering, Jaypee Institute of Information Technology, Noida, Uttar Pradesh, India
[b]Department of Physics, School of Basic Sciences and Research, Sharda University, Greater Noida, Uttar Pradesh,-India
[c]Department of Electronics Engineering, Ramswaroop Memorial Group of Professional Colleges, Lucknow, Uttar Pradesh, India

8.1 Introduction

Nanotechnology is one of the most growing fields for research with enormous applications in the real world. Green synthesis and green technology are branches of sustainable nanotechnology. Sustainability can be understood as a good correlation between the development and needs of the present as well as future generations without any compromise. Our scientific developments, future-oriented vision, and immense scientific achievements can lead us to visualize sustainable developments. Developments in nanotechnology toward green innovation lead us to green nanotechnology. Green nanotechnology refers to the use of nanotechnology to enhance the environmental sustainability of processes producing negative externalities. It refers to the use of the products of nanotechnology to enhance sustainability.

Nanotechnology provides decisive information about shape, characterization, design, structure of a material at the nanoscale. Hence, nanotechnology is subjected to reveal the structural features of a material that has a size of less than 100 nm. Such materials are said to be nanostructures, nanocomposites, and nanoparticle (NP) aggregates (Koduru et al., 2019; Sahu et al., 2019). The magnetic, electrical, optical properties, as well as surface activities and size-dependent functions of nanomaterials, make them different from other materials (Bhainsa & D'Souza, 2006; Guilger-Casagrande & Lima, 2019). During the last decades, for applications in medicine and pharmaceuticals, nanomaterials have been used for preparing NPs. These intelligent NPs have been used in cancer detection and its treatment, battery electrodes, carbon nanotubes, antimicrobial materials as well as in the food and textile industries (Gusseme, 2010; Sawle et al., 2008). On the other hand, the synthetic nanomaterials, whose production and applications were considered to very expensive and unfriendly to the environment, have a lot more applications and benefits (Lim et al., 2009). Because of the risks associated with side effects, the use of synthetic

Sustainable Nanotechnology for Environmental Remediation
DOI: https://doi.org/10.1016/B978-0-12-824547-7.00009-6

nanostructures in medicine is extremely limited. However, green synthesis is being used by scientists to avoid the side effects of such materials. Hence, green nanotechnology basically is the utilization of applications of nanotechnology for enhancing sustainability. Green technology is a set of new ideas developed by scientists for decreasing the risk and side effects to the environment as well as human health triggered by using the nanomaterials. Green technology also steps forward to replace existing products with nanoproducts that are more environment friendly (Dameron et al., 1989; Nair & Pradeep, 2002).

Along with several benefits, nanotechnology provides several hazards. The NPs are easily absorbed by the living systems and so they give rise to some byproducts that are risky. A developing branch "eco-nanotoxicology" addresses the effects of NPs on the earth as well as on its organisms (Corradini et al., 2010; Johnston et al., 2013). Green nanotechnology also addresses the issues related to shortage of global water and drinking water. Other green technology applications are addressing water treatment, renewable energy technologies, waste, environmental remediation, and sustainable production. The new challenges being faced by environmental science/engineering along with nanotechnology have envisioned a new branch known as green nanotechnology. The present chapter is focused on green nanotechnology, green synthesis, and their applications in human life.

Science and technology are complimentary to each other these days. There are a lot more drastic changes being witnessed by environmental sciences. In the same way, nanotechnology and green nanotechnology are achieving new millstones every day. It is demand of present era that the nanotechnology shall be redefined with scientific thoroughness. However, to maintain environmental standards and ecological diversity, the scientific community is in need of developing new vision toward newer innovations. In present time, when sustainable development is the uppermost requirement of the time, nanotechnology can help in overcoming most of the sustainability issues. Green nanotechnology includes the production as well as processing of nanomaterials, green chemistry, green engineering, and environmental applications. In other words, we can say that green nanotechnology is the utilization of nanotechnology for enhancing environmental sustainability (Fig. 8.1).

The upmost aim of producing green nanomaterials is to engage natural sources for solving problems related to environmental science. For example, NPs, which are synthesized magnetite, are being applied for removing toxic chlorinated organic solvents as well as arsenic from water (Makarov et al., 2014). Green nanotechnology can be very much useful in addressing environmental problems for example stopping use of nonrenewable resources, reducing the production of greenhouse gases, cutting down the utilization of fossil fuels, and by reducing the pollution caused by fuel combustion. One of the very important aspects of nanotechnology is the production of nanomembranes and nanocatalysts. The application of nanomembranes and nanocatalysts is to separate products from their matrix and to increase the efficiency of products, respectively

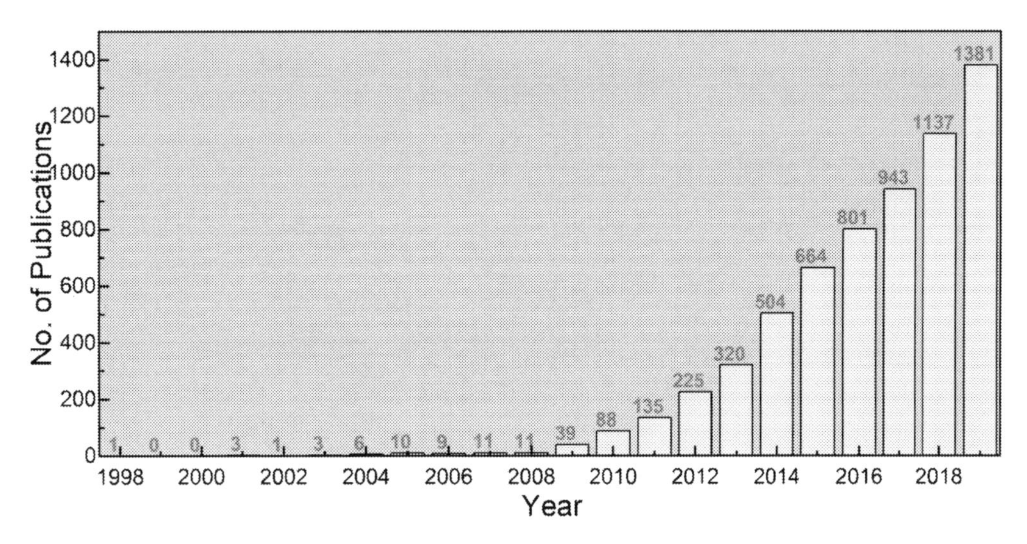

Figure 8.1 Progress of synthesis of nanostructures during 1998–2018.

(Mohammadinejad et al., 2015). Hence, by green nanotechnology, the synthetic nanotechnology can be efficiently replaced and several environmental issues can be addressed by producing nanomaterials and nanoproducts without harming human kind and the environment.

There is lot of literature available presenting the considerable pro effects and benefits of green science and green nanotechnology for example the use of green nanotechnology has substantially reduced the hazardous waste released to the environment by incorporating several research disciplines including investigations of green solvents, biomaterials, nanocatalysts and hazardous materials, etc. Sustainability, being a global concern, is correlated to global climate change and so the sustainable development has been promoted in recent technological systems. In present time, nanotechnology is capable of influencing and improving the sustainability of all industrial sectors.

Fig. 8.2 provides a detailed description of the different techniques used for the nanomaterial synthesis. These techniques may broadly be divided into three categories namely physical techniques, chemical techniques. The techniques that are neither purely physical nor chemical can be grouped in a separate group. Among these techniques, chemical route for the synthesis of nanomaterial is more popular. Especially the sol-technique is used by many researchers. It is not only bit easy to perform, works effectively, and does not require highly sophisticated instrumentation.

8.2 Factors affecting the characteristics of green nanomaterials

The characteristics of the green NPs depend upon the technique through which it is synthesized. The process involved during the synthesis, the chemical reagents used, the

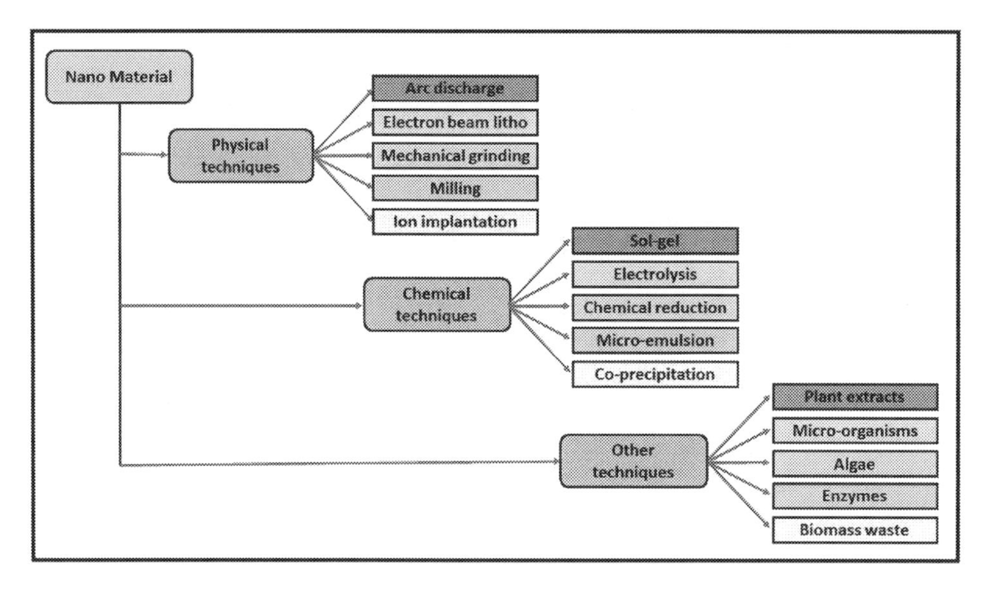

Figure 8.2 Classification of nanomaterial synthesis techniques.

solvent properties, its pH, temperature, pressure, and other parameters significantly affect the NP characteristics. A brief description of these parameters that affect the nanomaterial characteristics has been discussed in the following section.

8.2.1 pH of solvent

pH of the medium used for the synthesis of the NP is seen to alter the characteristics of the green nanocomposite. In a study conducted by Armendariz et al., it was found that the binding trend of oat (Avena sativa) adsorption capacity was highly dependent upon the pH of the medium in which the experiment was carried out. Oat sample in the study was used to reduce Au(III) in a medium of different pH ranging from 2 to 6. Studies showed the most absorption occurred with pH 3 (Armendariz et al., 2004). Another study performed by Namita and Soam showed that the geometry of the silver NPs can be altered by the pH of the solution used for the synthesis of the NPs (Namita & Soam, 2012). Similar results of pH affecting the characteristics are reported in few other articles available in the literature (Gamez, 2001; Gardea-Torresdey et al., 2000). All these studies confirm that while synthesizing a nanomaterial, the pH of the solution in which the reactions are performed is really significant.

8.2.2 Reaction temperature

Temperature is another important factor that influences the characteristics of the synthesized NP. Different synthesis techniques require different temperatures for the nanomaterial synthesis. Changing the temperature leads to a change in the characteristics of the synthesized material (Latif et al., 2018). For example, synthesis of nanomaterial through biological process generally is performed somewhere around 100 ^{0}C and very high temperature is not required in the nanomaterial synthesis through the biological route. A very high temperature may lead to the breaking of weaker bonds and formation of few newer bonds thus changing the whole composition of the newly developed material. The similar effects may be seen the chemical route where the temperature requirement (less than 350 ^{0}C) is little bit higher compared to the biological route. Here also higher temperature may cause the secondary reactions resulting in the formation of nanomaterials with modified properties (Patra & Baek, 2014; Rai et al., 2006).

8.2.3 Pressure during synthesis

Pressure is another physical parameter that significantly affects the characteristics of the synthesized nanomaterials. However, there are very few studies discussing the effects of pressure on the characteristics of the nano-sized material. On applying the pressure, particle size and shape gets altered (Abhilash & Pandey, 2012). It is a well-known fact the change in particle size and shape can affect the properties of the nanomaterial. A change in size of nanomaterial produces a changes the surface area of the material. It further can affect the porosity and pore volume of the nanomaterial also. In a study, Quang et al. observed that the reduction rate of the biological ions was affected by the pressure and it was found to be highest at the ambient pressure (Huy et al., 2013). Excess pressure on the nanomaterial can create the stress in the layers of the nanomaterial. This stress can break the layers and may lead to the deformation in the crystal structure and thus electrical, optical, and mechanical properties can get affected significantly.

8.2.4 Reaction time

The characteristics of the synthesized nanomaterials can be influenced by the time used for the incubation of the reaction. By changing the reaction time not only the chemical properties but the physical properties can also be altered. Darroudi et al. studied the effect of reaction time on the formation of silver NPs. The study revealed that the UV adsorption was highly influenced with the reaction time. As the reaction time was increased the UV adsorption also increased (Darroudi et al., 2011). Longer reaction time allows the slow reactions also to get completed and thus can change the chemical behavior. Moreover, the aggregation of the NPs may also be caused by a longer reaction time. These aggregation and agglomeration may affect the morphology of the nanomaterial (R., 2011). It may sometimes cause the change in the grain structure.

8.2.5 Other factors

Other than these factors the characteristics of the synthesized nanomaterial also depend upon the particle size, the catalyst, method adopted for the synthesis, etc. In the biological route, the plant, microorganism, or the biomass waste type influences the characteristics of nanomaterial. The moisture content of the precursor, its chemical composition, reactivity, etc. are some factors that affect the properties of synthesized nanomaterial. Few reports have claimed that by changing the adsorbate and activity of the catalyst morphological and surface characteristics of the nano materil can be altered significantly (Ajayan, 2004; Somorjai & Y. Park, 2008). Kuchibhatla et al. reported that on changing the environment from wet to dry, the crystalline nature of zinc sulfide (ZnS) changed suddenly. Another study conducted by Kuchibhatla et al. showed that the chemical behavior of cerium nitrate changes in the presence of peroxide (Kuchibhatla et al., 2012).

8.3 Characterization techniques of nanomaterials

Nanomaterials are very unique materials that exhibit the properties that are quite different from the bulk counterpart. The thermal, electrical, structural, mechanical, and other behavior of the nano-sized materials are different compared to their bulk sized materials. Therefore, it is highly important to characterize these nanomaterials prior to using it for some applications. The characteristics of the nanomaterials such as composition, particle size and shape, morphology, surface area, rheology, thermal conductivity, thermal diffusivity, heat capacity, electrical conductivity, mechanical strength, etc. should be known for an effective and efficient utilization of the nanomaterial. These characteristics of the nanomaterials can be studied and analyzed with different characterization techniques. The following sections will provide a brief description of few important characterization techniques. Table 8.1 gives a description of these techniques and the properties that can be determined of a nanomaterial. Please note that the properties that can be determined from these characterization techniques are not limited to the description provide in Table 8.1.

8.4 Particle size and shape

Particle size is one of the primary concerns while developing an NP. Particle size and its distribution can alter the material property up to a very high extent. Many physical and chemical properties of the material are highly sensitive to the particle size. During the particle size determination it is customary to evaluate the particle shape, agglomeration (if any), and the extent of particle agglomeration. Crystal structure and lattice parameters are also important to properties that should be known prior to use of the NP.

X-ray diffraction (XRD) is the most commonly used technique for the study of crystal structure of any particle. X-rays of known wavelength are allowed to fall on the sample and the intensity of reflected beam is noted as a function of angle. The X-ray scattered from different planes of the sample is scattered at different angles. With the help of the Scherrer equation corresponding to the most intense peak and its broadening the lattice parameters can be determined. Moreover, by noting the positions of the peak, a graph

Table 8.1 Few important characterization techniques used for the characterization of the nanomaterials.

Characterization technique (Abbreviated form)	Characterization technique (Expanded form)	Properties that can be determined
XRD	X-ray diffraction	Crystal structure, grain size, lattice parameter, elemental analysis
XPS	X-ray photoelectron spectroscopy	Elemental composition, impurity detection, compositional analysis
FT-IR	Fourier transform infra-red	Chemical bonding, functional groups present, nature of bonds
NMR	Nuclear magnetic resonance	Structure of organic molecules, electronic configuration, atomic structure,
BET surface area	Brunauer–Emmett–Teller surface area	Adsorption–desorption curve, surface area, porosity, pore size
SEM	Scanning electron microscopy	Morphology, grain boundary, distribution of particle, particle shape
TGA	Thermogravimetric analysis	Thermal stability, thermal decomposition rate, volatile component, moisture content
DSC	Differential scanning microscopy	Heat capacity, latent heat, thermal conductivity, phase transition temperature
Electron tomography	-	Crystal structure, lattice parameter, 3D structure of nanomaterials
AFM	Atomic force microscopy	Adhesion strength, hydro affinity, Topographic imaging
DLS	Dynamic light scattering	Hydro dynamical size, particle agglomeration, size distribution
VSM	Vibrating scan magnetometry	Magnetic moment, retentivity, coercivity, magnetization

Note: Properties determined from these characterization techniques are not limited to the description provide here.

between "intensity vs 2θ" can be plotted. One of the major advantages of the XRD is that the powdered samples of the NPs can be analyzed. These powdered samples analyzed at 360^0 represent a statistical average of all the particles. Instrumental broadening and lattice strain of grains are the two main reasons for the peak broadening (Upadhyay et al., 2016). The actual size of the crystal is smaller than the predicted by the Scherrer formula. The difference in these sizes is approximately 50–55 nm but it depends on the instrument and the sample itself. X-ray crystallography has helped in understanding the crystal structure in a better way and most of the times it is the most trusted characterization tool for the determination of size and shape of the sample.

8.5 Morphological characterization

Scanning electron microscopy (SEM) is one of the most widely used techniques to study the morphology of a given sample. The technique is based on developing an image of the sample with the help of the electron scattered from the sample. The SEM provides a 3D image of the sample that talks a lot about it. In SEM, the secondary electrons are used for the imagining purpose. The energy of these secondary electrons is often very less. It is typically of the order of 50 eV. Because of the low energy, only those secondary electrons can come out which have been originated at surface or in a very close layer from the top surface (Golding, 2016). In a morphological analysis, physical shape and size of the nanomaterials are studied. Moreover, the analysis of any crack, ridges, or any other type of dislocation in structure is also observed in the morphological analysis. Surface morphologies can be studied with the help of SEM. However, a background noise sometimes creates a problem but it can be countered using proper tools and techniques. The SEM image not only provides information about the surface morphologies but the particle size and shape can also be estimated. It also informs about the grains and their structures and the size. With the help of SEM one also can note the agglomeration of the particles at some places. The shape of the particle is also clearly visible in the SEM image, which also can be useful for many applications (Freyre-Fonseca et al., 2016).

Properties of the material are highly dependent on its size. The properties of the nanomaterials are very sensitive to the particle size. Even a small change in the size offers a drastic deviation in the characteristics of the nanomaterial. Particle size distribution of a nanomaterial is a less discussed area and is generally not discussed in research articles. Very few studies have been observed that discuss the particle size variation for the nanomaterial used in their respective study. The reason being that in most of the cases the nanomaterial is purchased from a supplier and the particle size is fixed with a very little variation in particle size. Particle size variation is difficult to measure for the NPs as the conventional techniques used for the bulk samples cannot be used for these nano-sized particles.

Particle size distribution however can be estimated with the help of SEM. With the help of SEM images the size of the different particles can be observed.

8.6 Thermal characterization

Thermal characterization of nanomaterials is useful to get the information regarding the glass transition temperature, heat capacity, and thermal stability. Thermo-gravimetric analysis (TGA), differential scanning calorimetry (DSC), differential thermos grapery are few basic characterization techniques employed to perform the thermal analysis. Thermal analysis of any sample is must and in most cases especially the materials that needs to undergo heating action (due to flow of current, chemical reaction, electrolysis, etc.) (Bannov et al., 2020; Glasscott et al., 2019).

TGA is the most common method to perform the thermal analysis. In this method, the sample is mounted on a sample holder under the influence of liquid nitrogen. A small heat pulse is provided to the sample and the heating rate is made constant. It produces a rise in temperature and the sample starts evaporating and the mass of the sample lowers. The residual mass at different time and temperature is recorded for the analysis. TGA can be used to obtain the moisture content in a sample and is widely used to estimate the ash present especially in biomass-based materials. Thermal stability of the material can be determined by noting down the temperature at a definite residue remaining. Thermophysical properties of the sample can be estimated by performing the DSC test. In this test, a fixed small heat pulse is given to the ample. This small heat pulse offers a localized temperature change, and localized temperature rise is monitored. DSC can measure the heat capacity of the sample directly and by some indirect methods thermal conductivity and thermal diffusivity can also be determined (Dey & Tripathi, 2010; Nam et al., 2018).

8.7 Elemental analysis

X-ray photoelectron spectroscopy (XPS) is one of the most commonly used techniques for the elemental analysis. The technique can be applied on a wide range of materials and is capable of extracting more quantitative and qualitative information such as its elemental composition, chemical state, etc. The technique is based on the photoelectric effect. The kinetic energy of the scattered electron is the characteristic of the element through which it has ejected. By noting the kinetic energy of the scattered electron the identity of the element can be known. XPS technique is a suitable technique for the surface studies also. In XPS, the analysis of the surface at a depth of somewhere between 5 and 10 nm is performed. Information about the spatial distribution can be obtained by scanning the sample with the X-ray beam. XPS can also provide information of surface layers and the structure of thin films. This information is very important and is required for many practical applications where surface or thin film composition is crucial (Baer & Thevuthasan, 2010; Mather, 2009).

XRD can also be helpful to determine the elements or compounds present in the sample. The "intensity vs 2θ" curve of XRD shows peak on intensity axis at different 2θ values. These peaks are a kind of fingerprint of the elements and compounds. The existence of peak at a particular 2θ value advocates the presence of a particular element or compound in sample.

8.8 Surface area of nanomaterials

Surface area of the material is one of the prime properties of the material especially if it is going to be used in the drug delivery, energy storage, pollutant removal, soil procurement, and many more applications. The large surface area of the material is always found to be suitable for all these abovementioned applications. Surface area determinations of materials are mostly based on the adsorption of some inert gas (generally N_2) over the sample. The amount of the adsorbed gas is a measure of the surface area of the material. The inert gas is allowed to pass through the sample material under varying pressure conditions and allowed to form a monolayer over the surface of the material. Brunauer–Emmett–Teller (BET) equation, which requires the number of gas molecules required and the cross-section area of a molecule, is used to determine the surface area of the sample. The BET equation (Eq. 8.3) measures the BET surface area of the sample and the adsorption-desorption curve can also determine the pore volume and micro- and mesopores available in the sample (Mubarak et al., 2014; Nizamuddin et al., 2016).

$$\frac{1}{V\left[\left(\frac{P_o}{P}\right)-1\right]}=\frac{c-1}{V_m c}\left(\frac{P}{P_o}\right)+\frac{1}{V_m c} \tag{8.1}$$

where

V is the adsorbed gas quantity,

V_m is the monolayer gas adsorbed quantity,

P is the equilibrium gas pressure of adsorbate,

P_o is the saturation pressure of adsorbate,

c is a constant given by $c = e^{\frac{E_1 - E_L}{RT}}$.

E_1 is heat of adsorption of first layer,

E_L is heat of adsorption of second layer.

A plot of $\frac{1}{V\left[\left(\frac{P_o}{P}\right)-1\right]}$ against the value $\left(\frac{P}{P_o}\right)$ is called the adsorption isotherm and is used to estimate the BET surface area. The curve mostly maintains a linearity between the region of $0.05 < \left(\frac{P}{P_o}\right) < 0.35$. The slope (S) and the intercept (I) of this plot are determined within the linear range. The volume of inert gas adsorbed in monolayer and constant c can be calculated by Equations (8.2) and (8.3).

$$V_m = \frac{1}{A+I} \tag{8.2}$$

$$V_m = \frac{1}{A+I} \tag{8.3}$$

Once the values of V_m and c are known, total surface area and the specific surface area can finally be calculated using the Equations (8.4) and (8.5).

$$S_{Total} = \frac{V_m N s}{V} \tag{8.4}$$

$$S_{BET} = \frac{S_{Total}}{a} \tag{8.5}$$

where

N is Avogadro number,

s is adsorption cross section of adsorbing material,

V is molar volume of adsorbate gas,

a is mass of adsorbent.

This method is widely used to determine the surface area of the solid materials. The information of surface area is important for many applications including construction, purification, agricultural, energy storage, etc. For example, the cement quality and the cement highly depend on the surface area and porous materials used as ingredients. Changing the surface area of the components used in cement can significantly affect the hardening and curing of the concrete. Similarly, in water purification activated carbon with large BET surface area is highly important. If the surface area of the activated carbon is lesser, then the pollutant removal will be smaller.

8.9 Electrical conductivity of nanomaterials

Electrical conductivity of the nanomaterials is highly important especially when this is used in fuel cells and electro catalytic applications. In electro-catalysis process, the electrical conductivity of the nanomaterial and that of the underlying substrate is crucial. Conventional electrical conductivity methods used for the bulk materials may be employed to measure the electrical conductivity of nano-sized materials (Glasscott et al., 2019). However, for these materials synthesized at atomic scale, some modified and improved testing techniques are also necessary. The small size of nanomaterials makes it completely different from the bulk-sized materials and it may bring the quantum mechanics into the picture. Because of the smaller size, the interaction and the bonding in

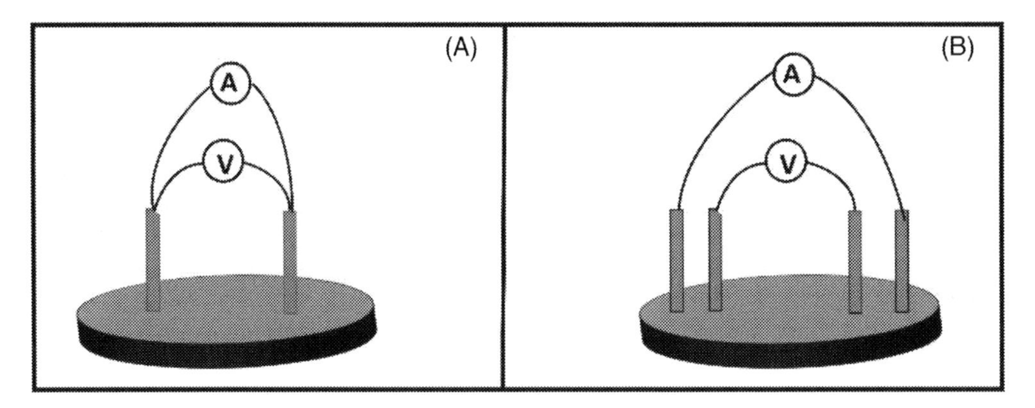

Figure 8.3 Schematic of (A) two-probe method and (B) four-probe method for the determination of electrical conductivity.

the nanomaterials are quite different compared to that of the bulk materials (Zienkiewicz-Strzałka, 2019). With the reduction in size, the energy bands are shifted and thus the band gaps between the energy levels are not same as in the bulk material.

Density of states of any materials can be directly linked with the help of the electrical conductivity of material. Earlier scanning tunnel microscope was used to determine it. Two-probe and four-probe methods are used to measure the electrical conductivity. In two-probe method, a small current is allowed to pass through the nano-sized material. The voltage developed across the material is noted, which may further be used to calculate resistivity of material using ohm's law. Once the resistivity is know the determination of conductivity can be made without much problem. It appears to be quite simple, effective, and nonerroneous technique to measure the electrical conductivity of the sample. However, once we use this method some practical problems arise which lead to the inaccuracy in the measurement of the resistivity. For example in this method the same probes are used to supply the current (i.e., used as source) and for the measurement of developed voltage (i.e., sensing response). Because of this some extra resistance of the voltage-measuring device is added up and the displayed voltage across the sample becomes inaccurate. In nanomaterials even a very small deviation from the actual value may result in larger inaccuracy on the conductivity value so to avoid it four-probe method is used. In the four-probe method, two probes are used to deliver the current to the specimen and two other probes measure the developed voltage. Here two separate source and sensing probes reduce the added resistance up to a significant level. The removal of contact resistance improves the accuracy of measurement and thus the accuracy in the determination of the electrical conductivity increases. Fig. 8.3 offers a schematic description of the two-probe and four-probe methods for the determination of electrical conductivity of a material. Note that in four-probe method, the current is supplied from the outer probes, awhile the developed voltage is recorded through the inner probes.

8.10 Characterization of magnetic nanomaterials

Magnetic nanomaterials are highly important materials and have a wide range of applications. Other than the conventional applications these magnetic nanomaterials have a lot of applications in biomedical engineering, purification industry, heavy metal removal, etc. This is the reason in last two decades the magnetic nanomaterials have been studied a lot and have enjoyed the attention of many researchers. These magnetic nanomaterials are synthesized and characterized to determine the magnetic behavior. The properties of the magnetic nanomaterials depend upon the method adopted to synthesize it. Basic magnetic properties generally of interest are the coercivity, retentivity, and magnetization of the material.

Direct current magnetometry is one of the most commonly used techniques to study the magnetic properties of a material. In this technique, the sample is kept in a constant field. Lorentz force displaces the sample physically and the displacement is recorded. This displacement can further be translated into the magnetic moment of the sample. VSM (vibrating scanning microscopy) is based on this direct current magnetometry technique. VSM determines the hysteresis curve for the sample under investigation. This hysteresis curve can further be analyzed to determine the required quantities such as coercivity, retentivity, and saturation magnetic field (Sandler et al., 2019).

Recently a new technique has been developed AGM (alternating gradient magnetometer) that can determine these properties relatively fast and with more accuracy. In this technique the magnetic sample is kept in a static magnetic field. Small varying-current coils are used to slightly modify this static magnetic field. This actually produces a magnetic field gradient and thus a varying force on the magnetic sample. The sample under this force is displaced from its original position. A piezo element is attached with the sample holder and can note the displacement with a very high accuracy. AGM can determine the magnetic properties more precisely compared to the VSM and has the advantage that it can be coupled with other devices such as superconducting quantum interference device (SQUID), etc. However, this AGM technique cannot be used to study the magnetic properties of the soft materials (Mourdikoudis et al., 2018).

Hall effect is another method that can determine the magnetic properties of the magnetic material. This method is popular because of its simplicity and ease in performing the experiment. The small probe size in this experiment makes it suitable to be used in many devices easily (Saima et al., 2019; Zhang et al., 2019). This method can determine the magnetic properties of a large variety of magnetic materials even with materials that do have very complex geometries. However, one issue with this method is that the calibration of the probe needs to be done for each sample. It makes the measurement a bit lengthy. The Hall effect method can work at low temperatures also which is its another advantage but the probes are sensitive to temperature change so the temperature needs to be stabilized during the experiment. A small variation in the temperature may result a large inaccuracy in the final result.

8.11 Applications of nanomaterials

8.11.1 In purification

Nanomaterials are widely used in the purification industry. The nano-sized carbon that is also porous in nature can be used to selectively remove the heavy metals such as Hg, Pb, As, Sb, etc., from the water (Tripathi et al., 2016, 2017). Carbon nanotubes are very effective in removing the chemicals such as phenol or dye from a chemical solution. This capability of the removal of selective participants from water or any other chemical is highly useful for the water purification, chemical industry, and agricultural industry as well. Nanofibers are derived from various agricultural products such as coconut shell, palm shell, wheat, rice husk, etc. (Thines et al., 2016). These biomass waste–derived char has shown a great improvement in the pollutant removal capability if the nano-sized activated carbon is used in place of micro-sized activated carbon (Dehghani et al., 2020; Karri et al., 2017; Karri & Sahu, 2018). The use of magnetic biochar with nano-sized carbon nanotubes mixed with it is another advancement in the materials used for the purification process (Hairuddin et al., 2019; Mubarak et al., 2014).

8.11.2 In electronics and computers

Electronic devices nowadays are equipped with so many features and more and more applications are added in very small-sized devices. These small-sized devices require the components, such as capacitors, resistors, transistors, etc., to be small in size while keeping the performance to be at the higher end. Nanomaterials can help and make these components work effectively even on reducing its size significantly. Various issues such as heat removal and thermal mismatch can also be taken care of by using suitable nanomaterials. These kinds of circuits are very popular for the computer chips and microprocessors. Use of these nanomaterials however can increase the cost of the device but it can be compensated with improved performance, compact design, and better reliability and efficiency.

8.11.3 In energy storage applications

Nanomaterials have a lot of applications in energy storage. These materials can be used in supercapacitors, batteries, or fuel cells. Moreover these nanomaterials such as TiO_2 are effectively being used as latent heat–based thermal energy storage systems. The use of nanomaterials in the batteries or supercapacitors can improve the performance of these devices in terms of charge storage capacity. Moreover, it has been observed that the charging–discharging cycles can also be improved if one uses the nanomaterial as an electrode material. Various studies have shown that the use of nanostructured carbon has enhanced the charge storage capacity of the supercapacitors. Moreover the charging–discharging cycles also is seen to be improved on using the nano-sized carbon

(Thines et al., 2016). Various biomass-derived char treated with a suitable filler have been found to enhance the performance of supercapacitors/batteries (Ghosh et al., 2019; Vijayakumar et al., 2019).

8.11.4 In biomedical

In biomedical applications nanomaterials based on gold and silver NPs are of great use. These nanomaterials are used in various different forms such as diagnostic, nanomedicines, carrier for drug delivery, tissue engineering, etc. These nanomaterials are used for the joints replacements, replacing the bones with a material, pacemakers, etc. Titanium is the most widely used material for the replacement of the organs. The nano-sized titanium can enhance the mechanical strength of the replacement keeping the weight to be small (Bi et al., 2019; Roy et al., 2003). Different polymeric compounds that use the nanofillers are also used for the special applications in magnetic drug delivery system where a drug is taken to a very specific position by means of the external magnetic field. Ornosil NPs are also used extensively while dealing with the cancer patients. It is used as a career and the dye kept within it, which does not get exposed to the other parts of the body (Sinani et al., 2003). It significantly decreases the adverse effects of the medication. It also minimizes the needed drug dose for therapy, which lessens the negative impact of the medicine on the patient.

8.11.5 In sensors

Nanomaterials are used to develop sensors with high sensitivity and better accuracy. These sensors are generally be used for the medical diagnostics, food safety, environmental pollution level diagnostics, water purity, moisture control, etc. (Omata et al., 2004). Compared to the normal materials the use of nanomaterial makes the sensing process more sensitive and can be controlled more efficiently. Various metal oxide–based nanomaterials, such as TiO_2, $BaTiO_3$, $NaNO_3$, polymer-based nanomaterials (Shahzad et al., 2017), hydrides, etc. can be used for sensing the various physical and chemical properties. The nanomaterials can also be used to determine the presence of a particular gas or chemical. Smoke sensors are a prominent example of how nanomaterials are used to detect smoke in any enclosed area (Chen et al., 2018).

8.11.6 Other applications

Other than the applications mentioned above, the nanomaterials can be used in a lot of applications. Truly speaking, these nanomaterials have their potential in almost all the areas. These nanomaterials are highly effective in solar cell applications. Nanostructured II-VI group and III-V group elements are of the great interest as they have a wide band gap and can enhance the efficiency of the solar cells up to a significant level (Razika, 2015). The nanomaterials have a wide range of applications in agriculture as well. The nanomaterials can be used to remove the unwanted materials present in the soil. Nanomaterial-based agrochemicals can be used for the high yield of the agriculture products. Nanotechnology can be used for the breeding and for the poultry industry also. Moving toward the manufacturing, the nanomaterials find themselves in a variety of applications because the mechanical strength, ductility, and stress-loading capacity of the nanomaterials are quite high. Hence these materials can be used to design and develop light weight structures with high mechanical strength.

8.12 Conclusions

Green nanomaterials, generally synthesized by means of the organic wastes, can be synthesized by various routes. The characteristics of these green nanomaterials highly depend on the processing techniques and the conditions in which these are prepared. By changing these conditions, the characteristics of these green nanomaterials can easily be changed. The characteristics mainly depend upon the pH value of the solution, catalyst used in the process, time for which the reaction took place, temperature used during the process, etc. For the effective utilization of these nanomaterials the properties of the synthesized nanomaterials must be known. Among these properties morphological study, elemental study, thermal analysis, and surface area analysis are most commonly used analysis. Scanning electron microscope (SEM) and transmission electron microscopes (TEM) are the most extensively characterization techniques for the morphological study. It finds the shape, size, distribution, grain conditions, and many more properties of the sample. XRD and energy dispersive spectroscopy (EDS) techniques are widely used to study the compositional analysis of the sample. It also can determine any impurity present in the sample. Particle size can also be determined using XRD analysis. TGA is used for the thermal analysis, which gives an idea of the thermal stability of the synthesized nanomaterial. However, it cannot be generalized that different applications demand different types of characterization. For example, thermal conductivity analysis is important for the material that is to be employed as insulator or heat exchanger, electrical concavity test is must for the material used in electrical circuitry–related applications. These nanomaterials have a lot of applications in various fields such as agriculture, sensing applications, energy storage, biomedical, etc. Hence use of the nanomaterials in many other fields is increasing day by day.

References

Abhilash, Pandey, B.D, 2012. Synthesis of zinc-based nanomaterials: a biological perspective. IET Nanobiotechnol. 6 (4), 144–148. https://doi.org/10.1049/iet-nbt.2011.0051.

Ajayan, P.M., 2004. How does a nanofibre grow? Nature 427 (6973), 402–403. https://doi.org/10.1038/427402a.

Armendariz, V., Herrera, I., Peralta-Videa, J.R., Jose-Yacaman, M., Troiani, H., Santiago, P., Gardea-Torresdey, J.L., 2004. Size controlled gold nanoparticle formation by *Avena sativa* biomass: use of plants in nanobiotechnology. J. Nanopart. Res. 6 (4), 377–382. https://doi.org/10.1007/s11051-004-0741-4.

Baer, D.R., Thevuthasan, S., 2010. Characterization of thin films and coatings. In: Martin, P.M. (Ed.), Handbook of Deposition Technologies for Films and Coatings. Elsevier Inc, Amsterdam, pp. 749–864.

Bannov, A.G., Popov, M.V., Kurmashov, P.B., 2020. Thermal analysis of carbon nanomaterials: advantages and problems of interpretation. J. Therm. Anal. Calorim. 142 (1), 349–370. https://doi.org/10.1007/s10973-020-09647-2.

Bhainsa, K.C., D'Souza, S.F, 2006. Extracellular biosynthesis of silver nanoparticles using the fungus *Aspergillus fumigatus*. Colloids Surf. B 47 (2), 160–164. https://doi.org/10.1016/j.colsurfb.2005.11.026.

Bi, Z., Kong, Q., Cao, Y., Sun, G., Su, F., Wei, X., Li, X., Ahmad, A., Xie, L., Chen, C.-M., 2019. Biomass-derived porous carbon materials with different dimensions for supercapacitor electrodes: a review. J. Mater. Chem. A 7 (27), 16028–16045. https://doi.org/10.1039/C9TA04436A.

Chen, C., Wang, X., Li, M., Fan, Y., Sun, R., 2018. Humidity sensor based on reduced graphene oxide/lignosulfonate composite thin-film. Sens. Actuators B 255, 1569–1576. https://doi.org/10.1016/j.snb.2017.08.168.

Corradini, E., de Moura, M.R., Mattoso, L.H.C., 2010. A preliminary study of the incorporation of NPK fertilizer into chitosan nanoparticles. Exp. Polym. Lett. 4 (8), 509–515. https://doi.org/10.3144/expresspolymlett.2010.64.

Dameron, C.T., Reese, R.N., Mehra, R.K., Kortan, A.R., Carroll, P.J., Steigerwald, M.L., Brus, L.E., Winge, D.R., 1989. Biosynthesis of cadmium sulphide quantum semiconductor crystallites. Nature 338 (6216), 596–597. https://doi.org/10.1038/338596a0.

Darroudi, M., Ahmad, M.B., Zamiri, R., Zak, A.K., Abdullah, A.H., Ibrahim, N.A., 2011. Time-dependent effect in green synthesis of silver nanoparticles. Int. J. Nanomed. 6 (1), 677–681. https://doi.org/10.2147/IJN.S17669.

Dehghani, M.H., Karri, R.R., Yeganeh, Z.T., Mahvi, A.H., Nourmoradi, H., Salari, M., Zarei, A., Sillanpää, M., 2020. Statistical modelling of endocrine disrupting compounds adsorption onto activated carbon prepared from wood using CCD-RSM and DE hybrid evolutionary optimization framework: comparison of linear vs non-linear isotherm and kinetic parameters. J. Mol. Liq. 302, 112526. https://doi.org/10.1016/j.molliq.2020.112526.

Dey, T.K., Tripathi, M., 2010. Thermal properties of silicon powder filled high-density polyethylene composites. Thermochim. Acta 502 (1–2), 35–42. https://doi.org/10.1016/j.tca.2010.02.002.

Freyre-Fonseca, V., Téllez-Medina, D.I., Medina-Reyes, E.I., Cornejo-Mazón, M., López-Villegas, E.O., Alamilla-Beltrán, L., Ocotlán-Flores, J., Chirino, Y.I., Gutiérrez-López, G.F., 2016. Morphological and physicochemical characterization of agglomerates of titanium dioxide nanoparticles in cell culture media. J. Nanomater. 2016,, 5937932. https://doi.org/10.1155/2016/5937932.

Gamez, G., 2001. Chemical processes involved in Au (III) binding and bioreduction by Alfalfa biomass. In: In: Proc. Hazardous Waste Research: Environmental Changes and Solutions to Resource Development, Production, and Use.

Gardea-Torresdey, J.L., Tiemann, K.J., Gamez, G., Dokken, K., Cano-Aguilera, I., Furenlid, L.R., Renner, M.W., 2000. Reduction and accumulation of gold(III) by *Medicago sativa alfalfa* biomass: X-ray absorption spectroscopy, pH, and temperature dependence. Environ. Sci. Technol. 34 (20), 4392–4396. https://doi.org/10.1021/es991325m.

Ghosh, S., Santhosh, R., Jeniffer, S., Raghavan, V., Jacob, G., Nanaji, K., Kollu, P., Jeong, S.K., Grace, A.N., 2019. Natural biomass derived hard carbon and activated carbons as electrochemical supercapacitor electrodes. Sci. Rep. 9 (1), 16315. https://doi.org/10.1038/s41598-019-52006-x.

Glasscott, M.W., Pendergast, A.D., Choudhury, M.H., Dick, J.E., 2019. Advanced characterization techniques for evaluating porosity, nanopore tortuosity, and electrical connectivity at the single-nanoparticle level. ACS Appl. Nano Mater. 2 (2), 819–830. https://doi.org/10.1021/acsanm.8b02051.

Golding, C.G., 2016. The scanning electron microscope in microbiology and diagnosis of infectious disease. Sci. Rep. 6 (1), 26516.

Guilger-Casagrande, M., Lima, R.d., 2019. Synthesis of silver nanoparticles mediated by fungi: a review. Front. Bioeng. Biotechnol. 7,, 287. https://doi.org/10.3389/fbioe.2019.00287.

Gusseme, 2010. Biogenic silver for disinfection of water contaminated with viruses. Appl. Environ. Microbiol. 76 (4), 1082–1087.
Hairuddin, M.N., Mubarak, N.M., Khalid, M., Abdullah, E.C., Walvekar, R., Karri, R.R., 2019. Magnetic palm kernel biochar potential route for phenol removal from wastewater. Environ. Sci. Pollut. Res. 26 (34), 35183–35197. https://doi.org/10.1007/s11356-019-06524-w.
Huy, T., Quy, N., Le, A.-T., 2013. Silver nanoparticles: Synthesis, properties, toxicology, applications and perspectives. Adv. Nat. Sci. Nanosci. Nanotechnol.,, 4,.
Johnston, C.W., Wyatt, M.A., Li, X., Ibrahim, A., Shuster, J., Southam, G., Magarvey, N.A., 2013. Gold biomineralization by a metallophore from a gold-associated microbe. Nat. Chem. Biol. 9 (4), 241–243. https://doi.org/10.1038/nchembio.1179.
Karri, R.R., Jayakumar, N.S., Sahu, J.N., 2017. Modelling of fluidised-bed reactor by differential evolution optimization for phenol removal using coconut shells based activated carbon. J. Mol. Liq. 231, 249–262. https://doi.org/10.1016/j.molliq.2017.02.003.
Karri, R.R., Sahu, J.N., 2018. Process optimization and adsorption modeling using activated carbon derived from palm oil kernel shell for Zn (II) disposal from the aqueous environment using differential evolution embedded neural network. J. Mol. Liq. 265, 592–602. https://doi.org/10.1016/j.molliq.2018.06.040.
Koduru, J.R., Karri, R.R., Mubarak, N.M.Inamuddin, S., Thomas, R.K., Mishra, A., Siri, A.M. (Eds.), 2019. Smart materials, magnetic graphene oxide-based nanocomposites for sustainable water purification. Sustainable Polymer Composites and Nanocomposites 759–781. Springer https://doi.org/10.1007/978-3-030-05399-4_26 .
Kuchibhatla, S.V.N.T., Karakoti, A.S., Baer, D.R., Samudrala, S., Engelhard, M.H., Amonette, J.E., Thevuthasan, S., Seal, S., 2012. Influence of aging and environment on nanoparticle chemistry: Implication to confinement effects in nanoceria. J. Phys. Chem. C 116 (26), 14108–14114. https://doi.org/10.1021/jp300725s.
Latif, M.S., Kormin, F., Mustafa, M.K., Mohamad, I.I., Khan, M., Abbas, S., Ghazali, M.I., Shafie, N.S., Abu Bakar, M.F., Sabran, S.F., Mohamad Fuzi, F.Z., 2018. Effect of temperature on the synthesis of *Centella asiatica* flavonoids extract-mediated gold nanoparticles: UV-visible spectra analyses. AIP Conference Proceedings 2016 (1), 020071 https://doi.org/10.1063/1.5055473.
Lim, B., Jiang, M., Camargo, P.H.C., Cho, E.C., Tao, J., Lu, X., Zhu, Y., Xia, Y., 2009. Pd-Pt bimetallic nanodendrites with high activity for oxygen reduction. Science 324 (5932), 1302–1305. https://doi.org/10.1126/science.1170377.
Makarov, V.V., Love, A.J., Sinitsyna, O.V., Makarova, S.S., Yaminsky, I.V., Taliansky, M.E., Kalinina, N.O., 2014. Green nanotechnologies: synthesis of metal nanoparticles using plants. Acta Naturae 6 (20), 35–44. https://doi.org/10.32607/20758251-2014-6-1-35-44.
Mather, R.R., 2009. Surface modification of textiles by plasma treatments. In: Wei, Q. (Ed.), Surface Modification of Textiles. Elsevier Inc, Amsterdam, pp. 296–317.
Mohammadinejad, R., Karimi, S., Iravani, S., Varma, R.S., 2015. Plant-derived nanostructures: types and applications. Green Chem. 18 (1), 20–52. https://doi.org/10.1039/c5gc01403d.
Mourdikoudis, S., Pallares, R.M., Thanh, N.T.K., 2018. Characterization techniques for nanoparticles: comparison and complementarity upon studying nanoparticle properties. Nanoscale 10 (27), 12871–12934. https://doi.org/10.1039/c8nr02278j.
Mubarak, N.M., Kundu, A., Sahu, J.N., Abdullah, E.C., Jayakumar, N.S., 2014. Synthesis of palm oil empty fruit bunch magnetic pyrolytic char impregnating with $FeCl_3$ by microwave heating technique. Biomass Bioenergy 61, 265–275. https://doi.org/10.1016/j.biombioe.2013.12.021.
Nair, B., Pradeep, T., 2002. Coalescence of nanoclusters and formation of submicron crystallites assisted by *Lactobacillus* strains. Cryst. Growth Des. 2 (4), 293–298. https://doi.org/10.1021/cg0255164.
Nam, H., Choi, W., Genuino, D.A., Capareda, S.C., 2018. Development of rice straw activated carbon and its utilizations. J. Environ. Chem. Eng. 6 (4), 5221–5229. https://doi.org/10.1016/j.jece.2018.07.045.
Namita, S., Soam, P., 2012. Factors affecting the geometry of silver nanoparticles synthesis in *Chrysosporium Tropicum* and *Fusarium Oxysporum*. Am. J. Nanotechnol., 2, pp. 112–121.
Nizamuddin, S., Mubarak, N.M., Tiripathi, M., Jayakumar, N.S., Sahu, J.N., Ganesan, P., 2016. Chemical, dielectric and structural characterization of optimized hydrochar produced from hydrothermal carbonization of palm shell. Fuel 163, 88–97. https://doi.org/10.1016/j.fuel.2015.08.057.

Omata, S., Murayama, Y., Constantinou, C.E., 2004. Real time robotic tactile sensor system for the determination of the physical properties of biomaterials. Sens. Actuators A 112 (2), 278–285. https://doi.org/10.1016/j.sna.2004.01.038.

Patra, J.K., Baek, K.H., 2014. Green nanobiotechnology: factors affecting synthesis and characterization techniques. J. Nanomater. 2014,, 417305. https://doi.org/10.1155/2014/417305.

R., B.D, 2011. Surface characterization of nanoparticles. J. Surf. Anal. 17 (3), 163–169. https://doi.org/10.1384/jsa.17.163.

Rai, A., Singh, A., Ahmad, A., Sastry, M., 2006. Role of halide ions and temperature on the morphology of biologically synthesized gold nanotriangles. Langmuir 22 (2), 736–741. https://doi.org/10.1021/la052055q.

Razika, T.-I., 2015. Nanomaterials in solar cells. In: Aliofkhazraei, M., Makhlouf, A.S.H. (Eds.), Handbook of Nanoelectrochemistry. Springer Science and Business Media LLC, Berlin, pp. 1–18.

Roy, I., Ohulchanskyy, T.Y., Pudavar, H.E., Bergey, E.J., Oseroff, A.R., Morgan, J., Dougherty, T.J., Prasad, P.N., 2003. Ceramic-based nanoparticles entrapping water-insoluble photosensitizing anticancer drugs: a novel drug-carrier system for photodynamic therapy. J. Am. Chem. Soc. 125 (26), 7860–7865. https://doi.org/10.1021/ja0343095.

Sahu, J.N., Karri, R.R., Zabed, H.M., Shams, S., Qi, X., 2019. Current perspectives and future prospects of nano-biotechnology in wastewater treatment. Sep. Purif. Rev. 50 (2), 139–158. https://doi.org/10.1080/15422119.2019.1630430.

Saima, G., Bahadar, K.S., Ur, R.I., Ali, K.M., I., K.M, 2019. A comprehensive review of magnetic nanomaterials modern day theranostics. Front. Mater. 6,, 179. https://doi.org/10.3389/fmats.2019.00179.

Sandler, S.E., Fellows, B., Thompson Mefford, O., 2019. Best practices for characterization of magnetic nanoparticles for biomedical applications. Anal. Chem. 91 (22), 14159–14169. https://doi.org/10.1021/acs.analchem.9b03518.

Sawle, B.D., Salimath, B., Deshpande, R., Bedre, M.D., Prabhakar, B.K., Venkataraman, A., 2008. Biosynthesis and stabilization of Au and Au-Ag alloy nanoparticles by fungus, *Fusarium semitectum*. Sci. Technol. Adv. Mater. 9 (3), 035012. https://doi.org/10.1088/1468-6996/9/3/035012.

Shahzad, F., Zaidi, S.A., Koo, C.M., 2017. Highly sensitive electrochemical sensor based on environmentally friendly biomass-derived sulfur-doped graphene for cancer biomarker detection. Sens. Actuators B 241, 716–724. https://doi.org/10.1016/j.snb.2016.10.144.

Sinani, V.A., Koktysh, D.S., Yun, B.G., Matts, R.L., Pappas, T.C., Motamedi, M., Thomas, S.N., Kotov, N.A., 2003. Collagen coating promotes biocompatibility of semiconductor nanoparticles in stratified LBL films. Nano Lett. 3 (9), 1177–1182. https://doi.org/10.1021/nl0255045.

Somorjai, G.A., Park, Y., J., 2008. Colloid science of metal nanoparticle catalysts in 2D and 3D structures. Challenges of nucleation, growth, composition, particle shape, size control and their influence on activity and selectivity. Top. Catal. 126–135. https://doi.org/10.1007/s11244-008-9077-0.

Thines, K.R., Abdullah, E.C., Ruthiraan, M., Mubarak, N.M., Tripathi, M., 2016. A new route of magnetic biochar based polyaniline composites for supercapacitor electrode materials. J. Anal. Appl. Pyrolysis 121, 240–257. https://doi.org/10.1016/j.jaap.2016.08.004.

Tripathi, M., Mubarak, N.M., Sahu, J.N., Ganesan, P., 2017. Overview on synthesis of magnetic bio char from discarded agricultural biomass. In: Thakur, V.K., Thakur, M.K., Kessler, M.R. (Eds.), Handbook of Composites from Renewable Materials, 1–8. Wiley,, NJ, pp. 435–460 Vols.

Tripathi, M., Sahu, J.N., Ganesan, P., 2016. Effect of process parameters on production of biochar from biomass waste through pyrolysis: a review. Renew. Sustain. Energy Rev. 55, 467–481. https://doi.org/10.1016/j.rser.2015.10.122.

Upadhyay, S., Parekh, K., Pandey, B., 2016. Influence of crystallite size on the magnetic properties of Fe_3O_4 nanoparticles. J. Alloys Compd. 678, 478–485. https://doi.org/10.1016/j.jallcom.2016.03.279.

Vijayakumar, M., Bharathi Sankar, A., Sri Rohita, D., Rao, T.N., Karthik, M., 2019. Conversion of biomass waste into high performance supercapacitor electrodes for real-time supercapacitor applications. ACS Sustain. Chem. Eng. 7 (20), 17175–17185. https://doi.org/10.1021/acssuschemeng.9b03568.

Zhang, Q., Yang, X., Guan, J., 2019. Applications of magnetic nanomaterials in heterogeneous catalysis. ACS Appl. Nano Mater. 2 (8), 4681–4697. https://doi.org/10.1021/acsanm.9b00976.

Zienkiewicz-Strzałka, M., 2019. Silver nanoparticles on chitosan/silica nanofibers: characterization and antibacterial activity. Int. J. Mol. Sci. 21 (1), 66.

CHAPTER 9

Chemical and physical properties of nanoparticles and hybrid materials

Renuka Gupta[a], Heena Chauhan[a], Vinod Kumar Garg[b] and Navish Kataria[a]
[a]Department of Environmental Sciences, J.C. Bose University of Science and Technology, YMCA, Faridabad, Haryana, India
[b]Department of Environmental Science and Technology, Central University of Punjab, Bathinda, Punjab, India

9.1 Introduction

Nanotechnology is a multidisciplinary branch of science that deals with the synthesis, characterization, exploitation, and exploration of nanoscale materials (Goyal, 2017). The word nanotechnology is formed from two Greek words "*Nano*" means "dwarf" and "*tekhnologia*" means "systematic treatment." Therefore, nanotechnology means providing the systematic treatment or solution to the problems by the usage of nano or dwarf particles. These particles must have at least one dimension in 1–100 nm range to be nanoparticle (NP) (Khan et al., 2019). Nanotechnology is undergoing unpredictable growth and is being practiced in almost every field, including pharmaceutical, environmental remediation, astronomical science, electrical and electronics, information technology, robotics, human health, wellness, etc. (Saleh, 2016). Scientists and researchers are continuously working on designing and development of novel micro/nano-sized materials in different forms namely nanocomposites, nanohybrids, metal–organic frameworks, doped metal oxides, nanocarbon, functionalized and surface modified nanomaterials, etc. The nanohybrids are a combination of inorganic units of hybrids held together by chemical bonding (De Clippel et al., 2013). NPs and nanohybrids have distinct and beneficial properties than the precursor bulk materials due to their small size and surface area to volume ratio. There are numerous research studies that have established magnificent catalytic properties exhibited by NPs (Lee et al., 2010). Gold and silver NPs have good antimicrobial properties (Sahu et al., 2019a; Sahu et al., 2019b; Yoksan & Chirachanchai, 2010). NPs of barium ferrite and nanohyrids of iron and cobalt have significant magnetic properties. Nanocomposites of polyurethane-acrylate coatings containing NPs of alumina and silica and copper oxide NPs have very good thermal properties (Nazari & Riahi, 2011; Sow et al., 2011) and zeta potential (Kamshad et al., 2019). These properties of nanomaterials vary with shape, composition, and structure and make them valuable materials for applications in diverse fields such as nanoadsorbents in wastewater treatment (Khan, Mubarak, Khalid, et al., 2020; Lingamdinne et al., 2019;

Sustainable Nanotechnology for Environmental Remediation
DOI: https://doi.org/10.1016/B978-0-12-824547-7.00024-2

Ruthiraan et al., 2019), phytoremediation (Joseph et al., 2019), semiconductors (Yan et al., 2020; Zhu et al., 2019), gas sensing (Zhang et al., 2019), and many more.

9.2 Historical progress in NPs and hybrid materials

Richard Feynman was the first to enroot the origin of nanotechnology during his lecture on "Plenty of Room at the bottom" in 1959. The term "nanotechnology" was first used in 1974 by Norio Taniguchi. Later in 1986, K. Eric Drexler in his book "Engines of Creation: The Coming Era of Nanotechnology" further publicized the term "nanotechnology."

Some major milestones in the history of nanotechnology are as follows:

- 1857—Michael Faraday prepared nanocolloids of gold.
- 1931—M. Knoll and E. Ruska invented the first electronic microscope.
- 1959—R. Feynman delivered his talk on "There is Plenty of Room at the Bottom," indicating researchers and scientists to reduce the size of materials that would originate robust materials.
- 1985—R.F. Smalley, R.F. Curl, and H.W. Kroto synthesized fullerene, which consists of 60 carbon atoms (C-60).
- 1986—K. Eric Drexler wrote a book, namely "Engines of creation: The Coming Era of Nanotechnology."
- 1991—S. lijima invented "carbon nanotubes."
- 2000—D.M. Eigler, with Fe atoms onto a copper substrate, illustrated "quantum mirage."

The list is unending, and nanotechnology has been creating many more milestones in the development of nanodevices for storing data, semiconductors, nanobiochips, nanobiorobotics, and much more. The treatment of cancer like disease is one of the significant achievements of nanotechnology.

9.3 NPs and hybrid materials

9.3.1 Classification of NPs and hybrid materials

The classification of NPs can be done based on size, morphology, and chemical and physical characteristics (Khan et al., 2019). The classification of NP is given in Fig. 9.1.

The classification of nanoparticles is based on dimensions as given in Fig. 9.2.

1. 0D—No dimension higher than 1–100 nm nanorange for example, nanocluster and nanosphere.
2. 1D—The nanoparticles must have at least one dimension in the nanorange. Nanoplatelets, nanorods, nanotubes, and nanofibers are considered in 1D nanoparticles.
3. 2D—The nanoparticles must exhibit at least two dimensions in the nanorange. Nanosheets, nanofilms, and whiskers are categorized into 2D nanoparticles.

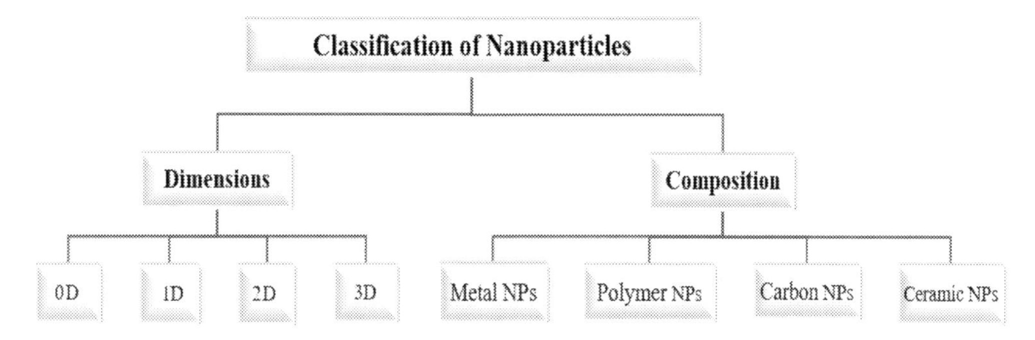

Figure 9.1 Classification of nanoparticles

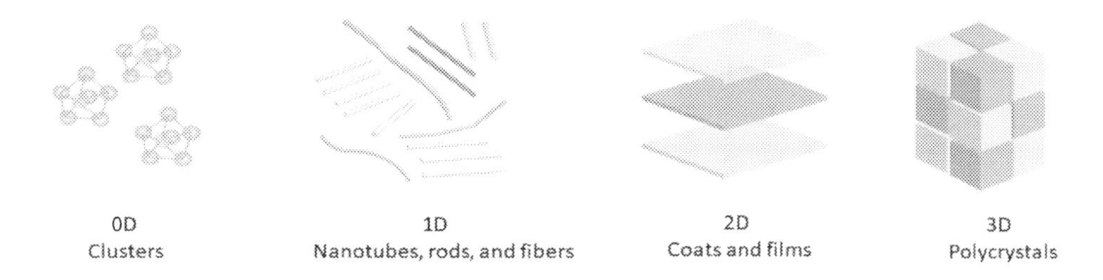

Figure 9.2 Nanomaterials of different dimensions.

4. 3D—The nanoparticles must have all three dimensions in the nanorange. Nanogranules, nanoclays, and equiaxed nanoparticles are considered in 3D nanoparticles.
   Classification of NPs based on chemical and physical characteristics is as follows:
1. *Metal NPs*: These are composed of pure metals and compounds such as oxides, sulfides, fluorides, and chlorides. CuO, ZnO, GaAs are few examples of metal NPs.
2. *Polymer NPs*: These NPs are composed of polymers. The macromolecules consisting of the same monomeric units attached to form chain-like structures are known as polymers. Polymers include chitosan, biomolecules, DNAs, etc.
3. *Carbon NPs*: These NPs are composed of carbon atoms and their derivatives. Carbon NPs are carbon nanotubes (CNTs), graphene oxide (GO), fullerenes.
4. *Ceramic NPs*: Ceramic is composed of inorganic compounds of metals, nonmetals, or metalloids. The NPs composed of ceramics are termed as ceramic NPs. Hydroxyapatite, silica (SiO_2), titanium oxide (TiO_2), and zirconia (ZrO_2) are examples of ceramic NPs.

9.3.2 Synthesis of NPs and hybrid materials

Nanomaterials are synthesized by two approaches: top-down approach and bottom-up approach. The "top-down" approach is a destructive type approach where the

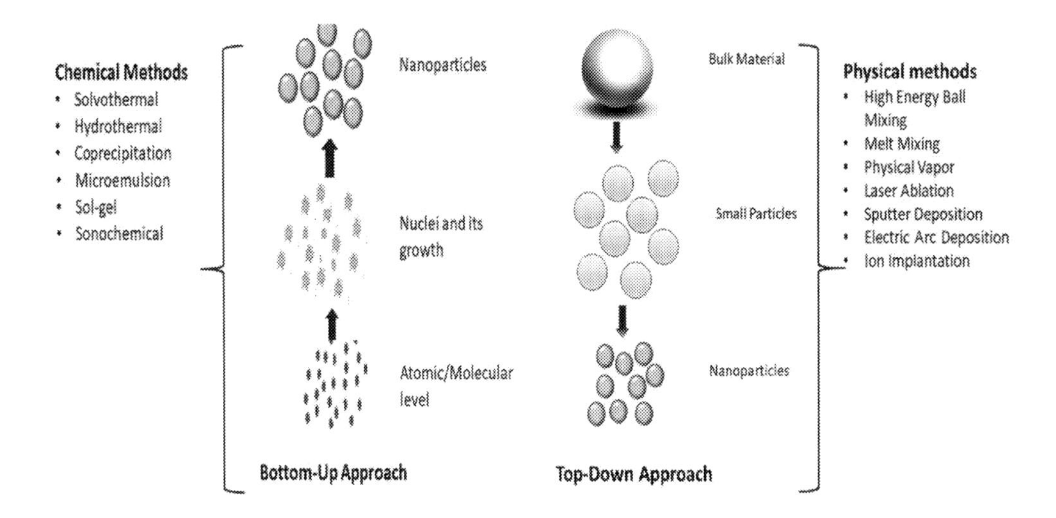

Figure 9.3 Synthesis approach of nanomaterials and nanohybrid.

precursor is broken down into smaller pieces and then further to nanomaterials, whereas in "bottom-up" approach, atoms or molecules adhere together to form nanomaterials (Sanganabhatla & SunderR, 2017). Nanomaterials are synthesized by various physical, chemical, biological, and hybrid methods (Dhand et al., 2015). Physical methods are based on the "top-down" approach, and chemical methods are based on the "bottom-up" approach (Fig. 9.3). The biological processes use specific biomolecules, fungi, bacteria, or biomembranes such as chitosan for the synthesis of nanomaterials. Hybrid methods include chemical, physical, or biological methods all together to synthesize NPs. Among all these methods, chemical methods are more appropriate as they require simple techniques and instrumentation. Further, nanomaterials of different shapes can easily be obtained, and nanomaterials can be prepared as a film, powder, or liquid easily.

Physical methods: These include high-energy ball mixing for preparing powder form, melt mixing for viscous liquid polymers, physical vapor deposition to make thin films, laser ablation, sputter deposition, electric arc deposition, and ion implantation.

Chemical methods: These include solvothermal, hydrothermal, coprecipitation, the Langmuir–Blodgett method, micro-emulsion, sol-gel sonochemical, and microwave synthesis.

Biological methods/green synthesis: These include synthesis of nanomaterials using microorganisms (fungi, bacteria, yeast, virus, etc.), use of plant extracts, biowastes, proteins template–based nanomaterials, DNA template–based nanomaterials, and bio-based membranes.

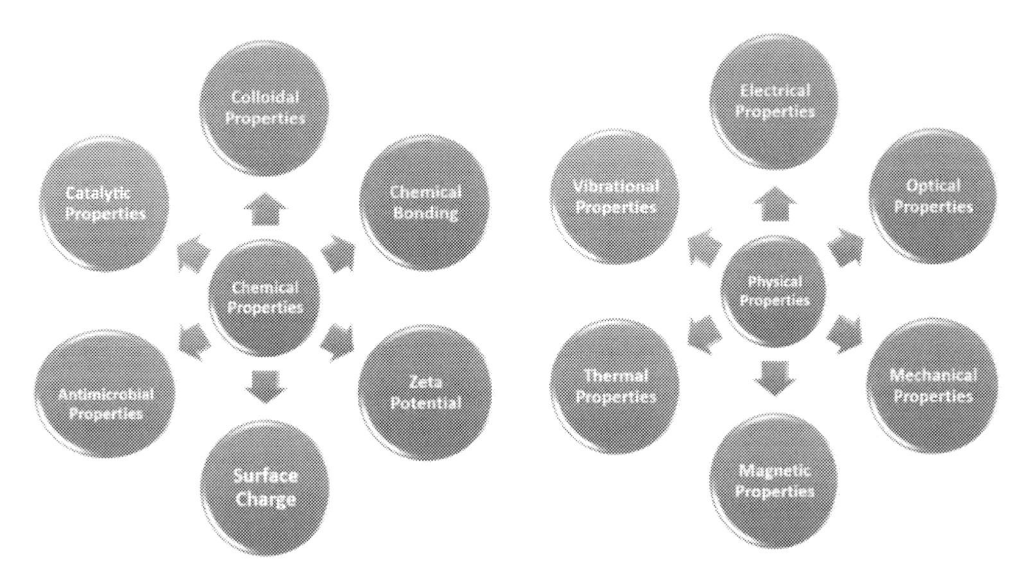

Figure 9.4 Chemical and physical properties of nanoparticles and hybrid materials.

9.3.3 Characterization of NPs and hybrid materials

The branch of nanotechnology, "Nanometrology," deals with the characterization of nanohybrids and NPs (Ikhmayies, 2014). Generally nanomaterials are characterized by morphology, composition, structure, physical and chemical properties. The techniques used for the characterization involve X-ray diffraction spectroscopy, Fourier-transform infrared spectroscopy, nuclear magnetic resonance, Brunauer-Emmett-Teller (BET) surface area analyzer, inductively coupled plasma mass spectrometry, thermogravimetric analysis, transmission electron microscopy, electron diffraction, high-resolution scanning electron microscopy, and various other spectroscopic methods.

9.4 Properties of NPs and hybrid materials

NPs and nanohybrid materials exhibit unique properties relative to bulk matter that transmits novel and advanced characteristics (Khan, Mubarak, Khalid, et al., 2020; Lingamdinne et al., 2019; Ruthiraan et al., 2019). The properties differ due to particle size, shape, surface charge, aspect ratio, composition and crystalline structure of nano and hybrid material (Garg & Kataria, 2016; Gatoo et al., 2014). The different chemical and physical properties enhance the affinity of NPs toward a variety of matters and their potential to be used in diverse science and engineering fields (Fig. 9.4).

9.4.1 Chemical properties of NPs and hybrid materials

The chemical properties of nano and hybrid materials define the chemical structure that becomes evident during, or after a chemical reaction. Chemical properties influence the internal structure of nano and hybrid materials. Important chemical properties of nanomaterials are given next.

9.4.1.1 Chemical bonding

Chemical bonding is a key aspect in synthesis of NPs and nanohybrids. Nature and type of chemical bonding influence several other properties such as melting point, boiling point, solubility, hardness, magnetic, and thermal properties of particles (Wu et al., 2018). Wei et al. (2012) while working on iron oxide (Fe_3O_4) NPs with acid modifiers (oleic acid and sodium citrate) reported that modifications in NPs could decrease the saturation magnetization due to surface charge. The modified magnetic NPs have dispersion capability, which reduces the surface charge and bipolar attraction and so could be used as biomedical materials. Zhao et al., (2018) worked on $MoSe_2$ nanosheets loaded with SnO_2NPs and demonstrated that charge accumulates at $MoSe_2$ and SnO_2 interface due to Se-O bond formation. Uniform dispersion of SnO_2 on the basal plane of $MoSe_2$ prevented aggregation, migration, and detachment of $MoSe_2$ during sodiation and enhanced cyclability. Hossain and Shimizu (2019) during covalent immobilization of gold NPs on graphene concluded that change in nanostructures chemical bonding can make them usable in biosensors and ultrasensitive chemical detectors. The epitaxial graphene on silicon-carbon (SiC) had functionalized –SH groups that were further decorated by gold NPs (AuNPs) that then bonded covalently with graphene by –S-Au bonds and became immobilized on to the surface of graphene and could react with molecules such as hexanedithiol and pentachlorobeneze thiol that made them apt for capturing biomolecules.

9.4.1.2 Zeta potential

The electrostatic potential at the double electrical layer of nanomaterials in solution is termed as zeta potential. It is used to predict the colloidal stability of nanomaterials. If the zeta potential value is low, then nanomaterials coagulate due to attraction of the particles. If zeta potential value is high, then there is no coagulation of the NPs due to electric repulsion. For neutral, the zeta potential lies between –10 mV to +10 mV, whereas for cationic and anionic, the zeta potential is higher than +30 mV or less than –30 mV. The zeta potential of metal oxide nanomaterials, dispersed in water, varies with change in pH and concentration. For example, the zeta potential of naked titanium oxide (TiO_2) increases from 6.7 to 8.2 mV at pH=6 as the concentration of particles changes from 0.5 to 5 mg/L (Wang et al., 2013). Hassan et al. (2018) worked on block-copolymer nanomaterial surfaces with conical nanopillars of varying height (60, 120, and 200 nm) in a hexagonal crystal with a similar spacing of approx. 50 nm over large areas. The nanostructure surface modified with silicon substrate displayed superhydrophobic properties and superhydrophilic properties. The key finding was that the surface conductivities could reduce the zeta potential magnitude and its dependencies on electrolyte concentration and pH. Khatami et al. (2019) worked on super-paramagnetic iron oxide nanomaterials fabricated by stevia plant extracts, which act as stabilizing and reducing agents. The fabricated nanohybrids have a small size and are stable due to biomembrane coating. The zeta potential value was –41.1 mV and developed opposing forces between nanomaterials and prevented aggregation.

9.4.1.3 *Surface charge*

The transfer of a species onto another species accompanied by adsorption and activation is because of the species' surface charge. Thus, the surface charge is an important property for tuning active surface sites and the catalytic performance of the species. Grover et al. (2013) reported that titania-based nanocatalysts such as sodium titanate nanotubes (TNT) and titania nanorods (TNR) had superior surface properties. Rice-like pure anatase-TNR particles (W = 8–13 nm and L = 81–134 nm) and straw-like hollow orthorhombic-TNT particles (W = 9–12 nm and L = 82–115 nm) were obtained by hydrothermal treatment. The result suggested that an alteration in the surface charge of titanium oxide depends on the method of preparation and morphology of NPs. The effect of nanohybrids' surface charge on the biocatalytic reaction kinetics of an enzyme by using plasmonic nanohybrids as an abiotic system model has been studied by Tadepalli et al. (2017). The results showed that the enzyme activity of bio-nanoconjugates and influence on surface charge proposes a possible correlation between biomolecular structure at the interface and functionality of bio-nanoconjugates. The surface charge transfer doping technique can modulate carrier concentration by extracting carrier charge between the surface dopant and semiconductor to achieve positive doping in a nondestructive manner due to work-function difference. This technique is commonly used for low-dimensional NPs that possess high surface area and single crystalline structure.

9.4.1.4 *Colloidal properties*

The NPs and hybrid materials have a tendency to form colloids, which results in enhancement of various properties such as uptake of materials onto surface, hydration, etc. Björnström et al. (2004) studied the accelerating effect of colloidal silica on the hydration process of Ca_3SiO_5 cement. Here colloidal silica acted as a nanocatalyst that accelerated the formation of calcium-silicate-hydrate binding phase and dissolution of Ca_3SiO_5 cement. Sara & Ahmad (2017) worked on polysaccharide-based nanocolloids as carriers in delivering contents to intracellular targets, which proved to be extremely efficient. The FeONPs, Fe_2O_3NPs, and Fe_3O_4NPs have been used for decoration of anodized titanium oxide nanotubes. These nanocolloids showed outstanding effective water splitting and irradiation activity (Kupracz et al., 2020).

9.4.1.5 *Antimicrobial properties*

Antimicrobial properties are those properties of the materials that have the potential to kill microbes (Sahu et al., 2019a). In the era of increasing multidrug resistance, in which microbes have developed resistance for many antibodies, it has become challenging to fight against infectious diseases and cure patients, which has led to severe mortality and morbidity. In the same concern, NPs tending to kill or stop microbial growth are alternatives to antibiotics. Usman et al. (2013) prepared high-purity metallic chitosan-copper NPs and tested their antimicrobial activities for several fungal and bacterial species. The result demonstrated that @ 0.2% concentration, these nanomaterials were highly effective. These chitosan-copper NPs are suitable to form different antimicrobial materials for biomedical and pharmaceutical applications. The antimicrobial properties of iron NPs fabricated GO have been reported by Ain et al. (2020). The study on magnetic GO revealed excellent antimicrobial properties against enterobacter agglomerans and *Escherichia coli*. Magnetic GO has potential for effective disinfection of contaminated drinking water (Lingamdinne et al., 2019).

9.4.1.6 Catalytic properties

Nanohybrids and NPs also have catalytic properties. These nanomaterials can either promote the reaction or demote the reaction. Luska et al. (2016) studied the catalytic properties of iron-ruthenium immobilized on supported ionic liquid phase (FeR-uNPs@SILP) NPs. Based on the results it was inferred that partial replacement of ruthenium by iron enhanced the catalytic properties of nanomaterial. Hassan et al. (2018) synthesized SnO_2/TiO_2 and TiO_2 with varying SnO_2 content ranging from 0 to 20 wt%. The catalytic properties were examined by photodegradation of methylene blue, rhodamine B, and synthesis of xanthene. The result showed that the synthesis of xanthene was improved when SnO_2 was doped TiO_2. The catalytic properties of silver nanomaterials (synthesized using *Cuminum cyminum* extract) were tested for the degradation of methylene blue, methyl red, rhodamine-B dyes, and p-nitrophenol. These silver NPs proved a very efficient nanocatalyst and degraded methylene blue, methyl red, rhodamine-B dyes in 2 min, while p-nitrophenol dye was degraded in 16 min (Choudhary et al., 2018).

9.4.2 Physical properties of NPs and hybrid materials

The physical properties of NPs and nanohybrids depend on the arrangement, physical system, and their changes between momentary states. Physical properties can be intensive properties, that is, these properties are not dependent on size, amount, or extent, and properties that reflect additive relationships. Both chemical and physical properties of NPs and hybrid materials are underlying and one property influences the other property.

9.4.2.1 Electrical properties

The electrical properties include conductivity and resistivity of the materials. The electrical properties depend on the morphological characteristics of nanomaterials. There is a difference in electric properties of bulk materials and nanomaterials; for instance, in case of bulk matter, the conductivity does not depend on cross-section area and diameter, whereas in case of CNTs conductivity changes with width. Cu_2O and CuO electronic structures, to evaluate their semiconducting nature, have been studied by Gupta et al. (2018). Direct and indirect bands were predicted during analysis; the direct band values were 2.01 eV and 1.17 eV for Cu_2O and CuO, respectively. According to Janas and Koziol (2016), the deposition of metal NPs such as Au, Pt, Pd on CNTs increased their resistivity up to 200%. The increase in resistivity is due to the addition of resistivity of CNT and metal NPs. The addition of TiO_2 of varying percentages from 0 to 3 mol% to SnO_2 resistors affected their electrical properties. The residual voltage ratio KR$\frac{1}{4}$2.4, nonlinear coefficient, $\alpha\frac{1}{4}50$, withstanding surge current density, 5 kA/cm^2, breakdown electric field, Eb$\frac{1}{4}$3 kV/cm by 0.25% mol of TiO_2, it was also found that above 0.25% it had an adverse effect such as increase in grain resistivity and homogeneity of sample is lost (Shahraki et al., 2015). Omri et al. (2018) worked on Cu-doped ZnONPs prepared by the sol-gel method. With the doping of Cu, there is an increase in crystalline structure and a decrease in activation energy, which enhances AC conductivity.

9.4.2.2 Optical properties

Optical property is the photoluminescence effect of NPs, which is dependent on the morphological characteristics and dopant. It can be determined by surface plasmon resonance. Koushki et al. (2019) in the study of Na-doped lead zirconate titanate found that optical band is indirect and reduced and have a probable change of reverse saturable absorption at 532 nm wavelength due to sodium acceptor ions; moreover, it increased extinction and adsorption coefficient in violet range. It decreased the permeability and refractive index. Bazta et al. (2019) studied the optical properties of n-type thin-film ZnO doped with yttrium (Y) with varying proportions. The thin films showed 83% transmittance in the visible region. Further, the band energy, E_g, was found to increase from 3.12 to 3.18 eV at 5% doping and ZnO-Y thin films can be used in optoelectronic devices. Aydın et al. (2019) observed the effect on ZnO thin-film properties after doping with Al. The result showed a direct band pass and changed the forbidden energy range. The thin film had characterization for ammonia (NH_3) gas sensing and was used in diagnosis, environmental protection, and cooling systems.

9.4.2.3 Mechanical properties

Mechanical properties define the strength, weight, and resistance to corrosion of particles. NPs have better mechanical strength compared to bulk particles. A dynamic mechanical analyzer determines these properties. The method of synthesis can also affect the mechanical properties of NPs. For instance, graphene/Sn-Ag-CuNPs were prepared by mechanical mixing and ball mining, respectively. The NPs of different proportions, SAC/0.03Ag-GNSS, SAC/0.1Ag-GNSS, and SAC/0.05GNSs, were developed. It was found that the ball milling method was more useful to enhance creep resistance and improve Young's modulus and hardness (Han et al., 2019). The mechanical properties of a composite laminate are improved by embedding metals with fibers (Chavhan & Wankhade, 2020). The *in-situ* free polymerization of nanocomposite with lignin as cross-linking junction enhances the synergic effect of nanocomposite network structure and reversible energy-dissipating mechanism of strong H-bond between polymeric chains for better mechanical strength. The mass ratio of polymer to lignin plays an essential role in determining the nanostructure; for robust hydrogen, it is 3.42 (Chen et al., 2019). Badawi et al. (2017) studied the mechanical properties of Fe_2O_3/chitosan films prepared by casting technique. During the work, an increase in glass transition temperature (T_g), storage modulus, and stiffness of Fe_2O_3/chitosan nanofilms compared to pure chitosan sheet were observed at different mass ratios.

9.4.2.4 Magnetic properties

Magnetic properties of NPs are dependent on chemical structure and method of synthesis (Lingamdinne et al., 2019, 2018). In general, most NPs show super para-magnetic properties that are caused due to thermal instability. The size of NPs can affect magnetic behavior. Duru et al. (2019) studied Zn0.95−xMgxNi0.050 NPs to observe the changes in ferromagnetic nature employing mild doping with $Mg^{2+}$and Ni^{2+} ions. The result showed that particle size in a range of 5–15 nm showed the best ferromagnetism. Monodispersed ferrite with nickel NPs in different proportions was synthesized by thermal deposition. One of the samples was grown by acetylacetonate to obtain a larger particle size, which showed an enhanced response in the magnetic hyperthermia experiment. Further, it was observed that Ni^{2+} ions play a vital role in the magnetic behavior of the samples, affecting Geff values, saturation magnetization, and anisotropy constant (Ghanbari et al., 2017). The heating capacity measured by magnetic hyperthermia decreases with higher Ni percentage. Cytotoxicity assays observations proved that these NPs are not cytotoxic up to 0.5 mg/mL concentration in Vero cells and can potentially be used as MRI contrast agents. These NPs can be alternatives to active agents used in diagnosis or therapy (Lasheras et al., 2016).

9.4.2.5 Thermal properties

Thermal properties of a matter are defined as the properties related to the conductivity of heat. These properties are exhibited by NPs when heat energy is passed through them. The composition, size, morphology, and cation distribution can influence the thermal properties of the NPs. For instance, $Co_xFe_3\ xO_4$ nanopowders were synthesized with varying concentrations and crystal structure and showed variation in thermal properties (Ajroudi et al., 2014). The thermal properties influence the thermal energy storage (TES) capacity of materials used for high-temperature solar systems. Conventionally, chloride salts are used as TES media, but with the advancement in nanotechnology, the chloride salts/NPs were prepared for TES. The chloride salt in the molar ratio of $MgCl_2$ (51): KCl (22): NaCl (27) was used as base salts, and CuO, ZnO, and Al_2O_3NPs were dispersed at

0.7 wt% to form viscous composites phase change materials. The results indicated that alumina synthesized were the optimal additives to improve TES of chloride salts (Han et al., 2019). The thermal properties of epoxy nanohybrids with hyperbranched aromatic polyamide-grated alumina (Al_2O_3) have been studied by Yu et al. (2011). The transmission electron microscopy characterization showed that Al_2O_3 forms a thin uniform layer in the epoxy matrix and improved interfacial interaction between alumina and epoxy matrix. Thus, the thermal stability showed an apparent increase compared to the epoxy matrix and Al_2O_3NPs. Oleyaei et al. (2016) researched on potato starch and TiO_2 nanomaterials to know their potential as food packaging materials. The outcome of research suggested that thin film of nanocomposites was stiffer with higher thermal stability and improved functional properties. With 1 wt% of TiO_2, nanomaterial could be regarded as optimum concentration to make starch-TiO_2 blend films.

9.4.2.6 *Vibrational properties*

The knowledge of the vibrational spectrum of NPs and nanohybrids is important because it is the signature of their morphology, and can be used to investigate thermodynamical, mechanical, and other physical properties. The vibrational properties are dependent on the size, material, and shape of nanohybrid and NPs (Sauceda et al., 2012). A vibrational spectroscopy of thiolate-protected gold clusters to understand structure revealed that the observed spectra was sensitive to the Au-S interface but was strongly affected by the ligands attached. Abdelouhab et al. (2019) synthesized zinc oxide nanostructures by hydrothermal method and studied their structural, morphological, and vibrational properties. The Raman spectroscopy showed a Raman shift and the influence of caustic bases and precursors on the intensity of mode peaks. Zinc oxide synthesized with hydrothermal methods is cost efficient, simple, and eco-friendly and are demanded products in photocatalysis.

9.5 Application of NPs and hybrid materials

Nanotechnology is growing like a huge tree and is spreading its roots in different and diverse fields. Some of them are mentioned next.

9.5.1 Agriculture and food industries

Nanotechnology has transformed numerous domains of agriculture and food industry involving food processing, preserving, packaging, storing, transporting, and other safety aspects of food (He et al., 2019). Nanotechnology provides green, sustainable, and efficient alternative methods for pest control in agriculture. The formulation of nanostructure-based pesticides and insecticides, bioconjugated NPs for active release of water and nutrients could enhance yield, transfer of NP-mediated gene, or DNA development of a persistent insect variety of crop (Rai & Ingle, 2012). Nanoherbicides are being synthesized to overcome perennial weed management and exhausting weed-seed banks. Nanofertilizers are efficient and sustainable to increase soil fertility compared to traditional ones due to their smaller size and more surface area. Different NPs such as silver, gold, copper, iron, and carbon-coated are used for root elongation, increased germination, and plants (Chhipa, 2019). Nanosensors or biosensors are used for monitoring soil quality and pest in agriculture, thus maintaining the health of the soil and crop.

9.5.2 Diagnosis

Nanobiotechnology is one of the promising scientific and technological areas. The invention of biocompatible devices and materials, engineered on the nanoscale that could interact with biological cells and molecules, provides imaging functions, specific diagnoses, and therapeutic (Kawadkar et al., 2011). NPs could be synthesized to carry therapeutics to specific cellular and subcellular compartments to provide enhanced imaging for the detection of strokes. NPs could be engineered to release its payload in response to distinct extracellular processes occurring around clot and in ischemic penumbra (Landowski et al., 2020).

9.5.3 Cancer therapy

According to a survey of The Indian Council of Medical Research (ICMR), around 2.25 million people of India are living with cancer disease and over 1.15 million new cancer patients are registered every year. Cancer has a high morbidity and mortality rate. So, it is highly required to treat this type of disease, and nanotechnology has come forward with best possible results. NPs such as polymeric NPs, liposomes, micelles, and solid lipid NPs are used as carriers of drug delivery in treatment of breast cancer (Tran et al., 2020). Xiao et al. (2019) studied the efficiency of silver NPs in treatment of colon cancer cells by single cell approach showing inhibition effects on cells. The silver NPs could kill topography and cytoskeleton structures, faltering cell membrane nanostructures, and thus increasing membrane roughness and depress cell membrane adhesion property and cell stiffness. Silver NPs altered all phenotypes or function of the cell in a dose-dependent manner. Thus, the research provided a paradigm for revealing destroying mechanisms of silver NPs against colon cancer cells. The RNA-based nanotherapeutics are used for enhancement of cancer immunotherapy and immunomodulation. The NPs effectively deliver RNA to immune cells or target tumors for triggering antitumor immune response. The RNA-mediated immunomodulation with chemotherapy, immune checkpoint blockade immunotherapy, and photodynamic therapy had bright future for clinical use of cancer treatment (Lin et al., 2020). It is believed that nanotechnology has immense potential to be used for cancer treatment.

9.5.4 Dentistry

Nanotechnology is used as nanodentistry in the dental field. NPs are used in toothpaste and other oral healthcare services that are less stressful to dental surgeons. NPs are used for polishing enamel surfaces, dental fillings, and also as implant materials (Priyadarsini et al., 2018). Sivolella et al. (2012) studied the use of silver NPs for reduction of bacteria adhesion and prevented biofilm formation. Silver NPs are used in alveolar bone surgery devices with promising results. Scaffolds and membranes for bone regeneration containing silver NPs have potential to reduce incidences of postoperative bacterial contaminations and prevent or delays peri-implantitis. Nitrogen-doped titanium oxide photocatalyst NPs are synthesized by the simple solvothermal method and are used to prevent the growth of *Streptococcus* mutants' biofilms. The study revealed that the nitrogen-doped titanium oxide resins are more feasible antibacterial agents against oral carcinogenic biofilms. It was also suggested that further optimization of materials and NPs are required to produce dental materials capable of preventing secondary caries (Esteban Florez et al., 2018). Nanodentistry attracts people as it is cost-effective and time saving.

9.5.5 Sensors

Nanomaterials are potential candidates for chemical and biological sensor application because of a high degree of selectivity and sensitivity compared to bulk counterparts due to higher surface area to volume ratio. Semiconductor metal oxide nanomaterials such as zinc oxide, tin oxide, tungsten oxide, titanium oxide, etc. are better materials for chemical sensing compared to other techniques or gas-sensing materials. Nanowires have a diameter comparable to biomolecules that make them good candidates as biosensors (Patra et al., 2008). Nanobiotechnology has efficiency to enable smart plant sensors that could communicate with electronic devices for enhancing plant productivity. Nanomaterials translate the chemical plants signal to digital signals that could be monitored by standoff electronic devices (Giraldo et al., 2019). The modification of graphene and GONPs leads to improvement of sensitivity, reduced fouling, and electrocatalytic behavior. These nanomaterial-based sensors are used in biosensing and chemical detection resulting in more sensitive healthcare instruments and prevent environmental problems (Beitollai et al., 2018).

9.5.6 Pollution remediation

Environmental pollution remediation depends on various technologies such as chemical reactions, photocatalysis, absorption, adsorption, and filtration used for removal of contaminants from air, water, and soil. The effectiveness and enhanced properties of nanomaterials make them particularly suitable for these processes due to high surface area to volume ratio which, in turn, increases reactivity (Kaur et al., 2020; Khan, Mubarak, Khalid, et al., 2020; Ruthiraan et al., 2019). The three main categories of nanomaterials, inorganic/metal based, carbon based, and polymer based, are extensively used for the removal of dyes, heavy metals, organophosphorus compounds, chlorinated organic compounds, and halogenated herbicides (Guerra et al., 2018). Mukherjee et al. (2016) worked on zero-valent iron oxide (nZVI) NPs as a tool for remediation of contaminated land and groundwater. nZVI has high reactivity compared to ZVI, cost effective, and is capable of treating a broad range of contaminants including organic pollutants (azo dyes, halogenated organic compounds, and pharmaceutical waste), and inorganic pollutants. NPs of gold, silver, silicon, CNTs are used for detection of low levels of pesticide concentrations. NPs of iron/nickel, titanium oxide, zinc oxide, and nanocomposites of graphene showed rapid degradation of pesticides (Rawtani et al., 2018).

9.5.7 Pharmaceutical and drug delivery

Nanotechnology has massive scope in pharmaceutical industries, which includes antimicrobial applications, burn dressing, health products, medical devices, and scaffolds. Nanoscale metal-organic frameworks have emerged as an important class of biomedical-relevant nanomaterials due to their multifunctionality, biocompatibility, and high porosity. Nanomaterials are used for improving drug delivery, biosensing, and biomedical imaging (Kuangda et al., 2018; Mustafa et al., 2017). Among various NPs, metal NPs of zinc, iron, gold, and silver have considerable applications in the biomedical field, not only because of surface area to volume ratio but also due to several medicinal properties (Kalia & Saini, 2017). Lavanya et al. (2020) conducted a comparative study on silver NPs and gelatin-stabilized NPs. The study suggested that gelatin-established NPs can be preferred over silver NPs as the later are comparatively more toxic than gelatin-stabilized silver NPs. Titanium dioxide (TiO_2) NPs are bright and have high refractive index (η=2.4), which make them suitable for pharmaceutical industries (Waghmode et al., 2019). Nithya and Sundrarajan (2020) observed the antibacterial and anticancer activities of cerium-oxide, gold-loaded cerium oxide, silver-loaded cerium oxide, and silver-gold-loaded cerium oxide NPs synthesized through hydrothermal method using *Justicia adhatoda* extract.

9.5.8 Cement industry

Nanotechnology has improved the performance of cement and concrete and led to the development of sustainable, novel, and advanced cement-based composites with unique electrical and mechanical properties. The addition of nanodimensions to concrete and cement accelerated the hydration process and the rate of hydration reaction increased with specific additional functionalities (Singh et al., 2017). CNTs exhibit unique properties, high thermal conductivity, ranging from ultra-high strength through unusual electric behavior, and ability to store NPs inside its tubes. The application of CNTs ranges from composite materials with high strength structural components to heat transfer technology. Concrete and cement CNT composites have strong potential, as CNTs act as a near-ideal reinforcing materials having a diameter similar in scale to the layers in calcium-silicate hydrate. The use of nanomaterials increases durability and strength of cement and concrete.

9.5.9 Textile industry

The textile industry is one of the largest industries worldwide producing a lot of waste. Nanotechnology has a wide use in different processes in this industry (Jun et al., 2020; Khan, Mubarak, Tan, et al., 2020). Nanomaterials have been used for wrinkle resistancy, antibacterial properties, odor control, antistatic properties, water repellence, lightweight, and strength to the cloth (Yetisen et al., 2016). Tin oxide (TiO_2) and zinc oxide (ZnO) NPs can absorb UV radiation. In the textile industry, a thin layer of TiO or ZnO is deposited on the surface of cloth which provides UV protection (Miśkiewicz, 2018). Shahidi and Moazzenchi (2018) worked on applications of CNTs in the textile industry. The results revealed that CNT has distinct properties and can be used in smart textiles, nanoreactors, fire retardants, electronic textiles, scaffolds, and flexible sensors.

9.6 Future prospective

Nanotechnology is spreading its roots to all fields as it is novel, cost-effective, sustainable, and time saving. Though nanotechnology is showing tremendous growth, still there is scope of development. The fabrications or modifications of nanomaterials enhance their mechanical, magnetic, thermal, optical, and other physiochemical properties that make them better candidates with promising results. Solid waste management is a challenge for urban local bodies as well as policymakers. Nanotechnology can be used for effective and sustainable solid waste management. The air quality index (AQI) is poor to very poor in many cities of the world due to failure to control pollution. In this field also there is a requirement of some NPs, which could improve the AQI. There is a requirement of some air sensors also which can detect the ultra-low concentrations of toxic and carcinogenic gases present in the atmosphere. Although NPs are highly efficient to remove the contaminants from the environment but due to nonbiodegradability of inorganic NPs, they stay in the environment for a longer time, so, there is a need to develop eco-friendly products. Some mediators/filters should be developed which could limit the emission of harmful gases from industries and automobiles. Nanofilters can be developed for the removal of contaminants from water. Some nanomaterials-based techniques can be developed for cleaning the underground aquifers, rivers, lakes, etc. To stop the overexploitation of nonrenewable resources, it is required to make efficient technologies by which renewable sources of energy could be increased. The improvement in pharmaceuticals and diagnosis have great scope. It can be used for developing nanomedicines for curing different diseases. Nanotechnology can be used for constructing infrastructure that could last long. Nanotechnology has a bright future in artificial intelligence. It could be used for making human robots. It can also be used for preparing electronic with less consumption of electricity, having high durability and that could be easily recycled. Moreover, nanotechnology can be used for developing processors to enhance the speed of computers, laptops, etc. It can be used for making chips that could store more data. There is a lot of need for nanotechnology in agriculture as the population is increasing and for the same the food production should be increased without the usage of excessive fertilizers and pesticides that are harmful to both soil and human health. There is a need of biosensors that could detect the deficiency of specific nutrient in the plant and could help in predicting droughts and rains in particular areas. For security reasons, there is a need of developing effective and accurate weapons with sensors that could detect enemy in the area. Nanotechnology can be used for preparing lightweight bulletproof jackets. There is need of develop lightweight and fuel-efficient spacecraft with better performance and durability. There is a need of develop strong satellites so that communication can be better. Nanomaterials may be toxic in nature so, there is a requirement of developing nontoxic nanomaterials as well to study the toxic effects of those nanomaterials that are widely in use.

9.7 Conclusion

This chapter suggested that NPs and hybrid materials have magnificent chemical and physical properties that make them a potential candidate to use in different fields. Nanomaterials of the same material prepared using different methods may have different properties. Although several techniques are being used to characterize the nanomaterials, there is a requirement to develop simple, feasible, cost effective, and easy to use techniques compared to currently available techniques. It is also evident from the available literature that nanomaterials are having wide applications in diverse fields such as electronics, pharmaceuticals, textile, health sciences, cement industry, pollution management, etc. But nanomaterials should be used cautiously as these may get released in different environment matrices which in turn may prove harmful for nontarget flora and fauna in a long run.

References

Abdelouhab, A.Z., Djouadi, D., Chelouche, A., Hammiche, L., Touam, T., 2019. Effects of precursors and caustic bases on structural and vibrational properties of ZnO nanostructures elaborated by hydrothermal method. Solid State Sci. 93–99. https://doi.org/10.1016/j.solidstatesciences.2019.01.001.

Ain, Q.U., Farooq, M.U., Jalees,, M.I., 2020. Application of magnetic graphene oxide for water purification: Heavy metals removal and disinfection. J. Water Proc. Eng. 33, 101044 https://doi.org/10.1016/j.jwpe.2019.101044.

Ajroudi, L., Mliki, N., Bessais, L., Madigou, V., Villain, S., Leroux, C., 2014. Magnetic, electric and thermal properties of cobalt ferrite nanoparticles. Mater. Res. Bull. 59, 49–58. https://doi.org/10.1016/j.materresbull.2014.06.029.

Aydın, H., Yakuphanoglu, F., Aydın, C., 2019. Al-doped ZnO as a multifunctional nanomaterial: structural, morphological, optical and low-temperature gas sensing properties. J. Alloys Compd. 773, 802–811. https://doi.org/10.1016/j.jallcom.2018.09.327.

Badawi, A., Ahmed, E.M., Mostafa, N.Y., Abdel-Wahab, F., Alomairy, S.E., 2017. Enhancement of the optical and mechanical properties of chitosan using Fe_2O_3 nanoparticles. J. Mater. Sci. Mater. Electron. 28 (15), 10877–10884. https://doi.org/10.1007/s10854-017-6866-x.

Bazta, O., Urbieta, A., Piqueras, J., Fernández, P., Addou, M., Calvino, J.J., Hungría, A.B., 2019. Influence of yttrium doping on the structural, morphological and optical properties of nanostructured ZnO thin films grown by spray pyrolysis. Ceram. Int. 45 (6), 6842–6852. https://doi.org/10.1016/j.ceramint.2018.12.178.

Beitollai, H., Safaei, M., Tajik, S., 2018. Application of graphene and graphene oxide for modification of electrochemical sensors and biosensors: a review. Int. J. Nano Dimens. 10 (2), 125–140.

Björnström, J., Martinelli, A., Matic, A., Börjesson, L., Panas, I., 2004. Accelerating effects of colloidal nano-silica for beneficial calcium-silicate-hydrate formation in cement. Chem. Phys. Lett. 392 (1–3), 242–248. https://doi.org/10.1016/j.cplett.2004.05.071.

Chavhan, G.R., Wankhade, L.N., 2020. Improvement of the mechanical properties of hybrid composites prepared by fibers, fiber-metals, and nano-filler particles-a review, 27, pp. 72–82.

Chen, Y., Zheng, K., Niu, L., Zhang, Y., Liu, Y., Wang, C., Chu, F., 2019. Highly mechanical properties nanocomposite hydrogels with biorenewable lignin nanoparticles. Int. J. Biol. Macromol. 128, 414–420. https://doi.org/10.1016/j.ijbiomac.2019.01.099.

Chhipa, H., 2019. Applications of nanotechnology in agriculture. Methods in Microbiology, 46. Academic Press Inc, MA, pp. 115–142.

Choudhary, M.K., Kataria, J., Sharma, S, 2018. Evaluation of the kinetic and catalytic properties of biogenically synthesized silver nanoparticles. J. Clean. Prod. 198, 882–890 https://doi.org/10.1016/j.jclepro.2018.09.015\.

De Clippel, F., Dusselier, M., Van De Vyver, S., Peng, L., Jacobs, P.A., Sels, B.F., 2013. Tailoring nanohybrids and nanocomposites for catalytic applications. Green Chem. 15 (6), 1398–1430. https://doi.org/10.1039/c3gc37141g.

Dhand, C., Dwivedi, N., Loh, X.J., Jie Ying, A.N., Verma, N.K., Beuerman, R.W., Lakshminarayanan, R., Ramakrishna, S., 2015. Methods and strategies for the synthesis of diverse nanoparticles and their applications: a comprehensive overview. RSC Adv. 5 (127), 105003–105037. https://doi.org/10.1039/c5ra19388e.

Duru, I.P., Ozugurlu, E., Arda, L., 2019. Size effect on magnetic properties of Zn0.95−xMgxNi0.05O nanoparticles by Monte Carlo simulation. Ceram. Int. 45 (5), 5259–5265. https://doi.org/10.1016/j.ceramint.2018.11.223.

Esteban Florez, F.L., Hiers, R.D., Larson, P., Johnson, M., O'Rear, E., Rondinone, A.J., Khajotia, S.S., 2018. Antibacterial dental adhesive resins containing nitrogen-doped titanium dioxide nanoparticles. Mater. Sci. Eng. C 93, 931–943. https://doi.org/10.1016/j.msec.2018.08.060.

Garg, V.K., Kataria, N., 2016. Nanomaterial-based sorbents for the removal of heavy metal ions from water. In: Gautam, R.K., Chattopadhyaya, M.C. (Eds.), Advanced Nanomaterials for Wastewater Remediation. CRC Press,, FL, pp. 179–200.

Gatoo, M.A., Naseem, S., Arfat, M.Y., Mahmood Dar, A., Qasim, K., Zubair, S., 2014. Physicochemical properties of nanomaterials: implication in associated toxic manifestations. Biomed. Res. Int. 2014. https://doi.org/10.1155/2014/498420.

Ghanbari, F., Arab, A., Shishe Bor, M., Mardaneh, M.R., 2017. Magnetic properties of La/Ni-substituted strontium hexaferrite nanoparticles prepared by coprecipitation at optimal conditions. J. Electron. Mater. 46 (4), 2112–2118. https://doi.org/10.1007/s11664-016-5140-y.

Giraldo, J.P., Wu, H., Newkirk, G.M., Kruss, S., 2019. Nanobiotechnology approaches for engineering smart plant sensors. Nat. Nanotechnol. 14 (6), 541–553. https://doi.org/10.1038/s41565-019-0470-6.

Goyal, R.K., 2017. Nanomaterials and Nanocomposites: Synthesis, Properties, Characterization Techniques, and Applications. CRC Press, FL, pp. 1–332.

Grover, I.S., Singh, S., Pal, B., 2013. The preparation, surface structure, zeta potential, surface charge density and photocatalytic activity of TiO_2 nanostructures of different shapes. Appl. Surf. Sci. 280, 366–372. https://doi.org/10.1016/j.apsusc.2013.04.163.

Guerra, F.D., Attia, M.F., Whitehead, D.C., Alexis, F., 2018. Nanotechnology for environmental remediation: materials and applications. Molecules 23 (7), 1760. https://doi.org/10.3390/molecules23071760.

Gupta, D., Meher, S.R., Illyaskutty, N., Alex,, Z.C., 2018. Facile synthesis of Cu_2O and CuO nanoparticles and study of their structural, optical and electronic properties. J. All. Comp. 743, 737–745 https://doi.org/10.1016/j.jallcom.2018.01.181.

Han, Y.D., Gao, Y., Zhang, S.T., Jing, H.Y., Wei, J., Zhao, L., Xu, L.Y., 2019. Study of mechanical properties of Ag nanoparticle-modified graphene/Sn-Ag-Cu solders by nanoindentation. Mater. Sci. Eng. A 761,, 138051. https://doi.org/10.1016/j.msea.2019.138051.

Hassan, S.M., Ahmed, A.I., Mannaa, M.A., 2018. Structural, photocatalytic, biological and catalytic properties of SnO_2/TiO_2 nanoparticles. Ceram. Int. 44 (6), 6201–6211. https://doi.org/10.1016/j.ceramint.2018.01.005.

He, X., Deng, H. & Hwang, H. m. (2019). The current application of nanotechnology in food and agriculture. J. Food Drug Anal., *27*(1), 1–21. https://doi.org/10.1016/j.jfda.2018.12.002

Hossain, M.Z., Shimizu, N., 2019. Covalent immobilization of gold nanoparticles on graphene. J. Phys. Chem. C 123 (6), 3512–3516. https://doi.org/10.1021/acs.jpcc.8b09619.

Ikhmayies, S.J., 2014. Characterization of nanomaterials. JOM 66 (1), 28–29. https://doi.org/10.1007/s11837-013-0826-6.

Janas, D., Koziol, K.K.K., 2016. The influence of metal nanoparticles on electrical properties of carbon nanotubes. Appl. Surf. Sci. 376, 74–78. https://doi.org/10.1016/j.apsusc.2016.02.233.

Joseph, L., Jun, B.M., Jang, M., Park, C.M., Muñoz-Senmache, J.C., Hernández-Maldonado, A.J., Heyden, A., Yu, M., Yoon, Y., 2019. Removal of contaminants of emerging concern by metal-organic framework nanoadsorbents: a review. Chem. Eng. J. 369, 928–946. https://doi.org/10.1016/j.cej.2019.03.173.

Jun, L.Y., Karri, R.R., Mubarak, N.M., Yon, L.S., Bing, C.H., Khalid, M., Jagadish, P., Abdullah, E.C., 2020. Modelling of methylene blue adsorption using peroxidase immobilized functionalized Buckypaper/polyvinyl alcohol membrane via ant colony optimization. Environ. Pollut. 259, 113940. https://doi.org/10.1016/j.envpol.2020.113940.

Kalia, V.C., Saini, A.K., 2017. Metabolic engineering for bioactive compounds: strategies and processes. In: Kalia, V.C., Saini, A.K. (Eds.), Metabolic Engineering for Bioactive Compounds: Strategies and Processes. Springer, Singapore, pp. 1–412.

Kamshad, M., Jahanshah Talab, M., Beigoli, S., Sharifirad, A., Chamani, J., 2019. Use of spectroscopic and zeta potential techniques to study the interaction between lysozyme and curcumin in the presence of silver nanoparticles at different sizes. J. Biomol. Struct. Dyn. 37 (8), 2030–2040. https://doi.org/10.1080/07391102.2018.1475258.

Kaur, M., Mubarak, N.M., Chin, Lai Fui, B., Khalid, M., Rao Karri, R., Walvekar, R., Abdullah, E., C., Amri Tanjung, F., 2020. Extraction of reinforced epoxy nanocomposite using agricultural waste biomass, 943 https://doi.org/10.1088/1757-899x/943/1/012021.

Kawadkar, J., Chauhan, M.K., Maharana, M., 2011. Nanobiotechnology: application of nanotechnology in diagnosis, drug discovery and drug development. Asian J. Pharma. Clin. Res. 4 (1), 23–28. http://www.ajpcr.com/Vol4Issue1/218.pdf.

Khan, F.S.A., Mubarak, N.M., Khalid, M., Walvekar, R., Abdullah, E.C., Mazari, S.A., Nizamuddin, S., Karri, R.R., 2020. Magnetic nanoadsorbents' potential route for heavy metals removal—a review. Environ. Sci. Pollut. Res. 27 (19), 24342–24356. https://doi.org/10.1007/s11356-020-08711-6.

Khan, F.S.A., Mubarak, N.M., Tan, Y.H., Karri, R.R., Khalid, M., Walvekar, R., Abdullah, E.C., Mazari, S.A., Nizamuddin, S., 2020. Magnetic nanoparticles incorporation into different substrates for dyes and heavy metals removal—a review. Environ. Sci. Pollut. Res. 27 (35), 43526–43541. https://doi.org/10.1007/s11356-020-10482-z.

Khan, I., Saeed, K., Khan, I., 2019. Nanoparticles: properties, applications and toxicities. Arab. J. Chem. 12 (7), 908–931. https://doi.org/10.1016/j.arabjc.2017.05.011.

Khatami, M., Alijani, H.Q., Fakheri, B., Mobasseri, M.M., Heydarpour, M., Farahani, Z.K., Khan, A.U., 2019. Super-paramagnetic iron oxide nanoparticles (SPIONs): greener synthesis using Stevia plant and evaluation of its antioxidant properties. J. Clean. Prod. 208, 1171–1177. https://doi.org/10.1016/j.jclepro.2018.10.182.

Koushki, E., Baedi, J., Tasbandi, A., 2019. Sodium doping effect on optical permittivity, band gap structure, nonlinearity and piezoelectric properties of PZT nano-colloids and nanostructures. J. Electron. Mater. 48 (2), 1066–1073. https://doi.org/10.1007/s11664-018-6834-0.

Kuangda, L., Theint, A., Nining, G., Ralph, W., Wenbin, L., 2018. Nanoscale metal-organic frameworks for therapeutic, imaging, and sensing applications. Adv. Mater., 1707634. https://doi.org/10.1002/adma.201707634.

Kupracz, P., Coy, E., Grochowska, K., Karczewski, J., Rysz, J., Siuzdak, K., 2020. The pulsed laser ablation synthesis of colloidal iron oxide nanoparticles for the enhancement of TiO_2 nanotubes photo-activity. Appl. Surf. Sci. 530. https://doi.org/10.1016/j.apsusc.2020.147097.

Landowski, L.M., Niego, B., Sutherland, B.A., Hagemeyer, C.E., Howells, D.W., 2020. Applications of nanotechnology in the diagnosis and therapy of stroke. Semin. Thromb. Hemost. 46 (5), 592–605. https://doi.org/10.1055/s-0039-3399568.

Lasheras, X., Insausti, M., Gil De Muro, I., Garaio, E., Plazaola, F., Moros, M., De Matteis, L., M. De La Fuente, J., Lezama, L., 2016. Chemical synthesis and magnetic properties of monodisperse nickel ferrite nanoparticles for biomedical applications. J. Phys. Chem. C 120 (6), 3492–3500. https://doi.org/10.1021/acs.jpcc.5b10216.

Lavanya, K., Kalaimurugan, D., Shivakumar, M.S., Venkatesan, S., 2020. Gelatin stabilized silver nanoparticle provides higher antimicrobial efficiency as against chemically synthesized silver nanoparticle. J. Cluster Sci. 31 (1), 265–275. https://doi.org/10.1007/s10876-019-01644-2.

Lee, Y., Garcia, M.A., Frey Huls, N.A., Sun, S., 2010. Synthetic tuning of the catalytic properties of Au-Fe_3O_4 nanoparticles. Angew. Chem. Int. Ed. 49 (7), 1271–1274. https://doi.org/10.1002/anie.200906130.

Lin, Y.X., Wang, Y., Blake, S., Yu, M., Mei, L., Wang, H., Shi, J., 2020. RNA nanotechnology-mediated cancer immunotherapy. Theranostics 10 (1), 281–299. https://doi.org/10.7150/thno.35568.

Lingamdinne, L.P., Koduru, J.R., Chang, Y.-Y., Karri, R.R., 2018. Process optimization and adsorption modeling of Pb(II) on nickel ferrite-reduced graphene oxide nano-composite. J. Mol. Liq. 250, 202–211. https://doi.org/10.1016/j.molliq.2017.11.174.

Lingamdinne, L.P., Koduru, J.R., Karri, R.R., 2019. A comprehensive review of applications of magnetic graphene oxide based nanocomposites for sustainable water purification. J. Environ. Manage. 231, 622–634. https://doi.org/10.1016/j.jenvman.2018.10.063.

Luska, K.L., Bordet, A., Tricard, S., Sinev, I., Grünert, W., Chaudret, B., Leitner, W., 2016. Enhancing the catalytic properties of ruthenium nanoparticle-SILP catalysts by dilution with iron. ACS Catal. 6 (6), 3719–3726. https://doi.org/10.1021/acscatal.6b00796.

Miśkiewicz, P., 2018. Nanotechnology in textile industry. Nanotechnology in a Nutshel 100, 74–85.

Mukherjee, R., Kumar, R., Sinha, A., Lama, Y., Saha, A.K., 2016. A review on synthesis, characterization, and applications of nano zero valent iron (nZVI) for environmental remediation. Crit. Rev. Environ. Sci. Technol. 46 (5), 443–466. https://doi.org/10.1080/10643389.2015.1103832.

Mustafa, F., Hassan, R.Y.A., Andreescu, S., 2017. Multifunctional nanotechnology-enabled sensors for rapid capture and detection of pathogens. Sensors (Switzerland) 17 (9), 2121. https://doi.org/10.3390/s17092121.

Nazari, A.H., Riahi, S., 2011. Effects of CuO nanoparticles on microstructure, physical, mechanical and thermal properties of self-compacting cementitious composites. J. Mater. Sci. Technol. 27 (1), 81–92. https://doi.org/10.1016/S1005-0302(11)60030-3.

Nithya, P., Sundrarajan, M., 2020. Ionic liquid functionalized biogenic synthesis of Ag–Au bimetal doped CeO_2 nanoparticles from *Justicia adhatoda* for pharmaceutical applications: antibacterial and anti-cancer activities. J. Photochem. Photobiol. B 202, 111706. https://doi.org/10.1016/j.jphotobiol.2019.111706.

Oleyaei, S.A., Zahedi, Y., Ghanbarzadeh, B., Moayedi, A.A., 2016. Modification of physicochemical and thermal properties of starch films by incorporation of TiO_2 nanoparticles. Int. J. Biol. Macromol. 89, 256–264. https://doi.org/10.1016/j.ijbiomac.2016.04.078.

Omri, K., Bettaibi, A., Khirouni, K., El Mir, L., 2018. The optoelectronic properties and role of Cu concentration on the structural and electrical properties of Cu doped ZnO nanoparticles. Physica B 537, 167–175. https://doi.org/10.1016/j.physb.2018.02.025.

Patra, M.K., Manzoor, K., Manoth, M., Negi, S.C., Vadera, S.R., Kumar, N., 2008. Nanotechnology applications for chemical and biological sensors. Def. Sci. J. 58 (5), 636–649. https://doi.org/10.14429/dsj.58.1686.

Priyadarsini, S., Mukherjee, S., Mishra, M., 2018. Nanoparticles used in dentistry: a review. J. Oral Biol. Craniofac. Res. 8 (1), 58–67. https://doi.org/10.1016/j.jobcr.2017.12.004.

Rai, M., Ingle, A., 2012. Role of nanotechnology in agriculture with special reference to management of insect pests. Appl. Microbiol. Biotechnol. 94 (2), 287–293. https://doi.org/10.1007/s00253-012-3969-4.

Rawtani, D., Khatri, N., Tyagi, S., Pandey, G., 2018. Nanotechnology-based recent approaches for sensing and remediation of pesticides. J. Environ. Manage. 206, 749–762. https://doi.org/10.1016/j.jenvman.2017.11.037.

Ruthiraan, M., Mubarak, N. M., Abdullah, E. C., Khalid, M., Nizamuddin, S., Walvekar, R. & Karri, R. R. (2019). An overview of magnetic material: preparation and adsorption removal of heavy metals from wastewater In: Abd-Elsalam K., Mohamed M., Prasad R. (eds) Magnetic Nanostructures. Nanotechnology in the Life Sciences. Springer, Cham (pp. 131–159). https://doi.org/10.1007/978-3-030-16439-3_8

Sahu, J.N., Karri, R.R., Zabed, H.M., Shams, S., Qi, X., 2019a. Current perspectives and future prospects of nano-biotechnology in wastewater treatment Sep. Purif. Rev. 50, 139–158.

Sahu, J.N., Zabed, H., Karri, R.R., Shams, S., Qi, X., Thomas, S., Grohens, Y., Pottathara, Y.B., 2019b. Chapter 11 - Applications of nano-biotechnology for sustainable water purification. In: Thomas, S., Grohens, Y., Pottathara, Y.B. (Eds.), Industrial Applications of Nanomaterials: Micro and Nano Technologies. Elsevier, Amsterdam, pp. 313–340.

Saleh, T.A., 2016. Nanomaterials for pharmaceuticals determination. Bioenergetics 5, 226. https://doi.org/10.4172/2167-7662.1000226.

Sanganabhatla, D., SunderR, S., 2017. Methods of synthesis of nanoparticles. Int. J. Adv. Eng. Nanotechnol. 8, 2347–6389.

Sara, S., Ahmad, Y.K., 2017. Overviews on the cellular uptake mechanism of polysaccharide colloidal nanoparticles. J. Cell. Mol. Med. 1668–1686. https://doi.org/10.1111/jcmm.13110.

Sauceda, H.E., Mongin, D., Maioli, P., Crut, A., Pellarin, M., Fatti, N.D., Vallée, F., Garzón, I.L., 2012. Vibrational properties of metal nanoparticles: Atomistic simulation and comparison with time-resolved investigation. J. Phys. Chem. C 116 (47), 25147–25156. https://doi.org/10.1021/jp309499t.

Shahidi, S., Moazzenchi, B., 2018. Carbon nanotube and its applications in textile industry–a review. J. Text. Inst. 109 (12), 1653–1666. https://doi.org/10.1080/00405000.2018.1437114.

Shahraki, M.M., Bahrevar, M.A., Mirghafourian, S.M.S., 2015. The effect of TiO_2 addition on microstructure and electrical properties of SnO_2 varistors prepared from nanomaterials. Ceram. Int. 41 (5), 6920–6924. https://doi.org/10.1016/j.ceramint.2015.01.146.

Singh, N.B., Kalra, M., Saxena, S.K., 2017. Nanoscience of cement and concrete, 4, pp. 5478–5487.

Sivolella, S., Stellini, E., Brunello, G., Gardin, C., Ferroni, L., Bressan, E., Zavan, B., 2012. Silver nanoparticles in alveolar bone surgery devices. J. Nanomater. 2012. https://doi.org/10.1155/2012/975842.

Sow, C., Riedl, B., Blanchet, P., 2011. UV-waterborne polyurethane-acrylate nanocomposite coatings containing alumina and silica nanoparticles for wood: mechanical, optical, and thermal properties assessment. J. Coat. Technol. Res. 8 (2), 211–221. https://doi.org/10.1007/s11998-010-9298-6.

Tadepalli, S., Wang, Z., Liu, K., Jiang, Q., Slocik, J., Naik, R.R., Singamaneni, S., 2017. Influence of surface charge of the nanostructures on the biocatalytic activity. Langmuir 33 (26), 6611–6619.

Tran, P., Lee, S.E., Kim, D.H., Pyo, Y.C., Park, J.S., 2020. Recent advances of nanotechnology for the delivery of anticancer drugs for breast cancer treatment. J. Pharma. Investig. 50 (3), 261–270. https://doi.org/10.1007/s40005-019-00459-7,

Usman, M.S., El Zowalaty, M.E., Shameli, K., Zainuddin, N., Salama, M., Ibrahim, N.A., 2013. Synthesis, characterization, and antimicrobial properties of copper nanoparticles. Int. J. Nanomed. 8, 4467–4479. https://doi.org/10.2147/IJN.S50837.

Waghmode, M.S., Gunjal, A.B., Mulla, J.A., N., P.N., N., N.N, 2019. Studies on the titanium dioxide nanoparticles: biosynthesis, applications and remediation. SN Appl. Sci. 1 (4), 310. https://doi.org/10.1007/s42452-019-0337-3.

Wang, N., Hsu, C., Zhu, L., Tseng, S., Hsu, J.P., 2013. Influence of metal oxide nanoparticles concentration on their zeta potential. J. Colloid Interf. Sci. 407, 22–28. https://doi.org/10.1016/j.jcis.2013.05.058.

Wei, Y., Han, B., Hu, X., Lin, Y., Wang, X., Deng, X., 2012. Synthesis of Fe_3O_4 nanoparticles and their magnetic properties. Procedia Eng., 27, pp. 632–637.

Wu, C., Huang, M., Luo, D., Jiang, Y., Yan, M., 2018. SiO_2 nanoparticles enhanced silicone resin as the matrix for Fe soft magnetic composites with improved magnetic, mechanical and thermal properties. J. Alloys Compd. 741, 35–43. https://doi.org/10.1016/j.jallcom.2017.12.322.

Xiao, H., Chen, Y., Alnaggar, M., 2019. Silver nanoparticles induce cell death of colon cancer cells through impairing cytoskeleton and membrane nanostructure. Micron 126. https://doi.org/10.1016/j.micron.2019.102750.

Yan, J., Feng, W., Kim, J.Y., Lu, J., Kumar, P., Mu, Z., Wu, X., Mao, X., Kotov, N.A., 2020. Self-assembly of chiral nanoparticles into semiconductor helices with tunable near-infrared optical activity. Chem. Mater. 32 (1), 476–488. https://doi.org/10.1021/acs.chemmater.9b04143.

Yetisen, A.K., Qu, H., Manbachi, A., Butt, H., Dokmeci, M.R., Hinestroza, J.P., Skorobogatiy, M., Khademhosseini, A., Yun, S.H., 2016. Nanotechnology in textiles. ACS Nano 10 (3), 3042–3068. https://doi.org/10.1021/acsnano.5b08176.

Yoksan, R., Chirachanchai, S., 2010. Silver nanoparticle-loaded chitosan-starch based films: fabrication and evaluation of tensile, barrier and antimicrobial properties. Mater. Sci. Eng. C 30 (6), 891–897. https://doi.org/10.1016/j.msec.2010.04.004.

Yu, J., Huang, X., Wang, L., Peng, P., Wu, C., Wu, X., Jiang, P., 2011. Preparation of hyperbranched aromatic polyamide grafted nanoparticles for thermal properties reinforcement of epoxy composites. Polym. Chem. 2 (6), 1380–1388. https://doi.org/10.1039/c1py00096a.

Zhang, D., Yang, Z., Li, P., Zhou, X., 2019. Ozone gas sensing properties of metal-organic frameworks-derived In_2O_3 hollow microtubes decorated with ZnO nanoparticles. Sens. Actuators B 301,, 127081. https://doi.org/10.1016/j.snb.2019.127081.

Zhao, X., Zhao, Y., Liu, Z., Yang, Y., Sui, J., Wang, H.E., Cai, W., Cao, G., 2018. Synergistic coupling of lamellar $MoSe_2$ and SnO_2 nanoparticles via chemical bonding at interface for stable and high-power sodium-ion capacitors. Chem. Eng. J. 354, 1164–1173. https://doi.org/10.1016/j.cej.2018.08.122.

Zhu, S., Chen, C., Li, Z., 2019. Magnetic enhancement and magnetic signal tunability of (Mn, Co) co-doped SnO_2 dilute magnetic semiconductor nanoparticles. J. Magn. Magn. Mater. 471, 370–380. https://doi.org/10.1016/j.jmmm.2018.09.106.

SECTION 2

Environmental Remediation Applications and Future Prospective

CHAPTER 10

Use of nanotechnology for wastewater treatment: potential applications, advantages, and limitations

Wajid Umar[a], Muhammad Zia ur Rehman[b], Muhammad Umair[b], Muhammad Ashar Ayub[b], Asif Naeem[c], Muhammad Rizwan[d], Husnain Zia[b] and Rama Rao Karri[e]

[a]School of Environmental Science, Hungarian University of Agriculture and Life Sciences, Gödöllő, Hungary
[b]Institute of Soil and Environmental Sciences, Faculty of Agriculture, University of Agriculture Faisalabad, Pakistan
[c]Institute of Plant Nutrition and Soil Science, Christian-Albrechts-Universität zu Kiel, Kiel, Germany
[d]Department of Environmental Sciences & Engineering, Government College University Faisalabad, Faisalabad, Pakistan
[e]Petroleum and Chemical Engineering, Faculty of Engineering, Universiti Teknologi Brunei, Brunei Darussalam

10.1 Introduction

Water is the most vital constituent of the life of all living organism, and it is expected that nearly 800 million people worldwide do not have access to good quality water for domestic use (Vallino et al., 2020). Water shortage and its quality regulation have become an important issue for sustainable development because the available water resources are not free from anthropogenic influences of organic (viruses/bacteria, pesticides, anhydrous phase liquids) and inorganic pollutants (salts, metals, and radioactive materials). Steady growth in population coupled with industry development (and expansion) have led to addition of new pollutants in soil-plant-environment continuum (primarily via water pollution) raising concerns for human health (Korashy et al., 2017; Sharma et al., 2016; Wu et al., 2019; Kumari et al., 2019). Contaminated water has a major impact on community health, water ecosystems, environmental sustainability, and societal economy and social wellbeing. For instance, it is analyzed that scarce water supply, along with poor hygiene is responsible for around 30,000 deaths in a day around the world (Mishra et al., 2020). Eighty percent of these cases are found in rural areas and are most common in infants. Today, the demand for hygienic and pure water is growing quickly all around the globe. However, the commonly used treatment methods are not sufficient to deliver the essential quantity of good quality water. Therefore, it is important to look for new energy-saving technology for water purification.

Among the available technologies, the application of nanomaterials (NMs) in the water treatment industry has become very important for the rapid removal of contaminants by adsorption or decomposition processes. For example, NMs are used

Sustainable Nanotechnology for Environmental Remediation
DOI: https://doi.org/10.1016/B978-0-12-824547-7.00002-3

in the wastewater treatment industry to remove dyes, phenols, and toxic trace metals. Another added advantage of magnetic NMs is its easy recovery from the mixture of reaction through an external magnetic field and the ability to reuse it several times (Kefeni et al., 2017). However, appropriate concerns are not addressed, some nanoparticles (NP) may be toxic and represent a new category of contaminants that are harmful to the environment and human health (Ying et al., 2017). From few years magnetic NPs (MNPs) also have gained attention because of their special superparamagnetic properties, the capability of high adsorption, and the relationship between size and surface area (Tshikovhi et al., 2020). Specifically, transition metal oxides with spinel structures called ferrites are one of the most important MNPs (Amiri et al., 2019). According to their magnetic properties and crystalline structures, ferrites are classified as spinel (MFe_2O_4, where M = Co, Mn, Zn, Fe, Ni, etc.), garnet ($M_3Fe_5O_{12}$, where M = rare cations), hexaferrite ($SrFe_{12}O_{19}$ and $BaFe_{12}O_{19}$), and orthoferrite ($MFeO_3$, M = rare earth cations). Among them, special attention was paid to spinel ferrite NPs (SFNPs). This is because of their simple chemical composition, excellent magnetic properties, and wide application in numerous fields, including water and sewage treatment, biomedicine, catalysts, and electronic devices (Ciocarlan et al., 2018).

The wastewater being generated by the urban community and industry is considered a major transporter of pollutants and must be manage and disposed off properly. The need to control the harmful pollutants present in the wastewater (organic/inorganic pollutants) and complex compounds has attracted direct attention for research. The use of effective technologies for the treatment of these compounds is one of the main ways to enhance water quality. In general, the purpose of this chapter is to summarize the application of NMs in water treatment including adsorption, obtaining analytes, membrane modification, photocatalysis, pathogen disinfection, and removal of metal ions. The properties, advantages, and disadvantages of different NMs and their role in the water treatment process were also studied and presented schematically that are understandable to a wide range of readers.

10.2 Nature of pollutants present in wastewater

Water pollution with organic and inorganic pollutants has turned out to be the most important environmental concern that threatens human life globally. In this regard, it is necessary to pay consideration to certain technologies and methods for their elimination from wastewater. Mainly there are three classes of pollutants present in water and wastewater, including inorganic, organic, and biological pollutants and are hazardous for the environment and living beings.

10.2.1 Inorganic pollutants

Industrial waste, sediments, heavy metals, and nutrients come under the category of inorganic contaminants. Arsenic comes under one of the most lethal inorganic pollutants, and along with other heavy metals such as lead, chromium, cadmium, etc. also pose serious threats to the environment and human beings (Rahman, 2020). Other inorganic ionic components such as F^-, SO_4^{-2}, NO^{3-}, Cl^-, etc. also reduce the quality of water at higher concentrations (Umar et al., 2016). Mostly inorganic pollutants occur naturally but their concentrations have gone raised owing to unmanaged human activities in the environment such as mining, industrialization, and agricultural activities (Masindi and Muedi, 2018). Often inorganic contaminants are not decomposable and their buildup in the human body can cause serious diseases (El-Kady and Abdel-Wahhab, 2018). Wastewater from production activities is characterized by an abundance of toxic and carcinogenic pollutants. In recent years, the restoration of the environment has been concentrated on the utilization of highly effective adsorbents to draw out metals and metalloids that subsist in wastewater (Frezzini et al., 2018).

10.2.2 Organic pollutants

Organic pollutants comprise hydrocarbons, pesticides, polychlorinated biphenyl, polycyclic aromatic hydrocarbon, pharmaceuticals, and surfactants (Alharbi et al., 2018; Frezzini et al., 2018; Izanloo et al., 2019; Kumari & Gupta, 2020). These contaminants show a wide range of properties (solubility, volatility, polarity) even occupying the same groups. Some of these organic pollutants are highly bio-accumulative, persistent, susceptible to far-off transport, and potentially toxic (Liu et al., 2020; Umar et al., 2016, 2019). Organic contaminants such as pesticides, pharmaceutical residues, paints, and other wastes cause environmental problems resulting from sewage spills, water runoff, industrial wastes, and agricultural activities (Castro-Jiménez et al., 2015; Hageman et al., 2015; Jaspers et al., 2013). Wastewater from treatment plants also contains millions of pathogenic and nonpathogenic bacteria per milliliter, including coliforms, *streptococci, staphylococci*, spore-forming anaerobic bacteria, and many other health-hazardous organisms (Frezzini et al., 2018; Karpińska & Kotowska, 2019; Liu et al., 2020). The persistence of microbial pathogens in treated water sources poses a serious health risk in general. Despite great boost in water treatment practices, waterborne diseases are yet a real-time prejudice to health worldwide. There are several reports on the presence of various organic pollutants and pathogenic microbes such as *Salmonella, Vibrios*, and *Listeria* in samples of treated and raw water collected from the inlets (Saingam et al., 2020).

10.3 Agricultural consumption and toxicities associated with wastewater irrigation

In most of the semi-arid and arid regions, it is becoming desirable for farmers to practice wastewater sources because of the situation of scarcity of freshwater (Ahmad et al., 2018; Qadir et al., 2010; Murtaza et al., 2010; Raja et al., 2015). Thus, the deficiency of appropriate and easy alternatives has made crop irrigation with the wastewater an effective preference. The reuse of the major share of the total municipal water has been done by the agriculture sector (Mireles et al., 2004; Angelakis & Snyder, 2015; Miller-Robbie et al., 2017), which makes the total consumption of agriculture water to reach up to 70% (Scotland, 2001). The studies proved that >20 million ha of crop around the globe use partially treated or untreated industrial/municipal wastewater for irrigation (Qishlaqi et al., 2008; Yucui & Yanjun, 2019). Currently, wastewater crop irrigation is acquiring a lot of consideration. In developed countries, the facilities and implementation of technologies for wastewater treatment have led to the increased application of treated wastewater (Chen et al., 2013).

The contamination of approximately 5500 bn m^3 of water/year is happening because of the worldwide discharge of the wastewater that reaches to 400 bn m^3/year (Yuan et al., 2019; Yucui & Yanjun, 2019). Different countries show a difference in the quality and nature of the used wastewater. The studies showed that around 44 countries are reusing reclaimed water for irrigation purposes that amounts to 15 million m^3/day (Winpenny et al., 2010). Also, the untreated wastewater usage in periurban and urban agriculture is estimated to be around 11% of irrigated croplands around the globe (Thebo et al., 2017). It is also estimated that the food produced by wastewater irrigation has been consumed by 10% of the total population of the world ("World Health Organization," 2006). No doubt, the utilization of wastewater for crop growth in low-income countries is poorly regulated and there is a poor understanding of economic and environmental issues (Qadir et al., 2010). In many poor countries, the sewage water has been used for irrigation despite the fact that it is illegal (Huibers & Lier, 2005). But in middle-income countries, the usage of wastewater is being practiced after treatment. All of these wastewater sources are leading to mass food contamination, which is a highly alarming situation regarding health risks. Several studies reporting the food chain contamination with wastewater irrigation have been summarized in Table 10.1.

In the agriculture sector, the inundation of crops with unprocessed wastewater has varied potential in relation of the negative and the positive impacts mainly on the production of crops, the human well-being, and soil productivity (Khalid et al., 2017, 2018). The risks to the environment and health might be due to the chemical constituents and unwanted contents that are present in wastewater. High nutrient contents, high total dissolved and suspended solids, and potentially toxic elements (PTEs) in wastewater are the main cause of negative impact on crops irrigated with wastewater (Khalid et al., 2017, 2018; Shahid et al., 2015). A large amount of salt persistence in wastewater that accumulates in roots can also pose an effect on the productivity and quality of the soil. The continued usage of salt-rich wastewater can affect the productivity of soil by deteriorating soil structure (Khalid et al., 2017). Recently, soil salinization because of irrigation with wastewater has been testified extensively. The heavy metal pollutants metal ions that are a considerable part of wastewater have a major effect on crop productivity (Shahid et al., 2015, 2017). In accordance with the remarks concluded by Mireles et al. (2004), the

Table 10.1 General mechanism of the degradation of organic pollutants through photocatalyst.

Crops grown with wastewater	Pollutants	Food contamination level	References
Brassica juncea L., *Lactuca sativa* L., *Spinacia oleracea* L., *Brassica napus*, *Raphanus sativus* L., *Brassica oleracea* L., and *Zea mays*	Cr, Ni, Pb, Cu, Zn, Cd	Moderate to high	Khan et al., 2008a
Raphanus sativus, *Spinacia oleracea*, *Petroselinum crispum*, *Mentha viridis*, *Coriandrum sativum*, and *Brassica oleracea* var botrytis	Cr, Ni, Pb, Cu, Zn	High	Gupta et al., 2012
Brassica oleracea and *Lactuca sativa*	Bisphenol A, naproxen, diclofenac sodium and 4-nonylphenol	High	Dodgen et al., 2013
Allium cepa, *Solanum lycopersicum* *Allium sativum*, and *Solanum melongena*	Fe, Cr, Pb, Ni, Co, Zn, Cu	High	Amin et al., 2013
Brassica rapa, *Brassica oleracea var. botrytis*, *Raphanus sativus*, *Solanum melongena*, *Spinacia oleracea*, *Nelumbium nelumbo*, *Trigonella foenum-graecum*, and *Coriandrum sativum*	Fe, Mn, Cu, Zn	Low to moderate	Arora et al., 2008

(continued on next page)

Table 10.1 General mechanism of the degradation of organic pollutants through photocatalyst—cont'd

Crops grown with wastewater	Pollutants	Food contamination level	References
Abelmoschus esculentus	Pb, Cd, Cr, Zn, Cu,	High	Balkhair & Ashraf, 2016
Lactuca sativa, Spinacia oleracea, Brassica oleracea var. capitata, *Allium cepa, Solanum tuberosum, Solanum lycopersicum,* and *Capsicum annuum*	Cu, Cd, Mn, Zn, Cr	Moderate to high	Cheshmazar et al., 2018
Vegetables	Sb, As, Cr, Hg, Mo, Ni, and Se	Moderate to high	Genthe et al., 2018
Raphanus sativus L., *Coriandrum sativum* L., *Solanum tuberosum* L., *Brassica oleracea, Daucus carota* L., *Brassica oleracea capitata Beta vulgaris* L., *Brassica rapa* L., *Spinacia oleracea* L., *Allium sativum* L., *Brassica campestris* L.,	Zn, Pb, Cr, Cd, Ni, Cu, Co, and, Mn	High	Mahmood & Malik, 2014
Brassica juncea, Brassica napus	Cu, Zn, Cd, Ni, Pb, Cr	High	Mapanda et al., 2007

(continued on next page)

Table 10.1 General mechanism of the degradation of organic pollutants through photocatalyst—cont'd

Crops grown with wastewater	Pollutants	Food contamination level	References
Crops: *(Trifolium alexandrinum, Sorghum bicolor, Zea mays, Oryza sativa, Triticum aestivum, Medicago sativa, Saccharum officinarum)* Vegetables: *(Brassica oleracea var. botrytis, Spinacia oleracea, Brassica juncea, Lagenaria siceraria, Brassica oleracea)*	Cr, Mn, Pb, Zn, Ni	High	Raja et al., 2015
Brassica pekinensis L., *Ipomoea aquatica Forsk, Lactuca sativa, Allium sativum* L., *Ipomoea batatas Lam, Brassica oleracea* L. var. Caulorapa DC	Cd, As, Zn, Cu Sb, Pb	High	Yuan et al., 2019
Triticum aestivum	Cu, Cr, Mn, Zn, Fe, Pb, Cd, Ni	High	Hassan et al., 2013
Allium cepa L., *Solanum melongena* L., *Daucus carota* L., *Triticum aestivum* L., *Lycopersicum esculentum* L., *Coriandrum sativum* L., *Capsicum annum* L., *Allium sativum* L., *Spinacia oleracea* L., *Mentha spicata* L., *Raphanus sativus* L., *Abelmoschus esculentus* L.	Cr, Ni, Pb, Mn, Cd	High	Khan et al., 2013

(continued on next page)

Table 10.1 General mechanism of the degradation of organic pollutants through photocatalyst—cont'd

Crops grown with wastewater	Pollutants	Food contamination level	References
Lactuca sativa L.	Polycyclic aromatic hydrocarbons: (naphthalene, acenapthylene, acenaphthylene, fluorine, benzo(b)fluoranthene, phenanthrene, benzo(a)pyrene, anthracene, benzo(k)fluoranthene, indeno(1,2,3-cd)pyrene, fluoranthene, benzo(a)anthracene, crycene, pyrene, dibenzo(ah)anthracene, benzo(ghi)perylene) Heavy metals: Ni, Pb, Cu, Cr, Cd	High	Khan et al., 2008a
Triticum aestivum, *Spinacia oleracea*, *Lactuca sativa* and *Apium graveolens*	Ni, Pb, Cd, Zn, Cr	Moderate	Qishlaqi et al., 2008

(continued on next page)

Table 10.1 General mechanism of the degradation of organic pollutants through photocatalyst—cont'd

Crops grown with wastewater	Pollutants	Food contamination level	References
Brassica nigra, Raphanus sativus and Colocasia esculentum	Fe, Cd, Mn, Cr, Pb	High	Gupta et al., 2012
Triticum aestivum	Cd, Mn	High	Ahmad et al., 2019
Triticum aestivum	Cd, Pb, Cu, Cr, Co	Moderate to high	Ahmad et al., 2018
Solanum lycopersicum L., *Solanum melongena, Cucurbita pepo, Piper nigrum, Brassica oleracea* var. *capitata, Lactuca sativa, Petroselinum crispum, Eruca vesicaria* ssp. *sativa, Solanum tuberosum,* and *Daucus carota* sp. *sativus*	Carbamazepine, acesulfame, caffeine	High	Riemenschneider et al., 2016
Brassica oleracea var. *botrytis, Phaseolus vulgaris, Beta vulgaris,* and *Spinacia oleracea*	Pb, Cu, Zn, Ni, Cd, Cr	High	Chopra and Pathak, 2015
Solanum tuberosum L., *Brassica oleracea* L., *Raphanus sativus* L., *Beta vulgaris* L., *Brassica nigra* L., and *Allium sativum* L., *Brassica oleracea* L. var. *capitata* L.	Cu, Cr, Zn, Ni, Cd, Pb, Co	High	Verma et al., 2015

agricultural sites in Mexico (D.F, Hidalgo and Tlahuac, Mixquiahuala) are deteriorated and unable to support plant growth due to wastewater irrigation for more than 50 years.

Wastewater irrigation of crops poses risks for human health including pathogens, organic and inorganic pollutants' exposure to farmers and consumers (Khalid et al., 2017). Human beings can expose to PTEs directly by accidental ways such as during working processes, by flooding actions due to heavy rains, by using wastewater for domestic activities, and during recreational activities (Fuhrimann et al., 2016). The source of many diseases owing to the viruses, nematodes, bacteria, and protozoa is also due to wastewater (Uyttendaele et al., 2015). In many low-income countries, wastewater is being released into surface water tables with no or little treatment (Qadir et al., 2010). Discharge of untreated industrial and municipal wastewater in other water bodies such as seas and oceans is a cause of the fast-growing deoxygenated dead zones. The studies found that the discharge of wastewater in water bodies has affected around 245,000 km^2 of oceanic ecosystems along with fisheries, food chains, and livelihoods (Corcoran, 2010). It is indicated by the recent international data that the countries where crop irrigation has been carried out by untreated wastewater have alarmingly growing records of sanitation and wastewater-related diseases. Intake of contaminated water and wastewater-fed crops and fish leads to indirect exposure (Mok & Hamilton, 2014). Humans are more prone to get exposed to these toxic compounds in the case of PTEs such as drinking water that is contaminated, inhaling dust, or atmospheric inhalation. However, the main reason for human exposure (>90%) to PTEs is with the food consumption contaminated with PTEs (Zia et al., 2017).

The escalation in the unrestrained utilization of the effluent-containing water without treatment for crop growth purposes in most areas has expand the possibility of human exposure to PTEs (Singh et al., 2010). The consumption of highly adulterated PTE diet may lead to very serious health issues connected with the etiology of many health problems such as heart problems, kidney, nervous system, blood, and bone diseases as revealed by clinical studies (Shahid et al., 2015, 2017). The usage of vegetables contaminated with PTE is the reason for the nutrient imbalance in humans that may cause many health problems such as immunological disorders, disability associated with malnutrition, upper gastrointestinal cancer, impaired psychosocial abilities, and intrauterine growth retardation (Khalid et al., 2018). Pb and Cd are the PTEs that have the capability of being mutagenic, carcinogenic, and teratogenic; the elevated accumulation of Pb and Cd in comestible parts of the plants are recognized as the etiology of gastrointestinal cancer (Järup, 2003). Furthermore, it is reported that Pb can also cause renal infection, reproductive system dysfunction, increase in blood pressure, and improper hemoglobin synthesis (Pourrut et al., 2013). Infants and children are most vulnerable to the contaminants of wastewater (Mara & Sleigh, 2010). In the early nineteenth century, the legislation regarding the utilization of wastewater for inundation of comestible crops and risks associated with health came into existence. Throughout that period in periurban fields, the utilization of wastewater for crop inundation purposes brought catastrophic epidemics of various waterborne diseases (Cotruvo et al., 2004). All these issues of health have resulted in establishing some legislation at a worldwide level, such as Britain's Public Health Act, regarding wastewater into the soil and the release of rainwater in the river (Jaramillo & Restrepo, 2017). The International Office of Public Hygiene has been established to implement sanitary controls alongside the borders (Vilar & Bernabeu-Mestre, 2011).

In the early 1950s, in many cities, the problem of diseases caused by wastewater had resulted in the growth of subversive sewage systems around the world. Furthermore, the chain of sanitary awareness workshops, scientific conferences, and educational seminars based on viable environment improvement is the result of the international environment and health movement, which is endorsed and sponsored by many European countries. Taking into consideration the associated risks of health and environment by using the wastewater in agriculture, the guidelines were made by World Health Organization (WHO) in 1973 regarding the "Reuse of effluents: methods of wastewater treatment and health safeguards." However, those rules were further restructured and reorganized in the year 1989 and 2006 by keeping epidemiological studies into consideration. Many parameters are now included in these restructured and updated guidelines such as health risk assessments, etc. Food contamination with wastewater irrigation is live and imminent threat as described in Table 10.1.

10.4 Characteristics making NPs pertinent for wastewater treatment

Characteristics of NMs such as higher reactivity, higher surface area, hydrophobic, hydrophilic, and electrostatic interaction, antimicrobial properties, and adjustable pore volume make NMs a suitable entity for the photocatalysis and adsorption catalyst in the removal of pollutants (Alharbi et al., 2018; Frezzini et al., 2018; Izanloo et al., 2019; M. Kumari & Gupta, 2020). There are three main categories of NMs such as; NPs, nanocomposites, and carbonaceous NMs. These NMs can be subject to use in the processing of ground as well as wastewater, etc.

NPs make a diversified group of molecules and atoms of metals and metal oxides. It was reported that most of these atoms are nonsaturated and exist on the facets of NPs, this property leads to the higher reactivity of NPs with other atoms (Mireles et al., 2004; Angelakis & Snyder, 2015; Miller-Robbie et al., 2017). That is the reason that nowadays NPs have been used in several environmental cleanup activities such as pesticides, radioactive elements, dyes, salts, nitrate, hydrocarbons, and heavy metals (HMs) (Cheng et al., 2018; Dhawan et al., 2018; Fan et al., 2017; Fato et al., 2019; Hosseini et al., 2018).

Since the 1950s the metal oxides NPs such as Fe_2O_3, CuO, CeO_2, ZnO, WO_3, TiO_2, etc. have been used in several catalytic applications (P. Jackson et al., 2013; S. D. Jackson & Hargreaves, 2008), to convert harmful contaminants to harmless compounds by decomposition (Bhatia et al., 2017; Kanakaraju et al., 2014). The extensive use of metal oxide NPs as a cocatalyst or photocatalyst is due to higher stability and porosity, abundance, light absorption efficiency, unique electronic structure, and charge transfer properties (Arya et al., 2019). MNPs are extensively used in the remediation of contaminated water by attacking the physical attributes of contaminants via their premier magnetic characteristics (Piao et al., 2012). MNPs possess a lower volume to surface ratio, facile separation from aqueous solution, and metal loaded magnetic adsorbents. Due to these properties, MNPs are considered as cost-effective, efficient, and a high potential technique for the processing of contaminated water in the existence of an exterior magnetic field (Bhimrao et al., 2020; Piao et al., 2012).

To augment the effectiveness of NPs composite, NMs have been derived by the saturation of these metal and metal oxide NPs into spongy supportive materials such as polymers, etc. (Fan et al., 2017; Hua et al., 2012; Ihsanullah et al., 2016). Combining several NPs to form a composite material and to evaluate the positive changes in the structure, properties, and activity of these composite NMs have gained significance nowadays. Bimetallic nanocomposites can attract photo-induced carriers for better separation of charge and can provide more redox-active site availability (Chen et al., 2013). The attainability of two types of surfaces enhances the catalytic reactivity and increases the surface area (Ataee-Esfahani et al., 2010). It is obvious from the reviewed works that nanocomposites of Ni/Fe, Fe/Cu, Pt/Fe, Ag/Fe, Pd/Fe have been used successfully for the processing of water spoiled with different types of organic and inorganic pollutants and heavy metals. Nanocomposites of semiconductor NPs coated with metals (e.g., Pd, Au, Pt, Ag) showed higher photocatalytic activity toward the pollutants such as phenols, heavy metals, dyes, formic acid, humic acid, oxalic acid, and urban wastewater (Pan & Xu, 2013).

Characteristics of carbonaceous NMs such as high facets' ratio, hydrophobicity, electrical attributes, adsorption behavior, higher mechanical strength, easy application, and separation characteristics make them one of the suitable materials for the processing of wastewater (Mittal et al., 2015). Fossil fuel stuff such as methane and ethane can be the source of these materials (Al-Jumaili et al., 2017). Fullerene NMs such as C_{60} can act as a photosensitizer because it can produce reactive oxygen species (ROS) by giving energy or electron to ground state oxygen. The produced ROS can act as an oxidizing mediator for the degradation of existing organic contaminants and the disinfection from microbes (Chae et al., 2009). Significant utilization of carbon nanotubes (CNTs) in the wastewater treatment is due to its exceptional properties such as higher surface area, higher attachment capacity, higher chemical and thermal stability, hollow structure, and the interaction of CNTs with pollutants (Fan et al., 2017; Hua et al., 2012; Ihsanullah et al., 2016).

10.5 Production technology of NPs

It is possible to synthesize different nanostructures by using a different process. The structure that has at least one dimension less than 100 nm is known as a nanostructure. Different methods have been used to synthesize liquid dispersed and dry NPs. NPs can be formed by the aggregation of atoms or by the reduction of micro-sized particles. Ultrasound, Inert gas condensation, template synthesis, coprecipitation, microwave, microemulsion, sputtering, hydrothermal synthesis, spark discharge, laser ablation, sol-gel, and biological synthesis are certain techniques for NPs production (Rane et al., 2018).

10.5.1 Coprecipitation

Nucleation, agglomeration, and growth processes have occurred simultaneously in a precipitation technique. Precipitation-related reactions show the following characteristics: products of precipitation are insoluble that are typically formed under saturation conditions. Many fine particles are formed under the process of nucleation, which is the main step in precipitation. Other processes such as ostwald ripening and aggregation also have a substantial influence on the properties and size of the NPs (Rane et al., 2018). Condition of supersaturation is required to induce precipitation (Sagadevan et al., 2018). The synthesis of different NPs such as ZnO, Co_3O_4 and MNPs has been reported by this method (Adam et al., 2018; Huang et al., 2018; Janjua, 2019).

10.5.2 Microemulsion

To prepare the inorganic NPs, microemulsion is considered as one of the ideal methods, but the process of formation of NP in microemulsions is not well defined. However, researchers have proposed a mechanism for the production of NPs by this method. In this method, exchange of reagents occurs due to the collision of water particles with reagents during the mixing process. The change of the reagent is very rapid and leads to the precipitation reaction in the NPs, after that the coagulation and nucleation growth of particles occurs and leads to the formation of NPs surrounded with water or by surfactant materials (Rivera-Rangel et al., 2018). Size-controlled synthesis of silver and starch NPs was carried out by this method (Chin et al., 2014; Solanki & Murthy, 2011).

10.5.3 Ultrasound

Nowadays, ultrasound has been used commonly for the preparation of NPs. In this method, ultrasonic cavitation has been formed when the liquid solutions are irradiated with ultrasonic waves. This process leads to several changes such as high pressure, cooling rate, and temperature and provides suitable conditions for chemical reactions. Ultrasound examination is a great way to prepare NPs with controlled morphology (Hajnorouzi and Afzalzadeh, 2019). The synthesis of AgNMs (Kumar et al., 2014), and gold NPs (Bhosale et al., 2017) was carried out by this method.

10.5.4 Hydrothermal synthesis

Hydrothermal synthesis of individual crystals of atoms solely depends on the absolute solubility of subjected minerals in the presence of hot water and elevated pressure. The NPs urged crystal growth takes place in a device comprising an iron pressure-maintaining vessel known as autoclave, in which the raw material is fed alongside with the water. At the adjacent corners of the crystal growth controlling chamber, a temperature gradient is maintained to monitor the hot tip–mediated dissolution of nutrients and the cold tip–driven seeds multiplication (Caramazana et al., 2018).

10.5.5 Inert gas condensation

This technique is frequently known for the prompt preparation of metal-based NPs. In inert gas condensation, metals evaporate at very high pressure in a vacuum chamber surrounded by argon or helium gas. The high energy containing excited metal atoms when tend to drop their kinetic energy owing to collisions with the pertaining medium gas molecules results in the formation of fine particles. The small particles then grow as a result of coagulation, eventually forming nanocrystals (Zheng and Branicio, 2020).

10.5.6 Sputtering

The explosion of atoms with energetic particles on the surface of a material is called evacuation. Fluctuation is a rapid transfer process that forms a cathode/target that drives atoms with bombarded ions. The pulverized atoms move until they reach the substrate, where they form the desired layer (Tan et al., 2019). Gunnarsson et al. (2015) described the production of TiO_2NPs by the sputtering method. The synthesis of a nanocatalyst (Pd, Pt) was also carried out by this method (Orozco-Montes et al., 2019).

10.5.7 Microwave-assisted technique

This technology has been used in nanotechnological and biochemical processes. Chemical reactions are usually faster and have a higher conductivity than traditional methods of convection heating, and the byproducts are less in number. Microwave reactors withstand high temperatures and pressures, perfectly control the reaction mixture, and exhibit reaction-to-reaction repeatability. While using a microwave technique, a high-degree separation of NPs growth and nucleation processes has been provided in an engineering way at the time of the start of the reaction at room temperature. By microwave heating, the front material of the NPs can be activated selectively, which is imperative for scaling up. Microwave synthesis can selectively heat solvent or precursor molecules to prepare NMs (Cao et al., 2017).

10.5.8 Laser ablation

The elimination of material from an exterior by laser irradiation is known as laser ablation. At low laser flow, sublimation and evaporation of material are carried out due to the absorption of laser energy. At high laser current, the material is usually converted to plasma (Moura et al., 2017). The production of metal NPs was reported by the laser ablation method (Mansoureh and Parisa, 2018).

10.5.9 Sol-gel

By the sol-gel method, small molecules can be used to produce fine size solid particles. In this method, gel-like material is formed by mixing two different solutions, whose morphologies vary from continuous polymer networks to discrete particles. To continue production, you need to follow the following steps. Make the solutions of the required materials. After that polycondensation and hydrolysis reactions will occur to create a transparent and stable soil system. Particles in colloidal suspension gradually accumulate as the sols kept undisturbed for a certain period, forming gels with the structure of three dimensions. After that, processes such as sintering and drying will lead to the formation of nano- and microstructures (Salavati-Niasari et al., 2016).

10.5.10 Template synthesis

Template synthesis is the green and environment-friendly method for the preparation of NMs. To prepare inorganic NPs (monodispersed), this method is considered as one of the promising techniques in which singular cavities of cellular materials are used as hosts to permit the NPs synthesized. The role of the model is double. It permits the structure to be as large as possible and plays the skeletal role in arranging the various functions, active components, and different interfaces of the device (Xin et al., 2018).

10.5.11 Biological synthesis

This method is based on the green chemistry approach. This method combines biotechnology and nanotechnology. The rate of NPs synthesis is slow in this process and the NPs are not monodispersed (Rauwel et al., 2015). Nucleation is dependent on the concentration of various components involved in the synthesis of macromolecules, leading to a drop in the polydispersity of the NPs and consequently, the synthesis rate (Rane et al., 2018). Optimization of several processes is needed to tackle the persistent issues involved in the NPs formation process such as microbial culture improvements, extraction methods enhancement; and a comprehensive methodology such as photobiological technique can be practiced. To improve the properties of NPs and to boost the potential of NPs synthesis, mechanisms such as molecular, cellular, and biochemical needed to be studied. The potential of biological materials for the synthesis of NPs is still needed to be explored due to the diversity in microbes and plants (Singh et al., 2016).

10.6 Nanotechnological processes in wastewater treatment

NMs possess some exceptional characteristics such as the smaller size, higher surface area, higher mobility in solution, high reactivity (Wu et al., 2019), dispersibility, porosity characters, hydrophobicity, hydrophilicity, and strong mechanical properties (Daer et al., 2015; Tang et al., 2014). Recently, some advanced developments occur in NMs to treat the polluted water such as nanosorbents, nanomotors, nanomembranes, nanophotocatalyst, etc.

10.6.1 Nanophotocatalyst

Two Greek words "Photo" and "catalysis" combined to form the word "photocatalysis," which means the degradation of compounds in the existence of light (Yaqoob et al., 2020). For photocatalysis, there is no consensus definition in the world of science (Saravanan et al., 2017). However, the term photocatalysis is used to describe a procedure to stimulate or activate the material by light (visible/sunlight/UV). The major difference between the photocatalyst and the thermocatalyst is that the prior is activated by light energy photons while the latter is activated by heat (Gomes et al., 2019). Nanophotocatalyst is commonly used to treat contaminated water, because due to shape-dependent features and higher surface ratio they can enhance catalytic activity (Chen et al., 2019).

The response of NPs is different than bulk materials due to exceptional properties of surface and different quantum effects (Yaqoob et al., 2020). It also increases their optical, magnetic, electric, mechanical properties, and chemical reactivity (Ong et al., 2018). Gómez-Pastora et al. (2017) reported that a nanophotocatalyst increased the production of oxidizing species on the material surface which expand the oxidation ability and enhanced the degradation of the pollutant from the polluted water.

Environmental pollutants such as chlorpyrifos, azo dyes, nitroaromatics, organochlorine pesticides, etc. are mostly treated by semiconductors, zero valance metal base, and some bimetallic NPs (Loeb et al., 2018; Samanta et al., 2016). Several researchers reported the use of TiO_2 nanotubes to remove the pollutants (organic pollutants i.e., chlorinated ethene, toluene, Congo red, azo dyes, dichlorophenol trichlorobenzene, phenol aromatic) from polluted water (Bhatia et al., 2017; Liang et al., 2019a). However, the most commonly used nanophotocatalyst are Al_2O_3, ZnO, SiO_2, TiO_2, etc. (Ali et al., 2019; Bhanvase et al., 2017). From all of these, TiO_2 is considered a very good photocatalyst due to reasons such as low cost, nontoxic properties, easy availability, and chemical stability. Naturally, it occurs in three different states brookite, rutile, and anatase; among these anatase, is supposed to be a good photocatalyst material (Yamakata and Vequizo, 2019). From other materials, ZnONPs are also used as a photocatalyst to treat the wastewater and can be reused effectively (Das et al., 2018; Hassan and Elhadidy, 2019). Li et al. (2014) examined the effectiveness of composite NMs CdS/TiO_2 as a photocatalyst to degrade the dimethyl sulfoxide (reference material) under the visible light. NMs doped with iron can be recycled and reused easily because of their ferromagnetism ability (Phokha et al., 2016; Serrà et al., 2020). Berekaa (2016) described the elimination of *Escherichia coli* from contaminated water by using the photocatalytic activity of Pd-incorporated ZnONPs. However, hard work has been focused on metal oxides to enhance their photocatalytic abilities by modifying them with other materials such as metal ions and metals

(Umar et al., 2019), dye sanitizers, and carbonaceous materials (Malik et al., 2014). However, further modification is need.

10.6.2 Nano- and micromotors

During the recent past, there are huge developments in the field of nanotechnology and provide different approaches to treat the water. Nowadays, nano/micromotors are introduced that can take energy from different resources and convert that energy into machine-driven force (Yaqoob et al., 2020). They have the ability to accomplish distinct goals through several processes. These advanced motors have exciting and significant applications and can be operated by energy or without energy sources (electric field, magnetic field, acoustics) (Jurado-Sánchez and Wang, 2018). Specific control movement, self-mix ability, high speed, and power are some characteristics of these motors. For the sustainability and stability of the environment, it is essential to get rid of the pollutants from contaminated resources (Moo and Pumera, 2015). Increased demand for purified and clean water globally enhanced water purification and treatment activities. Previously different traditional approaches have been used to treat the contaminated groundwater, sediments wastewater, freshwater, etc. (Yaqoob et al., 2020). Traditional methods are insufficient by diffusion and demand external agitation to enhance water treatment. Due to competencies and their self-propulsion, nano/micromotors can overcome the diffusion boundary through energetic mixing (Yaqoob et al., 2020). The self-propulsion ability of nano/micromotors significantly enhances the water treatment efficiency by integration with materials nano/microstructure that is, greater workability and surface area (Chi et al., 2018; Pacheco et al., 2019). Moreover, nano/micromotors can serve as programmed cleaners in microscale detentions, where traditional approaches do not work efficiently. These motors can be effectively used in the effective remediation of wastewater contamination (Table 10.2).

10.6.3 Nanomembranes

An exclusive type of membrane composed of different nanofibers to remove the undesirable small size particles from aqueous media is known as nanomembranes (Figs. 10.1 and 10.2). In this procedure, the development takes place at very high exclusion speed and it can also serve as a pretreatment method for reverse osmosis (Jhaveri and Murthy, 2016). Ultrafiltration (UF), reverse osmosis, and nanofiltration can be done by a water porous membrane. There are three categories of nanomembranes; inorganic, organic, and organic–inorganic hybrids. Organic nanomembranes are made up of polymeric materials that is cellulose acetates, polyethersulfone, cellulose nitrates, polyacrylonitrile, polyvinyl alcohol, biomacromolecules, etc. (Soyekwo et al., 2014; Wei et al., 2017). However, inorganic nanomembranes are made up of metals, metal oxides (Facciotti et al., 2014; Kanakaraju et al., 2014), CNTs (Brady-Estévez et al., 2010; Rahaman et al., 2012), and graphene (Zhang et al., 2019). When inorganic material is

Table 10.2 Nano/micromotors used in wastewater treatment.

Nanomotors	Fuel	Target pollutant	Mechanism	Reference
TiO_2/Au NPs/Pt nanomotors	H_2O_2	Superorganic mixture	Photocatalysis	Liu et al., 2020
Zero-valent iron (ZVI)/Pt	H_2O_2	Methylene blue	Fenton reaction	Pourrahimi et al., 2018
Au/TiO_2/Mg microspheres	———	Methyl paraoxon and bis-p-nitrophenyl phosphate (b-NPP)	Photocatalysis	Pourrahimi et al., 2015
GO_x-Ni/Pt	H_2O_2	Lead (Pb)	Adsorption	Zhang et al., 2019
Fe0 janus nanomotors	Citric acid	Azo dyes	Fenton reaction	Zhang et al., 2017
Nanomotor poly(3,4-ethylenedioxythiophene) (PEDOT)/Pt (bilayer)	H_2O_2	Nerve agents of organophosphorus	neutralization by oxidative reactions	Soler and Sánchez, 2014
Ni/Au/PEDOT/Pt microsubmarine coated with alkanethiols	H_2O_2	Oil	Hydrophobicity	Gao et al., 2015
Si janus micromotor functionalized by biotin	H_2O_2	Rhodamine B	Adsorption of charge	Ge et al., 2019
3D printed self-driven thumb-sized motors (TSM)	—	Oil	Adsorption mechanism	Yu et al., 2017

(continued on next page)

Table 10.2 Nano/micromotors used in wastewater treatment—cont'd

Nanomotors	Fuel	Target pollutant	Mechanism	Reference
CoNi/Pt nanorods	Borohydride	Methylene blue, rhodamine B, and 4-nitrophenol	Degradation	Garciá-Torres et al., 2017
Carbon-based motors with Pt coating	H_2O_2	Dyes, heavy metals and explosives of nitroaromatics	Active carbon adsorption	Li et al., 2011
Bi_2O_3-CoNi/BiOCl hybrid based microrobots	Ultraviolet light (UV)	Rhodamine B	Photocatalysis	Mushtaq et al., 2015
SiO_2-rGO-Pt	H_2O_2	PBDEs, 2,4 dichlorophene	Adsorption	Orozco et al., 2016
Au/Pt microtubes (DNA functionalized)	H_2O_2	$Hg+$	Adsorption	Wang et al., 2016
Pd-Ti/Cr/Fe (tubular microprojects)	Borohydride	4-nitrophenol	4-nitrophenol	Srivastava et al., 2016
Polystyrene Fe-Zn coreshell microparticles	4-nitrophenol	Rhodamine B	Fenton reaction	Gao et al., 2013

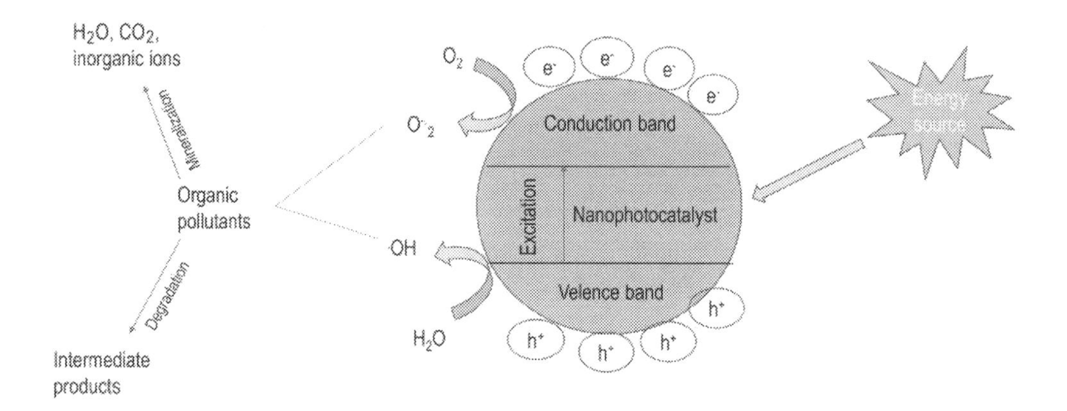

Figure 10.1 Schematic process of nanomembranes for complete removal of pollutants from wastewater including adsorption–desorption phenomenon.

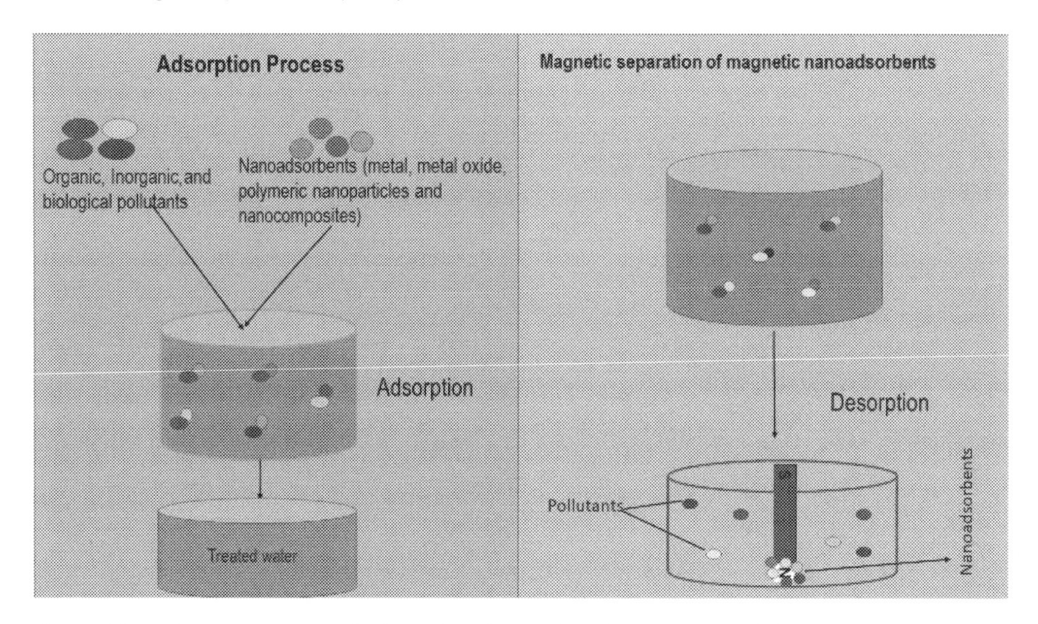

Figure 10.2 Schematic process of adsorption–desorption phenomenon in NPs for complete removal of pollutants from wastewater.

introduced in a polymetric system, it leads to the development of inorganic–organic hybrid nanomembranes (Goh et al., 2015).

Sun et al. (2013) developed a nanomembrane by using water channel protein aquaporin for water treatment. Aquaporin Z–based nanomembrane enhanced the water flux by 65% compared to the membrane without aquaporin Z and significantly reduced the concentration of $MgCl_2$ and NaCl (88.1% and 66.2%), respectively in water. In another study, Tijing et al. (2016) prepared a hydrophobic nanomembrane by using polyvinylidene fluoride-co-hexafluoropropylene and CNTs as nanofillers. It was reported that the water flux was improved by 33%–59% and the salt removal efficiency was recorded by >99.99%. Similarly, Liu et al. (2015) also prepared a nanomembrane for the desalination of the water by using zirconium (VI) based on a metal-organic framework. Test results revealed that the membrane significantly removes multivalent ions that is, Al^{+3} (99.3%), Ca^{+2} (86.3%), and Mg^{+2} (98%) from water. No decline in performance was observed during a test conducted up to 170 h using various saline solutions. In another study, Bolisetty et al. (2017) used the amyloid fibril–based membranes to remove the arsenic (arsenate) from contaminated water. Results showed about 99.6% removal of arsenic from water.

Due to small size and higher reactivity, NPs can easily penetrate into the body of human beings via food chains (drinking water) and may cause oxidative stress and other problems. So, to remove the NPs from water is of great importance. For this purpose, Zhang et al. (2016) prepared an ultra-thin nanomembrane using nanofibrils of cellulose. The pore diameter of the membrane varied from 2.5 to 12 nm and the thickness of the membrane was about 30 nm. Water flux of 1.14 L/h m^2 bar and significant removal of NPs was reported. In another study, Li et al. (2010) developed a CNT sponge-based nanomembrane. It was reported that CNT nanomembrane removes 100% of TiO_2 and AuNPs, about 80% of CdSNPs along with that rhodamine B and methyl orange molecules also removed completely from the water.

Organic contaminants such as oils, hydrocarbons, pharmaceuticals, pesticides, and phenols pose a threat to the aquatic organisms and may disturb the ecosystem. Several experiments have been carried out to remove the organic pollutants from water using nanomembranes. Karim et al. (2014) synthesized a composite nanomembrane using nanocrystals of cellulose in the matrix of chitosan (CS). The pore size of the membrane ranged from 10 to 13 nm. Results revealed a successful removal of rhodamine 6G (70%), victoria blue 2B (98%), and methyl violet 2B (84%). On the other hand, to separate the oil from water Lee and Baik (2010) prepared a nanomembrane by vertically aligned CNTs on steel mesh. The membrane has oleophilic and hydrophobic properties. The successful removal of low-viscosity diesel and high-viscosity lubricant oil was observed. Similarly, Li et al. (2014) used the immobilized Ag nanoclusters on aminated polyacrylonitrile (APAN) nanofibrous membrane for the exclusion of oil from water. Oil was significantly removed by APAN nanofibrous membrane in the hypersaline environment and also a broad range of pH. The removal of other organic pollutants using different nanomembranes was also reported such as methyl blue, direct red dye, dimethyl carbonate (Huang et al., 2014), ethanol, dyes, and humic acid (Gao et al., 2013).

Biological contaminants (viruses, bacteria, plankton, algae) are responsible for generically waterborne diseases. To clean the water from biological contaminants Zhang et al. (2009) prepared a TiO_2 nanowire-based UF nanomembrane. Pore size was reported to be around 20 nm. Membrane possesses different capabilities such as antibacterial, antifouling, photocatalytic oxidation, and concurrent separation. Yang et al. (2008) developed a highly selective, nanoporous membrane using copolymers to filter the viruses. The pore size diameter of the membrane was 17 nm and held high stability even at 2 bars. Human rhinovirus type 14, which has diameter of about 30 nm, was filtered through that membrane. Novel nanofibrous scaffolds significantly remove viruses and bacteria from the water (Sato et al., 2011). Similarly, Shayesteh et al. (2016) also developed a nanomembrane by coating ɤ-alumina and titania nanocrystals. Results showed that 100% removal of microorganisms and about 25% removal of ions by using that membrane.

10.6.4 Nanoadsorbents

Adsorption is widely used and examined techniques for the decontamination of aqueous media (Jiuhui, 2008). High efficiency and simplicity make it a favorable (Han et al., 2008). Preference is given to the adsorption technique compared to other available methods for the decontamination of contaminated water from a broad range of contaminants because of its effectiveness, easy to carry out, cost effectiveness, and flexibility (Kumari et al., 2019). NPs (organic/inorganic) that have high adsorption and isolating ability of a wide range of contaminants due to their high porosity, higher surface area, and higher active surfaces are known as nanoadsorbents (Santhosh et al., 2016). Nanoadsorbents work rapidly and efficiently and also have regeneration ability (Pacheco et al., 2006). It means that NPs can be used again after separation from the pollutant. For this purpose, desorption experiments were carried out by Yang and Xing (2007) for the regeneration of carbon NPs by separating adsorbed hydrocarbons. There are several types of nanoadsorbents used for the decontamination of the environment such as metal oxide-based, carbon-based, metal NPs, and composites of NPs (Kumari et al., 2019).

Nowadays, several types of pollutants are found in water which are categorized into three broad categories such as inorganic pollutants, organic pollutants, and biological pollutants. According to the literature, nanoadsorbents can adsorb inorganic, organic, and biological pollutants equally.

10.6.4.1 Inorganic pollutants

The track record of the adsorption technique for the removal of inorganic pollutants is quite impressive. Mostly, the oxides of titanium, iron, and aluminum are used for the adsorption of inorganic pollutants (Ali, 2012). Some scientists consider iron oxide NPs as very good adsorbents because these are easy to prepare and have a large surface area (Huang et al., 2007). Several metals and oxides of metals have been used for the adsorption of different heavy metal pollutants such as Cu, Cd, As, Cr, etc.

Due to groundwater contamination, arsenic (As) is considered as one of the harmful contaminants. People are consuming As-contaminated water all over the world

(Brammer and Ravenscroft, 2009). A study was conducted in Bangladesh to evaluate the As contamination in the groundwater. For this purpose, samples of water, urine, hairs, and nails were collected and the results demonstrated that about 900 villages have higher As concentration in groundwater than the permissible limits (Das et al., 1995). Several methods and techniques have been used by scientists to remove the As from groundwater. The adsorption technique proved effective in As removal. Martinez-Vargas et al. (2017) used the manganese ferrite and cobalt ferrite NPs for the adsorption of As. It was reported that cobalt ferrite NPs proved more effective and adsorbed a large amount of As. While the adsorption of As in the case of manganese ferrite NPs was decreased with the increase in manganese concentration in NPs. To evaluate the role of Fe oxide NPs in As biogeochemical cycling and the adsorption behavior (Dickson et al., 2017) used the hematite NPs to adsorb arsenate (As (+5)) and arsenite (As (+3)). Results revealed that hematite NPs showed higher adsorption capacity for As (+5) (4122 μg/g) compared to As (+3) (2899 μg/g) at equilibrium. Aggregation of hematite NPs was also reported during the adsorption process. It was also concluded from the Freundlich isotherm that hematite NPs showed higher affinity toward both As (+5) and As (+3) compared to hematite aggregates. In another study, (Hokkanen et al., 2015) used microfibrillated cellulose(MFC)–activated magnetite NPs (Magnetite NPs/MFC) for the adsorption of As (+5) from aqueous media and it was reported that magnetite NPs/MFC proved an exceptional adsorbent material because of its higher surface area, magnetic properties, and higher adsorption capacity. A comparative study to evaluate the effectiveness of the iron oxide NPs and iron oxide/alumina nanocomposites was carried out by Prathna et al. (2018) and it was reported that nanocomposites of iron oxide/alumina showed higher adsorption capacity for As (III) and As (V) compared to iron oxide NPs. A brief review of nonabsorbent role in management of these pollutants is given in Table 10.3.

Chromium (Cr) is used extensively in tanneries, pigment manufacturing, and mining. Two oxidation states of Cr exist in aqueous media: less toxic Cr^{+3} and highly toxic Cr^{+6}. Chromium (VI) is cancer causing for the living organisms (Depault et al., 2006). There is a dire need to develop methods to decontaminate the Cr-contaminated water. For this purpose, Al Nafiey et al. (2017) developed a graphene oxide (GO)/Co_3O_4 nanocomposite for Cr removal from water. It was reported that GO/Co_3O_4 nanocomposite showed a maximum adsorption capacity of 208.8 mg/g for Cr, which is quite high compared to other MNPs. Green synthesis of zero-valent iron NPs (NZVI) was carried out by Fazlzadeh et al. (2017) by using three plant extracts namely, *Urtica dioica*, *Rosa damascene*, and *Thymus vulgaris*. After characterization NZVI was applied for Cr removal from water. Results reported that 90% of Cr was removed in a contact time of 10 min and by increasing the contact time to 25 min leads to 100% removal of Cr from the water. Obaid (2019) reported that nanocomposite of magnetic carbon prepared by burning the straw proved a very good adsorbent material for Cr removal from water. Novel sandwiched nanocomposites GO/MnO_2/Fe_3O_4/PPy (GMFP) were prepared

Table 10.3 Nano adsorbents for the removal of pollutants from wastewater.

Contaminants	pH	Time of contact	Nanoadsorbents	Removal capacities	References
Arsenic (V)	2	5–600 min	Micro fibrillated modified Fe NPs	2.460 mg/g	Hokkanen et al., 2015
Cadmium (II)	5	50 min	Nano zero-valent iron (nZVI)/rGO	425.72 mg/g	Li et al., 2016
Arsenic (III)	2	4 h	Co and Mn ferrite NPs	24.81 and 24.11 mg/g	Martinez-Vargas et al., 2017
Chromium (VI)	2	10–30 min	nZVI	100%	Fazlzadeh et al., 2017
Arsenic (III and V)	6–8	8 h	Fe_2O_3	2899 and 4122 μg/g	Dickson et al., 2017
Lead (II)	6	12 h	Fe_3O_4/chitosan NPs	79.29 mg/g	Fan et al., 2017
Chromium (VI)	3	12 h	GO/CoO	208.8 mg/g	Al Nafiey et al., 2017
Cadmium (II)	5	30 min	ɤ-Al_2O_3 NPs	17.22 mg/g	Tabesh et al., 2018
Copper (II)	5	4 h	ɤ- Al_2O_3	31.3 mg/g	Fouladgar et al., 2015
Chromium (VI)	3	180 min	CoO NPs	15.62 mg/g	Gupta et al., 2016
Copper (II)	4	85 min	Modified henna with Fe_2O_3	99.11%	Davarnejad and Panahi, 2016
Nickel	8	35 min	Fe_3O_4 superparamagnetic	209.205–362.318 mg/g	Dickson et al., 2017
Cadmium (II)	7	60 min	Ascorbic acid stabilized nZVI	79.58%	Savasari et al., 2015
Nickel	10	30 min	MWCNTs and Al_2O_3 NPs	87.65% and 99.41%	Agarwal et al., 2016

(continued on next page)

Table 10.3 Nano adsorbents for the removal of pollutants from wastewater—cont'd

Contaminants	pH	Time of contact	Nanoadsorbents	Removal capacities	References
Lead (II)	7	24 h	Magnetic sulfonated NPs	108.9 mg/g	Chen et al., 2017
Nickel	5	6 h	Al-FeOOH	94.5 mg/g	Hokkanen et al., 2015
Lead (II)	—-	180 min	MgO NPs	2614 mg/m	Xiong et al., 2015
Copper (II)	4	90 min	ZnO/montmorillonite nanocomposites	——	Sani et al., 2017
Methyl orange	9	50 min	MWCNTs-NH_2	96 mg/g	Sadegh et al., 2019
Congo red	7	50 min	ZnO-MWCNTs	249.5 mg/g	Arabi et al., 2019
Diquat dibromide	6.5	300 min	O-MWCNTs-Fe_3O_4O-MWCNTs O-MWCNTs-k-carrageenan-Fe_3O_4	20.9 mg/g 58.3 mg/g 10.7 mg/g	Duman et al., 2019
Pefloxacin	6	240 min	O-MWCNTs	61.1 mg/g	Zhou et al., 2019
Acid orange	5	4 min	Silica NPs	230 mg/g	Rovani et al., 2018
Rhodamine B	5.8	40 min	Multi morphological mesoporous silica NPs	234.6 mg/g	Chen et al., 2018
Methylene blue	7.3	25 min	Fe_3O_4/montmorillonite	106.4 mg/g	Chang et al., 2016
Malathion	7	45 min	Magnetic/GO nanocomposites	43.3 mg/g	Kalantary et al., 2016
Metoprolol	7	15 min	k-carrageenan-Fe_3O_4@ SiO_2	447 mg/g	Soares et al., 2017
Crystal violet	7	140 min	Magnetic chitosan nanocomposite	333.3 mg/g	Massoudinejad et al., 2019
Diclofenac	4.5	60 min	Magnetic amine chitosan NPs	469.5 mg/g	Liang et al., 2019b
Natural organic matter	4.5–10	0–120 min	Magnetic NPs	98.7%	Kumari and Gupta, 2020
Natural organic matter	6.2	20 min	Activated carbon zeolite	90%	Wongcharee et al., 2020

from Fe_3O_4NPs, graphene oxide (GO), polypyrrole, and MnO_2 nanowires by Liu et al. (2018) and applied for Cr removal from water. It was observed that GMFP effectively adsorb Cr (VI), and adsorbed Cr (VI) was partially converted to Cr (III) (65%). It was concluded that GMFP has the potential to remove Cr (VI) from water efficiently and economically.

Cadmium (Cd) is nonbiodegradable and highly toxic metal (Keshavarz et al., 2013). When it penetrates the body, it gets accumulates there and its accumulation increases throughout life and leads to demineralization of bones and kidney damage (Kumari et al., 2019). To overcome the issue of water contamination by Cd, several scientists are working on the effective exclusion of Cd from contaminated water. For this purpose, Li et al. (2016) supported NZVI on reduced GO (rGO) by using the plasma reduction method (NZVI/rGO) and used for the removal of Cd from water. It was reported that NZVI/rGO has a maximum adsorption capacity of 425.7 mg/g for Cd (II) which is quite high. The process of adsorption was very rapid, equilibrium was reached in just 50 min. Gusain et al. (2019) synthesized a nanocomposite (MoS_2/SH-MWCNT) of multiwalled CNTs (MWCNTs) thiol-functionalized (SH-MWCNT) and molybdenum sulfide (MoS_2). The adsorption process followed pseudo-second-order kinetic reaction and the Freundlich adsorption isotherm. High adsorption capacity (66.6 mg/g) was observed for Cd (II). It was also reported that metal/sulfur complex formation contributed significantly to metal removal. Samadi et al. (2019) synthesized zeolite/PEG(polyethylene glycol)/GO nanocomposites by the sol-gel method and used it for the removal of heavy metals Pb and Cd from water. The maximum adsorption capacity was observed for Cd (50.2 mg/g) and Pb (49.6 mg/g).

The use of heavy metals increases day by day with an increase in industries related to chemical manufacturing. Lead (Pb) is one of the harmful metals that may cause several blood-related diseases (Chen et al., 2011). According to Luo et al. (2014), the acceptable limit of Cd is 0.010 mg/L for drinking water. Fan et al. (2017) synthesized the Fe_3O_4/CSNPs (Fe_3O_4/CSNPs). CS can chelate metals strongly, which ultimately increased the adsorption capacity of Fe_3O_4/CSNPs. Results reported that the maximum Pb (II) adsorption capacity of Fe_3O_4/CSNPs was 79.24 mg/g and the used material showed higher renewability even after five adsorption-desorption cycles. Safatian et al. (2019) reported that hydroxyapatite NPs synthesized from eggshells using microwave irradiation proved very effective adsorbent material for Pb removal from contaminated water. Novel synthesis of nanoadsorbents from agricultural waste (Husk of *Oryza sativa* (OSH)) was carried out by Kaur et al. (2020) and it was concluded that OSH nanoadsorbent material was effective in Pb removal from water. Additionally, it was cost effective and eco-friendly as well.

Copper (Cu) is one of the vital micronutrients for the body of human being but is required in very small quantities. Excessive intake of Cu can cause several health problems such as vomiting, nausea etc. Many scientists used the adsorption technique for the elimination of Cu from water. Guo et al. (2019) used citric acid and N-isopropyl acrylamide (NIPAM)-based nanocomposites *N*-isopropylacrylamide-*co*-citric acid (PNCA) for the removal of Cu and Pb from water and it was reported that PNCA significantly removed about 80% of Cu and Pb from water. Hybrid and novel nanoadsorbents were prepared by Plohl et al. (2019) for the elimination of Cu^{+2} from water-containing media by linking derived polyethyleneimine to magnetic silica nanocomposites. The adsorption capacity of 143 mg/g was observed for Cu^{+2} and showed great usability potential. García-Díaz et al. (2018) introduced new nanoadsorbents (carbon nanofibers) for the effective elimination of Cu^{+2} from wastewater. It was noted that the pH of the solution has a significant effect on Cu^{+2} adsorption and maximum adsorption was observed at pH 10. Altuntas et al. (2017) reported that 96% of Cu^{+2} was removed from water by the addition of 200 mg/L of nano zero-valent iron (nZVI) supported on activated carbon (AC).

In the above conversation, the elimination of single metal by using specific adsorbent was focused. Several studies promote the multimetal adsorption by a single adsorbent. In this regard, Fato et al. (2019) conducted a study to remove multimetal ions (Ni^{+2}, Cd^{+2}, Pb^{+2}, Cu^{+2}) simultaneously from river water by using ultrafine mesoporous Fe_3O_4NPs (UFMNPs). It was observed that UFMNPs showed the removal efficiencies of 84, 86, 80, and 54% for Cu^{+2}, Pb^{+2}, Cd^{+2}, and Ni^{+2}, respectively. In another study, (Manjuladevi and Sri, 2017) used the AC prepared from *Cucumis melo* peel for the adsorption of heavy metal cations (Pb^{+2}, Cd^{+2}, Cr^{+6}, Ni^{+2}) from industrial wastewater. Results showed that the adsorption process was dependent on pH, the dose of adsorbent, concentration of metal, and contact time. It was concluded that AC from C. melo peel could be a cost-effective and eco-friendly source to remove metals from wastewater. (Sun et al., 2018) used three different Fe_3O_4-based nanoadsorbents namely, Fe_3O_4, Fe_3O_4-GO, and Fe_3O_4-NH_2 to adsorb several metal cations including Ca^{+2}, Mn^{+2}, Cu^{+2}, As^{+5}, Cd^{+2}, Al^{+3}, Hg^{+2}, Mg^{+2}, Pb^{+2}, Zn^{+2}, and Fe^{+3}. It was reported that all three types of nanoadsorbents significantly adsorb metal cations. It was also concluded that all three nanoadsorbents have different affinities for different metal cations.

10.6.4.2 Organic pollutants

Numerous organic contaminants such as pesticides, dyes, and hydrocarbons (Alharbi et al., 2018) have been removed from wastewater by using adsorption techniques. The industries that are using a huge number of coloring agents that is, paper, painting, textile, printing, rubber, etc. are the main source of dyes in water bodies. Nowadays, scientists are working on the development of more efficient adsorbents for the removal of organic pollutants.

Asfaram et al. (2015) synthesized nanoadsorbent by loading ZnS:CuNPs on AC (ZnS:CuNP-AC) for the removal of auramine-O (AO) from aqueous media. The removal efficiency of AO by ZnS: CuNP-AC was 99.76% and the highest adsorption ability was 183.15 mg/g. It was concluded that ZnS:CuNP-AC has a promising potential to be used as an adsorbent for the removal of dyes. A novel nanoadsorbent has been developed by Lee et al. (2019) by immobilizing CS-coated Fe_3O_4NPs on the dextran gel column. For testing, a solution of Evans blue (EB) dye was passed through the column. As a result, efficient removal of EB with a maximum adsorption capacity of 243.9 mg/g was observed. To further evaluate the adsorption potential, adsorption experiments with acid red 1, acid yellow 25, Cy5.5, and green 25 were carried out. It was reported that all the dyes were removed efficiently. It was also concluded that the adsorbent material showed higher adsorption capacity for more negatively charged dyes. In another study, Soares et al. (2017) used magnetic-carrageenan/SiO_2 hybrid nanoadsorbent for the removal of methylene blue (MB). The maximum adsorption capacity was recorded at 530 mg/g. It was also stated that the hybrid nanoadsorbent has great regeneration potential just by rinsing with a KCl solution. It was reported that the removal efficiency of MB even after six consecutive adsorption/desorption batches was 97%, which shows the robustness and reusable potential of this hybrid nanoadsorbent.

Contamination of water bodies by pesticides is another big issue that may pose some serious issues for living beings. Industries (textile, food, cosmetics, etc.), agriculture, forestry, and domestic activities are the main source of pesticide introduction in the environment. Izanloo et al. (2019) used novel bifunctional nanoadsorbents for the removal of organic and inorganic pollutants from water. Significant removal of 2,4-dichlorophenoxyacetic acid (2,4 D) was observed. It was reported that pH of the aqueous media played an important role in the adsorption process. Malathion, which is an organophosphorus pesticide, was removed from water by using MWCNTs as an adsorbent. It was concluded that 100% removal of malathion is possible by using MWCNTs (Dehghani et al., 2017). CS-ZnONPs composite nanoadsorbent was used by Dehaghi et al. (2014) for the removal of Permethrin pesticide. Results reported that 99% of permethrin was removed when 0.5 g adsorbent was used and pH was adjusted to 7.0 at room temperature.

The pharmaceutical drugs are used for the cure of diseases but when accumulated in the environment may pose some threats to living beings. When accumulated in the environment, they categorized as organic pollutants. Recently, MWCNTs sandwiched between GO nanosheets were used for the efficient removal of tetracycline hydrochloride (TCH). The adsorption efficiency of adsorbent was recorded to be 99.23% for TCH (Yang et al., 2018). In another study, Nassar et al. (2019) used silica NPs synthesized from rice husks for the removal of Ciprofloxacin from polluted water. Adsorption capacity 54.1 mg/g was observed at pH 7.0 after 24 h. Metal oxide NPs are also used effectively for the removal of pollutants from wastewater. Fakhri and Adami (2014) used the MgONPs for the removal of antibiotics (cephalosporins) from aqueous media. The maximum adsorption capacity by MgONPs was recorded to be 500 mg/g at pH 9.0 in just 10 min. Table 10.3 summarizes ability of nanoadsorbents for removal of pollutants from water.

10.6.5 Biological contaminants

There are three major categories of biological contaminants namely natural organic matter, biological toxins, and microorganisms. Free-living microbes and human pathogens come under the category of microbial contaminants (Srinivasan et al., 2008). Most of the conventional water treatment systems are unable to filter cyanobacterial contaminants (Al Momani et al., 2008). Numerous adsorbents such as activated carbon (AC) have decent exclusion efficacies but on the other hand, several factors affect the process of adsorption (Wang et al., 2007). For the removal of Gram-positive bacterium and Gram-negative bacterium such as *Staphylococcus aureus*, *E. coli* from water, Fe_3O_4NPs are synthesized by two different approaches (top-down, bottom-up). The NPs synthesized by the top-down approach were denoted by Fe_3O_4-A, while the NPs synthesized by the bottom-up approach were denoted as Fe_3O_4-B. Results reported that both types of NPs efficiently remove Gram-negative and Gram-positive bacteria, while the Fe_3O_4-BNPs showed higher adsorption capacity compared to Fe_3O_4-ANPs (Darabdhara et al., 2017). Similarly, Zhan et al. (2014) used amine-modified Fe_3O_4NPs for the removal of disease-causing bacteria. Fe_3O_4-SiO_2-NH_2 were prepared for the adsorption of a broad range of bacteria such as, *E. coli*, *S. aureus*, *Pseudomonas aeruginosa*, *Salmonella*, *Bacillus subtilis*, Virus (Poliovirus-1), and bacteriophage f2. Results reported a 97.39% nonspecific removal efficiency. To remove the biological toxin (peptide-toxin) RNA-GO nanosheets were used by Hu et al. (2012). Results reported a significant removal of peptide toxin by RNA-GO nanosheets but the removal efficiency was reduced at extreme temperature, pH, natural organic matter (OM), and ionic strength. Dixon et al. (2011) used the UF and powder AC (PAC) for the removal of cyanobacterial toxins from natural blooms in Australia. It was reported that UF alone removed the cyanobacterial cells while the combination of UF and PAC was used for the removal of extracellular toxins.

10.7 Advantage of NPs over conventional treatments

According to a report by WHO presented recently, about 783 million people are suffering due to the poor quality of drinking water. It was also reported that about 1.87 million deaths of children are due to waterborne diseases (Van Der Heijden, 2018). In underdeveloped countries, the use of conventional treatment methods is not suitable due to high expenses, maintenance problems, etc. In many developing countries people collect and store water from one place to another for drinking purposes. There are chances of the contamination of water during collection and transport. The use of ceramic water filters (CWFs) is an efficient method to treat such problems. Nowadays, CWFs are modified and impregnated with silver NPs (Adeleye et al., 2016). By using CWFs, filtered and clean water is produced but the cleaning of water at a higher level is not possible by this method. Using silver NMs will remove the pathogens because of the antimicrobial properties of silver (Pandey et al., 2017). There are several pollutants, such as substances based on organics, present in trace amounts in water, and pose serious threats to human life. Flocculation and chlorination are used normally to remove water pollutants. There are drawbacks to these conventional treatments such as low efficiency and production of sludge in water, which also pose several threats to the environment (Umar et al., 2013). To reduce the contamination of water by sludge, nanomembranes have been used which stop the passage of any solid particles into the water. Due to these benefits, nanomembranes are considered the most suitable technique but still some improvements are needed to make it perfect. Photodegradation of several pollutants such as pharmaceutical drugs, pesticides, dyes, etc. is reported by several researchers by using NP-doped and NP-undoped metals and metal oxide. It was reported that NP-doped TiO_2 gives more prominent results in photodegradation compared to undoped TiO_2 (Umar et al., 2016). Compared to traditional techniques, nanomechanics offer several advantages. The small particles can be used for the *in-situ* and *ex-situ* decontamination and can lower the overall price of the treatment compared to conventional techniques (Safdar et al., 2017). Nanophotocatalysis is an important technique to degrade contaminants at 25 °C (Tahir et al., 2019). The advantages of nanophotocatalysis are low toxicity, cost effectiveness, higher stability, eco-friendliness, etc. (Ciambelli et al., 2019). Nanoadsorption is also an effective strategy to remove pollutants from water. Compared to conventional sorbents, nanoadsorbents remove a significantly higher amount of pollutants due to higher surface area, higher adsorption rate, and short equilibrium time (Vunain et al., 2016).

10.8 Limitations associated with nanotechnology

The impact of NMs on the aqueous environment decides the commercialization of NPs for the wastewater treatment. To evaluate the risk of NMs to health, different studies such as life cycle analysis, pathways, dispersal of NPs, assessment of technology, and toxicity tests have been conducted. The results of these studies will provide a better understanding of the behavior of a different type of NPs in the aqueous environment.

10.8.1 Potential ecotoxicity of nano-based materials and processes related to water applications

There is no direct effect of NPs present in water on humans, but there is a way through which humans can uptake NPs and that is by consuming fish, so it is important to evaluate the effects of NPs on the aquatic population. It is reported that NPs such as TiO_2, Ag, CNTs in water negatively affect the aquatic animals (Asghari et al., 2012; Jackson et al., 2013). Ecotoxicological effects and other limitations associated with NMs in wastewater treatment have been summarized in Table 10.4.

Table 10.4 Limitations for Nanomaterials application in water treatment.

Nanomaterial type		Limitation in use		Ecotoxicology	Reference
Nanoadsorbents	Polymeric Nanoadsorbents	Low adsorption potential, complex production and high regeneration cost	Pan et al., 2008	–	–
	Carbon nanotubes	Require supporting matrix and high production cost	Chatterjee and Deopura, 2002	Severe effect to human cells on exposure	Cui et al., 2005; Bottini et al., 2006
Nanometals and Nanometal oxides	Nano titanium dioxide	Requires UV activation due to big band gap energies	Etacheri et al., 2015	Effect daphnia, gills of fish and other aquatic organisms including algae and plants	(Environmental Protection Agency, 2010)
	Nano zinc oxide	Requires UV activation due to big band gap energies	Dimapilis et al., 2018	Potential effects on aquatic and public health	Beegam et al., 2016
	Nano iron oxide	Need modification, recovery and separation is difficult	Das et al., 2017	Phytotoxic	Chen et al., 2016
	Nano zero-valent iron	Need surface modification and stabilization	Stefaniuk et al., 2016	Can be a carrier for other pollutants, translocate over long distances, form sediments at bottoms of water reservoirs, and accumulate in living beings	Ghosh et al., 2017; Stefaniuk et al., 2016

(continued on next page)

Table 10.4 Limitations for Nanomaterials application in water treatment—cont'd

Nanomaterial type		Limitation in use		Ecotoxicology	Reference
	Nano zero-valent zinc	Limited to the degradation of CCl_4	Tratnyek et al., 2010	–	–
	Nanosilver	Limited durability	Gehrke et al., 2015	Adverse effects on environment and humans health	Rezvani et al., 2019
	Magnetic nanoparticles	Stabilization is required	Zhu et al., 2018	Human health concerns	Liu et al., 2013
Nanomembranes	Aquaporin-based membranes	Mechanically weak	(Gehrke et al., 2015	-	-
	Self-assembling membranes	Limited availability, not applicable at large scales			
	Nanofiltration membranes	Membrane blockage and fouling			
	Nanocomposite membranes	Require resistant material, chances of nanoparticles release	Tesh and Scott, 2014	–	–
	Nanofiber membranes	Low strength, fouling and pores blockage	Al-Husaini et al., 2019; Environmental Protection Agency, 2010	–	–

10.8.1.1 Harmful effects on aquatic organisms

An extensive study presenting the effects of TiO_2 on plants, bacteria, invertebrates, algae, and fish was published by US Environmental Protection Agency (EPA) in 2010 (EPA, 2010; Gehrke et al., 2015). In that study, effects on environment and organisms by different types of TiO_2NPs and different modes of entry were evaluated. It was reported that nano TiO_2 reduced the reproduction of daphnia and also led to behavioral changes, pathological variations in intestines, gills, and respiratory disturbance as well (EPA, 2010). Harmful effects on algae were also reported due to nano TiO_2 that depend mainly on the concentration and type of nano TiO_2.

The exposure time of NPs is another important factor that defines the toxicity level of NPs. As it was reported that nano TiO_2 caused severe toxicity effects to daphnia when the exposure time exceeds 72 h, while the toxicity effects were mild within 48 h of exposure (Zhu et al., 2010). Jiang et al. (2014) evaluated the effect of nano- and micro-sized AgNPs on *Spirodela polyrhiza* and it was reported that AgNPs significantly increased the ROS production and enhance the activity of superoxide dismutase (SOD), catalase (CAT), malonaldehyde, peroxidase (POD), antioxidant glutathione, while the micro-sized Ag particles did not show any effect on the activities of antioxidant enzymes. From the results, it was concluded that AgNPs have the potential to cause oxidative stress in aquatic organisms. Similarly, Bai et al. (2010) evaluated the effects of ZnONPs on zebrafish embryos. It was reported that ZnONPs at 50 and 100 ppm concentration killed the embryos of zebrafish, while the concentration up to 25 ppm reduced the hatching of embryos. It was observed by Blinova et al. (2010) that the amount of organic matter dissolved in water significantly changed the lethal concentration of ZnONPs and CuONPs in protozoa and crustacea. Quantum dots of CdSe/ZnS showed toxicity to daphnia at higher concentrations in the aquatic environment (Pace et al., 2010). Quantum dots of CdTe showed higher toxicity potential for green algae (*Chlamydomonas reinhardtii*) compared to nano TiO_2 (Wang et al., 2008). Similarly, quantum dots of CdTe showed DNA damage and oxidative stress to mussels (*Elliptio complanata*) living in freshwater (Gagné et al., 2008). It was reported that Gasterosteus aculeatus liver showed hepatocellular nuclear pleomorphism when exposed to quantum dots of CdS (Sanders et al., 2008).

Na^+ K^+-ATPase activity was increased in the intestines and gills of rainbow trout when treated with CNTs (single-walled). It caused respiratory dysfunction due to precipitation in the gills (Smith et al., 2007). The toxic effect of CNTs (double-walled) was also reported in amphibian larvae (Mouchet et al., 2008). Daphnia exposed to fullerene NPs showed oxidative damage (Klaper et al., 2009). It was also reported that C_{60} NPs lead to the apoptotic and necrotic death of cells in daphnia (Gaiser et al., 2011).

10.8.1.2 Fate and removal of NMs in water

NPs in the environment can come from nonpoint sources such as clothes, washing machines, and other NPs-based materials and point sources such as wastewater treatment, production facility, landfills, etc. (Nowack and Bucheli, 2007). For example, a considerable amount of AgNPs can be released to the environment during the washing of clothes depending on the washing conditions. Properties of NPs such as solubility in water, chemical reactivity such as redox potential, and binding to other components significantly affect their mobility. These properties and surface charge on NPs may increase the immobilization of NPs. After the contact with water, NPs adsorb and accumulate immediately due to a higher specific surface area. In an aqueous system, surface characteristics of polymers enhance the agglomeration of NPs and also enhance the interaction with different proteins (Mouchet et al., 2008). For example, the surface characteristics of polyethyleneimine (polymeric NPs) enhance the agglomeration in the water system. In adsorbed form or as agglomerates NPs can stay for a longer period in sediments in water bodies such as rivers, lakes etc. In this state, NPs are termed as inactive but an alteration in the ecological circumstances such as reduction in pH and increase in temperature can enhance the dissolution of NPs. The stability of metal oxide NPs is limited due to higher reactivity (Matlochova et al., 2013). However, several groups of scientists are working to enhance the stability of metal oxide NPs.

Degradation of NPs in an aqueous system can occur in different ways based on the specific conditions of the environment. Reactivity of NPs can be changed by photoreaction via sunlight depending on the water turbidity and chemical properties of NPs (Environmental Protection Agency, 2007). There is a possibility of the degradation of NPs through abiotic and biotic pathways. It was reported that few metallic and organic NMs transform anaerobic conditions (Nurmi et al., 2005).

During wastewater treatment, different processes such as sedimentation, flocculation, and filtration have been implemented in municipal wastewater treatment plants. Normally, biological treatment is adopted for the removal of NPs such as by interaction with the microbiological community that can enhance the sedimentation, incorporation, and degradation of NPs. The early agglomeration step is necessary to enhance the settling rate of NPs, during the sedimentation process because bigger particles will settle faster (US-EPA, 2012). For example, aggregates of C_{60} can be removed easily by the process described by (Hyung and Kim, 2009).

10.9 Future perspectives

In the domains of water reutilization and wastewater processing, nanotechnology can part a vibrant role. It can overcome the existing concerns and hazards regarding the production of harmful byproducts by conventional water treatment techniques. Unique properties of NMs such as catalysis, high facets ratio, antimicrobial activities, better surface attributes, and higher conductivity can effectively be used in high fluorescence for the detection and desalination by capacitive deionization method. Water processing can be supported by making the use of freely available solar energy via engineering NMs as a visible light photocatalyst. Water quality degradation, when it moves through the distribution networks, can also be addressed by nanotechnology. Nanotechnology-based water processing systems working with solar driven energy can be developed and may be best for catastrophe areas (Baruah et al., 2012). Nanotechnology has the ability to exploit substitute water sources for agriculture and drinking purposes by keeping the use of energy minimum. It can be a highly important technology for the developing world that is facing severe problems of water contamination. Nowadays, it is necessary to adopt innovative water treatment technologies that have high performance. In the future, most likely the water purification systems will be based on nanotechnology that will be cost effective, easy to operate, and efficient in the removal of a wide variety of water pollutant.

10.10 Summary

This chapter concludes the use of wastewater in different sectors and potential implications associated with the use of wastewater without treatment. Globally, a huge amount of wastewater has been produced and it contains different types of organic, inorganic, and biological contaminants. The contaminated wastewater affects plants, humans, animals directly or indirectly in a negative way. Potential toxic elements present in wastewater also pose serious threats to the soil microbial community and also play a role in soil degradation. Different treatment approaches have been discussed in this chapter. The removal of pollutants by the adsorption method is an old traditional technique. But the use of nanoadsorbents with higher surface area and higher reactivity has increased the efficiency of this method to manifolds. The use of nanomembrane filtration is a very efficient technique. However, it is an energy-intensive and expensive technique. Pollutants in wastewater can also be degraded by photocatalysis techniques using nanocatalysts. This technique is important for organic contaminants. Photocatalysis also has some drawbacks that reduced the application of this technique on a large scale. The use of antimicrobial NMs for example AgNPs is also a common technique to decontaminate the water from microbes. However, NMs have the potential to solve the problem of water contamination along with that NMs can also cause toxicity to the environment. Anyhow, it is assumed that the NMs will provide a practical solution to deal with contaminated wastewater in the near future.

References

Adam, R.E., Pozina, G., Willander, M., Nur, O., 2018. Synthesis of ZnO nanoparticles by co-precipitation method for solar driven photodegradation of Congo red dye at different pH. Photon. Nanostruct, 32, 11–18. https://doi.org/10.1016/j.photonics.2018.08.005

Adeleye, A.S., Conway, J.R., Garner, K., Huang, Y., Su, Y., Keller, A.A., 2016. Engineered nanomaterials for water treatment and remediation: costs, benefits, and applicability. Chem. Eng. J. 286, 640–662. https://doi.org/10.1016/j.cej.2015.10.105.

Agarwal, S., Tyagi, I., Gupta, V.K., Dehghani, M.H., Jaafari, J., Balarak, D., Asif, M., 2016. Rapid removal of noxious nickel (II) using novel γ-alumina nanoparticles and multiwalled carbon nanotubes: kinetic and isotherm studies. J. Mol. Liq. 224, 618–623. https://doi.org/10.1016/j.molliq.2016.10.032.

Ahmad, K., Nawaz, K., Khan, Z.I., Nadeem, M., Wajid, K., Ashfaq, A., Munirl, B., Memoonas, H., Sanas, M., Shaheen, F., Kokab, R., Rehmans, S.U., Ullah, M.F., Mehmood, N., Muqadas, H., Aslam, Z., Shezadi, M., Noorkals, I.R., Bask, H., Shads, H.A., Batool, F., Iqbal, S., Munk, M., Sohail, M., Sher, M., Ullahll, S., Ugulu, I., Dogan, I.Y., 2018. Effect of diverse regimes of irrigation on metals accummulation in wheat crop: an assessessment-dire need of the day. Fresenius Environ. Bull. 27 (2), 846–855.

Ahmad, K., Wajid, K., Khan, Z.I., Ugulu, I., Memoona, H., Sana, M., Nawaz, K., Malik, I.S., Bashir, H., Sher, M., 2019. Evaluation of Potential Toxic Metals Accumulation in wheat irrigated with wastewater. Bull. Environ. Contam. Toxicol. 102 (6), 822–828. https://doi.org/10.1007/s00128-019-02605-1.

Al Momani, F., Smith, D.W., El-Din, M.G., 2008. Degradation of cyanobacteria toxin by advanced oxidation processes. J. Hazard. Mater. 150 (2), 238–249.

Al Nafiey, A., Addad, A., Sieber, B., Chastanet, G., Barras, A., Szunerits, S., Boukherroub, R., 2017. Reduced graphene oxide decorated with Co_3O_4 nanoparticles (rGO-Co_3O_4) nanocomposite: a reusable catalyst for highly efficient reduction of 4-nitrophenol, and Cr (VI) and dye removal from aqueous solutions. Chem. Eng. J. 322, 375–384.

Alharbi, O.M.L., Basheer, A.A., Khattab, R.A., Ali, I., 2018. Health and environmental effects of persistent organic pollutants. J. Mol. Liq. 263, 442–453. https://doi.org/10.1016/j.molliq.2018.05.029.

Al-Husaini, I.S.Y., Lau, A.R.M., Ismail, W.J., Al-Abri, A.F., Al-Ghafri, M.Z., Wirzal, B.N.M.D.H., 2019. Fabrication of polyethersulfone electrospun nanofibrous membranes incorporated with hydrous manganese dioxide for enhanced ultrafiltration of oily solution. Sep. Purif. Technol. 212, 205–214.

Ali, I., 2012. New generation adsorbents for water treatment. Chem. Rev. 112 (10), 5073–5091.

Ali, I., AlGhamdi, K., Al-Wadaani, F.T., 2019. Advances in iridium nano catalyst preparation, characterization and applications. J. Mol. Liq., 104461278 doi:10.1016/J.MOLLIQ.2019.02.050.

Al-Jumaili, A., Alancherry, S., Bazaka, K., Jacob, M.V., 2017. Review on the antimicrobial properties of Carbon nanostructures. Materials 10 (9), 1066. https://doi.org/10.3390/ma10091066.

Altuntas, K., Debik, E., Kozal, D., Yoruk, I.I., 2017. Adsorption of copper metal ion from aqueous solution by nanoscale zero valent iron (nZVI) supported on activated carbon. Periodicals of Engineering and Natural Sciences 5 (1). http://dx.doi.org/10.21533/pen.v5i1.77.

Amin, N.U., Hussain, A., Alamzeb, S., Begum, S., 2013. Accumulation of heavy metals in edible parts of vegetables irrigated with waste water and their daily intake to adults and children, District Mardan, Pakistan. Food Chem. 136 (3–4), 1515–1523. https://doi.org/10.1016/j.foodchem.2012.09.058.

Angelakis, A.N., Snyder, S.A., 2015. Wastewater treatment and reuse: past, present, and future. Water (Switzerland), 7(9), 4887–4895. https://doi.org/10.3390/w7094887

Amiri, M., Salavati-Niasari, M., Akbari, A., 2019. Magnetic nanocarriers: evolution of spinel ferrites for medical applications. Adv. Colloid Interface Sci. 265, 29–44.

Arabi, S.M.S., Lalehloo, R.S., Olyai, M.R.T.B., Ali, G.A., Sadegh, H., 2019. Removal of congo red azo dye from aqueous solution by ZnO nanoparticles loaded on multiwall carbon nanotubes. Physica E 106, 150–155.

Arora, M., Kiran, B., Rani, S., Rani, A., Kaur, B., Mittal, N., 2008. Heavy metal accumulation in vegetables irrigated with water from different sources. Food Chem. 111 (4), 811–815. https://doi.org/10.1016/j.foodchem.2008.04.049.

Arya, S., Prerna, Singh, A., Kour, R, 2019. Comparative study of CuO, CuO@Ag and CuO@Ag:La nanoparticles for their photosensing properties. Mater. Res. Exp. 6 (11), 116313. https://doi.org/10.1088/2053-1591/ab49ab.

Asfaram, A., Ghaedi, M., Agarwal, S., Tyagi, I., Gupta, V.K., 2015. Removal of basic dye Auramine-O by ZnS: Cu nanoparticles loaded on activated carbon: optimization of parameters using response surface methodology with central composite design. RSC Adv. 5 (24), 18438–18450.

Asghari, S., Johari, S.A., Lee, J.H., Kim, Y.S., Jeon, Y.B., Choi, H.J., Yu, I.J., 2012. Toxicity of various silver nanoparticles compared to silver ions in *Daphnia magna*. Journal of nanobiotechnology 10 (1), 14.

Ataee-Esfahani, H., Wang, L., Nemoto, Y., Yamauchi, Y., 2010. Synthesis of bimetallic Au@Pt nanoparticles with Au core and nanostructured Pt shell toward highly active electrocatalysts. Chem. Mater. 22 (23), 6310–6318. https://doi.org/10.1021/cm102074w.

Bai, W., Zhang, Z., Tian, W., He, X., Ma, Y., Zhao, Y., Chai, Z., 2010. Toxicity of zinc oxide nanoparticles to zebrafish embryo: a physicochemical study of toxicity mechanism. J. Nanopart. Res. 12 (5), 1645–1654.

Balkhair, K.S., Ashraf, M.A., 2016. Field accumulation risks of heavy metals in soil and vegetable crop irrigated with sewage water in western region of Saudi Arabia. Saudi J. Biol. Sci. 23 (1), S32–S44. https://doi.org/10.1016/j.sjbs.2015.09.023.

Baruah, S., Jaisai, M., Dutta, J., 2012. Development of a visible light active photocatalytic portable water purification unit using ZnO nanorods. Catalysis Science & Technology 2 (5), 918–921.

Beegam, A., Prasad, P., Jose, J., Oliveira, M., Costa, F.G., Soares, A.M.V.M., Gonçalves, P.P., Trindade, T., Kalarikkal, N., Thomas, S., Pereira, M.de L., 2016. Environmental Fate of Zinc Oxide Nanoparticles: Risks and Benefits. In: Larramendy, M., SoloneskiIn, S. (Eds.), Toxicology - New Aspects to This Scientific Conundrum. Intechopen, Croatia, pp. 81–111.

Berekaa, M.M., 2016. Nanotechnology in wastewater treatment; influence of nanomaterials on microbial systems. Int. J. Curr. Microbiol. App. Sci. 5 (1), 713–726.

Bhanvase, B.A., Shende, T.P., Sonawane, S.H., 2017. A review on graphene–TiO_2 and doped graphene–TiO_2 nanocomposite photocatalyst for water and wastewater treatment. Environ. Technol. Rev. 6 (1), 1–14.

Bhatia, D., Sharma, N.R., Singh, J., Kanwar, R.S., 2017. Biological methods for textile dye removal from wastewater: a review. Crit. Rev. Environ. Sci. Technol. 47 (19), 1836–1876. https://doi.org/10.1080/10643389.2017.1393263.

Bhimrao, K.P., Haribhau, K.K., Gangadhar, D.U., pandurang, M.B., Jagannath, T.U., 2020. Sol-gel fabricated transition metal Cr^{3+}, Co^{2+} doped lanthanum ferric oxide (LFO-$LaFeO_3$) thin film sensors for the detection of toxic, flammable gases: a comparative study. Mater. Sci. Res. India, pp. 70–83.

Bhosale, M.A., Chenna, D.R., Bhanage, B.M., 2017. Ultrasound assisted synthesis of gold nanoparticles as an efficient catalyst for reduction of various nitro compounds. ChemistrySelect 2 (3), 1225–1231.

Blinova, I., Ivask, A., Heinlaan, M., Mortimer, M., Kahru, A., 2010. Ecotoxicity of nanoparticles of CuO and ZnO in natural water. Environ. Pollut. 158 (1), 41–47.

Bolisetty, S., Reinhold, N., Zeder, C., Orozco, M.N., Mezzenga, R., 2017. Efficient purification of arsenic-contaminated water using amyloid–carbon hybrid membranes. Chem. Commun. 53 (42), 5714–5717.

Bottini, M., Bruckner, S., Nika, K., Bottini, N., Bellucci, S., Magrini, A., Bergamaschi, A., Mustelin, T., 2006. Multi-walled carbon nanotubes induce T lymphocyte apoptosis. Toxicol. Lett. 160, 121–126.

Brady-Estévez, A.S., Schnoor, M.H., Kang, S., Elimelech, M., 2010. SWNT− MWNT hybrid filter attains high viral removal and bacterial inactivation. Langmuir 26 (24), 19153–19158.

Brammer, H., Ravenscroft, P., 2009. Arsenic in groundwater: a threat to sustainable agriculture in South and South-east Asia. Environ. Int. 35 (3), 647–654.

Cao, Q.Y., Jiang, R., Liu, M., Wan, Q., Xu, D., Tian, J., Huang, H., wen, Y., Zhang, X., Wei, Y., 2017. Microwave-assisted multicomponent reactions for rapid synthesis of AIE-active fluorescent polymeric nanoparticles by post-polymerization method. Mater. Sci. Eng. C 80, 578–583.

Caramazana, P., Dunne, P., Gimeno-Fabra, M., McKechnie, J., Lester, E., 2018. A review of the environmental impact of nanomaterial synthesis using continuous flow hydrothermal synthesis. Curr. Opin. Green Sustain. Chem. CURR OPIN GREEN SUST 12, 57–62.

Castro-Jiménez, J., Dachs, J., Eisenreich, S.J., 2015. Atmospheric deposition of POPs: implications for the chemical pollution of aquatic environments. In: Zeng, Eddy Y. (Ed.), Comprehensive Analytical Chemistry, 67. Elsevier B.V., pp. 295–322.

Chae, S.R., Hotze, E.M., Wiesner, M.R., 2009. Possible applications of fullerene nanomaterials in water treatment and reuse. In: Street, A., Sustich, R., Duncan, J., Savage, N. (Eds.), Nanotechnology Applications for Clean Water. Elsevier Inc, Amsterdam, pp. 167–177.

Chang, J., Ma, J., Ma, Q., Zhang, D., Qiao, N., Hu, M., Ma, H., 2016. Adsorption of methylene blue onto Fe_3O_4/activated montmorillonite nanocomposite. Appl. Clay Sci. 119, 132–140.

Chatterjee, A., Deopura, B.L., 2002. Carbon nanotubes and nanofibre: An overview. Fibers Polym. 3, 134–139. https://doi.org/10.1007/BF02912657.

Chen, K., He, J., Li, Y., Cai, X., Zhang, K., Liu, T., Liu, J., 2017. Removal of cadmium and lead ions from water by sulfonated magnetic nanoparticle adsorbents. J. Colloid Interface Sci. 494, 307–316.

Chen, L.F., Liang, H.W., Lu, Y., Cui, C.H., Yu, S.H., 2011. Synthesis of an attapulgite clay@ carbon nanocomposite adsorbent by a hydrothermal carbonization process and their application in the removal of toxic metal ions from water. Langmuir 27 (14), 8998–9004.

Chen, W., Liu, Q., Tian, S., Zhao, X., 2019. Exposed facet dependent stability of ZnO micro/nano crystals as a photocatalyst. Appl. Surf. Sci. 470, 807–816.

Chen, Z., Ngo, H.H., Guo, W., 2013. A critical review on the end uses of recycled water. Crit. Rev. Environ. Sci. Technol. 43 (14), 1446–1516. https://doi.org/10.1080/10643389.2011.647788.

Chen, X., O'Halloran, J., Jansen, M.A.K., 2016. The toxicity of zinc oxide nanoparticles to *Lemna minor* (L.) is predominantly caused by dissolved Zn. Aquat. Toxicol. 176, 46–53.

Chen, J., Sheng, Y., Song, Y., Chang, M., Zhang, X., Cui, L., Meng, D., Zhu, H., Shi, Z., Zou, H., 2018. Multimorphology mesoporous silica nanoparticles for dye adsorption and multicolor luminescence applications. ACS Sustain. Chem. Eng. 6 (3), 3533–3545.

Cheng, Y.H., Chou, W.C., Yang, Y.F., Huang, C.W., How, C.M., Chen, S.C., Chen, W.Y., Hsieh, N.H., Lin, Y.J., You, S.H., Liao, C.M., 2018. Response to "Letter to the editor re: Cheng YH, Chou WC, Yang YF, et al. Environ Sci Pollut Res (2018). https://doi.org/10.107/s11356-017-0875-4. Environ. Sci. Pollut. Res. 25 (33), 33836–33839. https://doi.org/10.1007/s11356-018-3178-5.

Cheshmazar, E., Arfaeinia, H., Enayat, S., 2018. Dataset for effect comparison of irrigation by wastewater and ground water on amount of heavy metals in soil and vegetables : accumulation transfer factor and health risk assessment. Data Brief 18, 1702-1710

Chi, Q., Wang, Z., Tian, F., You, J.A., Xu, S., 2018. A review of fast bubble-driven micromotors powered by biocompatible fuel: Low-concentration fuel, bioactive fluid and enzyme. Micromachines 9 (10), 537.

Chin, S.F., Azman, A., Pang, S.C., 2014. Size controlled synthesis of starch nanoparticles by a microemulsion method. Journal of Nanomaterials 2014, 763736. https://doi.org/10.1155/2014/763736.

Chopra, A.K., Pathak, C., 2015. Accumulation of heavy metals in the vegetables grown in wastewater irrigated areas of Dehradun, India with reference to human health risk. Environ. Monit. Assess. 187 (7), 445. https://doi.org/10.1007/s10661-015-4648-6.

Ciambelli, P., La Guardia, G., Vitale, L., 2019. Nanotechnology for green materials and processes. Studies in Surface Science and Catalysis, 179. Elsevier, pp. 97–116.

Ciocarlan, R.G., Seftel, E.M., Mertens, M., Pui, A., Mazaj, M., Tusar, N.N., Cool, P., 2018. Novel magnetic nanocomposites containing quaternary ferrites systems Co0. 5Zn0. 25M0. 25Fe2O4 (M= Ni, Cu, Mn, Mg) and TiO2-anatase phase as photocatalysts for wastewater remediation under solar light irradiation. Mater. Sci. Eng.: B. 230, 1–7.

Corcoran, E., 2010. Sick Water?: the Central Role of Wastewater Management in Sustainable Development: A Rapid Response Assessment. UNEP (United Nations Environment Programme), UN-HABITAT, Nairobi, Kenya.

Cotruvo, J.A., Dufour, A., Rees, Bartram, J., Carr, R., Cliver, D.O, Craun, G.F., Fayer, R., Gannon, V.P., 2004. Waterborne zoonoses. Iwa Publishing, London, UK.

Cui, D.X., Tian, F.R., Ozkan, C.S., Wang, M., Gao, H.J., 2005. Effect of single wall carbon nanotubes on human HEK293 cells. Toxicol. Lett. 155, 73–85.

Daer, S., Kharraz, J., Giwa, A., Hasan, G.W., 2015. Recent applications of nanomaterials in water desalination: a critical review and future opportunities. Desalination 367, 37–48.

Darabdhara, G., Boruah, P.K., Hussain, N., Borthakur, P., Sharma, B., Sengupta, P., Das, M.R., 2017. Magnetic nanoparticles towards efficient adsorption of gram positive and gram negative bacteria: An investigation of adsorption parameters and interaction mechanism. Colloids Surf. A 516, 161–170.

Das, D., Chatterjee, A., Mandal, B.K., Samanta, G., Chakraborti, D., Chanda, B., 1995. Arsenic in ground water in six districts of West Bengal, India: the biggest arsenic calamity in the world. Part 2. Arsenic concentration in drinking water, hair, nails, urine, skin-scale and liver tissue (biopsy) of the affected people. Analyst 120 (3), 917–924.

Das, P., Ghosh, S., Ghosh, R., Dam, S., Baskey, M., 2018. Madhuca longifolia plant mediated green synthesis of cupric oxide nanoparticles: a promising environmentally sustainable material for waste water treatment and efficient antibacterial agent. J. Photochem. Photobiol. B 189, 66–73. https://doi.org/10.1016/j.jphotobiol.2018.09.023.

Das, R., Vecitis, C.D., Schulze, A., Cao, B., Ismail, A.F., Lu, X., Chen, J., Ramakrishna, S., 2017. Recent advances in nanomaterials for water protection and monitoring. Chem. Soc. Rev. 46, 6946–7020. https://doi.org/10.1039/c6cs00921b.

Davarnejad, R., Panahi, P., 2016. Cu (II) removal from aqueous wastewaters by adsorption on the modified Henna with Fe_3O_4 nanoparticles using response surface methodology. Sep. Purif. Technol. 158, 286–292.

Dehaghi, S.M., Rahmanifar, B., Moradi, A.M., Azar, P.A., 2014. Removal of permethrin pesticide from water by chitosan–zinc oxide nanoparticles composite as an adsorbent. Journal of Saudi Chemical Society 18 (4), 348–355.

Dehghani, M.H., Niasar, Z.S., Mehrnia, M.R., Shayeghi, M., Al-Ghouti, M.A., Heibati, B., Yetilmezsoy, K., 2017. Optimizing the removal of organophosphorus pesticide malathion from water using multi-walled carbon nanotubes. Chem. Eng. J. 310, 22–32.

Depault, F., Cojocaru, M., Fortin, F., Chakrabarti, S., Lemieux, N., 2006. Genotoxic effects of chromium (VI) and cadmium (II) in human blood lymphocytes using the electron microscopy in situ end-labeling (EM-ISEL) assay. Toxicol. in Vitro 20 (4), 513–518.

Dhawan, A., Singh, Kumar, A, 2018. Nanobiotechnology: Human Health and the Environment. CRC Press, Boca Raton, FL.

Dickson, D., Liu, G., Cai, Y., 2017. Adsorption kinetics and isotherms of arsenite and arsenate on hematite nanoparticles and aggregates. J. Environ. Manage. 186, 261–267.

Dimapilis, E.A.S., Hsu, C.S., Mendoza, R.M.O., Lu, M.C., 2018. Zinc oxide nanoparticles for water disinfection. Sustain. Environ. Res. 28 (2), 47–56.

Dixon, M.B., Richard, Y., Ho, L., Chow, C.W., O'Neill, B.K., Newcombe, G., 2011. A coagulation–powdered activated carbon–ultrafiltration–Multiple barrier approach for removing toxins from two Australian cyanobacterial blooms. J. Hazard. Mater. 186 (2–3), 1553–1559.

Dodgen, L.K., Li, J., Parker, D., Gan, J.J., 2013. Uptake and accumulation of four PPCP/EDCs in two leafy vegetables. Environ. Pollut. 182, 150–156. https://doi.org/10.1016/j.envpol.2013.06.038.

Duman, O., Özcan, C., Polat, T.G., Tunç, S., 2019. Carbon nanotube-based magnetic and non-magnetic adsorbents for the high-efficiency removal of diquat dibromide herbicide from water: OMWCNT, OMWCNT-Fe3O4 and OMWCNT-κ-carrageenan-Fe3O4 nanocomposites. Environ. Pollut. 244, 723–732.

El-Kady, A.A., Abdel-Wahhab, M.A., 2018. Occurrence of trace metals in foodstuffs and their health impact. Trends Food Sci. Technol. 75, 36–45.

EPA (Environmental Protection Agency, U.S.), 2007. Nanotechnology white paper.

EPA (Environmental Protection Agency), 2010. RTP Division. Nanomaterial case studies: nanoscale titanium dioxide in water treatment and in topical sunscreen. 2010; EPA/600/R-09/057F. Available from: http://cfpub.epa.gov/ncea/cfm/recordisplay.cfm?deid=230972.

US-EPA, 2012. Nanomaterial Case Study: Nanoscale Silver in Disinfectant Spray.

Etacheri, V., Di Valentin, C., Schneider, J., Bahnemann, D., Pillai, S.C., 2015. Visible-light activation of TiO_2 photocatalysts: Advances in theory and experiments. J. Photochem. Photobiol. C Photochem. Rev. https://doi.org/10.1016/j.jphotochemrev.2015.08.003.

Facciotti, M., Boffa, V., Magnacca, G., Jørgensen, L.B., Kristensen, P.K., Farsi, A., Yue, Y., 2014. Deposition of thin ultrafiltration membranes on commercial SiC microfiltration tubes. Ceram. Int. 40 (2), 3277–3285.

Fakhri, A., Adami, S., 2014. Adsorption and thermodynamic study of Cephalosporins antibiotics from aqueous solution onto MgO nanoparticles. J. Taiwan Inst. Chem. Eng. 45 (3), 1001–1006.

Fan, H.L., Zhou, S.F., Jiao, W.Z., Qi, G.S., Liu, Y.Z., 2017. Removal of heavy metal ions by magnetic chitosan nanoparticles prepared continuously via high-gravity reactive precipitation method. Carbohydr. Polym. 174, 1192–1200. https://doi.org/10.1016/j.carbpol.2017.07.050.

Fato, F.P., Li, D.W., Zhao, L.J., Qiu, K., Long, Y.T., 2019. Simultaneous removal of multiple heavy metal ions from river water using ultrafine mesoporous magnetite nanoparticles. ACS Omega 4 (4), 7543–7549. https://doi.org/10.1021/acsomega.9b00731.

Fazlzadeh, M., Rahmani, K., Zarei, A., Abdoallahzadeh, H., Nasiri, F., Khosravi, R., 2017. A novel green synthesis of zero valent iron nanoparticles (NZVI) using three plant extracts and their efficient application for removal of Cr (VI) from aqueous solutions. Adv. Powder Technol. 28 (1), 122–130.

Fouladgar, M., Beheshti, M., Sabzyan, H., 2015. Single and binary adsorption of nickel and copper from aqueous solutions by γ-alumina nanoparticles: equilibrium and kinetic modeling. J. Mol. Liq. 211, 1060–1073.

Frezzini, M.A., Giuliano, A., Treacy, J., Canepari, S., Massimi, L., 2018. Food waste materials appear efficient and low-cost adsorbents for the removal of organic and inorganic pollutants from wastewater. Interface Sci 316, 298–309.

Fuhrimann, S., Winkler, M.S., Kabatereine, N.B., Tukahebwa, E.M., Halage, A.A., Rutebemberwa, E., Medlicott, K., Schindler, C., Utzinger, J., Cissé, G., 2016. Risk of intestinal parasitic infections in people with different exposures to wastewater and fecal sludge in Kampala, Uganda: a cross-sectional study. PLoS Negl. Trop. Dis. 10 (3), e0004469. https://doi.org/10.1371/journal.pntd.0004469.

Gagné, F., Auclair, J., Turcotte, P., Fournier, M., Gagnon, C., Sauvé, S., Blaise, C., 2008. Ecotoxicity of CdTe quantum dots to freshwater mussels: impacts on immune system, oxidative stress and genotoxicity. Aquatic. Toxicol. 86 (3), 333–340.

Gaiser, B.K., Biswas, A., Rosenkranz, P., Jepson, M.A., Lead, J.R., Stone, V., Fernandes, T.F., 2011. Effects of silver and cerium dioxide micro-and nano-sized particles on *Daphnia magna*. J. Environ. Monit. 13 (5), 1227–1235.

Gao, W., D'Agostino, M., Garcia-Gradilla, V., Orozco, J., Wang, J., 2013. Multi-fuel driven janus micromotors. Small 9 (3), 467–471.

Gao, W., Dong, R., Thamphiwatana, S., Li, J., Gao, W., Zhang, L., Wang, J., 2015. Artificial micromotors in the mouse's stomach: A step toward in vivo use of synthetic motors. ACS nano 9 (1), 117–123.

García-Díaz, I., López, F.A., Alguacil, F.J., 2018. Carbon Nanofibers: A New Adsorbent for Copper Removal from. Wastewater. Metals 8 (11), 914.

Ge, H., Chen, X., Liu, W., Lu, X., Gu, Z., 2019. Metal-Based Transient Micromotors: From Principle to Environmental and Biomedical Applications. Chem. Asian J. 14 (14), 2348–2356.

Gehrke, I., Geiser, A., Somborn-Schulz, A., 2015. Innovations in nanotechnology for water treatment. Nanotechnol. Sci. Appl. 8, 1–17.

Genthe, B., Kapwata, T., Le Roux, W., Chamier, J., Wright, C.Y., 2018. The reach of human health risks associated with metals/metalloids in water and vegetables along a contaminated river catchment: South Africa and Mozambique. Chemosphere 199, 1–9. https://doi.org/10.1016/j.chemosphere.2018.01.160.

Ghosh, I., Mukherjee, A., Mukherjee, A., 2017. In planta genotoxicity of nZVI: Influence of colloidal stability on uptake, DNA damage, oxidative stress and cell death. Mutagenesis 32 (3), 371–387.

Goh, P.S., Ng, B.C., Lau, W.J., Ismail, A.F., 2015. Inorganic nanomaterials in polymeric ultrafiltration membranes for water treatment. Separation & Purification Reviews 44 (3), 216–249.

Gomes, J., Domingues, E., Quinta-Ferreira, R.M., Martins, R.C., 2019. N–TiO2 photocatalysts: a review of their characteristics and capacity for emerging contaminants removal. water 11 (2), 373. doi:10.3390/w11020373.

Gómez-Pastora, J., Dominguez, S., Bringas, E., Rivero, M.J., Ortiz, I., Dionysiou, D.D., 2017. Review and perspectives on the use of magnetic nanophotocatalysts (MNPCs) in water treatment. Chem. Eng. Sci. 310, 407–427.

Gunnarsson, R., Helmersson, U., Pilch, I., 2015. Synthesis of titanium-oxide nanoparticles with size and stoichiometry control. J. Nanopart. Res. 17 (9), 353.

Guo, Y., Zhang, X., Sun, X., Kong, D., Han, M., Wang, X., 2019. Nanoadsorbents Based on NIPAM and Citric Acid: Removal Efficacy of Heavy Metal Ions in Different Media. ACS omega 4 (10), 14162–14168.

Gupta, V.K., Chandra, R., Tyagi, I., Verma, M., 2016. Removal of hexavalent chromium ions using CuO nanoparticles for water purification applications. J. Colloid Interface Sci. 478, 54–62.

Gupta, N., Khan, D.K., Santra, S.C., 2012. Heavy metal accumulation in vegetables grown in a long-term wastewater-irrigated agricultural land of tropical India. Environ. Monit. Assess. 184 (11), 6673–6682. https://doi.org/10.1007/s10661-011-2450-7.

Gusain, R., Kumar, N., Fosso-Kankeu, E., Ray, S.S., 2019. Efficient removal of Pb (II) and Cd (II) from industrial mine water by a hierarchical MoS_2/SH-MWCNT nanocomposite. ACS omega 4 (9), 13922–13935.

Hageman, K.J., Bogdal, C., Scheringer, M., 2015. Long-range and regional atmospheric transport of POPs and implications for global cycling. In: Zeng, Eddy Y. (Ed.), Comprehensive Analytical Chemistry, 67. Elsevier B.V., Amsterdam, pp. 363–387.

Hajnorouzi, A., Afzalzadeh, R., 2019. A novel technique to generate aluminum nanoparticles utilizing ultrasound ablation. Ultrason. Sonochem. 58, 104655. doi:10.1016/J.ULTSONCH.2019.104655.

Han, R., Ding, D., Xu, Y., Zou, W., Wang, Y., Li, Y., Zou, L., 2008. Use of rice husk for the adsorption of congo red from aqueous solution in column mode. Bioresour. Technol. 99 (8), 2938–2946.

Hassan, A.F., Elhadidy, H., 2019. Effect of Zr^{+4} doping on characteristics and sonocatalytic activity of TiO_2/carbon nanotubes composite catalyst for degradation of chlorpyrifos. J. Phys. Chem. Solids 129, 180–187.

Hassan, N.U., Mahmood, Q., Waseem, A., Irshad, M., Faridullah, Pervez, A, 2013. Assessment of heavy metals in wheat plants irrigated with contaminated wastewater. Polish J. Environ. Stud. 22 (1), 115–123.

Hokkanen, S., Repo, E., Lou, S., Sillanpää, M., 2015. Removal of arsenic (V) by magnetic nanoparticle activated microfibrillated cellulose. Chem. Eng. J. 260, 886–894.

Hosseini, S.M., Tosco, T., Ataie-Ashtiani, B., Simmons, C.T., 2018. Non-pumping reactive wells filled with mixing nano and micro zero-valent iron for nitrate removal from groundwater: vertical, horizontal, and slanted wells. J. Contam. Hydrol. 210, 50–64. https://doi.org/10.1016/j.jconhyd.2018.02.006.

Hua, M., Zhang, S., Pan, B., Zhang, W., Lv, L., Zhang, Q., 2012. Heavy metal removal from water/wastewater by nanosized metal oxides: a review. J. Hazard. Mater. 211–212, 317–331. https://doi.org/10.1016/j.jhazmat.2011.10.016.

Huang, H., Chen, T., Liu, X., Ma, H., 2014. Ultrasensitive and simultaneous detection of heavy metal ions based on three-dimensional graphene-carbon nanotubes hybrid electrode materials. Anal. Chim. Acta 852, 45–54.

Huang, W.J., Cheng, B.L., Cheng, Y.L., 2007. Adsorption of microcystin-LR by three types of activated carbon. J. Hazard. Mater. 141 (1), 115–122.

Huang, G., Lu, C.H., Yang, H.H., 2018. Magnetic nanomaterials for magnetic bioanalysis. In: Wang, X., Cheng, X. (Eds.), Novel Nanomaterials for Biomedical, Environmental and Energy Applications. Elsevier, Amsterdam, pp. 89–109.

Huibers, F.P., Lier, 2005. Use of wastewater in agriculture: the water chain approach. Irrig. Drain., 54, pp. S3–S9.

Hyung, H., Kim, J.H., 2009. Dispersion of C60 in natural water and removal by conventional drinking water treatment processes. Water Res. 43 (9), 2463–2470.

Ihsanullah, Abbas, A., Al-Amer, A., M., Laoui, T., Al-Marri, M.J., Nasser, M.S., Khraisheh, M., Atieh, M.A, 2016. Heavy metal removal from aqueous solution by advanced carbon nanotubes: critical review of adsorption applications. Sep. Purif. Technol. 157, 141–161. https://doi.org/10.1016/j.seppur.2015.11.039.

Izanloo, M., Esrafili, A., Jafari, A.J., Farzadkia, M., Behbahani, M., Sobhi, H.R., 2019. Application of a novel bi-functional nanoadsorbent for the simultaneous removal of inorganic and organic compounds: equilibrium, kinetic and thermodynamic studies. J. Mol. Liq. 273, 543–550. https://doi.org/10.1016/j.molliq.2018.10.013.

Jackson, P., Jacobsen, N.R., Baun, A., Birkedal, R., Kühnel, D., Jensen, K.A., Vogel, U., Wallin, H., 2013. Bioaccumulation and ecotoxicity of carbon nanotubes. Chem. Cent. J. 7 (1), 154. https://doi.org/10.1186/1752-153X-7-154.

Jackson, S.D., Hargreaves, 2008. Metal Oxide Catalysis, 2 Volume Set, 1. Wiley, NJ.

Janjua, M.R.S.A., 2019. Synthesis of Co_3O_4 nano aggregates by co-precipitation method and its catalytic and fuel additive applications. Open Chem. 17 (1), 865–873. https://doi.org/10.1515/chem-2019-0100.

Jaramillo, M.F, Restrepo, I., 2017. Wastewater reuse in agriculture: a review about its limitations and benefits. Sustainability (Switzerland) 9 (10). https://doi.org/10.3390/su9101734.

Järup, L., 2003. Hazards of heavy metal contamination. Br. Med. Bull. 68, 167–182. https://doi.org/10.1093/bmb/ldg032.

Jaspers, V., Megson, D., O'Sullivan, G, 2013. POPs in the terrestrial environment. In: O'Sullivan, G., Sandau, C. (Eds.), Environmental Forensics for Persistent Organic Pollutants. Elsevier B.V, Amsterdam, pp. 291–356.

Jhaveri, J.H., Murthy, Z.V.P., 2016. A comprehensive review on anti-fouling nanocomposite membranes for pressure driven membrane separation processes. Desalination 379, 137–154.

Jiang, H.S., Qiu, X.N., Li, G.B., Li, W., Yin, L.Y., 2014. Silver nanoparticles induced accumulation of reactive oxygen species and alteration of antioxidant systems in the aquatic plant Spirodela polyrhiza. Environ. Toxicol. Chem. 33 (6), 1398–1405.

Jiuhui, Q.U., 2008. Research progress of novel adsorption processes in water purification: a review. Journal of environmental sciences 20 (1), 1–13.

Jurado-Sánchez, B., Wang, J., 2018. Micromotors for environmental applications: A review. Environ. Sci.: Nano. 5 (7), 1530–1544.
Kalantary, R.R., Azari, A., Esrafili, A., Yaghmaeian, K., Moradi, M., Sharafi, K., 2016. The survey of Malathion removal using magnetic graphene oxide nanocomposite as a novel adsorbent: thermodynamics, isotherms, and kinetic study. Desalin. Water Treat. 57 (58), 28460–28473.
Kanakaraju, D., Glass, B.D., Oelgemöller, M., 2014. Titanium dioxide photocatalysis for pharmaceutical wastewater treatment. Environ. Chem. Lett. 12 (1), 27–47. https://doi.org/10.1007/s10311-013-0428-0.
Karim, Z., Mathew, A.P., Grahn, M., Mouzon, J., Oksman, K., 2014. Nanoporous membranes with cellulose nanocrystals as functional entity in chitosan: removal of dyes from water. Carbohydr. Polym. 112, 668–676.
Karpińska, J., Kotowska, U., 2019. Removal of organic pollution in the water environment. Water (Switzerland) 11 (10), 2017. https://doi.org/10.3390/w11102017.
Kaur, M., Kumari, S., Sharma, P., 2020. Removal of Pb (II) from aqueous solution using nanoadsorbent of *Oryza sativa* husk: Isotherm, kinetic and thermodynamic studies. Biotechnology Reports 25, e00410.
Kefeni, K.K., Mamba, B.B., Msagati, T.A., 2017. Application of spinel ferrite nanoparticles in water and wastewater treatment: a review. Sep. Purif. Technol. 188, 399–422.
Keshavarz, A., Parang, Z., Nasseri, A., 2013. The effect of sulfuric acid, oxalic acid, and their combination on the size and regularity of the porous alumina by anodization. Journal of Nanostructure in Chemistry 3 (1), 34.
Khalid, S., Shahid, M., Bibi, I., Sarwar, T., Shah, A.H., Niazi, N.K., 2018. A review of environmental contamination and health risk assessment of wastewater use for crop irrigation with a focus on low and high-income countries. Int. J. Environ. Res. Public Health 15 (5), 895.
Khalid, S., Shahid, M., Dumat, C., Niazi, N.K., Bibi, I., Bakhat, Gul, H., F.S., Abbas, G., Murtaza, B., Javeed, H.M.R, 2017. Influence of groundwater and wastewater irrigation on lead accumulation in soil and vegetables: implications for health risk assessment and phytoremediation. Int. J. Phytoremediat. 19 (11), 1037–1046. https://doi.org/10.1080/15226514.2017.1319330.
Khan, S., Aijun, L., Zhang, S., Hu, Q., Zhu, Y.G., 2008a. Accumulation of polycyclic aromatic hydrocarbons and heavy metals in lettuce grown in the soils contaminated with long-term wastewater irrigation. J. Hazard. Mater. 152 (2), 506–515. https://doi.org/10.1016/j.jhazmat.2007.07.014.
Khan, S., Cao, Q., Zheng, Y.M., Huang, Y.Z., Zhu, Y.G., 2008b. Health risks of heavy metals in contaminated soils and food crops irrigated with wastewater in Beijing, China. Environ. Pollut. 152 (3), 686–692. https://doi.org/10.1016/j.envpol.2007.06.056.
Khan, M.U., Malik, R.N., Muhammad, S., 2013. Human health risk from heavy metal via food crops consumption with wastewater irrigation practices in Pakistan. Chemosphere 93 (10), 2230–2238. https://doi.org/10.1016/j.chemosphere.2013.07.067.
Klaper, R., Crago, J., Barr, J., Arndt, D., Setyowati, K., Chen, J., 2009. Toxicity biomarker expression in daphnids exposed to manufactured nanoparticles: changes in toxicity with functionalization. Environ. Pollut. 157 (4), 1152–1156.
Korashy, H.M., Attafi, I.M., Famulski, K.S., Bakheet, S.A., Hafez, M.M., Alsaad, A.M., Al-Ghadeer, A.R.M., 2017. Gene expression profiling to identify the toxicities and potentially relevant human disease outcomes associated with environmental heavy metal exposure. Environ. Pollut. 221, 64–74.
Kumari, M., Gupta, S.K., 2020. A novel process of adsorption cum enhanced coagulation-flocculation spiked with magnetic nanoadsorbents for the removal of aromatic and hydrophobic fraction of natural organic matter along with turbidity from drinking water. J. Clean. Prod. 244, 118899.
Kumari, P., Alam, M., Siddiqi, W.A., 2019. Usage of nanoparticles as adsorbents for waste water treatment: an emerging trend. Sustain. Mater. Technol. 22, e00128. https://doi.org/10.1016/j.susmat.2019.e00128
Lee, C., Baik, S., 2010. Vertically-aligned carbon nano-tube membrane filters with superhydrophobicity and superoleophilicity. Carbon 48 (8), 2192–2197.
Lee, S.Y., Shim, H.E., Yang, J.E., Choi, Y.J., Jeon, J., 2019. Continuous flow removal of anionic dyes in water by chitosan-functionalized Iron oxide nanoparticles incorporated in a dextran gel column. Nanomaterials 9 (8), 1164.
Li, J., Chen, C., Zhu, K., Wang, X., 2016. Nanoscale zero-valent iron particles modified on reduced graphene oxides using a plasma technique for Cd (II) removal. J. Taiwan Inst. Chem. Eng. 59, 389–394.

Li, H., Gui, X., Zhang, L., Wang, S., Ji, C., Wei, J., Cao, A., 2010. Carbon nanotube sponge filters for trapping nanoparticles and dye molecules from water. Chem. Commun. 46 (42), 7966–7968.

Li, X., Xia, T., Xu, C., Murowchick, J., Chen, X., 2014. Synthesis and photoactivity of nanostructured CdS–TiO_2 composite catalysts. Catal. Today. 225, 64–73.

Liang, X., Cui, S., Li, H., Abdelhady, A., Wang, H., Zhou, H., 2019a. Removal effect on stormwater runoff pollution of porous concrete treated with nanometer titanium dioxide. Transp. Res. D Transp. Environ. 73, 34–45.

Liang, X.X., Omer, A.M., Hu, Z.H., Wang, Y.G., Yu, D., Ouyang, X.K., 2019b. Efficient adsorption of diclofenac sodium from aqueous solutions using magnetic amine-functionalized chitosan. Chemosphere 217, 270–278.

Liu, J., Chen, H., Shi, X., Nawar, S., Werner, J.G., Huang, G., Ye, M., Weitz, D.A., Solovev, A.A., Mei, Y., 2020. Hydrogel microcapsules with photocatalytic nanoparticles for removal of organic pollutants. Environ. Sci. Nano 7 (2), 656–664. https://doi.org/10.1039/c9en01108k.

Liu, X., Demir, N.K., Wu, Z., Li, K., 2015. Highly water-stable zirconium metal–organic framework UiO-66 membranes supported on alumina hollow fibers for desalination. J. Am. Chem. Soc. 137 (22), 6999–7002.

Liu, G., Gao, J., Ai, H., Chen, X., 2013. Applications and potential toxicity of magnetic iron oxide nanoparticles. Small 9 (9–10), 1533–1545.

Liu, W, Yang, L, Xu, S, Chen, Y, Liu, B, Li, Z, Jiang, C, 2018. Efficient removal of hexavalent chromium from water by an adsorption–reduction mechanism with sandwiched nanocomposites. RSC Adv. 8 (27), 15087–15093.

Loeb, S.K., Alvarez, P.J., Brame, J.E., Cates, E.L., Choi, W., Crittenden, J., Dionysiou, D.D., Li, Q., Li-Puma, J., Quan, X., Sedlak, D.L., 2018. The technology horizon for photocatalytic water treatment: sunrise or sunset? Environ. Sci. Technol. 53 (6), 2937–2947.

Luo, S., Lu, T., Peng, L., Shao, J., Zeng, Q., Gu, J.D., 2014. Synthesis of nanoscale zero-valent iron immobilized in alginate microcapsules for removal of Pb (II) from aqueous solution. J. Mater. Chem. A 2 (37), 15463–15472.

Mahmood, A., Malik, R.N., 2014. Human health risk assessment of heavy metals via consumption of contaminated vegetables collected from different irrigation sources in Lahore, Pakistan. . Arab. J. Chem. 7 (1), 91–99. https://doi.org/10.1016/j.arabjc.2013.07.002.

Malik, A., Hameed, S., Siddiqui, M.J., Haque, M.M., Umar, K., Khan, A., Muneer, M., 2014. Electrical and optical properties of nickel-and molybdenum-doped titanium dioxide nanoparticle: improved performance in dye-sensitized solar cells. J. Mater. Eng. Perform. 23 (9), 3184–9192.

Manjuladevi, M., Sri, M.O., 2017. Heavy metals removal from industrial wastewater by nano adsorbent prepared from cucumis melopeel activated carbon. Journal of Nanomedicine Research 5, 1–4.

Mansoureh, G., Parisa, V., 2018. Synthesis of metal nanoparticles using laser ablation technique. Emerging Applications of Nanoparticles and Architecture Nanostructures. Elsevier, pp. 575–596.

Mapanda, F., Mangwayana, E.N., Nyamangara, J., Giller, K.E., 2007. Uptake of heavy metals by vegetables irrigated using wastewater and the subsequent risks in Harare, Zimbabwe. Phys. Chem. Earth 32 (15–18), 1399–1405. https://doi.org/10.1016/j.pce.2007.07.046.

Mara, D., Sleigh, A., 2010. Estimation of Ascaris infection risks in children under 15 from the consumption of wastewater-irrigated carrots. J. Water Health 8 (1), 35–38. https://doi.org/10.2166/wh.2009.136.

Martinez-Vargas, S., Martínez, A.I., Hernández-Beteta, E.E., Mijangos-Ricardez, O.F., Vázquez-Hipólito, V., Patiño-Carachure, C., López-Luna, J., 2017. Arsenic adsorption on cobalt and manganese ferrite nanoparticles. J. Mater. Sci. 52 (11), 6205–6215.

Masindi, V., Muedi, K.L., 2018. Environmental contamination by heavy metals. Heavy metals. intechopen 115–132.

Massoudinejad, M., Rasoulzadeh, H., Ghaderpoori, M., 2019. Magnetic chitosan nanocomposite: Fabrication, properties, and optimization for adsorptive removal of crystal violet from aqueous solutions. Carbohydr. Polym. 206, 844–853.

Matlochova, A., Placha, D., Rapantová, N., 2013. The application of nanoscale materials in groundwater remediation. Polish journal of environmental studies 22 (5), 1401–1410.

Miller-Robbie, L., Ramaswami, A., Amerasinghe, P., 2017. Wastewater treatment and reuse in urban agriculture: exploring the food, energy, water, and health nexus in Hyderabad, India. Environ. Res. Lett. 12 (7), 075005. https://doi.org/10.1088/1748-9326/aa6bfe.

Mireles, A., Solıs, C., Andrade, E., Lagunas-Solar, M., Pina, C., Flocchini, R.G., 2004. Heavy metal accumulation in plants and soil irrigated with wastewater from Mexico city. Nucl. Instrum. Methods Phys. Res. Sect. B 219-220, 187–190.

Mishra, B.K., Chakraborty, S., Kumar, P., Saraswat, C., 2020. Water Quality Restoration and Reclamation. Sustainable Solutions for Urban Water Security. Springer, Cham, pp. 59–81.

Mittal, G., Dhand, V., Rhee, K.Y., Park, S.J., Lee, W.R., 2015. A review on carbon nanotubes and graphene as fillers in reinforced polymer nanocomposites. J. Ind. Eng. Chem. 21, 11–25. https://doi.org/10.1016/j.jiec.2014.03.022.

Mok, H.F., Hamilton, A.J., 2014. Exposure factors for wastewater-irrigated Asian vegetables and a probabilistic rotavirus disease burden model for their consumption. Risk Anal. 34 (4), 602–613. https://doi.org/10.1111/risa.12178.

Moo, J.G., Pumera, M., 2015. Chemical energy powered nano/micro/macromotors and the environment. Chem. Eur. J. 21 (1), 58–72.

Mouchet, F., Landois, P., Sarremejean, E., Bernard, G., Puech, P., Pinelli, E., Gauthier, L., 2008. Characterisation and in vivo ecotoxicity evaluation of double-wall carbon nanotubes in larvae of the amphibian *Xenopus laevis*. Aquatic. Toxicol. 87 (2), 127–137.

Moura, C.G, Pereira, R.S.F., Andritschky, M., Lopes, A.L.B., de Freitas Grilo, J.P., do Nascimento, R.M., Silva, F.S., 2017. Effects of laser fluence and liquid media on preparation of small Ag nanoparticles by laser ablation in liquid. Opt. Laser Technol. 97, 20–28.

Murtaza, G., Ghafoor, A., Qadir, M., Owens, G., Aziz, M.A., Ziz, M.H., Siafullah, 2010. Disposal and use of sewage on agricultural lands in Pakistan: a review. Pedosphere 20 (1), 23–34. https://doi.org/10.1016/S1002-0160(09)60279-4.

Nassar, M.Y., Ahmed, I.S., Raya, M.A., 2019. A facile and tunable approach for synthesis of pure silica nanostructures from rice husk for the removal of ciprofloxacin drug from polluted aqueous solutions. J. Mol. Liq. 282, 251–263.

Nowack, B., Bucheli, T.D., 2007. Occurrence, behavior and effects of nanoparticles in the environment. Environ. Pollut. 150 (1), 5–22.

Nurmi, J.T., Tratnyek, P.G., Sarathy, V., Baer, D.R., Amonette, J.E., Pecher, K., Driessen, M.D., 2005. Characterization and properties of metallic iron nanoparticles: spectroscopy, electrochemistry, and kinetics. Environ. Sci. Technol. 39 (5), 1221–1230.

Obaid, S.A., 2019. Removal chromium (VI) from water by magnetic carbon nano-composite made by burned straw. Journal of Physics: Conference Series, 1234. IOP Publishing.

Ong, C.B., Ng, L.Y., Mohammad, A.W., 2018. A review of ZnO nanoparticles as solar photocatalysts: synthesis, mechanisms and applications. Renew. Sust. Energ. Rev. 81, 536–551.

Orozco-Montes, V., Bigarre, J., Lecas, T., Brault, P., Caillard, A., 2019. PdPt nanocatalyst synthesis via liquid medium sputtering. ISPC hal-02394305. https://hal.archives-ouvertes.fr/hal-02394305/document.

Pace, H.E., Lesher, E.K., Ranville, J.F., 2010. Influence of stability on the acute toxicity of CdSe/ZnS nanocrystals to *Daphnia magna*. Environ. Toxicol. Chem. 29 (6), 1338–1344.

Pacheco, M., López, M.A., Jurado-Sánchez, B., Escarpa, A., 2019. Self-propelled micromachines for analytical sensing: a critical review. Anal. Bioanal. Chem. 411, 6561–6573.

Pacheco, S., Medina, M., Valencia, F., Tapia, J., 2006. Removal of inorganic mercury from polluted water using structured nanoparticles. J. Environ. Eng. 132 (3), 342–349.

Pan, B., Lin, D., Mashayekhi, H., Xing, B., 2008. Adsorption and hysteresis of bisphenol A and 17O ± -ethinyl estradiol on carbon nanomaterials. Environ. Sci. Technol. 42, 5480–5485.

Pan, X., Xu, Y.J., 2013. Defect-mediated growth of noble-metal (Ag, Pt, and Pd) nanoparticles on TiO_2 with oxygen vacancies for photocatalytic redox reactions under visible light. J. Phys. Chem. C 117 (35), 17996–18005. https://doi.org/10.1021/jp4064802.

Pandey, N., Shukla, S.K., Singh, N.B., 2017. Water purification by polymer nanocomposites: an overview. Nanocomposites 3 (2), 47–66.

Phokha, S., Klinkaewnarong, J., Hunpratub, S., Boonserm, K., Swatsitang, E., Maensiri, S., 2016. Ferromagnetism in Fe-doped MgO nanoparticles. J. Mater. Sci.: Mater. Electron. 27 (1), 33–39.

Piao, X., Ming, Z.G., Lian, H.D., Ling, F.C., Shuang, H., Hua, Z.M., Cui, L., Zhen, W., Chao, H., Xin, X.G., Feng, L.Z., 2012. Use of iron oxide nanomaterials in wastewater treatment: a review. Sci. Total Environ. 424, 1–10. https://doi.org/10.1016/j.scitotenv.2012.02.023.

Plohl, O., Finšgar, M., Gyergyek, S., Ajdnik, U., Ban, I., Zemljič, L.F., 2019. Efficient Copper Removal from an Aqueous Anvironment using a Novel and Hybrid Nanoadsorbent Based on Derived-Polyethyleneimine Linked to Silica Magnetic Nanocomposites. Nanomaterials 9 (2), 209.

Pourrahimi, A.M., Liu, D., Ström, V., Hedenqvist, M.S., Olsson, R.T., Gedde, U.W., 2015. Heat treatment of ZnO nanoparticles: new methods to achieve high-purity nanoparticles for high-voltage applications. J. Mater. Chem. A 3 (33), 17190–17200.

Pourrahimi, A.M., Villa, K., Ying, Y., Sofer, Z., Pumera, M., 2018. $ZnO/ZnO_2/Pt$ Janus Micromotors Propulsion Mode Changes with Size and Interface Structure: Enhanced Nitroaromatic Explosives Degradation under Visible Light. ACS Appl. Mater. Interfaces. 10 (49), 42688–42697.

Pourrut, B., Shahid, M., Douay, F., Dumat, C., Pinelli, E., 2013. Molecular mechanisms involved in lead uptake, toxicity and detoxification in higher plants. In: Gupta, D., Corpas, F., Palma, J. (Eds.), Heavy Metal Stress in Plants. Springer-Verlag, Berlin Heidelberg, pp. 121–147.

Prathna, T.C., Sharma, S.K., Kennedy, M., 2018. Application of iron oxide and iron oxide/alumina nanocomposites for arsenic and fluoride removal: A comparative study. International Journal of Theoretical and Applied Nanotechnology 6 (1), 1–4.

Qadir, M., Wichelns, D., Raschid-Sally, L., McCornick, P.G., Drechsel, P., Bahri, A., Minhas, P.S., 2010. The challenges of wastewater irrigation in developing countries. Agric. Water Manage. 97 (4), 561–568. https://doi.org/10.1016/j.agwat.2008.11.004.

Qishlaqi, A., Moore, F., Forghani, G., 2008. Impact of untreated wastewater irrigation on soils and crops in Shiraz suburban area, SW Iran. Environ. Monit. Assess. 141 (1–3), 257–273. https://doi.org/10.1007/s10661-007-9893-x.

Rahaman, M.S., Vecitis, C.D., Elimelech, M., 2012. Electrochemical carbon-nanotube filter performance toward virus removal and inactivation in the presence of natural organic matter. Environ. Sci. Technol. 46 (3), 1556–1564.

Raja, S., Cheema, H.M.N., Babar, S., Khan, A.A., Murtaza, G., Aslam, U., 2015. Socio-economic background of wastewater irrigation and bioaccumulation of heavy metals in crops and vegetables. Agric. Water Manage. 158, 26–34. https://doi.org/10.1016/j.agwat.2015.04.004.

Rane, A.V., Kanny, K., Abitha, V.K., Thomas, S., 2018. Methods for synthesis of nanoparticles and fabrication of nanocomposites. In: Bhagyaraj, S.M., Oluwafemi, O.S., Kalarikkal, N., Thomas, S. (Eds.), Synthesis of Inorganic Nanomaterials. Elsevier, Amsterdam, pp. 121–139.

Rauwel, P., Rauwel, E., Ferdov, S., Singh, M.P., 2015. A Review on the Green Synthesis of Silver Nanoparticles and Their Morphologies Studied via TEM. Adv. Mater. Sci. Eng., 682749 doi:10.1155/2015/682749.

Rezvani, E., Rafferty, A., McGuinness, C., Kennedy, J., 2019. Adverse effects of nanosilver on human health and the environment. Acta Biomater. 94, 145–159.

Riemenschneider, C., Al-Raggad, M., Moeder, M., Seiwert, B., Salameh, E., Reemtsma, T., 2016. Pharmaceuticals, their metabolites, and other polar pollutants in field-grown vegetables irrigated with treated municipal wastewater. J. Agric. Food Chem. 64 (29), 5784–5792. https://doi.org/10.1021/acs.jafc.6b01696.

Rivera-Rangel, R.D., González-Muñoz, M.P., Avila-Rodriguez, M., Razo-Lazcano, T.A., Solans, C., 2018. Green synthesis of silver nanoparticles in oil-in-water microemulsion and nano-emulsion using geranium leaf aqueous extract as a reducing agent. Colloids Surf. A 536, 60–67. https://doi.org/10.1016/j.colsurfa.2017.07.051.

Rovani, S., Santos, J.J., Corio, P., Fungaro, D.A., 2018. Highly pure silica nanoparticles with high adsorption capacity obtained from sugarcane waste ash. ACS omega 3 (3), 2618–2627.

Sadegh, H., Ali, G.A.M., Agarwal, S., Gupta, V.K., 2019. Surface modification of MWCNTs with carboxylic-to-amine and their superb adsorption performance. International Journal of Environmental Research 13 (3), 523–531.

Safatian, F., Doago, Z., Torabbeigi, M., Shams, H. R., Ahadi, N., 2019. Lead ion removal from water by hydroxyapatite nanostructures synthesized from egg sells with microwave irradiation. Applied Water Science 9 (4), 108.

Safdar, M., Simmchen, J., Jänis, J., 2017. Correction: Light-driven micro-and nanomotors for environmental remediation. Environmental Science: Nano 4 (11), 2235.

Sagadevan, S., Chowdhury, Z.Z., Rafique, R.F., 2018. Preparation and characterization of nickel ferrite nanoparticles via co-precipitation method. Mater. Res. 21 (2). https://doi.org/10.1590/1980-5373-mr-2016-0533.

Saingam, P., Li, B., Yan, T., 2020. Fecal indicator bacteria, direct pathogen detection, and microbial community analysis provide different microbiological water quality assessment of a tropical urban marine estuary. Water Res. 185, 116280.

Salavati-Niasari, M., Soofivand, F., Sobhani-Nasab, A., Shakouri-Arani, M., Fall, A.Y, Bagheri, S., 2016. Synthesis, characterization, and morphological control of ZnTiO3 nanoparticles through sol-gel processes and its photocatalyst application Adv. Powder Technol. 27, 2066–2075.

Samadi, S., Shalmani, M.M., Zakaria, S.A., 2019. Removal of heavy metals from Tehran south agricultural water by Zeolite N.P./PEG/GO nano-composite. J. Water Environ. Nanotechnol. 4 (2), 157–166.

Samanta, H.S., Das, R., Bhattachajee, C., 2016. Influence of nanoparticles for wastewater treatment-a short review. Austin. Chem. Eng. 3 (3), 1036.

Sanders, M. B., Sebire, M., Sturve, J., Christian, P., Katsiadaki, I., Lyons, B. P., Feist, S. W., 2008. Exposure of sticklebacks (*Gasterosteus aculeatus*) to cadmium sulfide nanoparticles: biological effects and the importance of experimental design. Marine environmental research 66 (1), 161–163.

Sani, H. A., Ahmad, M. B., Hussein, M. Z., Ibrahim, N. A., Musa, A., Saleh, T. A., 2017. Nanocomposite of ZnO with montmorillonite for removal of lead and copper ions from aqueous solutions. Process Safety and Environmental Protection 109, 97–105.

Santhosh, C., Velmurugan, V., Jacob, G., Jeong, S. K., Grace, A. N., Bhatnagar, A., 2016. Role of nanomaterials in water treatment applications: a review. Chemical Engineering Journal 306, 1116–1137.

Saravanan, R., Gracia, F., Stephen, A., 2017. Basic principles, mechanism, and challenges of photocatalysis. Nanocomposites for Visible Light-Induced Photocatalysis. Springer, Berlin/Heidelberg, Germany, pp. 19–40.

Sato, A., Wang, R., Ma, H., Hsiao, B. S., Chu, B., 2011. Novel nanofibrous scaffolds for water filtration with bacteria and virus removal capability. Journal of electron microscopy 60 (3), 201–209.

Savasari, M., Emadi, M., Bahmanyar, M. A., Biparva, P., 2015. Optimization of Cd (II) removal from aqueous solution by ascorbic acid-stabilized zero valent iron nanoparticles using response surface methodology. Journal of Industrial and Engineering Chemistry 21, 1403–1409.

Scotland, D., 2001. Alternative techniques to meet the world's water scarcity. Master thesis "Water Quality and Management". Mémoire du DESS «Qualité et Gestion de l'Eau», Faculté des Sciences. Faculty of Sciences, Amiens.

Serrà, A., Zhang, Y., Sepúlveda, B., Gómez, E., Nogués, J., Michler, J., Philippe, L., 2020. Highly reduced ecotoxicity of ZnO-based micro/nanostructures on aquatic biota: Influence of architecture, chemical composition, fixation, and photocatalytic efficiency. Water Res. 169, 115210.

Shahid, M., Khalid, S., Abbas, G., Shahid, N., Nadeem, M., Sabir, M., Aslam, M., Dumat, C., 2015. Heavy metal stress and crop productivity. In: Hakeem, K.R. (Ed.), Crop Production and Global Environmental Issues. Springer International Publishing,, NY, pp. 1–25.

Shahid, M., Rafiq, M., Niazi, N.K., Dumat, C., Shamshad, S., Khalid, S., Bibi, I., 2017. Arsenic accumulation and physiological attributes of spinach in the presence of amendments: an implication to reduce health risk. Environ. Sci. Pollut. Res. 24 (19), 16097–16106. https://doi.org/10.1007/s11356-017-9230-z.

Sharma, A., Kaur, M., Katnoria, J.K., Nagpal, A.K., 2016. Heavy metal pollution: A global pollutant of rising concern. Toxicity and waste management using bioremediation. IGI Global, pp. 1–26.

Shayesteh, M., Samimi, A., Shafiee Afarani, M., Khorram, M., 2016. Synthesis of titania–γ-alumina multilayer nanomembranes on performance-improved alumina supports for wastewater treatment. Desalination and Water Treatment 57 (20), 9115–9122.

singh, P., Kim, Y.J., Zhang, D., Yang, D.C., 2016. Biological synthesis of nanoparticles from plants and microorganisms. Trends Biotechnol 34 (7), 588–599.

Singh, A., Sharma, R.K., Agrawal, M., Marshall, F.M., 2010. Health risk assessment of heavy metals via dietary intake of foodstuffs from the wastewater irrigated site of a dry tropical area of India. Food Chem. Toxicol. 48 (2), 611–619. https://doi.org/10.1016/j.fct.2009.11.041.

Smith, C. J., Shaw, B. J., Handy, R. D., 2007. Toxicity of single walled carbon nanotubes to rainbow trout,(*Oncorhynchus mykiss*): respiratory toxicity, organ pathologies, and other physiological effects. Aquatic toxicology 82 (2), 94–109.

Soares, S. F., Simões, T. R., Trindade, T., Daniel-da-Silva, A. L., 2017. Highly efficient removal of dye from water using magnetic carrageenan/silica hybrid nano-adsorbents. Water, Air, & Soil Pollution 228 (3), 87.

Solanki, J.N., Murthy, Z.V.P., 2011. Controlled size silver nanoparticles synthesis with water-in-oil microemulsion method: a topical review. Ind. Eng. Chem. Res. 50 (22), 12311–12323. https://doi.org/10.1021/ie201649x.

Soler, L., Sánchez, S., 2014. Environmental monitoring and water remediation. Nanoscale 6 (13), 7175–7182.

Soyekwo, F., Zhang, Q. G., Deng, C., Gong, Y., Zhu, A. M., Liu, Q. L., 2014. Highly permeable cellulose acetate nanofibrous composite membranes by freeze-extraction. Journal of Membrane Science 454, 339–345.

Srinivasan, S., Harrington, G. W., Xagoraraki, I., Goel, R., 2008. Factors affecting bulk to total bacteria ratio in drinking water distribution systems. Water research 42 (13), 3393–3404.

Stefaniuk, M., Oleszczuk, P., Ok, Y.S., 2016. Review on nano zerovalent iron (nZVI): From synthesis to environmental applications. Chem. Eng. J. 287, 618–632.

Sun, G., Chung, T. S., Jeyaseelan, K., Armugam, A., 2013. Stabilization and immobilization of aquaporin reconstituted lipid vesicles for water purification. Colloids and Surfaces B: Biointerfaces 102, 466–471.

Sun, M., Li, P., Jin, X., Ju, X., Yan, W., Yuan, J., Xing, C., 2018. Heavy metal adsorption onto graphene oxide, amino group on magnetic nanoadsorbents and application for detection of Pb (II) by strip sensor. Food and agricultural immunology 29 (1), 1053–1073.

Tabesh, S., Davar, F., Loghman-Estarki, M. R., 2018. Preparation of γ-Al_2O_3 nanoparticles using modified sol-gel method and its use for the adsorption of lead and cadmium ions. Journal of Alloys and Compounds 730, 441–449.

Tahir, M. B., Kiran, H., Iqbal, T., 2019. The detoxification of heavy metals from aqueous environment using nano-photocatalysis approach: a review. Environmental Science and Pollution Research 26 (11), 10515–10528.

Tan, M., Wang, Y., Taguchi, A., Abe, T., Yang, G., Wu, M., Tsubaki, N., 2019. Highly-dispersed Ru nanoparticles sputtered on graphene for hydrogen production. Int. J. Hydrog. Energy. 44 (14), 7320–7325.

Tang, W.W., Zeng, G., Gong, J.L., Liang, J., Xu, P., Zhang, C., Haung, B.B., 2014. Impact of humic/fulvic acid on the removal of heavy metals from aqueous solutions using nanomaterials: a review. Sci. Total Environ. 468, 1014–1027. doi:10.1016/j.scitotenv.2013.09.044.

Tesh, S.J., Scott, T.B., 2014. Nano-composites for water remediation: A review. Adv. Mater. 26, 6056–6068.

Thebo, A.L., Drechsel, P., Lambin, E.F., Nelson, K.L., 2017. A global, spatially-explicit assessment of irrigated croplands influenced by urban wastewater flows. Environ. Res. Lett. 12 (7), 074008.

Tijing, L. D., Woo, Y. C., Shim, W. G., He, T., Choi, J. S., Kim, S. H., Shon, H. K., 2016. Superhydrophobic nanofiber membrane containing carbon nanotubes for high-performance direct contact membrane distillation. Journal of Membrane Science 502, 158–170.

Tratnyek, P.G., Salter, A.J., Nurmi, J.T., Sarathy, V., 2010. Environmental Applications of Zerovalent Metals: Iron vs Zinc. Nanoscale Materials in Chemistry: Environmental Applications, 1045. ACS Symposium Series 1045, pp. 165–178.

Tshikovhi, A., Mishra, S.B., Mishra, A.K., 2020. Nanocellulose-based composites for the removal of contaminants from wastewater. Int. J. Biol. Macromol. 152, 616–632.

Umar, K., Aris, A., Ahmad, H., Parveen, T., Jaafar, J., Majid, Z.A., Reddy, A.V.B., Talib, J., 2016. Synthesis of visible light active doped TiO_2 for the degradation of organic pollutants—methylene blue and glyphosate. J. Anal.Sci. Technol 7 (1), 29. https://doi.org/10.1186/s40543-016-0109-2.

Umar, K., Haque, M. M., Muneer, M., Harada, T., Matsumura, M., 2013. Mo, Mn and La doped TiO_2: synthesis, characterization and photocatalytic activity for the decolourization of three different chromophoric dyes. Journal of alloys and compounds 578, 431–438.

Umar, K., Ibrahim, M.N.M., Ahmad, A., Rafatullah, M., 2019. Synthesis of Mn-doped TiO_2 by novel route and photocatalytic mineralization/intermediate studies of organic pollutants. Res. Chem. Intermed. 45 (5), 2927–2945. https://doi.org/10.1007/s11164-019-03771-x.

Uyttendaele, M., Jaykus, L.A., Amoah, P., Chiodini, A., Cunliffe, D., Jacxsens, L., Holvoet, K., Korsten, L., Lau, M., McClure, P., Medema, G., Sampers, I., Jasti, Rao, 2015. Microbial hazards in irrigation water: standards, norms, and testing to manage use of water in fresh produce primary production. Comprehens. Rev. Food Sci. Food Saf. 14 (4), 336–356. https://doi.org/10.1111/1541-4337.12133.

Vallino, E., Ridolfi, L., Laio, F., 2020. Measuring economic water scarcity in agriculture: a cross-country empirical investigation. Environmental Science & Policy 114, 73–85.

Van Der Heijden, A.E.D.M., 2018. Developments and challenges in the manufacturing, characterization and scale-up of energetic nanomaterials–A review. Chemical Engineering Journal 350, 939–948.

Verma, P., Agrawal, M., Sagar, R., 2015. Assessment of potential health risks due to heavy metals through vegetable consumption in a tropical area irrigated by treated wastewater. Environ. Syst. Decis. 35 (3), 375–388. https://doi.org/10.1007/s10669-015-9558-1.

Vilar, J.L.B., Bernabeu-Mestre, J., 2011. La salud y el Estado: El movimiento sanitario internacional y la administración española (851–1945). Publicacions de la Universitat de València, València, Spain.

Vunain, E., Mishra, A. K., Mamba, B. B., 2016. Dendrimers, mesoporous silicas and chitosan-based nanosorbents for the removal of heavy-metal ions: a review. International journal of biological macromolecules 86, 570–586.

Wang, H., Ho, L., Lewis, D. M., Brookes, J. D., Newcombe, G., 2007. Discriminating and assessing adsorption and biodegradation removal mechanisms during granular activated carbon filtration of microcystin toxins. Water research 41 (18), 4262–4270.

Wang, J., Zhang, X., Chen, Y., Sommerfeld, M., Hu, Q., 2008. Toxicity assessment of manufactured nanomaterials using the unicellular green alga *Chlamydomonas reinhardtii*. Chemosphere 73 (7), 1121–1128.

Wei, G., Su, Z., Reynolds, N. P., Arosio, P., Hamley, I. W., Gazit, E., Mezzenga, R., 2017. Self-assembling peptide and protein amyloids: from structure to tailored function in nanotechnology. Chemical Society Reviews 46 (15), 4661–4708.

Winpenny, J., Heinz, I., Koo-Oshima, S., Salgot, M., Collado, J., Hernandez, F., Torricelli, R., 2010. The wealth of waste: the economics of wastewater use in agriculture. Water Rep. 35 https://www.fao.org/3/i1629e/i1629e.pdf.

Wongcharee, S., Aravinthan, V., Erdei, L., 2020. Removal of natural organic matter and ammonia from dam water by enhanced coagulation combined with adsorption on powdered composite nano-adsorbent. Environmental Technology & Innovation 17, 100557.

World Health Organization, 2006. Guidelines for the Safe Use of Wasterwater Excreta and Greywater, 1.

Wu, Y., Pang, H., Liu, Y., Wang, X., Yu, S., Fu, D., Wang, X., 2019. Environmental remediation of heavy metal ions by novel-nanomaterials: a review. Environ. Pollut. 246, 608–620.

Xin, Y., Huang, Y., Lin, K., Yu, Y., Zhang, B., 2018. Self-template synthesis of double-layered porous nanotubes with spatially separated photoredox surfaces for efficient photocatalytic hydrogen production. Sci. Bull. 63 (10), 601–608.

Xiong, C., Wang, W., Tan, F., Luo, F., Chen, J., Qiao, X., 2015. Investigation on the efficiency and mechanism of Cd (II) and Pb (II) removal from aqueous solutions using MgO nanoparticles. Journal of hazardous materials 299, 664–674.

Yamakata, A., Vequizo, J.J.M., 2019. Curious behaviors of photogenerated electrons and holes at the defects on anatase, rutile, and brookite TiO_2 powders: A review. J. Photochem. Photobiol. C: Photochem. Rev. 40, 234–243.

Yang, G. H., Bao, D. D., Zhang, D. Q., Wang, C., Qu, L. L., Li, H. T., 2018. Removal of antibiotics from water with an all-carbon 3D nanofiltration membrane. Nanoscale research letters 13 (1), 146.

Yang, S. Y., Park, J., Yoon, J., Ree, M., Jang, S. K., Kim, J. K., 2008. Virus filtration membranes prepared from nanoporous block copolymers with good dimensional stability under high pressures and excellent solvent resistance. Advanced Functional Materials 18 (9), 1371–1377.

Yaqoob, A.A., Parveen, T., Umar, K., Mohamad Ibrahim, M.N., 2020. Role of nanomaterials in the treatment of wastewater: a review. Water. 12 (2), 495. doi:10.3390/w12020495.

Ying, Y., Ying, W., Li, Q., Meng, D., Ren, G., Yan, R., Peng, X., 2017. Recent advances of nanomaterial-based membrane for water purification. Appl. Mater. Today. 7, 144–158.

Yuan, Y., Xiang, M., Liu, C., Theng, B.K.G., 2019. Chronic impact of an accidental wastewater spill from a smelter, China: a study of health risk of heavy metal(loid)s via vegetable intake. Ecotoxicol. Environ. Saf. 182, 109401.
Yucui, Z., Yanjun, S., 2019. Wastewater irrigation: past, present, and future. WIREs 6 (3), e1234.
Zhan, S., Yang, Y., Shen, Z., Shan, J., Li, Y., Yang, S., Zhu, D., 2014. Efficient removal of pathogenic bacteria and viruses by multifunctional amine-modified magnetic nanoparticles. Journal of hazardous materials 274, 115–123.
Zhang, Q. G., Deng, C., Soyekwo, F., Liu, Q. L., Zhu, A. M., 2016. Sub-10 nm Wide Cellulose Nanofibers for Ultrathin Nanoporous Membranes with High Organic Permeation. Advanced Functional Materials 26 (5), 792–800.
Zhang, Q., Dong, R., Wu, Y., Gao, W., He, Z., Ren, B., 2017. Light-driven Au-WO_3@ C Janus micromotors for rapid photodegradation of dye pollutants. ACS Appl. Mater. Interfaces. 9 (5), 4674–4683.
Zhang, B., Huang, G., Wang, L., Wang, T., Liu, L., Di, Z., Liu, X., Mei, Y., 2019. Rolled-Up Monolayer Graphene Tubular Micromotors: Enhanced Performance and Antibacterial Property. Chem. Asian J. 14 (14), 2479–2484.
Zhang, X., Zhang, T., Ng, J., Sun, D. D., 2009. High-performance multifunctional TiO2 nanowire ultrafiltration membrane with a hierarchical layer structure for water treatment. Advanced Functional Materials 19 (23), 3731–3736.
Zheng, K., Branicio, P.s., 2020. Synthesis of metallic glass nanoparticles by inert gas condensation. Phys. Rev. Mater 4 (7), 076001.
Zhou, Y., He, Y., Xiang, Y., Meng, S., Liu, X., Yu, J., Luo, L., 2019. Single and simultaneous adsorption of pefloxacin and Cu (II) ions from aqueous solutions by oxidized multiwalled carbon nanotube. Science of the Total Environment 646, 29–36.
Zhu, X., Chang, Y., Chen, Y., 2010. Toxicity and bioaccumulation of TiO_2 nanoparticle aggregates in *Daphnia magna*. Chemosphere 78 (3), 209–215.
Zhu, N., Ji, H., Yu, P., Niu, J., Farooq, M.U., Akram, M.W., Udego, I.O., Li, H., Niu, X., 2018. Surface modification of magnetic iron oxide nanoparticles. Nanomaterials 8 (10), 810.
Zia, M.H., Watts, M.J., Niaz, A., Middleton, D.R.S., Kim, A.W., 2017. Health risk assessment of potentially harmful elements and dietary minerals from vegetables irrigated with untreated wastewater, Pakistan. Environ. Geochem. Health 39 (4), 707–728. https://doi.org/10.1007/s10653-016-9841-1.

Further Reading

Gao, W., D'Agostino, M., Garcia-Gradilla, V., Orozco, J., Wang, J., 2013. Multi-fuel driven janus micromotors. Small 9 (3), 467–471.
García-Torres, J., Serrà, A., Tierno, P., Alcobé, X., Vallés, E., 2017. Magnetic propulsion of recyclable catalytic nanocleaners for pollutant degradation. ACS Appl. Mater. Interfaces 9 (28), 23859–23868.
Li, M., Wang, Q., Shi, X., Hornak, L.A., Wu, N., 2011. Detection of mercury (II) by quantum dot/DNA/gold nanoparticle ensemble based nanosensor via nanometal surface energy transfer. Anal. Chem. 83 (18), 7061–7065.
Mushtaq, F., Guerrero, M., Sakar, M.S., Hoop, M., Lindo, A.M., Sort, J., Chen, X., Nelson, B.J., Pellicer, E., Pané, S., 2015. Magnetically driven Bi_2O_3/BiOCl-based hybrid microrobots for photocatalytic water remediation. J. Mater. Chem. A. 3 (47), 23670–23676.
Orozco, J., Mercante, L.A., Pol, R., Merkoçi, A., 2016. Graphene-based Janus micromotors for the dynamic removal of pollutants. J. Mater. Chem. A. 4 (9), 3371–3378.
Srivastava, S.K., Guix, M., Schmidt, O.G., 2016. Wastewater mediated activation of micromotors for efficient water cleaning. Nano Letters 16 (1), 817–821.
Wang, H., Khezri, B., Pumera, M., 2016. Catalytic DNA-functionalized self-propelled micromachines for environmental remediation. Chem. 1 (3), 473–481.
Yu, F., Hu, Q., Dong, L., Cui, X., Chen, T., Xin, H., Liu, M., Xue, C., Song, X., Ai, F., Li, T., 2017. 3D printed self-driven thumb-sized motors for in-situ underwater pollutant remediation. Scientific Reports 7 (1), 1–6.

CHAPTER 11

Green biocomposite materials for sustainable remediation application

Shalu Rawat and Jiwan Singh
Department of Environmental Science, Babasaheb Bhimrao Ambedkar University, Lucknow, Uttar Pradesh, India

11.1 Introduction

Water, the second most essential thing after air, has become scarce nowadays due to overexploitation of water resources and water pollution because of the increasing global population (Biswal et al., 2020; Kumar et al., 2020). A continuous increase in environmental pollution from previous few decades has critically created a shortage of clean water and threatened human health. Water resources are severely affected by the effluent discharge of various industries. The main pollutants of a global concern are heavy metals and various organic compounds (natural and synthetic) even a trace amount of these pollutants in the water system can cause a serious environmental and human health issues. A huge amount of wastewater containing suspended solids, dyes, heavy metals, phenolic compounds, microorganisms, and other toxic compounds is generated from different industrial sources every year (Bhatti et al., 2017; Tang et al., 2020). Sustainable development has now become a requirement to give appropriation to that nation. When sustainable development is contributed with the credit, it is simply granted as the nation's progress as accustomed in the indices of sustainable development. In the recent, integrated environmental management meeting the topic of sustainable development was specially focused. New and advanced techniques for the treatment of air, water, and soil developed considering sustainable development indices can be successfully efficient. Therefore, development of a water treatment process keeping sustainable development in mind is an important aspect for the betterment of the environment. The fundamental steps to conserve water resources are management and recycling of wastewater, which also support water resource economy (Eugene et al., 2019; Ghorbal et al., 2019; Gwenzi et al., 2017).

The green technologies can bring different ecologically efficient methods that can be used for the clean-up and management of the environment in a sustainable way (Ghorbal et al., 2019). Currently researchers are focusing on the fabrication of an advance material with multifunctional properties such as high adsorption capability, catalytic activity and higher redox potential, stability, reusability, biocompatibility, and environmentally benign. Biocomposites are the materials that are being explore to achieve

Sustainable Nanotechnology for Environmental Remediation
DOI: https://doi.org/10.1016/B978-0-12-824547-7.00022-9

these properties and reduce environmental hazards due to which they are considered as efficient and capable materials (Maqbool et al., 2019). Composites are the materials composed of two or more material whose properties differ from their parent materials for preparation of biocomposites at least one part of composites needs to be of biological origin (Fowler et al., 2006). Preparation of advanced materials by green chemistry for the next generation is gaining focus because of their eco-efficient and sustainable behavior. In green synthesis microbial extract, plant extract, or biopolymer is used for formation of biocomposites. This process is environmental friendly as it replaces hazardous chemical compounds used in conventional chemical synthesis and materials prepared in this green way have a wide area of biological applications (Garza-Cervantes et al., 2020). The functional and high-performing materials developed using biomass as a raw material are biocompatible and biologically degradable and considered to be promising materials that can replace the conventional materials (Paulo et al., 2018; Tran et al., 2016). These biodegradable green materials play a noteworthy part in sustainable development due to having several disposing options with less adverse environmental impact (Sarasini & Fiore, 2018)

Over the last few years, nanotechnology has been investigated tremendously and has become a highly advance and effective technique with a wide area of application. In the water treatment, nanotechnology is now applied as an alternative for conventional chemical agents and it also meets the requirement of clean water with a comparatively lesser treatment cost (Bhati & Rai, 2017; Mohan et al., 2018). The synthesis of metallic nanoparticles by the use of biopolymers as an ideal stabilizer gives advantage of their comparatively lesser cost, more metallic interaction, and biogenic nature (Baran, 2019). The development and application of metal-organic frameworks (MOFs) is another attracting topic. These MOFs are composed of metal ions and their cluster with organic compounds. They have high porosity, flexibility, and also provide chance for modification post synthesis and this high porosity and high surface area provide them a wide area of application such as in gas adsorption, catalysis, luminescence, energy generation, gas storage, and environmental remediation (Wang et al., 2018; Zhu et al., 2019).

Carbon materials are also been extensively explore for the preparation of composites such as graphene oxide, fullerenes, carbon nanotube (CNT), activated carbon, and reduced graphene oxide because of their high specific surface area, electrical conductivity, charge carrying capacity, and good transparency (Dehghani et al., 2020; Lingamdinne et al., 2019, 2018; Ranjith et al., 2019). Characteristics, physical and chemical properties of graphene allow it be used in various premises and composite preparation (Ma et al., 2019). The main aim o study is to discussed preparation of green biocomposites and green nanobiocomposites using different biomasses and their application in sustainable water remediation techniques.

11.2 Green biocomposites

Due to a constant increase in human population, environmental pollution has become more persistent and also global energy demand is increasing day by day; hence new technologies should be developed to remediate environment in an environmentally safe and sustainable way (Ngo et al., 2019). The synthesis of modern-generation materials need to be industrial and ecological efficient, sustainable, and greener. Materials prepared via biological feedstock are biodegradable and biocompatible in nature that is a good alternative of the conventional materials (Sayfa et al., 2021). Composites prepared via green route are beneficiary over chemical ones in many ways which are their better mechanical performance, high durability, low cost, environmentally sustainability, renewability, recyclability, and biodegradability (Al-Oqla & Omari, 2017). Biocomposites can be produce using, wood waste, vegetable waste, crop byproducts, etc. as biological part. Usage of plant-based material for the synthesis of environment-friendly biocomposites is a novel technique in this era that will help in overcoming the increasing environmental threat. A large amount of biomass is produced in agriculture and food- and beverage-producing industries. These waste materials give us opportunity to use this biomass as raw material for the preparation of green composites that are widely and easily available as well as they have lower cost and environmentally convenient, hence their use is sustainable (Maqbool et al., 2019). Nonedible crop products and some other biologically renewable materials make the availability of raw material for sustainable and renewable biocomposite preparation almost unlimited.

11.3 Green nanobiocomposites

Nanomaterials possess specific physical, chemical, and biological properties that differentiate them from their bulk-sized or micro or macro-sized parent material because of their ultimate smaller size. Specific properties of nanoparticles include high surface to volume ratio, higher free surface energy, highly reactive and large number of active surface sites, economically favorable, and ease of availability (Lu & Astruc, 2020). The applicability of the nanoparticles has been expanded to a wider area (biological, medical, and environmental remediation) by preparation of nanocomposite (Sahu et al., 2019a, 2019b). Polymers and carbonaceous material are mainly used in the incorporation of nanoparticles for nanocomposite preparation. The preparation of nanocomposites increases the efficiency of nanoparticles comparatively (Taghipour et al., 2019). The nanocomposites are being prepared by combining organic bio-based compounds with inorganic salts; these green nanocomposites are used for adsorption, photocatalysis, sensing, ion exchange, and for other purposes (Ahmad et al., 2020). The biopolymer and transitional metals easily form complexes because of polymer functional groups (Baran, 2019).

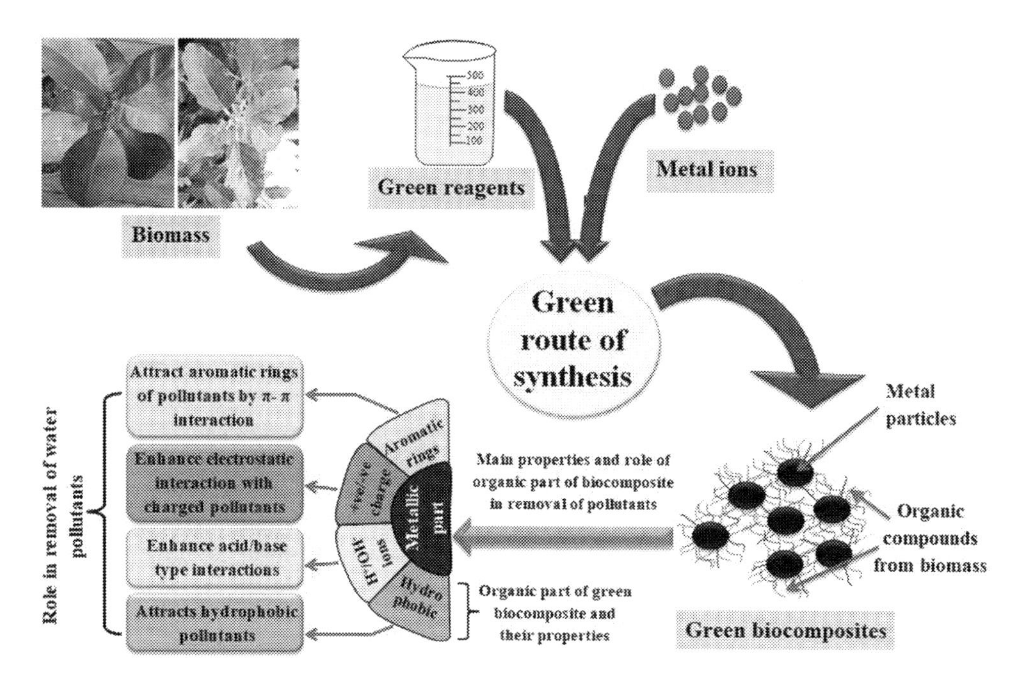

Figure 11.1 A general route for preparation of green biocomposite and the role of biochemicals in enhancement of composite efficiency for removal of water pollutant.

The greener approach to material preparation helps to lead toward environmental clean-up in a sustainable manner. The green nanocomposites are developed to increase pollutant interaction/reactivity, stability, areas of application, and to reduce negative environmental impact of nanoparticles (Kausar, 2020), Fig. 11.1 represents the preparation of composites using biomass and role of bio-based chemicals in improving composite's removal efficiency of pollutants. The advanced properties of the green nanocomposites encourage researchers to explore nanocomposites' potential in environmental remediation. Furthermore, the impregnation or attachment of nanocomposite on bulk material provides them mechanical properties important for their use in continuous process, which is needed for the treatment of industrial and domestic wastewater. The modification of nanocomposite in this way reduces their toxicity caused by free particles and agglomeration (Malakootian et al., 2020).

11.4 Application for sustainable wastewater remediation

Biocomposites have a wide area of application in environmental remediation and their capabilities have been extensively explored for in water pollution and yet their modification and enhancement of efficiency is under process. Water pollutants can be mainly categorized into three categories: inorganic, organic, and microorganisms. Consumption of water contaminated with pathogenic microbes causes several waterborne diseases in humans (Wang et al., 2018; Zhu et al., 2019). The inorganic category mainly includes contamination of water with heavy metals that persist for longer period of time and enter the food chain (Anawar & Strezov, 2019). The organic category includes several toxic and recalcitrant compounds such as phenols, pesticides, dyes, antibiotics, polyaromatic hydrocarbons (PAHs), and polychlorinated biphenyls. Dyes are one of the most noticeable organic pollutants released from industries such as textile, painting, printing, etc., and heavy metals and toxic chemical released from pharmaceutical industries are also a major water polluter that create health threat to humans even in trace amount (Thangavel et al., 2019). Green biocomposites in environmental remediation are known to be applied widely for the remediation of water pollution by mainly using their adsorptive, catalytic, and antimicrobial properties (Zhu et al., 2019).

11.4.1 Adsorptive removal of pollutants

Adsorption is an easier and efficient technique for the removal of organic and inorganic pollutants from the water due to its high adaptability, less environmental toxicity, and cost efficiency (Anastopoulos et al., 2018). Several metal, natural minerals, organic adsorbents such as metal and metal oxide nanoparticles, graphene, CNTs, activated carbon, clay, etc. are being used for the treatment of water pollutant (Dehghani et al., 2020; Lingamdinne et al., 2019, 2018). Nano-sized iron and its oxides are one of the most extensively used adsorbents and have proven to be highly effective in the area of adsorption (Gutierrez et al., 2017); and combining the nano-sized iron with organic components makes them more stable as they reduce their oxidation or dissolution in aqueous medium and enhance recyclability. Several studies have reported successful adsorption of various pollutants by iron based composite materials like in a study, researchers (Gautam et al., 2018) fabricated nano-sized iron particles coated with the organic compounds of tea via green synthesis method. These green iron nanocomposites were applied for the adsorption of phenol red dye from its aqueous solution. The nanocomposites were efficient adsorbent and removed 95% phenol red dye from the solution, the nanocomposites were also recycled for three cycles and the adsorption efficiency decreased with every cycle. (Sahu et al., 2019) in their study similarly utilized waste of sweet lime fruit for the preparation of iron nanocomposite by green synthesis with their particle size ranging from 86 to 113 nm. The nanocomposites were tested for the adsorption of methylene blue from an aqueous solution. The prepared nanocomposites were successful for the removal of dye and showed maximum adsorption efficiency 97% at 45 °C. Carbon and metal composites are another widely explored composite material for the remediation of environment. Biochar, another widely used cost-effective adsorbent prepared by thermal decomposition of biomass at higher temperature, is a solid material rich in carbon, and in comparison with other carbonaceous materials it is of low cost. Adsorption efficiency of biochar is being modified through different methods, modification of biochar by coating or impregnation formation of composites by combining with metal oxides, clay minerals, polymers, enhances its surface functional properties, coating also results in a change of surface charge on biochar that helps in achievement of better adsorption efficiencies (Qin et al., 2020). Mishra et al. (2019) prepared carbon and iron oxide composite for the adsorption of phenol and para-nitrophenol from their aqueous solutions. The composite material was prepared by the pyrolysis of waste corn husk biomass along with ferric chloride at 250 °C and 500 °C separately, and adsorption capacity of both the composite materials was compared. The results showed that the adsorbent synthesized at 500 °C had higher adsorption efficiency than that synthesized at 250 °C which was mainly due to the differences in the properties of both the composites; the one prepared at higher temperature was more crystalline and has lesser particles size and higher Iodine number. Similarly Nidhi et al. (2020) and Ibrahim et al. (2019) also prepared a carbon and iron composite from the waste biomass of cauliflower and sweet lime at two different temperatures and found the composite prepared at higher temperature to be more efficient for the adsorption of phenols (phenol and para-nitrophenol) and fluoride, respectively (Hairuddin et al., 2019; Karri et al., 2017). The activated carbon prepared from the waste biomass of cauliflower removed about 95% phenol and 100% para-nitrophenol and other synthesized from sweet lime removed 86% fluoride from its aqueous solution. MOFs are the advanced materials that comprise metal ions with organic cluster, they have a wide area of application, they are successfully synthesized through eco-friendly routes and utilized for adsorption of water pollutants. Shi et al. (2018) prepared magnetic and recyclable metal organic composite (Cu-MOF/Fe_2O_3) for the adsorption of Pb^{+2} and malachite green dye. The adsorption capacity of the composite was 113.6 and 219.0 mg/g for malachite green and Pb^{+2}, respectively. In an another study by Mahmoodi et al. (2019), an activated carbon and chromium modified nanocomposite was synthesized and used for the adsorptive removal of acid green 25 and reactive yellow 186 and the nanocomposites were recyclable also with less than 9% change in adsorption capacity after five cycles. Table 11.1 shows outcomes of some work conducted by various researchers for the adsorption of various water pollutants by using composites as adsorbent.

11.5 Catalytic degradation of pollutants

Catalytic degradation is a green chemical process with promising efficiency to completely degrade toxic organic pollutants, it generally utilizes involvement of highly reactive chemical species such as hydroxyl and sulfate radicals (Jin et al., 2018). Several green synthesized composite materials have been used till now for the degradation of different

Table 11.1 Adsorptive removal of some pollutants from water using different composite materials.

Biocomposites	Pollutants removed	Efficiency	References
Magnesium oxide embedded nitrogen self-doped biochar	Lead	893 mg/g	Ling et al., 2017
Novel magnetic porous composite ($CoFe_2O_4$-SiO_2 (CF-S))	Cu^{+2}, Mn^{+2}, Pb^{+2} and Cd+2 (in a mixed solution)	429.18 mg/g at 25 °C, 500.00 mg/g at 35 °C, 523.56 mg/g at 45 °C, 555.51 mg/g at 55 °C	Ren et al., 2017
Chitosan-lysozyme biocomposites	Methyl orange (MO) and Cr(VI)	435 mg/g for MO and 216 mg/g for Cr(VI)	Rathinam et al., 2018
Lignin-based biocomposites	Pb	344.83 mg/g	Mohammadabadi and Javanbakht, 2020
Polyaniline (PAN) and almond shell biocomposites	Cr(IV) and orange G dye (OG)	335.25 mg/g for Cr (VI) and 190.98 mg/g for OG	Hsini et al., 2020
Cellulose and clay composite	Drimarine yellow HF-3GL direct dye	48.96 mg/g	Kausar et al., 2020
Chitosan carbon based bio-nanocomposites	Phenol	409 mg/g	Soni et al., 2017
Magnetic Schiff's base-chitosan glyoxal/fly ash/Fe_3O_4 biocomposites	Reactive orange 16	112.5 mg/g	Malek et al., 2020
Montmorillonite sodium alginate nanocomposites	Polyaromatic hydrocarbons	1.20 mg/g (Ace), 0.90 mg/g (Flu), and 2.50 mg/g (Phe)	Dai et al., 2020
Chitosan beads modified with FeO and TiO_2 nanoparticles	Naphthalene	33.1 mg/g	Patinõ-Ruiz et al., 2020

(continued on next page)

Table 11.1 Adsorptive removal of some pollutants from water using different composite materials—cont'd

Biocomposites	Pollutants removed	Efficiency	References
Chitosan (CS) and lamb thigh bone powder (KT) composite (CS-KT) and CS-KT-metal oxide (CS-KT-M) composite	Radioactive Cesium ion (Cs-137)	CS-KT—2.52×10^{-4} and 2.74×10^{-4} mol/g, respectively, at 25 and 50 °C CS-KT-M—4.17×10^{-4} and 5.58×10^{-4} mol/g, respectively, at 25 and 50 °C	Işık et al., 2021
Enteromorpha prolifera–based novel magnetic composite	Radioactive cobalt (^{60}Co) and nickel (^{63}Ni)	135 mg/g (Co) and 137 mg/g (Ni)	Zhong et al., 2020
Biocomposites of polypyrrole and barley husk (PPY/BH), PAN/BH, and sodium alginate (Na-Alg/BH) with cellulose	2-4, dichlorophenol	31.95 mg/g for PPY/BH, 43.48 mg/g for PAN/BH, and 9.542 mg/g for Na-Alg/BH	Bhatti et al., 2020
Magnetic sugarcane bagasse (MBo) and magnetic peanut shell (MPSo) composites	Carbofuran and iprodione	For carbofuran 179 and 889.3 mg/g and for iprodione 119 and 2.76 mg/g for MBo and MPSo, respectively.	Toledo-Jaldin et al., 2020
MnO and CuO impregnated Sesbania bispinosa biochar (SBC) composite	Arsenic	SBC/Cuo= 12.47 mg/g and SBC/MnO = 7.34 mg/g	Imran et al., 2021

toxic organic pollutants such as dyes, PAHs, pesticides, and phenols. The organic compounds are attacked by the highly reactive radical species that are generated by a catalyst. The radical generation can be achieved by irradiation of light and ultrasonic waves. For this purpose, several composite materials have been prepared by combing reactive metallic nanoparticles to the organic matrix and have been successfully used to catalyze organic pollutants completely. Here are some studies explained in brief. Nasrollahzadeh et al. (2018) prepared green nanocomposites from eggshell modified with Fe_3O_4 and Cu by the use of Orchis mascula L. leaf extract. The nanocomposites were found highly effective for the catalytic degradation of 4-nitrophenol, methylene blue, methyl orange, Congo red, and rhodamine B from water. The use of waste eggshells and the facile synthesis of nanocomposites make them environmentally friendly and economically viable. Aksu Demirezen et al. (2019) degraded an antibiotic amoxicillin by green synthesized iron nanocomposites by extract of Ceratonia siliqua pericarp. The prepared nanocomposites were spherical with an average particle size of 7 nm. A study conducted by Ismail et al. (2018) fabricated a catalytic composite by loading of copper-silver, copper-nickel, and nickel on the ginger powder for the reductive removal of 2-nitrophenol, 4-nitrophenol, methyl orange, Congo red, and rhodamine B. The fabricated catalysts had high stability and reusability. Graphene-based biocomposite is used over a decade for the degradation of water pollutants. An excellent catalytic activity is observed to be possessed by a graphene-based catalyst because of direct charge transfer to graphene surface from the metallic nanomaterial that separates electron–hole pair and additionally generates hydroxyl radicals. The degradation process also facilitates electrostatic attraction between aromatic rings of graphene and organic pollutants (Boruah et al., 2017). Nanocomposites were synthesized by combining reduced graphene oxide with silver and iron oxide nanoparticles using peel extract of *Punica granatum* by green synthesis method. These magnetically separable nanocomposites were used for the catalytic degradation of 4-nitrophenol, methylene blue, methyl green, and methyl orange. The study concluded the nanocomposite to be highly effective for the degradation of pollutants within few minutes (Adyani & Soleimani, 2019).Raj Pant et al. (2013) in their study prepared a green Ag/ZnO/reduced graphene oxide composite via one-step hydrothermal synthesis. The green composite was used for the photocatalytic degradation of methylene blue and reactive blue 5 dyes. In water purification, photocatalytic degradation is a promising technique because of its energy and cost-efficient and environmental benign nature (Mou et al., 2019). It utilizes photoexcitation of electrons from a catalyst's surface that generates different reactive radicals subsequently which initiates degradation of pollutants. In a study, green cellulosic composites in combination with bismuth (Bi) and bromide (Br), which was BiOBr/BiOI/cellulose composite, were prepared for which waste paper pulp was used for the extraction of cellulose. Rhodamine B dye, fluorescein dye, and tetracycline hydrochloride were photocatalytically degraded using these synthesized cellulosic composites (Du et al., 2019). El-Shabasy et al. (2019) prepared

Ag/Ag_2O nanoparticles from the extract of *Capsicum annum L* coupled with P25 (80% anatase and 20% rutile phase coupled TiO_2) with an average particle size of 11.4 nm. The synthesized nanomaterial degraded more than 98% methylene blue dye and 2,4-dinitroaniline under solar irradiation within one minute, it had great stability and reusability. Yadav et al. (2019) prepared a photo-catalyst bismuth oxybromide/oxyiodide from green synthesis using the leaf extract of *Azadirachta indica*. A higher amount of methyl orange dye and amoxicillin was degraded under the irradiation of visible light using the synthesized photocatalysts. Table 11.2 shows various studies that include preparation of green biocomposites for the removal of various types of water pollutants.

11.6 Mechanisms involved in removal of pollutants from water

There are several mechanisms involved in the removal of pollutants by biocomposites; the removal takes place by the phenomenon of adsorption or catalytic degradation. Adsorption of the pollutants mainly includes their physically or chemically accumulation over the surface of the adsorbent (Pai et al., 2020). Adsorption depends on the surface morphology and surface properties of the biocomposites that include surface area, pore volume, and active reaction sites. The available literature shows that the adsorption of inorganic pollutants such as heavy metals mainly include electrostatic attractions among the charged pollutants and adsorbent surfaces, some other mechanism are responsible for the adsorption of inorganic compounds such as ion-exchange and complex formation. Deng et al. (2019) studied the decontamination of Pb^{+2} by carbonaceous nanochlorapatites that showed 90.37% removal in the decontamination process was chemically controlled mechanism primarily; the main mechanisms behind the removal of Pb^{+2} was reported to be surface precipitation and electrostatic attraction. Another study reported adsorption of Pb^{+2} by the use of a Tb-based MOF. The adsorbent was highly efficient having adsorption capacity 547 mg/g and was capable even after five cycles of recycling. Inner sphere complex (C-/=N····Pb) between Pb^{+2} and the nitrogenous group of the adsorbent was found to be the main mechanism of adsorption (H. Zhu et al., 2019). Wu et al., (2019) in their study performed adsorption of nitrate and phosphate from water by a novel organic modified aluminum-manganese bimetal oxide composite. Adsorption capacity of the adsorbent for nitrate and phosphate was observed to be 19.45 mg/g and 33.16 mg/g, respectively; after adsorption the spectroscopic analysis reveals participation of different mechanism for the adsorption of both the adsorbates. Nitrate adsorption was mainly based on ligand exchange with chloride ions and electrostatic interactions with amine groups, while the phosphate adsorption was governed by surface complexation and ion-exchange with sulfate. The adsorption of organic compounds involves several other mechanisms along with electrostatic attraction, which are chemical bonding, π-π interactions, hydrogen bonding, and others. Zhou et al., (2019)

Table 11.2 Some biocomposites and their application for the removal of various pollutants.

Biocomposites	Applications	Pollutant removed	Efficiency	References
Silica fumes based MCM-41 and cobalt oxide	Photocatalytic degradation	Omethoate pesticide	100%	Mohamed et al., 2020
ZnO-doped CuO@Alginate Bionanocomposites	Photodegradation	p-nitrophenol	98.38%	Hasan et al., 2020
Co_3O_4/ZnO nanocomposite	Photo-degradation	Methylene blue and Rhodamine B	80% and 90%, respectively	Hassanpour et al., 2017
$AgZnFe_2O_4$@RGO	Adsorption and photocatalytic degradation	Methylene blue, rhodamine b and methyl orange		Mady et al., 2017
Layered 1T-MoS_2/RGO	Catalytic degradation	4-nitrophenol		Meng et al., 2017
TiO_2@nitrogen doped carbon	Photocatalytic degradation	Methylene blue	90%	Atchudan et al., 2018
Polycrystalline NiO/MgO composite	Photocatalytic degradation	Methylene blue and methylene orange	73–87%	Fuku et al., 2018
Chitosan Fe_2O_3	Sorption	Pb^{+2} and Cd^{+2}		Ahmad and Mirza, 2018
Pd/Fe_3O_4 nanocomposites	Catalytic reduction	Cr^{+6}, 4-nitrophenol and 2,4-dinitrophenylhydrazine	~100%	Nasrollahzadeh et al., 2018
Ag-Cu_2O/rGO	Catalytic degradation	Methyl orange	90%	Sharma et al., 2018
nZVI/powdered activated carbon	Adsorption	Cr^{+6}	46.49 mg/g	Khosravi et al., 2018
Polythiophene/SiO_2	Adsorption	Zn^{+2}, Pb^{+2}, and Cu^{+2}	90%	Chen et al., 2019
Bentonite/chitosan@ cobalt oxide	Oxidative degradation and adsorption	Congo red dye and Cr^{+6}	303 mg/g for Congo red and Cr^{+6}	Abukhadra et al., 2019
Magnetic Egg Shell Membrane (Fe_3O_4@ESM)-Zeolitic imidazolate frameworks (ZIF 67)combinely reffered as ZIF-67@Fe_3O_4@ESM	Adsorption	Cu^{+2} and basic red 18	344.82 mg/g and 250.81 mg/g, respectively	Mahmoodi et al., 2019
nZVI-supported magnetic biochar composite	Oxidation/adsorption	Malachite green dye	515.77 mg/g	Eltaweil et al., 2020

conducted a study for the adsorption of ciprofloxacin and sparfloxacin by a novel Fe_3O_4/graphene oxide/citrus peel derived bionanocomposite with adsorption capacity of 283.44 mg/g (ciprofloxacin) and 502.37 mg/g (sparfloxacin). The study showed primary mechanisms for adsorption were π-π electron donor acceptor interactions, hydrogen bonding, hydrophobic interactions, and electrostatic attractions. According to the study conducted by Elessawy et al. (2020) the adsorption of methylene blue and acid blue 25 on green synthesized magnetic fullerenes nanocomposite was primarily governed by electrostatic attraction, π-π stacking, and hydrogen bonding. Djelad et al. (2019) studied the adsorptive removal of crystal violet onto the alginate-whey green biocomposite, and the main mechanisms for adsorption of dye were electrostatic interactions, strong hydrogen bonding, and chemical bonding between dye and adsorbent surface chemical groups. The oxidative degradation of an organic pollutant by a biocomposite material takes place by the generation of a highly reactive radicals. Hydroxyl radical–mediated degradation of the pollutants is the primary method; and the generation of hydroxyl radical is done by Fenton's reaction, ultrasonication, or photocatalysis. Besides hydroxyl radical now sulfate radical has also been used in oxidative degradation process that is generated by persulfate (peroxodisulfate) (Ike et al., 2018). The most common catalytic degradation mechanism is Fenton reaction that utilizes $^{\circ}OH$ radical by the excitation of H_2O_2 by ferrous ions. Lan et al. (2020) found contribution of three free radicals, $^{\circ}OH$, $^{\circ}O_2$ and $hv^{+}{}_{vb}$, in the degradation of an azo dye by a novel nanocomposite. The photocatalytic degradation involves light and the photocatalytic semiconductor. Photocatalysis can generate several reactive species such as OH°, hole superoxide radical, hydrogen radical, and singlet oxygen. A general photocatalytic mechanism involves four main processes: the first is photoexcitation of a catalyst and generation of an electron and hole pair, the second is the ionization of water and formation of $^{\circ}OH$ radical. The next is ionosorption of oxygen by reacting with the free electron generated in the first process and form anionic superoxide radical ($^{\circ}O_2{}^{-}$). Finally the superoxide radical get protonated and form hydroxy peroxyl radical ($^{\circ}HO_2$) and H_2O_2 subsequently that gives $^{\circ}OH$ radicals (Ajmal et al., 2014). Sharma et al., (2018) proposed two possible mechanisms for the photocatalytic degradation of crystal violet dye by titanium dioxide and graphene-oxide green nanocomposite. According to the first mechanism graphene-oxide generate photoexcited electrons on light irradiation and these electrons generate superoxide radicals. However according to the second mechanism the dye molecules get photoexcited and initiate photocatalysis.

11.6.1 Antimicrobial activity

According to the World Health Organization (WHO) about 2 million people die every year around the world due to waterborne diarrhea. A lot of pathogenic microorganisms are found in domestic wastewater; however, any type of water resource can possibly get contaminated with pathogens; thus, the elimination of these pathogens is important before its use for consumption (Dalal et al., 2020). The development of nanocomposites helps nanoparticles to expand their applicability in several areas such as microbiology, pharmaceuticals, and drug delivery system; and natural polymer usage in nanocomposite synthesis makes them biocompatible (Sahu et al., 2019a, 2019b; Shehabeldine & Hasanin, 2019). The biodegradable and eco-friendly polymer composite materials with antimicrobial capacity can serve as biodegradable food packaging materials that will reduce food spoilage and will help in increasing the shelf-life of the packed food. Polyethylene, polypropylene, polyvinyl chloride, and polyethylene terephthalate are mostly used in packaging of food materials that are nonbiodegradable and cause nuisance to the environment. The antimicrobial polymer composites can be produced either by immobilization of antimicrobial agents (such as metal nanoparticles) on the biological polymer or by their direct combination (Marrez et al., 2019). The bio-based composite materials as antimicrobial agents with biocompatibility and excellent efficiency are the new generation demand. A nanocomposite comprising cellulose, amino acid (L-Phenyl alanine and L-Tryptophan) having spherical shape and average particle size 72 (L-Phenyl alanine composite) and 44.37 nm (L-Tryptophan composite) was fabricated. These nanocomposites were used for antimicrobial activity against *Escherichia coli* (NCTC 10416), *Pseudomonas aeruginosa* (NCID-9016) (Gram-negative bacteria), *Streptococcus aureus* (NCTC-7447), *Bacillus subtilis* (NCID-3610), and a unicellular fungi *Candida albicans* (NCCLS 11). The results of this study suggested that nanocomposite L-tryptophan composites were better as they require comparatively less time to kill all the subjected microorganisms (Hasanin & Moustafa, 2020). He et al. (2017) biosynthesized a novel silver nanocomposite with the help of sericin obtained from silkworms. Sericin was used to reduce silver ions. The antibacterial activities of the prepared composites were tested against *S. aureus*, according to the results only 20 mg/L dosage was enough to resist the microbial growth, while 100 mg/L was required for their killing. Bimetallic and polymetallic composite materials are also gaining attention as they help increasing composite properties such as combination of antimicrobial metallic nanoparticle with magnetic nanomaterial result in formation of composite with antimicrobial and magnetic properties. Aqueous extract of *Carum carvi L.* seeds was used by Heydari et al., (2019) for the biosynthesis of Fe_3O_4 nanoparticles, Cu nanoparticles, and Fe_3O_4/Cu nanocomposites with average particle size 25, 37, and 62 nm, respectively. The study concluded the Fe_3O_4 nanoparticles with no antimicrobial activity became active upon the formation of Fe_3O_4/Cu nanocomposites and also the composite showed in increase in antimicrobial activity when compared to that of Cu nanoparticles. Silver and iron oxide nanocomposites (Ag@Fe_2O_3) green synthesized form the leaf extract of the *Adhatoda vasica* were applied for the antimicrobial functions against Gram-positive (*S. aureus*) and Gram-negative (*E. coli*) bacteria and a fungus *C. albicans*. The composites showed excellent antimicrobial activity for both classes of bacteria at 20 μg/mL amount while they also showed high antifungal activity against *C. albicans* with 60 μg/mL dosage. Further Ag@Fe_2O_3 composites were also used for the anticancer activity against MDA-MB-231human breast cancer (adenocarcinoma) cell lines and a visible apoptosis was seen (IC50 value ~135.29±2.9 μg/mL) (Kulkarni et al., 2017). Two-dimensional nanosheets such as graphene oxide and graphitic phase C_3N_4 (g-C_3N_4) in addition with metal nanoparticles have a wide application including antibacterial activity. Liu et al. (2016) prepared g-C_3N_4 nanosheet decorated with silver nanoparticles synthesized by grape seed extract. The antimicrobial activity of the nanocomposites was investigated against *E. coli, P. aeruginosa, S. aureus,* and *B. subtilis*. Grape seed extract and g-C_3N_4 were also tested and showed no visible antimicrobial activity, while the silver nanoparticles (AgNPs) and nanosheet decorated with AgNPs had significant antimicrobial activity. The g-C_3N_4/AgNPs nanocomposites showed better efficiency because of more stability and dispersion of AgNPs by their decoration on nanosheet. Sadhukhan et al. (2019) synthesized cadmium oxide nanoparticles in an efficient and green way, which were decorated on reduced graphene oxide to prepare nanocomposites. The composites were applied for the antibacterial activity against some Gram-positive (*Listeria monocytogenes, S. aureus* and *B. subtilis*) and Gram-negative (*E. coli, Salmonella typhimurium* and *Klebsiella pneumonia*) bacteria. The nanocomposites and CdO nanoparticles represented good antibacterial activity in comparison with reduced graphene oxide alone and commercially available antibiotic streptomycin. Gold nanoparticles (AuNPs) have shown antibacterial activities for both Gram-positive and Gram-negative bacteria; however their efficiency get affected by low dispersion and aggregation for which researcher coat them on polymers composites form for their better activity and this makes them biocompatible also. Tran et al., (2016) in their study fabricated gold nanocomposites with the help of cellulose, wool keratin, and chloroauric acid ($HAuCl_4$) in a one-pot method. The antibacterial activity of the composite material was tested against antibiotic-resistant bacterial species S. aureus (ATCC 33591) and *Enterococcus faecalis* (ATCC 51299) that result in their 98% and 97% growth inhibition, respectively. Plasmonic hybrids Ag@AgCl are gaining attention due to their high photocatalytic activity and antibacterial activity with low cytotoxicity. The hybridization of Ag@AgCl and ZnO increases the photocatalytic activity because electron transfer rate of ZnO is increased with immense amount of Ag. Nanocomposite Ag@AgCl/ZnO was prepared by green synthesis method using pectin as a matrix and the nanocomposites were utilized for the photocatalytic antibacterial activity against *S. aureus* and *E. coli*, under the illumination of visible light (Yu et al., 2019).

11.6.2 Biocomposites in membrane filtration

Membrane filtration technology of water purification is a widely acceptable technique among several owing to its easier fabrication, operation, and higher separation efficiency (Kamari & Shahbazi, 2020). The characteristics features selectivity and permeability of any membrane are dependent on the microstructure and chemistry of the material used in membrane fabrication. The traditional polymeric membranes encounter some disadvantages such as fouling and permeability and selectivity interchangeable relationship. Membrane fouling is a serious problem that requires cleanup, high-energy, and sometimes replacement of membrane. Immobilization of biocomposites/nanobiocomposites is now a new trend in preparation of composite materials in the field of green chemistry, engineering science, and biomedicine (Sahu et al., 2019a, 2019b). An introduction of inorganic nanomaterials and organic polymer membrane combination has attracted the focus of researchers these days (Wang et al., 2018; Zhu et al., 2019). The process of immobilization of composites deduct the need of removal of used composite material after usage, decreases the concern of ecotoxicity, and also reduces the problem of agglomeration of particles (Kim et al., 2018). The modification of polymer membrane by the addition of efficient composite or nanocomposite or nanomaterial improves the membrane properties. Decoration of membrane with nanomaterials overcome fouling problem, this results in the formation of "advanced composite membrane" that have better properties in several terms than the conventional membranes, the nanomaterial also provides strength to the membrane. TiO_2, zeolite, AgNPs, and CNTs are the most commonly used materials to modify membranes. The modifications not only help in mitigating the problem of membrane fouling but also enhance the permeability and selectivity of the membrane (Sinha et al., 2020). The membrane filtration is now a more advanced technique due to the introduction of nanomaterial in their fabrication. The functionality of membrane can be enhanced by changing properties of the attached nanomaterials such as change in their functional groups. The functional groups on nanomaterials enhance the selectivity, affinity, and reactivity toward the pollutants that improve the quality of the membrane ultimately (Saleh et al., 2019). Table 11.3 shows the various studies on modification and preparation of membrane by the use of green composite materials immobilization for the treatment of water.

11.7 Future perspectives and challenges

Biocomposites are suitable materials for environmental remediation and they have wide and interesting application in treatment of polluted water. Till now most of the research related to these materials has been done at a laboratory scale only, and for scaling up their use at a commercial level some issues need to be considered. The main issue is the management of the used and exhausted material, separation of composites from water stream, their regeneration and recycling is a tedious task. Besides, their biological nature disrupts their structural and function properties on storage. In environment degradable part of metal organic composites gets degraded and the metallic part free in the environment can pose toxic effect on environment. Immobilization over the polymer

Table 11.3 Immobilization of biocomposites for various applications.

Immobilization base material used	Immobilized composites	Applications	References
Cellulose-chitosan composite membrane	AgNPs	Antimicrobial activity against S. aureus and E. coli	Chook et al., 2017
Cotton fabric	$CuO/ViBO_4$	Antibacterial activity against E. coli	Ran et al., 2019
Alginate beads	NiFe nanoparticles	Tetracycline (87% removed)	Ravikumar et al., 2020
Chitosan	ZnO/CuO	Photodegradation of fast green dye	Alzahrani, 2018
Polyethersulfone membrane	$Fe_3O_4@SiO_2\text{-}NH_2$	Nanofiltration of Cd^{+2} and methyl red dye	Kamari and Shahbazi, 2020
Silver nanocomposites with Salai guggal (Sg) semi interpenetrating network using acrylic acid (AA) and acrylamide (AAm)	Sg-cl-poly-AAm-AA-Ag^0	Antimicrobial activity against S. aureus, E. coli, P. aeruginosa, and B. aureus	Sharma et al., 2018
Carbon black/ chitosan fibers	Co, Ag, and Cu (monometallic) and Co + Cu and Co + Ag (bimetallic)	Catalytic reduction of p-nitrophenol and azo dyes and antimicrobial activity against E. coli	Ali et al., 2017

for a membrane reduces the need of separation of composites after use. However this also has its own challenges such as an improper adhesion of composites, which leads their mobilization in water, loss of some matrix properties over their decoration with composites, need to increase their efficiency to remove a large range of water pollutants, and the life of the membrane. Additionally recovery of valuable materials from polluted water by membranes, easy recycling, and management of membrane at the end and overall economically efficient for the public are some areas that still need to be undertaken for further research.

11.8 Conclusion

The development of green multifunctional and highly efficient materials is important because of their biodegradability, biocompatibility, cost efficiency, and the demand to protect environment. This chapter provides information about application of green biocomposite in water treatment processes. Composites are prepared by combining two or more inorganic–inorganic, organic–organic, or inorganic–organic materials together to get unique and better properties. Preparation using biomass feed-stock makes the process more ecological and cost beneficial. Nano-sized biocomposites have emerged as a suitable material for the new-generation environmental remediation process. The biocomposites have been proved to be very efficient in water treatment. They have shown good adsorptive and catalytic activities, a range of pollutants have been removed from water till now using biocomposites including heavy metals, dyes, phenols, pesticides, etc. They have also shown good antibacterial, antifungal, and anticancer properties. Membrane filtration is among the advanced water treatment technologies and which is now in its advanced stage by their decoration with nanocomposite materials; it deducts the need of composite separation from water stream after their use and provides a continuous and convenient water treatment process. The nanomaterials offer their characteristics properties to the membrane owing to their ultra-small size, high surface to volume ration, reactivity, magnetic and optical properties. Decoration of composite materials increases the quality of membrane providing them better reactivity, adsorptive capacity, permeability, and also resist fouling. Green biocomposites can serve as material for future because renewable, recyclable, biodegradable, and sustainable material is the one that can make a real difference in present and future as well.

References

Abukhadra, M. R., Adlii, A., Bakry, B.M., 2019. Green fabrication of bentonite/chitosan@ cobalt oxide composite (BE/CH@ Co) of enhanced adsorption and advanced oxidation removal of Congo red dye and Cr (VI) from water. Int. J. Biol. Macromol. 126, 402–413.

Adyani, S.H., Soleimani, E., 2019. Green synthesis of Ag/Fe_3O_4/RGO nanocomposites by *Punica granatum* peel extract: catalytic activity for reduction of organic pollutants. Int. J. Hydrogen Energy 44 (5), 2711–2730 https://doi.org/10.1016/j.ijhydene.2018.12.012.

Ahmad, D.O., Ahmad, M.M., Ahmed, T.M.I., & Adil, H.A.(2020).Bionanocomposites in water treatment In: K.M.Zia, F.Jabeen, M.N.Anjum, and S.Ikram (eds.). Bionanocomposites: Green Synthesis and Applications (pp. 505–518). Elsevier BV, Amsterdam. https://doi.org/10.1016/b978-0-12-816751-9.00019-2

Ahmad, R., Mirza, A., 2018. Facile one pot green synthesis of Chitosan-Iron oxide (CS-Fe2O3) nanocomposite: Removal of Pb (II) and Cd (II) from synthetic and industrial wastewater. J. Clean. Prod. 186, 342–352.

Ajmal, A., Majeed, I., Malik, R.N., Idriss, H., Nadeem, M.A., 2014. Principles and mechanisms of photocatalytic dye degradation on TiO_2 based photocatalysts: a comparative overview. RSC Adv., 4 (70), 37003–37026 https://doi.org/10.1039/c4ra06658h.

Aksu Demirezen, D., Yıldız, Y.Ş., Demirezen Yılmaz, D., 2019. Amoxicillin degradation using green synthesized iron oxide nanoparticles: kinetics and mechanism analysis. Environ. Nanotechnol. Monit. Manage. 11,, 100219 https://doi.org/10.1016/j.enmm.2019.100219.

Ali, F., Khan, S.B., Kamal, T., Anwar, Y., Alamry, K.A., Asiri, A.M., 2017. Bactericidal and catalytic performance of green nanocomposite based-on chitosan/carbon black fiber supported monometallic and bimetallic nanoparticles. Chemosphere 188, 588–598.

Al-Oqla, F.M., Omari, M.A., 2017. Sustainable biocomposites: Challenges, potential and barriers for development. Green Energy Technol. 9783319466095, 13–29 https://doi.org/10.1007/978-3-319-46610-1_2.

Alzahrani, E., 2018. Chitosan membrane embedded with ZnO/CuO nanocomposites for the photodegradation of fast green dye under artificial and solar irradiation. Analytical chemistry insights 13, 1177390118763361.

Anastopoulos, I., Mittal, A., Usman, M., Mittal, J., Yu, G., Núñez-Delgado, A., Kornaros, M., 2018. A review on halloysite-based adsorbents to remove pollutants in water and wastewater. J. Mol. Liq. 269, 855–868. https://doi.org/10.1016/j.molliq.2018.08.104.

Anawar, H.M., Strezov, V., 2019. Synthesis of biosorbents from natural/agricultural biomass wastes and sustainable green technology for treatment of nanoparticle metals in municipal and industrial wastewater. In: Mishra, A.K., Anawar, H.M.D., Drouiche, N. (Eds.), Emerging and Nanomaterial Contaminants in Wastewater: Advanced Treatment Technologies. Elsevier, Amsterdam, pp. 83–104.

Atchudan, R., Edison, T.N.J.I., Perumal, S., Vinodh, R., Lee, Y.R., 2018. In-situ green synthesis of nitrogen-doped carbon dots for bioimaging and TiO2 nanoparticles@ nitrogen-doped carbon composite for photocatalytic degradation of organic pollutants. J. Alloys Compd. 766, 12–24.

Baran, T., 2019. Bio-synthesis and structural characterization of highly stable silver nanoparticles decorated on a sustainable bio-composite for catalytic reduction of nitroarenes. J. Mol. Struct. 1182, 213–218. https://doi.org/10.1016/j.molstruc.2019.01.057.

Bhati, M., Rai, R., 2017. Nanotechnology and water purification: Indian know-how and challenges. Environ. Sci. Pollut. Res. 24 (30), 23423–23435. https://doi.org/10.1007/s11356-017-0066-3.

Bhatti, H.N., Jabeen, A., Iqbal, M., Noreen, S., Naseem, Z., 2017. Adsorptive behavior of rice bran-based composites for malachite green dye: isotherm, kinetic and thermodynamic studies. J. Mol. Liq. 237, 322–333. https://doi.org/10.1016/j.molliq.2017.04.033.

Bhatti, Haq Nawaz, Mahmood, Zofishan, Kausar, Abida, Yakout, Sobhy M., Shair, Omar H., Iqbal, Munawar, 2020. Biocomposites of polypyrrole, polyaniline and sodium alginate with cellulosic biomass: Adsorption-desorption, kinetics and thermodynamic studies for the removal of 2,4-dichlorophenol. Int. J. Biol. Macromol. 153, 146–157. https://doi.org/10.1016/j.ijbiomac.2020.02.306.

Biswal, S.K., Panigrahi, G.K., Sahoo, S.K., 2020. Green synthesis of Fe_2O_3-Ag nanocomposite using *Psidium guajava* leaf extract: an eco-friendly and recyclable adsorbent for remediation of Cr(VI) from aqueous media. Biophys. Chem. 263, 106392. https://doi.org/10.1016/j.bpc.2020.106392.

Boruah, P.K., Sharma, B., Karbhal, I., Shelke, M.V., Das, M.R., 2017. Ammonia-modified graphene sheets decorated with magnetic Fe_3O_4 nanoparticles for the photocatalytic and photo-Fenton degradation of phenolic compounds under sunlight irradiation. J. Hazard. Mater. 325, 90–100. https://doi.org/10.1016/j.jhazmat.2016.11.023.

Chen, J., Zhu, J., Wang, N., Feng, J., Yan, W., 2019. Hydrophilic polythiophene/SiO2 composite for adsorption engineering: Green synthesis in aqueous medium and its synergistic and specific adsorption for heavy metals from wastewater. Chem. Eng. J. 360, 1486–1497.

Chook, S.W., Chia, C.H., Zakaria, S., Neoh, H.M., Jamal, R., 2017. Effective immobilization of silver nanoparticles on a regenerated cellulose–chitosan composite membrane and its antibacterial activity. New J. Chem. 41 (12), 5061–5065.

Dai, W-J., Wu, P., Liu, D., et al., 2020. Adsorption of polycyclic aromatic hydrocarbons from aqueous solution by organic montmorillonite sodium alginate nanocomposites. Chemosphere 251, 126074.

Dalal, U., Gupta, V.K., Reddy, S.N., Navani, N.K., 2020. Eradication of water borne pathogens using novel green nano Ag-biocomposite of *Citrus Limetta* peels. J. Environ. Chem. Eng. 8 (2), 103534. https://doi.org/10.1016/j.jece.2019.103534.

Dehghani, M.H., Karri, R.R., Yeganeh, Z.T., Mahvi, A.H., Nourmoradi, H., Salari, M., Zarei, A., Sillanpää, M., 2020. Statistical modelling of endocrine disrupting compounds adsorption onto activated carbon prepared from wood using CCD-RSM and DE hybrid evolutionary optimization framework: comparison of linear vs non-linear isotherm and kinetic parameters. J. Mol. Liq. 302, 112526. https://doi.org/10.1016/j.molliq.2020.112526.

Deng, R., Huang, D., Zeng, G., Wan, J., Xue, W., Wen, X., Liu, X., Chen, S., Li, J., Liu, C., Zhang, Q., 2019. Decontamination of lead and tetracycline from aqueous solution by a promising carbonaceous nanocomposite: interaction and mechanisms insight. Bioresour. Technol. 283, 277–285. https://doi.org/10.1016/j.biortech.2019.03.086.

Djelad, A., Mokhtar, A., Khelifa, A., Bengueddach, A., Sassi, M., 2019. Alginate-whey an effective and green adsorbent for crystal violet removal: kinetic, thermodynamic and mechanism studies. Int. J. Biol. Macromol. 139, 944–954. https://doi.org/10.1016/j.ijbiomac.2019.08.068.

Du, M., Du, Y., Feng, Y., Li, Z., Wang, J., Jiang, N., Liu, Y., 2019. Advanced photocatalytic performance of novel BiOBr/BiOI/cellulose composites for the removal of organic pollutant. Cellulose 26 (9), 5543–5557. https://doi.org/10.1007/s10570-019-02474-1.

Elessawy, N.A., El-Sayed, E.M., Ali, S., Elkady, M.F., Elnouby, M., Hamad, H.A., 2020. One-pot green synthesis of magnetic fullerene nanocomposite for adsorption characteristics. J. Water Process Eng. 34, 101047. https://doi.org/10.1016/j.jwpe.2019.101047.

El-Shabasy, R., Yosri, N., El-Seedi, H., Shoueir, K., El-Kemary, M., 2019. A green synthetic approach using chili plant supported $Ag/Ag_2O@P_{25}$ heterostructure with enhanced photocatalytic properties under solar irradiation. Optik 192,, 162943. https://doi.org/10.1016/j.ijleo.2019.162943.

Eltaweil, A.S., Mohamed, H.A., Abd El-Monaem, El-Subruiti, G.M., 2020. Mesoporous magnetic biochar composite for enhanced adsorption of malachite green dye: Characterization, adsorption kinetics, thermodynamics and isotherms. Adv. Powder Tech. 31 (3), 1253–1263.

Eugene, E.A., Phillip, W.A., Dowling, A.W., 2019. Data science-enabled molecular-to-systems engineering for sustainable water treatment. Curr. Opin. Chem. Eng. 26, 122–130. https://doi.org/10.1016/j.coche.2019.10.002.

Fowler, P.A., Hughes, J.M., Elias, R.M., 2006. Biocomposites: technology, environmental credentials and market forces. J. Sci. Food Agric. 86 (12), 1781–1789. https://doi.org/10.1002/jsfa.2558.

Fuku, X., Matinise, N., Masikini, M., Kasinathan, K., Maaza, M., 2018. An electrochemically active green synthesized polycrystalline NiO/MgO catalyst: use in photo-catalytic applications. Mater. Res. Bull. 97, 457–465.

Garza-Cervantes, J.A., Mendiola-Garza, G., de Melo, E.M., Dugmore, T.I.J., Matharu, A.S., Morones-Ramirez, J.R., 2020. Antimicrobial activity of a silver-microfibrillated cellulose biocomposite against susceptible and resistant bacteria. Sci. Rep. 10 (1), 7281. https://doi.org/10.1038/s41598-020-64127-9.

Gautam, A., Rawat, S., Verma, L., Singh, J., Sikarwar, S., Yadav, B.C., Kalamdhad, A.S., 2018. Green synthesis of iron nanoparticle from extract of waste tea: an application for phenol red removal from aqueous solution. Environ. Nanotechnol. Monit. Manage. 10, 377–387. https://doi.org/10.1016/j.enmm.2018.08.003.

Ghorbal, A., Sdiri, A., Elleuch, B., 2019. Green approaches for materials, wastes, and effluents treatment. Environ. Sci. Pollut. Res 26 (32), 32675–32677. https://doi.org/10.1007/s11356-019-06848-7.

Gutierrez, A.M., Dziubla, T.D., Hilt, J.Z., 2017. Recent advances on iron oxide magnetic nanoparticles as sorbents of organic pollutants in water and wastewater treatment. Rev. Environ. Health 32 (1–2), 111–117. https://doi.org/10.1515/reveh-2016-0063.

Gwenzi, W., Chaukura, N., Noubactep, C., Mukome, F.N.D., 2017. Biochar-based water treatment systems as a potential low-cost and sustainable technology for clean water provision. J. Environ. Manage. 197, 732–749. https://doi.org/10.1016/j.jenvman.2017.03.087.

Hairuddin, M.N., Mubarak, N.M., Khalid, M., Abdullah, E.C., Walvekar, R., Karri, R.R., 2019. Magnetic palm kernel biochar potential route for phenol removal from wastewater. Environ. Sci. Pollut. Res. 26 (34), 35183–35197. https://doi.org/10.1007/s11356-019-06524-w.

Hasan, I., Shekhar, C., Sharfan, I.I.B., Khan, R.A., Alsalme, A., 2020. Ecofriendly Green Synthesis of the ZnO-Doped CuO@Alg Bionanocomposite for Efficient Oxidative Degradation of p-Nitrophenol. ACS Omega 5 (49), 32011–32022.

Hasanin, M.S., Moustafa, G.O., 2020. New potential green, bioactive and antimicrobial nanocomposites based on cellulose and amino acid. Int. J. Biol. Macromol. 144, 441–448. https://doi.org/10.1016/j.ijbiomac.2019.12.133.

Hassanpour, M., Safardoust-Hojaghan, H., Salavati-Niasari, M., 2017. Degradation of methylene blue and Rhodamine B as water pollutants via green synthesized Co_3O_4/ZnO nanocomposite. J. Mol. Liq. 229, 293–299.

He, H., Tao, G., Wang, Y., Cai, R., Guo, P., Chen, L., Zuo, H., Zhao, P., Xia, Q., 2017. In situ green synthesis and characterization of sericin-silver nanoparticle composite with effective antibacterial activity and good biocompatibility. Mater. Sci. Eng. C 80, 509–516. https://doi.org/10.1016/j.msec.2017.06.015.

Heydari, R., Koudehi, M.F., Pourmortazavi, S.M., 2019. Antibacterial activity of Fe_3O_4/Cu nanocomposite: green synthesis using *Carum carvi* L. seeds aqueous extract. ChemistrySelect 4 (2), 531–535. https://doi.org/10.1002/slct.201803431.

Hsini, A., Essekri, A., Aarab, N., Laabd, M., Ait Addi, A., Lakhmiri, R., Albourine, A., 2020. Elaboration of novel polyaniline@ Almond shell biocomposite for effective removal of hexavalent chromium ions and Orange G dye from aqueous solutions. Environ. Sci. Pollut. Res. 27 (13).

Ibrahim, M., Siddique, A., Verma, L., Singh, J., Koduru, J.R., 2019. Adsorptive removal of fluoride from aqueous solution by biogenic iron permeated activated carbon derived from sweet lime waste. Acta Chim. Slov. 66 (1), 123–136. https://doi.org/10.17344/acsi.2018.4717.

Ike, I.A., Linden, K.G., Orbell, J.D., Duke, M., 2018. Critical review of the science and sustainability of persulphate advanced oxidation processes. Chem. Eng. J. 338, 651–669. https://doi.org/10.1016/j.cej.2018.01.034.

Imran, Muhammad, Iqbal, Muhammad Mohsin, Iqbal, Jibran, et al., 2021. Synthesis, characterization and application of novel MnO and CuO impregnated biochar composites to sequester arsenic (As) from water: Modeling, thermodynamics and reusability. J. Hazard. Mater., 123338. https://doi.org/10.1016/j.jhazmat.2020.123338.

Işık, B., Keçeli, A.E., Gürdağ, G., Gonul, K., 2021. Radioactive cesium ion removal from wastewater using polymer metal oxide composites. J. Hazard. Mater. 403, 123652. doi:10.1016/j.jhazmat.2020.123652.

Ismail, M., Khan, M.I., Khan, S.B., Khan, M.A., Akhtar, K., Asiri, A.M., 2018. Green synthesis of plant supported Cu–Ag and Cu–Ni bimetallic nanoparticles in the reduction of nitrophenols and organic dyes for water treatment. J. Mol. Liq. 260, 78–91. https://doi.org/10.1016/j.molliq.2018.03.058.

Jin, Q., Zhang, S., Wen, T., Wang, J., Gu, P., Zhao, G., Wang, X., Chen, Z., Hayat, T., Wang, X., 2018. Simultaneous adsorption and oxidative degradation of bisphenol A by zero-valent iron/iron carbide nanoparticles encapsulated in N-doped carbon matrix. Environ. Pollut. 243, 218–227. https://doi.org/10.1016/j.envpol.2018.08.061.

Kamari, S., Shahbazi, A., 2020. Biocompatible Fe_3O_4@SiO_2-NH_2 nanocomposite as a green nanofiller embedded in PES–nanofiltration membrane matrix for salts, heavy metal ion and dye removal: long–term operation and reusability tests. Chemosphere 243, 125282. https://doi.org/10.1016/j.chemosphere.2019.125282.

Kamari, S., Shahbazi, A., 2020. Biocompatible Fe_3O_4@SiO_2-NH_2 nanocomposite as a green nanofiller embedded in PES–nanofiltration membrane matrix for salts, heavy metal ion and dye removal: Long–term operation and reusability tests. Chemosphere 243, 125282. https://doi.org/10.1016/j.chemosphere.2019.125282.

Karri, R.R., Jayakumar, N.S., Sahu, J.N., 2017. Modelling of fluidised-bed reactor by differential evolution optimization for phenol removal using coconut shells based activated carbon. J. Mol. Liq. 231, 249–262. https://doi.org/10.1016/j.molliq.2017.02.003.

Kausar, A., 2020. Progress in green nanocomposites for high-performance applications. Mater. Res. Innov. 25 (1), 53–65. https://doi.org/10.1080/14328917.2020.1728489.

Kim, J.H., Joshi, M.K., Lee, J., Park, C.H., Kim, C.S., 2018. Polydopamine-assisted immobilization of hierarchical zinc oxide nanostructures on electrospun nanofibrous membrane for photocatalysis and antimicrobial activity. J. Colloid Interf. Sci. 513, 566–574. https://doi.org/10.1016/j.jcis.2017.11.061.

Kulkarni, S., Jadhav, M., Raikar, P., Barretto, D.A., Vootla, S.K., Raikar, U.S., 2017. Green synthesized multifunctional Ag@Fe_2O_3 nanocomposites for effective antibacterial, antifungal and anticancer properties. New J. Chem. 41 (17), 9513–9520. https://doi.org/10.1039/c7nj01849e.

Kumar, S., Bera, R., Das, N., & Koh, J. (2020). Chitosan-based zeolite-Y and ZSM-5 porous biocomposites for H2 and CO2 storage. Carbohydr. Polym., 232., 115808 https://doi.org/10.1016/j.carbpol.2019.115808.

Lan, J., Sun, Y., Huang, P., Du, Y., Zhan, W., Zhang, T.C., Du, D., 2020. Using Electrolytic Manganese Residue to prepare novel nanocomposite catalysts for efficient degradation of Azo Dyes in Fenton-like processes. Chemosphere 252, 126487. https://doi.org/10.1016/j.chemosphere.2020.126487.

Ling, L.-L., Liu, W.-J., Zhang, S., Jiang, H., 2017. Magnesium Oxide Embedded Nitrogen Self-doped Biochar Composites: Fast and High-Efficiency Adsorption of Heavy Metals in an Aqueous Solution. Env. Sci. Tech. 51, 10081–10089. https://doi.org/10.1016/j.chemosphere.2020.126487.

Lingamdinne, L.P., Koduru, J.R., Chang, Y.-Y., Karri, R.R., 2018. Process optimization and adsorption modeling of Pb(II) on nickel ferrite-reduced graphene oxide nano-composite. J. Mol. Liq. 250, 202–211. https://doi.org/10.1016/j.molliq.2017.11.174.
Lingamdinne, L.P., Koduru, J.R., Karri, R.R., 2019. A comprehensive review of applications of magnetic graphene oxide based nanocomposites for sustainable water purification. J. Environ. Manage. 231, 622–634. https://doi.org/10.1016/j.jenvman.2018.10.063.
Liu, C., Wang, L., Xu, H., Wang, S., Gao, S., Ji, X., Xu, Q., Lan, W., 2016. One pot green synthesis and the antibacterial activity of g-C_3N_4/Ag nanocomposites. Mater. Lett. 164, 567–570. https://doi.org/10.1016/j.matlet.2015.11.072.
Lu, F., & Astruc, D. (2020). Nanocatalysts and other nanomaterials for water remediation from organic pollutants. Coordinat.Chem. Rev., 408, 213180. https://doi.org/10.1016/j.ccr.2020.213180
Ma, L., Zhou, M., He, C., Li, S., Fan, X., Nie, C., Luo, H., Qiu, L., Cheng, C., 2019. Graphene-based advanced nanoplatforms and biocomposites from environmentally friendly and biomimetic approaches. Green Chem. 21 (18), 4887–4918. https://doi.org/10.1039/c9gc02266j.
Mady, Amr Hussein, Lara Baynosa, Marjorie, Dirk Tuma, Jae-Jin, Shim, 2017. Facile microwave-assisted green synthesis of Ag-ZnFe2O4@ rGO nanocomposites for efficient removal of organic dyes under UV-and visible-light irradiation. Appl. Catal. B. 203, 416–427.
Mahmoodi, N.M., Taghizadeh, M., Taghizadeh, A., Abdi, J., Hayati, B., Shekarchi, A.A., 2019. Bio-based magnetic metal-organic framework nanocomposite: ultrasound-assisted synthesis and pollutant (heavy metal and dye) removal from aqueous media. Appl. Surf. Sci. 480, 288–299. https://doi.org/10.1016/j.apsusc.2019.02.211.
Malakootian, M., Nasiri, A., Amiri Gharaghani, M., 2020. Photocatalytic degradation of ciprofloxacin antibiotic by TiO_2 nanoparticles immobilized on a glass plate. Chem. Eng. Commun. 207 (1), 56–72. https://doi.org/10.1080/00986445.2019.1573168.
Malek, N.N.A., Jawad, A.H., Abdulhameed, A.S., Ismail, K., Hameed, B.H., 2020. New magnetic Schiff's base-chitosan-glyoxal/fly ash/Fe_3O_4 biocomposite for the removal of anionic azo dye: An optimized process. Int. J. Biol. Macromol. 146, 530–539. doi:10.1016/j.ijbiomac.2020.01.020.
Maqbool, M., Bhatti, H.N., Sadaf, S., Zahid, M., Shahid, M., 2019. A robust approach towards green synthesis of polyaniline-Scenedesmus biocomposite for wastewater treatment applications. Mater. Res. Exp. 6 (5), 055308. https://doi.org/10.1088/2053-1591/ab025d.
Marrez, D.A., Abdelhamid, A.E., & Darwesh, O. M. (2019). Eco-friendly cellulose acetate green synthesized silver nano-composite as antibacterial packaging system for food safety. Food packaging and shelf life, 20, 100302. https://doi.org/10.1016/j.fpsl.2019.100302
Mishra, S., Yadav, S.S., Rawat, S., Singh, J., Koduru, J.R., 2019. Corn husk derived magnetized activated carbon for the removal of phenol and para-nitrophenol from aqueous solution: interaction mechanism, insights on adsorbent characteristics, and isothermal, kinetic and thermodynamic properties. J. Environ. Manage. 246, 362–373. https://doi.org/10.1016/j.jenvman.2019.06.013.
Mohamed, A.S., Abukkhadra, M.R., Abdallah, E.A., El-Sherbeeny, A.M., Mahmoud, R. K., 2020. The photocatalytic performance of silica fume based Co3O4/MCM-41 green nanocomposite for instantaneous degradation of Omethoate pesticide under visible light. J. Photochem. Photobiol. A. Chem. 392, 112434.
Mohammadabadi, S.I., Javanbakht, V., 2020. Lignin extraction from barley straw using ultrasound-assisted treatment method for a lignin-based biocomposite preparation with remarkable adsorption capacity for heavy metal. Int. J. Biol. Macromol. 164, 1133–1148.
Mohan, V.B., Lau, K.T., Hui, D., Bhattacharyya, D., 2018. Graphene-based materials and their composites: a review on production, applications and product limitations. Compos. B: Eng 142, 200–220. https://doi.org/10.1016/j.compositesb.2018.01.013.
Mou, Z., Zhang, H., Liu, Z., Sun, J., Zhu, M., 2019. Ultrathin BiOCl/nitrogen-doped graphene quantum dots composites with strong adsorption and effective photocatalytic activity for the degradation of antibiotic ciprofloxacin. Appl. Surf. Sci. 496,, 143655. https://doi.org/10.1016/j.apsusc.2019.143655.
Nasrollahzadeh, M., Issaabadi, Z., Sajadi, S.M., 2018. Green synthesis of Pd/Fe_3O_4 nanocomposite using *Hibiscus tiliaceus* L. extract and its application for reductive catalysis of Cr(VI) and nitro compounds. Sep. Purif. Technol. 197, 253–260. https://doi.org/10.1016/j.seppur.2018.01.010.

Ngo, H.H., Guo, W., Boopathy, R., 2019. Editorial overview: green technologies for environmental remediation. Curr. Opin. Environ. Sci. Health 12, A1–A3. https://doi.org/10.1016/j.coesh.2019.11.003.

Nidhi, Y., Dhruv, N.M., Shalu, R., Jiwan, S., 2020. Adsorption and equilibrium studies of phenol and para-nitrophenol by magnetic activated carbon synthesised from cauliflower waste. Environ. Eng. Res. (742) 742–752. https://doi.org/10.4491/eer.2019.238.

Pai, S., M Kini, S., Selvaraj, R., Pugazhendhi, A, 2020. A review on the synthesis of hydroxyapatite, its composites and adsorptive removal of pollutants from wastewater. J. Water Process Eng. 38,, 101574. https://doi.org/10.1016/j.jwpe.2020.101574.

Patinõ-Ruiz, D.A., De Avíla, G., Alarcoń-Suesca, C., Gonzaléz-Delgado, A.D., Herrera, A., 2020. Ionic cross-linking fabrication of chitosan-based beads modified with FeO and TiO2 nanoparticles: Adsorption mechanism toward naphthalene removal in seawater from cartagena bay area. ACS omega 5 (41), 26463–26475.

Paulo, P., Hugo, C., Hafiz, S., Marco, L., 2018. Natural fibre composites and their applications: a. review. J. Compos. Sci. 2 (4), 66. https://doi.org/10.3390/jcs2040066.

Qin, C., Wang, H., Yuan, X., Xiong, T., Zhang, J., Zhang, J., 2020. Understanding structure-performance correlation of biochar materials in environmental remediation and electrochemical devices. Chem. Eng. J. 382,, 122977. https://doi.org/10.1016/j.cej.2019.122977.

Raj Pant, H., Pant, B., Kim, Joo, H., Amarjargal, A., Hee Park, C., Tijing, L., D., Kyo Kim, E., Sang Kim, C., 2013. A green and facile one-pot synthesis of Ag-ZnO/RGO nanocomposite with effective photocatalytic activity for removal of organic pollutants. Ceram. Int. 39 (5), 5083–5091. https://doi.org/10.1016/j.ceramint.2012.12.003.

Ran, J., Chen, H., Bai, X., et al., 2019. Immobilizing $CuO/BiVO_4$ nanocomposite on PDA-templated cotton fabric for visible light photocatalysis, antimicrobial activity and UV protection. Applied Surface Science 493, 1167–1176. doi:10.1016/j.apsusc.2019.07.137.

Ranjith, R., Renganathan, V., Chen, S.M., Selvan, N.S., Rajam, P.S., 2019. Green synthesis of reduced graphene oxide supported TiO_2/Co_3O_4 nanocomposite for photocatalytic degradation of methylene blue and crystal violet. Ceram. Int. 45 (10), 12926–12933. https://doi.org/10.1016/j.ceramint.2019.03.219.

Rathinam, K., Singh, S.P., Arnusch, C.J., Kasher, R., 2018. An environmentally-friendly chitosan-lysozyme biocomposite for the effective removal of dyes and heavy metals from aqueous solutions. Carbohydr. Polym. 199, 506–515. doi:10.1016/j.carbpol.2018.07.055.

Ravikumar, K. V. G., Kubendiran, H., Ramesh, K., et al., 2020. Batch and column study on tetracycline removal using green synthesized NiFe nanoparticles immobilized alginate beads. Environmental Technology & Innovation 17, 100520.

Ren, C., Ding, X., Li, W., Wu, H., Yang, H., 2017. Highly Efficient Adsorption of Heavy Metals onto Novel Magnetic Porous Composites Modified with Amino Groups. J. Chem. Eng. Data. doi:10.1021/acs.jced.7b00198.

Sadhukhan, S., Ghosh, T.K., Roy, I., Rana, D., Bhattacharyya, A., Saha, R., Chattopadhyay, S., Khatua, S., Acharya, K., Chattopadhyay, D., 2019. Green synthesis of cadmium oxide decorated reduced graphene oxide nanocomposites and its electrical and antibacterial properties. Mater. Sci. Eng. C 99, 696–709. https://doi.org/10.1016/j.msec.2019.01.128.

Sahu, J.N., Karri, R.R., Zabed, H.M., Shams, S., Qi, X., 2019a. Current perspectives and future prospects of nano-biotechnology in wastewater treatment. Sep. Purif. Rev. 50 (2), 139–158. https://doi.org/10.1080/15422119.2019.1630430.

Sahu, J.N., Zabed, H., Karri, R.R., Shams, S., Qi, X., Thomas, S., Grohens, Y., Pottathara, Y.B., 2019b. Chapter 11 - Applications of nano-biotechnology for sustainable water purification. Editor(s): Sabu Thomas, Yves Grohens, Yasir Beeran Pottathara, In Micro and Nano Technologies, Industrial Applications of Nanomaterials, Elsevier, 2019, pp. 313–340, ISBN 9780128157497, https://doi.org/10.1016/B978-0-12-815749-7.00011-6.

Sahu, N., Rawat, S., Singh, J., Karri, R.R., Lee, S., Choi, J.S., Koduru, J.R., 2019. Process optimization and modeling of methylene blue adsorption using zero-valent iron nanoparticles synthesized from sweet lime pulp. Appl. Sci. (Switzerland) 9 (23), 5112. https://doi.org/10.3390/app9235112.

Saleh, T.A., Parthasarathy, P., Irfan, M., 2019. Advanced functional polymer nanocomposites and their use in water ultra-purification. Trends Environ. Anal. Chem. 24, e00067. https://doi.org/10.1016/j.teac.2019.e00067.

Sarasini, F., Fiore, V., 2018. A systematic literature review on less common natural fibres and their biocomposites. J. Cleaner Prod. 195, 240–267. https://doi.org/10.1016/j.jclepro.2018.05.197.

Sayfa, B., Saima, S., Suhail, S., Zain, K.M.Ahmed, S. (Ed.), 2021. Advanced application of green materials in environmental remediation. Applications of Advanced Green Materials 481–502. https://doi.org/10.1016/b978-0-12-820484-9.00019-2.

Sharma, M., Behl, K., Nigam, S., Joshi, M., 2018. TiO_2-GO nanocomposite for photocatalysis and environmental applications: a green synthesis approach. Vacuum 156, 434–439. https://doi.org/10.1016/j.vacuum.2018.08.009.

Shehabeldine, A., Hasanin, M., 2019. Green synthesis of hydrolyzed starch–chitosan nano-composite as drug delivery system to gram negative bacteria. Environ. Nanotechnol., Monit. Manage. 12,, 100252. https://doi.org/10.1016/j.enmm.2019.100252.

Shi, Z., Xu, C., Guan, H., Li, L., Fan, L., Wang, Y., Liu, L., Meng, Q., Zhang, R., 2018. Magnetic metal organic frameworks (MOFs) composite for removal of lead and malachite green in wastewater. Colloids Surf. A 539, 382–390. https://doi.org/10.1016/j.colsurfa.2017.12.043.

Sinha, R.S., Chinomso, I.A.O., Carlos, B.J., 2020. Polymer-based membranes and composites for safe, potable, and usable water: a survey of recent advances. Chem. Afr. https://doi.org/10.1007/s42250-020-00166-z.

Soni, U., Bajpai, J., Singh, S.K., Bajpai, A.K., 2017. Evaluation of chitosan-carbon based biocomposite for efficient removal of phenols from aqueous solutions. J. Water Proc. Eng. 16, 56–63.

Taghipour, S., Hosseini, S.M., Ataie-Ashtiani, B., 2019. Engineering nanomaterials for water and wastewater treatment: review of classifications, properties and applications. New J. Chem. 43 (21), 7902–7927. https://doi.org/10.1039/c9nj00157c.

Tang, F., Yu, H., Yassin Hussain Abdalkarim, S., Sun, J., Fan, X., Li, Y., Zhou, Y., Chiu Tam, K., 2020. Green acid-free hydrolysis of wasted pomelo peel to produce carboxylated cellulose nanofibers with super absorption/flocculation ability for environmental remediation materials. Chem. Eng. J. 395,, 125070. https://doi.org/10.1016/j.cej.2020.125070.

Thangavel, S., Raghavan, N., & Venugopal, G. (2019). Magnetically separable iron oxide-based nanocomposite photocatalytic materials for environmental remediation In: A. Pandikumar & K. Jothivenkatachalam, (Eds.), Photocatalytic Functional Materials for Environmental Remediation. John Wiley & Sons Inc., NJ, (pp. 243–265) https://doi.org/ https://doi.org/10.1002/9781119529941.ch8

Toledo-Jaldin, H.P., Sánchez-Mendieta, V., Blanco-Flores, A., López-Téllez, G., Vilchis-Nestor, A.R., Martín-Hernández, O., 2020. Low-cost sugarcane bagasse and peanut shell magnetic-composites applied in the removal of carbofuran and iprodione pesticides. Environ. Sci. Poll. Res. 27 (8), 7872–7885.

Tran, C.D., Prosencyes, F., Franko, M., Benzi, G., 2016. Synthesis, structure and antimicrobial property of green composites from cellulose, wool, hair and chicken feather. Carbohydr. Polym. 151, 1269–1276. https://doi.org/10.1016/j.carbpol.2016.06.021.

Wang, S., McGuirk, C.M., d'Aquino, A., Mason, J.A., Mirkin, C.A., 2018. Metal–organic framework nanoparticles. Adv. Mater. 30 (37), 1800202. https://doi.org/10.1002/adma.201800202.

Wu, K., Li, Y., Liu, T., Huang, Q., Yang, S., Wang, W., Jin, P., 2019. The simultaneous adsorption of nitrate and phosphate by an organic-modified aluminum-manganese bimetal oxide: adsorption properties and mechanisms. Appl. Surf. Sci. 478, 539–551. https://doi.org/10.1016/j.apsusc.2019.01.194.

Yadav, M., Garg, S., Chandra, A., Hernadi, K., 2019. Fabrication of leaf extract mediated bismuth oxybromide/oxyiodide (BiOBrxI1−x) photocatalysts with tunable band gap and enhanced optical absorption for degradation of organic pollutants. J. Colloid Interf. Sci. 555, 304–314. https://doi.org/10.1016/j.jcis.2019.07.090.

Yu, N., Peng, H., Qiu, L., Wang, R., Jiang, C., Cai, T., Sun, Y., Li, Y., Xiong, H., 2019. New pectin-induced green fabrication of Ag@AgCl/ZnO nanocomposites for visible-light triggered antibacterial activity. Int. J. Biol. Macromol. 141, 207–217. https://doi.org/10.1016/j.ijbiomac.2019.08.257.

Zhong, Q.Q., Zhao, Y.Q., Shen, L., et al., 2020. Single and binary competitive adsorption of cobalt and nickel onto novel magnetic composites derived from green macroalgae. Environ. Engin. Sci. 37 (3), 188–200.

Zhou, Y., Cao, S., Xi, C., Li, X., Zhang, L., Wang, G., Chen, Z., 2019. A novel Fe_3O_4/graphene oxide/citrus peel-derived bio-char based nanocomposite with enhanced adsorption affinity and sensitivity of ciprofloxacin and sparfloxacin. Bioresour. Technol. 292, 121951. https://doi.org/10.1016/j.biortech.2019.121951.

Zhu, H., Yuan, J., Tan, X., Zhang, W., Fang, M., Wang, X., 2019. Efficient removal of Pb 2+ by Tb-MOFs: identifying the adsorption mechanism through experimental and theoretical investigations. Environ. Sci. Nano 6 (1), 261–272. https://doi.org/10.1039/c8en01066h.

Zhu, L., Meng, L., Shi, J., Li, J., Zhang, X., Feng, M., 2019. Metal-organic frameworks/carbon-based materials for environmental remediation: a state-of-the-art mini-review. J. Environ. Manage 232, 964–977. https://doi.org/10.1016/j.jenvman.2018.12.004.

CHAPTER 12

Advanced green nanocomposite materials for wastewater treatment

Jai Kumar[a], Abdul Sattar Jatoi[b], Shaukat Ali Mazari[b], Esfand Yar Ali[b], Nazia Hossain[c], Rashid Abro[b], Nabisab Mujawar Mubarak[d] and Nizamuddin Sabzoi[c]

[a]State Key Laboratory of Organic-Inorganic Composites, Key Laboratory of Electrochemical Process and Technology for Materials, Beijing University of Chemical Technology, Beijing, China
[b]Department of Chemical Engineering, Dawood University of Engineering and Technology, Karachi, Pakistan
[c]School of Engineering, RMIT University, Melbourne, Australia
[d]Department of Chemical Engineering, Faculty of Engineering and Science, Curtin University, Sarawak, Malaysia

12.1 Introduction

Water is fundamental for survival. Apart from human needs water is also a primary need for industries. However, industrialization causes generation of wastewater, which tends to further adulterate large water bodies. This endangers the quality of drinking water leads to the difficulties for survival of human beings on the earth. Till now, organic, inorganic, microorganisms, and heavy metals are the well-known pollutants for degrading the quality of water. It has been reported by the UN, the shortage of drinking quality water is the most challenging issues of the twenty-first century (Gitis & Hankins, 2018; Hodges et al., 2018). In addition, the report of the world health organization (WHO) shows that around 1.7 million individuals died and 4 billion heath cases have been reported due to the shortages of quality water (Briggs et al., 2016). Thus, adulteration of the large water bodies may lead to environmental problems and raise concerns over the shortage of quality drinking water. So, purification of water is essential for alleviating both water pollution and drinking quality water depletion.

Among the currently available nanomaterial approaches for water purification, four types of purification technologies have been widely developed, as shown in Fig. 12.1.

Among them, for water purification metal oxides and zero-valent metal nanoparticles, nanocomposites, and carbon nanotubes (CNTs) are the most common and are found helpful for treating the wastewater containing the dyes, heavy metals, chlorinated organic compounds, pesticide, and polycyclic aromatic hydrocarbons, and bacteria (Fahad Khan Saleem et al., 2021; Khan Saleem Ahmed et al., 2021). These processes are relatively short and simple but are reported to be unsafe and costly due to aggregation phenomena, limited light adsorption of TiO_2 nanoparticles and ZnO nanoparticles, and the high cost of a CNT. Also, the difficulty in separation of the degraded system is one of

Sustainable Nanotechnology for Environmental Remediation
DOI: https://doi.org/10.1016/B978-0-12-824547-7.00015-1

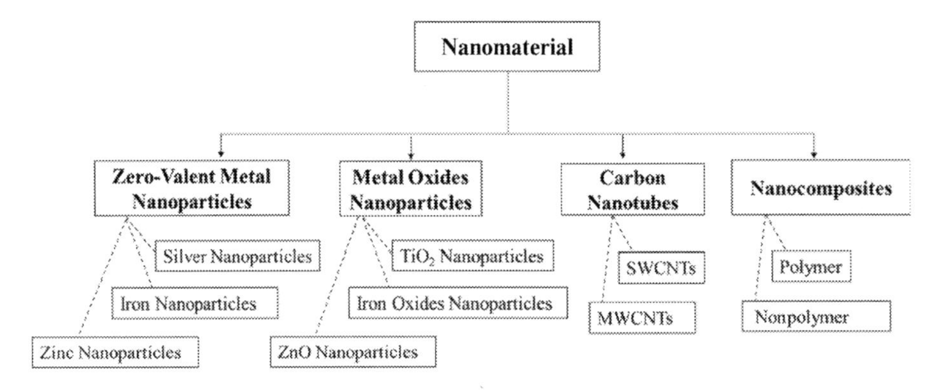

Figure 12.1 Types of nanomaterial for wastewater treatment.

the drawbacks of these processes. In contrast, another water purification approach is the use of nanocomposite materials. For the synthesis of nanocomposite compounds, these materials are getting significant consideration for water treatment in view of their high porosity, active surface sites, catalytic degradation ability, and accessibility to separation. Recently, many research attempts have been carried out the synthesis of nanocomposite materials for water treatment (Lingamdinne et al., 2019; Narayana et al., 2022). Kim et al. (2003) synthesized a thin-film composite material via interfacial polymerization phenomena, and then dipped the prepared thin-film composite into Ti_2O collide solution. In their work, they found that in the field of water treatment the prepared nanocomposite material has a major role. Zhang et al. (2010) achieved the activity of a visible light photocatalyst near >400 nm spectrometer via synthesis of graphene oxide (GO) nanocomposite material that is TiO_2 (P25)-GO. Although the synthesized nanocomposite materials are nontoxic, low in cost, and have durable stability; however as a result of the use of chemicals the nanocomposite are still facing problems.

Currently, the synthesis and application of green nanocomposites are gaining significant attention from water purification technologies. The waste and biomass-based nanocomposite materials are considered as the promising candidate for wastewater treatment owing to their green nature and relatively much lower cost. So far, several biomass materials such as Cassia fistula (golden shower), aloe vera (Kirti et al., 2018), wheat straw (Ahmad et al., 2015), biochar (Tan et al., 2016), agar (Kamal, 2019), red algae (Chen et al., 2016),etc. have been used for the synthesis of nanocomposites for wastewater treatment. Ahmad et al. (2015) prepared nanocomposite materials via a combination of wheat straw with the nanoparticle of hydrous zirconium oxide. They found that the developed nanocomposite of wheat straw-nitrogen-zirconium (Ws-N-Zr) has excellent potential for the elimination of phosphate from contaminated water because of its comparatively high level of coexisted sulfate ions. Kirti et al. (2018) investigated the application of green biomass–based synthesized nanocomposite materials. In their

Table 12.1 Common synthesis methods for nanocomposite materials.

Method	Advantage	Disadvantage	Model	Reference
Spinning	Good contact	High cost	Ethylene vinyl acetate/graphene oxide	F. Hussain et al., 2006
Sol-gel	High compatibility	High reactivity High cost	Polyethylene-octene elastomer	B. Arash et al., 2015
Phase separation	Excellent functionalization	Uncontrolled materialization	Carbon nanotube/cellulose acetate	Cope, C.O, 2014
In-situ polymerization	Free matrix agglomeration High liquid phase processing	High cost, Long process time	Carbon nanotube/polymer	Rahmat, M and Hubert, P., 2011
Electrochemical synthesis	Moderate conditions (i.e., low temperature and pressure)	High safety needed	Au/γ-Fe_2O_3/WO_3	Lin, K. et al., 2009

study, there were two types of mixtures: one biomass (i.e., C. fistula and aloe vera) and the other chemical (i.e., n-octanol). From the experimental result, they found that biomass-based synthesized nanocomposite material revealed excellent decontaminations properties. Besides, n-octanol-based synthesized nanocomposite material found to be economical and can reduce the synthesis cost to less than five times in comparison to relevant studies indicating the proposed research has a great economical advantage in wastewater treatment. It has been observed that cellulose-based materials are more thermally stable and crystalline, which is suitable for nanocomposites (Chen et al., 2016).

This chapter reports synthesis and characterization of biomass-based green nanocomposite materials and their applications for wastewater treatment. Furthermore, discussions have also been extended on feedstocks of material, process conditions, specific contaminant removal, material efficiency, and recyclability and contaminant removal mechanisms.

12.2 Waste and biomass-derived nanocomposites

12.2.1 Synthesis

The nanocomposite materials can be synthesized by a number of methods for example solution casting, melt compounding, in-situ polymerization, sol-gel, phase separation, and electrochemical synthesis. The distinct advantage of these methods and their synthesized compounds are listed in Table 12.1. For the preparation of nanoparticle-based materials, there are four typical processes namely chemical precipitation, ion exchange, membrane filtration, and adsorption. Among all these methods, coprecipitation and adsorption processess are getting more attention because of their high reactivity, simple operation, and low process time and cost. To get biomass-derived nanocomposite materials, initially, nanocomposite materials such as Ti_2O and Fe_3O_4 are prepared via a suitable process, and then prepared nanoparticles are mixed with pretreated biomass materials. In the end, the mixture is washed and dried. For example, Sun et al. (2015) synthesized xylan/poly (acrylic acid) magnetic nanocomposite hydrogel adsorbent in two-step: initially, they used chemical precipitation method for the synthesis of Fe_3O_4 nanoparticle and then, the prepared Fe_3O_4 nanoparticles were mixed with pretreated wheat straw (i.e., the source for xylan/poly). After the reaction, the black hydrogel slurry type material was obtained, which was then socked and dried according to the mature procedure. And finally, biomass-derived composite material was collected for wastewater treatment application. Furthermore, as shown in Fig. 12.2, Batool et al. (2020) prepared biomass-derived nanocomposite material from the peels of grapefruit and SnO_2 nanoparticles. In their work, they obtained SnO_2 nanoparticles from a commercial source and the biomass was pretreated to grapefruit-derived carbon material (GPC). To obtain GPC, initially, peels were removed manually from grapefruit and dried for 12 days, and then ground powder was obtained from dried peels and again dried at 80 °C for 12 h for the complete removal of moisture and to attain a uniform weight. Finally, the moisture free peel powder was put in a tube furnace under an inert atmosphere and calcined at 500 °C for 60 min, and then black solid material was taken out and sieved to a 20 μm mesh size after grinding. Finally, to get biomass-based nanocomposite material, SnO_2 nanoparticles and GPC were processed in a ball mill for 12 h at 1000 rpm.

12.2.2 Characterization of biomass-derived nanocomposite

Thermal analysis of biomass-derived nanocomposite is considered as an essential characterization, in which a suitable thermal kinetics model of prepared material can be investigated. In addition, for improving the thermal stability of nonpolymeric material in the polymeric matrix, thermal analysis plays a vital role. The thermal properties of biomass-derived nanocomposite are commonly analyzed by different thermal techniques such as thermogravimetric, thermomechanical analysis (Mohan et al., 2014), differential thermal analysis, differential scanning calorimetry, and dynamic mechanical analysis. These analyzer techniques can help to provide physical and chemical phenomena of biomass-derived nanocomposite, that is, absorption, phase transfer, desorption, chemisorption, thermal decomposition, and so on (Gan et al., 2018).

Surface investigation of biomass-derived nanocomposite is also considered as an important analysis, in which surface properties such as porosity, homogeneity, surface roughness, and combine the strength of prepared material can be studied. Along with that, the surface images of biomass-derived nanocomposite are also helping to inform the crystalline mismatch and compatibility. The image of surface measurement

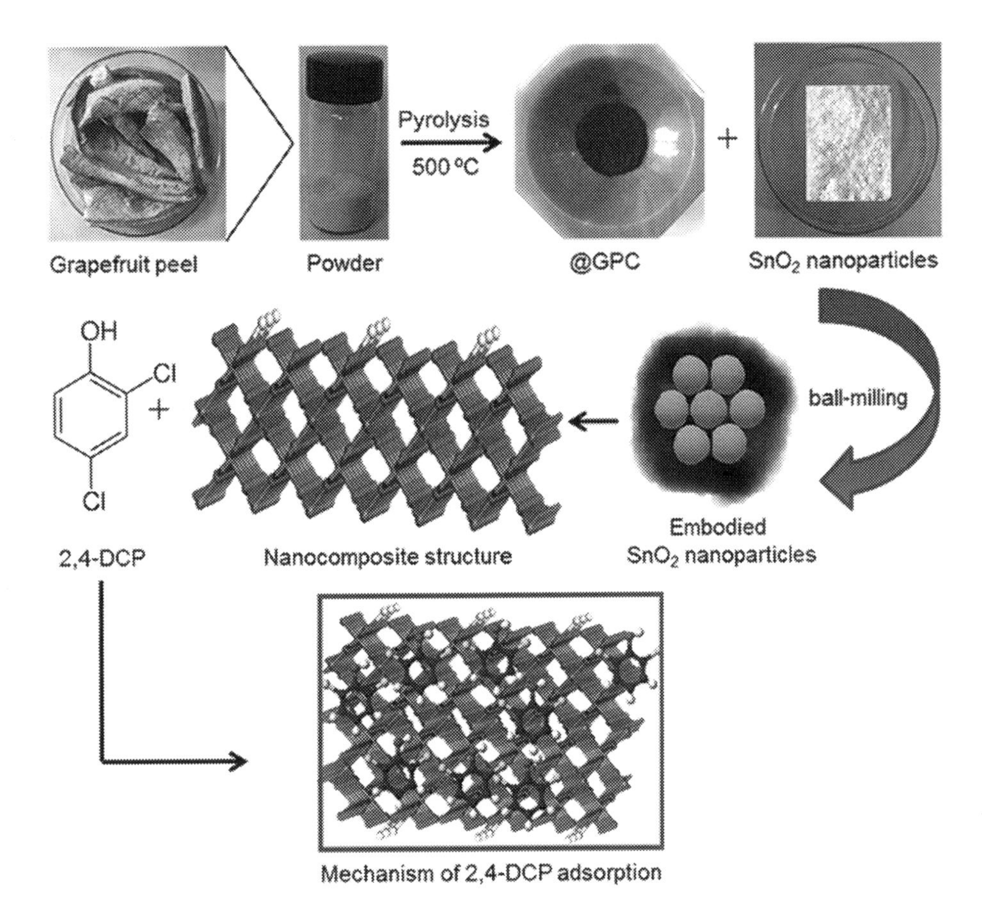

Figure 12.2 Design and characterization of a biomass template/SnO_2 nanocomposite for enhanced adsorption of 2,4-dichlorophenol. From Batool, S., Idrees, M., Ahmad, M., Ahmad, M., Hussain, Q., Iqbal, A., & Kong, J. (2020). Environ. Res., 181. https://doi.org/10.1016/j.envres.2019.108955 Schematic presentation of the SnO_2@GPC nanocomposite proposed by Batool et al., 2020.

is commonly captured by different electronic scanning techniques such as scanning tunneling microscopy, scanning emission morphology (SEM), atomic force spectroscopy (AFM), scanning probe microscopy, and transmission electron microscopy (TEM). These techniques differ according to their scanning strength. For example, The AFM is used for investigating the roughness of the material, while TEM helps to provide the quality of the nanocomposite material such as internal structure, structural defect, and distribution of various phase (Liang et al., 2020; Liou & Wu, 2010; Qiu et al., 2017).

To observe the crystalline structure of biomass-derived nanocomposite material, sometimes X-rays diffraction technique is also carried out, which is commonly abbreviated as X-ray power diffraction (XRD). In XRD techniques, X-rays penetrate at different angles on the sample and provide the different patterns of diffracted peaks associated with material crystallinity and its unit cell structure (Veerakumar et al., 2017). Notably, the XRD pattern is divided into two patterns according to their wide and small angle diffraction. The unit cell and crystalline structure of nanocomposite material can be observed corresponding to TEM images in a wide-angle X-ray diffraction pattern, while in small X-ray diffraction, the structure of the material can be seen in the order of 10 Å or larger. On the other hand, to find the interaction between biomass and nanoparticle, Fourier transform infrared (FT-IR) and Raman spectra techniques are usually performed, in which the strong level of interaction can be observed by different wave transmission numbers (Zhang et al., 2010).

12.2.3 Properties of biomass-derived nanocomposites

For water purification, adsorption, catalytic, physical, morphological, and mechanical properties are deemed necessary for any material. To get the desired properties of nanocomposite material, adding nanoparticles to the material matrix plays a vital role. Over the past few days, biomass-derived nanocomposite-based adsorbent material has attained consideration in the field of wastewater treatment owing to the existence of high surface area of nanoparticles in the composite lattice. The use of biomass-derived nanocomposite-based adsorbent material was found to have high potential for removing the dye, toxic metal ions, heavy metals, organic pollutants, and microorganisms from wastewater (Chen et al., 2016). Hari et al. synthesized biomass-based nanoadsorbent for the removal of fluoride found in wastewater (Paudyal et al., 2013). During their work, they prepared an adsorbent from waste biomass, that is, dried orange juice residue (DOJR) with Zr(IV) nanoparticles. They believed, the adsorbent DOJR-Zr(IV) has high potential to remove the fluoride from wastewater. Besides, the depth service time and Thomas model are also well fitted on the designed process. The adsorption and desorption phenomena between the DOJR loaded Zr(IV), and fluoride ions are illustrated in Fig. 12.3.

Catalytic properties of nanocomposites are very important for degradation of dyes and organic pollutants from wastewater (Vinayagam et al., 2018). In wastewater treatment, the nanocomposite material catalysts are used for the removal of several pollutants such as toxic organic compounds, dyes, and nitrogen containing compounds. An example of biomass-derived nanocomposite catalytic material was proposed by Kamal (2019). In this study, the author synthesized an agar biopolymer hydrogel based CuO nanoparticles catalyst and used it for the reduction of three types of nitroarenes such as 2,6-dinitrophenol (2,6-DNP), 2-nitrophenol (2-NP), and 4-nitrophenol (4-NP). The obtained experimental results revealed that an agar-CuO-based nanocomposite catalyst material could reduce the nitroarenes with a conversation rate of around 90%. Besides, the

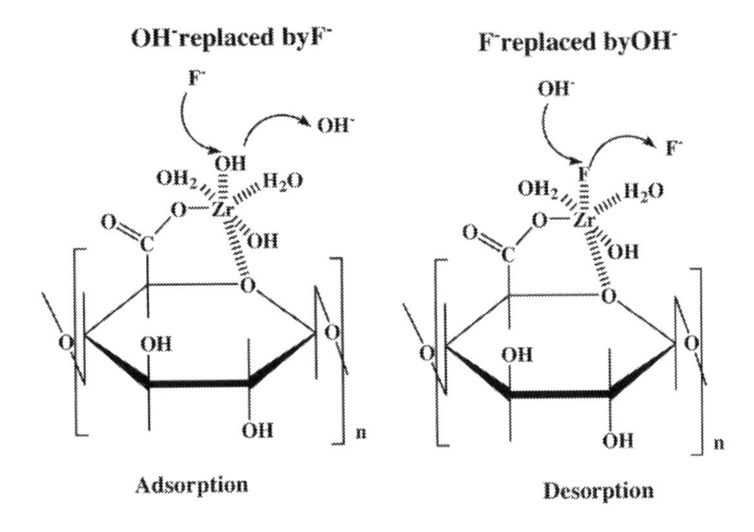

Figure 12.3 . Adsorptive removal of fluoride from aqueous medium using a fixed bed column packed with Zr(IV) loaded dried orange juice residue. From Paudyal, H., Pangeni, B., Inoue, K., Kawakita, H., Ohto, K., & Alam, S. (2013) Bioresour. Technol., 146, 713–720. *https://doi.org/10.1016/j.biortech.2013.07.014* Adsorption and desorption phenomena between DOJR loaded Zr(IV) and fluoride ions.

prepared agar-CuO-based nanocomposite catalyst can be recycled five times with a high conversion rate. Similarly, several other biomass-based nanocomposites have displayed favorable catalytic properties for degradation of dyes, organic pollutants, and metallic reduction (Atchudan et al., 2018; Gupta et al., 2020).

The mechanical strength of composite materials is considered as one of the main parameters during the synthesis of nanocomposite material because processability, strength, stability, and a lifetime of a nanocomposite depend on their mechanical properties (Lin et al., 2013). The interaction between biomass and nanoparticles in a composite not only influences mechanical behavior but also enhances the value-added mechanical qualities, which may not be found in an individual composite material. Besides, the anisotropy is also well-known characteristics for biomass-derived nanocomposite materials, which supports the mechanical properties in nanocomposites material (Chen et al., 2018). At the moment, carbon-based nanocomposite material getting close attention because carbon material itself owns thermal, mechanical, and exceptional electrical properties, along with that, the presence of ultra-mechanical properties in carbon material such as high tensile strength, and high stiffness makes this material suitable for the preparation of nanocomposite material. Considering the high mechanical properties of carbon materials, scientists are paying much attention to the biomass-derived carbon materials for nanocomposites. Osman et al. prepared a CNT for heavy metal removal application (Osman et al., 2019). In their research, they derived carbon from waste potato peels, and then consequently, multi-walled nanoparticles were prepared through carbon derived from potato peels waste by mixing it up with melamine a nitrogen-based material, and iron precursor, $C_6H_{12}Fe_2O_{18}$. Also, Batool et al. synthesized the biomass-derived nanocomposite materials from the peels of grapefruit (Batool et al., 2020). In their work, initially, the prepared carbon material from peels of grapefruit was heated at 500 °C under an inert atmosphere, and then derived carbon was mixed with SnO_2 nanoparticles. Their results revealed that the presence of carbon material increases the influences of mechanical behavior and increases Young's modulus as well as the tensile strength of the nanocomposite substance. The discussion demonstrates that utilization of waste biomass as carbon material can help to construct the green biomass-derived nanocomposite materials with higher mechanical strength, low cost, and eco-friendly process for wastewater application.

12.3 Wastewater treatment by using biomass-derived nanocomposites

Recently, substantial research efforts have been made to synthesize biomass-derived nanocomposites for the elimination of environmental contaminants, which appears to be beneficial for carbon sequestration and water pollution control (Ahmad et al., 2015). Due to the high availability of waste and biomass raw materials, their reduced price, and good chemical/physical surface properties of biochar, they have great potential to adsorb pollutants from water and degrade them catalytically (Tan et al., 2015; Vinayagam et al., 2018). The removal ability of biochar for different pollutants primarily relies on chemical and physical properties that are influenced through raw material feedstocks, synthesis techniques, and by adding nanoparticles. It is reported that raw biomass-based biochar from aqueous solutions has reduced capacity to absorb contaminants (Agnihotri et al., 2006; Yao et al., 2013), mostly for higher content of polluted water. Furthermore, it is hard to separate ground biochar from the aqueous solutions because of the small particle size (Chen et al., 2011; Quan et al., 2014). To overcome the above disadvantages, lots of experiments have been conducted to examine the engineered biomass-derived nanocomposites with improved structure and surface properties (Zhang et al., 2013). The synthesis of nanocomposites based on biochar is not just a technique to advance the chemical or physical properties of biochar, yet. In addition, it obtains a new composite materials and conjoin the pros of biochar with additional nanomaterials. As of late, several matrix-based nanocomposites have been enhanced for the purification of wastewater (Ahmad et al., 2014; Awual, 2016; Awual et al., 2021, 2014). Compared with other nanocomposites, using biomass as a substrate to produce nanocomposites has many advantages. First of all, biochar is abundant and a low-cost raw material, mainly derived from solid waste and agricultural biomass (Agnihotri et al., 2006; Sohi, S.P, 2012). In addition, the production of biochar is cheap and has low-energy requirements, typically formed at a comparatively low temperature (<800°C). Furthermore, biomass thermochemical treatment could produce biofuels couple with biochar. Consequently, biochar-based nanomaterials synthesis can attain four comprehensive objectives, comprising waste management, pollutant removal, energy production, and carbon sequestration.

12.3.1 Water contamination and their impact

Heavy metals are famed as metallic elements having a comparatively higher density than water (Fergusson, 1990). Toxic metals such as copper and lead are released into the atmosphere from industrial and concrete sources. Metals introduced into the marine atmosphere tend to accumulate within the sediments (Wang, 1987; Yadav et al., 2019). Metallic pollutants exist in urbanized estuaries by inputs from various sources, including urban runoff, industrial wastewater, shipping activities, and wastewater from sewage treatment plants, and can enter the aquatic environment through events such as low-dose continuous inflow and overflow (McBride, 1995). The largest metal input is lead due to its frequency and the frequent increase in lead content in contaminated waterways. Lead is a nonessential metal. It is a substance accumulated by most aquatic organisms and not regulated by them. It has a toxic effect at a lower concentration than many other metal pollutants. Heavy metals such as nickel (Ni), chromium (Cr), mercury (Hg), cadmium (Cd), thallium (Tl), and lead are harmful in nature as metal complexes as well as in the elemental form (Agnihotri et al., 2006). Due to their soluble nature related to the aqueous environments, they can easily be absorbed by living beings. The previous studies have spotted heavy metals in the liver, gills, and muscle tissues of several fish in the polluted marine ecosystem (Sobhanardakani et al., 2011). Just in case heavy metals become part of the food chain, they will be going to eventually accumulate within the human body (Barakat, 2011). Because these metals are widely utilized in various industries, they have effect on human beings working in these types of industries. Heavy metals exceeding the allowable limit will usually cause an adverse impact on humans, organisms, and the environment as well (Gholamreza & Mehrzad, 2016). Heavy metals that are allowed in food samples, their safety limits are associated with lower risks of human health. The level of toxicity for certain selected metals to the human body follows the following order: Co < Al < Cr < Pb < Ni < Zn < Cu < Cd < Hg (Gholamreza & Mehrzad, 2016). The damaging impacts of heavy metals on the human body depend on their rate of emission, dosage, and exposure time. In current decades, certain heavy metals have taken more and more consideration, including Cd, Hg, and Pb (Valavanidis & Vlachogianni, 2010). The contrary impact of mercury and mercury compounds on human health includes probable carcinogens; causing damage to the kidneys, brain, and lungs; damage to the growing fetus; high heart rate or hypertension; diarrhea and vomiting; eye irritation and skin rashes. In drinking water, the limit of mercury according to US Environmental Protection

Agency (EPA) is about 2 ppb. The suggested WHO safe limits regarding Hg in soils for agriculture as well as in wastewater are 0.05 and 0.0019 ppm, respectively (Onuegbu et al., 2013). In children, chronic cadmium toxicity includes damage to the kidneys, respiratory system, cardiovascular system, and bones similarly the growth of lung, kidney, prostate and gastric cancer (Lim et al., 2010; Satarug & Moore, 2004). People's exposure to cadmium includes smoking, consuming polluted food, and working in cd-contaminated workplaces and major processing industries that process metal. It was found according to a study carried out in Iran that in canned fish samples the cadmium content was much higher than the maximum permissible limit, which was caused by the release of pollutants rich in heavy metals into the aquatic ecosystem (Soheil, 2017). The US EPA's regulatory restriction of cadmium in drinking water is 5 ppb or 0.005 ppm (Gholamreza & Mehrzad, 2016). The WHO suggested safe limitations of cadmium in both soil for agriculture and wastewater is 0.003 ppm (Soheil, 2017). Intake of polluted dust and aerosols or ingestion of polluted water and food may cause exposure to lead. Lead poisoning can cause damage to the liver, kidneys, heart, brain, nervous system, and bones. The first signs of exposure to lead poisoning may include drowsiness, headache, irritability, and memory loss. It can also cause anemia and disturbances in the synthesis of hemoglobin. Long-term exposure to minimal levels of lead can reduce the intelligence of children. It was reported by the International Agency for Research on Cancer (IARC) that lead may be a human carcinogen. The permissible limit of lead according to US EPA in drinking water is 15 ppb. Safe limits of lead recommended by WHO in soils used for agriculture and wastewater and are 0.1 and 0.01 ppm, respectively. Chromium is used widely in electroplating, metallurgy, and the manufacture of pigments, paints, preservatives, paper, and pulp, etc. (Jaishankar et al., 2014). The insertion of chromium into the atmospheres is usually via fertilizer and wastewater (Ghani, 2011). Zn, Ca, Pb, and Sr chromates such as hexavalent chromium compounds are extremely toxic, water soluble, and carcinogenic (Wolińska et al., 2013). Additionally, chromium compounds have been related with ulcers that are slow-healing. There are reports as well that chromate-based compounds can cause damage to cells (Matsumoto et al., 2006; O'Brien et al., 2001). World Health Organization recommends that the safety limits of hexavalent chromium in agricultural soil and wastewater to be 0.1 and 0.05 ppm, respectively. Al is a pure form of odorless, soft blue-white metal, which when exposed to air will oxidize. The sources of Al include optical glasses, electronic equipment, mercury lamps, semiconductors, etc. Human beings are subjected to Al by means of inhalation, ingestion, and skin contact. The lethal toxicity of Al ranges from 6 up to 40 mg/kg. Some of the common health repurcursions include anorexia, poisoning, vomiting, abdominal pain, gastrointestinal bleeding, polyneuropathy, renal failure, alopecia, skin rash, mood changes, seizures, autonomic dysfunction, cardiotoxicity, and coma. The endorsed safety limit in China, for drinking water, is about 0.0001 ppm (Jiang et al., 2018). The document recommended by WHO does not give a safe limit of Al in wastewater and agricultural soils. Ni a silvery metal used to make, electronics, stainless steel, and coins. About 150,000 to 180,000 tons of nickel is estimated to be released into the environment each year worldwide. Nickel is exposed to human beings via air, water, and food. Past studies have indicated that compared to inhalation and skin routes, ingestion of nickel-contaminated dust is the primary route for residents to ingest heavy metals. After exposure to nickel, the nickel content of a person's urine and tissues may increase. The adverse influences of Ni on human health can comprise allergies, dermatitis, cancer of the respiratory tract and organ diseases. The safe limits recommended by WHO for nickel in agricultural soils and wastewater are 0.05 and 0.02 ppm, respectively. Wastewater coming from factories may have heavy metals that, by time, will gather in soil sediments alongside sewage canals and in the organisms inhabiting these canals (Jaishankar et al., 2014). It is often possible to expose people to polluted wastewater, especially in densely populated urban regions or when it is reused for different agricultural activities. However, prior studies have indicated that reusing wastewater efficiently is a foremost challenge for several countries around the globe.

The other major water pollutants include organic and inorganic compounds. It has been found that inorganic and organic chemicals can stimulate the production of free radicals (Yadav et al., 2019). Concerns about persistent organic pollutants (POPs) are confirmed. These substances generally exist in the ocean at very low concentrations. Due to their persistence, they accumulate in the tissues of aquatic organisms at high concentrations. Genotoxic compounds (such as POPs) can induce mutagenesis in aquatic organisms. Pesticides enter waterways with agricultural and urban waste. These toxic compounds are known to hydrolyze rapidly in the environment (Walker et al., 1997). Pollutants are currently dispersed in the environment (Bharagava et al., 2018). Pesticides are any substance or mixture of substances intended to prevent or destroy, including insecticides, herbicides, fungicides and various other substances used to control pests (Bonelli et al., 2017). They persist in the environment for a long time and can accumulate and transfer from one type to another throughout the food chain (Jones & De Voogt, 1999; S. & P., 1999).

12.3.2 Catalytic degradation of contaminants by using biomass-derived nanocomposites

Due to the increase in the world's population and the continuous development of industry and agriculture, the world has observed an increasing trend of organic pollutants in water sources. Wastewater containing high concentrations of the recyclable waste products can be handled by different biological treatment processes. However, wastewater coming from a number of industries for example pharmaceutical (Olama et al., 2018), textile (Touati et al., 2016), and agricultural (Kushniarou et al., 2019) frequently comprises toxic pollutants having low biodegradability. In municipal wastewater, groundwater and surface water treatment plants, widespread utilization of these toxic pollutants and partial removal from wastewater are common (Lee et al., 2017). Wastewater contains complex persistent organic pollutants; if detected they have a low ratio of biological oxygen demand (BOD) to chemical oxygen demand (COD) (Zhang et al., 2014).

Over the past two decades, due to existence of semiconductor oxides for example zinc oxide (ZnO), titanium dioxide (TiO_2), zinc sulfide (ZnS), cadmium sulfide (CdS), iron oxide (Fe_2O_3), nickel, etc., the catalytic degradation of organic pollutant has been common. Along with cheaper supports and materials such as biomass-derived nanocomposites these catalysts have reduced cost, reduced toxicity, and high chemical and physical stability, which can mineralize and degrade organic pollutants into harmless inorganic anions, water (H_2O), and carbon dioxide (CO_2) (Szczepanik, 2017). A large amount of the semiconductors unfortunately have wider energy bandgap (E_g) for instance TiO_2 (3.2 eV) (Szczepanik, 2017) and ZnO (3.37 eV) (Qi et al., 2017). For these semiconductors, catalytic efficiency is usually obtained by sufficient light absorption. The absorption of light can give excitation to electrons (e) and they jump to the conduction band from the valence band, leaving holes (h +) in the valence band to initiate a reaction known as photo-redox. It means that stronger ultraviolet rays should be used, which only make up 3%–5% of natural sunlight (Uyguner-Demirel et al., 2017). In past years, people have given attention to a greater extent using waste biomass as carbonaceous materials to produce biochar or activated carbon (AC), because it solves both problem waste disposal as well as reduction in cost of materials (Tan et al., 2017). These favorable conditions suggest using alternative carbon materials. Therefore, understanding the growth of activated biochar as a newfound potentially profitable and ecologically friendly carbonaceous substance is necessary to optimize and extend conditions of the process.

Vinayagam et al. developed zinc oxide/waste biomass AC nanocomposite (ZnO/WBAC) through hydrothermal process (Vinayagam et al., 2018). The synthesized nanocomposite was characterized by using FT-IR, XRD, UV-Vis Diffuse Reflectance Spectroscopy (UV-DRS), High-Resolution Scanning Electron Microscopy (HRSEM), and Energy Dispersive X Ray Analysis (EDAX). The ZnO/WBAC nanocomposite exhibited photocatalytic degradation of orange G dye up to 95% under-optimized conditions. Further, study reports that neutral pH conditions are suitable for photocatalytic activity of ZnO/WBAC nanocomposite. However, the activity decreases with the addition of external electrolytes. Biochar (Arash et al., 2015) incorporated Zn-Co-layered double hydroxide (LDH) (Zn-Co-LDH@BC) nanocomposite degraded one of the pharmaceutical water pollutants, the Gemifloxacin, up to 92.7%. Li et al. fabricated carbon nanofibers/gold nanoparticles (CNFs/Au) nanocomposites involving a two-step method, noncovalent functionalization of CNFs with polysiloxane, followed by noncovalent assembly of Au nanoparticles on the functionalized CNFs (Li et al., 2015). The nanocomposite was used to degrade one of the strongest organic pollutant, 4-nitrophenol. The catalytic degradation reduced initial concentration of 4-nitrophenol by 11% in 4 h, which is higher than the individual activities of CNFs or Au nanoparticles alone. Synthesis of nanocomposites of hemp-derived AC with zero-valent iron has been reported for the application of pulping effluent. Study reported that the synthesized nanocomposite removed COD of the pulping effluent by 87.74% (Mo et al., 2020). Khataee et al. synthesized TiO_2-biochar (TiO_2-BC) nanocomposites through sol-gel method for sonocatalytic degradation of reactive blue 69 (RB69) (Khataee et al., 2017). The study reports that optimized condition for the use of TiO_2-BC composite are at 7 pH, 1.5 g/L TiO_2-BC dosage, 20 mg/L RB69 initial concentration, and 300 W ultrasonic power. It is hypothesized that initially RB69 is oxidized to aromatic intermediates and aliphatics, which are attacked by hydroxyl OH radicals to convert them to H_2O and CO_2. Fig. 12.4 presents a detailed scheme of synthesis of TiO_2-BC, its characterization, and sonocatalytic application for degradation of RB69.

Furthermore, studies also indicate the use of biomass-based materials as green capping agents to synthesize nanocomposites and nanoparticles for wastewater treatment applications. For example, Ebrahimzadeh et al. synthesized $Fe_3O_4/SiO_2/Cu_2O$–Ag nanocomposites through extract of crataegus pentagyna fruit by using a facile and rapid sonochemical process (Ebrahimzadeh et al., 2019). The prepared nanocomposite showed desirable magnetic characteristics, morphology, and size for photocatalytic catalysis. The synthesized nanocomposite was successfully used for photocatalytic degradation of methylene blue (MB) and rhodamine B under visible light irradiation. Similarly, Mohammod et al. reported synthesis of NiO nanoparticles through *Annona muricata* L. fruit peel waste (Aminuzzaman, 2021). The synthesized nanoparticles were characterized through XRD, Energy Dispersive X-Ray Analysis (EDX), Raman spectroscopy, X-ray photoelectron spectroscopy (XPS), UV–vis spectroscopy, FT-IR spectroscopy, and Field Emission Scanning Electron Microscopy (FESEM) and applied on photocatalytic degradation of crystal violet dye. Authors reported 99.0% photodegradation of crystal violet dye at 105 min of illumination.

12.3.3 Adsorption of organic and inorganic pollutants by using biomass-derived nanocomposites

Biochar-based nanocomposites exhibit a strong affinity for organic pollutants. Various biochar-based nanocomposites adsorption capacities related to the organic contaminants are reported in various studies, such as methylene blue, crystal violet, (Inyang et al., 2014, 2015; Zhou et al., 2013) phenanthrene phenol (Agnihotri et al., 2006; Yao, Gao, Chen, & Yang, 2013), p-nitrotoluene (p-NT) (B. Chen et al., 2011), sulfapyridine (Inyang et al., 2015), naphthalene (NAPH), etc. The adsorption capacities of various nanocomposites are tabulated in Table 12.2. The interaction between organic pollutants and biochar functional groups can involve several adsorption mechanisms, including p–p interaction, hydrogen bonding, electrostatic attraction, and hydrophobicity interaction. Furthermore, nanoparticles coating on the carbon materials can improve adsorption of

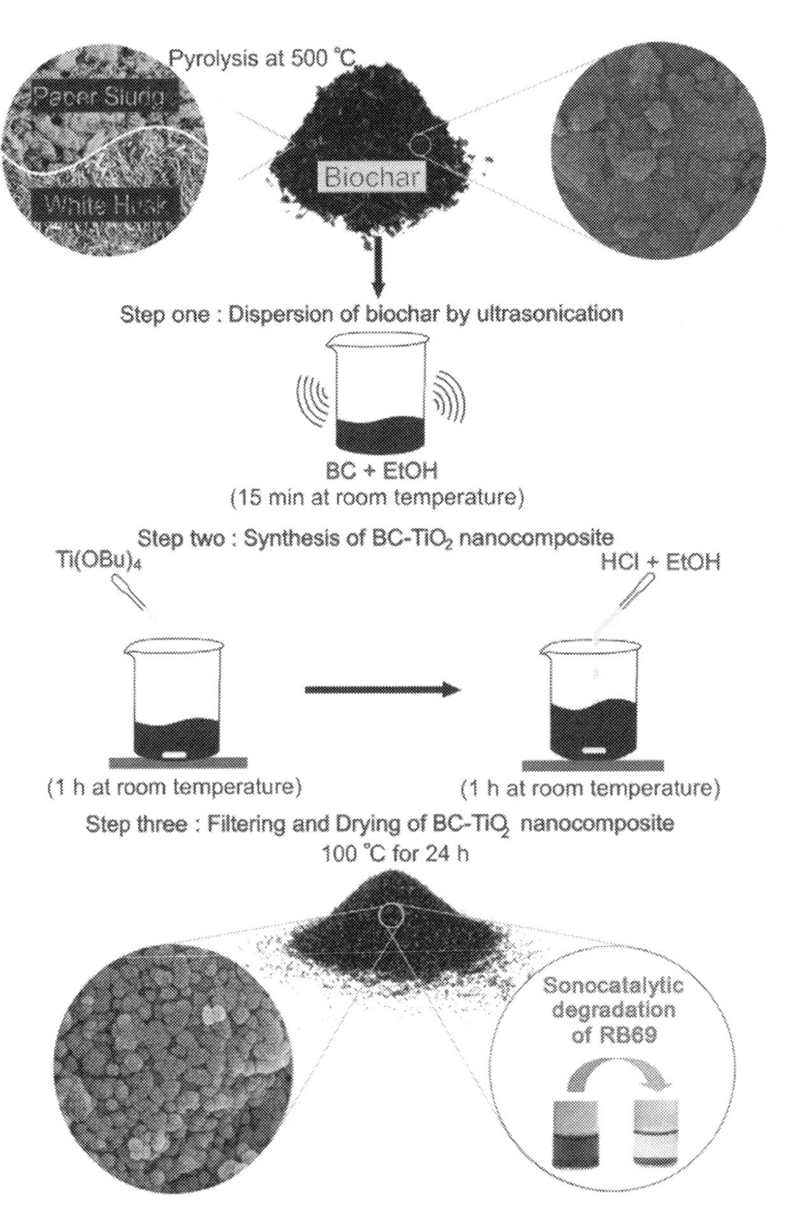

Figure 12.4 Sonocatalytic degradation of an anthraquinone dye using TiO_2-biochar nanocomposite. From Khataee, A., Kayan, B., Gholami, P., Kalderis, D., & Akay, S. (2017). Ultrasonics Sonochemistry, 39, 120–128. *https://doi.org/10.1016/j.ultsonch.2017.04.018* Synthesis of TiO_2-biochar nanocomposite, its characterization and sonocatalytic degradation of reactive blue 69 dye.

Table 12.2 Details of various nanocomposites for adsorption of heavy metals.

Feedstock	Nanomaterials	Contaminants	Q_{max} (mg/g)	BET surface area (m^2/g)	Adsorbent dosage (g/L)	Isotherm	Reference
Kans grass	Fe_3O_4	As(III)	2.004 (L)	31.45		L	Baig, S.A et al., 2014
Hickory chips	Iron hydroxide particles	As(V)	2.16 (L)	16.0	2	T	Hu, X et al., 2015
Cotton Wood	c-Fe_2O_3 particles	As(V)	3.147 (L)		2	L	M. Zhang et al., 2013
Kans grass	Fe_3O_4	As(V)	3.132 (L)	31.45		L	Baig, S.A et al., 2014
Loblolly pine wood	c-Fe_2O_3 particles	As(V)	0.4287 (L)	193.1	2.5	L	Wang et al., 2015
Pine bark waste	Magnetic $CoFe_2O_4$	Cd(II)	17.793 (L–F)			L–F	Reddy, D.H.K and Lee, S.M., 2014
Walnut shell	Chitosan	Cr(VI)	124.42 (L)	82.09, N_2	2	F	Huang, X., 2016
Rice husks	Iron oxide	As(V)	1.46 (L)	220, N_2		L	Cope, C.O, 2014
Hickory wood	MnO_x	Cd(II)	28.104 (L)	205, N_2	2	R–P	Wang et al., 2015
Rice husks	Iron oxide	As(V)	1.15 (L)	77.3, N_2		L	Cope, C.O, 2014
Eichhornia crassipes	Chitosan and c-Fe_2O_3	Cr(VI)	151.62 (L)	90.78, N_2	1	L	Fan, R., 2015
Sugarcane bagasse	ZnO	Cr(VI)	84.87 (L)	21.28, N_2	2	F	Chao Gan et al., 2015
Wheat straw	Graphene	Hg(II)	0.853 [115]	17.3	0.36	BET	Tang, J. et al., 2015
Corn straws	MnO_x	Cu(II)	160.3 (L)		5	L	Z. Song et al,. 2014
Hickory wood	MnO_x	Cu(II)	34.205 (L)	205, N_2	2	R–P	Wang et al., 2015
Hickory wood	MnO_x	Pb(II)	153.109 (L)	205, N_2	2	R–P	Wang et al., 2015
Bamboo	Chitosan	Pb(II)	14.3 (L)	166.9, N_2	2	L	Y. Zhou et al., 2013
Hickory chips	Carbon nanotubes (CNTs)	Pb(II)	31.05 (L–F)	359, N_2	2	L–F	M. Inyang et al., 2015
Hickory wood	MnOx	Pb(II)	153.109 (L)	205, N_2	2	R–P	Wang et al., 2015

organic pollutants present in water bodies (Sun et al., 2015). Other related functional nanoparticles (e.g., chitosan, ZnS, CNT, graphene, graphene oxide, nanocomposites, and LDH) coated on biochar surface can occupy a large area, additional functional groups, and have high efficiency of biochar, higher thermal stability, and higher biochar efficiency. The mechanism is closely related to the nature of the coated functional group. It is described that higher adsorption of organic pollutants by graphene-coated biochar is primarily ascribed to a strong interaction between molecules of organic pollutant molecules and graphene sheets on surface of biochar by p–p interactions. Similarly, the p–p interaction and hydrophobic interaction between organic pollutants on CNTs and graphite/hydrophobic sites are the main adsorption mechanisms that cause the removal of organic pollutants.

Ai and Li designed and synthesized biomass-derived carbon is supported on montmorillonite nanocomposites (MMT@C) through hydrothermal process (Ai & Li, 2013). The developed nanocomposites were applied for the adsorption of MB. The adsorption capacity determined through Langmuir equation was 194.2 mg/g. Inyang et al. fabricated hybrid multiwalled CNT (C)-coated biochar nanocomposites (Inyang et al., 2014). The study used untreated hickory (HC) and bagasse (Arash et al., 2015) biochars with CNTs for synthesis of biochar composites. At 1% being the limit for activating agent and remainder biomass, the highest removal efficiencies were found as 47% removal for HC—1% and 64% removal for BC—1%. In the process of removing inorganic pollutants, nanomaterials have two main effects on biochar: improving the surface area of biochar and the impregnation of nanoparticles on the surface of the biochar, which will considerably increase the adsorption of inorganic water pollutants (Jung & Ahn, 2016). Studies have shown that colloidal or nanoscale nanomaterials dispersed on biochar are the main reaction sites that bind to inorganic pollutants (Jung et al., 2015; Zhang & Gao, 2013). A number of interactions between inorganic contaminants and biochar matrix nanoparticles to explain potential adsorption mechanisms have been proposed, including precipitation, electrostatic attraction, anion exchange, and hydrogen bonding (Yao, Gao, Chen, Zhang, et al., 2013), through prevention of the accumulation of contaminants in the biochar matrix (Devi & Saroha, 2015).

Earlier, Zhang et al. (2012) synthesized MgO-biochar nanocomposites from magnesium chloride hexahydrate ($MgCl_2 \cdot 6H_2O$) and various biomass feedstocks such as sugar beet tailings (SBTs), sugarcane bagasse, cottonwoods (CWs), pine woods (PWs), and peanut shells (PSs). The synthesis was carried out by using pyrolysis process. The MgO content varied from 8.3 to 26.1 for various biomass feedstocks. For the removal of phosphate and nitration from aqueous solutions the synthesized nanocomposites were used. Study suggests that highest content of phosphate removed was up to 66.7% by sugar beet tailings-based MgO (SBT-MgO) biomass nanocomposite, whereas peanut shells (PS)-based nanocomposite removed the highest percentage of nitrate up to 11.7%. Mg-enriched plant tissues as the Mg source for synthesis of Mg-based biomass composites can also be used. Another study reported the synthesis of clay-biochar nanocomposites, where MMT and/or Kaolin powder used as clay feedstocks and biomass feedstocks included bamboo, bagasse, and HC chips. The composite was used for removal of MB. The nanocomposites of bagasse and MMT, which contained 13.13% of MMT, exhibited a highest percentage removal of MB up to 84.3% (Yao et al., 2014). Kirti et al. (2018) synthesized multifunctional superparamagnetic nanocomposites by using *C. fistula* and aloe vera as the biomass precursors and ferric nitrate as the iron nanoparticles source. The synthesized composites were used for the removal of Congo red and MB. Both the nanocomposites exhibited comparable performance for dye removal, which was above 80%.

12.3.4 Heavy metal removal by using biomass-derived nanocomposites

For wastewater purification, the removal of heavy metals through biochar-based nanocomposites has always got a significant interest in research society. For various biochar nanocomposites, the adsorption capacity for Pb(II) is in the range of 4.913 to 367.65 mg/g, while for the Langmuir of As(V) maximum capacity is in the ranges of 0.4287–3.147 mg/g. Magnetic ZnS/biochar composites have the best performance and their synthesis involves the deposition of nanocrystals of ZnS on surface of magnetic biochar (Yan et al., 2015). To illustrate equilibrium of adsorption of heavy metal on biochar nanocomposites and to analyze experimental data, a number of isothermal models have been widely used, among them the Freundlich, Langmuir, Temkin, Langmuir–Freundlich, Redlich–Peterson, and Brunauer-Emmett-Teller (BET) are most popular and used widely. To define equilibrium of heavy metal adsorption through biochar nanocomposites, all these equations can be used according to the facts by the data collected, (Table 12.2), the Freundlich, Langmuir, and Langmuir–Freundlich are used more frequently among them. Likewise, to analyze the adsorption of pollutants by means of biochar-based nanocomposites, certain kinetic models such as pseudo-first-order, pseudo-second-order, etc., have been used. The outcomes of kinetic and isothermal studies may deviate considerably subjected to target heavy metals and properties of the biochar-based nanocomposites as well as how they are synthesized. In addition, various procedures are required for the adsorption of heavy metals using several biochar-based nanocomposites. Their functions are manifested mainly in the following viewpoints: impact on the biochar's surface functional groups; either worsen or improve biochar pore properties, and act themselves as an active site. Relative to the original biochar, after the introduction of nanomaterials, the number of functional groups on the surface of the biochar can increase (e.g., carboxyl, COOH; and carboxylate, R–COO−, and hydroxyl, –OH) (Song et al., 2014). Improved functional groups can provide more available binding sites for heavy metals. These functional groups can enhance the interaction between biochar and heavy metals by forming surface complexes. Cation-p bonds, electrostatic attraction, and ion exchange contribute to higher adsorption capacity. Nanomaterials can also increase the surface area of biochar, which can promote the adsorption process. However, several research reports claim that nanomaterials have ability to strongly fill and obstruct pores and shrink volume of biochar and pore area. This indicates that the impact of nanomaterials on pore properties of biochar possibly will not be core determinant of heavy metal adsorption of biochar. Magnetic iron oxide (for instance c-Fe_2O_3, $CoFe_2O_4$, and Fe_3O_4), nanometal oxide or nanometal hydroxide (for instance ZnO, MnOx, MgO, CaO, Al_2O_3, AlOOH, and $Mg(OH)_2$), and further functional nanoparticles (e.g., chitosan, CNTs, ZnS NC, graphene, LDH, and GO) diffused upon the surface of biochar be able to react with heavy metals, causing biochar for heavy metals to have higher capacity of adsorption. For example, some studies report that an improved adsorption efficiency of several heavy metals' removal is ascribed to development of surface complexes with magnetic iron oxides and nanometal oxides/hydroxides. The analysis of Pb(II) adsorption on coated biochar as well as chitosan-modified biochar shows the adsorption of lead on biochar surface occurs due to interaction with the amine-functional group of chitosan and complexing with CNTs. Oxygen-containing functional groups generally containing only the functional groups and nanoparticles upon the top of biochar, can function as core sites for adsorption of heavy metal ions in aqueous solution. This kind of functional material may be used to simultaneously coat different functional nanoparticles on the biochar to play a synergistic effect on the removal of heavy metals. By depositing ZnS (NC) nanocrystals on the magnetic biochar in a polyol solution, biochar/ZnS magnetic composite material has been successfully synthesized, which has excellent superparamagnetic separation and separation performance. As compared to reported magnetic biochar the adsorption capacity of Pb(II) is about 10 times higher. Among them in this composite material, the coating of magnetic iron oxide and ZnS NCs will induce the magnetic properties on the biochar and the adsorption sites of nanoparticles. Bringing together the advantages of biochar, chitosan, and c-Fe_2O_3, a magnetic field was successfully prepared that was chitosan modified. Some of the characteristics of biochar are low operating cost, easy availability of biomass resources, and easy access to groups. It has strong magnetism and strong adsorption of Cr (VI) capacity. Cui et al. (2019) reported synthesis of biomass-derived magnetic nanocomposites for As(III) removal. Study reported the maximum As(III) removal of up to 16.23 mg/g. Table 12.2 presents feedstocks for various nanocomposites and their adsorption capacities for removal of heavy metals.

12.4 Conclusions

Advanced green nanocomposite materials have emerged for the applications of wastewater treatment. Several biomasses such as kans grass, HC chips, cotton wood, pine wood, loblolly pine woone bark waste, walnut shell, rice husks, sugarcane bagasse, wheat straw, corn straws, and several others have been used as the green feedstocks for advanced nanocomposite synthesis. In combination with biomass, metallic as well as nonmetallic elements or compounds have been used as the nanoparticles for the development of the nanocomposites. Surface functional groups of biochar are greatly impacted by the combination of biomass and nanoparticles, as a result improving or worsening the pore properties of biochar and acting themselves as an active site. Several studies reported the comparative effectiveness of various biomass-derived nanocomposites. The major applications include removal of organic and inorganic water contaminants and heavy metals through adsorption processes and catalytic degradation of organic pollutants. More specifically, common applications of nanocomposites are removal of dyes, heavy metals, chlorinated organic compounds, pesticide and polycyclic aromatic hydrocarbons, and bacteria.

References

Agnihotri, S., Mota, J.P.B., Rostam-Abadi, M., Rood, M.J., 2006. Adsorption site analysis of impurity embedded single-walled carbon nanotube bundles. Carbon 44 (12), 2376–2383. https://doi.org/10.1016/j.carbon.2006.05.038.

Ahmed, F.K.S., Mubarak, N.M., Hua Tan, Y., Khalid, M., Rama Rao, K., Rashmi, W., Ezzat Chan, A., Sabzoi, N., Mazari, S.A., 2021. A comprehensive review on magnetic carbon nanotubes and carbon nanotube-based buckypaper-heavy metal and dyes removal. J. Hazard. Mater., 125375.

Ahmad, A., Mohd-Setapar, S.H., Chuong, C.S., Khatoon, A., Wani, W.A., Kumar, R., Rafatullah, M., 2015. Recent advances in new generation dye removal technologies: novel search for approaches to reprocess wastewater. RSC Adv. 5 (39), 30801–30818. https://doi.org/10.1039/c4ra16959j.

Ahmad, M., Rajapaksha, A.U., Lim, J.E., Zhang, M., Bolan, N., Mohan, D., Vithanage, M., Lee, S.S., Ok, Y.S., 2014. Biochar as a sorbent for contaminant management in soil and water: a review. Chemosphere 99, 19–33. https://doi.org/10.1016/j.chemosphere.2013.10.071.

Ai, L., Li, L., 2013. Efficient removal of organic dyes from aqueous solution with ecofriendly biomass-derived carbon@montmorillonite nanocomposites by one-step hydrothermal process. Chem. Eng. J. 223, 688–695. https://doi.org/10.1016/j.cej.2013.03.015.

Aminuzzaman, M., 2021. Biosynthesis of NiO nanoparticles using soursop (*Annona muricata* L.) fruit peel green waste and their photocatalytic performance on crystal violet dye. J. Cluster Sci. 32, 949–958.

Arash, B., Park, H.S., Rabczuk, T., 2015. Mechanical properties of carbon nanotube reinforced polymer nanocomposites: a coarse-grained model. Compos. B: Eng. 80, 92–100. https://doi.org/10.1016/j.compositesb.2015.05.038.

Atchudan, R., Edison, T.N.J.I., Perumal, S., Vinodh, R., Lee, Y.R., 2018. In-situ green synthesis of nitrogen-doped carbon dots for bioimaging and TiO_2 nanoparticles@nitrogen-doped carbon composite for photocatalytic degradation of organic pollutants. J. Alloys Compd. 766, 12–24. https://doi.org/10.1016/j.jallcom.2018.06.272.

Awual, M.R., 2016. Assessing of lead(III) capturing from contaminated wastewater using ligand doped conjugate adsorbent. Chem. Eng. J. 289, 65–73. https://doi.org/10.1016/j.cej.2015.12.078.

Awual, M.R., Yaita, T., Shiwaku, H., Suzuki, S., 2015. A sensitive ligand embedded nano-conjugate adsorbent for effective cobalt(II) ions capturing from contaminated water. Chem. Eng. J. 276, 1–10.

Awual, M.R., Yaita, T., Taguchi, T., Shiwaku, H., Suzuki, S., Okamoto, Y., 2014. Selective cesium removal from radioactive liquid waste by crown ether immobilized new class conjugate adsorbent. J. Hazard. Mater. 278, 227–235. https://doi.org/10.1016/j.jhazmat.2014.06.011.

Baig, S.A, Zhu, J., Muhammad, N., Sheng, T., Xu, X., 2014. Effect of synthesis methods on magnetic Kans grass biochar for enhanced As (III, V) adsorption from aqueous solutions.. Biomass and Bioenergy 71.

Barakat, M.A., 2011. New trends in removing heavy metals from industrial wastewater. Arab. J. Chem. 4 (4), 361–377. https://doi.org/10.1016/j.arabjc.2010.07.019.

Batool, S., Idrees, M., Ahmad, M., Ahmad, M., Hussain, Q., Iqbal, A., Kong, J., 2020. Design and characterization of a biomass template/SnO_2 nanocomposite for enhanced adsorption of 2,4-dichlorophenol. Environ. Res. 181,, 108955. https://doi.org/10.1016/j.envres.2019.108955.

Bharagava, R.N., Saxena, G., Mulla, S.I., Patel, D.K., 2018. Characterization and identification of recalcitrant organic pollutants (ROPs) in tannery wastewater and its phytotoxicity evaluation for environmental safety. Arch. Environ. Contamin. Toxicol. 75 (2), 259–272. https://doi.org/10.1007/s00244-017-0490-x.

Bonelli, M.G., Ferrini, M., Manni, A., 2017. Artificial neural networks to evaluate organic and inorganic contamination in agricultural soils. Chemosphere 186, 124–131. https://doi.org/10.1016/j.chemosphere.2017.07.116.

Briggs, A.M., Cross, M.J., Hoy, D.G., Sànchez-Riera, L., Blyth, F.M., Woolf, A.D., March, L., 2016. Musculoskeletal health conditions represent a global threat to healthy aging: a report for the 2015 World Health Organization world report on ageing and health. Gerontologist 56, S243–S255. https://doi.org/10.1093/geront/gnw002.

Chao G., Liu, Y., Tan, X., Wang, S, Zeng, G., Heng, B., Li, T., Jiang, Z, Liu, W., et al., 2015. Effect of porous zinc–biochar nanocomposites on Cr (VI) adsorption from aqueous solution. RSC Advances 5 (44). doi:10.1039/C5RA04416B.

Chen, B., Chen, Z., Lv, S., 2011. A novel magnetic biochar efficiently sorbs organic pollutants and phosphate. Bioresour. Technol. 102 (2), 716–723. https://doi.org/10.1016/j.biortech.2010.08.067.

Chen, L., Ji, T., Mu, L., Shi, Y., Brisbin, L., Guo, Z., Khan, M.A., Young, D.P., Zhu, J., 2016. Facile synthesis of mesoporous carbon nanocomposites from natural biomass for efficient dye adsorption and selective heavy metal removal. RSC Adv. 6 (3), 2259–2269. https://doi.org/10.1039/c5ra19616g.

Chen, T., Zhang, J., Shi, P., Li, Y., Zhang, L., Sun, Z., He, R., Duan, T., Zhu, W., 2018. *Thalia dealbata* inspired anisotropic cellular biomass derived carbonaceous aerogel. ACS Sustain. Chem. Eng. 6 (12), 17152–17159. https://doi.org/10.1021/acssuschemeng.8b04528.

Cope, C.O, Webster, D.S., Sabatini, D.A., 2014. Arsenate adsorption onto iron oxide amended rice husk char. Sci. Total Environ. 488.

Cui, J., Jin, Q., Li, Y., Li, F., 2019. Oxidation and removal of As(III) from soil using novel magnetic nanocomposite derived from biomass waste. Environ. Sci. Nano 6 (2), 478–488. https://doi.org/10.1039/c8en01257a.

Devi, P., Saroha, A.K., 2015. Simultaneous adsorption and dechlorination of pentachlorophenol from effluent by Ni-ZVI magnetic biochar composites synthesized from paper mill sludge. Chem. Eng. J. 271, 195–203. https://doi.org/10.1016/j.cej.2015.02.087.

Ebrahimzadeh, M.A., Mortazavi-Derazkola, S., Zazouli, M.A., 2019. Eco-friendly green synthesis and characterization of novel $Fe_3O_4/SiO_2/Cu_2O$–Ag nanocomposites using *Crataegus pentagyna* fruit extract for photocatalytic degradation of organic contaminants. J. Mater. Sci. 30 (12), 10994–11004. https://doi.org/10.1007/s10854-019-01440-8.

Fan, R., Luo, J., Yan, S., Wang, T., Liu, L., Gao, Y., Zhang, Z., 2015. Use of water hyacinth (Eichhornia crassipes) compost as a peat substitute in soilless growth media. Compost Science & Utilization 23 (4). https://doi.org/10.1080/1065657X.2015.1046614.

Fergusson, J.E., 1990. The Heavy Elements: Chemistry, Environmental Impact and Health Effects. Pergamon Press, Oxford, UK.

Gan, S., Zakaria, S., Chia, C.H., Kaco, H., 2018. Effect of graphene oxide on thermal stability of aerogel bio-nanocomposite from cellulose-based waste biomass. Cellulose 25 (9), 5099–5112. https://doi.org/10.1007/s10570-018-1946-5.

Ghani, A.G.A., 2011. Effect of chromium toxicity on growth, chlorophyll and some mineral nutrients of *Brassica juncea* L. Egypt. Acad.J. Biol. Sci. Bot. 9–15. https://doi.org/10.21608/eajbsh.2011.17007.

Gholamreza, M., Mehrzad, M., 2016. Examination of the level of heavy metals in wastewater of Bandar Abbas wastewater treatment plant. Open J. Ecol. 55–61. https://doi.org/10.4236/oje.2016.62006.

Gitis, V., Hankins, N., 2018. Water treatment chemicals: trends and challenges. J. Water Process Eng. 25, 34–38. https://doi.org/10.1016/j.jwpe.2018.06.003.

Gupta, K., Kaushik, A., Tikoo, K.B., Kumar, V., Singhal, S., 2020. Enhanced catalytic activity of composites of $NiFe_2O_4$ and nano cellulose derived from waste biomass for the mitigation of organic pollutants. Arab. J. Chem. 13 (1), 783–798. https://doi.org/10.1016/j.arabjc.2017.07.016.

Hodges, B.C., Cates, E.L., Kim, J.H., 2018. Challenges and prospects of advanced oxidation water treatment processes using catalytic nanomaterials. Nat. NanoTechnol. 13 (8), 642–650. https://doi.org/10.1038/s41565-018-0216-x.

Hu, X., Ding, Z., Zimmerman, A.R., Wang, S., Gao, B., 2015. Batch and column sorption of arsenic onto iron-impregnated biochar synthesized through hydrolysis. Water Res. 68.

Huang, X., Liu, Y., Liu, S., Tan, X., Ding, Y., Zeng, G., Zhou, Y., Zhang, M., Wang, S., Zheng, B., 2016. Effective removal of Cr (VI) using β-cyclodextrin–chitosan modified biochars with adsorption/reduction bifuctional roles. RSC Adv. 6 (1). doi:10.1039/C5RA22886G.

Hussain, F., Hojjati, M., Okamoto, M., Gorga, R.E., 2006. Review article: polymer-matrix nanocomposites, processing, manufacturing, and application: an overview. J. Compos. Mater. 40 (17), 1511–1575. https://doi.org/10.1177/0021998306067321.

Inyang, M., Gao, B., Zimmerman, A., Zhang, M., Chen, H., 2014. Synthesis, characterization, and dye sorption ability of carbon nanotube-biochar nanocomposites. Chem. Eng. J. 236, 39–46. https://doi.org/10.1016/j.cej.2013.09.074.

Inyang, M., Gao, B., Zimmerman, A., Zhou, Y., Cao, X., 2015. Sorption and cosorption of lead and sulfapyridine on carbon nanotube-modified biochars. Environ. Sci. Pollut. Res. 22 (3), 1868–1876. https://doi.org/10.1007/s11356-014-2740-z.

Jaishankar, M., Tseten, T., Anbalagan, N., Mathew, B.B., Beeregowda, K.N., 2014. Toxicity, mechanism and health effects of some heavy metals. Interdiscip. Toxicol. 7 (2), 60–72. https://doi.org/10.2478/intox-2014-0009.

Jiang, Y., Xie, H., Zhang, H., Xie, Z., Cao, Y., 2018. Dissolved heavy metals distribution and risk assessment in the Le'an river subjected to violent mining activities. Polish J. Environ. Stud. 27 (4), 1559–1572. https://doi.org/10.15244/pjoes/77033.

Jones, K.C., De Voogt, P, 1999. Persistent organic pollutants (POPs): State of the science. Environ. Pollut., 100, pp. 209–221.

Jung, K.W., Ahn, K.H., 2016. Fabrication of porosity-enhanced MgO/biochar for removal of phosphate from aqueous solution: application of a novel combined electrochemical modification method. Bioresour. Technol. 200, 1029–1032. https://doi.org/10.1016/j.biortech.2015.10.008.

Jung, K.W., Jeong, T.U., Hwang, M.J., Kim, K., Ahn, K.H., 2015. Phosphate adsorption ability of biochar/Mg-Al assembled nanocomposites prepared by aluminum-electrode based electro-assisted modification method with $MgCl_2$ as electrolyte. Bioresour. Technol. 198, 603–610. https://doi.org/10.1016/j.biortech.2015.09.068.

Kamal, T., 2019. Aminophenols formation from nitrophenols using agar biopolymer hydrogel supported CuO nanoparticles catalyst. Polymer Testing 77, 105896. https://doi.org/10.1016/j.polymertesting.2019.105896.

Khan Saleem Ahmed, F., Mubarak, N.M., Mohammad, K., Walvekar, R., Abdullah, E.C., Awais, A., Karri, R.R., Harshini, P., 2021. Functionalized multi-walled carbon nanotubes and hydroxyapatite nanorods reinforced with polypropylene for biomedical application. Sci. Rep. 11 (1), 1–10.

Khataee, A., Kayan, B., Gholami, P., Kalderis, D., Akay, S., 2017. Sonocatalytic degradation of an anthraquinone dye using TiO_2-biochar nanocomposite. Ultrason. SonoChem. 39, 120–128. https://doi.org/10.1016/j.ultsonch.2017.04.018.

Kim, S.H., Kwak, S.Y., Sohn, B.H., Park, T.H., 2003. Design of TiO_2 nanoparticle self-assembled aromatic polyamide thin-film-composite (TFC) membrane as an approach to solve biofouling problem. J. Membr. Sci. 211 (1), 157–165. https://doi.org/10.1016/S0376-7388(02)00418-0.

Kirti, S., Bhandari, V.M., Jena, J., Bhattacharyya, A.S., 2018. Elucidating efficacy of biomass derived nanocomposites in water and wastewater treatment. J. Environ. Manage. 226, 95–105. https://doi.org/10.1016/j.jenvman.2018.08.028.

Kushniarou, A., Garrido, I., Fenoll, J., Vela, N., Flores, P., Navarro, G., Hellín, P., Navarro, S., 2019. Solar photocatalytic reclamation of agro-waste water polluted with twelve pesticides for agricultural reuse. Chemosphere 214, 839–845. https://doi.org/10.1016/j.chemosphere.2018.09.180.

Lee, C.M., Palaniandy, P., Dahlan, I., 2017. Pharmaceutical residues in aquatic environment and water remediation by TiO_2 heterogeneous photocatalysis: a review. Environ. Earth Sci. 76 (17). https://doi.org/10.1007/s12665-017-6924-y.

Li, J., Ni, Y.h., Liu, C., Zhang, L., 2015. Noncovalent assembly of the carbon nanofibers/Au nanocomposite and its application in 4-nitrophenol reduction. J. Cluster Sci. 26 (5), 1547–1556. https://doi.org/10.1007/s10876-015-0847-0.

Liang, G., Li, Y., Yang, C., Zi, C., Zhang, Y., Hu, X., Zhao, W., 2020. Production of biosilica nanoparticles from biomass power plant fly ash. Waste Manage. 105, 8–17. https://doi.org/10.1016/j.wasman.2020.01.033.

Lim, S.R., Lam, C.W., Schoenung, J.M., 2010. Quantity-based and toxicity-based evaluation of the U.S. Toxics Release Inventory. J. Hazard. Mater. 178 (1–3), 49–56. https://doi.org/10.1016/j.jhazmat.2010.01.041.

Lin, S., Huang, J., Chang, P.R., Wei, S., Xu, Y., Zhang, Q., 2013. Structure and mechanical properties of new biomass-based nanocomposite: castor oil-based polyurethane reinforced with acetylated cellulose nanocrystal. Carbohydr. Polym. 95 (1), 91–99. https://doi.org/10.1016/j.carbpol.2013.02.023.

Lin, K., Li, C., Tao, W., Huang, J., Wu, Q., Liu, Z., Liu, Z., Wang, D., Liu, X., et al., 2009. Electrochemical Synthesis and Electro-Optical Properties of Dibenzothiophene/Thiophene Conjugated Polymers With Stepwise Enhanced Conjugation Lengths. Frontiers in Chemistry 8.

Lingamdinne, L.P., Koduru, J.R., Karri, R.R., 2019. A comprehensive review of applications of magnetic graphene oxide based nanocomposites for sustainable water purification. J. Environ. Manage. 231, 622–634.

Liou, T.H., Wu, S.J., 2010. Kinetics study and characteristics of silica nanoparticles produced from biomass-based material. Ind. Eng. Chem. Res. 49 (18), 8379–8387. https://doi.org/10.1021/ie100050t.

Matsumoto, S.T., Mantovani, M.S., Malaguttii, M.I.A., Dias, A.L., Fonseca, I.C., Marin-Morales, M.A., 2006. Genotoxicity and mutagenicity of water contaminated with tannery effluents as evaluated by the micronucleus test and comet assay using the fish *Oreochromis niloticus* and chromosome aberrations in onion root-tips. Genetics Mol. Biol. 29 (1), 148–158. https://doi.org/10.1590/S1415-47572006000100028.

McBride, M.B., 1995. Toxic metal accumulation from agricultural use of sludge: are US EPA regulations protective? J. Environ. Quality 24 (1), 5–18. https://doi.org/10.2134/jeq1995.00472425002400010002x.

Mo, L., Zhou, S., Yang, S., Gong, J., Li, J., 2020. Hemp-derived activated carbon supported zero-valent iron as a heterogeneous Fenton catalyst for the treatment of pulping effluent. BioResources 15 (3), 4996–5011. https://doi.org/10.15376/biores.15.3.4996-5011.

Mohan, D., Sarswat, A., Ok, Y.S., Pittman, C.U., 2014. Organic and inorganic contaminants removal from water with biochar, a renewable, low cost and sustainable adsorbent - a critical review. Bioresour. Technol. 160, 191–202. https://doi.org/10.1016/j.biortech.2014.01.120.

Olama, N., Dehghani, M., Malakootian, M., 2018. The removal of amoxicillin from aquatic solutions using the TiO_2/UV-C nanophotocatalytic method doped with trivalent iron. Appl. Water Sci. 8 (4), 97.

Onuegbu, T., Umoh, E., Onwuekwe, I., 2013. Physico—chemical analysis of effluents from Jacbon chemical industries limited, makers of bonalux emulsion and gloss paints. Int. J. Sci. Technol. 2, 169–173.

Osman, A.I., Blewitt, J., Abu-Dahrieh, J.K., Farrell, C., Al-Muhtaseb, A.H., Harrison, J., Rooney, D.W., 2019. Production and characterisation of activated carbon and carbon nanotubes from potato peel waste and their application in heavy metal removal. Environ. Sci. Pollut. Res. 26 (36), 37228–37241. https://doi.org/10.1007/s11356-019-06594-w.

Narayana, P.L., Lingamdinne, L.P., Karri, R.R., Devanesan, S., AlSalhi, M.S., Reddy, N.S., Chang, Y.Y., Koduru, J.R., 2022. Predictive capability evaluation and optimization of Pb (II) removal by reduced graphene oxide-based inverse spinel nickel ferrite nanocomposite. Environ. Res. 204, 112029.

O'Brien, T., Xu, J., Patierno, S.R., 2001. Effects of glutathione on chromium-induced DNA crosslinking and DNA polymerase arrest. Mol. Cell. BioChem. 222 (1–2), 173–182. https://doi.org/10.1023/A:1017918330073.

Paudyal, H., Pangeni, B., Inoue, K., Kawakita, H., Ohto, K., Alam, S., 2013. Adsorptive removal of fluoride from aqueous medium using a fixed bed column packed with Zr(IV) loaded dried orange juice residue. Bioresour. Technol. 146, 713–720. https://doi.org/10.1016/j.biortech.2013.07.014.

Qi, K., Cheng, B., Yu, J., Ho, W., 2017. Review on the improvement of the photocatalytic and antibacterial activities of ZnO. J. Alloys Compd, 727, 792–820. https://doi.org/10.1016/j.jallcom.2017.08.142.

Qiu, H., Liang, C., Yu, J., Zhang, Q., Song, M., Chen, F., 2017. Preferable phosphate sequestration by nano-La(III) (hydr)oxides modified wheat straw with excellent properties in regeneration. Chem. Eng. J. 315, 345–354. https://doi.org/10.1016/j.cej.2017.01.043.

Quan, G.Z., Luo, G.C., Mao, A., Liang, J.T., Wu, D.S., 2014. Evaluation of varying ductile fracture criteria for 42CrMo steel by compressions at different temperatures and strain rates. Sci. World J.. 2014 https://doi.org/10.1155/2014/579328 .

Rahmat, M., Hubert, P., 2011. Carbon nanotube–polymer interactions in nanocomposites: a review. Compos. Sci. Technol. 72 (1).

Reddy, D.H.K, Lee, S.M., et al., 2014. Magnetic biochar composite: facile synthesis, characterization, and application for heavy metal removal. Colloids Surf. A: Physicochemical and Engineering Aspects 454. https://doi.org/10.1016/j.colsurfa.2014.03.105.

S., D., P., M.W, 1999. Accumulation of potentially toxic elements in plants and their transfer to human food chain. J. Environ. Sci. Health 681–708. Part B https://doi.org/10.1080/03601239909373221 .

Satarug, S., Moore, M.R., 2004. Adverse health effects of chronic exposure to low-level cadmium in foodstuffs and cigarette smoke. Environ. Health Perspect. 112 (10), 1099–1103. https://doi.org/10.1289/ehp.6751.

Sobhanardakani, S., Tayebi, L., Farmany, A., 2011. Toxic metal (Pb, Hg and As) contamination of muscle, gill and liver tissues of *Otolithes rubber, Pampus argenteus, Parastromateus niger, Scomberomorus commerson* and *Onchorynchus mykiss*. World Appl. Sci. J. 14 (10), 1453–1456. http://idosi.org/wasj/wasj14(10)11/4.pdf.

Soheil, S., 2017. Tuna fish and common kilka: health risk assessment of metal pollution through consumption of canned fish in Iran. J. Consumer Protect.Food Saf. 157–163. https://doi.org/10.1007/s00003-017-1107-z.

Sohi, S.P., 2012. Carbon storage with benefits. Science 338 (6110), 1034–1035. https://doi.org/10.1126/science.1225987.

Song, Z., Lian, F., Yu, Z., Zhu, L., Xing, B., Qiu, W., 2014. Synthesis and characterization of a novel MnO_x-loaded biochar and its adsorption properties for Cu^{2+} in aqueous solution. Chem. Eng. J. 242, 36–42. https://doi.org/10.1016/j.cej.2013.12.061.

Sun, P., Hui, C., Khan, R.A., Du, J., Zhang, Q., Zhao, Y.H., 2015. Efficient removal of crystal violet using Fe_3O_4-coated biochar: the role of the Fe_3O_4 nanoparticles and modeling study their adsorption behavior. Sci. Rep. 5, 12638. https://doi.org/10.1038/srep12638.

Szczepanik, B., 2017. Photocatalytic degradation of organic contaminants over clay-TiO_2 nanocomposites: a review. Appl. Clay Sci. 141, 227–239. https://doi.org/10.1016/j.clay.2017.02.029.

Tan, X.f., Liu, S.b., Liu, Y.g., Gu, Y.l., Zeng, G.m., Hu, X.j., Wang, X., Liu, S.h., Jiang, L.h., 2017. Biochar as potential sustainable precursors for activated carbon production: multiple applications in environmental protection and energy storage. Bioresour. Technol. 227, 359–372. https://doi.org/10.1016/j.biortech.2016.12.083.

Tan, X.f., Liu, Y.g., Gu, Y.l., Xu, Y., Zeng, G.m., Hu, X.j., Liu, S.b., Wang, X., Liu, S.m., Li, J., 2016. Biochar-based nano-composites for the decontamination of wastewater: a review. Bioresour. Technol. 212, 318–333. https://doi.org/10.1016/j.biortech.2016.04.093.

Tan, X., Liu, Y., Zeng, G., Wang, X., Hu, X., Gu, Y., Yang, Z., 2015. Application of biochar for the removal of pollutants from aqueous solutions. Chemosphere 125, 70–85. https://doi.org/10.1016/j.chemosphere.2014.12.058.

Tang, J., Lv, H., Gong, Y., Huang, Y., 2015. Preparation and characterization of a novel graphene/biochar composite for aqueous phenanthrene and mercury removal. Bioresour. Technol. 196. https://doi.org/10.1016/j.biortech.2015.07.047.

Touati, A., Hammedi, T., Najjar, W., Ksibi, Z., Sayadi, S., 2016. Photocatalytic degradation of textile wastewater in presence of hydrogen peroxide: effect of cerium doping titania. J. Ind. Eng. Chem. 35, 36–44. https://doi.org/10.1016/j.jiec.2015.12.008.

Uyguner-Demirel, C.S., Birben, N.C., Bekbolet, M., 2017. Elucidation of background organic matter matrix effect on photocatalytic treatment of contaminants using TiO_2: a review. Catal. Today 284, 202–214. https://doi.org/10.1016/j.cattod.2016.12.030.

Valavanidis, A., Vlachogianni, 2010. Metal pollution in ecosystems. Ecotoxicology studies and risk assessment in the marine environment. Science advances on Environment, Toxicology & Ecotoxicology issues 15784.

Veerakumar, P., Thanasekaran, P., Lin, K.C., Liu, S.B., 2017. Biomass derived sheet-like carbon/palladium nanocomposite: an excellent opportunity for reduction of toxic hexavalent chromium. ACS Sustain. Chem. Eng. 5 (6), 5302–5312. https://doi.org/10.1021/acssuschemeng.7b00645.

Vinayagam, M., Ramachandran, S., Ramya, V., Sivasamy, A., 2018. Photocatalytic degradation of orange G dye using ZnO/biomass activated carbon nanocomposite. J. Environ. Chem. Eng. 6 (3), 3726–3734. https://doi.org/10.1016/j.jece.2017.06.005.

Walker, J.M., Southworth, R.M., Rubin, A.B., 1997. US Environmental Protection Agency regulations and other stakeholder activities affecting the agricultural use of by-products and wastes. In: Rechcig, J.E., MacKinnon, H.C. (Eds.), Agricultural Uses of By-Products and Wastes. American Chemical Society, Washington.

Wang, W., 1987. Root elongation method for toxicity testing of organic and inorganic pollutants. Environ. Toxicol. Chem. 6 (5), 409–414. https://doi.org/10.1002/etc.5620060509.

Wang, S., Gao, B., AndrewZimmerman, R., Li, Y., Ma, L., WillieHarris, G., Kati, W.M., 2015. Removal of arsenic by magnetic biochar prepared from pinewood and natural hematite. Bioresour. Technol. 175, 391–395.

Wang, H., Gao, B., Wang, S., Fang, J., Xue, Y., Kai, Y., 2015. Removal of Pb (II), Cu (II), and Cd (II) from aqueous solutions by biochar derived from KMnO4 treated hickory wood. Bioresour. Technol. 197, 356–362.

Wolińska, A., Stepniewska, Z., Wlosek, R., 2013. The influence of old leather tannery district on chromium contamination of soils, water and plants. Nat. Sci., 5, pp. 253–258.

Yadav, A., Raj, A., Purchase, D., Ferreira, L.F.R., Saratale, G.D., Bharagava, R.N., 2019. Phytotoxicity, cytotoxicity and genotoxicity evaluation of organic and inorganic pollutants rich tannery wastewater from a common effluent treatment plant (CETP) in Unnao district, India using *Vigna radiata* and *Allium cepa*. Chemosphere 224, 324–332. https://doi.org/10.1016/j.chemosphere.2019.02.124.

Yan, L., Kong, L., Qu, Z., Li, L., Shen, G., 2015. Magnetic biochar decorated with ZnS nanocrytals for Pb (II) removal. ACS Sustain. Chem. Eng. 3 (1), 125–132. https://doi.org/10.1021/sc500619r.

Yao, Y., Gao, B., Chen, J., Yang, L., 2013. Engineered biochar reclaiming phosphate from aqueous solutions: mechanisms and potential application as a slow-release fertilizer. Environ. Sci. Technol. 47 (15), 8700–8708. https://doi.org/10.1021/es4012977.

Yao, Y., Gao, B., Chen, J., Zhang, M., Inyang, M., Li, Y., Alva, A., Yang, L., 2013. Engineered carbon (biochar) prepared by direct pyrolysis of Mg-accumulated tomato tissues: characterization and phosphate removal potential. Bioresour. Technol. 138, 8–13. https://doi.org/10.1016/j.biortech.2013.03.057.

Yao, Y., Gao, B., Fang, J., Zhang, M., Chen, H., Zhou, Y., Creamer, A.E., Sun, Y., Yang, L., 2014. Characterization and environmental applications of clay-biochar composites. Chem. Eng. J. 242, 136–143. https://doi.org/10.1016/j.cej.2013.12.062.

Zhang, H., Lv, X., Li, Y., Wang, Y., Li, J., 2010. P25-graphene composite as a high performance photocatalyst. ACS Nano 4 (1), 380–386. https://doi.org/10.1021/nn901221k.

Zhang, M., Gao, B., 2013. Removal of arsenic, methylene blue, and phosphate by biochar/AlOOH nanocomposite. Chem. Eng. J. 226, 286–292. https://doi.org/10.1016/j.cej.2013.04.077.

Zhang, M., Gao, B., Yao, Y., Inyang, M., 2013. Phosphate removal ability of biochar/MgAl-LDH ultra-fine composites prepared by liquid-phase deposition. Chemosphere 92 (8), 1042–1047. https://doi.org/10.1016/j.chemosphere.2013.02.050.

Zhang, M., Gao, B., Yao, Y., Xue, Y., Inyang, M., 2012. Synthesis of porous MgO-biochar nanocomposites for removal of phosphate and nitrate from aqueous solutions. Chem. Eng. J. 210, 26–32. https://doi.org/10.1016/j.cej.2012.08.052.

Zhang, T., Wang, X., Zhang, X., 2014. Recent progress in TiO_2-mediated solar photocatalysis for industrial wastewater treatment. Int. J. Photoenergy 2014,, 607954. https://doi.org/10.1155/2014/607954.

Zhou, Y., Gao, B., Zimmerman, A.R., Fang, J., Sun, Y., Cao, X., 2013. Sorption of heavy metals on chitosan-modified biochars and its biological effects. Chem. Eng. J. 231, 512–518. https://doi.org/10.1016/j.cej.2013.07.036.

CHAPTER 13

Application of green nanocomposites in removal of toxic chemicals, heavy metals, radioactive materials, and pesticides from aquatic water bodies

Emmanuel Ikechukwu Ugwu[a], Rama Rao Karri[b], Chidozie Charles Nnaji[c,d], Juliana John[e], V.C Padmanaban[f], Amina Othmani[g], Eberechukwu Laura Ikechukwu[h] and Wasim M.K. Helal[i]

[a]Department of Civil Engineering, College of Engineering and Engineering Technology, Michael Okpara University of Agriculture Umudike, Nigeria
[b]Petroleum and Chemical Engineering, Universiti Teknologi, Brunei
[c]Department of Civil Engineering, University of Nigeria, Nsukka, Nigeria
[d]Faculty of Engineering and Built Environment, University of Johannesburg, South Africa
[e]Department of Civil Engineering, National Institute of Technology Tiruchirapalli, Tamil Nadu, India
[f]Department of Biotechnology, Kamaraj College of Engineering & Technology, Tamil Nadu, India
[g]Department of Chemistry, Faculty of Sciences, University of Monastir, Monastir, Tunisia
[h]Department of Biochemistry, College of Natural Sciences, Michael Okpara University of Agriculture Umudike, Umuahia, Nigeria
[i]Department of Mechanical Engineering, Faculty of Engineering, Kafrelsheikh University, Kafrelsheikh, Egypt

13.1 Introduction

Water is an essential compound needed for the sustenance of life. However, the available water resources on the earth are being depleted owing to pollution. Access to clean and reliable water has been considered one of the basic human needs and the most challenging issues being faced in the twenty-first century. In developing countries, about 80% of the illnesses are associated with a poor water supply and hygienic conditions, with about 443 million days lost yearly from the school calendar because of water-related diseases (Corcoran et al., 2010). Water pollution results when toxic substances penetrate into the water and change the water quality (Mehtab et al., 2017). Water pollution emanates from the point and nonpoint sources with sources such as urbanization, industrial, and agricultural processes (Bassem, 2020). Rapid industrialization has brought about the degradation of water quality through the deposition of contaminants into the aquatic water bodies (Qu et al., 2013). Water pollution could constitute a significant risk to

Sustainable Nanotechnology for Environmental Remediation
DOI: https://doi.org/10.1016/B978-0-12-824547-7.00018-7

human health, animals, and aquatic water bodies; thus, about 3.1% of deaths are caused by the consumption of unsafe and bad quality water (Walha et al., 2007).

Pollutants' impact could differ by form and source. For example, dyes, heavy metals, and other organic contaminants have been pinpointed as carcinogens, whereas pharmaceutical, hormonal, beauty products, and cosmetic wastes are referred to as obstructive endocrine chemicals (Adeogun et al., 2016). The pollutants present a serious challenge to humans and the water bodies, whereas significant population growth stimulates climate change (Palmate et al., 2017). For example, several anthropogenic activities and the emission of greenhouse gases into the atmosphere by industries significantly cause global warming, planetary temperature increase, and air quality reduction. More so, they destroy fish, mollusks, seaweeds, marine birds, crustaceans, and other marine organisms that can be used as food for man (Mehtab et al., 2017). Such pollutants, which penetrate into the water bodies via diverse sources, primarily due to human activities, have become a key problem for researchers due to their numerous environmental hazards. Among all the water contaminants, pesticides, radioactive materials, heavy metals, and toxic chemicals are considered as the contaminants of priority due to their toxicity (Amoatey & Baawain, 2019). Thus, there is a need to discuss these contaminants in the subsequent sections. Though numerous reviews have been carried out on removing some emerging pollutants from aquatic water bodies, not sufficient research has been done to provide a detailed overview on the removal of toxic chemicals, heavy metals, radioactive materials, and pesticides from aquatic water bodies using green nanocomposites. In this chapter, recent studies on the pollutants' occurrence and their removal techniques using green nanocomposites are highlighted.

13.2 Major emerging pollutants

Pollutants are substances that create adverse effects once discharged into the environment. They could induce mild or severe effects depending on their concentration. Over time such pollutants persist in the atmosphere. Thus, they create issues whenever their concentration rises above the maximum absorbance in the environment (Chaudhry & Malik, 2017). The major factors that trigger water pollution are the disposal of industrial and domestic wastes, water tank leakages, sewage dumping close to water sources, and the emission of nuclear wastes and particulate matter (Mehtab et al., 2017). The sources and the major pollutants are listed in Table 13.1.

13.2.1 Heavy metals

Heavy metals are metallic elements with densities five times greater than water (5.0 g/cm^3). Heavy metals in wastewater are of a great concern due to their carcinogenic and hazardous nature compared to other contaminants such as dyes, pesticides, inorganic and

Table 13.1 Point source of major water pollutants.

Sources of water pollutants	Major pollutants
Agriculture	Pesticides, heavy metals, drug residues, nitrates and ammonia from manures
Wastewater and Sewage	Dyes, heavy metals, pesticides, pharmaceutical residues, oils, salts, chemicals, grease, and debris
Oil pollution–oil spills	Long-chain hydrocarbons
Radioactive waste	Radioactive materials
Plastics	Polyethene (PE) and polypropylene (PP), cellulose acetate (CA), polyethylene terephthalate, per-and polyfluoroalkyl substances (PFAS)

organic substances (Sushovan et al., 2018). A geometric increase in the industrial usage of metals and their discharge in water bodies has unavoidably brought about an increase of metallic substances in water bodies (Olukanni et al., 2014). These metals accumulate in the soil, air, and water, thus interfering with the ecosystem's normal functioning. Heavy metals accumulated in the environment can be transferred to humans by consuming contaminated water and accidental ingestion of soil, inhalation of polluted air, and consumption of plants grown on contaminated soil. Elevated levels of heavy metals have been detected in plants' edible parts (Vongdala et al., 2019). Studies have reported a gradual build-up of heavy metals in the environment and their attendant human and ecological hazards. The major sources of heavy metal release globally are mining, manufacturing, waste disposal, pesticide/fertilizer application, and rock weathering (Zhou et al., 2020). Toxicity, nonbiodegradability and persistence in the environment are the three most critical factors that make heavy metals a serious environmental concern (Zhongchen et al., 2019). To avert the detrimental effect of such wastes on humans and the environment, abatement measures should be put in place (Ugwu & Agunwamba, 2020) . Chromium occurs as both a natural trivalent ion [(Cr(III)] and a hexavalent ion [Cr(IV)] formed during industrial processes. Even though hexavalent chromium is far more dangerous because the trivalent type harms plants solely, owing to its oxidizing capacity and high permeability, the hexavalent type harms animals and plants. Its presence in aqueous ecosystems has adverse effects on humans and marine organisms (Karri et al., 2020). Zinc is ranked as the fourth important trace element (Ugwu & Agunwamba, 2020). The health implications of zinc-polluted water for man and other organisms are of a serious concern due to its mobility and nonbiodegradability (Zamani et al., 2017). Zn(II) permeates the food chain, thereby inducing long-lasting detrimental health effects (Karri & Sahu, 2018). Copper is a naturally occurring element (Yargiç et al., 2015), is a prevalent heavy metal, and it has been extensively utilized in areas such as manufacturing, mechanical, electrical, and public safety (Jurado-López et al., 2017). It is often considered an essential element and a pollutant in potable water (Ugwu & Agunwamba, 2020). It is detrimental to the pancreas, heart, kidney, skin, and liver when consumed in higher doses (Gupta & Gogate, 2016). Nickel has an unprecedented toxic effect among all heavy metals, thereby posing catastrophic environmental and human health damage. Thus, massive concentrations of nickel severely cause health problems (Aravind et al., 2015). Exposure to nickel could cause skin reactions, pneumonic fibrosis, and respiratory cancer (Hernández Rodiguez et al., 2018). Arsenic (As) is present as a mobile heavy metal in basically all the world regions, and it has been in use since ancient times as toxins and medications. Its presence in the atmosphere is related to natural processes and human activities. Like most other heavy metals, As contamination may occur across numerous pathways that include feeding and various sources such as air, soil, and water (Baker et al., 2018). As in an inorganic form is more harmful than the organic type, based on free radical formation and oxidative stress responsible for cell damage and death (Chen & Olsen, 2016). As occurs as arsenate and arsenite oxidation forms in groundwater (Ali, 2018). Arsenite and arsenate are hazardous heavy metals that adversely affect water quality (Bhatti et al., 2020). Mercury is among the most harmful substances affecting public health and the environment (Karagianni et al., 2020). Mercury could be accumulated in the human bodies via the skin, bronchi, and alimentary canal, thereby causing chronic and acute diseases such as bronchitis, high blood pressure, damage to the nervous system, cancer, and atrophy testis (Xiao et al., 2016). Lead is commonly utilized in various industrial applications, including electroplating, printing, and processing dyes, batteries, and paints (Rasoulzadeh et al., 2020). Lead is a highly toxic heavy metal that poses serious threats to the environment and public health, resulting in massive degenerative illnesses (Bhattacharya, 2019). Most human lead exposure mainly occurs by drinking water, airborne particulates, industrial processes, lead-based paints, gasoline, and fossil fuels (Xiao et al., 2016). Due to its high toxicity, Cadmium is considered a carcinogen. Its toxicity is solely focused on its inhibitory effect on cell differentiation, regrowth, and apoptosis, most of which come in contact with DNA repair (Rahimzadeh et al., 2017). Cadmium causes lung cancer, urinary protein excretion, demineralization, and kidney failure (Ekanem, 2017).

13.2.2 Textile dyes

Textile is among the oldest industries, yet it has a dominant presence globally, contributing to about one-third of global total export revenues and providing profitable employment opportunities to millions of people. The textile industry is unique among some of the world's industries, making a significant contribution to the country's total industrial output, exports, tax revenues, and GDP. The textile industry consists mainly of small-scale, unintegrated rolling, entwining, and finishing and garments-making enterprises. This particular industry structure is totally an endowment of government policies that facilitated labor-intensive, small-scale enterprises but segregated toward larger firms. The textile industry equally directly connects to the regional economy and the performance of key fiber crops and art, employing millions of craftsmen and farmers in the rural and suburban regions. It was projected that one in six rural households relies explicitly or implicitly on this industry. Yet this industry is among the main contributors to environmental contamination as it releases vast amounts of textile wastewater that contains recalcitrant dye substances that have proven to be toxic to so many living organisms (Abhijith et al., 2012). Among the high-polluting sectors, the textile dyeing industry known for dyeing different fibers forms is the most notable one. It is projected that about 10%–50% of dyes used during the textile dyeing are discarded and eventually leading to contamination of the environment with approximately one million tons of such substances and reducing azo dyes lead to high mutagenic aromatic compounds, even more mutagenic than that of the original dye. The effluent's colloidal structure induces turbidity. It inhibits the sunlight penetration required for photosynthesis. It disrupts the air-water network for the oxygen transfer process. Depleting dissolved oxygen from the water is perhaps the most extreme consequence of textile waste as dissolved oxygen is important for aquatic life. This also hampers water self-purification.

Furthermore, whenever this effluent is permitted to circulate in the fields, it clutters the soil pores culminating in soil productivity loss (Ashraf et al., 2013). Soil texture gets strengthened, and root penetration is obstructed. If the wastewater flows in drainage systems and rivers, it affects the drinking water quality in hand pumps, rendering it unsuitable for human consumption.

13.2.3 Pesticides

These are conglomerates of agents such as fungicides, herbicides, insecticides, avicides, and rodenticides used to control certain unwanted living organisms' intrusion into farms, gardens, orchards, etc. They have been of great help in controlling crop pests and diseases, thereby increasing crop yield and decreased insect-borne diseases (Rekha et al., 2006). However, its misuse leads to detrimental effects on man and the environment. The overuse and misuse of pesticides affect the ecosystem by intruding into the food web, thereby leading to the pollution of the surface and groundwater and the air and soil (Abhilash & Singh, 2009). Though pesticides are primarily applied to agricultural soils, they are easily leached into groundwater and surface water bodies and can also be airborne by attachment to a particulate matter. Besides, even when moderately applied, residues of these chemicals are left behind after serving their purpose. These substances are different from other chemical substances because they are applied intentionally in the environment as they are made to tackle certain living species. They are of diverse levels of toxicity (Corsini et al., 2008). Certain pesticides have been classified as mutagenic, teratogenic, and carcinogenic (Ghorab & Khalil, 2016). Pesticides made from organophosphorus compounds are known to reduce fertility in human beings by reducing testosterone level, causing behavioral changes in children, and inducing immune problem (Ghorab & Khalil, 2016). The mobility of pesticides in the environment is facilitated by their solubility in water and organic solvents. They may decompose under certain environmental conditions to release their metabolites that may also be toxic (Székács et al., 2015).

13.2.4 Radioactive materials

Radioactive materials (radionuclides) emit alpha, beta, and gamma rays that can penetrate living tissues and cause severe damage (Gortam, 2018). Radionuclides exist in the environment (water, air, and soil) in minute and harmless quantities. Anthropogenic activities associated with nuclear power plants and nuclear weapons testing have caused a spike in the overall concentration of radionuclides in the environment. Radionuclides can also be emitted from uranium mining, geological repository, fuel processing plant, hospital activities (treatment and diagnostics), and nuclear accidents. These wastes are hazardous to humans and the environment. Radioactive wastes from nuclear weapons and defensive activities constitute a great risk of environmental protection for future and present generations (Rao, 2001). Notably, the Great East Japan Earthquake that ravaged the Fukushima Daiichi Nuclear Power Plant has remained a constant source of release of radionuclides into the Pacific Ocean (Kobza & Geremek, 2017). The effects of radioactive wastes can be short and long ranges. The short-range effects are ephemeral, and they include subcutaneous bleeding, loss of nails and hairs, distortion of blood cells and metabolism; however, the range effects can last for months or years (Mondal, 2015).

13.3 Conventional water treatment strategies

The geometric increase of contaminants in aquatic water bodies requires the development of cost-effective techniques for their treatment. Decontamination of the contaminants to harmless chemicals and removing the polluted water is required for a clean environment (Rasalingam et al., 2014). Owing to this, several treatment techniques such as chemical precipitation (Zhu et al., 2007), ultrafiltration (Chao et al., 2018), coagulation (Lee et al., 2006), microfiltration (Strathmann, 2001), reverse osmosis (Walha et al., 2007), nanofiltration (Walha et al., 2007), and adsorption (Rafatullah et al., 2010) have been used for contaminant removal from aquatic water bodies. Even though these treatment techniques are effective in contaminant removal, each technique has its advantages and shortcomings as shown in Table 13.2. Therefore, it is pertinent to highlight the principles adopted in these conventional techniques for contaminant removal.

13.3.1 Chemical-based coagulation and precipitation

Chemical precipitation is used in water and wastewater treatment, whereby the materials dissolved in water change to solid particles (Wang & Chen, 2006, 2006). It involves converting the soluble chemical contaminant into a precipitate (solid form) using a precipitating agent through a sequence of neutralization, precipitation, coagulation/flocculation, solid-liquid separation, and dewatering (Dahman, 2017). It is used to remove polar constituents dissolved in water through ionization, which aims to reduce solubility. It is most widely used in the metal plating industry, where metals such as cadmium, chromium,

Table 13.2 Advantages and limitations of various adsorbents toward adsorption process.

S. No	Treatment methodology	Advantages	Limitations
1	Activated carbon	Removes a wide variety of dyes	Cost-intensive regeneration process
2	Cucurbituril	Good sorption capacity for various dyes	Very expensive
3	Peat	Effective adsorbent due to cellular structure. No activation required	Low surface area than activated carbon
4	Wood chips	Effective adsorbent due to cellular structure. Good adsorption capacity for acid dyes	Long retention time. Huge quantities of wood chips are required
5	Silica gel	Effective for the removal of basic dyes	Side reactions prevent the commercial application
6	Ion exchange	Regeneration with low loss of adsorbents	Not effective adsorbent for all dyes

copper, lead, zinc, and nickel need to be removed from the solution for possible recovery (Wang & Chen, 2006). In the chemical precipitation technique, chemicals are added to the water, followed by the separation process to remove the water's precipitates. The separation process normally takes place in a clarifier. However, it can also take place in ceramic, other membranes, or by filtration. The technique can equally be used to remove contaminants from water bodies or pit lakes where the precipitated solutes tend to sink down to the bottom of the pool. The technique is specifically applicable for the treatment of large volumes of radioactive wastes with low concentrations. It can equally be used in the treatment of several wastes that contain high salt concentration. But, its effectiveness is reduced due to high concentration of salt and thereby making it difficult to undergo secondary treatment technologies such as ion exchange and thus, not meeting the stipulated limit for effluent discharges. The coagulation-flocculation technique refers to the stabilization and aggregation of colloidal particles to form flocs. To remove colloidal particles from water or wastewater, many techniques are utilized, and among all the techniques, coagulation-flocculation is an imperative unit process for achieving this purpose. In water and wastewater treatment, the coagulation-flocculation technique has been practiced from time immemorial, using different types of materials. The coagulation-flocculation treatment technique is a significant improvement in the water/wastewater treatment process. The coagulation-flocculation technique aims to agglomerate colloids and fine particles into particles of larger intensity to reduce the soluble inorganic and organic contaminants, natural organic matter, and turbidity. The technique consists of two phases; speedy stirring of destabilized coagulant into the aquatic water bodies to be treated through quick mixing and the agglomeration of flocculants into larger particles called flocs through gentle excitation (Spellman, 2004). The coagulation-flocculation technique is generally used in different industries due to its low-energy consumption, ease of operation, and simplicity in design (Almubaddal et al., 2009). It can also be used as both preliminary treatment and tertiary treatment process and can be used as the basic treatment technique due to its versatility (Szyguła et al., 2009). However, excess sludges are generated during its operation. Thus, there is a need for innovative technologies to take care of the shortcomings. The majority of the studies reported high removal of contaminants using the electrocoagulation method; thus, less information is available on plant-based coagulants to remove pesticides, radioactive materials, heavy metals, and toxic chemicals.

13.3.2 Surface-based removal of micropollutants through adsorption

Adsorption is a mass transport of substances from a fluid or vapor phase to a strong interface connected by chemical or physical relationships (Kurniawan et al., 2006). It could be a physicochemical treatment technique for removing undesirable and regularly harmful particles. Amidst this process, the heavy metals get connected to the adsorbents in a reversible-like reaction. Adsorption can be chemical or physical. Physical adsorption occurs when electrostatic forces exist between the metals and the adsorbents, while in the case of chemical adsorption, chemical interactions are involved. Among all the techniques, adsorption is higher efficiency than other methods due to its ease of operation, capacity for recovery, cost-effectiveness, and applicability in large-scale usage (Gopal Reddi et al., 2017). The removal of pesticides, radioactive materials, heavy metals, and toxic chemicals using the adsorption process has been achieved through its physical, chemical, and biological characteristics due to its disintegration properties in an aqueous medium (Pradhan et al., 2017).

Various adsorbent groups are categorized into natural materials such as zeolites (Ansari et al., 2014), clays (Nadine & F., 2005), and siliceous substances (Phan et al., 2000). These natural resources are readily available, as well as cost effective. They equally have considerable capacity to modify the adsorption performance (Chakraborty et al., 2020). Agricultural waste adsorbents based on the activated carbon is commonly used to remove heavy metal ions. Commercial activated carbon is effective for the adsorption of heavy metals; however, it is costly and limited. For the adsorption of heavy metals from water, it is important to develop relatively cheap and commonly available adsorbents. Agricultural waste is readily available worldwide and could be used for activated carbon production. Its potential for adsorption is lower than the commercial activated carbon, but the cost is quite low. It is quite environmentally safe, cost effective, and extremely effective in removing heavy metals from water. It comprises starch, sugar, lignin, hemicelluloses, and cellulose (Chakraborty et al., 2020). Agricultural wastes contain numerous functional groups such as keto, aldehyde, amine groups, and so on.

Such characteristics increase its efficiency for eliminating harmful contaminants (Jain et al., 2016). Some agricultural waste materials have been proposed as adsorbents to remove heavy metals because of the excellent adsorption performance. Wan Ngah & Hanafiah (2008) documented the usage of chemically modified plant waste–based adsorbent for the adsorption of heavy metals from effluents. Overall, the chemically modified adsorbents exhibited greater potential for adsorption than the unmodified ones. The adsorption of hazardous substances from aqueous systems through biological matter is called bioadsorption, while the materials are called bioadsorbents. Such bioadsorbents and their products are composed of many binding sites that could attach the ions in the heavy metals. Examples of bioadsorbents are bacteria (Mousavi et al., 2019), fungus (Li et al., 2009), yeasts (Wang & Chen, 2006), peat (Sun & Yang, 2003), chitosan, and chitin (Dursun & Kalayci, 2005).

13.3.3 Electrochemically based removal of recalcitrants

Electrocoagulation is one of the most widely used electrochemical techniques for wastewater treatment. During, electrocoagulation process, electrochemical coagulants, which possess the capacity to remove different wastewater contaminants, are produced (Genawi et al., 2020). Ideally, a conventional electrocoagulation cell should be made up of two electrodes, cathode and anode, with an external connection to a power source immersed wholly in the polluted water. The most widely used electrodes for electrocoagulation are made of aluminum and iron materials (Moussa et al., 2017). Electrocoagulation has different advantages over other techniques such as low operation cost, minimal sludge generation, and high removal efficiency. Besides, its chemical requirement is low, leading to a reduction in sludge quantity (Adhoum et al., 2004). The electrocoagulation technique has been successfully used in the treatment of electroplating effluents (Adhoum et al., 2004), textile effluents (Zaroual et al., 2006), and polishing effluents. However, it has some drawbacks, such as the fact that it requires a lot of labor to maintain and clean the electrodes. Also, the electrodes are susceptible to tear and wear, which attracts an extra cost of replacement.

13.3.4 Membrane-based removal of contaminants

Membrane filtration techniques such as reverse osmosis, ultrafiltration, microfiltration, and nanofiltration are pressure-driven filtration techniques regarded as some of the modern exceedingly viable processes (Walha et al., 2007). They are well known as alternative techniques for removing colossal amounts of biodegradable contaminants (Bodzek et al., 2006). Treatment of water/wastewater using membrane filtration techniques is cost effective and feasible. It can be a way better option for the conventional treatment frameworks because their high effectiveness in removing contaminants meets the high environmental guidelines (Owen et al., 1995). Microfiltration and ultrafiltration are almost identical to the conventional filtration process. They are applicable for removing colloidal particles, suspended solids, substances of high molecules, viruses, bacteria, and so on, and the mechanism is sieve-effect-controlled (Kocurek et al., 2014). The reverse osmosis and nanofiltration follow the same principle of operation. However, the filtration capacity for nanofiltration is ordinarily significantly lower. This filtration technique can filter mainly monovalent particles. Thus with the nanofiltration technique having lower proficiency than the reverse osmosis, removal of polyvalent particles is almost comparable, with the mechanism being diffusion-controlled (Kocurek et al., 2014). Typically, the reverse osmosis process is the opposite of normal osmosis, which is the solvent movement from an area of low concentration to that of higher concentration through a semipermeable membrane. By and large, membranes consisting of polyamide are utilized to decontaminate wastewater-containing chromium (Hintermeyer et al., 2008).

The reverse osmosis method has been effectively utilized to purify electroplating flush waters, not just to meet the wastewater discharge standards but also to recover the deposited metallic salt solutions for further uses. However, its fundamental demerits are high cost of equipment and costly monitoring framework, excess generation of sludge, and high energy demand (Bogeshwaran et al., 2014). Nanofiltration and reverse osmosis have demonstrated a very viable filtration technique for removing emerging pollutants (Yoon et al., 2006). Reverse osmosis is moderately more viable than nanofiltration, but higher energy utilization in reverse osmosis makes it less appealing than nanofiltration. Even though nanofiltration-based membrane processes are very viable in removing colossal loads of contaminants (Bolong et al., 2009), innovative techniques are required to treat some recalcitrant wastes' heavy metals, particularly chromium, as they are nonbiodegradable. The major drawback of the membrane process is its restrained lifetime in achieving the membrane fouling process. Also, the cost of intermittent substitution required for the process limits its usage.

13.3.5 Biological treatment

In this process, the wastes are decomposed to harmless inorganic solids through aerobic or anaerobic decomposition with bacteria's aid (Durai & Rajasimman, 2011) . The use of biological treatment, especially microbial remediation of hexavalent chromium, has gained major attention recently. This is because it is economical, sustainable, and effective (Fernandes et al., 2009) . This has an advantage over the physiochemical techniques as most of those methods have shortcomings ranging from unsuitability, nonenvironmental friendliness, and low efficiency (Bharagava & Mishra, 2018). Besides, they are not economically viable at low concentrations but rather at high concentrations (Mamais et al., 2016). However, it has some demerits, such as the microorganisms' inability to remediate the contaminants from synthetic solution to perform well under the real

Table 13.3 Merits and demerits of different water purification techniques.

Methods	Merits	Demerits
Adsorption	Ease of design and operation. Efficiency in the removal of toxic contaminants	Regeneration of adsorbents. A large volume of effluents
Chemical precipitation	Inexpensive and simple operation.	Enormous generation of sludge. Expensive sludge disposal.
Coagulation/flocculation	Metal selective and easy removal as they form large agglomerates.	Increases metal ions concentration. Generation of sludge
Electro-Fenton process	Effective removal of several wastes	Generation of sludge
Oxidation Processes	A rapid removal of toxic contaminants	The high cost of energy and scaling up
Electrochemical methods	Metal selective and easy to use.	Scaling up
Membrane filtration	Low consumption of compounds and minimal generation of sludge.	Implementation and maintenance cost
Ion exchange	Selective to metals	Implementation and maintenance cost
Biological treatment	Easy removal of metal ions, biomass loaded with heavy metals can be incinerated.	The problem of early saturation and limited potential for biological process improvement

case scenario, such as with industrial wastewater (Bhattacharya, 2019). As the wastewater contains different pollutants, there may be competition between the microbes and the pollutants. Thus to ensure an efficient biological treatment technique, effective reducing microbes should be used for recalcitrant contaminants. Table 13.1 shows the merits and demerits of different wastewater purification techniques (Abdi & Kazemi, 2015; Ahmaruzzaman, 2011; Staicu et al., 2015) .

13.3.6 Merits and demerits of different water purification techniques

Selecting an appropriate treatment method depends mainly on some range of factors, such as the type and quality of waste, the effluent heterogeneity, the appropriate amount of removal, and economic conditions (Rajasulochana & Preethy, 2016). The merits and demerits of different water purification techniques are given in Table 13.3. The key reason for water treatment is to mitigate the impact of water contamination and preserve the environment and human life by promoting and protecting water supplies from disease transmission. These can be achieved by several treatment methods, which may be offsite or on-site treatment techniques.

Several physicochemical methods are used to overcome the problems posed by these emerging contaminants. However, they have been unable to tackle the emerging microcontaminants to meet environmental standards (Amin et al., 2014). Given this, efforts are being made to overcome the shortcomings. The effective material required for adequate water purification is expected to possess high porosity, sequestration capacity, regenerability, and economical nature (Malaviya & Singh, 2011). To meet this, requirement nanotechnology is adopted. Nanotechnology is geared toward developing materials with high hydrophilic, porous, and mechanical strength properties for efficient water purification (Daer et al., 2015). Nanomaterials are materials ranging in size from 1 to 100 nm. Reducing the size of a material at a nanoscale helps in improving its properties to a greater extent. This is because as the material's size is reduced, its surface area tends to increase (Saxena et al., 2020). Nanomaterials with a high surface area can be of great application in water purification. But its usage is limited due to aggregation/agglomeration of masses (Ray & Shipley, 2015). Agglomeration of nanoparticles reduces the potential enhancement of nanoparticle's mechanical properties (Ashraf et al., 2013).

Nevertheless, agglomeration can be reduced by converting nanomaterials to nanocomposites (Pandey et al., 2017). The study of green nanocomposites in water purification is expected to improve the quality of aquatic water bodies. Thus, we have summarized the green nanocomposites' application to remove toxic chemicals, heavy metals, pesticides, and radioactive materials to solve the water crisis. Green nanocomposites have the potential for improving water quality. However, there are some limitations to be addressed, which are highlighted in this chapter. Also, the advantages and disadvantages of different types of green nanocomposites are given in Table 13.4. Moreover, several water purification techniques using green nanocomposites and future studies' perspectives are discussed in this chapter.

13.4. Green nanocomposites for water pollution

A nanocomposite is a material composed of one or two different materials of dimension around 10^{-9} m. Nanocomposites present small particles with a high aspect ratio and large surface area. This property allows a better adhesion with the matrix surface. Also, they present good mechanical properties and better transparency compared to the parent material (Camargo et al., 2009). With their abundance and unique properties, green nanocomposites have attracted great attention compared to polymers. This hybrid combination gives them not only intermediate properties between the mineral and the organic, but also new behavior such as stiffness, toughness, chemical resistance, low density, and produce low CO_2 emission, low cost of acquisition barrier properties, thermal stability, and biodegradability. Various mineral and organic fillers have been used to assess these nanocomposites such as silica nanoparticles and carbon nanotubes metallic particles (Uyama et al., 2003).

Table 13.4 Merits and demerits of emerging water purification methods.

S. no	Method	Merits	Demerits
1	Photocatalysis: The process in which a catalyst harnesses the radiation and uses the energy to break down the organic compounds and several other compounds in the effluent results in degradation.	Complete mineralization with shorter detention times. Effective for a small number of colored dyes	Cost-intensive process
2	Sonication: Ultrasonic frequencies (>20 kHz) are used to agitate the particles in the sample resulting in the breakdown of the compounds.	Effective in integrated systems	Relatively new method and awaiting for large scale application
3	Enzymatic treatment: Enzymes accelerate chemical reactions. Thus implying enzymes in the process helps in the degradation of certain specific compounds.	Effective for specifically selected compounds	Enzyme isolation and purification is a tedious process
4	Redox mediators: These are chemicals with electrochemical activity	Easily available and enhances the process by increasing the electron transfer rate	Concentration of redox mediator may give an antagonistic effect. Depends on the biological activity of the system

13.4.1 Cellulose-based nanocomposites

Cellulose is one of the most abundant polymers produced by two major sources: plants and microbes. Plant biomass, including wood, cotton, dried hemp, etc., are important plant sources of celluloses, whereas few bacteria and algae species contribute to their microbial production. Both plant cellulose (PC) as well as microbial cellulose (MC) are abundantly present in nature and possess interesting properties such as high mechanical strength, high biocompatibility, high purity, water holding capacity, and nanofibrous architecture (40–70 nm) (Kaiyan & N., 2014). Their nanofibrous structure has larger surface areas in their porous structures and has high thermal stability. The hydroxyl groups present in the cellulose structure make them potential functional candidates to remove pollutants from contaminated waters through metal chelation and hydrogen bonding. PC contains hemicellulose or lignin and cellulose. Their nanofiber extraction from plant sources can be carried out using chemical and mechanical treatments, whereas MC does not contain them. Many techniques such as particle deposition, resin infiltration, electrospinning, dispersion and casting, and *in-situ* self-assembly are used to fabricate these nanocomposites.

Considering these properties, many nanocomposites and nanofibers are being synthesized using these celluloses and applied in various fields. These celluloses can act either as matrices to load filler materials or as reinforcement during nanocomposites fabrication. Engineering them in the form of nanocomposites and using them as reinforcements generally improve their strength and other additional properties such as antimicrobial properties and can be employed in wider applications. When Ahmed et al. (2011) added different amounts of cellulose whiskers as reinforcements to green Polyvinyl alcohol (PVA) nanocomposites, 5 wt% of cellulose whiskers promoted their orientation and mechanical strength. Also, it filled their interfibrillar voids within the nanocomposites. Pretreating them by chemical modifications such as silylation, mercerization, and so on or by bacterial modifications or by grafting polymers with them can modify their surface properties and mechanical properties to a greater extent. Anupama et al. (2010) fabricated a green nanocomposite by using thermoplastic starch and cellulose nanofibrils (CNFs) obtained from wheat straw and found that improved barrier properties and mechanical properties were observed when the concentrations of nanofiber and nanofiller concentrations were increased. Iman et al. (2013) also found that the strength and properties of green jute based cross-linked soy flour nanocomposite was found to have improved when cellulose whiskers and nanoclay were used as fillers for the nanocomposite material. Using such green nanocomposites would be environmentally safe, cost-effective, and readily biodegradable and would be a solution while moving toward sustainable and a greener environment (Kalia et al., 2011).

13.4.2 Application of cellulose-based nanocomposites in water remediation

Cellulose and regenerated cellulose can be used in many applications (Fig. 13.1). They are in water remediation, such as removing toxic chemicals, heavy metals, radioactive materials, and pesticides from aquatic water bodies. Viscose rayon and lyocell rayon, and cellulose-derived acetate fiber can be successfully used for the removal of toxic compounds from wastewater. The modification of natural cellulose can enhance their properties, which allow them to be performing candidates used for water remediation. The chemical modification can be used for the reinforcement of these cellulosic fibers. In this context, a chemically modified cellulose filter paper with ethylene diamine tetraacetic acid was successfully applied for Ag(I), Pb(II), Cd(II), Ni(II), Zn(II), Sn(II), and Cu(II) remediation. The use of this performing chemically modified cellulose has shown a good ability for the removal of metals from aqueous solution with 90%–95% removal efficiency (Halluin et al., 2015). Table 13.5 presents some methods used for the

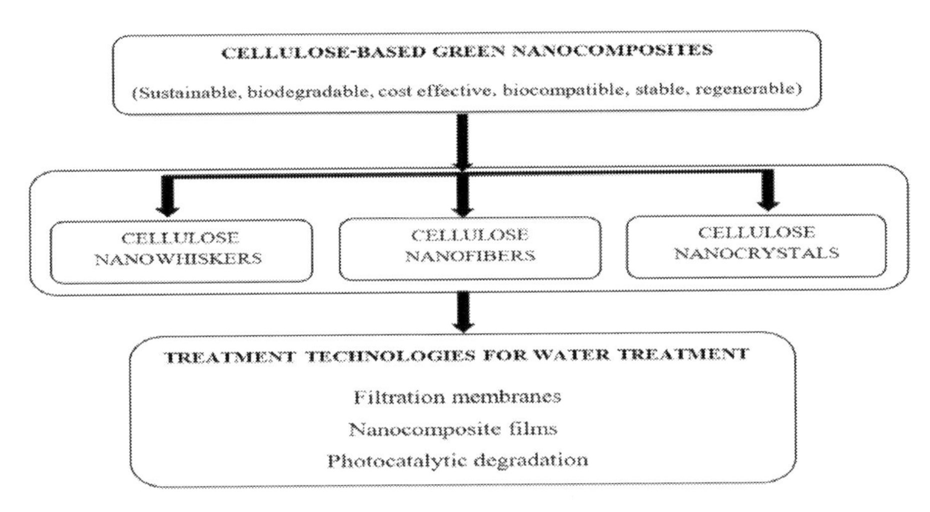

Figure 13.1 Cellulose-based green nanocomposites.

Table 13.5 Method used for the chemical modification of cellulosic fibers and their use for water remediation.

Method applied	Pathway of modification and main achievement	References
Esterification	The formation of ester bonds with the hydroxyl of cellulose can take place using carboxylic acid, alkyds, alketonedimers, or acid chloride. The esterification can enhance the hydrophobicity nature of cellulose surface.	Ashori et al., 2014; Hubbe et al., 2015
Oxidation of cellulose	During this process, the functional groups of cellulose are attacked by the oxidizing media	Abd Rahman et al., 2018
Alkaline treatment	It presents cheap and performing surface treatment. During the alkaline treatment, the wax and lignin covering the surface of the fiber can be removed, which enhances the mechanical properties of the used fibers.	Mohanty et al., 2001; Li et al., 2015
Grafting or graft polymerization	This method enhances the properties of cellulose and increases its functionality through the combination of two or more polymers in one physical unit.	Roy et al., 2009; Kang et al., 2007

chemical modification of cellulosic fibers and their use for water remediation. Different forms of cellulose such as CNFs and cellulose nanocrystals can be applied for water treatment applications, and they have been discussed in Table 13.6. Due to their high strength and mechanical properties, cellulosic nanofibrils (CNFs) act as a bio modifier to improve their strength when employed as reinforcements in nanocomposites. When

Table 13.6 Forms of cellulose and their roles in different green nanocomposites.

Type of nanocomposite	Form of cellulose	Application	Results	Reference
Green PVA nanocomposite	Cellulose whiskers	As reinforcements	Improved orientation and mechanical strength	Ahmed 2011
Thermoplastic starch/cellulose nanocomposite	Cellulose nanofibrils	As reinforcements and fillers	Improved barrier and mechanical strength	Kaushik et al., 2010
Green jute based cross-linked soy flour nanocomposite	Cellulose whiskers	As fillers	Improved mechanical strength	Iman, 2013
Thermoplastic starch-based nanocomposite	Cellulose nanofibrils	As reinforcements	Increased flexural and tensile properties	Lomelí-Ramírez, 2018
Green nanocomposite membrane	Carboxymethylcellulose	As cross-linker	Removal of 546 mg/g of crystal violet dye and 781 mg/g of Cadmium from aqueous solutions	Saber-Samandari 2016

1% of chemically modified CNF was added as reinforcement to a thermoplastic starch-based nanocomposite, an increase in flexural as well as tensile properties was observed (Lomelí-Ramírez et al., 2018) . Cellulose green nanocomposites, when cast in the form of membranes, can provide large surface areas for adsorbing different micropollutants and so they can be used successfully for treating waters in water treatment plants. When Saber-Samandari et al. (2016) synthesized a green nanocomposite membrane using carboxy cellulose, they showed high sorption capacities toward dyes and heavy metals and removed them at a rate of 546 mg/g and 781 mg/g for dye and cadmium from aqueous solutions. These nanocomposite membranes can act as antimicrobial filters, water filters, air filters, filters for removing heavy metals from waters, pollutant-based sensors, and a catalyst for performing catalytic degradation of pollutants from waters (Mazhar et al., 2016). As chitosan has good film-forming abilities, Fernandes et al. (2009) blended chitosan with different proportions of bacterial cellulose to synthesize green nanocomposite films and found that due to their high levels of biodegradability and thermal stabilities, they can be employed widely in water treatment applications. Immobilizing photocatalytic nanoparticles within these green nanocomposites can improve their adsorptive properties and make them more selective and recyclable during contaminant removal from waters. A porous cellulose-based nanocomposite film with functional photocatalytic nanoparticles (TiO_2 and Fe_3O_4) was prepared by Wittmar et al. (2017) using nonsolvent-induced phase separation technique and as they exhibited excellent magnetic as well as photocatalytic properties, they can be successfully employed for treating wastewaters as well industrial waters in future. Liu et al. (2018) prepared a green nanocomposite by sandwiching polydopamine onto cellulose fibers and adhering TiO_2 nanoparticles onto them and when used for treating contaminated waters, they showed excellent adsorption and photocatalytic removal of methylene blue (15 mg/g) as well as Pb(II) ions (20 mg/g) from waters, proving them as environment-friendly candidates for wastewater treatment plants in future. A metallic nanocomposite of silver nanoparticle supported on cellulose acetate filter paper as well as TiO_2 was synthesized by Albukhari et al. (2019) and when applied for decontamination of wastewaters, it was found that these catalysts promoted the reduction of various organic pollutants such as nitrophenols and dyes and they can be used as solutions for wastewater treatment.

13.5. Future perspective

Fabricating innovatively engineered nanocomposites using greener technologies would be a sustainable way for synthesizing nanocomposites in large-scale levels, and as they are highly specific, they can be employed for effective removal of a wide range of pollutants from water treatment plants. Some of the future perspectives of green nanocomposites are shown in Fig. 13.2.

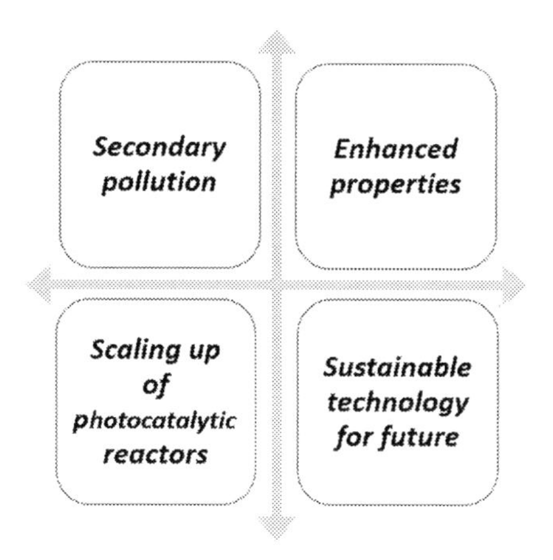

Figure 13.2 Future perspectives on nanocomposites-based water treatment.

13.5.1 Enhanced properties

Producing green nanocomposites in different forms such as nanomembranes, nanocatalysts, nanohybrids, etc. and enhancing their properties by novel functionalization and antibacterial properties may also help in decontaminating a wide range of pollutants from waters. Enhancing their stability and reusability characteristics by various processes can also help in extending their lifetime and can be used repeatedly for a long number of cycles when applied in large-scale levels. Conducting toxicity studies on such produced nanocomposite materials may ensure their level of biosafety and toxicity to the public and may pave a way to try them out in industrial applications.

13.5.2 Secondary pollution

As secondary pollution is one of the main drawbacks faced while using nanocomposites, focusing future studies on developing green nanocomposites that could overcome this problem would be very much beneficial. Developing biofiber-based nanocomposites may be one of the solutions to reduce such secondary pollutions as they are environment friendly and are readily biodegradable.

13.5.3 Scaling up of photocatalytic reactors

Synthesizing photocatalysts by these greener methods may reduce the usage of toxic chemicals as in chemical reactions and would be a novel green route of synthesis. Applying these green synthesized nanocomposites as catalysts in photocatalytic reactions may enhance their photocatalytic efficiencies as well as surface areas and can be reused for a longer period of time in water treatment systems. They may also have a reduced electron–hole recombination rate and high electron conductivities than usual catalysts and can be used as anode materials in electrochemical applications for selective redox reactions of pollutant removal. Producing nanocomposites that would offer higher rates of photodegradation with reduced use of photocatalysts may be beneficial in terms of large-scale applications. Focusing on cost-analysis aspects, careful monitoring of each step of green nanocomposite production can help in reducing their production costs to a greater level and comparing them with different production methods may help in analyzing their market potentiality. As all the components of green synthesis are from renewable sources, they are comparatively a very much cost-effective process than other ones.

13.5.4 Sustainable technology for future

Biogenic nanoparticles, one of the potential candidates produced using greener methods, are inexpensive ones and can readily offer sustainable contaminant removal from waters. Another important candidate is the green clay-polymer nanocomposites, and they are eco-friendly in nature; and due to their high adsorption capacities, they can be applied for effective removal of heavy metals and other persistent pesticides from wastewaters. Green solvents such as ionic liquids can also be used for synthesizing these nanocomposites as they are more reactive and environment friendly than other organic solvents and their desired properties can be easily tailored by varying their compositions. Thus, green nanocomposites can play a major role in moving toward a safe, sustainable future because of their interesting integrated properties and cost-effectiveness features.

13.6 Conclusions

Heavy metals, toxic chemicals, radioactive materials and pesticides present grave environmental challenges, especially with respect to drinking water. Their presence in water has been linked to the myriads of human health issues. Several conventional methods such as ion exchange, reverse osmosis, forward osmosis, coagulation/flocculation, electrocoagulation, ultrafiltration, and microfiltration are commonly used for the removal of these contaminants from water. However, they present some drawbacks ranging from cost to adverse environmental effects. Nanocomposites present a very promising option for the treatment of water contaminated with heavy metals, toxic chemicals, pesticides and radionuclides because of their unique surface properties and high structural integrity. Besides, they are environmentally friendly, cost-effective, and highly efficient.

References

Abd Rahman, N.S., Yhaya, M.F., Azahari, B., Ismail, W.R., 2018. Utilisation of natural cellulose fibres in wastewater treatment. Cellulose 25 (9), 4887–4903.

Abdi, O., Kazemi, M., 2015. A review study of biosorption of heavy metals and comparison between different biosorbents. J. Mater. Environ. Sci. 6 (5), 1386–1399. http://www.jmaterenvironsci.com/Document/vol6/vol6_N5/164-JMES-1454-2015-Abdi.pdf.

Abhijith, B.D., Ramesh, M., Poopal, R.K., 2012. Sublethal toxicological evaluation of methyl parathion on some haematological and biochemical parameters in an Indian major carp *Catla catla*. Compar. Clin. Pathol. 21 (1), 55–61. https://doi.org/10.1007/s00580-010-1064-8.

Abhilash, P.C., Singh, N., 2009. Pesticide use and application: an Indian scenario. J. Hazard. Mater. 165 (1–3), 1–12. https://doi.org/10.1016/j.jhazmat.2008.10.061.

Adeogun, A.O., Ibor, O.R., Adeduntan, S.D., Arukwe, A., 2016. Intersex and alterations in reproductive development of a cichlid, *Tilapia guineensis*, from a municipal domestic water supply lake (Eleyele) in Southwestern Nigeria. Sci. Total Environ. 541, 372–382. https://doi.org/10.1016/j.scitotenv.2015.09.061.

Adhoum, N., Monser, L., Bellakhal, N., Belgaied, J.E., 2004. Treatment of electroplating wastewater containing Cu^{2+}, Zn^{2+} and Cr(VI) by electrocoagulation. J. Hazard. Mater. 112 (3), 207–213. https://doi.org/10.1016/j.jhazmat.2004.04.018.

Ahmaruzzaman, M., 2011. Industrial wastes as low-cost potential adsorbents for the treatment of wastewater laden with heavy metals. Adv. Colloid Interface Sci. 166 (1–2), 36–59. https://doi.org/10.1016/j.cis.2011.04.005.

Ahmed, J.U., Jun, A., Yasuo, G., 2011. Toward \strong\ green nanocomposites: polyvinyl alcohol reinforced with extremely oriented cellulose whiskers. Biomacromolecules 12, 617–624.

Albukhari, S.M., Muhammad, I., Kalsoom, A., Danish, E.Y., 2019. Catalytic reduction of nitrophenols and dyes using silver nanoparticles @ cellulose polymer paper for the resolution of waste water treatment challenges. Colloids Surf. A PhysicoChem. Eng. Asp. 577, 548–561. https://doi.org/10.1016/j.colsurfa.2019.05.058.

Ali, I., 2018. Microwave assisted economic synthesis of multi walled carbon nanotubes for arsenic species removal in water: batch and column operations. J. Mol. Liquids 271, 677–685. https://doi.org/10.1016/j.molliq.2018.09.021.

Almubaddal, F., Alrumaihi, K., Ajbar, A., 2009. Performance optimization of coagulation/flocculation in the treatment of wastewater from a polyvinyl chloride plant. J. Hazard. Mater. 161 (1), 431-438.

Amin, M.T., Alazba, A.A., Manzoor, U., 2014. A review of removal of pollutants from water/wastewater using different types of nanomaterials. Adv. Mater. Sci. Eng. 2014, 825910. https://doi.org/10.1155/2014/825910.

Amoatey, P., Baawain, M.S., 2019. Effects of pollution on freshwater aquatic organisms. Water Environ. Res. 91 (10), 1272–1287. https://doi.org/10.1002/wer.1221.

Ansari, M., Aroujalian, A., Raisi, A., Dabir, B., Fathizadeh, M., 2014. Preparation and characterization of nano-NaX zeolite by microwave assisted hydrothermal method. Adv. Powder Technol. 25 (2), 722–727. https://doi.org/10.1016/j.apt.2013.10.021.

Anupama, K., Mandeep, S., Gaurav, V., 2010. Green nanocomposites based on thermoplastic starch and steam exploded cellulose nanofibrils from wheat straw. Carbohydr. Polym. 82 (2), 337–345.

Aravind, J., Lenin, C., Nancyflavia, C., Rashika, P., Saravanan, S., 2015. Response surface methodology optimization of nickel (II) removal using pigeon pea pod biosorbent. Int. J. Environ. Sci. Technol. 12 (1), 105–114. https://doi.org/10.1007/s13762-013-0391-0.

Ashori, A., Babaee, M., Jonoobi, M., Hamzeh, Y., 2014. Solvent-free acetylation of cellulose nanofibers for improving compatibility and dispersion. Carbohydr. Polym. 102, 369–375.

Ashraf, M.A., Hussain, M., Mahmood, K., Wajid, A., Yusof, M., Alias, Y., Yusoff, I., 2013. Removal of acid yellow-17 dye from aqueous solution using eco-friendly biosorbent. Desalin. Water Treat. 51 (22–24), 4530–4545. https://doi.org/10.1080/19443994.2012.747187.

Baker, B.A., Cassano, V.A., Carolyn, M., 2018. Arsenic exposure, assessment, toxicity, diagnosis, and management. J. Occupation. Environ. Med. 60 (12), e634–e639. https://doi.org/10.1097/jom.0000000000001485.

Bassem, S.M., 2020. Water pollution and aquatic biodiversity. Biodivers. Int. J. 4 (1), 10–16.

Bharagava, R.N., Mishra, S., 2018. Hexavalent chromium reduction potential of *Cellulosimicrobium* sp. isolated from common effluent treatment plant of tannery industries. Ecotoxicol. Environ. Saf. 147, 102–109. https://doi.org/10.1016/j.ecoenv.2017.08.040.

Bhattacharya, S., 2019. Probiotics against alleviation of lead toxicity: recent advances. Interdiscip. Toxicol. 12 (2), 89–92. https://doi.org/10.2478/intox-2019-0010.

Bhatti, Z.A., Qureshi, K., Maitlo, G., Ahmed, S., 2020. Study of PAN fiber and iron ore adsorbents for arsenic removal. Civil Eng. J. (Iran) 6 (3), 548–562. https://doi.org/10.28991/cej-2020-03091491.

Bodzek, M., Łobos-Moysa, E., Zamorowska, M., 2006. Removal of organic compounds from municipal landfill leachate in a membrane bioreactor. Desalination 198 (1–3), 16–23. https://doi.org/10.1016/j.desal.2006.09.004.

Bogeshwaran, K., Manikandan, G.N., Akila, S., Bibiana, M., Gayathri, R., 2014. Efficient methods for the removal of chromium from textile effluents – a review. Int. J. ChemTech Res. 6 (9), 4324–4327. http://sphinxsai.com/2014/RTBCE/5/(4324-4327)%20014.pdf.

Bolong, N., Ismail, A.F., Salim, M.R., Matsuura, T., 2009. A review of the effects of emerging contaminants in wastewater and options for their removal. Desalination 239, 229–246. https://doi.org/10.1016/j.desal.2008.03.020.

Camargo, P.H.C., Satyanarayana, K.G., Wypych, F., 2009. Nanocomposites: synthesis, structure, properties and new application opportunities. Mater. Res. 12 (1), 1–39. https://doi.org/10.1590/S1516-14392009000100002.

Chakraborty, R., Asthana, A., Singh, A.K., Jain, B., Susan, A.B.H., 2020. Adsorption of heavy metal ions by various low-cost adsorbents: a review. Int. J. Environ. Anal. Chem. https://doi.org/10.1080/03067319.2020.1722811.

Chao, G., Shuili, Y., Yufei, S., Zhengyang, G., Wangzhen, Y., Liumo, R., 2018. A review of ultrafiltration and forward osmosis: application and modification, 128 https://doi.org/10.1088/1755-1315/128/1/012150.

Chaudhry, F., Malik, M., 2017. Factors affecting water pollution: a review. J. Ecosyst. Ecogr 7, 225. https://doi.org/10.4172/2157-7625.1000225.

Chen, A.Y.Y., Olsen, T., 2016. Chromated copper arsenate–treated wood: a potential source of arsenic exposure and toxicity in dermatology. Int. J. Women's Dermatol. 2 (1), 28–30. https://doi.org/10.1016/j.ijwd.2016.01.002.

Corcoran, E., Nellemann, C., Baker, E., Bos, R., Osborn, D., Savelli, H., 2010. Sick Water? The Central Role of Wastewater Management in Sustainable Development: A Rapid Response Assessment. UNEP (United Nations Environment Programme, UN-HABITAT, GRID-Arendal. www.grida.no.

Corsini, E., Liesivuori, J., Vergieva, T., Van Loveren, H., Colosio, C., 2008. Effects of pesticide exposure on the human immune system. Hum. Exp. Toxicol. 27 (9), 671–680. https://doi.org/10.1177/0960327108094509.

Daer, S., Kharraz, J., Giwa, A., Hasan, S.W., 2015. Recent applications of nanomaterials in water desalination: a critical review and future opportunities. Desalination 367, 37–48. https://doi.org/10.1016/j.desal.2015.03.030.

Dahman, Y., 2017. Nanotechnology and Functional Materials for Engineers. Elsevier Inc,, Amsterdam, pp. 1–268.

Durai, G., Rajasimman, M., 2011. Biological treatment of tannery wastewater - a review. J. Environ. Sci. Technol. 4 (1), 1–17. https://doi.org/10.3923/jest.2011.1.17.

Dursun, A.Y., Kalayci, C.S., 2005. Equilibrium, kinetic and thermodynamic studies on the adsorption of phenol onto chitin. J. Hazard. Mater. 123 (1–3), 151–157. https://doi.org/10.1016/j.jhazmat.2005.03.034.

Ekanem S.E., 2017. Adsorption of Copper ion from Aqueous Solution onto Thermally Treated Eggshell: Kinetic and Equilibrium Study (B. Eng Thesis). Department of Chemical & Petroleum Engineering, Ado-Ekiti, Nigeria.

Fernandes, S.C.M., Oliveira, L., Freire, C.S.R., Silvestre, A.J.D., Neto, C.P., Gandini, A., Desbriéres, J., 2009. Novel transparent nanocomposite films based on chitosan and bacterial cellulose. Green Chem. 11 (12), 2023–2029. https://doi.org/10.1039/b919112g.

Genawi, N.M., Ibrahim, M.H., El-Naas, M.H., Alshaik, A.E., 2020. Chromium removal from tannery wastewater by electrocoagulation: optimization and sludge characterization. Water, 12, p. 1374.

Ghorab, M., Khalil, M., 2016. The effect of pesticides pollution on our life and environment. J. Pollut. Eff. Cont 4, 2. https://doi.org/10.4172/2375-4397.1000159.

Gopal Reddi, M.R., Gomathi, T., Saranya, M., Sudha, P.N., 2017. Adsorption and kinetic studies on the removal of chromium and copper onto chitosan-g-maliec anhydride-g-ethylene dimethacrylate. Int. J. Biol. Macromol. 104, 1578–1585. https://doi.org/10.1016/j.ijbiomac.2017.01.142.

Gortam, H., 2018. Radioactive pollution: an overview. Holist. Approach Environ. 8, 48–65.

Gupta, H., Gogate, P.R., 2016. Intensified removal of copper from waste water using activated watermelon based biosorbent in the presence of ultrasound. Ultrason. SonoChem. 30, 113–122. https://doi.org/10.1016/j.ultsonch.2015.11.016.

Halluin, M., Rull-Barrull, J., Bretel, G., Labrugère, C., Grognec, L., Felpin, E., X, F, 2015. Chemically modified cellulose filter paper for heavy metal remediation in water. ACS Sustain. Chem. Eng. 5, 1965–1973.

Hernández Rodiguez, M., Yperman, J., Carleer, R., Maggen, J., Daddi, D., Gryglewicz, G., Van der Bruggen, B., Falcón Hernández, J., Otero Calvis, A., 2018. Adsorption of Ni(II) on spent coffee and coffee husk based activated carbon. J. Environ. Chem. Eng. 6 (1), 1161–1170. https://doi.org/10.1016/j.jece.2017.12.045.

Hintermeyer, B.H., Lacour, N.A., Perez Padilla, A., Tavani, E.L., 2008. Separation of the chromium(III) present in a tanning wastewater by means of precipitation, reverse osmosis and adsorption. Latin Am. Appl. Res. 38 (1), 63–71. http://www.scielo.org.ar/scielo.php?script=sci_arttext&pid=S0327-07932008000100008&lng=en&nrm=iso&tlng=en.

Hubbe, M.A., Rojas, O.J., Lucian A. Lucia, L.A., 2015. Surface modification: Review, BioResources 10 (3), 6095–6206. 6108

Iman, M., Bania, K.K., Maji, T.K., 2013. Green jute-based cross-linked soy flour nanocomposites reinforced with cellulose whiskers and nanoclay. Ind. Eng. Chem. Res. 52 (21), 6969–6983. https://doi.org/10.1021/ie400609t.

Jain, C.K., Malik, D.S., Yadav, A.K., 2016. Applicability of plant based biosorbents in the removal of heavy metals: a review. Environ. Process. 3 (2), 495–523. https://doi.org/10.1007/s40710-016-0143-5.

Jurado-López, B., Vieira, R.S., Rabelo, R.B., Beppu, M.M., Casado, J., Rodríguez-Castellón, E., 2017. Formation of complexes between functionalized chitosan membranes and copper: a study by angle resolved XPS. Mater. Chem. Phys. 185, 152–161. https://doi.org/10.1016/j.matchemphys.2016.10.018.

Kaiyan, Q., N., N.A, 2014. A review of fabrication and applications of bacterial cellulose based nanocomposites. Polym. Rev. 598–626. https://doi.org/10.1080/15583724.2014.896018.

Kalia, S., Dufresne, A., Cherian, B.M., Kaith, B.S., Avérous, L., Njuguna, J., Nassiopoulos, E., 2011. Cellulose-based bio- and nanocomposites: a review. Int. J. Polym. Sci. https://doi.org/10.1155/2011/837875.

Karagianni, E., Xenidis, A., Papassiopi, N., 2020. Enhanced Hg removal from aqueous streams by sulfurized activated carbon products: equilibrium and kinetic studies. Water Air Soil Pollut. 231 (6), 262. https://doi.org/10.1007/s11270-020-04606-x.

Karri, R.R., Sahu, J.N., 2018. Process optimization and adsorption modeling using activated carbon derived from palm oil kernel shell for Zn (II) disposal from the aqueous environment using differential evolution embedded neural network. J. Mol. Liquids 265, 592–602. https://doi.org/10.1016/j.molliq.2018.06.040.

Karri, R.R., Sahu, J.N., Meikap, B.C., 2020. Improving efficacy of Cr (VI) adsorption process on sustainable adsorbent derived from waste biomass (sugarcane bagasse) with help of ant colony optimization. Ind. Crops Prod. 143, 111927. https://doi.org/10.1016/j.indcrop.2019.111927.

Kaushik, A., Mandeep, S., Gaurav, V., 2010. Green nanocomposites based on thermoplastic starch and steam exploded cellulose nanofibrils from wheat straw. Carbohydrate Polymers 82 (2), 337–345.

Kobza, J., Geremek, M., 2017. Do the pollution related to high-traffic roads in urbanised areas pose a significant threat to the local population? Environ. Monit. Assess. 189 (1), 33. https://doi.org/10.1007/s10661-016-5697-1.

Kocurek, P., Kolomazník, K., Bařinová, M., 2014. Chromium removal from wastewater by reverse osmosis. WSEAS Trans. Environ. Dev. 10 (1), 358–365. http://www.wseas.org/multimedia/journals/environment/2014/a185715-241.pdf.

Kurniawan, T.A., Chan, G.Y.S., Lo, W.H., Babel, S., 2006. Physico-chemical treatment techniques for wastewater laden with heavy metals. Chem. Eng. J. 118 (1–2), 83–98. https://doi.org/10.1016/j.cej.2006.01.015.

Kang P.H., Jeun J.P., Chung B.Y., Kim J.S., Nho Y.C., 2007. Preparation and characterization of glycidyl methacrylate (GMA) grafted kapok fiber by using radiation induced grafting technique. J. Ind. Eng. Chem. Seoul. 13, 956.

Lee, J.W., Choi, S.P., Thiruvenkatachari, R., Shim, W.G., Moon, H., 2006. Evaluation of the performance of adsorption and coagulation processes for the maximum removal of reactive dyes. Dyes Pigments 69 (3), 196–203. https://doi.org/10.1016/j.dyepig.2005.03.008.

Li, X., Xu, Q., Han, G., Zhu, W., Chen, Z., He, X., Tian, X., 2009. Equilibrium and kinetic studies of copper(II) removal by three species of dead fungal biomasses. J. Hazard. Mater. 165 (1–3), 469–474. https://doi.org/10.1016/j.jhazmat.2008.10.013.

Li, Q., Li, X., Wageh, S., Al-Ghamdi, A.A., Yu, J., 2015. CdS/graphene nanocomposite photocatalysts. Adv. Energy Mater. 5 (14), 1500010. https://doi.org/10.1002/aenm.201500010.3516P.

Liu, R., Dai, L., Si, C.L., 2018. Mussel-inspired cellulose-based nanocomposite fibers for adsorption and photocatalytic degradation. ACS Sustainable Chem. Eng. 6 (11), 15756–15763. https://doi.org/10.1021/acssuschemeng.8b04320.

Lomelí-Ramírez, M.G., Valdez-Fausto, E.M., Rentería-Urquiza, M., Jiménez-Amezcua, R.M., Hernández, Anzaldo, J., Torres-Rendon, J., G., García Enriquez, S., 2018. Study of green nanocomposites based on corn starch and cellulose nanofibrils from Agave Tequilana Weber. Carbohydr. Polym. 201, 9–19. https://doi.org/10.1016/j.carbpol.2018.08.045.

Malaviya, P., Singh, A., 2011. Physicochemical technologies for remediation of chromium-containing waters and wastewaters. Crit. Rev. Environ. Sci. Technol. 41 (12), 1111–1172. https://doi.org/10.1080/10643380903392817.

Mamais, D., Noutsopoulos, C., Kavallari, I., Nyktari, E., Kaldis, A., Panousi, E., Nikitopoulos, G., Antoniou, K., Nasioka, M., 2016. Biological groundwater treatment for chromium removal at low hexavalent chromium concentrations. Chemosphere 152, 238–244. https://doi.org/10.1016/j.chemosphere.2016.02.124.

Mazhar, U.-I., Muhammad, U., Shaukat, K., Tahseen, K., Salman, U.-I., Nasrullah, S., Joong, P., 2016. Recent advancement in cellulose based nanocomposite for addressing environmental challenges. Recent Pat. NanoTechnol. 169–180. https://doi.org/10.2174/1872210510666160429144916.

Mehtab, H., Muhammad, F.M., Asma, J., Sidra, A., Nayab, A., Sharon, Z., Jaweria, H., 2017. Water pollution and human health. Environ. Risk Assess. Remediat https://doi.org/10.4066/2529-8046.100020.

Mohanty, J.K., Misra, S.K., Nayak, B.B., 2001. Sequential leaching of trace elements in coal: A case study from Talcher coalfield, Orissa J. Geo. Soc. India 58, 441–447.

Mondal, P., 2015. Essay on Radioactive Pollution: Sources, Effects and Control of Radioactive Pollution https://www.yourarticlelibrary.com/pollution/essay-on-radioactive-pollution-sources-effects-and-control-of-radioactive-pollution/23271. Accessed 20 August 2020.

Mousavi, S.M., Hashemi, S.A., Babapoor, A., Savardashtaki, A., Esmaeili, H., Rahnema, Y., Mojoudi, F., Bahrani, S., Jahandideh, S., Asadi, M., 2019. Separation of Ni (II) from industrial wastewater by Kombucha Scoby as a colony consisted from bacteria and yeast: kinetic and equilibrium studies. Acta Chim. Slovenica 66 (4), 865–873. https://doi.org/10.17344/acsi.2019.4984.

Moussa, D.T., El-Naas, M.H., Nasser, M., Al-Marri, M.J., 2017. A comprehensive review of electrocoagulation for water treatment: potentials and challenges. J. Environ. Manage. 186, 24–41. https://doi.org/10.1016/j.jenvman.2016.10.032.

Nadine, K., F., M.D, 2005. Extraction of selected heavy metals using modified clays. J. Environ. Sci. Health A 40 (3), 601–608. https://doi.org/10.1081/ESE-200046606.

Olukanni, D.O., Agunwamba, J.C., Ugwu, E.I., 2014. Biosorption of heavy metals in industrial wastewater using microorganisms (*Pseudomonas aeruginosa*). Am. J. Sci. Ind. Res. 5 (2), 81–87.

Owen, G., Bandi, M., Howell, J.A., Churchouse, S.J., 1995. Economic assessment of membrane processes for water and waste water treatment. J. Membr. Sci. 102 (C), 77–91. https://doi.org/10.1016/0376-7388(94)00261-V.

Palmate, S.S., Pandey, A., Dheeraj, K., Pandey, R.P., Mishra, S.K., 2017. Climate change impact on forest cover and vegetation in Betwa Basin, India. Appl. Water Sci. 103–114. https://doi.org/10.1007/s13201-014-0222-6 .

Pandey, N., Shukla, S.K., Singh, N.B., 2017. Water purification by polymer nanocomposites: an overview. Nanocomposites 3 (2), 47–66. https://doi.org/10.1080/20550324.2017.1329983.

Phan, T.N.T., Bacquet, M., Morcellet, M., 2000. Synthesis and characterization of silica gels functionalized with monochlorotriazinyl β-cyclodextrin and their sorption capacities towards organic compounds. J. Incl. Phenom 38 (1–4), 345–359. https://doi.org/10.1023/a:1008169111023.

Pradhan, P., Costa, L., Rybski, D., Lucht, W., Kropp, J.P., 2017. A systematic study of sustainable development goal (SDG) interactions. earth's. Future 5 (11), 1169–1179. https://doi.org/10.1002/2017EF000632.

Qu, X., Alvarez, P.J.J., Li, Q., 2013. Applications of nanotechnology in water and wastewater treatment. Water Res. 47 (12), 3931–3946. https://doi.org/10.1016/j.watres.2012.09.058.

Rafatullah, M., Sulaiman, O., Hashim, R., Ahmad, A., 2010. Adsorption of methylene blue on low-cost adsorbents: a review. J. Hazard. Mater. 177 (1–3), 70–80. https://doi.org/10.1016/j.jhazmat.2009.12.047.

Rahimzadeh, M.R., Rahimzadeh, M.R., Kazemi, S., Moghadamnia, A.A., 2017. Cadmium toxicity and treatment: an update. Casp. J. Int. Med. 8 (3), 135–145. https://doi.org/10.22088/cjim.8.3.135.

Rajasulochana, P., Preethy, V., 2016. Comparison on efficiency of various techniques in treatment of waste and sewage water – a comprehensive review. Resource-Efficient Technol. 175–184. https://doi.org/10.1016/j.reffit.2016.09.004.

Rao, K.R., 2001. Radioactive waste: the problem and its management. Curr. Sci. 81 (12), 1534–1546.

Rasalingam, S., Peng, R., Koodali, R.T., 2014. Removal of hazardous pollutants from wastewaters: applications of TiO_2-SiO_2 mixed oxide materials. J. NanoMater. https://doi.org/10.1155/2014/617405.

Rasoulzadeh, H., Dehghani, M.H., Mohammadi, A.S., Karri, R.R., Nabizadeh, R., Nazmara, S., Kim, K.H., Sahu, J.N., 2020. Parametric modelling of Pb(II) adsorption onto chitosan-coated Fe_3O_4 particles through RSM and DE hybrid evolutionary optimization framework. J. Mol. Liquids 297. https://doi.org/10.1016/j.molliq.2019.111893.

Ray, P.Z., Shipley, H.J., 2015. Inorganic nano-adsorbents for the removal of heavy metals and arsenic: a review. RSC Adv., 5 (38), 29885–29907. https://doi.org/10.1039/c5ra02714d.

Rekha, Naik, S., N., Prasad, R, 2006. Pesticide residue in organic and conventional food-risk analysis. J. Chem. Health Saf. 13 (6), 12–19. https://doi.org/10.1016/j.chs.2005.01.012.

Roy, D., Semsarilar, M., Guthrie, J.T., Perrier, S, 2009. Cellulose modification by polymer grafting: a review. Chem. Soc. Rev. 38, 2046–2064.

Saber-Samandari, S., Saber-Samandari, S., Heydaripour, S., Abdouss, M., 2016. Novel carboxymethyl cellulose based nanocomposite membrane: synthesis, characterization and application in water treatment. J. Environ. Manage. 166, 457–465. https://doi.org/10.1016/j.jenvman.2015.10.045.

Saxena, R., Saxena, M., Lochab, A., 2020. Recent progress in nanomaterials for adsorptive removal of organic contaminants from wastewater. Chem. Select 5 (1), 335–353. https://doi.org/10.1002/slct.201903542.

Spellman, F.R., 2004. Mathematic Manual for Water and Wastewater Treatment Plant Operators. CRC Press, Boca Raton, FL.

Staicu, L.C., Van Hullebusch, E.D., Oturan, M.A., Ackerson, C.J., Lens, P.N.L., 2015. Removal of colloidal biogenic selenium from wastewater. Chemosphere 125, 130–138. https://doi.org/10.1016/j.chemosphere.2014.12.018.

Strathmann, H., 2001. Membrane separation processes: current relevance and future opportunities. AIChE J. 47 (5), 1077–1087. https://doi.org/10.1002/aic.690470514.

Sun, Q., Yang, L., 2003. The adsorption of basic dyes from aqueous solution on modified peat-resin particle. Water Res. 37 (7), 1535–1544. https://doi.org/10.1016/S0043-1354(02)00520-1.

Sushovan, S., Somnath, N., Susmita, D., 2018. Application of RSM and ANN for optimization and modeling of biosorption of chromium(VI) using cyanobacterial biomass. Applied Water Science. https://doi.org/10.1007/s13201-018-0790-y.

Székács, A., Mörtl, M., Darvas, B., 2015. Monitoring pesticide residues in surface and ground water in Hungary: surveys in 1990-2015. J. Chem. 2015, 717948. https://doi.org/10.1155/2015/717948.

Szygułа, A., Guibal, E., Palacín, M.A., Ruiz, M., Sastre, A.M., 2009. Removal of an anionic dye (acid blue 92) by coagulation-flocculation using chitosan. J. Environ. Manage. 90 (10), 2979–2986. https://doi.org/10.1016/j.jenvman.2009.04.002.

Ugwu, E.I., Agunwamba, J.C., 2020. A review on the applicability of activated carbon derived from plant biomass in adsorption of chromium, copper, and zinc from industrial wastewater. Environ. Monit. Assess. 192 (4), 240. https://doi.org/10.1007/s10661-020-8162-0.

Uyama, H., Kuwabara, M., Tsujimoto, T., Nakano, M., Usuki, A., Kobayashi, S., 2003. Green nanocomposites from renewable resources: plant oil-clay hybrid materials. Chem. Mater. 15 (13), 2492–2494. https://doi.org/10.1021/cm0340227.

Vongdala, N., Tran, H.D., Xuan, T.D., Teschke, R., Khanh, T.D., 2019. Heavy metal accumulation in water, soil, and plants of municipal solid waste landfill in Vientiane, Laos. Int. J. Environ. Res. Public Health 16 (1). https://doi.org/10.3390/ijerph16010022.

Walha, K., Amar, R.B., Firdaous, L., Quéméneur, F., Jaouen, P., 2007. Brackish groundwater treatment by nanofiltration, reverse osmosis and electrodialysis in Tunisia: performance and cost comparison. Desalination 207 (1–3), 95–106. https://doi.org/10.1016/j.desal.2006.03.583.

Wan, Ngah, S., W., Hanafiah, M.A.K.M, 2008. Removal of heavy metal ions from wastewater by chemically modified plant wastes as adsorbents: a review. Bioresour. Technol. 3935–3948. https://doi.org/10.1016/j.biortech.2007.06.011.

Wang, J., Chen, C., 2006. Biosorption of heavy metals by Saccharomyces cerevisiae: A review. Biotechnol. Adv. 24 (5), 427–451. https://doi.org/10.1016/j.biotechadv.2006.03.001.

Wittmar, A.S.M., Fu, Q., Ulbricht, M., 2017. Photocatalytic and Magnetic Porous Cellulose-Based Nanocomposite Films Prepared by a Green Method. ACS Sustain. Chem. Eng. 5 (11), 9858–9868. https://doi.org/10.1021/acssuschemeng.7b01830.

Xiao, X.F., Yang, N., Wang, Z.L., Huang, Y.Q., 2016. Determination of trace mercury(II) in wastewater using on-line flow injection spectrophotometry coupled with supported liquid membrane enrichment. Anal. Methods 8 (3), 582–586. https://doi.org/10.1039/c5ay02725j.

Yargiç, A.S., Yarbay Şahin, R.Z., Özbay, N., Önal, E., 2015. Assessment of toxic copper(II) biosorption from aqueous solution by chemically-treated tomato waste. J.Clean. Prod. 88, 152–159. https://doi.org/10.1016/j.jclepro.2014.05.087.

Yoon, Y., Westerhoff, P., Snyder, S.A., Wert, E.C., 2006. Nanofiltration and ultrafiltration of endocrine disrupting compounds, pharmaceuticals and personal care products. J. Membr. Sci. 270 (1–2), 88–100. https://doi.org/10.1016/j.memsci.2005.06.045.

Zamani, S.A., Yunus, R., Samsuri, A.W., Salleh, M.A.M., Asady, B., 2017. Removal of zinc from aqueous solution by optimized oil palm empty fruit bunches biochar as low cost adsorbent. Bioinorg. Chem. Appl. 2017, 7914714. https://doi.org/10.1155/2017/7914714.

Zaroual, Z., Azzi, M., Saib, N., Chainet, E., 2006. Contribution to the study of electrocoagulation mechanism in basic textile effluent. J. Hazard. Mater. 131 (1–3), 73–78. https://doi.org/10.1016/j.jhazmat.2005.09.021.

Zhongchen, H., Jianwu, L., Hailong, W., Zhengqian, Y., Xudong, W., Yongfu, L., Dan, L., Zhaoliang, S., 2019. Soil contamination with heavy metals and its impact on food security in China. J. Geosci. Environ. Protect. 7, 168–183. https://doi.org/10.4236/gep.2019.75015.

Zhou, Q., Yang, N., Li, Y., Ren, B., Ding, X., Bian, H., Yao, X., 2020. Total concentrations and sources of heavy metal pollution in global river and lake water bodies from 1972 to 2017. Glob. Ecol. Conserv. 22, e00925. https://doi.org/10.1016/j.gecco.2020.e00925.

Zhu, M.X., Lee, L., Wang, H.H., Wang, Z., 2007. Removal of an anionic dye by adsorption/precipitation processes using alkaline white mud. J. Hazard. Mater. 149 (3), 735–741. https://doi.org/10.1016/j.jhazmat.2007.04.037.

CHAPTER 14

Functionalized green carbon-based nanomaterial for environmental application

Oscar M. Rodríguez-Narvaez[a], Daniel A. Medina-Orendain[b] and Lorena N. Mendez-Alvarado[c]

[a]Dirección de investigación y soluciones tecnológicas, Centro de Innovación Aplicada en Tecnologías Competitivas (CIATEC), Omega 201, Col. Industrial Delta, León, Guanajuato, C.P. 37545, México.
[b]Departamento de Química y Bioquímica, Tecnológico Nacional de México-Instituto Tecnológico de Tepic, Av. Tecnológico # 2595, Lagos del Country, 63175, Tepic, Nayarit, México
[c]División de ciencias naturales y exactas, Departamento de Ingeniería Química, Universidad de Guanajuato, Noria Alta s/n, 36050, Guanajuato, Guanajuato, México

14.1 Introduction

Over the past decade, the advanced oxidation processes (AOPs) have shown high efficiency in organic contaminant degradation (Rodriguez-Narvaez et al., 2017; Pacheco-Álvarez et al., 2018; Ramírez-Sánchez et al., 2017). However, as several AOPs use metals as catalysts, there is an increasing concern about the environmental and human health risks of using these materials (e.g., metal oxides and metal salts) (Rodríguez-Narváez et al., 2019). To prevent the release of metal catalysts into the environment, different solid matrices (e.g., silicate and carbon-based materials) have been used as support materials, and metal-free catalysts have been investigated (Rodríguez-Narváez et al., 2019; Rodriguez-Narvaez et al., 2019; Choi et al., 2010). Among different metal-free catalysts and support materials developed, carbon-based materials (e.g., graphene, biochar, carbon nanotubes) have shown considerable relevance because their low-cost and environmentally friendly nature (Zhang et al., 2019; Mohan et al., 2014; Tan et al., 2015).

Carbon-based materials are produced by a physicochemical treatment (e.g., oxidation and pyrolysis) from carbon feedstock (Zhang et al., 2019; Chee H. Chia and Downie, 2019; Chen et al., 2018). A wide variety of carbon-based materials (e.g., graphene, biochar, hydrochar) have been reported with different physical and chemical properties (e.g., adsorption capacity, photo-activity) (Zhao et al., 2017; Zanias et al., 2020; Xiao et al., 2018; Khataee et al., 2017). However, some carbon-based materials do not possess the characteristics required for their efficient use in AOPs. Therefore, functionalization of these materials has been developed to improve their chemical and physical properties (e.g., high adsorption, photoactivity) for their application in AOPs (Rodriguez-Narvaez et al., 2019; fei Tan et al., 2016; Rajapaksha et al., 2016).

Sustainable Nanotechnology for Environmental Remediation
DOI: https://doi.org/10.1016/B978-0-12-824547-7.00005-9

Different carbon-based material synthesis methods, functionalization, and applications for AOPs have been reported. Therefore, this work aims to undertake an in-depth review of the current state-of-the-art in functionalization processes of carbon-based materials applied to AOPs and identify existing knowledge gaps for future research lines.

14.2 Carbon-based material classification

Carbon is one of the most abundant elements on our planet and stands out as feedstock for the production of different materials (Zhang et al., 2019; Manyà, 2012; Chen et al., 2010). Moreover, as carbon has several allotropic forms, different carbon-based materials can be generated depending on the sysnthesis method (e.g., diamond, graphite, biochar, graphene) (Rodriguez-Narvaez et al., 2019; Chen et al., 2018; Wang et al., 2018). Therefore, carbon-based materials are usually classified according to the synthesis and functionalization methods.

14.2.1 Graphene

Graphene has a honeycomb-shaped (hexagonal) crystal lattice with one atom thickness and sp^2 hybridization that produces a two-dimensional (2-D) structure. This structure causes that each carbon atom has a free electron in the last valence shell, which creates an electron cloud throughout the graphene lamina, generating a special electronic structure (Yadav and Dixit, 2017; Lim et al., 2018). Moreover, graphene's electronic properties give rise to nonconventional carbon characteristics, such as mechanical (e.g., high hardness, resistance, flexibility) and thermal features, as well as the ability to maintain high electric current densities (Ramos-Galicia et al., 2013; Ray, 2015).

14.2.1.1 Synthesis methods

Graphene synthesis can be performed by physical or chemical methods. Physical methods involve breaking a bond between the adjacent graphene layers that are detached layer by layer from the graphite with the use of mechanical, thermal, or electrical force (e.g., arc discharge method, thermal reduction, micromechanical exfoliation) (Choi et al., 2010; Yi and Shen, 2015; Qin et al., 2016). Chemical methods, however, involve the oxidation of graphite by adding functional groups between the layers to separate them. Once the separation is produced, an exfoliation and reduction process is carried out to eliminate functional groups to generate graphite oxide (GO) layers (Cao and Zhang, 2015; Kumar et al., 2019), and finally, graphene. Table 14.1 shows the different graphite chemical oxidation processes. Among them, the Hummers method is considered an easy and low-cost methodology based on graphite oxidation by potassium permanganate ($KMnO_4$) in concentrated sulfuric acid (H_2SO_4) and sodium nitrate ($NaNO_3$) (De Silva et al., 2017; Amir Faiz et al., 2020). Nevertheless, its main drawback is that temperature

Table 14.1 Graphite chemical oxidation processes.

Method	Oxidant	Solvent	Oxidation time (days)	Temperatures (°C)[a]	Reference
Brodie	$KClO_3$	HNO_3	3–5	40–60	(Hermanová et al., 2015)
Staudenmaier	$KClO_3$	HNO_3/H_2SO_4	4	Room temperature	(Ong et al., 2012)
Hummers	$NaNO_3/KMnO_4$	H_2SO_4	0.25	20–35 and 70–98[b]	(Yoo and Park, 2019)
Hummers Modified	$KMnO_4$	H_2SO_4/H_3PO_4	0.3	35–40	(Zaaba et al., 2017, Alam et al., 2017)

[a] When oxidants are added, a 0 °C temperature should be used.
[b] A two-step process is used at different temperatures.

must be controlled because an exothermic reaction is produced (Shao et al., 2012; Dimiev and Tour, 2014). Therefore, method modifications have been reported using acid mixtures (e.g., H_2SO_4/H_3PO_4) and other oxidants (e.g., $KMnO_4$, $K_2S_2O_8$, P_2O_5) to avoid the temperature disadvantage (Shao et al., 2012; Dimiev and Tour, 2014; Lee et al., 2019).

14.2.1.2 Graphene functionalization

Graphene oxide functionalization to improve its chemical and physical properties can be performed by different methods. The chemical reduction of graphene oxide has been reported as one of the most used for graphene oxide functionalization because it can be achieved using different techniques (e.g., thermal reduction, microwave assistance, photoreduction), even at environmental conditions, and it does not require special equipment (Dong et al., 2017; Zhang et al., 2018; Tai et al., 2019). However, as most reduction methods, it uses toxic reagents (e.g., hydrazine and boron sodium hydride). Therefore, some studies have reported non-toxic solvents (e.g., organic acids, plant extracts, microorganisms, sugars, antioxidants, amino acids, proteins) used to overcome the disadvantage of using toxic reagents (De Silva et al., 2017).

Graphene functionalization with non-metals and metals significantly modifies graphene, providing chemical and physical improvement (e.g., increased availability of active sites and improved conductivity and optical properties) (Xu et al., 2017; Tabrizian et al., 2019; Wu et al., 2019). Hence, different non-metal and metal functionalization methods were developed. For non-metal addition, one of the most reported techniques is the solvothermal/hydrothermal method because of its low-temperature range (110–180 °C), which is used for nitrogen-doped graphene production (Jagannatham et al., 2015; Somu et al., 2017). For metal addition, however, photoactivation is the most common. It is carried out by producing a graphene oxide, a metal feedstock (e.g., iron), and short-chain alcohol (e.g., methanol, ethanol, or isopropanol) solution, which is irradiated with ultraviolet (UV) radiation afterward (Kecsenovity et al., 2013).

14.2.2 Carbon nanotubes

Carbon nanotubes (CNTs) are graphene sheets with a tubular form and sp^2 and sp^3 intermediate hybridization. CNTs' main characteristics are affected by the graphene sheets synthesis method and morphology (e.g., number of concentric layers, tube diameter, roll-up) (Segawa et al., 2016). According to the number of walls, CNTs are classified into single- or multi-walled. Monolayer or single-walled CNTs (SWCNTs) are a single graphene layer with 0.71–3 nm diameter, while multiwalled CNTs (MWCNTs) are several concentric graphene layers, consisting of 2–50 cylindrical graphene layers with interlayer spacing equal to graphene. Their internal and external diameters range between 2–10 and 15–30 nm, respectively (Kobashi et al., 2019).

CNTs' roll-up has been reported as the main parameter as it defines helicity, impacting CNT properties. As for SWCNTs, three types are produced, depending on the helicity: armchair, zigzag, and chiral forms. Armchair and zigzag CNTs are straight nanotubes, whereas the chiral form is a twisted structure. As MWCNTs are a collection of SWCNT structures, different helicity-dependent forms of SWCNTs, and the quantitative presence of each of them in MWCNT structures will affect MWCNT chemical and physical properties (Kumar et al., 2018).

14.2.2.1 Synthesis methods

The electric arc-discharge, laser ablation technique, chemical vapor deposition (CVD), and hydrothermal methods stand out among the several synthesis methodologies. The electric arc-discharge uses high temperature (3,000 °C) to generate cathode decomposition, providing carbon atoms for CNT production in the anode. A potential of 20–30 V is applied to 5–20 μm diameter graphite electrodes with 1 mm separation under inert atmosphere (e.g., nitrogen, helium, or argon) at a moderate pressure range (50–700 mbar) (Sharma et al., 2015). Both SWCNTs and MWCNTs can be produced using the arc-discharge method, but for SWCNTs, metal catalysts (e.g., cobalt, iron, or nickel) should be used on the cathode (Kecsenovity et al., 2013; Mohammed et al., 2017; Bhongade et al., 2019; Manawi et al., 2018).

In the laser ablation method, a graphite plate, with 98.8% graphite and 1.2% cobalt/nickel, is placed in the center of a tube furnace at 1,200 °C, under an inert atmosphere (e.g., argon or helium), and exposed to controlled laser pulses. As graphite plate carbon vaporizes, the vaporized atoms are transported by an inert gas to the tube furnace outlet, where they are condensed to generate CNTs with 10–20 nm diameter and 100 μm length (Das et al., 2016).

Laser ablation and arc-discharge methods have one main technical drawback: no CNTs with specific characteristics can be produced (e.g., length, diameter, orientation, and purity). For this reason, other synthesis methods have been developed, among which the CVD method stands out for its capacity to produce CNTs with specific characteristics

(Saifuddin et al., 2013). The CVD method uses carbon feedstock (e.g., methane, ethane, acetylene, or aromatic compounds), which is carried into a tube furnace by an inert gas (e.g., Ar, N_2, or He) with continuous flow, and temperature 500–1200 °C and atmospheric pressure, to be decomposed on a catalytic surface (e.g., silica gels, alumina, zeolites, or copper matrices). The released carbon atoms are gradually cooled down at room temperature and deposited on the tube walls, used as nucleation sites for CNT generation (Kecsenovity et al., 2013; Mohammed et al., 2017; Bhongade et al., 2019). Also, reports in which a metal catalyst was mixed with the inert gas and the carbon feedstock showed an improvement in reaction time, bigger CNTs' length, continuous reaction operation, and a large-scale application (Manawi et al., 2018).

14.2.2.2 CNT functionalization

CNT functionalization has been thoroughly studied, and metal and non-metal dopings (e.g., iron and nitrogen atoms, respectively) are the most frequently reported (Xu et al., 2018; Ma et al., 2010). Among the functionalization methods, the reports of Duan et al. (2015) and Cheng et al. (2016) are interesting because their use of novel non-metal and metal doping. Duan et al. (2015) used melanin as a nitrogen source to dope SWCNTs at 700 °C, achieving high chemical stability. Cheng et al. (2016) performed an impregnation-precipitation method using iron and sulfur, generating iron/sulfur-modified CNTs.

14.2.3 Biochar

Biochar is a highly porous carbon-based material produced from different feedstocks (e.g., agricultural and garden residues) under a low-oxygen atmosphere with significant aromatization and antidecomposition properties (Wang et al., 2017; Clurman et al., 2020). Also, it has been reported to have a high specific surface area and a wide variety of negative surface functional groups (e.g., hydroxyl, carboxyl, and amino), which make it a multifunctional adsorbent (Rodriguez-Narvaez et al., 2017; Tan et al., 2015; Inyang et al., 2016).

14.2.3.1 Synthesis methods

Biomass conversion technologies have been developed, using biological (e.g., anaerobic digestion, hydrolysis, and fermentation) and thermal (e.g., combustion, pyrolysis, liquefaction, Torre-faction, and gasification) treatments (Czernik and Bridgwater, 2004; Mohan et al., 2006). However, most of these methods are used for biomass conversion to liquid and gas fuel. Table 14.2 shows the thermochemical processes in which carbon-based adsorbents are produced (Brewer et al., 2009). Although several thermochemical processes are reported to generate different carbon-based adsorbent materials, biochar is mainly produced by pyrolysis. It is a thermal treatment in which high temperatures at low- or free-oxygen atmosphere are used to decompose organic matter with irreversible

Table 14.2 Thermochemical processes, reactions, and residence time (modified) (Mohan et al. 2014).

Thermochemical processes	Temperature range (°C)	Heating rate, °C/min	Pressure	Residence time
Slow pyrolysis	350–800	<10	Atmospheric	Second–hours
Fast pyrolysis	400–600	≈60,000	Vacuum – atmospheric	Seconds
Gasification	700–1500	10–1,000	Atmospheric – elevated	Seconds–minutes
Hydrothermal carbonization	175–250	<10	-	Hours
Torrefaction	200–300	<10	Atmospheric	Minute–hours

changes. Conventional pyrolysis methods can be classified into slow and fast pyrolysis, differentiated by thermal conditions used (i.e., temperature, heating rate, and vapor residence time) (Mohan et al., 2014).

Fast pyrolysis is a low-efficiency and high-cost thermochemical treatment because it requires pretreatment of feedstock to obtain a low humidity percentage (i.e., less than 10%) and 1–2 mm particle size. Also, a rapid heating rate (see Table 14.2) has to be used to reach 400–500 °C in seconds, and only 10% of biochar is generated from the total feedstock (Brewer et al., 2009; Lima et al., 2010). However, when slow or conventional pyrolysis is used, a more efficient biochar production is reported because the conversion rate distribution into biochar in solid, liquid, and gas is 35%, 30%, and 35%, respectively (Antal and Grønli, 2003). Also, a low heating rate up to 500 °C and short residence times are required (see Table 14.2).

14.2.3.2 Biochar functionalization

The biochar functionalization process can be mainly divided into three methods: chemical modification, physical modification, and the use as a support matrix for nanomaterials (Rajapaksha et al., 2016). Chemical modifications are performed via one- or two-step processes. The one-step method is a simultaneous carbonization-activation process in the presence of an activating reagent (e.g., acid, oxidant, or alkaline). The two-step process is biomass carbonization, followed by activation in a mixed solution using an activating reagent (Qian et al., 2015; Azargohar and Dalai, 2008). Most of the reports on chemical modifications used acids, hydroxyls, and oxidants (Jin et al., 2014; Xue et al., 2012; Li et al., 2014). However, some studies also added specific functional groups (i.e., amine and carboxyl) to the biochar surface to provide it with a predominant functional group and achieve a more homogenous surface (Fig. 14.1) (Macoveanu and Gavrilescu, 2010; Li et al., 2016).

Biochar physical modifications are performed by thermal methods and partial gasification of the carbon-based material with a gas (e.g., steam, carbon dioxide, and air) at 800–1000 °C temperature (Nandi et al., 2012; Bouchelta et al., 2008; Maciá-Agulló et al., 2004). Physical methods are usually used for scale-up processes because they are simple

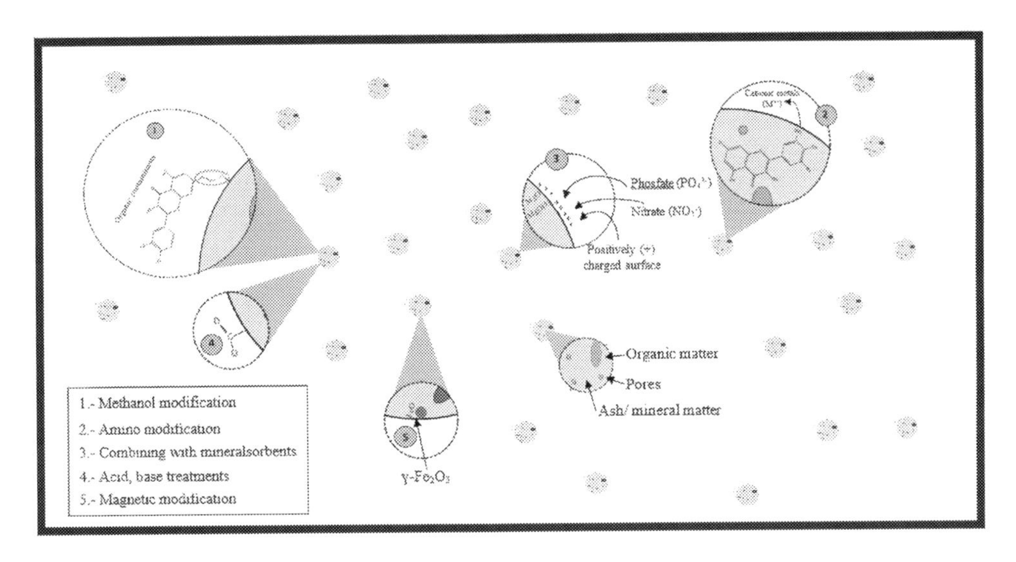

Figure 14.1 Biochar scheme using different functionalization methods (modified from Rajapaksha et al., 2016).

and economically feasible, despite their low efficiency (Rajapaksha et al., 2016; Mestre et al., 2015; Sun et al., 2016).

Several studies immobilized nano-sized materials into biochar surfaces to increase physical and chemical characteristics are reported. Furthermore, different classification methods were reported depending on the immobilized nano-sized materials, developed material properties, and application (Rodriguez-Narvaez et al., 2019; Rajapaksha et al., 2016). Among various research works, that of Rodriguez-Narvaez et al. (2019) should be highlighted as it deals with the types of nanomaterials used to functionalize the biochar: (1) nanometal oxide/hydroxide-biochar (N-MO/HB) composites; (2) magnetic biochar composites; and (3) functional-nanoparticle-coated biochar; it also reports how the different materials were used for soil remediation, water treatment, and CO_2 capture.

14.2.4 Hydrochar

Hydrochar is a carbon-based material produced by hydrothermal carbonization (HTC) at 180–250 °C (see Table 14.2). This method has recently emerged as a highly cost-effective product with significant advantages such as low-energy consumption, high yield, and low ash content (Zhang et al., 2019). Hydrochar has been reported to have low porosity and low specific surface area, and to contain polar surface functional groups, which suggests low adsorption capacity, particularly for non-polar organic matter (Nizamuddin et al., 2019). Therefore, opportunities for improving this material are highlighted.

14.2.4.1 Synthesis methods

Despite being produced in the same way as biochar in terms of thermochemical treatment, hydrochar does not follow the biochar-specific conditions (e.g., slow heat rate and oxygen-free atmosphere). Therefore, the number of studies on hydrochar synthesis is increasing (Wang et al., 2018; Mihajlović et al., 2016). Hydrochar is synthesized using HTC in which the biomass is mixed together with a solution (e.g., water) to generate a solid/liquid mass ratio; then, it is inserted and sealed in a reactor, which is heated at low temperatures (Table 14.2). HTC will generate an autogenic pressure inside the reactor, producing a char– water–slurry mixture, in which the solid part is the hydrochar (Zhang et al., 2019).

Hydrochar synthesis method is modified by using other assisted methods (e.g., electrochemical and microwave). Among different HTC-assisted methods, microwave-assisted synthesis stands out because it uses high-frequency electromagnetic radiation (e.g., 300 MHz-300 GHz) with a dipole rotation and ionic conduction to induce heat, which reduces process rates, energy consumption, and improves dewaterability (Dai et al., 2017; Shao et al., 2019; Cao et al., 2019; Kannan et al., 2017).

14.2.4.2 Hydrochar functionalization

Hydrochar functionalization can be performed in the same way as biochar, but some reports have highlighted the use of biological assistance for that purpose (Hua et al., 2020; Kaushik et al., 2014). Biological modification for hydrochar aging has been carried out by adding microorganisms to a hydrochar solution to produce anaerobic fermentation (Hua et al., 2020). Further, enzymatic-assisted hydrolysis as a pretreatment was performed to improve biomass decomposition and its solubilization prior to the HTC treatment (Kaushik et al., 2014).

14.2.5 Carbon dots

Carbon quantum dots (CQDs) are fluorescent graphene spherical or semispherical nanoparticles with a 1–20 nm average size and high carbon proportions. CQDs have sp^2 and sp^3 hybridized carbon atoms and different surface functional groups (e.g., hydroxyl, carboxyl, carbonyl, ether, and epoxy), which generate high thermal stability, optical photostability, and specific surface area (Fernando et al., 2015; Sagbas and Sahiner, 2018; Chen et al., 2019).

14.2.5.1 Synthesis methods

CQD synthesis has been developed from different carbon feedstocks (e.g., coal, graphite, citric acid, and glycerol) and natural products (e.g., aloe, mushroom, watermelon peel, and hair fiber) through inexpensive and easy-to-do processes (Nurunnabi et al., 2013; Xu et al., 2015). Two main CQD synthesis methods are "top-down" and "bottom-up."

Table 14.3 "Bottom-up" CQDs synthesis methods using organic and natural source (modified) (Das et al., 2018).

Synthesis method	Carbon feedstock	Solvent/additive	Reaction conditions, temperature/time
Carbonization	Watermelon	Water	220 °C/2 h
	Citric acid	Water/polyethyleneimine	200 °C/20 min
	Sweet red pepper	Water	180 °C/5 h
	Albumin bovine serum	Water/concentrated sulfuric acid	50°C/2 h
Microwave	Sucrose	Water/phosphoric acid	100 W[a]/220 s
	Cashew gum	Water	800 W[a]/30–40 min
Hydrothermal/ solvothermal	Grass	Water	180 °C/3 h
	Pomelo peel	Water	200 °C/3 h
	Orange juice	Ethanol	120 °C/150 min
	Aloe	Water	180 °C/11 h

[a] Microwave power.

The "top-down" technique is done through material fragmentation by acid oxidation, arc discharge, ablation with laser, or ultrasonic exfoliation, whereas the "bottom-up" methods include carbonization, hydrothermal, microwave, and thermolysis methods (Wang and Hu, 2014; Wang et al., 2019).

The "bottom-up" processes are clearly preferred to the others because, as shown in Table 14.3, they use organic feedstocks (e.g., watermelon) and are therefore easy, cheap, and eco-friendly methods (Rai et al., 2017; Zhou et al., 2012). Also, hydrothermal/solvothermal synthesis of CQDs has been widely studied and compared with other treatments, being an easy-to-do method that achieves uniform particle size distribution (Sagbas and Sahiner, 2018; Singh et al., 2020; Das et al., 2018).

14.2.5.2 Carbon dots functionalization

CQD surface area has several functional groups, allowing the functionalization of different functional groups (e.g., amines, phosphorous, sulfur-containing heteroatoms, and metals) to improve chemical and physical characteristics (Wu et al., 2017; Park et al., 2016). Therefore, CQD functionalization focuses on the desired material properties, such as, increased visible light adsorption and specific surface area, and reduced charge recombination. For example, Han et al. (2020) functionalized CQDs with phenylhydrazine, benzoic anhydride, and 2-bromo-1-phenylethanone to increase luminol and hydrogen peroxide chemiluminescent activity through a singlet oxygen mechanism.

14.2.6 Nanodiamonds

Diamond nanocrystals or nanodiamonds (NDs) are a carbon allotropic form with a cubic crystalline structure, which has unique sp^3 hybridization in tetrahedral junction units

Table 14.4 NDs synthesis methods (modified) (Kaur and Badea, 2013).

Synthesis method	Carbon feedstock	Reaction condition
Ball milling technique	Microscale synthetic Natural diamond crystal	Environmental conditions; in presence ceramic beads
Chemical vapor deposition	Hydrocarbons	Temperature = 300 °C; Ar or H_2 atmosphere
High-pressure high temperature	Microdiamond	Pressure = 7–10 GPa; Temperature = 1500–2200 °C; Metal catalysis with Fe or Ni
Detonation	Carbon-rich explosive mixture	Autogenerated temperature and pressure; Temperature = 3000–4000 °C; Pressure = 20–30 GPa; Oxygen deficient atmosphere

and can be modified from a zero-dimensional (0-D) group to a three-dimensional (3-D) frame with versatile face-centered terminations. Therefore, NDs have high chemical stability, specific surface area, low toxicity, thermal stability, refraction coefficient, thermal conductivity, biocompatibility properties, and alkaline and acidic media resistance (Duan et al., 2019).

14.2.6.1 Synthesis methods

Table 14.4 shows as main ND synthesis methods ball milling technique (BMT) and high-pressure-high-temperature (HPHT) method, in which microscale synthetic or natural diamond crystals are used as the main source. BMT is performed at environmental conditions with ceramic beds used as catalysts (Khan et al., 2016), whereas HPHT method is carried out at high pressures and temperatures (i.e., 7–10 GPa and 1500–2200 °C, respectively) and catalyzed by metals (i.e., Fe and Ni) (Shenderova and McGuire, 2015; Stehlik et al., 2015).

On the other hand, CVD method generates polycrystalline NDs in situ by growing them on a metallic or silicon substrate, using different hydrocarbons as source (e.g., methane, ethane, propane), where hydrogen stabilizes carbon atoms during the ND nucleation and growth (Butler and Sumant, 2008; Liu et al., 2018). The detonation method, however, uses a carbon-rich explosive mixture (e.g., trinitrotoluene, octogen, and hexogen), placed into closed chambers, which produces an explosion that generates instantaneously high pressures (20–30 GPa) and temperatures (3000–4000 °C) in an oxygen-free atmosphere. The supersaturated carbon vapor is condensed into liquid drops, and these drops are then crystallized into 3–5 nm diamond particles as the pressure and temperature drop (Duan et al., 2019; Nunn et al., 2017).

14.2.6.2 ND functionalization

NDs require a purification process due to soot and non-diamond carbon production and the presence of metal residues throughout the entire process. Therefore, a deagglomeration and fractionization posttreatment (DFPT) is carried out to achieve homogeneous size distribution. This posttreatment uses the oxidation of mineral acids (concentrated sulfuric, nitric, and perchloric acid) and/or air at 425–450 °C temperature (Pichot et al., 2008; Rouhani et al., 2016). The latter process is also used as a functionalization process due to its ability to generate hydroxyl, carbonyl, and carboxylic groups on the ND surface (Krueger and Lang, 2012; Reina et al., 2019).

Nitrogen doping is reported as an important functionalization method, in which a thermal treatment at 600–700 °C, and an NH_3/Ar atmosphere is used as nitrogen feedstock (Kausar, 2018). Further, when melamine is used as nitrogen-rich feedstock, oxygen and nitrogen functionalization are employed simultaneously (Duan et al., 2016).

14.3 AOPs application

14.3.1 Photocatalysis

Organic contaminant degradation using photocatalysts has been deeply studied, because its great advantage of using light irradiation on photocatalyst activation (Wang et al., 2018; Rueda-Marquez et al., 2020; Li et al., 2018). However, most photocatalysts are nanoparticles that are not easily recovered after being added to wastewater (McKee and Filser, 2016). Therefore, further research is needed to develop low-risk nanoparticles and/or photocatalyst insertion into a low-cost solid matrix (e.g., carbon-based materials) (Hoseini et al., 2017; Zhang et al., 2020; Li et al., 2016; Zhang et al., 2020). As mentioned before, a carbon-based material is important as a solid matrix for photocatalysts and as a photocatalyst (Rodriguez-Narvaez et al., 2019; Park et al., 2016; Li et al., 2016; Li et al., 2021). Therefore, Table 14.5 shows different carbon-based materials that have been reported for photocatalysis.

CQDs have attracted increasing interest within the research community because of their important electro- and photo-chemical characteristics (Wang and Hu, 2014) and photocatalyst efficiency (Chen et al., 2019; Wang et al., 2017). Table 14.5 shows CQDs with and without functionalization have high efficiency for organic contaminant degradation by photocatalysis, highlighting the reports of Wang et al. (2017) and Markad et al. (2017), where the authors functionalized CQDs and used for organic contaminant photocatalytic degradation. Wang et al. (2017) reported indomethacin degradation improved after 80 min (i.e., 20 vs. 90%) when nitrogen (i.e., graphitic carbon nitride, g-C_3N_4) was doped into CQDs. While Markad et al. (2017) reported an 80% increase in

Table 14.5 Organic contaminant degradation using carbon-based materials for photocatalysis.

Carbon-based Material	Functionalization	Contaminant	Removal (%)	Wavelength (nm)	Notes	Reference
Carbon dots	None	Diethyl phthalate	100	365	Diethyl phthalate $[DEP]_0 = 20$ mg/L; $[CDs]_0 = 0–500$ mg/L; Power = 300 W; Intensity: 5.7×10^{-5} Einstein/cm^2 s	(Chen et al., 2019)
Carbon dots	$g-C_3N_4$	Indomethacin	92	420	Indomethacin $[IDM]_0 = 4$ mg/L; $[CDs]_0 = 1$ g/L; Power = 350 W	(Wang et al., 2017)
Carbon dots	Graphene/TiO_2	Methyl blue	10–100	None reported	Methyl blue $[MB]_0 = 3$ mg/L; $[CDs]_0 = 0.2$ g/L; Power =50 W; Intensity = 100 W/m^2	(Markad et al., 2017)
Biochar	Bismuth	Estrone	36–95	350–700	$[Estrone]_0 = 2.8$ mg/L; $[Biochar]_0 = 1$ g/L; Potential = 500 W	(Zhu et al., 2020)
Biochar	TiO_2	Acid orange 7	90	380–480	$[AO7]_0 = 20$ mg/L; $[Biochar]_0 = 0.1$ g/L; Power =15 W; Irradiation intensity = 10 mW/cm^2	(Silvestri et al., 2019)
Hydrochar	Ag_3PO_4	Sulfamethoxazole	12–98	420	Sulfamethoxazole $[SMX]_0 =$ 1 mg/L; $[Hydrochar]_0 =$ 0.1 g/L; Power = 300 W; Irradiation intensity = 180 mW/cm^2	(Zhou et al., 2019)

(continued on next page)

Hydrochar	None	Sulfadimidine	72	300–750	$[SM2]_0$ = 250 μg/L; $[Hydrochar]_0$ = 100 mg/L; Power = 40 W; irradiation intensity = 3.82 mW/cm^2	(Chen et al., 2017)
Graphene	TiO_2/Co_3O_4	Oxytetracycline congo red	91 91	400	Oxytetracycline $[OTC]_0$ = 10 mg/L; congo red $[CR]_0$ = 10 mg/L; $[Graphene]_0$ = g/L; Power =300 W; Intensity = 100 mW/cm^2	(Jo et al., 2017)
Graphene	g-C_3 $N_4/MnFe_2$	Metronidazole	78	400	Metronidazole $[MNZ]_0$ = 20 mg/L; $[Graphene]_0$ = 1 g/L; Potential = 300 W	(Wang et al., 2017)
Graphitic carbon nitride	Phosphorous	p-hydroxybenzoic acid	5 – 77	420	p-hydroxybenzoic acid $[HBA]_0$ = 1 mg/L; $[g\text{-}C_3N_4]_0$ = 0.5 g/L; Power =300 W; Intensity = 1 sun	(Wang et al., 2019)
Activated carbon	Phosphorous/Fe	Methylene blue	22–44	400–700	$[MB]_0$ = 2–20 mg/L; $[AC]_0$ = 50.4 mg/L; Power = 300 W; Irradiation intensity = 105 W/m^2	(Matos et al., 2019)

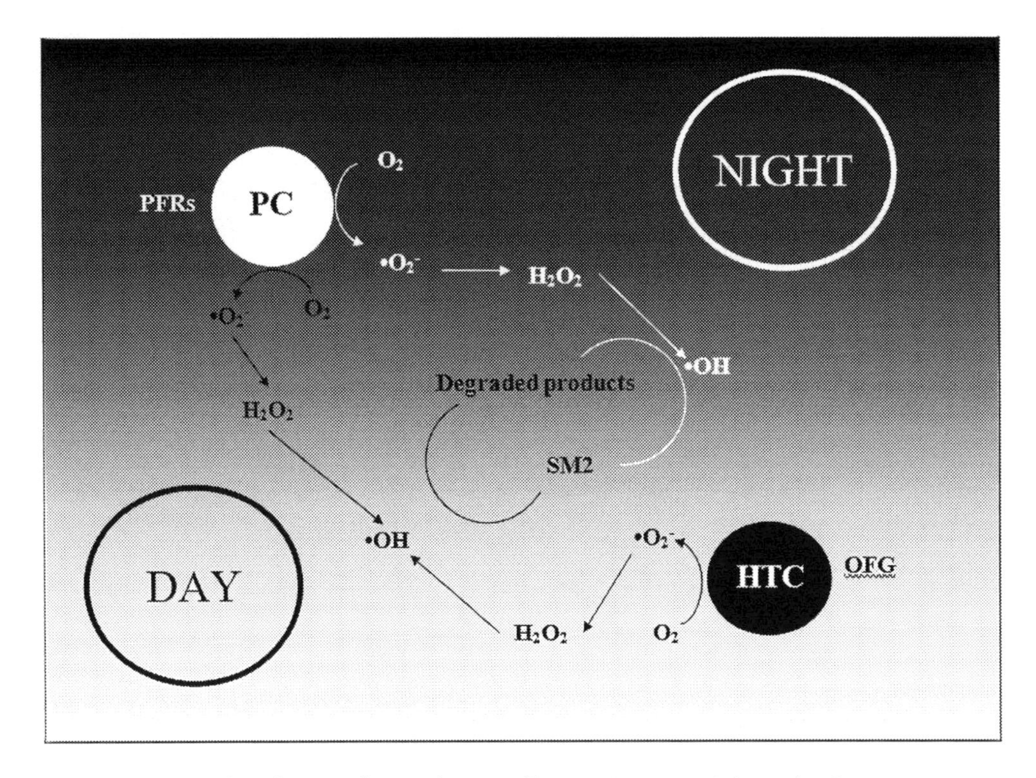

Figure 14.2 Hydrochar photocatalytic and Fenton-like reaction synergistic mechanism.

methyl blue degradation (10 vs. 90% degradation) when a ternary graphene/CQDs/TiO_2 composite was synthesized.

Photocatalysis using carbon-based materials without functionalization has been previously reported, such as CQDs, graphene, and CNTs (Wang and Hu, 2014; Ge et al., 2019; Li et al., 2016). However, these materials were reported to have an important disadvantage in full-scale production (Emiru and Ayele, 2017; Ali et al., 2019). Therefore, low-cost and high-volume production of carbon-based materials for photocatalysts was found to be a research gap. Even so, hydrochar stands out (Table 14.5) as an efficient photocatalyst because black carbon fractions on hydrochar can directly generate singlet oxygen ($^1O_2$) and superoxide (O_2^-) under sunlight, producing hydrogen peroxide that can be decomposed into hydroxyl radical (Fig. 14.2) (Fu et al., 2016; Fang et al., 2014). However, Table 14.5 shows that only Chen et al. (2017) used hydrochar as a photocatalyst, which achieved 60% more sulfadimidine degradation when hydrochar was exposed to sunlight than under dark conditions (70% vs. 10% degradation, respectively).

Table 14.5 shows that some carbon-based materials (e.g., CNTs, biochar, and hydrochar) have been widely used as a support material for photocatalysts (e.g., bismuth oxide and plasmonic materials), in which Zhu et al., (2020) report is highlighted, because the authors reported that 100% estrone degradation was achieved after 60 min and an estrone degradation mechanism was developed, in which biochar contribution for photocatalysis was improved (Eqs. 14.1 to 14.8) when biochar was doped with bismuth. However, there is an important information gap regarding the use of materials doped with bismuth oxide. Different bismuth oxide shapes, embedded in a carbon-based material, generate different photochemical properties that affect its photocatalytic activity and degradation efficiency (Fang and Shangguan, 2019; Mendez-Alvarado et al., 2020).

$$\text{biochar} + hv \rightarrow \text{biochar}^{+} + \text{biochar}(e^{-}) \tag{14.1}$$

$$\text{biochar}(e^{-}) + O_2 \rightarrow O_2^{\bullet -} \tag{14.2}$$

$$(Bi_2O_3/Bi) + hv \rightarrow Bi(h^{+} + e^{-}) \tag{14.3}$$

$$Bi(e^{-}) + O_2 \rightarrow O_2^{\bullet -} \tag{14.4}$$

$$O_2^{\bullet -} + H^{+} \rightarrow HO_2^{\bullet} \tag{14.5}$$

$$2HO_2^{\bullet} \rightarrow H_2O_2 + O_2 \tag{14.6}$$

$$H_2O_2 + O_2^{\bullet -} \rightarrow OH^{-} + O_2 + {}^{\bullet}OH \tag{14.7}$$

$$Bi(h^{+}) + H_2O \rightarrow {}^{\bullet}OH + OH^{-} \tag{14.8}$$

Table 14.5 also shows some heterojunction-based photocatalysts supported on carbon-based materials for organic degradation (Wang et al., 2017; Matos et al., 2019). Among them, plasmonic nanostructures (e.g., silver and gold) stand out because surface plasmons of these materials are used to improve their optical properties (Sánchez-Cid et al., 2019) and enhance photocatalytic activity through plasmonic energy transfer from metal nanostructure to the semiconductor, which decreases the band gap and extends light absorption range (Kale et al., 2014; Wu, 2018). Therefore, plasmonic materials (e.g., Ag_3PO_4) supported on hydrochar were used for organic contaminant degradation (Zhou et al., 2019). Zhou et al. (2019) showed that they improved sulfamethoxazole degradation compared to hydrochar alone (10 vs. 100%, respectively). Also, the authors implied that 100% sulfamethoxazole degradation could be achieved without using high hydrochar and Ag_3PO_4 composition ratios (i.e., 0.1), allowing the possible scale-up process. Further research may be carried out using of other plasmonic materials as photocatalysts coupled with different semiconductors and doped with other carbon-based materials.

14.3.2 Fenton-like reactions

Fenton-like reactions have been extensively studied because of their high organic contaminant degradation efficiency, which is based on a transition metal (e.g., cobalt, iron, or manganese) for oxidant decomposition (e.g., hydrogen peroxide, peroxymonosulfate, and percarbonate). Nevertheless, important human health and environmental risks have been underlined (Ghanbari and Moradi, 2017; Rodríguez-Narváez et al., 2018; Huang and Zhang, 2019; Rodriguez-Narvaez et al., 2020; Rodríguez-Narvaez et al., 2020). Therefore, metal immobilization in a solid matrix (e.g., silicate- and carbon-based materials) (Mortazavian et al., 2019; Wei et al., 2020; Fang et al., 2017) and metal-free catalyst production for Fenton-like reactions have been developed (Xiao et al., 2018; Yu et al., 2020).

Carbon-based materials (e.g., biochar, hydrochar, and CNTs) are mainly used as a solid matrix for metal (e.g., iron, cobalt, zero-valent iron) immobilization for Fenton-like reactions (Yu et al., 2019; Duan et al., 2017; Pu et al., 2014; Yi et al., 2019; Barrios-Bermúdez et al., 2020). However, metal addition to the carbon-based material imposes an environmental and human health risk as this material can leach the metal (Rodríguez-Narváez et al., 2019; Ghanbari and Moradi, 2017). Therefore, carbon-based material has been recently used with and without functionalization as a metal-free catalyst (or carbo-catalyst) for Fenton-like reaction (Table 14.6). Table 14.6 shows that phenols and dyes are mainly used as target contaminants. However, there is an important information gap on the degradation of other contaminants used (e.g., emerging contaminants and microplastics). Table 14.6 also shows that peroxymonosulfate (PMS) is mainly used as an oxidant, indicating a further research line to develop on the variation of oxidants used for free radical production (e.g., peroxyacetic acid and percarbonate).

Although metal-free oxidant decomposition is carried out using sulfate-based oxidants (e.g., PMS and persulfate) (Duan et al., 2016), only PMS/CNT decomposition mechanism has been developed (Eqs. 14.9 to 14.12) (Xiao et al., 2018; Lee et al., 2016). Therefore, further development of oxidant decomposition mechanisms is needed to establish different decomposition pathways and elucidate the double carbon-oxygen bond behavior when in contact with other oxidants (e.g., percarbonate and persulfate).

$$C-\pi^{+}+\mathrm{HSO}_5^{-}\rightarrow C-\pi+\mathrm{OH}^{-}+\mathrm{SO}_4(^{\bullet}-) \tag{14.9}$$

$$C-\pi^{+}+\mathrm{HSO}_5^{-}\rightarrow C-\pi+\mathrm{H}^{-}+\mathrm{SO}_5(^{\bullet}-) \tag{14.10}$$

$$C=C-\mathrm{O}^{+}\mathrm{HSO}_5^{-}\rightarrow C=C-\mathrm{O}^{+}\mathrm{OH}^{-}+\mathrm{SO}_4(^{\bullet}-) \tag{14.11}$$

$$C=C-\mathrm{O}^{+}\mathrm{HSO}_5^{-}\rightarrow C=C=\mathrm{O}+\mathrm{H}^{+}+\mathrm{SO}_5(^{\bullet}-) \tag{14.12}$$

Carbon-based materials' functionalization with non-metals (e.g., sulfur and nitrogen) is reported as a more efficient for oxidant decomposition compared to unmodified materials (Duan et al., 2015; Han et al., 2019; Ren et al., 2020). Therefore, several

Table 14.6 Organic contaminant degradation using carbon-based materials for Fenton-like reactions.

Carbon-based material	Oxidant	Contaminant	Removal, %	Notes	Reference
Carbon nanotubes	Peroxymonosulfate Peroxydisulfate	Phenol	100	[Phenol] = 20 mg/L; [CNTs] = 0.2 g/L; [PMS] = 2 g/L; Peroxydisulfate [PDS] = 1.48 g/L	(Sun et al., 2014)
Carbon nanotubes	Peroxymonosulfate Peroxydisulfate	Phenol	50–100	[Phenol] = 0.1 mM; [CNTs] = 0.1 g/L; [PDS] = 1 mM	(Lee et al., 2015)
Carbon nanotubes	Peroxymonosulfate	Phenol	40–100	[Phenol] = 0.0.53–0.212 mM; [CNTs] = 0.1 g/L; [PMS] = 2.5–12 mM	(Hou et al., 2019)
Graphene	Peroxymonosulfate	Diclofenac	30–100	[Diclofenac] = 1–10 mg/L; [Graphene] = 10–250 mg/L; [PMS] = 0.1–1 mM	(Solís et al., 2020)
Graphene	Peroxydisulfate	Rhodamine B	70–100	[Diclofenac] = 1–10 mg/L; [Graphene] = 10–250 mg/L; [PMS] = 0.1–1 mM	(Ren et al., 2018)
Biochar	Hydrogen peroxide	2-chlorobiphenyl	100	2-chlorobiphenyl [2-CB] = 10.6 μM; [Biochar] = 1 g/L; [H_2O_2] = 10 mM	(Fang et al., 2014)
Biochar	Hydrogen peroxide	Tetracycline	80–100	[Tetracycline] = 67.5 μM; [Biochar] = 0.5 g/L; [H_2O_2] = 5 mM	(Huang et al., 2019)
Biochar	Persulfate	Sulfamethoxazole	30– 80	[SMX] = 0.25–1.5 g/L; [Biochar] = 100– 300 mg/L; Persulfate [PS] = 0.25–1.5 mg/L	(Magioglou et al., 2019)
Biochar	Peroxydisulfate	Sulfadiazine	20–100	[Sulfadiazine] = 1–50 μM; [Biochar] = 0.5–3 g/L; [PS] = 250–1500 mg/L	(Wang et al., 2019)
Porous carbon	Peroxymonosulfate	Phenol	40–100	[Phenol] = 20 mg/L; [Catalyst] = 0.05–0.3 g/L; [PMS] = 0.3–3.3 mM	(Wang et al., 2017)

(continued on next page)

Table 14.6 Organic contaminant degradation using carbon-based materials for Fenton-like reactions—cont'd

Carbon-based material	Oxidant	Contaminant	Removal, %	Notes	Reference
Graphitic carbon nitride	Peroxymonosulfate	Several dyes	10–100	[Dye] = 50 mg/L; [g-C_3N_4] = 1 g/L; [PMS] = 0.118 g/L	(Dong et al., 2016)
Graphitic carbon nitride	Peroxymonosulfate	Rhodamine B	20–100	[Rhodamine B] = 10 mg/L; [g-C_3N_4] = 0.25–1 g/L; [PMS] = 0.1–1 g/L	(Lin and Zhang, 2017)
Cubic mesoporous carbon	Persulfate	Phenol	30–100	[Phenol] = 20 mg/L; [Catalyst] = 0.2 g/L; [Persulfate] = 6.5 mM	(Duan et al., 2018)
Carbon black	Persulfate	Diclofenac	60–100	[Diclofenac] = 0.5 – 4 mg/L; [g-C_3N_4] = 25–75 mg/L; [Persulfate] = 25 – 200 mg/L	(Dimitriadou et al., 2019)
Carbon fibers	Peroxyacetic acid	Reactive brilliant red X-3B	35–100	Reactive brilliant red X-3B [RRX-3B] = 50 μM; [Fibers] = 1–3 g/L; Peroxyacetic acid [PAA] = 5 mM	(Zhou et al., 2015)
Carbon fibers	Peroxymonosulfate	Reactive brilliant red X-3B	30	[RRX-3B] = 50 μM; [Fibers] = 2 g/L; [PAA] = 1 mM	(Huang et al., 2016)
Nanodiamond	Peroxymonosulfate	4-chlorophenol	20–100	[4-C P] = 0.16 mM; [Nanodiamond] = 0.1 g/L; [PMS] = 0.25 mM	(Shao et al., 2018)
Nanodiamond	Peroxydisulfate	Several organic contaminants	20–100	[Contaminant] = 0.01 mM; [Nanodiamond] = 0.1 g/L; [PDS] = 1 mM	(Lee et al., 2016)
Nanodiamond	Persulfate	Phenol	30–100	[Phenol] = 20 mg/L; [Nanodiamond] = 0.2 g/L; [Persulfate] = 6.5 mM	(Duan et al., 2016)

studies using non-metal functionalization were performed and highlighted that non-metal feedstock affected efficiency performance (Wang et al., 2017; Chen and Carroll, 2016). However, a lack of information on the diversification of nitrogen feedstock and its impact on contaminant degradation efficiency was shown. Also, another non-metal used for carbon-based materials' functionalization is sulfur, which was reported to improve organic pollutant degradation (Cheng et al., 2016; Pu et al., 2014; Ren et al., 2018). Nonetheless, there is a lack of information on how sulfur affects carbon-based materials and their organic contaminant degradation performance because there are no reports in which it is used alone.

14.3.3 Electro-oxidation

Electro-oxidation (EC), has been used for organic contaminant degradation because of its high degradation efficiency and no reagents required (Särkkä et al., 2015). The main factor for organic contaminant degradation efficiency is the type of electrode used because of its effect on electron transfer (Särkkä et al., 2015; Dominguez et al., 2018). Therefore, a search for electrodes with a broader potential window, high life-time, mechanical stability, application in continuous systems, and resistance toward passivation is needed (Tajik et al., 2020). Carbon-based materials are promising for electrodes as they are reported to have high current densities, wide potential range, and long-term stability (Li, 2020; Li et al., 2019; Cen et al., 2020; Castañeda et al., 2017). Table 14.7 shows carbon-based materials used as electrodes for electro-oxidation.

As shown in Table 14.7, a few of these materials (i.e., graphene, CNTs, and carbon felt) are used for carbon-based electrodes in electro-oxidation. Thus, an important research line using different carbon-based materials (e.g., biochar and hydrochar) should be developed because biochar and hydrochar electrodes have proved to have high-efficiency performance as supercapacitors (Chen et al., 2015; Thines et al., 2016; Yu et al., 2015). Also, although boron-doped diamond (BDD) is widely used for electro-oxidation (Pacheco-Álvarez et al., 2018; Garcia-Segura et al., 2015; Peralta-Hernández et al., 2012), the authors kept BDD materials out of this review due to their high cost and complex synthesis and functionalization (Cornejo et al., 2020). ND electrodes, however, represent a good option because of their low-cost feedstock and processes compared to BDD electrodes.

Electro-oxidation coupled with other treatments (e.g., Fenton-like process and ultrasound) and carbon-based materials as electrodes were also reported (Akbarpour et al., 2016; Khataee et al., 2013; Peralta-Hernandez et al., 2020; Soltani and Mashayekhi, 2019; Liu et al., 2019). The work of Soltani and Mashayekhi (2019 is particularly important because it reported that acetaminophen EC degradation, using carbon cloth coated with black carbon as a cathode, was improved when it was coupled with ultrasound (US) treatment (i.e., 40 vs. 75%, respectively). According to the authors, the US treatment decomposes *in-situ* hydrogen peroxide into hydroxyl radicals (Eq. 14.13) and the coupling

Table 14.7 Organic contaminant degradation using carbon-based materials for electrooxidation.

Carbon-based Material	Contaminant	Cathode/Anode	Removal, %	Notes	Reference
Carbon nanotubes	Basic Yellow 28	CNTs-PTFE (Polytetrafluoroethylene)/ Ti-RiO$_2$	39	$[BY28]_0$ = 30 mg/L; Cathode dimension = 25 mm diameter and 0.6 mm thickness; Anode dimension = 7 × 6 × 1 mm; J = 40.76 mA/cm^2; Flow rate = 10 L/h	(Akbarpour et al., 2016)
Carbon nanotubes	Acid Blue 92	CNTs-PTFE/ Boron-doped diamond	37–96	$[AB92]_0$ = 20 mg/L; Cathode dimension = 25 mm diameter and 0.6 mm thickness; Anode dimension = 10 mm × 5 cm × 2 mm; I = 0.1 A; Flow rate = 10 L/h	(Khataee et al., 2013)
Carbon nanotubes	Direct blue 129	CNTs-PTFE/ Ti-RiO$_2$	37	$[DB129]_0$ = 60 mg/L; Cathode dimension = 25 mm diameter and 0.6 mm thickness; Anode dimension = 7 × 6 × 1 mm; I = 0.2 A; Flow rate = 10 L/h	(Khataee et al., 2014)
Graphene	Methylene blue acid red 14	GO-GE (Graphite electrode)/ Graphite rGO-GE/ Graphite	96–100 98–100	$[MB]_0$ = 10–50 mg/L; $[AR14]_0$ = 25–200 mg/L; Cathode dimension = 3 × 50 × 130 mm Anode dimension = 3 × 50 × 130 mm; I = 0.02 – 0.22 A	(Akerdi et al., 2017)
Graphene	Rhodamine B 2,4-dichlorophenol	Graphene/Pt	63–98 97.6	$[RhB]_0$ = 4.79 mg/L; Cathode dimension = 1 × 1 × 0.5 cm; Anode dimension = 1 cm^2; J = 40 mA/cm^2	(Zhao et al., 2016)

(continued on next page)

Graphene Carbon nanotubes/ Graphene	Dimethyl Phthalate	CNT/Pt Graphene-CNT/Pt Graphene/Pt N-Graphene-CNT/Pt	6.8–57.8 58.6–76.2 17.8–46 100	$[DMP]_0$ = 50 mg/L; Cathode dimension = 2 × 2 × 0.2 cm; Anode dimension = 1 × 2 cm Applied current = -0.2 V to −0.5 V	(Liu et al., 2016)
Carbon nanotubes/ graphene	Methyl blue	Fe-CNT-Graphene/Pt	99	$[MB]_0$ = 10 mg/L; Cathode dimension = 8 cm^2; Anode dimension = 8 cm^2; I = 15 mA	(Chen et al., 2016)
Carbon felt	Sulfanilamide	Carbon felt/Pt	78–93	$[Sulfanilamide]_0$ = 0.103 g/L, Cathode dimension = 15 × 4 × 0.5 cm; Anode dimension = 3 cm diameter and 5 cm height. I = 100–1000 mA	(El-Ghenymy et al., 2014)
Carbon felt	Sulfanilamide	Carbon felt/boron-doped diamond	97–99	$[Sulfanilamide]_0$ = 0.103 g/L, Cathode dimension = 15 cm × 4 cm × 0.5 cm Anode dimension = 25 cm^2; I = 100 –1000 mA	(El-Ghenymy et al., 2014)
Carbon-toped compounds	Rhodamine B	g-C_3N_4-ACF/Pt	91[a]	Total organic carbon $[TOC]_0$ = 42 mg/L; Cathode dimension = 40 cm^2; Anode dimension = 5 cm^2; J = 3 mA/cm^2	(He et al., 2018)
Carbon-toped compounds	Dimethyl phthalate (DMP)	Fe_3O_4@OMC (Ordered mesoporous carbon)/CA (Carbon aerogel)/BDD	93	$[DMP]_0$ = 50 mg/L Cathode dimension = 1.5 × 2 × 0.3 cm J = 50 mA/cm^2; Flow rate = 400 mL/min	(Wang et al., 2016)

[a] Total organic carbon removal.

technology generates "electrohydraulic cavitation," in which non-converted oxygen molecules (O_2) pass through the carbon-based electrode and react with hydrogen radicals ($H^{\bullet}$) to generate the hydroxyl radical ($^{\bullet}OH$). The US is also involved in generating oxygen radicals ($O^{\bullet}$) (Eqs. 14.14 to 14.15).

$$H_2O_2 + US \rightarrow {}^{\bullet}OH \tag{14.13}$$

$$H^{\bullet} + O_2 \rightarrow {}^{\bullet}OH + O \tag{14.14}$$

$$O_2 + US \rightarrow 2O^{\bullet} \tag{14.15}$$

Oxidant decomposition by electro-assistance has been reported as a Fenton-like reaction, where no catalyst addition is required because it is mainly performed by electron transfer. Nevertheless, metal can be released from the electrode, generating metal contamination (Jaafarzadeh et al., 2017; Kim et al., 2020; Ghalebizade and Ayati, 2020). Therefore, electro-oxidation using carbon-based materials is an alternative technology because, as Liu et al. (2019) reported, carbamazepine (CBZ) was degraded by PMS decomposition using an activated carbon fiber (ACF) cathode and platinum (Pt) anode. The authors determined that four main CBZ degradation pathways were followed: (1) OH electrode-generated on Pt surface, (2) CBZ oxidation by electro-oxidation, (3) sulfate radical production by PMS decomposition by ACF, and (4) PMS decomposition by electron transfer. However, important information gaps are shown in relation to different oxidants used and oxidant decomposition mechanisms, as in a Fenton-like reaction.

Different ways to produce carbon-based electrodes have also been reported, among which carbon paste offers easy electrode production. This process is based on mixing carbon-based material (i.e., electrode feedstock) with a polymer (e.g., parafilm drops) to paste it, after which the solid is packed into an electrode body (e.g., glass tube) (Tajik et al., 2020). Another way is electrodeposition, where the carbon-based material suspended in a solution and an inert surface electrode (e.g., titanium or graphite) are connected to a power supply. Once the system is set up, different current values are passed through the system in short periods of time (i.e., seconds) to generate electrodeposition, and finally, the calcination process is performed (Espinola-Portilla et al., 2017; Walsh and Low, 2016).

14.4 Conclusion

As highlighted in the present work, significant research on carbon-based materials and their application in AOPs has been performed in recent years. Their application for Fenton-like reaction and electro-oxidation processes is especially useful because both processes proved sustainable and efficient performance without the drawbacks of metal catalysts or electrodes. Furthermore, it has been shown that carbon-based materials' synthesis methods have been improved by using easy synthesis and non-toxic reagents. However, there are still several knowledge gaps to fill related to the synthesis and application of different carbon-based materials.

- There have been no reports on using AOPs with carbon-based materials for water disinfection, although the carbon-based materials have been reported as effective bacterial adsorbents (Pajerski et al., 2020).
- There is still a lack of knowledge about the photocatalysis process using carbon-based materials because even though several photocatalysts have been reported with high organic degradation efficiency, few of them have been doped on carbon-based materials.
- Carbon-based materials have been reported to be efficient photocatalysts. However, for some materials as hydrochar, photochemical properties and how to improve the functionalization have not been discussed.
- Carbon-based materials as metal-free catalysts for oxidant decomposition are shown as an important line of research to develop because there are reports on high organic degradation, but few on different oxidants and carbon-based materials.
- Few carbon-based materials have been used as electrodes; therefore, important drawbacks on EC properties and the electron transfer mechanism are shown.
- The electro-oxidation process with and without coupling is shown as a possible scale-up process because, despite little information on carbon-based electrodes, several studies have been reported on EC reactors showing excellent performance in the mineralization of persistent organic pollutants.
- AOPs using carbon-based materials are presented in the scientific literature as efficient in degrading organic contaminants. However, a limited number of scale-up reports have been undertaken.

Acknowledgment

O.M. Rodriguez-Narvaez would like to thank CONACyT and the Universidad Autónoma de Nayarit for the scholarship. The authors are also grateful to Dubravka Suznjevic for her editorial review.

References

Akbarpour, A., Khataee, A., Fathinia, M., Vahid, B., 2016. Development of kinetic models for photoassisted electrochemical process using Ti/RuO_2 anode and carbon nanotube-based O_2-diffusion cathode. Electrochim. Acta. 187, 300–311. https://doi.org/10.1016/j.electacta.2015.11.014.

Akerdi, A.G., Es'Haghzade, Z., Bahrami, S.H., Arami, M., 2017. Comparative study of GO and reduced GO coated graphite electrodes for decolorization of acidic and basic dyes from aqueous solutions through heterogeneous electro-Fenton process. J. Environ. Chem. Eng. 5, 2313–2324. https://doi.org/10.1016/j.jece.2017.04.028.

Alam, S.N., Sharma, N., Kumar, L., 2017. Synthesis of graphene oxide (GO) by modified Hummers method and its thermal reduction to obtain reduced graphene oxide (rGO)*. Graphene 06, 1–18. https://doi.org/10.4236/graphene.2017.61001.

Ali, S., Rehman, S.A.U., Luan, H.Y., Farid, M.U., Huang, H., 2019. Challenges and opportunities in functional carbon nanotubes for membrane-based water treatment and desalination. Sci. Total Environ. 646, 1126–1139. https://doi.org/10.1016/j.scitotenv.2018.07.348.

Amir Faiz, M.S., Che Azurahanim, C.A., Raba'ah, S.A., Ruzniza, M.Z., 2020. Low cost and green approach in the reduction of graphene oxide (GO) using palm oil leaves extract for potential in industrial applications. Results Phys 16, 102954. https://doi.org/10.1016/j.rinp.2020.102954.

Antal, M.J., Grønli, M., 2003. The art, science, and technology of charcoal production. Ind. Eng. Chem. Res. 42, 1619–1640. https://doi.org/10.1021/ie0207919.

Azargohar, R., Dalai, A.K., 2008. Steam and KOH activation of biochar: experimental and modeling studies. Micropor. Mesopor. Mater 110, 413–421. https://doi.org/10.1016/j.micromeso.2007.06.047.

Barrios-Bermúdez, N., González-Avendaño, M., Lado-Touriño, I., Cerpa-Naranjo, A., Rojas-Cervantes, M.L., 2020. Fe-Cu doped multiwalled carbon nanotubes for Fenton-like degradation of paracetamol under mild conditions. Nanomaterials 10, 11–13. https://doi.org/10.3390/nano10040749.

Bhongade, T., Gogaram, D.M.Gautam, Vijayakumar, R.P., 2019. Synthesis of MWCNTs using waste toner powder as carbon source by chemical vapor deposition method. Fuller. Nanotub. Car. N. 27, 864–872. https://doi.org/10.1080/1536383X.2019.1652169.

Bouchelta, C., Medjram, M.S., Bertrand, O., Bellat, J.P., 2008. Preparation and characterization of activated carbon from date stones by physical activation with steam. J. Anal. Appl. Pyrolysis. 82, 70–77. https://doi.org/10.1016/j.jaap.2007.12.009.

Brewer, C.E., Schmidt-Rohr, K., Satrio, J.A., Brown, R.C., 2009. Characterization of biochar from fast pyrolysis and gasification systems. Environ. Prog. Sustain. Energy. 28, 386–396. https://doi.org/10.1002/ep.10378.

Butler, J.E., Sumant, A.V., 2008. The CVD of nanodiamond materials. Chem. Vap. Deposition 14, 145–160. https://doi.org/10.1002/cvde.200700037.

Cao, L., Yu, I.K.M., Cho, D.W., Wang, D., Tsang, D.C.W., Zhang, S., Ding, S., Wang, L., Ok, Y.S., 2019. Microwave-assisted low-temperature hydrothermal treatment of red seaweed (*Gracilaria lemaneiformis)* for production of levulinic acid and algae hydrochar. Bioresour. Technol. 273, 251–258. https://doi.org/10.1016/j.biortech.2018.11.013.

Cao, N., Zhang, Y., 2015. Study of reduced graphene oxide preparation by Hummers' method and related characterization. J. Nanomater. 2015, 168125. https://doi.org/10.1155/2015/168125.

Castañeda, L.F., Walsh, F.C., Nava, J.L., Ponce de León, C., 2017. Graphite felt as a versatile electrode material: properties, reaction environment, performance and applications. Electrochim. Acta. 258, 1115–1139. https://doi.org/10.1016/j.electacta.2017.11.165.

Cen, B., Li, K., Lv, C., Yang, R., 2020. A novel asymmetric activated carbon electrode doped with metal-organic frameworks for high desalination performance. J. Solid State Electrochem. 24, 687–697. https://doi.org/10.1007/s10008-020-04510-8.

Chee H. Chia, P.M., Downie, Adriana, 2019. Characteristics of biochar: physical and structural properties. In: Lehman, J., Joseph, S. (Eds.), Biochar for Environmental Management. Taylor & Francis, London, UK, pp. 1–2.

Chen, D., Tang, L., Li, J., 2010. Graphene-based materials in electrochemistry. Chem. Soc. Rev. 39, 3157. https://doi.org/10.1039/b923596e.

Chen, H., Carroll, K.C., 2016. Metal-free catalysis of persulfate activation and organic-pollutant degradation by nitrogen-doped graphene and aminated graphene. Environ. Pollut. 215, 96–102. https://doi.org/10.1016/j.envpol.2016.04.088.

Chen, M.D., Wumaie, T., Li, W.L., Song, H.H., Song, R.R., 2015. Electrochemical performance of cotton stalk based activated carbon electrodes modified by MnO_2 for supercapacitor. Mater. Technol. 30, A2–A7. https://doi.org/10.1179/1753555714Y.0000000241.

Chen, N., Huang, Y., Hou, X., Ai, Z., Zhang, L., 2017. Photochemistry of hydrochar: reactive oxygen species generation and sulfadimidine degradation. Environ. Sci. Technol. 51, 11278–11287. https://doi.org/10.1021/acs.est.7b02740.

Chen, W., Yang, X., Huang, J., Zhu, Y., Zhou, Y., Yao, Y., Li, C., 2016. Iron oxide containing graphene/carbon nanotube based carbon aerogel as an efficient E-Fenton cathode for the degradation of methyl blue. Electrochim. Acta. 200, 75–83. https://doi.org/10.1016/j.electacta.2016.03.044.

Chen, X., Da Oh, W., Lim, T.T., 2018. Graphene- and CNTs-based carbocatalysts in persulfates activation: material design and catalytic mechanisms. Chem. Eng. J. 354, 941–976. https://doi.org/10.1016/j.cej.2018.08.049.

Chen, X., Fang, G., Liu, C., Dionysiou, D.D., Wang, X., Zhu, C., Wang, Y., Gao, J., Zhou, D., 2019. Cotransformation of carbon dots and contaminant under light in aqueous solutions: a mechanistic study. Environ. Sci. Technol. 53, 6235–6244. https://doi.org/10.1021/acs.est.8b07124.

Cheng, X., Guo, H., Zhang, Y., Liu, Y., Liu, H., Yang, Y., 2016. Oxidation of 2,4-dichlorophenol by non-radical mechanism using persulfate activated by Fe/S modified carbon nanotubes. J. Colloid Interface Sci. 469, 277–286. https://doi.org/10.1016/j.jcis.2016.01.067.

Choi, W., Lahiri, I., Seelaboyina, R., Kang, Y.S., 2010. Synthesis of graphene and its applications: a review. Crit. Rev. Solid State Mater. Sci. 35, 52–71. https://doi.org/10.1080/10408430903505036.

Clurman, A.M., Rodríguez-Narvaez, O.M., Jayarathne, A., De Silva, G., Ranasinghe, M.I., Goonetilleke, A., Bandala, E.R., 2020. Influence of surface hydrophobicity/hydrophilicity of biochar on the removal of emerging contaminants. Chem. Eng. J. 402, 126277. https://doi.org/10.1016/j.cej.2020.126277.

Cornejo, O.M., Murrieta, M.F., Castañeda, L.F., Nava, J.L., 2020. Characterization of the reaction environment in flow reactors fitted with BDD electrodes for use in electrochemical advanced oxidation processes: a critical review. Electrochim. Acta. 331, 135373. https://doi.org/10.1016/j.electacta.2019.135373.

Czernik, S., Bridgwater, A.V., 2004. Overview of applications of biomass fast pyrolysis oil. Energy Fuels 18, 590–598. https://doi.org/10.1021/ef034067u.

Dai, L., He, C., Wang, Y., Liu, Y., Yu, Z., Zhou, Y., Fan, L., Duan, D., Ruan, R., 2017. Comparative study on microwave and conventional hydrothermal pretreatment of bamboo sawdust: hydrochar properties and its pyrolysis behaviors. Energy Convers. Manag. 146, 1–7. https://doi.org/10.1016/j.enconman.2017.05.007.

Das, R., Bandyopadhyay, R., Pramanik, P., 2018. Carbon quantum dots from natural resource: a review. Mater. Today Chem. 8, 96–109. https://doi.org/10.1016/j.mtchem.2018.03.003.

Das, R., Shahnavaz, Z., Ali, M.E., Islam, M.M., Abd Hamid, S.B., 2016. Can we optimize arc discharge and laser ablation for well-controlled carbon nanotube synthesis? Nanoscale Res. Lett. 11. https://doi.org/10.1186/s11671-016-1730-0.

De Silva, K.K.H., Huang, H.H., Joshi, R.K., Yoshimura, M., 2017. Chemical reduction of graphene oxide using green reductants. Carbon 119, 190–199. https://doi.org/10.1016/j.carbon.2017.04.025.

Dimiev, A.M., Tour, J.M., 2014. Mechanism of graphene oxide formation. ACS Nano 8, 3060–3068. https://doi.org/10.1021/nn500606a.

Dimitriadou, S., Frontistis, Z., Petala, A., Bampos, G., Mantzavinos, D., 2019. Carbocatalytic activation of persulfate for the removal of drug diclofenac from aqueous matrices. Catal. Today. 355, 937–944. https://doi.org/10.1016/j.cattod.2019.02.025.

Dominguez, C.M., Oturan, N., Romero, A., Santos, A., Oturan, M.A., 2018. Lindane degradation by electrooxidation process: effect of electrode materials on oxidation and mineralization kinetics. Water Res 135, 220–230. https://doi.org/10.1016/j.watres.2018.02.037.

Dong, H., Wei, M., Li, J., Fang, J., Gao, L., Li, X., Xu, A., 2016. Catalytic performance of supported g-C3N4 on MCM-41 in organic dye degradation with peroxymonosulfate. RSC Adv 6, 70747–70755. https://doi.org/10.1039/c6ra15721a.

Dong, L., Yang, J., Chhowalla, M., Loh, K.P., 2017. Synthesis and reduction of large sized graphene oxide sheets. Chem. Soc. Rev. 46, 7306–7316. https://doi.org/10.1039/c7cs00485k.

Duan, L., Zhou, X., Liu, S., Shi, P., Yao, W., 2017. 3D-hierarchically structured Co_3O_4/graphene hydrogel for catalytic oxidation of orange II solutions by activation of peroxymonosulfate. J. Taiwan Inst. Chem. Eng. 76, 101–108. https://doi.org/10.1016/j.jtice.2017.04.019.

Duan, X., Ao, Z., Li, D., Sun, H., Zhou, L., Suvorova, A., Saunders, M., Wang, G., Wang, S., 2016. Surface-tailored nanodiamonds as excellent metal-free catalysts for organic oxidation. Carbon 103, 404–411. https://doi.org/10.1016/j.carbon.2016.03.034.

Duan, X., Ao, Z., Zhou, L., Sun, H., Wang, G., Wang, S., 2016. Occurrence of radical and nonradical pathways from carbocatalysts for aqueous and nonaqueous catalytic oxidation. Appl. Catal. B Environ. 188, 98–105. https://doi.org/10.1016/j.apcatb.2016.01.059.

Duan, X., Su, C., Zhou, L., Sun, H., Suvorova, A., Odedairo, T., Zhu, Z., Shao, Z., Wang, S., 2016. Surface controlled generation of reactive radicals from persulfate by carbocatalysis on nanodiamonds. Appl. Catal. B Environ. 194, 7–15. https://doi.org/10.1016/j.apcatb.2016.04.043.

Duan, X., Sun, H., Tade, M., Wang, S., 2018. Metal-free activation of persulfate by cubic mesoporous carbons for catalytic oxidation via radical and nonradical processes. Catal. Today. 307, 140–146. https://doi.org/10.1016/j.cattod.2017.04.038.

Duan, X., Sun, H., Wang, Y., Kang, J., Wang, S., 2015. N-doping-induced nonradical reaction on single-walled carbon nanotubes for catalytic phenol oxidation. ACS Catal 5, 553–559. https://doi.org/10.1021/cs5017613.

Duan, X., Tian, W., Zhang, H., Sun, H., Ao, Z., Shao, Z., Wang, S., 2019. Sp2/sp3 framework from diamond nanocrystals: a key bridge of carbonaceous structure to carbocatalysis. ACS Catal 9, 7494–7519. https://doi.org/10.1021/acscatal.9b01565.

El-Ghenymy, A., Rodríguez, R.M., Brillas, E., Oturan, N., Oturan, M.A., 2014. Electro-Fenton degradation of the antibiotic sulfanilamide with Pt/carbon-felt and BDD/carbon-felt cells. Kinetics, reaction intermediates, and toxicity assessment. Environ. Sci. Pollut. Res. 21, 8368–8378. https://doi.org/10.1007/s11356-014-2773-3.

Emiru, T.F., Ayele, D.W., 2017. Controlled synthesis, characterization and reduction of graphene oxide: a convenient method for large scale production. Egypt. J. Basic Appl. Sci. 4, 74–79. https://doi.org/10.1016/j.ejbas.2016.11.002.

Espinola-Portilla, F., Navarro-Mendoza, R., Gutiérrez-Granados, S., Morales-Muñoz, U., Brillas-Coso, E., Peralta-Hernández, J.M., 2017. A simple process for the deposition of TiO_2 onto BDD by electrophoresis and its application to the photoelectrocatalysis of acid blue 80 dye. J. Electroanal. Chem. 802, 57–63. https://doi.org/10.1016/j.jelechem.2017.08.041.

Fang, G., Gao, J., Liu, C., Dionysiou, D.D., Wang, Y., Zhou, D., 2014. Key role of persistent free radicals in hydrogen peroxide activation by biochar: implications to organic contaminant degradation. Environ. Sci. Technol. 48, 1902–1910. https://doi.org/10.1021/es4048126.

Fang, J., Li, J., Gao, L., Jiang, X., Zhang, J., Xu, A., Li, X., 2017. Synthesis of OMS-2/graphite nanocomposites with enhanced activity for pollutants degradation in the presence of peroxymonosulfate. J. Colloid Interface Sci. 494, 185–193. https://doi.org/10.1016/j.jcis.2017.01.049.

Fang, W., Shangguan, W., 2019. A review on bismuth-based composite oxides for photocatalytic hydrogen generation. Int. J. Hydrogen Energy. 44, 895–912. https://doi.org/10.1016/j.ijhydene.2018.11.063.

fei Tan, X., guo Liu, Y., ling Gu, Y., Xu, Y., ming Zeng, G., jiang Hu, X., bo Liu, S., Wang, X., mian Liu, S., Li, J., 2016. Biochar-based nano-composites for the decontamination of wastewater: a review. Bioresour. Technol. 212, 318–333. https://doi.org/10.1016/j.biortech.2016.04.093.

Fernando, K.A.S., Sahu, S., Liu, Y., Lewis, W.K., Guliants, E.A., Jafariyan, A., Wang, P., Bunker, C.E., Sun, Y.P., 2015. Carbon quantum dots and applications in photocatalytic energy conversion. ACS Appl. Mater. Interf. 7, 8363–8376. https://doi.org/10.1021/acsami.5b00448.

Fu, H., Liu, H., Mao, J., Chu, W., Li, Q., Alvarez, P.J.J., Qu, X., Zhu, D., 2016. Photochemistry of dissolved black carbon released from biochar: reactive oxygen species generation and phototransformation. Environ. Sci. Technol. 50, 1218–1226. https://doi.org/10.1021/acs.est.5b04314.

Garcia-Segura, S., Vieira Dos Santos, E., Martínez-Huitle, C.A., 2015. Role of sp3/sp2 ratio on the electrocatalytic properties of boron-doped diamond electrodes: a mini review. Electrochem. Commun. 59, 52–55. https://doi.org/10.1016/j.elecom.2015.07.002.

Ge, J., Zhang, Y., Park, S.-J., 2019. Recent advances in carbonaceous photocatalysts with enhanced photocatalytic performances: a mini review. Materials (Basel) 12. https://doi.org/10.3390/ma12121916.

Ghalebizade, M., Ayati, B., 2020. Investigating electrode arrangement and anode role on dye removal efficiency of electro-peroxone as an environmental friendly technology. Sep. Purif. Technol. 251, 117350. https://doi.org/10.1016/j.seppur.2020.117350.

Ghanbari, F., Moradi, M., 2017. Application of peroxymonosulfate and its activation methods for degradation of environmental organic pollutants: review. Chem. Eng. J. 310, 41–62. https://doi.org/10.1016/j.cej.2016.10.064.

Han, C., Duan, X., Zhang, M., Fu, W., Duan, X., Ma, W., Liu, S., Wang, S., Zhou, X., 2019. Role of electronic properties in partition of radical and nonradical processes of carbocatalysis toward peroxymonosulfate activation. Carbon 153, 73–80. https://doi.org/10.1016/j.carbon.2019.06.107.

Han, W., Li, D., Zhang, M., Ximin, H., Duan, X., Liu, S., Wang, S., 2020. Photocatalytic activation of peroxymonosulfate by surface-tailored carbon quantum dots. J. Hazard. Mater. 395, 122695. https://doi.org/10.1016/j.jhazmat.2020.122695.

He, Z., Chen, J.J., Chen, Y., Makwarimba, C.P., Huang, X., Zhang, S., Chen, J.J., Song, S., 2018. An activated carbon fiber-supported graphite carbon nitride for effective electro-Fenton process. Electrochim. Acta. 276, 377–388. https://doi.org/10.1016/j.electacta.2018.04.195.

Hermanová, S., Zarevúcká, M., Bouša, D., Pumera, M., Sofer, Z., 2015. Graphene oxide immobilized enzymes show high thermal and solvent stability. Nanoscale 7, 5852–5858. https://doi.org/10.1039/c5nr00438a.

Hoseini, S.N., Pirzaman, A.K., Aroon, M.A., Pirbazari, A.E., 2017. Photocatalytic degradation of 2,4-dichlorophenol by Co-doped TiO_2 (Co/TiO_2) nanoparticles and Co/TiO_2 containing mixed matrix membranes. J. Water Process Eng. 17, 124–134. https://doi.org/10.1016/j.jwpe.2017.02.015.

Hou, J., Xu, L., Han, Y., Tang, Y., Wan, H., Xu, Z., Zheng, S., 2019. Deactivation and regeneration of carbon nanotubes and nitrogen-doped carbon nanotubes in catalytic peroxymonosulfate activation for phenol degradation: variation of surface functionalities. RSC Adv 9, 974–983. https://doi.org/10.1039/C8RA07696K.

Hua, Y., Zheng, X., Xue, L., Han, L., He, S., Mishra, T., Feng, Y., Yang, L., Xing, B., 2020. Microbial aging of hydrochar as a way to increase cadmium ion adsorption capacity: process and mechanism. Bioresour. Technol. 300, 122708. https://doi.org/10.1016/j.biortech.2019.122708.

Huang, D., Luo, H., Zhang, C., Zeng, G., Lai, C., Cheng, M., Wang, R., Deng, R., Xue, W., Gong, X., Guo, X., Li, T., 2019. Nonnegligible role of biomass types and its compositions on the formation of persistent free radicals in biochar: insight into the influences on Fenton-like process. Chem. Eng. J. 361, 353–363. https://doi.org/10.1016/j.cej.2018.12.098.

Huang, J., Zhang, H., 2019. Mn-based catalysts for sulfate radical-based advanced oxidation processes: a review. Environ. Int. 133, 105141. https://doi.org/10.1016/j.envint.2019.105141.

Huang, Z., Bao, H., Yao, Y., Lu, J., Lu, W., Chen, W., 2016. Key role of activated carbon fibers in enhanced decomposition of pollutants using heterogeneous cobalt/peroxymonosulfate system. J. Chem. Technol. Biotechnol. 91, 1257–1265. https://doi.org/10.1002/jctb.4715.

Inyang, M.I., Gao, B., Yao, Y., Xue, Y., Zimmerman, A., Mosa, A., Pullammanappallil, P., Ok, Y.S., Cao, X., 2016. A review of biochar as a low-cost adsorbent for aqueous heavy metal removal. Crit. Rev. Environ. Sci. Technol. 46, 406–433. https://doi.org/10.1080/10643389.2015.1096880.

Jaafarzadeh, N., Ghanbari, F., Alvandi, M., 2017. Integration of coagulation and electro-activated $HSO_5-$to treat pulp and paper wastewater. Sustain. Environ. Res. 27, 223–229. https://doi.org/10.1016/j.serj.2017.06.001.

Jagannatham, M., Sankaran, S., Prathap, H., 2015. Electroless nickel plating of arc discharge synthesized carbon nanotubes for metal matrix composites. Appl. Surf. Sci. 324, 475–481. https://doi.org/10.1016/j.apsusc.2014.10.150.

Jin, H., Capareda, S., Chang, Z., Gao, J., Xu, Y., Zhang, J., 2014. Biochar pyrolytically produced from municipal solid wastes for aqueous As(V) removal: adsorption property and its improvement with KOH activation. Bioresour. Technol. 169, 622–629. https://doi.org/10.1016/j.biortech.2014.06.103.

Jo, W.K., Kumar, S., Isaacs, M.A., Lee, A.F., Karthikeyan, S., 2017. Cobalt promoted TiO_2/GO for the photo-catalytic degradation of oxytetracycline and congo red. Appl. Catal. B Environ. 201, 159–168. https://doi.org/10.1016/j.apcatb.2016.08.022.

Kale, M.J., Avanesian, T., Christopher, P., 2014. Direct photocatalysis by plasmonic nanostructures. ACS Catal 4, 116–128. https://doi.org/10.1021/cs400993w.

Kannan, S., Gariepy, Y., Raghavan, G.S.V., 2017. Optimization and characterization of hydrochar produced from microwave hydrothermal carbonization of fish waste. Waste Manag 65, 159–168. https://doi.org/10.1016/j.wasman.2017.04.016.

Kaur, R., Badea, I., 2013. Nanodiamonds as novel nanomaterials for biomedical applications: drug delivery and imaging systems. Int. J. Nanomedicine. 8, 203–220. https://doi.org/10.2147/IJN.S37348.

Kausar, A., 2018. Properties and applications of nanodiamond nanocomposite. Am. J. Nanosci. Nanotechnol. Res. 6, 46–54.

Kaushik, R., Parshetti, G.K., Liu, Z., Balasubramanian, R., 2014. Enzyme-assisted hydrothermal treatment of food waste for co-production of hydrochar and bio-oil. Bioresour. Technol. 168, 267–274. https://doi.org/10.1016/j.biortech.2014.03.022.

Kecsenovity, E., Fejes, D., Reti, B., Hernadi, K., 2013. Growth and characterization of bamboo-like carbon nanotubes synthesized on Fe-Co-Cu catalysts prepared by high-energy ball milling. Phys. Status Solidi Basic Res. 250, 2544–2548. https://doi.org/10.1002/pssb.201300075.

Khan, M., Shahzad, N., Xiong, C., Zhao, T.K., Li, T., Siddique, F., Ali, N., Shahzad, M., Ullah, H., Rakha, S.A., 2016. Dispersion behavior and the influences of ball milling technique on functionalization of detonated nano-diamonds. Diam. Relat. Mater. 61, 32–40. https://doi.org/10.1016/j.diamond.2015.11.007.

Khataee, A., Akbarpour, A., Vahid, B., 2014. Photoassisted electrochemical degradation of an azo dye using Ti/RuO_2 anode and carbon nanotubes containing gas-diffusion cathode. J. Taiwan Inst. Chem. Eng. 45, 930–936. https://doi.org/10.1016/j.jtice.2013.08.015.

Khataee, A., Kayan, B., Kalderis, D., Karimi, A., Akay, S., Konsolakis, M., 2017. Ultrasound-assisted removal of acid red 17 using nanosized Fe3O4-loaded coffee waste hydrochar. Ultrason. Sonochem. 35, 72–80. https://doi.org/10.1016/j.ultsonch.2016.09.004.

Khataee, A.A., Khataee, A.A., Fathinia, M., Vahid, B., Joo, S.W., 2013. Kinetic modeling of photoassisted-electrochemical process for degradation of an azo dye using boron-doped diamond anode and cathode with carbon nanotubes. J. Ind. Eng. Chem. 19, 1890–1894. https://doi.org/10.1016/j.jiec.2013.02.037.

Kim, C., Ahn, J.Y., Kim, T.Y., Hwang, I., 2020. Mechanisms of electro-assisted persulfate/nano-Fe0 oxidation process: roles of redox mediation by dissolved Fe. J. Hazard. Mater. 388, 121739. https://doi.org/10.1016/j.jhazmat.2019.121739.

Kobashi, K., Ata, S., Yamada, T., Futaba, D.N., Okazaki, T., Hata, K., 2019. Classification of commercialized carbon nanotubes into three general categories as a guide for applications. ACS Appl. Nano Mater. 2, 4043–4047. https://doi.org/10.1021/acsanm.9b00941.

Krueger, A., Lang, D., 2012. Functionality is key: recent progress in the surface modification of nanodiamond. Adv. Funct. Mater. 22, 890–906. https://doi.org/10.1002/adfm.201102670.

Kumar, R., Sahoo, S., Joanni, E., Singh, R.K., Tan, W.K., Kar, K.K., Matsuda, A., 2019. Recent progress in the synthesis of graphene and derived materials for next generation electrodes of high performance lithium ion batteries. Prog. Energy Combust. Sci. 75, 100786. https://doi.org/10.1016/j.pecs.2019.100786.

Kumar, S., Nehra, M., Kedia, D., Dilbaghi, N., Tankeshwar, K., Kim, K.H., 2018. Carbon nanotubes: a potential material for energy conversion and storage. Prog. Energy Combust. Sci. 64, 219–253. https://doi.org/10.1016/j.pecs.2017.10.005.

Lee, H., Il Kim, H., Weon, S., Choi, W., Hwang, Y.S., Seo, J., Lee, C., Kim, J.H., 2016. Activation of persulfates by graphitized nanodiamonds for removal of organic compounds. Environ. Sci. Technol. 50, 10134–10142. https://doi.org/10.1021/acs.est.6b02079.

Lee, H.H.J., Lee, H.H.J., Jeong, J., Lee, J., Park, N.B., Lee, C., 2015. Activation of persulfates by carbon nanotubes: oxidation of organic compounds by nonradical mechanism. Chem. Eng. J. 266, 28–33. https://doi.org/10.1016/j.cej.2014.12.065.

Lee, X.J., Hiew, B.Y.Z., Lai, K.C., Lee, L.Y., Gan, S., Thangalazhy-Gopakumar, S., Rigby, S., 2019. Review on graphene and its derivatives: synthesis methods and potential industrial implementation. J. Taiwan Inst. Chem. Eng. 98, 163–180. https://doi.org/10.1016/j.jtice.2018.10.028.

Li, G., 2020. Direct laser writing of graphene electrodes. J. Appl. Phys. 127, 10901. https://doi.org/10.1063/1.5120056.

Li, G., Mo, X., Law, W.C., Chan, K.C., 2019. Wearable fluid capture devices for electrochemical sensing of sweat. ACS Appl. Mater. Interf. 11, 238–243. https://doi.org/10.1021/acsami.8b17419.

Li, J.H., Lv, G.H., Bai, W.B., Liu, Q., Zhang, Y.C., Song, J.Q., 2016. Modification and use of biochar from wheat straw (*Triticum aestivum* L.) for nitrate and phosphate removal from water. Desalin. Water Treat. 57, 4681–4693. https://doi.org/10.1080/19443994.2014.994104.

Li, X., Xie, J., Jiang, C., Yu, J., Zhang, P., 2018. Review on design and evaluation of environmental photocatalysts. Front. Environ. Sci. Eng., 12, p. 14.

Li, X., Yu, J., Wageh, S., Al-Ghamdi, A.A., Xie, J., 2016. Graphene in photocatalysis: a review. Small 12, 6640–6696. https://doi.org/10.1002/smll.201600382.

Li, Y., Shao, J., Wang, X., Deng, Y., Yang, H., Chen, H., 2014. Characterization of modified biochars derived from bamboo pyrolysis and their utilization for target component (furfural) adsorption. Energy Fuels 28, 5119–5127. https://doi.org/10.1021/ef500725c.

Li, Z., Meng, X., Zhang, Z., 2018. Recent development on MoS2-based photocatalysis: a review. J. Photochem. Photobiol. C Photochem. Rev. 35, 39–55. https://doi.org/10.1016/j.jphotochemrev.2017.12.002.

Lim, J.Y., Mubarak, N.M., Abdullah, E.C., Nizamuddin, S., Khalid, Inamuddin, M., 2018. Recent trends in the synthesis of graphene and graphene oxide based nanomaterials for removal of heavy metals—a review. J. Ind. Eng. Chem. 66, 29–44. https://doi.org/10.1016/j.jiec.2018.05.028.

Lima, I.M., Boateng, A.A., Klasson, K.T., 2010. Physicochemical and adsorptive properties of fast-pyrolysis bio-chars and their steam activated counterparts. J. Chem. Technol. Biotechnol. 85 (11), 1515–1521. https://doi.org/10.1002/jctb.2461.

Lin, K.Y.A., Zhang, Z.Y., 2017. Metal-free activation of Oxone using one-step prepared sulfur-doped carbon nitride under visible light irradiation. Sep. Purif. Technol. 173, 72–79. https://doi.org/10.1016/j.seppur.2016.09.008.

Liu, T., Wang, K., Song, S., Brouzgou, A., Tsiakaras, P., Wang, Y., 2016. New electro-Fenton gas diffusion cathode based on nitrogen-doped graphene@carbon nanotube composite materials. Electrochim. Acta. 194, 228–238. https://doi.org/10.1016/j.electacta.2015.12.185.

Liu, Y., Tzeng, Y.K., Lin, D., Pei, A., Lu, H., Melosh, N.A., Shen, Z.X., Chu, S., Cui, Y., 2018. An ultrastrong double-layer nanodiamond interface for stable lithium metal anodes. Joule 2, 1595–1609. https://doi.org/10.1016/j.joule.2018.05.007.

Liu, Z., Ding, H., Zhao, C., Wang, T., Wang, P., Dionysiou, D.D., 2019. Electrochemical activation of peroxymonosulfate with ACF cathode: kinetics, influencing factors, mechanism, and application potential. Water Res 159, 111–121. https://doi.org/10.1016/j.watres.2019.04.052.

Ma, P.C., Siddiqui, N.A., Marom, G., Kim, J.K., 2010. Dispersion and functionalization of carbon nanotubes for polymer-based nanocomposites: a review. Compos. A Appl. Sci. Manuf. 41, 1345–1367. https://doi.org/10.1016/j.compositesa.2010.07.003.

Maciá-Agulló, J.A., Moore, B.C., Cazorla-Amorós, D., Linares-Solano, A., 2004. Activation of coal tar pitch carbon fibres: physical activation vs. chemical activation. Carbon, 42, pp. 1367–1370.

Macoveanu, M., Gavrilescu, M., 2010. Environmental engineering and management journal. Environ. Eng. Manag. J. 9, 473–480. http://omicron.ch.tuiasi.ro/EEMJ/(accessed. August 4, 2020.

Magioglou, E., Frontistis, Z., Vakros, J., Manariotis, I.D., Mantzavinos, D., 2019. Activation of persulfate by biochars from valorized olive stones for the degradation of sulfamethoxazole. Catalysts 9, 419. https://doi.org/10.3390/catal9050419.

Manawi, Y.M., Ihsanullah, A.Samara, Al-Ansari, T., Atieh, M.A., 2018. A review of carbon nanomaterials' synthesis via the chemical vapor deposition (CVD) method. Materials 11 (5), 822. https://doi.org/10.3390/ma11050822.

Manyà, J.J., 2012. Pyrolysis for biochar purposes: a review to establish current knowledge gaps and research needs. Environ. Sci. Technol. 46, 7939–7954. https://doi.org/10.1021/es301029g.

Markad, G.B., Kapoor, S., Haram, S.K., Thakur, P., 2017. Metal free, carbon-TiO_2 based composites for the visible light photocatalysis. Sol. Energy. 144, 127–133. https://doi.org/10.1016/j.solener.2016.12.025.

Matos, J., Poon, P.S., Montaña, R., Romero, R., Gonçalves, G.R., Schettino, M.A., Passamani, E.C., Freitas, J.C.C., 2019. Photocatalytic activity of P-Fe/activated carbon nanocomposites under artificial solar irradiation. Catal. Today. 356, 226–240. https://doi.org/10.1016/j.cattod.2019.06.020.

McKee, M.S., Filser, J., 2016. Impacts of metal-based engineered nanomaterials on soil communities. Environ. Sci. Nano. 3, 506–533. https://doi.org/10.1039/c6en00007j.

Mendez-Alvarado, L.N., Medina-Ramirez, A., Manriquez, J., Navarro-Mendoza, R., Fuentes-Ramirez, R., Peralta-Hernandez, J.M., 2020. Synthesis of microspherical structures of bismuth oxychloride (BiOCl) towards the degradation of reactive orange 84 dye with sunlight. Mater. Sci. Semicond. Process. 114, 105086. https://doi.org/10.1016/j.mssp.2020.105086.

Mestre, A.S., Tyszko, E., Andrade, M.A., Galhetas, M., Freire, C., Carvalho, A.P., 2015. Sustainable activated carbons prepared from a sucrose-derived hydrochar: remarkable adsorbents for pharmaceutical compounds. RSC Adv 5, 19696–19707. https://doi.org/10.1039/c4ra14495c.

Mihajlović, M., Petrović, J., Stojanović, M., Milojković, J., Lopičić, Z., Koprivica, M., Lačnjevac, Č., 2016. Hydrochars, perspective adsorbents of heavy metals: a review of the current state of studies. Zast. Mater. 57, 488–495. https://doi.org/10.5937/zasmat1603488m.

Mohammed, I.A., Bankole, M.T., Abdulkareem, A.S., Ochigbo, S.S., Afolabi, A.S., Abubakre, O.K., 2017. Full factorial design approach to carbon nanotubes synthesis by CVD method in argon environment. South Afr. J. Chem. Eng. 24, 17–42. https://doi.org/10.1016/j.sajce.2017.06.001.

Mohan, D., Pittman, C.U., Steele, P.H., 2006. Pyrolysis of wood/biomass for bio-oil: a critical review. Energy and fuels 20, 848–889. https://doi.org/10.1021/ef0502397.

Mohan, D., Sarswat, A., Ok, Y.S., Pittman, C.U., 2014. Organic and inorganic contaminants removal from water with biochar, a renewable, low cost and sustainable adsorbent—a critical review. Bioresour. Technol. 160, 191–202. https://doi.org/10.1016/j.biortech.2014.01.120.

Mortazavian, S., Jones-Lepp, T., Bae, J.H., Chun, D., Bandala, E.R., Moon, J., 2019. Heat-treated biochar impregnated with zero-valent iron nanoparticles for organic contaminants removal from aqueous phase: material characterizations and kinetic studies. J. Ind. Eng. Chem. 76, 197–214. https://doi.org/10.1016/j.jiec.2019.03.041.

Nandi, M., Okada, K., Dutta, A., Bhaumik, A., Maruyama, J., Derks, D., Uyama, H., 2012. Unprecedented CO_2 uptake over highly porous N-doped activated carbon monoliths prepared by physical activation. Chem. Commun. 48, 10283–10285. https://doi.org/10.1039/c2cc35334b.

Nizamuddin, S., Qureshi, S.S., Baloch, H.A., Siddiqui, M.T.H., Takkalkar, P., Mubarak, N.M., Dumbre, D.K., Griffin, G.J., Madapusi, S., Tanksale, A., 2019. Microwave hydrothermal carbonization of rice straw: optimization of process parameters and upgrading of chemical, fuel, structural and thermal properties. Materials (Basel) 12 (3), 403. https://doi.org/10.3390/ma12030403.

Nunn, N., Torelli, M., McGuire, G., Shenderova, O., 2017. Nanodiamond: a high impact nanomaterial. Curr. Opin. Solid State Mater. Sci. 21, 1–9. https://doi.org/10.1016/j.cossms.2016.06.008.

Nurunnabi, M., Khatun, Z., Huh, K.M., Park, S.Y., Lee, D.Y., Cho, K.J., Lee, Y.K., 2013. In vivo biodistribution and toxicology of carboxylated graphene quantum dots. ACS Nano 7, 6858–6867. https://doi.org/10.1021/nn402043c.

Ong, B.K., Poh, H.L., Chua, C.K., Pumera, M., 2012. Graphenes prepared by Hummers, Staudenmaier and Hofmann methods for analysis of TNT-based nitroaromatic explosives in seawater. Electroanalysis 24, 2085–2093. https://doi.org/10.1002/elan.201200474.

Pacheco-Álvarez, M.O., Rodríguez-Narváez, O.M., Wrobel, K., Navarro-Mendoza, R., Nava-Montes De Oca, J.L., Peralta-Hernández, J.M., 2018. Improvement of the degradation of methyl orange using a TiO_2/BDD composite electrode to promote electrochemical and photoelectro-oxidation processes. Int. J. Electrochem. Sci. 13, 11549–11567. https://doi.org/10.20964/2018.12.70.

Pajerski, W., Duch, J., Ochonska, D., Golda-Cepa, M., Brzychczy-Wloch, M., Kotarba, A., 2020. Bacterial attachment to oxygen-functionalized graphenic surfaces. Mater. Sci. Eng. C. 113, 110972. https://doi.org/10.1016/j.msec.2020.110972.

Park, Y., Yoo, J., Lim, B., Kwon, W., Rhee, S.W., 2016. Improving the functionality of carbon nanodots: doping and surface functionalization. J. Mater. Chem. A. 4, 11582–11603. https://doi.org/10.1039/c6ta04813g.

Peralta-Hernández, J.M., Méndez-Tovar, M., Guerra-Sánchez, R., Martínez-Huitle, C.A., Nava, J.L., nchez, R., Marted, C.A.nez-Huitle, Nava, J.L., Peralta-Hernel, J.M.ndez, Me9, M.ndez-Tovar, Guerra-Se1, R.nchez, Marted, C.A.nez-Huitle, Nava, J.L., 2012. A brief review on environmental application of boron doped diamond electrodes as a new way for electrochemical incineration of synthetic dyes. Int. J. Electrochem. 2012, 154316. https://doi.org/10.1155/2012/154316.

Peralta-Hernandez, J.M., Pacheco-Alvarez, M., Picos, R., Rodriguez-Narvaez, O.M., 2020. Progress in the preparation of TiO_2 films at boron-doped diamond toward environmental applications. In: Hussain, C.M., Mishra, A.K. (Eds.), Handbook of Smart Photocatalytic Mater. Elsevier, Amsterdam, pp. 197–224.

Pichot, V., Comet, M., Fousson, E., Baras, C., Senger, A., Normand, F.Le, Spitzer, D., 2008. An efficient purification method for detonation nanodiamonds. Diam. Relat. Mater. 17, 13–22. https://doi.org/10.1016/j.diamond.2007.09.011.

Pu, M., Ma, Y., Wan, J., Wang, Y., Huang, M., Chen, Y., 2014. Fe/S doped granular activated carbon as a highly active heterogeneous persulfate catalyst toward the degradation of Orange G and diethyl phthalate. J. Colloid Interf. Sci. 418, 330–337. https://doi.org/10.1016/j.jcis.2013.12.034.

Qian, K., Kumar, A., Zhang, H., Bellmer, D., Huhnke, R., 2015. Recent advances in utilization of biochar. Renew. Sustain. Energy Rev. 42, 1055–1064. https://doi.org/10.1016/j.rser.2014.10.074.

Qin, B., Zhang, T., Chen, H., Ma, Y., 2016. The growth mechanism of few-layer graphene in the arc discharge process. Carbon 102, 494–498. https://doi.org/10.1016/j.carbon.2016.02.074.

Rai, S., Singh, B.K., Bhartiya, P., Singh, A., Kumar, H., Dutta, P.K., Mehrotra, G.K., 2017. Lignin derived reduced fluorescence carbon dots with theranostic approaches: nano-drug-carrier and bioimaging. J. Lumin. 190, 492–503. https://doi.org/10.1016/j.jlumin.2017.06.008.

Rajapaksha, A.U., Chen, S.S., Tsang, D.C.W., Zhang, M., Vithanage, M., Mandal, S., Gao, B., Bolan, N.S., Ok, Y.S., 2016. Engineered/designer biochar for contaminant removal/immobilization from soil and water: potential and implication of biochar modification. Chemosphere 148, 276–291. https://doi.org/10.1016/j.chemosphere.2016.01.043.

Ramírez-Sánchez, I.M., Tuberty, S., Hambourger, M., Bandala, E.R., 2017. Resource efficiency analysis for photocatalytic degradation and mineralization of estriol using TiO_2 nanoparticles. Chemosphere 184, 1270–1285. https://doi.org/10.1016/j.chemosphere.2017.06.046.

Ramos-Galicia, L., Mendez, L.N., Martínez-Hernández, A.L., Espindola-Gonzalez, A., Galindo-Esquivel, I.R., Fuentes-Ramirez, R., Velasco-Santos, C., 2013. Improved performance of an epoxy matrix as a result of combining graphene oxide and reduced graphene. Int. J. Polym. Sci. 2013, 493147. https://doi.org/10.1155/2013/493147.

Ray, S.C., 2015. Application and uses of graphene oxide and reduced graphene oxide. In: Roy, S.C. (Ed.), Applications of Graphene and Graphene-Oxide based Nanomaterials: A volume in Micro and Nano Technologies. Elsevier, Amsterdam, pp. 39–55.

Reina, G., Zhao, L., Bianco, A., Komatsu, N., 2019. Chemical functionalization of nanodiamonds: opportunities and challenges ahead. Angew. Chem. Int. Ed. 131, 18084–18095. https://doi.org/10.1002/ange.201905997.

Ren, W., Nie, G., Zhou, P., Zhang, H., Duan, X., Wang, S., 2020. The intrinsic nature of persulfate activation and N-doping in carbocatalysis. Environ. Sci. Technol. 54, 6438–6447. https://doi.org/10.1021/acs.est.0c01161.

Ren, X., Feng, J., Si, P., Zhang, L., Lou, J., Ci, L., 2018. Enhanced heterogeneous activation of peroxydisulfate by S, N co-doped graphene via controlling S, N functionalization for the catalytic decolorization of dyes in water. Chemosphere 210, 120–128. https://doi.org/10.1016/j.chemosphere.2018.07.011.

Ren, X., Guo, H., Ma, X., Hou, G., Chen, L., Xu, X., Chen, Q., Feng, J., Si, P., Zhang, L., Ci, L., 2018. Improved interfacial floatability of superhydrophobic and compressive S, N co-doped graphene aerogel by electrostatic spraying for highly efficient organic pollutants recovery from water. Appl. Surf. Sci. 457, 780–788. https://doi.org/10.1016/j.apsusc.2018.06.289.

Rodriguez-Narvaez, O.M., Pacheco-Alvarez, M.O.A., Wróbel, K., Páramo-Vargas, J., Bandala, E.R., Brillas, E., Peralta-Hernandez, J.M., 2020. Development of a Co^{2+}/PMS process involving target contaminant degradation and PMS decomposition. Int. J. Environ. Sci. Technol. 17, 17–26. https://doi.org/10.1007/s13762-019-02427-y.

Rodriguez-Narvaez, O.M., Peralta-Hernandez, J.M., Goonetilleke, A., Bandala, E.R., 2017. Treatment technologies for emerging contaminants in water: a review. Chem. Eng. J. 323, 361–380. https://doi.org/10.1016/j.cej.2017.04.106.

Rodriguez-Narvaez, O.M., Peralta-Hernandez, J.M., Goonetilleke, A., Bandala, E.R., 2019. Biochar-supported nanomaterials for environmental applications. J. Ind. Eng. Chem. 78, 21–33. https://doi.org/10.1016/j.jiec.2019.06.008.

Rodríguez-Narváez, O.M., Pérez, L.S., Yee, N.G., Peralta-Hernández, J.M., Bandala, E.R., 2019. Comparison between Fenton and Fenton-like reactions for l-proline degradation. Int. J. Environ. Sci. Technol. 16, 1515–1526. https://doi.org/10.1007/s13762-018-1764-1.

Rodríguez-Narvaez, O.M., Rajapaksha, R.D., Ranasinghe, M.I., Bai, X., Peralta-Hernández, J.M., Bandala, E.R., 2020. Peroxymonosulfate decomposition by homogeneous and heterogeneous Co: kinetics and application for the degradation of acetaminophen. J. Environ. Sci. 93, 30–40. https://doi.org/10.1016/j.jes.2020.03.002.

Rodríguez-Narváez, O.M., Serrano-Torres, O., Wrobel, K., Brillas, E., Peralta-Hernandez, J.M., 2018. Production of free radicals by the Co 2þ/Oxone system to carry out diclofenac degradation in aqueous medium. Water Sci. Technol. 78, 2131–2140. https://doi.org/10.2166/wst.2018.489.

Rouhani, P., Govindaraju, N., Iyer, J.K., Kaul, R., Kaul, A., Singh, R.N., 2016. Purification and functionalization of nanodiamond to serve as a platform for amoxicillin delivery. Mater. Sci. Eng. C. 63, 323–332. https://doi.org/10.1016/j.msec.2016.02.075.

Rueda-Marquez, J.J., Levchuk, I., Fernández Ibañez, P., Sillanpää, M., 2020. A critical review on application of photocatalysis for toxicity reduction of real wastewaters. J. Clean. Prod. 258, 120694. https://doi.org/10.1016/j.jclepro.2020.120694.

Sagbas, S., Sahiner, N.Khan, A., Jawaid, M., Inamuddin, Asiri, A.M. (Eds.), 2018. Carbon dots: Preparation, properties, and application. Nanocarbon and Its Composites. Preparation, Properties and Applications 651–676. https://doi.org/10.1016/B978-0-08-102509-3.00022-5.

Saifuddin, N., Raziah, A.Z., Junizah, A.R., 2013. Carbon nanotubes: a review on structure and their interaction with proteins. J. Chem. 2013. https://doi.org/10.1155/2013/676815.

Sánchez-Cid, P., Jaramillo-Páez, C., Navío, J.A., Martín-Gómez, A.N., Hidalgo, M.C., 2019. Coupling of Ag_2CO_3 to an optimized ZnO photocatalyst: advantages vs. disadvantages. J. Photochem. Photobiol. A Chem. 369, 119–132. https://doi.org/10.1016/j.jphotochem.2018.10.024.

Särkkä, H., Bhatnagar, A., Sillanpää, M., 2015. Recent developments of electro-oxidation in water treatment—a review. J. Electroanal. Chem. 754, 46–56. https://doi.org/10.1016/j.jelechem.2015.06.016.

Segawa, Y., Yagi, A., Matsui, K., Itami, K., 2016. Design and synthesis of carbon nanotube segments. Angew. Chem. Int. Ed. 55, 5136–5158. https://doi.org/10.1002/anie.201508384.

Shao, G., Lu, Y., Wu, F., Yang, C., Zeng, F., Wu, Q., 2012. Graphene oxide: the mechanisms of oxidation and exfoliation. J. Mater. Sci. 47, 4400–4409. https://doi.org/10.1007/s10853-012-6294-5.

Shao, P., Tian, J., Yang, F., Duan, X., Gao, S., Shi, W., Luo, X., Cui, F., Luo, S., Wang, S., 2018. Identification and regulation of active sites on nanodiamonds: establishing a highly efficient catalytic system for oxidation of organic contaminants. Adv. Funct. Mater. 28, 1705295. https://doi.org/10.1002/adfm.201705295.

Shao, Y., Long, Y., Wang, H., Liu, D., Shen, D., Chen, T., 2019. Hydrochar derived from green waste by microwave hydrothermal carbonization. Renew. Energy. 135, 1327–1334. https://doi.org/10.1016/j.renene.2018.09.041.

Sharma, R., Sharma, A.K., Sharma, V., 2015. Synthesis of carbon nanotubes by arc-discharge and chemical vapor deposition method with analysis of its morphology, dispersion and functionalization characteristics. Cogent. Eng 2. https://doi.org/10.1080/23311916.2015.1094017.

Shenderova, O.A., McGuire, G.E., 2015. Science and engineering of nanodiamond particle surfaces for biological applications (Review). Biointerphases 10, 030802. https://doi.org/10.1116/1.4927679.

Silvestri, S., Gonçalves, M.G., Da Silva Veiga, P.A., Matos, T.T.D.S., Peralta-Zamora, P., Mangrich, A.S., 2019. TiO_2 supported on *Salvinia molesta* biochar for heterogeneous photocatalytic degradation of acid orange 7 dye. J. Environ. Chem. Eng. 7, 102879. https://doi.org/10.1016/j.jece.2019.102879.

Singh, H., Bamrah, A., Khatri, M., Bhardwaj, N., 2020. One-pot hydrothermal synthesis and characterization of carbon quantum dots (CQDs). Mater. Today Proc 28, 1891–1894. https://doi.org/10.1016/j.matpr.2020.05.297.

Solís, R.R., Mena, I.F., Nadagouda, M.N., Dionysiou, D.D., 2020. Adsorptive interaction of peroxymonosulfate with graphene and catalytic assessment via non-radical pathway for the removal of aqueous pharmaceuticals. J. Hazard. Mater. 384, 121340. https://doi.org/10.1016/j.jhazmat.2019.121340.

Soltani, R.D.C., Mashayekhi, M., 2019. Ultrasonically facilitated electrochemical degradation of acetaminophen using nanocomposite porous cathode and Pt anode. Chem. Biochem. Eng. Q. 33, 35–42. https://doi.org/10.15255/CABEQ.2018.1446.

Somu, C., Karthi, A., Singh, S., Karthikeyan, R., Dinesh, S., Ganesh, N., 2017. Synthesis of various forms of carbon nanotubes by AC arc discharge methods—comprehensive review. Int. Res. J. Eng. Technol. 04, 344–354.

Stehlik, S., Varga, M., Ledinsky, M., Jirasek, V., Artemenko, A., Kozak, H., Ondic, L., Skakalova, V., Argentero, G., Pennycook, T., Meyer, J.C., Fejfar, A., Kromka, A., Rezek, B., 2015. Size and purity control of

HPHT nanodiamonds down to 1 nm. J. Phys. Chem. C. 119, 27708–27720. https://doi.org/10.1021/acs.jpcc.5b05259.

Sun, H., Kwan, C.K., Suvorova, A., Ang, H.M., Tadé, M.O., Wang, S., 2014. Catalytic oxidation of organic pollutants on pristine and surface nitrogen-modified carbon nanotubes with sulfate radicals. Appl. Catal. B Environ. 154–155, 134–141. https://doi.org/10.1016/j.apcatb.2014.02.012.

Sun, Y., Zhang, J.P., Guo, F., Zhang, L., 2016. Hydrochar preparation from black liquor by CO_2 assisted hydrothermal treatment: optimization of its performance for Pb^{2+} removal. Korean J. Chem. Eng. 33, 2703–2710. https://doi.org/10.1007/s11814-016-0152-0.

Tabrizian, P., Ma, W., Bakr, A., Rahaman, M.S., 2019. pH-sensitive and magnetically separable Fe/Cu bimetallic nanoparticles supported by graphene oxide (GO) for high-efficiency removal of tetracyclines. J. Colloid Interface Sci. 534, 549–562. https://doi.org/10.1016/j.jcis.2018.09.034.

Tai, X.H., Chook, S.W., Lai, C.W., Lee, K.M., Yang, T.C.K., Chong, S., Juan, J.C., 2019. Effective photoreduction of graphene oxide for photodegradation of volatile organic compounds. RSC Adv 9, 18076–18086. https://doi.org/10.1039/c9ra01209e.

Tajik, S., Beitollahi, H., Nejad, F.G., Safaei, M., Zhang, K., Van Le, Q., Varma, R.S., Jang, H.W., Shokouhimehr, M., 2020. Developments and applications of nanomaterial-based carbon paste electrodes. RSC Adv 10, 21561–21581. https://doi.org/10.1039/d0ra03672b.

Tan, X., Liu, Y., Zeng, G., Wang, X., Hu, X., Gu, Y., Yang, Z., 2015. Application of biochar for the removal of pollutants from aqueous solutions. Chemosphere 125, 70–85. https://doi.org/10.1016/j.chemosphere.2014.12.058.

Thines, K.R., Abdullah, E.C., Ruthiraan, M., Mubarak, N.M., Tripathi, M., 2016. A new route of magnetic biochar based polyaniline composites for supercapacitor electrode materials. J. Anal. Appl. Pyrolysis. 121, 240–257. https://doi.org/10.1016/j.jaap.2016.08.004.

Walsh, F.C., Low, C.T.J., 2016. A review of developments in the electrodeposition of tin-copper alloys. Surf. Coatings Technol. 304, 246–262. https://doi.org/10.1016/j.surfcoat.2016.06.065.

Wang, B., Gao, B., Fang, J., 2017. Recent advances in engineered biochar productions and applications. Crit. Rev. Environ. Sci. Technol. 47, 2158–2207. https://doi.org/10.1080/10643389.2017.1418580.

Wang, F., Chen, P., Feng, Y., Xie, Z., Liu, Y., Su, Y., Zhang, Q., Wang, Y., Yao, K., Lv, W., Liu, G., 2017. Facile synthesis of N-doped carbon dots/g-C_3N_4 photocatalyst with enhanced visible-light photocatalytic activity for the degradation of indomethacin. Appl. Catal. B Environ. 207, 103–113. https://doi.org/10.1016/j.apcatb.2017.02.024.

Wang, G., Chen, S., Quan, X., Yu, H., Zhang, Y., 2017. Enhanced activation of peroxymonosulfate by nitrogen doped porous carbon for effective removal of organic pollutants. Carbon 115, 730–739. https://doi.org/10.1016/j.carbon.2017.01.060.

Wang, H., Guo, W., Liu, B., Wu, Q., Luo, H., Zhao, Q., Si, Q., Sseguya, F., Ren, N., 2019. Edge-nitrogenated biochar for efficient peroxydisulfate activation: an electron transfer mechanism. Water Res 160, 405–414. https://doi.org/10.1016/j.watres.2019.05.059.

Wang, S., He, F., Zhao, X., Zhang, J., Ao, Z., Wu, H., Yin, Y., Shi, L., Xu, X., Zhao, C., Wang, S., Sun, H., 2019. Phosphorous doped carbon nitride nanobelts for photodegradation of emerging contaminants and hydrogen evolution. Appl. Catal. B Environ. 257, 117931. https://doi.org/10.1016/j.apcatb.2019.117931.

Wang, T., Zhai, Y., Zhu, Y., Li, C., Zeng, G., 2018. A review of the hydrothermal carbonization of biomass waste for hydrochar formation: process conditions, fundamentals, and physicochemical properties. Renew. Sustain. Energy Rev. 90, 223–247. https://doi.org/10.1016/j.rser.2018.03.071.

Wang, W., Tadé, M.O., Shao, Z., 2018. Nitrogen-doped simple and complex oxides for photocatalysis: a review. Prog. Mater. Sci. 92, 33–63. https://doi.org/10.1016/j.pmatsci.2017.09.002.

Wang, X., Feng, Y., Dong, P., Huang, J., 2019. A mini review on carbon quantum dots: preparation, properties, and electrocatalytic application. Front. Chem. 7, 1–9. https://doi.org/10.3389/fchem.2019.00671.

Wang, X., Wang, A., Ma, J., 2017. Visible-light-driven photocatalytic removal of antibiotics by newly designed C_3N_4@$MnFe_2O_4$-graphene nanocomposites. J. Hazard. Mater. 336, 81–92. https://doi.org/10.1016/j.jhazmat.2017.04.012.

Wang, Y., Hu, A., 2014. Carbon quantum dots: synthesis, properties and applications. J. Mater. Chem. C. 2, 6921–6939. https://doi.org/10.1039/c4tc00988f.

Wang, Y., Zhao, H., Zhao, G., 2016. Highly ordered mesoporous Fe_3O_4@carbon embedded composite: high catalytic activity, wide pH range and stability for heterogeneous electro-Fenton. Electroanalysis 28, 169–176. https://doi.org/10.1002/elan.201500488.

Wei, J., Liu, Y., Zhu, Y., Li, J., 2020. Enhanced catalytic degradation of tetracycline antibiotic by persulfate activated with modified sludge bio-hydrochar. Chemosphere 247, 125854. https://doi.org/10.1016/j.chemosphere.2020.125854.

Wu, F., Su, H., Wang, K., Wong, W.K., Zhu, X., 2017. Facile synthesis of N-rich carbon quantum dots from porphyrins as efficient probes for bioimaging and biosensing in living cells. Int. J. Nanomedicine. 12, 7375–7391. https://doi.org/10.2147/IJN.S147165.

Wu, N., 2018. Plasmonic metal-semiconductor photocatalysts and photoelectrochemical cells: a review. Nanoscale 10, 2679–2696. https://doi.org/10.1039/c7nr08487k.

Wu, Y., Pan, W., Li, Y., Yang, B., Meng, B., Li, R., Yu, R., 2019. Surface-oxidized amorphous Fe nanoparticles supported on reduced graphene oxide sheets for microwave absorption. ACS Appl. Nano Mater. 2, 4367–4376. https://doi.org/10.1021/acsanm.9b00809.

Xiao, R., Luo, Z., Wei, Z., Luo, S., Spinney, R., Yang, W., Dionysiou, D.D., 2018. Activation of peroxymonosulfate/persulfate by nanomaterials for sulfate radical-based advanced oxidation technologies. Curr. Opin. Chem. Eng. 19, 51–58. https://doi.org/10.1016/j.coche.2017.12.005.

Xu, H., Yang, X., Li, G., Zhao, C., Liao, X., 2015. Green Synthesis of fluorescent carbon dots for selective detection of tartrazine in food samples. J. Agric. Food Chem. 63, 6707–6714. https://doi.org/10.1021/acs.jafc.5b02319.

Xu, J., Cao, Z., Zhang, Y., Yuan, Z., Lou, Z., Xu, X., Wang, X., 2018. A review of functionalized carbon nanotubes and graphene for heavy metal adsorption from water: preparation, application, and mechanism. Chemosphere 195, 351–364. https://doi.org/10.1016/j.chemosphere.2017.12.061.

Xu, T., Xue, J., Zhang, X., He, G., Chen, H., 2017. Ultrafine cobalt nanoparticles supported on reduced graphene oxide: efficient catalyst for fast reduction of hexavalent chromium at room temperature. Appl. Surf. Sci. 402, 294–300. https://doi.org/10.1016/j.apsusc.2017.01.114.

Xue, Y., Gao, B., Yao, Y., Inyang, M., Zhang, M., Zimmerman, A.R., Ro, K.S., 2012. Hydrogen peroxide modification enhances the ability of biochar (hydrochar) produced from hydrothermal carbonization of peanut hull to remove aqueous heavy metals: batch and column tests. Chem. Eng. J. 200–202, 673–680. https://doi.org/10.1016/j.cej.2012.06.116.

Yadav, R., Dixit, C.K., 2017. Synthesis, characterization and prospective applications of nitrogen-doped graphene: a short review. J. Sci. Adv. Mater. Dev. 2, 141–149. https://doi.org/10.1016/j.jsamd.2017.05.007.

Yi, M., Shen, Z., 2015. A review on mechanical exfoliation for the scalable production of graphene. J. Mater. Chem. A. 3, 11700–11715. https://doi.org/10.1039/c5ta00252d.

Yi, Y., Tu, G., Zhao, D., Tsang, P.E., 2019. Pyrolysis of different biomass pre-impregnated with steel pickling waste liquor to prepare magnetic biochars and their use for the degradation of metronidazole. Bioresour. Technol. 289, 121613. https://doi.org/10.1016/j.biortech.2019.121613.

Yoo, M.J., Park, H.B., 2019. Effect of hydrogen peroxide on properties of graphene oxide in Hummers method. Carbon. 141, 515–522. https://doi.org/10.1016/j.carbon.2018.10.009.

Yu, J., Zhu, Z., Zhang, H., Chen, T., Qiu, Y., Xu, Z., Yin, D., 2019. Efficient removal of several estrogens in water by Fe-hydrochar composite and related interactive effect mechanism of H_2O_2 and iron with persistent free radicals from hydrochar of pinewood. Sci. Total Environ. 658, 1013–1022. https://doi.org/10.1016/j.scitotenv.2018.12.183.

Yu, J., Zhu, Z., Zhang, H., Di, G., Qiu, Y., Yin, D., Wang, S., 2020. Hydrochars from pinewood for adsorption and nonradical catalysis of bisphenols. J. Hazard. Mater. 385, 121548. https://doi.org/10.1016/j.jhazmat.2019.121548.

Yu, S., Liu, D., Zhao, S., Bao, B., Jin, C., Huang, W., Chen, H., Shen, Z., 2015. Synthesis of wood derived nitrogen-doped porous carbon-polyaniline composites for supercapacitor electrode materials. RSC Adv 5, 30943–30949. https://doi.org/10.1039/c5ra01949d.

Zaaba, N.I., Foo, K.L., Hashim, U., Tan, S.J., Liu, W.W., Voon, C.H., 2017. Synthesis of graphene oxide using modified Hummers method: solvent influence. Procedia Eng., 184, pp. 469–477.

Zanias, A., Frontistis, Z., Vakros, J., Arvaniti, O.S., Ribeiro, R.S., Silva, A.M.T., Faria, J.L., Gomes, H.T., Mantzavinos, D., 2020. Degradation of methylparaben by sonocatalysis using a Co–Fe magnetic carbon xerogel. Ultrason. Sonochem. 64, 105045. https://doi.org/10.1016/j.ultsonch.2020.105045.

Zhang, L., Li, L., Wang, Y., Yu, X., Han, B., 2020. Multifunctional cement-based materials modified with electrostatic self-assembled CNT/TiO_2 composite filler. Constr. Build. Mater. 238, 117787. https://doi.org/10.1016/j.conbuildmat.2019.117787.

Zhang, P., Li, Z., Zhang, S., Shao, G., 2018. Recent advances in effective reduction of graphene oxide for highly improved performance toward electrochemical energy storage. Energy Environ. Mater. 1, 5–12. https://doi.org/10.1002/eem2.12001.

Zhang, S., Zhu, X., Zhou, S., Shang, H., Luo, J., Tsang, D.C.W., 2019. Hydrothermal Carbonization for Hydrochar Production and Its Application. Elsevier Inc., Amsterdam https://doi.org/10.1016/b978-0-12-811729-3.00015-7.

Zhang, X., Gao, B., Fang, J., Zou, W., Dong, L., Cao, C., Zhang, J., Li, Y., Wang, H., 2019. Chemically activated hydrochar as an effective adsorbent for volatile organic compounds (VOCs). Chemosphere 218, 680–686. https://doi.org/10.1016/j.chemosphere.2018.11.144.

Zhang, Z., Yi, G., Li, P., Zhang, X., Fan, H., Zhang, Y., Wang, X., Zhang, C., 2020. A minireview on doped carbon dots for photocatalytic and electrocatalytic applications. Nanoscale 12, 13899–13906. https://doi.org/10.1039/d0nr03163a.

Zhang, Z., Zhu, Z., Shen, B., Liu, L., 2019. Insights into biochar and hydrochar production and applications: a review. Energy 171, 581–598. https://doi.org/10.1016/j.energy.2019.01.035.

Zhao, Q., Mao, Q., Zhou, Y., Wei, J., Liu, X., Yang, J., Luo, L., Zhang, J., Chen, H., Chen, H., Tang, L., 2017. Metal-free carbon materials-catalyzed sulfate radical-based advanced oxidation processes: a review on heterogeneous catalysts and applications. Chemosphere 189, 224–238. https://doi.org/10.1016/j.chemosphere.2017.09.042.

Zhao, X., Liu, S., Huang, Y., 2016. Removing organic contaminants by an electro-Fenton system constructed with graphene cathode. Toxicol. Environ. Chem. 98, 530–539. https://doi.org/10.1080/02772248.2015.1123495.

Zhou, F., Lu, C., Yao, Y., Sun, L., Gong, F., Li, D., Pei, K., Lu, W., Chen, W., 2015. Activated carbon fibers as an effective metal-free catalyst for peracetic acid activation: implications for the removal of organic pollutants. Chem. Eng. J. 281, 953–960. https://doi.org/10.1016/j.cej.2015.07.034.

Zhou, J., Sheng, Z., Han, H., Zou, M., Li, C., 2012. Facile synthesis of fluorescent carbon dots using watermelon peel as a carbon source. Mater. Lett. 66, 222–224. https://doi.org/10.1016/j.matlet.2011.08.081.

Zhou, L., Cai, M., Zhang, X., Cui, N., Chen, G., Zou, G.Y., 2019. Key role of hydrochar in heterogeneous photocatalytic degradation of sulfamethoxazole using Ag_3PO_4-based photocatalysts. RSC Adv 9, 35636–35645. https://doi.org/10.1039/c9ra07843f.

Zhu, N., Li, C., Bu, L., Tang, C., Wang, S., Duan, P., Yao, L., Tang, J., Dionysiou, D.D., Wu, Y., 2020. Bismuth impregnated biochar for efficient estrone degradation: the synergistic effect between biochar and Bi/Bi_2O_3 for a high photocatalytic performance. J. Hazard. Mater. 384, 121258. https://doi.org/10.1016/j.jhazmat.2019.121258.

CHAPTER 15

Photocatalytic applications of biogenic nanomaterials

Erick R. Bandala
Division of Hydrologic Sciences, Desert Research Institute, Las Vegas, Nevada, USA

15.1 Introduction

Ensuring water security—defined as the reliable availability of an acceptable quantity and quality of water for health, livelihoods, and production—is one of the most ambitious and cross-linked objectives included in the Sustainable Development Goals proposed by the United Nations as the world's shared plan to end extreme poverty, reduce inequality, and protect the planet by 2030 (Bandala & Goonetilleke, 2016). The challenge of ensuring acceptable quality of water is paramount and continues growing with the increased emerging of novel groups of pollutants, with a mounted list of known and a terrifying potential for unknown effects on human health and aquatic ecosystems (Zeidman et al., 2020). In the last few years, the need for developing more efficient water treatment processes able to remove emerging contaminants has increased (Rodriguez-Narvaez et al., 2020) with the consequence of a wide variety of novel methodologies being reported to avoid the release of poorly treated wastewater effluents into the environment (Chiliquinga et al., 2020; Medrano-Rodríguez et al., 2020; Ortiz-Marin et al., 2020). Among the different treatment technologies, advanced oxidation processes (AOPs) have been identified as high cost-efficient alternative for water treatment, in some cases capable to over perform conventional water treatment technologies (Rodríguez-Narvaez et al., 2020). The most accepted definition of AOPs involves the *in-situ* production of hydroxyl radicals (HO•) using either homogeneous or heterogeneous processes, and a wide variety of materials and processes have been reported for efficient promoting AOPs in water treatment experiments (Mortazavian et al., 2020). The use of engineered nanomaterials (ENMs) with specific designed features such as specific chemical and physical properties to promote AOPs and their application to improve water treatment process has showed a significant increase in the interest from the scientific community in the last decade, along with many other uses in the building, electronics, computer, pharmaceutical and medical industries (Bandala and Berlli, 2019). Nevertheless, ENMs have also recently found with some environmental drawbacks such as the involvement of toxic chemicals in their synthesis procedures or the potential for generation of hazardous byproducts (Villasenor et al., 2018). In the last few years, the search for novel, green,

Sustainable Nanotechnology for Environmental Remediation
DOI: https://doi.org/10.1016/B978-0-12-824547-7.00023-0

and ecologically viable methodologies to synthesize NMs has become a significant topic of research driven by environmental stewardship and the need to reduce the environmental footprint of restoration processes (Rodriguez-Narvaez et al., 2019). Using biogenic synthetic protocols to produce NMs has been suggested to be a nontoxic, ecofriendly alternative to chemical processes by replacing toxic chemicals with natural products (Villasenor et al., 2018). These natural products can act as reducing, capping, and stabilizing agents during the synthesis of the materials (Vaseghi et al., 2018). Biogenic synthesis uses natural products, such as polyphenols, vitamins, amino acids, carbohydrates, biopolymers, and natural surfactants from plants, bacteria, algae, fungi, or yeast (Santos et al., 2019). It is less time and resource intensive than conventional methods, and it is capable of producing smaller sized NMs (de Barros et al., 2018). A wide variety of biogenic sources have been used to synthesize NMs and many of them have interesting advantages to conventional chemical processes, such as reduced defective surface, higher production rate, and lower energy requirements (Srivastava, 2019). This chapter presents a critical review of application of biogenic NMs (BNMs) in the photocatalytic degradation of organic contaminants in water, identifies the current knowledge gaps, and provides recommendations for future research directions.

15.2 Fundamentals of photocatalysis

Photocatalysis is a specific AOP defined as the acceleration of a photoreaction in the presence of a catalyst (Ramírez-Sánchez et al., 2017). When radiation of the proper wavelength impinges the surface of the catalyst involved in the photocatalysis process—also known as photocatalyst— the photocatalyst will absorb photons with energy greater or equal to the band gap (BG) energy, inducing generation of charge carriers (e.g., electron-hole pairs) that allow reaction with surface-adsorbed chemical species (e.g., water molecule, molecular oxygen, organic molecule) and/or production of hydroxyl (HO •) radicals (Ramírez-Sánchez et al., 2019). Fig. 15.1 shows the most currently accepted mechanism for photocatalytic activation of titanium dioxide (TiO_2), the most widely studied photocatalyst for water treatment applications (Ramírez-Sánchez & Bandala, 2018).

In agreement with the density functional theory, the valence band (VB) and conduction band (CB) of pure TiO_2 are composed of mainly O2p and Ti3d orbitals, respectively, and the Fermi level is located in the middle of the BG, which indicates that VB is fully filled while CB is empty. When interacting with photons of energy equal or higher than the BG energy (3.2 eV in the case of TiO_2), photoexcitation of the semiconductor promotes electrons from VB to CB creating a charge vacancy (or hole (h^+)) in the VB. The h^+ in the VB is able to react with hydroxide ion to form HO • or can also be filled by donor-absorbed organic molecules. Photo-generated electrons in the CB can be transferred to electron acceptors and also produce HO • (see Fig 15.1). Several different studies report BNMs with interesting photocatalytic activity. For example, biogenic metal

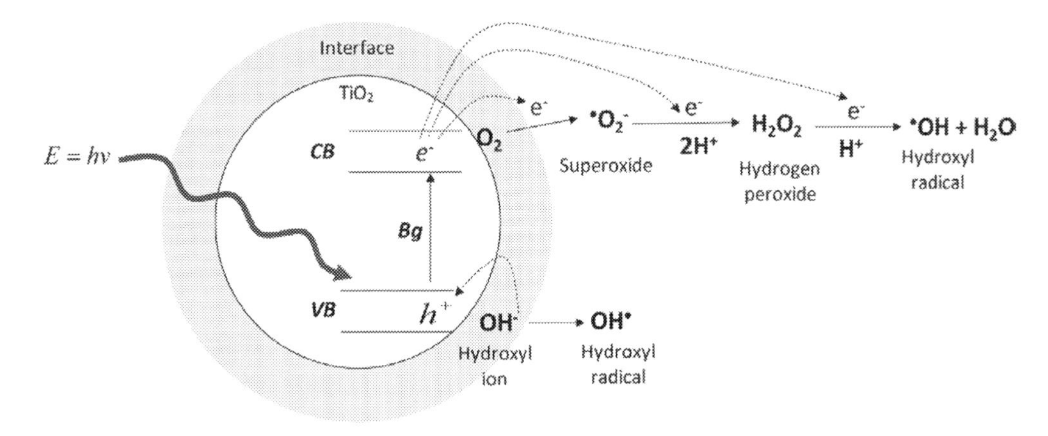

Figure 15.1 Schematic representation of the photocatalytic process (Modified from Ramírez-Sánchez, I. M., Tuberty, S., Hambourger, M., & Bandala, E. R. (2017). Resource efficiency analysis for photocatalytic degradation and mineralization of estriol using TiO_2 nanoparticles. Chemosphere, 184, 1270–1285. https://doi.org/10.1016/j.chemosphere.2017.06.046).

nanoparticles such as silver nanoparticles (AgNPs) in colloidal water have been found able to interact with photons to generate excited electrons at their surface. The dissolved oxygen molecules have been suggested able to accept the excited electrons from the AgNP surface becoming oxygen anion radicals that will interact with unsaturated bonds in organic molecules and transform it into simpler organic structures (Ramírez-Sánchez et al., 2017).

Because the surface nature of photocatalysis, it is reasonable to consider looking into the role played by the different capping agents generated during the biogenic production of NMs on the efficiency of the process as an interesting topic for further research. In agreement with other studies, charge carriers are produced on, or will migrate to, the surface of the NM after generation through interaction with the appropriate wavelength radiation. Once generated, charge carriers will react with oxygen and/or water molecules adsorbed on the surface of the photocatalyst to produce reactive oxygen species, capable of participating in redox processes that will led to the degradation of organic contaminants (Ramírez-Sánchez & Bandala, 2018). The presence of capping agents in the surface of BNMs may compete with oxygen or water molecules, interact with charge carriers, and prevent the photocatalytic process. Fig. 15.2A shows an overview of Ag nanoparticles obtained using *Fusarium oxysporum* (strain 551) where the capping layer is clearly observed as shown in Fig. 15.2B.

15.3 Photocatalytic activity in BNMs

Besides TiO_2, many other semiconductor-based materials have been reported with photocatalytic properties including ZnO, SnO_2, CuO, among many others

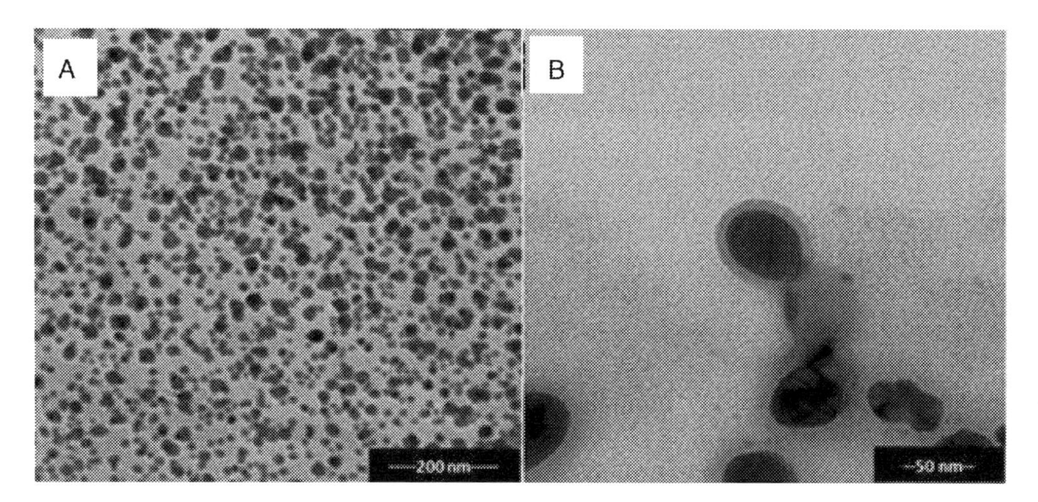

Figure 15.2 (A) Ag biogenic nanoparticles obtained from *Fusarium oxysporum* (strain 551), (B) Ag BNMs zoom showing the capping layer produced.

Table 15.1 Main advantages and disadvantages of conventional and biogenic NMs.

Materia	Advantages	Disadvantages
Conventional NMs	Chemically simpler Production cost and yield, well-known synthesis methods, standard reliable/scalable production, production conditions replicable	Unstable toxic, environmental fate/health effects unknown
Biogenic NMs	Recyclable sources eco-friendly synthesis, stabilized by capping agents	Environmental fate unknown, unknown production mechanisms, high variability, unknown life cycle

(Choudhary et al., 2019). The use of NMs of biogenic origin, however, is significantly less common. Table 15.1 lists a comparison of the main advantages and disadvantages of conventional and BNMs (Bandala et al., 2020). As shown, both conventional and biogenic NMs possess significant disadvantages that require attention and suggest potential opportunities for future research to fill the knowledge gaps detected. One of the major disadvantages found for BNMs is the sensible absence of information related with their environmental fate and health effects, once released into the environment independently of their proposed application. This lack of information is a significant knowledge gap worth attention.

The same principle explained previously applies for the mechanisms of most of the BNMs reported with photocatalytic activity, and the only difference in many of the cases is the BG energy for the different semiconducting materials, which made them suitable for applications with different energy sources. In some cases, for example, the addition of noble metals nanoparticles has been reported to serve to contribute in stimulating wide-band-gap semiconductors that are able to shift their BG energy to the red, allowing to operate within the visible radiation spectrum (Choudhary et al., 2019). In this regard, the production of doped semiconductors by biogenic green synthesis has gained great attention as a very attractive way to produce BNMs with photocatalytic properties able to absorb radiation other than ultraviolet which has been stated as a way to improve their performance (Kadam et al., 2020). The synthesis procedures reported for preparation of photocatalytic BNMs are quite variable and have been discussed with accuracy in other chapters of this book. Nevertheless, it is worth to mention that besides the shift in the BG energy reported for different biogenic semiconductors, other interesting features observed include their small particle size, and significant Brunauer-Emmett-Teller (BET) specific surface area (Fatimah et al., 2019). Some BNMs are reported to produce photo-generated charge carrier species (e.g., electrons, and holes) when interacting with either UV or visible radiation. The charge carrier species may recombine in microsecond after generation, but they also interact with organic compounds generating redox processes (Kumar Bachheti et al., 2020). To our best knowledge, however, the role played by other chemical species attached to the surface of the BNMs is unknown (Bandala et al., 2020). For example, chemical species attached to the surface of BNMs might inhibit or delay charge carrier recombination. The lack of experimental evidence for the roles played by the additional chemical species of biogenic origin on the surface of BNM photocatalyst in their photocatalytic activity is a significant knowledge gap. The effect of these species needs to be understood to know whether any differences exist between conventional and biogenic photocatalysts and to identify opportunities to improve the performance of BNMs for water treatment applications. Some examples of synthetic methodologies for the generation of BNMs with photocatalytic activity are summarized in Table 15.2. As shown, very few studies are available with information on the photocatalytic activity of BNMs. The lack of information identified from Table 15.2 is also highlighted by the fact that the full set of application of the BNMs with photocatalytic activity was only used for dye degradation. Despite the high interest and relevant impact of dyes in the environment, other water contaminants would be desirable to be included in the test. The lack of information is identified as a significant knowledge gap worth exploration.

15.4 Application of BNMs for photocatalytic degradation of organic contaminants

Table 15.3 shows a recollection of different BNMs recently reported with photocatalytic activity, their main characteristics, and applications for water treatment AOPs. As shown, almost all the organic contaminants included in the last studies are dyes, with very few exceptions such as chlorpyrifos and phenol which are considered pesticides included in the conventional organic contaminant list. It is worth noticing the variety of biogenic sources and BNM types are described in Table 15.1, including semiconductors and

Table 15.2 Synthesis methods, characteristics, and applications of photocatalytic BNMs

Synthetic Method	BNM morphology	Application(s)	Reference
Camellia sinensis methanol extract mixed with $CdSO_4$	CdS quantum dots (QDs)	The CdS QDs were used to decorate nanotubes which showed photocatalytic activity	Khan et al., 2020
Bacterial intracellular synthesis	Particle size range 22–39 nm, spherical shape, protein coating	Photocatalytic degradation of dyes using UV radiation	Noman et al., 2020
Pomegranate seed molasses mixed with Zn precursor	Particle size range 5–50 nm, active under visible radiation	Photocatalytic degradation of dyes	Essawy et al., 2020
Extract of Ficus benghalensis was added into SeS_2 solution	Average particle size 64 nm, highly stable SeNPs	Photocatalytic degradation of dyes using UV radiation	Tripathi et al., 2020
CuO mixed with Tamarindus indica extracts	5–10 nm QDs and 50–100 nm NPs were generated	Photocatalytic degradation of dyes using UV radiation	Zaman et al., 2020
Ludwigia octovalvis extract was mixed with $AgNO_3$	Particle size range 28–50 nm, polyhedral shape	Photocatalytic degradation of dyes	Kannan et al., 2020
Eucalyptus globulus extract is mixed with $Ce(NO_3)_3$ at pH>7.0	Average particle size 13.7 nm, pore diameter 7.8 nm	Photocatalytic degradation of dyes using UV radiation	Balaji et al., 2020

combined metallic composites (Kalathil et al., 2012, 2013; Shayegan Mehr et al., 2018). In most of the cases, the biogenic synthesis starts mixing a precursor salt with the biogenic material during a wide variety of reaction times, and calcination temperatures (ranging from 180 °C to 500 °C) with few studies reporting, for example, the control of the calcination atmosphere (Ijaz et al., 2017; Kalathil et al., 2012; Naghikhani et al., 2018; Shayegan Mehr et al., 2018; Somasundaram & Rajan, 2018). The resulting BNM (e.g., a metallic oxide) may be produced; however, even in the absence of the biogenic material and no study was found where a control performing the synthesis in the absence of biogenic material was included to demonstrate their significance. The absence of quality control during BNM synthesis processes is identified as a significant knowledge gap because including the proper quality control/assurance in the experimental design is basic for demonstrating the significance of the synthetic process. Some other examples highlighting the need for quality control in the synthesis of BNMs can be cited, for example, the use of electrochemically active biofilm to reduce the BG energy in Degussa P25 TiO_2 (Kalathil et al., 2013). Changes in TiO_2 color and BG energy changes (e.g., from $E_g = 3.1$ eV to $E_g = 2.85$ eV) were reported using nitrogen gas to ensure anoxic

Table 15.3 Biogenic nanomaterials used for the photocatalytic degradation of organic contaminants.

BNM photocatalyst	Main characteristics	Applications	Reference
Ag-ZnO heterojunctions from Fennel vulgare	E_g: 3.08–3.19 eV for Ag–ZnO 1–3 wt%, respectively, particle size 6–10 nm	Rhodamine B and chlorpyrifos degradation using solar radiation	Choudhary et al., 2019
TiO_2/SiO_2 composite from Gigantochloa apus leaves	E_g: 3.21 eV, particle size: 6–13 nm, pore volume 0.02–0.4 cc, BET: 21–289 m^2/g	Methylene blue (MB) degradation	Fatimah et al., 2019
TiO_2-AgNPs from carambola fruit extract	E_g: 2.24 eV, pore size: 3.7–10 nm, pore volume: 0.13–0.25 cm^3/g, BET: 30–55 m^2/g.	Phenol degradation	Panchal et al., 2020
MnONPs from Cinnamomum verum	Particle size: 50–100 nm, spherical morphology with tendency to aggregate	MB degradation	Rajeswari et al., 2018
ZnONPs from Coriandrum sativum leafs	E_g: 3.4 eV, particle size 24 nm, activated using UV lamp and solar radiation	Yellow 186 dye degradation	Kamran et al., 2019
CuONPs from Tamarindus indica L.	Particle size: 5–10 nm using fruit, 50–100 using leaf extract	Rhodamine B degradation	Zaman et al., 2020
NiONPs from Limonia acidissima	Average particle size: 20 nm, 91% Evans blue degradation achieved using solar radiation after 150 min.	Evans blue	Fatimah et al., 2019
$CuFe_2O_4$-CuONPs from different plant extracts	Particle size: 40–50 nm dye degradation of 93%, 82%, and 86% achieved in 60 min using visible light.	Acid black 1, acid brown 14, and blue 9 dyes	Naghikhani et al., 2018
Iron oxide NPs from Gigantochloa apus leaves	E_g: 2.27 eV, particle size: 10–40 nm, Fe content 5–10%wt., combined hematite/magnetite nanostructure	Rhodamine B degradation	Choudhary et al., 2019

conditions in the reaction mixture. Nitrogen-doped TiO_2 is very well-known to possess a narrower semiconductor BG and being activate to visible radiation (Castillo-Ledezma et al., 2011; Castillo-Ledezma et al., 2014). Nevertheless, the study failed considering that using nitrogen may generate the observed narrowing in the BG and the role of the electrochemically active biofilm in the process remains in unclear (Kalathil et al., 2012). The use of Ag-ZnO heterojunctions produced from *Foeniculum vulgare* for rhodamine B and chlorpyrifos is an example of application of biogenic deposition of metal nanoparticles in semiconductors (see Table 15.3). Casting noble metal nanoparticles on the surface of semiconductors is reported by photodeposition, colloidal deposition, and deposition precipitation (Singh et al., 2019). The biogenic processes for noble metal insertion within the crystal lattice of semiconductors are unknown and no information is available on the mechanistic processes involved. This is a significant knowledge gap because the lack of information about the mechanisms impedes the improvement of the processes, scaling up, or optimization. For example, lipids, polyphenols, or carbohydrates are suggested to shape the Ti^{4+} ions into specific patterns that eventually produce small anatase-TiO_2 crystallographic planes (Kannan et al., 2020) but no experimental evidence is included to support these statements. Another study by Kadam et al. (2020) claimed using fennel seed extract as a precipitating agent for a deposition/precipitation procedure to insert Ag nanoparticles into ZnO (Choudhary et al., 2019) but no evidence is provided on the way the extract served as a precipitating agent or what specific chemical component within the extract was acting as the precipitating agent. In general, several biogenic processes use extraction with polar solvents for a significant amount of time, which most likely results in a complex mixture of secondary metabolites, which makes almost impossible to understand the synthesis of BNMs (Ngoepe et al., 2018). Therefore, a systematic phytochemical study of the plant extracts used for biogenic synthesis of NMs is highly desirable and a pending research avenue worth exploring. After identifying the biogenic sources that successfully synthesize NMs it will be easier to determine the compounds playing a major role in the synthetic process that will significantly reduce the perception of empiricism created by a lack of proper characterization in several of the studies currently available. Most of the tested organic compounds shown in Table 15.3 are dyes, with only a few other pollutants included (Choudhary et al., 2019; Das et al., 2018) probably because dyes are particularly easy to follow during discoloration processes. The lack of information on the effectivity of BNMs for photocatalytic degradation of organic contaminants other than dyes is another significant knowledge gap. Several different organic contaminants besides dyes have been identified as public concern related to their biological activity, persistence, or bioaccumulation, as well as several other undesirable characteristics (García-Alonso et al., 2019; Mortazavian et al., 2020; Rodriguez-Narvaez et al., 2017). For example, a recent study reported the use of biogenic manganese oxides and bio-palladium (bio-Pd) to remove recalcitrant pharmaceuticals from sewage water and using bio-Pd via catalytic reduction to sequester contrast mediums. In both cases, however, the BNMs were used as adsorbents, sequestrants, oxidants/reducers, catalysts, or electrocatalysts. We consider that assessing the capabilities of BNMs to generate photocatalytic processes to remove antibiotics, pharmaceuticals, pesticides, or polyfluoroalkyl substances, identify reaction mechanisms, byproducts, and reaction kinetics are pending research avenues worth exploring.

15.5 Challenges and perspectives

Obtaining NMs by biological means possess several advantages such as involving one-step, low-cost, energy-efficient, and easy to scale-up processes (David & Moldovan, 2020). To ensure the broad use of BNMs as a real alternative to conventional NMs, however, requires facing some feasibility challenges including, for example, generating accurate information about the characteristics of materials, identifying biocapping agents and their impact, improving BNM purification processes, generating standardized synthetic procedures, estimating production costs and optimizing yields, and assessing process scaling up (Grasso et al., 2020). The lack of information about the characterization of biocapping agents is a significant challenge for determining BNM feasibility. For example, phytochemicals are frequently suggested as reducing and stabilizing agents, but most studies do not identify these phytochemicals or their role as capping agents for the synthesized materials (Shaheen & Ahmad, 2020). Other biocapping agents, such as amino acids, have attracted attention because their physical properties may influence their interactions with other compounds (Basnet et al., 2020). This is another significant knowledge gap because there is an urgent need to define the mechanisms involved in their synthesis. Production scale up is another interesting feasibility challenge for the full-scale application of BNMs. Some studies have suggested that using cell cultures as a biogenic source for BNM production is comparable to the bioproduction of pharmaceuticals (Srivastava, 2019). The use of living cells to produce compounds of interest prevents the need for extraction and purification steps, ensures structural integrity, and allows production in continuous cell culturing and the associated downstream process. Nonetheless, controlling BNM properties remains the biggest challenge and, despite glycolipids and proteins have shown control on BNMs' size/shape, the biochemical pathways and biomolecules involved remain unknown and generating knowledge on this topic remains a pending research task (Chellamuthu et al., 2019; Verma et al., 2019).

15.6 Conclusions

This chapter reviewed the use of BNMs to promote photocatalytic degradation of organic contaminants and highlights the potential of BNMs for water treatment. The following were our main findings:

- The use of BNMs for the photocatalytic degradation of organic contaminants was found mostly limited to single or combined metallic or semiconducting nanoparticles tested mainly for the photocatalytic degradation of dyes in synthetic water.

- Different knowledge gaps related to a lack of control over the synthesis processes used to generate the BNMs tested as photocatalysts and some apparent flaws in the preparation methods were found. These flaws may produce artifact results and overstatements of the photocatalytic capability of the tested materials.
- A need for more accurate quality assurance/quality control was identified to generate comparable results, allow fairer comparisons among BNMs, and better assess their performance against conventional NMs.
- Several challenges were found to making BNMs feasible for engineered use. Those challenges include achieving uniformity in the material characteristics, improving material characterization, enhancing BNM purification processes, and standardizing synthetic procedures, production costs, and yields.
- Only few studies were found reporting BNM production scaling up or final quality standardization, which limits their potential for real-scale application. All these knowledge gaps were discussed in this paper and some suggestions for future research directions were included.

References

Balaji, S., Mandal, B., Reddy, V., Sen, D., 2020. Biogenic Ceria Nanoparticles (CeO_2 NPs) for Effective Photocatalytic and Cytotoxic Activity. Bioengineering 7 (1), 26. doi:10.3390/bioengineering7010026. In this issue.

Bandala, E.R., Berlli, M., 2019. Engineered nanomaterials (ENMs) and their role at the nexus of food, energy, and water. Mater. Sci. Technol. 2 (1), 29–40. doi:10.1016/j.mset.2018.09.004, In this issue.

Bandala, E.R., Goonetilleke, A., 2016. Emerging Contaminants in urban storm water : challenges and perspectives for sustainable water use, Brief for GSDR 2016 Update. The United Nations 1–4.

Bandala, E.R., Stanisic, D., Tasic, L., 2020. Biogenic nanomaterials for photocatalytic degradation and water disinfection: a review. Environ. Sci. Water Res. Technol. 6 (12), 3195–3213. https://doi.org/10.1039/d0ew00705f.

Basnet, P., Samanta, D., Chanu, T.I., Jha, S., Chatterjee, S., 2020. Glycine-A bio-capping agent for the bioinspired synthesis of nano-zinc oxide photocatalyst. J. Mater. Sci. 31 (4), 2949–2966. https://doi.org/10.1007/s10854-019-02839-z.

Castillo-Ledezma, J., Lopez-Malo, A., Pelaez, M., Dionysiou, D.D., Bandala, E.R., 2014. Modeling the enhanced photocatalytic solar disinfection of *Escherichia coli* using nitrogen-doped TiO_2. J. Surf. Interfaces Mat. 2 (4), 334–342. doi:10.1166/jsim.2014.1067.

Castillo-Ledezma, J.H., Salas, J.L.S., López-Malo, A., Bandala, E.R., 2011. Effect of pH, solar irradiation, and semiconductor concentration on the photocatalytic disinfection of *Escherichia coli* in water using nitrogen-doped TiO_2. Eur. Food Res. Technol. 233 (5), 825–834. https://doi.org/10.1007/s00217-011-1579-5.

Chellamuthu, P., Tran, F., Silva, K.P.T., Chavez, M.S., El-Naggar, M.Y., Boedicker, J.Q., 2019. Engineering bacteria for biogenic synthesis of chalcogenide nanomaterials. Microb. Biotechnol. 12 (1), 161–172. https://doi.org/10.1111/1751-7915.13320.

Chiliquinga, M., Espinoza-Montero, P.J., Rodríguez, O., Picos, A., Bandala, E.R., Gutiérrez-Granados, S., Peralta-Hernández, J.M., 2020. Simultaneous electrochemical generation of ferrate and oxygen radicals to blue BR dye degradation. Processes 8 (7), 753. https://doi.org/10.3390/PR8070753.

Choudhary, M.K., Kataria, J., Bhardwaj, V.K., Sharma, S., 2019. Green biomimetic preparation of efficient Ag-ZnO heterojunctions with excellent photocatalytic performance under solar light irradiation: a novel biogenic-deposition-precipitation approach. Nanoscale Adv. 1 (3), 1035–1044. https://doi.org/10.1039/c8na00318a.

Das, S., Das, A., Maji, A., Beg, M., Singha, A., Hossain, M., 2018. A compact study on impact of multiplicative Streblus asper inspired biogenic silver nanoparticles as effective photocatalyst, good antibacterial agent and interplay upon interaction with human serum albumin. J. Mol. Liquids 259, 18–29. https://doi.org/10.1016/j.molliq.2018.02.111.

David, L., Moldovan, B., 2020. Green synthesis of biogenic silver nanoparticles for efficient catalytic removal of harmful organic dyes. Nanomaterial. 10 (2), 202. https://doi.org/10.3390/nano10020202.

de Barros, C.H.N., Cruz, G.C.F., Mayrink, W., Tasic, L., 2018. Bio-based synthesis of silver nanoparticles from orange waste: effects of distinct biomolecule coatings on size, morphology, and antimicrobial activity. Nanotechnol., Sci. Appl. 11, 1–14. https://doi.org/10.2147/NSA.S156115.

Essawy, A., Alsohaimi, I., Alhumaimess, M., Hassan, H., Kamel, M., 2020. Green synthesis of spongy Nano-ZnO productive of hydroxyl radicals for unconventional solar-driven photocatalytic remediation of antibiotic enriched wastewater. J. Environ. Manage. 271, 110961. doi:10.1016/j.envman.2020.110961. In this issue.

Fatimah, I., Prakoso, N.I., Sahroni, I., Musawwa, M.M., Sim, Y.L., Kooli, F., Muraza, O., 2019. Physicochemical characteristics and photocatalytic performance of TiO_2/SiO_2 catalyst synthesized using biogenic silica from bamboo leaves. Heliyon 5 (11), e02766. https://doi.org/10.1016/j.heliyon.2019.e02766.

García-Alonso, J.A., Sulbarán-Rangel, B.C., Bandala, E.R., del Real-Olvera, J., 2019. Adsorption and kinetic studies of the removal of ciprofloxacin from aqueous solutions by diatomaceous earth. Desalin. Water Treat. 162, 331–340. https://doi.org/10.5004/dwt.2019.24313.

Grasso, G., Zane, D., Dragone, R., 2020. Microbial nanotechnology: challenges and prospects for green biocatalytic synthesis of nanoscale materials for sensoristic and biomedical applications. NanoMater. 10 (1), 11. https://doi.org/10.3390/nano10010011.

Ijaz, F., Shahid, S., Khan, S.A., Ahmad, W., Zaman, S., 2017. Green synthesis of copper oxide nanoparticles using *Abutilon indicum* leaf extract: antimicrobial, antioxidant and photocatalytic dye degradation activities. Trop. J. Pharma. Res. 16 (4), 743–753. https://doi.org/10.4314/tjpr.v16i4.2.

Kadam, A.N., Salunkhe, T.T., Kim, H., Lee, S.W., 2020. Biogenic synthesis of mesoporous N–S–C tri-doped TiO_2 photocatalyst via ultrasonic-assisted derivatization of biotemplate from expired egg white protein. Appl. Surf. Sci. 518, 146194. https://doi.org/10.1016/j.apsusc.2020.146194.

Kalathil, S., Khan, M.M., Ansari, S.A., Lee, J., Cho, M.H., 2013. Band gap narrowing of titanium dioxide (TiO_2) nanocrystals by electrochemically active biofilms and their visible light activity. Nanoscale 5 (14), 6323–6326. https://doi.org/10.1039/c3nr01280h.

Kalathil, S., Khan, M.M., Banerjee, A.N., Lee, J., Cho, M.H., 2012. A simple biogenic route to rapid synthesis of Au@TiO_2 nanocomposites by electrochemically active biofilms. J. Nanoparticle Res. 14 (8), 1051. https://doi.org/10.1007/s11051-012-1051-x.

Kamran, U., Bhatti, H., Iqbal, M., Jamil, S., Zahid, M., 2019. Biogenic synthesis, characterization and investigation of photocatalytic and antimicrobial activity of manganese nanoparticles synthesized from Cinnamomum verum bark extract. J. Mol. Struct. 1179, 532–539. doi:10.1016/j.molstruc.2018.11.006. In this issue.

Kannan, K., Radhika, D., Nikolova, M.P., Sadasivuni, K.K., Mahdizadeh, H., Verma, U., 2020. Structural studies of bio-mediated NiO nanoparticles for photocatalytic and antibacterial activities.. Inorgan. Chem. Commun. 113, 107755. https://doi.org/10.1016/j.inoche.2019.107755.

Khan, Y., Ahmad, A., Ahmad, N., Mir, F., Schories, G., 2020. Biogenic synthesis of a green tea stabilized PPy/SWCNT/CdS nanocomposite and its substantial applications, photocatalytic degradation and rheological behavior. Nanoscale Adv. 2 (4), 1634–1645, doi:10.1039/D0NA00029A. In this issue.

Kumar Bachheti, R., Fikadu, A., Bachheti, A., Husen, A., 2020. Biogenic fabrication of nanomaterials from flower-based chemical compounds, characterization and their various applications: a review. Saudi J. Biol. Sci. 27 (10), 2551–2562. https://doi.org/10.1016/j.sjbs.2020.05.012.

Medrano-Rodríguez, F., Picos-Benítez, A., Brillas, E., Bandala, E.R., Pérez, T., Peralta-Hernández, J.M., 2020. Electrochemical advanced oxidation discoloration and removal of three brown diazo dyes used in the tannery industry. J. Electroanal. Chem. 873, 114360. https://doi.org/10.1016/j.jelechem.2020.114360.

Mortazavian, S., Bandala, E.R., Bae, J.H., Chun, D., Moon, J., 2020. Assessment of p-nitroso dimethylaniline (pNDA) suitability as a hydroxyl radical probe: investigating bleaching mechanism using immobilized zero-valent iron nanoparticles. Chem. Eng. J. 385, 123748. https://doi.org/10.1016/j.cej.2019.123748.

Naghikhani, R., Nabiyouni, G., Ghanbari, D., 2018. Simple and green synthesis of $CuFe_2O_4$–CuO nanocomposite using some natural extracts: photo-degradation and magnetic study of nanoparticles. J. Mater. Sci. 29 (6), 4689–4703. https://doi.org/10.1007/s10854-017-8421-1.

Ngoepe, N.M., Mbita, Z., Mathipa, M., Mketo, N., Ntsendwana, B., Hintsho-Mbita, N.C., 2018. Biogenic synthesis of ZnO nanoparticles using *Monsonia burkeana* for use in photocatalytic, antibacterial and anticancer applications. Ceram. Int. 44 (14), 16999–17006. https://doi.org/10.1016/j.ceramint.2018.06.142.

Noman, M., Shahid, M., Ahmed, T., Niazi, M., Song, F., Mazoor, I., 2020. Use of biogenic copper nanoparticles synthesized from a native Escherichia sp. as photocatalysts for azo dye degradation and treatment of textile effluent. Environ. Pollut. 257, 113514. doi:10.1016/j.envopol.2019.113514. In this issue.

Ortiz-Marin, A.D., Amabilis-Sosa, L.E., Bandala, E.R., Guillén-Garcés, R.A., Treviño-Quintanilla, L.G., Roé-Sosa, A., Moeller-Chávez, G.E., 2020. Using sequentially coupled UV/H_2O_2-biologic systems to treat industrial wastewater with high carbon and nitrogen contents. Process Saf. Environ. Protect 137, 192–199. https://doi.org/10.1016/j.psep.2020.02.020.

Panchal, P., Paul, D., Sharma, A., Choundhary, P., Meena, P., Nehra, S., 2020. Biogenic mediated Ag/ZnO nanocomposites for photocatalytic and antibacterial activities towards disinfection of water. J. Colloid Interface Sci. 563, 370380. doi:10.1016/j.jcis.2019.12.079. In this issue.

Rajeswari, P., Tiwari, B., Ram, S., Pradhan, D., 2018. A biogenic TiO_2-C-O nanohybrid grown from a Ti^4+-polymer complex in green tissues of chilis, interface bonding, and tailored photocatalytic properties. J. Mater. Sci. 53, 3131–3148. doi:10.1007/s10853-017-1763-5. In this issue.

Ramírez-Sánchez, I.M., Bandala, E.R., 2018. Photocatalytic degradation of estriol using iron-doped TiO_2 under high and low UV irradiation. Catalysts 8 (12). https://doi.org/10.3390/catal8120625.

Ramírez-Sánchez, I.M., Máynez-Navarro, O.D., Bandala, E.R., 2019. Degradation of emerging contaminants using Fe-doped TiO_2 under UV and visible radiation. In: Prasad, R. (Ed.), Nanotechnology in the Life Sciences. Springer Science and Business Media B.V., Dordecht, pp. 263–285. https://doi.org/10.1007/978-3-030-02381-2_12.

Ramírez-Sánchez, I.M., Tuberty, S., Hambourger, M., Bandala, E.R., 2017. Resource efficiency analysis for photocatalytic degradation and mineralization of estriol using TiO_2 nanoparticles. Chemosphere 184, 1270–1285. https://doi.org/10.1016/j.chemosphere.2017.06.046.

Rodriguez-Narvaez, O.M., Pacheco-Alvarez, M.O.A., Wróbel, K., Páramo-Vargas, J., Bandala, E.R., Brillas, E., Peralta-Hernandez, J.M., 2020. Development of a Co^{2+}/PMS process involving target contaminant degradation and PMS decomposition. Int. J. Environ. Sci. Technol. 17 (1), 17–26. https://doi.org/10.1007/s13762-019-02427-y.

Rodriguez-Narvaez, O.M., Peralta-Hernandez, J.M., Goonetilleke, A., Bandala, E.R., 2017. Treatment technologies for emerging contaminants in water: a review. Chem. Eng. J. 323, 361–380. https://doi.org/10.1016/j.cej.2017.04.106.

Rodriguez-Narvaez, O.M., Peralta-Hernandez, J.M., Goonetilleke, A., Bandala, E.R., 2019. Biochar-supported nanomaterials for environmental applications. J. Ind. Eng. Chem. 78, 21–33. https://doi.org/10.1016/j.jiec.2019.06.008.

Rodríguez-Narvaez, O.M., Rajapaksha, R.D., Ranasinghe, M.I., Bai, X., Peralta-Hernández, J.M., Bandala, E.R., 2020. Peroxymonosulfate decomposition by homogeneous and heterogeneous Co: kinetics and application for the degradation of acetaminophen. J. Environ. Sci. 93, 30–40. https://doi.org/10.1016/j.jes.2020.03.002.

Santos, L.M., Stanisic, D., Menezes, U.J., Mendonça, M.A., Barral, T.D., Seyffert, N., Azevedo, V., Durán, N., Meyer, R., Tasic, L., Portela, R.W., 2019. Biogenic silver nanoparticles as a post-surgical treatment for Corynebacterium pseudotuberculosis infection in small ruminants. Front. Microbiol. 10, 824. https://doi.org/10.3389/fmicb.2019.00824.

Shaheen, I., Ahmad, K.S., 2020. Chromatographic identification of "green capping agents" extracted from *Nasturtium officinale* (Brassicaceae) leaves for the synthesis of MoO_3 nanoparticles. J.Separation Sci. 43 (3), 598–605. https://doi.org/10.1002/jssc.201900840.

Shayegan Mehr, E., Sorbiun, M., Ramazani, A., Taghavi Fardood, S., 2018. Plant-mediated synthesis of zinc oxide and copper oxide nanoparticles by using *Ferulago angulata* (schlecht) boiss extract and comparison of their photocatalytic degradation of Rhodamine B (RhB) under visible light irradiation. J. Mater. Sci. Mater. Electron. 29 (2), 1333–1340. https://doi.org/10.1007/s10854-017-8039-3.

Singh, J., Kaur, S., Kaur, G., Basu, S., Rawat, M., 2019. Biogenic ZnO nanoparticles: a study of blueshift of optical band gap and photocatalytic degradation of reactive yellow 186 dye under direct sunlight. Green Process. Synth. 8 (1), 272–280. https://doi.org/10.1515/gps-2018-0084.

Somasundaram, G., Rajan, J., 2018. Effectual role of *Abelmoschus esculentus* (Okra) extract on morphology, microbial and photocatalytic activities of CdO tetrahedral clogs. J. Inorg. Organometal. Polym. Mater. 28 (1), 152–167. https://doi.org/10.1007/s10904-017-0695-5.

Srivastava, S.K., 2019. Microbial membrane-supported catalysts: a paradigm shift in clean energy and greener production. ACS Sustain. Chem. Eng. 7 (24), 19321–19331. https://doi.org/10.1021/acssuschemeng.9b05084.

Tripathi, R., Hammed, P., Rao, R., Shrivastava, N., Mittal, J., Mohapatra, S., 2020. Biosynthesis of highly stable fluorescent selenium nanoparticles and the evaluation of their photocatalytic degradation of dye. Bio. Nano. Science. 10, 389–396. doi:doi:10.1007/s12668-020-00718-0. In this issue.

Vaseghi, Z., Nematollahzadeh, A., Tavakoli, O., 2018. Green methods for the synthesis of metal nanoparticles using biogenic reducing agents: a review. Rev. Chem. Eng. 34 (4), 529–559. https://doi.org/10.1515/revce-2017-0005.

Verma, M., Sharma, S., Dhiman, K., Jana, A.K., 2019. Microbial production of nanoparticles: mechanisms and applications. In: Prassad, R. (Ed.), Microbial Nanobionics. Nanotechnology in the Life Sciences, pp. 159–176. https://doi.org/10.1007/978-3-030-16383-9_7.

Villasenor, D., Pedavoah, M.-M., Bandala, E.R., 2018. Plant materials for the synthesis of nanomaterials: greener sources. In: Martinez, K.B., Kharissova, L. (Eds.), Handbook of Ecomaterials. Springer, Cham, pp. 1–9.

Zaman, M., Poolla, R., Singh, P., Gudipati, T., 2020. Biogenic synthesis of CuO nanoparticles using Tamarindus indica L. and a study of their photocatalytic and antibacterial activity. Environ. Nanotechnol. Monit. Manag. 14, 100346. doi:10.1016/j.enmm.2020.100346. In this issue.

Zeidman, A.B., Rodriguez-Narvaez, O.M., Moon, J., Bandala, E.R., 2020. Removal of antibiotics in aqueous phase using silica-based immobilized nanomaterials: a review. Environ.Technol. Innov., 20. https://doi.org/10.1016/j.eti.2020.101030.

CHAPTER 16

Synthesis and photocatalytic applications of Cu_xO/ZnO in environmental remediation

Deborah L. Villaseñor-Basulto[a], Erick R. Bandala[b], Irwing Ramirez[c] and Oscar M. Rodriguez-Narvaez[d]

[a]Departamento de Química. División de Ciencias Naturales y Exactas, Campus Guanajuato, Universidad de Guanajuato, Guanajuato, Gto. México

[b]Division of Hydrologic Sciences. Desert Research Institute, Las Vegas Nevada, USA

[c]School of Engineering and Innovation, The Open University, Milton Keynes, MK76AA UK

[d]Dirección de investigación y soluciones tecnológicas, Centro de Innovación Aplicada en Tecnologías Competitivas (CIATEC), Omega 201, Col. Industrial Delta, León, Guanajuato, C.P. 37545, México

16.1 Introduction

Population rise and industrial development have led to water pollution, emerging antibiotic-resistant bacteria, and faster spread of diseases (Al-Badaii & Shuhaimi-Othman, 2015; Cheng et al., 2020; Manisalidis et al., 2020). These issues are seriously worsening in many regions globally, threatening the environment and human health. Presently, nanotechnology has demonstrated to be a promising alternative for water treatment and disinfection (Gao et al., 2020). Nanoparticles (NPs) are favorable materials in environmental applications as NPs have a large specific surface areas, high photostability, tunable electronic band structure, and cost effectiveness (Ruthiraan et al., 2019; Khan et al., 2020). NPs based on semiconductors are commonly used in photocatalysts for their unique electronic structure, and a bandgap (E_g) that can absorb UV-visible light (Long et al., 2020).

Photocatalysis begins when an semiconductor absorbs a photon with higher energy (hv) than its E_g. Then electrons (e^-) can transfer from the valence band (VB) to the conduction band (CB), and electron/hole (e^-/h^+) pairs are generated. When e^-/h^+ pairs are efficiently separated, they may induce redox reactions on the surface.

Semiconductors such as SiC, TiO_2, MnO_2, Fe_2O_3, CuO, ZnO, ZnS, ZrO_2, $SrTiO_3$, Nb_2O_5, MoO_3, MoS_2, CdS, CdTe, CdSe, SnO_2, WO_3, PbS, $BiWO_6$, $BiVO_4$, BiOBr, BiOCl, Bi_2O_3, and CeO_2 are promising in photocatalysis (Karthikeyan et al., 2020; Ramchiary et al., 2020; Riente & Noël, 2019). However, the high e^-/h^+ recombination rate impairs its photocatalytic performance in most of these catalysts. Consequently, various strategies have been examined (e.g., loading, doping, compositing, or coupling) to take this limitation away. An instance of these strategies is coupling various

Sustainable Nanotechnology for Environmental Remediation
DOI: https://doi.org/10.1016/B978-0-12-824547-7.00026-6

semiconductors to make a heterojunction that may decrease the e^-/h^+ recombination rate. Parul et al. reviewed the progress of semiconductor heterojunction in environmental applications and stood out that a better understanding should be given to confirm the heterojunction's electron migration mechanisms (Parul et al., 2020). In particular, ZnO heterojunctions have withdrawn more attention because of their different applications in sensing, CO_2 reduction, and water treatment. Given the significance of this kind of heterojunction, Goktas and Goktas summarized ZnO-based heterojunction progress and shown ZnO/CuO heterojunction could perform one of the highest photocatalysts performance of ZnO-based heterojunction (Goktas & Goktas, 2021). Although the literature has been reviewed on the basis of synthesis of ZnO and Copper oxides (Cu_xO) based heterojunction, a better understanding of the methods and mechanisms to improve ZnO functionality when joined with Cu_xO in environmental remediation is needed for its promising applications (Goktas & Goktas, 2021; Raizada et al., 2020; Susmita Das & Vimal Chandra Srivastava, 2018).

Cu_xOs are efficient catalysts with E_g between 1.7 and 2.2 eV (Heinemann et al., 2013), Cu_xOs included in this group are cuprous oxide (Cu_2O), copper peroxide (CuO_2), cupric oxide (CuO), and copper (III) oxide (Cu_2O_3). Among Cu_xOs, Cu_2O and CuO are the most stable and studied forms (Rydosz, 2018). Between these two stable forms, CuO has an E_g between 1.2 and 2.1 eV and is a nontoxic and abundant material (Dhineshbabu et al., 2016). On the other hand, Cu_2O has an E_g between 2.0 and 2.2 eV and is considered an efficient catalyst (Heinemann et al., 2013). So the coupling between ZnO and Cu_xOs creates a p–n heterojunction, which can boost the photocatalytic activity because p–n heterojunction facilitates e^-/h^+ separation (Oliveira et al., 2020), which notably expands efficient adsorption toward the visible-light region (Zhang, 2013). This chapter looks into Cu_xO/ZnO coupling and reviews the last 10 years on the synthesis, electron transfer mechanism, and environmental application for wastewater treatment and disinfection.

16.2 Synthesis of Cu_xO/ZnO

The synthesis of Cu_xO/ZnONPs and their composites has drawn more attention as these NPs have excellent photocatalytic activity, electrical, magnetic, and optical properties, and environment-friendly nature (Das and Srivastava, 2015). The synthesis of Cu_xO/ZnONPs can be grouped into physical, chemical, and biological methods (Fig. 16.1).

A chemical synthesis is an approach from the bottom to up. The most frequent chemical methods used in Cu_xO/ZnO synthesis include precipitation (Sepulveda-Guzman et al., 2009), sol-gel (Ristić et al., 2005), hydrothermal (Zhou et al., 2008), solvothermal (Xu et al., 2009), chemical vapor deposition (CVD) methods (Jeong & Aydil, 2009), photolysis and photochemical deposition, and chemical combustion. Still, the less used methods have been electron beam evaporation, laser ablation, and ultrasonic

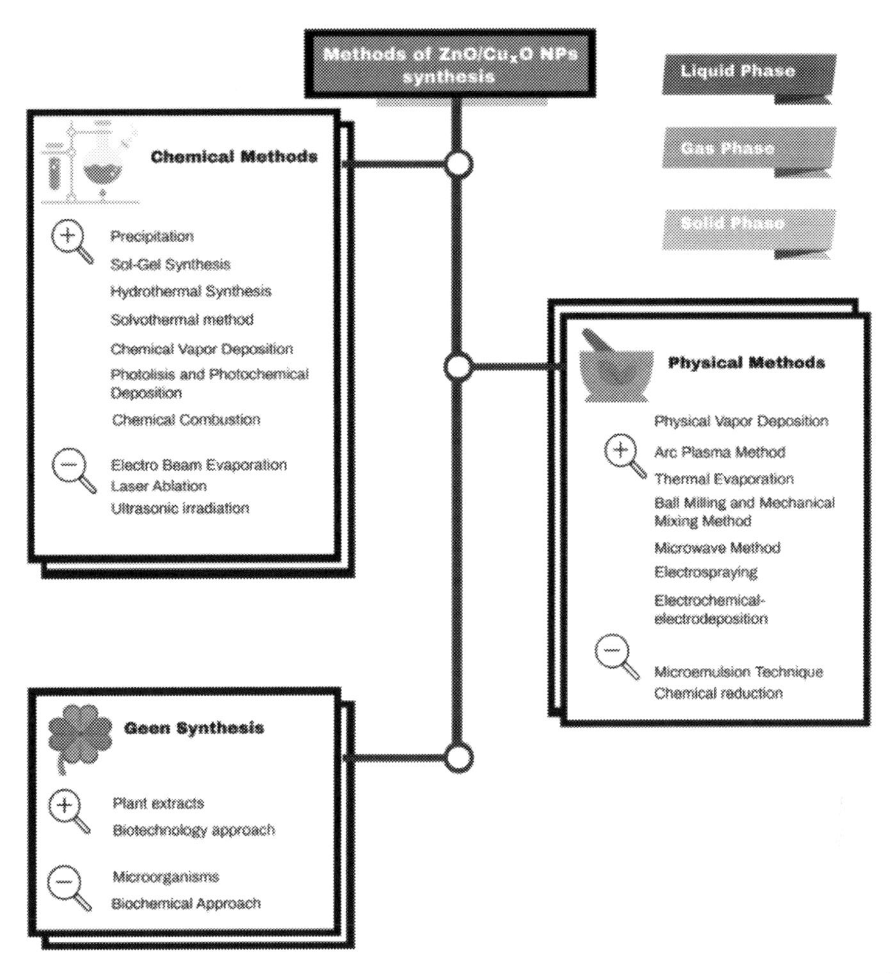

Figure 16.1 Photocatalytic process in semiconductors and e^-/h^+ pairs generation. Where $O_2^{\bullet-}$ is superoxide anion; •OH is hydroxyl radicals; OH− hydroxyl ion; CB is conduction band; VB is valence band; E_g is bandgap energy; hv is photon energy; and E_f is Fermi level.

irradiation. Meanwhile, the physical process is a top-down strategy that starts with a bulk counterpart, depleting and generating NPs. The most frequent physical methods reported in the literature for Cu_xO/ZnO synthesis have been arc plasma, ball milling, electrospinning, electrochemical, microwave, photolysis, thermal evaporation, ultrasonic irradiation, and vapor deposition methods. On the other hand, green synthesis is an emerging and eco-friendly synthesis method for Cu_xO/ZnO in which nontoxic and inexpensive precursors and solvents are used (Iravani, 2011; Yuvakkumar et al., 2014). For example, plant extracts and eggshell (ES) templating have been frequently reported in Cu_xO/ZnO synthesis by biological methods.

For all synthesis approaches, suitable synthetic parameters are critical for controlling specific crystallographic directions and interfacial interactions (Keis et al., 2001). In general, the synthesis outcomes can be grouped into 3-D structures, 2-D), and 1-D nanostructures. Some examples of 3-D nanostructures are coniferous urchin-like, dandelion, flower, and snowflakes. For instance, the 2-D nanostructures are pellets, plates, and sheets. Meanwhile, 1-D nanostructure includes belts, combs, helixes, needles, ribbons, rings, rods, springs, tubes, and wires (Keis et al., 2001).

16.2.1 Chemical synthesis methods

16.2.1.1 Precipitation method

An effective precipitation synthesis relies on a suitable precursor selection (Król et al., 2017). In general, the precipitation method begins with reactions between metals and OH^-. Then, the aggregation and formation of stable colloid suspension takes place in an aqueous solution (Hu et al., 2003). Different morphologies of Cu_xO/ZnONPs were obtained by controlling calcination temperature, pH, solution concentration, and washing medium during the precipitation processes (Table 16.1).

Three-dimensional flower-leaf ZnO-CuO nanocomposite resulted from managing media and using different types and amounts of precursors (Li & Wang, 2010). And calcination temperature was a significant factor to control the photocatalytic activity of spherical CuO/ZnO nanocomposites (Liu et al., 2008). The photocatalytic performance first increased and then decreased as the calcination temperature increased; such an effect was due to different particle size formation (Liu et al., 2008). Further, photocatalysis and absorption in visible light improved when Zn^{2+}/Cu^{2+} molar ratio was 10:2 (Liu et al., 2019), and Cu_2O/ZnO ratio was 0.78 (Wang et al., 2019), respectively.

The precipitation method's advantages are that it is simple, and morphology and composition can be easily controlled using harmless solvents at a low temperature. But due to precipitation reactions, impurities may precipitate along with NPs (Rane et al., 2018).

16.2.1.2 Sol-gel method

The five steps in the sol-gel method are hydrolysis, polycondensation, aging, drying, and thermal decomposition (Król et al., 2017; Wang et al., 2019). Table 16.2 shows CuO/ZnO nanostructures and Cu_2O/ZnO nanocomposites produced by the sol-gel method. The medium, aging method, and calcination time had an essential effect on Cu_xO/ZnO nanostructures (Giahi et al., 2013; Lavín et al., 2019). It was reported that 500 °C calcination temperature was optimal for synthesizing CuO/ZnO uniform structures to obtain an average particle size smaller than 60 nm (Nguyen, 2020), and formed spherical-like particles in the range 35–45 nm (Habibi & Karimi, 2014). Besides, a modification in

Table 16.1 Synthesis of Cu_xO/ZnO by precipitation method.

Composite	Precursor	Reaction medium	Washing medium	Calcination temperature	Reference
CuO-ZnO	Metallic Cu (0.064 g) + ZnO (4.05 g)	NH_4HCO_3 + ammonia	DI	350 °C, 450 °C, 500 °C, 550 °C and 650 °C	(Hassanpour et al., 2017; Liu et al., 2008)
CuO/ZnO	$ZnCl_2$ (0.2726 g) + $CuSO_4$ (0.2497 g)	NaOH + DI (controlled media)	DI + absolute alcohol	80 °C, 12 h	(Li and Wang, 2010)
CuO/ZnO	$Cu(II)Cl_2$ (0.68 g) + $ZnSO_4$ (0.87 g)	NaOH + DI (controlled media)	Centrifuged + washed with DI	80 °C, 12 h.	(Das and Srivastava, 2015)
CuO/ZnO	$ZnC_4H_6O_4$ + $C_2H_4CuO_2$ (mole ratio 1:1)	NaOH + deionized water	DI	500 °C, 3 h	(Yakout and El-Sayed, 2019)
CuO/ZnO	$CuSO_4$ (0.01 mol) + $Zn(Ac)_2$ (0.04 mol)	200 mL ethylene glycol, 50 mmol urea + 200 mL of EG + 52 mmol of tetramethylammo-nium hydroxide	DI	460 °C, 4 h	(Zhang, 2013)
Cu_2O/ZnO	Cu(Ac)2 (0, 0.028, to 0.112 mol/L), ZnO (0.3 g)	EtOH + HAC + glucose solution + NaOH + DE + absolute EtOH	EtOH + DI	60 °C, 8 h	(Wang et al., 2019)
CuO/ZnO	Zn^{2+}/Cu^{2+} ratio 10:1, 10:2, 10:3, 10:4, 10:5, and 10:6	Cetyltrimethylammonium bromide by ultrasonic + NaOH	EtOH + DI	500 °C, 2 h	(Liu et al., 2019)

DI - Distilled water; DE - Deionized water; EtOH - Ethanol; HAC - Acetic Acid.

Table 16.2 Synthesis of Cu_xO/ZnO by sol-gel method.

Composite	Precursor	Medium	Aging	Calcination temperature	Reference
10% CuO Doped ZnO	$Zn(NO_3)_2$ (5.3 g) + $CuSO_4$ (0.5 g)	Ethylene glycol + H_2O + citric acid	Kept 24 h under dark conditions	500 °C, 4 h.	(Giahi et al., 2013)
CuO/ZnO mixed oxide nanocrystalline	$CuCO_3$ (0.5 g) + $ZnCO_3$ (1 g)	Glacial acetic acid + DI + monoethanolamine + ethylene glycol	Aged at room temperature	450 °C, 2 h	(Habibi and Karimi, 2014)
ZnO/CuONP	$Zn(NO_3)_2$ (0.5 g) + $Cu(NO_3)_2$ (0.4 g)	Deionized water+ EtOH l+ polyvinyl alcohol	Sonicated 53 kHz, 80 °C	500 °C, 8 h	(Lavín et al., 2019)
Cu_2O/ZnO composite	$Zn(NO_3)$ (23.7 g) + $Cu(NO_3)$ (4.8 g)	DI+ oxalic acid+ ethylene glycol	80 °C, 2 h	Dried at 200 °C, 2 h. Calcinated in pure N_2 flow (3 L/h1) at 450, 500, and 550 °C	(Nguyen, 2020)

DI - Distilled water; EtOH - Ethanol.

the way ZnO was added on copper oxide generated photocatalytic performance changes (Giahi M. et al., 2013; Lavín et al., 2019).

Advantages of this method are faster nucleation and growth and that it is a suitable method for large-scale production. Meanwhile, its main disadvantage relies on high precursor costs, long reaction times, and solvents used in synthesis may harm the environment and humans (Król et al., 2017; Rane et al., 2018).

16.2.1.3 Hydrothermal method

The hydrothermal method occurs in an aqueous medium (water, alcohol, organic solvent, or inorganic solvent) in a closed vessel at temperature and pressure above the boiling point and 1 bar (Nunes et al., 2019). Table 16.3 shows CuO/ZnO nanocomposites synthesized by the hydrothermal method that reported interactions and the molar ratio between ZnO and CuO that causes modifications of the electronic structure, and consequently, the photocatalytic activity.

For instance, Cu_2O/tetrapod-like ZnO whisker (T-ZnOW) improved the photocatalytic performance for its 3-D nanostructure and p–n heterojunction (Liu et al., 2013). Three-dimensional flower-like CuO/ZnO composite confirmed intimate p–n heterojunction for Cr(VI) reduction (Yu et al., 2015). The CuO/ZnO synthesis with a molar ratio of 1:3 showed the highest photocatalytic degradation (Salehi et al., 2017). When the atomic ratio of Cu/Zn was 0.52, the visible emission intensity was relatively largest (Lu et al., 2017), and a better reflectance spectrum was obtained when a Cu: Zn molar rate was 1:3 (Lu et al., 2017b).

The merit of this technique is that the morphology can be adjusted through operational parameters such as solvents, precursors, reaction time, and temperature (Mahendiran et al., 2019). However, some disadvantages are safety issues during autoclaves operation (Rane et al., 2018).

16.2.1.4 Chemical vapor deposition

In CVD synthesis, a substrate is exposed to volatile precursors that react and decompose into a film. Besides, variants of CVD synthesis are atomic layer deposition (ALD), spray pyrolysis (SP), ultrasonic SP (USP), and chemical combustion. In CVD, the precursor type, concentrations, solvents, velocities, temperatures, and carrier gas fluxes are optimized (Sáenz-Trevizo et al., 2016). Table 16.4 shows nanostructures (e.g., nanowires, thin nanofilms, plasmon-induced nanocomposites, and nanosheets) synthesized by CVD and their variants.

CuO/ZnO nanocomposite was obtained with a crystalline size between 15 and 18 nm and wurtzite structure for ZnO and cubic phase for CuO (Renuka et al., 2017). USP technique formed a nanocomposite thin film of N-doped ZnO-CuO with a crystalline size of about 18 nm with aggregated crumpled-shape morphology

Table 16.3 Hydrothermal method in the synthesis of Cu_xO/ZnO nanocomposites.

Composite	Precursor	Medium	Reaction conditions	Washing medium	Reference
CuO/ZnO nanocomposites	Zn^{2+}/Cu^{2+} ratio 10:3, 10:4, 10:5, and 10:6	DI + hexamethylenetetramine (HMT) ($Zn(NO_3)_2$: HMT = 1:2)	90 °C/3 h/ Teflon-lined stainless (TSA)	N.D.	(Xu et al., 2017)
CuO/ZnO nanocomposites	Zn^{2+}/Cu^{2+} ratio 1:3, 2:3, 3:3, 3:2, and 3:1	Urea (>99.0%)+ DE	180 °C/10 h/TSA	Centrifuged + washed with DE + absolute EtOH	(Lu et al., 2017a)
CuO nanosheets/ZnO nanorods	Cu^{2+}/Zn^{2+} ratio 1:10, 1:2, 1:1, and 2:1	NaOH + DE	50–180 °C/12 h/ TSA	Centrifuged + washed with DE+ absolute EtOH	(Lu et al., 2017b)
CuO/ZnO	$Zn(NO_3)_2$ (0.04–0.05 M) + $CuH_{12}N_2O_{12}$ (0.04–0.01 M)	NaOH	100–200 °C/19 h/ teflon chamber	DE	(Taufique et al., 2018)
CuO/ZnO p–n junction nanocomposite	$Cu(CH_3COO)_2$ (1.167 mmol) + $Zn(NO_3)_2$ (2.141 mmol)	0.5 M NaOH	180 °C/24 h/ stainless-steel autoclave	DE+ absolute EtOH	(Lashgari et al., 2017)
Urchin-like CuO/ZnO nanocomposites	$Zn(CH_3COO)_2$ (0.5 mmol) + $Cu(NO_3)_2$ (1.5 mmol)	Glutamine (Gln) + urea + DE	120 °C/11 h/TSA	DE + absolute EtOH	(Fang et al., 2018)
CuO/ZnO nanocomposites	Zn $(CH_3COO)_2$ (2.54 g) + $Cu(CH_3COO)_2$ (0.76 g)	NaOH + DE	140 °C/12 h/TSA	DE + absolute EtOH	(Mahendiran et al., 2019)

(*continued on next page*)

CuO/ZnO composite	$ZnC_4H_6O_4$ (15 mmol) + $Cu(CH_3COO)_2$ (10 mmol)	NaOH + DE	150 °C/1.5 h/ autoclave	DI + absolute EtOH	(Sodeifian and Behnood, 2020)
Def-ZnO@Cu_2O *	Zn $(CH_3COO)_2$ + $CuSO_4$	Absolute EtOH + ammonium hydroxide	200 °C/8 h/ autoclave	DI	(Heo, 2020)
CuO/ZnO heterojunction	ZnO + $Cu(CH_3COO)_2$	Toluene	120 °C/6 h/ Vigreux condenser	Filtration EtOH	(Prabhu et al., 2019)
CuO/ZnO nanocomposites	ZnO (0.5 g) + $CuSO_4$ (0.2 g)	DE + hexam-ethylenete-tramine (0.3–0.5 g)	120–130 °C/12–24 h/TSA	DE	(Ma et al., 2016)
CuO/ZnO nanocomposites	ZnO nanostructures (40 mL) + $Cu(CH_3COO)_2$ (0.4 mmol)	DE	120 °C/8 h/TSA	N.D.	(Chang et al., 2013)
ZnO quantum dots/CuO nanosheets composites	$ZnC_4H_6O_4$ (200 mg) + CuO nanosheets (500 mg)	KOH + DI	180 °C/10 h/TSA	DI + absolute EtOH	(Fakhri et al., 2017)
Cu_2O spherical particles deposited on the surface of T-ZnOW	T-ZnOW (3.00 g) + $Cu(NO_3)_2$ (0.45 g)	0.1 g poly(vinyl pyrrolidone)	0–200 °C then cold 180 °C leave 30 min/ oil bath	DI + EtOH	(Liu et al., 2013)

(continued on next page)

Table 16.3 Hydrothermal method in the synthesis of Cu_xO/ZnO nanocomposites—cont'd

Composite	Precursor	Medium	Reaction conditions	Washing medium	Reference
CuO/ZnO nanocomposite	ZnO (1, 2, and 3 mol) + CuO (1 mol)	NaOH + trimethylamine	150 °C/12 h/ teflon reactor	DE	(Salehi et al., 2017)
Cu_2O/T-ZnOW nanocompound	T-ZnOW (3.00 g) + $Cu(NO_3)_2$ (0.15, 0.30, 0.45, 0.60, 0.75, 0.90 g)	0.06 g poly(vinyl pyrrolidone) + EG	0 to 180 °C leave 30 min/oil bath	DI + absolute EtOH	(Liu et al., 2013)
CuO/ZnO p–n heterojunction	$Cu(CH_3COO)_2$ (0, 0.1, 0.2, 0.4, and 0.5 mmol) + ZnONPs (100 mg)	Toluene	120 °C/6 h/N_2 atmosphere	Filtration + EtOH	(Yendrapati, 2019)
CuO@ZnO core-shell nanocomposites	0.4, 2, 10, and 50% CuO according to the ratio of initial molar concentrations of copper to zinc	DI-ultrasonicated + ammonia solution	70 °C/1 h/teflon autoclave	Centrifuged + washed	(Mansournia and Ghaderi, 2017)
3-D CuO/ZnO p–n junction	CuO microflowers coated with the ZnO seeds + $Zn(NO_3)_2$ (0.025 M)	$C_6H_{12}N_4$	95 °C/10 h/teflon autoclave	DI + absolute EtOH	(Yu et al., 2015)

N.D. - No Data reported; DI - Distilled water; DE - Deionized water; EtOH - Ethanol.

Table 16.4 Chemical vapor in Cu_xO/ZnO synthesis.

Technique	Composite type	Precursor type	Solvents	Temperatures	Reference
ALD	ZnO-coated CuO nanowires	CuO nanowires + Diethyl zinc	DI + precursor pulse time of 1.5 and 1.0 s	150 °C + background pressure of 50 mTorr	(Wang et al., 2015)
USP	N-doped ZnO-CuO nanocomposite thin film	$Zn(CH_3COO)_2$ (0.1 M) + $Cu(CH_3COO)_2$ (0.03 M)	DE + ammonium acetate	450 °C + Compressed air as a carrier gas with flow rate of 10 mL/min	(Rahemi Ardekani et al., 2018)
	Nano-sized ZnO/CuO nanocomposite	$Zn(NO_3)_2$ + $Cu(NO_3)_2$	Double DI + oxalyl dihydrazide as fuel	450 °C	(Renuka et al., 2017)
Chemical combustion synthesis	ZnO-CuO heterostructure	$Cu(NO_3)_2$ (10 mL, 5 mL and 2.5 mL 0.1 M) + $Zn(NO_3)_2$ (50 mL 0.1 M)	Sucrose as fuel	250 °C	(Bajiri et al., 2019)
	Copper plasmon-induced Cu-doped ZnO-CuO nanophotocatalysts	Different concentrations $Cu(NO_3)_2$ + $Zn(NO_3)_2$	DE + nitrate molar ratios of ethylene glycol as fuel	350 °C	(Abbasi et al., 2020)

Abbreviations: DI, distilled water; DE, deionized water.

(Rahemi Ardekani et al., 2018). CuO nanowires together with ZnO islands nanostructures resulted from ALD with few pulsed cycles, where ALD cycle numbers and ZnO concentration impacted crystalline orientation and surface defects (Wang et al., 2015). ZnO/CuO cotton-like porous structure was obtained by combustion synthesis (Bajiri et al., 2019).

The CVD method allows the control of crystal structure and surface morphology and easy scale-up, but the hazard of precursor gases needs special handling, and multilayer deposition is difficult (Jamkhande et al., 2019).

16.2.2 Physical methods

Several types of physical methods and their combination with chemical methods have been examined to produce Cu_xO/ZnO nanocomposites. Physical methods reported in Cu_xO/ZnO synthesis are microwave heating, physical vapor deposition, arc plasma, thermal evaporation, ball milling, electrospinning, and electrochemical deposition (Table 16.5).

The microwave heating, coupled with other methods, makes the NPs synthesis a user-friendly task to control morphology (Mohammadi et al., 2018). The microwave method alone formed CuO/ZnO spherical nanocomposite (Hassanpour et al., 2017) and ZnO/CuO composites with quasi-sphere-shaped wurtzite structures (Li and Lin, 2015). Besides, the microwave method produced CuO/ZnO nanocomposites using urea–nitrate combustion synthesis (Sherly et al., 2015). The merit of microwave-assisted synthesis is that it provides high reaction yields and has demonstrated reproducibility in reactions, crucial for scalability (Rane et al., 2018).

The arc plasma is a method in which an anode and a cathode are used in an arcing chamber saturated with gas. Cathode and anode come together to close the circuit, and just before this moment, an arc is produced, increasing temperature and bringing about ionized gas (Agrawal, 2001). CuO/ZnO heterostructured films were synthesized by the arc plasma method (Yu et al., 2019). CuO/ZnO nanoflowers were synthesized by the low-voltage liquid plasma discharge method using a brass wire cathode and a platinum foil as an anode in which the products were controlled by the applied voltage (Li et al., 2019). Some important features of the arc plasma method are that it can produce high purity NPs of about 10 nm, but the low production rates are a disadvantage (Rane et al., 2018).

In thermal evaporation, the material is vaporized at an elevated temperature and is condensed under certain conditions to form the desired product (Yang et al., 2005). Thermal evaporation method assisted CuO/ZnO film formation by evaporation and annealing of ZnO and Cu (Kuriakose et al., 2015). An advantage of thermal evaporation is that a thin layer can be easily obtained at a low cost (Dai et al., 2003).

Table 16.5 Synthesis of Cu_xO/ZnO nanocomposites assisted by microwave.

Technique	Composite type	Precursor type	Solvents	Microwave conditions	Temperature	Reference
Microwave method + urea–nitrate combustion synthesis	ZnO/CuO nanocomposites	$Zn(NO_3)_2$ + $Cu(NO_3)_2$ (1:1, 2:1, and 1:2)	DEr + urea	2.45 GHz, 850 W, 7 min	Spontaneous combustion	(Sherly et al., 2015)
Microwave method + pot solution combustion synthesis	ZnO/CuO nanocomposites	$Zn(NO_3)_2$ + $Cu(NO_3)_2$ (2:1)	DE + urea	2.45 GHz, 850 W, 7 min	Spontaneous combustion	(Sherly et al., 2016)
Microwave method	CuO/ZnO hallow spherical nanocomposite	$Cu(CH_3CO_2)_2$ (0.1 g) + $Zn(CH_3CO_2)_2$ (0.1 g)	Ethylene glycol	900 W, 10 min	Calcined at 400 °C, 2 h	(Hassanpour et al., 2017)
Microwave method	ZnO/CuO composites	$Zn(CH_3CO_2)_2$ + $Cu(CH_3CO_2)_2$ (99.9:0.1, 98.0:2.0, and 95.0:5.0)	DE + EtOH	20 min	Annealed at 400 °C, 2 h	(Ruan et al., 2020)

Abbreviations : DE, Deionized water; EtOH, Ethanol.

Ball milling grinds fine particles into homogenous powders. Circularity, roundness, and aspect ratio rely on the type of mill equipment (Moosakazemi et al., 2017). Through this method, p-CuO/n-ZnO nanocomposites were synthesized, presenting catalytic differences for both less (Sapkota & Mishra, 2013) and high milling hours (Chabri et al., 2016). Besides, agate mortar was for producing CuO/ZnO nanocomposites, and optimal CuO/ZnO ratio was obtained with a low concentration of CuO (Silva et al., 2015). The merit of this method is that it can be used for the large-scale production of high purity NPs. However, high energy and long milling periods could demerit extensive uses (Jamkhande et al., 2019).

Electrospinning, similar to electrospraying, is used to produce fibers, which involves an electrohydrodynamic process to generate fibers (Xue et al., 2019). CuO/ZnO nanofibers (NFs) were fabricated by one-step electrospinning by a polymer precursor and annealing in air, giving different CuO concentrations (Naseri et al., 2017). In general, electrospinning has shown the advantage of self-assembly of nanostructures (Zhang and Yu, 2014).

Electrochemical deposition is developed in an electrolyte solution at the metal interface to be deposited and a metal substrate (Singaravelan & Bangaru Sudarsan Alwar, 2015). This method synthesized monoclinic CuO within hexagonal wurtzite ZnO nanocomposite, and results showed a decrease in maximum reflectance and an increase of E_g (Ternes et al., 2007). Also, voltammetry was used to deposit the Cu_2ONPs onto ZnO nanorod arrays used in antibacterial applications (Wang et al., 2012). The advantage of electrochemical deposition is that NPs are directly grown on the substrate (Jamkhande et al., 2019).

16.2.3 Green synthesis

16.2.3.1 Plant extracts

The Cu_xO/ZnO synthesis can use plants using interesting phytochemical compounds in plant extracts (Table 16.6). For example, Zn hyperaccumulating plants formed ZnONPs (Król et al., 2017). Both CuONPs and CuO/ZnO nanocomposite were synthesized by Melissa Officinalis L. powdered dried leaves extract (Bordbar et al., 2018). Also, silver-doped ZnO/CuONPs were synthesized by leaves of Sida rhombifolia, forming the wurtzite hexagonal phase for ZnO and monoclinic for CuO (Babu & Antony, 2019). However, these compounds differ in plant location and type (Villaseñor-Basulto et al., 2019).

16.2.3.2 Biotemplating (biogenic approach)

Biological structures can provide materials with remarkable optical, electronic, and catalytic properties (Table 16.7). Biotemplating lets us prepare materials with complex structures taking advantage of nature's ability to generate unique morphologies (Magnabosco et al., 2020).

Table 16.6 Green synthesis of Cu_xO/ZnO nanocomposites using plant extracts.

	Pant extraction		Metal oxide NPs synthesis					
Composite type	**Plant part**	**Reaction condition**	**Plant extract**	**Precursor concentration**	**Reaction condition**	**Posttreatment**	**Calcination temperature**	**Reference**
CuO/ZnO nanocomposite	10 g of *Melissa Officinalis* L. leaves	Double DI water; 100 °C; 440 min	50 mL	ZnONPs (1.0 g)/ $CuCl_2$ (0.3 g)	75 °C	Centrifuged/ washed with water/EtOH	N.D.	(Bordbar et al., 2018)
ZnO/CuONPs	*Sida Rhombifolia* leaves	Water; 80 °C; 30 min	N.D.	$Zn(NO_3)_2$/ $Cu(NO_3)_2$ (5 g)	60 °C; 30 min	Centrifuged/ dried	400 °C, 3 h	(Babu and Antony, 2019)

Abbreviations: N.D.: No data reported; DI, distilled water; DE, deionized water; EtOH, ethanol.

Table 16.7 Green synthesis used in synthesis of Cu_xO/ZnO nanocomposites.

	Eggshell preparation			Metal oxide NPs				
Composite type	Eggshell treatment	Reaction condition	Posttreatment	Metal oxide NPs method	Precursor concentration/ conditions	Reaction condition	Posttreatment	Reference
ZnO-CuO/ES and 1:1 ZnO-CuO catalyst	Crushed and grinded	Water; 80 °C; 24 h	Oven-dried at 100 °C until constant weight	Electro-chemical synthesis	Pt cathode + Zn/Cu alternately as anode (120 mA/cm^2)	N,N-dimethyl-formamide (10 mL) + methylam-monium perchlorate (0.1 M), naphthalene (6 mmol) + ES (15 g)	[Calcinated]= 550 °C for 3 h	(Khairol, 2019)
CuO/ ZnO/ES	Ground into powder, was sieved 200-mesh	6% NaOH (0.1 M); 80 °C; 0.5 h	Dried at 60 °C overnight	Impregnation and calcination	$Cu(NO_3)$+ $Zn(NO_3)2$ (1:1)	Eggshell (2 g) + Cu:Zn media for 24 h	[Dried]=60 °C for 10 h [Calcinated]= 600 °C for 2 h	(Zhang et al., 2019)
3-D fibrous networks ZnO–CuO compos-ite	Eggs shell membranes	Metal salt precursors (0.3 M); Room temperature; 24 h	[Substrate]= Si/SiO_2 Room temperature; 12 h; [calcinated]=550 °C for 3 h		$Cu(NO_3)2$ + $Zn(NO_3)2$ (1:1) or $Cu(CH_3COO)_2$ + $Zn(CH_3COO)_2$	Impregnated eggshell membrane + Cu:Zn media	N.D.	(Preda et al., 2020)

N.D.: no data reported.

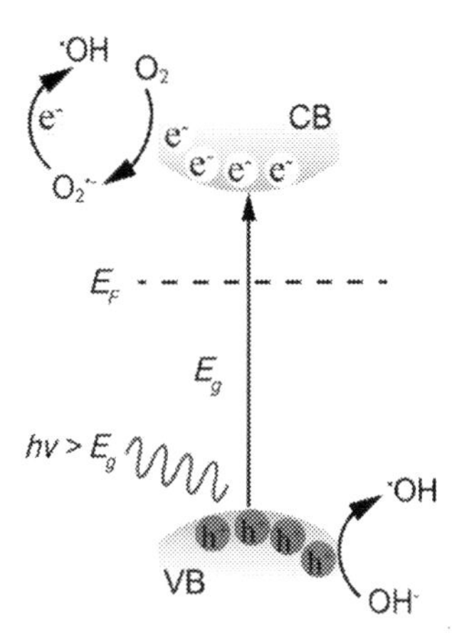

Figure 16.2 Classification of methods for Cu_xO/ZnO synthesis.

For instance, the ES as a biological template produced hierarchical self-assembled structures formed during calcination (Zhang et al., 2019). Further, ES was used as a biotemplate in CuO/ZnO nanocomposites (Preda et al., 2020) and used in electrochemical synthesis, and results showed a better photocatalytic performance than ZnO/ES or CuO/ES alone (Fahmi Khairol et al., 2019a, 2019b). Some CuO/ZnO/ES composites were found to have photocatalytic and antibacterial activities (Zhang et al., 2019). Further, using $ES/Si/SiO_2$ substrate resulted in a 3-D fibrous network biomorphic CuO/ZnO composite. The structure was hexagonal wurtzite for ZnO and the monoclinic structure for CuO and resulted in an E_g of 1.56 eV.

In general, the green approach's advantages are that precursors are abundantly available, and the process is flexible and easy to scale-up. However, extraction processes and heating may increase the production cost of NPs (Jamkhande et al., 2019).

16.3 Photocatalytic activity of Cu_xO/ZnO

When a semiconductor absorbs a photon with higher energy than its E_g, electrons (e^-) can transfer from the VB to the CB, and electron/hole (e^-/h^+) pairs are generated. So e^- can reduce O_2, which can turn into reactive oxygen species (ROS), for example, superoxide anion ($O_2^{\bullet -}$), hydrogen peroxide (H_2O_2), and hydroxyl radical ($^\bullet OH$). Meanwhile, holes (h^+) can oxidize both hydroxyl anions (OH^-) and electron donors, bringing about $^\bullet OH$ and subproducts, respectively (Fig. 16.2).

ZnO is an intrinsically n-type semiconductor with a direct E_g of 3.15 eV (Valles-Pérez et al., 2020; Zhao et al., 2019). Both Cu_2O and CuO are p-type semiconductors

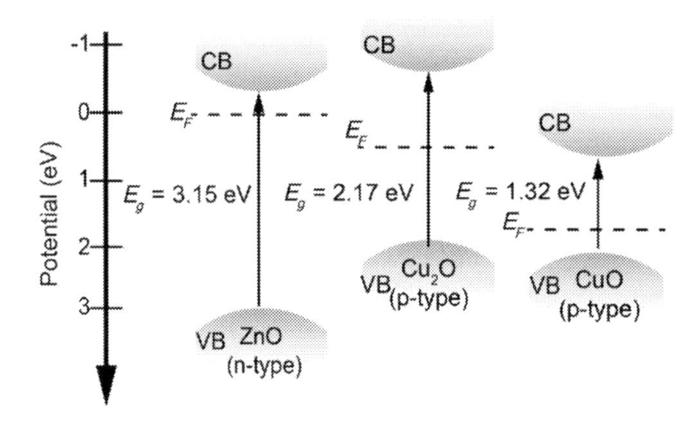

Figure 16.3 Energy band diagrams for ZnO, Cu_2O, CuO before contact.

with a direct E_g of 2.17 eV (Zhao et al., 2019) and an indirect E_g of 1.32 eV (Valles-Pérez et al., 2020), respectively. Fig. 16.3 depicts the band structure of pure compounds before contact. ZnO CB is higher than CuO CB but lower than Cu_2O CB, and the E_f of ZnO is in a higher position than both CuO and Cu_2O.

When n-type and p-type semiconductors are brought together, the E_f of individual materials reaches the same energy level, resulting in charge transfer and forming a depletion layer (Nadargi et al., 2020). This synergy brings about substantial enhancement in photocatalytic activity compared with their single-component system. The synthesis of materials with p–n heterojunction is discussed in the previous section. This heterojunction can be formed by CuO/ZnO, Cu_2O/ZnO, and a combination of them in Cu_xO/ZnO, which E_g is 2.61 eV (Mardikar et al., 2020), 2.17 eV (Kramm et al., 2012), and between 2 and 3.3 eV (Fuku et al., 2016), respectively. In general, p−n heterojunction construction is a practical approach to enhance the photocatalytic activity under the visible light region. The photocatalytic activity improvement may rely on efficient e^-/h^+ separation and the enhanced utilization of photons in the visible-light region (Xu et al., 2017).

16.3.1 Cu_2O/ZnO and CuO/ZnO

The coupling of ZnO and CuO generates a heterojunction that minimizes the recombination rate in both semiconductors (Singh & Soni, 2020). The heterojunction brings about that ZnO loses electrons, and CuO gathers electrons, so the E_f of the two semiconductors can reach an alignment (Han et al., 2020). Two possible electron-hole migration mechanisms have been suggested to explain the CuO/ZnO activity under light, that is, electron-hole transfer mechanism and the Z-scheme.

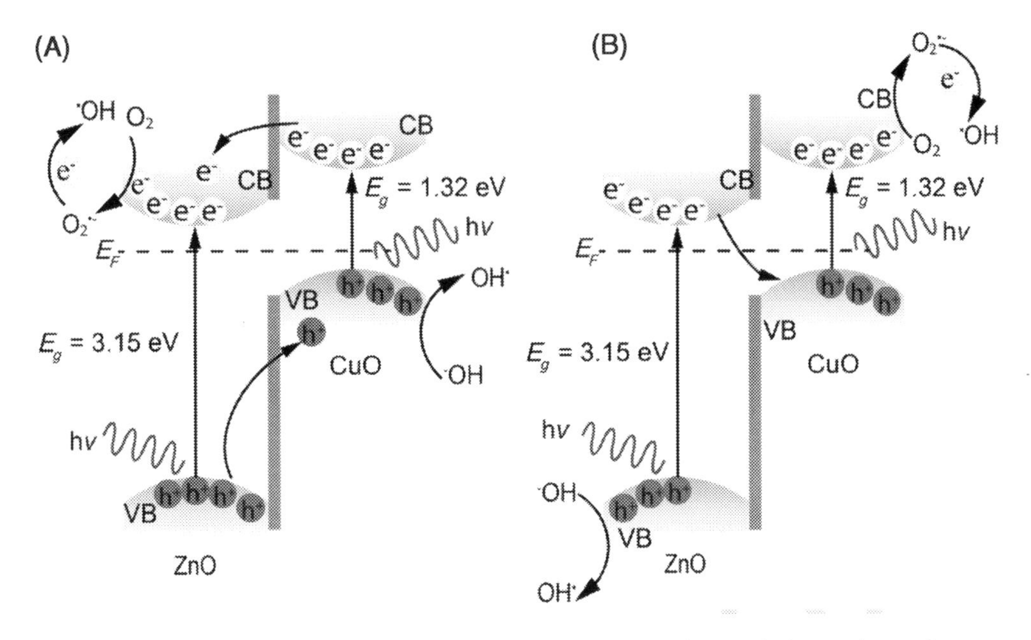

Figure 16.4 Electron-hole migration mechanisms of CuO/ZnO. (A) Electron-hole transfer mechanism; (B) the Z-scheme electron transfer mechanism.

Fig. 16.4A shows the electron-hole migration mechanism that occurs when UV radiation impinges the CuO/ZnO heterostructure. First, e^- in VB are excited to the CB, resulting in the formation of e^-/h^+ pairs (Fang et al., 2018; Fang et al., 2020). Then, when many p–n heterojunctions are formed due to the combination of p-type and n-type semiconductors, the e^- in the CuO CB can quickly move to the ZnO CB, whereas the holes can migrate from ZnO VB to CuO VB (Abbasi et al., 2020; Mardikar et al., 2020). In consequence, photocatalytic behavior improves because the lifetime of the e^-/h^+ pair increases (Sahu et al., 2020).

In the Z-scheme (Fig. 16.4B), the e^- moves from the ZnO CB to the CuO VB by electrostatic attraction, resulting in that e^- could long stay in the CuO CB; consequently, holes remain in the ZnO VB (Ruan et al., 2020). The Z-scheme has recently been discussed based on density functional theory, confirming the heterojunction reduces the electron-hole recombination rate (Oliveira et al., 2020).

In both mechanisms suggested in Fig. 16.4, e^- can react with dissolved O_2 leading to $O_2^{\bullet -}$ according to Equations (16.1) and (16.2). Then $O_2^{\bullet -}$ can yield H_2O_2 according to Equations (16.3) to (16.5) to produce $^{\bullet}OH$. The h^+ could oxidize adsorbed organic molecules or trap e^- form OH^- from H_2O molecules to produce $^{\bullet}OH$ (Eq. 16.6)

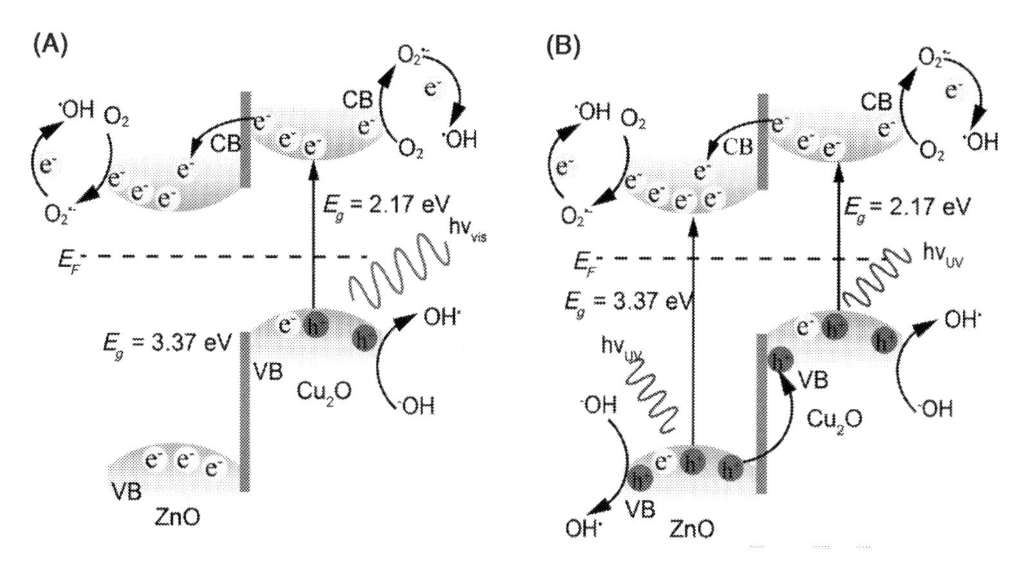

Figure 16.5 Electron-hole migration mechanism of Cu_2O/ZnO stimulated by visible light (A) and UV light (B).

(Mardikar et al., 2020).

$$CuO/ZnO + h\nu \rightarrow e^- + {h^+}_{p-CuO} \tag{16.1}$$

$$e^- + O_2 \rightarrow {O_2}^{\bullet -} \tag{16.2}$$

$${O_2}^{\bullet -} + H_2O \rightarrow {HO_2}^{\bullet} + OH^- \tag{16.3}$$

$${HO_2}^{\bullet} + H_2O \rightarrow H_2O_2 + OH^{\bullet} \tag{16.4}$$

$$H_2O_2 \rightarrow 2\,OH^{\bullet} \tag{16.5}$$

$$h^+ + OH^- \rightarrow OH^{\bullet} \tag{16.6}$$

Likewise, the p–n heterojunction in Cu_2O/ZnO (E_g of 1.9 eV) improves photocatalytic performance under visible light (Abdolhoseinzadeh & Sheibani, 2020). Two photocatalytic mechanisms have been suggested depending on whether visible or ultraviolet radiation stimulates Cu_2O/ZnO (Hong et al., 2017). Fig. 16.5A shows the mechanism when visible radiation stimulates Cu_2O/ZnO. Under visible light, Cu_2O can absorb visible light photons (hv_{vis}) to produce e^-/h^+ pairs. Then, the e^- in Cu_2O CB may directly produce $O_2^{\bullet -}$, or be transferred to ZnO CB, giving rise to $O_2^{\bullet -}$. Meanwhile, h^+ in Cu_2O VB can react with OH^- to produce $^{\bullet}OH$ (Eq. 16.6)) or oxidize organic matter. Fig. 16.5B shows the Z-scheme when UV radiation stimulates Cu_2O/ZnO. The e^- may

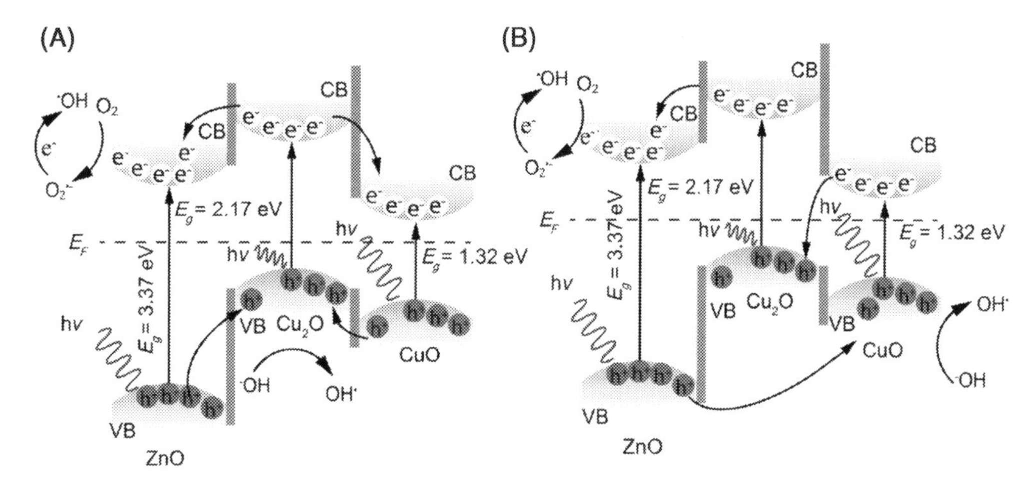

Figure 16.6 Electron-hole migration mechanism of $ZnO/Cu_2O/CuO$. (A) Electron-hole transfer; (B) the Z-scheme, showing relative potential energy for the $ZnO/Cu_2O/CuO$ and including electron flows between each bands.

migrate from the Cu_2O CB to the ZnO CB, and the h^+ may migrate from the ZnO VB to the Cu_2O VB, which reduces the recombination rate and prolongs the lifetime of e^-/h^+ pairs (Han et al., 2020). Both mechanisms could follow Equations (16.1) to (16.6) to generate $^{\bullet}OH$.

16.3.2 $ZnO/Cu_2O/CuO$ composites

$ZnO/Cu_2O/CuO$ composite nanostructure is an interesting material for its optical and catalytic properties. CB and VB of ZnO and CuO remain at relatively lower energy positions than the CB and VB of Cu_2O, so an electron-hole transfer mechanism and the Z-scheme electron-hole transfer have been suggested to describe the performance of $ZnO/Cu_2O/CuO$ (Nandi & Das, 2020). The electron-hole transfer mechanism (Fig. 16.6A) indicates that the e^- can separate from Cu_2O CB to CB of both ZnO and CuO after photon absorption. The h^+ from the VBs of both CuO and ZnO could transfer to the Cu_2O (Nandi & Das, 2020).

On the other hand, the Z-scheme mechanism suggests the e^- excited from the CuO may recombine to Cu_2O VB. Then the e^- from Cu_2O CB can be efficiently used without recombination process or move to ZnO CB. The h^+ in the ZnO VB could transfer to the higher potential in Cu_2O VB (Yoo et al., 2020). In both ways, the e^- and h^+ transfers might delay the recombination rate and gain a longer life for their effective utilization (Nandi and Das, 2020). In both mechanisms in Fig. 16.6, the mechanism could follow Equations (16.1) to (16.6) to generate ROS.

16.4 Cu_xO/ZnO in water treatment applications

Cu_xO/ZnO in water treatment has been studied for inorganic and organic contaminant removal. The Cu_xO/ZnO is advantageous in photocatalyst and adsorption for its porosity, specific area, and E_g (Ma et al., 2016; Chauhan, 2019). However, adsorption applications have a disadvantage because the pollutant is only transferred to a different phase. Table 16.8 shows the photocatalytic applications of Cu_xO/ZnO on organic and inorganic removal from water.

There is a lack of information for inorganic contaminants removal (e.g., metals and metalloids) because only a few studies on arsenite and chromium (VI) removal have been performed (Samad et al., 2016; Yu et al., 2015). Arsenite adsorption increased when the catalyst dose increased as more active sites were available on the material (Samad et al., 2016). Nanoflowers of 3D CuO/ZnO improved the reduction of chromium (VI) under visible light ($\lambda > 420$ nm) (Yu et al., 2015). The improvement was suggested due to the formation of p–n heterojunction that reduced e^-/h^+ recombination rate.

Cu_xO/ZnO materials have been used for different organic pollutant degradation (e.g., dyes, emerging contaminants). Some reports have been focused on dye degradation (e.g., phenol-based compounds) and only a few on emerging contaminants (i.e., Lidocaine HCl, Bisphenol A, Sulfamethazine, amoxicillin (AMX), 2,4-dichlorophenol) (Belaissa et al., 2016; Giahi M. et al., 2013; Naseri et al., 2017; Yu et al., 2019; Sherly et al., 2016).

Methylene blue (MB) is frequently used in dyeing, office supplies, active pharmaceutical substance, and therapeutic applications (Ginimuge & Jyothi, 2010). Therefore, MB is a potential wastewater pollutant because MB cannot be degraded through conventional water treatment processes (Hou et al., 2018). Photocatalytic oxidation is considered an effective way of degrading MB (Mansournia & Ghaderi, 2017; Nguyen & Juang, 2019). In particular, photocatalysis using CuO/ZnO hollow spheres, NFs, 2-D nanohybrids, and hollow disk-like nanoparticles have demonstrated to be effective for MB degradation under complete spectrum (UV-Visible radiation) (Naseri et al., 2017; Pawar et al., 2015; Sahu et al., 2020; Sodeifian & Behnood, 2020; Zhang et al., 2012), and also was confirmed MB mineralization over CuO/ZnO composite under visible radiation ($\lambda > 420$ mn) (Minh, 2019). Reusability tests of CuO/ZnO showed a reduction in MB degradation over four continuous cycles owing to the passivation of the active surface sites and the destruction of heterojunction via multiple washing (Chang et al., 2013; Joshi et al., 2017; Tuncel & Ökte, 2019). Evaluation of carrier scavengers during photocatalytic MB degradation confirmed that h^+ and $^\bullet OH$ were the main species responsible for MB degradation over CuO/ZnO (Naseri et al., 2017). Besides, Cu_xO/ZnO degraded rhodamine B (RhB), Congo red (CR), direct Blue 7, methyl Orange, and acid black under UV-visible radiation (Table 16.8), also the CuO/ZnO nanorods showed stability,

Table 16.8 Cu_xO/ZnO in water treatment applications.

Synthesis method	CuO/ZnO nanostructure	Catalyst load (g/L)	Contaminant	Initial concentration (mg/L)	Region and irradiation source	Removal	Reference
Mechanical mixing	Nanoparticles	0.67	Arsenite	30	UV: Black light (352 nm)	100% in 11.5 h	(Samad et al., 2016)
Hydrothermal growth	3-D flower-like composite	Sheet of 1 cm × 3.5 cm	Chromium (VI)	20	Visible: 300 W Xenon lamp with a 420 nm cut-off filter (λ> 420 nm)	100% in 800 min	(Yu et al., 2015)
Hydrothermal-deposition	Nanorod	1	Phenol	20	UV-visible: 55 W compact fluorescent lamp (14,500 lux)	100% in 300 min	(Lam et al., 2014)
Electrodeposition	Films on aluminum substrate	Sheet of 3 cm × 2 cm	Phenol	5	UV-visible: sunlight	40–70% in 300 min	(Paz et al., 2013)
Wet chemical	2-D nanohybrids	1	4-NP	0.7	UV-visible: sunlight	100% in 120 min	(Sahu et al., 2020)
Wet impregnation	Core shell	1	2-, 3-, and 4-NP, 2,4-, 2,5-, and 2,6- NP	30	UV-visible: sunlight (800 ± 100 × 102 lux)	99% in 180 min	(Qamar et al., 2015)
Combustion	ZnO–CuO	0.5	2, 4-dichlorophenol	50	Visible: 300 W tungsten–halogen lamp	82% in 250 min	(Sherly et al., 2015)

(continued on next page)

Table 16.8 Cu_xO/ZnO in water treatment applications—cont'd

Synthesis method	CuO/ZnO nanostructure	Catalyst load (g/L)	Contaminant	Initial concentration (mg/L)	Region and irradiation source	Removal	Reference
Sol-gel method	Nanoparticles	0.48	Lidocaine HCl	30	UV-visible: 400-W mercury lamp (354 nm)	93% in 360 min and 7 mM H_2O_2	(Giahi et al., 2013)
Chemical route	Dense grains	0.125 - 2	AMX	10 - 100	UV-Visible: sunlight (109 mW/cm^2) Visible: 200 W tungsten lamp	40–95 in	(Belaissa et al., 2016)
One-step electrospinning	Nanofibers	0.4	MB	20	UV-Visible: 300 W Xenon lamp (800 ± 50 lx)	5–90% in 120 min	(Naseri et al., 2017)
Wet chemical method	2-D nanohybrids	1	MB	1.6	UV-Visible: Sunlight	93.8% in 20 min	(Sahu et al., 2020)
Hydrothermal	Hollow spheres	0.25	MB	20	UV-Visible: 800 W Xenon-arc lamp	100% in 90 min	(Zhang et al., 2012)
Coprecipitation and impregnation	Nanoparticles	1	MB	0.81–32.7	UV-Visible: black fluorescent lamp Philips 15 W (320–440 nm, 4.7 × 1015 photon s-1)	80–99% in 100 min	(Tuncel and Ökte, 2019)

(continued on next page)

Thermal decomposition	Nanocomposites	1	MB, MeO, and textile wastewater	95.9	Visible: 250 W projection lamp using acetone as cut-off filter	20–90% in 120 min	(Saravanan et al., 2013)
Coprecipitation	Nanocomposite	0.25	MeO	20	UV-Visible: 500 W Xenon lamp	100% in 120 min	(Zhang, 2013)
Chemical	Hollow disk-like		MB, RhB	100	Visible: 100 W Halogen lamp (10 mW cm)	90–99% in 150 min	(Pawar et al., 2015)
Coprecipitation	Nanorods	0.8	RhB	10	400 W metal halide lamp	20–90% in 120 min	(Nandi and Das, 2020)
	nanorods	1.3	RhB, CR	15 μM	Visible: Tungsten incandescent lamp (150 mW cm-2)	100% in 50 min	(Senthil Kumar et al., 2017)
Hydrothermal	nanocomposite	0.5–2.5	DB71	20–60	UV-Visible: sunlight	89.58% in 177.13 min	(Salehi et al., 2017)

Methylene Blue (MB); Methyl Orange (MeO); Rhodamine B (RhB); 4-nitrophenol (4NP); amoxicillin (AMX); Congo Red (CR); Direct Blue 71 dye (DB71).

reusability, and an excellent degradation against the mixed dye solution of CR and RhB (Senthil Kumar et al., 2017).

Nanohybrids of 2-D CuO-ZnO had better activity than CuO for 4-nitrophenol degradation under sunlight, in which photocatalytic activity was suggested due to p–n heterojunctions at CuO and ZnO inhibited charge carrier recombination (Sahu et al., 2020). For this material, the pH had an essential role in the process as pH affected electronic bands and surface charge potential (Sahu et al., 2020). The CuO/ZnO arrangement showed degradation and mineralization of phenol under solar radiation (Paz et al., 2013).

Qamar et al. noticed that the increase in the optical absorption of CuO/ZnO did not increase photocatalytic activity (Qamar et al., 2015). And Lam et al. suggested CuO/ZnO heterojunction did not show appreciable improvement under UV–Vis radiation for the increase of recombination rate on the CuO (Lam et al., 2014). Sapkota et al. reported when CuO concentration for ZnO/CuO increased, the degradation rate increased, reaching a maximum. However, the photocatalytic degradation rate decreased when CuO content increased after the maximum (Naseri et al., 2017; Sapkota & Mishra, 2013). If CuO grows excessively in CuO/ZnO, more recombination centers may form, accelerating the recombination rate (Liu et al., 2019).

The use of CuO/ZnO to degrade emerging contaminants, for example, lidocaine (local anesthetic), sulfamethazine (antibiotic), and AMX (antibiotic), has been reported. Optimal lidocaine degradation was obtained when 0.48 g/L CuO/ZnONPs and 7 mM H_2O_2 were used under 400 W high-pressure mercury vapor. The ZnO/CuO led to sulfamethazine degradation under solar light (complete spectrum) and visible radiation, achieving 85% and 26% antibiotic degradation in 240 min, respectively (Yu et al., 2015). The removal and mineralization of AMX over CuO/ZnO were accomplished after 4 h under the complete spectrum (solar light). Therefore, CuO/ZnO could become a promising material for pharmaceutical removal in wastewater treatment (Belaissa et al., 2016).

As shown in Table 16.8, CuO/ZnO accomplished degradation-mineralization of dyes and pharmaceuticals on a lab scale, but few real industrial effluent applications have been performed. Saravanan et al. used CuO/ZnO nanocomposite for textile wastewater degradation under visible light and confirmed the mineralization of pollutants (Saravanan et al., 2013). Also, CuO/ZnO photocatalyst effectively degraded chlortetracycline in wastewater from aquaculture under visible light (20 W fluorescent light) (Liu et al., 2019). The lack of applications in real wastewater shows a critical information gap in using Cu_xO/ZnO.

Coupled applications have been performed by Yu et al. (2015) and Sharma et al. (2020), who reported that ZnO/CuO materials could be used in organic pollutants degradation and heavy metals reduction at the same time. The mechanism suggested in

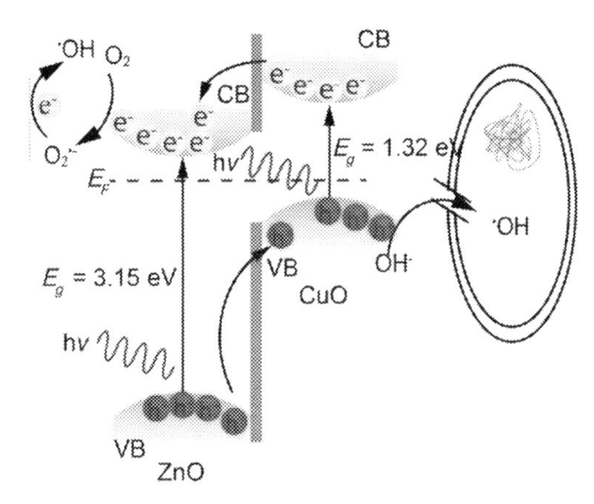

Figure 16.7 Hypothetical antibacterial mechanism of ZnO/CuO.

this coupled application h^+ produces the free radicals (i.e., $^\bullet OH$) for RhB degradation, while the electrons (e^-) reduces the chromium (VI) to chromium (III).

16.5 Cu_xO/ZnO in water disinfection

The photocatalytic disinfection mechanism lay on ROS effects on microorganisms, that is, ROS may attack the cell membrane. Then membrane permeability could increase, or cell membrane could break down (Deng et al., 2020). $^\bullet OH$ was demonstrated to be the ROS species that can quickly and efficiently degrade many bacteria components because of its high redox potential and nonselective reactivity (Ireland et al., 1993). Specifically, CuO/ZnO follows the electron-hole transfer mechanism suggested in Fig. 16.4 to generate $^\bullet OH$, and schematic generalization on its antibacterial mechanism is depicted in Fig. 16.7.

Cu_2O/ZnO composite and CuO impregnated on Fe-doped ZnO showed antibacterial against Escherichia coli under visible light (Gupta et al., 2018; Hong et al., 2017). The ROS generation by Cu_2O/ZnO under visible radiation may follow the mechanism suggested in Fig. 16.5A. The CuO/ZnO/ES composites exhibited significant antibacterial activity under UV-visible radiation (300 W xenon arc lamp) for the inactivation of *E.* coli and Staphylococcus aureus attributed to Cu^{2+} and Zn^{2+} ions released and the ROS generation (Zhang et al., 2019). Zhang et al. noticed using scanning electron microscope images that *E. coli* and *S. aureus'* cell membranes were damaged, which allowed cytoplasm to flow out (Zhang et al., 2019).

When microorganisms were exposed to ultraviolet radiation, mesoporous silica (SBA)/CuZnO displayed high photocatalytic activity against E. coli and S. aureus (Qiu et al., 2020). Qiu et al. identified that the bactericidal effect was due to H_2O_2

generation by SBA/CuZnO under ultraviolet radiation, confirming the formation of ROS (Qiu et al., 2020).Also, CuO/ZnO and forest-like TiO_2 NF/ZnO and forest-like TiO_2/ZnO/CuONPs completely inactivated E. coli in the solution under solar light simulator and solar light, respectively (Bai et al., 2012; Liu, 2012). Furthermore, ZnO quantum dots/CuO nanosheets composites had antibacterial activity against Enterococcus faecalis and Micrococcus luteus due to electrostatic interaction between nanostructures and cells membrane (Fakhri et al., 2017).

16.6 Conclusion and perspectives

The Cu_xO/ZnO has driven photocatalytic reactions for environmental remediation. The hydrothermal chemical method has been the most used for Cu_xO/ZnO synthesis because of a user-friendly approach. Further, it was demonstrated that the Cu ratio in Cu_xO/ZnO is a critical factor to enhance photocatalysis, controlling the recombination rate of electron/hole pairs. Although the coupling of Cu_xO/ZnO has been extensively studied, a broader understanding of the relationship between electron/hole transfer mechanisms and photon energy is needed to improve the photocatalytic processes in UV-visible regions and full-spectrum under solar radiation. Moreover, Cu_xO/ZnONPs and their composites have been efficiently used in adsorption, inorganic and organic contaminant removal, and disinfection, but a significant information gap is about scalability to real wastewater treatment and scale up.

In summary, Cu_xO/ZnONPs and their composites may come as an emerging alternative material for water purification and disinfection due to adsorption-photocatalytic-disinfection activity. So this material opens new possibilities to treat complex pollutant matrices and explore their high potential from biomedical to construction applications.

Acknowledgments

This chapter was supported by the University of Texas System and the Consejo Nacional de Ciencia y Tecnología (CONACYT) through the ConTex Postdoctoral Fellowship Program and Ph.D. Scholarship. The opinions expressed are those of the authors and do not represent the views of these funding agencies. The authors thank the University of Texas at Austin for granting financial resources, access to facilities, and training to conduct this work.

References

Abbasi, E., Haghighi, M., Shokrani, R., Shabani, M., 2020. Copper plasmon-induced Cu-doped ZnO-CuO double-nanoheterojunction: in-situ combustion synthesis and photo-decontamination of textile effluents. Mater. Res. Bull. 129, 110880. https://doi.org/10.1016/j.materresbull.2020.110880.

Abdolhoseinzadeh, A., Sheibani, S., 2020. Enhanced photocatalytic performance of Cu_2O nano-photocatalyst powder modified by ball milling and ZnO. Adv. Powder Technol. 31 (1), 40–50. https://doi.org/10.1016/j.apt.2019.09.035.

Agrawal, K.C., 2001. Introduction. In Agrawal, K.C. (Ed.). Industrial Power Engineering Handbook (pp. xv–xvi). Butterworth-Heinemann, UK. https://doi.org/10.1016/B978-075067351-8/50078-7.
Al-Badaii, F., Shuhaimi-Othman, M., 2015. Water pollution and its impact on the prevalence of antibiotic-resistant *E. coli* and total coliform bacteria: a study of the Semenyih River, Peninsular Malaysia. Water Qual. Expo. Health 7 (3), 319–330. https://doi.org/10.1007/s12403-014-0151-5.
Babu, A.T., Antony, R., 2019. Green synthesis of silver doped nano metal oxides of zinc & copper for antibacterial properties, adsorption, catalytic hydrogenation & photodegradation of aromatics. J. Environ. Chem. Eng. 7 (1), 102840. https://doi.org/10.1016/j.jece.2018.102840.
Bai, H., Liu, Z., Sun, D.D., 2012. Solar-light-driven photodegradation and antibacterial activity of hierarchical $TiO_2/ZnO/CuO$ Material. ChemPlusChem 77 (10), 941–948. https://doi.org/10.1002/cplu.201200131.
Bajiri, M.A., Hezam, A., Namratha, K., Viswanath, R., Drmosh, Q.A., Bhojya Naik, H.S., Byrappa, K., 2019. $CuO/ZnO/g\text{-}C_3N_4$ heterostructures as efficient visible light-driven photocatalysts. J. Environ. Chem. Eng. 7 (5), 103412. https://doi.org/10.1016/j.jece.2019.103412.
Belaissa, Y., Nibou, D., Assadi, A.A., Bellal, B., Trari, M., 2016. A new hetero-junction p-CuO/n-ZnO for the removal of amoxicillin by photocatalysis under solar irradiation. J. Taiwan Inst. Chem. Eng. 68, 254–265. https://doi.org/10.1016/j.jtice.2016.09.002.
Bordbar, M., Negahdar, N., Nasrollahzadeh, M., 2018. Melissa Officinalis L. leaf extract assisted green synthesis of CuO/ZnO nanocomposite for the reduction of 4-nitrophenol and rhodamine B. Sep. Purif. Technol., 191, 295–300. https://doi.org/10.1016/j.seppur.2017.09.044.
Chabri, S., Dhara, A., Show, B., Adak, D., Sinha, A., Mukherjee, N., 2016. Mesoporous CuO–ZnO p–n heterojunction based nanocomposites with high specific surface area for enhanced photocatalysis and electrochemical sensing. Catal. Sci. Technol. 6 (9), 3238–3252. https://doi.org/10.1039/C5CY01573A.
Chang, T., Li, Z., Yun, G., Jia, Y., Yang, H., 2013. Enhanced photocatalytic activity of ZnO/CuO nanocomposites synthesized by hydrothermal method. Nano-Micro Lett. 5 (3), 163–168. https://doi.org/10.1007/BF03353746.
Chauhan, et al., 2019. Green synthesis of CuO nanomaterials and their proficient use for organic waste removal and antimicrobial application. Environ. Res. 168, 85–95. doi:10.1016/j.envres.2018.09.024.
Cheng, D., Ngo, H.H., Guo, W., Chang, S.W., Nguyen, D.D., Liu, Y., Wei, Q., Wei, D., 2020. A critical review on antibiotics and hormones in swine wastewater: water pollution problems and control approaches. J. Hazard. Mater. 387, 121682. https://doi.org/10.1016/j.jhazmat.2019.121682.
Dai, Z.R., Pan, Z.W., Wang, Z.L., 2003. Novel nanostructures of functional oxides synthesized by thermal evaporation. Adv. Funct. Mater. 13 (1), 9–24. https://doi.org/10.1002/adfm.200390013.
Das, S., 2015. Hierarchical nanostructured ZnO-CuO nanocomposite and its photocatalytic activity. J. Nano Res. 35, 21–26. doi:10.4028/www.scientific.net/JNanoR.35.21.
Das, Susmita, Srivastava, Vimal Chandra, 2018. An overview of the synthesis of CuO-ZnO nanocomposite for environmental and other applications. Nanotechnol. Rev. 7 (3), 267–282. https://doi.org/10.1515/ntrev-2017-0144.
Deng, Y., Li, Z., Tang, R., Ouyang, K., Liao, C., Fang, Y., Ding, C., Yang, L., Su, L., Gong, D., 2020. What will happen when microorganisms "meet" photocatalysts and photocatalysis? Environ. Sci. Nano 7 (3), 702–723. https://doi.org/10.1039/C9EN01318K.
Dhineshbabu, N.R., Rajendran, V., Nithyavathy, N., Vetumperumal, R., 2016. Study of structural and optical properties of cupric oxide nanoparticles. Appl. Nanosci. 6 (6), 933–939. https://doi.org/10.1007/s13204-015-0499-2.
Fahmi Khairol, N., Sapawe, N., Danish, M., 2019a. Effective photocatalytic removal of different dye stuffs using ZnO/CuO-incorporated onto eggshell templating, 19, pp. 1255–1260.
Fahmi Khairol, N., Sapawe, N., Danish, M., 2019b. Photocatalytic study of ZnO-CuO/ES on degradation of congo red, 19, pp. 1333–1339.
Fakhri, A., Azad, M., Tahami, S., 2017. Degradation of toxin via ultraviolet and sunlight photocatalysis using ZnO quantum dots/CuO nanosheets composites: preparation and characterization studies. J. Mater. Sci. Mater. Electron. 28 (21), 16397–16402. https://doi.org/10.1007/s10854-017-7550-x.
Fang, H., Guo, Y., Wu, T., Liu, Y., 2018. Biomimetic synthesis of urchin-like CuO/ZnO nanocomposites with excellent photocatalytic activity. New J. Chem. 42 (15), 12779–12786. https://doi.org/10.1039/C8NJ02052C.

Fang, W., Yu, L., Xu, L., 2020. Preparation, characterization and photocatalytic performance of heterostructured CuO–ZnO-loaded composite nanofiber membranes. Beilstein J. Nanotechnol. 11, 631–650.

Fuku, X., Kaviyarasu, K., Matinise, N., Maaza, M., 2016. Punicalagin green functionalized Cu/Cu2O/ZnO/CuO nanocomposite for potential electrochemical transducer and catalyst. Nanoscale Res. Lett. 11 (1), 386. https://doi.org/10.1186/s11671-016-1581-8.

Gao, Y., Qian, K., Xu, B., Li, Z., Zheng, J., Zhao, S., Ding, F., Sun, Y., Xu, Z., 2020. Recent advances in visible-light-driven conversion of CO_2 by photocatalysts into fuels or value-added chemicals. Carbon Resour. Convers. 3, 46–59. https://doi.org/10.1016/j.crcon.2020.02.003.

Giahi, M., Badalpoor, N., Habibi, S., Taghavi, H., 2013. Synthesis of CuO/ZnO nanoparticles and their application for photocatalytic degradation of lidocaine HCl by the trial-and-error and Taguchi methods. Bull. Kor. Chem. Soc. 34 (7), 2176–2182. https://doi.org/10.5012/BKCS.2013.34.7.2176.

Ginimuge, P.R., Jyothi, S.D., 2010. Methylene blue: revisited. J. Anaesthesiol. Clin. Pharmacol. 26 (4), 517–520. https://pubmed.ncbi.nlm.nih.gov/21547182.

Goktas, S., Goktas, A., 2021. A comparative study on recent progress in efficient ZnO based nanocomposite and heterojunction photocatalysts: a review. J. Alloys Compd. 863, 158734. https://doi.org/10.1016/j.jallcom.2021.158734.

Gupta, R., Eswar, N.K., Modak, J.M., & Madras, G. (2018). Ag and CuO impregnated on Fe doped ZnO for bacterial inactivation under visible light. Photoactive mixed metal oxides for solar fuel and environmental processes, 300, 71–80. https://doi.org/10.1016/j.cattod.2017.05.032.

Habibi, M.H., Karimi, B., 2014. Preparation of nanostructure CuO/ZnO mixed oxide by sol–gel thermal decomposition of a $CuCO_3$ and $ZnCO_3$: TG, DTG, XRD, FESEM and DRS investigations. J. Ind. Eng. Chem. 20 (3), 925–929. https://doi.org/10.1016/j.jiec.2013.06.024.

Han, W., Luo, Z., Zhao, Y., Li, J., Li, P., 2020. Facile solution synthesis of quasi-0D/2D ZnO/CuO and ZnO/CuO/NiO composites for improved photocatalytic performance. Phys. Status Solidi A 217 (9), 1900904. https://doi.org/10.1002/pssa.201900904.

Hassanpour, M., Safardoust-Hojaghan, H., Salavati-Niasari, M., Yeganeh-Faal, A., 2017. Nano-sized CuO/ZnO hollow spheres: synthesis, characterization and photocatalytic performance. J. Mater. Sci. Mater. Electron. 28 (19), 14678–14684. https://doi.org/10.1007/s10854-017-7333-4.

Heinemann, M., Eifert, B., Heiliger, C., 2013. Band structure and phase stability of the copper oxides Cu_2O, CuO, and Cu_4O_3. Phys. Rev. B 87 (11), 115111. https://doi.org/10.1103/PhysRevB.87.115111.

Heo, et al., 2020. Self-assembled electron-rich interface in defected ZnO:rGO-Cu:Cu_2O, and effective visible light-induced carbon dioxide photoreduction. Appl. Catal. B Environ. 266. doi:10.1016/j.apcatb.2020.118648.

Hong, W., Meng, M., Liu, Q., Gao, D., Kang, C., Huang, S., Zhou, Z., Chen, C., 2017. Antibacterial and photocatalytic properties of Cu_2O/ZnO composite film synthesized by electrodeposition. Res. Chem. Intermed. 43 (4), 2517–2528. https://doi.org/10.1007/s11164-016-2777-3.

Hou, C., Hu, B., Zhu, J., 2018. Photocatalytic degradation of methylene blue over TiO_2 pretreated with varying concentrations of NaOH. Catalysts 8 (12), 575. https://doi.org/10.3390/catal8120575.

Hu, Z., Oskam, G., Searson, P.C., 2003. Influence of solvent on the growth of ZnO nanoparticles. J. Colloid Interf. Sci. 263 (2), 454–460. https://doi.org/10.1016/S0021-9797(03)00205-4.

Iravani, S., 2011. Green synthesis of metal nanoparticles using plants. Green Chem 13 (10), 2638–2650. https://doi.org/10.1039/C1GC15386B.

Ireland, J.C., Klostermann, P., Rice, E.W., Clark, R.M., 1993. Inactivation of *Escherichia coli* by titanium dioxide photocatalytic oxidation. Appl. Environ. Microbiol. 59 (5), 1668. http://aem.asm.org/content/59/5/1668.abstract.

Jamkhande, P.G., Ghule, N.W., Bamer, A.H., Kalaskar, M.G., 2019. Metal nanoparticles synthesis: an overview on methods of preparation, advantages and disadvantages, and applications. J. Drug Deliv. Sci. Technol. 53, 101174. https://doi.org/10.1016/j.jddst.2019.101174.

Jeong, S., Aydil, E.S., 2009. Heteroepitaxial growth of Cu_2O thin film on ZnO by metal organic chemical vapor deposition. J. Cryst. Growth 311 (17), 4188–4192. https://doi.org/10.1016/j.jcrysgro.2009.07.020.

Joshi, S., C., B., R., K., Jones, L.A., Mayes, E.L.H., Ippolito, S.J., Sunkara, M.V, 2017. Modulating interleaved ZnO assembly with CuO nanoleaves for multifunctional performance: perdurable CO_2 gas sensor and visible light catalyst. Inorg. Chem. Front. 4 (11), 1848–1861. https://doi.org/10.1039/C7QI00474E.

Karthikeyan, C., Arunachalam, P., Ramachandran, K., Al-Mayouf, A.M., Karuppuchamy, S., 2020. Recent advances in semiconductor metal oxides with enhanced methods for solar photocatalytic applications. J.Alloys Compd. 828, 154281. https://doi.org/10.1016/j.jallcom.2020.154281.

Keis, K., Vayssieres, L., Rensmo, H., Lindquist, S.-E., Hagfeldt, A., 2001. Photoelectrochemical properties of nano- to microstructured ZnO electrodes. J. Electrochem. Soc. 148 (2), A149. https://doi.org/10.1149/1.1342165.

Khairol, et al., 2019. Photocatalytic Study of ZnO-CuO/ES on Degradation of Congo Red. Materials Today: Proceedings 19 (4), 1333–1339. doi:10.1016/j.matpr.2019.11.146.

Khan, F.S.A., Mubarak, N.M., Khalid, M., Walvekar, R., Abdullah, E.C., Mazari, S.A., Nizamuddin, S., Karri, R.R., 2020. Magnetic nanoadsorbents' potential route for heavy metals removal—a review. Environ. Sci. Pollut. Res. 27 (19), 24342–24356. https://doi.org/10.1007/s11356-020-08711-6.

Kramm, B., Laufer, A., Reppin, D., Kronenberger, A., Hering, P., Polity, A., Meyer, B.K., 2012. The band alignment of Cu_2O/ZnO and Cu2O/GaN heterostructures. Appl. Phys. Lett. 100 (9), 094102. https://doi.org/10.1063/1.3685719.

Król, A., Pomastowski, P., Rafińska, K., Railean-Plugaru, V., & Buszewski, B. (2017). Zinc oxide nanoparticles: Synthesis, antiseptic activity and toxicity mechanism., Adv Colloid Interface Sci. 249, 37–52. https://doi.org/10.1016/j.cis.2017.07.033.

Kuriakose, S., Avasthi, D.K., Mohapatra, S., 2015. Effects of swift heavy ion irradiation on structural, optical and photocatalytic properties of ZnO–CuO nanocomposites prepared by carbothermal evaporation method. Beilstein J. Nanotechnol. 6, 928–937.

Lam, S.-M., Sin, J.-C., Abdullah, A.Z., & Mohamed, A.R. (2014). Transition metal oxide loaded ZnO nanorods: preparation, characterization and their UV–vis photocatalytic activities. separation and purification technology, 132, 378–387. https://doi.org/10.1016/j.seppur.2014.05.043.

Lashgari, M., Elyas-Haghighi, P., Takeguchi, M., 2017. A highly efficient pn junction nanocomposite solar-energy-material [nano-photovoltaic] for direct conversion of water molecules to hydrogen solar fuel. Sol. Energy Mater. Sol. Cells 165, 9–16. https://doi.org/10.1016/j.solmat.2017.02.028.

Lavín, A., Sivasamy, R., Mosquera, E., & Morel, M.J. (2019). High proportion ZnO/CuO nanocomposites: synthesis, structural and optical properties, and their photocatalytic behavior. Surf. Interf., 17, 100367. https://doi.org/10.1016/j.surfin.2019.100367.

Li, B., Wang, Y., 2010. Facile synthesis and photocatalytic activity of ZnO–CuO nanocomposite. Superlattices Microst. 47 (5), 615–623. https://doi.org/10.1016/j.spmi.2010.02.005.

Li, C.J., Cao, X., Li, W.H., Zhang, B.W., Xiao, L.Q., 2019. Co-synthesis of CuO-ZnO nanoflowers by low voltage liquid plasma discharge with brass electrode. J. Alloys Compd. 773, 762–769. https://doi.org/10.1016/j.jallcom.2018.09.250.

Li, S.-W., Lin, A.Y.-C., 2015. Increased acute toxicity to fish caused by pharmaceuticals in hospital effluents in a pharmaceutical mixture and after solar irradiation. Chemosphere 139, 190–196. https://doi.org/10.1016/j.chemosphere.2015.06.010.

Liu, H., Wang, J., Fan, X.M., Zhang, F.Z., Liu, H.R., Dai, J., Xiang, F.M., 2013. Synthesis of Cu_2O/T-ZnOW nanocompound and characterization of its photocatalytic activity and stability property under UV irradiation. Mater. Sci. Eng. B 178 (2), 158–166. https://doi.org/10.1016/j.mseb.2012.10.041.

Liu, J., Yu, X., Wang, L., Guo, M., Zhu, W., Tian, S., 2019. Photocatalytic degradation of chlortetracycline hydrochloride in marine aquaculture wastewater under visible light irradiation with CuO/ZnO. Water Sci. Technol. 80 (7), 1249–1256. https://doi.org/10.2166/wst.2019.372.

Liu, et al., 2012. Hierarchical CuO/ZnO membranes for environmental applications under the irradiation of visible light. Int. J. Photoenergy doi:10.1155/2012/804840.

Liu, Z.-L., Deng, J.-C., Deng, J.-J., Li, F.-F., 2008. Fabrication and photocatalysis of CuO/ZnO nano-composites via a new method. Mater. Sci. Eng. B 150 (2), 99–104. https://doi.org/10.1016/j.mseb.2008.04.002.

Long, Z., Li, Q., Wei, T., Zhang, G., Ren, Z., 2020. Historical development and prospects of photocatalysts for pollutant removal in water. J. Hazard. Mater. 395, 122599. https://doi.org/10.1016/j.jhazmat.2020.122599.

Lu, P., Zhou, W., Li, Y., Wang, J., Wu, P., 2017a. Abnormal room temperature ferromagnetism in CuO/ZnO nanocomposites via hydrothermal method. Appl. Surf. Sci. 399, 396–402. https://doi.org/10.1016/j.apsusc.2016.12.113.

Lu, P., Zhou, W., Li, Y., Wang, J., Wu, P., 2017b. CuO nanosheets/ZnO nanorods synthesized by a template-free hydrothermal approach and their optical and magnetic characteristics. Ceram. Int. 43 (13), 9798–9805. https://doi.org/10.1016/j.ceramint.2017.04.159.

Ma, X., Zhang, B., Cong, Q., He, X., Gao, M., Li, G., 2016. Organic/inorganic nanocomposites of ZnO/CuO/chitosan with improved properties. Mater. Chem. Phys. 178, 88–97. https://doi.org/10.1016/j.matchemphys.2016.04.074.

Magnabosco, G., Papiano, I., Aizenberg, M., Aizenberg, J., Falini, G., 2020. Beyond biotemplating: multi-scale porous inorganic materials with high catalytic efficiency. Chem. Commun. 56 (23), 3389–3392. https://doi.org/10.1039/D0CC00651C.

Mahendiran, M., Mathen, J.J., Racik, K.M., Madhavan, J., Raj, M.V.A., 2019. Facile synthesis of n-ZnO @ p-CuO nanocomposite for water purification enhanced decolorization of methyl orange. J. Mater. Sci. Mater. Electron. 30 (17), 16099–16109. https://doi.org/10.1007/s10854-019-01980-z.

Manisalidis, I., Stavropoulou, E., Stavropoulos, A., & Bezirtzoglou, E. (2020). Environmental and health impacts of air pollution: a review. Front. Public Health, 8, 14. https://doi.org/10.3389/fpubh.2020.00014.

Mansournia, M., Ghaderi, L., 2017. CuO@ZnO core-shell nanocomposites: novel hydrothermal synthesis and enhancement in photocatalytic property. J. Alloys Compd. 691, 171–177. https://doi.org/10.1016/j.jallcom.2016.08.267.

Mardikar, S.P., Kulkarni, S., Adhyapak, P.V., 2020. Sunlight driven highly efficient degradation of methylene blue by CuO-ZnO nanoflowers. J. Environ. Chem. Eng. 8 (2), 102788. https://doi.org/10.1016/j.jece.2018.11.033.

Minh, et al., 2019. Synthesis of porous octahedral ZnO/CuO composites from Zn/Cu-based MOF-199 and their applications in visible-light-driven photocatalytic degradation of dyes. J. Nanomater doi:10.1155/2019/5198045.

Mohammadi, E., Aliofkhazraei, M., Hasanpoor, M., Chipara, M., 2018. Hierarchical and complex ZnO nanostructures by microwave-assisted synthesis: morphologies, growth mechanism and classification. Crit. Rev. Solid State Mater. Sci. 43 (6), 475–541. https://doi.org/10.1080/10408436.2017.1397501.

Moosakazemi, F., Tavakoli Mohammadi, M.R., Mohseni, M., Karamoozian, M., Zakeri, M., 2017. Effect of design and operational parameters on particle morphology in ball mills. Int. J. Miner. Process. 165, 41–49. https://doi.org/10.1016/j.minpro.2017.06.001.

Nadargi, D.Y., Tamboli, M.S., Patil, S.S., Dateer, R.B., Mulla, I.S., Choi, H., Suryavanshi, S.S., 2020. Microwave-epoxide-assisted hydrothermal synthesis of the CuO/ZnO heterojunction: a highly versatile route to develop H_2S gas sensors. ACS Omega 5 (15), 8587–8595. https://doi.org/10.1021/acsomega.9b04475.

Nandi, P., Das, D., 2020. ZnO-CuxO heterostructure photocatalyst for efficient dye degradation. J. Phys. Chem. Solids 143, 109463. https://doi.org/10.1016/j.jpcs.2020.109463.

Naseri, A., Samadi, M., Mahmoodi, N.M., Pourjavadi, A., Mehdipour, H., Moshfegh, A.Z., 2017. Tuning composition of electrospun ZnO/CuO nanofibers: toward controllable and efficient solar photocatalytic degradation of organic pollutants. J. Phys. Chem. C 121 (6), 3327–3338. https://doi.org/10.1021/acs.jpcc.6b10414.

Nguyen, et al., 2020. Fabrication of Cu_2O-ZnO nanocomposite by the sol-gel technique and its antibacterial activity. J. Biochem. Technol. 11 (1), 18–24. https://jbiochemtech.com/en/article/fabrication-of-cu2o-zno-nanocomposite-by-the-sol-gel-technique-and-its-antibacterial-activity.

Nguyen, C.H., Juang, R.-S., 2019. Efficient removal of methylene blue dye by a hybrid adsorption–photocatalysis process using reduced graphene oxide/titanate nanotube composites for water reuse. J. Ind. Eng. Chem. 76, 296–309. https://doi.org/10.1016/j.jiec.2019.03.054.

Nunes, D., Pimentel, A., Santos, L., Barquinha, P., Pereira, L., Fortunato, E., Martins, R., Nunes, D., Pimentel, A., Santos, L., Barquinha, P., Pereira, L., Fortunato, E., Martins, R., et al., 2019. Synthesis, design, and morphology of metal oxide nanostructures. In: Nues, D., et al. (Eds.), Metal Oxide Nanostructures: Synthesis, Properties and Applications,. Elsevier,, Amsterdam, pp. 21–57.

Oliveira, M.C., Fonseca, V.S., Andrade Neto, N.F., Ribeiro, R.A.P., Longo, E., de Lazaro, S.R., Motta, F.V., Bomio, M.R.D., 2020. Connecting theory with experiment to understand the photocatalytic activity of CuO–ZnO heterostructure. Ceram. Int. 46 (7), 9446–9454. https://doi.org/10.1016/j.ceramint.2019.12.205.

Parul, Kaur, K., Badru, R., Singh, P., P., Kaushal, S, 2020. Photodegradation of organic pollutants using heterojunctions: a review. J. Environ. Chem. Eng. 8 (2), 103666. https://doi.org/10.1016/j.jece.2020.103666.

Pawar, R.C., Choi, D.-H., Lee, J.-S., Lee, C.S., 2015. Formation of polar surfaces in microstructured ZnO by doping with Cu and applications in photocatalysis using visible light. Mater. Chem. Phys. 151, 167–180. https://doi.org/10.1016/j.matchemphys.2014.11.051.

Paz, D.S., Foletto, E.L., Bertuol, D.A., Jahn, S.L., Collazzo, G.C., da Silva, S.S., Chiavone-Filho, O., Nascimento, C.A.O., 2013. CuO/ZnO coupled oxide films obtained by the electrodeposition technique and their photocatalytic activity in phenol degradation under solar irradiation. Water Sci. Technol. 68 (5), 1031–1036. https://doi.org/10.2166/wst.2013.345.

Prabhu, Y.T., Rao, V.N., Shankar, M.V., Sreedhar, B., Pal, U., 2019. The facile hydrothermal synthesis of CuO@ZnO heterojunction nanostructures for enhanced photocatalytic hydrogen evolution. New J. Chem. 43 (17), 6794–6805. doi:10.1039/c8nj06056h.

Preda, N., Costas, A., Enculescu, M., Enculescu, I., 2020. Biomorphic 3D fibrous networks based on ZnO, CuO and ZnO–CuO composite nanostructures prepared from eggshell membranes. Mater. Chem. Phys. 240, 122205. https://doi.org/10.1016/j.matchemphys.2019.122205.

Qamar, M.T., Aslam, M., Ismail, I.M.I., Salah, N., Hameed, A., 2015. Synthesis, characterization, and sunlight mediated photocatalytic activity of CuO Coated ZnO for the removal of nitrophenols. ACS Appl. Mater. Interf. 7 (16), 8757–8769. https://doi.org/10.1021/acsami.5b01273.

Qiu, S., Zhou, H., Shen, Z., Hao, L., Chen, H., Zhou, X., 2020. Synthesis, characterization, and comparison of antibacterial effects and elucidating the mechanism of ZnO, CuO and CuZnO nanoparticles supported on mesoporous silica SBA-3. RSC Adv 10 (5), 2767–2785. https://doi.org/10.1039/C9RA09829A.

Rahemi Ardekani, S., Sabour Rouhaghdam, A., Nazari, M., 2018. N-doped ZnO-CuO nanocomposite prepared by one-step ultrasonic spray pyrolysis and its photocatalytic activity. Chem. Phys. Lett. 705, 19–22. https://doi.org/10.1016/j.cplett.2018.05.052.

Raizada, P., Sudhaik, A., Patial, S., Hasija, V., Parwaz Khan, A.A., Singh, P., Gautam, S., Kaur, M., Nguyen, V.-H., 2020. Engineering nanostructures of CuO-based photocatalysts for water treatment: current progress and future challenges. Arab. J. Chem. 13 (11), 8424–8457. https://doi.org/10.1016/j.arabjc.2020.06.031.

Ramchiary, A., Mustansar Hussain, C., Mishra, A.K., 2020. Metal-oxide semiconductor photocatalysts for the degradation of organic contaminants. In: Hussain, C.M., Mishra, A.K. (Eds.), , Handbook of Smart Photocatalytic Materials: Environment, Energy, Emerging Applications, and Sustainability. Elsevier, Amsterdam, pp. 23–38.

Rane, A.V., Kanny, K., Abitha, V.K., Thomas, S., Mohan Bhagyaraj, S., Oluwafemi, O.S., Kalarikkal, N., Thomas, S., 2018. Methods for synthesis of nanoparticles and fabrication of nanocomposites. In: Bhagyaraj, S.M., Oluwafemi, O.S., Kalarikkal, N., Thomas, S. (Eds.), Synthesis of Inorganic Nanomaterials: Advances and Key Technologies: A volume in Micro and Nano Technologies. Woodhead Publishing, UK, pp. 121–139.

Renuka, L., Anantharaju, K.S., Vidya, Y.S., Nagaswarupa, H.P., Prashantha, S.C., Nagabhushana, H., 2017. Synthesis of sunlight driven ZnO/CuO nanocomposite: characterization, optical, electrochemical and photocatalytic studies, 4(11, Part 3), pp. 11782–11790.

Riente, P., Noël, T., 2019. Application of metal oxide semiconductors in light-driven organic transformations. Catal. Sci. Technol. 9 (19), 5186–5232. https://doi.org/10.1039/C9CY01170F.

Ristić, M., Musić, S., Ivanda, M., Popović, S., 2005. Sol–gel synthesis and characterization of nanocrystalline ZnO powders. J. Alloys Compd. 397 (1), L1–L4. https://doi.org/10.1016/j.jallcom.2005.01.045.

Ruan, S., Huang, W., Zhao, M., Song, H., Gao, Z., 2020. A Z-scheme mechanism of the novel ZnO/CuO n-n heterojunction for photocatalytic degradation of acid orange 7. Materi. Sci. Semiconduct. Process. 107, 104835. https://doi.org/10.1016/j.mssp.2019.104835.

Ruthiraan, M., Mubarak, N.M., Abdullah, E.C., Khalid, M., Nizamuddin, S., Walvekar, R., Karri, R.R.Abd-Elsalam, K.A., Mohamed, M.A., Prasad, R. (Eds.), 2019. An overview of magnetic material: preparation and adsorption removal of heavy metals from wastewater. Magnetic Nanostructures 131–159. https://doi.org/10.1007/978-3-030-16439-3_8.

Rydosz, A., 2018. The use of copper oxide thin films in gas-sensing applications. Coatings, 8, p. 425.

Sáenz-Trevizo, A., Amézaga-Madrid, P., Pizá-Ruiz, P., Antúnez-Flores, W., Ornelas-Gutiérrez, C., Miki-Yoshida, M., 2016. Efficient and durable ZnO core-shell structures for photocatalytic applications in aqueous media. Mater. Sci. Semiconduct. Process. 45, 57–68. https://doi.org/10.1016/j.mssp.2016.01.018.
Sahu, K., Bisht, A., Kuriakose, S., Mohapatra, S., 2020. Two-dimensional CuO-ZnO nanohybrids with enhanced photocatalytic performance for removal of pollutants. J. Phys. Chem. Solids 137, 109223. https://doi.org/10.1016/j.jpcs.2019.109223.
Salehi, K., Bahmani, A., Shahmoradi, B., Pordel, M.A., Kohzadi, S., Gong, Y., Guo, H., Shivaraju, H.P., Rezaee, R., Pawar, R.R., Lee, S.-M., 2017. Response surface methodology (RSM) optimization approach for degradation of direct blue 71 dye using CuO–ZnO nanocomposite. Int. J. Environ. Sci. Technol. 14 (10), 2067–2076. https://doi.org/10.1007/s13762-017-1308-0.
Samad, A., Furukawa, M., Katsumata, H., Suzuki, T., Kaneco, S., 2016. Photocatalytic oxidation and simultaneous removal of arsenite with CuO/ZnO photocatalyst. J. Photochem. Photobiol. A Chem. 325, 97–103. https://doi.org/10.1016/j.jphotochem.2016.03.035.
Sapkota, B.B., Mishra, S.R., 2013. A simple ball milling method for the preparation of p-CuO/n-ZnO nanocomposite photocatalysts with high photocatalytic activity. J. Nanosci. Nanotechnol 13 (10), 6588–6596. https://doi.org/10.1166/jnn.2013.7544.
Saravanan, R., Karthikeyan, S., Gupta, V.K., Sekaran, G., Narayanan, V., Stephen, A., 2013. Enhanced photocatalytic activity of ZnO/CuO nanocomposite for the degradation of textile dye on visible light illumination. Mater. Sci. Eng. C 33 (1), 91–98. https://doi.org/10.1016/j.msec.2012.08.011.
Senthil Kumar, P., Selvakumar, M., Ganesh Babu, S., Induja, S., Karuthapandian, S, 2017. CuO/ZnO nanorods: an affordable efficient p-n heterojunction and morphology dependent photocatalytic activity against organic contaminants. J. Alloys Compd. 701, 562–573. https://doi.org/10.1016/j.jallcom.2017.01.126.
Sepulveda-Guzman, S., Reeja-Jayan, B., de la Rosa, E., Torres-Castro, A., Gonzalez-Gonzalez, V., Jose-Yacaman, M., 2009. Synthesis of assembled ZnO structures by precipitation method in aqueous media. Mater. Chem. Phys. 115 (1), 172–178. https://doi.org/10.1016/j.matchemphys.2008.11.030.
Sharma, M., Poddar, M., Gupta, Y., Nigam, S., Avasthi, D.K., Adelung, R., Abolhassani, R., Fiutowski, J., Joshi, M., Mishra, Y.K., 2020. Solar light assisted degradation of dyes and adsorption of heavy metal ions from water by CuO–ZnO tetrapodal hybrid nanocomposite. Mater. Today Chem. 17, 100336. https://doi.org/10.1016/j.mtchem.2020.100336.
Sherly, E.D., Vijaya, J.J., Kennedy, L.J., 2015. Visible-light-induced photocatalytic performances of ZnO–CuO nanocomposites for degradation of 2,4-dichlorophenol. Chin. J. Catal. 36 (8), 1263–1272. https://doi.org/10.1016/S1872-2067(15)60886-5.
Sherly, Eluvathingal Devassy, Vijaya, J.J., Kennedy, L.J., Meenakshisundaram, A., Lavanya, M, 2016. A comparative study of the effects of CuO, NiO, ZrO_2 and CeO_2 coupling on the photocatalytic activity and characteristics of ZnO. Kor. J. Chem. Eng. 33 (4), 1431–1440. https://doi.org/10.1007/s11814-015-0285-6.
Silva, W.J.C.D., Silva, M.R.D., Takashima, K., 2015. Preparation and characterization of ZnO/CuO semiconductor and photocatalytic activity on the decolorization of direct red 80 azodye. J. Chil. Chem. Soc. 60, 2749–2751.
Singaravelan, R., Bangaru Sudarsan Alwar, S., 2015. Electrochemical synthesis, characterisation and phytogenic properties of silver nanoparticles. Appl. Nanosci. 5 (8), 983–991. https://doi.org/10.1007/s13204-014-0396-0.
Singh, J., Soni, R.K., 2020. Controlled synthesis of CuO decorated defect enriched ZnO nanoflakes for improved sunlight-induced photocatalytic degradation of organic pollutants. Appl. Surf. Sci. 521, 146420. https://doi.org/10.1016/j.apsusc.2020.146420.
Sodeifian, G., Behnood, R., 2020. Hydrothermal synthesis of N-doped GQD/CuO and N-doped GQD/ZnO nanophotocatalysts for MB dye removal under visible light irradiation: evaluation of a new procedure to produce N-doped GQD/ZnO. J. Inorg. Organomet. Polym. Mater. 30 (4), 1266–1280. https://doi.org/10.1007/s10904-019-01232-x.
Taufique, M.F.N., Haque, A., Karnati, P., Ghosh, K., 2018. ZnO–CuO nanocomposites with improved photocatalytic activity for environmental and energy applications. J. Electron. Mater. 47 (11), 6731–6745. https://doi.org/10.1007/s11664-018-6582-1.

Ternes, T.A., Bonerz, M., Herrmann, N., Teiser, B., Andersen, H.R., 2007. Irrigation of treated wastewater in Braunschweig, Germany: an option to remove pharmaceuticals and musk fragrances. Chemosphere 66 (5), 894–904. https://doi.org/10.1016/j.chemosphere.2006.06.035.

Tuncel, D., Ökte, A.N., 2019. ZnO@CuO derived from Cu-BTC for efficient UV-induced photocatalytic applications, 328, pp. 149–156.

Valles-Pérez, B.Y., Badillo-Ávila, M.A., Torres-Delgado, G., Castanedo-Pérez, R., Zelaya-Ángel, O., 2020. Photocatalytic activity of ZnO + CuO thin films deposited by dip coating: coupling effect between oxides. J. Sol-Gel Sci. Technol. 93 (3), 517–526. https://doi.org/10.1007/s10971-020-05223-0.

Villaseñor-Basulto, D.L., Pedavoah, M.-M., Bandala, E.R., 2019. Plant materials for the synthesis of nanomaterials: greener sources. In: Martínez, L.M.T., Kharissova, O.V., Kharisov, B.I. (Eds.), Handbook of Ecomaterials. Springer International Publishing,, NY, pp. 105–121.

Wang, W.-N., Wu, F., Myung, Y., Niedzwiedzki, D.M., Im, H.S., Park, J., Banerjee, P., Biswas, P., 2015. Surface engineered CuO nanowires with ZnO islands for CO_2 Photoreduction. ACS Appl. Mater. Interf. 7 (10), 5685–5692. https://doi.org/10.1021/am508590j.

Wang, X., Zhang, Y., Wang, Q., Dong, B., Wang, Y., Feng, W., 2019. Photocatalytic activity of Cu_2O/ZnO nanocomposite for the decomposition of methyl orange under visible light irradiation. Sci. Eng. Compos. Mater. 26 (1), 104–113 https://doi.org/doi:10.1515/secm-2018-0170.

Wang, Y., She, G., Xu, H., Liu, Y., Mu, L., Shi, W., 2012. Cu_2O nanoparticles sensitized ZnO nanorod arrays: electrochemical synthesis and photocatalytic properties. Mater. Lett. 67 (1), 110–112. https://doi.org/10.1016/j.matlet.2011.09.046.

Xu, Linhua, Zhou, Y., Wu, Z., Zheng, G., He, J., Zhou, Y., 2017. Improved photocatalytic activity of nanocrystalline ZnO by coupling with CuO. J. Phys. Chem. Solids 106, 29–36. https://doi.org/10.1016/j.jpcs.2017.03.001.

Xu, Linping, Hu, Y.-L., Pelligra, C., Chen, C.-H., Jin, L., Huang, H., Sithambaram, S., Aindow, M., Joesten, R., Suib, S.L., 2009. ZnO with different morphologies synthesized by solvothermal methods for enhanced photocatalytic activity. Chem. Mater. 21 (13), 2875–2885. https://doi.org/10.1021/cm900608d.

Xue, J., Wu, T., Dai, Y., Xia, Y., 2019. Electrospinning and electrospun nanofibers: methods, materials, and applications. Chem. Rev. 119 (8), 5298–5415. https://doi.org/10.1021/acs.chemrev.8b00593.

Yakout, S.M., El-Sayed, A.M., 2019. Enhanced ferromagnetic and photocatalytic properties in Mn or Fe doped p-CuO/n-ZnO nanocomposites. Adv. Powder Technol. 30 (11), 2841–2850. https://doi.org/10.1016/j.apt.2019.08.033.

Yang, Z.X., Zhu, F., Zhou, W.M., & Zhang, Y.F. (2005). Novel nanostructures of β-Ga2O3 synthesized by thermal evaporation. Phys. E Low-Dimens. Syst. Nanostructures, 30(1), 93–95. https://doi.org/10.1016/j.physe.2005.07.011.

Yendrapati, et al., 2019. Controlled addition of Cu/Zn in hierarchical CuO/ZnO p-n heterojunction photocatalyst for high photoreduction of CO_2 to MeOH. J. CO_2 Util 31, 207–214. doi:10.1016/j.jcou.2019.03.012.

Yoo, H., Kahng, S., Hyeun Kim, J., 2020. Z-scheme assisted ZnO/Cu_2O-CuO photocatalysts to increase photoactive electrons in hydrogen evolution by water splitting. Sol. Energy Mater. Sol. Cells 204, 110211. https://doi.org/10.1016/j.solmat.2019.110211.

Yu, J., Zhuang, S., Xu, X., Zhu, W., Feng, B., Hu, J., 2015. Photogenerated electron reservoir in hetero-p–n CuO–ZnO nanocomposite device for visible-light-driven photocatalytic reduction of aqueous Cr(VI). J. Mater. Chem. A 3 (3), 1199–1207. https://doi.org/10.1039/C4TA04526B.

Yu, Z., Moussa, H., Liu, M., Schneider, R., Wang, W., Moliere, M., & Liao, H. (2019). Development of photocatalytically active heterostructured MnO/ZnO and CuO/ZnO films via solution precursor plasma spray process. Surf. Coat. Technol., 371, 107–116. https://doi.org/10.1016/j.surfcoat.2019.02.053.

Yuvakkumar, R., Suresh, J., Nathanael, A.J., Sundrarajan, M., Hong, S.I., 2014. Novel green synthetic strategy to prepare ZnO nanocrystals using rambutan (*Nephelium lappaceum* L.) peel extract and its antibacterial applications. Mater. Sci. Eng. C 41, 17–27. https://doi.org/10.1016/j.msec.2014.04.025.

Zhang, C., Yin, L., Zhang, L., Qi, Y., Lun, N., 2012. Preparation and photocatalytic activity of hollow ZnO and ZnO–CuO composite spheres. Mater. Lett. 67 (1), 303–307. https://doi.org/10.1016/j.matlet.2011.09.073.

Zhang, C.-L., Yu, S.-H., 2014. Nanoparticles meet electrospinning: recent advances and future prospects. Chem. Soc. Rev. 43 (13), 4423–4448. https://doi.org/10.1039/C3CS60426H.

Zhang, D., 2013. Photobleaching of pollutant dye catalyzed by p-n junction ZnO-CuO photocatalyst under UV-visible light activation. Russ. J. Phys. Chem. A 87 (1), 137–144. https://doi.org/10.1134/S0036024413010068.

Zhang, X., He, X., Kang, Z., Cui, M., Yang, D.-P., Luque, R., 2019. Waste eggshell-derived dual-functional CuO/ZnO/eggshell nanocomposites: (photo)catalytic reduction and bacterial inactivation. ACS Sustain. Chem. Eng. 7 (18), 15762–15771. https://doi.org/10.1021/acssuschemeng.9b04083.

Zhao, Y., Yin, H.-B., Fu, Y.-J., Wang, X.-M., Wu, W.-D., 2019. Energy band alignment at Cu_2O/ZnO heterojunctions characterized by in situ X-ray photoelectron spectroscopy. Chin. Phys. B 28 (8), 087301. https://doi.org/10.1088/1674-1056/28/8/087301.

Zhou, Y., Wu, W., Hu, G., Wu, H., Cui, S., 2008. Hydrothermal synthesis of ZnO nanorod arrays with the addition of polyethyleneimine. Mater. Res. Bull. 43 (8), 2113–2118. https://doi.org/10.1016/j.materresbull.2007.09.024.

CHAPTER 17

Phytogenic-mediated nanoparticles for the management of water pollution

Abdul Rehman[a], Shama Sehar[b], Adnan Younis[c], Muhammad Anees[a], Riaz Muhammad[d], Kashif Latif[e] and Iffat Naz[f]

[a]Department of Microbiology, Kohat University of Science & Technology (KUST), Kohat, Khyber Pakhtunkhwa, Pakistan
[b]College of Science, University of Bahrain, Sakhir, Kingdom of Bahrain
[c]Department of Physics, College of Science, University of Bahrain, Sakhir, Kingdom of Bahrain
[d]Government College Peshawar, Khyber Pakhtunkhwa, Pakistan
[e]Department of Microbiology, Quaid-i-Azam University, Islamabad, Pakistan
[f]Department of Biology, Deanship of Education Services, Scientific Unit, Qassim University, Kingdom of Saudi Arabia (KSA)

17.1 Introduction

Fresh and clean water being the utmost necessity of life is getting contaminated day by day due to a rapid growth in population, urbanization, and a continuous increase in an industrial discharge into underground and above-ground water streams without any prior treatment. In addition, this contaminated wastewater not only poses serious threats to terrestrial and aquatic life, disturb the natural ecosystem, flora, fauna, and fisheries, etc. but also narrows socioeconomic and agricultural growth/production worldwide (Sheng et al., 2019). According to a survey report conducted by WHO in 2015, about 3.4 million people (mostly children less than 5 years) die per year in developing countries due to unhygienic conditions, poor sanitation, and waterborne diseases (Serrà et al., 2020; Water & Supply, 2015). Moreover, in developing countries, about 50% of the community has no access to desired standard sanitation (Azizi et al., 2020). Food and Agriculture Organization of the United Nations reported that in many parts of Asia, more than 800 million of the population face severe water shortage that is possibility of fresh and potable water revenue would be less than 1500 m^3 per person per year, which might be expected to reach more than 2.5 billion of the population in 2025 (Connor et al., 2017).

Thus, keeping in mind the importance and continuous increase in the demand for fresh and clean water, it is an urgent need of the time to protect available water resources from being contaminated as well as to further purify already affected and polluted wastewater, especially in the developing and underdeveloped countries. Besides, a large number of wastewater treatment plants and projects are required to be installed to further purify different types of wastewater originating from domestic, agricultural, and industrial sectors. The complex wastewater treatment process can be divided into three sections.

Sustainable Nanotechnology for Environmental Remediation
DOI: https://doi.org/10.1016/B978-0-12-824547-7.00006-0

that is (1) Plain effluent reuse, where the effluent of one treatment system will be the influent of other systems with or without supplementary treatment; (2) programmed usage of treated water, where the untreated wastewater first goes to the treatment facility and then after treatment, discharge for further reuse; (3) haphazard water reuse, where the wastewater mixes into naturally occurring surface water reservoirs, acting as a source of water for other usages (Diaz-Elsayed et al., 2019; Voulvoulis, 2018).

17.1.1 Wastewater and its treatment

Wastewater is generally defined as a mixture of various liquids, polluted with attenuated contents released from the household, recreational, trading, and industrial sectors (Varma et al., 2020). In addition, it may also sometimes contain flooded as well as unclean surface water (Abu Hasan et al., 2020). Depending on the sources of production, wastewater has been classified as industrial or municipal wastewater. As the name indicates, industrial wastewater is a type of water generated during manufacturing processes going on in industrial division, while the later one is a type of sewage spawned from the domestic sites including houses, commercial areas, and institutes (Hairuddin et al., 2019; Karri et al., 2020; Lingamdinne et al., 2019). Chemistry of both is quite different from each other that is domestic wastewater mostly contains carbonaceous compounds such as carbohydrates, proteins, oil and greases, etc.; whereas each industry produced its own individual complex harmful wastewater mostly depending on the activities performed within industries (Abu Hasan et al., 2020; Varma et al., 2020). Table 17.1 (Yaqoob et al., 2020) illustrates the origins and adverse effects of various different kinds of pollutants in the used water (Yaqoob et al., 2020). Generally, wastewater contains 99.6% water and 0.4% solids (Voulvoulis, 2018). The solid part of wastewater is further divided into biodegradable organics (70%) and inorganics (30%) portion. The biodegradable organics in wastewater are usually composed of proteins, carbohydrates (starch, cellulose), fats, oils and grease, etc. whereas the inorganic portion of wastewater is composed of sediments, salts, and metals (Kim et al., 2019).

Treatment of wastewater is one of the ultimate processes to eradicate the possible toxicity of the compounds present in wastewater. In this process, multifarious carbonaceous compounds, either by physicochemical or biological means, are converted into simpler composites that are nontoxic and safe to the aquatic ecosystem (Abu Hasan et al., 2020). On the other hand, if untreated waste effluent is discharged as such to the environment, it may lead to the emission of putrid gases, reduction of dissolved oxygen contents from the desired level as required by aquatic organisms, eutrophication of the lakes and streams due to the presence of excessive nutrients that may facilitate the growth of unnecessary aquatic plants and algae. Moreover, it contains pathogenic microorganisms belonging to the family *Enterobacteriaceae* and is responsible for the variety of waterborne diseases in living beings. Because of these reasons, wastewater treatment must be performed to produce an effluent suitable for agricultural or aquaculture reuse or to produce an effluent

Table 17.1 The origins and adversative effects of different kinds of contaminants in water.

Contaminants	Origin	Adverse Effects	Reference
Organic	Detergents, soaps fruits, and vegetable wastes from recreational centers insecticides, fungicides, and herbicides	Carcinogenic in nature. Affect growth of both aquatic and terrestrial flora and fauna. Used as a common source of nutrient for pathogenic bacteria.	Yaqoob et al. (2020)
Inorganic	Inorganic salts, heavy metals, trace elements, and Mineral acids	Affect health of both aquatic as well as terrestrial flora and fauna.	Yaqoob et al. (2020)
Agricultural	Insecticides, fungicides, herbicides, chemicals used in irrigation process	Affect soil fertility when used in high concentration. Directly affect the freshwater resources.	Yaqoob et al. (2020)
Pathogenic Organisms	Fecal coliform bacteria from domestic wastewater and other pathogenic organisms from soil sources.	Caused deadly waterborne infections such as cholera, diarrhea, etc.	Yaqoob et al. (2020)
Industrial	Chemicals used in industries during manufacturing process such as textile dyes, ethers, aldehydes, etc.	Caused water, soil, and air pollution. Adverse effect on public health.	Yaqoob et al. (2020)
Macroscopic	Marine debris, sediments, and suspended solids	Plastic pollution. Clogging drainage pipelines.	Yaqoob et al. (2020)

that can be safely discharged into the ground or surface water reservoirs (Sahu, Karri, et al., 2019; Sahu, Zabed, et al., 2019).

Treatment of wastewater can be achieved through physical and chemical means. Physical processes involve screening by screen bars to remove larger suspended solids and particulate matters, mixing, flocculation, sedimentation, and physical filtration, while precipitation, coagulation, flocculation, adsorption and oxidation are the examples of chemical unit processes used in sewage treatment (Aziz & Ali, 2019; Dehghani et al., 2020; Hairuddin et al., 2019; Khan et al., 2020). Although these processes are effective in the removal of contaminants, they may have additional environmental and economic disadvantages (Hazirah et al., 2014). The physical methods used for wastewater treatment give demerits related with membrane fouling problems, limited lifetime of treatment units, and the rate of periodic replacement must be counted in any economic feasibility analysis. Although chemical processes such as coagulation, flocculation, and oxidation are successful in the removal of pollutants, these approaches are costly, requiring prolonged detention time, sludge retention time, and hence also create a disposal problem. Therefore, biological wastewater treatment methods and technologies are gaining much more attention in recent few years as they are not the only environment and budget friendly, demands fewer operational facilities and infrastructure but also proved to have superior treatment efficiencies in comparison to physical and chemical treatment methods (Magwaza et al., 2020). The primary goal of biological wastewater treatment technologies is to eradicate biodegradable carbonaceous compounds in wastewater along with nutrients such as phosphate, sulfate, nitrate, and nitrite by means of microorganisms.

These different treatment units provide diverse levels of treatment that is preliminary, primary, secondary, and tertiary wastewater treatment. In the primary wastewater treatment, substances such as dead animals, tree branches, wood, plastics, and rags that cause damage to the treatment design and operational process are removed by a physical mean. Besides, suspended solids, organic and inorganic substances such as oils and greases, and other organic matters are removed by physical sedimentation process, thus reducing Biochemical Oxygen Demand (BOD_5) up to 10%–25% (Abu Hasan et al., 2020; Bolognesi et al., 2020). The secondary wastewater treatment involves oxidation of carbonaceous compounds present in the wastewater (coming from primary treatment unit) through biological means under both aerobic and anaerobic conditions. In these treatment units, microorganisms play a vital role to oxidize organic compounds present in wastewater and hence produce an effluent that fulfills standard limits. The tertiary treatment is an advanced wastewater treatment system used to further purify effluent of secondary treatment units. In the tertiary treatment process, excessive nutrients and organic compounds have been removed up to some extent but this method is mostly used to kill and destroy pathogenic microorganisms from wastewater by chemical disinfection, flocculation, sedimentation, and chlorination methods (Kour et al., 2019; Mirajkar et al., 2019).

The existence of well-designed and well-operated wastewater treatment plants is very less in developing countries due to a lack of funds, professional expertise, and well-established pollution control laws and legislations. However, a wide range of consolidated wastewater treatment facilities are available in developing countries, which include stabilization ponds, activated sludge, constructed wetlands, and trickling filter systems as shown in Table 17.2. Stabilization ponds are aerobic in nature and primarily remove organic compounds present in the wastewater by their aerobic oxidation process. Moreover, it has been reported that most of these ponds removed 80%–85% BOD_5, 40%–45% ammonia, and 70%–95% indicator organisms from wastewater. However, the main drawback of stabilization ponds is that they required larger space and extended retention time due to the continuous mode of aeration (Kaetzl et al., 2019). On the other side, the

Table 17.2 Conventional wastewater treatment facilities with their possible merits and demerits.

Treatment System	Merits	Demerits
Septic tank	Low cost, simple maintenance	Low efficiency
Leaching field system	Low cost, high efficiency	Limitations with local soil conditions.
Aerobic biological treatment unit	High efficiency, application on small scale	High initial capital cost. Need operational control and routine periodic maintenance.
Membrane bioreactor		
Lagoons	Cost effective, simple maintenance	Odor problem. Required large land area.
Constructed wetland	Cost-effective, simple maintenance, high efficiency	Large area required for installation. Easily affected by seasonal variations.
Trickling filter system	High efficiency, simple maintenance, low space required	Proprietary treatment units are required. Require technical personnel and routine periodic inspection.

activated sludge process is also a suspended growth process in which microbial suspension is used for the removal of carbonaceous compounds, nitrogen, and phosphorus, but there are certain shortcomings of this process that is removal efficiency of pathogenic indicators is not constant and also settling of suspended solids are inappropriate which in turn enhanced treatment cost (Zhang et al., 2019).

Constructed wetlands are ecological contrived systems that consume the natural processes comprising wetland foliage, territories, and the accompanying microbial biofilm to treat wastewaters (Zhuang et al., 2019). It usually provides two types of treatment systems: free water surface system and subsurface flow system. The main downsides of constructed wetlands are that they require large land, continuous monitoring of operating units, and sludge removal. As compared to the other conventional treatment methods, fixed-film bioreactors, that is, trickling filter provides extended retention time due to high active biomass concentration and thus bear high organic loading rates (Bandeira et al., 2020). Moreover, it demands less space, energy, operational and maintenance cost, which makes them an attractive option for the treatment of all types of wastewater especially in the developing countries (Kwon et al., 2020). However, the only drawback of attached growth system is that void spaces among filter media are easy to clog due to climate changes and production of excess biomass, therefore time to time washing of the filter media is required (Capodaglio, 2017).

Nanomaterials are called "a wonder of the modern world." In new modern nanotechnology, different sizes and shapes of nanoparticles have been fabricated using different reducing agents that is plants, microbial cells, and their products (Sahu, Karri, et al., 2019; Sahu, Zabed, et al., 2019). Nowadays, nanoparticle innovation can be utilized to treat diverse sorts of microbial infections as well as for the remediation of different environmental pollutants (Venkateasan et al., 2017). Plants extracts are mostly preferred for the green synthesis of nanoparticles because chances of toxicity are less compared to chemically synthesized or biosynthesized (using microbial cells) nanoparticles (Patra & Baek, 2015). The phytogenic-mediated nanoparticles have great advantages such as simple, cost-effective, biocompatible, eco-friendly, easily available, and have wide application compared to chemically synthesized, physically synthesized, and biosynthesized nanoparticles.

In the recent years, the green synthesis of nanoparticles has gained much attention due to its cost effectiveness, less time consumption, nontoxic nature, and environmental friendliness compared to physicochemical synthesis where toxic compounds are used in the start-up stage of nanoparticle's fabrication. However, the reducing and capping agents (polyphenols, amino acids, alkaloids, terpenoids, proteins, carboxyl, polysaccharides) used in the green synthesis are from different plant sources that are safe and eco-friendly (Devatha et al., 2016; Lingamdinne et al., 2019). The main goal of green synthesis is to use plant-based resources as capping and stabilizing agents under controlled laboratory conditions to minimize experimental as well as environmental risks (Zhou et al., 2016). To handle surfactants and avoid clumping of nanoparticles, capping agents play a significant role in the fabrication of nanoparticles, and in this way, they maintain proper shape, size, and configuration of synthesized nanoparticles. In addition, capping agents obtained from plants such as (polyphenols, amino acids, alkaloids, terpenoids, proteins, carboxyl, polysaccharides) are easily recovered from the solution using the filtration and fractionation process. However, in chemical synthesis, capping agents are attached permanently with the surface of nanoparticles, reducing not only the surface area of nanoparticles but also hard to degrade and hence increasing environmental risks. This property makes phytogenic-mediated nanotechnology superior to other physical and chemical technologies. Besides antimicrobial potential, phytogenic-mediated nanoparticles are also used nowadays for the management of water pollution and detoxification of other environmental pollutants such as heavy metals and dyes, because nanomaterials are excellent adsorbents and catalysts due to their larger specific surface areas and high reactive nature (Bhateria & Singh, 2019). A general illustration regarding phytogenic-mediated nanoparticle preparation, reaction mechanisms, and applications is shown in Fig. 17.1.

17.2 Overview of nanotechnology and phytogenic-mediated nanoparticles

Nanotechnology is the field of material science that deals in the synthesis of particles having size below 100 nm (Shafey, 2020). Moreover, these particles have different chemical properties that is why they have been extensively used in different fields such as optics, mechanics, photoelectrochemical sciences, medical sciences, chemical industry,

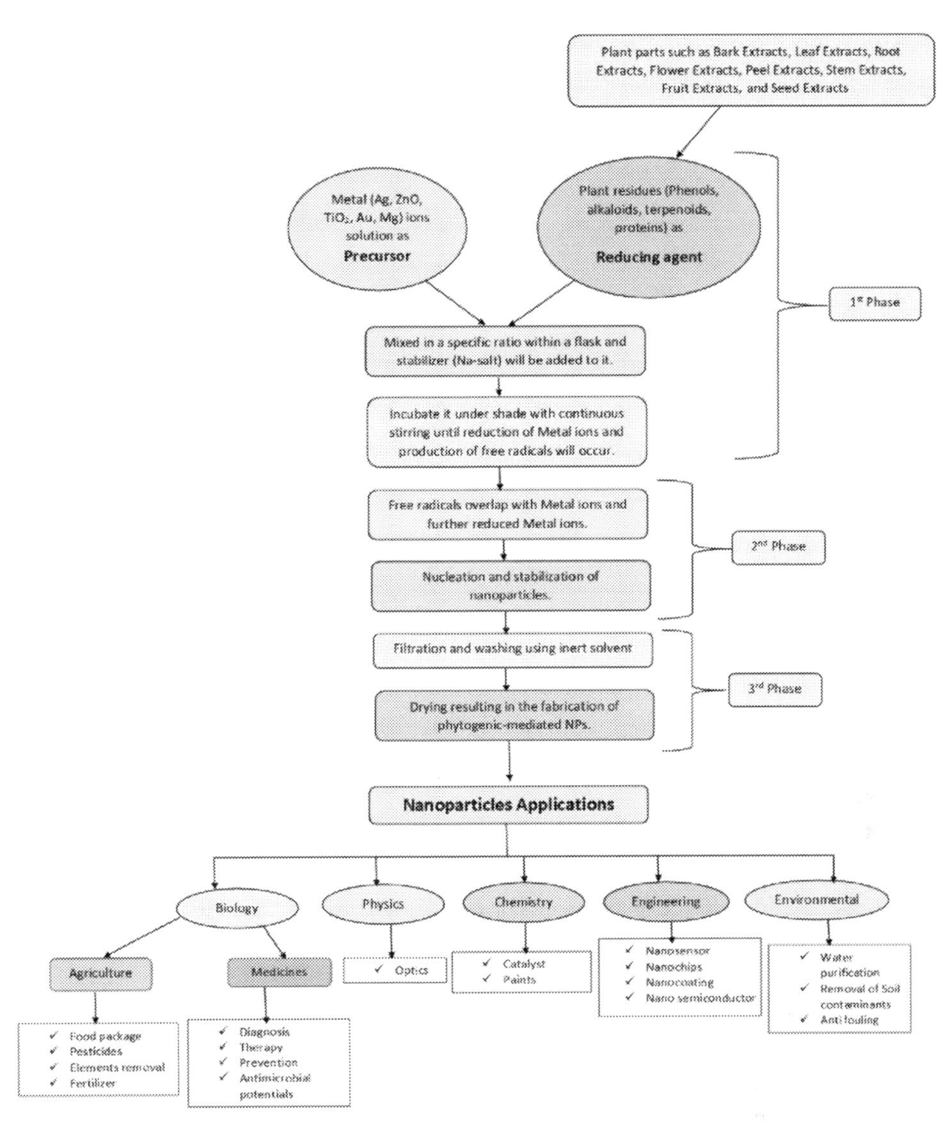

Figure 17.1 General illustration regarding phytogenic–mediated nanoparticle preparation, reaction mechanisms, and applications.

catalysis, drug-gene delivery, energy sciences, electronics, nonlinear optical sciences, and environmental remediation (Rolim et al., 2019; Shafey, 2020). Nanoparticles are extensively used in the formulation of different products such as pigments, ceramics, lubricants, adhesives tapes, ointment, and batteries (Rolim et al., 2019). Nanoparticles had a wide range of applications in the environmental sector especially in the remediation of water and air pollution, because they act as an excellent adsorbent, catalyst, and sensor due to their reactive nature (mainly due to the high density of coordinated atoms at the surface, edges and vortices) (Khan et al., 2020; Lingamdinne et al., 2019). Besides, due to their smaller size, they offered greater surface area for different pollutants present in the wastewater to be adsorbed on its surface. Due to these properties, nanomaterials are used in small amounts to scan larger volumes of pollutants present in the wastewater. Moreover, these adsorbed pollutants are easily removed after treatment by applying centrifugal or magnetic force. Different methods such as chemical and biological methods are utilized to synthesize nanoparticles. Chemical methods for the synthesis of nanoparticles include precipitation, microemulsification, hydrothermal approach, thermal method, sol-gel method. These chemical methods are not only utilizing expensive salts, harsh reducing agents, and organic solvents but also demands high energy, temperature and pressure, and releases toxic chemicals. However, on the other hand, biological methods for synthesizing nanoparticles are not safe, cost effective but also environment friendly (Gautam et al., 2019).

Biological synthesis or "green synthesis" is an easy, safe, reliable, and environment-friendly technology that employs naturally occurring substances such as plant extracts, microorganisms (yeast, fungi, and bacteria), sugars, biodegradable polymers, and biological particles such as enzyme, protein, peptides, polysaccharide, etc. in the synthesis of nanoparticles. These naturally occurring components function as reducing as well as capping agents (Dhall & Self, 2018; Sangsefidi et al., 2017). Phytogenic-mediated synthesis of nanoparticles is ecofriendly, safe, and beneficial approach as plants are easily accessible to produce nanoparticles on a large scale. In addition, the rate of production is faster compared to other biological models such as algae, fungi, and bacteria (Koli et al., 2018; Ravichandran et al., 2019). Further, nanoparticles synthesized from plants are nontoxic and biocompatible semiconductor used in the biological fields that is, nanomedicine due to their extensive antimicrobial properties and in the remediation of vast variety of environmental pollutants (Cai et al., 2018; Patel et al., 2017). In addition, phytogenic-mediated nanomaterials including chitosan, silver nanoparticles (AgNPs), photocatalytic titanium dioxide (TiO_2), zinc oxide nanoparticles (ZnONPs), and carbon nanotubes had also been proved to have strong antimicrobial properties against vast varieties of microbial pathogens (Chen et al., 2019; Das et al., 2017; Deshmukh et al., 2019; Zeng et al., 2017). Beside this, phytogenic-mediated nanomaterials are also used for the detection of pesticides and heavy metals such as cadmium, copper, lead, mercury, arsenic, etc. from wastewater due to their enhanced redox and photocatalytic properties (Han et al., 2017; Bhateria & Singh, 2019). Phytogenic-mediated nanoparticles are used for the management of water pollution because nanomaterials are excellent adsorbents and catalysts due to their large specific surface areas and high reactive nature. Moreover, the flexibility of nanomaterials in solution is high and even a small number of nanomaterials are enough to degrade toxic chemical and microbial pollutants present in wastewater. Furthermore, plants provide ecofriendly, safe, and beneficial approach for the synthesis of

Table 17.3 Plant used in the synthesis of phytogenic-mediated nanoparticles.

Name of plant	Plant parts used	Nanoparticles synthesized	References
Acalypha indica	Leaf extract	CeO_2NPs	Dhall and Self, 2018
Aloe vera	Leaf extract	CeO_2NPs	Thakur et al. (2019)
Gloriosa superba	Leaf extract	CeO_2NPs	Thakur et al. (2019)
Hibiscus sabdariffa	Flower extract	CeO_2NPs	Rajeshkumar and Naik (2018)
Olea europaea	Leaf extract	CeO_2NPs	Maqbool et al. (2016)
Calendula officinalis	Seed extract	AgNPs	Ahmad et al. (2019)
Diospyros paniculata	Root extracts	AgNPs	Rao et al. (2016)

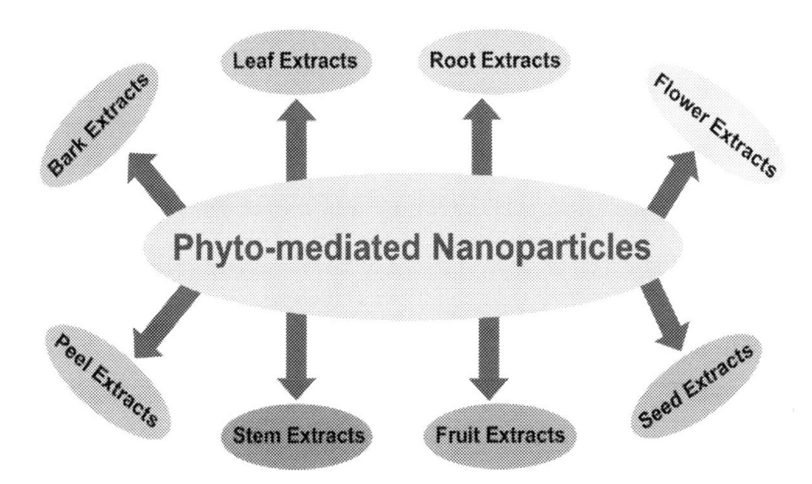

Figure 17.2 Parts of plants used for synthesis of phyto-mediated nanoparticles.

nanoparticles because plants are easily accessible to produce nanoparticles at a commercial scale.

A wide variety of different plants are reported in recent years to synthesize nanoparticles, some of them are listed in Table 17.3. Different parts of plants can be incorporated in biosynthesis of phytogenic-mediated nanoparticles such as seeds, leaves, fruits, flowers, bark, stem, roots, and peel extract as shown in Fig. 17.2. These plant extracts are rich in different phytochemicals (carboxylic acid, phenolic acids, alkaloids, flavonoids, tannins, vitamins, amino acids, saponins, inositol, and terpenes) that may act as both reducing and stabilization agent (Ahmad et al., 2019, 2019; Dhall & Self, 2018, 2018; Maqbool et al., 2016, 2016; Rajeshkumar & Naik, 2018, 2018; Rao et al., 2016, 2016; Thakur et al., 2019). Some of phytogenic-mediated nanoparticles used in the management of water pollution are summarized in Table 17.4 (Dong et al., 2019; Gomes et al., 2019; Jorge de Souza et al., 2019; Lu et al., 2016; Ray & Shipley, 2015; Tao et al., 2020; Yaqoob et al., 2020).

Table 17.4 Phytogenic-mediated nanoparticles used in the management of water pollution.

Nanoparticles (NPs)	Target analytes	Treatment mechanism	Positive aspects	References
Silver nanoparticles (AgNPs)	Pathogenic bacteria, viruses, fungi	Adhere to the cell wall and cell membrane of organisms and bring structural changes and ultimately lead to death.	Applied in water and wastewater disinfection process.	de Souza et al., (2019)
Zero-valent metal nanoparticles (Fe, Zn, Al, and Ni NPs)	Toxic organic compounds, organic dyes, pesticides	Adsorption, reduction, precipitation, and oxidation mechanisms	Achieved pilot-scale and full-scale applications at real water contaminated field sites	Dong et al. (2019)
Metal-oxide nanoparticle (TiO_2-NPs)	Organic pollutants and heavy metals	Electrochemical photolysis and photocatalysis	Completely degrade chlorinated organic compounds, polycyclic aromatic hydrocarbons, phenols, pesticides.	Lu et al. (2016)
Metal-oxide nanoparticle (ZnO-NPs)	Organic as well as microbial pollutants	Photocatalysis and oxidation-reduction mechanism	Degrade organic compounds present in wastewater and also eradicate pathogenic organisms from the treated effluent.	Gomez-Solis et al. (2015)
Iron-oxide nanoparticles	Heavy metals	Adsorption and magnetic properties	Cheap material in sensors, catalysis, and environmental applications	Tao et al. (2020)
Carbon-based nanoparticles	Organic pollutants, dyes	Adsorption, mechanical, optical, electrical, and chemical catalysis	Applied in drinking water purification process	Ray and Shipley (2015)
Dendrimers	Organic pollutants and heavy metals	Adsorption and entrapping mechanism	*In-situ* water remediation, harmless to handle	Yaqoob et al. (2020)
Polymeric nanoparticles	Arsenic and toxic chemicals	Adsorption, mechanical, and chemical catalysis	Applied in drinking water purification process	Yaqoob et al. (2020)

17.2.1 Factors affecting fabrication of phytogenic-mediated nanoparticles

There are certain factors that need to be considered during fabrication of phytogenic-mediated nanoparticles, and some of them are described next.

17.2.1.1 Incubation temperature

Temperature is an important factor during the fabrication of nanoparticles as it regulates the morphology, adsorbent, and magnetic properties of synthesized nanoparticles. Researchers investigated that magnetization of metal nanoparticles synthesized from the plant sources increased by increasing reaction temperature and incubation time (Karade et al., 2018). The same research group also reported that temperature affects the morphology and configuration of nanoparticles. However, another research group investigated the effect of temperature on the magnetic and adsorbent properties of nanoparticles synthesized from the different plant constituents (including glucose, gluconic acids as capping and reducing agents) and also reported that by increasing temperature, the magnetic and adsorbent properties of phytogenic-mediated nanoparticles greatly reduced (Yew et al., 2020). However, they also reported that phytogenic-mediated nanoparticles displayed superparamagnetic properties at room temperature that further enhanced the adsorbent property of synthesized nanoparticles. In the wastewater clarification process, this superparamagnetic property played a vital role by enhancing the contaminants adsorption on the surfaces of nanoparticles. Besides this, it also canceled the effects of external magnetic forces and hence reduced the operational cost of the treatment facility.

17.2.1.2 Nature of plant and plant parts

Plants contain different types of active compounds such as tannins, terpenoids, alkaloids, and flavonoids that can be used as reducing agents during the fabrication of phytogenic-mediated nanoparticles (Rolim et al., 2019). The nature of the plant and plant parts played a significant role in the fabrication of phytogenic-mediated nanoparticles due to the presence of different active metabolites (Khan et al., 2020; Ravichandran et al., 2019). Spectroscopic and chromatographic techniques are normally used to investigate the concentration of active compounds present in the plant extracts (used as a reducing agent in nanomaterial fabrication). They furthermore concluded that the size, shape, adsorbent, magnetic properties, and chemical reactivity of the phytogenic-mediated nanoparticles vary distinctly due to the presence of different active metabolites in different plants. Moreover, it also depends on the nature of plant extract, their antioxidant activity, and experimental conditions such as incubation temperature, nature of the solvents, metal ions concentration, and pH of the solution. Therefore, the selection of plant or plant parts is an essential factor for the large-scale production of phytogenic-mediated nanoparticles.

17.2.1.3 pH

Although pH has no direct effect on aquatic as well as terrestrial life, but it has been considered as one of the important parameters to define the quality of phytogenic-mediated nanoparticles with respect to their structural configuration, adsorbent, and magnetic properties (Saif et al., 2016). They also reported that the pH of the solution decreased during the fabrication of phytogenic-mediated metal nanoparticles. Besides this, no other information is available in the literature reflecting the effect of pH on the fabrication of nanoparticles using plant extracts as a reducing agent.

17.3 Nanoparticles-based wastewater treatment technologies

17.3.1 Nanophotocatalysts

The term photocatalyst is a Greek word and it has two parts : "photo" means light and "catalyst" means agent used to catalyze a chemical reaction. Overall, scientifically photocatalysts are defined as substances that decompose compounds in the presence of light (Saravanan et al., 2017). Photocatalysts are the stable substances that change the rate of reaction during the chemical transformation process. They are different from thermal catalysts that usually require heat energy as activation energy while photocatalysts require energetic radiation as kinetic energy (Gomes et al., 2019). Nanophotocatalysts are nanoparticles fabricated from different plant sources and exhibit photocatalytic properties. Furthermore, they are commonly applied in the wastewater clarification process because they offer a greater surface area for the adsorption of pollutants present in the wastewater and exhibit maximum magnetic, chemical reactivity, and optical properties (Chen et al., 2019a; Das et al., 2017; Deshmukh et al., 2019; Zeng et al., 2017). As a result, its pollutants degradative ability is more compared to the conventional wastewater treatment technologies. Phytogenic-mediated metal nanoparticles are commonly reported in the literature where they were used as nanophotocatalyst to eradicate toxic dyes, carbonaceous compounds, and nitro compounds from the polluted water (Bhaskar Reddy et al., 2015; Samanta et al., 2016). Different researchers reported that TiO_2-based nanotubes significantly removed toxic compounds such as azo dyes, textile dyes, Congo red dye, and aromatic compounds from the industrial wastewater (Bhatia et al., 2017; Linhares et al., 2018). Further, the commonly reported substantial metal oxide nanophotocatalysts include SiO_2, ZnO, TiO_2, and Al_2O_3 (Bhanvase et al., 2017). All of these metal oxide nanophotocatalysts played a significant role in the reduction of different toxic chemicals as well as microbial pollutants from the waste effluent using photo-oxidation properties. But among them, TiO_2 is one of the excellent photocatalysts because of numerous characteristics such as cost effectiveness, toxic-free property, chemical stability, and easily accessible (Bishoge et al., 2018; Di Mauro et al., 2017). It was also reported that nanophotocatalysis occurred in two states heterogeneous and homogenous nanophotocatalyst. Among them, heterogeneous nanophotocatalyst are most applied in photoreactor for wastewater clarification as it offers higher chemical reactivity and stability compared to homogenous-based nanophotocatalysts (Berekaa, 2016; Gupta et al., 2020). That is why phytogenic-mediated nanoparticles having heterogeneous nature are suitable to be used as nanophotocatalyst for the management of water pollution.

17.3.1.1 Advantages of nanophotocatalysts

- Nanophotocatalysts are temperature-independent catalysts and play a significant role in the reduction of toxic chemical and microbial pollutants from the industrial as well as domestic wastewater, respectively, even at room temperature.
- Nanophotocatalysts are cost effective, eco-friendly, chemically inert, and easily accessible to be fabricated from different sources.
- The flexibility of nanophotocatalysts in solution is high and even a small amount of nanomaterial is enough to degrade chemical and microbial pollutants from the wastewater.

17.3.1.2 Disadvantages of nanophotocatalysts

- Only TiO_2 nanophotocatalyst exhibits maximum light stability compared to ZnO and other metal-based nanophotocatalysts (Bishoge et al., 2018).
- Modifications are required in the fabrication of phytogenic-mediated nanophotocatalysts to achieve maximum photocatalytic activities for a longer period.
- Sometimes, they offer toxicity and recovery issue especially with ZnO nanophotocatalysts fabricated through a chemical method and these issues limit their application at a pilot scale.

17.3.1.3 Future perspective of nanophotocatalysts

- Researchers are now focusing to fabricate nanophotocatalysts from plant-based sources to minimize their toxicity during wastewater treatment processes.
- They also try to improve structural morphology of the nanophotocatalyst to enhance their adsorptive properties, magnetic properties, and chemical reactivity especially during environmental remediation processes.

17.3.2 Nanosorbents

Nanosorbents are extensively applied in the laboratory-scale treatment of wastewater owing to their high pollutants sorption and adsorption capabilities (Karri et al., 2020; Khan et al., 2020; Lingamdinne et al., 2019). However, researchers are now focusing to fabricate nanosorbents from the plant-based sources to perform wastewater treatment process on a commercial scale. Most of the carbon-based nanosorbents such as C-black, graphite, etc. were reported in the literature for the management of water pollution; however, metal oxides and polymeric nanosorbents were also reported in the literature for environmental remediation (Dong et al., 2019; Gomes et al., 2019; Jorge de Souza et al., 2019; Lu et al., 2016a; Ray & Shipley, 2015; Tao et al., 2020; Yaqoob et al., 2020). In addition, a combination of different metals such as Ag/C, Ag/TiO_2, and C/TiO_2 are also used nowadays in the fabrication of nanosorbent because these metals exhibit efficient capability to reduced toxicity of the polluted water (Salim & Ho, 2015). Carbon-based nanosorbent fabricated from different plants (Soyabean, Cucumber, Tomato and *Tobbaco callus*) exhibits maximum surface area and thus offers maximum adsorption of carbonaceous pollutants from the wastewater. Moreover, carbon-based nanosorbents hydrophobic in nature also become more stabilized compared to other metal oxides nanosorbents. On the other hand, polymeric nanosorbents such as dendrimers reduced the level of organic pollutants from the wastewater. Moreover, they are biocompatible, easily degradable, eco-friendly, and removed organic pollutants up to 99% but the only disadvantage is that they are used only one time for the management of water pollution (Fuwad et al., 2019; Peng et al., 2017; Yang et al., 2018). Zeolite nanosorbent containing several nanoparticles such as Cu, Ag ions, etc. is also employed for the heavy metals reduction and removal of microbial pollutants from the wastewater (Giwa et al., 2017; Qi et al., 2019). Further, phytogenic-mediated magnetic nanosorbents are extensively used nowadays in conventional wastewater treatment facilities for the removal of almost all types of pollutants. These magnetic nanosorbents are easily recovered after being applied in the wastewater treatment process, either by magnetic forces or ion-exchange processes and hence reduced the treatment cost (Perez et al., 2019; Sahebi et al., 2019).

17.3.2.1 Advantages of nanosorbents

- Nanosorbents are easily fabricated from different sources and exhibit larger surface area, chemical reactivity and adsorptive properties due to their nano-sized range.
- They offer maximum reduction properties for all types of pollutants compared to conventional wastewater treatment technologies.
- Carbon-based and phytogenic-mediated magnetic nanoparticles are reproducible nanosorbents and can reduce the treatment cost of environmental remediation process.

17.3.2.2 Disadvantages of nanosorbents

- Carbon-based nanosorbents sometimes exhibit health risks such as allergy, digestive problems, arthritis, etc. due to their toxicity which largely depends on their structural configuration, nature of solvents, and surfactants used in their fabrication.
- Polymeric nanosorbents such as graphene, fullerene, and carbon based derivatives are excellent in their performance to reduce heavy metals and microbial pollutants up to 99% (Sahebi et al., 2019) but they are used only one time for the management of water pollution that is nonreproducible and thus does not offer cost-effective treatment process.

17.3.2.3 Future perspective of nanosorbents

- Researchers are nowadays working on commercialization of nansorbents for a large-scale application in the wastewater treatment as well as environmental remediation processes.
- They are also working on their structural morphology so that a more stable nanosorbents would be synthesized in near-future with enhanced chemical and adsorptive properties and exhibit less toxicity.
- Some of the research work on graphite nanosorbents is in progress as graphite proved to be biocompatible, durable, and eco-friendly. Besides it also offers maximum adsorptive properties. Therefore, it can be safely applied in the wastewater treatment processes to remove toxic pollutants from the wastewater.

17.3.3 Nanomembranes

Nanomembranes are made up of nanofibers from different plant sources and provide larger surface area as well as chemical reactivity to remove carbonaceous, nitrogenous, and microbial pollutants from the wastewater (Jun, Karri, Mubarak, et al., 2020; Jun, Karri, Yon, et al., 2020). They are mostly used in the pretreatment process where larger suspended particles are removed from the wastewater, but nanomembranes due to their reactive nature also offer removal of dissolved organic contaminants from the sewage water (Jhaveri and Murthy, 2016). The membrane contains small pores of uniform mesh size, facilitating the transport of water while retaining larger debris present in it. They have a composite layer arrangement that is made up of carbon that is, graphite or carbon nanotubes surrounded by a polymeric matrix. Nowadays, different types of nanomembranes are fabricated by using different nanomaterials including TiO_2, zeolite, Ag, Cu, ZnO, etc., and used widely in both applied as well as environmental sectors. They are also used in the hybrid attached growth wastewater treatment facilities, where they perform a noteworthy role in the decrease of different environmental pollutants from the wastewater (Linhares et al., 2018; Saleh et al., 2019; Zhu et al., 2018). However, there are few limitations of the nanomembrane filter such as its clogging and fouling. Therefore, super hydrophilic nanoparticles are suggested to be used in the fabrication of nanomembranes because they prevent these fouling and clogging issues when dealing with high strength wastewater. Furthermore, it was reported that phytogenic-mediated metal oxide nanoparticles are very effectively solving the issues of membranes clogging and fouling via offering greater porosity and hydrophilicity (Giwa et al., 2017; Y. Qi et al., 2019). In addition, the efficacy of nanomembranes with respect to the degradation of toxic pollutants are further improved by the introduction of nanophotocatalysts in the composite layer of nanomembranes for example TiO_2 nanophotocatalysts are incorporated into the nanofibers, where they played role in the deactivation of microbial pollutants and also degrade organic pollutants present in the wastewater (Gopalakrishnan et al., 2018).

This enhanced the application of nanomembrane at an industrial scale to minimize the level of toxic chemical compounds in industrial effluents. Moreover, a modification in the structural configuration of nanomembranes with nanophotocatalysts enhanced their application due to high selectivity, chemical reactivity, absorptivity, permeability, and reduces fouling and clogging issues (Abdullah et al., 2019).

17.3.3.1 Advantage of nanomembranes

- Nanofiber membranes are commonly used in ultrafiltration process where they offer excellent permeability, porosity, and offer bactericidal activities.
- Nanocomposite membrane provides high hydrophilicity, permeability, chemical reactivity, and reduces clogging and fouling issue especially when dealing with high strength wastewater.
- Aquaporin-based nanomembrane improved ionic selectivity and permeability during treatment of industrial wastewater.
- Nanomembranes maintain uniform nanopores thus providing maximum efficiency during filtration of different environmental pollutants. These nanomembranes also retain small size pathogenic organisms and thus reduce the toxicity of wastewater.
- Nanofiltration membranes have charge-based repulsion properties and also provide better selectivity during wastewater treatment process.

17.3.3.2 Disadvantage of nanomembranes

- Membrane fouling and clogging of nanomembrane especially when it deals with high strength wastewater are serious issues, but these issues largely depend on the structural configuration and the nature of nanomaterials used in the fabrication of nanomembranes.
- Another demerit of nanomembrane is stability because it does not persist for a longer period of time, which limits their application at a commercial scale.
- To enhance the stability of nanomembranes, expensive hydrophilic nanomaterials must be incorporated that makes them expensive. Besides, the incorporation of such hydrophilic nanomaterials may affect their selectivity and chemical reactivity during wastewater treatment at a commercial scale.

17.4 Applications of nanomaterials in management of water pollution

Engineered phytogenic-mediated nanoparticles proved to be effective for the management of water pollution. Some of the applications of phytogenic-mediated nanoparticles in wastewater treatment are summarized in Table 17.4.

- Nanomaterials such as TiO_2 integrated with metal oxide played a significant role in the reduction of arsenic from the drinking water (Ersan et al., 2017). These heterostructures help in reduction of heavy metals present in the drinking water as well as wastewater due to their adsorptive and photooxidative properties.

- Magnetic and carbon nanotubes were used as a sensor to monitor water quality with respect to toxic pollutants because these nanomaterials had stable morphology and profound optical, electrical, and chemical properties. These nanomaterials are mostly equipped with conventional wastewater treatment facilities to detect minute changes in the quality of treated effluent before allowing it to be discharged into the running water streams (Kumar et al., 2018).
- Recently, conventional wastewater treatment facilities are equipped with AgNPs due to their antimicrobial properties and thus used to eradicate microbial pollution from the wastewater as well as treated effluents (Pandey et al., 2017).
- Researchers reported that polymeric nanomaterials displayed substantial activity in the reduction of carbonaceous as well as nitrogenous pollution from the wastewater (Santhosh et al., 2016).
- Nanocomposite and nanosorbents provide high hydrophilicity, permeability, chemical reactivity, and mostly used in the ultrafiltration process to retain small size pathogenic organisms, thus helps in reducing toxicity of wastewater (Tate et al., 2000).
- Some of the nanoparticles such as AgNPs, TiO_2NPs and carbon-based nanoparticles displayed strong antibacterial and oxidative properties and thus commonly used in the purification of drinking water (Prathna et al., 2018).
- Nanomembranes played a significant role in the elimination of toxic organic-based pollutants as well as minimizing the level of pathogenic organisms due to their strong adsorptive and chemical reactivity and thus routinely installed in the laboratory-scale wastewater treatment facilities to get maximum efficiency of the reactors (Rozman et al., 2019).
- Phytogenic-mediated metal oxide nanoparticles are also used as photocatalysts to remove toxic compounds, organic pesticides, dyes, and pharmaceutical products from industrial waste effluents. Because these nanomaterials possess unique properties as they are smaller in size, offer larger surface area, have high magnetic and chemical properties (Lu et al., 2016b; Nasrollahzadeh et al., 2020a; Salim & Ho, 2015).

17.5 Conclusions and future outlooks

Herein we concluded that plant and plant-based products/compounds proved to be eco-friendly, safe, and beneficial approaches for the large-scale fabrication of nanoparticles. Moreover, phytogenic-mediated nanoparticles had a wide range of applications in the environmental sector especially in the remediation of water pollution, because they act as an excellent adsorbent, catalyst, and sensor due to their small size, greater surface area, and reactive nature. Further, the reducing and capping agents (polyphenols, amino acids, alkaloids, terpenoids, proteins, carboxyl, polysaccharides) used in the green synthesis of phytogenic-mediated nanoparticles are safe, eco-friendly, and also possess strong chemical reactivity. Moreover, phytogenic-mediated nanophotocatalysts are temperature-independent catalysts play a significant role in the reduction of toxic chemicals and microbial pollutants from industrial as well as domestic wastewater. Phytogenic-mediated nanomembranes maintain uniform nanopores thus providing maximum efficiency during filtration of different environmental pollutants. Besides, phytogenic-mediated AgNPs, TiO_2NPs, and carbon-based nanoparticles displayed strong antibacterial and oxidative properties. Further, it is suggested that experimental conditions including incubation temperature, nature of the solvents, pH of the solution, nature of the plant parts must be considered during fabrication of phytogenic-mediated nanoparticles.

References

Abdullah, N., Yusof, N., Lau, W.J., Jaafar, J., Ismail, A.F., 2019. Recent trends of heavy metal removal from water/wastewater by membrane technologies. J. Ind. Eng. Chem. 76, 17–38. https://doi.org/10.1016/j.jiec.2019.03.029.

Abu Hasan, H., Muhammad, M.H., Ismail, N.I, 2020. A review of biological drinking water treatment technologies for contaminants removal from polluted water resources. J. Water Process Eng. 33, 101135. https://doi.org/10.1016/j.jwpe.2019.101035.

Ahmad, S., Munir, S., Zeb, N., Ullah, A., Khan, B., Ali, J., Bilal, M., Omer, M., Alamzeb, M., Salman, S.M., Ali, S., 2019. Green nanotechnology: a review on green synthesis of silver nanoparticles — an ecofriendly approach. Int. J. Nanomed. 14, 5087–5107. https://doi.org/10.2147/IJN.S200254.

Aziz, S., Ali, S., 2019. Treatment of synthetic dairy wastewater using disposed plastic materials as trickling filter media: optimization and statistical analysis by RSM. Adv. Environ. Biol. 13 (10), 1–6.

Azizi, S., Nosrati, H., Danafar, H., 2020. Simple surface functionalization of magnetic nanoparticles with methotrexate-conjugated bovine serum albumin as a biocompatible drug delivery vehicle. Appl. Organomet. Chem. 34 (4), e5479. https://doi.org/10.1002/aoc.5479.

Bandeira, F.O., Alves, P.R.L., Hennig, T.B., Schiehl, A., Cardoso, E.J.B.N., Baretta, D., 2020. Toxicity of imidacloprid to the earthworm Eisenia andrei and collembolan *Folsomia candida* in three contrasting tropical soils. J. Soils Sediments 20 (4), 1997–2007. https://doi.org/10.1007/s11368-019-02538-6.

Berekaa, M.M., 2016. Nanotechnology in wastewater treatment; influence of nanomaterials on microbial systems. Int. J.Curr. Microbiol.Appl. Sci. 713–726. https://doi.org/10.20546/ijcmas.2016.501.072.

Bhanvase, B., Shende, T., Sonawane, SH., 2017. A review on graphene–TiO_2 and doped graphene–TiO_2 nanocomposite photocatalyst for water and wastewater treatment. Environ. Technol. Rev. 6 (1), 1–4.

Bhaskar Reddy, A.V., Jaafar, J., Abdul Majid, Z., Aris, A., Umar, K., Talib, J., Madhavi, G, 2015. Relative efficiency comparison of carboxymethyl cellulose (CMC) stabilized Fe0and Fe0/Ag nanoparticles for rapid degradation of chlorpyrifos in aqueous solutions. Digest J. Nanomater. Biostruct. 10 (2), 331–340. http://www.chalcogen.ro/331_Reddy.pdf.

Bhateria, R., Singh, 2019. A review on nanotechnological application of magnetic iron oxides for heavy metal removal. J. Water Process Eng. 31, 100845.

Bhatia, D., Sharma, N.R., Singh, J., Kanwar, R.S., 2017. Biological methods for textile dye removal from wastewater: A review. Crit. Rev. Environ. Sci. Technol. 47 (19), 1836–1876. https://doi.org/10.1080/10643389.2017.1393263.

Bishoge, O.K., Zhang, L., Suntu, S.L., Jin, H., Zewde, A.A., Qi, Z., 2018. Remediation of water and wastewater by using engineered nanomaterials: a review. J. Environ. Sci. Health 53 (6), 537–554. https://doi.org/10.1080/10934529.2018.1424991.

Bolognesi, S., Cecconet, D., Callegari, A., Capodaglio, A.G., 2020. Bioelectrochemical treatment of municipal solid waste landfill mature leachate and dairy wastewater as co-substrates. Environ. Sci. Pollut. Res. 28, 24639–24649. https://doi.org/10.1007/s11356-020-10167-7.

Cai, Z., Dwivedi, A.D., Lee, W.N., Zhao, X., Liu, W., Sillanpää, M., Zhao, D., Huang, C.H., Fu, J., 2018. Application of nanotechnologies for removing pharmaceutically active compounds from water: development and future trends. Environ. Sci. Nano 5 (1), 27–47. https://doi.org/10.1039/c7en00644f.

Capodaglio, A.G., 2017. Integrated, decentralized wastewater management for resource recovery in rural and peri-urban areas. Resources 6 (2), 22. https://doi.org/10.3390/resources6020022.

Chen, W., Liu, Q., Tian, S., Zhao, X., 2019. Exposed facet dependent stability of ZnO micro/nano crystals as a photocatalyst. Appl. Surf. Sci. 470, 807–816. https://doi.org/10.1016/j.apsusc.2018.11.206.

Connor, R., Renata, A., Ortigara, C., Koncagül, E., Uhlenbrook, S., Lamizana-Diallo, B., Zadeh, S., Qadir, M., Kjellén, M., Sjödin, J., Hendry, 2017. UN world water development report 2017. Wastewater: the untapped resource. The United Nations World Water Development Report.

Das, R., Vecitis, C.D., Schulze, A., Cao, B., Ismail, A.F., Lu, X., Chen, J., Ramakrishna, S., 2017. Recent advances in nanomaterials for water protection and monitoring. Chem. Soc. Rev. 46 (22), 6946–7020. https://doi.org/10.1039/c6cs00921b.

Dehghani, M.H., Karri, R.R., Yeganeh, Z.T., Mahvi, A.H., Nourmoradi, H., Salari, M., Zarei, A., Sillanpää, M., 2020. Statistical modelling of endocrine disrupting compounds adsorption onto activated carbon prepared from wood using CCD-RSM and DE hybrid evolutionary optimization framework: comparison of linear vs non-linear isotherm and kinetic parameters. J. Mol. Liq. 302, 112526. https://doi.org/10.1016/j.molliq.2020.112526.

Deshmukh, S.P., Patil, S.M., Mullani, S.B., Delekar, S.D., 2019. Silver nanoparticles as an effective disinfectant: a review. Mater. Sci. Eng. C 97, 954–965. https://doi.org/10.1016/j.msec.2018.12.102.

Devatha, C.P., Thalla, A.K., Katte, S.Y., 2016. Green synthesis of iron nanoparticles using different leaf extracts for treatment of domestic waste water. J. Clean. Prod. 139, 1425–1435. https://doi.org/10.1016/j.jclepro.2016.09.019.

Dhall, A., Self, W., 2018. Cerium oxide nanoparticles: a brief review of their synthesis methods and biomedical applications. Antioxidants 7 (8), 97. https://doi.org/10.3390/antiox7080097.

Di Mauro, A., Cantarella, M., Nicotra, G., Pellegrino, G., Gulino, A., Brundo, M.V., Privitera, V., Impellizzeri, G., 2017. Novel synthesis of ZnO/PMMA nanocomposites for photocatalytic applications. Sci. Rep. 7, 40895. https://doi.org/10.1038/srep40895.

Diaz-Elsayed, N., Rezaei, N., Guo, T., Mohebbi, S., & Zhang, Q. (2019). Wastewater-based resource recovery technologies across scale: a review. resources, conservation and recycling, 145, 94–112. https://doi.org/10.1016/j.resconrec.2018.12.035

Dong, H., Li, L., Lu, Y., Cheng, Y., Wang, Y., Ning, Q., Wang, B., Zhang, L., Zeng, G., 2019. Integration of nanoscale zero-valent iron and functional anaerobic bacteria for groundwater remediation: a review. Environ. Int. 124, 265–277. https://doi.org/10.1016/j.envint.2019.01.030.

Ersan, G., Apul, O.G., Perreault, F., Karanfil, T., 2017. Adsorption of organic contaminants by graphene nanosheets: a review. Water Res. 126, 385–398. https://doi.org/10.1016/j.watres.2017.08.010.

Fuwad, A., Ryu, H., Malmstadt, N., Kim, S.M., Jeon, T.J., 2019. Biomimetic membranes as potential tools for water purification: preceding and future avenues. Desalination 458, 97–115. https://doi.org/10.1016/j.desal.2019.02.003.

Gautam, P.K., Singh, A., Misra, K., Sahoo, A.K., Samanta, S.K., 2019. Synthesis and applications of biogenic nanomaterials in drinking and wastewater treatment. J. Environ. Manage. 231, 734–748. https://doi.org/10.1016/j.jenvman.2018.10.104.

Giwa, A., Hasan, S.W., Yousuf, A., Chakraborty, S., Johnson, D.J., Hilal, N., 2017. Biomimetic membranes: a critical review of recent progress. Desalination 420, 403–424. https://doi.org/10.1016/j.desal.2017.06.025.

Gomes, J., Lincho, J., Domingues, E., Quinta-Ferreira, R., Martins, R.C., 2019. N–TiO_2 photocatalysts: a review of their characteristics and capacity for emerging contaminants removal. Water 11 (2), 373.

In: Gopalakrishnan, I., Samuel, R., Sridharan, K., 2018. Nanomaterials-based adsorbents for water and wastewater treatments. In: Sridharan, K. (Ed.), Emerging Trends of Nanotechnology in Environment and Sustainability. Springer, Cham, pp. 89–98.

Gupta, A., Tandon, M., Kaur, A., 2020. Role of metallic nanoparticles in water remediation with special emphasis on sustainable synthesis: a review. Nanotechnol. Environ. Eng. 5 (3), 27. https://doi.org/10.1007/s41204-020-00092-y.

Hairuddin, M.N., Mubarak, N.M., Khalid, M., Abdullah, E.C., Walvekar, R., Karri, R.R., 2019. Magnetic palm kernel biochar potential route for phenol removal from wastewater. Environ. Sci. Pollut. Res. 26 (34), 35183–35197. https://doi.org/10.1007/s11356-019-06524-w.

Han, J., Wang, M., Hu, Y., Zhou, C., Guo, R., 2017. Conducting polymer-noble metal nanoparticle hybrids: synthesis mechanism application. Prog. Polym. Sci. 70, 52–91. https://doi.org/10.1016/j.progpolymsci.2017.04.002.

Hazirah, N., Nurhaslina, C., Ku Halim, K., 2014. Effect of Interaction Time on Immobilization Process of Bambusa Heterostachya as Lactobacillus Matrix. J. Environ. Sci. Toxicol. Food Technol 8, 6–10.

Jhaveri, J.H., Murthy, Z.V.P., 2016. A comprehensive review on anti-fouling nanocomposite membranes for pressure driven membrane separation processes. Desalination 379, 137–154. https://doi.org/10.1016/j.desal.2015.11.009.

Souza, Jorge de, A., T., Souza, Rosa, L., R., Franchi, L.P, 2019. Silver nanoparticles: an integrated view of green synthesis methods, transformation in the environment, and toxicity. Ecotoxicol. Environ. Saf. 171, 691–700. https://doi.org/10.1016/j.ecoenv.2018.12.095.

Jun, L.Y., Karri, R.R., Mubarak, N.M., Yon, L.S., Bing, C.H., Khalid, M., Jagadish, P., Abdullah, E.C., 2020. Modelling of methylene blue adsorption using peroxidase immobilized functionalized Buckypaper/polyvinyl alcohol membrane via ant colony optimization. Environ. Pollut. 259, 113940. https://doi.org/10.1016/j.envpol.2020.113940.

Jun, L.Y., Karri, R.R., Yon, L.S., Mubarak, N.M., Bing, C.H., Mohammad, K., Jagadish, P., Abdullah, E.C., 2020. Modeling and optimization by particle swarm embedded neural network for adsorption of methylene blue by jicama peroxidase immobilized on buckypaper/polyvinyl alcohol membrane. Environ. Res. 183, 109158. https://doi.org/10.1016/j.envres.2020.109158.

Kaetzl, K., Lübken, M., Uzun, G., Gehring, T., Nettmann, E., Stenchly, K., Wichern, M., 2019. On-farm wastewater treatment using biochar from local agroresidues reduces pathogens from irrigation water for safer food production in developing countries. Sci. Total Environ. 682, 601–610. https://doi.org/10.1016/j.scitotenv.2019.05.142.

Karade, V.C., Dongale, T.D., Sahoo, S.C., Kollu, P., Chougale, A.D., Patil, P.S., Patil, P.B., 2018. Effect of reaction time on structural and magnetic properties of green-synthesized magnetic nanoparticles. J. Phys. Chem. Solids 120, 161–166. https://doi.org/10.1016/j.jpcs.2018.04.040.

Karri, R.R., Sahu, J.N., Meikap, B.C., 2020. Improving efficacy of Cr (VI) adsorption process on sustainable adsorbent derived from waste biomass (sugarcane bagasse) with help of ant colony optimization. Ind. Crops Prod. 143, 111927. https://doi.org/10.1016/j.indcrop.2019.111927.

Khan, F.S.A., Mubarak, N.M., Tan, Y.H., Karri, R.R., Khalid, M., Walvekar, R., Abdullah, E.C., Mazari, S.A., Nizamuddin, S., 2020. Magnetic nanoparticles incorporation into different substrates for dyes and heavy metals removal—a review. Environ. Sci. Pollut. Res. 27 (35), 43526–43541. https://doi.org/10.1007/s11356-020-10482-z.

Kim, H.C., Park, S.H., Noh, J.H., Choi, J., Lee, S., Maeng, S.K., 2019. Comparison of pre-oxidation between O_3 and O_3/H_2O_2 for subsequent managed aquifer recharge using laboratory-scale columns. J. Hazard. Mater. 377, 290–298. https://doi.org/10.1016/j.jhazmat.2019.05.099.

Koli, S.H., Mohite, B.V., Suryawanshi, R.K., Borase, H.P., Patil, S.V., 2018. Extracellular red Monascus pigment-mediated rapid one-step synthesis of silver nanoparticles and its application in biomedical and environment. Bioprocess. Biosyst. Eng. 41 (5), 715–727. https://doi.org/10.1007/s00449-018-1905-4.

Kour, D., Rana, K.L., Kaur, T., Yadav, N., Yadav, A.N., Rastegari, A.A., Saxena, A.K., 2019. Microbial biofilms: Functional annotation and potential applications in agriculture and allied sectors. In: Yadav, M.K., Singh, B.P. (Eds.), New and Future Developments in Microbial Biotechnology and Bioengineering: Microbial Biofilms Current Research and Future Trends in Microbial Biofilms. Elsevier, Amsterdam, pp. 283–301.

Kumar, V., Kumar, P., Pournara, A., Vellingiri, K., Kim, K.H., 2018. Nanomaterials for the sensing of narcotics: challenges and opportunities. TrAC Trends Anal. Chem. 106, 84–115. https://doi.org/10.1016/j.trac.2018.07.003.

Kwon, G., Nam, J.H., Kim, D.M., Song, C., Jahng, D., 2020. Growth and nutrient removal of chlorella vulgaris in ammonia-reduced raw and anaerobically-digested piggery wastewaters. Environ. Eng. Res. 25 (2), 135–146. https://doi.org/10.4491/eer.2018.442.

Lingamdinne, L.P., Koduru, J.R., Karri, R.R., 2019. A comprehensive review of applications of magnetic graphene oxide based nanocomposites for sustainable water purification. J. Environ. Manage 231, 622–634. https://doi.org/10.1016/j.jenvman.2018.10.063.

Linhares, A.M.F., Grando, R.L., Borges, C.P., da Fonseca, F.V., 2018. Technological prospection on membranes containing silver nanoparticles for water disinfection. Recent Pat. Nanotechnol. 12 (1), 3–12. https://doi.org/10.2174/1872210511666170920144342.

Lu, H., Wang, J., Stoller, M., Wang, T., Bao, Y., Hao, H., 2016. An overview of nanomaterials for water and wastewater treatment. Adv. Mater. Sci. Eng. 2016, 4964828. https://doi.org/10.1155/2016/4964828.

Magwaza, S.T., Magwaza, L.S., Odindo, A.O., Mditshwa, A., 2020. Hydroponic technology as decentralised system for domestic wastewater treatment and vegetable production in urban agriculture: a review. Sci. Total Environ. 698, 134154. https://doi.org/10.1016/j.scitotenv.2019.134154.

Maqbool, Q., Nazar, M., Naz, S., Hussain, T., Jabeen, N., Kausar, R., Anwaar, S., Abbas, F., Jan, T., 2016. Antimicrobial potential of green synthesized CeO_2 nanoparticles from *Olea europaea* leaf extract. Int. J. Nanomed. 11, 5015–5025. https://doi.org/10.2147/IJN.S113508.

Mirajkar, S.J., Dalvi, S.G., Ramteke, S.D., Suprasanna, P., 2019. Foliar application of gamma radiation processed chitosan triggered distinctive biological responses in sugarcane under water deficit stress conditions. Int. J. Biol. Macromol. 139, 1212–1223. https://doi.org/10.1016/j.ijbiomac.2019.08.093.

Nasrollahzadeh, M., Sajjadi, M., Iravani, Varma, RS, 2020. Green-synthesized nanocatalysts and nanomaterials for water treatment: current challenges and future perspectives. J. Hazard. Mater 401, 123401.

Pandey, N., Shukla, S.K., Singh, N.B., 2017. Water purification by polymer nanocomposites: an overview. Nanocomposites 3 (2), 47–66. https://doi.org/10.1080/20550324.2017.1329983.

Patel, K., Bharatiya, B., Mukherjee, T., Soni, T., Shukla, A., Suhagia, B.N., 2017. Role of stabilizing agents in the formation of stable silver nanoparticles in aqueous solution: characterization and stability study. J. Dispers. Sci. Technol. 38 (5), 626–631. https://doi.org/10.1080/01932691.2016.1185374.

Patra, J.K., Baek, K.H., 2015. Novel green synthesis of gold nanoparticles using *Citrullus lanatus* rind and investigation of proteasome inhibitory activity, antibacterial, and antioxidant potential. Int. J. Nanomed. 10, 7253–7264. https://doi.org/10.2147/IJN.S95483.

Peng, F., Xu, T., Wu, F., Ma, C., Liu, Y., Li, J., Zhao, B., Mao, C., 2017. Novel biomimetic enzyme for sensitive detection of superoxide anions. Talanta 174, 82–91. https://doi.org/10.1016/j.talanta.2017.05.028.

Perez, T., Pasquini, D., De Faria Lima, A., Rosa, E.V., Sousa, M.H., Cerqueira, D.A., De Morais, L.C., 2019. Efficient removal of lead ions from water by magnetic nanosorbents based on manganese ferrite nanoparticles capped with thin layers of modified biopolymers. J.Environ. Chem. Eng. 7 (1), 102892. https://doi.org/10.1016/j.jece.2019.102892.

Prathna, T.C., Sharma, S.K., Kennedy, M., 2018. Nanoparticles in household level water treatment: an overview. Sep. Purif. Technol. 199, 260–270. https://doi.org/10.1016/j.seppur.2018.01.061.

Qi, L., Liu, Z., Wang, N., Hu, Y., 2018. Facile and efficient in situ synthesis of silver nanoparticles on diverse filtration membrane surfaces for antimicrobial performance. Appl. Surf. Sci. 456, 95–103. https://doi.org/10.1016/j.apsusc.2018.06.066.

Qi, Y., Zhu, L., Shen, X., Sotto, A., Gao, C., & Shen, J. (2019). Polythyleneimine-modified original positive charged nanofiltration membrane: removal of heavy metal ions and dyes. Sep. Purif. Technol., 222, 117–124. https://doi.org/10.1016/j.seppur.2019.03.083

Rajeshkumar, S., Naik, P., 2018. Synthesis and biomedical applications of cerium oxide nanoparticles – a review. Biotechnol. Rep. 17, 1–5. https://doi.org/10.1016/j.btre.2017.11.008.

Rao, N.H., Lakshmidevi, N., Pammi, S.V.N., Kollu, P., Ganapaty, S., Lakshmi, P., 2016. Green synthesis of silver nanoparticles using methanolic root extracts of *Diospyros paniculata* and their antimicrobial activities. Mater. Sci. Eng. C 62, 553–557. https://doi.org/10.1016/j.msec.2016.01.072.

Ravichandran, V., Vasanthi, S., Shalini, S., Shah, S.A.A., Tripathy, M., Paliwal, N., 2019. Green synthesis, characterization, antibacterial, antioxidant and photocatalytic activity of *Parkia* speciosa leaves extract mediated silver nanoparticles. Results Phys. 15, 102565. https://doi.org/10.1016/j.rinp.2019.102565.

Ray, P.Z., Shipley, H.J., 2015. Inorganic nano-adsorbents for the removal of heavy metals and arsenic: a review. RSC Adv., 5 (38), 29885–29907. https://doi.org/10.1039/c5ra02714d.

Rolim, W.R., Pelegrino, M.T., de Araújo Lima, B., Ferraz, L.S., Costa, F.N., Bernardes, J.S., Rodigues, T., Brocchi, M., Seabra, A.B., 2019. Green tea extract mediated biogenic synthesis of silver nanoparticles: characterization, cytotoxicity evaluation and antibacterial activity. Appl. Surf. Sci. 463, 66–74. https://doi.org/10.1016/j.apsusc.2018.08.203.

Rozman, N., Tobaldi, D., Cvelbar, U., Puliyalil, H., Labrincha, J., Legat, A., Škapin, S., 2019. Hydrothermal synthesis of rare-earth modified titania: Influence on phase composition, optical properties, and photocatalytic activity. Materials 12 (5), 713.

Sahebi, S., Sheikhi, M., Ramavandi, B., 2019. A new biomimetic aquaporin thin film composite membrane for forward osmosis: characterization and performance assessment. Desalin. Water Treat. 148, 42–50. https://doi.org/10.5004/dwt.2019.23748.

Sahu, J.N., Karri, R.R., Zabed, H.M., Shams, S., Qi, X., 2019. Current perspectives and future prospects of nano-biotechnology in wastewater treatment. Sep. Purif. Rev 50 (2), 139–158. https://doi.org/10.1080/15422119.2019.1630430.

Sahu, J.N., Zabed, H., Karri, R.R., Shams, S., Qi, X., Thomas, S., Grohens, Y., Pottathara, Y.B., 2019. Applications of nano-biotechnology for sustainable water purification. In: S. Thomas, Y. Grohens and Y.B. Pottathara (Eds.), Micro and Nano Technologies. Elsevier, Amsterdam, pp. 313–340.

Saif, S., Tahir, A., Chen, Y., 2016. Green synthesis of iron nanoparticles and their environmental applications and implications. Nanomaterials 6 (11), 209. https://doi.org/10.3390/nano6110209.

Saleh, T.A., Parthasarathy, P., Irfan, M., 2019. Advanced functional polymer nanocomposites and their use in water ultra-purification. Trends Environ. Anal. Chem. 24. https://doi.org/10.1016/j.teac.2019.e00067.

Salim, W., Ho, W.W., 2015. Recent developments on nanostructured polymer-based membranes. Curr. Opin. Chem. Eng. 8, 76–82. https://doi.org/10.1016/j.coche.2015.03.003.

Samanta, H., Das, R., Bhattachajee, C., 2016. Influence of nanoparticles for wastewater treatment-a short review. Austin Chem. Eng. 3 (3), 1031–1036.

Sangsefidi, F.S., Nejati, M., Verdi, J., Salavati-Niasari, M., 2017. Green synthesis and characterization of cerium oxide nanostructures in the presence carbohydrate sugars as a capping agent and investigation of their cytotoxicity on the mesenchymal stem cell. J. Clean. Prod. 156, 741–749. https://doi.org/10.1016/j.jclepro.2017.04.114.

Santhosh, C., Velmurugan, V., Jacob, G., Jeong, S.K., Grace, A.N., Bhatnagar, A., 2016. Role of nanomaterials in water treatment applications: a review. Chem. Eng. J. 306, 1116–1137. https://doi.org/10.1016/j.cej.2016.08.053.

Saravanan, R., Gracia, F., Stephen, A., 2017. Basic principles, mechanism, and challenges of photocatalysis. In: Khan, M.M., Pradhan, D., Sohn, Y. (Eds.), Nanocomposites for Visible Light-Induced Photocatalysis. Springer, Cham, pp. 19–40.

Serrà, A., Zhang, Y., Sepúlveda, B., Gómez, E., Nogués, J., Michler, J., Philippe, L., 2020. Highly reduced ecotoxicity of ZnO-based micro/nanostructures on aquatic biota: influence of architecture, chemical composition, fixation, and photocatalytic efficiency. Water Res. 169. https://doi.org/10.1016/j.watres.2019.115210.

Shafey, A.M.E., 2020. Green synthesis of metal and metal oxide nanoparticles from plant leaf extracts and their applications: a review. Green Process. Synth. 9 (1), 304–339. https://doi.org/10.1515/gps-2020-0031.

Sheng, Y., Wei, Z., Miao, H., Yao, W., Li, H., Zhu, Y., 2019. Enhanced organic pollutant photodegradation via adsorption/photocatalysis synergy using a 3D g-C_3N_4/TiO_2 free-separation photocatalyst. Chem. Eng. J. 370, 287–294. https://doi.org/10.1016/j.cej.2019.03.197.

Tao, Q., Bi, J., Huang, X., Wei, R., Wang, T., Zhou, Y., Hao, H., 2020. Fabrication, application, optimization and working mechanism of Fe_2O_3 and its composites for contaminants elimination from wastewater. Chemosphere 263, 127889.

Tate, J., Burton, A., Boschi-Pinto, Parashar, World Health Organization–Coordinated Global Rotavirus Surveillance Network, 2000. Global, regional, and national estimates of rotavirus mortality in children < 5 years of age. Clin. Infect. Dis., 62, pp. 96–105.

Thakur, N., Manna, P., Das, J., 2019. Synthesis and biomedical applications of nanoceria, a redox active nanoparticle. J. Nanobiotechnol. 17 (1), 84. https://doi.org/10.1186/s12951-019-0516-9.

Varma, S., Tayade, J., Kinjal, J., Pradyuman, A., Atindra, D., Vimal, D., 2020. Photocatalytic degradation of pharmaceutical and pesticide compounds (PPCs) using doped TiO_2 nanomaterials: a review. Water-Energy Nexus 3, 46–61. https://doi.org/10.1016/j.wen.2020.03.008.

Venkateasan, A., Prabakaran, R., Sujatha, V., 2017. Phytoextract-mediated synthesis of zinc oxide nanoparticles using aqueous leaves extract of *Ipomoea pes-caprae* (L).R.br revealing its biological properties and photocatalytic activity. Nanotechnol. Environ. Eng. 2 (1), 8. https://doi.org/10.1007/s41204-017-0018-7.

Voulvoulis, N. (2018). Water reuse from a circular economy perspective and potential risks from an unregulated approach. Curr. Opin. Environ. Sci. Health, 2, 32–45. https://doi.org/10.1016/j.coesh.2018.01.005
Water, W.J.Supply, 2015. Sanitation Monitoring Programme, World Health Organization. Progress on sanitation and drinking water: 2015 Update and MDG Assessment. World Health Organization.
Xu, C., Chen, W., Gao, H., Xie, X., Chen, Y., 2020. Cellulose nanocrystal/silver (CNC/Ag) thin-film nanocomposite nanofiltration membranes with multifunctional properties. Environ. Sci. Nano 7 (3), 803–816. https://doi.org/10.1039/c9en01367a.
Yang, Z., Ma, X.H., Tang, C.Y., 2018. Recent development of novel membranes for desalination. Desalination 434, 37–59. https://doi.org/10.1016/j.desal.2017.11.046.
Yaqoob, A.A., Parveen, T., Umar, K., Ibrahim, M.N.M., 2020. Role of nanomaterials in the treatment of wastewater: a review. Water (Switzerland) 12 (2). https://doi.org/10.3390/w12020495.
Yew, Y.P., Shameli, K., Miyake, M., Ahmad Khairudin, N.B.B., Mohamad, S.E.B., Naiki, T., Lee, K.X., 2020. Green biosynthesis of superparamagnetic magnetite Fe_3O_4 nanoparticles and biomedical applications in targeted anticancer drug delivery system: a review. Arab. J. Chem. 13 (1), 2287–2308. https://doi.org/10.1016/j.arabjc.2018.04.013.
Zeng, X., Wang, Z., Wang, G., Gengenbach, T.R., McCarthy, D.T., Deletic, A., Yu, J., Zhang, X., 2017. Highly dispersed TiO_2 nanocrystals and WO_3 nanorods on reduced graphene oxide: Z-scheme photocatalysis system for accelerated photocatalytic water disinfection. Appl. Catal. B 218, 163–173. https://doi.org/10.1016/j.apcatb.2017.06.055.
Zhang, M., Gu, J., Liu, Y., 2019. Engineering feasibility, economic viability and environmental sustainability of energy recovery from nitrous oxide in biological wastewater treatment plant. Bioresour. Technol. 282, 514–519. https://doi.org/10.1016/j.biortech.2019.03.040.
Zhou, H., Pan, H., Xu, J., Xu, W., Liu, L., 2016. Acclimation of a marine microbial consortium for efficient Mn(II) oxidation and manganese containing particle production. J. Hazard. Mater. 304, 434–440. https://doi.org/10.1016/j.jhazmat.2015.11.019.
Zhuang, L.L., Yang, T., Zhang, J., Li, X., 2019. The configuration, purification effect and mechanism of intensified constructed wetland for wastewater treatment from the aspect of nitrogen removal: a review. Bioresour. Technol. 293, 122086. https://doi.org/10.1016/j.biortech.2019.122086.

CHAPTER 18

Magnetic nanoparticles and their application in sustainable environment

Megha Singh, Shikha Dhiman, Nitai Debnath and Sumistha Das
Amity Institute of Biotechnology, Amity University Haryana, Gurugram, Haryana, India

18.1 Introduction

Nanotechnology refers to the branch of science that deals with particles of very small size. By definition, nanoparticles (NPs) are entities of any shape with at least one dimension in the range of 1×10^{-9} and 1×10^{-7} m. Several natural objects fall under this size range. For example, nucleic acids, many of the proteins floating in the human body are in nano-dimensions. Similarly, volcanic eruption generates a huge amount of suspended NPs. Technically, these are a thousand times smaller than a cell. NPs exhibit multiple specialized features in comparison to their bulk counterparts. With the advancement in the field of nanoscience and technology, myriads of nanomaterials are being applied in several facets of human lives. Among them, magnetic NPs (MNPs) offer immense potential almost in all sectors of medical science (Sahu et al., 2019).

MNPs are specialized because of their unique response and tunable properties under magnetic fields (Lingamdinne et al., 2019, 2018). These engineered nanostructures are mostly composed of two components: one is a magnetic core (ferrites, iron, nickel, cobalt) another is a nonmagnetic chemical constituent. Iron as core is preferred over other magnetic metallic components, to design MNPs because of their relatively less toxic profile. For this reason, iron oxide–based materials are used in the field of disease and diagnostics because of their special features (Vallabani and Singh, 2018). The movement of molecules with mass and charge renders these special magnetic properties to such nanomaterials. According to the published literature, these properties are tunable with size. So in nanodomain, MNPs exhibit even better and more beneficial physical properties in respect to their bulk counterparts. For this reason, MNPs have huge industrial applications. Starting from catalysts to synthetic dyes MNPs are very useful. Similarly, separation technology has reached a new dimension with the help of MNPs.

With an ever-increasing load of pollution in the environment, there is always a need for technologies to remove environmental contaminants. Different filtration and adsorption mechanisms are still inadequate to address all kinds of pollutants from the air,

Sustainable Nanotechnology for Environmental Remediation
DOI: https://doi.org/10.1016/B978-0-12-824547-7.00007-2

soil, water, etc. For example, the magnetic feature is very much useful for the selective separation of environmental contaminats (Khan et al., 2020; Lingamdinne et al., 2020; Ruthiraan et al., 2019). Similarly, due to magnetic properties detection of agricultural, heavy metal and other unwanted waste ions are also easily identified. Even MNPs are very efficient in detecting very minute quantities of toxic substances present in drinking water. The use of MNPs has excellent potential in wastewater treatment. These can eliminate contaminants and pathogens either as stand-alone treatment agents or by incorporating them into biological membranes and combining them with existing treatment methods (Sahu et al., 2019). For example, Mounting NPs on a magnetic core–like nanomagnetite have recently been used to remove Cd from water (Devi et al., 2017). Unique properties of NPs can be assembled on the nanomagnetite core to create a multifunctional nanocomposite material that can be collected and recycled through magnetic separation. Another way is to use covalent bonding, wherein graphene oxide obtained via the modified Hummers method can be further functionalized with magnetic Fe_3O_4NPs through ionic liquid (IL) such as BF4 [Bmim] to form core-shell structured Fe_3O_4@Graphene Oxide (GO) nanospheres for optimal cadmium extraction from wastewater (Alvand and Shemirani, 2016; Lingamdinne et al., 2019, 2018). Some researchers have produced magnetic graphene oxide nanocomposites for multidimensional applications, including energy storage, water treatment, and drug delivery (Chung et al., 2013).

Apart from this, biological pathogens are another threat to environment that poses major public health issues as well. Such biological pathogens can be detected and removed from water bodies with the help of MNPs. Apart from this, due to possibility of tunability in properties with engineered MNPs, scientists are synthesizing MNP with high stability, specificity, and other beneficial properties. For example, for detection of disease in medical applications MNPs are having huge potential (Khan et al., 2020; Lingamdinne et al., 2019; Williams, 2017).

Magnetic resonance imaging has taken a new shape with advanced nanostructures and their unique features. Similarly in therapeutics as well, MNPs are playing a huge role in delivery of drug, generation of controlled heat in uncontrollably growing tissues, and providing other therapeutic benefits. Even interaction of biomolecules with MNPs has shown interesting observations such as weak and strong hydrophilic, hydrophobic, covalent, and electrostatic interactions. All such interactions create interesting nanobioconjugates that traverse the majority of organs and tissues with excellent specificity and target accession. Such magnetic nanomaterials help to detect various bioactivity ongoing in-vitro and in-vivo conditions (Kudr et al., 2017). Bioimaging and detection is another emerging field with MNPs. Apart from these, MNPs are also good for tissue engineering and allied applications (Ortega & Reguera, 2019). Almost all the sectors of medical science and plant science are now very much including MNPs for various applications. Delivery of biomolecules such as hormones and nucleic acids are delivered to provide improved features in crop growth and yield by using an MNP-based transfection system. For all these purposes, the safe and biocompatible approach of MNP synthesis is very popular. Keeping this in interest, much research is ongoing for the sustainable use of green sources for MNP synthesis. Plant leaf extract, metabolic products are now used as source, reducing, and capping material for MNP synthesis.

But the major disadvantage with MNPs comes from their sensitivity to air and water which can lead to corrosion and rusting (Ali et al., 2016)). For this reason, selection of an appropriate surface coating material is of much importance. For example, iron cores can have non porous coatings. Iron alloys such as FePt and FeAu, are very common and widely used as core materials for MNPs. The coatings can be of various types, for example, polymer, inorganic, or semiconductors depending on the application. Both natural and synthetic polymers can be used but as natural polymers lack mechanical strength, synthetic polymers are preferred. Coating silica on iron oxide particles is difficult as the amorphous structure does not allow silica to form a homogeneous layer on the surface of the iron oxide. Gold is one of the most commonly used materials for coating.

Several experimental and review articles are already published based on the synthesis, characterization, and application of MNPs in different areas (Bedanta et al., 2013). In most cases, MNPs gained more attention in material science, specifically in the removal of toxicants, sensing, and other approaches where magnetic field induction was the priority. Also, significant MNPs were found in medical science as biosensors, magnetic field–based detectors, imaging, and drug delivery purposes. Few more references are also evident where natural sources of material and environment-safe approaches are considered to synthesize MNPs. This is attributable to the production of biosafe MNPs for expanded use in the medical and agricultural sectors. Limited reports are cited on the possibility of MNPs in agriculture use and their impact. This chapter contains summarized information of all the possible MNPS with their detailed synthesis, properties, and implication in ensuring a sustainable environment.

18.2 Physical properties of MNPs

Particles such as protons, electrons, and other charged ions have both mass and electrical charges thus endure magnetic effects due to oscillation. And these spinning charged particles create dipoles and are called magnetons. The alignment of this magneton is known as the magnetic domain of ferromagnetic materials. The MNPs range from 1 to 100 nm in size and can display super-paramagnetic behavior (Belanova et al., 2018). Thermal effects cause the superparamagnetic nature of the MNPs. These particles have zero coercivity and have no hysteresis. In the absence of an externally applied magnetic field, it is a ferromagnetic material in bulk, which has nonzero coercivity and at nanodomain, the shrunk structure further develops the coercivity that goes up or induces level increase which reaches its peak as it is becoming a single magnetic domain. Larger particles have more domains. When they shrink to just one domain, the coercivity falls to zero and then it makes the transition to be called a superparamagnetic material. If an external magnetic field is applied in this state, it will magnetize the NPs with much greater magnetic susceptibility. No magnetization is found when the field is eliminated. This property can be very beneficial in multiple applications. The physicochemical properties of MNPs include factors such as shape, size, charge, solution stability, and zeta potential, and the coating of the NPs. These factors are interrelated as the charge of the surface or surface charge affects the zeta potential, whereas the zeta potential helps in actuating the stability factor of the solution of NPs. NPs when synthesized can turn out to be negative, positive, or neutral. And due to the differences in the clearance speed and biodistribution of positively and negatively charged NPs, mostly the neutral or negatively charged NPs are preferred and considered as the optimal surface charge. The bare surface of magnetic iron oxide NPs attains agglomeration due to strong magnetic attractions. To avoid this problem stabilizing groups such as carboxylates, inorganic compounds, and polymeric compounds are used to adorn the stability of the particle suspended. These stabilizers are proven to be quite incompatible when used in biomedical applications. In this situation, solution stability is directed further by the zeta potential. Zeta potential is a bulk property with parameters such as the charge, ionic content, and strength of the solution of the NPs. For the biosafe application of the MNPs, zeta potential should be below -25 mV or above +25 mV. Shape and size play a very crucial role as larger surface areas are preferred due to their superior deposition of tumor, and the correct size of the NPs is important as it helps in the rate of cellular uptake for biomedical applications in vitro and in vivo (Belanova et al., 2018).

18.3 Chemical synthesis of MNPs

There have been numerous methods manifested for the synthesis of MNPs. The methods for synthesis are picked precisely considering their size, shape, stability, for enhancing their magnetic properties in relevance to their applications. Earlier, the synthesis of MNPs played a vital role in enhancing the physical and chemical properties concerning their bulk counterparts and then testing their behavior. The synthesis of MNPs is primarily done keeping in mind the large-scale production with cost effectiveness (Devi et al., 2017). Due to the usage of metals (mostly oxides of iron, cobalt, nickel) as core materials in combination with numerous other metals such as copper, zinc, barium, and strontium as a magnetic part of the particles engineered as the surface coating is done to make them more stable. This is for improving their solubility and other important functions such as target specificity, resistance to corrosion, toxicity, oxidation, and agglomeration. The core material and its surface coating with modifications are done by focusing on the desired use and functionality of the MNPs. Some functional ligands (such as bifunctional linkers using the avidin-biotin technique) of desired properties are too attached to the

NP for improving its functions and specific qualities (Kudr et al., 2017). Another example can be the synthesis of MNPs with iron-based cores that can be easily manufactured by immobilization of any of the oxides of iron such as iron (II) oxalate (FeO) or iron (III) citrate (FeC) on magnetite NPs. They are further stabilized with polyethylene glycol (PEG) (Neamtu et al., 2018). Here we will be discussing a few of the techniques commonly approached for the production of stable, cost-effective, and desired MNPs in different physical and chemical routes.

Different bottom-up and top-down approaches are commonly used for the synthesis of MNPs. The bottom-up approach is primarily focused on physical and mechanical paths; therefore has a higher probability of generating NPs with fewer irregularities, and has a more homogeneous chemical composition than the top-down approach. The other difference between these two approaches is that in the bottom-up approach crystal surfaces add to the substrate, while in the top-down method the crystal surfaces are taken away from the substrate surface to form the NPs. Chemical methods are preferred and used more frequently over the other methods as these are cheaper while manufacturing the desired MNPs in large proportions. NPs ranging from nanometers to micrometers in size can be regulated and achieved more conveniently by the chemical method. The synthesis of monodisperse NPs is commenced by injecting reagents inwards the hot surfactant solution and after this procedure, further steps include the aging and selection process for a suitable size. However, in this method initially, a small burst of nucleation is achieved then their growth is slowed to produce monodisperse particles. This controlled growth is essential for obtaining stable MNPs and their sizes can be controlled by manipulating reaction parameters such as temperature, time, reagents concentrations, and stabilizing surfactants (Devi et al., 2017).

18.3.1 Microemulsion method

Microemulsion also plays a vital role in the synthesis of MNPs. In this approach, a thermodynamically stable isotropic dispersion, "microemulsion" is formed with the help of two immiscible liquids. By integrating an interfacial layer of surfactant molecules into a mixture, the microdomain of both or one of the liquids (e.g., mixtures of oil, water, and surfactant, in addition to a co-surfactant) is made stable. This important technique gives rise to MNPs production (Devi et al., 2017).

18.3.2 Wet precipitation and coprecipitation

This is one of the oldest methods used for the preparation of MNPs. It is a simple process implemented by careful control of the pH of an iron salt solution. In this process, a fine suspension with a particle size of 5 nm of iron oxide is formed. In another approach, a complex method of coprecipitation has been used as the testing method for analyzing the existed oxides of iron such as iron (II) or iron (III) ions in an aqueous solution. With a stoichiometric solution of the two metal ions such as magnetite (Fe_3O_4), ferrites including $CoFe_2O_4$, $NiFe_2O_4$ can also be prepared by coprecipitation (Majidi et al., 2016).

18.3.3 Chemical vapor condensation

This technique for the preparation of MNPs produces extremely high-quality NPs. For this process, specialized facilities are needed. It is a simple process using a few volatile metal compounds that when heated in an inert gas atmosphere form metal NP as a result of decomposition. The size distribution comparison of iron NPs synthesized from pentacarbonyl iron vaporized in a condensation system with and without a chilling unit showed that rapid cooling allowed for the preparation of smaller iron particles with narrower particle sizes (Iqbal et al., 2017).

18.3.4 Thermal decomposition and reduction

This chemical approach for the synthesis of MNPs focuses on the phenomenon of decomposition in which metal oxy-salts (such as acetates, carbonates, and nitrates) are heated to a certain level so that after their heating the phenomenon of decomposition occurs in which the oxy-salts decompose to form metal oxides. Let us understand this phenomenon with the help of an example where iron (III) nitrate decomposes to form iron (III) oxide in the following equation:

$$4Fe(NO_3)_3(s) \rightarrow 2Fe_2O_3(s) + 12NO_2(g) + 3O_2(g) \tag{18.1}$$

Metal oxide NPs further reduce under a reducing gas to metal by heating them to a particular temperature. The most commonly used gases are hydrogen (H_2) or carbon monoxide (CO) in the following equation:

$$MO + H_2 \rightarrow M + H_2O \tag{18.2}$$

$$MO + CO \rightarrow M + CO_2 \tag{18.3}$$

Except for alkaline and alkaline earth metals, this method of reduction applies to generally most metal oxides.

18.3.5 Liquid phase reduction

In this approach magnetic or nonmagnetic metal oxides are reduced by liquid-phase reduction to magnetic metal or metal alloy with powerful reducing agents such as $NaBH_4$ and $LiAlH_4$. Due to the great solubility of $NaBH_4$ in both water and methanol, it is quite popular in this practice. Metal oxides reduction using $NaBH_4$ is shown in the following equations:

Chemical reaction E°/V vs normal hydrogen electrode (NHE) at 25 °C

$$Fe^{2+} + 2e^- \rightarrow Fe - 0.440$$

$$BH^{-4} + 4OH^- \rightarrow BO^{-2} + 2H_2O + 2H_2 + 4e^- - 1.73$$

$$N_2H_4 + 4OH^- \rightarrow N_2 + 4H_2O + 4e^- - 1.16$$

Most of the hydrides are difficult to handle and are moisture sensitive. This method of liquid phase reduction is more advantageous over other methods.

18.3.6 Polyol method

In this method, certain precursor compounds such as nitrates, acetates, and oxides are diffused or dissolved in a diol, such as diethylene glycol or ethylene glycol (EG), and the reaction mixture is heated between a temperature of 180 °C and 199 °C to reflux. In this reaction, an intermediate is formed after the metal precursors were solubilized in the diol and reduced further to form metal nuclei that further form metal particles. This approach also has been used in the buildup of nanocrystalline powder such as Ni, Cu, Fe, Co, Pd, Ru, Rh, Ag, W, Sn, Re, Pt, Au, (Fe, Cu), (Ni, Cu), (Co, Cu), and (Co, Ni) with the help of various salt precursors. For example, sodium hydroxide and Fe (II) chloride when reacted with PEG or EG, precipitation occurs at a temperature between 80 °C and 100 °C. Polyol being a nonaqueous solvent reduces the issue of hydrolysis of fine metallic particles; a phenomenon very common in the aqueous solution (Majidi et al., 2016).

18.3.7 Sol–gel reactions

Nanostructured metal oxides can be easily prepared with the help of the sol–gel method. This method is established on condensation and hydroxylation of molecular precursors in solution and initiating a “sol "of nanometric particles. Additional steps such as condensation and inorganic polymerization lead to wet gel that is a Network of 3-D metal oxide. These reactions are mostly performed at room temperature but for finding out the final crystalline state extra heat treatments are required. The sol–gel method incorporates condensation and hydrolysis of metal alkoxides. Due to metal alkoxide endurance in hydrolysis, they are considered to be good precursors. When hydrolysis takes place, alkoxide replaces hydroxide group from water, and the generation of free alcohol is evident. There are few factors to be contemplated in this method, such as solvent type, precursor temperature, pH, catalysts, additives, and mechanical agitation. These factors further affect growth, kinetics, hydrolysis, and condensation reactions. Magnetite NPs synthesis using metalloorganic precursors can be developed through this method. Majidi et al. (2016) successfully synthesized NPs, primarily of α-Fe_2O_3, by this method and studied the magnetic characterization.

18.3.8 Sonochemical synthesis

Sonochemical is much less time-consuming than the other preparation techniques. It has been used across the broad generation of materials with unusual properties. There is an ultrasound with physiochemical effects that arises from the acoustic cavitation, due to the generation of the implosive collapse of bubbles in the liquid medium and growth. The effect of the implosive collapse generates a localized hot spot within the gas phase of the bubble collapse, either through shock wave formation or adiabatic compression. The conditions in hotspots present here have been determined to have transient temperatures and pressure, respectively, of 5000 K, and 1800 atm. A new phase was formed due to these extreme conditions and the agglomeration shear effect, which, in turn, are beneficial in the preparation of highly uniform stable NPs. Sonochemically formed Fe_3O_4NPs showed higher saturation, crystallinity, and magnetization than generally formed during the coprecipitation method (Abu-Dief et al., 2016).

18.3.9 Hydrothermal method

The hydrothermal method is also known as the solvothermal method. It is an approach that deals with the preparation of MNPs and ultra-fine powders. Various crystals of different materials can be grown successfully through this. Hydrothermal method accommodates numerous wet-chemical technologies. This wet-chemical technology method can crystallize a material inside a sealed container. At high vapour pressure ranging from 0.3 to 4 Mpa and any aqueous solution from high temperature range of 130°C to 250°C. Several studies have been conducted to identify adequate ligands for the synthesis of monodisperse and high-quality nanocrystals smaller than 10 nm in a hydrophilic environment. The synthesis of such material with a hydrophilic surface still poses a challenge. For example, Fe_3O_4NPs that have a diameter of 27 nm can easily be prepared by using a hydrothermal method in the active participation of sodium bis (2-ethylhexyl) sulfosuccinate, a surfactant. For the preparation of Fe_3O_4 powder that has a diameter of 40 nm, the hydrothermal method is reported to be used for 6 h at 140 °C. This method was stated to have a saturation magnetization of 85.8 emu/g, much lower than bulk Fe_3O_4 (92 emu/g) (Majidi et al., 2016).

18.4 Green synthesis of MNPs

18.4.1 Iron MNPs

Fe_3O_4 (ferric oxide) has unique physiochemical properties and is quite capable of facilitating the cellular and molecular level of biological interactions. Because of these reasons they are being considered as the successful severing agents in the field of MRI in nanomedicine. Fe_3O_4 (ferric oxide) belongs to the class of metal oxides having photocatalytic and photo-oxidizing capacities (Satishkumar et al., 2018).

18.4.1.1 Green synthesis of superparamagnetic Fe_3O_4 MNPs using alpha-D-glucose and gluconic acid

This method of green synthesis of superparamagnetic Fe_3O_4 using alpha-D-glucose and gluconic acid is well accomplished, mild, and renewable. Because of this reason, it shows low toxicity and high biocompatibility. Alpha-D-glucose is used as a reducing agent in this method, while gluconic acid acts as a stabilizer and dispersant. Iron oxides such as (Fe_3O_4 and g-Fe_2O_3) are the most popular, worked upon, and analyzed materials for MNPs in various fields due to their elevated biocompatibility and benevolent magnetic properties. These materials are multitasking due to their ferrimagnetic nature with magnetic transitions above 800 K. Reducing agents can pose as the biggest concern in the preparation process of super-paramagnetic nano-Fe_3O_4. Widely used reductants such as hydrazine, sodium borohydride ($NaBH_4$), dimethylformamide and carbon monoxide (CO) are highly reactive and can cause potential biological risks to the environment. The synthesis of Fe_3O_4 MNPs is done without the use of any additional stabilizer and dispersant in this method. First, 5 mL 0.05 M of glucose solution is added in 5 mL of an aqueous solution containing 0.3M of Fe^{3+} under mechanical stirring for some time at 80 °C. Then the mixed solution is added drop-wise to 50 mL 1M of $NH_3.H_2O$ solution and is kept at a constant pH of 10, under vigorous mechanical stirring (2000 rpm) for 30 min at 60 °C. The suspended material becomes black and is cooled down to room temperature. Then, the precipitated powder is collected from the suspension. After the separation, the precipitated powder is washed three times with doubly distilled water and is washed various times with ethanol too with the help of a magnetic field applied. The supernatant solution was removed thoroughly from the precipitated powder followed by the decantation step. In the final step, magnetic Fe_3O_4NPs were dried in a vacuum oven for 12 h at 60 °C and are kept stored in a stoppered bottle for further applications. By application of controlled partial reduction method of Fe^{3+} to Fe^{2+} using a-D-glucose as the reducing agent with the help of gentle heating in the solution through a redox reaction, superparamagnetic Fe_3O_4NPs can be retrieved. This method of synthesis of Fe_3O_4NPs is facile, simple, and can be executed at a gentle temperature in solution and obtains Fe_3O_4NPs, that can be used as a magnetic carrier and also later as templates that can form polyaniline microspheres, which might have broad applications in drug delivery and removal of waste in biomedical field (Ashraf et al., 2020).

18.4.1.2 Green synthesis of magnetic Fe_3O_4NPs using Couroupita guianensis Aubl. fruit extract

A novel simple method of formulating magnetic CGFe_3O_4NPs was developed in accomplice with active constituents of C. guianensis. This sustainable green establishment has been instilled in the manufacture of magnetic Fe_3O_4NPs with the aid of aqueous fruit extract of edible C. guianensis (CGFE). It is a safe, environmentally friendly route that promotes low-energy consumption and the prevention of environmental contaminations with a sustainable approach. For the synthesis of $CGFe_3O_4NPs$, initially well-scrapped (outer covering removed) and chopped pieces of cannonball fruit were collected and dried off for 8–10 days under a shade. With the help of a commercial electrical stainless-steel kitchen blender, the dried fruit samples were finely grounded into a powdered form, and with the help of the decoction method the extract was prepared. For the maximum extraction, 20 g of fruit powder was autoclaved and mixed in 400 mL of sterile distilled water and are boiled for 20 min at 60 °C. For the elimination of the rubble formed during the process, the mixture was filtered using Whatman no. 1 filter paper. The filtered extract was refrigerated for further analysis. The aftermath of the procedure above concluded that after adding CGFE into the precursor $FeCl_3$ solution it forms $CGFe_3O_4NPs$ and also apprehensions were made that there was a color change in the reaction medium. After the addition, the precursor material $FeCl_3$ showed golden yellow coloration, whereas NPs synthesized exhibits color shift from dark brown to light red. Reduction of $FeCl_3$ using bioactive phytomolecules is a mechanism through which the hydroxyl and amine groups of the bioactive molecules aim to bind with Fe^{3+} and further they form ferric hydroxide in the plant extracts and also there are other bioactive materials too which partially reduce it for the production of Fe_3O_4 particles. Through this ecofriendly green synthesis, production of well-characterized MNPs exhibits good magnetic properties, high stability, crystalline spherical–shaped Fe_3O_4 in co-ordination with active biocompound's functional groups, and biocompatibility which led them for multiple specialized applications (Satishkumar et al., 2018).

18.4.1.3 *Green synthesis of Fe_3O_4 MNPs using watermelon rinds*

Bio-inspired synthesis of Fe_3O_4 MNPs using watermelon rinds is another well-received, novel, nontoxic, and biodegradable approach. In this green synthesis method, watermelon rind extract is used as a solvent and the reduction agent present is coated. In the synthesis of Fe_3O_4 MNPs, some developments such as X-ray diffraction (XRD), transmission electron microscopy (TEM), Fourier transforming infrared (FTIR) spectroscopy, and vibrating sample magnetometer (VSM) techniques showed some breakthrough results. It is a very convenient and rational method for the synthesis of small-sized and highly stable NPs with a 2–20 nm size distribution. For the eco-friendly preparation of Fe_3O_4 MNPs, a common reaction procedure is carried out. 2.26 g $FeCl_3.6H_2O$ and 6.46 g sodium acetate were added to 30 mL of freshly processed watermelon rind powder extract, and the mixture is then vigorously stirred for 3 h at 80 °C. After 3 h, the resultant solution is black and homogenous, displaying the formation of Fe_3O_4 MNPs solution and is left to cool at room temperature. An external magnetic field is applied through the black product for its isolation and is later cleaned with the help of ethanol and is dried in a vacuum for 12 h at 90 °C. After this, it is stored in a stop-cocked bottle for further applications. A solution of ethyl acetoacetate (1, 1.0 mmol), aldehyde (2, 1.2 mmol), and urea (3, 1.2 mmol) was prepared in ethanol (5.0 mL) for the synthesis of compounds.

To this mixture of solutions, 5 mmol Fe_3O_4NPs were added and further this reaction mixture is heated for 10–12 h in regression. Using thin layer chromatography (TLC) the solvents are filtered in a reduced pressure evaporator after the reaction has reached its final potential. With the aid of column chromatography on silica gel, the resulting residue is further purified to assist the final product. Studies such as XRD and FTIR disclosed the high degree crystalline, monophasic nature of FeNPs coated with surface capping agent (mostly proteins). The particles are more in the reduced and stabilized form in the solution, respectively. These MNPs demonstrate amazing magnetic response activity as they exhibit a higher 14.2 emu/g saturation magnetization (Ms). They can further be used as inorganic catalysts for the high yield production of 2-oxo-1,2,3,4-tetrahydro pyrimidine derivatives. It is a proficient prospect to synthesis magnetic iron NPs, as it is an economical and easy-to-execute method that produces zero contamination (Prasad et al., 2016).

18.4.2 Cobalt MNPs

18.4.2.1 Green synthesis of $CoFe_2O_4$NPs using okra plant gel

Using biological agents, such as okra plant gel, cobalt ferrite NPs are synthesized. With the aid of the heating methods, the plant component can be used as a reducing agent. Here, okra plant extract is used as a reduction agent for the synthesis of cobalt ferrite NPs due to the advantageous properties such as eco-friendly nature, cost effectiveness, and simple method of large-scale development. Another important beneficial property to use okra gel comes from its antimicrobial property against the number of pathogenic fungi and bacteria (Kombaiah et al., 2018). The NPs synthesized using this approach are outlined by various methods for analyzing the shape, size, crystallinity, magnetic, optical, and antimicrobial properties. The XRD pattern showed that the NPs synthesized by this process had crystalline single-phase structures having an average size of 45–55 nm. Ferrites are considered to be one of the most efficient magnetic materials. It can be further be subdivided into normal and inverse spinel structures. Cobalt ferrite belonging to the inverse spinel structures has metal ions occupying two sublattices and shows strong magnetostriction. For the preparation of the plant extract, 5 g okra (Abelmoschus esculentus) crop portion was collected and washed thoroughly without using any toxic organic compounds. Afterward, the supernatant was removed, and the plant gel was finely chopped (inner part) and then mixed in 30 mL of deionized water. The mixture was homogenized using a magnetic stirrer for an hour at 32 °C. The solution was then filtered to obtain the extract for further use. The synthesis of $CoFe_2O_4$NPs can be performed through two heating methods: conventional heating and microwave heating. Cobalt nitrate ($(NO_3)_2 \cdot 6H_2O$) and ferric nitrate ($Fe(NO_3)_3 \cdot 9H_2O$) were heated in a 1:2 molar ratio dissolved in double-distilled water. The mixture was stirred thoroughly to obtain a clear solution. At room temperature, the okra (A. esculentus) plant extract was added drop-wise to the reaction mixture with vigorous stirring for several hours once a clear solution was obtained. The solution is then moved to a hot air oven for 3 h at 180 °C for drying and the powder is thoroughly grounded with the aid of pestle and mortar and then sintered in a muffle furnace at 1000 °C for 3 h. The finished product was then rinsed with ethanol and was labeled conventional heating method (CHM). In the microwave method, after the vigorous stirring of the metal nitrates and plant extract, the obtained clear solution was transferred to a silica crucible. The mixture was then irradiated at 850 W for 15 min at a frequency of 2.54 GHz in a domestic microwave oven. Because of this, first, the solution begins to boil resulting in dehydration and later a large number of gases are produced due to decomposition. The powder was grounded using mortar and pestle and washed with ethanol, and finally dried in a hot air oven at 100 °C for 1 h. This final product is labeled microwave heating method (MHM). Within 15 min, the single-phase crystallinity was formed after combustion caused under the microwave irradiation without the calcination treatment. Finally, MHM demonstrated superior optical, structural, magnetic, and antimicrobial activities in comparison to CHM (Kombaiah et al., 2018).

18.4.3 Nickel MNPs

18.4.3.1 Green synthesis of nickel NPs using Ocimum sanctum

Nickel NPs are synthesized using an eco-friendly biological agent such as *O. sanctum* leaf extract. These plant extracts are used because of their bio-safe nature, easy availability, and being great as a reducing and capping agent, unlike dye-mediated chemical route of nickel NP synthesis. The discharge of these harmful dyes and chemicals in the water bodies causes low penetration of sunlight, photosynthesis in water, and also increases the toxicity for the aquatic beings by producing carcinogens posing threat to environmental sustainability. On the other hand, nickel oxide is known as a strong adsorbent for its special chemical and magnetic properties. Diatomite and polyvinyl alcohol/titanium oxide modified with nickel oxide NPs were tailored to adsorb dyes, aromatic compounds, and even heavy metals from aqueous solutions. Nickel oxide NPs-modified diatomite is very a well-known discovered choice for removing basic red 46 dye from aqueous solution. The synthesis of nickel NPs was done by collecting fresh leaves of O. sanctum and was washed thoroughly with both tap water and distilled water and then it was left to shade drying for 3 days at 25(± 1) °C. With the help of a mechanical grinder, the leaves were powdered well and were sieved using a 15-mesh sieve and then the powder was washed with 2% HCl solution and then dried well for further applications. The extract of leaf was produced by dissolving 1 g of leaf powder in 50 mL of distilled water and shaken well for 2 h and the precipitates were filtered out. The filtrate was used for the synthesis of nickel NPs. Nickel NPs are synthesized by mixing aqueous 1 mmol/L $Ni(NO_3)_2$ with 10 mL of O. sanctum leaf extract. After mixing them well the contents were vigorously stirred at 60 °C for 3 h. Thereafter the solution was freeze-dried for about 24 h for obtaining the dried powdered form of NPs. The final powder was stored at 4 °C until further application (Pandian et al., 2015).

18.4.3.2 Green synthesis of nickel NPs using **Desmodium gangeticum** *aqueous root extract*

The synthesis of NPs has increased over time and also its applications. The preparation done chemically can be hazardous to the environment because of which a more safe, nontoxic, eco-friendly alternative biological agent Desmodium gangeticum (DG) aqueous root extract is used for the synthesis of nickel NPs without any stabilizing and reducing agent. Green synthesized NiNPs also show their reduced size and improved monodispersity. The green synthesis also possesses reliable antioxidant and antibacterial activity of synthesized nickel NPs with this method. DG is a plant derived from the Fabaceae family. They are mainly found in India, China, Africa, and Australia. This plant is well known for its great medicinal benefits. This plant is quite effective against various critical disease conditions such as jaundice, rheumatism, puerperal fever, paralysis, edema, filaria, ischemia-reperfusion injury. For preparing the aqueous extract of DG (Linn) roots were collected from the botanical garden of St. Berchmans College, Changanassery, Kerala, India. The roots are collected and are washed thoroughly under tap water and were dried in the shade. Roots were then cut into small pieces and using a pulverizer was ground coarsely. The extraction of Soxhlet is done to obtain the crude aqueous extract and the sample obtained is then vaporized and lyophilized for further applications. Green synthesis of nickel NPs is done by a simple reduction method. Nickel chloride solution (1M) was mixed with crude aqueous DG extract of 1 mg/mL concentration in three different ratios of 1:2, 1:5, and 1:10 in which part 1 was of the precursor ($NiCl_2$) and 2, 5, and 10 parts were of crude aqueous DG extract. Afterward, this mixture was heated at 80 °C and was thoroughly stirred for 45 min approximately until the color turned grey from brown. The prepared colloidal NPs were then dried off, grounded to obtain a fine powder, and were stored for further applications. The NPs synthesized through this method showed to have colloidal stability, magnetic properties, and antioxidant properties (Sudhasreea et al., 2014). All the chemical and green synthesis routes for MNP synthesis are summarized in Fig. 18.1.

18.5 Applications of MNPs

18.5.1 Application of MNPs in pollution sensing

Detection of pollutants and other contaminants from soil and water sample is a primary challenge for a sustainable environment (Jiang et al., 2018). To address these, multiple studies have already taken into account different optically active nanomaterial magnetic iron oxide NPs reported being sensing $\leq$1 ppb of Cd, Cu, Pb, etc. Using the coprecipitation method Au-Fe_3O_4 MNPs were synthesized by Yang et al. (2013) to

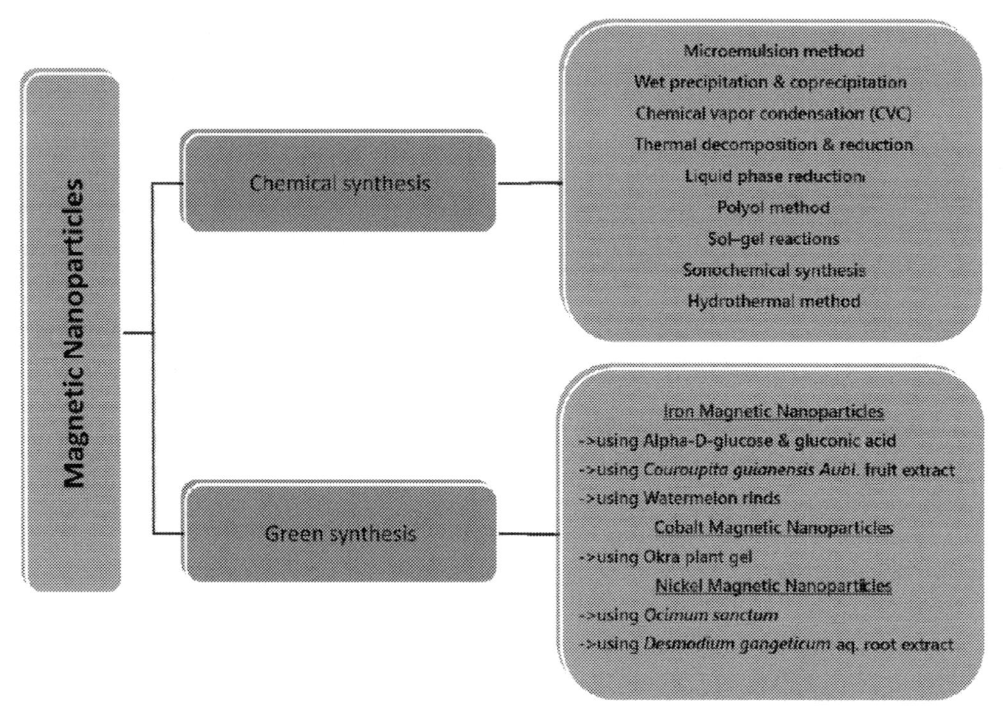

Figure 18.1 Different chemical and green synthesis routes for magnetic nanoparticle synthesis.

very efficiently detect H_2O_2 pollution. Such easy to synthesize nanosensors are very useful to detect Cr^{3+} in a very narrow spectrum of 0.2–0.6 nmol/L (Prakash et al., 2012). Gan et al. (2010) reported a very innovative approach in developing magnetic nanostructure imprinted with carbon electrode to detect dimethoate in cabbage. Juhong et al. (2016) reported the fabrication of a novel magnetic nanoconjugate of magnetic Fe_3O_4@SiO_2/Au complex that was proved to be very sensitive in detecting nitrite ions. The detection of microbial contamination in ultra-small quantity is very important for constant quality assessment from contaminated soil and water sample. For example, rapid detection of *Escherichia coli* at a very low concentration ranging from 10 to 100 CFU/mL was reported by Li et al. (2019). The author showed that positively charged magnetic iron oxide NPs were found to be efficient in such a detection mechanism. Another set of MNPs functionalized whole cell bioreceptors (ADPWH_recA) was reported to detect soil pathogen at the coal cinder site with huge reproducibility (Jia et al., 2016). Although much research has already been reported to ensure high hope with MNPs to be the future trend of MNP-based sensors for the detection of different environmental pollutants, still much exploration in the said field is expected to come up with a better prospect and rapid evaluation.

18.5.2 Application of MNPs in soil and groundwater remediation

Remediation is an approach in which contaminants are removed from contaminated soil, groundwater, and aquifers by using numerous methods so that the toxic load is reduced. Huge agricultural and industrial waste harms soil and groundwater leading to its contamination. There can be different types of contaminations depending on nature, such as chemical, physical, biological, radioactive waste, etc. Multiple technologies have been introduced to extract and clean the affected areas. The increased surface area of nanomaterials has opened the avenue of nanoscale technology toward successful application in environmental pollution detection and remediation. Specifically, MNPs have garnered enough attraction due to adsorption properties much better than conventional sorbent beds. There have been numerous methods used for treating contaminated soil and groundwater. One such method is by using reactive NPs such as Iron, due to its magnetic properties, which leads to the degradation of contaminants to a much lesser harmful product through the process of reduction. Through a direct reaction, degradation of contaminant takes place when interacted with Fe^0 particle or by reacting with Fe^0 corrosion products specifically with Fe^{2+} and hydrogen. These reactions may take place in both aerobic and anaerobic conditions.

$$Fe^0 + 2H_2O \rightarrow Fe^{2+} + H_2 + 2OH$$

$$2Fe^0 + O_2 + 2H_2O \rightarrow 2Fe^{2+} + 4OH$$

The degradation process can be done on harmful organic contaminants such as chlorinated organic solvents (e.g., CCl_4 and C_2Cl_4) reduced to many harmless substances such as CO_2 and H_2O. The treatment of a wide range of substances such as fertilizers (NO_3^-), toxic metals, and pesticides (lindane, DDT) is possible using this methodology. Absorption is a nonreactive approach through which the sequestration of contaminants occurs. In abundant cases, where Fe^0 with metallic cations declines to produce certain insoluble products; in such cases, the removal is through adsorption. Nano-scale Zero-valent iron (nZVI) particles are of great interest recently due to the years of research on them. nZVI particles used in remediation applications are mostly in a size range between tens and hundreds of nanometers acting as environmental colloids. As they are nanodomain, nZVI has a high specific area and with more active sites on the surface which in turn leads to high reactivity due to which a greater range of degradation reactions take place and a faster degradation rate is anticipated for contaminants thereby they produce much lesser harmful byproducts. This type of treatment can be done both *in situ* (nZVI injections directly introduced into the contaminated area) and *ex-situ* (the contaminated material is extracted away to a treatment plant instead of the affected area for the treatment). One backdrop of nZVI treatment comes from its laboratory-based studies failing to mimic natural environments of contaminants which lead to overestimation of performance over longer timescales where contaminant release may occur (Pratt et al., 2014). Wang et al. (2016) designed a novel magnetic chitosan-coated graphene oxide imprint on Pb^{2+} to selectively remove Pb from the aqueous body. Here the sorbent followed Langmuir adsorption isotherm with 90% adsorption and stability even after applying it for five times.

18.5.3 Application of MNPs in removal of biological waste from water

Polluted water is a major health risk for a sustainable environment. Huge industrial, agricultural, pharmaceutical, and other metal, nonmetals, organic nonorganic chemical discharge is a reason to make water safe for drinking, food, and feed development purposes. To make water potable or at least agriculturally suitable, researchers are pursuing multiple efforts. But still, due to hospital sewage, laboratory sewage, and many other factors, pathogenic biomass growth is now an ever-increasing problem. This issue is also being addressed by a group of scientists worldwide. A unique magnetic binary nanocomposites of Ag@Fe_3O_4 (Magnetite) and γ-Fe_2O_3(Maghemite)@Ag in the size of 70 nm and 20–40 nm were reported to be very efficient as antimicrobial and antifungal agent reported against 10 bacterial strains (Prucek et al., 2011). This kind of material is in wide use for medicinal and disinfectant purposes. But such studies give impetus to the researchers to deeply experiment on MNPs for their potential role in eradicating microbial contamination from water and soil sources with the aim of a sustainable environment. Sharma et al. (2015) reported magnetic graphene-carbon nanotube iron composites to be very instrumental in the removal of both Gram-positive and Gram-negative bacterial pathogens (e.g., *E. coli* and *Staphylococcus aureus*) from wastewater. According to Jin et al. (2014), MNPs conjugated with amino acid can efficiently clean water from biological contaminants (Gram-positive and Gram-negative). Coliform bacteria from soil and water can also be removed using magnetic iron oxide NPs in the size range of 6–40 nm diameters (Singh et al., 2011). Rana et al. (2016) prepared Ni-doped ZnONPs using a wet chemical method and studied their biofunctionality which exhibited very good antimicrobial efficiency against Gram-negative (*Shigella dysenteriae*, *Vibrio cholerae*, and *E. coli*) using the agar well diffusion technique. This composite also promises wide scope in the removal of aquatic pathogenic waste. Zhang et al. (2017) also demonstrated the facile synthesis of POHABA (poly-N,N'-[(4,5-dihydroxy-1,2-phenylene)bis(methylene)]bisacrylamide)-coated Fe_3O_4MNP to adsorb wastewater pathogen. The conjugate was found to kill 100% bacterial contamination from water (both Gram-positive and Gram-negative). Chitosan oligosaccharide–functionalized iron oxide NPs were synthesized using a chemical method to remove very deadly waterborne pathogenic Entamoeba histolytica cysts responsible for a very dangerous health condition known as amoebiasis (Shukla et al., 2015). Amine-functionalized magnetic Fe_3O_4–SiO_2–NH_2 nanomaterial was synthesized and experimented on its antimicrobial potential. The analysis was carried out to assess the potential of nanocomposite against a wide spectrum of microbes including *bacteriophage f2*, *Poliovirus* -1, and pathogenic bacteria such as *Pseudomonas aeruginosa*, *S. aureus*, *Salmonella*, *E. coli* O157:H7, and *Bacillus subtilis* (Zhan et al., 2014). The efficiency of this MNP was more than 97% for both viruses and bacteria. The efforts are already in practice to remove pathogenic waste from water with the help of MNPs. Much research is successful in lab-scale and needs industry collaboration to lead toward a large-scale treatment plant.

18.5.4 Dye adsorption through MNPs

Organic compounds such as dyes are often very toxic and possess very high color strength which can impact photosynthesis activity in the marine ultimately leading toward disruption in the microbial activity of organisms that are found in the affected areas. Dyes used in the paper and textile industries are very toxic. Few examples of dyes are methylene blue, carcinogenic Congo red, acid black, and basic Fuchsin which are easily found in the water bodies due to industrial waste dumping. Wang et al. (2015) investigated, for Congo red, the adsorption capabilities of ferrite NPs, and found $CoFe_2O_4$ to be the most proficient adsorbent which displayed the maximum adsorption capacity of 245 mg/g. Electrostatic absorption as the primary removal mechanism was suggested. Congo red from aqueous samples can also be removed successfully by g-Fe_2O_3NPs. Polyacrylic acid–bound iron oxide NPs and ZnO-coated $ZnFe_2O_4$NPs under the UV radiation were found the most effective for the removal of methylene blue. Graphene sheet–supported Fe_3O_4NPs composite showed promising results for the removal of both Congo red and methylene blue dyes. Various cationic dyes such as alkali blue 6B, methylene blue, crystal violet, and basic Fuchsin, a magenta dye are few dyes that can be removed by the use of the modified Fe_3O_4NPs from affected areas (Pratt et al., 2014). Wang et al. (2014) synthesized magnetic iron NPs using green technology where they used tea and eucalyptus leaf extract. These MNPs were found to be effectively removing nitrate contaminants from aqueous solutions. Similarly, Soliemanzadeh et al. (2016) reported pH-dependent removal of phosphorous from wastewater using MNP. All these examples are enough to put big insight into the science of dye adsorption using MNPs leading to reduced pollution levels to the environment (Jun et al., 2020; Khan et al., 2020; Lau et al., 2020).

18.5.5 Application of MNPs for the removal of heavy metal ions from wastewaters

MNPs have shown promising results in the removal of heavy metals from wastewaters. The extraction of heavy metals from water bodies is necessary as contaminants such as heavy metals are highly toxic to the environment. Researchers demonstrated that MNPs being cost-effective, safe, and easy to use posed to be a better option than the other methods such as reverse osmosis, membrane separation, and ion exchange in the removal of toxic pollutants from industrial waters. MNPs with 10 nm as average diameter were synthesized through the coprecipitation method. Now, these iron-based NPs were coated by cationic resin, and then the products (Fe_3O_4-T-RH) were characterized by XRD and TEM. Atomic spectrometry indicated toxic metals such as cadmium, copper, and nickel can be removed efficiently through MNPs. Magnetic separation helped in the removal efficiency which was tested on a column. Results proved fast heavy metal ions adsorption on MNPs and demonstrated a strong possibility for the removal of toxic materials from wastewaters efficiently. Maghemite (γ-Fe_2O_3) NPs have excellent magnetic properties, which make them very attractive for heavy metal removal from wastewater (Yang et al., 2019). Such materials with a surface area of 79.35 m^2/g can remove toxic lead and copper ions from wastewater. In another approach, Madrakrian et al. (2012) reported the synthesis of mercapto-ethylamino monomer-modified maghemite nanocomposite and shown their high removal capacity of lead, cadmium, and many other toxicants from water. Huang et al. (2018) also developed a unique core-shell magnetic nanostructure where amino-decorated Zr-based metal-organic frameworks (Zr-MOFs) used as shell material over $Fe_3O_4@SiO_2$ core material exhibited high adsorption capacity of lead and methylene blue dye to pollutant from water. The aforementioned examples show enormous potential of MNPs as heavy metal adsorbents from water because of their unique magnetic properties, surface modulation, and encapsulation abilities (Khan et al., 2020; Ruthiraan et al., 2019).

18.5.6 Application of MNPs in detection and removal of contaminants from water samples

MNPs demonstrated an efficient approach for easy detection of low concentration pesticides (0.001–0.008 mg/L) in contaminated water bodies. In this approach C_{18}-modified $Fe_3O_4@SiO_2$NPs incorporated in gas-phase chromatography–mass spectrometry cause solid-phase coupled magnetic extraction, which helps in regulating the pesticide residue concentration in contaminated water samples. Fe_3O_4NPs implanted with hydrous aluminum oxide have an affinity toward fluoride ions due to which fluoride concentration could be reduced to 0.3 mg/L from 20 mg/L, meeting the WHO standard quality for drinking water (Sun et al., 2018). Apart from pesticides, pharmaceutical waste is a major contributor to water pollution. A huge amount of layover from the pharmaceutical industry makes water polluted with harsh chemical waste. These water bodies are equally unhealthy for the aquatic ecosystem and human use. In this scenario, researchers have paid much attention to clean with the help of MNPs. For example, gallium acid–coated MNPs are reported to be highly efficient in removing antiinflammatory drug meloxicam (nonsteroid) from the aqueous system through photocatalysis-mediated degradation (Nadim et al., 2015). The excessive load of pharmaceutical and personal care products (PCCPs) is also a huge contributor to the aquatic ecosystem. This challenge has been taken by Salem Attia et al. (2013) by designing zeolite-coated MNP that can eradicate almost 95% PCCP load from water using pH-dependent adsorption strategy experimented with different column and batch studies. A similar study from Hayasi et al. (2017) showed the rapid removal of three pharmaceuticals (ceftriaxone sodium, diclofenac sodium, and atenolol) waste efficiently from water using poly (styrene-2-acrylamido-2-methyl propane sulfonic acid) coated MNPs with high adsorption capacity.

18.5.7 Application of MNPs in sustainable agriculture

Magnetic nanostructures of Fe_3O_4 have huge potential to increase soil fertility by beneficially modulating the carbon, nitrogen, and phosphorous cycle in the soil microenvironment. It was evident from the report came from He et al. (2011), where scientists have shown that efficient control over soil microbes Mesoporous Silica nanoparticles (MSNs) can be employed for increasing soil fertility through modulation of the nutrient cycle. Another way of MSNs in sustainable agriculture relies on a smart detection system for plant health and growth evaluation. Researchers have developed MSNs-based agrochemical delivery system and also utilized for localization studies in plant tissue. Even MSNs are reported to take a pivotal role in identifying and mitigating plant infections. MSNs have also been shown to increase plant growth by working as a fertilizer better than conventional bulk-sized chemical fertilizers. For example, Rui et al. (2016) proved that plant growth and biomass yield in every aspect can be improved by iron-based MSN as fertilizer. The author described that this improvement was due to stimulation by MSN at the hormonal and enzymatic levels both. Such magnetic nanostructures are very useful to detect soil and water contaminant and thus can be utilized in irrigation largely. The effect of MNPs in crop growth and biomass production is also an important parameter of sustainable agriculture. Zia-ur-Rehman et al., 2018 reported the effect of magnetic iron oxide NPs on germination time, rate, and root biomass of wheat crops. Iron deficiency is a major problem in crop growth resulting in multiple diseases. Magnetic iron NPs can supplement the iron deficiency in plants compared to micronutrient fertilizer. Rui et al. (2016) showed that iron oxide NPs can increase iron uptake and translocation in Peanut (Arachis hypogaea).

18.5.8 Application of MNP in crop protection, growth, and yield

Sustainable environmental effect is the cumulative effect of safe and justified use of land and aquatic systems with utmost care to the ecosystem. In this endeavor, long-term strategies to ensure plant growth and yield need to be maintained. There is already a huge impact of growing industrialization and deforestation resulting in the huge loss of the green ecosystem. The global demand for greenery worldwide is of utmost expectation but still, we are lagging. Here in this chapter, we will explain a few strategies to ensure plant growth, protection, and yield using MNPs for a sustained green environment. The presence of carbon-coated MNPs in a plant system under an applied magnetic field is also reported (Tang et al., 2013). This concept can be utilized very much to sense material/pesticide within a plant. Such MNPs can also be utilized as a smart delivery system to improve plant growth by transferring beneficial biomolecules such as hormones, nucleic acids, enzymes, pesticides, or fertilizers. As cobalt-, manganese-, and nickel-based magnetic iron oxide NPs can be utilized for the transfection of desires feature in terms of nucleic acid (Wegmann and Scharr 2018). It is also reported that surface-modified MNPs can enhance gene delivery of the HVJ-E vector. Similarly, the successful transfer of plant gene and development of transgenic cotton plant was carried out using MNP in the presence of a magnetic field namely magnetofection (Zhao et al., 2017). Hao et al. (2013) reported the use of gold-based MNPs to transfer genes to canola protoplasts. MNP-based viral and nonviral gene delivery systems are useful because of easy delivery, target specificity, low-cost stable delivery, and hazardless penetration (Jat et al., 2020). A high load of pesticides in the soil makes soil quality inappropriate for agriculture. Detection of overloaded pesticides, fertilizers and other pollutants is of utmost importance for crop growth and development. In this regard, it is needless to mention about the huge sequestration of organophosphorus-based pesticide in soil. Dzudzevi Cancar et al. (2016) developed polymer conjugated core-shell MNPs for the detection of acetylcholinesterase-based organophosphorus pesticide. Urease-immobilized MNPs were developed for the successful detection of atrazine-based pesticides from the environment (Braham et al., 2013). Similarly, Sun et al. (2018) fabricated cube-like Fe_3O_4@SiO_2@Au@Ag MNPs for pesticide detection using Surface Enhanced Raman Spectroscopy (SERS) technology. Aflatoxin a metabolic byproduct of microbes is a very powerful food contaminant to multiple crop health due to various factors such as inappropriate environmental conditions, poor weather situation, and unfavorable storage conditions. Identification and remediation of such mycotoxin are very much important. In this regard, Urusov et al. (2014) developed MNP-based immunoassay techniques to detect aflatoxin b1. The removal through adsorption of such contaminants is also possible through MNP. For example, magnetic carbon nanoconjugates can remove aflatoxin b1 very effectively (Zahoor et al., 2018). Possible applications of MNPs for sustainable development are schematically represented in Fig. 18.2.

18.6 Conclusion

The field of nanoscience and technology is rapidly spreading over all the sectors of human welfare. Because of extraordinary features compared to bulk counterparts, nanomaterials are popular for broad spectrum applications. This chapter specifically focused on chemical and green synthesis of MNPs along with their applications in development of a sustainable environment. MNPs can be synthesized using a wide range of methodologies,

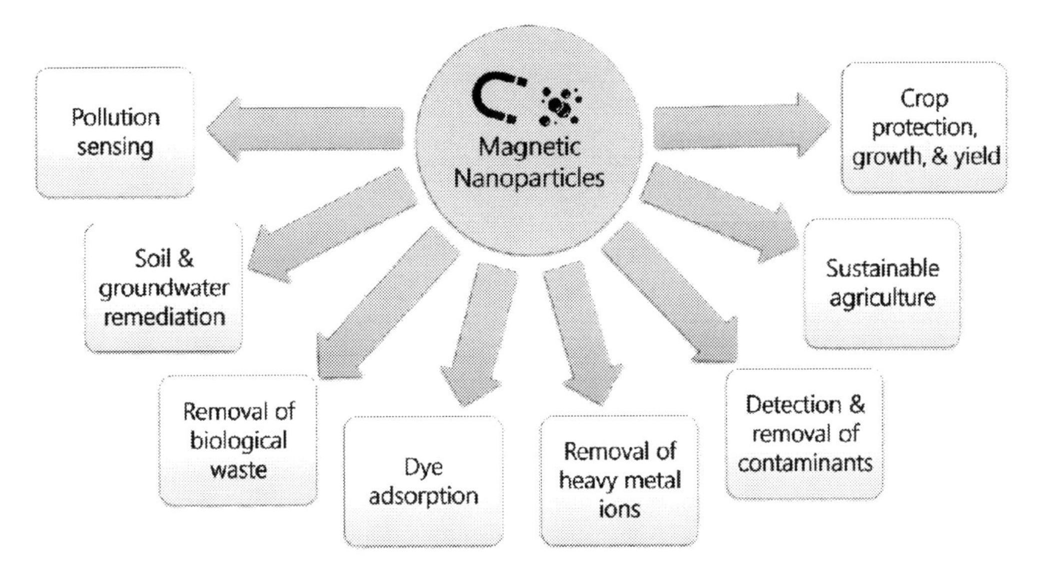

Figure 18.2 Applications of magnetic nanoparticles for sustainable environment.

including ecologically responsible methods, using natural resources as the source, or using nonhazardous reactants that open up their probable induction toward every biomedical and agricultural application. In this chapter, we have outlined the chemical and physical methods to synthesize MNPs. But recently, the synthesis of NPs using natural substances and green methods is a promising area in green nanotechnology. These bio-inspired methods eliminate the need to use hazardous, toxic, and expensive chemical substances and materials. Using biomass and/or cell-free extract of the organisms results in the biosynthesis of MNPs of different shapes and sizes. The sizes and morphologies of the produced NPs depend on the type and concentration of biomolecules produced, varied growth and reaction conditions, precursor solutions utilized, and reaction time. Biosynthesis methods are still under development, and encounter issues such as stability, aggregation, and the regulation of crystal growth, shape, size, and size distribution. The isolation and purification of produced NPs and their separation by size and shape are crucial aspects of bio-based synthesis methods. The properties of NPs can be regulated by modifying the critical reaction parameters that affect the growth status of organisms, their cellular activities, and enzyme processes. The method of plant-mediated synthesis is relatively recent but can alter hazardous chemical synthesis procedures. Tuning the surface energy of farmed nuclei, their growth, and synthesis of desired-sized MNPs are some challenges that remain yet to be solved.

Unique properties such as superparamagnetism, ferromagnetism, etc. make these NPs very attractive for multiple applications. Many of the MNPs are having extraordinary adsorption capacity. This has been utilized for the removal of unwanted waste such as physical, chemical, biological, and pharmaceutical toxicants from soil and water. Similarly, with tunable magnetic efficiency, these MNPs have already been proved to be good detecting material. MNP-based sensors for detection of overloaded pesticide, fertilizer, microbial pathogen, harmful dyes, antibiotic toxins, heavy metal ions are very popular and well-established methodology. For sustainable environment, maintenance of the ecosystem is very important. In this context, huge environmental pollution and faulty industrialization have reduced the greenery in the last five decades. To address this problem, planned agricultural practices, improvement in crop growth and yield, smart pollution detection and management practices are of utmost importance. MNP-based sensors and adsorbents are very advantageous in maintaining the soil and water quality. Our chapter also emphasized much on MNP-based agricultural practices which in turn has opened a new horizon in developing sustainable agriculture and the environment in broader. MNPs can deliver plant genetic material, detect and adsorb heavy fertilizer load, and itself can increase soil fertility by ensuring improved micronutrient availability. When we discuss various applications of MNPs a critical view on the probable toxic nature of MNPs needs to be addressed. Multiple studies have shown that MNPs have toxic effect on organs and tissues. Based on such studies detailed bio-safety analysis is designed for the usage of MNPs in medical sector. Toxicity profiling in terms of its detrimental effect on ecosystem and environmentally beneficial organisms is not significantly studied so far. In this context, it is very important to analyze all these MNPs for their concentration-dependent toxicity studies. All these measures would be very advantageous in the future to make strategies with novel MNPs for their wide spread application in the development and maintenance of a sustainable environment.

References

Abu-Dief, A., Hamdan, SK., 2016. Functionalization of magnetic nano particles: synthesis, characterization and their application in water purification. Am. J. NanoSci. 2 (3), 26–40.

Ali, A., Zafar, H., Zia, M., ul Haq, I., Phull, A.R., Ali, J.S., Hussain, A., 2016. Synthesis, characterization, applications, and challenges of iron oxide nanoparticles. Nanotechnol. Sci. Appl. 9, 49–67. https://doi.org/10.2147/NSA.S99986.

Alvand, M., Shemirani, F., 2016. Fabrication of Fe_3O_4@graphene oxide core-shell nanospheres for ferrofluid-based dispersive solid phase extraction as exemplified for Cd(II) as a model analyte. Microchim. Acta 183 (5), 1749–1757. https://doi.org/10.1007/s00604-016-1805-8.

Ashraf, H., Anjum, T.Microwave, 2020. Assisted green synthesis and characterization of silver nanoparticles using *Melia azedarach* for the management of *Fusarium* wilt in tomato. Front. Microbiol. 11, 238.

Bedanta, S., Barman, A., Kleemann, W., Petracic, O., Seki, T., 2013. Magnetic nanoparticles: a subject for both fundamental research and applications. J. Nanomater. 2013,, 952540. https://doi.org/10.1155/2013/952540.

Belanova, A.A., Gavalas, N., Makarenko, Y.M., Belousova, M.M., Soldatov, A.V., Zolotukhin, P.V., 2018. Physicochemical properties of magnetic nanoparticles: implications for biomedical applications in vitro and in vivo. Oncol. Res. Treat. 41 (3), 139–143. https://doi.org/10.1159/000485020.

Braham, Y., Barhoumi, H., Maaref, A., 2013. Urease capacitive biosensors using functionalized magnetic nanoparticles for atrazine pesticide detection in environmental samples. Anal. Methods 5 (18), 4898–4904. https://doi.org/10.1039/c3ay40579f.

Chung, C., Kim, Y.K., Shin, D., Ryoo, S.R., Hong, B.H., Min, D.H., 2013. Biomedical applications of graphene and graphene oxide. Acc. Chem. Res. 46 (10), 2211–2224. https://doi.org/10.1021/ar300159f.

Devi, V., Selvaraj, M., Selvam, P., Kumar, A.A., Sankar, S., Dinakaran, K., 2017. Preparation and characterization of CNSR functionalized Fe_3O_4 magnetic nanoparticles: an efficient adsorbent for the removal of cadmium ion from water. J. Environ. Chem. Eng. 5 (5), 4539–4546. https://doi.org/10.1016/j.jece.2017.08.036.

Dzudzevic Cancar, H., Soylemez, S., Akpinar, Y., Kesik, M., GÃker, S., Gunbas, G., Toppare, L., 2016. A novel acetylcholinesterase biosensor: core-shell magnetic nanoparticles incorporating a conjugated polymer for the detection of organophosphorus pesticides. ACS Appl. Mater. Interf. 8 (12), 8058–8067. https://doi.org/10.1021/acsami.5b12383.

Gan, N., Yang, X., Xie, D., Wu, Y., Wen, W., 2010. A disposable organophosphorus pesticides enzyme biosensor based on magnetic composite nano-particles modified screen printed carbon electrode. Sensors 10 (1), 625–638. doi:10.3390/s100100625.

Hao, Y., Yang, X., Shi, Y., Song, S., Xing, J., Marowitch, J., Chen, J., 2013. Magnetic gold nanoparticles as a vehicle for fluorescein isothiocyanate and DNA delivery into plant cells. Botany 91 (7), 457–466. https://doi.org/10.1139/cjb-2012-0281.

Hayasi, M., Saadatjo, N., 2017. Preparation of magnetic nanoparticles functionalized with poly (styrene-2-acrylamido-2-methyl propanesulfonic acid) as novel adsorbents for removal of pharmaceuticals from aqueous solutions. Adv. Polym. Tech. 37 (6), 1941–1953.

He, S, Feng, Y, Ren, H, Zhang, Y, Gu, N, Lin, X, 2011. The impact of iron oxide magnetic nanoparticles on the soil bacterial community. J Soils Sediments 11, 1408–1417. doi:10.1007/s11368-011-0415-7.

Huang, L., He, M., Chen, B., Hu, B., 2018. Magnetic Zr-MOFs nanocomposites for rapid removal of heavy metal ions and dyes from water. Chemosphere 199, 435–444. https://doi.org/10.1016/j.chemosphere.2018.02.019.

Iqbal, A., Iqbal, K., Li, B., Gong, D., Qin, W., 2017. Recent advances in iron nanoparticles: preparation, properties, biological and environmental application. J. Nanosci. Nanotechnol. 17 (7), 4386–4409. doi:10.1166/jnn.2017.14196.

Jat, S.K., Bhattacharya, J., Sharma, M.K., 2020. Nanomaterial based gene delivery: a promising method for plant genome engineering. J. of Mater. Chem. B 8 (19), 4165–4175. https://doi.org/10.1039/d0tb00217h.

Jia, J., Li, H., Zong, S., Jiang, B., Li, G., Ejenavi, O., Zhang, D., 2016. Magnet bioreporter device for ecological toxicity assessment on heavy metal contamination of coal cinder sites. Sens. Actuat. B 222, 290–299. https://doi.org/10.1016/j.snb.2015.08.110.

Jiang, B., Lian, L., Xing, Y., Zhang, N., Chen, Y., Lu, P., Zhang, D., 2018. Advances of magnetic nanoparticles in environmental application: environmental remediation and (bio)sensors as case studies. Environ. Sci. Pollution Res. 25 (31), 30863–30879. https://doi.org/10.1007/s11356-018-3095-7.

Jin, Y., Liu, F., Shan, C., Tong, M., Hou, Y., 2014. Efficient bacterial capture with amino acid modified magnetic nanoparticles. Water Res. 50, 124–134. https://doi.org/10.1016/j.watres.2013.11.045.

Jun, L.Y., Karri, R.R., Mubarak, N.M., Yon, L.S., Bing, C.H., Khalid, M., Jagadish, P., Abdullah, E.C., 2020. Modelling of methylene blue adsorption using peroxidase immobilized functionalized Buckypaper/polyvinyl alcohol membrane via ant colony optimization. Environ. Pollut. 259, 113940. https://doi.org/10.1016/j.envpol.2020.113940.

Juhong, C., Shintaro, P., Lili, H., R., N.S, 2016. Highly sensitive and selective detection of nitrite ions using Fe_3O_4@SiO_2/Au magnetic nanoparticles by surface-enhanced Raman spectroscopy. Biosens. Bioelectron. 85, 726–733. https://doi.org/10.1016/j.bios.2016.05.068.

Khan, F.S.A., Mubarak, N.M., Tan, Y.H., Karri, R.R., Khalid, M., Walvekar, R., Abdullah, E.C., Mazari, S.A., Nizamuddin, S., 2020. Magnetic nanoparticles incorporation into different substrates for dyes and heavy metals removal—a review. Environ. Sci. Pollut. Res. 27 (35), 43526–43541. https://doi.org/10.1007/s11356-020-10482-z.

Kombaiah, K., Vijaya, J.J., Kennedy, L.J., Bououdina, M., Ramalingam, R.J., Al-Lohedan, H.A., 2018. Okra extract-assisted green synthesis of $CoFe_2O_4$ nanoparticles and their optical, magnetic, and antimicrobial properties. Mater. Chem. Phys. 204, 410–419. https://doi.org/10.1016/j.matchemphys.2017.10.077.

Kudr, J., Haddad, Y., Richtera, L., Heger, Z., Cernak, M., Adam, V., Zitka, O., 2017. Magnetic nanoparticles: from design and synthesis to real world applications. Nanomaterials 7 (9), 243.

Lau, Y.J., Karri, R.R., Mubarak, N.M., Lau, S.Y., Chua, H.B., Khalid, M., Jagadish, P., Abdullah, E.C., 2020. Removal of dye using peroxidase-immobilized Buckypaper/polyvinyl alcohol membrane in a multi-stage filtration column via RSM and ANFIS. Environ. Sci. Pollut. Res. 27 (32), 40121–40134. https://doi.org/10.1007/s11356-020-10045-2.

Lingamdinne, L.P., Koduru, J.R., Chang, Y.-Y., Karri, R.R., 2018. Process optimization and adsorption modeling of Pb(II) on nickel ferrite-reduced graphene oxide nano-composite. J. Mol. Liq. 250, 202–211. https://doi.org/10.1016/j.molliq.2017.11.174.

Lingamdinne, L.P., Koduru, J.R., Karri, R.R., 2019. A comprehensive review of applications of magnetic graphene oxide based nanocomposites for sustainable water purification. J. Environ. Manage. 231, 622–634. https://doi.org/10.1016/j.jenvman.2018.10.063.

Lingamdinne, L.P., Vemula, K.R., Chang, Y.-Y., Yang, J.-K., Karri, R.R., Koduru, J.R., 2020. Process optimization and modeling of lead removal using iron oxide nanocomposites generated from bio-waste mass. Chemosphere 243, 125257. https://doi.org/10.1016/j.chemosphere.2019.125257.

Li, Z., Ma, J., Ruan, J., Zhuang, X., 2019. Using positively charged magnetic nanoparticles to capture bacteria at ultralow concentration. Nanoscale Res. Lett. 14. https://doi.org/10.1186/s11671-019-3005-z.

Madrakian, T., Afkhami, A., Zolfigol, M.A., Ahmadi, M., Koukabi, N., 2012. Application of modified silica coated magnetite nanoparticles for removal of iodine from water samples. Nano-Micro Lett. 4 (1), 57–63. https://doi.org/10.3786/nml.v4i1.p57-63.

Majidi, S., Sehrig, F.Z., Farkhani, S.M., Goloujeh, M.S., Akbarzadeh, A., 2016. Current methods for synthesis of magnetic nanoparticles. Artific. Cells, Nanomed. Biotechnol. 44 (2), 722–734. https://doi.org/10.3109/21691401.2014.982802.

Nadim, A.H., Al-Ghobashy, M.A., Nebsen, M., Shehata, M.A., 2015. Gallic acid magnetic nanoparticles for photocatalytic degradation of meloxicam: Synthesis, characterization and application to pharmaceutical wastewater treatment. RSC Adv. 5 (127), 104981–104990. https://doi.org/10.1039/c5ra20281g.

Neamtu, M., Nadejde, C., Hodoroaba, V.D., Schneider, R.J., Verestiuc, L., Panne, U., 2018. Functionalized magnetic nanoparticles: synthesis, characterization, catalytic application and assessment of toxicity. Sci. Rep. 8 (1), 6278. https://doi.org/10.1038/s41598-018-24721-4.

Pandian, C.J., Palanivel, R., Dhananasekaran, S., 2015. Green synthesis of nickel nanoparticles using *Ocimum sanctum* and their application in dye and pollutant adsorption. Chin. J. Chem. Eng. 23 (8), 1307–1315. https://doi.org/10.1016/j.cjche.2015.05.012.

Prakash, A., Chandra, S., Bahadur, D., 2012. Structural, magnetic, and textural properties of iron oxide-reduced graphene oxide hybrids and their use for the electrochemical detection of chromium. Carbon 50 (11), 4209–4219. https://doi.org/10.1016/j.carbon.2012.05.002.

Prasad, C., Gangadhara, S., Venkateswarlu, P., 2016. Bio-inspired green synthesis of Fe_3O_4 magnetic nanoparticles using watermelon rinds and their catalytic activity. Appl. NanoSci. (Switzerland) 6 (6), 797–802. https://doi.org/10.1007/s13204-015-0485-8.

Pratt, A., 2014. Environmental applications of magnetic nanoparticles. In: Binns, C. (Ed.), Frontiers of Nanoscience: Nanomagnetism: Fundamentals and Applications (Vol. 6. Elsevier Ltd, United Kingdom, pp. 259–307.

Prucek, R., Tuček, J., Kilianová, M., Panáček, A., Kvítek, L., Filip, J., et al., 2011. The targeted antibacterial and antifungal properties of magnetic nanocomposite of iron oxide and silver nanoparticles. Biomaterials 32 (21), 4704–4713. doi:10.1016/j.biomaterials.2011.03.039.

Rana, S.B., Singh, R.P.P., 2016. Investigation of structural, optical, magnetic properties and antibacterial activity of Ni-doped zinc oxide nanoparticles. J. Mater. Sci. Mater. Electron. 27 (9), 9346–9355. https://doi.org/10.1007/s10854-016-4975-6.

Rui, M., Ma, C., Hao, Y., Guo, J., Rui, Y., Tang, X., Zhu, S., 2016. Iron oxide nanoparticles as a potential iron fertilizer for peanut (*Arachis hypogaea*). Front. Plant Sci. 7 (2016), 815. https://doi.org/10.3389/fpls.2016.00815.

Ruthiraan, M., Mubarak, N.M., Abdullah, E.C., Khalid, M., Nizamuddin, S., Walvekar, R., Karri, R.R., 2019. An overview of magnetic material: preparation and adsorption removal of heavy metals from wastewater. Magnetic Nanostructures. Nanotechnology in the Life Sciences 131–159. https://doi.org/10.1007/978-3-030-16439-3_8.

Sahu, J., Karri, R., Zabed, H., Shams, S., Qi, X., 2019. Current perspectives and future prospects of nano-biotechnology in wastewater treatment. Sep. Purif. Rev. 50 (2), 139–158. doi:10.1080/15422119.2019.1630430.

Salem Attia, T.M., Hu, X.L., Yin, D.Q., 2013. Synthesized magnetic nanoparticles coated zeolite for the adsorption of pharmaceutical compounds from aqueous solution using batch and column studies. Chemosphere 93 (9), 2076–2085. https://doi.org/10.1016/j.chemosphere.2013.07.046.

Sathishkumar, G., Logeshwaran, V., Sarathbabu, S., Jha, P.K., Jeyaraj, M., Rajkuberan, C., Sivaramakrishnan, S., 2018. Green synthesis of magnetic Fe_3O_4 nanoparticles using *Couroupita guianensis* Aubl. fruit extract for their antibacterial and cytotoxicity activities. Artific.Cells Nanomed. Biotechnol. 46 (3), 589–598. https://doi.org/10.1080/21691401.2017.1332635.

Sharma, V.K., McDonald, T.J., Kim, H., Garg, V.K., 2015. Magnetic graphene-carbon nanotube iron nanocomposites as adsorbents and antibacterial agents for water purification. Adv. Colloid Interf. Sci. 225, 229–240. https://doi.org/10.1016/j.cis.2015.10.006.

Shukla, S., Arora, V., Jadaun, A., Kumar, J., Singh, N., Jain, V.K., 2015. Magnetic removal of entamoeba cysts from water using chitosan oligosaccharide-coated iron oxide nanoparticles. Int. J. Nanomed. 10, 4901–4917. https://doi.org/10.2147/IJN.S77675.

Singh, S., Barick, K.C., Bahadur, D., 2011. Surface engineered magnetic nanoparticles for removal of toxic metal ions and bacterial pathogens. J. Hazard. Mater. 192 (3), 1539–1547. https://doi.org/10.1016/j.jhazmat.2011.06.074.

Soliemanzadeh, A., Fekri, M., Bakhtiary, S., Mehrizi, M.H., 2016. Biosynthesis of iron nanoparticles and their application in removing phosphorus from aqueous solutions. Chem. Ecol. 32 (3), 286–300. https://doi.org/10.1080/02757540.2016.1139091.

Sudhasreea, S., Banua, Brindhab, P., Kuriana, GA, 2014. Synthesis of nickel nanoparticles by chemical and green route and their comparison in respect to biological effect and toxicity. Toxicol. Environ. Chem. 96 (5), 743–754.

Sun, M., Zhao, A., Wang, D., Wang, J., Chen, P., Sun, H., 2018. Cube-like Fe_3O_4@SiO_2@Au@Ag magnetic nanoparticles: a highly efficient SERS substrate for pesticide detection. Nanotechnology 29 (16), 165302. https://doi.org/10.1088/1361-6528/aaae42.

Tang, S.C.N., Lo, I.M.C., 2013. Magnetic nanoparticles: essential factors for sustainable environmental applications. Water Res. 47 (8), 2613–2632. doi:10.1016/j.watres.2013.02.039.

Urusov, A.E., Petrakova, A.V., Vozniak, M.V., Zherdev, A.V., Dzantiev, B.B., 2014. Rapid immunoenzyme assay of aflatoxin B1 using magnetic nanoparticles. Sensors (Switzerland) 14 (11), 21843–21857. https://doi.org/10.3390/s141121843.

Vallabani, N.V.S., Singh, S., 2018. Recent advances and future prospects of iron oxide nanoparticles in biomedicine and diagnostics. 3 Biotech 8 (6), 279. https://doi.org/10.1007/s13205-018-1286-z.

Wang, T., Lin, J., Chen, Z., Megharaj, M., Naidu, R., 2014. Green synthesized iron nanoparticles by green tea and eucalyptus leaves extracts used for removal of nitrate in aqueous solution. J. Clean. Prod. 83, 413–419. https://doi.org/10.1016/j.jclepro.2014.07.006.

Wang, Y., Li, L.L., Luo, C., Wang, X., Duan, H., 2016. Removal of Pb^{2+} from water environment using a novel magnetic chitosan/graphene oxide imprinted Pb^{2+}. Int. J. Biol. Macromol. 86, 505–511. https://doi.org/10.1016/j.ijbiomac.2016.01.035.

Wang, Y., Shi, L., Gao, L., Wei, Q., Cui, L., Hu, L., Du, B., 2015. The removal of lead ions from aqueous solution by using magnetic hydroxypropyl chitosan/oxidized multiwalled carbon nanotubes composites. J. Colloid Interf. Sci. 451, 7–14. https://doi.org/10.1016/j.jcis.2015.03.048.

Wegmann, M., Scharr, M., 2018. Synthesis of magnetic iron oxide nanoparticles. In: Deigner, H-P., Kohl, M. (Eds.), Precision Medicine: Tools and Quantitative Approaches. Elsevier Inc, Germany, pp. 145–181.

Williams, H.M., 2017. The application of magnetic nanoparticles in the treatment and monitoring of cancer and infectious diseases. BioSci. Horizons Int. J. Student Res. 10, hzx009. https://doi.org/10.1093/biohorizons/hzx009.

Yang, J., Hou, B., Wang, J., Tian, B., Bi, J., Wang, N., Huang, 2019. Nanomaterials for the removal of heavy metals from wastewater. Nanomaterials 9 (3), 424.

Yang, X., Xiao, F., Lin, W., Wu, F., Chen, u Z, 2013. A novel H_2O_2 biosensor based on Fe_3O_4-Au magnetic nanoparticles coated horseradish peroxidase and graphene sheets-Nafion film modified screen-printed carbon electrode. Electrochim Acta 109, 750–755.

Zahoor, M., Ali Khan, F., 2018. Adsorption of aflatoxin B1 on magnetic carbon nanocomposites prepared from bagasse. Arab. J. Chem. 11 (5), 729–738. https://doi.org/10.1016/j.arabjc.2014.08.025.

Zhan, S., Yang, Y., Shen, Z., Shan, J., Li, Y., Yang, S., Zhu, D., 2014. Efficient removal of pathogenic bacteria and viruses by multifunctional amine-modified magnetic nanoparticles. J. Hazard. Mater. 274, 115–123. https://doi.org/10.1016/j.jhazmat.2014.03.067.

Zhang, Z., Xing, D., Zhao, X., Han, X., 2017. Controllable synthesis Fe_3O_4@POHABA core-shell nanostructure as high-performance recyclable bifunctional magnetic antimicrobial agent. Environ. Sci. Pollut. Res. 24 (23), 19011–19020. doi:10.1007/s11356-017-9535-y.

Zhao, X., Meng, Z., Wang, Y., Chen, W., Sun, C., Cui, B., Cui, H., 2017. Pollen magnetofection for genetic modification with magnetic nanoparticles as gene carriers. Nat. Plants 3 (12), 956–964. https://doi.org/10.1038/s41477-017-0063-z.

Zia-ur-Rehman, M., Naeem, A., Khalid, H., Rizwan, M., Ali, S., Azhar, M., et al., 2018. Responses of Plants to Iron Oxide Nanoparticles. In: Tripathi, D.K., et al. (Eds.), Nanomaterials in Plants, Algae, and Microorganisms: Concepts and Controversies, 1. Elsevier Inc, Pakistan, pp. 221–238.

CHAPTER 19

Future development, prospective, and challenges in the application of green nanocomposites in environmental remediation

Pranta Barua[a], Nazia Hossain[b], MTH Sidddiqui[b], Sabzoi Nizamuddin[b], Shaukat Ali Mazari[c] and Nabisab Mujawar Mubarak[d]

[a]Department of Electronic Materials Engineering, Kwangwoon University, Seoul-01891, South Korea
[b]School of Engineering, RMIT University, Melbourne VIC, Australia
[c]Department of Chemical Engineering, Dawood University of Engineering and Technology, Karachi, Pakistan
[d]Petroleum, and Chemical Engineering, Faculty of Engineering, Universiti Teknologi Brunei, Bandar Seri Begawan, Brunei Darussalam

19.1 Introduction

In the era of advanced technology, synthetic and environment-friendly nanopolymers are being used in different fields of industries (Lett et al., 2019; Li & Wang, 2015). Environmentally sound polymers are sustainable, dissolvable, low carbon emitting materials, and such kinds of polymers are called green nanocomposites (Mohanty et al., 2002; Yu et al., 2007). Scientific efforts are upgraded to make sustainable, eco-friendly, and zero-pollutant nanocomposites that can be dominant in different fields of industries related to the composite material. A recent study shows that the inclusion of nanoparticles (NPs) with base polymer helps to improve the properties that can be utilized in construction automation and medical fields. The properties changed fundamentally are mechanical (e.g., quality, versatile modulus, and dimensional solidness), thermo-mechanical, and penetrability (e.g., gases, water, and hydrocarbons). Others are warm soundness and temperature of warmth contortion, protection from fire and emanation of smoke, retardancy of concoction exercises, the presence of surface, physical weight, and electrical conduction (Njuguna et al., 2008). The nanocomposite is a class of composite that is reinforced to a particular dimension in nanometers. They are much flexible and more sustainable than conventional composites because of the maximum interfacial adhesion.

The required critical factors of these nanocomposite materials are dependent on nanofiller content, nanofiller processing technique, and matrix/filler interfacial interaction or link (Magendran et al., 2019). Green polymeric nanocomposites build a different variety of earnings; however, their proper usage in ultra-tech applications is still formidable. Sustainable natural polymer matrices comprise cellulose, natural rubber,

Sustainable Nanotechnology for Environmental Remediation
DOI: https://doi.org/10.1016/B978-0-12-824547-7.00027-8

Table 19.1 Synthetic and natural polymer matrices.

Synthetic polymer	Structure
Polyhydroxyalkanoate (PHA)	
Natural rubber	
Polycaprolactone (PCL)	
Polylactic acid (PLA)	
Poly (vinyl alcohol)	
Polyethylene glycol	

starch, polyesters, chitin, and such as polyhydroxyalkanoate (PHA). Additionally, a vast area of green synthetic polymers is also commercially harnessed, such as polylactic acid (PLA), polycaprolactone (PCL), polyethylene glycol, poly (vinyl alcohol), and others (Jun et al., 2020; Karri et al., 2018; Lau et al., 2020; Desa et al., 2014). Along with this, environment-friendly bio-based polyamide has been plasticized and made using a green softening agent such as H_2O, oil from soybean seeds, and glycerol (He et al., 2016). Here PLA is a thermoplastic polymer with exclusive mechanical properties. PLA can sustain in microbial/bacterial attacks, and by this attack, the longevity is not declined. In drug delivery, tissue engineering, biomedical engineering, PCL is now used worldwide. Moreover, in packaging and bioimplant, PHA and PCL are being applied (Aimer et al., 2016; Guerra et al., 2018; Hardy et al., 2016; Kalita & Netravali, 2017; Tiwari & Ed, 2012). The chemical configurations of these polymers are provided in Table 19.1.

Recent studies show significant growth in the research of green nanocomposite (Lingamdinne et al., 2019a, 2019b, 2020). In 2015, research on the green combination of silver/polystyrene (PS) nanocomposite and NPs in the antibacterial process had come to a fruitful outcome. The key finding of this project is that the use of AgNPs/PS nanocomposite has a unique property that is antibacterial activity can be utilized for the food packaging process (Awad et al., 2015). In 2019, an experiment was performed so that toxic Hg (II) metal ion can be removed from the aqueous environment. The result of this project finds that curcumin-based magnetic nanocomposite that is green and eco-friendly can be applied thoroughly to remove Hg (II) ion from aqueous solution. This study has added a noteworthy improvement in the aqueous environment, which will help marine lives to exist in a nontoxic environment (Naushad et al., 2019). It is known that silver amalgam is widely used for the decayed dental structure. But this amalgam is losing its popularity due to the adverse effect of mercury such as disposal of mercury as a threat for the environment, the toxicity of mercury, secondary caries, and others. Progressed research in 2018 extended that amalgamation of titanium NPs in an eco-accommodating and manageable approach to be used as fillers which will be in half-breed structure. The physical quality, compressive quality, versatile modulus, and flexural quality and different attributes increment than before, after the polymerization shrinkage starts to decrease as a result of filler volume part increments. Titanium NPs are blended in with bond specialist 3-aminopropyltriethoxysilane (APTES) and strengthened with a grid that is incorporated from natural materials to acquire dental therapeutic nanocomposite material by applying the light-relieving strategy. Annona squamosa or organic product strip concentrate of Citrus limon is utilized as a dissolvable for the union of NPs (Chaughule, 2018). Bio-stable materials in the region of recuperative dentistry have been improved, and antimicrobial attributes have the inborn probability for the treatment of both tooth weakening and tooth fixation.

Nanocellulose (NC) is now being used as a green nanocomposite that has exceptional qualities compared to other polymer composites. Much more modified mechanical properties can be achieved for these materials. The inalienable property of these nanocomposites, for the most part, arranged under the name NC, is legitimized for broad utilitarian nanomaterials (Nin et al., 2012). Nonetheless, it is a bio-based strengthening nanofiller that is accessible in the market and has improved enthusiasm for the course of the most recent 20 years altogether (A. et al., 2010; Dufresne, 2012; J. et al., 2011). The future possibility of green composites extends to the search for reinforcements that are eco-friendly and renewable. Cellulose is the most available and unlimited renewable fiber in nature. From agricultural waste, these nanofibers can be extracted and used for making nanocomposite polymer. Table 19.2 shows extracted natural fiber and their properties based on which green nanocomposite is prepared (Panthapulakkal et al., 2006; Reddy & Yang, 2005, 2007, 2008, 2009a, 2009b; Sain & Panthapulakkal, 2006; Rai et al., 2006).

The use of green nanocomposites from renewable sources over synthetic sources has taken a lot of attraction worldwide. As they are economically sustainable, eco-friendly, and zero-carbon-emitting materials, which will bring a permanent solution *in situ* of plastic usage that is harmful to our environment (A. et al., 2010a; Dufresne, 2012; J. et al., 2011). Green nanocomposite has been widely researched and is now a hot topic because

Table 19.2 Different natural fibers and their properties (Panthapulakkal, Zereshkian, & Sain, 2006; Reddy & Yang, 2005, 2007, 2008, 2009a, 2009b; Sain & Panthapulakkal, 2006; Rai et al., 2006).

Fiber source	Percentage of cellulose content	Tensile strength, MPa	Elastic modulus, GPa	Percentage of elongation
Banana	NA	779	32	2
Coconut husk	NA	220	6	1.5–2.5
Velvet leaf	69	325–500	18–38	1.6–2.6
Jute	NA	400–800	10–30	1.8
Switch grass leaves	61.2	715	31	2.2
Switch grass stem	68.2	351	9.1	6.8
Soybean straw	85	351	12	3.9
Wheat straw	47.5–63.14	58.7–146	3–8	NA
Corn stocks	52	286	16.5	2.2

of different prerequisites in this industrial revolution. Recent technology demands the synthesis of materials from biodegradable polymers, storage of fossil fuels, biodegradability, and reduction of carbon dioxide from the atmosphere. Moreover, agricultural waste and forest residue can also be invested and utilized to form green nanocomposite that is cost effective and available too. All these reasons above described drive to green nanocomposite and their polymers (Rai et al., 2012).

19.2 Global potentiality and opportunity of green nanocomposites

The global demand for green nanocomposite is uplifting because of its exceptional properties. From a recent study, it is found that the rate of growth is around 18.5% that is developed for 2020–2027 in nanocomposite production. The nanocomposite market is dominating the market of materials in developed countries and the Asia Pacific, because of having different manufacturers and industries China is taking the lead for nanocomposite production. So, it is predictable that the recent outcome of green nanopolymer research inspires the developed and developing countries to proceed with their comprehensive technology on green nanocomposite. Another examination guarantees that the polymer nanocomposite market will hit US$ 31.8 billion by 2026. The polymer nanocomposites market development is constrained by strong government guidelines for the utilization of nanocomposites in the bundling and food and drink industry, elevating interest for nanocomposites in end-use businesses, and developing interest expanded mechanical and physical properties. Additionally, expected up-and-coming applications from building and likely divisions, for example, bundling, car and transportation, and food and drink in rising nations likewise underpin the development of the worldwide polymer nanocomposites market. The polymer nanocomposites are additionally used in the car business to bolder the toughness of segments, for example, headlamp covers, motor spread, tires, and inside and outer parts (Adelere & Lateef, 2016a, 2016b; Ali et al., 2019).

In Pakistan, a study on nanocarbon-based nanocomposites shows that it is a green polymer. The types of nanocarbon building blocks are carbon nanotube, graphene, fullerene, and nanodiamond and others. Polymer/nanocarbon nanocomposites have electrical, mechanical, thermal, photovoltaic, and air/water pollution control mechanisms in a very unique and exceptional form (Huang et al., 2014; Sánchez-González et al., 2005a, 2005b; Siddiqui et al., 2018, 2019)[37]. For the treatment of textile dye, green synthesis of silver has experimented as a nanocomposite in India. Aqueous of fresh Tulsi leaves (*Ocimum tenuiflorum*) is extracted, and then as a reducing agent, it is mixed with 95 mL 1 mmol silver to form the composite. This nanocomposite was fixed to show a productive activity and more excellent performance compared to the soil that is acting as an adsorbent to remove reactive turquoise blue dye bound to some reaction conditions.

Additionally, as adsorbent, Ag-nanocomposites improved in obtaining the removal of reactive turquoise blue dye from the effluent solution that is near about 96.8% (Baruwati et al., 2009). The latest study in China shows how green nanocomposite can improve aircraft construction. In green aviation, it is a dominant fact to synthesize biologically derived highly manufactured resins to alternate the conventional epoxies to utilize in structural applications. Rosin acid (Lett et al., 2019; Li & Wang, 2015) and itaconic acid (Ma et al., 2013) are then synthesized, and prepared epoxy resins have also been fabricated to produce green composites. Furthermore, from renewable daidzein, an inherently flame-resistant epoxy resin (diglycidyl ether of daidzein) has been prepared despite having a mixture of any flame-retardant element (Xiao-Su et al., 2018; Zhang et al., 2017). In Saudi Arabia, research is conducted in 2015 on PS, where green silver NPs (AgNPs) are amalgamated, and orange peel extract is utilized to apply as a reducing agent for silver nitrate salt. The significant outcome of this experiment is using a secured AgNPs/PS nanocomposite, which carries an unusual antibacterial activity and has a future usage in the packaging of food. Such kind of green synthesis is the potential future of green nanopolymers/composites (Adelere & Lateef, 2016a; Ali et al., 2019; Ashori, 2008). Research in Brazil reports that multifunctional green nanocomposite is developed based on poly (furfuryl alcohol) resin that is perfluoroalkoxy alkanes (PFA) bio-resin, obtained from sugarcane bagasse fortified with various graphite nanosheets. It has unique and better mechanical and electrical properties than pure PFA resin (Stieven et al., 2018). Another study in Kairo demands that they have developed a silver-polyvinyl alcohol nanocomposite film using Beta-D glucose as a reducing agent and can be used for wound dressing and is considered as a bio-friendly polymer (Badawy, 2014). An experiment is performed by the collaboration of Iraq, Iran, and the United Kingdom in 2019. They found that ZnO/SiO_2 nanocomposite can be synthesized from pomegranate seed extract as a reducing/stabilizing agent and is coated by natural xanthan polymer. It is a green process that is simple and economical, and this nanocomposite occupies the surfactant and polymer flooding in enhancing oil recovery (Ali et al., 2019).

From the research, as mentioned earlier, findings in different developed and developing countries, it is assumed that the potentiality of green nanocomposite is increasing rapidly. Its global demand in automobiles, food packaging, textile, dentistry, fabrics has a significant impact on the environment and economy. In the future world of polymer science, green nanocomposite will dominate than any other classes of polymers.

19.3 Green polymer nanocomposites

The polymer of nanocomposites comprises a polymer or copolymer that has nano-based materials divided into the matrix of the polymer. These can be of a variety of shapes (e.g., platelets, fibers, spheroids), though at best, one dimension should be in the limit of 1–50 nm (Harito et al., 2019). The polymer is viewed as green when it contains emanation nonpartisan and condition inviting properties, for example, boundlessness and degradability. Biodegradation characterizes corruption of a polymer; normally, that suggests an alteration in the concoction structure, declining mechanical and auxiliary attributes, and changing into different materials that apply to the earth (Jamshidian et al., 2010). Different degradable polymers have been used to integrate green composites, and a portion of these are recorded next.

19.3.1 Thermoplastic starch-based composite

A high volume of water or plasticizer such as glycerol, sorbitol is needed to make a film dependent on starch. These kinds of materials having applications in mechanical and thermal energy are referred to as thermoplastic starches (TPSs) (Peelman et al., 2013). TPS shows low properties of the mechanical formation such as weak tensile strength and maximum deformations, which are barriers to applying it in packaging or films. The application of fortified agents in the matrix of starch is a successful attempt to remove these limitations and numerous classes of biodegradable reinforcements, such as cellulosic fibers, whiskers, and nanofibers, have been used to build up more modified and cheap starch biocomposites. The ever uplifting improvement and application of TPS are assumed to take a lead role in reducing the more considerable amount of synthetic plastic wastes globally (S. Ma et al., 2013).

TPS is a compound that is gained through the modification of structure functioning in the starch granule developed with low content of water and the activity of shear force and temperature when plasticizers are present that cannot be evaporated simply during the processing (Bastioli, 2001). Up to the current time, it is entrenched that the strategies, for instance, expulsion, infusion trim, and the film projects, are utilized to measure starch-based materials that are probably going to be utilized for normal polymers. A simple and better method for making sheets or movies by expulsion is the utilization of a twin-screw extruder with a cut, or level film bite the dust. Twin-screw expulsion is the settled strategy and is considered for its straightforward method of taking care of, the delayed season of home, better huge scope shear, and better-loosened up the temperature control (Walenta et al., 2001). Improved consistency and low properties of the stream of starch-based materials show difficulties in infusion shaping, other than the deficiency of reasonable boundaries make it hard to depict the most basic handling conditions (Walenta et al., 2001). The strategy of projecting for the amalgamation of starch films alludes to the readiness of scattering, gelatinization at 95 °C, shaping in acrylic, or teflon plates. The drying time frame is 24 h at around 40–75 °C. Glycerol is a typical plastifier that is being utilized in the planning of starch films. The readied starch film may hold a thickness somewhere in the range of 0.02 and 0.10 mm (A.J.F et al., 2001; Averous & Boquillon, 2004a, 2004b; Bangyekan et al., 2006).

In previous years, many types of research have been conducted based on the change of the starch to produce a better thermoplastic compound. Different types of TPS derived from bio-based polymers dependent on TPS are commercially available with a certain extent of achievement, by various companies such as Mater Bi in Italy, Carghill-Down in the United States, and many more countries such as Spain, Germany, France, Japan, Denmark, and Canada (A.J.F et al., 2001; Averous & Boquillon, 2004a, 2004b; Bangyekan et al., 2006a). An examination was led to the change of starch by photograph actuated cross-connecting. Composite movies have been set up from the fluid scatterings of starch tangled with microcrystalline cellulose utilizing glycerol going about as a plasticizer and illuminated in the scope of bright (UV) light applying sodium benzoate as photograph sensitizer through projecting (Kumar & Singh, 2008). Another experiment outcome first reported that TPS could be used in the synthesis of composites via melt intercalation in a twin-screw extruder. The composites were produced with regular starch derived from corn and remonstrated with glycerin and fortified with hydrated kaolin. The original copy shows vital improvement in the elasticity from 5 to 7.5 MPa for the composite with 50 phr creation of dirt (A.J.F et al., 2001). From reference (Pandey & Singh, 2005), it is found that the addition of glycerol/plasticizer has been sequenced that had a significant impact on change, such as the formation of the nanocomposite. The research was conducted by the varying sequence of plasticizers to observe the thermal, mechanical properties of modified composites using the solution method. Few studies reported that without modification by reinforcing nanofibers with TPS, a sharp rise in tensile strength and improvement in thermal properties must be possible (Guimarães et al., 2010; Kaith et al., 2010; Kaushik et al., 2010; Liu et al., 2010; Lu et al., 2006; X. F. Ma et al., 2007; Sharif et al., 2019; Svagan, 2008; Rai et al., 2006).

19.3.2 PLA nanocomposite

PLA is the rapidly and broadly explored perishable thermoplastic polyester that has the property of renewability. It can replace polymers from resources of fossil fuel. However, specific characteristics such as unconfined degradation, few thermal properties, gas permeability, and extreme brittleness may be an obstacle for the broad-scale usage of PLA. With the recent improvement in advanced nanotechnology, PLA nanocomposites are considered unique and beyond the ordinary compound. These materials have an excellent future for utilization such as medical applications, tissue cultures, and food packaging (Sharif et al., 2019).

Other than low-density polyethylene (LDPE), PLA has a superior bright light obstruction property, yet somewhat subpar contrasted with cellophane, PS, and polyethylene terephthalate (PET). PLA films have mechanical properties contrasted with PET and superior to properties of PS. Lower dissolving and glass progress temperatures of PLA recognize with PS. The glass change temperature of the PLA fluctuates with time. In the event that mugginess is between 10% and 95% and capacity temperatures are 5–40 °C, at that point these conditions would not significantly affect the progress temperature of PLA, which can be depicted by water sorption esteems that are low and the cutoff is between 1000 ppm and Aw=1 (Auras et al., 2004). It decreases carbon discharge by expending itself and has the property of noteworthy vitality sparing mode. The existence cycle for polylactic corrosive polymers is drawn in Fig. 19.1.

Ogata et al. detailed that they arranged nanocomposites by using PLA that is naturally changed that is made by dissolving the polymer in chloroform (in hot temperature) in the habitation of dimethyl distearyl ammonium altered montmorillonite (MMT) (2C18MMT) (Ogata et al., 1997). This modification will help to increase the Modulus of Young of the hybrid nanocomposite. Different studies have come to the conclusion that mechanical, thermal, bending properties are improved in PLA-based nanocomposite than before (Sinha, Okamoto, et al., 2002; Sinha, Yamada, et al., 2002; Sinha et al., 2003, 2003, 2003; Sinha Ray et al., 2002). PLA is applied as the structure with cellulose stubbles misleadingly mixed in with anionic surfactant in different wholes such as 5, 10, and 20 wt% as support. The disturbed materials were shot out in three divisions with pelletizing between the first and the ensuing advances. Conveyed nanocomposites were pressure planned and shown the inflexibility and protracting at break decreased for this composite stood out from PLA (Lee et al., 2008). Thermal debasement, thermal progress, morphological, and mechanical properties of the PLA composites were assessed in Lu et al., (2006). The malleable modulus of the composite improved from 62.5% to 169.5% in spite of the past examinations where any imperative change was recorded. PLA light obstruction properties are determined and have gone to a correlation with the apparent and splendid light transmission paces of standard business PS, poly(ethylene terephthalate), low thickness polyethylene, and cellophane films in the range 190–800 nm (Awad et al., 2015). Table 19.3 shows the profiles for PLA and other standard business films using the CIELAB system that is utilized to depict concealing and choose model shades. Here, three factors are referenced that is softness L^* running from 0 to 100 nm. Variable a^* changes from green (negative) to red (positive) whether or not variable b^* changes from

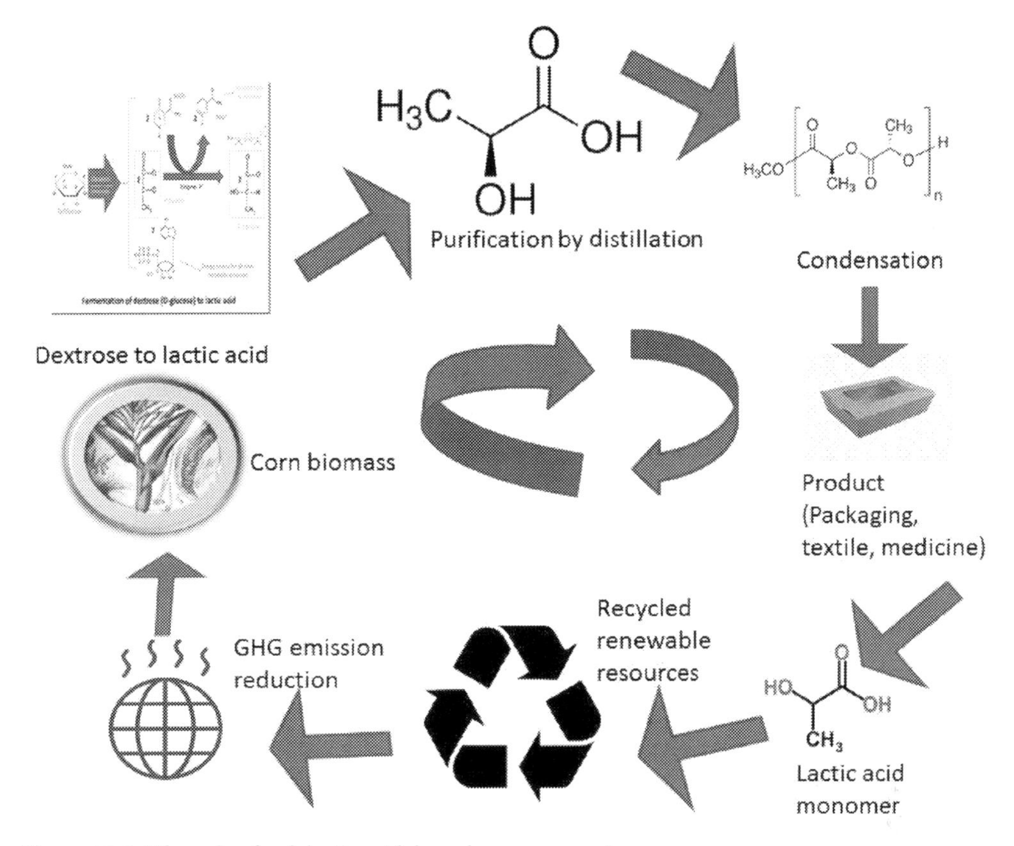

Figure 19.1 Life cycle of polylactic acid–based nanocomposite.

Table 19.3 Values of $L^*a^*b^*$ of PLA (98% L-lactic acid) and commercial sample polymers.

	L^*	a^*	b^*	ΔE^*	YI
PLA	90.64 ± 0.21	-0.99 ± 0.01	-0.50 ±0.04	-	4.67
Cellophane	88.94 ± 0.07	−0.83 ± 0.12	1.45 ± 0.03	2.59	6.30
Polystyrene	89.80 ± 0.67	−0.72 ± 0.02		1.33	4.32
LDPE	89.73 ± 0.09			0.93	4.67
PET	88.10 ± 0.22			2.61	5.71

blue (negative) to yellow (confirmed). They are expressed in various measures, ASTM D 6290-98 and DIN 6176-03 (ASTM-D6290-98e1, n.d.; "Colorimetric Calculation of Color Differences with the DIN 99 Formula," n.d.).

From Table 19.3, the color difference between two points can be calculated using Equations (19.1) and (19.2).

$$\Delta E * ab = \left[(\Delta L*)^2 + (\Delta a*)^2 + (\Delta b*)^2\right]^{\frac{1}{2}} \tag{19.1}$$

Likewise, the yellowness file (YI) can be determined utilizing the tristimulus qualities, X and Y; $YI = 100(CxX \sim CzZ)/Y$, ... Eq. (19.2) where $Cx = 1.2769$ and $Cz = 1.0592$ according to ASTM D 6290-98. The higher L value of PLA depicts a slight increase in lightness. The negative value of a signifies a green symbol in the film. Negative b value mentions more blue color than other sample polymers.

19.3.3 Cellulose-based composite

Cellulose-fiber-reinforced polymer composites are now accessible as they show tremendous improvement in biodegradability, minimal effort, low thickness, nonabrasive, nontoxicity, combustible properties. Numerous studies and experiments have been conducted around the globe on the use of cellulose strands as a strengthening material for the combination of different kinds of composites. Nevertheless, weak interfacial bond, corrupting dissolving point, and water affectability make it hard to use in all segments. Pretreatments of the cellulose fibers can improve the obstacles in the fiber surface; for example, concoction functionalization can boycott the dampness retention procedure and increment the unpleasantness of surface (Kalia et al., 2009).

Cellulose is the most widely recognized kind of living everyday biomass and has an alternate point of view of utilizations in different modern zones (Crawford, 1981). Anselm Payen first referenced the revelation of cellulose in each plant divider in 1838 (Payen, 1838). Cellulose is a long-chain polymer with repeating units of D-glucose, an essential sugar. It is made in an essentially unadulterated structure in cotton fiber. Also, in wood, plant leaves, and stalks, its existence is found with combining various materials, for instance lignin and hemicelluloses. Be that as it may, a few microorganisms are likewise dependable to create cellulose moreover. In 2004, experiments were done to extract cellulose nanofiber from wheat straw and soy hulls resulting in pure cellulose by chemomechanical technique (Alemdar & Sain, 2008a, 2008b). Cellulose nanofibers were extracted from soybean stock also. A process flow of the chemomechanical method is drawn in Fig. 19.2 (Wang & Sain, 2007).

Among different latest studies, in 2017, a study experimented on an increase of cellulose concentration. The 5 wt% sample shows high density, whereas 13 wt% sample improves to reduce the density of cellulose polymer. There is a gradual change in the reinforcing effect while increasing the concentration of cellulose. Such kind of polymer can be used in lightweight structure design or different load-bearing applications (Rajapaksha et al., 2017). Another research shows an inspection of the mechanical and optical properties of cellular-based nanocomposite film. Expansion of toluene diisocyanate as a coupling operator and avicel filaments as strengthening components give films with the most elevated mechanical attributes. Glass change temperature T_g of the considerable number of materials is beneath the room temperature while examined and that the T_g has step by step expanded with cross-connecting and presentation of avicel

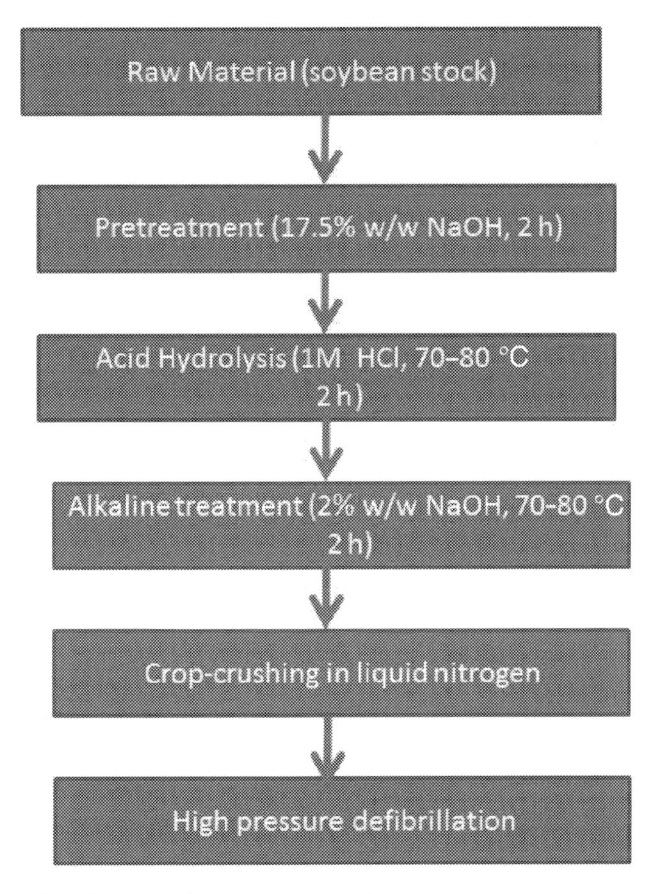

Figure 19.2 Process flow diagram of chemomechanical technique.

(Borges et al., 2001). Many studies (Aouda et al. 2011; Nakayama et al., 2004) announced the first nanocomposite hydrogels was obtained from bacterial cellulose and gelatin. Bacterial cellulose is set up by microorganisms and gives some improvement. Exceptional property, such as high mechanical quality, the high furthest reaches of water digestion, high crystallinity, and a super fine and significantly unadulterated fiber (10–100 nm) orchestrate structure. Another examination revealed a basic, brisk, and low-esteem method for the formation of nanocomposites by direct soaking dry polyacrylamide-methylcellulose (PAM-MC) hydrogels into cellular nanocrystals (CNC) liquid suspensions (Aouda et al. 2011; Nakayama et al., 2004).

19.3.4 Plant oil–based nanocomposites

Vegetable oils, including consumable to drying oils, are a potential wellspring of polymer mix that can be changed to highlight different kinds of functionalities that will incite new materials with a sweeping extent of basic to utilitarian properties. Greasy oil iotas isolated from vegetable oil can be artificially improved through hydrolysis or transesterification or by reacting/modifying unsaturation available in the unsaturated fat chains. A lot of studies, tests are performed on this plant oil–based nanocomposite by changing various properties, synthetic detailing, and others (Mosiewicki & Aranguren, 2013).

Desroches' brief discussion on vegetable oil–inferred bio-polyurethanes (PUs) features a point by point survey of different manufactured courses and included a proficient rundown of business bio-based polyols that can be used in the creation of PUs (Desroches et al., 2012). Another report from Japan created green nanocomposites by a corrosive catalyzed relieving of epoxidized plant oils within sight of organophilic dirt. The composite has extensive flexible properties. Researchers first ensured the biodegradability of this extracted polymer. They expect it can be considered for a unique and new type of biodegradable plastic and new coating materials that can be synthesized from cheap and available renewable resources (Pandey & Singh, 2005). In another study, NC has been separated from oil palm frond leaves utilizing fading, antacid treatment, and acid hydrolysis. The X-ray diffractogram determined the crystallinity index is 70.2% that indicates this NC can be considered as nanofiller that is NC-reinforced nanocomposite for efficient synthesis of butyl butyrate. The yield percentage of butyl butyrate is 76.3% for 4 h (Nursyafiqah et al., 2017). An experiment for surface coating applications is performed using oil palm–based PU/clay nanocomposites. Nanocomposites are blended from PU with earth fill varieties of 3%, 5%, and 8% by weight of the ideal blend of 40 g. The thermal endurance test using thermogravimetric analysis highlighted that the support of dirt fillers in PU exhibited particular and broadened results for better thermal stability in nanocomposite materials than flawless polymers (Satriananda et al., 2018). Other outcomes of different studies have been outlined in Table 19.4. In Table 19.4, some recent findings of polymer nanocomposite derived from plant oil polymers and improvements (if any) are highlighted (Guimarães et al., 2010; Kaith et al., 2010; Kaushik et al., 2010; D. Liu et al., 2010; Lu et al., 2006; X. F. Ma et al., 2007; Sharif et al., 2019; Svagan, 2008; Rai et al., 2006).

The blending of, in any event, two polymers to obtain a polymer that is biodegradable has gotten interested due to its' phenomenal properties (Willett & Shogren, 2002). These polymers have been tried for their degradability and are considered for application in various polymers. Starch/PLA mixes, polybutylenes succinate/cellulose, acetic acid derivation mixes, starch/adjusted polyester mixes, PCL/polyvinyl alcohol mixes, and TPS/polyesteramide mixes have been integrated (Jun et al., 2020; Ke & Sun, 2001; Lau et al., 2020; Uesaka et al., 2000). There are different types of the polymer-polymer blend, and they are as follows (A. et al., 2010b; Adeniyi et al., 2016a; Ali et al., 2019):

1. *Homologous polymer blend*: a mixture of the fraction of two or more polymers.
2. *Miscible polymer mix*: Miscible mixes are like the atomic level. For this kind of combination, Gibbs free vitality of blending is $\Delta G_m \approx \Delta H_m \leq 0$ where ΔH_m is the enthalpy of blending, and this is a result of explicit connections.

Table 19.4 Recent findings of polymer nanocomposite derived from plant oil polymers (Chandrashekhara, Sundararaman, Flanigan, & Kapila, 2005; Z. Liu, Erhan, & Xu, 2005; Lu & Larock, 2007; Merlini, Soldi, & Barra, 2011; Miyagawa, Mohanty, Drzal, & Misra, 2004; Pukanszky, Tudos, Jancar, & Kolarik, 1989; Rai et al., 2006; Wik, Aranguren, & Mosiewicki, 2011).

Polymer matrix	Fiber/particle	Percentage	Other improvement
Epoxidized soybean oil (ESO) (10% replacement of epoxy) + epoxy amine	Glass fiber	63	Lubricity, 21% reduction of the pulling force with 10% epoxy-ESO replacement
Low saturated soy oil	Glass fiber	52	-
Linseed oil (DCDPROM Cross-linked)	Glass fiber	40	
Conjugated soy oil (St-divinylbenzene (DVB) cross-linked)	Wheat straw	75	
Conjugated soy oil (DVB cross-linked)	Corn stover	80	
Acrylated epoxidized soybean oil (AESO) (St cross-linked)	Keratin fibers (from chicken feathers)	30	Low dielectric constant and expansion coefficient
Castor oil (PU)	Banana fiber	15(vol.) untreated 5.10 _ 107 (T) 4.80 _ 106 (T) NaOH treat	
Castor oil (waterborne PU)	Nanocellulose	4	Elongation remainsP300%. Nanocellulose favors phase separation
Conjugated low saturated soy oil (St-DVB cross-linked)	Nanoclay (reactive clay by using a vinyl modifier)	2	Improved barrier properties and thermal degradation behavior

3. *Immiscible polymer blend*: Immiscible polymer blends are phase separated with:

$$\Delta G_m \approx \Delta H_m \geq 0 \tag{19.2}$$

Here ΔG_m is Gibb's free energy, and ΔH_m is the enthalpy heat of mixing.

4. *Compatible polymer blend*: In this sort, a little division of one of the mixing segment is disintegrated in the other part, so the mix shows excellent stage morphology and improved properties. Another examination revealed a survey on thermoplastic/thermoset mixes in a thermoplastic grid (Adeniyi et al., 2016a). Thermoplastics and thermosets construct a profoundly different mix due to their most significant contrasts in extremity and high interfacial pressures. These mixes can assume a job to decrease cost and guarantee economical procedure capacity. Thermoplastics additionally improve the grip, sway quality, printability, paintability of thermosets (Tabaddor et al., 2000).

An experimental study (Ashori, 2008) reported a brief discussion on wood-plastic nanocomposite that is also a popular polymer blend. The term wood plastic composite (WPC) characterizes any composites that bear plant fiber and thermosets or thermoplastics. Similar to different sinewy materials, plant strands are acclimated to strengthen plastics because of relatively high quality and firmness, low cost, less thickness, helpless CO_2 outflow, biodegradability, and inexhaustible. Plant filaments as fillers and fortifications for polymers are as of now the quickest developing kind of polymer added substances. As automakers are intending to make each part either recyclable or biodegradable, there still is, by all accounts, some degree for green composites dependent on biodegradable polymers and plant filaments. From a specialized perspective, these bio-based composites will increment mechanical quality and acoustic execution, decline material weight and fuel utilization, lower creation cost, improve traveler wellbeing and break evidence execution under extraordinary temperature changes, and improve biodegradability for the auto inside parts.

19.4 Production mechanism of green nanocomposite polymer

The conventional and well-known methods used for the production of polymer nanocomposite are solution blending, melt mixing, and *in-situ* polymerization. Other methods are melt intercalation, the direct mixture of polymer and particulates, template synthesis, sol–gel process, hydrothermal synthesis, chemical vapor deposition; microwave synthesis, polymerized complicated method. These methods are applied to the union of the green nanocomposite, likewise (Sahay et al., 2014).

Arrangement mixing is most surely understood as the strategy for divulging various properties to the center material. This technique just alludes to the blending of two sections in an answer followed by electro-turning, of which only a single part arrangement is required to be electro-spinnable. In various points of view, mixing is applied in situations where the ideal material is not fit to orchestrate filaments along these lines, a partner, and the electro-spinnable polymer is utilized. Colloids that are not solvent, for example, carbon nanotubes, silica particles, hydroxyapatite, have been suspended in arrangement and electrospun to orchestrate strands (Lim et al., 2006). Different materials are polymers and solvent salts. Three necessary basic arrangement mixing strategies are chiefly examined consistently; the initial one is the blending materials solvent in a typical dissolvable; second is that there are no solvents exist that can break up both. The third is one material insoluble. Given the one of kind properties of mixing, reconciliation of the three situations might be applied by and large in fiber turning (Nie et al., 2009).

Melt blending is the proposed strategy for surrounding earth/polymer nanocomposites of a thermoplastics and elastomeric polymeric matrix (Nie et al., 2009). Generally, the polymer is disintegrated and gotten together with the ideal proportion of the intercalated earth using Banbury or an extruder. Condense blending acted inside sight of latent gas, for instance, argon, nitrogen, or neon. Then again, the polymer can be mixed in the wake of drying with the intercalant, and thereafter can be warmed in a blender and subject to shear satisfactory to gather the ideal earth polymer nanocomposites. Mellow blending has varying advantages over *in-situ* intercalative polymerization of polymer game plan intercalation. Mollify blending is earth persuading record of the nonattendance of normal solvents (Utracki, 2004).

In-situ polymerization techniques allude to the blending of nanomaterial in a flawless monomer (or numerous monomers) or an answer of monomer, in the event with polymerization within sight of the scattered nanomaterials. Multiple investigations on *in-situ* polymerization techniques guarantee that in the blended nanocomposites, between the framework and nanomaterial covalent linkages happen. Be that as it may, *in-situ* polymerization is applied to accomplish noncovalent nanocomposites dependent on various polymers, for example, phycoerythrin (PE), poly methyl methacrylate (PMMA), and so on. A high level of dispersion of gadolinium (GD) is obtained compared to the melt mixing technique (Fim et al., 2010; Jang et al., 2009; Potts et al., 2011; Sinha Ray & Okamoto, 2003). For *in-situ* polymerizations, some specific conditions should be followed accordingly and they are the use of low viscosity prepolymers (pressure less than 1 Pascal), short duration of polymerization, utilization of polymers with extensive mechanical properties, and no formation of side products that can hamper polymerization synthesis (Advani & Hsiao, 2012).

Sol–gel is a framework for mixing strong materials from little particles. In this substance strategy, the sol (or plan) rapidly propels toward the advancement of a gel-like diphasic structure containing a liquid stage and a significant stage, the morphologies of which limit from discrete particles to perpetual polymer frameworks. Ultrafine and uniform terminated powders can be set up by precipitation. These powders of single and multisegment creation can be conveyed on a nanoscale particle size for dental and biomedical applications (Ugo & Moretto, 2007). Among various blend measures, chemical vapor deposition (CVD) onto temporary metal substrates is a modest, simple, and accessible cycle that is being utilized for graphene. The cycle is included in the carbon submersion of a changing metal upon prologue to a hydrocarbon gas at high temperature. As the substrate helps with cooling the dissolvability of carbon in the change of metal reductions and a dainty film of carbon rush from the surface. Different hydrocarbons such as methane, ethylene, acetylene, and benzene have been rotted on various change metal substrates, for instance, Ni, Cu, Co, Au, and Ru (Coraux et al., 2008; Kwon et al., 2009; Reina et al., 2008; Sutter et al., 2008).

Distinctive ongoing investigations have utilized different techniques to incorporate green nanocomposite effectively. Expulsion, trailed by infusion shaping, has been utilized (Averous & Boquillon, 2004a; Avérous & Le Digabel, 2006). The cycle shows exceptional property: better shedding of strands in the network. Liquefy expulsion is another preparing technique that is likewise utilized. In this technique, the expulsion speed assumes an urgent job in warm conduct, spiral extension proportion, polymer froth compressibility. Another is film stacking that packs a polymer film and fiber for quite a while.

Moreover, other researchers used different moldings such as compression molding, injection molding, direct molding, solution casting after gelatinization, and other steps. Molding improves the tensile strength of composites. In *in-situ* intercalative solution, polymerization enhances the conversion of nanofillers in monomer(s), followed by bulk or solution polymerization. The nanofillers are constantly extemporized by practical gatherings to ascend the association between the polymers and the nanofillers or to get a decent scattering in the polymer framework. It is accounted for that, numerous points of interest are engaged with this technique, for example, straightforward dealing with and better execution of the completed items (Guan & Hanna, 2006; Huskić & Žigon, 2007; Lei et al., 2007; D. Liu et al., 2010; Majdzadeh-Ardakani & Sadeghi-Ardakani, 2010; Wei et al., 2010; Zabihzadeh, 2010; Zou et al., 2008).

19.5 Functional properties of green nanocomposite

Green nanocomposites exhibit improved functional behaviors that distinguish the difference between standard composite and green nanocomposite (Lingamdinne et al., 2019b). Useful behaviors are thermal insulation, vibration, damping, biodegradation. Thermal insulation is the most significant characteristic of green nanocomposite. A recent study shows that polyvinyl chloride (PVC)/SiO_2 nanocomposite exhibits better thermal insulation quality. In the power cable connection, PVC is used as insulating material. But some drawbacks are the obstacles for this extensive thermal insulation property. Under overload or short circuit conditions, PVC-insulated cables are carried out to thermal aging. This stress leads to the rapid declining effect of primary product characteristics. The degradation brings an irrecoverable change that leads to reduce their lifetime. But after using silica with PVC, this weak point can be overcome. The addition of SiO_2 NPs to PVC plays a lead role in the improvement in its dielectric properties when the warm maturing process if we contrast with the unadulterated PVC (M. et al., 2019).

PU foams are commercially chosen available insulation material. It has excellent thermal protecting properties, low dampness-fume porousness, and high protection from water assimilation, generally high mechanical quality, and small thickness. Now, this foam is synthesized from bio-based materials or bio-waste such as walnut shells, cellulose from agricultural/forest residue, diatomite, and others. This experiment finds this green nanocomposite can be best suited for thermal insulation (Ferdinando et al., 2020).

Another insulation material is synthesized by the assembly of graphene oxide (GO) (Lingamdinne et al., 2019a) and polyimide. This nanocomposite shows lightweight, thermally superinsulation, and flexible. The thermal conductivity is much low as for 0.012 W/mK in the radial direction. It is highly efficient to use as a thermal insulator in aerospace, wearable device, and building construction (Yuyang et al., 2019). Sprayed PU is a more improved insulating material that is recently fabricated. Different studies found out in high voltage outdoor applications; nanocomposites can be applied thoroughly as a thermal insulator than conventional insulating material. For that fact, how this nanocomposite should be synthesized is highlighted in detail by different conduction experiments on particular properties such as functionalized nanofillers, degree of dispersion, AC dielectric breakdown, and DC dielectric breakdown as a function of environmental factors, more aging of these nano and hybrid composites. The studies suggest that such kind of nanocomposite should be fabricated industrially (Becker et al., 2005; Casciola & Pic, 2005; Isaias et al., 2009; Mark, 1996; Purkansky, 2005; Mi et al., 2013).

19.6 Application of green nanocomposite in the engineering field

The polymer-based materials have been integrated with hybrid nanomaterials that have exclusive properties; surface characteristics such as high thermal conductivity has been utilized in the engineering field for the development of advanced nanoelectronics or photonics (S. Y. Lee et al., 2008). Recently, as conductive-based polymers are also widely considered for practical solutions for energy storage applications that are linked with advanced areas including fuel cells, batteries, development of electrochemical supercapacitors or ultracapacitors (Chen et al., 2016). However, specific polymers that have higher conductivity values are being utilized for the energy-related applications include polythiophene, polyaniline, and polypyrrole-based composites. It is of utmost importance to consider features along with the thermal conductivity, including crystallinity, the orientation of polymer chains, chain structure, and direction of ordered domains that are significant for their applications (Thomassin et al., 2013). One illustration is the polyamide-based GO nanocomposites that not only offer lightweight but also provide higher thermal conductivity that can be utilized in aerospace, constructional engineering, and the development of smart devices. One exciting field that is emerged with the development and rapid growth of smart electronics is electromagnetic interferences (EMIs) that can be represented by the electromagnetic pollution being created by electrical operations consuming energy for gigahertz from electronic devices and advanced telecommunication system that never existed before that requires solutions for EMI shielding composites and polymer-based carbon nanocomposites that have been considered as efficient shield materials (Sánchez-González et al., 2005a). For this application, carbon black on integration with polymer-based materials facilitates characteristics including higher modulus, increased hardness, increased electrical conductivity along with higher viscosity, and, therefore, can be sufficient in reducing electromagnetic pollution (Siddiqui et al., 2019). They are, however, fabricating magnetic-based conductive composites such as magnetic carbon nanocomposites claimed to have increased the efficiency for concerned application attributed by achieving higher stable structures, enhancing conductivity as well as linking with higher saturation magnetization and coercivity that can enhance overall shielding effectiveness (Huang et al., 2014; Sánchez-González et al., 2005a, 2005b; Siddiqui et al., 2018a, 2019).

On the other hand, magnetoelectric/polymer nanocomposite formulated a new type of advanced material using a combination of zinc, iron, cobalt/polyvinylidene fluoride trifluroethylene that exhibited extraordinary piezoelectric composite that shows performance at room temperature (Guo et al., 2015; Peyghambarian et al., 2004). Recently, multiwalled carbon nanotubes/polymer hybrid nanocomposite showed incredible enhancement for dielectric features by providing an effective conductive network (Huang et al., 2014). Also, bistriarylamine-based composite claimed to be applied for the field of photorefractive devices in the field of optical development (Tugrul Seyhan et al., 2008). Considering the engineering advanced field, there are outstanding applications for polymer-based materials being carried out that provided convenient fabrication of the materials covering areas such as catalysis, solar cell, development of unique characteristics materials, water purification, and advanced optics (J. J. Huang et al., 2014; Sánchez-González et al., 2005a, 2005b; Siddiqui et al., 2018a, 2019).

19.7 Prospective strategies for improvement in green nanocomposite

Among different nanocomposites, nanofibers are now prevalent in the industrial revolution. They are now used in packaging applications. For industrial production, NC powder is preferred. To improve the quality of NC, two modifications are applied, and they are physical modification, which implies coating of the CNC surfaces with surfactants having polar heads and long hydrophobic tails and another one is a chemical modification that is grafting of the cellular surface. In the physical change, by using surfactants, stable suspension can be obtained. A well-efficient nanocellular composite can be obtained using the dispersion of CNCs within polymer matrices. In chemical modification, grafting is the best solution to achieve the goal. The help of the "grafting can obtain grafting of polymer matrices on the surface of NC onto" or "grafting from" approaches (Arrieta et al., 2016; Habibi, 2014; Wu, 2009; Zini & Scandola, 2011).

Nanobiotechnology is the latest concept applied to transform metals into nano-sizes. Different variables such as the strategy used for union, pH, temperature, pressure, time, molecule size, pore size, condition, and nearness impact a lot of the properties of the orchestrated NPs and their determinations and applications. Besides, the portrayal of the combined NPs is required for their possible use in different medication conveyance and biomedical applications. The creation of NPs utilizing traditional strategies brings about poisonous and destructive natural risks (Arrieta et al., 2016; Mohanty et al., 2002; Thakur et al., 2013). In recent years, different eco-friendly syntheses of NP production are introduced that do not produce toxic waste or hazards. For the most part, three fundamental advances are utilized for the blend of NPs by methods for an organic framework: the decision of dissolvable medium used, the conclusion of an eco-friendly and earth benevolent diminishing operator, and the resolution of a nontoxic material as a topping specialist to balance out the integrated NPs. Like NP, size can be controlled by altering the pH of solution media. Temperature also plays a pivotal role in the biosynthesis of NPs. In the biochemical process, the temperature can be limited to less than 350 °C, whereas in conventional methods, the temperature needs to be higher than 350 °C. Moreover, pressure, time, particle size, and other physical parameters can be redesigned using nanobiotechnology (Mohanty et al., 2002).

Another study focuses on the improvement of the mechanical properties of the nanocomposite. They reported that to accomplish this complex task, a different way of thinking than a traditional one is needed. They proposed classification of different models on the improvement of mechanical properties. An extraordinary property of nanocomposite can be mentioned as lifting in the particular surface territory (ST), along with the development of atomic associations at the nanoscale. Improvement of lattice sturdiness the various powerless network controlled properties of composite covers is the specific objective of nanoalteration (Daniel & Astruc, 2004).

An ongoing report on green union of NPs exhibited that green blend is the watchword for the mix of NPs by plants or their metabolites. This inventive adjustment is astoundingly redressing and decreases the harmful quality raised by the expectedly coordinated NPs. Green techniques for combination are common as their capability to diminish the poisonousness of NPs. In addition, the utilization of nutrients, amino acids, plant extricates assumes a lead job these days (Xiangqian et al., 2011). NPs can be set up from agro-waste such as Cocos nucifera coir, corn cob, common item seeds and strips, wheat and rice grain, palm oil, etc. These blends are improved with biomolecules such as flavonoids, phenolic, and proteins that can be applied as a reductive administrator for the association of NPs (Kharissova et al., 2013).

The green blend of Ag and palladium nanocircles, nanowires, and nanoposts by using supplement B2 (as reducing and garnish administrators) has been highlighted additionally. The supplement B2 is being utilized as the diminishing master for the mixture of the nanowires and nanorods. This is a novel philosophy in the field of green nanotechnology that propels the usage of standard administrators in the movement of this field, similar to their effect on different tumor cells ("Factors Affecting the Geometry of Silver Nanoparticles Synthesis in Chrysosporium Tropicum and Fusarium Oxysporum," 2011). A scientist combined AgNPs stable in dull spot for 8 months without using cell culture supernatants of psychrophilic microorganisms Pseudomonas antarctica, Pseudomonas proteolytica, Pseudomonas meridiana, Arthrobacter kerguelensis, Arthrobacter gangotriensis, Bacillus indicus, and Bacillus cecembensis (Rai et al., n.d.). There are more ongoing studies that are trying to modify green nanocomposite syntheses such as metal synthesis action from the green combination, improvement of chemical property of plant and phytochemicals, algae, yeast, fungi, and others that are bio-based reformation to improve the efficiency of synthesis (Adelere & Lateef, 2016a; Baruwati et al., 2009; Quaresimin et al., 2012; Rai et al., 2006).

19.8 Environmental impact of green nanocomposite

Recently cellulose derived from plants for producing green composites is instantly utilized in vast applications because of the economic, renewable, and eco-friendly nature, considering the organic waste materials from plants that can be suitable for clean energy processing (Nadagouda & Varma, 2006). It is also a fact that natural fibers utilized to be converted into green composites have been better than synthetic fibers due to their exclusive benefits due to having lower cost and density and higher abundance and mechanical properties such as modulus (Shivaji et al., 2011). History was made in 2003 when the first-ever green composite emerged as a commercial product in the form of a spare tire composed of PLA-based bioplastic-reinforced kenaf fibers (Ahmed et al., 2018). Recently, composite produced using renewable polylactide and green coconut fiber showed fabrication of copolymer material having excellent thermal properties, and the biodegradation rate was significantly increased (Bakir et al., 2018). Rastogi et al. (2017) fabricated pine needle–based nanocomposites using fiber that has shown better mechanical and physiochemical properties revealing better interfacial compatibility as that made of synthetic composites.

Furthermore, thermosetting green composites such as natural oils are cost-effective and renewable sources composed of triglycerides of fatty acids. They can polymerize using chemical upgrade using the allylic carbons' double bonds using reaction with haloacids to functionalize into esters (Ahmed et al., 2018). Therefore, using biodegradable polymers and bio-based matrix copolymers or renewable resources such as biomass-based derivatives is a solution that offers sustainable, eco-efficient advantages to secure productions that are presently controlled entirely on petroleum-derived unsustainable products in the market (Bakir et al., 2018; Kou & Varma, 2012; Rastogi et al., 2017). Considering all the edges with the renewable sources and using them as alternate resources using synthetic polymers produced using petroleum products is a win-win approach for reducing pollution, reducing cost, and offering materials that are widely available naturally can become an essential program for future synthesis bearing in mind the environmental aspects by releasing green gases and global warming.

19.9 Conclusions

The green nanocomposite is now playing a significant role in the green synthesis of materials. In the near future, this type of composite will dominate other obvious material. Because of the industrial revolution, as green house gas (GHG) emission is uplifting day by day, such kinds of materials should have to be chosen that will play a vital role in carbon emission reduction. The review summarizes the crucial facts that show a sustainable future of green nanocomposite. To accomplish the green synthesis of materials from the engineering perspective, these green nanocomposite materials will be in the leading position, among other elements. If scientists and researchers pay great attention to green nanocomposite and financial investment is adequately supported by industrialists, then green nanocomposite will be the future of upcoming decades.

References

A., B., A., B.L., T., P., H., Y., K., A., M., N., N., N.A., A., M., J., S., S., B.A, Eichhorn, S.J., Dufresne, A., Aranguren, M., Marcovich, N.E., Capadona, J.R., Rowan, S.J., Weder, C., Thielemans, W., Roman, M., Renneckar, S., Gindl, W., Veigel, S., Keckes, J., Yano, H., Abe, K., Nogi, M., Nakagaito, A.N., Mangalam, A., Simonsen, J., Benight, A.S., Bismarck, A., Berglund, L.A., Peijs, T., 2010. Review: current international research into cellulose nanofibres and nanocomposites. J. Mater. Sci. 1–33. https://doi.org/10.1007/s10853-009-3874-0.

A.J.F, de Carvalho, A.A.S, Curvelo, J.A.M, Agnelli, 2001. A first insight on composites of thermoplastic starch and kaolin. Carbohydr. Polym. 189–194. https://doi.org/10.1016/s0144-8617(00)00315-5.

Adelere, I.A., Lateef, A., 2016. A novel approach to the green synthesis of metallic nanoparticles: the use of agro-wastes, enzymes, and pigments. Nanotechnol. Rev. 5 (6), 567–587. https://doi.org/10.1515/ntrev-2016-0024.

Adeniyi, A., Agboola, O., Sadiku, E.R., Durowoju, M.O., Olubambi, P.A., Reddy, Babul, A., Ibrahim, I., D., Kupolati, W.K, 2016. Thermoplastic-thermoset nanostructured polymer blends. In: Thomas, S., Shanks, R., Chandrasekharakurup, S. (Eds.), Design and Applications of Nanostructured Polymer Blends and Nanocomposite Systems. Elsevier Inc, Amsterdam, pp. 15–38.

Advani, S.G., Hsiao, K.T., 2012. Manufacturing techniques for polymer matrix composites (PMCs). Manufacturing Techniques for Polymer Matrix Composites (PMCs). Elsevier Inc, Amsterdam, pp. 491–497.

Ahmed, A.A., Hamzah, H., Maaroof, M., 2018. Analyzing formation of silver nanoparticles from the filamentous fungus *Fusarium oxysporum* and their antimicrobial activity. Turkish J. of Biol. 42 (1), 54–62. https://doi.org/10.3906/biy-1710-2.

Aimer, M., Klemm, E., Langanke, B., Gehrke, H., Stubenrauch, C., 2016. Reactive extraction of lactic acid by using tri-n-octylamine: structure of the ionic phase. Chem. Eur. J. 22 (10), 3268–3272. https://doi.org/10.1002/chem.201503799.

Alemdar, A., Sain, M., 2008a. Biocomposites from wheat straw nanofibers: morphology, thermal and mechanical properties. Compos. Sci. Technol. 68 (2), 557–565. https://doi.org/10.1016/j.compscitech.2007.05.044.

Alemdar, A., Sain, M., 2008b. Isolation and characterization of nanofibers from agricultural residues—wheat straw and soy hulls. Bioresour. Technol. 99 (6), 1664–1671. https://doi.org/10.1016/j.biortech.2007.04.029.

Ali, J.A., Kolo, K., Mohammed Sajadi, S., Manshad, A.K., Stephen, K.D., 2019. Green synthesis of ZnO/SiO_2 nanocomposite from pomegranate seed extract: coating by natural xanthan polymer and its characterisations. Micro Nano Lett. 14 (6), 638–641. https://doi.org/10.1049/mnl.2018.5617.

Aouada, F.A., de Moura, M.R., Orts, W.J., Mattoso, L.H.C., 2011. Preparation and characterization of novel micro- and nanocomposite hydrogels containing cellulosic fibrils. J. Agric. Food Chem. 59 (17), 9433–9442.

Arrieta, M.P., Fortunati, E., Burgos, N., Peltzer, M.A., López, J., Peponi, L., 2016. Nanocellulose-based polymeric blends for food packaging applications. In: Puglia, D., Fortunati, E., Kenny, J.M. (Eds.), Multifunctional Polymeric Nanocomposites Based on Cellulosic Reinforcements. Elsevier Inc, pp. 205–252.

Ashori, A., 2008. Wood-plastic composites as promising green-composites for automotive industries!. Bioresour. Technol. 99 (11), 4661–4667. https://doi.org/10.1016/j.biortech.2007.09.043.

ASTM-D6290-98e1, 2021. Standard Test Method for Color Determination of Plastic Pellets 8.03, pp. 892–896.

Auras, R., Harte, B., Selke, S., 2004. An overview of polylactides as packaging materials. Macromol. Biosci. 4 (9), 835–864. https://doi.org/10.1002/mabi.200400043.

Averous, L., Boquillon, N., 2004. Biocomposites based on plasticized starch: thermal and mechanical behaviours. Carbohydr. Polym. 56 (2), 111–122. https://doi.org/10.1016/j.carbpol.2003.11.015.

Avérous, L., Le Digabel, F., 2006. Properties of biocomposites based on lignocellulosic fillers. Carbohydr. Polym. 66 (4), 480–493. https://doi.org/10.1016/j.carbpol.2006.04.004.

Awad, M.A., Mekhamer, W.K., Merghani, N.M., Hendi, A.A., Ortashi, K.M.O., Al-Abbas, F., Eisa, N.E., 2015. Green synthesis, characterization, and antibacterial activity of silver/polystyrene nanocomposite. J. of Nanomater. 2015, 943821. https://doi.org/10.1155/2015/943821.

Badawy, S.M., 2014. Green synthesis and characterisations of antibacterial silver-polyvinyl alcohol nanocomposite films for wound dressing. Green Process. Synth. 3 (3), 229–234. https://doi.org/10.1515/gps-2014-0022.

Bakir, E.M., Younis, N.S., Mohamed, M.E., El Semary, N.A., 2018. Cyanobacteria as nanogold factories: chemical and anti-myocardial infarction properties of gold nanoparticles synthesized by *Lyngbya majuscula*. Marine Drugs 16 (6), 217. https://doi.org/10.3390/md16060217.

Bangyekan, C., Aht-Ong, D., Srikulkit, K., 2006. Preparation and properties evaluation of chitosan-coated cassava starch films. Carbohydr. Polym. 63 (1), 61–71. https://doi.org/10.1016/j.carbpol.2005.07.032.

Baruwati, B., Polshettiwar, V., Varma, R.S., 2009. Glutathione promoted expeditious green synthesis of silver nanoparticles in water using microwaves. Green Chem. 11 (7), 926–993. https://doi.org/10.1039/b902184a.

Bastioli, C., 2001. Global status of the production of biobased packaging materials. Starch/Staerke 53 (8), 351–355. https://doi.org/10.1002/1521-379X(200108)53:8<351::AID-STAR351>3.0.CO;2-R.

Becker, O., Simon, G.P., Dusek, K., 2005. Inorganic Polymeric Nanocomposites and Membranes. Springer-Verlag, Berlin.

Borges, J.P., Godinho, M.H., Martins, A.F., Trindade, A.C., Belgacem, M.N., 2001. Cellulose-based composite films. Mech. Compos. Mater. 37 (3), 257–264. https://doi.org/10.1023/A:1010650803273.

Casciola, M., Pic, M., 2005. Nafion–Zirconium Phosphate Nanocomposite Membranes with High Filler Loadings: Conductivity and Mechanical Properties. Macromol. Symp.

Chandrashekhara, K., Sundararaman, S., Flanigan, V., Kapila, S., 2005. Affordable composites using renewable materials. Mater. Sci. Eng. A 412 (1–2), 2–6. https://doi.org/10.1016/j.msea.2005.08.066.

Chaughule, R., 2018. Green and rapid synthesis of size controlled TiO2 nanoparticles used as fillers in light curing dental nanocomposite resins. J. of Nanomed. & Nanotechnol. 9. https://doi.org/10.4172/2157-7439-C3-073.

Chen, H., Ginzburg, V.V., Yang, J., Yang, Y., Liu, W., Huang, Y., Du, L., Chen, B., 2016. Thermal conductivity of polymer-based composites: fundamentals and applications. Prog. Polym. Sci. 59, 41–85. https://doi.org/10.1016/j.progpolymsci.2016.03.001.

Colorimetric calculation of color differences with the DIN 99 Formula. (2021). In DIN-6176.

Coraux, J., N'Diaye, A.T., Busse, C., Michely, T., 2008. Structural coherency of graphene on Ir(111). Nano Lett. 8 (2), 565–570. https://doi.org/10.1021/nl0728874.

Crawford, R.L., 1981. Lignin Biodegradation and Transformation. John Wiley and Sons, NY.

Daniel, M.C., Astruc, D., 2004. Gold nanoparticles: assembly, supramolecular chemistry, quantum-size-related properties, and applications toward biology, catalysis, and nanotechnology. Chem. Rev. 104 (1), 293–346. https://doi.org/10.1021/cr030698+.

Desa, M.S.Z.M., Hassan, A., Arsad, A.B., Mohammad, N.N.B., 2014. Mechanical properties of poly(lactic acid)/multiwalled carbon nanotubes nanocomposites. Mater. Res. Innov. 18 (Supp 6), S6-14–S6-17. https://doi.org/10.1179/1432891714Z.000000000924.

Desroches, M., Escouvois, M., Auvergne, R., Caillol, S., Boutevin, B., 2012. From vegetable oils to polyurethanes: synthetic routes to polyols and main industrial products. Polym. Rev. 52 (1), 38–79. https://doi.org/10.1080/15583724.2011.640443.

Dufresne, A., 2012. Nanocellulose: From Nature to High Performance Tailored Materials. De Gruyter, Berlin.

Ferdinando, D.L.B., Chiara, S., Letizia, V., Pietro, C., Andrea, M., Laura, B., Simona, L., Francesca, C., Salvatore, I., C., L.G, 2020. Greener nanocomposite polyurethane foam based on sustainable polyol and natural fillers: investigation of chemico-physical and mechanical properties. Materials, 13, p. 211.

Fim, F.D.C., Guterres, J.M., Basso, N.R.S., Galland, G.B., 2010. Polyethylene/graphite nanocomposites obtained by in situ polymerization. J. Polym. Sci. A Polym. Chem. 48 (3), 692–698. https://doi.org/10.1002/pola.23822.

Guan, J., Hanna, M.A., 2006. Selected morphological and functional properties of extruded acetylated starch-cellulose foams. Bioresour. Technol. 97 (14), 1716–1726. https://doi.org/10.1016/j.biortech.2004.09.017.

Guerra, A.J., Cano, P., Rabionet, M., Puig, T., Ciurana, J., 2018. Effects of different sterilization processes on the properties of a novel 3D-printed polycaprolactone stent. Polym. Adv. Technol. 29 (8), 2327–2335. https://doi.org/10.1002/pat.4344.

Guimarães, J.L., Wypych, F., Saul, C.K., Ramos, L.P., Satyanarayana, K.G., 2010. Studies of the processing and characterization of corn starch and its composites with banana and sugarcane fibers from Brazil. Carbohydr. Polym. 80 (1), 130–138. https://doi.org/10.1016/j.carbpol.2009.11.002.

Guo, Q., Xue, Q., Sun, J., Dong, M., Xia, F., Zhang, Z., 2015. Gigantic enhancement in the dielectric properties of polymer-based composites using core/shell MWCNT/amorphous carbon nanohybrids. Nanoscale 7 (8), 3660–3667. https://doi.org/10.1039/c4nr05264a.

Habibi, Y., 2014. Key advances in the chemical modification of nanocelluloses. Chem. Soc. Rev. 43 (5), 1519–1542. https://doi.org/10.1039/c3cs60204d.

Hardy, J.G., Palma, M., Wind, S.J., Biggs, M.J., 2016. Responsive biomaterials: advances in materials based on shape-memory polymers. Adv. Mater. 28 (27), 5717–5724. https://doi.org/10.1002/adma.201505417.

Harito, C., Bavykin, D.V., Yuliarto, B., Dipojono, H.K., Walsh, F.C., 2019. Polymer nanocomposites having a high filler content: synthesis, structures, properties, and applications. Nanoscale 11 (11), 4653–4682. https://doi.org/10.1039/c9nr00117d.

He, M., Wang, Z., Wang, R., Zhang, L., Jia, Q., 2016. Preparation of bio-based polyamide elastomer by using green plasticizers. Polymer 8 (7), 257. https://doi.org/10.3390/polym8070257.

Huang, J.J., Zhong, Z.F., Rong, M.Z., Zhou, X., Chen, X.D., Zhang, M.Q., 2014. An easy approach of preparing strongly luminescent carbon dots and their polymer based composites for enhancing solar cell efficiency. Carbon 70, 190–198. https://doi.org/10.1016/j.carbon.2013.12.092.

Huskić, M., Žigon, M., 2007. PMMA/MMT nanocomposites prepared by one-step in situ intercalative solution polymerization. Eur. Polym. J. 43 (12), 4891–4897. https://doi.org/10.1016/j.eurpolymj.2007.09.009.

Isaias, R., Shesha, J., Edward, C., Mario, G., Leonardo, S., 2009. Erosion resistance and mechanical properties of silicone nanocomposite insulation. IEEE Trans. Dielectr. Electr. Insul. 16 (1), 52–59. https://doi.org/10.1109/TDEI.2009.4784551.

Jamshidian, M., Tehrany, E.A., Imran, M., Jacquot, M., Desobry, S., 2010. Poly-lactic acid: production, applications, nanocomposites, and release studies. Compr. Rev. Food Sci. Food Saf. 9 (5), 552–571. https://doi.org/10.1111/j.1541-4337.2010.00126.x.

Jang, J.Y., Kim, M.S., Jeong, H.M., Shin, C.M., 2009. Graphite oxide/poly(methyl methacrylate) nanocomposites prepared by a novel method utilizing macroazoinitiator. Compos. Sci. Technol. 69 (2), 186–191. https://doi.org/10.1016/j.compscitech.2008.09.039.

Jun, L.Y., Karri, R.R., Mubarak, N.M., Yon, L.S., Bing, C.H., Khalid, M., Jagadish, P., Abdullah, E.C., 2020. Modelling of methylene blue adsorption using peroxidase immobilized functionalized Buckypaper polyvinyl alcohol membrane via ant colony optimization. Environ. Pollut. 259, 113940. https://doi.org/10.1016/j.envpol.2020.113940.

Kaith, B.S., Jindal, R., Jana, A.K., Maiti, M., 2010. Development of corn starch based green composites reinforced with *Saccharum spontaneum* L fiber and graft copolymers—evaluation of thermal, physicochemical and mechanical properties. Bioresour. Technol. 101 (17), 6843–6851. https://doi.org/10.1016/j.biortech.2010.03.113.

Kalia, S., Kaith, B.S., Kaur, I., 2009. Pretreatments of natural fibers and their application as reinforcing material in polymer composites—a review. Polym. Eng. Sci. 49 (7), 1253–1272. https://doi.org/10.1002/pen.21328.

Kalita, D., Netravali, A.N., 2017. Thermoset Resin Based Fiber Reinforced Biocomposites. In: Mittal, K.L., Bahners, T. (Eds.), Textile Finishing: Recent Developments and Future Trends. Wiley Blackwell, NJ, pp. 425–484.

Karri, R.R., Tanzifi, M., Tavakkoli Yaraki, M., Sahu, J.N., 2018. Optimization and modeling of methyl orange adsorption onto polyaniline nano-adsorbent through response surface methodology and differential evolution embedded neural network. J. Environ. Manage. 223, 517–529. https://doi.org/10.1016/j.jenvman.2018.06.027.

Kaushik, A., Singh, M., Verma, G., 2010. Green nanocomposites based on thermoplastic starch and steam exploded cellulose nanofibrils from wheat straw. Carbohydr. Polym. 82 (2), 337–345. https://doi.org/10.1016/j.carbpol.2010.04.063.

Ke, T., Sun, X., 2001. Effects of moisture content and heat treatment on the physical properties of starch and poly(lactic acid) blends. J. Appl. Polym. Sci. 81 (12), 3069–3082. https://doi.org/10.1002/app.1758.

Kharissova, O.V., Dias, H.V.R., Kharisov, B.I., Pérez, B.O., Pérez, V.M.J., 2013. The greener synthesis of nanoparticles. Trends Biotechnol. 31 (4), 240–248. https://doi.org/10.1016/j.tibtech.2013.01.003.

Kou, J., Varma, R.S., 2012. Beet juice-induced green fabrication of plasmonic AgCl/Ag nanoparticles. ChemSusChem 5 (12), 2435–2441. https://doi.org/10.1002/cssc.201200477.

Kumar, A.P., Singh, R.P., 2008. Biocomposites of cellulose reinforced starch: improvement of properties by photo-induced crosslinking. Bioresour. Technol. 99 (18), 8803–8809. https://doi.org/10.1016/j.biortech.2008.04.045.

Kwon, S.Y., Ciobanu, C.V., Petrova, V., Shenoy, V.B., Bareño, J., Gambin, V., Petrov, I., Kodambaka, S., 2009. Growth of semiconducting graphene on palladium. Nano Lett. 9 (12), 3985–3990. https://doi.org/10.1021/nl902140j.

Lau, Y.J., Karri, R.R., Mubarak, N.M., Lau, S.Y., Chua, H.B., Khalid, M., Jagadish, P., Abdullah, E.C., 2020. Removal of dye using peroxidase-immobilized Buckypaper/polyvinyl alcohol membrane in a multi-stage filtration column via RSM and ANFIS. Environ. Sci. Pollut. Res 27, 40121–40134. https://doi.org/10.1007/s11356-020-10045-2.

Lee, S.Y., Kang, I.A., Doh, G.H., Yoon, H.G., Park, B.D., Wu, Q., 2008. Thermal and mechanical properties of wood flour/talc-filled polylactic acid composites: effect of filler content and coupling treatment. J. Thermoplast. Compos. Mater. 21 (3), 209–223. https://doi.org/10.1177/0892705708089473.

Lei, Y., Wu, Q., Yao, F., Xu, Y., 2007. Preparation and properties of recycled HDPE/natural fiber composites. Compos. A 38 (7), 1664–1674. https://doi.org/10.1016/j.compositesa.2007.02.001.

Lett, A., Sagadevan, J., Prabhakar, J., et al., 2019. Exploring the binding effect of a seaweed-based gum in the fabrication of hydroxyapatite scaffolds for biomedical applications. Mater. Res. Innov. 24, 75–81.

Li, u L., Wang, J., 2015. The influence of curing age and mix proportion on the uniaxial compressive properties of green high-performance fibre-reinforced cementitious composites. Mater. Res. Innov. 19, 8–624.

Lim, J.M., Moon, J.H., Yi, G.R., Heo, C.J., Yang, S.M., 2006. Fabrication of one-dimensional colloidal assemblies from electrospun nanofibers. Langmuir 22 (8), 3445–3449. https://doi.org/10.1021/la053057d.

Lingamdinne, L.P., Koduru, J.R., Karri, R.R., 2019a. A comprehensive review of applications of magnetic graphene oxide based nanocomposites for sustainable water purification. J. Environ. Manage. 231, 622–634. https://doi.org/10.1016/j.jenvman.2018.10.063.

Lingamdinne, L.P., Koduru, J.R., Karri, R.R., 2019b. Green synthesis of iron oxide nanoparticles for lead removal from aqueous solutions. Key Eng. Mater. 805, 122–127. https://doi.org/10.4028/www.scientific.net/KEM.805.122.

Lingamdinne, L.P., Vemula, K.R., Chang, Y.Y., Yang, J.K., Karri, R.R., Koduru, J.R., 2020. Process optimization and modeling of lead removal using iron oxide nanocomposites generated from bio-waste mass. Chemosphere 243,, 125257. https://doi.org/10.1016/j.chemosphere.2019.125257.

Liu, D., Zhong, T., Chang, P.R., Li, K., Wu, Q., 2010. Starch composites reinforced by bamboo cellulosic crystals. Bioresour. Technol. 101 (7), 2529–2536. https://doi.org/10.1016/j.biortech.2009.11.058.

Liu, Z., Erhan, S.Z., Xu, J., 2005. Preparation, characterization and mechanical properties of epoxidized soybean oil/clay nanocomposites. Polymer 46 (23), 10119–10127. https://doi.org/10.1016/j.polymer.2005.08.065.

Lu, Y., Larock, R.C., 2007. Fabrication, morphology and properties of soybean oil-based composites reinforced with continuous glass fibers. Macromol. Mater. Eng. 292 (10–11), 1085–1094. https://doi.org/10.1002/mame.200700150.

Lu, Y., Weng, L., Cao, X., 2006. Morphological, thermal and mechanical properties of ramie crystallites—reinforced plasticized starch biocomposites. Carbohydr. Polym. 63 (2), 198–204. https://doi.org/10.1016/j.carbpol.2005.08.027.

M., H.M., M., A.-E.A., A., E.R., A., I.M, 2019. Performance of PVC/SiO_2 nanocomposites under thermal ageing. Appl. Nanosci. 11, 2143–2151. https://doi.org/10.1007/s13204-018-00941-y.

Ma, S., Liu, X., Jiang, Y., Tang, Z., Zhang, C., Zhu, J., 2013. Bio-based epoxy resin from itaconic acid and its thermosets cured with anhydride and comonomers. Green Chem. 15 (1), 245–254. https://doi.org/10.1039/c2gc36715g.

Ma, X.F., Yu, J.G., Wang, N., 2007. Fly ash-reinforced thermoplastic starch composites. Carbohydr. Polym. 67 (1), 32–39. https://doi.org/10.1016/j.carbpol.2006.04.012.

Magendran, S.S., Khan, F.S.A., Mubarak, N.M., Khalid, M., Walvekar, R., Abdullah, E.C., Nizamuddin, S., Karri, R.R., 2019. Synthesis of organic phase change materials by using carbon nanotubes as filler material. Nano-Struct. Nano-Objects 19, 100361. https://doi.org/10.1016/j.nanoso.2019.100361.

Majdzadeh-Ardakani, K., Sadeghi-Ardakani, S., 2010. Experimental investigation of mechanical properties of starch/natural rubber/clay nanocomposites. Digest J. Nanomater. BioStruct. 5 (2), 307–316. http://www.chalcogen.infim.ro/307_Majdzadeh-1.pdf.

Mark, J.E., 1996. Ceramic-reinforced polymers and polymer-modified ceramics. Polym. Eng. Sci. 36 (24), 2905–2920. https://doi.org/10.1002/pen.10692.

Merlini, C., Soldi, V., Barra, G.M.O., 2011. Influence of fiber surface treatment and length on physico-chemical properties of short random banana fiber-reinforced castor oil polyurethane composites. Polym. Test. 30 (8), 833–840. https://doi.org/10.1016/j.polymertesting.2011.08.008.

Mi, Y.N., Liang, G., Gu, A., Zhao, F., Yuan, L., 2013. Thermally conductive aluminum nitride–multiwalled carbon nanotube/cyanate ester composites with high flame retardancy and low dielectric loss. Ind. Eng. Chem. Res. 52 (9), 3342–3353.

Miyagawa, H., Mohanty, A., Drzal, L.T., Misra, M., 2004. Effect of clay and alumina-nanowhisker reinforcements on the mechanical properties of nanocomposites from biobased epoxy: A comparative study. Ind. Eng. Chem. Res. 43 (22), 7001–7009. https://doi.org/10.1021/ie049644w.

Mohanty, A.K., Misra, M., Drzal, L.T., 2002. Sustainable Bio-Composites from renewable resources: opportunities and challenges in the green materials world. J. Polym. Environ. 10 (1–2), 19–26. https://doi.org/10.1023/A:1021013921916.

Moon., R.J., Ashlie, M., John, N., John, S., Jeff, Y, 2011. Cellulose nanomaterials review: structure, properties and nanocomposites. Chem. Soc. Rev. 3941. https://doi.org/10.1039/c0cs00108b.

Mosiewicki, M. A., & Aranguren, M. I. (2013). A short review on novel biocomposites based on plant oil precursors. Eur. Polym. J. 49 (6), 1243–1256. https://doi.org/10.1016/j.eurpolymj.2013.02.034.

Nadagouda, M.N., Varma, R.S., 2006. Green and controlled synthesis of gold and platinum nanomaterials using vitamin B2: density-assisted self-assembly of nanospheres, wires and rods. Green Chem. 8 (6), 516–518. https://doi.org/10.1039/b601271j.

Nakayama, A, Kakugo, A., Gong, J.P., Takai, M., Erata, T., Kawano, S., 2004. High mechanical strength double-network hydrogel with bacterial cellulose. Adv. Funct. Mater. 14 (11), 1124–1128.

Naushad, M., Ahamad, T., AlOthman, Z.A., Al-Muhtaseb, A.H., 2019. Green and eco-friendly nanocomposite for the removal of toxic Hg(II) metal ion from aqueous environment: adsorption kinetics & isotherm modelling. J. Mol. Liq. 279, 1–8. https://doi.org/10.1016/j.molliq.2019.01.090.

Nie, H., Ho, M.L., Wang, C.K., Wang, C.H., Fu, Y.C., 2009. BMP-2 plasmid loaded PLGA/HAp composite scaffolds for treatment of bone defects in nude mice. Biomater. 30 (5), 892–901. https://doi.org/10.1016/j.biomaterials.2008.10.029.

Nin, L., Huang, J., Dufresne, A., 2012. Preparations, properties and applications of polysaccharide nanocrystals in advanced functional nanomaterials: a review. Nanoscale 4, 3274–3294. https://doi.org/10.1039/C2NR30260H.

Njuguna, J., Pielichowski, K., Desai, S., 2008. Nanofiller-reinforced polymer nanocomposites. Polym. Adv. Technol. 19 (8), 947–959. https://doi.org/10.1002/pat.1074.

Nursyafiqah, E., Sheela, C., Nursyafreena, A., Arafat, M.N., Abdul, R.F.I., Joazaizulfazli, J., Abdul, W.R., 2017. Structure and properties of oil palm-based nanocellulose reinforced chitosan nanocomposite for efficient synthesis of butyl butyrate. Carbohydr. Polym. 176,, 281–292. https://doi.org/10.1016/j.carbpol.2017.08.097.

Ogata, N., Jimenez, G., Kawai, H., Ogihara, T., 1997. Structure and thermal/mechanical properties of poly(*l*-lactide)-clay blend. J. Polym. Sci. B Polym. Phys. 35 (2), 389–396. https://doi.org/10.1002/(SICI)1099-0488(19970130)35:2<389::AID-POLB14>3.0.CO;2-E.

Pandey, J.K., Singh, R.P., 2005. Green nanocomposites from renewable resources: effect of plasticizer on the structure and material properties of clay-filled starch. Starch/Staerke 57 (1), 8–15. https://doi.org/10.1002/star.200400313.

Panthapulakkal, S., Zereshkian, A., Sain, M., 2006. Preparation and characterization of wheat straw fibers for reinforcing application in injection molded thermoplastic composites. Bioresour. Technol. 97 (2), 265–272. https://doi.org/10.1016/j.biortech.2005.02.043.

Payen, A., 1838. Mémoire sur la composition du tissu propre des plantes et du ligneux. Comptes Rendus 7, 1052–1056.

Peelman, N., Ragaert, P., De Meulenaer, B., Adons, D., Peeters, R., Cardon, L., Van Impe, F., Devlieghere, F., 2013. Application of bioplastics for food packaging. Trends Food Sci. Technol. 32 (2), 128–141. https://doi.org/10.1016/j.tifs.2013.06.003.

Pérez-Pacheco, E., Canto-Pinto, J.C., Moo-Huchin, V.M.Poletto, M. (Ed.), 2021. Thermoplastic starch (TPS)-cellulosic fibers composites: mechanical properties and water vapor barrier: a review. Composites from Renewable and Sustainable Materials. https://doi.org/10.5772/65397.

Peyghambarian, N., Fuentes-Hernandez, C., Yamamoto, M., Cammack, K., Matsumoto, K., Walker, G.A., Barlow, S., Kippelen, B., Meredith, G., Marder, S.R., 2004. Bistriarylamine polymer-based composites for photorefractive applications. Adv. Mater. 16 (22), 2032–2036. https://doi.org/10.1002/adma.200400102.

Potts, J.R., Dreyer, D.R., Bielawski, C.W., Ruoff, R.S., 2011. Graphene-based polymer nanocomposites. Polymer 52 (1), 5–25. https://doi.org/10.1016/j.polymer.2010.11.042.

Pukanszky, B., Tudos, F., Jancar, J., Kolarik, 1989. The possible mechanisms of polymer–filler interaction in polypropylene–$CaCO_3$ composites. J. Mater. Sci. Lett. 8, 1040.

Purkansky, B., 2005. Effect of molecular interactions on the miscibility and structure of polymer blends. Eur. Polym. J. 41, 727–736.

Quaresimin, M., Salviato, M., Zappalorto, M., 2012. Strategies for the assessment of nanocomposite mechanical properties. Compos. B Eng. 43 (5), 2290–2297. https://doi.org/10.1016/j.compositesb.2011.12.012.

Rai, A., Singh, A., Ahmad, A., & Sastry, M. (2006) Role of halide ions and temperature on the morphology of biologically synthesized gold nanotriangles. Langmuir 22, 736–741.

Rajapaksha, L.D., Saumyadi, H.A.D., Samarasekara, A.M.P.B., Amarasinghe, D.A.S., Karunanayake, L., 2017. Development of cellulose based light weight polymer composites. In: In Proc. 3rd International Moratuwa Engineering Research Conference, MERCon 2017. Institute of Electrical and Electronics Engineers Inc, pp. 182–186.

Rastogi, A., Zivcak, M., Sytar, O., Kalaji, H.M., He, X., Mbarki, S., Brestic, M., 2017. Impact of metal and metal oxide nanoparticles on plant: a critical review. Front. Chem. 5, 78. https://doi.org/10.3389/fchem.2017.00078.

Rai, S.S., Bousmina, M., 2012. Biodegradable polymer/layered silicate nanocomposites. In: Mai, Y-W, Yu, Z-Z (Eds.), Polymer Nanocomposites. Woodhead Publishing, UK, pp. 57–129.

Reddy, N., Yang, Y., 2005. Structure and properties of high quality natural cellulose fibers from cornstalks. Polymer 46 (15), 5494–5500. https://doi.org/10.1016/j.polymer.2005.04.073.

Reddy, N., Yang, Y., 2007. Natural cellulose fibers from switchgrass with tensile properties similar to cotton and linen. Biotechnol. Bioeng. 97 (5), 1021–1027. https://doi.org/10.1002/bit.21330.

Reddy, N., Yang, Y., 2008. Characterizing natural cellulose fibers from velvet leaf (Abutilon theophrasti) stems. Bioresour. Technol. 99 (7), 2449–2454. https://doi.org/10.1016/j.biortech.2007.04.065.

Reddy, N., Yang, Y., 2009a. Natural cellulose fibers from soybean straw. Bioresour. Technol. 100 (14), 3593–3598. https://doi.org/10.1016/j.biortech.2008.09.063.

Reddy, N., Yang, Y., 2009b. Properties of natural cellulose fibers from hop stems. Carbohydr. Polym. 77 (4), 898–902. https://doi.org/10.1016/j.carbpol.2009.03.013.

Reina, A., Jia, X., Ho, J., Nezich, D., Son, H., Bulovic, V., Dresselhaus, M., Kong, J., 2008. Large area, few layer graphene films on arbitrary substrates by chemical vapor deposition. Nano Lett 9 (1), 30–35.

Sahay, R., Reddy, V.J., Ramakrishna, S., 2014. Synthesis and applications of multifunctional composite nanomaterials. Int. J. Mech. Mater. Eng. 9 (1), 25. https://doi.org/10.1186/s40712-014-0025-4.

Sain, M., Panthapulakkal, S., 2006. Bioprocess preparation of wheat straw fibers and their characterization. Ind. Crops Prod. 23 (1), 1–8. https://doi.org/10.1016/j.indcrop.2005.01.006.

Sánchez-González, J., MacÍas-García, A., Alexandre-Franco, M.F., Gómez-Serrano, V, 2005. Electrical conductivity of carbon blacks under compression. Carbon 43 (4), 741–747. https://doi.org/10.1016/j.carbon.2004.10.045.

Satriananda, Riza, M., Mulyati, S., Mulana, F, 2018. Polyurethane/clay nanocomposites from palm oil for surface-coating applications. Polym. Renew. Resour. 9 (3–4), 103–110. https://doi.org/10.1177/2041247918800243.

Sharif, A., Mondal, S., Hoque, M.E., 2019. Polylactic acid (PLA)-based nanocomposites: processing and properties. In: Sanyang, M.L., Jawaid, M. (Eds.), Bio-based Polymers and Nanocomposites: Preparation, Processing, Properties & Performance. Springer International Publishing, pp. 233–254.

Shivaji, S., Madhu, S., Singh, S., 2011. Extracellular synthesis of antibacterial silver nanoparticles using psychrophilic bacteria. Process Biochem. 46 (9), 1800–1807. https://doi.org/10.1016/j.procbio.2011.06.008.

Siddiqui, M.T.H., Nizamuddin, S., Baloch, H.A., Mubarak, N.M., Al-Ali, M., Mazari, S.A., Bhutto, A.W., Abro, R., Srinivasan, M., Griffin, G., 2019. Fabrication of advance magnetic carbon nano-materials and their potential applications: a review. J. Environ. Chem. Eng. 7 (1), 102812. https://doi.org/10.1016/j.jece.2018.102812.

Siddiqui, M.T.H., Nizamuddin, S., Baloch, H.A., Mubarak, N.M., Dumbre, D.K., Inamuddin, Asiri, A., M., Bhutto, A.W., Srinivasan, M., Griffin, G.J, 2018. Synthesis of magnetic carbon nanocomposites by hydrothermal carbonization and pyrolysis. Environ. Chem. Lett. 16 (3), 821–844. https://doi.org/10.1007/s10311-018-0724-9.

Sinha Ray, S., Okamoto, K., Yamada, K., Okamoto, M., 2002. Novel porous ceramic material via burning of polylactide/layered silicate nanocomposite. Nano Lett. 2 (4), 423–425. https://doi.org/10.1021/nl020284g.

Sinha Ray, S., Okamoto, M., 2003. Polymer/layered silicate nanocomposites: a review from preparation to processing. Prog. Polym. Sci. (Oxford) 28 (11), 1539–1641. https://doi.org/10.1016/j.progpolymsci.2003.08.002.

Sinha, R.S., Okamoto, M., Yamada, K., Ueda, K., 2002. New biodegradable polylactide/layered silicate nanocomposites: preparation, characterization and materials properties. Macromolecules 35, 659–660.

Sinha, R.S., Okamoto, M., Yamada, K., Ueda, K., 2003. New polylactide/layered silicate nanocomposites: concurrent improvement of materials properties and biodegradability. Polymer 44, 857–866.

Soni, N., Prakash, S., 2011. Factors affecting the geometry of silver nanoparticles synthesis in *Chrysosporium Tropicum* and *Fusarium Oxysporum*. Am. J. Nanotechnol. 2 (1), 112–121. https://doi.org/10.3844/ajnsp.2011.112.121.

Stieven, M.L., Sizuka, O.S., Faria, D.M., Larissa, do A.M.T., Santos, da S.F., Roberto, P.F., Cerqueira, R.M, 2018. Multifunctional green nanostructured composites: preparation and characterization. Mater. Res. Exp. 5, 055010. https://doi.org/10.1088/2053-1591/aabf66.

Sutter, P.W., Flege, J.I., Sutter, E.A., 2008. Epitaxial graphene on ruthenium. Nat. Mater. 7 (5), 406–411. https://doi.org/10.1038/nmat2166.

Svagan, A., 2008. Bio-inspired cellulose Nanocomposites and foams based on starch matrix. Ph.D. thesis. KTH Chem. Sci. Eng. Stockholm, Sweden.

Tabaddor, P.L., Aloisio, C.J., Bair, H.E., Plagianis, C.H., Taylor, C.R., 2000. Thermal analysis characterization of a commercial thermoplastic/thermoset adhesive. J. Therm. Anal. Calorim. 59 (1), 559–570.

Thakur, V.K., Singha, A.S., Thakur, M.K., 2013. Fabrication and physico-chemical properties of high-performance pine needles/green polymer composites. Int. J. Polym. Mater. Polym. Biomater. 62 (4), 226–230. https://doi.org/10.1080/00914037.2011.641694.

Thomassin, J.M., Jérôme, C., Pardoen, T., Bailly, C., Huynen, I., Detrembleur, C., 2013. Polymer/carbon based composites as electromagnetic interference (EMI) shielding materials. Mater. Sci. Eng. R Rep. 74 (7), 211–232. https://doi.org/10.1016/j.mser.2013.06.001.

Tiwari, A., Srivastava, R.B., 2012. Biotechnology in Biopolymers: Developments, Applications & Challenging Areas. Smithers Rapra, UK.

Tugrul Seyhan, A., Tanoglu, M., Schulte, K., 2008. Mode I and mode II fracture toughness of E-glass non-crimp fabric/carbon nanotube (CNT) modified polymer based composites. Eng. Fract. Mech. 75 (18), 5151–5162. https://doi.org/10.1016/j.engfracmech.2008.08.003.

Uesaka, T., Nakane, K., Maeda, S., Ogihara, T., Ogata, N., 2000. Structure and physical properties of poly(butylene succinate)/cellulose acetate blends. Polym. 41 (23), 8449–8454. https://doi.org/10.1016/S0032-3861(00)00206-8.

Ugo, P., Moretto, L.M., 2007. Template deposition of metals. In: Zoski, C.G. (Ed.), Handbook of Electrochemistry. Elsevier, Amsterdam, pp. 678–709.

Utracki, L.A., 2004. Clay-Containing Polymeric Nanocomposites 1, 1–85957.

Walenta, E., Fink, H.P., Weigel, P., Ganster, J., Schaaf, E., 2001. Structure-property relationships of extruded starch, 2: Extrusion products from native starch. Macromol. Mater. Eng. 286 (8), 462–471. https://doi.org/10.1002/1439-2054(200108)286:8<462::AID-MAME462>3.0.CO;2-A.

Wang, B., Sain, M., 2007. Dispersion of soybean stock-based nanofiber in a plastic matrix. Polym. Int. 56 (4), 538–546. https://doi.org/10.1002/pi.2167.

Wei, L., Hu, N., Zhang, Y., 2010. Synthesis of polymer-mesoporous silica nanocomposites. Materials. 3 (7), 4066–4079. https://doi.org/10.3390/ma3074066.

Wik, V.M., Aranguren, M.I., Mosiewicki, M.A., 2011. Castor oil-based polyurethanes containing cellulose nanocrystals. Polym. Eng. Sci. 51 (7), 1389–1396. https://doi.org/10.1002/pen.21939.

Willett, J.L., Shogren, R.L., 2002. Processing and properties of extruded starch/polymer foams. Polymer 43 (22), 5935–5947. https://doi.org/10.1016/S0032-3861(02)00497-4.

Wu, C.S., 2009. Renewable resource-based composites of recycled natural fibers and maleated polylactide bioplastic: characterization and biodegradability. Polym. Degrad. Stab. 94 (7), 1076–1084. https://doi.org/10.1016/j.polymdegradstab.2009.04.002.

Xiangqian, L., Huizhong, X., Zhe-Sheng, C., Guofang, C., 2011. Biosynthesis of nanoparticles by microorganisms and their applications. J. Nanomater. 2011, 270974. https://doi.org/10.1155/2011/270974.

Xiao-Su, Y., Xvfeng, Z., Fangbo, D., Jianfeng, T., 2018. Development of bio-sourced epoxies for biocomposites. Aerospace 5 (2), 65. https://doi.org/10.3390/aerospace5020065.

Yu, L., Petinakis, S., Dean, K., Bilyk, A., Wu, D., 2007. Green polymeric blends and composites from renewable resources. Macromol. Sympos., 249–250, pp. 535–539.

Yuyang, Q., Qingyu, P., Yue, Z., Xu, Z., Zaishan, L., Xiaodong, H., Yibin, L., 2019. Lightweight, mechanically flexible and thermally superinsulating rGO/polyimide nanocomposite foam with an anisotropic microstructure. Nanoscale Adv. 1, 4895–4903. https://doi.org/10.1039/c9na00444k.

Zabihzadeh, S.M., 2010. Water uptake and flexural properties of natural filler/HDPE composites. BioResources 5 (1), 316–323. http://www.ncsu.edu/bioresources/BioRes_05/BioRes_05_1_0316_Zabihzadeh_Water_Uptake_Flex_Prop_Nat_HDP_E_Compos_804.pdf.

Zhang, X.F., Wu, Y.Q.Q.G., Wei, J.H., Tong, J.F., Yi, X.S., 2017. Curing kinetics and mechanical properties of bio-based composite using rosin-sourced anhydrides as curing agent for hot-melt prepreg. Sci. China Technol. Sci. 60 (9), 1318–1331. https://doi.org/10.1007/s11431-016-9029-y.

Zini, E., Scandola, M., 2011. Green composites: an overview. Polym. Compos. 32 (12), 1905–1915. https://doi.org/10.1002/pc.21224.

Zou, H., Wu, S., Shen, J., 2008. Polymer/Silica Nanocomposites: preparation, characterization, properties, and applications. Chem. Rev. 108 (9), 3893–3957. https://doi.org/10.1021/cr068035q.

SECTION 3

Miscellaneous Applications

CHAPTER 20

Nanotechnology for biosensor applications

Bhanu Shrestha
Department of Electronics Engineering, Kwangwoon University, Seoul, Korea

20.1 Introduction

Biosensors can be defined as capable of providing qualitative and quantitative analytical information with the help of a transducer. In other words, it is an analytical device that converts a biological response into a measurable electrical signal. It can be fabricated in nano-scaled integrated circuit technology. Nanotechnology is a one-billionth of a meter or it can be comparable to ten times the diameter of a hydrogen atom (Sahu et al., 2019a, 2019b). We can also compare this scale to our hairs which are 80,000 nm on average. On other hand, the biosensor can be represented by Fig. 20.1 which shows analytes, recognition elements, transducers, and signal processors with a display unit. Basically, bioreceptor, electrochemical interface, transducers, signal processor (amplifier), and display are typical components of a biosensor which is shown in Fig. 20.1. Each component will be briefly described as follows:

Analyte: It is a substance that needs to be analyzed and characterized. In biosensor, the analyte is captured by the recognition element (i.e., receptor). All the biological agents can be a sample of analytes such as blood, serum, glucose, etc. and this is also termed as a target.

Recognition elements: The recognition element is called a bioreceptor that can specifically recognize the analytes. It is capable of detecting the target analyte. We can use the target analytes, for example, enzymes, cells antibodies, DNA/RNA, bacteria, tissue, organelle, etc.

Transducer: It is a device that can convert bio-signal to an electrical signal. This is also called signalization. Various types of components can be used as transducers such as acoustic waves, Silicon wafer, thermistor, Field-Effect Transistor (FET), graphite electrode, etc. These are the part of biosensors and can be used for various detection purposes.

Signal processor with a display unit: A very small bio-signal must be amplified to have a readable or measurable signal. Such a signal can be achieved by using a signal processor and the signal can be sent to the display unit that gives meaningful parameters for analysis.

Sustainable Nanotechnology for Environmental Remediation
DOI: https://doi.org/10.1016/B978-0-12-824547-7.00013-8

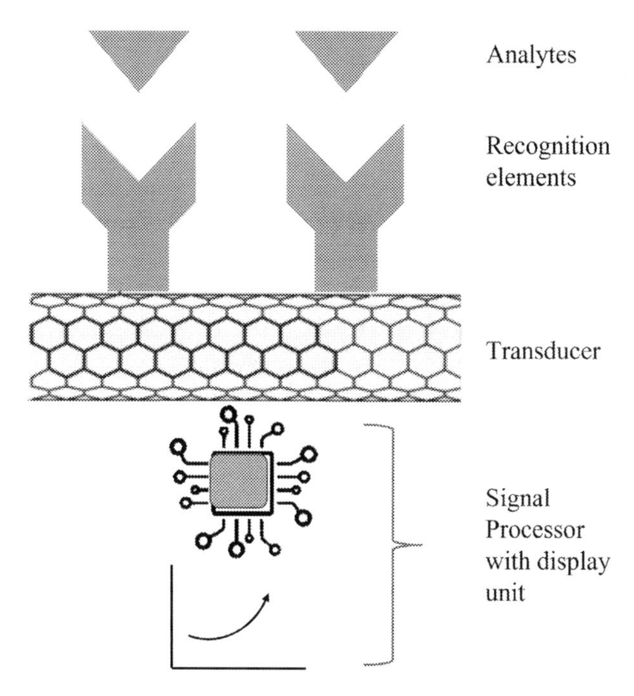

Figure 20.1 Schematic representation of biosensors.

To fabricate nanobiosensors, various biomaterials are used such as carbon nanotubes (CNTs), graphene, gold (Au), silver (Ag), semiconductors, etc. Significant properties of these biomaterials are used for sensing purposes with various technologies. Therefore, obviously, there are different types of biosensors such as electromechanical biosensor, nanoshell biosensor, nanowire biosensor, optical biosensor, nanotube-based biosensor, electrical biosensor, viral biosensor, etc. These biosensors are briefly described with their different applications. The characteristics of a biosensor are selectivity, reproducibility, stability, sensitivity, and linearity. These characteristics are equally important in making biosensor for different purposes (Chao et al., 2016; Jianrong et al., 2004; Thiébaut & Gordon, 2019).

20.2 Biomaterials in nanoscales

Biomaterials are used to fabricate biosensors for various applications. Biosensors in the nanoscale are getting challenging for precise and accurate device performance. This technology is composed of nanotechnology and biosensor, which is interestingly being popular among the related researchers. The nano-scaled biosensor can detect various analytes such as DNA, RNA, antibodies, bacteria, etc.(Ozer et al., 2019). In this technology, nanomaterials are used in nano-size and it is principally single unit sized between 1 and

100 nm. These nanomaterials are used to make biosensors by using microfabrication technology from which the biosensor can be fabricated on the nano-scale. These materials are being commercialized due to the use for various purposes. It is often categorized on how many of their dimensions meet in the nano-size. If three external dimensions of the nanoobject are in the nanoscale, then it is called a nanoparticle. If two external dimensions are in the nano-scale and it can include nanotube and nanorods. If one external dimension and two other dimensions are different within the nano-scale, it is called nanoribbon. Besides, there are also metal-based nanoparticles such as quantum dots (QDs), nanowires, and nanorods which are also known as inorganic nanomaterials. As organic nanomaterials, we have organic solar cells, OLEDs, etc. Broadly, these particles can not only be used in biomedical applications but also in the field of textile, biomedical, health care, food agriculture, industrial, electronics, environments, renewable energy, etc. (Li et al., 2015).

Thus, the nanostructure can be categorized as a 1-D structure, 2-D structure, and bulk structured material. The cylindrical confinement is engineered in a 1-D pattern and it gives mechanical supports and prevents detachment from a chain of atoms. For example, a CNT is a natural semi 1-D nanostructure, nanowires are 1-D, and so on. These can be used for various purposes in fabricating biosensors. However, 2-D structures are crystalline in nature consisting of a single layer of atoms with two dimensions. For example, graphene is considered a 2-D material that was discovered in 2004 (Katsnelson, 2007). As a bulk structure material, we can take nanocomposite, nanocrystals, nanofilm, and nanotextured surfaces.

These nanomaterials can be synthesized with bottom-up and top-down methods, of which the characteristic size is 1–100 nm as described above. In the bottom-up method, atoms/molecules are arrayed in nanoscale and the sources can be in the form of gases, liquids, or solids. In this method, a chaotic and controlled process is involved. In a chaotic process, atoms/molecules of a source remain in a chaotic state to make an unstable state by suddenly changing the conditions. In this case, we need the clever manipulation of the parameters. But when it collapses from the chaotic state, it is not easy to control the sizes, whereas the controlled process involves the controlled transferring of atoms/molecules to the nanoparticles. In such nanoparticles formation processes, the distances from the atom to atom/molecules to molecules can be controlled on the nano-scale. We can take an example of it as chemical vapor deposition (CVD), molecular beam epitaxy (MBE), etc.

In the top-down methods, the mechanical force can be implemented to break the bulk materials into nano-sized particles. For example, ball milling. A short-pulsed laser can also be implemented to ablate the bulk materials or, we can create circuits on the silicon substrate by various kinds of etching processes.

A few popular and important nanomaterials are briefly described as follows: Recently, some nanomaterials such as gold nanoparticles and CNTs have been implemented as a nano-biosensor to have a precise and accurate response.

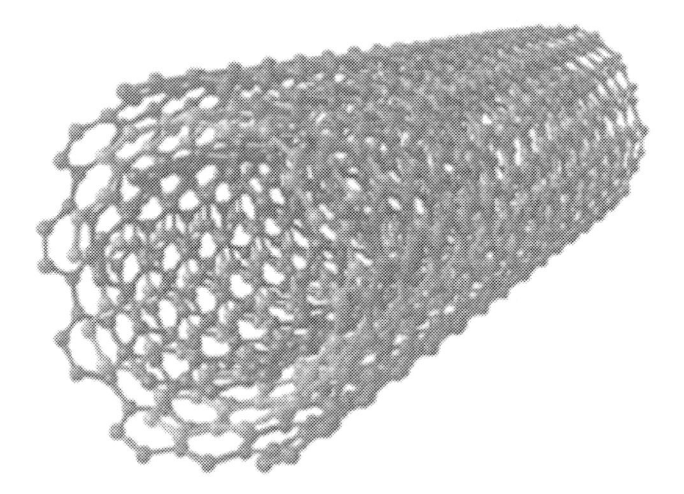

Figure 20.2 Carbon nanotubes with multiple walls.

20.2.1 Carbon nanotube

After the discovery of CNTs, it is being an intensive research interest among the researchers. This is due to the unique behavior of CNT with its electrical conductivity, good biocompatibility, chemical stability, and mechanical strength. A. G. Mamalis et al. stated that CNTs are promising structure for the semiconductor industry and it is leading the miniaturization of the sensing devices. It has a property of stiffness and conducting electronics (Mamalis et al., 2004).Fig. 20.2 shows the multiwalled CNT.

20.2.2 Graphene

Graphene consists of a single layer of atoms in a 2-D lattice with electronic, thermal, and mechanical properties. Recently, graphene has been successfully implemented in many biomedical applications, such as analysis of DNA, enzyme activity, protein, and drug delivery (Shen et al., 2012). Especially, graphene oxide is being an attractive candidate in biological applications due to its unique characteristics (Lingamdinne et al., 2019, 2018). Thus, graphene can also be used for nanoelectronics (Shintaro, 2015). In addition, we can also pattern by a lithographic process at the micro-level to generate graphene interconnected patterned devices (Heer, 2007). Different types of graphene have existed in the form of nanomaterials and these types of nanomaterials are based on their production approaches such as CVD (Wang et al., 2009), shear exfoliation in liquid (Paton et al., 2014), MBE, atomic layer epitaxy, chemical synthesis, etc. After the production of graphene, it can be transferred by the wet or dry transfer process. After various kinds of further processes, graphene can be prepared for use (Bonaccorso et al., 2012).

20.2.3 Gold

Gold nanoparticles (AuNPs) have been extensively studied and implemented in nanotechnology thanks to their special properties and functionalities. The AuNP provides various platforms for assembling in nanoscale with antibodies, proteins, and nucleoid (Zhang et al., 2020). Particularly, AuNP and nanorods have simple synthesis with large surface area, facile conjugation to various biomolecules with strong ability in adsorption. Recently, the immobilization of DNA/RNA on AuNP-based nanobiosensors has been reported, especially for cancer detection (Du et al., 2010). Besides, it is also revealed that AuNP exhibits a strong plasmon band in the form of an aqueous solution depending upon their geometric shape and size. Thus, Au is also a promising nanomaterial for cellular and molecular biology. Sperling et al. reviewed that AuNPs are used in biology widely and identified that it is used not only in labeling and delivering but also in heating, and sensing until these studies (Sperling et al., 2008). Gold nanoparticles can be synthesized by gold nanospheres, gold nanorods, gold nanoshells, gold nanocages, etc. These nanoparticles are studied in various fields such as *in-vitro* and *in-vivo* imaging, drug delivery, and cancer treatment (Cai et al., 2008). AuNPs are also used in delivery applications such as drug and gene delivery (Ghosh et al., 2008).

20.2.4 Silver

Silver nanoparticles (AgNPs) are commonly implemented to have considerable attention in biological detection (Sahu, Zabed, et al., 2019). Due to their significant physicochemical properties including the surface plasmon resonance (SPR) and large effective scattering cross-section of individual silver nanoparticles, AgNPs are advantageous for electrochemical and SPR nanobiosensors. In addition, the hydrophobic Ag-AuNPs demonstrated a strong adsorption and conduction properties that lead to use in biosensors. Both Ag and Au can be used in biosynthesis such as the synthesis of nanoparticles by a microorganism (bacteria, etc.) (Barabadi, 2017). Silver nanoparticles can be synthesized by numerous methods. Among them, there are spark discharging, reduction of electrochemical, irradiating of the solution, and synthesis of cytochemical. Interestingly one investigation notified that when the germ cells are exposed to AgNPs, mitochondrion function of the cell was significantly decreased and morphology of the cell was changed (Chen & Schluesener, 2008). This study shows that mitochondrion can be a targeted factor for cytotoxicity due to AgNPs.

20.2.5 Fullerenes

The fullerenes are also used as nanomaterials. They are single- or double-bonded carbon allotrope to make a closed mesh and have almost 5–7 atoms in that mesh. Such kinds of carbon nanoparticles can be manufactured for various objectives but mass production results in environmental pollution that can cause serious brain damage as well. It is also

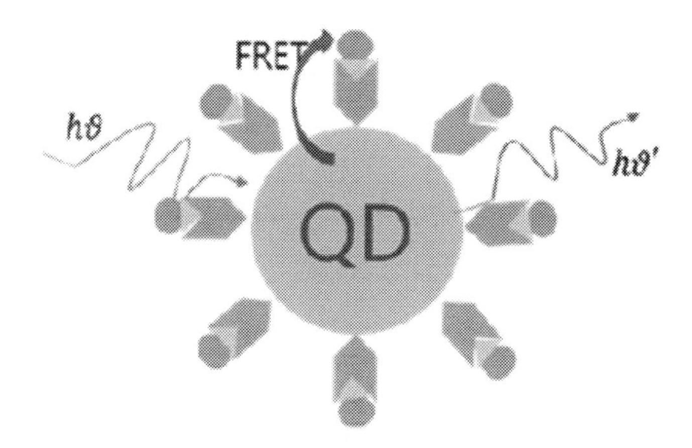

Figure 20.3 Symbolic representation of QD-based FRET.

toxic for culturing human cells. However, it can be used for medical applications such as anti-aging and anti-damage in cosmetics, drug delivery systems, sports goods industries, etc. Interesting thing is that the fullerenes can also be applied in the plant system. The concentration of fullerenes can be used for yielding the cereal as well as fruit crop which is a significant achievement of nanotechnology (Husen & Siddiqi, 2014).

20.2.6 Semiconductors

Semiconductors nanoparticles are very important materials for fabricating biosensors that have found wide application for the detection of analytes. The surface potential of semiconductor nanoparticles is responsible for the performance and characteristics of biosensors based on semiconductors. To detect the photonic biorecognition process, the tunable fluorescence properties of semiconductor nanoparticles have been used. They show tunable absorbance and fluorescence-based on their tunable size.

Elements from the group II-V, III-V, or IV-VI semiconductor nanocrystals having unique photophysical properties are called QDs of the semiconductor. This is also termed as 0-D materials. One of the most typical QDs is Cd-chalcogenide nanocrystals whose shell and a core is composed of ZnS and a centrosome of about 2–10 nm, respectively. The outer surface of the shell is coated by a polymer with a size of 10–15 nm. The centrosome of QDs is particularly composed of CdSe, CdTe, and CdS. A QD–based biosensors such as QD-based fluorescence resonance energy transfer (FRET) genosensor, QD-based FRET immuno-sensor, and QD-based bioluminescence resonance energy transfer (BRET) immuno-sensor, etc. are based on the various types of molecular beacon conjugated on QDs and transduction signals. The schematic principle of QD-based FRET, widely used in biological applications, is shown in Fig. 20.3 (Agasti. et al., 2010). Such type of sensor

Table 20.1 A review of biosensor applications of nanomaterials.

Nanomaterials	Applications	References
Carbon nanotube	Electrochemical transducer	Vamvakaki & Chaniotakis, (2007)
Graphene	FET transducer	(Peña-Bahamonde et al., 2018)
Nanoparticles (Au, Ag, etc.)	SPR transducer	(Pedersen & Duncan, 2005)
Quantum dots	Optical transducer	(Clapp et al., 2006)
Nanowires	Receptor for DNA, etc.	(Wang Z et al 2016)
Magnetic nanoparticles	Diagnostic magnetic resonance	(Haun et al 2010)
Fullerenes	Mediator between recognition site and the electrode (in electrochemical biosensor)	(Hwang et al., 2020)
Semiconductors	FET transducer	(Sapsford et al., 2006)

is currently widely used due to its high sensitivity, reproducibility, stability, and immune-resistance. The emissive fluorescent molecules can be promising acceptors. In the case of FRET, QDs are frequently used as donors. In the case of bioluminescence resonance energy transfer chemiluminescence response energy transfer (BRETCRET), they can also act as an energy donor (Kim, 2012).

Thus, we can see that the various types of nanomaterials are used in biosensors with different technologies. The main challenge in the biosensor is to capture efficient signal in the biological recognition process (Holzinger et al., 2014). A review of biosensor applications of nanomaterials is depicted in Table 20.1 (Vamvakaki & Chaniotakis, 2007; Peña-Bahamonde et al., 2018; Pedersen & Duncan, 2005; Clapp et al., 2006; X. Wang et al., 2009; Haun et al., 2010; Hwang et al., 2020; Sapsford et al., 2006).

20.3 Classification of nanobiosensor

Nanobiosensors can be classified in various ways. The classification is based on nanomaterials with biosensing operation. That means, it is not easy to classify because of its diversity. Therefore, it is classified in terms of its type of materials to be analyzed and transducer used. The name of the biosensor is based on the analyte we used. For example, if we screen any antigen or enzyme through biosensors, then it is known antigen or enzyme biosensors. If we think of it in terms of sensing mechanism, mainly electrochemical, and optical biosensors are named. Therefore, we can classify the nanobiosensor as shown in Fig. 20.4. Each type of sensors is briefly described as follows.

However, biosensors can be described in various aspects. We can classify the biosensor in terms of receptors and transducers as well. Antibody biosensors, DNA biosensors, cell biosensors, and viral biosensors can be included for the first classification, and for the second one, electrochemical biosensors, calorimetric biosensors, optical biosensors, and piezoelectric biosensors can be included. The electrochemical biosensor can be further

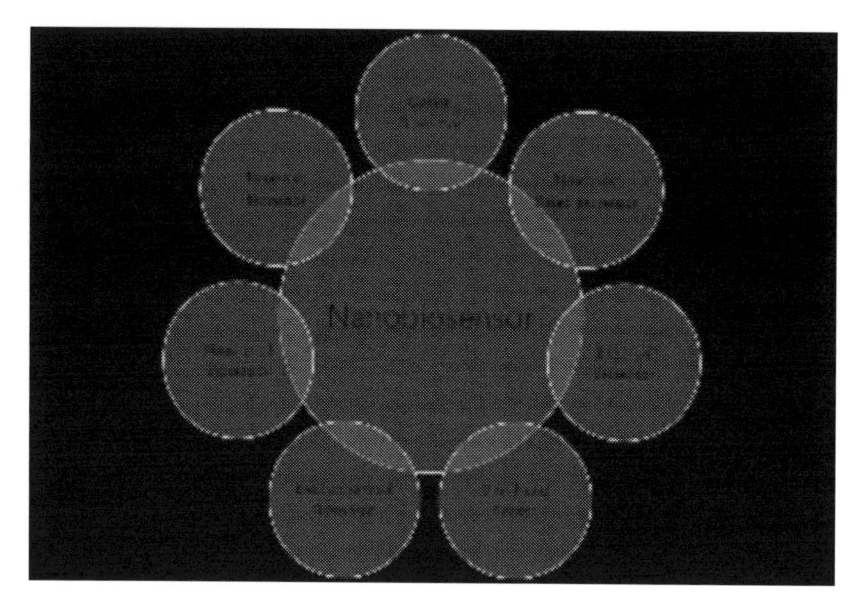

Figure 20.4 Classification of nanobiosensors.

classified into potentiometric biosensors, amperometric biosensors, and impedimetric biosensors, while piezoelectric biosensors can be further classified into surface acoustic wave and quartz crystal microbalance (QCM). In the same way, optical biosensors can be further classified into fluorescence biosensors, SPR and SPR imaging (SPRi) biosensors, flow cytometric biosensors, and chemiluminescence biosensors (Wu et al., 2019).

20.3.1 Electrochemical biosensors

Electrochemical biosensor used the biological molecules coated on a probing surface. The reaction of sensing molecules on the compound to be detected and the electrical signal produces with the quality measured. It has electrodes that change the chemical signal into an electrical signal. The electrochemical biosensor is mainly classified into potentiometric, amperometric, voltammetric, and impedimetric transducers. These are employed to detect various biological analytes such as DNA/RNA, proteins, etc.

The electrochemical biosensor is typical sensing devices based on transducing the biochemical to electrical signals. The main component, an electrode, in the biosensor can be employed as a support for the immobilization of biomolecules and electron movement. Because of various kinds of nanomaterials with the large surface area, loading capacity and mass transport of reactants can be enabled due to synergic effects to get good performance in the sense of analytical sensitivity (Rasouli et al., 2019).

Recently, novel biosensor designs are appearing which can play increasingly important roles in our life. Especially, rapid growth of biomaterials, in terms of availability and application of polymers associated with new sensing technologies, has led to remarkable innovations with significant improvements in functionality (Jung et al., 2020). Wei et al. (2009) and Nitin et al. (2007) stated various kinds of biosensors at the nanoscale such as amperometric and voltammetric biosensors, impedance biosensors, conductance nanosensors, and potentiometric nanosensors and their impact on the biosensing process.

20.3.2 Nanoshell biosensor

Nanoshell represents nanoshell particles for biosensing applications. Andronescu & Grumezescu (2017) state, "Nanoshells are gold-coated nanoscale beads that can absorb near-infrared light in a size-dependent fashion to generate intense heat that can be used to selectively kill target cells, for example, cancer cells." In the future, it can be used in detecting oral and dental cancer cells as well. For example, the size of Au-nanoshells ranges from 10 to 200 nm which are composed of a dielectric material coated by a thin gold shell. These materials have extra-ordinary properties such as optical, chemical, and physical properties and also can be candidates in the field of medical biosensing, cancer-detection, etc. (Erickson & Tunnell, 2009).

20.3.3 Nanowire biosensor

Nanowire biosensor is composed of nanowires using biological molecules such as DNA molecules, fibrin proteins, etc. It is a fibril-like 1-D nanostructure. Particularly, it is used in biomedical applications to detect possible diseases such as breast cancer. Due to its high sensitivities, it is becoming a potential research area for investigating various diseases. Next, we can also use a field-effect transistor (FET) based on nanowire in which doped channels and their gates can be changed with nanowires and receptors so that it can detect various kinds of biological agents such as proteins, DNA/RNA, viruses, etc. Practically silicon nanowire biosensors are fabricated with the top-down or bottom-up methods as described in Section 20.2. After fabrication of the biosensors, this can be influenced by various factors such as its dimensions, density of carriers, surface topology, and their mobility where the studies found experimentally that the smaller dimension has sharper sensitivities; however, for concern regarding analytes, the optimized amount of analytes can be allowed to have a minimum detectable signal (J. Zhang et al., 2020).

20.3.4 Optical biosensor

Optical fiber can also be used in fabricating an optical biosensor that is a compact analytic device with an integrated transducer. This uses characteristics of optical fiber such as absorption, scattering, and fluorescence. The sensor detects such characteristics of the sensing elements. The result can be obtained either by characteristics of optical fibers or by the refractive index of the interacting surface. For example, if the antibodies such as the biological elements are bound with a metal layer, the variation results can be obtained when the refractive index of the medium comes in contact with this layer. Such biosensor has the merit of nonelectric nature that allows analyzing more than two elements on a unit layer by applying the variation of the light wavelengths (Xueqing et al., 2009). Optical biosensors have clear-cut merits over conventional approaches such as high sensitivity, smaller dimensions, and cost-effectiveness. The merit of optical biosensors is the utilization in the drug discovery process.

20.3.5 Nanotube-based biosensor

The diameter of a nanotube is typically measured in nanometer and hence it is known as a nanotube and it is composed of carbon. A CNT exhibits significant electrical conductivity property. It can be designated as single or multi-walled (Wohlstadter et al., 2003) . Due to its unique features, it became attractive for biosensing applications. CNTs act as scaffolding for biomaterials on the surfaces and it combines physical, electrical, and optical properties that make excellent materials for the transduction of signals with analytes detection. Wang et al. (2003) used a multiwalled CNT-based biosensor for glucose detection and compared it with the glassy carbon-based biosensor. However, the single-walled CNT is not only used to destroy cancer cells using photothermal therapeutic agents but also in photoacoustic imaging. Because of its one dimension, it can exhibit strong resonance and high absorption. Jacob N. Wohlstadter et al. reported that CNT acts as an electrode as well as an immobilization phase in sensing devices based on electro-chemiluminescence. They used this CNT sensor to detect a fetoprotein. They have chosen the fetoprotein as a clinical experiment because they achieved both sensitivity and dynamic range which were desired requirements (J.N. Wohlstadter et al., 2003).

20.3.6 Electrical biosensor

Electrical biosensor can be used to detect various kinds of biological analytes with various methodologies. This biosensor can be easily integrated using a standard electronic microfabrication process in which capability in microfluidics develops and then the detection of proteins in a small analytical volume (Luo & Davis, 2013). Chung et al. used the electrical biosensor for detecting the circulating tumor. In their study, they used blood samples as a result of the immunomagnetic isolation process and detected circulating cancer cells by measurement of impedance (Yao-Kuang et al., 2011). To fabricate such a biosensor, an electrical transduction mechanism is essential.

20.3.7 Viral biosensor

Recently, pathogens such as coronavirus that threaten human life are becoming a great problem worldwide. Many human lives are lost day-by-day. Due to the pandemic of SARs, MERs, and recently spreading COVID-19, researchers are compelled to find ways to detect those infectious viruses. Thus, viral biosensors are used to detect the pathogens of viruses rapidly in contaminated water and food, surfaces, or in samples of the patients. For this purpose, polymerase chain reaction is used to detect DNA or RNA of the virus. Semiconductor materials such as silicon (Si) are also used to make a viral biosensor (Rossi et al., 2007) and novel universal immuno-sensor can be fabricated to detect the influenza virus as well (Nidzworski et al., 2014). Jiang et al research team investigated that the COVID-19 can be detected by graphene biosensors (Jiang et al., 2020). Many biomolecules such as nucleic acids, bacteriophages, etc. have significant recognition factors due to the biocompatibility and solubility of graphene.

Besides those nanobiosensors described above, radio frequency (RF) biosensor is also emerging sensing technology for various purposes. In this concept of RF biosensor, resonator or filter can be used which are very sensitive by nature and changes small parameters. To design such resonators or filters, we can use substrates not only Teflon or printed circuit board (PCB) but also semiconductors such as GaAs, Si, etc. The fabrication can be achieved by using various integrated circuit (IC) technologies such as omplementary metal oxide semiconductor (CMOS), monolithic microwave integrated circuit (MMIC), integrated passice device (IPD), etc. For example, a glucose biosensor has already been investigated by various researchers. A simple resonator used in RF biosensor is shown in Fig. 20.5. Kim et al. stated that the RF biosensors can be possibly detect the glucose level in serum using an RF resonator which acts as a biosensor (Kim, 2012).

In this biosensor, it is demonstrated that a certain amount of serum is dropped down on the surface of the RF resonator (i.e., biosensor) and it produced some variation of resonance frequency shifts with amplitude caused by the certain volume of analytes (i.e., serum). For the experimental process, different samples of serum with different volume and glucose solution with different volumes were taken. The results were extracted in terms of S-parameters and regression analysis. Similar research has been performed using interdigital capacitor structure as a biosensing device (Kim et al., 2015). Therefore, it is proved that we can also use RF devices such as resonators, filters, etc. for biosensing purpose.

20.4 Nanotechnology-based biosensor applications

Nanotechnology is assisting to considerably revolutionize many fields such as information technology, security, medicine, energy, food safety, environmental science, etc.

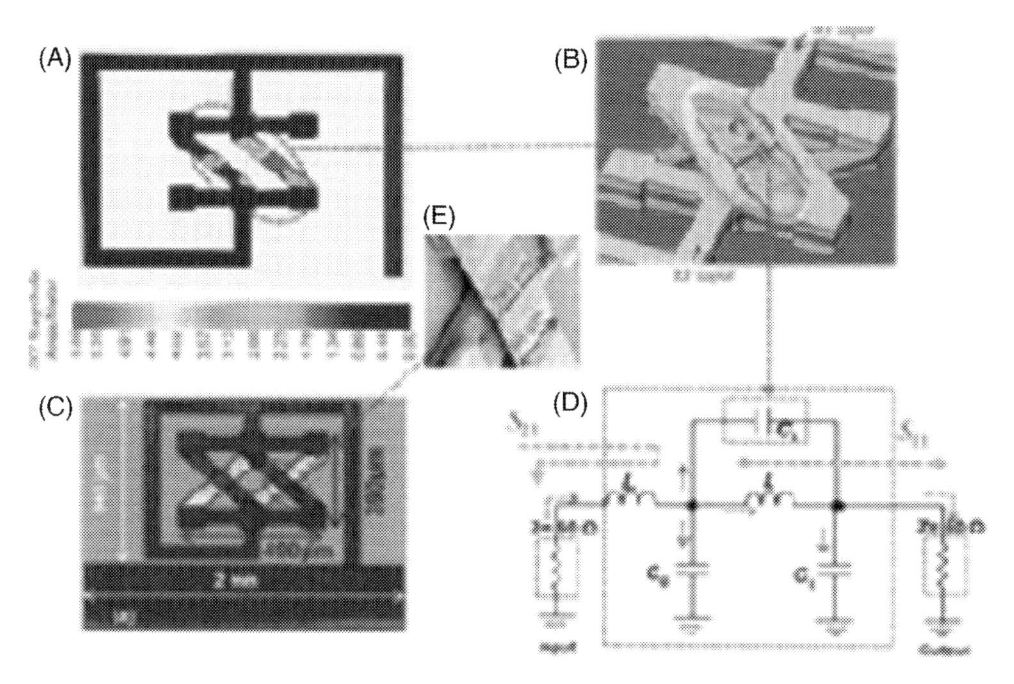

Figure 20.5 The resonator used as a biosensor. (A) The current density. (B, C) Fabricated biosensor. (D) The equivalent circuit. (E) The fabricated air-bridge type resonator.[Courtesy: RFIC Lab, Kwangwoon University, Korea]. From Kim, N. Y., Dhakal, R., Adhikari, K. K., Kim, E. S., & Wang, C. (2015). A reusable robust radio frequency biosensor using microwave resonator by integrated passive device technology for quantitative detection of glucose level. Biosensors Bioelectron., 67, 687–693. https://doi.org/10.1016/j.bios.2014.10.021.

(Wennersten et al., 2008). Due to the improvement of the performance and sensitivity of nanotechnology-based biosensors, several types of biosensors are developed. Many applications of nanotechnology-based biosensors are briefly described next.

20.4.1 Electronics and IT applications

The great contribution of nanotechology in advanced electronics and computing shows fast, small, and portable with a huge amount of data storage capacity. Particularly, the fabrication process of transistors is being smaller and smaller with nanotechnology. Various companies such as Intel, IBM, Samsung, etc. fabricated a nano-scaled transistor. Moreover, nano-scaled magnetoresitive random access memory (MRAM) for the computer booting process enhanced the system shutdown. Due to the advent of nanotechnology, QDs are used in ultra-high definition (UHD) TV to give highly vibrant colors with energy efficiency (Pandit et al., 2016). Many flexible electronic devices are developed that give ease of using those products such as wearable devices. Graphene is also used for making flexible electronic devices. There are also memory chips, hearing aids, a keyboard, and a case of smartphones coating with antibacterial materials, etc. which are used nanoparticles to produce them. In this way, nanotechnology is contributing to electronic and IT applications as well (Allsopp et al., 2007).

20.4.2 Biomedical and healthcare applications

The application of biosensors in nanoscale can be implemented to nanomedicine for biological solutions such as disease prevention, diagnosis, and treatment. For example, nanogold or nanogold particles are potential treatments for cancer and other diseases that have been already investigated clinically (Dreaden et al., 2011; Krzysztof et al., 2018). In treatment and diagnosis of atherosclerosis or buildup of plaque in arteries, nanotechnology can be implemented (S. et al., 2010). Therefore, for example, high-density lipoprotein (HDL) is the good cholesterol that helps to shrink plaque. Some researchers have studied about the number of therapeutics. We can also use the biosensor to detect diabetes, cancer-cell, including so many other disorders based on the serum analysis. The detection can help in diagnosing the related patients whether infected or not. The applications include the detection of glucose in diabetic patients, detection of urinary tract bacterial infections, detection of HIV-AIDS, and diagnosis of cancer. Besides bacteria, in most cases, fungi can also cause food spoilage and create severe health-related problems. So, fungal pathogen detection is also being an emerging solution for health-related problems.

Biosensor plays a significant role in tissue engineering as well. Due to the development of biomedical research, tissue engineering has emerged, and it is a practice of associating cells, scaffolds, and biomolecules into functional tissue. The purpose of tissue engineering is to integrate functional constructs that maintain, restore, or improve damaged tissues or whole organs. Therefore, small biomolecules such as glucose, hydrogen peroxide, adenosine, etc. are the significant candidates in configuring the cell culture, organ-specific on chips, and maintaining 3-D integrity. The numerous signals existing in the metabolic cells which have significant roles in transmuting and transfering. The signal may be physical or chemical (in any form) that varies with ionic concentration, protein content, pH, oxygen ingestion, etc. Such analytes can be monitored to have real-time insights into the cells. In addition, nanofiber can also be used for various purposes such as blood purification, air purification, water purification, etc. which has a size of less than 100 nm (Cho et al., 2020). In addition, in biomedical and healthcare applications, carbon dots have also been employed as photonic nanoagent due to its wide range of tunable optical and physicochemical properties (Jung et al., 2020).

20.4.3 Environmental applications

Biosensors can be used in the field of environmental pollution monitoring (Guerra et al., 2018) because many harmful environmental agents are existing and cause serious health hazards to humans and the ecosystem. To avoid such hazards to human health and to the surroundings, we need to remove the contaminants such as soil, water, air, etc. Therefore, to have a green environment, biosensing technologies are important for detecting such contaminants. The substantial advantages can be obtained when it is used to detect impurities in drinking water. For example, nitrates and phosphates are common in water as pollutants. And also, we use nanoparticles for cleaning industrial water pollutants. The heavy metals with toxic levels such as Ag, Cd, Hg, Cu, Ni, Co, etc. are also a serious problem for our health and these can be detected using biosensors (Carolin et al., 2017). Besides, the nano-scaled biosensors can also be implemented for detecting airborne contaminants and pathogens. In this case, phage-sensors are used for monitoring those pathogenic microbes (Rowe-Taitt et al., 2000).

20.4.4 Food industry applications

Fresh and healthy foods are essential for humans and these must be secured and safe. Therefore, several industries adopted specific methods to detect spoiled or damaged foods that cause serious health problems. We also need safe processed foods from industries. The biosensors with high sensitivity and fast response are necessary to determine the cause of spoilage. It can also be used to make the freshness to maintain the quality of food. Many types of nano-scaled biosensors can be used in the food industry particularly enzyme-based biosensors and immunosensors. Rashidi and Khosravi-Darini (Rashidi and Khosravi-Darani, 2011) stated that gases, pathogens, toxins can be detected using biosensors in packaged food. Therefore, nanobiosensors have played a significant role in protecting packaged food on an industrial scale. The heat, as well as mechanical resistance properties, can be improved due to the use of nanoparticles in the food packaging process and it assists to span the period of food protection. Thus, nanotechnology is a potential technology for the food industry application as well.

20.4.5 Energy applications

Nanobiosensors can also be used in energy applications to optimize various types of detections. Nanobiosensors can be regulated especially in the industrial operations and commercially further enhancement can be employed on an industrial scale. For instance, the separation of impurities can be employed in the metallurgical operation which is in the form of ores. We can use nanosensors to separate the selected impurities by trying out various types of biosensors with sensing enzymes. The sources of nanobiosensors can be fossil, nuclear fuels, solar energy, etc. These sources then converted into heat and electricity and a considerable amount of CO_2 emission can be enhanced by improving efficiency. To enhance efficiency, the nanostructured materials can be used by optimizing the layer design and by utilizing the wasted heat. Thus, such energy can be stored and distributed for practical purposes. One more thing is that this technology has already been implemented to develop novel batteries. These batteries are responsible for fast charging, high performance, and high power density. The windmill blade can be made with an epoxy-containing CNT that can be used efficiently in the energy-generating process due to its lighter and stronger capability of the windmill blade. Nanotechnology has also been used in the energy-harvesting process by utilizing piezoelectric nanowires used in the oven that can generate practical energy. In this way, lives are being convenient thanks to the development of nanotechnology and its applications.

20.5 Future trends

Nanotechnology is a promising technology that uses and develops multifunctional materials and we can expect to them lighter, safer, smarter, and more efficient vehicles, aircraft, spacecraft, and ships. Besides, nanotechnology helps in various ways to enhance transportation infrastructure. Nanotechnology has really become very essential for the development of biosensors. The overall mechanism of nanotechnology is being faster, smarter, and user friendly. The transduction mechanisms have been significantly improved with the use of nanomaterials and nanostructures like those of QDs, nanoparticles for enzyme immobilization, and hybrid nanostructure with multiple functionalities. In the future, it will be a more dynamic, versatile, and fast recognition system. The promising nanomaterials can be considered for combining chemical and biological sensors to make the overall process fast and easy to execute with better performance. We can imagine that most of the sensing or detecting processes have been more versatile, robust, and dynamic due to the use of nanomaterials. This is due to the tremendous demand for rapid and cost-effective sensing systems and they can be implemented for food and drink, defense and security, industries and environment, health and medicals, and so on. Therefore, it seems that the future of nanotechnology is bright with a huge project for the human as well. For example, if we see the recent trends of nanowire biosensors, the advancement of this technology is getting popular in the nanotechnology field due to its practical utility and suitable applications. Thus, recently nanowires biosensors are used in *in-vivo* sensing, integrating it with paper-based devices, self-powering, signal processing, data analysis, etc.

20.6 Conclusion

Nanobiosensor research focuses on developing innovative technologies that could make eminent contributions in human and animal disease marker detection with promising therapeutic compound identification and analysis. In the development of biosensors in nanoscale, nanotechnology's contribution plays a very important role. The sensitivity and the device performance are being enhanced by employing nanomaterials for the construction of biosensors. Various types of effects in nano-scaled are core topics for the study including QDs effects that are very unique and significant aspects in nanotechnology. We have studied the various kinds of biosensors such as electrochemical biosensors, nanoshell biosensor, nanowire biosensor, optical biosensor, nanotube-based biosensor, electrical biosensor, and viral biosensors and such kinds of biosensors can be applied in various fields such as electronics and IT applications, biomedical and healthcare applications, environmental applications, food industry applications, and energy applications. Due to technical enhancement of this technology, different types of biosensors have appeared for practical use and such a trend is increasing recently in the electromedical field. The advances in biosensors using nanotechnology have been increasing and generated a huge number of patents, publications, and projects. Therefore, there are many advantages of nanotechnology-based biosensors such as early detection of breast cancer, AIDS, high throughputs, smart and rapid sensing, cost-effective, user friendly, robust and repetitive, stable and portable, etc.

References

Agasti, S.S., Subinoy, R., Myoung-Hwan, P., Kyu, K.C., Chang-Cheng, Y., Rotello, V.M, 2010. Nanoparticles for detection and diagnosis. Adv. Drug Deliv. Rev. 62 (3), 316–328. https://doi.org/10.1016/j.addr.2009.11.004.

Allsopp, M., Walters, A., Santillo, D., 2007. Nanotechnologies and nanomaterials in electrical and electronic goods: A review of uses and health concerns. Technical Note 09/2007. Greenpeace Res. Laboratories London.

Andronescu, E., Grumezescu, A.M., 2017. Nanostructures for Oral Medicine. Elsevier, Amsterdam.

Barabadi, H., 2017. Nanobiotechnology: a promising scope of gold biotechnology. Cell. Mol. Biol. 63 (12), 3–4. https://doi.org/10.14715/cmb/2017.63.12.2.

Bonaccorso, F., Lombardo, A., Hasan, T., Sun, Z., Colombo, L., Ferrari, A.C., 2012. Production and processing of graphene and 2D crystals. Mater. Today 15 (12), 564–589.

Cai, W., Gao, T., Hong, H., Sun, J., 2008. Applications of gold nanoparticles in cancer nanotechnology. Nanotechnol., Sci. Appl. 1, 17–32.

Carolin, C.F., Kumar, P.S., Saravanan, A., Joshiba, G.J., Naushad, M., 2017. Efficient techniques for the removal of toxic heavy metals from aquatic environment: a review. J. Environ. Chem. Eng. 5 (3), 2782–2799.

Chao, J., Zhu, D., Zhang, Y., Wang, L., Fan, C., 2016. DNA nanotechnology-enabled biosensors. Biosensors Bioelectron. 76, 68–79. https://doi.org/10.1016/j.bios.2015.07.007.

Chen, X., Schluesener, H.J., 2008. Nanosilver: a nanoproduct in medical application. Toxicol. Lett. 176 (1), 1–12.

Cho, I.-H., Kim, D.H., Park, S., 2020. Electrochemical biosensors: perspective on functional nanomaterials for on-site analysis. Biomater. Res. 24 (1), 1–12.

Clapp, A.R., Medintz, I.L., Mattoussi, H., 2006. Förster resonance energy transfer investigations using quantum-dot fluorophores. Chem. Phys. Chem. 7 (1), 47–57.

Dreaden, E.C., Mac key, M.A., Huang, X., Kang, B., El-Sayed, M.A., 2011. Beating cancer in multiple ways using nanogold. Chem. Soc. Rev. 40 (7), 3391–3404. https://doi.org/10.1039/c0cs00180e.

Du, Y., Li, B., Wang, E., 2010. Analytical potential of gold nanoparticles in functional aptamer-based biosensors. Bioanal. Rev. 1 (2–4), 187–208.

Erickson, T.A., Tunnell, J.W., 2009. Gold nanoshells in biomedical applications. Mixed Metal Nanomater. 3, 1–44.

Ghosh, P., Han, G., De, M., Kim, C.K., Rotello, V.M., 2008. Gold nanoparticles in delivery applications. Adv. Drug Deliv. Rev. 60 (11), 1307–1315.

Guerra, F.D., Attia, M.F., Whitehead, D.C., Alexis, F., 2018. Nanotechnology for environmental remediation: materials and applications. Molecules 23 (7), 1760.
Haun, J.B., Yoon, T., Lee, H., Weissleder, R., 2010. Magnetic nanoparticle biosensors.. WIREs Nanomed. NanoBiotechnol. 2 (3), 291–304.
de Heer, W.A., Berger, C., Conrad, E., First, P., Murali, R., Meindl, J., 2007. Pionics: the emerging science and technology of graphene-based nanoelectronics. IEEE Int. Electron Devices Meet., 2007, pp. 199–202.
Holzinger, M., Le Goff, A., Cosnier, S., 2014. Nanomaterials for biosensing applications: a review. Front. Chem. 2, 63.
Husen, A., Siddiqi, K.S., 2014. Carbon and fullerene nanomaterials in plant system. J. NanoBiotechnol. 12 (1), 16.
Hwang, H.S., Jeong, J.W., Kim, Y.A., Chang, M., 2020. Carbon nanomaterials as versatile platforms for biosensing applications. Micromachines 11 (9), 814.
Jiang, Z., Feng, B., Xu, J., Qing, T., Zhang, P., Qing, Z., 2020. Graphene biosensors for bacterial and viral pathogens. Biosensors Bioelectron. 166, 112471.
Jianrong, C., Yuqing, M., Nongyue, H., Xiaohua, W., Sijiao, L., 2004. Nanotechnology and biosensors. Biotechnol. Adv. 22 (7), 505–518. https://doi.org/10.1016/j.biotechadv.2004.03.004.
Jung, C.Y., Jinhyun, K., Beum, P.C., 2020. Photonic carbon dots as an emerging nanoagent for biomedical and healthcare applications. ACS Nano 14 (6), 6470–6497. https://doi.org/10.1021/acsnano.0c02114.
Katsnelson, M.I., 2007. Graphene: carbon in two dimensions. Mater. Today 10 (1–2), 20–27.
Kim, N.Y., Dhakal, R., Adhikari, K.K., Kim, E.S., Wang, C., 2015. A reusable robust radio frequency biosensor using microwave resonator by integrated passive device technology for quantitative detection of glucose level. Biosensors Bioelectron. 67, 687–693. https://doi.org/10.1016/j.bios.2014.10.021.
Kim, Y.-P., 2012. Energy transfer-based multiplex analysis using quantum dots. In: A. Al-Ahmadi (Ed.). Quantum Dots: A Variety of New Applications, pp. 225–240.
Krzysztof, S., Michał, G., Barbara, K.-M., 2018. Gold nanoparticles in cancer treatment. Mol. Pharma. 16 (1), 1–23. https://doi.org/10.1021/acs.molpharmaceut.8b00810.
Li, G., Li, Y., Chen, G., He, J., Han, Y., Wang, X., Kaplan, D.L., 2015. Silk-based biomaterials in biomedical textiles and fiber-based implants. Adv. Healthc. Mater. 4 (8), 1134–1151.
Lingamdinne, L.P., Koduru, J.R., Chang, Y.-Y., Karri, R.R., 2018. Process optimization and adsorption modeling of Pb(II) on nickel ferrite-reduced graphene oxide nano-composite. J. Mol. Liquids 250, 202–211. https://doi.org/10.1016/j.molliq.2017.11.174.
Lingamdinne, L.P., Koduru, J.R., Karri, R.R., 2019. A comprehensive review of applications of magnetic graphene oxide based nanocomposites for sustainable water purification. J. Environ. Manage. 231, 622–634. https://doi.org/10.1016/j.jenvman.2018.10.063.
Luo, X., Davis, J.J., 2013. Electrical biosensors and the label free detection of protein disease biomarkers. Chem. Soc. Rev. 42 (13), 5944–5962. https://doi.org/10.1039/c3cs60077g.
Mamalis, A.G., Vogtländer, L.O.G., Markopoulos, A., 2004. Nanotechnology and nanostructured materials: trends in carbon nanotubes. Precision Eng. 28 (1), 16–30. https://doi.org/10.1016/j.precisioneng.2002.11.002.
Nidzworski, D., Pranszke, P., Grudniewska, M., Król, E., Gromadzka, B., 2014. Universal biosensor for detection of influenza virus. Biosensors Bioelectron. 59, 239–242. https://doi.org/10.1016/j.bios.2014.03.050.
Nitin, C., G., G.V., G., B.L., J., H.B., G., B.L, 2007. Functional one-dimensional nanomaterials: applications in nanoscale biosensors. Anal. Lett. 40 (11), 2067–2096. https://doi.org/10.1080/00032710701567170.
Ozer, T., Geiss, B.J., Henry, C.S., 2019. Chemical and biological sensors for viral detection. J. ElectroChem. Soc. 167 (3), 037523.
Pandit, S., Dasgupta, D., Dewan, N., Prince, A., 2016. Nanotechnology based biosensors and its application. Pharma Innov. 5 (6), 18 Part A).
Paton, K.R., Varrla, E., Backes, C., Smith, R.J., Khan, U., O'Neill, A., Boland, C., Lotya, M., Istrate, O.M., King, P., 2014. Scalable production of large quantities of defect-free few-layer graphene by shear exfoliation in liquids. Nat. Mater. 13 (6), 624–630.
Pedersen, D.B., Duncan, E., 2005. Surface plasmon resonance spectroscopy of gold nanoparticle-coated substrates: Use as an indicator of exposure to chemical warfare simulants. Defence Research and Development Suffield (Alberta).

Peña-Bahamonde, J., Nguyen, H.N., Fanourakis, S.K., Rodrigues, D.F., 2018. Recent advances in graphene-based biosensor technology with applications in life sciences. J. NanoBiotechnol. 16 (1), 1–17.
Rashidi, L., Khosravi-Darani, K., 2011. The applications of nanotechnology in food industry. Crit. Rev. Food Sci. Nutr. 51 (8), 723–730.
Rasouli, R., Barhoum, A., Bechelany, M., Dufresne, A., 2019. Nanofibers for biomedical and healthcare applications. MacroMol. BioSci. 19 (2), e1800256. https://doi.org/10.1002/mabi.201800256.
Rossi, A.M., Wang, L., Reipa, V., Murphy, T.E., 2007. Porous silicon biosensor for detection of viruses. Biosensors Bioelectron. 23 (5), 741–745. https://doi.org/10.1016/j.bios.2007.06.004.
Rowe-Taitt, C.A., Golden, J.P., Feldstein, M.J., Cras, J.J., Hoffman, K.E., Ligler, F.S., 2000. Array biosensor for detection of biohazards. Biosensors Bioelectron. 14 (10–11), 785–794.
Sahu, J.N., Karri, R.R., Zabed, H.M., Shams, S., Qi, X., 2019. Current perspectives and future prospects of nano-biotechnology in wastewater treatment. Sep. Purif. Rev. 50 (2), 139–158. https://doi.org/10.1080/15422119.2019.1630430.
Sahu, J.N., Zabed, H., Karri, R.R., Shams, S., Qi, X., 2019. Applications of nano-biotechnology for sustainable water purification. Micro and Nano Technologies, Thomas, S., Grohens, Y., Pottathara, Y.B. (Eds.). Elsevier, Amsterdam, pp. 313–340. https://doi.org/10.1016/B978-0-12-815749-7.00011-6.
Sapsford, K.E., Pons, T., Medintz, I.L., Mattoussi, H., 2006. Biosensing with luminescent semiconductor quantum dots. Sensors 6 (8), 925–953.
Shen, H., Zhang, L., Liu, M., Zhang, Z., 2012. Biomedical applications of graphene. Theranostics 2 (3), 283.
Shintaro, S., 2015. Graphene for nanoelectronics. Jpn. J. Appl. Phys. 54, 040102. https://doi.org/10.7567/JJAP.54.040102.
Sperling, R.A., Gil, P.R., Zhang, F., Zanella, M., Parak, W.J., 2008. Biological applications of gold nanoparticles. Chem. Soc. Rev. 37 (9), 1896–1908.
Thiébaut, B., Gordon, R., 2019. Nanotechnology and biosensors. Johnson Matthey Technol. Rev. 63 (2), 143–146. http://dx.doi.org/10.1595/205651319x15518685669774.
Vamvakaki, V., Chaniotakis, N.A., 2007. Carbon nanostructures as transducers in biosensors. Sensors Actuat. B Chem. 126 (1), 193–197.
Wang, S.G., Qing, Z., Ruili, W., Y, S.F., 2003. A novel multi-walled carbon nanotube-based biosensor for glucose detection. BioChem. BioPhys. Res. Commun. 311 (3), 572–576. https://doi.org/10.1016/j.bbrc.2003.10.031.
Wang, X., You, H., Liu, F., Li, M., Wan, L., Li, S., Li, Q., Xu, Y., Tian, R., Yu, Z., 2009. Large-scale synthesis of few-layered graphene using CVD. Chem. Vap. Depos. 15 (1-3), 53–56.
Wang, Z., Lee, S., Koo, K., Kim, K., 2016. Nanowire-based sensors for biological and medical applications. IEEE Trans. NanobioSci. 15 (3), 186–199.
Wei, D., Bailey, M.J.A., Andrew, P., Ryhänen, T., 2009. Electrochemical biosensors at the nanoscale. Lab Chip 9 (15), 2123–2131. https://doi.org/10.1039/b903118a.
Wennersten, R., Fidler, J., Spitsyna, A., 2008. Nanotechnology: a new technological revolution in the 21st Century. In: Advani, S.G., Hsiao, K-T (Eds.), Handbook of Performability Engineering. Springer, Cham, pp. 943–952.
Wohlstadter, Jacob N, Wilbur, J.L., Sigal, G.B., Biebuyck, H.A., Billadeau, M.A., Dong, L., Fischer, A.B., Gudibande, S.R., Jameison, S.H., Kenten, J.H., 2003. Carbon nanotube-based biosensor. Adv. Mater. 15 (14), 1184–1187.
Wu, Q., Zhang, Y., Yang, Q., Yuan, N., Zhang, W., 2019. Review of electrochemical DNA Biosensors For Detecting Food Borne Pathogens. Sensors 19 (22), 4916.
Xueqing, Z., Qin, G., Daxiang, C., 2009. Recent advances in nanotechnology applied to biosensors. Sensors 9 (2), 1033–1053. https://doi.org/10.3390/s90201033.
Yao-Kuang, C., Julien, R., Chuan, L.K., Min, L.H., Yi, L.P., Yanping, W.K., Cheong, T.K., HongMiao, J., Yu, C., 2011. An electrical biosensor for the detection of circulating tumor cells. Biosensors Bioelectron. 26 (5), 2520–2526. https://doi.org/10.1016/j.bios.2010.10.048.
Zhang, J., Mou, L., Jiang, X., 2020. Surface chemistry of gold nanoparticles for health-related applications. Chem. Sci. 11 (4), 923–936.

CHAPTER 21

Ultrasmall fluorescent nanomaterials for sensing and bioimaging applications

Jigna R. Bhamore[a,b], Tae-Jung Park[b] and Suresh Kumar Kailasa[a]
[a]Department of Chemistry, Sardar Vallabhbai National Institute of Technology, Surat – 395007, Gujarat, India
[b]Research Institute of Chem-Bio Diagnostic Technology, Department of Chemistry, Chung-Ang University, 84 Heukseok-ro, Dongjak-gu, Seoul 06974, Republic of Korea

21.1 Introduction

Over the past few decades, nanomaterials have attained significant interest in analytical and bioanalytical chemistry due to their outstanding optical properties (Rafique et al., 2019). The fabrication of nanomaterials-based sensor is one of the key applications in nanoanalytical science. Nanomaterials integrated analytical tools have proved to be rising and promising analytical platforms in assaying various molecules including pesticides, drugs, inorganic species, and biomolecules (Kailasa et al., 2008, 2018; Kailasa & Wu, 2015). In recent years, nanomaterials have been shown exciting applications in multidisciplinary science because of their size-dependent electrical, optical, and molecular properties (Aragay et al., 2012). Fluorescent nanomaterials exhibited outstanding features including color tenability, photophysical and photochemical properties. Furthermore, nanomaterials provide analyte-specific interaction due to the molecular assemblies on the surfaces of nanoparticles (NPs) (Kailasa & Rohit, 2017a, 2017b; Mehta et al., 2016; Rana et al., 2018). Especially, ultrasmall fluorescent nanomaterials (metal nanoclusters (NCs) and carbon dots (CDs)) have received considerable interest in analytical and bioanalytical sciences due to their enhancing analytical figures (selectivity, simplicity, and sensitivity). As a result, ultrasmall nanomaterials have been used as potential probes for various analytical applications (Jaque & Vetrone, 2012; Kasibabu, Bhamore, et al., 2015; Kasibabu, D'Souza, et al., 2015; Kasibabu, D'souza, et al., 2015). Furthermore, ultrasmall fluorescent nanomaterials such as metal NCs and CDs have evolved as a new class of fluorescent water-dispersible and biocompatible nature, offering impressive sensing and bioimaging applications These fluorescent metal NCs, as a new class of ultrasmall, water-soluble, stable, and biocompatible fluorophores, could complement or even replace conventional fluorescent probes, offering immense opportunities for advancements in sensor and biological label technologies (D'Souza, Deshmukh, Bhamore, et al., 2016; D'Souza, Deshmukh, Rawat, et al., 2016; D'souza et al., 2018; Kailasa et al., 2015; Li & Jin, 2013).

Sustainable Nanotechnology for Environmental Remediation
DOI: https://doi.org/10.1016/B978-0-12-824547-7.00003-5

In this chapter, we intend to summarize the recent reports on the synthesis, properties, and applications of ultrasmall fluorescent nanomaterials (metal NCs and CDs). This chapter divided into two main sections: one section illustrates the metal NCs synthetic approaches, properties and their analytical applications (sensing and bioimaging). The second section describes the synthetic approaches, properties, and applications (sensing and bioimaging) of CDs.

21.2 Metal NCs

Among ultrasmall nanomaterials, metal NCs have fascinated as exceptional and potential probes in identifying and quantifying molecular species at micro volume samples because of their molecule-like properties (Díez & Ras, 2011; Li & Jin, 2013). Metal NCs consist of few to hundred atoms and their size is equal to the Fermi wavelength of electrons. Thus, metal NCs exhibited tremendous optical, electrical, and molecular properties together with high photoluminescence (PL), excellent photostability, large stroke shift, brilliant biocompatibility, and subnanometer diameter. Because of all these properties, metal NCs are used as stimulating agents in various fields of science including therapeutics, bioimaging, biosensing and chemosensing, modern catalysis, and electronic devices (Díez & Ras, 2011). Luminescent metal NCs have been explored as a new group of fluorophores in recognizing a wide variety of molecules. As of now fluorescence-based analytical strategies have been performed using organic dye, fluorescence proteins, and semiconductor quantum dots (QDs) as probes. Organic fluorophores exist as a variety of chemical functional groups with typical spectral properties, but they are expensive, toxic, and needs tedious steps to fabricate pure fluorophore. Generally, metal NCs exhibit interband transitions between the ligand/d-band and the sp band or electronic transitions between highest occupied molecular orbital and lowest unoccupied molecular orbital. As a result, metal NCs exhibit fluorescence properties due to the metal core with its intrinsic quantization effects, and to the particle surface governed by the interaction between metal core and ligand chemistry.

21.2.1 Properties of metal NCs

21.2.1.1 Absorption and fluorescence

Optical properties of metal NPs are governed by surface plasmon resonance (SPR), in which collective oscillation of electrons occurs at the surface of NPs. As size decreased to less than 3 nm, metal NCs cannot hold SPR property and behave as a molecule due to the presence of ~25 atoms in each cluster. These few atom metal NCs display discrete energy levels and exhibit unique spectral properties, which are quite different from their large size nanomaterials. Further, definite metal NCs rely upon a distinctive absorption and can be recognized from each other by their absorption profiles (Muhammed et al., 2009). For example, glutathione (GSH)-capped Au_{25} NCs display a few trademark assimilations in the range of 400–1000 nm, which are ascribed from intraband (sp←sp) or interband advances (sp←d) of the mass gold. Negishi et al. (2005) demonstrated that the absorption peak of GSH- Au NCs is due to intraband change, which has tendency to blue-shift with diminishing core size of metal NCs. Due to their ultrasmall size (~3.0 nm), the distinct states are separated and expanded in each band, which leads blue shift in the absorption spectra. Similarly, the absorption peaks of AgNCs additionally show discrete electronic changes rather than collective plasmon excitations. For example, Petty and co-workers (Petty et al., 2004) announced fluorescent Ag_1—Ag_4 NCs with retention tops at 440 nm and 357 nm respectively. Discrete electronic changes were plainly demonstrated due to few atoms in the metal NCs, which display subatomic conduct.

21.2.1.2 Two-photon absorption

To energize the atom from the ground state to higher state, simultaneous absorption of photon of same or discrete frequencies is required, which is referred as two-photon absorption (TPA). TPA in near-infrared region increases the penetration depth and spatial resolution of metal NCs because of lower scattering and reduce the autofluorescence compared to one-photon excitation, which is well suited for bioimaging and photodynamic treatment (So et al., 2000). TPA properties of ultrasmall-sized Au NCs in hexane were investigated and found that the fabricated AuNCs exhibited two photons excited at fluorescence in the visible region when excited at 800 nm, which exhibit superior optical properties than that of other organic chromophores. Liu's group studied TPA properties of water dispersible 11-mercaptoundecanoic acid–capped AuNCs, which provides TPA cross-section of around 8761 GM (Hsiao et al., 2009). Patel and co-workers determined that the highest TPA cross-section for DNA encapsulated AgNCs, which is 35,000 GM for 660 nm, 34,000 GM for 680 nm, and 50,000 GM for 710 nm excitation wavelengths (Patel et al., 2008). This large TPA cross-section makes them efficient candidates for the multiphoton bioimaging of various cells and other optical applications.

21.2.1.3 Solvatochromic properties

Metal NPs display obvious solvatochromic impact because of their SPR. Recent investigations revealed that the metal NCs show comparable solvatochromic properties. Solvent-dependent PL properties of Au_8NCs were examined by Zhou et al. (2009) when exposed to various organic solvents that is, chloroform, tetrahydrofuran, and N,N-dimethylformamide. They have also investigated the solvatochromic effect of Au_8NCs capped by various ligands that is, amino acids, peptide, protein, and DNA and the solvatochromic properties of AuNCs were attributed to the surface functionalizations of AuNCs. The solvent effects on the optical properties of AgNCs were also studied by Diez's group (Díez et al., 2009). It was observed that the absorption and fluorescence properties of AgNCs were effectively tuned by selecting a suitable solvent. An around

70 nm stroke shift of absorption peak was observed by changing the solvent from water to methanol for AgNCs. DNA-encapsulated AgNCs exhibited solvent polarity–dependent absorption and fluorescence.

21.2.1.4 Fluorescence lifetime

With respect to the fluorescence lifetime of metal NCs, most of metal NCs display mono-exponential nanosecond fluorescence lifetimes. Through a several picoseconds segment and the other 2–3 ns part, some polymer- and peptide-synthesized metal NCs demonstrated biexponential lifetime decays. However, thiol-encapsulated metal NCs exhibited exceedingly short picoseconds fluorescence lifetimes (Udaya Bhaskara Rao & Pradeep, 2010). The ligands, that is, dihydrolipoic acid (DHLA) and liopic acid capped Au/AgNCs, exhibited submicrosecond fluorescence lifetime, these features make them as promising probes for sensitive and selective detection of various analytes and imaging of various cells (Shang et al., 2013). For example, liopic acid–encapsulated AgNCs exhibit 37 μs fluorescence lifetime when excited at 425 nm. Poly(methacrylic acid) (PMAA) capped-Ag NCs have also demonstrated the multiple lifetime segments, showing higher lifetime . These functionalized metal NCs exhibited unique spectral properties and successfully integrated with fluorescence spectroscopic tools for the analysis of various molecules with high sensitivity.

21.3 Synthesis of metal NCs

Recent years, much attention has been focused on the synthesis of metal NCs with high quantum yield (QY) and their applications in analytical sciences. For the synthesis of ultrasmall fluorescent NCs, selection of capping and stabilizing agent plays a key role to encapsulate and to reduce metal ions with atomically precise metal NCs, which allow them to exhibit unique fluorescence properties. In view of this, various synthetic routes have been developed for the fabrication of fluorescent metal NCs using various templates. The following sections will highlight the synthesis of metal NCs by using various organic ligands.

21.3.1 Mercapto group–containing molecules

Recently, various metal (Au, Ag, and Cu) NCs have been synthesized by using various mercapto group containing molecules as templates (Jin, 2010). For example, GSH was selected as a protective ligand for the controlled reduction of Au^{3+} ion, which leads to form atomically precise Au NCs (Link et al., 2002). Moreover, various metal NCs have been synthesized by using GSH as a template and studied their spectral characteristics. In addition, a series of other mercapto group–containing molecules such as tiopronin thiolate, thiolate alpha cyclodextrin and 3-mercaptopropionic have been utilized as effective ligands for the fabrication of ultrasmall fluorescent AuNCs (Paau et al., 2010; Wang et al., 2009). Adhikari and Banerjee used bidentate (DHLA) as a reducing and capping agent for the synthesis of AgNCs. In this, lipoic acid, and Ag^+ were reduced by using $NaBH_4$, and the particle size and fluorescence properties of AgNCs were strongly dependent on the mercapto ligand and Ag^+ ion ratio (Adhikari & Banerjee, 2010). In spite of the above specific one-step metal reduction protocols by thiols, the fluorescent metal NCs have also been synthesized by etching of metal NPs with thiols. By utilizing the excess amount of GSH, Abubaker et al. (2008) fabricated luminescent AuNCs from mercapto succinic acid-capped AuNPs by etching process. The etching strategy has considered as pH-dependent synthetic process and the maximum yield of Au_{25} and Au_8 NCs were achieved at pH 3. In etching process, two possible mechanisms were proposed to generate metal NCs that is, gold atoms are isolated from the NPs surface by excess amount of ligands, resulting to form Au^+-thiolate complex, which further leads to strong aurophilic interaction to form fluorescent AuNCs. In second, ligands were used to etch the surface of Au atom of AuNPs, yielding to develop small-sized luminescent particles. The evacuated atoms have ability to form Au^+-thiolate complex, which allows the reduction of Au^+ ion with atomically precise atoms. Furthermore, Muhammed et al. described an interfacial way for the synthesis of red luminescent Au_{23} NCs through etching process (Muhammed et al., 2009). Similarly, two-step "bottom-up" approach was developed for the synthesis of water-soluble mercaptobenzoic acids (MBA) protected AuNCs (Bertorelle et al., 2018). In this approach, three isomers of MBA (p/m/o-MBA) and Me_3NBH_3 were used as precursors. The $Au_{25}(MBA)_{18}$ isomers exhibited absorption bands at 690, 470, and 430 nm. In addition, GSH-capped Cu NCs were synthesized and used as a probe for fluorescence detection of Zn^{2+} ion via a fluorescence-enhanced mechanism (Lin et al., 2017).

21.3.2 Peptide and proteins

Peptides and proteins have been used as potential templates for the fabrication of fluorescent metal NCs with high QY. Protein have structural similarities with dendrimers and polymers, thereby encapsulating the metal atoms with a controlled reduction process, which results to form NCs with precisely modified atomic structures. Xie et al. developed a strategy for one-step synthesis of bright red fluorescent AuNCs by using bovine serum albumin (BSA) as a reducing and capping agent at physiological temperature (37 °C) (Xie et al., 2009). The metal NCs have been fabricated by a bio-mineralization process using proteins as a template. For this, Au^+ ion are effectively coordinated with BSA, where BSA acts as a reducing agent at basic media (pH 12.0). As a result, BSA-protected AuNCs were formed and used as probes for fluorescence detection of various molecules. Miao et al. adopted environment friendly method for the fabrication of BSA-capped CuNCs (Miao et al., 2018). Blue emitted lysozyme protected CuNCs were synthesized for cell imaging

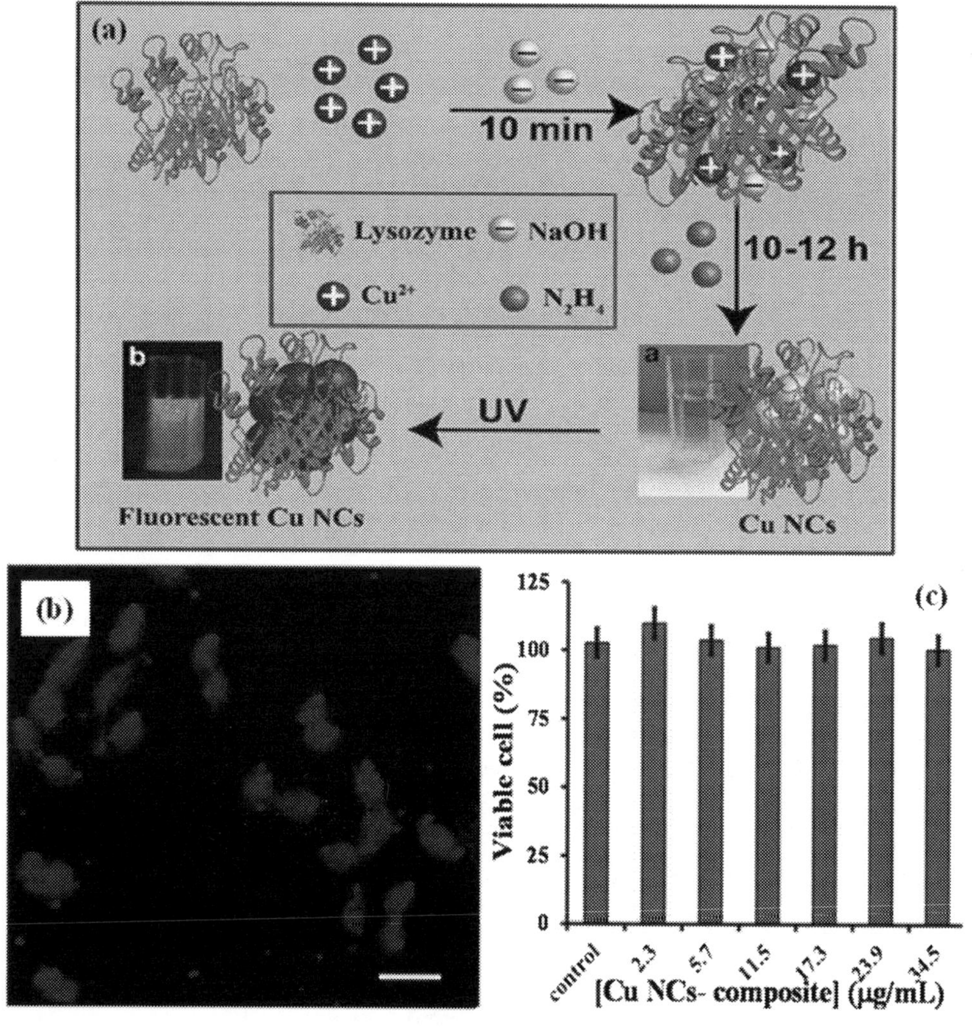

Figure 21.1 (A) Preparation of blue fluorescent Cu NCs using lysozyme as a ligand. (B) Fluorescence microscopic images of HeLa cells using lysozyme Cu NCs. (C) Evaluation of lysozyme-Cu NCs cytotoxicity on HeLa cells as measured by XTT assay after 24 h. From Ghosh, R., Sahoo, A. K., Ghosh, S. S., Paul, A., & Chattopadhyay, A. (2014). Blue-emitting copper nanoclusters synthesized in the presence of lysozyme as candidates for cell labeling. ACS Appl. Mater. Interf., 6(6), 3822–3828. https://doi.org/10.1021/am500040t.

applications as shown in Fig. 21.1 (Ghosh et al., 2014). Mohanty's group built-up a simple synthetic strategy for the fabrication of Au-AgNCs using BSA as a template (Mohanty et al., 2012). Moreover, horseradish peroxidase, insulin, pepsin, and transferrin have been utilized for the synthesis of metal NCs and used as promising probes for assaying of various molecules and imaging of various cells.

21.3.3 Nucleic acids

It is well known that metal ions and DNA have distinguished interactions, which allows developing DNA template–based metal NCs preparation. In this connection, Ag^{+} ions have high affinity to bind with cytosine moiety of single-stranded DNA, which yields to form AgNCs with controlled size as well as atoms. Petty et al. (2004) revealed the combination of short oligonucleotide-stabilized AgNCs using $NaBH_4$ as a reducing agent. Mass spectra of synthesized AgNCs revealed that each AgNC consisted with 1–4 Ag atoms bind with 12-base oligonucleotide. Atomically precise monodispersed Au NCs were obtained by etching of gold atoms using amino acids, protein, and DNA as templates via a sonication process. Under sonication, double-stranded DNA was used to etch Au nanorods for the synthesis of fluorescent Au_8NCs. The DNA molecules act as promising ligands to hold Au^{3+} ions for their controlled reduction to form AuNCs.

21.4 Applications of metal NCs

21.4.1 Metal NCs as probe for metal ions sensing

Due to their molecule-like properties, metal NCs have been used as promising probes for the fluorescence detection of various metal ions. The metal NCs–based analytical strategy has ability to recognize metal ions via either "fluorescence turn on" or "turn off" mechanisms. In the fluorescence "turn-off" system, fluorescence was quenched due to the metallophilic interaction between metal ion and metal NCs or to the aggregation of metal NCs by metal ions. In contrast, in fluorescence "turn-on" system, metal ion is deposited on the metal NCs surface or aggregates due to metal ions, yielding to enhance the fluorescence intensity. Due to the potential hazardous effect of metal ions (Hg^{2+}, Cu^{2+}, Pb^{2+}, Fe^{3+}, Cr^{3+}, and As^{3+}) on the biological system and environment, it is essential to detect metal ions at a trace level. To this, Xie et al., developed BSA-stabilized Au NCs system for the sensitive detection of Hg^{2+} ions (Xie et al., 2010). The fluorescence of ultrasmall Au NCs was effectively quenched by the addition Hg^{2+} ions due to metallophilic interaction (d^{10}–d^{10}) between Hg^{2+} ion and AuNCs. Deng's group developed a simple analytical strategy for fluorescence "turn-on" detection of Hg^{2+} ion using DNA duplex–stabilized Ag NCs as a probe (Deng et al., 2011). Zhou et al., fabricated lysozyme templated AgNCs for fluorescence sensing of Hg^{2+} ion, which is based on the quenching of AgNCs via $5d^{10}(Hg^{2+})$-$4d^{10}(Ag^{+})$ metallophilic interaction (R. Zhou et al., 2009). The fluorescence intensity of DNA-encapsulated Cu/Ag NCs was quenched by 3-mercaptopropionic acid; however, fluorescence was restored on the addition of Cu^{2+} ion, which facilitates to detect Cu^{2+} ion even at 2.7 nM (Su et al., 2010). Shang and Dong developed (PMAA) AgNCs-based sensitive fluorescence strategy for quantification of Cu^{2+} ion with a lower detection limit of 8 nM (Shang & Dong, 2008). A novel analytical strategy was developed for the detection of L-histidine (His) using

Cu^{2+}-assisted DNA-Ag NCs as a fluorescent probe (Zheng et al., 2015). The bimetallic Au/Ag NCs system was developed by Huang et al., for the detection of Fe^{3+} ion in living cell (Huang et al., 2016). Bian and coworkers developed on-site detection of Hg^{2+} and Pb^{2+} ion based on the AuNCs paper strips, where fluorescence was quenched due to Hg^{2+} ion and enhanced due to Pb^{2+} ion (Bian et al., 2017). In addition, a highly selective and sensitive AuNCs fluorescent sensor (turn on) was proposed by Roy's group and used as a promising probe for the detection of As^{3+} ion in water samples (Roy et al., 2012). Various metal NCs (Au, Cu, and Ag) have been applied as fluorescent probes for sensing of various metal ions including Pb^{2+}, Cd^{2+}, Al^{3+}, Zn^{2+}, Cu^{2+}, Co^{2+}, and Hg^{2+} ions from various sample matrices (R. Ghosh et al., 2014). Metal NCs have been synthesized using various proteins or biomolecules as templates, and exhibited good QY. The methods exhibited wider linear ranges and impressive detection limits for the above metal ions sensing, confirming the metal NCs acted as a promising fluorescent probe for sensing of trace-level metal ions with reduced sample preparations. For example, AuNCs were synthesized with bright green fluorescence using His and GSH as ligands as shown in Fig. 21.2. The synthesized green fluorescent AuNCs acted as a sensor for the detection of Pb^{2+} ion, showing good linearity (Fig. 21.2B,C).

21.4.2 Metal NCs as probes for sensing of organic molecules

Due to their significant roles including detoxification, metabolism process, and reversible redox reaction, the identification of biothiols such as homocysteine (Hcy), cysteine (Cys) and GSH plays important roles in estimating biochemical pathways in living organisms. In the early diagnosis of many diseases, detection of biothiols levels plays a critical role. A simple and sensitive system has been developed for the detection of trace-level Cys using PMAA-Ag NPs as a probe (Shang & Dong, 2009). The His-AuNCs system was successfully used to detect GSH in biofluids (Zhang et al., 2014). A nonfluorescent complex was formed between biothiols and DNA-encapsulated AgNCs, resulting to develop fluorescence "turn-off" method for simultaneous detection of Cys, Hcy, and GSH with detections limit of 4.0, 4.0, and 200 nM, respectively (Z. Huang et al., 2011). Recently, metal NCs (Pt and Au) have been synthesized via simple one-step reactions and used as probes for sensing of dopamine, spermine, deltamethrin, and GSH from various sample matrices (Bhamore, Jha, et al., 2019; Bhamore, Murthy, et al., 2019; Borse et al., 2020).

21.4.3 Metal NCs as probes for sensing of proteins

To develop protein sensor, particular receptors such as antibodies and aptamer are need to be attached on the surface of luminescent metal NCs, which can selectively bind with proteins. It is well known that thrombin is a coagulated protein in blood stream, which takes part in the conversion of soluble fibrinogen to insoluble strands and many

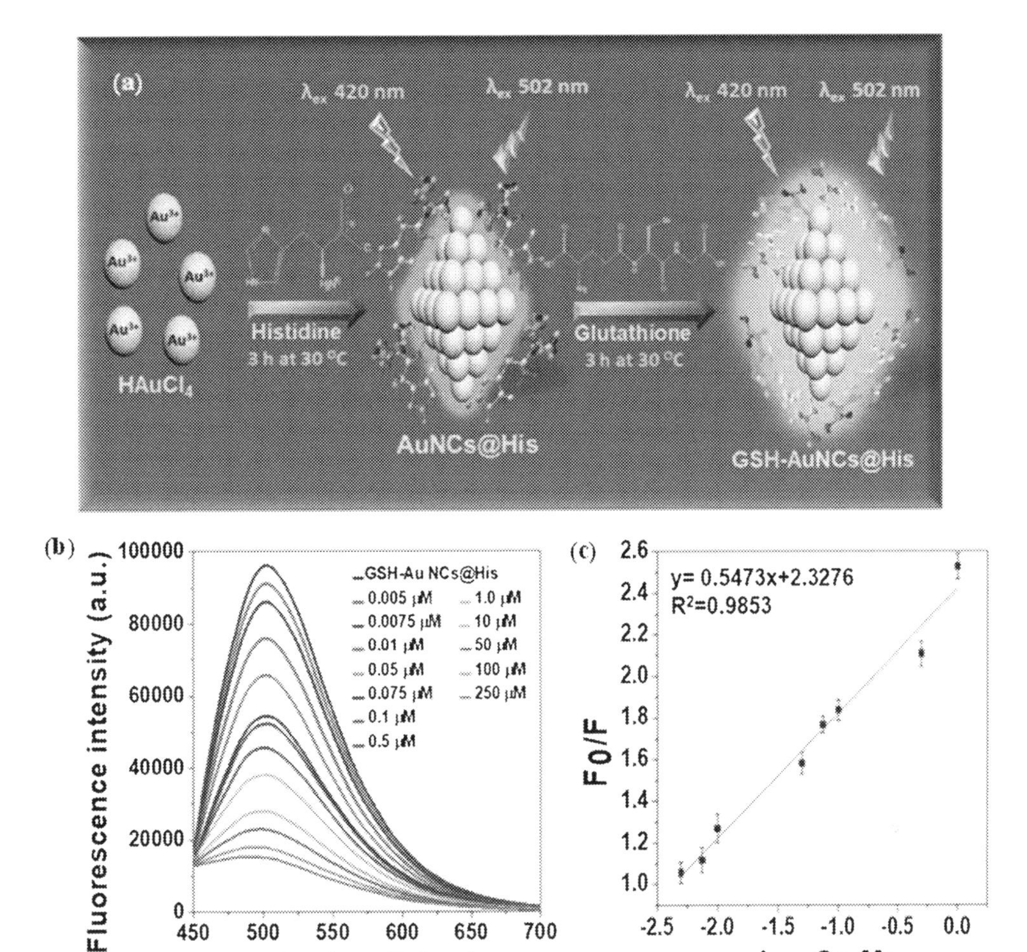

Figure 21.2 (A) Synthesis of AuNCs using GSH with His potential ligands. (B) Fluorescence emission spectra of AuNCs with the addition of Pb^{2+} ion (0.005–250 μM). (C) The calibration graph plotted in the range of 0.005–1.0 μM. Reprinted with permission From Bhamore, J. R., Gul, A. R., Chae, W. S., Kim, K. W., Lee, J. S., Park, H., Kailasa, S. K., & Park, T. J. (2020). One-pot fabrication of amino acid and peptide stabilized gold nanoclusters for the measurement of the lead in plasma samples using chemically modified cellulose paper. Sensors Actuat. B Chem., 322, 128603. https://doi.org/10.1016/j.snb.2020.128603.

other catalytic activities. Martinez's group proposed a simple analytical strategy for the sensitive detection of thrombin by using DNA-conjugated Ag NCs as a probe. DNA aptamer has strong binding affinity toward thrombin. The DNA has two active domains that is, one for strong coordination with thrombin, and the second one for the fabrication of AgNCs (Sharma et al., 2011). Li et al. developed a fluorescent "turn-on" system for the detection of human α-thrombin by using DNA-templated Ag NCs as a probe

(D'Souza, Deshmukh, Bhamore, et al., 2016; D'Souza, Deshmukh, Rawat, et al., 2016; D'souza et al., 2018; Kailasa et al., 2015; G. Li & Jin, 2013). It was observed that human α-thrombin has remarkably "turned-on" the red fluorescence of DNA-Ag NCs, showing good linearity between the fluorescence intensity and concentration of human α-thrombin (5 nM–4 μM) (Fig. 21.3A,B). The detection limit was 1 nM based on a signal-to-background ratio of 3. The fluorescent enhancement was based on the binding-induced DNA hybridization and luminescence enhancement of Ag NCs with G-rich DNA sequences. The ssDNA-binding protein (SSB) is an essential protein in cells for all living organisms. Lan et al., developed a DNA-based Cu/Ag NCs system for ultrasensitive detection of SSB in biofluids (Lan et al., 2011a). The SSB has strong affinity to bind with DNA template, resulting a change in the conformation of template, which yields to quench the fluorescence of DNA-stabilized Cu/Ag NCs.

21.4.4 Metal NCs as probes for sensing of nucleic acids

Arrangement of luminous metal NCs relies on various variables, including solvent, pH, and template. For example, the luminescence properties of DNA-stabilized AgNCs may shift incredibly contingent upon the nearby condition and nucleotide sequences, which allows to quantify the nucleic acid. By virtue of the profoundly sequence-dependent formation of fluorescent Ag NCs in hybridized DNA duplex platforms, Guo's group (Guo et al., 2010) built-up AgNCs-based fluorescent test for specific recognition of single-nucleotide alterations in DNA. Lan and co-workers developed one-pot synthetic approach for DNA-Ag NCs and used as a probe for selective and sensitive detection of DNA (Lan et al., 2011a). The particular DNA scaffold binds with the DNA sequence for the recognition of target DNA.

21.4.5 Metal NCs as probes for bioimaging applications

Metal NCs render as promising fluorescent probes for the biological imaging because of their attractive feature such as ultrasmall size, excellent biocompatibility, nontoxic behavior, and photostability. In addition, a large stroke shift of metal NCs can prevent the spectral cross-link thereby increasing the detection signal. Due to deeper tissue penetration and low fluorescence background, metal NCs attracted more attention. Various imaging tools have been used for the detection of cancer at early stage including magnetic resonance imaging, photoacoustic imaging, and X-ray computed tomography, which contains the distinctive advantage with inherent limitations. Up to now, various groups have reported massive work on fluorescent metal NCs–based biological imaging applications. The first imaging application of AgNCs reported by Makarava et al. (2005) was presented in combination with a fluorophore, thioflavin T (ThT). The DHLA-Au NCs were used as intracellular imaging probes for Hg^{2+} ion in living human cervical cancer cells (HeLa) cells, where intracellular fluorescence quenching was

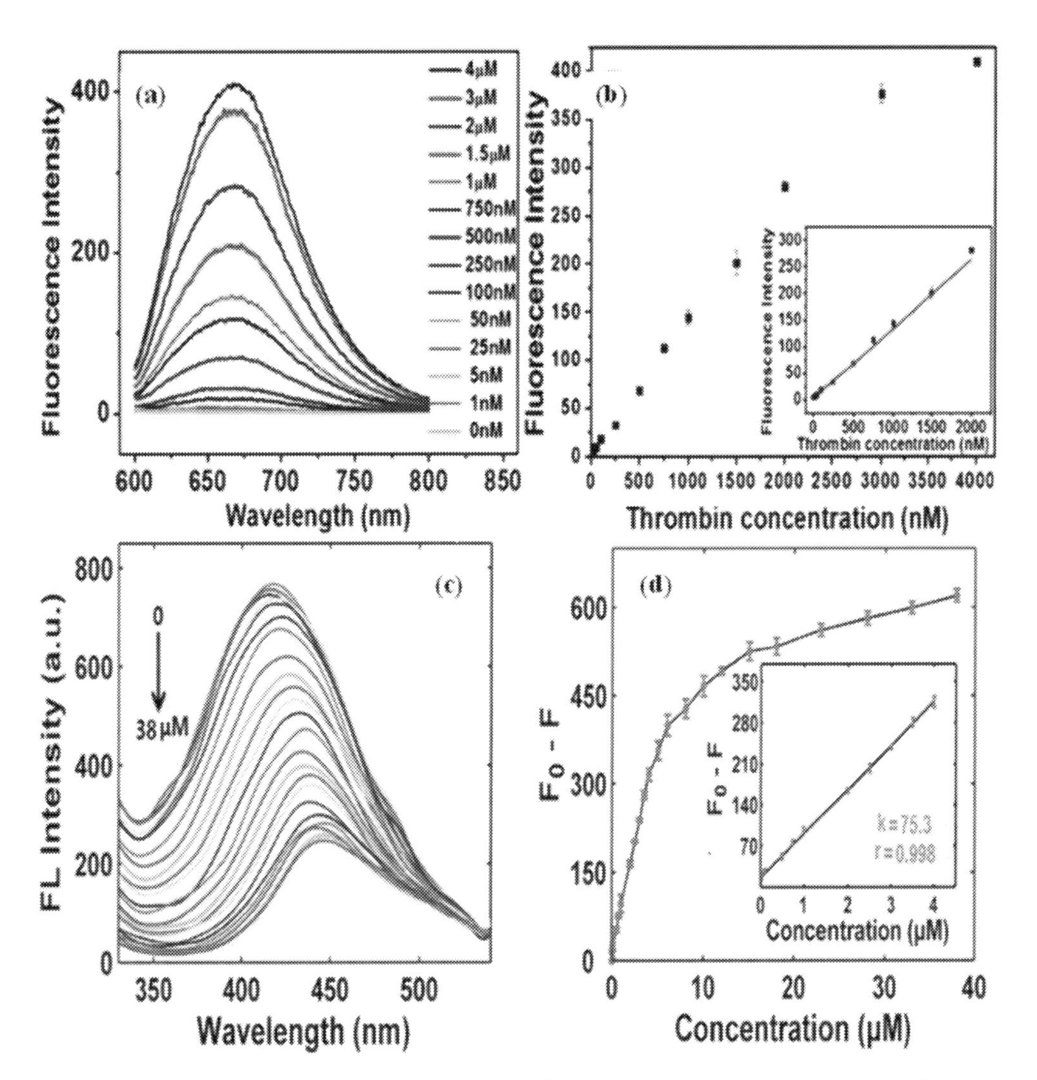

Figure 21.3 (A) Fluorescence spectra of DNA-Ag NCs for the detection of human α-thrombin (0–4 μM) via fluorescence turn-on mechanism. (B) Calibration graph was constructed between the emission at 669 nm and the concentration of human α-thrombin (5 nM–2 μM). Reprinted with permission from Ref. (J. Lee et al., 2012).Emission spectra of CuNCs upon the addition of nitrofurantoin (0–38.0 μM). (D) Calibration graph was constructed between the value of (F0-F) at 285 nm and nitrofurantoin. Reprinted with permission from Ref. (Y. Wang et. al., 2018) From Li, J., Zhong, X., Zhang, H., Le, X. C., & Zhu, J. J. (2012). Binding-induced fluorescence turn-on assay using aptamer-functionalized silver nanocluster DNA probes. Anal. Chem., 84(12), 5170–5174. https://doi.org/10.1021/ac3006268.

observed upon the addition of Hg^{2+} ion (Shang et al., 2012). Wang et al. observed that the red fluorescent peptide-Au NCs particularly target the nucleus of various cell line, which includes normal cells human gastric mucosa cells, HeLa, and human embryonic lung fibroblast cells (Paau et al., 2010; Z. Wang et al., 2009; Y. Wang et al., 2012). The AS14115-encapsulated Ag NCs have shown high affinity to bind with nucleolin of cancer cell thereby establishing a simple tool for the bioimaging of HeLa cells (Ai et al., 2012). Table 21.1 summarizes the analytical applications of metal NCs for sensing of various target analytes from various samples (X. Zhang et al., 2014).

For example, the emission spectra of adenosine-Cu NCs were gradually quenched with increasing concentration of nitrofurantoin, exhibiting good linearity between F_0-F values and concentration of nitrofurantoin (0.05–4.0 μM) (Fig. 21.3C,D), which achieved the limit of detetion (LOD) of 30 nM.

21.5 Carbon dots

CDs have chosen to be a new class of materials, which are quite different from NPs. The CDs are normally little oxygenous carbon NPs with spherical geometry and have size less than 10 nm. The CDs represent a fascinating class of photoluminescent 0-D carbon nanomaterials and exhibit tremendous applications in the field of analytical sciences and biomedical sciences. They possess physicochemical properties with multiple surface functionalities and low toxicity, allowing them as potential probes that can be effectively replaced toxic metal–based semiconductor QDs (Bhamore et al., 2016; Bhamore, Deshmukh, et al., 2018; Bhamore, Jha, Basu, et al., 2018; Bhamore, Gul, et al., 2020; Desai et al., 2018a, 2020; S. Ghosh et al., 2019; Kateshiya et al., 2020). Nowadays, the fluorescent CDs have been applied as potential candidates in various applications that is, bioimaging, sensing, and drug delivery (Ashrafizadeh et al., 2020; Bhamore et al., 2017; Bhamore, Jha, Park, et al., 2018; Bhamore, Jha, Singhal, et al., 2018; Bhamore, Jha, Park, et al., 2019; Bhamore, Park, et al., 2020; Desai et al., 2018b; Desai, Basu, et al., 2019; Desai, Jha, et al., 2019; Gupta et al., 2020; Kailasa, Ha, et al., 2019; Rawat et al., 2016; Mehta et al., 2017; Phan et al., 2019; Shahnawaz Khan et al., 2015; J. Zhang et al., 2016), because of their multifunctional groups, biocompatibility, low toxicity, and high PL. The formation of CDs was observed during the purging of single-walled carbon nanotubes via arc discharge process. Up to now, various synthetic methods have been used for the preparation of fluorescent CDs by using various precursors as carbon sources. Generally, the fluorescent CDs were synthesized by the following approaches that is, (1) top-down and (2) bottom-up approaches. The top-down approach deals with the breaking down of bigger carbon structures into ultrasmall particles using graphite, carbon soot, nanodiamonds, carbon nanotubes, activated carbon, and graphite oxide as carbon sources via laser ablation, arc discharge and electrochemical oxidation. On

Table 21.1 Overview of synthesis of metal NCs using various precursors and their properties and applications.

Metal NCs	Reducing agent	Size (nm)	Ex/Em (nm)	Applications	Reference
CuNCs	Double-strands DNA	~2	354/595	Biosensor for protein kinase activity	M. Wang et al., 2018
AgNCs	Lysozyme	–	365,490/ 445,640	Sensing of ascorbic acid	Mo et al., 2018
AuNCs	GSH	2.5±0.2	395/610	Sensing of iodide	W. Hou et al., 2018
AgNCs	Aptamer DNA	~1	585/650	Detection of Pb^{2+} ion	M. Zhang et al., 2018
CuNCs	Adenosine	2.4±0.3	285/417	Detection of nitrofurantoin	Y. Wang et al., 2018
CuNCs	Cysteamine	2.3±0.5	330/430	Detection of Al^{3+} ion	Boonmee et al., 2018
CuNCs	Ovalbumin	5.6 ± 1.1	370/440	Detection of L-lysine	Zhang & Wei, 2018
CuNCs	GSH	1.7 ± 0.3	365/442	Fe^{3+}sensing	H. Huang et al., 2017
AuNCs	Inositol	~1.95	370/470	Fe^{3+}sensing	Halawa et al., 2018
CuNCs	Dithiothreitol	~2.3	360/590	Al^{3+}sensing	X. Hu et al., 2017
AgNCs	Trisodium citrate dihydrate	~2	348/423	Hg^{2+}, Cu^{2+}, and Fe^{3+} sensing	Xiao et al., 2018
Au/AgNCs	Lysozyme	~1.75	420/665	Detection of ascorbic acid and acid phosphatase	Pang & Liu, 2017
CDs-AuNCs	BSA	~3–4	420/610	Dopamine sensing	He et al., 2018
AuNCs	GSH	< 2	–	$Fe^{3+}$and Cu^{2+} sensing	Zhao et al., 2016
AuNCs	GSH	1.04± 0.4	416/588	Sensing and imaging of Cu^{2+} and temperature in human cells and bacterial cells	L. Kong et al., 2016
AuNCs	N-acetyl-l-cysteine	~1–2	340/590	Detection of Hg^{2+} ion	Y. Zhang et al., 2016
AuNCs	β–Lactoglobulin	< 2	510/650	Detection of Hg^{2+} ion	J. Zhang et al., 2016
AuNCs	DNA templates	–	370/455	Detection of Hg^{2+} ion	Qing et al., 2016

(*continued on next page*)

Table 21.1 Overview of synthesis of metal NCs using various precursors and their properties and applications—cont'd

Metal NCs	Reducing agent	Size (nm)	Ex/Em (nm)	Applications	Reference
AuNCs	Mercaptan acids	1.8–2.2	300/ 590,615, 630, 630	Mo^{6+} and Hg^{2+} ion sensing	Y. Yang et al., 2017
AgNCs	poly(acrylic acid-co-maleic acid	2.5–4.7	500/600	–	Wu et al., 2018
AgNCs-N-doped CDs	$NaBH_4$	–	370/450	Determination of cysteine and GSH	S. Zhang et al., 2018
AgNCs	DNA-templated	~2–3	620/680	Detection of virulence genes	Shang et al., 2011

the other hand, the bottom-up approach deals with the synthesis of CDs from various molecular precursors such as carbohydrates, citrate, and polymer–silica nanocomposites through ignition/thermal treatment and microwave irradiation. Usually, the fluorescent CDs exhibit excitation-dependent PL because of surface defeats, quantum size impact, triplet carbenes at the zigzag edges (Boonmee et al., 2018; Halawa et al., 2018; He et al., 2018; Hou et al., 2018; Hu et al., 2017; H. Huang et al., 2017; Kong et al., 2016; Mo et al., 2018; Pang & Liu, 2017; Qing et al., 2016; Shang et al., 2011; M. Wang et al., 2018; Y. Wang et al., 2018; Wu et al., 2018; Xiao et al., 2018; Y. Yang et al., 2017; Zang et al., 2016; B. Zhang & Wei, 2018; M. Zhang et al., 2018; S. Zhang et al., 2018; Y. Zhang et al., 2016; Zhao et al., 2016), and radiative recombination of the excitons (Fowley et al., 2012). The PL properties of CDs play essential roles in chemical species sensing and bioimaging applications (Baker & Baker, 2010). Generally, the PL quenching is due to either collisional/dynamic quenching or static quenching. Both static and dynamic quenching requires subatomic contact between the fluorophore and quencher. In case of dynamic quenching, the quencher must diffuse with the fluorophore during the lifetime of the excited state. In static quenching, a complex is formed between the fluorophore and the quencher, which results to generate nonfluorescent complex. The quenching might be derived from various mechanisms including excited-state responses, molecular rearrangements, ground-state complex formation, and collisional quenching, energy/electron transfer and emission group destruction, respectively. The PL quenching mechanisms and the behavior of CDs have already been discussed in the literature (Jaschinski et al., 1999; Javey et al., 2003).

21.5.1 Properties of fluorescent CDs

21.5.1.1 Fluorescence nature

Recently, the fluorescent CDs have been considered as alternative to semiconductor QDs as the CDs exhibit the excellent fluorescence properties with high photostability and emission tenability. The excitation-dependent emission mechanism of CDs may be attributed to the radiative recombination of surface-confined electron and holes, which are generated from efficient photo-induced charge separations in the CDs (J. Xie et al., 2009). The surface passivation of CDs by the organic or other functional groups plays a critical role by providing more stability to surface sites, which result to exhibit a more effective radiative recombination (Sun & Lei, 2017). In CDs, isolated sp^2 hybridization between the graphene sheets with large distance could lead to radiative recombination and illuminate the bright blue fluorescence (Cao et al., 2011). It was confirmed that the multicolor emission of CDs arises not only from a radiative recombination between emissive trap sites but also from the different size of CDs (F. Wang et al., 2011).

21.5.1.2 QY of CDs

QY is another important parameter for the effective use of CDs in bioimaging and biosensing. The QY of CDs is mainly dependent on the synthesis, separation, and surface functionalities. The CDs without surface functionalities exhibit a bright fluorescence with generally very low QYs (Qu et al., 2012). Thus, effective surface passivation of CDs with the polymers, small organic molecules, and inorganic salts may dramatically improve the QY of CDs (W. Wang et al., 2014).

21.5.1.3 Stability and cytotoxicity of CDs

The stability of fluorescence CDs can be categorized into two ways that is, photostability and store-stability. In photostability, the fluorescence emission intensity remains stable during a long-time continuous excitation, while store stability revenue the fluorescence intensity of CDs remain after a long time period. The cytotoxicity is the most important parameter to use CDs as efficient fluorescent probes for bioimaging of live cells, tissues, and animals. Therefore, the extensive efforts have been devoted on the toxicity evaluation of CDs and the experimental data revealed that they are biocompatible in nature even at high concentrations (Bourlinos et al., 2008; Ding et al., 2014; Liu et al., 2007; J. Zhou et al., 2007; Zhu et al., 2009).

21.5.2 Synthetic routes for the preparation of CDs

Nowadays, significant advancements have been achieved in synthesizing fluorescent CDs using various molecules as precursors due to their phenomenal photostability, biocompatibility, low toxicity, high water dispersibility, high sensitivity and excellent selectivity, tunable fluorescence emission and excitation, high QY and large stokes shifts. Various synthetic approaches including laser ablation, electrochemical oxidation, combustion/microwave heating, and other synthesis method have been applied for the fabrication of fluorescent CDs (Liu et al., 2007). However, some of these strategies require either costly instrumentation or complex treatment forms. Therefore, there is an urgent need to build-up cost-effective technique for the facile synthesis of CDs, which meets the perspective of green chemistry. In this connection, a few techniques such as aqueous, solvothermal, and microwave synthesis strategies have been developed for the synthesis of fluorescent CDs.

21.5.2.1 Microwave method

Microwave is a sort of electromagnetic wave, which has the large wavelength range from 1.0 nm to 1.0 m. The microwave radiation provides rigorous energy to substrate to break the chemical bonds (J. Zhou et al., 2007). Importantly, the microwave method offers several advantages such as shorten reaction time, simultaneous and homogeneous heating, which allow to form CDs with uniform size. Microwave reactions are carried out by a microwave reactor to minimize the reaction time. A characteristic aspect of this method is that no surface passivation reagent is required. Moreover, the PL intensity of the synthesized CDs does not change at the pathological and physiological pH range of 4.5–9.5, and there is no photobleaching observed. Many works have been reported on the preparation of CDs by using a microwave reactor. Table 21.2 summarizes the use of microwave reactor for the synthesis of fluorescent CDs using various precursors and their optical properties. Chatzimitakos et al. (2018) prepared CDs using human nails by applying microwave heating at 400 W for 2 min. Mitra's group (Mitra et al., 2012) prepared fluorescent CDs by using poloxamer as a carbon source. Zhang et al. (2012) reported a microwave heating method for the preparation of fluorescent CDs using acrylic acid as a source. Chowdhury and co-worker (Chowdhury et al., 2012) synthesized fluorescent CDs from chitosan gel by microwave heating, and synthesized CDs exhibited blue fluorescence when irradiated under UV light. Many studies (Boonmee et al., 2018; Halawa et al., 2018; He et al., 2018; W. Hou et al., 2018; Hu et al., 2017; H. Huang et al., 2017; Kong et al., 2016; Mo et al., 2018; Pang & Liu, 2017; Qing et al., 2016; Shang et al., 2011; M. Wang et al., 2018; Y. Wang et al., 2018; Wu et al., 2018; Xiao et al., 2018; Y. Yang et al., 2017; Zang et al., 2016; B. Zhang & Wei, 2018; M. Zhang et al., 2018; S. Zhang et al., 2018; Y. Zhang et al., 2016; Zhao et al., 2016; Hou et al., 2013) have developed one-pot route for large-scale preparation of highly luminescent CDs powder by using triammonium citrate and phosphate as precursors by applying microwave heating at 750 W. Ramezani's group (Ramezani et al., 2018) presented one-step green synthetic approach for the preparation of CDs from quince fruit (*Cydonia oblonga*) powder as a carbon source via microwave irradiation (Fig. 21.4). Many studies (Bourlinos et al., 2008; Ding et al., 2014; H. Liu et al., 2007; J. Zhou et al., 2007; Zhu et al., 2009; Liu et al., 2011) have developed acid-driven microwave-assisted synthetic approach for the preparation of CDs from N, N-dimethylformamide. Q. Wang et al. (2012) reported a facile and green synthetic approach for the preparation of CDs through an eggshell membrane at 400 °C for 120 min. Zhai et al. (2012) described microwave-assisted pyrolysis for the synthesis of fluorescent CDs by using citric acid as a carbon source and used as a probe for biomedical

Table 21.2 Microwave synthesis of fluorescent CDs and their spectral characteristics and applications.

Source	Conditions		Size (nm)	UV-visible (nm)	Fluorescence (Ex/Em; nm)	Application	Reference
	Energy/ temperature	Time (min)					
Human nail	400 W	2	2.2	300	330/380	Bioimagining and biolabeling	Chatzimitakos et al., 2018
Poloxamer	450 W	4	5–20	285	380/442	Surface application	Mitra et al., 2012
Eggshell	400 °C	120	5	–	275/450	–	Q. Wang et al., 2012
Citric acid	–	–	2.2–3.0	240	360/460	Biomedical imaging Applications	Zhai et al., 2012
Amino acids	700 W	2.4	2 ± 0.4	299	360/440	Optoelectronic devices, bioassays, biomedical and bioimaging applications	Jiang et al., 2012
Poly(ethylene glycol)	900 W	10	1.4–4.0	245–325	325/445	Bio-labeling and bioimaging	Jaiswal et al., 2012
Calcium citrate, urea	800 W	5	2.18 ± 0.36	217,249, 272	330–385/420	Growth of fluorescent bean sprouts, and fingerprint detection tools	M. Xu et al., 2014
Lysin	750 W	4–6	5–10	208, 290	398/431	Bioimaging	Park et al., 2017
Shrimp eggs	180 °C	25	3.25 ± 1.06	295	280–500/420–520	Biological labeling and imaging	P. Y. Lin et al., 2014
Xylitol	700 W	2	4.65 ± 1.21	286	360/446	Biomedical applications	Kim et al., 2014

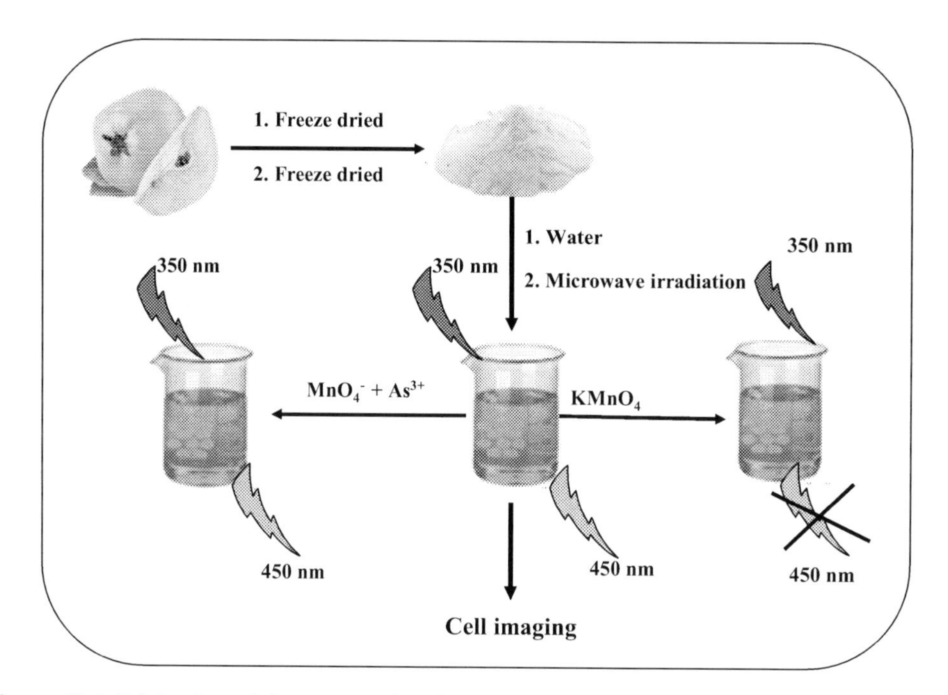

Figure 21.4 Fabrication of fluorescent CDs from quince fruit (*Cydonia oblonga*) via microwave irradiation. From Ramezani, Z., Qorbanpour, M., & Rahbar, N. (2018). Green synthesis of carbon quantum dots using quince fruit (*Cydonia oblonga*) powder as carbon precursor: application in cell imaging and As^{3+} determination. Colloids Surf. A Physicochem. Eng. Asp., 549, 58–66. https://doi.org/10.1016/j.colsurfa.2018.04.006.

imaging applications. Various organic compound such as amino acid (Jiang et al., 2012), polyethylene glycol (Jaiswal et al., 2012), calcium citrate (Xu et al., 2014), lysin (Park et al., 2017), shrimp eggs (L. Lin et al., 2017), and xylitol (Kim et al., 2014) have been successfully used for the preparation of fluorescent CDs via microwave heating.

21.5.2.2 Hydrothermal method

To synthesize nontoxic and eco-friendly CDs with low-cost materials, hydrothermal carbonization method was adopted for the fabrication of CDs using various natural resources as a carbon source. Generally, a solution of organic precursor or a natural source is conserved and reacted in a hydrothermal reactor at high temperature and pressure. Materials or natural resources such as chocolate (Y. Liu et al., 2016), corn flour (Wei et al., 2014), orange juice (Sahu et al., 2012), His (Dai et al., 2014), pork (Zhao et al., 2018), various fruits (Boonmee et al., 2018; Halawa et al., 2018; He et al., 2018; W. Hou et al., 2018; Hu et al., 2017; H. Huang et al., 2017; Kong et al., 2016; Mo et al., 2018; Pang & Liu, 2017; Qing et al., 2016; Shang et al., 2011; M. Wang et al., 2018; Y. Wang et al., 2018; Wu et al., 2018; Xiao et al., 2018; Y. Yang et al., 2017; Zang et al., 2016; B. Zhang & Wei, 2018; M. Zhang et al., 2018; S. Zhang et al., 2018; Y. Zhang et al., 2016; Q. Zhao et al., 2016), pure milk (W. Wang et al., 2014; D. Wang et al., 2014), *Saccharum officinarum* juice

(Mehta, Jha, & Kailasa, 2014), cucumber juice (C. Wang et al., 2014), and natural vegetables (J. Wang et al., 2014) have been used as carbon precursors in the hydrothermal method for the synthesis of fluorescent CDs. These developed CDs show blue or green fluorescence and efficiently act as probes for sensing and imaging applications. Similarly, numerous reports have described hydrothermal method for the synthesis of fluorescent CDs using organic molecules including citric acid (Song et al., 2014), mandelic acid (Boonmee et al., 2018; Halawa et al., 2018; He et al., 2018; W. Hou et al., 2018; Hu et al., 2017; H. Huang et al., 2017; Kong et al., 2016; Mo et al., 2018; Pang & Liu, 2017; Qing et al., 2016; Shang et al., 2011; M. Wang et al., 2018; Y. Wang, Chen, et al., 2018; Wu et al., 2018; Xiao et al., 2018; Y. Yang et al., 2017; Zang et al., 2016; B. Zhang & Wei, 2018; M. Zhang et al., 2018; S. Zhang et al., 2018; Y. Zhang et al., 2016; Q. Zhao et al., 2016), broccoli (Arumugam & Kim, 2018), carbon source (Shen et al., 2013), roseheart radish (W. Liu et al., 2017), glucose (Gong et al., 2014), formaldehyde (M. et al., 2014), and sucrose (Pan et al., 2014) as carbon sources. Further, used candle soot (LiQin et al., 2011), gum tragacanth and chitosan bio-polymers (Moradi et al., 2018), hongcaitai (L. S. Li et al., 2018), phytic acid (L. S. Li et al., 2018), alkoxysilanes (Chen et al., 2013), cetylpyridinium chloride (Kozák et al., 2013), and glycerine (Amjadi et al., 2015) were used as carbon sources for the synthesis of fluorescent CDs via hydrothermal treatment. Diao's group synthesized blue and green fluorescent CDs from *Syringa oblata* Lindl by the hydrothermal method (Diao et al., 2018) (Fig. 21.5). Table 21.3 illustrates the preparation of fluorescent CDs from various natural resources and their morphology and optical properties.

21.5.2.3 Carbonization method

Carbonization of various molecules/materials has proven to be a green approach for the fabrication of ultrasmall fluorescent CDs. Thus, eco- friendly CDs were successfully synthesized on a large scale and used as promising materials for various scientific applications. Jia et al. (2012) proposed a novel one-step reaction for the synthesis of CDs by using low-temperature aqueous heating. To this, aqueous solutions of ascorbic acid and copper acetate were heated at 90 °C for 5 h, which results to form CDs. De & Karak (2013) used banana juice as a precursor for the synthesis of fluorescent CDs. Mewada et al. (2013) synthesized fluorescent CDs by using *Trapa bispinosa* as a carbon source. Li's group (W. Li et al., 2013) used soya bean as a carbon source for preparation of fluorescent CDs. Further, several chemicals or natural resources such as calcium disodium ethylene diamine tetraacetate (X. Li et al., 2015), citric acid (Boonmee et al., 2018; Halawa et al., 2018; He et al., 2018; W. Hou et al., 2018; Hu et al., 2017; H. Huang et al., 2017; L. Kong et al., 2016; Mo et al., 2018; Pang & Liu, 2017; Qing et al., 2016; Shang et al., 2011; M. Wang et al., 2018; Y. Wang, Chen, et al., 2018; Wu et al., 2018; Xiao et al., 2018; Y. Yang et al., 2017; Zang et al., 2016; B. Zhang & Wei, 2018; M. Zhang et al., 2018; S. Zhang et al., 2018; Y. Zhang et al., 2016; Q. Zhao et al., 2016), watermelon peel (J. Zhou et al., 2012), Assam tea (Konwar et al., 2015), poly-ethylene glycol (Chen et al., 2014), and cyclodextrin (Boonmee et al., 2018; Halawa et al., 2018; He et al., 2018; W. Hou et al., 2018; X. Hu et al., 2017;

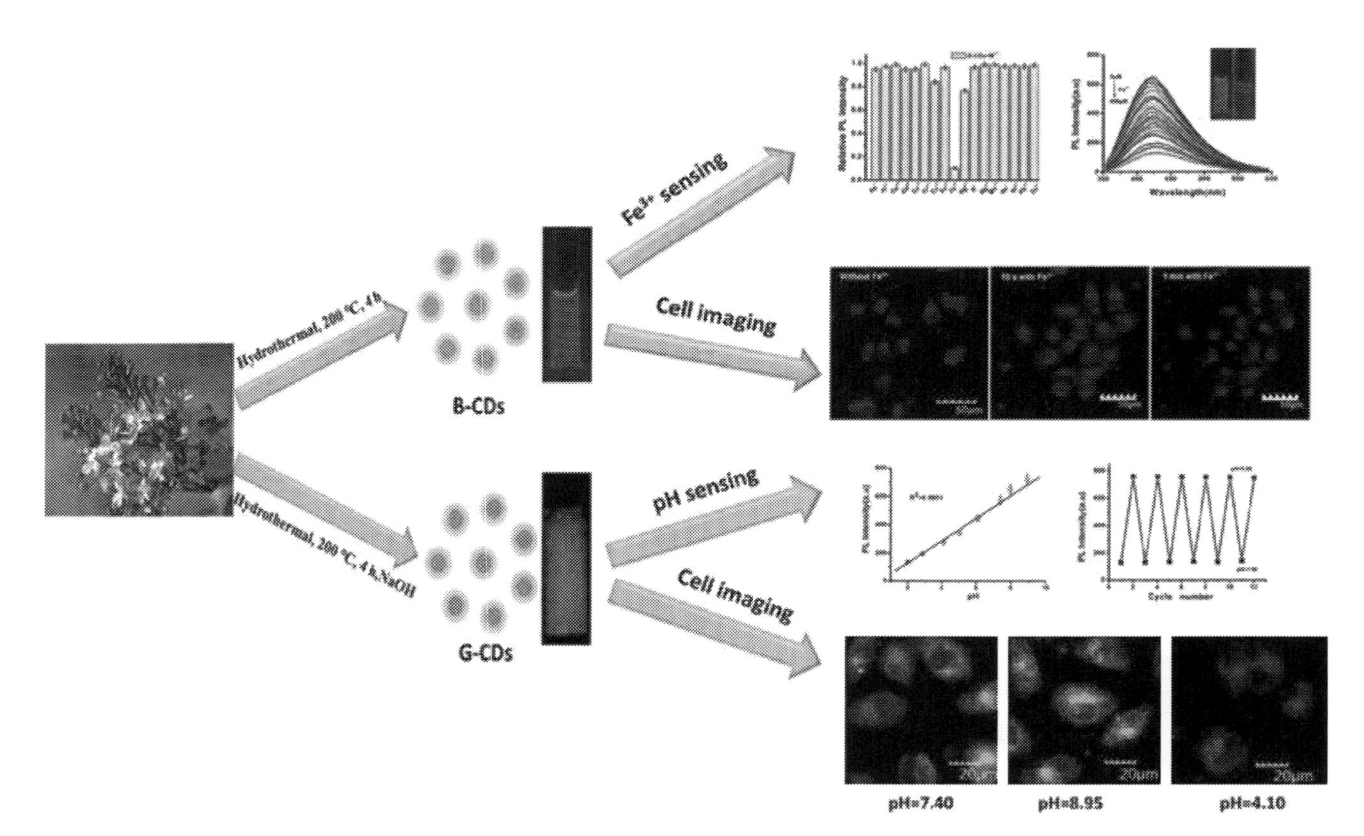

Figure 21.5 Schematic illustration for the preparation of blue and green fluorescent CDs from *Syringa oblata* Lindl via hydrothermal method and their applications in sensing and imaging . From Diao, H., Li, T., Zhang, R., Kang, Y., Liu, W., Cui, Y., Wei, S., Wang, N., Li, L., Wang, H., Nie, W., & Sun, T. (2018). Facile and green synthesis of fluorescent carbon dots with tunable emission for sensors and cells imaging. Spectrochim. Acta A Mol. Biomol. Spectrosc., 200, 226–234. https://doi.org/10.1016/j.saa.2018.04.029.

Table 21.3 Hydrothermal synthesis of fluorescent CDs using various natural resources as carbon sources.

Source	Condition		Size (nm)	UV-visible (nm)	Fluorescence (Ex/Em, nm)	Application	Reference
	Temperature (°C)	Time (h)					
Chocolate	200	8	6.41	–	280/354	Chemosensor for Pb2+ ion	Y. Liu et al., 2016
Corn flour	180	5	3.5	282	365/441	Biological application	Wei et al., 2014
Orange juice	120	2.5	1.5–4.5	288	390/455	Bioimaging agents	Sahu et al., 2012
Histidine	180	12	3–5	280	350/438	Melamine detection	Dai et al., 2014
Pork	200	10	3.5	235, 281	310/412	Uric acid detection	C. Zhao et al., 2018
Pure milk	180	8	5 ± 0.27	275	380/466	–	Wang et al., 2014
Saccharum officinarum juice	120	3	2.71	370	390/474	Cell biology. Sensor	Mehta et al., 2014
Cucumber juice	100–150	–	2.7–5.6	259, 283	370/450	Sensor	C. Wang et al., 2014
Natural vegetables	180	4	10	270	400/440–610	–	J. Wang et al., 2014
Citric acid and ethylene diamine	200	5	–	–	360/443	Sensor	Song et al., 2014
Mandelic acid and ethylenediamine	200	5	2.5	247,333	342/429	Detection of picric acid	Y. Wang et al., 2018
Broccoli	190	6	2–6	233, 288	355/450	Detection of Ag+ ion	Arumugam and Kim, 2018
Carbon source	140	12	5–30	–	330 –450/460 - 550	Biological labeling, imaging, catalysis, and chromatography	Shen et al., 2013
Rose heart radish	180	3	3.6	281.5	330/420	Fe^{3+} detection and cell imaging	W. Liu et al., 2017

(continued on next page)

Table 21.3 Hydrothermal synthesis of fluorescent CDs using various natural resources as carbon sources—cont'd

Source	Condition		Size (nm)	UV-visible (nm)	Fluorescence (Ex/Em, nm)	Application	Reference
	Temperature (°C)	Time (h)					
Glucose, serine	180	4	2.3	–	350/450	–	Gong et al., 2014
Formaldehyde	180	24	12.5± 4.03	–	360/440	Cell imaging applications	Algarra et al., 2014
Sucrose	180	5	–	554	380/550–850	Design of highly efficient Photocatalytic and potential photoelectric	Pan et al., 2014
Candle soot	200	12	3.1 ± 0.5	–	310/450	Optical sensing of metal ions	Qin, 2015
Gum Tragacanth and chitosan biopolymers	250	45	70–90	–	340/428	Biological imaging	Moradi et al., 2018
Hongcaitai	240	20	2.6	283	330/410	Detection of hypochlorite and mercuric ions and cell imaging	L. S. Li et al., 2018
Phytic acid and sodium citrate	160–240	4	1.5	238, 330	330/425	Cu^{2+} ion sensing	Y. Li et al., 2018

H. Huang et al., 2017; L. Kong et al., 2016; Mo et al., 2018; Pang & Liu, 2017; Qing et al., 2016; Shang et al., 2011; M. Wang et al., 2018; Y. Wang, Chen, et al., 2018; Wu et al., 2018; Xiao et al., 2018; Y. Yang et al., 2017; Zang et al., 2016; B. Zhang & Wei, 2018; M. Zhang et al., 2018; S. Zhang et al., 2018; Y. Zhang et al., 2016; Q. Zhao et al., 2016; Hu et al., 2014) have been used as a carbon sources for the synthesis of fluorescent CDs. Apart from these, there are some other carbon sources such as citric acid and GSH (Zhuo et al., 2015), L-glutamic acid (Niu & Gao, 2014), and ethylene glycol (Y. Liu et al., 2012) that were used as precursors for the preparation of fluorescent CDs and used as fluorescent probes for sensing and imaging applications. Walnut shells were used as precursors for the fabrication of photoluminescent CDs through carbonization method (Cheng et al., 2017). Fig. 21.6 depicts the preparation of ultrasmall fluorescent CDs using walnut shells as a carbon source and bioimaging applications (Cheng et al., 2017). Table 21.4 summarizes the synthesis of fluorescent CDs via carbonization method and their optical properties.

21.6 Applications of CDs

21.6.1 Bioimaging

Because of their unique fluorescence and biocompatible nature, QDs have been successfully replaced with CDs in bioimaging of various cells (Fowley et al., 2012; Mehta, Jha, & Kailasa, 2014; Mehta, Jha, Singhal, et al., 2014; Mehta et al., 2015; Tiwari et al., 2020; S. T. Yang et al., 2009). The ultrasmall fluorescent CDs have become popular probes to visualize biological systems both *in vitro* and *in vivo*. In general, the carbon cores of CDs themselves are not toxic and they can easily enter the cells. Lu et al. prepared luminescent N- and S- atoms doped CDs from dl-malic acid as a carbon precursor via hydrothermal method and used as probes for imaging of HeLa cells (Lu et al., 2015). Fig. 21.7 shows the fluorescence images of N- and S- atoms doped CDs incubated HeLa cells. HeLa cell was internalized by CDs and exhibited blue, green and red fluorescence when excited laser wavelengths at 405, 488, and 559 nm, respectively.

21.6.2 Drug delivery

The fluorescent CDs were synthesized quickly via many different low-cost and simple synthetic routes and served as promising agents in delivering various drugs molecules toward target cells (Bechet et al., 2008; X. Huang et al., 2013; Zeng et al., 2016). To confirm their biocompatible nature, *in-vivo* toxicity studies were carried out on mice, where mice was intravenously injected with CDs and tested after 4 weeks. These studies revealed that the as-prepared CDs did not show any significant toxicity on mice (Zeng et al., 2016). The fluorescent CDs exhibited good biocompatibility, which was confirmed by prothrombin time assays in plasma samples. In recent years, drug delivery system based on nanotechnology is progressively developed. Drug delivery system was broadly examined on AuNPs, but their toxicity concern limits their applications in clinical

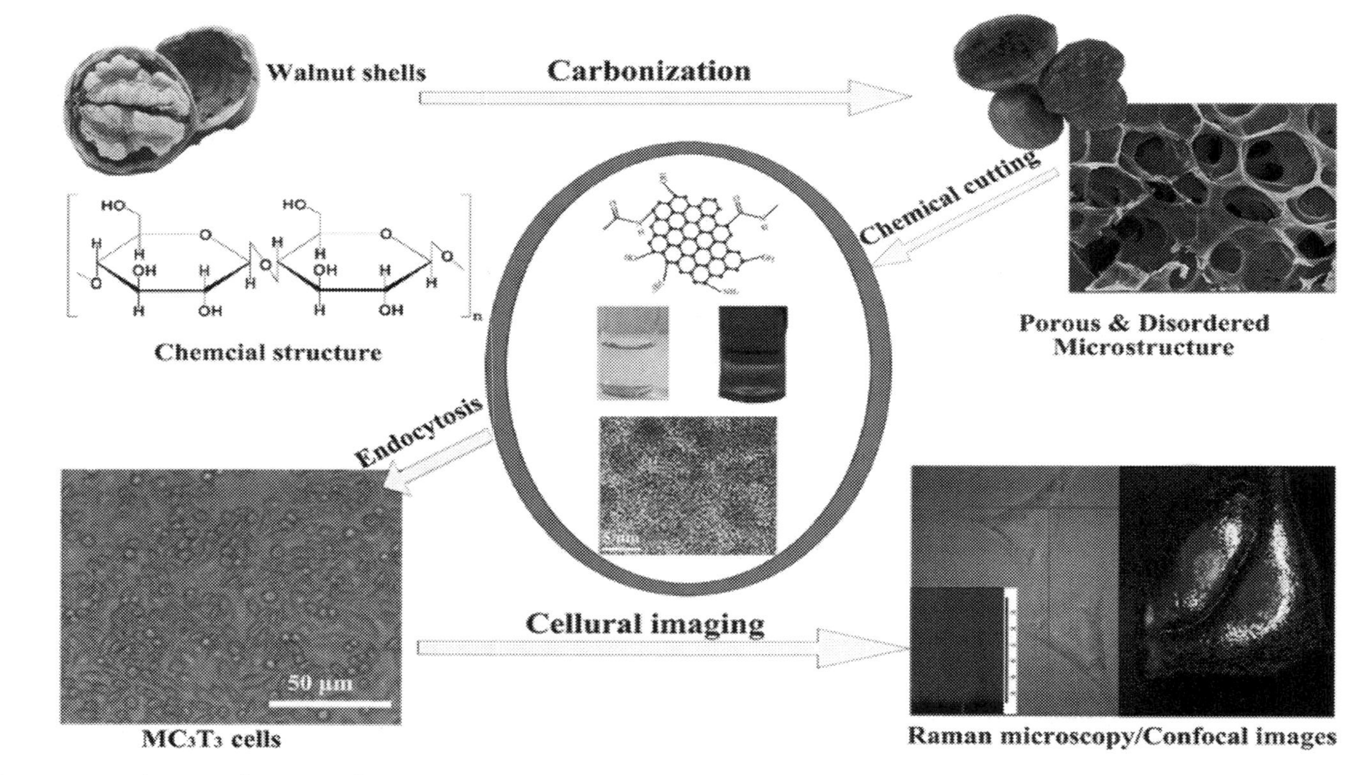

Figure 21.6 Schematic illustration for the preparation of fluorescent CDs from walnut shells by carbonization method and their applications in bioimaging (Cheng et al., 2017). From Cheng, C., Shi, Y., Li, M., Xing, M., & Wu, Q. (2017). Carbon quantum dots from carbonized walnut shells: Structural evolution, fluorescence characteristics, and intracellular bioimaging. Mater. Sci. Eng. C, 79, 473–480. https://doi.org/10.1016/j.msec.2017.05.094.

Table 21.4 Overview of carbonization of various chemicals or natural resources as carbon sources for synthesis of CDs and their optical characteristics.

Source	Condition		Size (nm)	UV-visible (nm)	Fluorescence (Ex/Em, nm)	Application	Reference
	Temperature (°C)	Time (h)					
Ascorbic acid	90	5	3.20 ±0.72	287	371/455	Biochemistry and diagnostics.	Jia et al., 2012
Banana juice	150	4	3	283	360/460	–	De and Karak, 2013
Trapa bispinosa	150	2	1 - 2.5	450–650	–	Biological application	Mewada et al., 2013
Soya bean grounds	–	–	3.0	260	360–500/440–560	Biomedical imaging	W. Li et al., 2013
Calcium disodium ethylene diamine tetraacetate	400	2	7.7	260–320	340/460	Bioimaging or biodiagnosis applications	X. Li et al., 2015
Citric acid	240	0.083	2.7	–	–	Bio applications	Q. Kong et al., 2014
Water-melon peel	220	2	2	–	310/490–580	Cell imaging	J. Zhou et al., 2012
Assam tea	200	8	2.7	–	380/459	Biomedical and industrial fields	Konwar et al., 2015
Polyethylene glycol	160	0.5–6	2.3–4.7	246, 310	360/450	Drug delivery, targeting therapy, and bioimaging	M. Chen et al., 2014
Cyclodextrin	70	4	1.7–3.3	–	420/510	Sensor	M. Hu et al., 2014
Citric acid and GSH	200	0.17	3	345	367/453, 482	Cell imaging	Zhuo et al., 2015
L-Glutamic acid	200	–	4.6	310	400/470	Drug detection	Niu and Gao, 2014
Ethylene glycol	140	6	1–4	233, 293, 350	280–370/400	Sensor	Y. Liu et al., 2012

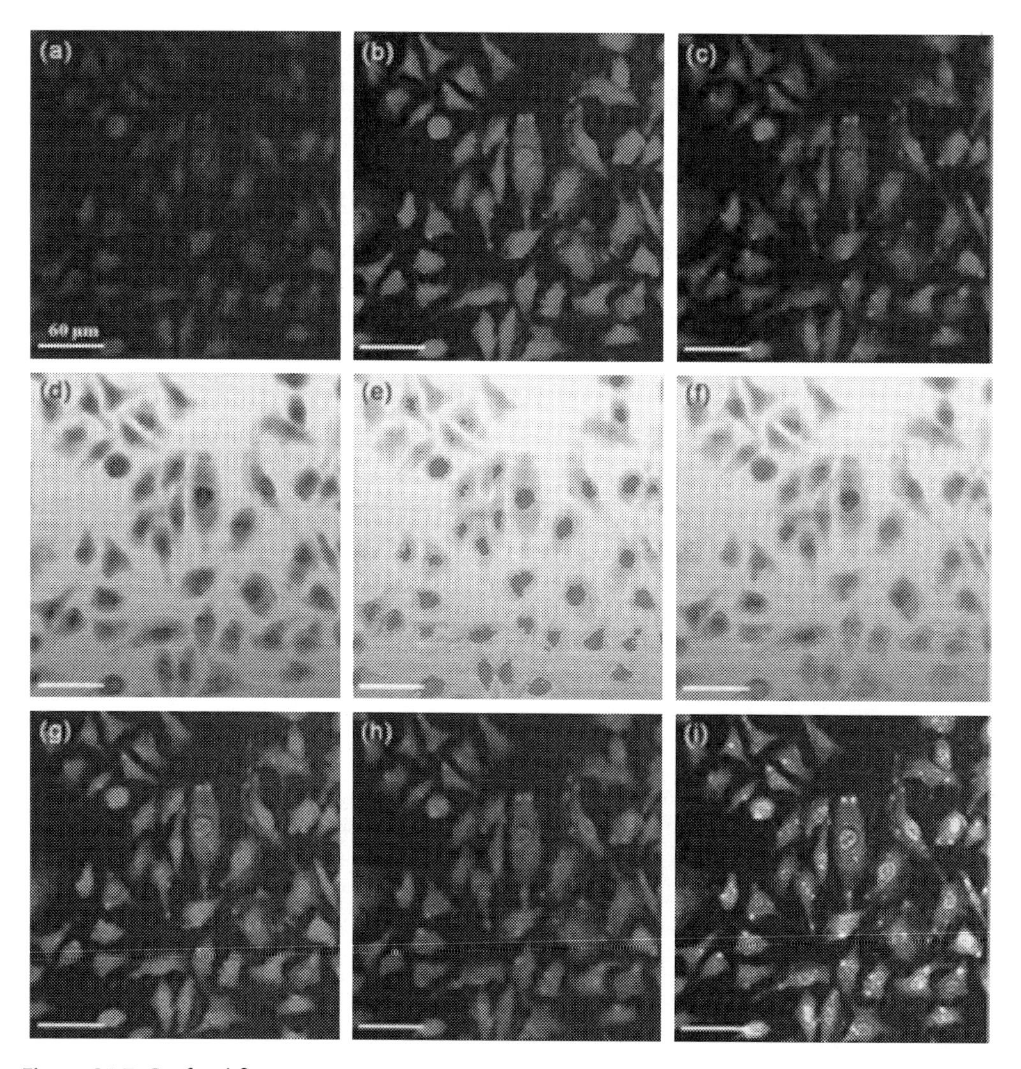

Figure 21.7 Confocal fluorescence microscopic images of *Pseudomonas aeruginosa* cells (A) without CDs and with CDs at wavelengths (B) 488 (green), and (C) 561 (red) nm without bright field. Confocal fluorescence laser microscopic images of *Fusarium avenaceum* cells (D) bright field without CDs and with CDs at excitation wavelengths of (E) 488 (green), and (F) 561 nm (red). The cells were incubated with CDs for 1.0–6.0 h at 37 °C. From Lu, W., Gong, X., Nan, M., Liu, Y., Shuang, S., & Dong, C. (2015). Comparative study for N and S doped carbon dots: Synthesis, characterization and applications for Fe3+ probe and cellular imaging. Anal. Chim. Acta, 898, 116–127. https://doi.org/10.1016/j.aca.2015.09.050.

therapy as well as the use of expensive chemicals in AuNPs synthesis. The fluorescent CDs have proved as good alternatives to AuNPs because of their easy preparation, and diverse functional groups, which make them potential drug vehicles to carry the target drugs toward target positions. For example, the fluorescent CDs were successfully used

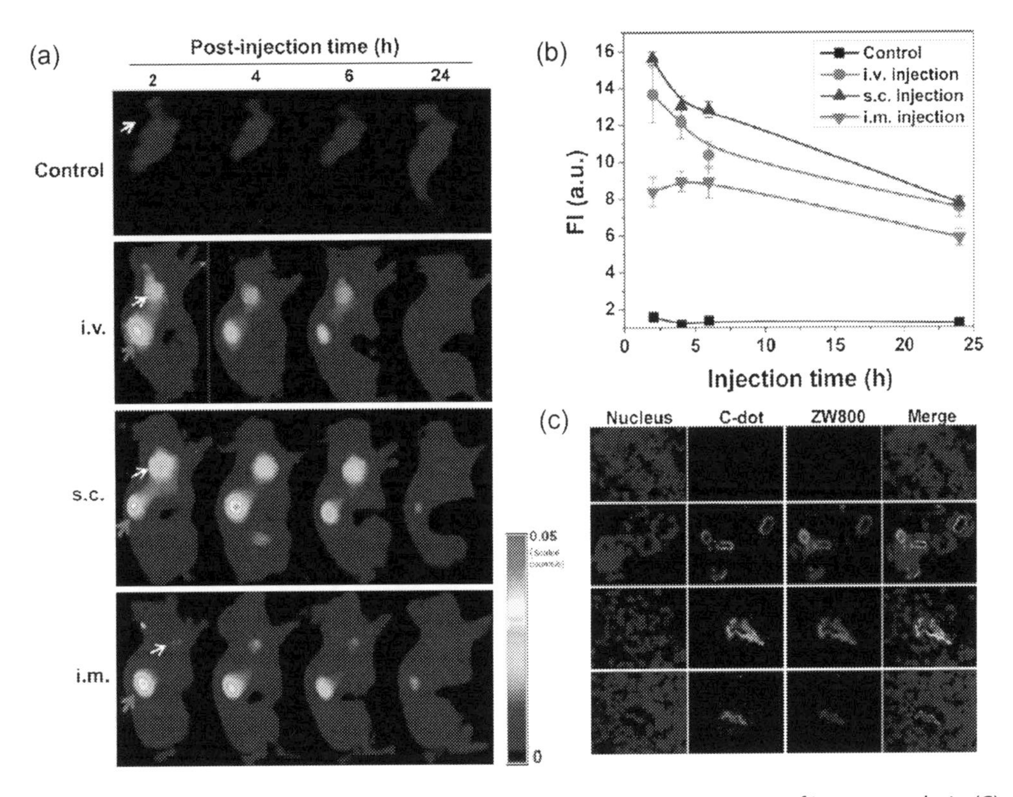

Figure 21.8 (A) Fluorescence images of tumor-bearing mice, (B) tumor region of interest analysis. (C) *Ex-vivo* fluorescence images derived from the emission of CDs and ZW800 were acquired to confirm tumor uptake of particles (Huang et. al., 2013) From Huang, X., Zhang, F., Zhu, L., Choi, K. Y., Guo, N., Guo, J., Tackett, K., Anilkumar, P., Liu, G., Quan, Q., Choi, H. S., Niu, G., Sun, Y. P., Lee, S., & Chen, X. (2013). Effect of injection routes on the biodistribution, clearance, and tumor uptake of carbon dots. ACS Nano, 7(7), 5684–5693. https://doi.org/10.1021/nn401911k.

for photodynamic therapy for the treatment of external tumors (Bechet et al., 2008). It involves the localization and accumulation of photosensitizer tumor tissue, following that they are irradiated with a particular wavelength. As a result, singlet oxygen species were formed, yielding to inhibit the biochemical pathways of cancer cells, which results the cell death. It was noticed that the CDs have high inhibition impact on Michigan Cancer Foundation-7 (MCF-7) and MDA-MB-231 cancer cells (X. Huang et al., 2013). The ultrasmall fluorescent CDs act as effective photosensitizers to absorb laser, causing to generate reactive oxygen species, which inhibits the regular function of cells. It was noticed that the circulation and uptake of CDs within the body rely on their surface coating and the route of administration. Huang et al. investigated the effect of the injection route on the distribution, clearance, and tumor uptake of CDs (Fig. 21.8) (X. Huang et al., 2013).

21.6.3 Sensing of metal ions

Apart from these application, ultrasmall fluorescent CDs act as potential probes for sensing of various chemical species including metal ions, drugs, and biomolecules via "fluorescence turn-off" and "turn-on" mechanisms. For example, Liu et al. described the use of bamboo leaves for the generation of fluorescent CDs and their uses as a fluorescent probe for sensitive and selective detection of Cu^{2+} ions in water samples (Y. Liu et al., 2014). Xu's group reported one-step hydrothermal method for the preparation of fluorescent CDs from potatoes as a carbon source (M. Xu et al., 2014). The synthesized fluorescent CDs were used as a sensing probe for label-free, sensitive, and selective detection of Fe^{3+} ions in environmental samples. Similarly, Huang et al., synthesized fluorescent nitrogen-doped CDs via one-pot hydrothermal treatment using strawberry juice and applied as a sensor for the detection of Hg^{2+} ions in environmental water samples (H. Huang et al., 2013). Cui et al., synthesized CDs and capped them with oligodeoxyribonucleotide and graphene oxide for effective sensing of mercury with lower detection limit (Cui et al., 2015). Furthermore, ultrasmall fluorescent CDs served as promising probes for the detection of heavy metals such as Hg^{2+} (Desai, Jha, et al., 2019), Cu^{2+} (Bhamore, Jha, Park, et al., 2018), Fe^{3+} (Abubaker et al., 2008; Kailasa, Ha, et al., 2019), Al^{3+} (Bhamore, Jha, Singhal, et al., 2018), Pb^{2+} (Wee et al., 2013), and Cr^{6+} (Zheng et al., 2013) in various sample matrices. These reports illustrated that the performance of fluorescence spectroscopy was significantly improved for the selective and sensitive detection of wide variety of molecules (biomolecules, organic molecules, and metal ions) at low volume of samples.

21.7 Summary

In this chapter, ultrasmall fluorescent nanomaterials (metal NCs and CDs) properties, synthesis, and their applications in sensing of various chemical species and bioimaging of cells are discussed. The size, doping of heteroatoms, and surface chemistry of metal NCs and CDs play a key role in tuning their fluorescence properties and analytical applications. Similarly, mercapto molecules, phytochemicals, peptides, and proteins are extensively used for the fabrication of various metal NCs, thereby controlling the growth of metal NCs with atomically precise manner, which allows them to exhibit different fluorescent properties. It was noticed that the doping of heteroatom into CDs has greatly improved the optical properties of CDs, allowing them to act as promising agents for sensing and bioimaging applications. Furthermore, ultrasmall fluorescent CDs have been synthesized by various synthetic routes including the top-down (Arc discharge, electrochemical oxidation and ultrasonic) and bottom-up (hydrothermal, microwave, solvothermal and carbonization) approaches using various chemical molecules and natural resources as precursors. The integration of ultrasmall fluorescent nanomaterials (metal NCs and CDs) with analytical techniques (fluorescence, UV-visible and confocal microscopy) has greatly improved the analytical performance of the above techniques, allowing them to detect trace-level target analytes with high selectivity and sensitivity at low volume of samples. The fabricated CDs acted as carriers for the delivery of drugs selectively at targeted molecules or cells. Overall, ultrasmall fluorescent nanomaterials (metal NCs and CDs) have proven to be ideal candidates for sensing of various trace-level chemical species and bioimaging of various cells.

Acknowledgments

This work was supported by the Department of Science and Technology (DST), Government of India (EMR/2016/002621/IPC). Authors thank the Director, SVNIT, Surat for providing necessary facilities to carry out this work.

References

Abubaker, H.M.M., Subramani, R., Sekhar, S.S., Kumar, P.S., Thalappil, P., 2008. Two distinct fluorescent quantum clusters of gold starting from metallic nanoparticles by pH-dependent ligand etching. Nano Res. 333–340. https://doi.org/10.1007/s12274-008-8035-2.

Adhikari, B., Banerjee, A., 2010. Facile synthesis of water-soluble fluorescent silver nanoclusters and HgII sensing. Chem. Mater. 22 (15), 4364–4371. https://doi.org/10.1021/cm1001253.

Ai, J., Guo, W., Li, B., Li, T., Li, D., Wang, E., 2012. DNA G-quadruplex-templated formation of the fluorescent silver nanocluster and its application to bioimaging. Talanta 88, 450–455. https://doi.org/10.1016/j.talanta.2011.10.057.

Algarra, M., Pérez-Martín, M., Cifuentes-Rueda, M., Jiménez-Jiménez, J., da Silva, J.E., Bandosz, T.J., Rodriguez-Castellon, E., Navarrete, J.L., Casado, J., 2014. Carbon dots obtained using hydrothermal treatment of formaldehyde. Cell imaging in vitro. Nanoscale 6 (15), 9071–9077.

Amjadi, M., Manzoori, J.L., Hallaj, T., 2015. A novel chemiluminescence method for determination of bisphenol A based on the carbon dot-enhanced HCO_3^-–H_2O_2 system. J. Lumin. 158, 160–164. https://doi.org/10.1016/j.jlumin.2014.09.045.

Aragay, G., Pino, F., Merkoçi, A., 2012. Nanomaterials for sensing and destroying pesticides. Chem. Rev. 112 (10), 5317–5338. https://doi.org/10.1021/cr300020c.

Arumugam, N., Kim, J., 2018. Synthesis of carbon quantum dots from Broccoli and their ability to detect silver ions. Mater. Lett. 219, 37–40. https://doi.org/10.1016/j.matlet.2018.02.043.

Ashrafizadeh, M., Mohammadinejad, R., Kailasa, S.K., Ahmadi, Z., Afshar, E.G., Pardakhty, A., 2020. Carbon dots as versatile nanoarchitectures for the treatment of neurological disorders and their theranostic applications: A review. Adv. Colloid Interface Sci. 278. https://doi.org/10.1016/j.cis.2020.102123.

Baker, S.N., Baker, G.A., 2010. Luminescent carbon nanodots: emergent nanolights. Angew. Chem. Int. Ed. 49 (38), 6726–6744. https://doi.org/10.1002/anie.200906623.

Bechet, D., Couleaud, P., Frochot, C., Viriot, M.L., Guillemin, F., Barberi-Heyob, M., 2008. Nanoparticles as vehicles for delivery of photodynamic therapy agents. Trends Biotechnol. 26 (11), 612–621. https://doi.org/10.1016/j.tibtech.2008.07.007.

Bertorelle, F., Russier-Antoine, I., Comby-Zerbino, C., Chirot, F., Dugourd, P., Brevet, P.F., Antoine, R., 2018. Isomeric Effect of mercaptobenzoic acids on the synthesis, stability, and optical properties of Au25(MBA)18 nanoclusters. ACS Omega 3 (11), 15635–15642. https://doi.org/10.1021/acsomega.8b02615.

Bhamore, J.R., Deshmukh, B., Haran, V., Jha, S., Singhal, R.K., Lenka, N., Kailasa, S.K., Murthy, Z.V.P., 2018. One-step eco-friendly approach for the fabrication of synergistically engineered fluorescent copper nanoclusters: sensing of Hg^{2+} ion and cellular uptake and bioimaging properties. New J. Chem. 42 (2), 1510–1520. https://doi.org/10.1039/c7nj04031h.

Bhamore, J.R., Gul, A.R., Chae, W.S., Kim, K.W., Lee, J.S., Park, H., Kailasa, S.K., Park, T.J., 2020. One-pot fabrication of amino acid and peptide stabilized gold nanoclusters for the measurement of the lead in plasma samples using chemically modified cellulose paper. Sensors Actuat. B Chem. 322, 128603. https://doi.org/10.1016/j.snb.2020.128603.

Bhamore, J.R., Jha, S., Basu, H., Singhal, R.K., Murthy, Z.V.P., Kailasa, S.K., 2018. Tuning of gold nanoclusters sensing applications with bovine serum albumin and bromelain for detection of Hg^{2+} ion and lambda-cyhalothrin via fluorescence turn-off and on mechanisms. Anal. BioAnal. Chem. 410 (11), 2781–2791. https://doi.org/10.1007/s00216-018-0958-1.

Bhamore, J.R., Jha, S., Mungara, A.K., Singhal, R.K., Sonkeshariya, D., Kailasa, S.K., 2016. One-step green synthetic approach for the preparation of multicolor emitting copper nanoclusters and their applications in chemical species sensing and bioimaging. Biosensors Bioelectron. 80, 243–248. https://doi.org/10.1016/j.bios.2016.01.066.

Bhamore, J.R., Jha, S., Park, T.J., Kailasa, S.K., 2018. Fluorescence sensing of Cu^{2+} ion and imaging of fungal cell by ultra-small fluorescent carbon dots derived from *Acacia concinna* seeds. Sensors Actuat. B Chem. 277, 47–54. https://doi.org/10.1016/j.snb.2018.08.149.

Bhamore, J.R., Jha, S., Park, T.J., Kailasa, S.K., 2019. Green synthesis of multi-color emissive carbon dots from *Manilkara zapota* fruits for bioimaging of bacterial and fungal cells. J. PhotoChem. Photobiol. B: Biol. 191, 150–155. https://doi.org/10.1016/j.jphotobiol.2018.12.023.

Bhamore, J.R., Jha, S., Singhal, R.K., Kailasa, S.K., 2017. Synthesis of water dispersible fluorescent carbon nanocrystals from *Syzygium cumini* fruits for the detection of Fe^{3+} ion in water and biological samples and imaging of *Fusarium avenaceum* Cells. J. Fluoresc. 27 (1), 125–134. https://doi.org/10.1007/s10895-016-1940-y.

Bhamore, J.R., Jha, S., Singhal, R.K., Murthy, Z.V.P., Kailasa, S.K., 2019. Amylase protected gold nanoclusters as chemo- and bio- sensor for nanomolar detection of deltamethrin and glutathione. Sensors Actuat. B Chem. 281, 812–820. https://doi.org/10.1016/j.snb.2018.11.001.

Bhamore, J.R., Jha, S., Singhal, R.K., Park, T.J., Kailasa, S.K., 2018. Facile green synthesis of carbon dots from *Pyrus pyrifolia* fruit for assaying of Al^{3+} ion via chelation enhanced fluorescence mechanism. J. Mol. Liquids 264, 9–16. https://doi.org/10.1016/j.molliq.2018.05.041.

Bhamore, J.R., Murthy, Z.V.P., Kailasa, S.K., 2019. Fluorescence turn-off detection of spermine in biofluids using pepsin mediated synthesis of gold nanoclusters as a probe. J. Mol. Liquids 280, 18–24. https://doi.org/10.1016/j.molliq.2019.01.132.

Bhamore, J.R., Park, T.J., Kailasa, S.K., 2020. Glutathione-capped Syzygium cumini carbon dot-amalgamated agarose hydrogel film for naked-eye detection of heavy metal ions. J. Anal. Sci. Technology 11 (1). https://doi.org/10.1186/s40543-020-00208-8.

Bian, R.X., Wu, X.T., Chai, F., Li, L., Zhang, L.Y., Wang, T.T., Wang, C.G., Su, Z.M., 2017. Facile preparation of fluorescent Au nanoclusters-based test papers for recyclable detection of Hg^{2+} and Pb^{2+}. Sensors Actuat. B Chem. 241, 592–600. https://doi.org/10.1016/j.snb.2016.10.120.

Boonmee, C., Promarak, V., Tuntulani, T., Ngeontae, W., 2018. Cysteamine-capped copper nanoclusters as a highly selective turn-on fluorescent assay for the detection of aluminum ions. Talanta 178, 796–804. https://doi.org/10.1016/j.talanta.2017.10.006.

Borse, S., Murthy, Z.V.P., Kailasa, S.K., 2020. Chicken egg white mediated synthesis of platinum nanoclusters for the selective detection of carbidopa. Optical Mater. 107, 110085. https://doi.org/10.1016/j.optmat.2020.110085.

Bourlinos, A.B., Stassinopoulos, A., Anglos, D., Zboril, R., Georgakilas, V., Giannelis, E.P., 2008. Photoluminescent carbogenic dots. Chem. Mater. 20 (14), 4539–4541. https://doi.org/10.1021/cm800506r.

Cao, L., Sahu, S., Anilkumar, P., Bunker, C.E., Xu, J., Fernando, K.A.S., Wang, P., Guliants, E.A., Tackett, K.N., Sun, Y.P., 2011. Carbon nanoparticles as visible-light photocatalysts for efficient CO_2 conversion and beyond. J. Am. Chem. Soc. 133 (13), 4754–4757. https://doi.org/10.1021/ja200804h.

Chatzimitakos, T., Kasouni, A., Sygellou, L., Leonardos, I., Troganis, A., Stalikas, C., 2018. Human fingernails as an intriguing precursor for the synthesis of nitrogen and sulfur-doped carbon dots with strong fluorescent properties: analytical and bioimaging applications. Sensors Actuat. B Chem. 267, 494–501. https://doi.org/10.1016/j.snb.2018.04.059.

Chen, M., Wang, W., Wu, X., 2014. One-pot green synthesis of water-soluble carbon nanodots with multicolor photoluminescence from polyethylene glycol. J. Mater. Chem. B 2 (25), 3937–3945. https://doi.org/10.1039/c4tb00292j.

Chen, P.C., Chen, Y.N., Hs u, P.C., Shih, C.C., Chang, H.T., 2013. Photoluminescent organosilane-functionalized carbon dots as temperature probes. Chem. Commun. 49 (16), 1639–1641. https://doi.org/10.1039/c3cc38486a.

Cheng, C., Shi, Y., Li, M., Xing, M., Wu, Q., 2017. Carbon quantum dots from carbonized walnut shells: structural evolution, fluorescence characteristics, and intracellular bioimaging. Mater. Sci. Eng. C 79, 473–480. https://doi.org/10.1016/j.msec.2017.05.094.

Chowdhury, D., Gogoi, N., Majumdar, G., 2012. Fluorescent carbon dots obtained from chitosan gel. RSC Adv., 2 (32), 12156–12159. https://doi.org/10.1039/c2ra21705h.

Cui, X., Zhu, L., Wu, J., Hou, Y., Wang, P., Wang, Z., Yang, M., 2015. A fluorescent biosensor based on carbon dots-labeled oligodeoxyribonucleotide and graphene oxide for mercury (II) detection. Biosensors Bioelectron. 63, 506–512. https://doi.org/10.1016/j.bios.2014.07.085.

Dai, H., Shi, Y., Wang, Y., Sun, Y., Hu, J., Ni, P., Li, Z., 2014. A carbon dot based biosensor for melamine detection by fluorescence resonance energy transfer. Sensors Actuat. B Chem. 202, 201–208. https://doi.org/10.1016/j.snb.2014.05.058.

De, B., Karak, N., 2013. A green and facile approach for the synthesis of water soluble fluorescent carbon dots from banana juice. RSC Adv., 3 (22), 8286–8290. https://doi.org/10.1039/c3ra00088e.

Deng, L., Zhou, Z., Li, J., Li, T., Dong, S., 2011. Fluorescent silver nanoclusters in hybridized DNA duplexes for the turn-on detection of Hg^{2+} ions. Chem. Commun. 47 (39), 11065–11067. https://doi.org/10.1039/c1cc14012d.

Desai, M.L., Basu, H., Saha, S., Singhal, R.K., Kailasa, S.K., 2019. Investigation of silicon doping into carbon dots for improved fluorescence properties for selective detection of Fe^{3+} ion. Optical Mater., 96,, p. 109374.

Desai, M.L., Basu, H., Saha, S., Singhal, R.K., Kailasa, S.K., 2020. One pot synthesis of fluorescent gold nanoclusters from *Curcuma longa* extract for independent detection of Cd^{2+}, Zn^{2+} and Cu^{2+} ions with high sensitivity. J. Mol. Liq 304, 112697–112706.

Desai, M.L., Jha, S., Basu, H., Singhal, R.K., Park, T.J., Kailasa, S.K., 2019. Acid oxidation of muskmelon fruit for the fabrication of carbon dots with specific emission colors for recognition of Hg^{2+} ions and cell imaging. ACS Omega 4 (21), 19332–19340. https://doi.org/10.1021/acsomega.9b02730.

Desai, M.L., Jha, S., Basu, H., Singhal, R.K., Sharma, P.K., Kailasa, S.K., 2018a. Chicken egg white and L-cysteine as cooperative ligands for effective encapsulation of Zn-doped silver nanoclusters for sensing and imaging applications. Colloids Surf. A PhysicoChem. Eng. Asp. 559, 35–42. https://doi.org/10.1016/j.colsurfa.2018.09.036.

Desai, M.L., Jha, S., Basu, H., Singhal, R.K., Sharma, P.K., Kailasa, S.K., 2018b. Microwave-assisted synthesis of water-soluble Eu^{3+} hybrid carbon dots with enhanced fluorescence for the sensing of Hg^{2+} ions and imaging of fungal cells. New J. Chem. 42 (8), 6125–6133. https://doi.org/10.1039/c7nj04835a.

Diao, H., Li, T., Zhang, R., Kang, Y., Liu, W., Cui, Y., Wei, S., Wang, N., Li, L., Wang, H., Nie, W., Sun, T., 2018. Facile and green synthesis of fluorescent carbon dots with tunable emission for sensors and cells imaging. SpectroChim. Acta A Mol. Biomol. Spectrosc. 200, 226–234. https://doi.org/10.1016/j.saa.2018.04.029.

Díez, I., Pusa, M., Kulmala, S., Jiang, H., Walther, A., Goldmann, A.S., Müller, A.H.E., Ikkala, O., Ras, R.H.A., 2009. Color tunability and electrochemiluminescence of silver nanoclusters. Angew. Chem. Int. Ed. 48 (12), 2122–2125. https://doi.org/10.1002/anie.200806210.

Díez, I., Ras, R.H.A., 2011. Fluorescent silver nanoclusters. Nanoscale 3 (5), 1963–1970. https://doi.org/10.1039/c1nr00006c.

Ding, C., Zhu, A., Tian, Y., 2014. Functional surface engineering of C-dots for fluorescent biosensing and in vivo bioimaging. Acc. Chem. Res. 47 (1), 20–30. https://doi.org/10.1021/ar400023s.

D'souza, S.L., Chettiar, S.S., Koduru, J.R., Kailasa, S.K., 2018. Synthesis of fluorescent carbon dots using *Daucus carota* subsp. sativus roots for mitomycin drug delivery. Optik 158, 893–900. https://doi.org/10.1016/j.ijleo.2017.12.200.

D'Souza, S.L., Deshmukh, B., Bhamore, J.R., Rawat, K.A., Lenka, N., Kailasa, S.K, 2016. Synthesis of fluorescent nitrogen-doped carbon dots from dried shrimps for cell imaging and boldine drug delivery system. RSC Adv. 6 (15), 12169–12179. https://doi.org/10.1039/c5ra24621k.

D'Souza, S.L., Deshmukh, B., Rawat, K.A., Bhamore, J.R., Lenka, N., Kailasa, S.K, 2016. Fluorescent carbon dots derived from vancomycin for flutamide drug delivery and cell imaging. New J. Chem. 40 (8), 7075–7083. https://doi.org/10.1039/c6nj00358c.

Fowley, C., Mc Caughan, B., Devlin, A., Yildiz, I., Raymo, F.M., Callan, J.F., 2012. Highly luminescent biocompatible carbon quantum dots by encapsulation with an amphiphilic polymer. Chem. Commun. 48 (75), 9361–9363. https://doi.org/10.1039/c2cc34962k.

Ghosh, R., Sahoo, A.K., Ghosh, S.S., Paul, A., Chattopadhyay, A., 2014. Blue-emitting copper nanoclusters synthesized in the presence of lysozyme as candidates for cell labeling. ACS Appl. Mater. Interf. 6 (6), 3822–3828. https://doi.org/10.1021/am500040t.

Ghosh, S., Bhamore, J.R., Malek, N.I., Murthy, Z.V.P., Kailasa, S.K., 2019. Trypsin mediated one-pot reaction for the synthesis of red fluorescent gold nanoclusters: Sensing of multiple analytes (carbidopa, dopamine, Cu2+, Co2+ and Hg2+ ions). SpectroChim. Acta A Mol. Biomol. Spectrosc. 215, 209–217. https://doi.org/10.1016/j.saa.2019.02.078.

Gong, A., Xu, W., Shao, Q., Li, J., Zhou, Z., Li, Z., Cao, M., Chen, J., Zhang, L., Liu, X., 2014. Nitrogen-doped carbon dots with heterogeneous multi-layered structures. RSC Adv., 4 (71), 37536–37541. https://doi.org/10.1039/c4ra06818a.

Guo, W., Yuan, J., Dong, Q., Wang, E., 2010. Highly sequence-dependent formation of fluorescent silver nanoclusters in hybridized DNA duplexes for single nucleotide mutation identification. J. Am. Chem. Soc. 132 (3), 932–934. https://doi.org/10.1021/ja907075s.

Gupta, D.A., Desai, M.L., Malek, N.I., Kailasa, S.K., 2020. Fluorescence detection of Fe^{3+} ion using ultra-small fluorescent carbon dots derived from pineapple (*Ananas comosus*): development of miniaturized analytical method. J. Mol. Struct. 1216. https://doi.org/10.1016/j.molstruc.2020.128343.

Halawa, M.I., Wu, F., Nsabimana, A., Lou, B., Xu, G., 2018. Inositol directed facile "green" synthesis of fluorescent gold nanoclusters as selective and sensitive detecting probes of ferric ions. Sensors Actuat. B Chem. 257, 980–987. https://doi.org/10.1016/j.snb.2017.11.046.

He, Y.S., Pan, C.G., Cao, H.X., Yue, M.Z., Wang, L., Liang, G.X., 2018. Highly sensitive and selective dual-emission ratiometric fluorescence detection of dopamine based on carbon dots-gold nanoclusters hybrid. Sensors Actuat. B Chem. 265, 371–377. https://doi.org/10.1016/j.snb.2018.03.080.

Hou, J., Yan, J., Zhao, Q., Li, Y., Ding, H., Ding, L., 2013. A novel one-pot route for large-scale preparation of highly photoluminescent carbon quantum dots powders. Nanoscale 5 (20), 9558–9561. https://doi.org/10.1039/c3nr03444e.

Hou, W., Chen, Y., Lu, Q., Liu, M., Zhang, Y., Yao, S., 2018. Silver ions enhanced AuNCs fluorescence as a turn-off nanoprobe for ultrasensitive detection of iodide. Talanta 180, 144–149. https://doi.org/10.1016/j.talanta.2017.12.047.

Hsiao, J.K., Chou, P.T., Chen, Y.C., Hsieh, C.C., Lin, Y.C., Wang, Y.H., Yang, M.J., Duan, H.S., Chen, B.S., Lee, J.F., 2009. Thiol-functionalized gold nanodots: two-photon absorption property and imaging in vitro. J. Phys. Chem. C 113 (50), 21082–21089. https://doi.org/10.1021/jp9080492.

Hu, M., Yang, Y., Gu, X., Hu, Y., Huang, J., Wang, C., 2014. One-pot synthesis of photoluminescent carbon nanodots by carbonization of cyclodextrin and their application in Ag+ detection. RSC Adv. 4 (107), 62446–62452. https://doi.org/10.1039/c4ra11491d.

Hu, X., Mao, X., Zhang, X., Huang, Y., 2017. One-step synthesis of orange fluorescent copper nanoclusters for sensitive and selective sensing of Al^{3+} ions in food samples. Sensors Actuat. B Chem. 247, 312–318. https://doi.org/10.1016/j.snb.2017.03.050.

Huang, H., Li, H., Feng, J.J., Feng, H., Wang, A.J., Qian, Z., 2017. One-pot green synthesis of highly fluorescent glutathione-stabilized copper nanoclusters for Fe^{3+} sensing. Sensors Actuat. B Chem. 241, 292–297. https://doi.org/10.1016/j.snb.2016.10.086.

Huang, H., Li, H., Feng, J.J., Wang, A.J., 2016. One-step green synthesis of fluorescent bimetallic Au/Ag nanoclusters for temperature sensing and in vitro detection of Fe^{3+}. Sensors Actuat. B Chem. 223, 550–556. https://doi.org/10.1016/j.snb.2015.09.136.

Huang, H., Lv, J.J., Zhou, D.L., Bao, N., Xu, Y., Wang, A.J., Feng, J.J., 2013. One-pot green synthesis of nitrogen-doped carbon nanoparticles as fluorescent probes for mercury ions. RSC Adv. 3 (44), 21691–21696. https://doi.org/10.1039/c3ra43452d.

Huang, X., Zhang, F., Zhu, L., Choi, K.Y., Guo, N., Guo, J., Tackett, K., Anilkumar, P., Liu, G., Quan, Q., Choi, H.S., Niu, G., Sun, Y.P., Lee, S., Chen, X., 2013. Effect of injection routes on the biodistribution, clearance, and tumor uptake of carbon dots. ACS Nano 7 (7), 5684–5693. https://doi.org/10.1021/nn401911k.

Huang, Z., Pu, F., Lin, Y., Ren, J., Qu, X., 2011. Modulating DNA-templated silver nanoclusters for fluorescence turn-on detection of thiol compounds. Chem. Commun. 47 (12), 3487–3489. https://doi.org/10.1039/c0cc05651k.

Jaiswal, A., Sankar Ghosh, S., Chattopadhyay, A., 2012. One step synthesis of C-dots by microwave mediated caramelization of poly(ethylene glycol). Chem. Commun. 48 (3), 407–409. https://doi.org/10.1039/c1cc15988g.

Jaque, D., Vetrone, F., 2012. Luminescence nanothermometry. Nanoscale, 4 (15), 4301–4326. https://doi.org/10.1039/c2nr30764b.

Jaschinski, O., Roth, S., Kertesz, M., Iqbal, Z., Barisci, J.N., Spinks, G.M., Wallace, G.G., Mazzoldi, A., De Rossi, D., Rinzler, A.G., 1999. Carbon nanotube actuators. Sci. 284 (5418), 1340–1344. https://doi.org/10.1126/science.284.5418.1340.

Javey, A., Guo, J., Wang, Q., Lundstrom, M., Dai, H., 2003. Ballistic carbon nanotube field-effect transistors. Nature 424 (6949), 654–657. https://doi.org/10.1038/nature01797.

Jia, X., Li, J., Wang, E., 2012. One-pot green synthesis of optically pH-sensitive carbon dots with upconversion luminescence. Nanoscale 4 (18), 5572–5575. https://doi.org/10.1039/c2nr31319g.

Jiang, J., He, Y., Li, S., Cui, H., 2012. Amino acids as the source for producing carbon nanodots: microwave assisted one-step synthesis, intrinsic photoluminescence property and intense chemiluminescence enhancement. Chem. Commun. 48 (77), 9634–9636. https://doi.org/10.1039/c2cc34612e.

Jin, R., 2010. Quantum sized, thiolate-protected gold nanoclusters. Nanoscale 2 (3), 343–362. https://doi.org/10.1039/b9nr00160c.

Kailasa, S.K., Bhamore, J.R., Koduru, J.R., Park, T.J., 2019. Carbon dots as carriers for the development of controlled drug and gene delivery systems. In: Grumezescu, A.M. (Ed.), Biomedical Applications of Nanoparticles. Elsevier, Cham, pp. 295–317.

Kailasa, S.K., D'Souza, S., Wu, H.F, 2015. Analytical applications of nanoparticles in MALDI-MS for bioanalysis. Bioanalysis 7 (17), 2265–2276. https://doi.org/10.4155/bio.15.149.

Kailasa, S.K., Ha, S., Baek, S.H., Phan, L.M.T., Kim, S., Kwak, K., Park, T.J., 2019. Tuning of carbon dots emission color for sensing of Fe^{3+} ion and bioimaging applications. Mater. Sci. Eng. C 98, 834–842. https://doi.org/10.1016/j.msec.2019.01.002.

Kailasa, S.K., Kiran, K., Wu, H.F., 2008. Comparison of ZnS semiconductor nanoparticles capped with various functional groups as the matrix and affinity probes for rapid analysis of cyclodextrins and proteins in surface-assisted laser desorption/ionization time-of-flight mass spectrometry. Anal. Chem. 80 (24), 9681–9688. https://doi.org/10.1021/ac8015664.

Kailasa, S.K., Koduru, J.R., Desai, M.L., Park, T.J., Singhal, R.K., Basu, H., 2018. Recent progress on surface chemistry of plasmonic metal nanoparticles for colorimetric assay of drugs in pharmaceutical and biological samples. TrAC Trends Anal. Chem. 105, 106–120. https://doi.org/10.1016/j.trac.2018.05.004.

Kailasa, S.K., Rohit, J.V., 2017a. Multi-functional groups of dithiocarbamate derivative assembly on gold nanoparticles for competitive detection of diafenthiuron. Sensors Actuat. B Chem. 244, 796–805. https://doi.org/10.1016/j.snb.2017.01.075.

Kailasa, S.K., Rohit, J.V., 2017b. Tuning of gold nanoparticles analytical applications with nitro and hydroxy benzylindole-dithiocarbamates for simple and selective detection of terbufos and thiacloprid insecticides in environmental samples. Colloids Surf. A PhysicoChem. Eng. Asp. 515, 50–61. https://doi.org/10.1016/j.colsurfa.2016.11.067.

Kailasa, S.K., Wu, H.F., 2015. Nanomaterial-based miniaturized extraction and preconcentration techniques coupled to matrix-assisted laser desorption/ionization mass spectrometry for assaying biomolecules. TrAC Trends Anal. Chem. 65, 54–72. https://doi.org/10.1016/j.trac.2014.09.011.

Kasibabu, B.S.B., Bhamore, J.R., D'souza, S.L., Kailasa, S.K., 2015. Dicoumarol assisted synthesis of water dispersible gold nanoparticles for colorimetric sensing of cysteine and lysozyme in biofluids. RSC Adv. 5 (49), 39182–39191. https://doi.org/10.1039/c5ra06814b.

Kasibabu, B.S.B., D'Souza, S.L., Jha, S., Kailasa, S.K, 2015. Imaging of bacterial and fungal cells using fluorescent carbon dots prepared from Carica papaya juice. J. Fluoresc. 25 (4), 803–810. https://doi.org/10.1007/s10895-015-1595-0.

Kasibabu, B.S.B., D'souza, S.L., Jha, S., Singhal, R.K., Basu, H., Kailasa, S.K., 2015. One-step synthesis of fluorescent carbon dots for imaging bacterial and fungal cells. Anal. Methods 7 (6), 2373–2378. https://doi.org/10.1039/c4ay02737j.

Kateshiya, M.R., Malek, N.I., Murthy, Z.V.P., Kailasa, S.K., 2020. Designing of glutathione-lactose derivative for the fabrication of gold nanoclusters with red fluorescence: sensing of Al3+and Cu2+ ions with two different mechanisms. Optical Mater. 100, 109704. https://doi.org/10.1016/j.optmat.2020.109704.

Kim, D., Choi, Y., Shin, E., Jung, Y.K., Kim, B.S., 2014. Sweet nanodot for biomedical imaging: carbon dot derived from xylitol. RSC Adv., 4 (44), 23210–23213. https://doi.org/10.1039/c4ra01723d.

Kong, L., Chu, X., Ling, X., Ma, G., Yao, Y., Meng, Y., Liu, W., 2016. Biocompatible glutathione-capped gold nanoclusters for dual fluorescent sensing and imaging of copper(II) and temperature in human cells and bacterial cells. MicroChim. Acta 183 (7), 2185–2195. https://doi.org/10.1007/s00604-016-1854-z.

Kong, Q., Zhang, L., Liu, J., Wu, M., Chen, Y., Feng, J., Shi, J., 2014. Facile synthesis of hydrophilic multi-colour and upconversion photoluminescent mesoporous carbon nanoparticles for bioapplications. Chem. Commun. 50 (99), 15772–15775. https://doi.org/10.1039/c4cc07121b.

Konwar, A., Gogoi, N., Majumdar, G., Chowdhury, D., 2015. Green chitosan-carbon dots nanocomposite hydrogel film with superior properties. Carbohydr. Polym. 115, 238–245. https://doi.org/10.1016/j.carbpol.2014.08.021.

Kozák, O., Datta, K.K.R., Greplová, M., Ranc, V., Kašlík, J., Zbořil, R., 2013. Surfactant-derived amphiphilic carbon dots with tunable photoluminescence. J. Phys. Chem. C 117 (47), 24991–24996. https://doi.org/10.1021/jp4040166.

Lan, G.Y., Chen, W.Y., Chang, H.T., 2011a. Characterization and application to the detection of single-stranded DNA binding protein of fluorescent DNA-templated copper/silver nanoclusters. Analyst 136 (18), 3623–3628. https://doi.org/10.1039/c1an15258k.

Lan, G.Y., Chen, W.Y., Chang, H.T., 2011b. One-pot synthesis of fluorescent oligonucleotide Ag nanoclusters for specific and sensitive detection of DNA. Biosensors Bioelectron. 26 (5), 2431–2435. https://doi.org/10.1016/j.bios.2010.10.026.

Li, G., Jin, R., 2013. Atomically precise gold nanoclusters as new model catalysts. Acc. Chem. Res. 46 (8), 1749–1758. https://doi.org/10.1021/ar300213z.

Li, J., Zhong, X., Zhang, H., Le, X.C., Zhu, J.J., 2012. Binding-induced fluorescence turn-on assay using aptamer-functionalized silver nanocluster DNA probes. Anal. Chem. 84 (12), 5170–5174. https://doi.org/10.1021/ac3006268.

Li, L.S., Jiao, X.Y., Zhang, Y., Cheng, C., Huang, K., Xu, L., 2018. Green synthesis of fluorescent carbon dots from Hongcaitai for selective detection of hypochlorite and mercuric ions and cell imaging. Sensors Actuat. B Chem. 263, 426–435. https://doi.org/10.1016/j.snb.2018.02.141.

Li, W., Yue, Z., Wang, C., Zhang, W., Liu, G., 2013. An absolutely green approach to fabricate carbon nanodots from soya bean grounds. RSC Adv., 3 (43), 20662–20665. https://doi.org/10.1039/c3ra43330g.

Li, X., Chang, J., Xu, F., Wang, X., Lang, Y., Gao, Z., Wu, D., Jiang, K., 2015. Pyrolytic synthesis of carbon quantum dots, and their photoluminescence properties. Res. Chem. Intermed. 41 (2), 813–819. https://doi.org/10.1007/s11164-013-1233-x.

Li, Y., He, X., Wang, C., Cao, Y., Li, Y., Yan, L., Liu, M., Lv, M., Yang, Y., Zhao, X., 2018. Controllable and eco-friendly synthesis of P-riched carbon quantum dots and its application for copper (II) ion sensing. Appl. Surf. Sci. 448, 589–598. https://doi.org/10.1016/j.apsusc.2018.03.246.

Lin, L., Hu, Y., Zhang, L., Huang, Y., Zhao, S., 2017. Photoluminescence light-up detection of zinc ion and imaging in living cells based on the aggregation induced emission enhancement of glutathione-capped copper nanoclusters. Biosensors Bioelectron. 94, 523–529. https://doi.org/10.1016/j.bios.2017.03.038.

Lin, P.Y., Hsieh, C.W., Kung, M.L., Chu, L.Y., Huang, H.J., Chen, H.T., Wu, D.C., Kuo, C.H., Hsieh, S.L., Hsieh, S., 2014. Eco-friendly synthesis of shrimp egg-derived carbon dots for fluorescent bioimaging. J. Biotechnol. 189, 114–119. https://doi.org/10.1016/j.jbiotec.2014.08.043.

Link, S., Beeby, A., FitzGerald, S., El-Sayed, M.A., Schaaff, T.G., Whetten, R.L., 2002. Visible to infrared luminescence from a 28-atom gold cluster. J. Phys. Chem. B 106 (13), 3410–3415. https://doi.org/10.1021/jp014259v.

LiQin, L., YuanFang, L., Lei, Z., Yue, L., ChengZhi, H., 2011. One-step synthesis of fluorescent hydroxyls-coated carbon dots with hydrothermal reaction and its application to optical sensing of metal ions. Sci. China Chem. 1342–1347. https://doi.org/10.1007/s11426-011-4351-6.

Liu, H., Ye, T., Mao, C., 2007. Fluorescent carbon nanoparticles derived from candle soot. Angew. Chem. Int. Ed. 46 (34), 6473–6475. https://doi.org/10.1002/anie.200701271.

Liu, S., Wang, L., Tian, J., Zhai, J., Luo, Y., Lu, W., Sun, X., 2011. Acid-driven, microwave-assisted production of photoluminescent carbon nitride dots from N,N-dimethylformamide. RSC Adv. 1 (6), 951–953. https://doi.org/10.1039/c1ra00249j.

Liu, W., Diao, H., Chang, H., Wang, H., Li, T., Wei, W., 2017. Green synthesis of carbon dots from rose-heart radish and application for Fe^{3+} detection and cell imaging. Sensors Actuat. B Chem. 241, 190–198. https://doi.org/10.1016/j.snb.2016.10.068.

Liu, Y., Liu, C.Y., Zhang, Z.Y., 2012. Synthesis of highly luminescent graphitized carbon dots and the application in the Hg^{2+} detection. Appl. Surf. Sci. 263, 481–485. https://doi.org/10.1016/j.apsusc.2012.09.088.

Liu, Y., Zhao, Y., Zhang, Y., 2014. One-step green synthesized fluorescent carbon nanodots from bamboo leaves for copper(II) ion detection. Sensors Actuat. B Chem. 196, 647–652. https://doi.org/10.1016/j.snb.2014.02.053.

Liu, Y., Zhou, Q., Li, J., Lei, M., Yan, X., 2016. Selective and sensitive chemosensor for lead ions using fluorescent carbon dots prepared from chocolate by one-step hydrothermal method. Sensors Actuat. B Chem. 237, 597–604. https://doi.org/10.1016/j.snb.2016.06.092.

Lu, W., Gong, X., Nan, M., Liu, Y., Shuang, S., Dong, C., 2015. Comparative study for N and S doped carbon dots: synthesis, characterization and applications for Fe^{3+} probe and cellular imaging. Anal. Chim. Acta 898, 116–127. https://doi.org/10.1016/j.aca.2015.09.050.

M., A., M., P.-M., M., C.-R., J., J.-J., G., E.da S.J.C., J., B.T., E., R.-C., T., L.N.J., J., C, 2014. Carbon dots obtained using hydrothermal treatment of formaldehyde. Cell imaging in vitro. Nanoscale 6,, 9071–9077. https://doi.org/10.1039/C4NR01585A.

Makarava, N., Parfenov, A., Baskakov, I.V., 2005. Water-soluble hybrid nanoclusters with extra bright and photostable emissions: a new tool for biological imaging. BioPhys. J. 89 (1), 572–580. https://doi.org/10.1529/biophysj.104.049627.

Mehta, V.N., Chettiar, S.S., Bhamore, J.R., Kailasa, S.K., Patel, R.M., 2017. Green synthetic approach for synthesis of fluorescent carbon dots for lisinopril drug delivery system and their confirmations in the cells. J. Fluoresc. 27 (1), 111–124. https://doi.org/10.1007/s10895-016-1939-4.

Mehta, V.N., Jha, S., Basu, H., Singhal, R.K., Kailasa, S.K., 2015. One-step hydrothermal approach to fabricate carbon dots from apple juice for imaging of mycobacterium and fungal cells. Sensors Actuat. B Chem. 213, 434–443. https://doi.org/10.1016/j.snb.2015.02.104.

Mehta, V.N., Jha, S., Kailasa, S.K., 2014. One-pot green synthesis of carbon dots by using *Saccharum officinarum* juice for fluorescent imaging of bacteria (*Escherichia coli*) and yeast (*Saccharomyces cerevisiae*) cells. Mater. Sci. Eng. C 38 (1), 20–27. https://doi.org/10.1016/j.msec.2014.01.038.

Mehta, V.N., Jha, S., Singhal, R.K., Kailasa, S.K., 2014. Preparation of multicolor emitting carbon dots for HeLa cell imaging. New J. Chem. 38 (12), 6152–6160. https://doi.org/10.1039/c4nj00840e.

Mehta, V.N., Rohit, J.V., Kailasa, S.K., 2016. Functionalization of silver nanoparticles with 5-sulfoanthranilic acid dithiocarbamate for selective colorimetric detection of Mn^{2+} and Cd^{2+} ions. New J. Chem. 40 (5), 4566–4574. https://doi.org/10.1039/c5nj03454j.

Mewada, A., Pandey, S., Shinde, S., Mishra, N., Oza, G., Thakur, M., Sharon, M., Sharon, M., 2013. Green synthesis of biocompatible carbon dots using aqueous extract of *Trapa bispinosa* peel. Mater. Sci. Eng. C 33 (5), 2914–2917. https://doi.org/10.1016/j.msec.2013.03.018.

Miao, Z., Hou, W., Liu, M., Zhang, Y., Yao, S., 2018. BSA capped bi-functional fluorescent Cu nanoclusters as pH sensor and selective detection of dopamine. New J. Chem. 42 (2), 1446–1456. https://doi.org/10.1039/c7nj03524a.

Mitra, S., Chandra, S., Kundu, T., Banerjee, R., Pramanik, P., Goswami, A., 2012. Rapid microwave synthesis of fluorescent hydrophobic carbon dots. RSC Adv., 2 (32), 12129–12131. https://doi.org/10.1039/c2ra21048g.

Mo, Q., Liu, F., Gao, J., Zhao, M., Shao, N., 2018. Fluorescent sensing of ascorbic acid based on iodine induced oxidative etching and aggregation of lysozyme-templated silver nanoclusters. Anal. Chim. Acta 1003, 49–55. https://doi.org/10.1016/j.aca.2017.11.068.

Mohanty, J.S., Xavier, P.L., Chaudhari, K., Bootharaju, M.S., Goswami, N., Pal, S.K., Pradeep, T., 2012. Luminescent, bimetallic AuAg alloy quantum clusters in protein templates. Nanoscale 4 (14), 4255–4262. https://doi.org/10.1039/c2nr30729d.

Moradi, S., Sadrjavadi, K., Farhadian, N., Hosseinzadeh, L., Shahlaei, M., 2018. Easy synthesis, characterization and cell cytotoxicity of green nano carbon dots using hydrothermal carbonization of *Gum Tragacanth* and chitosan bio-polymers for bioimaging. J. Mol. Liquids 259, 284–290. https://doi.org/10.1016/j.molliq.2018.03.054.

Muhammed, M.A.H., Verma, P.K., Pal, S.K., Kumar, R.C.A., Paul, S., Omkumar, R.V., Thalappil, P., 2009. Bright, NIR-emitting Au23 from Au25: characterization and applications including biolabeling. Chem. Eur. J. 15 (39), 10110–10120. https://doi.org/10.1002/chem.200901425.

Negishi, Y., Nobusada, K., Tsukuda, T., 2005. Glutathione-protected gold clusters revisited: bridging the gap between gold(I)-thiolate complexes and thiolate-protected gold nanocrystals. J. Am. Chem. Soc. 127 (14), 5261–5270. https://doi.org/10.1021/ja042218h.

Niu, J., Gao, H., 2014. Synthesis and drug detection performance of nitrogen-doped carbon dots. J. Lumin. 149, 159–162. https://doi.org/10.1016/j.jlumin.2014.01.026.

Paau, M.C., Lo, C.K., Yang, X., Choi, M.M.F., 2010. Synthesis of 1.4 nm α-cyclodextrin-protected gold nanoparticles for luminescence sensing of mercury(II) with picomolar detection limit. J. Phys. Chem. C 114 (38), 15995–16003. https://doi.org/10.1021/jp101571k.

Pan, J., Sheng, Y., Zhang, J., Wei, J., Huang, P., Zhang, X., Feng, B., 2014. Preparation of carbon quantum dots/TiO_2 nanotubes composites and their visible light catalytic applications. J. Mater. Chem. A 2 (42), 18082–18086. https://doi.org/10.1039/c4ta03528c.

Pang, S., Liu, S., 2017. Lysozyme-stabilized bimetallic gold/silver nanoclusters as a turn-on fluorescent probe for determination of ascorbic acid and acid phosphatase. Anal. Methods 9 (47), 6713–6718. https://doi.org/10.1039/c7ay02372c.

Park, S.Y., Thongsai, N., Chae, A., Jo, S., Kang, E.B., Paoprasert, P., Park, S.Y., In, I., 2017. Microwave-assisted synthesis of luminescent and biocompatible lysine-based carbon quantum dots. J. Ind. Eng. Chem. 47, 329–335. https://doi.org/10.1016/j.jiec.2016.12.002.

Patel, S.A., Richards, C.I., Hsiang, J.C., Dickson, R.M., 2008. Water-soluble Ag nanoclusters exhibit strong two-photon-induced fluorescence. J. Am. Chem. Soc. 130 (35), 11602–11603. https://doi.org/10.1021/ja804710r.

Petty, J.T., Zheng, J., Hud, N.V., Dickson, R.M., 2004. DNA-templated ag nanocluster formation. J. Am. Chem. Soc. 126 (16), 5207–5212. https://doi.org/10.1021/ja031931o.

Phan, L.M.T., Gul, A.R., Le, T.N., Kim, M.W., Kailasa, S.K., Oh, K.T., Park, T.J., 2019. One-pot synthesis of carbon dots with intrinsic folic acid for synergistic imaging-guided photothermal therapy of prostate cancer cells. BioMater. Sci. 7 (12), 5187–5196. https://doi.org/10.1039/c9bm01228a.

Qin, D., 2015. A facile approach for the fabrication of superhydrophobic surface with candle smoke particles. Physicochemical Problems of Mineral Processing 51.

Qing, T., He, X., He, D., Qing, Z., Wang, K., Lei, Y., Liu, T., Tang, P., Li, Y., 2016. Oligonucleotide-templated rapid formation of fluorescent gold nanoclusters and its application for Hg^{2+} ions sensing. Talanta 161, 170–176. https://doi.org/10.1016/j.talanta.2016.08.045.

Qu, S., Wang, X., Lu, Q., Liu, X., Wang, L., 2012. A biocompatible fluorescent ink based on water-soluble luminescent carbon nanodots. Angew. Chem. Int. Ed. 51 (49), 12215–12218. https://doi.org/10.1002/anie.201206791.

Rafique, R., Kailasa, S.K., Park, T.J., 2019. Recent advances of upconversion nanoparticles in theranostics and bioimaging applications. TrAC Trends Anal. Chem. 120. https://doi.org/10.1016/j.trac.2019.115646.

Ramezani, Z., Qorbanpour, M., Rahbar, N., 2018. Green synthesis of carbon quantum dots using quince fruit (*Cydonia oblonga*) powder as carbon precursor: Application in cell imaging and As^{3+} determination. Colloids Surf. A PhysicoChem. Eng. Asp. 549, 58–66. https://doi.org/10.1016/j.colsurfa.2018.04.006.

Rana, K., Bhamore, J.R., Rohit, J.V., Park, T.J., Kailasa, S.K., 2018. Ligand exchange reactions on citrate-gold nanoparticles for a parallel colorimetric assay of six pesticides. New J. Chem. 42 (11), 9080–9090. https://doi.org/10.1039/c8nj01294f.

Rawat, K.A., Singhal, R.K., Kailasa, S.K., 2016. Colorimetric and fluorescence "turn-on" methods for the sensitive detection of bromelain using carbon dots functionalized gold nanoparticles as a dual probe. RSC Adv. 6, 32025–32036. https://doi.org/10.1039/C6RA01575A.

Roy, S., Palui, G., Banerjee, A., 2012. The as-prepared gold cluster-based fluorescent sensor for the selective detection of As III ions in aqueous solution. Nanoscale 4 (8), 2734–2740. https://doi.org/10.1039/c2nr11786j.

Sahu, S., Behera, B., Maiti, T.K., Mohapatra, S., 2012. Simple one-step synthesis of highly luminescent carbon dots from orange juice: application as excellent bio-imaging agents. Chem. Commun. 48 (70), 8835–8837. https://doi.org/10.1039/c2cc33796g.

Shahnawaz Khan, M., Bhaisare, M.L., Pandey, S., Talib, A., Wu, S.M., Kailasa, S.K., Wu, H.F., 2015. Exploring the ability of water soluble carbon dots as matrix for detecting neurological disorders using MALDI-TOF MS. Int. J. Mass Spectrom. 393, 25–33. https://doi.org/10.1016/j.ijms.2015.10.007.

Shang, L., Dong, S., 2008. Silver nanocluster-based fluorescent sensors for sensitive detection of Cu(II). J. Mater. Chem. 18 (39), 4636–4640. https://doi.org/10.1039/b810409c.

Shang, L., Dong, S., 2009. Sensitive detection of cysteine based on fluorescent silver clusters. Biosensors Bioelectron. 24 (6), 1569–1573. https://doi.org/10.1016/j.bios.2008.08.006.

Shang, L., Dong, S., Nienhaus, G.U., 2011. Ultra-small fluorescent metal nanoclusters: synthesis and biological applications. Nano Today 6 (4), 401–418. https://doi.org/10.1016/j.nantod.2011.06.004.

Shang, L., Stockmar, F., Azadfar, N., Nienhaus, G.U., 2013. Intracellular thermometry by using fluorescent gold nanoclusters. Angew. Chem. Int. Ed. 52 (42), 11154–11157. https://doi.org/10.1002/anie.201306366.

Shang, L., Yang, L., Stockmar, F., Popescu, R., Trouillet, V., Bruns, M., Gerthsen, D., Nienhaus, G.U., 2012. Microwave-assisted rapid synthesis of luminescent gold nanoclusters for sensing Hg^{2+} in living cells using fluorescence imaging. Nanoscale 4 (14), 4155–4160. https://doi.org/10.1039/c2nr30219e.

Sharma, J., Yeh, H.C., Yoo, H., Werner, J.H., Martinez, J.S., 2011. Silver nanocluster aptamers: in situ generation of intrinsically fluorescent recognition ligands for protein detection. Chem. Commun. 47 (8), 2294–2296. https://doi.org/10.1039/c0cc03711g.

Shen, C., Yao, W., Lu, Y., 2013. One-step synthesis of intrinsically functionalized fluorescent carbon nanoparticles by hydrothermal carbonization from different carbon sources. J. Nanoparticle Res. 15 (10), 2019. https://doi.org/10.1007/s11051-013-2019-1.

So, P.T.C., Dong, C.Y., Masters, B.R., Berland, K.M., 2000. Two-photon excitation fluorescence microscopy. Annu. Rev. Biomed. Eng. 2 (2000), 399–429. https://doi.org/10.1146/annurev.bioeng.2.1.399.

Song, Y., Zhu, S., Xiang, S., Zhao, X., Zhang, J., Zhang, H., Fu, Y., Yang, B., 2014. Investigation into the fluorescence quenching behaviors and applications of carbon dots. Nanoscale 6 (9), 4676–4682. https://doi.org/10.1039/c4nr00029c.

Su, Y.T., Lan, G.Y., Chen, W.Y., Chang, H.T., 2010. Detection of copper ions through recovery of the fluorescence of DNA-templated copper/silver nanoclusters in the presence of mercaptopropionic acid. Anal. Chem. 82 (20), 8566–8572. https://doi.org/10.1021/ac101659d.

Sun, X., Lei, Y., 2017. Fluorescent carbon dots and their sensing applications. TrAC Trends Anal. Chem. 89, 163–180. https://doi.org/10.1016/j.trac.2017.02.001.

Tiwari, P., Kaur, N., Sharma, V., Mobin, S.M., 2020. A spectroscopic investigation of carbon dots and its reduced state towards fluorescence performance. J. PhotoChem. Photobiol. A 403, 112847.

Udaya Bhaskara Rao, T., Pradeep, T., 2010. Luminescent Ag7 and Ag8 clusters by interfacial synthesis. Angew. Chem. Int. Ed. 49 (23), 3925–3929. https://doi.org/10.1002/anie.200907120.

Wang, C., Sun, D., Zhuo, K., Zhang, H., Wang, J., 2014. Simple and green synthesis of nitrogen-, sulfur-, and phosphorus-co-doped carbon dots with tunable luminescence properties and sensing application. RSC Adv. 4 (96), 54060–54065. https://doi.org/10.1039/c4ra10885j.

Wang, D., Wang, X., Guo, Y., Liu, W., Qin, W., 2014. Luminescent properties of milk carbon dots and their sulphur and nitrogen doped analogues. RSC Adv., 4 (93), 51658–51665. https://doi.org/10.1039/c4ra11158c.

Wang, F., Chen, Y.H., Liu, C.Y., Ma, D.G., 2011. White light-emitting devices based on carbon dots' electroluminescence. Chem. Commun. 47 (12), 3502–3504. https://doi.org/10.1039/c0cc05391k.

Wang, J., Ng, Y.H., Lim, Y.F., Ho, G.W., 2014. Vegetable-extracted carbon dots and their nanocomposites for enhanced photocatalytic H_2 production. RSC Adv., 4 (83), 44117–44123. https://doi.org/10.1039/c4ra07290a.

Wang, M., Lin, Z., Liu, Q., Jiang, S., Liu, H., Su, X., 2018. DNA-hosted copper nanoclusters/graphene oxide based fluorescent biosensor for protein kinase activity detection. Anal. Chim. Acta 1012, 66–73. https://doi.org/10.1016/j.aca.2018.01.029.

Wang, Q., Liu, X., Zhang, L., Lv, Y., 2012. Microwave-assisted synthesis of carbon nanodots through an eggshell membrane and their fluorescent application. Analyst 137 (22), 5392–5397. https://doi.org/10.1039/c2an36059d.

Wang, W., Cheng, L., Liu, W., 2014. Biological applications of carbon dots. Sci. China Chem. 57 (4), 522–539. https://doi.org/10.1007/s11426-014-5064-4.

Wang, Y., Chang, X., Jing, N., Zhang, Y., 2018. Hydrothermal synthesis of carbon quantum dots as fluorescent probes for the sensitive and rapid detection of picric acid. Anal. Methods 10 (23), 2775–2784. https://doi.org/10.1039/c8ay00441b.

Wang, Y., Chen, T., Zhuang, Q., Ni, Y., 2018. Label-free photoluminescence assay for nitrofurantoin detection in lake water samples using adenosine-stabilized copper nanoclusters as nanoprobes. Talanta 179, 409–413. https://doi.org/10.1016/j.talanta.2017.11.014.

Wang, Y., Cui, Y., Zhao, Y., Liu, R., Sun, Z., Li, W., Gao, X., 2012. Bifunctional peptides that precisely biomineralize Au clusters and specifically stain cell nuclei. Chem. Commun. 48 (6), 871–873. https://doi.org/10.1039/c1cc15926g.

Wang, Z., Cai, W., Sui, J., 2009. Blue luminescence emitted from monodisperse thiolate-capped Au clusters. ChemPhysChem 10 (12), 2012–2015. https://doi.org/10.1002/cphc.200900067.

Wee, S.S., Ng, Y.H., Ng, S.M., 2013. Synthesis of fluorescent carbon dots via simple acid hydrolysis of bovine serum albumin and its potential as sensitive sensing probe for lead (II) ions. Talanta 116, 71–76. https://doi.org/10.1016/j.talanta.2013.04.081.

Wei, J., Zhang, X., Sheng, Y., Shen, J., Huang, P., Guo, S., Pan, J., Feng, B., 2014. Dual functional carbon dots derived from cornflour via a simple one-pot hydrothermal route. Mater. Lett. 123, 107–111. https://doi.org/10.1016/j.matlet.2014.02.090.

Wu, P.C., Chen, C.Y., Chang, C.W., 2018. The fluorescence quenching and aggregation induced emission behaviour of silver nanoclusters labelled on poly(acrylic acid-: Co -maleic acid). New J. Chem. 42 (5), 3459–3464. https://doi.org/10.1039/c7nj04399f.

Xiao, N., Dong, J.X., Liu, S.G., Li, N., Fan, Y.Z., Ju, Y.J., Li, N.B., Luo, H.Q., 2018. Multifunctional fluorescent sensors for independent detection of multiple metal ions based on Ag nanoclusters. Sensors Actuat. B Chem. 264, 184–192. https://doi.org/10.1016/j.snb.2018.02.177.

Xie, J., Zheng, Y., Ying, J.Y., 2009. Protein-directed synthesis of highly fluorescent gold nanoclusters. J. Am. Chem. Soc. 131 (3), 888–889. https://doi.org/10.1021/ja806804u.

Xie, J., Zheng, Y., Ying, J.Y., 2010. Highly selective and ultrasensitive detection of Hg^{2+} based on fluorescence quenching of Au nanoclusters by Hg^{2+}-Au^{+} interactions. Chem. Commun. 46 (6), 961–963. https://doi.org/10.1039/b920748a.

Xie, S.Y., Sun, Y.P., Meziani, M.J., Lu, F., Wang, H., Luo, P.G., Lin, Y., Harruff, B.A., Veca, L.M., Murray, D., 2007. Carbon dots for multiphoton bioimaging. J. Am. Chem. Soc. 129 (37), 11318–11319. https://doi.org/10.1021/ja073527l.

Xu, J., Zhou, Y., Liu, S., Dong, M., Huang, C., 2014. Low-cost synthesis of carbon nanodots from natural products used as a fluorescent probe for the detection of ferrum(III) ions in lake water. Anal. Methods 6 (7), 2086–2090. https://doi.org/10.1039/c3ay41715h.

Xu, M., He, G., Li, Z., He, F., Gao, F., Su, Y., Zhang, L., Yang, Z., Zhang, Y., 2014. A green heterogeneous synthesis of N-doped carbon dots and their photoluminescence applications in solid and aqueous states. Nanoscale 6 (17), 10307–10315. https://doi.org/10.1039/c4nr02792b.

Yang, S.T., Wang, X., Wang, H., Lu, F., Luo, P.G., Cao, L., Meziani, M.J., Liu, J.H., Liu, Y., Chen, M., Huang, Y., Sun, Y.P., 2009. Carbon dots as nontoxic and high-performance fluorescence imaging agents. J. Phys. Chem. C 113 (42), 18110–18114. https://doi.org/10.1021/jp9085969.

Yang, Y., Han, A., Li, R., Fang, G., Liu, J., Wang, S., 2017. Synthesis of highly fluorescent gold nanoclusters and their use in sensitive analysis of metal ions. Analyst 142 (23), 4486–4493. https://doi.org/10.1039/c7an01348e.

Zang, J., Li, C., Zhou, K., Dong, H., Chen, B., Wang, F., Zhao, G., 2016. Nanomolar Hg^{2+} detection using β-lactoglobulin-stabilized fluorescent gold nanoclusters in beverage and biological media. Anal. Chem. 88 (20), 10275–10283. https://doi.org/10.1021/acs.analchem.6b03011.

Zeng, Q., Shao, D., He, X., Ren, Z., Ji, W., Shan, C., Qu, S., Li, J., Chen, L., Li, Q., 2016. Carbon dots as a trackable drug delivery carrier for localized cancer therapy: in vivo. J. Mater. Chem. B 4 (30), 5119–5126. https://doi.org/10.1039/c6tb01259k.

Zhai, X., Zhang, P., Liu, C., Bai, T., Li, W., Dai, L., Liu, W., 2012. Highly luminescent carbon nanodots by microwave-assisted pyrolysis. Chem. Commun. 48 (64), 7955–7957. https://doi.org/10.1039/c2cc33869f.

Zhang, B., Wei, C., 2018. Highly sensitive and selective detection of Pb^{2+} using a turn-on fluorescent aptamer DNA silver nanoclusters sensor. Talanta 182, 125–130. https://doi.org/10.1016/j.talanta.2018.01.061.

Zhang, M., Qiao, J., Zhang, S., Qi, L., 2018. Copper nanoclusters as probes for turn-on fluorescence sensing of L-lysine. Talanta 182, 595–599. https://doi.org/10.1016/j.talanta.2018.02.035.

Zhang, P., Li, W., Zhai, X., Liu, C., Dai, L., Liu, W., 2012. A facile and versatile approach to biocompatible "fluorescent polymers" from polymerizable carbon nanodots. Chem. Commun. 48 (84), 10431–10433. https://doi.org/10.1039/c2cc35966a.

Zhang, S., Lin, B., Yu, Y., Cao, Y., Guo, M., Shui, L., 2018. A ratiometric nanoprobe based on silver nanoclusters and carbon dots for the fluorescent detection of biothiols. SpectroChim. Acta A Mol. Biomol. Spectrosc. 195, 230–235. https://doi.org/10.1016/j.saa.2018.01.078.

Zhang, X., Wu, F.G., Liu, P., Gu, N., Chen, Z., 2014. Enhanced fluorescence of gold nanoclusters composed of $HAuCl_4$ and histidine by glutathione: glutathione detection and selective cancer cell imaging. Small 10 (24), 5170–5177. https://doi.org/10.1002/smll.201401658.

Zhang, Y., Yan, M., Jiang, J., Gao, P., Zhang, G., Choi, M.M.F., Dong, C., Shuang, S., 2016. Highly selective and sensitive nanoprobes for Hg(II) ions based on photoluminescent gold nanoclusters. Sensors Actuat. B Chem. 235, 386–393. https://doi.org/10.1016/j.snb.2016.05.108.

Zhao, C., Jiao, Y., Hu, F., Yang, Y., 2018. Green synthesis of carbon dots from pork and application as nanosensors for uric acid detection. SpectroChim. Acta A: Mol. Biomol. Spectrosc. 190, 360–367. https://doi.org/10.1016/j.saa.2017.09.037.

Zhao, Q., Yan, H., Liu, P., Yao, Y., Wu, Y., Zhang, J., Li, H., Gong, X., Chang, J., 2016. An ultra-sensitive and colorimetric sensor for copper and iron based on glutathione-functionalized gold nanoclusters. Anal. Chim. Acta 948, 73–79. https://doi.org/10.1016/j.aca.2016.10.024.

Zheng, M., Xie, Z., Qu, D., Li, D., Du, P., Jing, X., Sun, Z., 2013. On-off-on fluorescent carbon dot nanosensor for recognition of chromium(VI) and ascorbic acid based on the inner filter effect. ACS Appl. Mater. Interf. 5 (24), 13242–13247. https://doi.org/10.1021/am4042355.

Zheng, X., Yao, T., Zhu, Y., Shi, S., 2015. Cu^{2+} modulated silver nanoclusters as an on-off-on fluorescence probe for the selective detection of l-histidine. Biosensors Bioelectron. 66, 103–108. https://doi.org/10.1016/j.bios.2014.11.013.

Zhou, J., Booker, C., Li, R., Zhou, X., Sham, T.K., Sun, X., Ding, Z., 2007. An electrochemical avenue to blue luminescent nanocrystals from multiwalled carbon nanotubes (MWCNTs). J. Am. Chem. Soc. 129 (4), 744–745. https://doi.org/10.1021/ja0669070.

Zhou, J., Sheng, Z., Han, H., Zou, M., Li, C., 2012. Facile synthesis of fluorescent carbon dots using watermelon peel as a carbon source. Mater. Lett. 66 (1), 222–224. https://doi.org/10.1016/j.matlet.2011.08.081.

Zhou, R., Shi, M., Chen, X., Wang, M., Chen, H., 2009. Atomically monodispersed and fluorescent sub-nanometer gold clusters created by biomolecule-assisted etching of nanometer-sized gold particles and rods. Chem. Eur. J. 15 (19), 4944–4951. https://doi.org/10.1002/chem.200802743.

Zhou, T., Huang, Y., Li, W., Cai, Z., Luo, F., Yang, C.J., Chen, X., 2012. Facile synthesis of red-emitting lysozyme-stabilized Ag nanoclusters. Nanoscale 4 (17), 5312–5315. https://doi.org/10.1039/c2nr31449e.

Zhu, H., Wang, X., Li, Y., Wang, Z., Yang, F., Yang, X., 2009. Microwave synthesis of fluorescent carbon nanoparticles with electrochemiluminescence properties. Chem. Commun. 34, 5118–5120. https://doi.org/10.1039/b907612c.

Zhuo, Y., Miao, H., Zhong, D., Zhu, S., Yang, X., 2015. One-step synthesis of high quantum-yield and excitation-independent emission carbon dots for cell imaging. Mater. Lett. 139, 197–200. https://doi.org/10.1016/j.matlet.2014.10.048.

CHAPTER 22

Synthesis of advanced carbon-based nanocomposites for biomedical application

Geoffrey S. Simate
School of Chemical and Metallurgical Engineering, University of the Witwatersrand, Johannesburg, South Africa

22.1 Introduction

From as early as the 1990s after the rediscovery (Monthioux and Kuznetsov, 2006) of carbon nanotubes (CNTs) by Iijima (1991), there had been countless research on various aspects of CNTs such as synthesis, properties, and applications. The CNTs are considered as a pillar for innovation in the fundamental science and technological applications (Setaro, 2017), and represent one of the most studied allotropes of carbon (Alshehri, 2016). The CNT materials are cylinder-shaped macromolecules (Balasubramanian and Burghard, 2005) characterized by having at least a single dimension with the size of a billionth of a meter (10^{-9} m) (Pokropivny, 2007). Their shape can be viewed as rolled hexagonal carbon networks (or graphene layers) that are capped by pentagonal carbon rings (Terrones, 2003). In other words, the CNTs have walls that are made up of a hexagonal lattice of carbon atoms analogous to the atomic planes of graphite, and are capped at their ends by one half of a fullerene-like molecule (Balasubramanian and Burghard, 2005). Eatemadi (2014) states further that a CNT is theoretically considered as a cylinder fabricated from rolled up graphene sheets.

In most cases, the CNTs can be categorized into two types depending on the number of graphene layers rolled up to make up the CNT (as depicted in Fig. 22.1), that is, single-walled CNTs (SWCNTs) and multiwalled CNTs (MWCNTs). Furthermore, depending on the way the graphene layer is wrapped into a cylinder, three different geometries can be formed for the SWCNTs: armchair, chiral, and zigzag (Eatemadi, 2014) as shown in Fig. 22.2 (Grobert, 2007). According to Eatemadi (2014), the structures and/or different geometries of the SWCNT are characterized by a pair of indices (n, m) that describe the chiral vector and have a direct effect on the electrical properties of nanotubes. It is also noted that the number of unit vectors in the honeycomb crystal lattice of graphene along two directions is determined by the integers n and m. Currently, there is a common understanding among scientists that when $m = 0$, the nanotubes are named zigzag nanotubes; when $n = m$, the nanotubes are named armchair nanotubes, and other states

Sustainable Nanotechnology for Environmental Remediation
DOI: https://doi.org/10.1016/B978-0-12-824547-7.00019-9

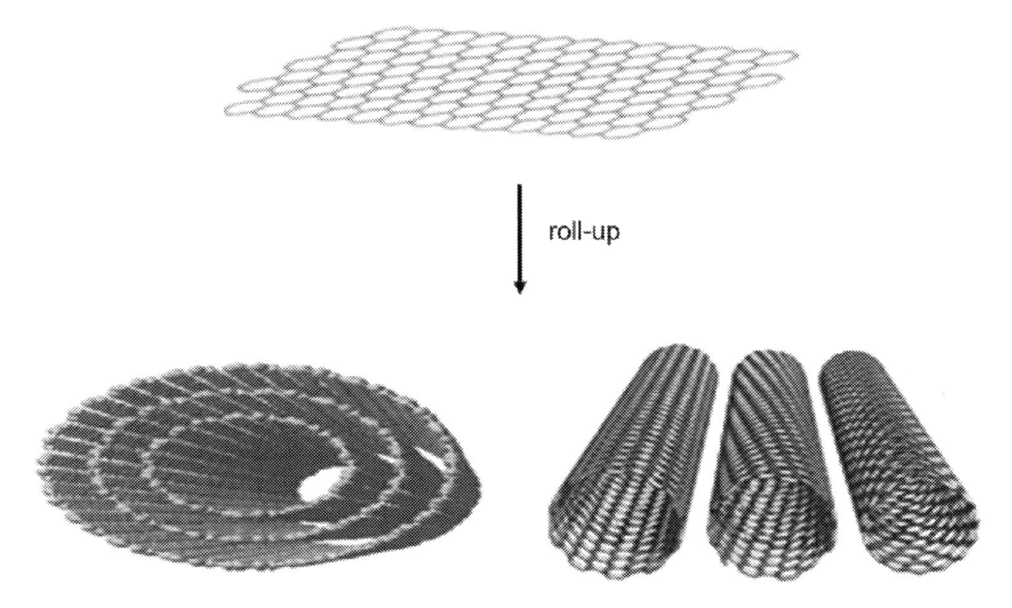

Figure 22.1 Models and schematic representation of multi-walled carbon nanotubes and single-walled carbon nanotubes.

are called chiral. Grobert (2007) also argues that in the zigzag structure, two opposite C-C bonds of each hexagon are parallel to the tube axis, whereas in the armchair structure the C-C bonds are perpendicular to the axis. In all other arrangements, the opposite C-C bonds lie at an angle to the tube axis, resulting in the so-called "helical nanotube" that is chiral.

As for MWCNTs, which were reported earlier than SWCNT (Iijima, 1991; Iijima and Ichihashi, 1993), there are two models that can be used to describe their structures: the Russian doll model and the Parchment model (Scoville, 1991; Iyuke and Simate, 2011; Eatemadi, 2014; Simate and Walubita, 2016). In the Russian doll model, a CNT contains another nanotube inside it and the inner nanotube has a smaller diameter than the outer nanotube, whereas in the Parchment model, a single graphene sheet is rolled around itself multiple times resembling a rolled up scroll of paper (Scoville, 1991). In MWCNTs, the nanotubes are typically bound together by strong van der Waals interaction forces and form tight bundles (Dai, 2002a; Simate and Walubita, 2016), whereas SWCNTs in many cases come together and form bundles in form of ropes (Eatemadi, 2014; Simate and Walubita, 2016). Chico et al. (1996) state that in a bundle structure, SWCNTs are hexagonally organized to form a crystal-like construction.

Extensive work has been carried out worldwide in recent years and the ongoing studies have indicated that CNTs are novel nanomaterials that have unique properties and potential to be developed into useful products at industrial scales, as they are attractive

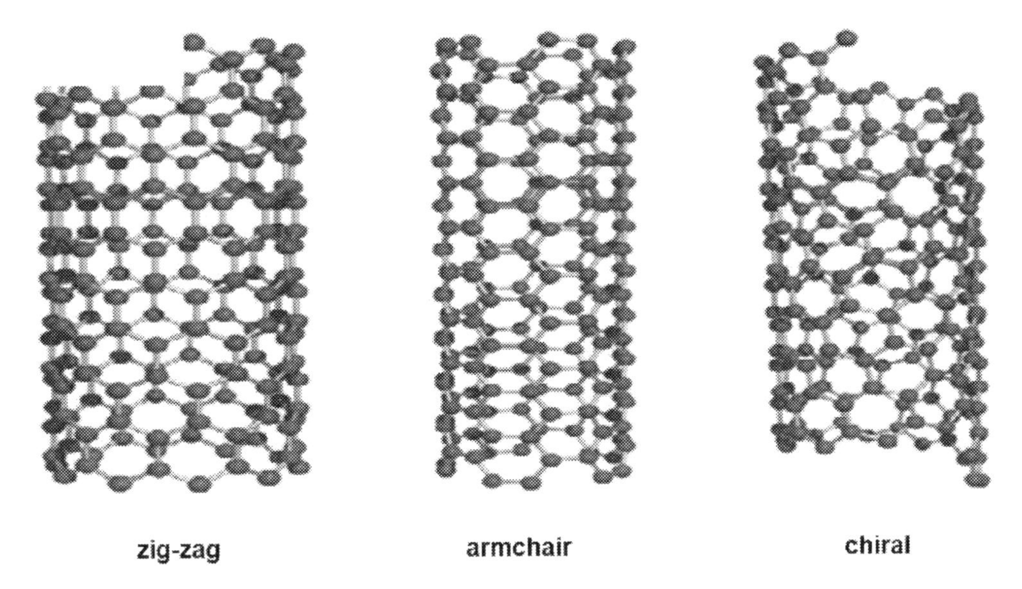

Figure 22.2 Different forms of singled-walled carbon nanotubes (Grobert, 2007).

for a variety of applications in various fields (Alshehri, 2016). The main focus of this chapter is to elucidate the numerous applications of CNT-based nanocomposites in the biomedical field. First, the chapter will give a snapshot of the synthesis of CNTs, in general. This will be followed by the synthesis of CNT-based nanocomposites. The functionalization of CNTs, which results in the improvement of their solubility and biocompatibility including altering their cellular interactions pathways, resulting in much reduced cytotoxic effects (Vardharajula et al., 2012) will also be discussed. The concluding remarks will follow after an in-depth discussion of various applications of CNT-based nanocomposites in biomedicine.

22.2 Synthesis of CNTs

It is well known that heating carbon black and graphite in a controlled flame environment can successfully produce CNTs (Iijima, 1991). Unfortunately, the nanotubes synthesized by this technique are irregular in size, shape, mechanical strength, quality, and purity due to uncontrollable natural environment (Singh et al., 2016). In the past decades, therefore, different techniques which mainly involve gas phase processes have been developed to produce CNTs in appropriate amounts (Dresselhaus et al., 2001; Agboola et al., 2007; Simate et al., 2010; Iyuke and Simate, 2011; Simate, 2012; Eatemadi, 2014; Lamberti, 2015; Alshehri, 2016; Simate and Walubita, 2016; Singh et al., 2016). However, the three very useful, widespread, and well-established methodologies for the production of CNTs include chemical vapor deposition (CVD), laser ablation, and arc discharge

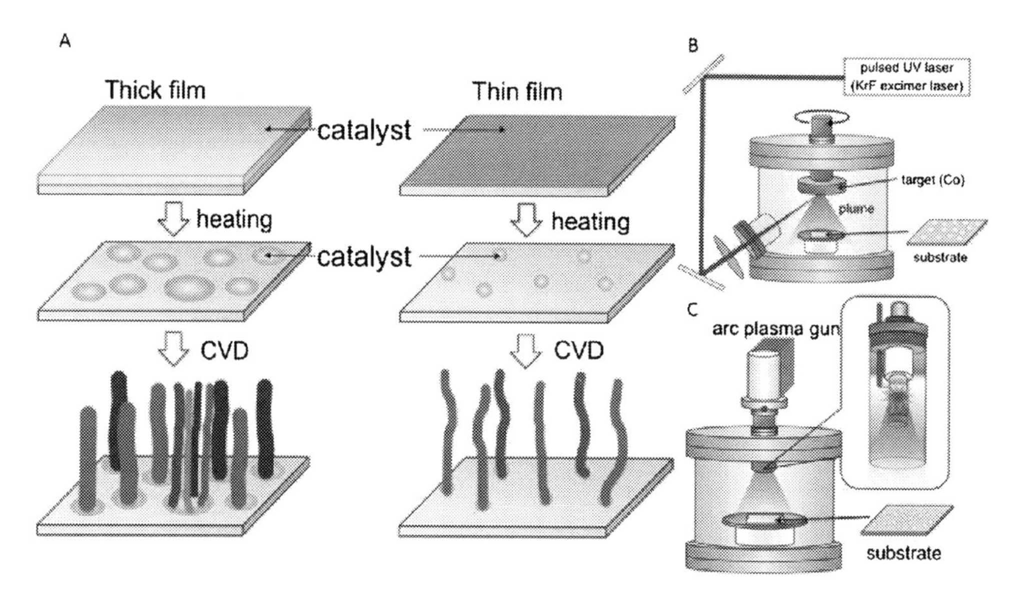

Figure 22.3 Schematic illustration showing different synthesis methods of carbon nanotubes. (A) Chemical vapor deposition (CVD) method used for carbon nanotubes synthesis. (B) Laser-ablation technique used in the synthesis of carbon nanotubes. (C) Carbon arc-discharge technique performed to synthesize carbon nanotubes (Alshehri et al., 2016).

(Jose et al., 1993; Chico et al., 1996; Ajayan and Ebbesen, 1997; Robertson, 2004; Agboola et al., 2007; Abbasi et al., 2014; Eatemadi, 2014; Simate and Walubita, 2016). Fig. 22.3 is a schematic illustration showing the three different methods for synthesizing CNTs. A review of the three techniques shows that two key requirements for the synthesis of CNTs include (1) a carbon source, and (2) a heat source for achieving the desired operating temperature (See and Harris, 2007). However, the type of nanotube that is produced also depends strongly on the absence or presence of a catalyst; MWCNTs are most commonly produced via noncatalytic means, whereas SWCNTs are usually the dominant products under catalytic growth conditions (Bernholc et al., 1998; Eatemadi, 2014). A summary of the three methodologies for CNT production is given in Table 22.1.

22.3 Synthesis of CNT-based nanocomposites

22.3.1 Introduction

Ajayan et al. (1997) define nanocomposites as materials consisting of more than one solid phase compositionally or structurally, where at least one dimension falls in the nanometer range. Similarly, Porwal and Saggar (2018) define nanocomposite materials as materials with more than one Gibbsian solid phase where at least one phase has dimensions in

Table 22.1 Summary of the three well-established techniques for producing carbon nanotubes.

Technique	General principles	References
Chemical vapor deposition	There are many different types of CVD such as catalytic chemical vapor deposition—either thermal or plasma-enhanced oxygen-assisted CVD, water-assisted CVD, microwave plasma-enhanced (MPECVD), radiofrequency CVD (RF-CVD), or hot-filament (HF-CVD). The technique simply involves the use of an energy source (such as plasma, a resistive or inductive heater, or furnace) to transfer energy to a gas-phase carbon source so as to produce fullerenes, CNTs, and other sp^2-like nanostructures. The technique can be applied both in the absence and presence of a catalyst substrate; the former being a homogeneous gas-phase process where the catalyst is in the gas-phase and the latter being a heterogeneous process that uses a supported catalyst. However, the CVD is currently considered as the standard technique for the synthesis of CNTs.	Iijima, 1991; Ebbesen and Ajayan, 1992; Ajayan et al., 1999; Dervishi et al., 2009; Iyuke and Simate, 2011; Eatemadi, 2014; Simate and Walubita, 2016
Laser ablation	In this technique, atomic carbon species are generated at ±1200 °C through laser irradiation of graphite. Ideally, in this methodology a quartz tube containing a block of pure graphite is heated inside a furnace at high temperatures in an argon atmosphere using high-power laser vaporization (yttrium-aluminum-garnet type). The aim of using laser is basically to vaporize the graphite within the quartz. The carbon species formed are swept by the flowing inert gas from the high-temperature zone and deposited on a conical water-cooled copper collector at the end of the apparatus where the soot-containing nanotubes is collected. This method produces MWCNTs when the vaporized carbon target is pure graphite, whereas the addition of transition metals (e.g., Co, Ni, Fe or Y) as catalysts to the graphite target results in the production of SWCNTs.	Journet and Bernier, 1998; Balasubramanian and Burghard, 2005; Paradise and Goswami, 2007; Agboola et al., 2007; Abbasi et al., 2014; Eatemadi, 2014; Simate and Walubita, 2016
Electric arc discharge	In this technique, CNTs are produced from carbon vapor generated by an electric arc discharge between two graphite electrodes (with or without catalysts), under an inert gas atmosphere of helium or argon. This technique uses higher temperatures (> 1700 °C) for CNT synthesis and typically produces CNTs of reasonable quality with fewer structural defects in comparison with other methods. The electric-arc process produces MWCNTs in the absence of catalysts, whereas SWCNTs are produced in the presence of catalysts.	Journet et al., 1997; Journet and Bernier, 1998; Lee et al., 2002; Agboola et al., 2007; Grobert, 2007; Simate and Walubita, 2016

Table 22.2 Categories of nanocomposites.

Matrix	Examples
Metal	Fe-Cr/Al_2O_3, Ni/Al_2O_3, Co/Cr, Fe/MgO, Al/CNT, Mg/CNT
Ceramic	Al_2O_3/SiO_2, SiO_2/Ni, Al_2O_3/TiO_2, Al_2O_3/SiC, Al_2O_3/CNT, SiO_2/Fe, $PbTiO_3$/PbZrO3, Al_2O_3/NdAlO3, Al_2O_3/$LnAlO_3$, $BaTiO_3$/SiC
Polymer	Thermoplastic/thermoset polymer/layered silicates, polyester/TiO_2, polymer/CNT, polymer/layered double hydroxides.

the nanometer range while the solid phase can exist in an amorphous, semicrystalline, or crystalline state. However, some researchers argue that even though one of the dimensions of the filler material is of the order of a nanometer, the final product does not have to be in nanoscale, but can be micro- or macroscopic in size (Hussain, 2006). In the recent past, the study of size effects on material properties has attracted enormous attention due to their scientific and industrial importance (Guisbiers et al., 2012). In fact, the material properties of nanostructures, for example, are different from the bulk materials due to the high surface area over volume ratio and possible appearance of quantum effects at the nanoscale (Guisbiers et al., 2012). Therefore, in nanocomposites, the nanophase (one or more) acts as a functional component in the composite that delivers new size-sensitive properties (Ajayan et al., 1997).

In general, nanocomposite materials can be classified into three different categories according to their matrix materials as shown in Table 22.2—metal, ceramic, and polymer (Camargo et al., 2009; Mikličanin et al., 2020). Metal matrix nanocomposites are basically materials consisting of a ductile metal or alloy matrix in which some nano-sized reinforcement material is implanted (Camargo et al., 2009). These materials combine metal and ceramic features, that is, ductility and toughness with high strength and modulus (Camargo et al., 2009). Therefore, these properties make metal matrix nanocomposites suitable constituents for the production of materials with high strength in shear/compression properties and high service temperature capabilities.

Ceramic nanocomposites comprise the matrix phase that can be either a ceramic or a glass material, while a nanophase can be nanoparticles (SiC, Si_3N_4), nanotubes (CNTs), nanoplatelets (graphene), or hybrids of the nanophase materials (Porwal and Saggar, 2018). Ceramic matrix composites have been developed to overcome the intrinsic brittleness and mechanical unreliability of monolithic ceramics, which are otherwise attractive for their high stiffness and strength (Chawla, 2003; Cho et al., 2009). Therefore, the addition of a nanophase to the ceramic matrix is very important because it helps to improve the mechanical properties of the ceramics (Porwal and Saggar, 2018). Ideally, the nanophases are expected to improve the mechanical properties of the nanocomposites either by deflecting crack propagations or acting as bridging elements to stop further crack growth. In addition, the nanophases also have the potential to improve properties such as wear

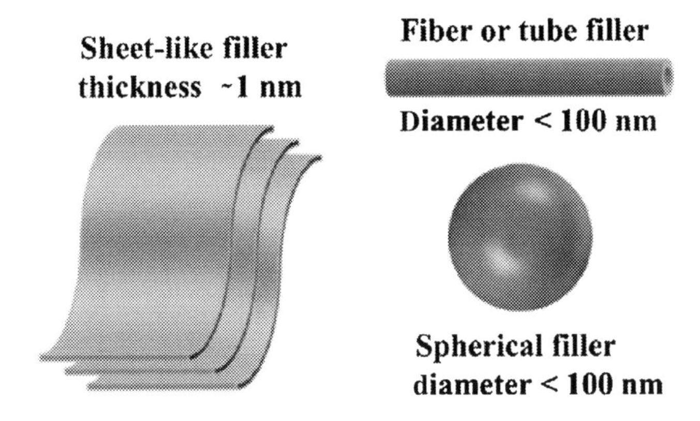

Figure 22.4 Schematic diagrams of three types of nanoscale fillers (Fu et al., 2019).

resistance, thermal shock resistance, bioactivity, electrical, and thermal conductivities of the nanocomposites (Sternitzke, 1997; Cho et al., 2009; Porwal et al., 2013; Porwal and Saggar, 2018).

Polymer nanocomposites are a group of hybrid materials that use polymer as the matrix and nanomaterial as the filler (also called nanofiller) (Abdulkadir et al., 2016). Das et al. (2018) consider polymer nanocomposites as a combination of a continuous phase of polymer and discontinuous phase of nanoparticles that show several advantages in mechanical, electric, and optical properties compared with individual components. Polymer nanocomposites can generally be grouped into three major types, depending on the dimensions of the dispersed nanoscale fillers (Fu et al., 2019). In the first type, the 2-D nanoscale fillers such as layered silicate, graphene, and many others such as MXene which are in the form of sheets of one to a few nanometer thick and of hundreds to thousands nanometers long are present in polymeric matrices (Du et al., 2012; Albdiry et al., 2013; Li et al., 2018; Fu et al., 2019). The corresponding polymer nanocomposites in the 2-D group can be categorized into the type of layered polymer nanocomposites (Fu et al., 2019). In the second type of polymer nanocomposites, two dimensions of the composite are in the nanometer scale and the third is larger thus forming an elongated 1-D structure (Fu et al., 2019). The nanoscale fillers in the second type of polymer nanocomposites that are used as reinforcing components to obtain materials with exceptional properties include nanofibers or nanotubes, for example, carbon nanofibers and carbon nanotubes (Calvert, 1999) or halloysite nanotubes (Liu et al., 2014). The third type is that of nanocomposites containing nanoscale fillers of three dimensions in the order of nanometers. These nanoscale fillers are iso-dimensional low aspect ratio nanoparticles such as quantum dots (Huang et al., 2016), spherical silica (Mark, 1996; Reynaud et al., 1999), and semiconductor nanoclusters (Herron and Thorn, 1998). Fig. 22.4 shows a representation of various nanoscale fillers.

Among the nanocomposites, CNT-based nanocomposites have attracted significant research interest in recent years owing to their important applications in various technological fields (Li et al., 2007; Camargo et al., 2009) including biomedical applications (Zhang et al., 2010; Lamberti et al., 2015). Some of the examples of CNT-based nanocomposites include polymer/CNTs, activated carbon/CNTs, metal oxide/CNTs, carbon fibers/CNTs, and many more (Ates et al., 2017).

22.3.2 Preparation of CNT-based nanocomposites

Many methods have been developed to prepare the CNT nanocomposites. However, the preparation of good quality nanocomposites including CNT-based nanocomposites using a proper processing method is critical to obtain nanocomposites with high performance characteristics (Fu et al., 2019). There is no universal technique for preparing all types of CNT-based nanocomposites. So far different processing techniques have been studied and/or employed for the preparation of CNT-based nanocomposites and other nanocomposites including noncovalent interaction, covalent reaction, electro and electroless plating, hydrothermal and solvothermal growth, electrochemical and electrophoresis deposition, photochemical reactions, and physical deposition and mixing. Other techniques, which are discussed in subsections 22.3.2.1–22.3.2.5 include melt blending, solution blending, *in-situ* polymerization, electrospinning, and layer-by-layer assembly. However, the subsections will focus mainly on the use of polymers because CNT-polymer nanocomposites have attracted increasing attention in recent years (Yellampalli, 2011) for biomedical applications (Wang et al., 2014). In addition, polymer is also considered to be a versatile material having many unique properties such as low density, reasonable strength, flexibility, easy processability, and many others (Choudhary and Gupta, 2011). The mechanical properties of several CNT-polymer composites produced by various techniques are summarized in Table 22.3.

22.3.2.1 Melt processing of CNT composites

In the melt processing technique, the polymer in a molten state and the nanotubes are mixed in a shear environment in a mixing device (typically a screw extruder) (Bhattacharya and Seong, 2013). In other words, the process involves heat treatment of the polymer and the CNTs in a mixing equipment (screw extruder or batch mixer). The mixer imparts shear and elongational stress to the process, thus helping to break CNT agglomerates apart and dispersing them uniformly in the polymer matrix. Ideally, the objective is to uniformly disperse the nanotubes in the polymer matrix for reinforcement (Bhattacharya and Seong, 2013). The compounded CNT-polymer composite can be further processed using other polymer-processing techniques such as injection molding, profile extrusion, blow molding, etc. (Bhattacharya and Seong, 2013). One of the advantages of melt processing is that it does not require the use of organic solvents during processing.

Table 22.3 Summary of the mechanical properties of various carbon nanotube composites processed using different techniques (Bhattacharya and Seong, 2013).

Polymer	Type of nanotube	Percent concentration of nanotubes	Processing method	Nanotube functionality	Modulus (GPa)	Tensile Strength (MPa)
High density polyethylene	MWCNT	1	Melt	Acid	1.2	28
Polystyrene	MWCNT	1	Melt	None	2	35
Polypropylene	MWCNT	1	Melt	None	1.5	26
Polyamide-12	SWCNT	0–15	Melt	None	2.4–13.2	–
Polyamide-6	SWCNT	2–12	Melt	None	3.0–4.18	–
Polyamide-6	MWCNT	0–2	Melt	Acid	2.0–3.0	35–54
Poly methyl methylacrylate	MWCNT	1-10	*In-situ* polymerization	None	–	47.2–71.5
Poly vinyl alcohol	SWCNT	0–0.8	Solution casting	Hydroxyl	2.4–4.3	74–07
Poly vinyl alcohol	MWCNT	1.5	Solution casting	Ferritin protein	7.2	–
Polystyrene	MWCNT	1	Solution casting	Chlorinated polypropylene	2.63	
Polyurethane	MWCNT	0–20	Solution Casting	Acid	0.05–0.42	7.6–21.3
Nylon 610	MWCNT	0–1.2	*In-situ* polymerization	Acid	0.9–1.4	36–54
Nylon 610	MWCNT	0–1.5	*In-situ* polymerization	Acid	0.9–2.4	36–52
Epoxy	SWCNT	0–4	*In-situ* polymerization	Acid	2.62–3.40	83–102

22.3.2.2 Solution processing of CNT composites

This technique is still the most popular method of producing composites, particularly at a laboratory scale. In solution processing technique, the polymer is dissolved in solution and the nanotubes are added thereafter (Bhattacharya and Seong, 2013). As the tubes are held together by van der Waals forces, they are separated and dispersed in solution using sonication or mechanical stirring. Ideally, the lower viscosity of the polymer in solution (as opposed to a melt) coupled with agitation by a mechanical stirrer or ultrasonication aids in the dispersion of the CNTs. Once adequate dispersion and homogeneity are obtained, the solvent is evaporated to yield the nanotube filled polymer.

22.3.2.3 In-situ polymerization technique

A variety of CNT-polymer composites have been prepared using *in-situ* polymerization (Bhattacharya and Seong, 2013). This process involves polymerizing vinyl monomers and CNTs. This process is very attractive for polymers that are thermally unstable (thereby making melt processing difficult) or are insoluble in solvents, and the technique can be used to produce both thermoset and thermoplastic materials (Bhattacharya and Seong, 2013). When using this process, it is possible to have a high nanotube loading (McClory et al., 2009; Bhattacharya and Seong, 2013), including the grafting of the polymer to the nanotube surface, which promotes interfacial adhesion between the polymer and the nanotube thus increasing its bulk properties (Bhattacharya and Seong, 2013). In fact, one of the advantages of this technique is that it allows the grafting of polymer molecules onto the walls of the nanotubes (Bhattacharya and Seong, 2013).

22.3.2.4 Electrospinning

This is an alternative technique for fabricating polymer/CNT composite fibers. The technique produces fibers with diameters ranging from microns to a few nanometers (Bhattacharya and Seong, 2013). It was originally applied to polymers, but the process can be applied to the production of glass, metal, and ceramic (Greiner and Wendorff, 2008; Bhattacharya and Seong, 2013).

The following is the description of a typical electrospinning system according to Bhattacharya and Seong (2013) and as illustrated in Fig. 22.5. The elements of a basic electrospinning unit include an electrode connected to a high-voltage power supply that is inserted into a syringe-like container containing the polymeric solution. Connected to the syringe is a capillary. The syringe-capillary set up can be mounted vertically, horizontally, or tilted at a defined angle. A grounded collector plate, which is connected to the other end of the electrode, is placed at a distance of 10–30 cm from the tip of the capillary.

The polymer solution at the end of the capillary upon the application of high voltage becomes charged. As the voltage is increased, a charge is induced on the surface of the

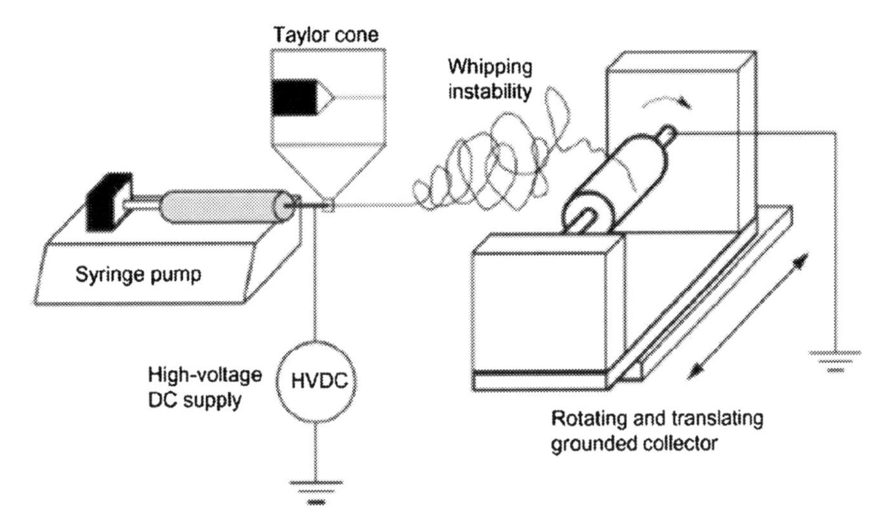

Figure 22.5 Schematic of a typical electrospinning system (Bhattacharya and Seong, 2013).

liquid. Mutual charge repulsion leads to development of force directly opposite to the surface tension. A jet is ejected from the suspended liquid meniscus at the end of the capillary when the applied electric field overcomes the surface tension of the liquid. A further increase in the electric field causes the hemispherical surface of the droplet at the tip of the capillary tube to elongate and form a conical shape known as the Taylor cone. When the repulsive electrostatic force overcomes the surface tension of the fluid, the charged jet is ejected from the tip of the Taylor cone. Within a few centimeters of travel from the tip, the discharged jet undergoes bending instability (Rayleigh instability) and begins to whip and splits into bundles of smaller fibers. In addition to bending instability, the jet undergoes elongation (strain $\sim 10^5$ and rate of strain $\sim 10^3$ s^{-1}) which causes it to become very long and thin (diameter in the range of nanometers to micrometers). The solvent evaporates, leading to the formation of skin and solidification of the fluid jet followed by the collection of solid charged polymer fibers on the collector, usually in the form of nonwoven fabric.

A number of CNT/polymer composites (mostly consisting of MWCNTs) have been successfully electrospun, making it a versatile fiber processing technique. The electrospinning technique allows the alignment of the CNTs along the fiber axis (Bhattacharya and Seong, 2013). The alignment of the CNTs in the polymer enhances the aspect ratio for reinforcing and increases the area for interfacial bonding (Kim et al., 2005; Bhattacharya and Seong, 2013). Many biologically functional molecules and cells often interact at the nanoscale level, making these electrospun matrices attractive for tissue engineering.

22.3.2.5 *Layer-by-layer assembly*

According to Bhattacharya and Seong (2013), the layer-by-layer technique is a powerful tool used to assemble multilayer and multimaterial thin films. The technique involves immersing a negatively (or positively) charged substrate in an oppositely charged polyelectrolyte that is adsorbed onto the substrate. After equilibrium is reached, the substrate is removed, rinsed, dried, and immersed in a negatively (or positively) charged polyelectrolyte solution. This process is repeated until the desired thickness is achieved. In simple terms, the layer-by-layer technique exploits the electrostatic attraction between oppositely charged species to induce the growth of a 1-D structure (Bhattacharya and Seong, 2013). The absorption of the polyelectrolyte is irreversible and charge overcompensation leads to charge reversal at the surface (Caruso et al., 1998; Bhattacharya and Seong, 2013). Different materials can be inserted between layers as long as they have the opposite charge. This enables the coating of various different shapes and sizes by uniformly layered materials with controllable thickness. The layer-by-layer assembly can also be performed on a colloidal substrate (Becker et al., 2010; Such et al., 2011; Bhattacharya and Seong, 2013).

22.3.3 Typical studies of synthesizing CNT-based nanocomposites

This section discusses a couple of examples of some of the methodologies used for fabricating various types of CNTs-based nanocomposites, namely ceramics, metals, and polymers. Basically, only a sample of the techniques used for fabricating nanocomposites are covered in the examples. Azqhandi et al. (2018) produced Ag-doped ZnO-loaded MWCNTs (Ag@ZnO/MWCNT) nanocomposite in three main stages. In the first stage, carboxylic acid–functionalized MWCNTs (MWCNT-COOH) were produced by sonication in strong acid (sulfuric acid–nitric acid) to introduce carboxyl (–COOH) groups on the CNT surface. The carboxylation stage basically involved taking an appropriate amount of MWCNT powder and adding it into 150 mL of H_2SO_4–HNO_3 (3:1 v/v) solution followed by direct sonication for 3 h at 40 °C. Thereafter, the mixture was cooled to room temperature followed by filtration and washing of the functionalized MWCNTs with deionized water until they were completely neutralized and were later dried at 80 °C for 5 h in a vacuum oven. The second step involved the deposition of ZnO and Ag nanoparticles onto functionalized MWCNTs. In this stage, 0.1 g of MWCNT-COOH was dispersed in an appropriate amount of deionized water in a bath sonicator for 1 h. A mixed solution of $Zn(NO_3)_2 \cdot 6H_2O$ and $AgNO_3$ in deionized water was then added and dispersed by magnetic stirring for 30 min to allow adsorption of cations (i.e., Ag^+ and Zn^{2+}) on the surface of MWCNTs. The content of Ag in Ag@ZnO/MWCNT nanocomposite was fixed to obtain a nanocomposite with 3 wt% Ag. The pH was adjusted by adding NH_4OH (1:1) until the pH of 10 was reached. Thereafter, a solution of oxalic acid (0.5 M) was added to the mixture and magnetically stirred for 1 h at 60 °C. The mixture was then cooled for 45 min until zinc oxalate and silver oxalate were formed. In the last step, the nanocomposites obtained from the previous step were centrifuged, washed several times, dried at 100 °C in a vacuum oven, and calcined at 500 °C for 10 h in an argon flow (20 mL/min). After this step, all of the silver oxalate and zinc oxalate were converted to metallic silver and zinc oxide.

As was shown by the scanning electron microscopy (SEM) micrographs that displayed the surface morphology of pure ZnO nanoparticles, Ag@ZnO, and Ag@ZnO/MWCNT nanocomposites, the results of a study by Azqhandi et al. (2018) indicated that the addition of MWCNTs to the nanocomposite reduced the particle size of Ag and ZnO nanoparticles as well as improved the distribution of smaller particles on the surface of MWCNTs. The study also showed that the specific surface area of Ag@ZnO nanocomposite was increased with the addition of Ag to ZnO nanoparticles. Previous studies also showed that doping of transition metals into ZnO structure creates smaller nanostructures with higher surface area (Samadi et al., 2016). In addition, loading of MWCNTs to the nanocomposite increased the specific surface area of the nanocomposite from 20.89 m^2/g for Ag@ZnO to 175.50 m^2/g for Ag@ZnO/MWCNT. The chemical composition of the prepared Ag@ZnO and Ag@ ZnO/MWCNT nanocomposites was determined using energy dispersive X-ray (EDX) analysis. The EDX dot-mapping analysis of synthesized Ag@ZnO/MWCNT nanocomposites showed the presence and distribution of the desired elements such as Ag, Zn, and O in the structure of samples.

Li et al. (2007) reported the production of various nanocomposites of metals, metal oxides, and CNTs via self-assembly. In one study, Li et al. (2007) produced TiO_2/CNTs nanocomposites using the as-received CNTs and the HNO_3-treated CNTs. The first step involved the dispersion of CNTs in 4–60 mL of toluene under ultrasonic conditions (in an ultrasonic water bath) for 0.5–2 h. Thereafter, 1–8 mL of the TiO_2 suspension (in toluene) was added, and the mixture was continuously sonicated for another 0.5–1 h. In the same study, TiO_2/CNTs nanocomposites were also produced under mechanical stirring conditions for a much longer process time of 12 h without any sonication. The resulting TiO_2/CNTs nanocomposite was washed thoroughly with toluene and ethanol solvents and dried in an electric oven at 60 °C for several hours. In the same study, Li et al. (2007) purged about 50 mg of Co_3O_4/CNTs sample with argon gas inside a quartz tube reactor at a flow rate of 70 mL/min to a temperature of 500 °C (the heating rate was 10°C/min). A stream of pure hydrogen was then introduced at a rate of 50 mL/min, and the sample was held at 500 °C for 2 h. The composite samples (Co/CoO/Co_3O_4/CNTs) were collected when the furnace temperature was decreased to room temperature.

Other CNT-based nanocomposite of Au/CNTs, Au/TiO_2/CNTs, TiO_2/Co_3O_4/CNTs were also produced by Li et al. (2007). The results of the study showed that it is possible to carry out step-by-step self-assembly approach for the preparation of binary, ternary, and quaternary CNTs-based nanocomposites using presynthesized nanoparticles as primary building units. The study demonstrated that good controls of particle shape, size, and distribution for the highly complex inorganic-organic nanocomposites and nanohybrids can be attained. For example, the content of overlayer components can be easily controlled by predetermining ratios of assembled components (i.e., overlayer/CNTs ratios) and assembling time. In general, a greater ratio of overlayer/CNTs and a longer processing time lead to a higher loading of the surface phase(s).

Béguin et al. (2004) used MWCNTs produced by catalytic decomposition of acetylene as a backbone for preparing a C/C composite by one-step carbonization of CNTs/polyacrylonitrile blends. Initially, the CNTs having a content ranging from 15 to 70 wt% were mixed with polyacrylonitrile in excess acetone so as to generate a slurry. Thereafter, the acetone was evaporated and the composite mixture was pressed at 1–2 tons/cm^2. The pellets were later carbonized at 700–900 °C for 30–420 min under nitrogen thus giving rise to C/C composites. The pure components of the composites, that is, CNTs and polyacrylonitrile, and the C/C composites were later characterized using SEM (Hitachi S 4200) and transmission electron microscopy (TEM Philips CM20). The chemical composition of the composites was determined by elemental analysis and by X-ray photoelectron spectroscopy (XPS, Escalab 250, VG Scientific). The porous texture was characterized by nitrogen adsorption at a temperature of 77 K (–196.15 °C) on an Autosorb 6 (Quanta Chrome) after degassing for 15 h at 200 °C. The results of the chemical composition by both elemental analysis and XPS showed that the final C/C composite was still rich in nitrogen (9.2 atomic% and 7.2 atomic%, respectively), demonstrating that polyacrylonitrile is an efficient nitrogen carrier. The study found that the amount of oxygen in the composite was quite high, probably due to its incorporation by the addition on the dangling bonds when the C/C composite was exposed to air after its formation. The SEM images of the CNTs/polyacrylonitrile (30/70 wt%) blend carbonized at 700 °C under nitrogen showed that the resulting C/C composite was homogeneous and similar to the carbonized pure polyacrylonitrile. The study concluded that the production of an optimally performing nanocomposite can be achieved by: (1) high amount of polyacrylonitrile that would favor a large gas evolution thus enabling the creating of pores; (2) a sufficient amount of CNTs, which prevents shrinkage of the composite during the carbonization of polyacrylonitrile and consequently assists in pores formation; and (3) a reasonable proportion of polyacrylonitrile to get the largest amount of residual nitrogen in the C/C composites.

Sharma and Kumar (2018) studied the fabrication and preparation of CNTs-based polymer nanocomposites using an injection molding machine. The MWCNTs reinforcements were mixed with pure polymer (high-density polyethylene) of an injection molding grade matrix. Ideally, the carbon nanofluid prepared with a given quantity of CNTs was mixed with polyethylene pellets. The mixture was heated and stirred continuously to create a uniform coating of CNTs on the polyethylene pellets. Thereafter, the fluid was evaporated and the pellets were kept in an oven for 4 h at 100 °C to remove any moisture present on the CNT-coated pellets. The CNT-coated pellets were later used as raw materials in an injection molding machine to form "dog bone shaped" CNT/high density polyethylene nanocomposite of 120 mm long and 2 mm thick. In the whole study by Sharma and Kumar (2018) different weight fractions of CNTs were mixed with pure high-density polyethylene and the improvements in the strength, toughness, Young's modulus, stiffness and many other properties were studied using universal testing machines, SEM, and optical microscopic. The results showed that the composites had higher mechanical properties compared to high-density polyethylene.

22.4 Functionalization of CNT-based nanocomposites for biomedical applications

22.4.1 Introduction

Section 22.3 discussed the use of CNTs as nanofillers particularly in polymer nanocomposites. Indeed, CNTs can greatly improve the mechanical strength of existing polymer materials and create a highly anisotropic nanocomposite (Sen et al., 2004; Urooj et al., 2016). Unfortunately, the as-produced CNTs are almost insoluble in any aqueous solution or organic solvents except by sonication that usually only leads to CNT dispersion (Tasis at al., 2006; Alshehri et al., 2016). The insoluble nature of the CNTs is due to strong intertube van der Waals and π-π interactions (Liu, 2008; Liu and Speranza, 2019). In other words, CNTs readily bundle together and it is very difficult to dissolve or disperse them in solution (Simate, 2012). Vardharajula et al. (2012) also argue that the smooth surface of carbon nanomaterials that make them to have no overhanging bonds renders them chemically inert and incompatible with almost all organic and inorganic solvents. The lack of solubility and the difficulty of manipulation in any solvent by the freshly prepared CNTs hinders and/or limits their practical applications downstream in fields such as biomedical (Matarrendona, 2003; Tasis at al., 2006; Vardharajula et al., 2012; Liu and Speranza, 2019). Ideally, the successful incorporation of CNTs into practical materials relies on the capability of breaking up the bundles into individual nanotubes and keeping them in homogeneous and stable suspensions (Matarrendona, 2003).

In view of the inherent nature of the as-produced CNTs, it is important to functionalize them not only to make them more soluble, but also to allow their integration into many organic, inorganic, and biological systems and applications (Tasis at al., 2006; Beg et al., 2011; Simate, 2012; Alshehri et al., 2016). In other words, CNTs need to be functionalized to obtain their optimal performance in various applications including biomedical applications (Mehra et al., 2008; Simate, 2012). Moreover, Sireesha et al. (2017) state that functionalization of CNTs provides them with biocompatibility and solubility. More specifically, several of the functionalization techniques have useful biomedical applications (Alshehri et al., 2016). In fact Liu et al. (2009) argue that surface functionalization is critical to the behavior of CNTs in biological systems.

22.4.2 Functionalization of CNTs for biomedical applications

Functionalization is the process of immobilizing molecules onto surfaces such as that of CNTs to impart the surfaces with specific functions such as biospecificity and/or catalytic activity (Zhou et al., 2019). The overall objective of functionalizing CNTs for biomedical applications is to increase their solubility or dispersion in biocompatible (aqueous) media, thereby reducing their toxic effects (Vardharajula et al., 2012). In addition, functionalization moieties can be introduced to CNTs such that they specifically interact with cell surface receptors that can guide their internalization (Alshehri et al., 2016). For example, the receptor-mediated targeting strategies can facilitate specific cell loading, thereby lowering the quantity of drugs needed in disease treatment. Indeed, functionalized CNTs display unique properties that enable a variety of medicinal applications, including the diagnosis and treatment of cancer, infectious diseases and central nervous system disorders, and applications in tissue engineering (Zhang et al., 2010); some of which are discussed in detail in Section 22.5.

The main techniques for the functionalization or modification of the CNT structures can be grouped into three categories (Tasis et al., 2006; Wu et al., 2010; Alshehri et al., 2016): (1) the covalent attachment of chemical groups through reactions onto the π-conjugated skeleton of CNT, (2) the noncovalent adsorption or wrapping of various functional molecules, and (3) the endohedral filling of their inner empty cavities. Interestingly, several of these functionalization techniques have useful biomedical applications (Alshehri et al., 2016). The three techniques are discussed in the subsequent subsections. However, the sections will focus mainly on the use of polymers because CNT-polymer nanocomposites have attracted increasing attention in recent years (Yellampalli, 2011) for biomedical applications (Wang et al., 2014). In addition, polymer is also considered to be a versatile material having many unique properties such as low density, reasonable strength, flexibility, easy processability, and many others (Choudhary and Gupta, 2011). Most importantly, compared to different ranges of nanofillers, CNTs have emerged as the most promising nanofiller for polymer composites due to their remarkable characteristics that include mechanical and electrical properties (Ishikawa et al., 2001; Kracke and Damaschke, 2000; Choudhary and Gupta, 2011). Beyou et al. (2013) also argue that the incorporation of CNTs into a polymer matrix is a very attractive way to combine the mechanical and electrical properties of individual nanotubes with the advantages of plastics.

22.4.2.1 Covalent approaches

Covalent functionalization is widely used to disperse CNTs so as to improve CNT biocompatibility and to bring biomedical functionality to CNTs (Liu and Speranza, 2019). The covalent functionalization process involves several chemical reactions, through which covalent chemical bonds are formed between CNTs and the entities for functionalization

(Liu and Speranza, 2019). As stated earlier, this section will focus mainly on the use of polymers.

The covalent reaction of CNTs with polymers is essential because the long polymer chains help to dissolve the tubes into a wide range of solvents even at a low degree of functionalization (Tasis et al., 2006). Two main methodologies exist for the covalent attachment of polymeric substances onto the surfaces of CNTs (Liu, 2005; Tasis et al., 2006; Kitano et al., 2007; Sahoo et al., 2010; Beyou et al., 2013), that is, "grafting to" and "grafting from" methods. The "grafting to" method begins with the synthesis of a polymer of a specific molecular weight followed by end group transformation (Beyou et al., 2013). Subsequently, the polymer chain is attached to the graphitic surface of CNTs. In other words "grafting to" method means that the readymade polymers with reactive end groups are reacted with the functional groups on the nanotube surfaces, that is, it is the reaction between the surface groups of nanotubes and readymade polymers (Liu, 2005). A disadvantage of this method is that the grafted polymer contents are limited because of high steric hindrance of macromolecules (Beyou et al., 2013). The "grafting from" is based on the covalent immobilization of the polymer precursors (or initiators) on the surface of the nanotubes and subsequent propagation of the polymerization in the presence of monomeric species (Tasis et al., 2006; Wu et al., 2011; Beyou et al., 2013). In other words "grafting from" method means the reactive groups are covalently attached to the nanotube surface and then the polymers graft from the reactive groups, that is, it is the reaction between the reactive groups on the surface of the nanotubes and monomers (Liu, 2005). Fig. 22.6 shows different forms of chemical functionalization processes.

It is noted, however, that covalent sidewall functionalization of nanotubes from sp^2 to sp^3 structure is both difficult and undesirable because of the loss of conjugation (Dai, 2002b). In other words, side wall covalent functionalization of CNTs usually involves the breaking of sp^2 bonding networks to form sp^3 binding on CNT walls (Liu and Speranza, 2019). This technique, therefore, generally requires harsh reaction conditions and leads to extreme changes in the properties of the CNTs (Liu and Speranza, 2019). Park et al. (2006) argue that because the covalent attachment of the surface modifiers involves the partial disruption of the sidewall sp^2 hybridization system, covalently modified CNTs inevitably lose some degree of their electrical and/or electronic performance properties. In addition, the covalent attachment of functional groups to the surface of nanotubes might introduce defects on the walls of the perfect structure of the nanotubes thus lowering the strength of the reinforcing component (Liu, 2005).

22.4.2.2 Noncovalent approaches

Noncovalent sidewall functionalization is the process of adsorption or wrapping of surfactants, polymers, or biopolymers onto the surface of CNTs (Liu and Speranza, 2019). In other words, in this method polymer chains are wrapped, or small molecules

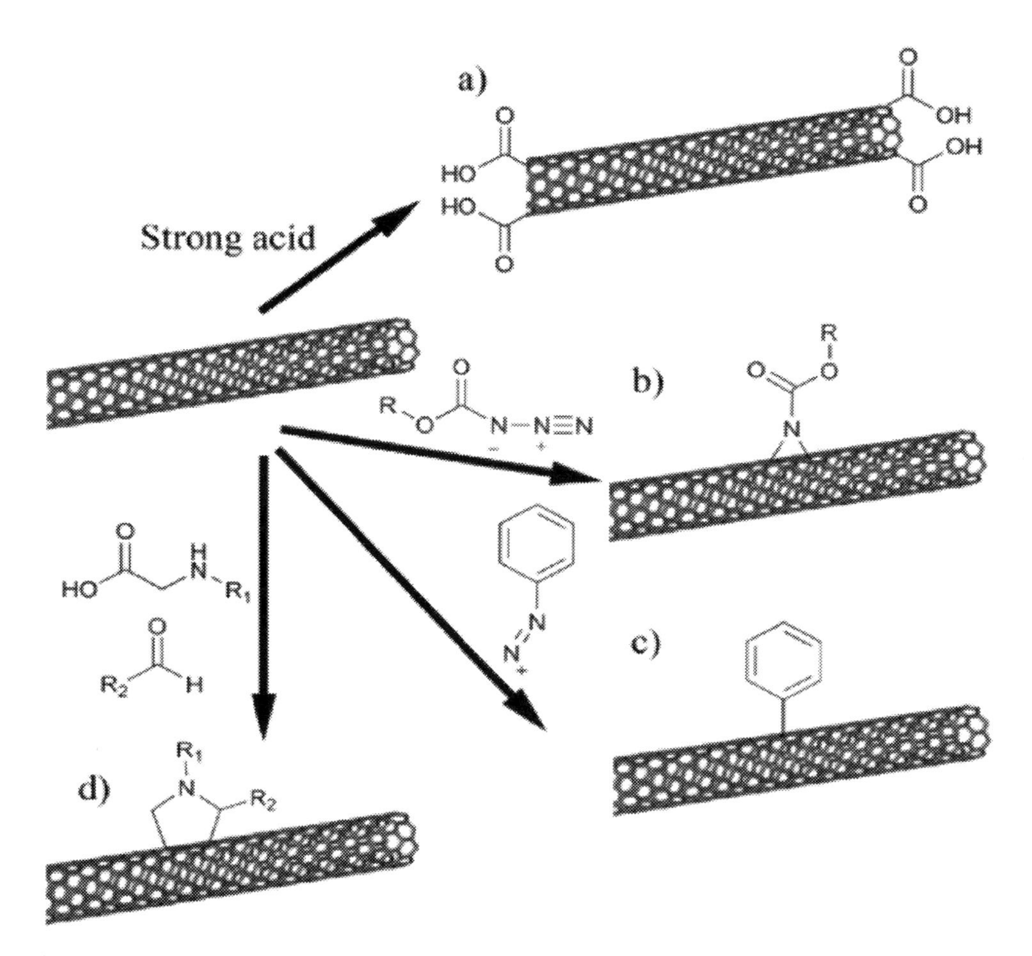

Figure 22.6 Common methods for chemical functionalization of carbon nanotubes. (A) oxidation by strong acids, (B) nitrene cycloaddition, (C) arylation using diazonium salts, and (D) 1,3-dipolar cycloadditions (Yang et al., 2007).

or biochemically active molecules are attached onto the sidewall of CNTs (Trojanowicz, 2006). The technique is based on weak interactions, for example, hydrogen bonding, $\pi-\pi$ stacking, electrostatic forces, van der Waals forces and hydrophobic interactions (Trojanowicz, 2006), and it is controlled by thermodynamics (O'Connell et al., 2001; Liu, 2005; Tasis et al., 2006). The solubility of CNTs in a solvent depends on the type and concentration of the molecules adsorbed on the tube surface (Liu, 2008). In principle, the nonpolar part of the surfactant, for example, will adsorbed on the CNT surface with the polar end sticking out to contact with the solvent used. Fig. 22.7 shows polymer wrapping model. Another type of noncovalent functionalization model is termed polymer attachment or adsorption (Dai et al., 2003).

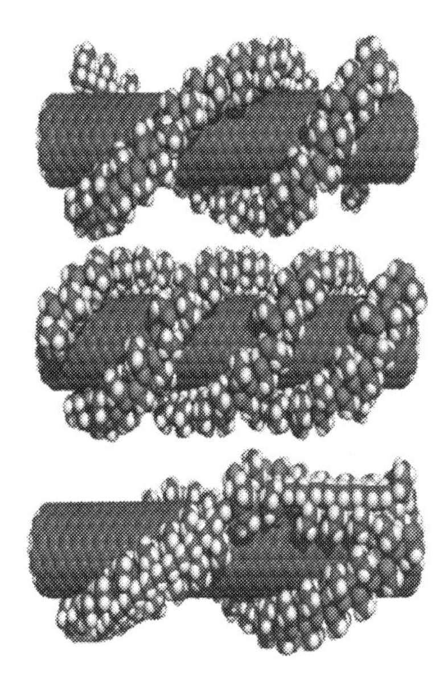

Figure 22.7 Polymer wrapping model of noncovalent functionalization (O'Connell et al., 2001).

Noncovalent functionalization of CNTs is particularly attractive because it offers the possibility of attaching chemical handles without affecting the electronic networks of the nanotubes (Tasis et al., 2006), thus their mechanical properties should not change (Goyanes et al., 2007). In other words, the technique is expected to bring less disturbance to the sp^2 structure of CNTs and, subsequently, preserve the mechanical and electrical properties of CNTs (Liu and Speranza, 2019). Electronic properties and sp^3 structures are useful for various postfunctionalization applications (Dai, 2002b). The other two advantages are its simple preparation procedure and, it is applicable to as-produced full-length CNTs (Liu, 2008; Liu and Speranza, 2019). Furthermore, most surfactants (or even polymers and biopolymers) used are easy to obtain and have already been used in pharmaceutical products (Lamprecht et al., 2011; Liu and Speranza, 2019). The disadvantage of noncovalent attachment is that the forces between the wrapping molecule and the nanotube might be weak, thus as a filler in a composite the efficiency of the load transfer might be low (Liu, 2005; Goyanes et al., 2007).

22.4.2.3 Endohedral filling

Open-ended CNTs provide internal cavities capable of accommodating molecules of suitable sizes (Tasis et al., 2006). In this regard, of particular interest to researchers is the use of CNTs in nanofluidic applications (Werder et al., 2001; Tasis et al., 2006). Nanofluidics is the study of fluid (gas, liquid) flow around and inside nanoscale systems

(Werder et al., 2001). Nanofluidics is envisioned as a key technology for designing engineering devices for biological applications, such as biomedical devices (e.g., nanoexplorers, cell manipulators, etc.) in which the dominant biomolecular transport process is carried out by natural and forced convection (Werder et al., 2001; Tasis et al., 2006). Other applications include the usage of CNT as nanopipets, as sieves for DNA sequencing applications, and (in arrays of CNTs) as acoustic sensors in the form of artificial stereocilia (Werder et al., 2001; Tasis et al., 2006).

It has been predicted that any liquid having a surface tension below approximately 180 m/Nm should be able to wet the inner cavity of nanotubes through an open end at atmospheric pressure (Dujardin et al., 1994; Tasis et al., 2006), whereas in the case of high surface tension, a highly pressurized liquid must be used to force it to enter inside the cavity (Tasis et al., 2006). This limit implies that typical pure metals will not be drawn into the inner cavity of nanotubes through capillarity, whereas water and organic solvents will (Dujardin et al., 1994).

22.5 Applications of CNT and their nanocomposites in biomedicine

For close to three decades there has been a wealth of research and new avenues that have been opened up for possible biomedical applications of CNTs and their composites. This section explores the development of CNTs and their composites for biomedical applications including drug delivery, cancer therapy, biosensing, bioimaging, bacterial inhibition, tissue engineering, and dental applications.

22.5.1 Drug delivery

Li et al. (2019) define drug delivery systems as methods by which drugs are delivered to desired tissues, organs, cells, and subcellular organs for drug release and absorption through a variety of drug carriers. Such systems are usually designed to (1) improve aqueous solubility and chemical stability of active agents, (2) increase pharmacological activity, and (3) reduce side effects (Li et al., 2019). Amongst the present drug delivery systems, CNTs and other nanoparticles have shown great potential as carriers in recent years. The encapsulation of drugs in CNTs and many other nanoparticles, including micelles, liposomes, dendrimers, nanocapsules, nanospheres and others, improves the therapeutic index and reduces the adverse side effects (Li et al., 2019).

There is no doubt the use of carbon-based nanostructures, such as CNTs, in biomedicine, and drug delivery, in particular, is increasingly attracting attention (Lacerda et al., 2007). In fact, Liu et al. (2009) state that the CNT-based drug delivery has shown promise in various *in-vitro* and *in-vivo* experiments including delivery of small interfering RNA (siRNA), paclitaxel, and doxorubicin. The unique hollow interior of CNTs as well as functional groups on the exterior surface allows for use of both covalent functionalization and physical entrapment as methods for functionalizing and loading drugs onto and inside CNTs (Loh et al., 2018). It is also noted that the external attachment of drugs onto CNTs is usually achieved by attaching molecules by amide, ester, or disulfide bonds (Kushwaha et al., 2013). This approach is used to employ a bond that is biologically cleaved either near the cell or more usefully, within the cell before releasing the payload (Kushwaha et al., 2013). Similarly, other scholars also argue that the location of the drug to be delivered by the CNT can be internal or external (Kushwaha et al., 2013). In addition, it is also argued that internalization or encapsulation relies on Van der Waals forces for insertion into the CNT and is best used for drugs that are sensitive to external environments and easily broken down (Hillebrenner et al., 2006; Kushwaha et al., 2013).

One of the key advantages of CNTs is their ability to translocate through plasma membranes thus allowing their use for the delivery of therapeutically active molecules in a manner that resembles cell-penetrating peptides (Lacerda et al., 2007). Similarly, Lu et al. (2009) state that CNTs have been pursued for their ability to penetrate into cells without the need for any external transporter system and the potentially high loading capacity. In fact, research has demonstrated that CNTs are effective carriers for shuttling and delivering various peptides, proteins, nucleic acids, and small molecular drugs into living cells (Kam and Dai, 2006; Lacerda et al., 2007; Prato et al., 2008; Lu et al., 2009). For example, in a study by Pastorin et al. (2006) a "double functionalization" strategy was employed to attach both fluorescein isothiocyanate (FITC, a fluorescent probe) and methotrexate (an anticancer drug) onto the sidewall of MWNTs via the 1,3-cycloaddition reaction of azomethine ylides. According to *in-vitro* experiments with Human Jurkat T lymphocytes cells, the results of the study showed that the nanotube-bound drugs were rapidly internalized into the cells and subsequently accumulated into the cell cytoplasm (Pastorin et al., 2006).

22.5.2 Cancer therapy

Cancer or malignant neoplasm is a genetic disorder that results from genetic or epigenetic alterations in the somatic cells (Chakraborty and Rahman, 2012). According to Feinberg et al. (2006), cancer is a highly heterogeneous complex disease that encompasses a group of disorders characterized by continuous indefinite growth. In simple terms, cancer is described as a range of diseases that can affect different organs and any part of the body, and the most common types of cancer are: breast in women; bowel (colorectal); prostate in men; skin melanoma; and lung (Cancer Council of Australia, 2017). According to Kushwaha et al., (2013) cancer is considered to be among the top three killers in modern society, next to heart and cerebrovascular diseases. Treatment for cancer is often successful if the cancer is found early, and the most common types of cancer treatment are surgery, chemotherapy, and radiotherapy (radiation therapy); whereas, in recent years, hormone (endocrine) therapy, targeted therapy, and immunotherapy are becoming more common for some types of cancers (Cancer Council of Australia, 2017). However, despite many advances in cancer-treatment techniques, it still remains a problem (Augustine et al., 2017). In fact a study by Chakraborty and Rahman (2012) outlined a number of difficulties in cancer treatment including the difficulty in targeting cancer stem cells (CSCs), drug resistance properties of CSCs that make them immune to anticancer drugs, lack of cancer epigenetic profiling and specificity of existing epidrugs, problems associated with cancer diagnosis making it difficult to treat, unavailability of effective biomarkers for cancer diagnosis and prognosis, existing chemotherapeutic drugs are toxic to all cells including cancer and normal cells, and metastatic nature of cancer pose a huge problem in cancer treatment.

To improve the survival rate of patients suffering from cancer, the availability of novel technologies for early diagnosis, monitoring, and therapy is essential (Augustine et al., 2017). With recent advances in the field of nanobiotechnology, the use of nanostructured materials such as CNTs for the development of cancer theranostic agents is receiving enormous research and practical interests (Gobbo et al., 2015; Crozals et al., 2016; Augustine et al., 2017). The CNTs have specifically been explored for cancer theranostics because of their unique physicochemical properties (Augustine et al., 2017) and research has shown that it is possible to engineer CNTs with highly efficient multifunction diagnostics and therapeutics agents (Bartelmess et al., 2015; Chen et al., 2015). Indeed, as discussed in Section 22.4.2, CNTs can be surface engineered (i.e., functionalized) to enhance their dispersibility in the aqueous phase or to provide the appropriate functional groups that can bind to the desired therapeutic material or the target tissue to elicit a therapeutic effect (Elhissi et al., 2012). For example, the CNTs can help the attached therapeutic molecule to penetrate through the target cell to treat diseases (Pantarotto et al., 2004a; Pantarotto et al., 2004b; Bianco et al., 2005) and an example of CNTs with a variety of functional groups relevant for cancer therapy is shown in Fig. 22.8 (Bhirde et al., 2009; Elhissi et al., 2012).

At the moment there are a variety of CNTs and other nanoparticle systems that are being explored for cancer therapeutics (Haley and Frenkel, 2008; Jabir et al., 2012). Some of the various approaches toward cancer treatment with CNTs and different other nanomaterials include, (1) targeted drug delivery via nanocarriers, (2) targeting tumor cells, (3) targeting the tumor microenvironment, and (4) targeting recurrent and drug-resistant cancers (Jabir et al., 2012). One good illustration of some of the approaches (e.g., tumor targeting) pertains to a study by McDevitt et al. (2007) who reported a successful multiple derivatization of CNTs with a monoclonal antibody used as a targeting ligand. In the study, a CNT-antibody conjugate was developed to specifically target the CD20 epitope on Human Burkitt lymphoma cells and simultaneously deliver a radionuclide (McDevitt et al., 2007).

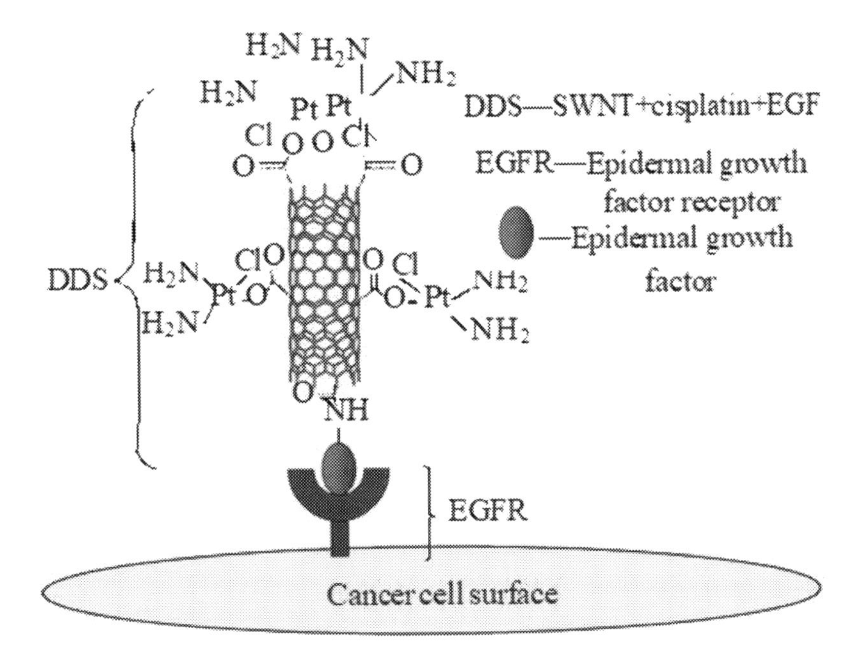

Figure 22.8 Single-walled carbon nanotube bioconjugated with cisplatin and epidermal growth factor (EGF), targeting cancer cell surface receptor overexpressing EGF receptors (EGFR) (Bhirde et al., 2009).

22.5.3 Biosensors

According to Balasubramanian and Burghard (2006), sensors are a class of devices whose use range from the detection of gas molecules to the real-time tracking of chemical signals in biological cells. In general, according to Balasubramanian and Burghard (2006), a sensor consists of an active sensing element and a signal transducer, and produces an electrical, optical, thermal, or magnetic output signal. While the sensing element is responsible for the selective detection of the analyte, the transducer converts a chemical event into an appropriate signal that can be used with or without amplification to determine the analyte concentration in a given test sample (Balasubramanian and Burghard, 2006).

In recent past, the detection of biomolecules has become a topical research issue in many areas of health care, clinical medicine, food safety, environmental monitoring, and homeland security (Yang et al., 2015). On the other hand, there has been significant effort to demonstrate the feasibility and potential use of CNTs as biosensors (Tîlmaciu and Morris, 2015; Yang et al., 2015; Kour et al., 2020) for various biomedical applications as shown in Fig. 22.9 (Pastorin et al., 2006; Kour et al., 2020). It must be noted that biosensors differ from classical chemical sensors in the following two ways: (1) the sensing element consists of a biological material such as proteins (e.g., cell receptors, enzymes, antibodies), oligo- or polynucleotides, microorganisms, or even whole biological tissues

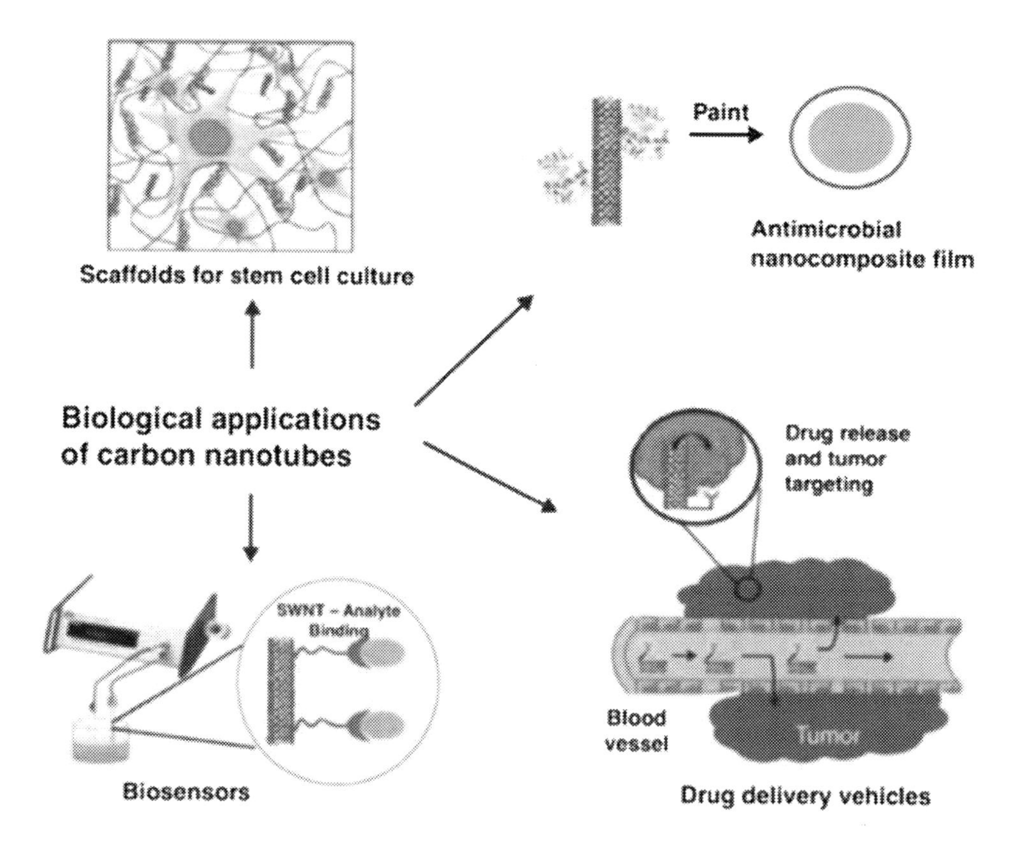

Figure 22.9 Carbon nanotubes as biosensors for different applications (Kour et al., 2020).

(Malhotra et al., 2005; Wilson and Gifford, 2005), and (2) the sensor is used to monitor biological processes or for the recognition of biomolecules (Balasubramanian and Burghard, 2006). Malhotra et al. (2005) define biosensors as analytical devices incorporating biological materials such as enzymes, tissues, microorganisms, antibodies, cell receptors or biologically derived materials or a biomimic component in intimate contact with a physicochemical transducer or transducing microsystems. As already discussed, transducers are the components that convert a biochemical signal into a quantifiable electrical signal (Fig. 22.9).

Rapid advances in biosensors have recently been reported due to the rapid growth in the development of nanomaterials such as CNTs (Tîlmaciu and Morris, 2015; Yang et al., 2015; Kour et al., 2020). Nanomaterials possess unique features that make them particularly attractive for biosensing applications (Tîlmaciu and Morris, 2015). In particular, CNT-based biosensors are recognized to be the next-generation building blocks for ultrasensitive and ultrafast biosensing systems (Yang et al., 2015). Moreover,

CNTs can serve as scaffolds for the immobilization of biomolecules on their surfaces, and combine several exceptional physical, chemical, electrical, and optical characteristics which make them one of the best suited materials for the transduction of signals associated with the recognition of analytes, metabolites, or disease biomarkers (Balasubramanian and Burghard, 2006; Tîlmaciu and Morris, 2015).

Interestingly, Kour et al. (2020) argue that the properties of CNTs can be customized to suit their potential as biosensors. For example, the antibodies and enzymes can be used to customize the features of CNTs in the electrochemical biosensors. The characteristics of CNTs as biosensors can also be tailored by the peptides and nucleic acids as they do have the inherent capability to be acquainted with bioelements or biomolecules. However, the method of analysis, that is, *in-vivo* or *in-vitro* determines the design of a biosensor. Ideally, the biomolecules are attached on the surface of CNTs to prepare the surface for a particular detection process.

A selection of typical studies of CNT biosensors include the following: (1) a glucose biosensor was fabricated by Gomathi et al. (2011) using phase separation method that employed MWCNT-grafted chitosan (CS)-nanowire (NW) to which glucose oxidase was attached to obtain the biosensor. The electrochemical detection of glucose was done by employing cyclic voltammetry and amperometry. The fabricated biosensor exhibited a high sensitivity of 5.03 μA/mM in a concentration range of 1–100 mM and a low response time to the detection of glucose. The MWCNTs-CS-NW facilitated the conduction of electrons between glucose oxidase and target molecules; and (2) Tsai et al. (2007) developed a biosensor that employed alcohol-dehydrogenase enzyme along with MWCNT and poly(vinyl alcohol) and reported a response time of about 8 s for ethanol detection. The electro oxidation of the reduced form of nicotinamide adenine dinucleotide (NAD), which is produced during the enzymatic activity, produces a current that is taken into account for generating the response by the biosensor. It must be noted that NAD exists in two forms: an oxidized and reduced form, abbreviated as NAD^+ and NADH, respectively.

22.5.4 Biomedical imaging

One of the most important steps in disease biology is early detection and diagnostic, which facilitates the design and choice of a better therapeutic approach (Raghavendra and Pullaiah, 2018; Han et al., 2019). In this regard, there are cellular and molecular diagnostic methods that serve as tools for enabling the understanding of a wide variety of clinically and genetically heterogeneous disorders with specific or overlapping phenotypes (Raghavendra and Pullaiah, 2018). Owing to computing power growth, current imaging techniques aim at providing complete visualization from the molecular scale to cellular, organ, tissue, lesion scales, and to the whole organism scale (Wallyn et al., 2019). There are five prime imaging techniques dedicated for preclinical and clinical biomedical applications. According to Wallyn et al. (2019), the most commonly used imaging modalities are X-ray computed tomography (X-ray CT); magnetic resonance imaging (MRI); optical imaging by fluorescence and bioluminescence; nuclear imaging, including positron emission tomography (PET) and single-photon emission computed tomography (SPECT); and ultrasound imaging. The existence of such a wide range of imaging techniques is mainly due to their respective abilities to reveal structural and/or functional information at different scales and accuracy levels.

Unfortunately, the traditional clinical imaging techniques have a number of disadvantages including a limitation in detection (Han et al., 2019; Wallyn et al., 2019). In addition, the current medical imaging contrast agents being used are mostly small molecules that exhibit fast metabolism and have nonspecific distribution and potential undesirable toxicities (Chronos et al., 1993; Petrik et al., 2005; Han et al., 2019). Therefore, nanomaterials, and in particular CNTs, have invigorated efforts of finding ways to improve biomedical detection and imaging due to their unique passive, active, and physical targeting properties (Han et al., 2019).

Indeed, the intrinsic properties of CNTs have been studied and exploited for the imaging and therapeutic purposes (Martincic and Tobias, 2014). For example, as a result of the strong optical absorption in the near-infrared region, SWCNTs have been used for photoacoustic imaging and photothermal therapy (Martincic and Tobias, 2014). Liu et al. (2009) also state that SWCNTs with various interesting intrinsic optical properties have been used as novel photoluminescence, Raman, and photoacoustic contrast agents for imaging of cells and animals. In addition, it is also observed that the emission range of SWCNTs (800–2000 nm) covers the biological tissue transparency window, thus allowing biological imaging (Liu et al., 2009; Martincic and Tobias, 2014).

Some of the studies involving the use of CNTs as imaging agents for MRI, PET, SPECT, and X-ray CT include the following as reviewed in an article by Hernández-Rivera et al. (2016): (1) PET: Ruggiero et al. (2010) developed an amine-functionalized SWCNT material that was conjugated to both the tumor neovascular-targeting antibody E4G10 and to one of two different metal chelates that contained either $^{225}Ac^{3+}$, an alpha emitter for radiotherapy ($t_{1/2} = 10$ d) chelated by 1,4,7,10- tetraazacyclododecane-1,4,7,10-tetraacetic acid (DOTA), or $^{89}Zr^{4+}$, a β^{+} emitter for PET imaging ($t_{1/2} = 78$ h) chelated by desferrioxamine B. The results showed that using the targeted, radiolabeled SWCNT material resulted in an approximately fivefold increase in specific activity, which improved the signal-to-noise ratio of the image (Ruggiero et al., 2010); (2) MRI: Sitharaman et al. (2005) reported the first CNT-based MRI contrast agents containing Gd^{3+} that was termed the "Gadonanotube" or GNT. The contrast agents consisted of ultrashort (20–80 nm) SWCNTs loaded with Gd^{3+} ion clusters (3–4% Gd by weight). The GNTs displayed 40–90 times superior relaxivity values compared to the clinically used MRI contrast agents Magnevist. Furthermore, the Gd^{3+} ion cluster were so tightly contained by the CNT platform such that the ions did not leak out of the CNTs under biological conditions (Hartman et al., 2008), which is critical for clinical applications given the high toxicity of free Gd^{3+} (Broome et al., 2008); and (3) X-ray CT: The use of CT in combination with a CNT-based contrast agent to track stem cells was studied by Rivera et al. (2013). X-ray contrast agents (or radiocontrast agents) are used to provide transient contrast enhancement in X-ray-based imaging modalities such as radiography, CT, and fluoroscopy and have also been investigated for cell labeling (Kim et al., 2017).

22.5.5 Bacterial inhibition

According to Yah and Simate (2015), microbial infections have become a global health burden due to emerging and resistant strains of viruses, bacteria, pathogenic fungi, and protozoa that are defying clinical treatment. Similarly, Mocan et al. (2017) also acknowledge that multidrug-resistant bacterial infections evolved via natural selection have increased alarmingly at a global level. Consequently, this has culminated into lengthy treatment, higher health expenditure, mortality risk, and low life expectancy (Tanwar et al., 2014; Yah and Simate, 2015). Therefore, it is imperative that novel antibiotics for the treatment of microbial infections are developed (Mocan et al., 2017). In other words, there is a need to seek new alternative and safer antimicrobial agents against the "super bugs" of viruses, bacteria, fungi, and protozoa (Yah and Simate, 2015).

With the development of biomedical nanomaterials, new antimicrobial agents have begun to emerge either as novel and/or augmenting the activities of the current conventional antimicrobials (Yah and Simate, 2015). In particular, a number of carbon-based nanomaterials such CNTs and their composites have been found to possess powerful bactericidal properties toward pathogenic microorganisms (Al-Jumaili et al., 2017). The bactericidal actions of CNTs, their composites, and other carbon nanostructures depend on a combination of physical and chemical mechanisms (Al-Jumaili et al., 2017). More specifically, the mechanism by which CNTs inactivate microorganisms is complex and depends on intrinsic properties of CNTs, for example, composition and surface modification, the nature of the target microorganisms, and the characteristics of the environment in which cell-CNT interactions take place (Al-Jumaili et al., 2017). Some scholars have suggested that the antimicrobial actions of nanoparticles including CNTs and their composites consist of the destruction of cell membranes, blockage of enzyme pathways, alterations of microbial cell wall, and nucleic materials pathway (Galdiero et al., 2011; Yah and Simate, 2015). Table 22.4 gives an overview of CNTs studies that show some of their antimicrobial mechanisms. It can be seen from the table that a range of mechanisms are used by CNTs to deactivate microorganisms.

22.5.6 Tissue scaffold reinforcements

A particular area that has generated enormous interest in the use of CNTs is tissue engineering (Edwards et al., 2009). According to Boccaccini et al. (2010), tissue

Table 22.4 Antimicrobial actions of carbon nanotubes (Mocan et al., 2017).

Type of nanoparticle	Size	Bacterial strain	Mechanism of action
Silver nanoparticles within vertically aligned MWCNTs	~ 200 nm	*Escherichia coli*	Bacterial cell membrane ruptured by direct contact
MWCNTs functionalized with lysine	Diameter of ~ 30 nm	Gram-negative and Gram-positive bacteria	Electrostatic adsorption of bacterial membrane due to positive charges of the lysine groups on CNT
SWCNTs; MWCNTs with different surface	Diameter of 15–30 nm groups to bacterial pathogens	Gram-negative and Gram-positive bacteria	Physical piercing; oxidative stress
Surfactant-modified MWCNTs	Diameter of ~ 15 nm	*S. mutans*	Cell membrane damage by direct contact with the dispersed CNTs
Pegylated SWCNTs-Ag (pSWCNTs-Ag)	Outer diameter of 1–5 nm; length of 1–2 μm	*Salmonella typhimurium*	Downregulations of some genes associated with metabolism and outer membrane integrity
SWNTs; short and long MWCNTs; functionalized MWCNTs (hydroxyl and carboxyl modification)	Length: SWCNTs 1–3 μm; short MWCNTs 0.5–2 μm; long MWCNTs >50 μm	*Lactobacillus acidophilus, Bifidobacterium adolescentis, E. coli, Enterococcus faecalis, Staphylococcus aureus* Outer diameter of 1.2 nm; length of 10–20 nm.	Diameter-dependent piercing; length-dependent wrapping

engineering endeavors to restore and regenerate damaged or diseased tissues by inducing the human body to heal itself using its intrinsic repair mechanisms. With an expanding range of tissue types being considered for tissue engineering, the demand for novel scaffold properties is growing (Edwards et al., 2009). More specifically, tissue engineering requires the development of biocompatible materials to fabricate scaffolds with optimized surface and structural compatibility (Ramakrishna et al., 2001; Boccaccini and Gerhardt, 2010). Biocompatibility is necessary so as to achieve optimal physicochemical matching of the scaffold/tissue interface and thus leading to efficient biomaterial integration in the human body (Ramakrishna et al., 2001; Boccaccini et al., 2010).

Research has shown that the compatibility on the interface involves chemical, physical, and topographical suitability of a scaffold surface in relation to a particular tissue (Ramakrishna et al., 2001; Boccaccini et al., 2010). On the other hand, structural compatibility is related to the mechanical properties of a biomaterial and to optimal load transmission that displays a minimal interfacial strain mismatch at the implant/tissue interface (Ramakrishna et al., 2001; Boccaccini et al., 2010).

It is noted by various scholars that an ideal scaffold for tissue regeneration should provide, (1) structural support to the newly formed tissue, including mechanical properties similar to those of the tissue repair site, (2) biodegradability so that it can be gradually replaced by native tissue, and (3) large porosity, high pore size and interconnectivity for tissue in-growth, vascularization, nutrient delivery, and waste removal (Karageorgiou and Kaplan, 2005; Rezwan et al., 2006; Boccaccini and Gough, 2007). In addition to possessing the correct properties—structural and chemical—it is also important for the scaffold to be electrically conductive, as electrical stimulation has been shown to be able to enhance cell differentiation, proliferation, and dendritic expansion (Schmidt et al., 1997; Supronowicz et al., 2002; Edwards et al., 2009). In this regard, and in view of their unique physical, mechanical and electronic properties, CNTs have been found to be valuable reinforcements or additives that can improve the properties and introduce novel functionalities to scaffolding materials (Haniu et al., 2012). More specifically, CNTs can be used as reinforcing agents aimed at imparting sufficient strength to (bio)polymers (Shi et al., 2007; Armentano et al., 2008; Boccaccini et al., 2010) or ceramic (Wang et al., 2007; Boccaccini et al., 2010) matrices. In addition, nanoscale filamentous CNTs have been compared to the dimensions of the biological scaffolding material, the extracellular matrix (Firkowska et al., 2008; Edwards et al., 2009), and, therefore, they have the ability to stimulate the desired cellular functions (Edwards et al., 2009). Indeed, a number of studies have shown that CNTs can favorably improve the mechanical properties of tissue engineering composites (Boccaccini et al., 2010). Therefore, there is no doubt that investigations and the usage of CNT and their composites in scaffolds for tissue engineering are on the increase in recent past.

22.5.7 Dental applications

Dental materials are used in different treatments such as plaque removal, caries treatment, aesthetic interventions, teeth reconstruction, and implants (do Nascimento and Chirani, 2015). Other scholars also state that a wide variety of materials are used by dentists for various treatments, and many factors need to be considered by both the dentist and the patient when choosing the optimal restorative material for each procedure, with the longevity of that particular restorative material being one of the most important factors (Kim et al., 2013; Schwass et al., 2013; Fernandes et al., 2015). Indeed, dental material could be made from diverse sources, from metal alloys to shape-memory polymers (do Nascimento and Chirani, 2015). However, there is no material that has been found to be ideal for any kind of dental application (Mitra et al., 2013; Turagam and Mudrakola, 2019), and some of them have a number of disadvantages (Khurram, et al., 2018; Priyadarsini et al., 2018; DBC, 2020). For examples, there are major toxicity and esthetics concerns related to silver amalgam that has been a dental restorative material for more than a century (Smart et al., 1995; Eley, 1997; Jones, 1998; Turagam and Mudrakola, 2019). Therefore, there is a continuing need for the development of better materials and methods for dental applications (Imazato, 2003; Akasaka, 2009; Padovani, 2015).

For a couple of decades, CNT and their composites have been found to be novel potential materials that may aid in the restoration of bone defects in dentistry (Akasaka, 2009; Martins-Júnior et al., 2013). Indeed, the CNTs are attracting considerable attention because of their unique physical properties and various potential applications (Akasaka, 2009). Research has shown that because of their unique properties, CNTs have generated interest regarding their use in dentistry as they can (1) improve the strength of composite materials and implants, (2) increase cell adhesion and proliferation, (3) effect nucleation of hydroxyapatite, and (4) provide protection against bacteria (Zhang et al., 2008).

Studies by Mendes et al. (2010) and Sá et al. (2013) have demonstrated that CNT composites have great potential use for bone regenerative procedures in dentistry, in both normal and adverse metabolic states. Other studies have also indicated that because of the excellent mechanical and electrical properties of CNTs and their composites such as heat stability, heat transmission efficiency, high strength, and lower density, they are used as a candidate for teeth filling (Priyadarsini et al., 2018). Functionalized SWCNTs have also been applied to dental composite resin-based matrix such as bisphenol A-glycidyl methacrylate and an inorganic filler such as silica to increase its tensile strength and Young's modulus so as to help in improving the longevity of composite restoration in the oral cavity (Bhattacharya and Seong, 2013). The addition of functionalized SWCNTs also increased the flexural strength of the CNT-reinforced composite resin significantly by absorbing more stress (Zhang et al., 2008; Bhattacharya and Seong, 2013). The CNTs have also been applied to the interface of dentin and composite resin to compensate for microleakage development in long-term use, which is a major cause of restoration failure (Bhattacharya and Seong, 2013). For example, once microleakage develops between the tooth and composite resin interface, it works as a nidus for bacterial colonization, thus secondary decay can develop (Bhattacharya and Seong, 2013). However, the coating of CNTs on teeth has shown the potential to provide protection against bacteria and initiate the nucleation of hydroxyapatite (a component of bone) on their surface (Akasaka et al., 2009; Bhattacharya and Seong, 2013).

22.5.8 Other biomedical applications

There are several other biomedical applications of CNTs that are closely related to the ones discussed in Sections 22.5.1–22.5.7. For example the use of CNTs for the delivery of proteins has been reported by many researchers (Klumpp et al., 2006; Wang et al., 2012). The CNTs can improve the penetration of protein into cells and also increase protein uptake by the cells (Karimi et al., 2015). In a study by Chen et al. (2008), a novel CNT-based delivery vector using poly(lactic-co-glycolic acid) (PLGA) was designed to internalize different proteins into cells. The carboxylated CNT preparation was functionalized with PLGA and then 5 μg of protein was added to the system. Chen et al. (2008) showed that this new system had a high transfection rate, reduced toxicity and a highly controlled drug release profile and was also stable for weeks at −20 °C. A variety of nanovectors including CNTs have also been explored as possible gene therapy vehicles (Wen et al., 2013). The unique properties of CNTs such as high length-to-diameter ratio, easy chemical functionalization, and good biocompatibility make them a suitable candidate for this purpose (Klumpp et al., 2006). Different types of nucleic acids such as micro-RNA (miRNA) (Dong et al., 2011), small-interfering (siRNA) (Wang et al., 2012), and plasmid DNA (pDNA) (Pruthi et al., 2012) can be bound to CNTs and transferred into mammalian cells. To minimize or eliminate the undesirable tissue reaction, strategies such as delivering/releasing anti-inflammatory drugs or neurotrophic factors to the vicinity of the implant have also been explored via CNTs (Luo et al., 2013). All these delivery systems (protein, gene, anti-inflammatory therapy, etc.) relate to drug delivery systems (Karimi et al., 2015).

22.6 Concluding remarks

This chapter has presented and shown that the integration of nanotechnology with medicine introduces a rare opportunity for producing novel materials that can be applied in the medical field. The CNTs, in particular, have inherent properties that can be modified and extensively used in medicine. More specifically, the physical properties of CNTs such as mechanical strength, electrical conductivity, and optical properties are great and unique characteristics for creating advanced biomaterials. This chapter has thoroughly discussed the most recent progress of CNTs in biomedical applications in the past decades. These applications are due to the extensive research efforts that have elevated CNTs as one of the most widely used class of nanomaterials.

The chapter started by giving a brief overview of the synthesis of CNTs via three very useful, widespread, and well-established methodologies—CVD, laser ablation, and arc discharge. The synthesis of CNTs-based nanocomposites and typical case studies were also given. Thereafter, the first critical aspect discussed was the functionalization of CNTs. Strategies such as covalent functionalization, noncovalent functionalization, and endohedral filling of empty cavities of CNTs for immobilizing various biomolecules such as proteins, enzymes, antibodies and nucleic acids onto CNTs were extensively discussed. The second and last critical aspect dealt with the potential biomedical applications of functionalized CNTs. Some of the successful biomedical applications of CNTs, in general, discussed in this chapter include drug delivery, cancer therapy, biosensors, biomedical imaging, bacterial inhibition, dental applications, tissue scaffold reinforcements, and others such as gene delivery, anti-inflammatory therapy, protein delivery that may be categorized as part of drug delivery.

With the continued research coupled with the high versatility of CNTs it is expected that CNTs will continue to be used in many facets of biomedical engineering. However, in all these studies, the difficulty that remains is due to the vast diverse surface and physiological properties of the CNTs, sizes, the various attached ligands, the bioaccumulation of the nanoparticles after biomedical applications, and the hypersensitivity reactions that may emerge during the course of action.

References

Abbasi, E., Sedigheh, F.A., Abolfazl, A., Morteza, M., Hamid, T.N., Younes, H., Kazem, N.K., Roghiyeh, P.A., 2014. Dendrimers: synthesis, applications, and properties. Nanoscale Res. Lett. 9 (1), 247–255.

Albdiry, M.T., Yousif, B.F., Ku, H., Lau, K.T., 2013. A critical review on the manufacturing processes in relation to the properties of anoclay/polymer composites. J. Composite Mater. 47, 1093–1115.

Abdulkadir, A., Sarker, T., He, Q., Guo, Z., Wei, S., 2016. Mössbauer spectroscopy of polymer nanocomposites. In: Thomas, S., Rouxel, D., Ponnamma, D. (Eds.), Spectroscopy of Polymer Nanocomposites. William Andrew Publishing, Norwich, USA, pp. 393–409.

Agboola, A.E., Pike, R.W., Hertwig, T.A., Lou, H.H., 2007. Conceptual design of carbon nanotube processes. Clean Technol. Environ. Policy 9, 289–311.

Ates, M., Eker, A.A., Eker, B., 2017. Carbon nanotube-based nanocomposites and their applications. J. Adhes. Sci. Technol. 31 (18), 1977–1997.

Ajayan, P.M., Ebbesen, T.W., 1997. Nanometre-size tubes of carbon. Rep. Prog. Phys. 60 (10), 1025–1062.

Ajayan, P.M., Redlich, P., Rühle, M., 1997. Structure of carbon nanotube-based nanocomposites. J. Microscopy 185 (2), 275–282.

Ajayan, P.M., Charlier, J.C., Rinzler, A.G., 1999. Carbon nanotubes: from macromolecules to nanotechnology. Proc. Natl. Acad. Sci. USA 96 (25), 14199–14200.

Akasaka, T., Nakata, K., Uo, M., Watari, F., 2009. Modification of the dentin surface by using carbon nanotubes. Bio-Med. Mater. Eng. 19, 179–185.

Al-Jumaili, A., Alancherry, S., Bazaka, K., Jacob, M.V., 2017. Review on the antimicrobial properties of carbon nanostructures. Materials 10(9), 1066.

Alshehri, R., Ilyas, A.M., Hasan, A., Arnaout, A., Ahmed, F., Memic, A., 2016. Carbon nanotubes in biomedical applications: factors, mechanisms, and remedies of toxicity. J. Med. Chem. 59, 8149–8167.

Armentano, I., Dottori, M., Puglia, D., Kenny, J.M., 2008. Effects of carbon nanotubes (CNTs) on the processing and in-vitro degradation of poly(DL-lactide-co-glycolide)/CNT films. J. Mater. Sci.: Mater. Med. 19 (6), 2377–2387.

Augustine, S., Singh, J., Srivastava, M., Sharma, M., Das, A., Malhotra, B.D., 2017. Recent advances in carbon based nanosystems for cancer theranostics. BioMater. Sci. 5, 901–952.

Azqhandi, M.A., Shekari, M., Ghalami-Choobar, B., 2018. Synthesis of carbon nanotube-based nanocomposite and application for wastewater treatment by ultrasonicated adsorption process. Appl. Organometal. Chem. 32 ((8), e4410.

Balasubramanian, K., Burghard, M., 2005. Chemically functionalized carbon nanotubes. Small 1 (2), 180–192.

Balasubramanian, K., Burghard, M., 2006. Biosensors based on carbon nanotubes. Anal. Bioanal. Chem. 385, 452–468.

Bartelmess, J., Quinn, S.J., Giordani, S., 2015. Carbon nanomaterials: multi-functional agents for biomedical fluorescence and Raman imaging. Chem. Soc. Rev. 44 (14), 4672–4698.

Becker, A.L., Johnston, A.P.R., Caruso, F., 2010. Layer-by-layer-assembled capsules and films for therapeutic delivery. Small 6, 1836–1852.

Beg, S., Rizwan, M., Sheikh, A.M., Hasnain, M.S., Anwer, K., Kohli, K., 2011. Advancement in carbon nanotubes: basics, biomedical applications and toxicity. J. Pharmacy Pharmacol. 63, 141–163.

Béguin, F., Szostak, K., Lillo-Rodenas, M., Frackowiak, E., 2004. Carbon nanotubes as backbones for composite electrodes of supercapacitors. AIP Conf. Proc. 723, 460–464.

Bernholc, J., Brabec, C., Buongiorno, N.M., Maiti, A., Roland, C., Yakobson, B.I., 1998. Theory of growth and mechanical properties of nanotubes. Appl. Phys. A 67, 39–46.

Beyou, E., Akbar, S., Chaumont, P., Cassagnau, P., 2013. Polymer nanocomposites containing functionalised multiwalled carbon nanotubes: a particular attention to polyolefin based materials. In: Suzuki, S. (Ed.), Syntheses and Applications of Carbon Nanotubes and Their Composites. Intech Open, Rijeka, Croatia.

Bhattacharya, M., Seong, W.J., 2013. Carbon nanotube-based materials—preparation, biocompatibility, and applications in dentistry. In: Subramani, K., Ahmed, W., Hartsfield, J.K. (Eds.), Nanobiomaterials in Clinical Dentistry. William Andrew Publishing, USA.

Bhirde, A.A., Patel, V., Gavard, J., Zhang, G., Sousa, A.A., Masedunskas, A., Leapman, R.D., Weigert, R., Gutkind, J.S., Rusling, J.F., 2009. Targeted killing of cancer cells in vivo and in vitro with EGF-directed carbon nanotube-based drug delivery. ACS Nano 3 (2), 307–316.

Bianco, A., Kostarelos, K., Prato, M., 2005. Applications of carbon nanotubes in drug delivery. Curr. Opin. Chem. Biol. 9 (6), 674–679.

Boccaccini, A.R., Gough, J.E., 2007. Tissue Engineering using Ceramics and Polymers. Woodhead Publishing, Cambridge.

Boccaccini, A.R., Gerhardt, L.C., 2010. Carbon nanotube composite scaffolds and coatings for tissue engineering applications. Key Eng. Mater. 441, 31–52.

Broome, D.R., 2008. Nephrogenic systemic fibrosis associated with gadolinium based contrast agents: a summary of the medical literature reporting. Eur. J. Radiol. 66, 230–234.

Calvert, P., 1999. Nanotube composites—a recipe for strength. Nature 399, 210–211.

Camargo, P.H.C., Satyanarayana, K.G., Wypych, F., 2009. Nanocomposites: synthesis, structure, properties and new application opportunities. Mater. Res. 12 (1), 1–39.

Cancer Council of Australia, 2017. Cancer: an overview https://www.cancerwa.asn.au/resources/2017-05-09-cancer-an-overview.pdf.

Caruso, F., Caruso, R.A., Mohwald, H., 1998. Nanoengineering of inorganic and hybrid hollow spheres by colloidal templating. Science 282, 1111–1114.

Chakraborty, S., Rahman, T., 2012. The difficulties in cancer treatment. https://www.researchgate.net/publication/235791893_The_difficulties_in_cancer_treatment. [Accessed 30 August 2020].

Chawla, K.K., 2003. Ceramic Matrix Composites. Springer, New York.

Chen, J., Chen, S., Xianrui, Z., Kuznetsova, L., Wong, S.S., Ojima, I., 2008. Functionalized single-walled carbon nanotubes as rationally designed vehicles for tumor-targeted drug delivery. J. Am. Chem. Soc. 130 (49), 16778–16785.

Chen, D., Dougherty, C.A., Zhu, K., Hong, H., 2015. Theranostic applications of carbon nanomaterials in cancer: focus on imaging and cargo delivery. J. Control. Release 210, 230–245.
Chico, L., Crespi, V.H., Benedict, L.X., Louie, S.G., Cohen, M.L., 1996. Pure carbon nanoscale devices: nanotube heterojunctions. Phys. Rev. Lett. 76 (6), 971–974.
Cho, J., Boccaccini, A.R., Shaffer, M.S.P., 2009. Ceramic matrix composites containing carbon nanotubes. J. Mater. Sci. 44 (8), 1934–1951.
Choudhary, V., Gupta, A., 2011. Polymer/carbon nanotube nanocomposites. In: Yellampalli, S. Ed. Carbon Nanotubes – Polymer Nanocomposites. IntechOpen, Rijeka, Croatia.
Chronos, N.A., Goodall, A.H., Wilson, D.J., Sigwart, U., Buller, N.P., 1993. Profound platelet degranulation is an important side effect of some types of contrast media used in interventional cardiology. Circulation 88, 2035–2044.
Crozals, G.D., Bonnet, R., Farre, C., Chaix, C., 2016. Nanoparticles with multiple properties for biomedical applications: a strategic guide. Nano Today 11, 435–463.
Dai, H., 2002a. Carbon nanotubes: opportunities and challenges. Surf. Sci. 500, 218–241.
Dai, H., 2002b. Carbon nanotubes: synthesis, integration, and properties. Acc. Chem. Res. 35 (12), 1035–1044.
Dai, L., He, P., Li, S., 2003. Functionalized surfaces based on polymers and carbon nanotubes for some biomedical and optoelectronic applications. Nanotechnology 14, 1081–1097.
Das, R., Pattanayak, A.J., Swain, S.K., 2018. Polymer nanocomposites for sensor devices. In: Jawaid, M., Khan, M.M. (Eds.), Polymer-Based Nanocomposites for Energy and Environmental Applications. Woodhead Publishing, Cambridge, pp. 205–218.
Dental Board of California (DBC), 2020. Dental materials—advantages and disadvantages. https://cfdds.com/images/dental_materials.pdf. [Accessed 5 September 2020].
Dervishi, E., Li, Z., Xu, Y., Saini, V., Biris, A.R., Lupu, D., Biris, A.S., 2009. Carbon nanotubes: synthesis, properties, and applications. Particulate Sci. Technol. 27 (2), 107–125.
do Nascimento, R.O., Chirani, N., 2015. Shape-memory polymers for dental applications. In: Yahia, L. (Ed.), Shape Memory Polymers for Biomedical Applications. Woodhead Publishing, Cambridge, pp. 267–280.
Dong, H., Ding, L., Yan, F., Ji, H., Ju, H., 2011. The use of polyethylenimine-grafted graphene nanoribbon for cellular delivery of locked nucleic acid modified molecular beacon for recognition of microRNA. Biomaterials 32 (15), 3875–3882.
Dresselhaus, M.S., Dresselhaus, G., Avouris, P. Ed. 2001. Carbon Nanotubes: Synthesis, Structure, Properties and Applications. Springer, New York.
Du, J.H., Cheng, H.M., 2012. The fabrication, properties, and uses of graphene/polymer composites. Macromol. Chem. Phys. 213, 1060–1077.
Dujardin, E., Ebbesen, T.W., Hiura, H., Tanigaki, K., 1994. Capillarity and wetting of carbon nanotubes. Science 265, 1850–1852.
Eatemadi, A., Daraee, H., Karimkhanloo, H., Kouhi, M., Zarghami, N., Akbarzadeh, A., Abasi, M., Hanifehpour, Y., Joo., S.W., 2014. Carbon nanotubes: properties, synthesis, purification, and medical applications. Nanoscale Res. Lett. 9 (393), 1–13.
Ebbesen, T.W., Ajayan, P.M, 1992. Large-scale synthesis of carbon nanotubes. Nature 358 (6383), 220–222.
Edwards, S.L., Werkmeister, J.A., Ramshaw, J.A.M., 2009. Carbon nanotubes in scaffolds for tissue engineering. Exp. Rev. Med. Dev. 6 (5), 499–505.
Eley, B.M., 1997. The future of dental amalgam: a review of the literature. Part 7: possible alternative materials to amalgam for the restoration of posterior teeth. Br. Dent. J. 183, 11–14.
Elhissi, A.M.A., Ahmed, W., Hassan, I.U., Dhanak, V.R., D'Emanuele, A., 2012. Carbon nanotubes in cancer therapy and drug delivery. J. Drug Deliv. 2012, 1–10.
Feinberg, A.P., Ohlsson, R., Henikoff, S., 2006. The epigenetic progenitor origin of human cancer. Nat. Rev. Genet. 7 (1), 21–33.
Fernandes, N., Vally, Z., Sykes, L., 2015. The longevity of restorations—a literature review. South Afr. Dent. J. 70, 410–413.
Firkowska, I., Godehardt, E., Giersig, M., 2008. Interaction between human osteoblast cells and inorganic two-dimensional scaffolds based on multiwalled carbon nanotubes: a quantitative AFM Study. Adv. Funct. Mater. 18 (23), 3765–3771.

Fu, S., Sun, Z., Huang, P., Li, Y., Hu, N., 2019. Some basic aspects of polymer nanocomposites: a critical review. Nano Mater. Sci. 1, 2–30.
Galdiero, S., Falanga, A., Vitiello, M., Cantisani, M., Marra, V., Galdiero, M., 2011. Silver nanoparticles as potential antiviral agents. Molecules 16, 8894–8918.
Gobbo, O.L., Sjaastad, K., Radomski, M.W., Volkov, Y., Mello, A.P., 2015. Magnetic nanoparticles in cancer theranostics. Theranostics 5 (11), 1249–1263.
Gomathi, P., Kim, M.K., Park, J.J., Ragupathy, D., Rajendran, A., Lee, S.C., Kim, J.C., Lee, S.H., Ghim, H.D., 2011. Multiwalled carbon nanotubes grafted chitosan nanobiocomposite: a prosperous functional nanomaterials for glucose biosensor application. Sens. Actuators B Chem. 155, 897–902.
Goyanes, S., Rubiolo, G.R., Salazar, A., Jimeno, A., Corcuera, M.A., Mondragon, I., 2007. Carboxylation treatment of multiwalled carbon nanotubes monitored by infrared and ultraviolet spectroscopies and scanning probe microscopy. Diam. Relat. Mater. 16, 412–417.
Guisbiers, G., Mejia, R.S., Deepak, F.L., 2012. Nanomaterial properties: size and shape dependencies. J. Nanomater. 2012, 180976.
Greiner, A., Wendorff, J.H., 2008. Functional self-assembled nanofibers by electrospinning. Adv. Polym. Sci. 219, 107–171.
Grobert, N., 2007. Carbon nanotubes - becoming clean. Mater. Today 10 (1), 28–35.
Haley, B., Frenkel, E., 2008. Nanoparticles for drug delivery in cancer treatment. Urologic Oncol. 26 (1), 57–64.
Han, X., Xu, K., Taratula, O., Farsad, K., 2019. Applications of nanoparticles in biomedical imaging. Nanoscale 11, 799–819.
Haniu, H., Saito, N., Matsuda, Y., Tsukahara, T., Usui, Y., Narita, N., Hara, K., Aoki, K., Shimizu, M., Ogihara, N., Takanashi, S., Okamoto, M., Kobayashi, S., Ishigaki, N., Nakamura, K., Kato, H., 2012. Basic potential of carbon nanotubes in tissue engineering applications. J. Nanomater. 2012, 343747.
Hartman, K.B., Laus, S., Bolskar, R.D., Muthupillai, R., Helm, L., Toth, E., Merbach, A.E., Wilson, L.J., 2008. Gadonanotubes as ultrasensitive pH-smart probes for magnetic resonance imaging. Nano Lett. 8, 415–419.
Hernández-Rivera, M., Zaibaq, N.G., Wilson, L.J., 2016. Toward carbon nanotube-based imaging agents for the clinic. Biomaterials. 101, 229–240.
Herron, N., Thorn, D.L., 1998. Nanoparticles: uses and relationships to molecular cluster compounds. Adv. Mater. 10, 1173–1184.
Hillebrenner, H., Buyukserin, F., Kang, M., Mota, M.O., Stewart, J.D., Martin, C.R., 2006. Corking nano test tubes by chemical self-assembly. J. Am. Chem. Soc. 128 (13), 4236–4237.
Huang, P., Shi, H.Q., Fu, S.Y., Xiao, H.M., Hu, N., Li, Y.Q., 2016. Greatly decreased redshift and largely enhanced refractive index of mono-dispersed ZnO-QD/silicone nanocomposites. J. Mater. Chem. 4, 8663–8669.
Hussain, F., Hojjati, M., Okamoto, M., Gorga, R.E., 2006. Polymer-matrix nanocomposites, processing, manufacturing, and application: An overview. J. Compos. Mater. 40, 1511–1575.
Iijima, S., 1991. Helical microtubules of graphitic carbon. Nature 354, 56–58.
Iijima, S, Ichihashi, T., 1993. Single-shell carbon nanotubes of 1-nm diameter. Nature 363, 603–605.
Imazato, S., 2003. Antibacterial properties of resin composites and dentin bonding systems. Dent. Mater. 19, 449–457.
Ishikawa, H., Fudetani, S., Hirohashi, M., 2001. Mechanical properties of thin films measured by nanoindenters. Appl. Surf. Sci. 178, 56–62.
Iyuke, S.E., Simate, G.S., 2011. Synthesis of carbon nanomaterials in a swirled floating catalytic chemical vapour deposition reactor for continuous and large scale production. In: Naraghi., M. (Ed.), Carbon Nanotubes – Growth and Applications. IntechOpen, Rijeka, pp. 35–58.
Jabir, N., Tabrez, S., Ashraf, G.M., Shakil, S., Damanhouri, G.A., Kamal, M.A., 2012. Nanotechnology-based approaches in anticancer research. Int. J. Nanomed. 7, 4391–4408.
Jones, D.W., 1998. A Canadian perspective on the dental amalgam issue. Br. Dent. J. 184, 581–586.
Jose, Y.M., Miki, Y.M., Rendon, L., Santiesteban, J.G., 1993. Catalytic growth of carbon microtubules with fullerene structure. Appl. Phys. Lett. 62 (2), 202–204.

Journet, C., Maser, W.K., Bernier, P., Loiseau, A., de la Chapelle, M.L., Lefrant, S., Deniard, P., Lee, R., Fischer, J.E., 1997. Large-scale production of single-walled carbon nanotubes by the electric-arc technique. Lett. Nat. 388, 756–758.

Journet, C., Bernier, P., 1998. Production of carbon nanotubes. Appl. Phys. A 67, 1–9.

Kam, N.W.S., Dai, H., 2006. Single walled carbon nanotubes for transport and delivery of biological cargos. Phys. Status Solidi B 243 (13), 3561–3566.

Karimi, M., Solati, N., Ghasemi, A., Estiar, M.A., Hashemkhani, M., Kiani, P., Mohamed, E., Saeidi, A., Taheri, M., Avci, P., Aref, A.R., Amiri, M., Baniasadi, F., Hamblin, M.R., 2015. Carbon nanotubes part II: a remarkable carrier for drug and gene delivery. Exp. Opin. Drug Deliv. 12 (7), 1089–1105.

Karageorgiou, V., Kaplan, D., 2005. Porosity of 3D biomaterial scaffolds and osteogenesis. Biomaterials. 26, 5474–5491.

Khurram, M., Zafar, K.J., Qaisar, A., Atiq, T., Khan, S.A., 2018. Restorative dental materials: A comparative evaluation of surface microhardness of three restorative materials when exposed to acidic beverages. Profess. Med. J. 25 (1), 140–149.

Kim, G.M., Michler, G.H., Potschke, P., 2005. Deformation processes of ultrahigh porous multiwalled carbon nanotubes/polycarbonate composite fibers prepared by electrospinning. Polymer 46 (18), 7346–7351.

Kim, K.L., Namgung, C., Cho, B.H., 2013. The effect of clinical performance on the survival estimates of direct restorations. Restor. Dent. Endod. 38 (1), 11–20.

Kim, J., Chhour, P., Hsu, J., Litt, H.I., Ferrari, V.A., Popovtzer, R., Cormode, D.P., 2017. Use of nanoparticle contrast agents for cell tracking with computed tomography. Bioconjug. Chem. 28 (6), 1581–1597.

Kitano, H., Tachimoto, K., Anraku, Y.J., 2007. Functionalization of single-walled carbon nanotube by the covalent modification with polymer chains. J. Colloid Interf. Sci. 306 (1), 28–33.

Klumpp, C., Kostarelos, K., Prato, M., Bianco, A., 2006. Functionalized carbon nanotubes as emerging nanovectors for the delivery of therapeutics. Biochim. Biophys. Acta 1758 (3), 404–412.

Kour, R., Arya, S., Young, S.J., Gupta, V., Bandhoria, P., Khosla, A., 2020. Review—Recent advances in carbon nanomaterials as electrochemical biosensors. J. Electrochem. Soc. 167, 1–23.

Kracke, B., Damaschke, B., 2000. Measurement of nanohardness and nanoelasticity of thin gold films with scanning force microscope. Appl. Phys. Lett. 77, 361–363.

Kushwaha, S.K.S., Ghoshal, S., Rai, A.K., Singh, S., 2013. Carbon nanotubes as a novel drug delivery system for anticancer therapy: a review. Braz. J. Pharma. Sci. 49, 629–643.

Lacerda, L., Raffa, S., Prato, M., Bianco, A., Kostarelos, K., 2007. Cell-penetrating CNTs for delivery of therapeutics. Nano Today 2 (6), 38–43.

Lamberti, M., Pedata, P., Sannolo, N., Porto, S., De Rosa, A., Caraglia, M., 2015. Carbon nanotubes: properties, biomedical applications, advantages and risks in patients and occupationally-exposed workers. Int. J. Immunopathol. Pharmacol. 28 (1), 4–13.

Lamprecht, C., Torin, H.J., Ivanova, V.M., Foldvari, M., 2011. Non-covalent functionalization of carbon nanotubes with surfactants for pharmaceutical applications — A critical mini-review. Drug Deliv. Lett. 1, 45–57.

Lee, S.J., Baik, H.K., Yoo, J., Han, J.H., 2002. Large scale synthesis of carbon nanotubes by plasma rotating arc discharge technique. Diam. Relat.Mater. 11, 914–917.

Li, J., Tang, S., Lu, L., Zeng, H.C., 2007. Preparation of nanocomposites of metals, metal oxides and carbon nanotubes via self-assembly. J. Am. Chem. Soc. 129, 9401–9409.

Li, X.Q., Wang, C.Y., Cao, Y., Wang, G.X., 2018. Functional MXene materials: progress of their applications. Chem. Asian J. 13, 2742–2757.

Li, C., Wang, J., Wang, Y., Gao, H., Wei, G., Huang, Y., Yu, H., Gan, Y., Wang, Y., Mei, L., Chen, H., Hu, H., Zhang, Z., Jin, Y., 2019. Recent progress in drug delivery – Annual review. Acta Pharma. Sin. B 9 (6), 1145–1162.

Liu, P., 2005. Modifications of carbon nanotubes with polymers. Eur. Polym. J. 41, 2693–2703.

Liu, R., 2008. The Functionalisation of Carbon Nanotubes Ph.D. thesis). University of New South Wales, Australia.

Liu, Z., Tabakman, S., Welsher, K., Dai, H.J., 2009. Carbon nanotubes in biology and medicine: in vitro and in vivo detection, imaging and drug delivery. Nano Res. 2, 85–120.

Liu, M.X., Jia, Z.X., Jia, D.M., Zhou, C.R., 2014. Recent advance in research on halloysite nanotubes-polymer nanocomposite. Prog. Polym. Sci. 39, 1498–1525.
Liu, W., Speranza, G., 2019. Functionalization of carbon nanomaterials for biomedical applications. J. Carbon Res. 5, 72.
Loh, K.P., Ho, D., Chiu, G.N.C., Leong, D.T., Pastorin, G., Chow, E.K., 2018. Clinical applications of carbon nanomaterials in diagnostics and therapy. Adv. Mater. 30 (1802368), 1–21.
Lu, F.S., Gu, L.R., Meziani, M.J., Wang, X., Luo, P.G., Veca, L.M., Cao, L., Sun, Y.P., 2009. Advances in bioapplications of carbon nanotubes. Adv. Mater. 21 (2), 139–152.
Luo, X.L., Matranga, C., Tan, S.S., Alba, N., Cui, X.Y.T., 2013. Carbon nanotube nanoreservior for controlled release of anti-inflammatory dexamethasone. Biomaterials. 32, 6316–6323.
Malhotra, B.D., Singhal, R., Chaubey, A., Sharma, S.K., Kumar, A., 2005. Recent trends in biosensors. Curr. Appl. Phys. 5, 92–97.
Mark, J.E., 1996. Ceramic-reinforced polymers and polymer-modified ceramics. Polym. Eng. Sci. 36, 2905–2920.
Martincic, M., Tobias, G., 2014. Filled carbon nanotubes in biomedical imaging and drug delivery. Exp. Opin. Drug Deliv. 12 (5), 1–19.
Martins-Junior, P.A., Alcantara, C.E., Resende, R.R., Ferreira, A.J., 2013. Carbon nanotubes: directions and perspectives in oral regenerative medicine. J. Dent. Res. 92, 575–583.
Matarredona, O., Rhoads, H., Li, Z., Harwell, J.H., Balzano, L., Resasco, D.E., 2003. Dispersion of single-walled carbon nanotubes in aqueous solutions of the anionic surfactant NaDDBS. J. Phys. Chem. B 107, 13357–13367.
McClory, C., Chin, S.J., McNally, T., 2009. Polymer/carbon nanotube composites. Aust. J. Chem. 62, 762–785.
McDevitt, M.R., Chattopadhyay, D., Kappel, B.J., Jaggi, J.S., Schiffman, S.R., Antczak, C., Njardarson, J.T., Brentjens, R., Scheinberg, D.A., 2007. Tumor targeting with antibody-functionalized, radiolabeled carbon nanotubes. J. Nucl. Med. 48 (7), 1180–1189.
Mehra, N.K., Jain, A.K., Lodhi, N., Raj, R., Dubey, V., Mishra, D., Nahar, M., Jain, N.K., 2008. Challenges in the use of carbon nanotubes for biomedical applications. Crit. Rev. Therap. Drug Carrier Syst. 25, 169–206.
Mendes, R.M., Silva, G.A., Caliari, M.V., Silva, E.E., Ladeira, L.O., Ferreira, A.J., 2010. Effects of single wall carbon nanotubes and its functionalization with sodium hyaluronate on bone repair. Life Sci. 87, 215–222.
Mikličanin, E.O., Badnjević, A., Kazlagić, A., Hajlovac., M., 2020. Nanocomposites: a brief review. Health Technol. 10, 51–59.
Mitra, S.B., Wu, D., Holmes, B.N., 2013. An application of nanotechnology in advanced dental materials. J. Am. Dent. Assoc. 134, 1382–1390.
Mocan, T., Matea, C.T., Pop, T., Mosteanu, O., Buzoianu, A.D., Suciu, S., Puia, C., Zdrehus, C., Iancu, C., Mocan, L., 2017. Carbon nanotubes as anti-bacterial agents. Carbon nanotubes as anti-bacterial agents. Cell. Mol. Life Sci. 74, 3467–3479.
Monthioux, M., Kuznetsov, V.L., 2006. Who should be given the credit for the discovery of carbon nanotubes? Carbon 44, 1621–1623.
O'Connell, M.J., Boul, P., Ericson, L.M., Huffman, C., Wang, Y., Haroz, E., Kuper, C., Tour, J., Ausman, K.D., Smalley, R.E., 2001. Reversible water-solubilisation of single-walled carbon nanotubes by polymers wrapping. Chem. Phys. Lett. 342, 265–271.
Padovani, G.C., Feitosa, V.P., Sauro, S., Ta, F.R., Durán, G., Paula, A.J., Durán, N., 2015. Advances in dental materials through nanotechnology: facts, perspectives and toxicological aspects. Trends Biotechnol. 33, 621–636.
Pantarotto, D., Briand, J.P., Prato, M., Bianco, A., 2004a. Translocation of bioactive peptides across cell membranes by carbon nanotubes. Chem. Commun. 10 (1), 16–17.
Pantarotto, D., Singh, R., McCarthy, D., Erhardt, M., Briand, J.P., Prato, M., Kostarelos, K., Bianco, A., 2004b. Functionalized carbon nanotubes for plasmid DNA gene delivery. Angew. Chem. Int. Ed. 43 (39), 5242–5246.
Paradise, M., Goswami, T., 2007. Carbon nanotubes—production and industrial applications. Mater. Des. 28, 1477–1489.
Park, H., Zhao, J., Lu, J.P., 2006. Effects of sidewall functionalization on conducting properties of single wall carbon nanotubes. Nano Lett. 6 (5), 916–919.

Pastorin, G., Wu, W., Wieckowski, S., Briand, J.P., Kostarelos, K., Prato, M., Bianco, A., 2006. Double functionalization of carbon nanotubes for multimodal drug delivery. Chem. Commun. 21 (11), 1182–1184.

Petrik, M., Weigel, C., Kirsch, M., Hosten, N., 2005. No detectable nephrotoxic side effect using a dimer, non-ionic contrast media in cerebral perfusion computed tomography in case of suspected brain ischemia. Rofo 177 (9), 1242–1249.

Pokropivny, V., Lohmus, R., Hussainova, I., Pokropivny, A., Vlassov, S., 2007. Introduction to nanomaterials and nanotechnology. https://www.researchgate.net/profile/Alex_Pokropivny/publication/299345334_Introduction_in_nanomaterials_and_nanotechnology/links/56f160e008ae1cb29a3d0c2d.pdf. [Accessed 15 August 2020].

Porwal, H., Grasso, S., Reece, M., 2013. Review of graphene-ceramic matrix composites. Adv. Appl. Ceram. 112 (8), 443–454.

Porwal, H., Saggar, R., 2018. Ceramic matrix nanocomposites. In: Beaumont, P.W.R., Zweben, C.H. (Eds.), Comprehensive Composite Materials II, 6. Elsevier, Netherlands, pp. 138–161.

Prato, M., Kostarelos, K., Bianco, A., 2008. Functionalized carbon nanotubes in drug design and discovery. Acc. Chem. Res. 41 (1), 60–68.

Priyadarsini, S., Mukherjee, S., Mishra, M., 2018. Nanoparticles used in dentistry: a review. J. Oral Biol. Craniofac. Res. 8, 58–67.

Pruthi, J., Mehra, N.K., Jain, N.K., 2012. Macrophages targeting of amphotericin B through mannosylated multiwalled carbon nanotubes. J. Drug Target. 20 (7), 593–604.

Raghavendra, P., Pullaiah, T., 2018. Advances in Cell and Molecular Diagnostics. Academic Press, Elsevier, Cambridge.

Ramakrishna, S., Mayer, J., Wintermantel, E., Leong, K.W., 2001. Biomedical applications of polymer-composite materials: a review. Compos. Sci. Technol. 61 (9), 1189–1224.

Reynaud, E., Gauthier, C., Perez, J., 1999. Nanophases in polymers. Rev. Métal. 96, 169–176.

Rezwan, K., Chen, Q.Z., Blaker, J.J., Boccaccini, A.R., 2006. Biodegradable and bioactive porous polymer/inorganic composite scaffolds for bone tissue engineering. Biomaterials. 27 (18), 3413–3431.

Rivera, E.J., Tran, L.A., Hern´andez-Rivera, M., Yoon, D., Mikos, A.G., Rusakova, I.A., Cheong, B.Y., Cabreira-Hansen, M.D., Willerson, J.T., Perin, E.C., Wilson, L.J., 2013. Bismuth@US-tubes as a potential contrast agent for X-ray imaging applications. J. Mater. Chem. B 1 (37), 4792–4800.

Robertson, J., 2004. Realistic applications of CNTs. Mater. Today 7 (10), 46–52.

Ruggiero, A., Villa, C.H., Holland, J.P., Sprinkle, S.R., May, C., Lewis, J.S., Scheinberg, D.A., McDevitt, M.R., 2010. Imaging and treating tumor vasculature with targeted radiolabeled carbon nanotubes. Int. J. NanoMed. 5, 783–802.

Sá, M., Andrade, V., Mendes, R., Caliari, M., Ladeira, L., Silva, E.E., Silva, G.A.B., Corrêa-Júnior, J.D., Ferreira, A.J., 2013. Carbon nanotubes functionalized with sodium hyaluronate restore bone repair in diabetic rat sockets. Oral Dis. 19, 484–493.

Sahoo, N.G., Rana, S., Cho, J.W.L.S.H., Chan, Li., 2010. Polymer nanocomposites based on functionalized carbon nanotubes. Prog. Polym. Sci. 35 (7), 837–867.

Samadi, M., Zirak, M., Naseri, A., Khorashadizade, E., Moshfegh, A.Z., 2016. Recent progress on doped ZnO nanostructures for visible-light photocatalysis. Thin Solid Films 605, 2–9.

Schmidt, C.E., Shastri, V.R., Vacanti, J.P., Langer, R., 1997. Stimulation of neurite outgrowth using an electrically conductingpolymer, 94, pp. 8948–8953.

Schwass, D.R., Lyons, K.M., Purten, D.G., 2013. How long will it last? The expected longevity of prosthodontic and restorative treatment. NZ Dent. J. 109 (3), 98–105.

Scoville, C., Cole, R., Hogg, J., 1991. Carbon nanotubes. https://courses.cs.washington.edu/courses/csep590a/08sp/projects/CarbonNanotubes.pdf. [Accessed 15 August 2020].

See, C.H., Harris, A.T., 2007. A review of carbon nanotube synthesis via fluidized-bed chemical vapor deposition. Ind. and Eng. Chem. Res. 46, 997–1012.

Sen, R., Zhao, B., Perea, D., Itkis, M.E., Hu, H., Love, J., Haddon, R.C., 2004. Preparation of single-walled carbon nanotube reinforced polystyrene and polyurethane nanofibers and membranes by electrospinning. Nano Lett. 4 (3), 459–464.

Setaro, A., 2017. Advanced carbon nanotube functionalisation. J. Phys. Condens. Matter 29 (42), 1–19.
Sharma, A., Kumar, R., 2018. Fabrication and characterization of carbon nanotubes-based polymer nanocomposites. Integr. Res. Adv. 5 (2), 42–75.
Shi, X., Sitharaman, B., Pham, Q.P., Liang, F., Wu, K., Billups, W.E., Wilson, L.J., Mikos, A.G., 2007. Fabrication of porous ultra-short single-walled carbon nanotube nanocomposite scaffolds for bone tissue engineering. BioMatererials. 28 (28), 4078–4090.
Simate, G.S., Iyuke, S.E., Ndlovu, S., Yah, C., Walubita, L.F., 2010. The production of carbon nanotubes from carbon dioxide: challenges and opportunities. J. Nat. Gas Chem. 19 (5), 453–460.
Simate, G.S., 2012. The treatment of brewery wastewater using carbon nanotubes synthesized from carbon dioxide carbon source. PhD Thesis, University of the Witwatersrand, South Africa.
Simate, G.S., Walubita, L.F., 2016. Synthesis, properties, and applications of carbon nanotubes in water and wastewater treatment. In: Gautam, R.K., Chattopadhyaya, M.C. (Eds.), Advanced Nanomaterials for Wastewater Remediation. CRC Press, Boca Raton, pp. 201–225.
Singh, B., Lohan, S., Sandhu, P.S., Jain, A., Mehta, S.K., 2016. Functionalized carbon nanotubes and their promising applications in therapeutics and diagnostics. In: Grumezescu, A.M. (Ed.), Nanobiomaterials in Medical Imaging: Applications of Nanobiomaterials. Elsevier, Amsterdam, pp. 455–478.
Sireesha, M., Babu, V.J., Ramakrishna, S., 2017. Functionalised carbon nanotubes in bio-world: applications, limitations and future directions. Mater. Sci. Eng. B 223, 43–63.
Sitharaman, B., Kissell, K.R., Hartman, K.B., Tran, L.A., Baikalov, A., Rusakova, I., Sun, Y., Khant, H.A., Ludtke, S.J., Chiu, W., Laus, S., Tóth, E., Helm, L., Merbach, A.E., Wilson, L.J., 2005. Superparamagnetic gadonanotubes are high-performance MRI contrast agents. Chem. Commun. 31, 3915–3917.
Smart, E.R., Macleod, R.I., Lawrence, C.M., 1995. Resolution of lichen-planus following removal of amalgam restorations in patients with proven allergy to mercury salts—a pilot study. Br. Dent. J. 178, 108–112.
Sternitzke, M., 1997. Structural ceramic nanocomposites. J. Eur. Ceram. Soc. 17 (9), 1061–1082.
Such, G.K., Johnston, A.P.R., Caruso, F., 2011. Engineered hydrogen-bonded polymer multilayers: from assembly to biomedical applications. Chem. Soc. Rev. 40, 19–29.
Supronowicz, P.R., Ajayan, P.M., Ullmann, K.R., Arulanandam, B.P., Metzger, D., Bizios, R.J., 2002. Novel current-conducting composite substrates for exposing osteoblasts alternating current stimulation. J. Biomed. Mater. Res. 59 (3), 499–506.
Tanwar, J., Das, S., Fatima, Z., Hameed, S., 2014. Multidrug resistance: an emerging crisis. Interdiscip. Perspect. Infect. Dis. 2014, 541340.
Tasis, D., Tagmatarchis, N., Bianco, A., Prato, M., 2006. Chemistry of carbon nanotubes. Chem. Rev. 106 (3), 1105–1136.
Terrones, M., 2003. Science technology of the twenty-first century: synthesis, properties, and applications of carbon nanotubes. Annu. Rev. Mater. Res. 33, 419–501.
Tîlmaciu, C.M., Morris, M.C., 2015. Carbon nanotube biosensors. Front. Chem. 3 (59), 1–21.
Trojanowicz, M., 2006. Analytical applications of carbon nanotubes: a review. Trends Anal. Chem. 25 (5), 480–489.
Tsai, Y.C., Huang, J.D., Chiu, C.C., 2007. Amperometric ethanol biosensor based on poly(vinyl alcohol)–multiwalled carbon nanotube–alcohol dehydrogenase biocomposite. Biosens. Bioelectron. 22, 3051–3056.
Turagam, N., Mudrakola, D.P., 2019. Advantages and limitations of CNT-polymer composites in medicine and dentistry. In: Saleh, H., El-Sheikh, S.M. (Eds.), Perspective of Carbon Nanotubes. Intech Open, Rijeka, Croatia, pp. 651–660.
Urooj, S., Singh, S.P., Pal, N.S., Lay-Ekuakille, A., 2016. Carbon-based nanomaterials in biomedical applications https://ieeexplore.ieee.org/stamp/stamp.jsp?tp=&arnumber=8521437.
Vardharajula, S., Ali, S.Z., Tiwari, P.M., Eroglu, E., Vig, K., Dennis, V.A., Singh, S.R., 2012. Functionalized carbon nanotubes: biomedical applications. Int. J. Nanomed. 7, 5361–5374.
Wallyn, J., Anton, N., Akram, S., Vandamme, T.F., 2019. Biomedical imaging: principles, technologies, clinical aspects, contrast agents, limitations and future trends in nanomedicines. Pharma. Res. 36 (78), 4–31.
Wang, X.P., Ye, J.D., Wang, Y.J., Chen, L., 2007. Reinforcement of calcium phosphate cement by bio-mineralized carbon nanotube. J. Am. Ceram. Soc. 90, 962–964.
Wang, T., Upponi, J.R., Torchilin, V.P., 2012. Design of multifunctional non-viral gene vectors to overcome physiological barriers: dilemmas and strategies. Int. J. Pharma. 427 (1), 3–20.

Wang, W., Zhu, Y., Liao, S., Li, J., 2014. Carbon nanotubes reinforced composites for biomedical applications. BioMed Res. Int. 2014, 518609.
Wen, S., Liu, H., Cai, H., 2013. Targeted and pH-responsive delivery of doxorubicin to cancer cells using multifunctional dendrimer-modified multi-walled carbon nanotubes. Adv. Healthc. Mater. 2 (9), 1267–1276.
Werder, T., Walther, J.H., Jaffe, R.L., Halicioglu, T., Noca, F., Koumoutsako, P., 2001. Molecular dynamics simulation of contact angles of water droplets in carbon nanotubes. Nano Lett. 1 (12), 697–702.
Wilson, G.S., Gifford, R., 2005. Biosensors for real-time in vivo measurements. Biosens. Bioelectron. 20 (15), 2388–2403.
Wu, H.C., Chang, X., Liu, L., Zhao, F., Zhao, Y., 2010. Chemistry of carbon nanotubes in biomedical applications. J. Mater. Chem. 20, 1036–1052.
Wu, X., Chen, X., Wang, J., Liu, J., Fan, Z., Chen, X., Chen, J., 2011. Functionalization of multiwalled carbon nanotubes with thermotropic liquid-crystalline polymer and thermal properties of composites. Ind. Eng. Chem. Res 50, 891–897.
Yah, C.S., Simate, G.S., 2015. Nanoparticles as potential new generation broad spectrum antimicrobial agents. DARU J. Pharma. Sci. 23, 43–57.
Yang, W., Thordarson, P., Gooding, J.J., Ringer, S.P., Braet, F., 2007. Carbon nanotubes for biological and biomedical applications. Nanotechnology 18 (41), 412001.
Yang, N., Chen, X., Ren, T., Zhang, P., Yang, D., 2015. Carbon nanotube based biosensors. Sens. Actuators B 207, 690−715.
Yellampalli, S., 2011. Carbon Nanotubes—Polymer Nanocomposites. IntechOpen, London, UK.
Zhang, F., Xia, Y., Xu, L., Gu, N., 2008. Surface modification and microstructure of single-walled carbon nanotubes for dental resin-based composites. J. Biomed. Mater. Res. B 86, 90–97.
Zhang, Y., Bai, Y., Yan, B., 2010. Functionalised carbon nanotubes for potential medicinal applications. Drug Discov. Today 15 (11-12), 428–453.
Zhou, Y., Fang, Y., Ramasamy, R.P., 2019. Non-covalent functionalization of carbon nanotubes for electrochemical biosensor development. Sensors 19, 392.

CHAPTER 23

Synthesis of metal oxide–based nanocomposites for energy storage application

Asim Ali Yaqoob[a], Akil Ahmad[b], Mohamad Nasir Mohd Ibrahim[a], Rama Rao Karri[c], Mohd Rashid[a] and Zahoor Ahamd[d]

[a]School of Chemical Sciences, Universiti Sains Malaysia, Penang, Malaysia
[b]Centre of Lipids Engineering and Applied Research, Universiti Teknologi Malaysia, Johor Bahru, Malaysia
[c]Petroleum and Chemical Engineering, Faculty of Engineering, Universiti Teknologi Brunei, Brunei Darussalam
[d]Department of Chemistry, Mirpur University of Science and Technology, Mirpur AJK, Pakistan

23.1 Introduction

The broad study on different energy storage resources and devices with their utilization is a developing question due to the high demand for energy consumption and less availability of fossil fuel sources. Rapidly reducing levels of fossil fuel resources with high environmental pollution impact are the most significant alarms of the twenty-first century. At present, the rate of consumption of natural fossil fuels is very high. According to a rough estimate, in the next 40 year, all known fossil fuels of the world due to the complete depletion of sources. Natural fossil fuel resources are a rich energy source, which carries approximately 30–50 MJ energy/kg (Block and Schmücker, 2016). The high rate of fossil fuel combustion produces the emission of N_2O, CH_4, CO_2, etc., in the air, capturing solar radiation. Natural solar radiation is significant for living organisms on the earth. However, due to extreme emissions (N_2O, CH_4, CO_2, etc.) gases, the earth's temperature is warmer, which disturbs the eco-life. NASA's Goddard institute revealed that the earth's average temperature increased by 0.8 °C, subsequently beginning the industrial revolution (Ray et al., 2018). This increase may appear minor, but the unpleasant circumstance is that a slightly extra temperature increment worldwide will produce the glaciers and Antarctic ice caps to melt, leading the sea close to intensification flooding. Therefore, to maintain human evolution, these problems have to be answered on priority.

To decrease fossil fuel deprivation while sustaining the green energy natural resources such as wind, solar, tidal, etc. These natural resources carry the capacity to encounter global energy supplies, but the alternating energy resources is an inevitable problem that expressively stimulates the study of different energy storage sources (Yaqoob et al., 2021a). Many energy-storing devices exist, such as supercapacitors, batteries, fuel cells, traditional capacitors, etc. Among all, supercapacitors as energy storage got much attention in the

Sustainable Nanotechnology for Environmental Remediation
DOI: https://doi.org/10.1016/B978-0-12-824547-7.00028-X

recent era due to higher power density, rapid charging period and long life span, while having reasonable energy density but less than batteries (Yan et al., 2020; Yaqoob et al., 2020a). The electrochemical energy systems include supercapacitors and batteries as the most commonly used energy storage sources, but the most prominent research topic is supercapacitors. Several researchers are studying different materials and their composite in supercapacitors and batteries to enhance the energy storage capacity to reduce the world energy crises (Yaqoob et al., 2020b; An et al., 2019).

In the modern era, most of the research found that metal oxide–based supercapacitors or batteries led to significant results due to their excellent properties such as high conductivity, less costly than pure metals, high mechanical, chemical, and thermal stability (Yaqoob et al., 2020c). The metal oxide–based materials well known as the most promising material to be utilized in several types of energy storage systems due to their productive assets, ecological geniality, informal accessibility and other exciting characteristics; for example, their various morphologies, compositions, constituents, high specific surface area, and large theoretic precise capacitance (Seok et al., 2019). Furthermore, they show a dominant part as electrode material in different electrochemical-based batteries and supercapacitors and offers a noticeable capacitance development through regulating and monitoring their faults and interfaces. However, the energy density has revealed an improvement to a positive degree; the lower electric conductive effect, irrepressible capacity growth, and inactive ions dispersion in the bulk stage have delayed their real-world applications (Mamulová Kutláková et al., 2015). Hereafter, it is significant to study and investigate the metal oxide–based material nanocomposite to enhance the quality of the electrochemical aspects of energy storage devices. However, the composition of metal oxide–based nanocomposites, fabrication of shape, electric conductivity, and potential oxygen vacancies improved the physiochemical behavior of metal oxide–based nanocomposites, concerning their electric conductivities, precise surface area, chemical, mechanical and thermal stability (Ravikumar et al., 2017; Yaqoob et al., 2020d).

The existence of binary cations in one crystal assembly could cause further electrons than sole metal oxide, which improves the electric conductivity of the composite of two metal oxides. For example, $NiCo_2O_4$ nanocomposite materials hold electric conductivity in two or more magnitude orders which are higher than corresponding metal oxide of NiO or Co_3O_4. Therefore, Ni doping significantly improves the electric conductivity of material (Huang et al., 2013). Furthermore, the effect of metal doping has the potential for corresponding metal oxide with additional redox responses. It decreases the impedance of charge transfer for supercapacitors. However, in contrast, innovative nano-range structured-based metal oxides are generally absorbent and can offer a high surface area, which is encouraging the penetration of the electrode materials. Additionally, the proper electroactive positions and high mechanical, thermal, and chemical stability of metal oxide promise a higher pseudo-capacitive concert and cyclical stability.

Similarly, the metal oxide/carbon-based composites are widely reported to enhance the electric properties to use the material in energy storage devices such as carbon fibers, nanotubes, graphite, graphene oxide, single-layer graphene, and shapeless carbonaceous materials, which have intensely enhanced electric conductivity and additional precise capacitance and their performances (Khan et al., 2015; Ibrahim et al., 2021). In explanation of unique aspects, the utilization of metal oxides composites has brought a major revolution for supercapacitors. It has progressively upgraded the battery scale's energy compactness without losing the conventional capacitor energy supply, which links the gap between capacitors and batteries. Compared with bulk metal oxide with lower precise surface area and capacitance behavior, the increasing tendency to enterprise and produce original permeable nanocomposites with carbonaceous materials is necessary to raise the surface area, electrical and chemical properties. The metal oxide material is used as an electrode in supercapacitors and batteries to store energy.

In this chapter, the most commonly used synthesis methods are comprehensively reviewed and discussed. The basic principles and the advanced progress in diverse supercapacitor systems also have been reviewed and elaborated. The role of different metal oxide–based nanocomposites in energy storage devices is summarized and presented.

23.2 Potential methods for synthesis of metal oxide nanocomposites

Among energy storage materials, the metal oxides such as ZnO, CuO, AgO, TiO_2, Al_2O_3, ZrO_2, CeO_2, etc. Nanocomposites have become one of the emerging research areas because metal oxides–based composite material is associated with a function of distinct nanomaterials as well as showing exclusive shared and synergetic electric properties compared to individual materials. Usually, the nanocomposites are simply synthesized through several methods, for example, deposition–precipitation, sol–gel method, hydrothermal, solvothermal, co-precipitation, impregnation, and so many others. Some most commonly used and standard methods to prepare the metal oxides–based nanocomposites are discussed next.

23.2.1 Coprecipitation/chemical precipitation/chemical method

The coprecipitation/chemical precipitation/chemical method is employed as one of the most well-known and cost-effective methods for preparing metal oxides nanocomposites. The residue is produced separately from the solution. This method is the simplest and most effective method for synthesizing metal oxides–based nanocomposites. The inorganic salts are mostly served as a precursor agent, liquified in solvents and water to get the homogenous-based ions solution; after that, the used salts' precursor starts precipitation in the form of oxalates or hydroxides when the serious species concentration is achieved tailed through growth and nucleation phases (Yaqoob et al., 2021a). This method depends on three phases: first, the liquid phase–based solution preparation in the presence of optimized chemical composition; second, the heat treatment directly affects the morphology (size, shape, ratio, porosity), crystallinity, and structure of nanomaterials (Khan et al., 2015). The third phase is the utilization of precipitation agents such as urea, NaOH, Na_2CO_3, etc., which might fluctuate the nanoparticles' sizes. The morphology of particles such as shape, size, porosity is significantly influenced by salt concentration, temperature, and pH of the solution. After producing the precipitation, washing and filtration are achieved through calcination to change the produced hydroxide in oxide forms with a crystal-like structure (de Oliveira Sousa Neto et al., 2019). The most commonly used precipitating agents are Na_2CO_3, NH_4OH, NH_3, NaOH, etc. (Kalantari et al., 2013). The utilization of surfactants is one of the regular exercises to evade accumulation, which also interferes with the nanocomposites' particle achieved through this method. This method provides the advantage of lower cost, easy handling, flexibility, optimized reaction environments, and morphology control. Although this method is extensively used for FeO nanocomposite preparation, it has also been known to be effective in formulating several further metal oxides for example, CuO, ZnO-SnO_2, AgO, TiO_2-ZrO_2, and ZnO-MnO_2 (Hamidreza et al., 2014). Yang et al., (2018) studied the formation of magnetic Fe_3O_4 deposition on carbon nanotubes' (CNTs) surface by using the coprecipitation method and achieved significant material that showed excellent electric properties. Furthermore, it seems to be expected to produce shell/core nanocomposite by using this method. Some more examples are shown in Table 23.1.

23.2.2 Sol–gel method

This method attracts much attention as a promising method for preparing nanorange materials due to their stable reaction environments and the structure of materials from molecular agents leading to a difference in materials properties. The resultant product obtained by this method is either colloidal powder or films (Yan et al., 2020). The general graphical representation of this material is shown in Fig. 23.1. The sol–gel method can make macro- as well as microstructures. The compositional ratio, shape, size, and structure of the ultimate product are significantly affected through different reaction parameters. The sol–gel method is also known as the exploited process for metal oxide and its nanocomposites preparation (Yaqoob et al., 2020e). The sol–gel method depends on condensation, hydrolysis, and metal precursors polymerization. Then, polymerization, the gel formation, is detected, made up of 3-D metal oxide systems. Therefore, the sol–gel process signifies an effective and multipurpose approach for synthesizing several diverse porous substances with diverse 3-D structural morphologies (Zhang and Chen, 2009). The stable optimization of the entire process can be determined through modification in several parameters such as time, nature of the material, salt precursor's concentration and quality, pH, nature of solvent and surfactants temperature in the form of solution

Table 23.1 List of synthesis of metal-oxide nanocomposites

Metal oxide nanocomposites	Preparation method	Precursor	Size of particle (nm)	Temperature of reaction	Reference
MgO/Al_2O_3	Coprecipitation	$MgCl_2$, NaOH	4–30	70	(Mirzaei and Neri, 2016)
ZnO/SnO_2	Coprecipitation	$ZnCl_2$, NaOH	30	105	(Nazari and Halladj, 2014)
ZnO/CuO	Coprecipitation	$CuCl_2$, $NaBH_4$	22	120	(Rao et al., 2017)
MgO/CuO	Coprecipitation	Mg $(NO_3)_2$, Cu $(NO_3)_2$ and glycine (NH_2CH_2COOH)	20	120	(Ravikumar et al., 2017)
ZnO/Fe_3O_4	Coprecipitation	Fe_3O_4, NaOH	40	90	(Ray et al., 2018)
NiO/CeO2/ZnO	Coprecipitation	$ZnCl_2$, NaOH	14–25	120	(Seok et al., 2019)
Ag/talc	Coprecipitation	$AgNO_3$, $NaBH_4$	11.3	–	(Sheng et al., 2019)
Ag/activated carbon	Coprecipitation	Soluble starch, NaOH	55	–	(Sultana et al., 2013)
$Zeolite/Fe_3O_4$	Coprecipitation	Zeolite	<100	70	(Thimmaiah et al., 2001)
ZnO/activated carbon	Coprecipitation	Hexamethylenetetramine ($(CH_2)_6N_4$; HMTA)	50–200	70	(Wang et al., 2011, Wu et al., 2010)
TiO_2/Al_2O_3	Sol–gel method	Titanyl sulfate	5–9	–	(Yan et al., 2020)
TiO_2/Fe_2O_3	Sol–gel method	TiO_2	4–10	100–2000	(Yang et al., 2005)
Ag/TiO_2	Sol–gel method	$AgNO_3$, titanium(IV) isopropoxide	13–20	550	(Yang et al., 2018)
$Al_2(OH)_3(VO_4)$	Hydrothermal method	V_2O_5, $Al(NO_3)_3.9H_2$	10	–	(Zhang and Chen, 2009)
Mixed phase TiO_2	Hydrothermal method	Titanium chloride, anatase powder	48	155	(Zhang et al., 2017)

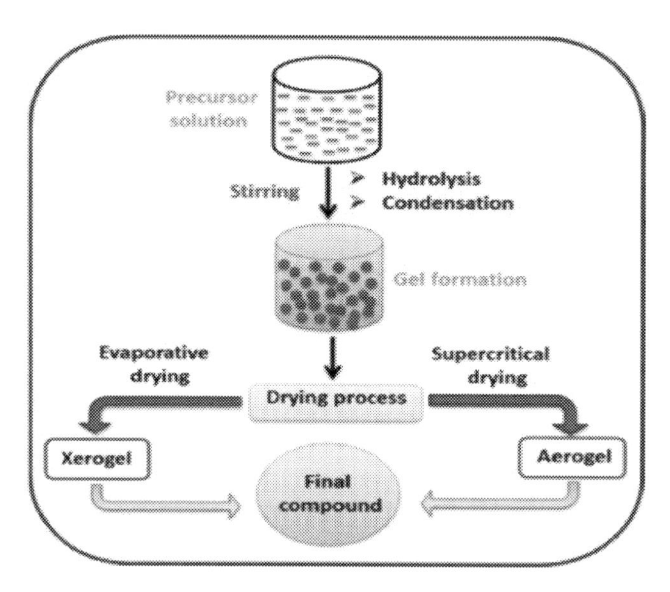

Figure 23.1 General synthesis route of metal oxide nanocomposite using the sol–gel method (Adapted from reference (Rao et al., 2017) with Elsevier permission). From Rao, B. G., Mukherjee, D., & Reddy, B. M. (Rao et al., 2017). Novel approaches for preparation of nanoparticles. In: Nanostructures for Novel Therapy: Synthesis, Characterization and Applications (1st ed., Vol. 2, pp. 1–36). Elsevier Inc., Amsterdam https://doi.org/10.1016/B978-0-323-46142-9.00001-3.

(Yang et al., 2005). The surfactants addition is well known as the most important factor that changes the surface morphological parameters and surface charges, excluding the crystal-based structure because it cannot be changed. Nonaqueous and aqueous solvents have been utilized to prepare polar and nonpolar nanocomposites material through this versatile process (Rao et al., 2017). In the case of aqueous, the sol–gel method, the oxygen is supplied from water to produce the metal oxide.

In contrast, in nonaqueous, oxygen is provided by adding external organic solvents such as ethers, ethyl, methyl, aldehydes, ketone, alcohols, etc. This method can be concisely described in six phases such as (1) the optimized stable metal-based precursor formation led to solution, (2) polycondensation reaction used to produce gel, (3) the age of produced gel is few hours which shows expulsion of organic solvent that is a production of solid mass and Ostwald ripening, (4) gel drying process, (5) surface stability and dehydration, and (6) thermal treatment at high temperature to achieve the crystalline nano-sized material (Mirzaei and Neri, 2016). An advantage is that this method is the preparation of a nano-range structure containing more than two compounds. The slow rate of kinetics reaction leads to good operational production of a product. One more advantage is that the entire reaction is carried out at lower or room temperature, which is easily maintainable. This process includes inorganic precursors that carry, in the presence of several chemical responses, the subsequent 3-D molecule network formation. The size range of prepared materials through this technique is mentioned in Table 23.1 (Umar et al., 2020).

23.2.3 Solvothermal/hydrothermal method

The solvothermal and hydrothermal methods are usually showed in steel-based pressure vessels such as autoclave to control temperature and pressure. The temperature synthesis is frequently reliant on the organic solvent employed in the method because it is usually raised directly above the solvent boiling point to influence the vapor pressure saturation (Mirzaei and Neri, 2016). Therefore, solvothermal/hydrothermal methods are considered as suitable and applied methods for metal oxide preparation because they permit the evasion of special tools, complex processes, and unwieldy preparation circumstances. This method has been effectively applied to synthesize diverse nanoparticle categories with precise morphology (shape and size) in previous studies. The synthesis of metal oxide nanocomposites using solvothermal/hydrothermal methods can include the processes of oxidation, thermolysis, and hydrolysis. According to the reported earlier literature, most metal oxide nanocomposites are prepared using these solvothermal/hydrothermal methods (Nazari and Halladj, 2014). The thermolysis synthesis route depends on the organometallic complex decomposition process of metal oxide through heating and improved structure pressure. Thimmaiah et al., (2001) produced a promising way to prepare metal oxide nanocomposites comprising more than two metallic atoms. They presented that the transition metal complex decomposition in toluene as a solvent in the presence of solvothermal synthesis.

Many metal oxide nanocomposites can be prepared through hydrolysis; meanwhile, it is predictable that several metals can be solvated through water shells. Furthermore, this way offers an informal way to adapt nanomaterials' size and shape due to a tremendous dielectric constant, which referred to it as an ideal solvent for many structural-directing molecules. Li and Wang, (2010) deliberated the variance among solvothermal/hydrothermal methods by making the $CuCo_2S_4$/rGO (reduced graphene oxide) nanocomposite in the presence of using thiourea as a solvent precursor, and reaction temperature was kept at 180 °C for 3 h. The experimental explanation is presented in Fig. 23.2. Several other metal oxides–based composites examples are summarized in Table 23.1.

23.3 Importance of metal-oxide nanocomposites in energy storage devices

The metal oxides–based materials are commonly used as electrodes in batteries and capacitors. This topic got much attention in the recent era, but some factors are limiting the metal oxides–based electrode performance due to lack of further study. For example,

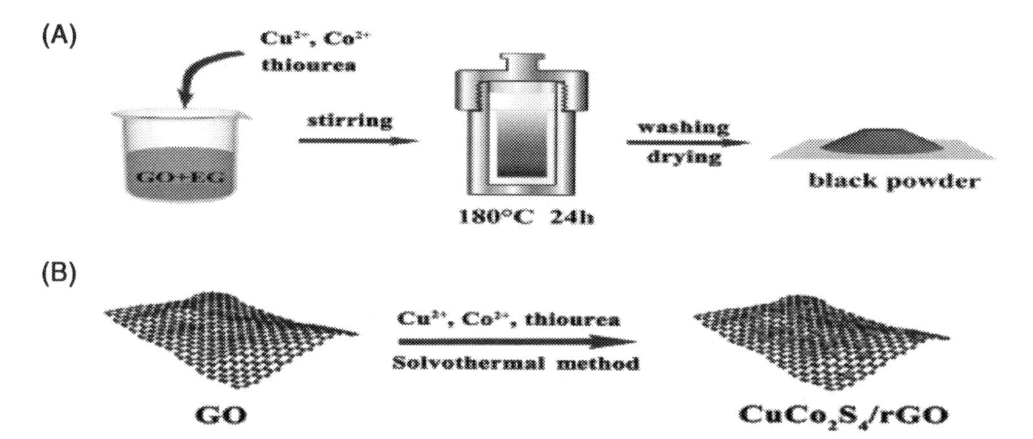

Figure 23.2 (A) Preparation of $CuCo_2S_4$/rGO oxide nanocomposite by hydrothermal method. (B) Preparation of $CuCo_2S_4$/rGO oxide nanocomposite by using the solvothermal method. (Adapted from reference Li & Wang, (Li and Wang, 2010) with Elsevier permission). Gong, Y., Zhao, J., Wang, H., & Xu, J. (Yang et al., 2018). $CuCo_2S_4$/reduced graphene oxide nanocomposites synthesized by one-step solvothermal method as anode materials for sodium ion batteries. Electrochim. Acta, 292, 895–902. https://doi.org/10.1016/j.electacta.2018.09.194.

in the case of pseudocapacitor, the metal oxide–based electrode was used, and their theoretical specific capacitance can be measured using the following equation (An et al., 2019):

$$C_t^s = (n \times F)/(M \times V) \quad (23.1)$$

where

n = No. of transferred electrons in redox reaction,

F = Faraday's constant,

M = Metal oxides molar mass,

V = Voltage.

Basically, the metal oxides with a low molar mass and high electrons transformation rate usually showed a high specific capacitance value. Therefore, diverse chemical compositions, particularly the metal oxide valence state and their different components in the composite form of two or more metal oxide materials, lead to significant electrochemical performance. Furthermore, the specific surface area is also a concern because, in the case of pseudocapacitance development, of the reactions carried out at the specific surface area of metal oxides-electrodes. So, a large and effective definite surface area would provide other electroactive positions for a pseudocapacitance reaction. The specific surface area of metal oxides is contingent mainly on the microstructures. The balanced model of a microstructure with an exclusive porous size and appropriate definite surface area would support attracting the metal oxide–based electrodes for capacitors (Yan et al., 2020). One more question that needs attention to address is electroconductivity of metal oxide–based electrodes. Most metal oxides typically have comparatively significant band gaps than pure metal strips and display a semiconductor behavior. Functioning as an electrode in the device, the lower electroconductivities are still limiting the capacitance capability. Also, capacitance, the metal oxide–based electrodes power density and energy density are two critical matters in related applications. These might be achieved by using the following equation:

$$E = CV^2/2 \tag{23.2}$$

$$P = V^2/4R \tag{23.3}$$

where

C = Farad capacitance
V = Nominal voltage
R = Equivalent series resistance
V = Nominal voltage

These factors play an active part in power and energy density, and it is mainly associated with specific electrolyte classifications in supercapacitors (Yaqoob et al., 2021b). The above discussion showed that the metal oxide–based electrodes limiting performance is credited to their microstructure, chemical constitution, electrolytes, and electroconductivities (Zhao et al., 2018). Then, we reread some recent advanced pioneering research papers to explain the further advanced growth of metal oxide electrodes for capacitors.

23.3.1 Various chemical compositions

Many investigations and studies are concentrating on the various chemical compositions of the composite, which depend on more than two metal oxides. The single-metal oxide appears unsuitable to entirely encounter the complete performances desired for good supercapacitor as electrodes (Khaja Hussain and Su Yu, 2019). For instance, He et al., (2018) studied the fabrication of $FeCo_2O_4/NiCo$ composite and used it as an electrode to show the unique specific capacitance, that is, 2426 F/g at 1 A/g, and super high rate competence with high capacitance retention (72.5%) at 20 A/g. These kinds of first-rate performances might be ascribed to the mixture of the two or more classes and rich electrochemical vigorous sites. Further, Zhang et al. (2017) also studied the fabrication of electrode using the $NiCo_2O_4/MnO_2$ nanocomposite and showing a tremendous actual capacitance value, that is, 13.9 F c/m at 4 mA c/m. The aqueous asymmetric expedient assembled to carry a higher energy density, that is 60.4 W h k/g and the power density was 950.1 W k/g. However, the variety of metal oxides can be improved through the diversity and metal ions, which currently shows high electrochemical activities. Another side, the synergistic influence among diverse species would increase the entire performances of various metal oxides by enabling these kinds of activities as diffusion and ion adsorption (Baek et al., 2011). The metal oxide–based electrodes chemical composition fluctuates distinctly, but entirely considering the comprehensive, synergistic outcome from diverse mechanisms to realistically model electrodes still needs more research and study.

23.3.2 Electroconductivity and microstructure

Nanomaterials are widely used due to their unique physicochemical properties. Also, the metal oxide–based electrodes microstructures also affect the overall electrochemical activities (Yaqoob et al., 2021c). Some essential factors, such as ion solid diffusion, electrolytes wettability, and specific surface area, depend on electrodes microstructures to a large degree. Several nanostructures-based metal oxides have been realistically achieved to increase the system performances. One-dimensional $NiCo_2O_4$ nano-sized wire is coated with rGO to provide the higher specific capacitance (1248 F/g) due to its exclusive synergistic influence on electrolytes. Similar, 2-D-based $MnMO_4$-H_2O/MnO_2 nanocomposite sheets owned a specific capacitance, that is, 3560.2 F/g with a significant power density (507.3 W kg) and energy density (45.6 W h kg). Also, 3-D resonating $NiCo_2O_4$ nano-spheres with an excellent surface area also achieved the capacitance of 1229 F/g special performance rate and virtuous cycle activities (Fig. 23.3A–C). (Sultana et al., 2013).

Diverse nanostructures with diverse dimensions hold their exclusive advantages, and a precise regulator for gaining superior nanostructures is a vital task. Metal oxides, with well-behaved microstructures consequent from metal-organic frameworks (MOFs), have progressively achieved a significant place in supercapacitors (Guan et al., 2017).For example, Guan et al., (2017) studied the multilayered of Ni-Co oxide of MOFs (Fig. 23.3D–I) with potentially high capacitance (1900 F/g), virtuous ability rate and super high driving stability, which must be directed toward the microstructural benefits that the volume modified the oxide nanograins as well-accommodated at the shell level. One more generally criticized matter of metal oxide electrodes is the electroconductivities. Certainly, metal oxides' inherent electroconductivities are contingent on the specific bandgap and are generally unacceptable than electrodes of supercapacitor such as carbon-based electrodes. Providentially, the application of element doping, compositing, and the intended formation of oxygen vacancy can be directed to advance the electroconductivities of metal oxide electrodes. The metal oxide electrodes compositing with high electroconductivities materials, is a general system. Later, a great 2-D transition family of metal nitrides and carbides, cooperatively raised to as MXenes, has considered substantial consideration due to the metal-based electric conductivities, which formed an innovative electroconductivity composite of metal oxides. For instance, the self-assembled MXene (Ti_3C_2TX)-alpha/Fe_2O_3 composites-derived electrodes exhibited a 405.4 F/g capacities value. The achieved capacity retention was 97.7% after 2000 cycles

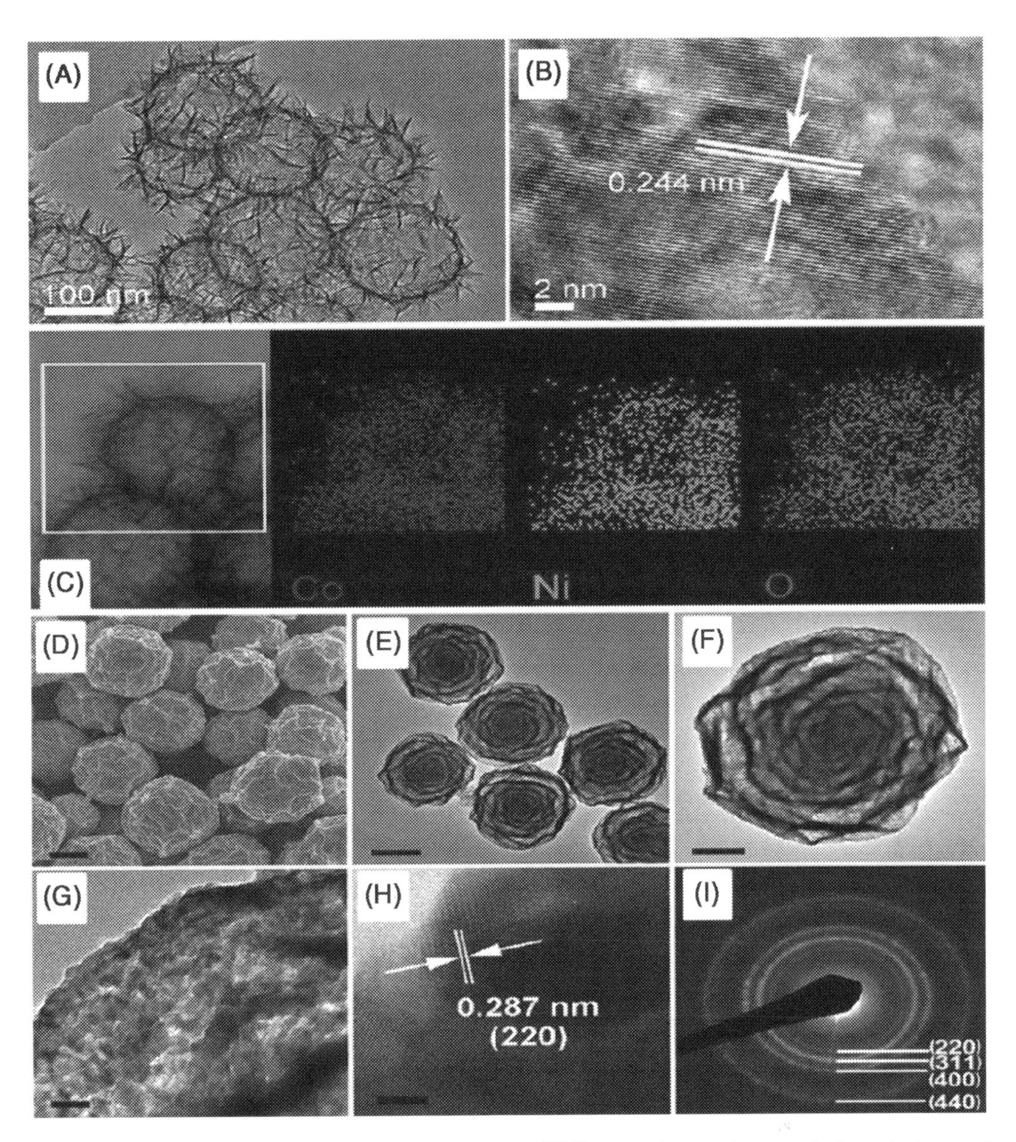

Figure 23.3 (A–C) Energy dispersive spectroscopy (EDS) mapping and transmission electron microscopy (TEM) of 3-D $NiCo_2O_4$ nanospheres, (D–I) scanning electron microscope (SEM)/TEM of multilayered Ni-Co oxide (Adapted from Sultana et al. (2013) with permissions).

(Gao et al., 2017). The composition of metal oxides with high electroconductivities, like carbon-based materials such as CNTs, graphite, graphene oxide, etc.) and many other electroconductivity polymers materials have a great future to explore further (Lu et al., 2017). Yang et al., (2018) studied the prepared $FeCo_2O_4$/shell nanowires/polypyrrole core, which showed a specific capacitance and displayed first-rate cycle activities.

The interfacial production among metal oxides versus electroconductive polymeric material is one of the common issues and the growth of a modest process to produce organic-based electroconductivities polymer materials and their powerful composite with prominent metal oxides (Yaqoob et al., 2021d). Above studies indicate that various chemical constitutions offer a parameter to find the possible novelty of metal oxides with a substantial role as electrodes in supercapacitors. Furthermore, the variety of metal oxides with specific functional classes can cover the lack of individual metal oxides. Though, for an assumed metal oxide, its microstructure and electroconductivities directly influenced its performance intensely (An et al., 2019). Therefore, the balanced model of metal oxides with superior microstructures and the improvement of the electroconductivities would promote them to achieve excellent performances.

23.4 Energy-based applications of metal-oxide nanocomposites

The literature observed that the metal oxides–based nanocomposites carried high electron storing capacity in electrochemical energy storage devices and served as electrode materials in devices such as supercapacitors and batteries (Umar et al., 2020). The individual role of metal oxide does not show high electrical conductivity as required to use at an industrial scale. Therefore, the nanocomposite idea is a novel approach to combine different materials to enhance material properties. Several studies were carried out to enhance the performance of electrochemical devices, and different methods were used to make nanocomposites more efficient, as mentioned earlier (Yan et al., 2020; Yaqoob et al., 2021c). Various kinds of metal oxides–based nanocomposites are used as energy storage devices, and some very common and familiar metal oxides–based nanocomposites are summarized next.

23.4.1 TiO_2-V_2O_5 nanocomposites

Among all the transition metal oxides, the vanadium oxides (V_2O_5) are well known and extensively studied as an ideal material for electrodes supercapacitor for energy storage application. The V_2O_5 possess excellent physiochemical properties such as biocompatibility, eco-friendly, easy synthesis, layered structures, large specific capacity, and show different oxidation states, but despite all, V_2O_5 shows less conductivity at an industrial scale. Furthermore, the mechanism of charge storage in supercapacitor might be based on material superficial properties. Therefore, nano-sized material is considered as a potential material for an electrode in energy storage devices. The nanomaterial has a higher specific surface area and good surface energy, but nanomaterial aggregation is still an emerging challenge. This efficiently rises the stress for electrolytic ions diffusion inside the nanorange materials at the electrode surface. So, the scientific community is currently studying the design and fabrication of 3-D, well-ordered, and mesoporous material to address this challenge. The four essential phases to regulate the charge storing behaviors are electron bounding among two or more nanoparticles; electron bonding inside the individual nanoparticle; electron bonding among active electrode substances and current accumulator; and finally, the proton diffusion in the interior part of nanoparticles. However, electron hopping and proton diffusion are well known and essential properties of nanomaterials, but the potential resistance can be reduced by piling on metal oxide. The resistance is produced due to the presence of the intraparticle electron bounding, respectively. Practically it has also been noticed that the stacking on a constant metal oxide distresses the rise of proton diffusion barrier inside the vigorous materials. This usually tracks active sites' damage due to the mixture of TiO_2 into V_2O_5 and makes Ti-O-V bonds as a replacement of metal-metal bonds. This can influence the entire development of electrochemical activities and natural constancy due to the droplet in the intraparticle electron bounding confrontation (Ray et al., 2018). Therefore, TiO_2-V_2O_5 nanocomposite is the most prominent material as an electrode for supercapacitor. Ray et al. (2018) studied the TiO_2-V_2O_5 nanocomposite as supercapacitors and reported a very significant result. They proved that the TiO_2-V_2O_5 nanocomposite-based electrode showed long-term stability and achieved a current density of 5.0 mA c/m^2.

23.4.2 Ni-Mn_2O_4 nanocomposites

The Ni-Mn_2O_4 composite material has recently acknowledged an excessive interest than other metal oxide materials because this composite exhibited high electrochemical performance, toxic-free character, and cost-effective material. As specified previously, supercapacitors are utilized as electrode materials with high electric conductivity and high electrochemical behavior, they are desired for attaining good high energy density and good power density. The Ni-Mn_2O_4 nanostructured composite material is more appropriate than conventional bulk substances for supercapacitor as electrodes. They provide a high ratio of surface to volume and briefer electron/ions transport stations. The Ni-Mn_2O_4 nanostructures composite becomes the goal of up-to-date study for their application in higher energy storage system performance.

Furthermore, the morphology of the nanostructures is seen to influence their electrochemical behavior (X. Dong et al., 2014; Y. C. Dong et al., 2013; Ibrahim et al., 2021; Yaqoob et al., 2020d). So, manipulating metal oxide nanocomposite structures of measured morphology with high electric conductivities is a task. Ray et al. (2018) studied the Ni-Mn_2O_4 nanocomposite material as an anode. The observed compactly crowded agglomerated globular Ni-Mn_2O_4 nanocomposite particle size was between 6 and 10 nm. The formation of densely Ni-Mn_2O_4 nanocomposite particles efficiently produces the porous high definite surface area, which improves the electrochemical activities of Ni-Mn_2O_4 nanocomposite-based electrodes due to the high interaction area of electrodes with electrolytes. They also confirmed the Ni-Mn_2O_4 nanostructures porous nature by using the Brumaire–Emmett–Teller (BET) for measuring the surface area. BET's adsorption and desorption results indicated that the Ni-Mn_2O_4 nanocomposite nature is nanocrystals and achieved surface area was 43.6 m^2/g with 13.3 nm pore size, which provide large active sites for electrochemical performance. This precise, definite surface area of Ni-Mn_2O_4 nanocomposite nanostructures can also offer a great connection among the electrodes and electrolyte solution following fast ion transmission at the border.

23.4.3 Iron oxide nanocomposites

The iron oxides, mostly Fe_3O_4 and Fe_2O_3, have found to be potential anode materials for energy generation devices due to lower cost, environmental friendliness, and high specific capacitance. However, their limited electric conductivity and significant volume expansion hinder their larger applications. Synthesizing nanostructures embedded with a highly conductive matrix enhance the challenges. Sheng et al. (2019) studied the fabrication of Fe_3O_4-graphene nanocomposite-based electrode by using the solvothermal method. The prepared Fe_3O_4-graphene nanocomposite-based electrode showed an outstanding energy density, that is, 87.6 W h k/g, larger cycling constancy, and the capacity retention was 93.1%. Furthermore, another research group, Zhao et al., (2018), proved that a clipped united Fe_2O_3 nano-needle array as anode showed good performances. The definite energy density was reported high at 3.5 kW k/g power density, and the capacitance retention was 86.6%.

23.4.4 Cobalt-based nanocomposites

Among various metal oxides, cobalt attracts attention as an electrode (cathode) in several electrochemical devices. The Co_3O_4 electrode is measured through dispersion in the electrochemical aqueous medium electrolyte and usually known as battery-based material for electrodes. The Co_3O_4 cathode electrode is energetically industrialized due to its significant redox reversibility and high-power capacitance. However, the cycling capability rate is delayed due to its lower electrical conductivities. So, substantial research has been focused on manipulating nanostructured Co_3O_4 composite materials to enhance electrochemical behavior. Wu et al. (2010) studied the Co_3O_4/N-doped carbon hollow spheres (Co_3O_4/NHCS) material with potential hierarchical nanostructure as a progressive supercapacitor electrode for a higher irregular behavior. This prepared electrode had shown an excellent energy density of 34.5 W-h-K/g at a power density of 753 W-K/g when used as a cathode in a supercapacitors device.

23.4.5 Manganese oxide–based nanocomposites

The manganese oxide (MnO) materials are widely studied as cathode electrode materials for electrochemical devices. Among all, Mn_3O_4 and MnO_2 have significant consideration for applications due to their cost effectiveness, large functioning potential space, and earth rich nature. Hard work has been engrossed on passage the hole among theoretic capacitance and tentative capacitance. Zhu et al. (2018) fabricate the core shell–based electrode comprising core MnO_2 and vastly associated shell-based MnO_2. A packed electrode carried a higher energy density with a maximum power density (17.6 kW k/g). Lately, numerous clusters have established that the supplement of a higher number of cations such as K^+ and Na^+ into the MnO_2 nanostructure can achieve improved electrochemical behavior, depending on the further redox reaction. Jabeen et al., (2017) fabricated the Na-MnO_2 nano-wall-based cathode and used carbon-Fe_3O_4 nanocomposite as an anode electrode, showing a first-rate energy density (81 Wh k/g) in a prolonged potential space.

23.4.6 Ruthenium oxide–based nanocomposites

The Ru-based electrode materials showed large theoretic species capacitance, excellent electrical conductivities, and adjustable redox reaction. The RuO_2 is known as the most prominent and significant material for electrode fabrication. Further, the RuO_2 has excellent corrosion resistance power toward both basic and acidic environments, capable of extensive applications in aqueous medium schemes with diverse aqueous medium electrolytes. The quartz phase (RuO_2) and amorphous hydrous (RuO_2 xH_2O) are two forms of RuO_2. They have shown great active sites and hence advanced capacitance. The amorphous hydrous was extensively used as a pseudocapacitive material for electrodes in acid-natured electrolytes in supercapacitors (Kong et al., 2017). However, one of the significant problems in RuO_2-based electrodes is the thoughtful accumulation throughout the cycling. Kong et al. (2017) studied the fabrication of exclusive nanostructure electrode material containing RuO_2 nanomaterials with diameters (1.9 nm) coated on graphene-based derivatives, which could efficiently constrain the accumulation phenomenon.

23.4.7 Carbon-based nanocomposites

Several carbon-based nanocomposites with metal oxide were applied to enhance the energy storage capacity in many electrochemical devices. The structural configuration is classified into 1-D, 2-D, and 3-D nanocomposites, as shown in Fig. 23.4. This figure also presents their properties and advantageous features of these classifications in terms of electrochemical performance of a 1-D nanostructured composite of carbon/metal-oxide that was widely considered due to their particular nanocomposite aspects. The 1-D nanomaterials considered mostly the carbon nanofibers (CNFs) or CNTs) characteristically form a nanocomposite via homogeneous doping of metal oxides around related material. As-prepared nanocomposites show enhanced features approaching their entire structure. CNTs with a higher properties ratio in a constant conductive system can enable the charged carriage because of the condensed interaction resistance with head-to-head nanomaterials higher than metal-oxide nanostructure. Zhao et al. (2018) studied

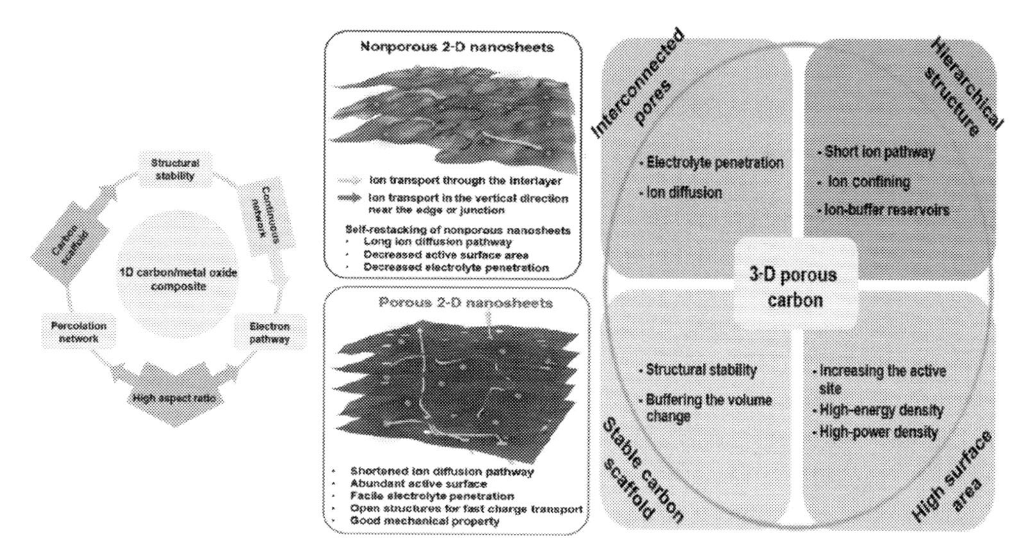

Figure 23.4 Overview of 1-D, 2-D and 3-D-based carbon/metal-oxide composite for electrochemical energy device.

$Co@Co_3O_4$/CNTs nanocomposite's electric conductivity to enhance the material properties by using an arc discharge method that is low-temperature oxidation. The single metal oxide (Co_3O_4) conductivity is 7.1×10^{-4} S/m, but the nanocomposite of $Co@Co_3O_4$/CNTs showed a highly remarkable conductivity value (7.1×10^{-4} S/m).

Similarly, the 2-D nanostructured carbon-metal-oxide composite involves the carbon nanosheet as an illustrative carbon-based 2-D nanostructure, and the graphene has an sp^2-hybridized atom in their structure. For example, Su et al. (Fan et al., 2011) studied the synthesis of graphene/Fe_3O_4 (GNS/Fe_3O_4) nanocomposite and used it in the lithium-ion battery as an anode. This kind of carbon-metal oxide composite permits the material (graphene) to carry an exposed porous structure and the material's porous surface (graphene), which makes it conceivable to obtain the ideal electrode because there is no contact between them. Wang et al. (2011) studied the 3-D structured-based carbon-metal oxide composite as electrode material for electrochemical capacitors. The prepared carbon shows a larger porous size, signifying its usage as a suitable electrode material for electrochemical devices. Some most prominent carbon-metal oxide composites are presented in Table 23.2 with their precursors and synthesis methods.

23.5 Conclusions and future perspective

The present review summarizes the importance and application of different metal oxide–based nanocomposites for electrochemical devices. Despite the enormous capacity and large power density of metal-oxide electrodes in the energy storage system, the

Table 23.2 List and Applications of carbon (graphene)-metal oxide supported nanocomposites.

Composite material	Precursors	Preparation method	Applications	Reference
TiO_2/reduced graphene	GO/$TiSO_4$	Hydrothermal method	Lithium-ion battery electrodes	(Qiu et al., 2014)
	GO/$TiSO_4$	Self-assembly method	Electrochemical system	(Wu et al., 2012)
	GO/titanium isopropoxide (hydrazine hydrate)	Hydrothermal method	Lithium-ion battery electrodes	(Fan et al., 2011)
ZrO_2/reduced graphene	GO–$ZrOCl_2$	Hydrothermal method	Electrochemical system	(Du et al., 2011)
	GO–$ZrO(NO_3)_2.3H_2O$	Hydrothermal method	Electrochemical system	(Ghosh et al., 2014)
V_2O_5/reduced graphene	GO–V_2O_5	Hydrothermal method	Lithium-ion cathode electrodes	(Cheng et al., 2013)
	GO–V_2O_5. nH_2O (H_2O_2)	Hydrothermal method	Cathode electrodes	(Nethravathi et al., 2013)
	GO–$(NH_4)_2V_6O_{16}$	Electrochemical deposition	Battery cathode electrodes	(Zhu et al., 2014)
	GO–vanadium oxy triisopropoxide	Solvothermal method	Supercapacitors	(Perera et al., 2013)
MnO_2/reduced graphene	GO–Mn $(CH_3COO)_2$	Electrodeposition	Supercapacitors	(Yan et al., 2010)
	GO–K_xMnO_2	Hydrothermal method	Supercapacitors	(Cheng et al., 2011)
	GO–$Mn(NO_3)_2$	Electrostatic coprecipitation method	Supercapacitors	(Wu et al., 2012)
MnO_2/reduced graphene/CNTs	GO–$Mn(NO_3)_2$–SWCNT (Single walled carbon nanotubes)	Electrodeposition	Supercapacitors	(Jin et al., 2013)
MnO_2/reduced graphene	GO–$KMNO_4$	Ultrafiltration method	Anode material	(Yu et al., 2011)
Mn_3O_4/reduced graphene	GO–MnO_2	Thermal method	Supercapacitors	(Wang et al., 2010)
	GO–$MnAc_2$.$4H_2O$	Hydrothermal method	Supercapacitors	(Kou et al., 2002)
	GO–$MnAc_2$.$4H_2O$	Solvothermal method	Supercapacitors	(Zhang et al., 2012a)
	GO–$Mn(CH_3COO)_2$	Hydrothermal method	Supercapacitors	(Wang et al., 2010)

(continued on next page)

Fe/reduced graphene	GO–$FeCl_3$. $6H_2O$	Hydrothermal method	Electrode material	(Guo et al., 2011)
Fe_3O_4/reduced graphene	GO–$FeCl_3$.$6H_2O$ (hydrazine hydrate)	Hydrothermal method	Electrode material	(Su et al., 2011)
	GO–Fe powder	Hydrothermal method	Electrode material	(Dong et al., 2014)
	GO–Fe(acac)3(benzyl alcohol)	Sol–gel method	Anode material	(Baek et al., 2011)
	GO–ferric citrate	Chemical method	Electrochemical system	(Dong et al., 2014)
	GO–$FeCl_3$.$6H_2O$, $FeCl_2$.$4H_2O$	Hydrothermal method	Anode material	(Shi et al., 2011)
$LiFePO_4$/reduced graphene	GO–H_3PO_4 and LiOH (H_2O_2)	Hydrothermal method	Cathode material	(Zhou et al., 2013)
	GO–$RuCl_3$	Sol–gel method	Supercapacitors	(Wang et al., 2011)
Co_3O_4/reduced graphene	GO–$CoCl_2$.$6H_2O$ ($NaBH_4$)	Hydrothermal method	Anode material	(Li et al., 2011)
	GO–$CoSO_4$.$7H_2O$	Sol–gel method	Anode material	(Tao et al., 2012)
	GO–$(C_2H_3O_2)_2Co$.$4H_2O$ (NH_2NH_2)	Chemical method	Anode material	(Kim et al., 2011)
CoO/reduced graphene	GO–$Co_4(CO)_{12}$	Ultrasonic method	Electrode material for battery	(Peng et al., 2012)
	GO–$Co(acac)_3$	Thermal method	Electrode material for battery	(Qi et al., 2013)
Ni $(OH)_2$/reduced graphene	GO–$Ni(C_2O_4)$. H_2O	Solid-state synthesis	Supercapacitor	(Sun and Lu, 2012)
	GO–$Ni(NO_3)_2$.$6H_2O$	Thermal method	Pseudo capacitor	(Zhu et al., 2011)
	GO–Ni foam	Thermal method	Supercapacitor	(Min et al., 2014)
	GO–$Ni(NO_3)_2$ ($NaBH_4$)	Chemical method	Electrode material	(Li et al., 2011)
Cu_2O/reduced graphene	GO–Cu $(CH_3COO)_2$(glucose)	Ultrasound assisted	Electrode material	(Zhang et al., 2012b)

electric conductivity is observed still insufficient in the case of the single metal oxide–based electrode. Therefore, different types of hybrid metal–metal oxide nanocomposites, carbon-metal oxides nanocomposites, etc., are receiving much attention to fabricating their electrodes for different devices to enhance the electrochemical activities. This ideal approach carried a synergistic effect on electrode material which may lead to an industrial scale in the near future. The metal oxide nanomaterials offer an excellent opportunity to enhance the supercapacitors application to facilitate large battery scale energy density. To achieve complete potentiality, the nanostructure composition and amount of metal oxides need to be adjusted by compositing with other materials such as metal oxides, conductive polymers and carbon-based derivatives such as CNTs and graphene derivatives. The metal oxide nanocomposites show an improved feature such as electronic conductivities, precise surface area, and the richness of vigorous electrochemical sites than bulk components, which are more important to achieve significant improvement in electrochemical aspects.

Furthermore, the innovative metal oxides nanocomposites with various features as anodes have shown attractive resistance and electrochemical aspects. Also, several metal oxides nanocomposites-based supercapacitor systems have shown significant improvements and progress at a large scale. Thus, these metal oxides nanocomposites supercapacitor devices can replace the traditional supercapacitors and conventional batteries to enhance energy storage. However, the development of metal oxide–based nanocomposite electrodes for supercapacitors is still at an early stage, and the inherent energy storage mechanism is not yet completely explored. Therefore, there is a need to extensively study the energy storage mechanism and identify potential metal oxide to further increase the energy storage to multifold.

Acknowledgments

The authors gratefully acknowledged Universiti Sains Malaysia, 11800 Penang Malaysia under the Research University Grant; 1001/PKIMIA/8011070).

Conflicts of interest

There are no conflicts to declare.

References

An, C., Zhang, Y., Guo, H., Wang, Y., 2019. Metal oxide-based supercapacitors: progress and prospectives. Nanoscale Adv. 1 (12), 4644–4658. https://doi.org/10.1039/c9na00543a.

Baek, S., Yu, S.-H., Park, S.-K., Pucci, A., Marichy, C., Lee, D.-C., Sung, Y.-E., Piao, Y., Pinna, N., 2011. A one-pot microwave-assisted non-aqueous solgel approach to metal oxide/graphene nanocomposites for Li-ion batteries. RSC Adv. 1, 1687–1690. https://doi.org/10.1039/c1ra00797a.

Block, T., Schmücker, M., 2016. Metal oxides for thermochemical energy storage: a comparison of several metal oxide systems. Solar Energy 126, 195–207. https://doi.org/10.1016/j.solener.2015.12.032.

Cheng, Q., Tang, J., Ma, J., Zhang, H., Shinya, N., Qin, L.-C., 2011. Graphene and nanostructured MnO_2 composite electrodes for supercapacitors. Carbon 49, 2917–2925. https://doi.org/10.1016/j.carbon.2011.02.068.

Cheng, J., Wang, B., Xin, H.L., Yang, G., Cai, H., Nie, F., Huang, H., 2013. Self-assembled V_2O_5 nanosheets/reduced graphene oxide hierarchical nanocomposite as a high-performance cathode material for lithium ion batteries. J. Mater. Chem. A 1 (36), 10814. https://doi.org/10.1039/c3ta12066j.

de Oliveira Sousa Neto, V., Freire, T.M., Saraiva, G.D., Muniz, C.R., Cunha, M.S., Fechine, P.B.A., Nascimento, R.F.D., 2019. Water treatment devices based on zero-valent metal and metal oxide nanomaterials. In: do Nascimento, R., Ferreira, O.P., De Paula, A., Neto, V.S. (Eds.), Nanomaterials Applications for Environmental Matrices: Water, Soil and Air. Elsevier, Amsterdam, pp. 187–225.

Dong, X., Li, L., Zhao, C., Liu, H.-K., Guo, Z., 2014. Controllable synthesis of RGO/Fe_xO_y nanocomposites as high-performance anode materials for lithium ion batteries. J. Mater. Chem. A 2 (25), 9844–9850. https://doi.org/10.1039/c4ta01804d.

Dong, Y.C., Ma, R.G., Hu, M.J., Cheng, H., Tsang, C.K., Yang, Q.D., Li, Y.Y., Zapien, J.A., 2013. Scalable synthesis of Fe_3O_4 nanoparticles anchored on graphene as a high-performance anode for lithium ion batteries. J. Solid State Chem. 201, 330–337. https://doi.org/10.1016/j.jssc.2012.12.021.

Du, D., Liu, J., Zhang, X., Cui, X., Lin, Y., 2011. One-step electrochemical deposition of a graphene-ZrO_2 nanocomposite: Preparation, characterization and application for detection of organophosphorus agents. J. Mater. Chem. 21 (22), 8032. https://doi.org/10.1039/c1jm10696a.

Fan, Y., Lu, H.-T., Liu, J.-H., Yang, C.-P., Jing, Q.-S., Zhang, Y.-X., Yang, X.-K., Huang, K.-J., 2011. Hydrothermal preparation and electrochemical sensing properties of TiO_2-graphene nanocomposite. Colloids Surf. B 83 (1), 78–82. https://doi.org/10.1016/j.colsurfb.2010.10.048.

Gao, W., Chen, D., Quan, H., Zou, R., Wang, W., Luo, X., Guo, L., 2017. Fabrication of hierarchical porous metal-organic framework electrode for aqueous asymmetric supercapacitor. ACS Sustain. Chem. Eng. 5 (5), 4144–4153. https://doi.org/10.1021/acssuschemeng.7b00112.

Ghosh, D., Giri, S., Mandal, M., Das, C.K., 2014. High performance supercapacitor electrode material based on vertically aligned PANI grown on reduced graphene oxide/$Ni(OH)_2$ hybrid composite. RSC Adv. 4 (50), 26094–26101. https://doi.org/10.1039/c4ra02653e.

Guan, B.Y., Kushima, A., Yu, L., Li, S., Li, J., Lou, X.W.D., 2017. Coordination polymers derived general synthesis of multishelled mixed metal-oxide particles for hybrid supercapacitors. Adv. Mater. 29 (17), 1605902. https://doi.org/10.1002/adma.201605902.

Guan, C., Zhao, W., Hu, Y., Lai, Z., Li, X., Sun, S., Zhang, H., Cheetham, A.K., Wang, J., 2017. Cobalt oxide and N-doped carbon nanosheets derived from a single two-dimensional metal-organic framework precursor and their application in flexible asymmetric supercapacitors. Nanoscale Horizons 2 (2), 99–105. https://doi.org/10.1039/c6nh00224b.

Guo, P., Zhu, G., Song, H., Chen, X., Zhang, S., 2011. Graphene-encapsulated iron microspheres on the graphene nanosheets. Phys. Chem. Chem. Phys. 13 (39), 17818. https://doi.org/10.1039/c1cp22378j.

Hamidreza, S., Ramin, S., Maryam, K., 2014. Study in synthesis and characterization of carbon nanotubes decorated by magnetic iron oxide nanoparticles. Int. Nano Lett. 129–135. https://doi.org/10.1007/s40089-014-0128-1.

He, X., Li, R., Liu, J., Liu, Q., chen, R.R., Song, D., Wang, J., 2018. Hierarchical $FeCo_2O_4$@NiCo layered double hydroxide core/shell nanowires for high performance flexible all-solid-state asymmetric supercapacitors. Chem. Eng. J. 334, 1573–1583. https://doi.org/10.1016/j.cej.2017.11.089.

Huang, L., Chen, D., Ding, Y., Feng, S., Wang, Z.L., Liu, M., 2013. Nickel-cobalt hydroxide nanosheets coated on $NiCo_2O_4$ nanowires grown on carbon fiber paper for high-performance pseudocapacitors. Nano Lett. 13 (7), 3135–3139. https://doi.org/10.1021/nl401086t.

Ibrahim, M.N.M., Yaqoob, A.A., Umar, K., 2021. Biomass-derived composite anode electrode: Synthesis, characterizations, and application in microbial fuel cells (MFCs). J. Environ. Chem. Eng. 9 (5), 106111. https://doi.org/10.1016/j.jece.2021.106111.

Jabeen, N., Hussain, A., Xia, Q., Sun, S., Zhu, J., Xia, H., 2017. High-performance 2.6 V aqueous asymmetric supercapacitors based on in situ formed Na0.5MnO_2 nanosheet assembled nanowall arrays. Adv. Mater. 29 (32), 1700804. https://doi.org/10.1002/adma.201700804.

Jin, Y., Chen, H., Chen, M., Liu, N., Li, Q., 2013. Graphene-patched CNT/MnO_2 nanocomposite papers for the electrode of high-performance flexible asymmetric supercapacitors. ACS Appl. Mater. Interf. 5 (8), 3408–3416. https://doi.org/10.1021/am400457x.

Kalantari, K., Ahmad, M.B., Shameli, K., Khandanlou, R., 2013. Synthesis of talc/Fe_3O_4 magnetic nanocomposites using chemical co-precipitation method. Int. J. Nanomed. 8, 1817–1823. https://doi.org/10.2147/IJN.S43693.

Khaja Hussain, S., Su Yu, J., 2019. Cobalt-doped zinc manganese oxide porous nanocubes with controlled morphology as positive electrode for hybrid supercapacitors. Chem. Eng. J. 361, 1030–1042. https://doi.org/10.1016/j.cej.2018.12.152.

Khan, M., Tahir, M.N., Adil, S.F., Khan, H.U., Siddiqui, M.R.H., Al-Warthan, A.A., Tremel, W., 2015. Graphene based metal and metal oxide nanocomposites: synthesis, properties and their applications. J. Mater. Chem. A 3 (37), 18753–18808. https://doi.org/10.1039/c5ta02240a.

Kim, H., Seo, D.-H., Kim, S.-W., Kim, J., Kang, K., 2011. Highly reversible Co_3O_4/graphene hybrid anode for lithium rechargeable batteries. Carbon 49 (1), 326–332. https://doi.org/10.1016/j.carbon.2010.09.033.

Kong, S., Cheng, K., Ouyang, T., Gao, Y., Ye, K., Wang, G., Cao, D., 2017. Facile electrodepositing processed of RuO_2-graphene nanosheets-CNT composites as a binder-free electrode for electrochemical supercapacitors. ElectroChim. Acta 246, 433–442. https://doi.org/10.1016/j.electacta.2017.06.019.

Kou, H.-Z., Gao, S., Li, C.-H., Liao, D.-Z., Zhou, B.-C., Wang, R.-J., Li, Y., 2002. Characterization of a soluble molecular magnet: unusual magnetic behavior of cyano-bridged Gd(III)−Cr(III) complexes with one-dimensional and nanoscaled square structures. Inorg. Chem. 41 (18), 4756–4762.

Li, B., Cao, H., Shao, J., Li, G., Qu, M., Yin, G., 2011a. Co_3O_4@graphene composites as anode materials for high-performance lithium ion batteries. Inorg. Chem. 50 (5), 1628–1632. https://doi.org/10.1021/ic1023086.

Li, B., Cao, H., Shao, J., Zheng, H., Lu, Y., Yin, J., Qu, M., 2011b. Improved performances of β-$Ni(OH)_2$@reduced-graphene-oxide in Ni-MH and Li-ion batteries. Chem. Commun. 47 (11), 3159. https://doi.org/10.1039/c0cc04507a.

Li, B., Wang, Y., 2010. Facile synthesis and photocatalytic activity of ZnO-CuO nanocomposite. Superlattices MicroStruct. 47 (5), 615–623. https://doi.org/10.1016/j.spmi.2010.02.005.

Lu, W.J., Huang, S.Z., Miao, L., Liu, M.X., Zhu, D.Z., Li, L.C., Duan, H., Xu, Z.J., Gan, L.H., 2017. Synthesis of MnO_2/N-doped ultramicroporous carbon nanospheres for high-performance supercapacitor electrodes. Chinese Chem. Lett. 28 (6), 1324–1329. https://doi.org/10.1016/j.cclet.2017.04.007.

Mamulová Kutláková, K., Tokarský, J., Peikertová, P., 2015. Functional and eco-friendly nanocomposite kaolinite/ZnO with high photocatalytic activity. Appl. Catal. B Environ. 162, 392–400. https://doi.org/10.1016/j.apcatb.2014.07.018.

Min, S., Zhao, C., Chen, G., Qian, X., 2014. One-pot hydrothermal synthesis of reduced graphene oxide/$Ni(OH)_2$ films on nickel foam for high performance supercapacitors. Electrochim. Acta 115, 155–164. https://doi.org/10.1016/j.electacta.2013.10.140.

Mirzaei, A., Neri, G., 2016. Microwave-assisted synthesis of metal oxide nanostructures for gas sensing application: a review. Sensors Actuat. B Chem. 237, 749–775. https://doi.org/10.1016/j.snb.2016.06.114.

Nazari, M., Halladj, R., 2014. Adsorptive removal of fluoride ions from aqueous solution by using sonochemically synthesized nanomagnesia/alumina adsorbents: an experimental and modeling study. J. Taiwan Inst. Chem. Engrs. 45 (5), 2518–2525. https://doi.org/10.1016/j.jtice.2014.05.020.

Nethravathi, C., Rajamathi, C.R., Rajamathi, M., Gautam, U.K., Wang, X., Golberg, D., Bando, Y., 2013. N-doped graphene–VO_2(B) nanosheet-built 3D flower hybrid for lithium ion battery. ACS Appl. Mater. Interf. 5 (7), 2708–2714. https://doi.org/10.1021/am400202v.

Peng, C., Chen, B., Qin, Y., Yang, S., Li, C., Zuo, Y., Liu, S., Yang, J., 2012. Facile ultrasonic synthesis of CoO quantum dot/graphene nanosheet composites with high lithium storage capacity. ACS Nano 6 (2), 1074–1081. https://doi.org/10.1021/nn202888d.

Perera, S.D., Liyanage, A.D., Nijem, N., Ferraris, J.P., Chabal, Y.J., Balkus, K.J., 2013. Vanadium oxide nanowire–graphene binder free nanocomposite paper electrodes for supercapacitors: A facile green approach. J. Power Sources 230, 130–137. https://doi.org/10.1016/j.jpowsour.2012.11.118.

Qi, Y., Zhang, H., Du, N., Yang, D., 2013. Highly loaded CoO/graphene nanocomposites as lithium-ion anodes with superior reversible capacity. J. Mater. Chem. A 1 (6), 2337. https://doi.org/10.1039/c2ta00929c.

Qiu, B., Xing, M., Zhang, J., 2014. Mesoporous TiO_2 nanocrystals grown in situ on graphene aerogels for high photocatalysis and lithium-ion batteries. J. Am. Chem. Soc. 136 (16), 5852–5855. https://doi.org/10.1021/ja500873u.

Ray, A., Roy, A., Sadhukhan, P., Chowdhury, S.R., Maji, P., Bhattachrya, S.K., Das, S., 2018. Electrochemical properties of TiO_2-V_2O_5 nanocomposites as a high performance supercapacitors electrode material. Appl. Surf. Sci. 443, 581–591. https://doi.org/10.1016/j.apsusc.2018.02.277.

Rao, B.G., Mukherjee, D., Reddy, B.M., 2017. Novel approaches for preparation of nanoparticles. In: Ficai,D., Grumezescu, A.M. (Eds.),Nanostructures for Novel Therapy: Synthesis,Characterization and Applications, Elsevier Inc., Amsterdam, 1st ed., Vol. 2, pp. 1--36.

Ravikumar, C.R., Kotteeswaran, P., Murugan, A., Raju, Bheema, V., Santosh, M., S., Nagaswarupa, H.P., Nagabhushana, H., Prashantha, S.C.,Kumar, Anil,M., R.,Gurushantha,K, 2017. Electrochemical studies of nano metal oxide reinforced nickel hydroxide materials for energy storage applications. Mater. Today Proc., 4(11), 12205--12214.

Seok, D., Jeong, Y., Han, K., Yoon, D.Y., Sohn, H., 2019. Recent progress of electrochemical energy devices: metal oxide-carbon nanocomposites as materials for next-generation chemical storage for renewable energy. Sustainability (Switzerland) 11 (13), 3694. https://doi.org/10.3390/su11133694.

Sheng, S., Liu, W., Zhu, K., Cheng, K., Ye, K., Wang, G., Cao, D., Yan, J., 2019. Fe_3O_4 nanospheres in situ decorated graphene as high-performance anode for asymmetric supercapacitor with impressive energy density. J. Colloid Interface Sci. 536, 235–244. https://doi.org/10.1016/j.jcis.2018.10.060.

Shi, W., Zhu, J., Sim, D.H., Tay, Y.Y., Lu, Z., Zhang, X., Sharma, Y., Srinivasan, M., Zhang, H., Hng, H.H., Yan, Q., 2011. Achieving high specific charge capacitances in Fe_3O_4/reduced graphene oxide nanocomposites. J. Mater. Chem. 21 (10), 3422. https://doi.org/10.1039/c0jm03175e.

Su, J., Cao, M., Ren, L., Hu, C., 2011. Fe_3O_4–graphene nanocomposites with improved lithium storage and magnetism properties. J. Phys. Chem. C 115 (30), 14469–14477. https://doi.org/10.1021/jp201666s.

Sultana, S., Rafiuddin, Khan, M., Z., Umar, K., Muneer, M, 2013. Electrical, thermal, photocatalytic and antibacterial studies of metallic oxide nanocomposite doped polyaniline. J. Mater. Sci. Technol. 29 (9), 795–800. https://doi.org/10.1016/j.jmst.2013.06.001.

Sun, Z., Lu, X., 2012. A solid-state reaction route to anchoring $Ni(OH)_2$ nanoparticles on reduced graphene oxide sheets for supercapacitors. Ind. Eng. Chem. Res. 51 (30), 9973–9979. https://doi.org/10.1021/ie202706h.

Tao, L., Zai, J., Wang, K., Zhang, H., Xu, M., Shen, J., Su, Y., Qian, X., 2012. Co_3O_4 nanorods/graphene nanosheets nanocomposites for lithium ion batteries with improved reversible capacity and cycle stability. J. Power Sources 202, 230–235. https://doi.org/10.1016/j.jpowsour.2011.10.131.

Thimmaiah, S., Rajamathi, M., Singh, N., Bera, P., Meldrum, F., Chandrasekhar, N., Seshadri, R., 2001. A solvothermal route to capped nanoparticles of γ-Fe_2O_3 and $CoFe_2O_4$. J. Mater. Chem. 11 (12), 3215–3221. https://doi.org/10.1039/b104070g.

Umar, K., Yaqoob, A.A., Ibrahim, M.N.M., 2020. Silver nanoparticles: various methods of synthesis, size affecting factors and their potential applications–a review. Appl. NanoSci. 1369–1378. https://doi.org/10.1007/s13204-020-01318-w.

Wang, B., Park, J., Wang, C., Ahn, H., Wang, G., 2010a. Mn_3O_4 nanoparticles embedded into graphene nanosheets: Preparation, characterization, and electrochemical properties for supercapacitors. Electrochim. Acta 55 (22), 6812–6817. https://doi.org/10.1016/j.electacta.2010.05.086.

Wang, H., Cui, L.-F., Yang, Y., Sanchez Casalongue, H., Robinson, J.T., Liang, Y., Cui, Y., Dai, H., 2010b. Mn_3O_4−graphene hybrid as a high-capacity anode material for lithium ion batteries. J. Am. Chem. Soc. 132 (40), 13978–13980. https://doi.org/10.1021/ja105296a.

Wang, H., Liang, Y., Mirfakhrai, T., Chen, Z., Casalongue, H.S., Dai, H., 2011. Advanced asymmetrical supercapacitors based on graphene hybrid materials. Nano Res. 4 (8), 729–736. https://doi.org/10.1007/s12274-011-0129-6.

Wu, Z., Ren, W., Wang, D., Li, F., Liu, B., Cheng, H., Li, Y., Zhao, N., Shi, C., Liu, E., He, C., 2010. High-energy MnO_2 nanowire/graphene and graphene asymmetric electrochemical capacitors. J. Phys. Chem. C. 4, 25226–25232.

Wu, Z.-S., Zhou, G., Yin, L.-C., Ren, W., Li, F., Cheng, H.-M., 2012. Graphene/metal oxide composite electrode materials for energy storage. Nano Energy 1 (1), 107–131. https://doi.org/10.1016/j.nanoen.2011.11.001.

Yan, J., Fan, Z., Wei, T., Qian, W., Zhang, M., Wei, F., 2010. Fast and reversible surface redox reaction of graphene–MnO_2 composites as supercapacitor electrodes. Carbon 48 (13), 3825–3833. https://doi.org/10.1016/j.carbon.2010.06.047.

Yan, Y., Wang, K., Clough, P.T., Anthony, E.J., 2020. Developments in calcium/chemical looping and metal oxide redox cycles for high-temperature thermochemical energy storage: a review. Fuel Process. Technol. 199, 106280. https://doi.org/10.1016/j.fuproc.2019.106280.

Yang, H., Shi, R., Zhang, K., Hu, Y., Tang, A., Li, X., 2005. Synthesis of WO_3/TiO_2 nanocomposites via sol-gel method. J. Alloys Compd. 398 (1–2), 200–202. https://doi.org/10.1016/j.jallcom.2005.02.002.

Yang, Z.F., Li, L.Y., Hsieh, C.T., Juang, R.S., 2018. Co-precipitation of magnetic Fe_3O_4 nanoparticles onto carbon nanotubes for removal of copper ions from aqueous solution. J. Taiwan Inst.Chem. Engrs. 82, 56–63. https://doi.org/10.1016/j.jtice.2017.11.009.

Yaqoob, A.A., Ahmad, H., Parveen, T., Ahmad, A., Oves, M., Ismail, I.M.I., Qari, H.A., Umar, K., Ibrahim, Mohamad, M., N, 2020a. Recent advances in metal decorated nanomaterials and their various biological applications: a review Front. Chem. 8, 341.

Yaqoob, A.A., Ibrahim, M.N.M., Rafatullah, M., Chua, Y.S., Ahmad, A., Umar, K., 2020b. Recent advances in anodes for microbial fuel cells: an overview. Materials 13 (9), 13092078. https://doi.org/10.3390/ma13092078.

Yaqoob, A.A., Ibrahim, M.N.M., Rodríguez-Couto, S., 2020c. Development and modification of materials to build cost-effective anodes for microbial fuel cells (MFCs): an overview. Bio Chem. Eng. J. 164, 107779. https://doi.org/10.1016/j.bej.2020.107779.

Yaqoob, A.A., Noor, N.H.B.M., Serrà, A., Ibrahim, M.N.M., 2020d. Advances and challenges in developing efficient graphene oxide-based ZnO photocatalysts for dye photo-oxidation. Nanomaterials 10 (5), 932. https://doi.org/10.3390/nano10050932.

Yaqoob, A.A., Parveen, T., Umar, K., Ibrahim, M.N.M., 2020e. Role of nanomaterials in the treatment of wastewater: a review. Water (Switzerland) 12 (2), 495. https://doi.org/10.3390/w12020495.

Yaqoob, A.A., Ibrahim, M.N.M., Ahmad, A., Vijaya Bhaskar Reddy, A., 2021a. Toxicology and environmental application of carbon nanocomposite. In: Jawaid, M., Ahmad, A., Ismail, N., Rafatullah, M. (Eds.), Green Energy and Technology. Springer Science and Business Media Deutschland GmbH, Switzerland, pp. 1–18.

Yaqoob, A.A., Ibrahim, M.N.M., Yaakop, A.S., Umar, K., Ahmad, A., 2021b. Modified graphene oxide anode: a bioinspired waste material for bioremediation of Pb^{2+} with energy generation through microbial fuel cells. Chem. Eng. J. 417, 128052. https://doi.org/10.1016/j.cej.2020.128052.

Yaqoob, A.A., Ibrahim, M.N.M., Umar, K., Bhawani, S.A., Khan, A., Asiri, A.M., Khan, M.R., Azam, Al Ammari, A.M., 2021c. Cellulose derived graphene/polyaniline nanocomposite anode for energy generation and bioremediation of toxic metals via benthic microbial fuel cells. Polymers 13 (1), 135.

Yaqoob, A.A., Umar, K., Adnan, R., Ibrahim, M.N.M., Rashid, M., 2021. Graphene oxide–ZnO nanocomposite: an efficient visible light photocatalyst for degradation of rhodamine. B. Appl. 11, 1291–1302.

Yu, A., Park, H.W., Davies, A., Higgins, D.C., Chen, Z., Xiao, X., 2011. Free-standing layer-by-layer hybrid thin film of graphene-MnO_2 nanotube as anode for lithium ion batteries. J. Phys. Chem. Lett. 2 (15), 1855–1860. https://doi.org/10.1021/jz200836h.

Zhang, H., Chen, G., 2009. Potent antibacterial activities of Ag/TiO_2 nanocomposite powders synthesized by a one-pot sol-gel method. Environ. Sci. Technol. 43 (8), 2905–2910. https://doi.org/10.1021/es803450f.

Zhang, X., Sun, X., Chen, Y., Zhang, D., Ma, Y., 2012a. One-step solvothermal synthesis of graphene/Mn_3O_4 nanocomposites and their electrochemical properties for supercapacitors. Mater. Lett. 68, 336–339. https://doi.org/10.1016/j.matlet.2011.10.092.

Zhang, Y., Wang, X., Zeng, L., Song, S., Liu, D., 2012b. Green and controlled synthesis of Cu_2O–graphene hierarchical nanohybrids as high-performance anode materials for lithium-ion batteries via an ultrasound assisted approach. Dalton Trans. 41 (15), 4316. https://doi.org/10.1039/c2dt12461k.

Zhang, S.W., Yin, B.S., Liu, C., Wang, Z.B., Gu, D.M., 2017. Self-assembling hierarchical $NiCo_2O_4/MnO_2$ nanosheets and MoO_3/PPy core-shell heterostructured nanobelts for supercapacitor. Chem. Eng. J. 312, 296–305. https://doi.org/10.1016/j.cej.2016.11.144.

Zhao, J., Li, Z., Yuan, X., Yang, Z., Zhang, M., Meng, A., Li, Q., 2018. A high-energy density asymmetric supercapacitor based on Fe_2O_3 nanoneedle arrays and $NiCo_2O_4/Ni(OH)_2$ hybrid nanosheet arrays grown on SiC nanowire networks as free-standing advanced electrodes. Adv. Energy Mater. 8 (12), 1702787. https://doi.org/10.1002/aenm.201702787.

Zhao, Y., Dong, W., Riaz, M.S., Ge, H., Wang, X., Liu, Z., Huang, F., 2018. "Electron-sharing" mechanism promotes Co@Co_3O_4/CNTs composite as the high-capacity anode material of lithium-ion battery. ACS Appl. Mater. Interf. 10 (50), 43641–43649. https://doi.org/10.1021/acsami.8b15659.

Zhou, G., Yin, L.-C., Wang, D.-W., Li, L., Pei, S., Gentle, I.R., Li, F., Cheng, H.-M., 2013. Fibrous hybrid of graphene and sulfur nanocrystals for high-performance lithium–sulfur batteries. ACS Nano 7 (6), 5367–5375. https://doi.org/10.1021/nn401228t.

Zhu, J., Chen, S., Zhou, H., Wang, X., 2011. Fabrication of a low defect density graphene-nickel hydroxide nanosheet hybrid with enhanced electrochemical performance. Nano Res. 5 (1), 11–19. https://doi.org/10.1007/s12274-011-0179-9.

Zhu, X., Song, X., Ma, X., Ning, G., 2014. Enhanced electrode performance of Fe_2O_3 nanoparticle-decorated nanomesh graphene as anodes for lithium-ion batteries. ACS Appl. Mater. Interf. 6 (10), 7189–7197. https://doi.org/10.1021/am500323v.

Zhu, S., Li, L., Liu, J., Wang, H., Wang, T., Zhang, Y., Zhang, L., Ruoff, R.S., Dong, F., 2018. Structural directed growth of ultrathin parallel birnessite on β-MnO_2 for high-performance asymmetric supercapacitors. ACS Nano 12 (2), 1033–1042. https://doi.org/10.1021/acsnano.7b03431.

CHAPTER 24

Engineered uses of nanomaterials for sustainable cementitious composites

Paul O. Awoyera[a], Mehmet Serkan Kırgız[b] and Adeyemi Adesina[c]
[a]Department of Civil Engineering, Covenant University, Ota, Nigeria
[b]İstanbul Üniversitesi-Cerrahpaşa, Avcılar, İstanbul, Turkey
[c]Civil and Environmental Engineering, University of Windsor, Windsor, Canada

24.1 Introduction

Cementitious composites, which are the most used construction material in the world, are evolving with current trends in the scientific world. The high demand for high-performance cementitious composites has resulted in the use of recent advances in science and engineering to improve the performance of the composites. The evolution of nanotechnology in recent times has opened a pathway for an effective way to enhance the performance of cementitious composites. One of the ways nanotechnology is used in the construction industry is by incorporating materials with particle size ranging between 1 and 100 nm in construction materials such as cementitious composites. These nanomaterials modify the nanostructure of cementitious composites resulting in a significant improvement in the performance compared to when additives such as mineral additives are used (Hakamy et al., 2015). The ability to achieve higher performance in cementitious composites with the use of nanomaterials also creates an avenue to reduce the required content of Portland cement (Mostafa et al., 2015). Reducing the content of Portland cement used to produce cementitious composites will result in a significant reduction in the cost and embodied carbon of cementitious composites (Meyer, 2009; Sivakrishna et al., 2019). The incorporation of nanomaterials such as nanosilica in cementitious composites can also be used to improve the early age strength of the composites when industrial waste products such as fly ash with low reactivity are incorporated as a sustainable alternative to partially replace Portland cement as the binder (Gengying, 2004; Lin et al., 2008; Sato & Beaudoin, 2011). Hence, nanotechnology advances the development and application of cementitious composites made with high volume supplementary cementitious materials (SCMs) as a replacement of Portland cement. In addition to the refinement of the microstructure and improvement of the durability and mechanical properties of cementitious composites, nanomaterials such as carbon nanotubes (CNT) and carbon nanofibers (CNF) offer alternative reinforcement

Sustainable Nanotechnology for Environmental Remediation
DOI: https://doi.org/10.1016/B978-0-12-824547-7.00004-7

to short fibers in these composites. The tensile strength and modulus of elasticity of CNT are several ten to hundred times that of the conventional steel used as reinforcement in cementitious composites (Siad et al., 2018; Walters et al., 1999). Also, these nanomaterials used as reinforcement (i.e., CNT and CNF) do not undergo corrosion. Hence, cementitious composites reinforced with nanomaterials would exhibit higher service life and would require lower maintenance. The reinforcement of cementitious composites on a nanoscale also proffers a way to mitigate early age cracking and enhances the interface between the cementitious matrix and the aggregates in the composite (Laila et al., 2010). Nanomaterials can also be used to improve functional properties such as electric conductivity and sensing capabilities of cement-based composites (Adeyemi, 2020). The use of nanomaterials in cementitious composites also embodied it with other beneficial properties such as self-cleaning, self-sensing (Han et al., 2014), self-disinfection, and high photocatalytic activity (Hashimoto et al., 2007). As the use of nanomaterials in construction applications is an emerging area, stakeholders in the construction industry must be aware of how these materials influence the performance of cementitious composites. Hence, this chapter presents a comprehensive overview of the engineered use of nanomaterials in cementitious composites.

24.1.1 Nanotechnology applications in construction—overview

Nanotechnology is generally used to describe processes and/materials that involve the utilization of a nanoscale. The advent of nanotechnology started in physical sciences but it has now evolved into applied science where it is used in various construction materials. The possible significant improvement in the performance of composites that can be achieved with the use of nanotechnology has made this a promising and emerging area in recent times. With cementitious composites being the most used building material in the world, one of the major areas where nanotechnology is being used in construction applications is as additives in cementitious composites. Several studies have shown that the incorporation of nanomaterials such as nanosilica can be used to improve the durability and mechanical properties of cementitious composites (Adeyemi, 2019; Al-Najjar et al., 2016; Zemei et al., 2016). These nanomaterials are able to enhance the performance of cementitious composites by accelerating the hydration reaction and densification of the nanostructure and microstructure of the composites. Nanomaterials are relatively smaller compared to other components used in cementitious composites. Due to the smaller particle sizes of nanomaterials, the kinetics of cementitious composites can be modified on a nanoscale. A representation of the size at which some nanomaterials are used compared to other components in cementitious composites depicted by Sobolev and Flores (2008) is presented in Fig. 24.1. Fig. 24.2 presents the effect of material size on its ability to fill the voids among the components in cementitious composites. It can be observed from Fig. 24.2 that the smaller the particle size of the material, the

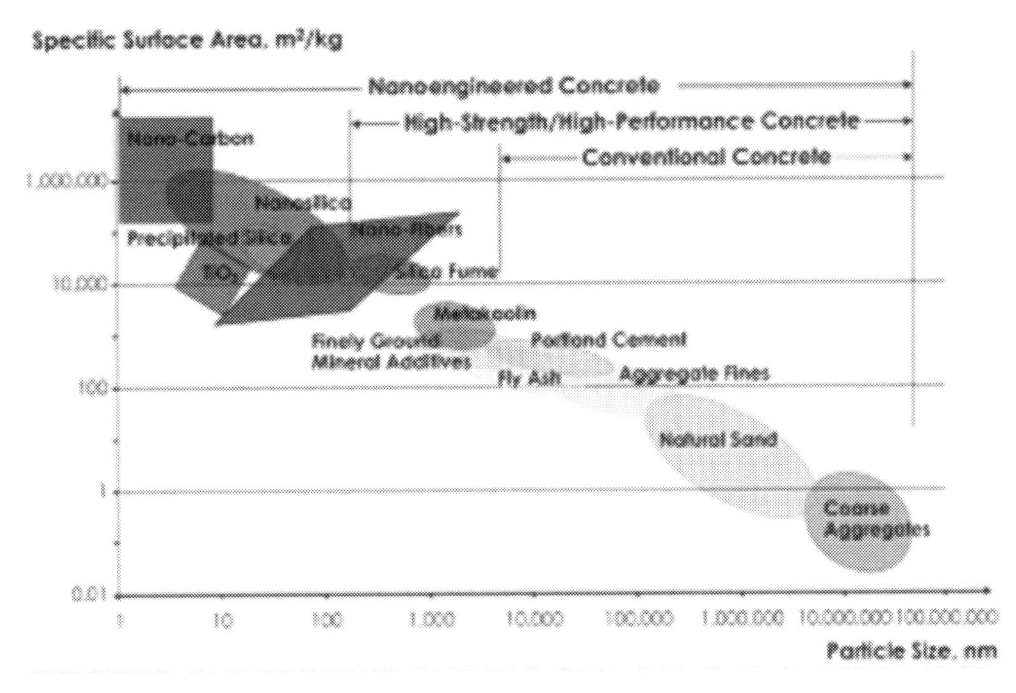

Figure 24.1 Particle size and specific surface area of nanomaterials and other components in cementitious composites (Sobolev, 2015).

more suitable it is to fill the voids among the components of the compositions. Also, the magnetite particles with particle size range of 20 nm to 100 nm were able to result in more densification compared to when magnetite particles of higher sizes (i.e., greater than 100 nm) were used. Hence, in addition to the chemical contribution of nanomaterials, the ability to fill nanovoids contributes to the enhancement in the performance exhibited by such composites. Nanotechnology also creates an avenue to solve many challenges related to the performance, sustainability, and safety of cement-based composites. With the current trends in the use of nanomaterials in construction materials, it is anticipated that the large-scale production and application of nanomaterials will revolutionize high-performance composites for future constructions.

24.1.2 Significance of the study

One of the promising areas that will revolutionize construction materials is nanotechnology. Hence, an adequate understanding of how this technology can be utilized safely while achieving desired properties is critical to its success. This study is of great significance because it proffers innovative ways on how nanomaterials can be used to improve the performance of construction materials such as cementitious composites. This study also explores how various types of nanomaterials interact and influence the resulting

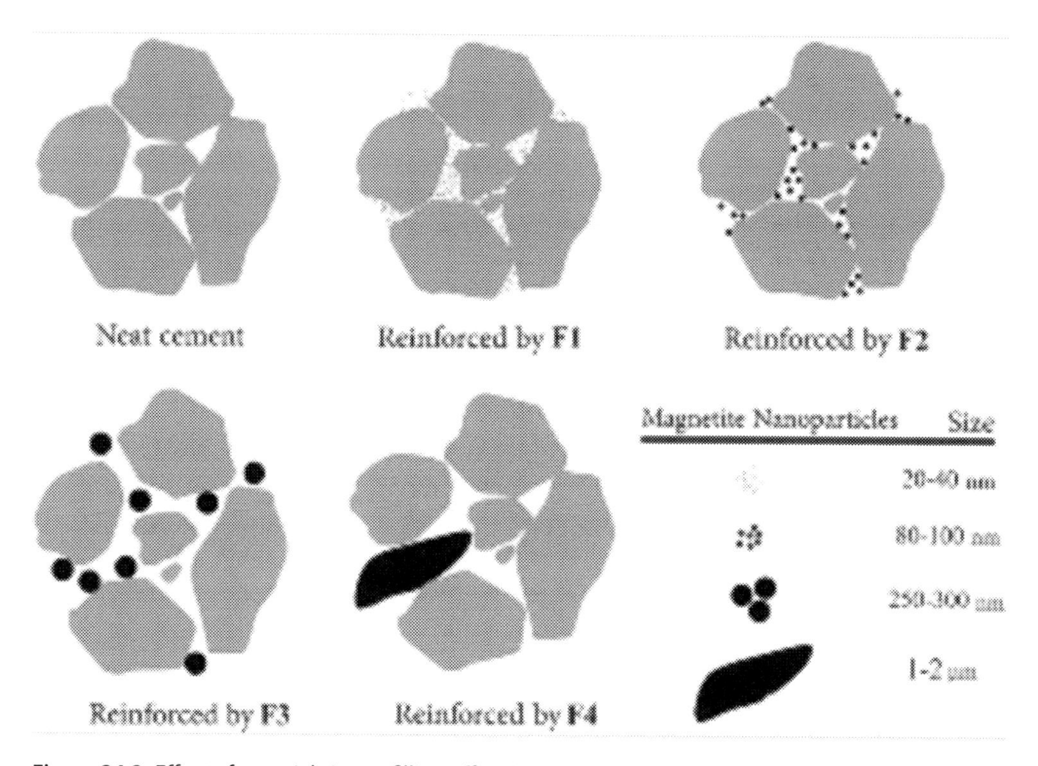

Figure 24.2 Effect of material size on filling effect in cementitious composites.

performance of these composites. Discussions presented in this study will be a very useful resource for stakeholders in the construction industry exploring innovative ways to improve the performance of cementitious composites. The limitations with the use of nanomaterials in construction materials discussed in this study will also gear more research and development in improving the application of nanomaterials in various construction materials.

24.2 Nanotechnology—the state of the art

24.2.1 Types and mechanisms of nanomaterials

The mechanism by which nanomaterials influence the performance of cement-based materials is dependent on the physical and chemical properties of the nanomaterials. As these properties differ between nanomaterials, their corresponding effect on the performance of cement-based composites is different. Generally, the mechanism of nanomaterials in cementitious composites evolves around hydration seeding effects and filling effects (Land & Stephan, 2012). Depending on the chemical composition of the nanomaterials, their mechanism to improve the performance of composites will range

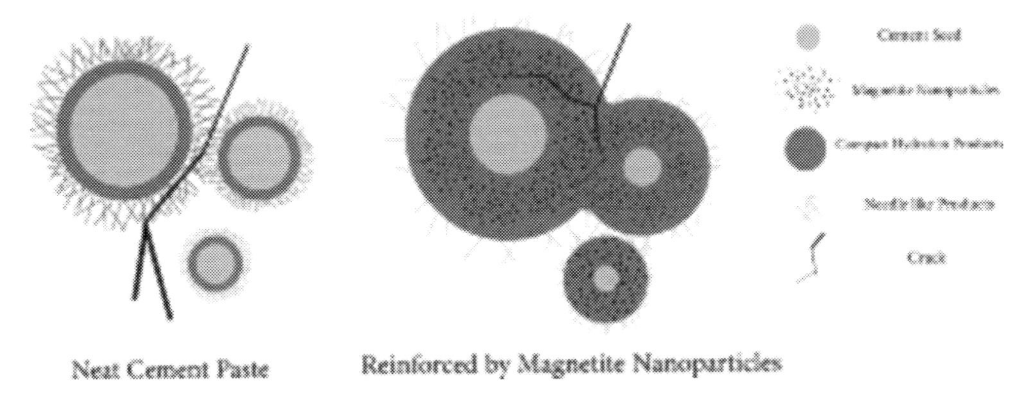

Figure 24.3 Seeding and filling effect of nanomagnetite.

in the form of acting as a nucleation site to the exhibition of pozzolanic properties. Nanomaterials acting as a nucleation site for the formation of hydration products exhibit what is called seeding effect in which these nanoparticles provide areas for the formation of additional products, which results in corresponding acceleration of the hydration process (Pengkun et al., 2013). The ability of these nanomaterials to act as nucleations sites in the cementitious matrix is a result of their higher surface area, which results in an increase in the rate of chemical reactivity (Li et al., 2018). For nanomaterials with pozzolanic characteristics and mostly metallic oxides, they react with the calcium hydroxide in the pore solution of the composites to form additional calcium silicate hydrates (C-S-H). The formation of this secondary C-S-H results in the densification of the microstructure and a corresponding enhancement of the mechanical and durability properties of the composites. Nanomaterials such as nanosilica have also been found to accelerate the initial polymerization of silicate coupled with the exhibition of pozzolanic characteristics and acting as nucleation sites (Bjornstrom et al., 2004). A schematic of the seeding and filling effect of nanomagnetite in cement paste is shown in Fig. 24.3 (Imanian et al., 2020). It can be observed from Fig. 24.3 that before the addition of the nanomaterials, the volume of hydration products is lower and there exists a bigger crack in the composite. However, there was a significant increase in the formation of hydration products and a lower crack formation when nanomagnetite was incorporated into the composites. Nanomaterials can exist in the form of powders or tubes that also affect their corresponding influence on the performance of cement-based composites. Hence, nanomaterials can be classified as 0-D, 1-D, and 2-D, which corresponds to nanoparticles, nanofibers, and nanofoils, respectively (Samuel et al., 2014). As mentioned earlier, the initial crack formation in cementitious composites can be significantly mitigated with the use of nanofibers. The ability of these nanofibers to be able to prevent/reduce these initial cracks is due to C-S-H which is the main hydration product being a nanomaterial itself

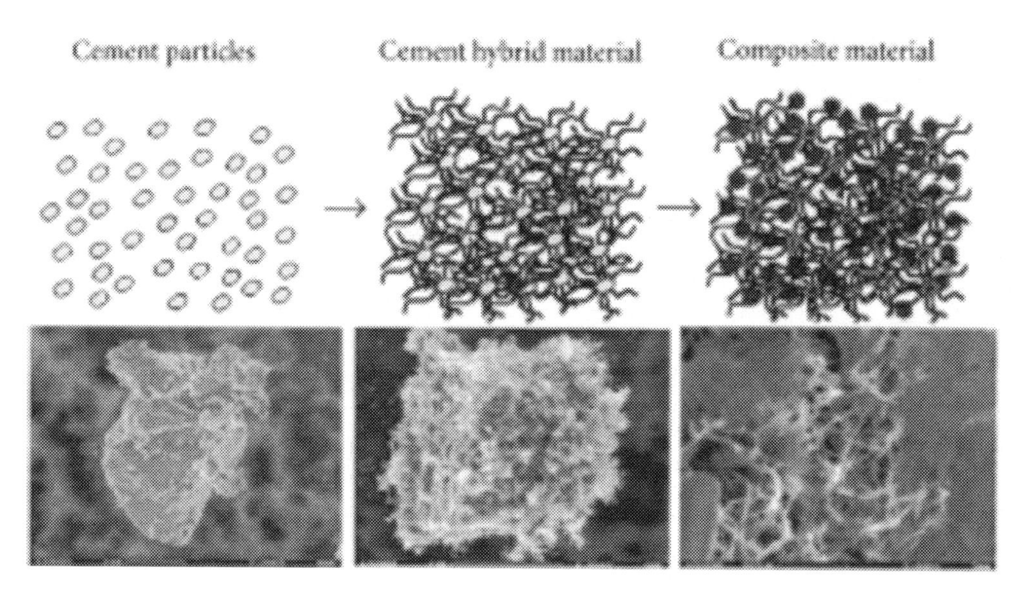

Figure 24.4 Schematic and scanning electron image of the mechanism of CNT and CNF.

and the incorporation of these fibers being able to improve its stiffness on a nanolevel. A schematic representation of the interaction between CNT/CNF and cement particles made by Imanian et al. (2020) is presented in Fig. 24.4. It can be observed from the figure that the CNT/CNF acts as a bridging agent between the cement particles and allows the formation of products around the nanomaterials resulting in an enhancement in the stiffness of the hydration products.

24.2.2 Potential risks in nano-enabled construction materials

With the beneficial prospects with the use of nanomaterials in construction materials such as cementitious composites come some potential risks. However, the current research and development trends in the use of nanomaterials are more focused on its application while its effect on the potential health and environmental risks is limited. As humans are always in contact with construction materials as these materials are used for the construction of our infrastructures, it is critical to ensure that these materials do not pose any potential risk to the health and safety of humans and other living organisms. Generally, the potential risk of nano-enabled materials can be classified into human exposure and environmental exposure (Tsuji et al., 2006). The major risk associated with the use of nanomaterials in construction materials can be associated with its super fine size that makes it possible to be ingested and absorb easily. The study by Maynard and Kuempel (2005) has shown that the toxicity of nanomaterials increases with their corresponding decreasing size. With nanomaterials containing various deleterious compounds that are detrimental to the health and safety of living organisms, their ingestion might result in various health hazards. The study by Grassian et al. (2007) showed that the inhalation of nano-titanium particles at a concentration of 8.8 mg/m^3 would cause lung inflammation. This observation corresponds to that reported by Aihong et al. (2008) and Pacurari et al. (2010). Nanomaterials from aluminum have high ignition sensitivity and are susceptible to inflammation and explosion (Vignes et al., 2012). Hence, it is critical to ensure that when such materials are used in cementitious composites; they do not pose any threat. The pro-inflammatory and pro-oxidative properties of some of the nanomaterials obtained from metal oxides make them susceptible to cause heart-related diseases in humans (Xiaoyan et al., 2012). Some nanomaterials have also been associated with damage to organs in humans (Liou et al., 2015; Enrico et al., 2015). The increasing sustainability awareness in the construction industry has also resulted in a significant amount of cementitious composites being recycled at the end of their service life and reused in new composites. Hence, it is paramount that these recycled composites containing nanomaterials be properly labeled to ensure their proper handling and reuse. In general, the potential risk of cementitious composites made with nanomaterials revolves around the handling and storage of nanomaterials, mixing process and application of the composites, and abrasive actions on the composites such as drilling, cutting, grinding, etc. (María et al., 2019). The limited knowledge of the potential risks of these nanomaterials when used for construction applications also poses a major risk to the users. Hence, it is recommended that more research should be carried out to identify the potential risks with the use of these materials.

24.3 Cementitious composites incorporating nanomaterials

24.3.1 Physical and mechanical characteristics of products

Comprehensive evaluation works demonstrated that the substitution of nanoparticle of cement for micro- and macroparticle of cement in mortar mixing leads to an increase in the physical and mechanical performance of the mortar prepared (Parang et al., 2014). Moreover, a broad field of use of different nanoparticles such as CNTs, nano SiO_2, nano TiO_2, nano Fe_2O_3, nano CuO, nano ZrO_2, nano ZnO_2, nano Al_2O_3, nano $CaCO_3$ (NC), and nano Cr_2O_3 has been recently evaluated in comprehensive research works. Most of the works pointed out that the presence of nanoparticles as admixtures in cementitious binders could enhance the physical and mechanical characteristics of cement-based materials (Gengying, 2004). As the most worked nanomaterial in cement-based materials is nano SiO_2 particle supplement, there are comprehensive research results for the nano SiO_2 in the science literature (Amin & Alireza, 2019). They studied the effect of the mechanical strength of concrete made by incorporation of two different sizes of nano SiO_2 particles (15 nm and 80 nm) with percentages of 0.5%, 1.0%, 1.5%, and 2.0% by cement weight. Table 24.1 summarizes compressive strength, splitting tensile strength,

Table 24.1 The compressive strength, splitting tensile strength, and bending moment strength of concrete containing nano SiO_2 particle at the days of 7, 28, and 90, and their improvement percent.

Type of test	Curing days	Nano SiO_2 particle 2% by weight of cement in size of 15 (nm) and 1.5% by weight of cement in size 80 (nm)
Compressive strength (MPa)	7 days	52.1
	28 days	63.7
	90 days	78.1
Improvement of compressive strength (%)	7 days	98.1
	28 days	83
	90 days	93.8
Splitting tensile strength (MPa)	7 days	4.9
	28 days	4.3
	90 days	5.9
Improvement of splitting tensile strength (%)	7 days	276.9
	28 days	186.7
	90 days	210.5
Bending moment strength (MPa)	7 days	7
	28 days	7.3
	90 days	7.8
Improvement of bending moment strength (%)	7 days	75
	28 days	73.8
	90 days	73.3

and bending moment strength of concrete containing nano SiO_2 particles at the days of 7, 28, and 90, and their improvement percent.

For instance, on substitution of cement with 1% of nanoparticle of SiO_2, both the bending moment strength and compressive strength of concrete were, respectively, measured to be increased by 4.2% and 12.3%; and the chloride permeability and porosity were, respectively, determined to be decreased by 18% and 6.9% (Mao-hua & Hui, 2011). There is plenty of research explaining the effect of other nanomaterials on the physical and mechanical characteristics of cement-based materials. These are summarized in the following. One of the studies regarding cement concrete prepared with the incorporating of 1% nano TiO_2 particle presented current results that the bending moment strength, split tensile strength, and compressive strength were, respectively, established to be increased by 25%, 67%, and 18%; and its water absorption was over 3.6% lower when compared to normal concrete (Nazari et al., 2010). However, the addition of a 5% nano TiO_2 particle significantly influenced the physical and mechanical characteristics of self-compacting mortar containing 25% fly ash. Its compressive strength and electrical resistivity, respectively, increased over 13% and 257%, and water absorption and chloride permeability, respectively, reduced over 11% and 58% (Rahmat et al., 2015). As the 0.3% CNT was added in concrete, the enhancement in the bending moment strength and the compressive strength were, respectively, determined to be over 21% and 23% (Li et al., 2018). The density of mortar containing 1% CNT was found to be 2.6% greater than the standard mortar (Arnon et al., 2010). The total porosity of mortar blended 0.5% CNT was established to be over 64% lesser than the standard mortar (Li et al., 2018). It was observed that the compressive strength, splitting tensile strength, and bending moment strength of concrete was, respectively, enhanced over 71.8%, 93.7%, and 76.1% by the addition of 4% nano Fe_2O_3 particle; and its water absorption and total specific pore volume was also, respectively, reduced by over 73.7% and 10.5% (Ali et al., 2012). In addition to the enhancement provided by the nanoparticles aforementioned, the nano CuO particle was also proved that it was an effective nanoparticle for activating the cement-based materials (Rahmat et al., 2015). The mortar containing 3% nano CuO particle, respectively, demonstrated over 60% lesser capillary permeability and over 9% lesser water absorption than that of normal mortar. Moreover, the mortar blended 4% nano CuO particle showed over 275% growth in the electrical resistivity and decreasing over 44% in the chloride permeability (Rahmat et al., 2015). The supplement of 1% NC particle increased the compressive strength and chloride penetration resistance more than over 17% and 20%, respectively, and decreased the water absorption and volume of permeable voids more than over 19% and 46%, respectively (Shaikh and Supit, 2014). These improvements could be mainly referred to as the specific surface area of nanoparticles. The great specific surface area of nanoparticle as admixture not only fills the tiny voids in the cement-based materials but also plays a key role in accelerating the hydration process resulting in more consumption of portlandite ($Ca(OH)_2$) and making greater quantity of C–S–H gel (Decheng et al., 2013). On the other hand, the following researchers proved that nano Al_2O_3 particle increased the compressive strength of cement-based materials (Jian et al., 2019). As the Al_2O_3 nanofiber was added with a dosage of 0.25% by cement weight, this addition improved the compressive strength of cement-based materials by up to 30% (Scott et al., 2019). Experimental findings have shown that the 7th-day and 28th-day compressive strength and bending moment strength of the cement mortar mixed with the nanoparticles of Fe_2O_3 and SiO_2 were higher than that of plain cement mortar in a paper published by Hui et al. (2004). Wengui et al. (2015) evaluated the physical and mechanical characteristics of ultrahigh-performance concrete (UHPC), which was mixed with nanoparticles of limestone (NC) and SiO_2; and type I plain Portland cement, fly ash, and silica fume were constituent mixing materials. The percentages of nano SiO_2 particle and NC by cement weight were, respectively, changed between 0.5%, 1.0%, 1.5%, and 2.0% and 2.0%, 3.0%, and 4.0%. The workability of the mixture was decreased with respect to the control concrete, and it was fixed when the percentage of NC was increased. The authors took into account their results that the small size of nanoparticle was located on the surface and left less water to contribute to the workability. Therefore, the compressive strength and splitting tensile strength of concrete containing both NC and nano SiO_2 were improved with respect to concrete without any nanoparticle and SCM additions (Wengui et al., 2015). The review effort would present the last study related to the effect of nano Fe_2O_3 particles on the physical and mechanical characteristics of cement-based materials. Fang et al. (2017) determined the mechanical characteristics of cement-based materials with different additions, percentage of nano Fe_2O_3 particles for example, 3%, 5%, and 10% by cement weight at water curing ages of 7, 14, and 28 days. As the nano Fe_2O_3 content was increased, the surface morphology was denser in the cement-based material. For all measurements and every percentage, the supplement of nano Fe_2O_3 particles developed the compressive strength of cement mortar when compared to the control mortar. For the best compressive strength gain, the authors suggested that there was a need of using 10% of nano Fe_2O_3 particle supplement.

24.3.2 Microscale features of products

There are many flaws in the microscale features of ordinary cement-based material, which would have an unfavorable effect on the physical and mechanical properties. Recently, many types of research have been conducted on the microscale features of cement modified through blending mineral supplements and nanoparticles, and have succeeded in important advances (He et al., 2012). With the development of nanotechnology, researchers understood that nanomaterials were practical materials with many perfect properties such as size effect, quantum effect, surface effect, and microscale effect (Norhasri et al., 2017). These properties can improve the microscale features of cement, and open new fields for cement research. Many researchers used such nanomaterials as nano SiO_2, nano TiO_2 (Jianping et al., 2015), NC (.), nano Al_2O_3 (Salim et al., 2014), and CNT (Florence & Konstantin, 2010) as an additive, and have found some remarkable results. For instance, nano SiO_2 particle has more advantages, as it has a greater pozzolanic effectivity than that of other nanomaterials. The nano SiO_2 particle has a retardation effect on the microscale feature of cement paste. Nevertheless, Zhu et al. (2015) and Yu et al. (2018) reported that nano SiO_2 particle developed the hydration process of cement and increased more C-S-H gel formation (He et al., 2012). Similar to Zhu et al. (2015) and Yu et al. (2018) results were reported that when a current nanoparticle employed in cement-based material is nano-silica, it accelerates the hydration of cement to generate more calcium-silicate-hydrate (C-S-H). Additionally, this nano SiO_2 particle led to the acceleration of hydration of cement because it acted as a seed for the nucleation of the C-S-H gel (Scott et al., 2019). As mentioned in Introduction, smaller cement particles result in better hydration process and create less large ettringite crystals and amorphous calcium hydrate stack (Parang et al., 2014). For transforming standard cement particles to cement nanoparticle, a planetary ball milling (PBM) instrument has to be used. The PBM granulates the particle of cement by using the steel balls that undergo a rotational motion within the container of PBM and/or the bowl of milling (Ramezan & Neitzert, 2012).

Another study with the scanning electron miscroscopy (SEM) monitoring of the microscale features among the cement mortars made of the nanoparticles and the plain cement revealed that the nanoparticles of Fe_2O_3 and SiO_2 reduced the portlandite ($Ca(OH)_2$) compound between cement hydrates (Xiaoyan et al., 2012). However, it was pointed out that the mixing of nanoparticle of SiO_2 led to the pozzolanic reaction effectively when compared to the mixing of silica fume. This mechanism could be examined by the fact that the nanoparticle of SiO_2 not only function as an enhancer in the microstructure of the mortar but also function as an activator in the reaction of pozzolanic (Byung-Wan et al., 2007). Moreover, it was taken into account that summarizing the most comprehensive research outcome on the use of nanoparticle for cement-based material was reported in the study performed by Sobolev and Flores (2008). The most current results regarding the microscale features are as follows: (1) The nanoparticle fills up microvoids between cement grains so that it could increase the density. (2) A well-dispersed nanoparticle stack boosters for crystallization between the cement hydrates, therefore it accelerates the hydration process of cement-based material (Sobolev and Flores, 2008).

Fang et al., (2017) did present another important study to measure the mechanical and microscale features of cement-based materials containing different supplements percentage of nano Fe_2O_3 particles, such as 3%, 5%, and 10% by cement weight, at ages of 7, 14, and 28 days.

With the increase in nano Fe_2O_3 particle supplement, Fang et al. (2017) revealed an important result that the microscale surface morphology of cement-based material was denser compared to reference cement-based material. Moreover, Liu et al. (2012) pointed that at early ages, the use of nano Al_2O_3 particle supplement could accelerate advances in the C–S–H gel formation, which improves the strength and microscale features of cement-based material (Jintao et al., 2015). Salim et al., (2014) supported aforementioned result explained by Liu et al. Their work examined the effect of the nano Al_2O_3 particle supplement on the microscale features of the cement paste hydrated at 7 days age. Plain Portland cement was replaced with nano Al_2O_3 powder in 2% and 4% by cement weight, and its water-to-cement ratio was kept constant at 0.4. At an early age, in light of their x-ray diffraction (XRD) analysis, Barbhuiya et al. (2014) found no novel mineral phase generated by the supplement of nano Al_2O_3 particle. As Liu et al. (2012) they also reported that the nano Al_2O_3 supplement led to the generation of dense microscale features with larger amorphous bulk of portlandite in the cement paste. Similar to Liu et al. (2012) and Barbhuiya et al. (2014), Decheng et al. (2013) examined the microscale features of concrete made of nano TiO_2 particle supplement as well as the mechanical properties of the cement paste. And their result supported the microscale result of Liu et al. (2012) and Barbhuiya et al. (2014). Ehsan et al. (2016) worked on the effects of nano Al_2O_3 particle and rice husk ash in polypropylene fiber reinforced cement mortar. This addition of nano Al_2O_3 particle generated a denser microstructure

Table 24.2 The properties of nanomaterial used for cement-based material in the different applications.

Properties	Type of nanomaterials	
	Nano SiO_2	Nano Cu
Molecular weight (g/mol)	60.1	–
Melting point (°C)	1750	–
Boiling point (°C)	2355	–
Density (g/cm^3)	2.65	–
Heat capacity (J/g)	937	–
Thermal conductivity (W/mK)	1.3	–
Purity	> 99	> 99
Surface area (m^2/g)	300	200

as well as 20% and 41% greater compressive and bending moment strength at 90 days, respectively.

24.3.3 Special properties of products

Nanoparticles could develop special properties of cement-based material such as an interfacial transition zone (ITZ) between the aggregate stack and the cement matrix, durability, electrical resistance, self-cleaning feature, and overall quality of the cement matrix. Therefore, this section would summarize the effect of nanomaterials on the above-mentioned feature of cement-based materials from the current scientific literature. In the study carried out by Hui et al. (2004), the nanoparticles of Fe_2O_3 and SiO_2 filled up pores in the ITZ between the aggregate stack and the cement matrix. In addition, it was taken into account that a summary of the most comprehensive research outcome on the use of nanoparticle for cement-based material was reported in the study published by Sobolev and Flores (2008). The most current results regarding the special properties are as follows: (1) The ITZ between the aggregate stack and the cement matrix is improved by the nanoparticles, leading to better bonding between the aggregate stack and cement matrix. (2) The shear, toughness, tensile, and bending moment strength of cement-based material is enhanced by the cracking arrest and interlocking effect between the slip planes provided by the nanoparticle. Table 24.2 reports the properties of nanomaterial used for cement-based material in the different applications

Additionally, many types of research were published related to the effect of nano SiO_2 particle on the special property (such as durability) of cement-based material, and the results pointed that a 4% of 120 (nm) nano SiO_2 particle tremendously enhanced the durability of cement-based material (M. Ehsan et al., 2016). Nano SiO_2 particle raised the heat of hydration of cement paste. Its effect on the heat of hydration was obvious that as the dosage of nano SiO_2 particle increased from 3% to 5%, the heat of hydration was increased from 0.002 W/g to 0.003 W/g in the second exothermic peak. Moreover, the formation of the second exothermic peak was nearly 2.5 h earlier

(Deyu et al., 2013). On the other hand, nano Al_2O_3 particle is another nanomaterial that is also currently used as an additive in the cement-based materials, and its specific surface area is very close to the nano SiO_2 particle. Zhou et al. (2019) also reported that the early reaction of nano Al_2O_3 particle with the hydration products of cement caused a greater ettringite formation at the initial hydration stance (Jin et al., 2019). SEM micrograph pointed out the formation of a much denser microstructure with the addition of nano Al_2O_3 particle when compared with plain cement paste. Xiaoyan et al. (2012) determined that NCNC particle has a tiny bit of influence on the water demand for standard consistency of cement. However, as the NC content was increased from 0 wt% to 2 wt% in cement paste mixing, its addition reduced both flow setting time of cement paste significantly. Another special property improved by the nano CuO particle was the thermal properties of cementitious materials. Nazari et al. (2010) showed the effect of nano CuO particle on the thermal properties of self-compacting concrete (SCC). Their results pointed out that the thermal properties of SCC increased with the addition of nano CuO particle, up to 4%. Mehrinejad et al. (2015) carried out a laboratory-based work on the durability of self-compacting mortar containing nano CuO particle and fly ash. They demonstrated that the addition of 4% nano CuO particle and 25% fly ash led to the best results for electrical resistivity and chloride permeability in the self-compacting mortar. Moreover, it can be inferred that the addition of up to 3% NC to mortar mixing is suggested by Nazari et al. (2010) because they reported results similar to Mehrinejad et al. (2015). In view of the ITZ, nanoparticles enhanced the contact field of aggregate, thus they could make a proper bonding structure between aggregate and cement paste, and result in an increase in copacity at any ITZ between aggregate and cement paste. It was also established that the addition of 3% nano CuO and 10% metakaolin improved a unit volume mass and quality conformity of cement-based materials. The results might strongly depend on the boosted cement hydration which incorporates surplus silicon and alumina content in the metakaolin and great free energy of nano CuO particle. Florence & Konstantin (2010) mixed Portland cement with nano SiO_2 particle to enhance both the rheological performance and the durability of cement-based materials; a nano SiO_2 content of 0.25% by weight of cement. Macquarie & Ahmed (2015) established the durability feature of concrete with the addition of nano SiO_2 and high volume fly ash. They used plain Portland cement and various series of mixtures with a water-to-cement ratio of 0.40. The addition of 4% nano SiO_2 reduced the water absorption of concrete when compared to concrete without nano SiO_2 particle. They studied the resistance of chloride penetration at the ages of 28 and 90 days, and the concrete mixing with 2% nano SiO_2 showed the lowest penetration value. Peng et al. (2019) evaluated the durability of concrete including nano SiO_2 particles and steel-based fiber. They prepared concrete samples with cement; 1%, 3%, 5%, 7%, and 9% of nano SiO_2 particles; and 0.5%, 1%, 1.5%, 2% and 2.5% of steel fiber. The tested properties were the resistance of carbonation and cracking, and resistance of permeability and freezing-thawing as the durability tests. As the nano SiO_2 particles dosage was increased from 1% to 7%, the number of cracks was decreased in the concrete. And the lowest number of crack was achieved with a 7% nano SiO_2 particles dosage, but this number is increased as the addition of nano SiO_2 particle is up to 9%. Based on their results, the addition of nano SiO_2 improved the durability of concrete as the nano SiO_2 particle content did not exceed 9% by weight of cement in the concrete mixing.

Rashad (2013) conducted a review study on the influences of nano Fe_2O_3 particle, nano Al_2O_3 particle, nano Fe_3O_4 particle, and nano clay particle on properties of cement such as hydration heat, water absorption, workability, setting time, and durability. In the light of results, it was concluded that the addition of inclusion of nano Fe_2O_3 in the cement matrix reduced the adsorption of water and heat growth as well as boosted the time of heat peak, and the workability of the concrete was decreased as the nano Fe_2O_3 particle content was increased. Nazari et al. (2010) reported workability measured from a concrete mixing, in which cement was partially substituted with 0.5%, 1%, 1.5%, and 2% nano Fe_2O_3 particle by cement weight and its water-to-binder ratio was constant as 0.4. The workability of concrete decreased as the nano Fe_2O_3 particle content is increased in the concrete mixing. In addition to the workability, K. Ali et al., (2012) presented two mathematical models using genetic programming and artificial neural networks to estimate the percentage of water adsorption and splitting tensile strength from concrete including nano Fe_2O_3 particle supplement. Another special feature provided by the addition of nano TiO_2 particle for cement-based material is self-cleaning property. The nano TiO_2 particle could let photocatalytic deterioration of pollutants such as VOCs, CO, NOx, aldehydes, and chlorophenols generated by the emissions of industries and automobiles. However, aging decreases the efficiency of the feature because of the carbonization of cement-based materials was encountered (Aslani, 2015). This feature depends on a nano TiO_2 thin film that actively releases oxygen on the surface of cement-based materials under the Ultra Violet light provided by sunlight. Therefore, its mission is to catalyze the deterioration of organic matters placed at the surface of cement-based materials coated by nano TiO_2 particle (Singh et al., 2017). The rainy weather, that could help prevent the makeup of dirt, also cleans the surface of cement-based materials. Moreover, nano TiO_2 particles could improve the resistance of water permeability of cement-based materials as well (Silvestre et al., 2016). Except for the contribution of nano TiO_2 particle on the physical and mechanical properties of cement-based materials, Li et al., (2018) evaluated the effect of various contents of nano TiO_2 particle on the hydration feature of cement paste under curing temperatures of 0, 5, 10, and 20 °C. They mixed natural river sand, plain Portland cement (type I), and nano TiO_2 nanoparticle dosages of 1%, 2%, 3%, 4%, and 5% by cement weight with the size of 15 nm to product mortar. For the mortar mixing, the water-to-binder ratio was constant as 0.5. The hydration degree of the mortars improved through the addition of the nano TiO_2 particle dosage. This is due to the nano TiO_2 particle supplying extra space for the formation of novel and existing hydration products. New hydration products and the C–S–H gels filled pores in cement paste cause a greater enhanced strength for mortar. Additionally, the nano TiO_2 particle, which has a very large surface area, lets an extra surface field to form novel hydration products. And, the nano TiO_2 particle makes a strong bond between hydration products that enhances the strength properties of the mortar (Li et al., 2018). Mostafa et al. (2013) worked on the special feature of high resistance in SCC made of the addition of fly ash and nano TiO_2 particles. They prepared the self-compacted concrete with 80% plain Portland cement, 15% fly ash, and 5% nano TiO_2 particle. As the 5% nano TiO_2 particle was added into the SCC, it enhanced the consistency and reduced the segregation in the SCC. Moreover, the nano TiO_2 particle decreased the water absorption and capillarity of SCC significantly. Zhiqiang et al., (2019) published an important article regarding the advances in concrete microstructure along with the nano TiO_2 particles supplement. Harmful gases in the air are decomposed by the catalyzer mission of the nano TiO_2 particles supplement as being in the studies conducted. In addition to the catalyzer effect of nano TiO_2 particle supplement, they also evaluated that change in temperature of cement paste containing nano TiO_2 particle could reduce cracks and boost the hydration reaction as evaluated by the durability of UHPC containing nano TiO_2 particle supplement. This addition of 1% nano TiO_2 particle enhanced the durability of UHPC as well as its mechanical properties. They examined the effects of 1% nano TiO_2 particle supplement on the dry shrinkage, carbonation resistance, freeze-thaw resistance, and resistance to chloride ingress of UHPC. The addition of nano TiO_2 particle allowed the UHPC a self-cleaning and photocatalytic behavior. In addition, the nano TiO_2 particle supplement decreased the capillary porosity in the UHPC. The review would present the last study related to the effect of nano Al_2O_3 particle on the chloride-bonding capacity of cement paste from Zhiqiang et al., (2019). The authors evaluated the effect of nano Al_2O_3 particle on the chloride-bonding capacity of cement paste. These samples were made of nano Al_2O_3 particles dosages of 0.5%, 1.0%, 3.0%, and 5%. They examined the chloride-bonding capacity using traditional equilibrium tests, in which the samples were, respectively, subjected to a NaCl solution at 0.05 mol/L, 0.1 mol/L, 0.3 mol/L, 0.5 mol/L, and 1.0 mol/L. Based on their results, the chloride-bonding capacity showed an increase of 37.2% at 0.05 mol/L NaCl solution with the addition of a 5% nano Al_2O_3 particle. Therefore, they suggested that the 5% addition of nano Al_2O_3 is an optimum percent to enhance the chloride-bonding capacity of cement paste in the NaCl medium.

24.4 Conclusion and future perspectives

In the quest for ensuring sustainable development and eco-friendly construction, various materials have been persistently investigated for use in the production of cementitious composites. In this chapter, discussions have been based on the engineered uses of nanomaterials for sustainable cementitious composites. The study has highlighted the properties of the products, and the advances in the use of nanomaterials, their effect on strength and microscale features. A further detailed evaluation of the composite developed with nano contents can be based on durability features, environmental impact, and life cycle assessments. The chapter sections have discussed various innovative ways of how nanomaterials can be used to improve the performance of construction materials such as cementitious composites. Moreover, the mechanism in which various types of nanomaterials interact and influence the resulting performance of composites has been relayed. Overall, the discussions presented in this study will be a very useful resource for stakeholders in the construction industry exploring innovative ways to improve the performance of cementitious composites. The limitations with the use of nanomaterials in construction materials discussed in this study will also gear more research and development in improving the application of nanomaterials in various construction materials.

References

Adeyemi, A., 2019. Durability enhancement of concrete using nanomaterials: an overview. Mater. Sci. Forum 967, 221–227. https://doi.org/10.4028/www.scientific.net/msf.967.221.

Adeyemi, A., 2020. Nanomaterials in cementitious composites: review of durability performance. J. Build. Pathol. Rehab 5, 21. https://doi.org/10.1007/s41024-020-00089-9.

Aihong, L., Kangning, S., Jiafeng, Y., Dongmei, Z., 2008. Toxicological effects of multi-wall carbon nanotubes in rats. J. Nanoparticle Res. 10, 1303–1307. https://doi.org/10.1007/s11051-008-9369-0.

Ali, K., Ali, N., Gholamreza, K., 2012. Effects of Fe_2O_3 nanoparticles on water permeability and strength assessments of high strength self-compacting concrete. J. of Mater. Sci. Technol 28 (1), 73–82. https://doi.org/10.1016/s1005-0302(12)60026-7.

Al-Najjar, Y., Yeşilmen, S., Majeed Al-Dahawi, A., Şahmaran, M., Yildirim, G., Lachemi, M., Amleh, L., 2016. Physical and chemical actions of nano-mineral additives on properties of high-volume fly ash engineered cementitious composites. ACI Mater. J. 113 (6), 791–801. https://doi.org/10.14359/51689114.

Amin, N., Alireza, N., 2019. Study on mechanical properties of ternary blended concrete containing two different sizes of nano-SiO_2. Compos. B Eng. 167, 20–24. https://doi.org/10.1016/j.compositesb.2018.11.136.

Arnon, C., Thanongsak, N., Watcharapong, W., Pincha, T., 2010. Compressive strength and microstructure of carbon nanotubes–fly ash cement composites. Mater. Sci. Eng. A 527 (4-5), 1063–1067. https://doi.org/10.1016/j.msea.2009.09.039.

Aslani, F., 2015. Nanoparticles in self-compacting concrete—a review. Mag. Concr. Res. 67 (20), 1084–1100. https://doi.org/10.1680/macr.14.00381.

Barbhuiya, S., Mukherjee, S., Nikraz, H., 2014. Effects of nano-Al_2O_3 on early-age microstructural properties of cement paste. Construction and Building Materials 52, 189–193. https://doi.org/10.1016/j.conbuildmat.2013.11.010.

Björnström, J., Martinelli, A., Matic, A., Börjesson, L., Panas, I., 2004. Accelerating effects of colloidal nano-silica for beneficial calcium–silicate–hydrate formation in cement. Chem. Phys. Lett. 393 (1-3), 242–248. https://doi.org/10.1016/j.cplett.2004.05.071.

Byung-Wan, J., Chang-Hyun, K., Ghi-ho, T., Jong-Bin, P., 2007. Characteristics of cement mortar with nano-SiO_2 particles. Constr. Build. Mater. 21 (6), 1351–1355. https://doi.org/10.1016/j.conbuildmat.2005.12.020.

Decheng, F., Ning, X., Chunwei, G., Zhen, L., Huigang, X., Hui, L., Xianming, S., 2013. Portland cement paste modified by TiO_2 nanoparticles: a microstructure perspective. Ind. Eng. Chem. Res. 52 (33), 11575–11582. https://doi.org/10.1021/ie4011595.

Deyu, K., Yong, S., Xiangfei, D., Yang, Y., Su, W., P., S.S, 2013. Influence of nano-silica agglomeration on fresh properties of cement pastes. Constr. Build. Mater. 43, 557–562. https://doi.org/10.1016/j.conbuildmat.2013.02.066.

Ehsan, M., Mehrinejad, K.M., Farzad, N., Maryam, M., Prabir, S., 2016. Polypropylene fiber reinforced cement mortars containing rice husk ash and nano-alumina. Constr. Build. Mater. 111, 429–439. https://doi.org/10.1016/j.conbuildmat.2016.02.124.

Enrico, B., Craig, P., Irina, G.C., Adriele, P.-M., 2015. The role of biological monitoring in nano-safety. Nano Today 10 (3), 274–277. https://doi.org/10.1016/j.nantod.2015.02.001.

Fang, Y., Sun, Y., Lu, M., Xing, F., Li, W., 2017. Mechanical and pressure-sensitive properties of cement mortar containing nano-Fe_2O_3 BT. In: Proc. 4th Annual International Conference on Material Engineering and Application.

Florence, S., Konstantin, S., 2010. Nanotechnology in concrete—a review. Constr. Build. Mater., 24, pp. 2060–2071.

Gengying, L., 2004. Properties of high-volume fly ash concrete incorporating nano-SiO_2. Cem. Concr. Res. 34 (6), 1043–1049. https://doi.org/10.1016/j.cemconres.2003.11.013.

Grassian, V.H., O'Shaughnessy, P.T., Adamcakova-Dodd, A., Pettibone, J.M., Thorne, P.S, 2007. Inhalation exposure study of titanium dioxide nanoparticles with a primary particle size of 2 to 5 nm. Environ. Health Perspect. 115 (3), 397–402. https://doi.org/10.1289/ehp.9469.

Hakamy, A., Shaikh, F.U.A., Low, I.M., 2015. Characteristics of nanoclay and calcined nanoclay-cement nanocomposites. Compos. B Eng. 78, 174–184. https://doi.org/10.1016/j.compositesb.2015.03.074.

Han, B., Yu, X., Ou, J., 2014. Self-Sensing Concrete in Smart Structures. Elsevier Inc., Amsterdam, pp. 1–385.

Hashimoto, K., Irie, H., Fujishima, A., 2007. TiO_2 photocatalysis: a historical overview and future prospects. Jpn. J. Appl. Phys., 44, p. 8269.

He, Z., Liu, J., Zhu, K., 2012. Influence of mineral admixtures on the short and long-term performance of steam-cured concrete. Energy Procedia 16, 836–841. https://doi.org/10.1016/j.egypro.2012.01.134.

Hui, L., Hui-gang, X., Jie, Y., Jinping, O., 2004. Microstructure of cement mortar with nano-particles. Compos. B Eng., 35, pp. 185–189.

Imanian, G.S., Maisam, J., Sadegh, S., Habibnejad, K.A., 2020. A comparative study on the mechanical, physical and morphological properties of cement-micro/nanoFe_3O_4 composite. Sci. Rep. 10, 2859. https://doi.org/10.1038/s41598-020-59846-y.

Jian, Z.B., Xing, X.D., Sun, P.C., 2019. The effect of nanoalumina on early hydration and mechanical properties of cement pastes. Constr. Build. Mater. 202, 169–176. https://doi.org/10.1016/j.conbuildmat.2019.01.022.

Jianping, Z., Chunhua, F., Haibin, Y., Zhanying, Z., Shah, S.P, 2015. Effects of colloidal nanoBoehmite and nanoSiO_2 on fly ash cement hydration. Constr. Build. Mater. 101, 246–251. https://doi.org/10.1016/j.conbuildmat.2015.10.038.

Jintao, L., Qinghua, L., Shilang, X., 2015. Influence of nanoparticles on fluidity and mechanical properties of cement mortar. Constr. Build. Mater. 101, 892–901. https://doi.org/10.1016/j.conbuildmat.2015.10.149.

Laila, R., James, B., Rouhollah, A., Jon, M., Taijiro, S., 2010. Cement and concrete nanoscience and nanotechnology. Materials 3 (2), 918–942. https://doi.org/10.3390/ma3020918.

Land, G., Stephan, D., 2012. The influence of nano-silica on the hydration of ordinary Portland cement. J. Mater. Sci. 47 (2), 1011–1017. https://doi.org/10.1007/s10853-011-5881-1.

Li, W., Hongliang, Z., Yang, G., 2018. Effect of TiO_2 nanoparticles on physical and mechanical properties of cement at low temperatures. Adv. Mater. Sci. Eng. 2018, 8934689. https://doi.org/10.1155/2018/8934689.

Lin, D.F., Lin, K.L., Chang, W.C., Luo, H.L., Cai, M.Q., 2008. Improvements of nano-SiO_2 on sludge/fly ash mortar. Waste Manage. 28 (6), 1081–1087. https://doi.org/10.1016/j.wasman.2007.03.023.

Liou, S.H., Tsai, C.S.J., Pelclova, D., Schubauer-Berigan, M.K., Schulte, P.A., 2015. Assessing the first wave of epidemiological studies of nanomaterial workers. J. Nanoparticle Res. 17 (10), 413. https://doi.org/10.1007/s11051-015-3219-7.

Liu, X., Chen, L., Liu, A., Wang, X., 2012. Effect of nano-$CaCO_3$ on properties of cement paste. Energy Procedia 16, 991–996. https://doi.org/10.1016/j.egypro.2012.01.158.

Macquarie, S.S.W., Ahmed, S.F.U., 2015. Durability properties of high volume fly ash concrete containing nano-silica. Mater. Struct. 48, 2431–2445. https://doi.org/10.1617/s11527-014-0329-0.

Mao-hua, Z., Hui, L., 2011. Pore structure and chloride permeability of concrete containing nano-particles for pavement. Constr. Build. Mater. 25, 608–616. https://doi.org/10.1016/j.conbuildmat.2010.07.032.

María, D.-S.B., Dolores, M.-A.M., Mónica, L.-A., 2019. Potential risks posed by the use of nano-enabled construction products: a perspective from coordinators for safety and health matters. J. Clean. Prod. 220, 33–44. https://doi.org/10.1016/j.jclepro.2019.02.056.

Maynard, A.D., Kuempel, E.D., 2005. Airborne nanostructured particles and occupational health. J. Nanoparticle Res. (7) 587–614. https://doi.org/10.1007/s11051-005-6770-9.

Mehrinejad, K.M., Ehsan, M., Ali, Y.M., Prabir, S., Mohammad, R.M., 2015. Effect of nano-CuO and fly ash on the properties of self-compacting mortar. Constr. Build. Mater. 94, 758–766. https://doi.org/10.1016/j.conbuildmat.2015.07.063.

Meyer, C., 2009. The greening of the concrete industry. Cem. Concr. Compos. 31 (8), 601–605. https://doi.org/10.1016/j.cemconcomp.2008.12.010.

Mostafa, J., Alireza, P., Fasihi, H.O., Davoud, J., 2015. Comparative study on effects of Class F fly ash, nano silica and silica fume on properties of high performance self compacting concrete. Constr. Build. Mater. 94, 90–104. https://doi.org/10.1016/j.conbuildmat.2015.07.001.

Mostafa, J., Mojtaba, F., Mohammad, F., 2013. Effects of fly ash and TiO_2 nanoparticles on rheological, mechanical, microstructural and thermal properties of high strength self-compacting concrete. Mech. Mater. 61, 11–27. https://doi.org/10.1016/j.mechmat.2013.01.010.

Nazari, A., Riahi, S., Riahi, S., Shamekhi, S.F., Khademno, A., 2010. Improvement the mechanical properties of the cementitious composite by using TiO2 Nanoparticles. J. Am. Sci. 6(4), 98–101.

Norhasri, M.S.M., Hamidah, M.S., Fadzil, A.M., 2017. Applications of using nano material in concrete: a review. Constr. Build. Mater. 133, 91–97. https://doi.org/10.1016/j.conbuildmat.2016.12.005.

Pacurari, M., Castranova, V., Vallyathan, V., 2010. Single- and multi-wall carbon nanotubes versus asbestos: are the carbon nanotubes a new health risk to humans? J. Toxicol. Environ. Health 73 (5–6), 378–395. https://doi.org/10.1080/15287390903486527.

Parang, S., Frisky, S., Aji, F.D., 2014. The effect of nano-cement content to the compressive strength of mortar. Procedia Eng. 95, 386–395. https://doi.org/10.1016/j.proeng.2014.12.197.

Peng, Z., Qingfu, L., Yuanzhao, C., Yan, S., Yi-Feng, L., 2019. Durability of steel fiber-reinforced concrete containing SiO_2 nano-particles. Materials 12 (13), 2184. https://doi.org/10.3390/ma12132184.

Pengkun, H., Shiho, K., Deyu, K., J., C.D., Jueshi, Q., P., S.S, 2013. Modification effects of colloidal nanoSiO_2 on cement hydration and its gel property. Compos. B Eng. 45 (1), 440–448. https://doi.org/10.1016/j.compositesb.2012.05.056.

Rahmat, M., Ehsan, M., Yasin, M.S., Maryam, N., 2015. An experimental investigation on the durability of self-compacting mortar containing nano-SiO_2, nano-Fe_2O_3 and nano-CuO. Constr. Build. Mater. 86, 44–50. https://doi.org/10.1016/j.conbuildmat.2015.03.100.

Ramezan, M., Neitzert, T., 2012. Mechanical milling of aluminum powder using planetary ball milling process. J. Achievements Mater. Manufactur. Eng. 55, 790–798.

Rashad, A.M., 2013. A synopsis about the effect of nano-Al_2O_3, nano-Fe_2O_3, nano-Fe_3O_4 and nano-clay on some properties of cementitious materials – a short guide for civil engineer. Mater. Des. 52, 143–157. https://doi.org/10.1016/j.matdes.2013.05.035.

Salim, B., Shaswata, M., Hamid, N., 2014. Effects of nano-Al_2O_3 on early-age microstructural properties of cement paste. Constr. Build. Mater. 52, 189–193. https://doi.org/10.1016/j.conbuildmat.2013.11.010.

Samuel, C., Zhu, P., G., S.J., Ming, W.C., Hui, D.W, 2014. Nano reinforced cement and concrete composites and new perspective from graphene oxide. Constr. Build. Mater. 73, 113–124. https://doi.org/10.1016/j.conbuildmat.2014.09.040.

Sato, T., Beaudoin, J.J., 2011. Effect of nano-$CaCO_3$ on hydration of cement containing supplementary cementitious materials. Adv. Cem. Res. 23 (1), 33–43. https://doi.org/10.1680/adcr.9.00016.

Scott, M., Ismael, F.-V., Konstantin, S., 2019. Ultra-high strength cement-based composites designed with aluminum oxide nano-fibers. Constr. Build. Mater. 220, 177–186. https://doi.org/10.1016/j.conbuildmat.2019.05.175.

Shaikh, F.U.A., Supit, S.W.M., 2014. Mechanical and durability properties of high volume fly ash (HVFA) concrete containing calcium carbonate ($CaCO_3$) nanoparticles. Constr. Build. Mater. 70, 309–321. https://doi.org/10.1016/j.conbuildmat.2014.07.099.

Siad, H., Lachemi, M., Sahmaran, M., Mesbah, H.A., Hossain, K.A., 2018. Advanced engineered cementitious composites with combined self-sensing and self-healing functionalities. Constr. Build. Mater. 176, 313–322. https://doi.org/10.1016/j.conbuildmat.2018.05.026.

Silvestre, J., Silvestre, N., De Brito, J, 2016. Review on concrete nanotechnology. Eur. J. Environ. Civil Eng. 20 (4), 455–485. https://doi.org/10.1080/19648189.2015.1042070.

Singh, N.B., Kalra, M., Saxena, S.K., 2017. Nanoscience of cement and concrete, 4, pp. 5478–5487.

Sivakrishna, A., Adesina, A., Awoyera, P.O., Kumar, K.R., 2019. Green concrete: a review of recent developments, 27, pp. 54–58.

Sobolev, K., 2015. Nanotechnology and nanoengineering of construction materials. In: Nanotechnology in Construction. Springer International Publishing. https://doi.org/10.1007/978-3-319-17088-6_1.

Sobolev, K., Flores, I., Hermosillo, R., Torres-Martínez, L.M., 2008. Nanomaterials and nanotechnology for high-performance cement composites. ACI Mater. J., 254. American Concrete Institute, p. 254.

Tsuji, J.S., Maynard, A.D., Howard, P.C., James, J.T., Lam, C.W., Warheit, D.B., Santamaria, A.B., 2006. Research strategies for safety evaluation of nanomaterials, part IV: risk assessment of nanoparticles. Toxicol Sci. 89 (1), 42–50. https://doi.org/10.1093/toxsci/kfi339.

Vignes, A., Muñoz, F., Bouillard, J., Dufaud, O., Perrin, L., Laurent, A., Thomas, D., 2012. Risk assessment of the ignitability and explosivity of aluminum nanopowders. Process Saf. Environ. Protect. 90 (4), 304–310. https://doi.org/10.1016/j.psep.2011.09.008.

Walters, D.A., Ericson, L.M., Casavant, M.J., Liu, J., Colbert, D.T., Smith, K.A., Smalley, R.E., 1999. Elastic strain of freely suspended single-wall carbon nanotube ropes. Appl. Phys. Lett. 74 (25), 3803–3805. https://doi.org/10.1063/1.124185.

Wengui, L., Zhengyu, H., Fangliang, C., Zhihui, S., P., S.S, 2015. Effects of nano-silica and nano-limestone on flowability and mechanical properties of ultra-high-performance concrete matrix. Constr. Build. Mater. 95, 366–374. https://doi.org/10.1016/j.conbuildmat.2015.05.137.

Xiaoyan, L., Lei, C., Aihua, L., Xinrui, W., 2012. Effect of Nano-$CaCO_3$ on properties of cement paste. Energy Procedia 16, 991–996. https://doi.org/10.1016/j.egypro.2012.01.158.

Yu, X., Kang, S., Long, X., 2018. Compressive strength of concrete reinforced by TiO_2 nanoparticles. In: AIP Conference Proceedings (Vol. 2036, No. 1, p. 030006). AIP Publishing LLC. https://doi.org/10.1063/1.5075659.

Zemei, W., Caijun, S., K.H., K., Shu, W, 2016. Effects of different nanomaterials on hardening and performance of ultra-high strength concrete (UHSC). Cem. Concr. Compos. 70, 24–34. https://doi.org/10.1016/j.cemconcomp.2016.03.003.

Zhiqiang, Y., Yun, G., Song, M., Honglei, C., Wei, S., Jinyang, J., 2019. Improving the chloride binding capacity of cement paste by adding nano-Al_2O_3. Constr. Build. Mater. 415–422. https://doi.org/10.1016/j.conbuildmat.2018.11.012.

Zhou, J., Zheng, K., Liu, Z., Fuqiang, H., 2019. Chemical effect of nano-alumina on early-age hydration of Portland cement. Cem. Concr. Res. 116, 159–167. https://doi.org/10.1016/j.cemconres.2018.11.007.

Zhou, J., Zheng, K., Liu, Z., Fuqiang, H., 2019. Chemical Effect of Nano-Alumina on Early-Age Hydration of Portland Cement. Cement and Concrete Research 116, 159–167. https://doi.org/10.1016/j.cemconres.2018.11.007.

Zhu, J., Feng, C., Yin, H., Zhang, Z., Surendra, P.S., 2015. Effects of colloidal nanoboehmite and NanoSiO_2 on fly ash cement hydration. Construct. Build. Mater. 101, 246–251. https://doi.org/10.1016/j.conbuildmat.2015.10.038.

CHAPTER 25

The carbon nanomaterials with abnormally high specific surface area for liquid adsorption

Alexander V. Melezhik, Elena A. Neskoromnaya, Alexander E. Burakov, Alexander V. Babkin, Irina V. Burakova and Alexey G. Tkachev
Tambov State Technical University, Tambov, Russian Federation

25.1 Introduction

Environmental pollution, especially pollution of water bodies, is becoming a serious threat to existence, and the problem attracts the attention of experts and activists in this area. Different countries and regional politicosocial blocs create administrative bodies to save the lives of the population by regulating or preventing environmental disturbance that happens as a result of improper use or inappropriate handling of water bodies. Concerned with the accelerating growth of the world's population, continuous industrialization and active manmade human activity in the various manufacturing sector have a detrimental negative effect on natural hydrogeosystems (Burakov et al., 2019; Kang et al., 2020). In addition, the intensive development of one of the main consumers of clean freshwater is an agro-industrial complex, and a significant increase of the wastewater at agricultural in general, made ecologists worried about the management of used water resources (Liang et al., 2021).

Millions tons of wastewater is generated every day by various industries (e.g., textiles, fertilizers and agrochemicals, tanning, etc.), and a significant amount is discharged into surrounding water bodies without any primary or secondary treatment (Kashif et al., 2017; Li et al., 2020). This wastewater contains a variety of inorganic (e.g., heavy metals such as mercury (Hg), cadmium (Cd), lead (Pb), nickel (Ni), zinc (Zn), chromium (Cr), cobalt (Co), and rare-earth elements) and organic pollutants (e.g., dyes contaminated by the textile and dyeing industries, pharmaceuticals, personal care products, surfactants, phenols, pesticides, etc.) that are toxic, carcinogenic, resistant to degradation, and able to bioaccumulate (Ali et al., 2019; Awad et al., 2020; Babkin et al., 2018; Huber et al., 2016; Khan, Mubarak, Khalid, et al., 2020; Khan, Mubarak, Tan, et al., 2020; Lingamdinne et al., 2019; Ruthiraan et al., 2019; Yi et al., 2017). A significant number of research prove that the impact of many toxic pollutants contained in polluted water on the human body underlies many physiological disorders, such as infant mortality, Alzheimer's disease,

Sustainable Nanotechnology for Environmental Remediation
DOI: https://doi.org/10.1016/B978-0-12-824547-7.00016-3

carcinogenicity, neurotoxicity, reproductive toxicity and metabolic toxicity, etc. (Roop et al., 2021; Sivaranjanee & Kumar, 2021). Consequently, polluted water treatment is an urgent problem as it has a detrimental effect on people, animals, and the environment (Tang et al., 2018).

Several treatment technologies are traditionally used to remove toxic pollutants from polluted waters, such as adsorption, advanced oxidation process, membrane separation, reverse osmosis, chemical precipitation, coagulation and flocculation, ion exchange, aerobic and anaerobic degradation, electrochemical treatment, biological treatment, etc. (Ali et al., 2018; Garba et al., 2020; Lu et al., 2020; Zhou et al., 2014, 2020). Unfortunately, many of these methods have significant drawbacks, such as overuse of chemicals that is difficult to dispose of it, etc. Among all of the purification methods mentioned above, adsorption is considered the most promising because of its high efficiency, simplicity of operation, profitability, and the possibility of implementation on a large scale (Yang et al., 2019; Zhou et al., 2018). The simplicity of the equipment design, the possibility for the adsorbent reusing, recovery of the adsorbate, and the availability of raw materials are the significant advantages of the method.

Since then, various adsorption materials, such as activated carbon (Dehghani et al., 2020; Karri & Sahu, 2018; Yazidi et al., 2020), clay minerals (Chauhan et al., 2020), polymers (Huang et al., 2020; Kaur et al., 2020), agricultural waste (Ali et al., 2019; Awad et al., 2020; Babkin et al., 2018; Huber et al., 2016; Khan, Mubarak, Khalid, et al., 2020; Khan, Mubarak, Tan, et al., 2020; L.P. Lingamdinne et al., 2019; Ruthiraan et al., 2019; Yi et al., 2017), zeolites (Hong et al., 2019) and others have been studied for effective removal of pollutants from aquatic environments. However, most of these adsorbents have low efficiency of adsorption and desorption or are not selective enough for the target pollutants.

Creating the new, highly efficient materials for the purification of polluted waters is necessary to meet increasingly restrictive emission regulations for environmental friendliness and the implementation of "green" technologies and processes. Commercially available products based on graphene structures, porous carbon, and its derivatives with a high specific surface area, such as graphene aerogels (Ali et al., 2019), mesoporous carbon (A. Burakov et al., 2019; Kang et al., 2020), and composite materials based on carbon nanotubes (CNTs) (Worsley et al., 2012), are the most attractive option for the technological development of remediation processes of wastewater, high-quality water treatment, and solutions to other environmental problems. Such materials are the most attractive for sorption due to their unique physical and chemical characteristics (Burakova et al., 2018; Nasrollahzadeh et al., 2021).

The purpose of the study is to create a new effective sorption material with a broad spectrum of action (for removing organic and inorganic compounds) based on graphene structures and to study the sorption characteristics of the developed sorbent.

25.2 Materials and methods

25.2.1 Materials synthesis

The following starting materials were used in the present study: (1) urotropin $C_6H_{12}N_4$ (hexamethylenetetramine, HMTA), analytical grade; (2) sulfuric acid (95% H_2SO_4), analytical grade; (3) oleum, technical grade, with the content of 24% free SO_3. Diluted oleum containing 5% SO_3 was prepared by mixing calculated amounts of sulfuric acid and 24% oleum; (4) Multiwalled CNTs called Taunit-M were produced by "NanoTechCenter" (Tambov). Multi-walled carbon nanotubes (MWCNTs) were purified from the rest of the catalyst by treatment with hydrochloric acid; (5) potassium hydroxide, analytical grade, was in the form of granules with a KOH content of 85% (the rest is water).

The carbon-containing substance synthesis of a supposed cumulene structure with the code name "polycumulen" was carried out by analogy with the soluble substance synthesis of a supposed cumulene structure described in Melezhik et al. (2017) and Tkachev et al. (2018). However, in the present research, the heat treatment temperature was increased to 180-200 °C for deeper polymerization, which resulted in an insoluble product. The synthesis was carried out as follows. Forty-five milliliters of 5% oleum was placed into a 2-L beaker of heat-resistant glass under the exhaust ventilation, and while cooling in a bath with cold water, at continuous stirring, 30 g of HMTA were gradually added. Then the beaker was placed in a thermostatic jacket, provided with thermal insulation and a thermocouple, and slowly heated on the electric cooker. When the temperature reached 110-120 °C, an exothermic reaction took place with foaming of the reaction mixture due to the release of a certain amount of gas (analysis showed that it was mainly sulfur dioxide). The reaction mixture was kept for 2 h at a temperature of 180-200 °C. After cooling to room temperature, the product was poured with water, crushed, and washed on the filter to a neutral pH, and then dried at 110 °C to a constant weight. The amount of produced substance was 17.91 g per one synthesis with the specified load. For the production of the material, synthesis was repeated several times. The energy dispersive analysis showed the presence of 11.1% of oxygen and 15.9% of sulfur in this substance, in addition to carbon (73.0% of the mass). Further, this substance will be conventionally denoted as C-110.

The synthesis of a composite containing CNTs was carried out in a similar way. However, in the presence of CNTs, the foaming of the reaction mixture was less, which allowed the loading to be tripled (135 mL of 5% oleum, 90 g of HMTA). The initial mixing of HMTA with oleum was carried out with cooling, as described above. In this case, the reaction mixture partially crystallizes, probably due to the formation of the HMTA hydrosulfate at this stage. Then the beaker with the reaction mixture was heated until 60–70°C, the reaction mixture melted, then the viscous clear liquid was formed. In this liquid, keeping the temperature in the specified range, 12 g of purified CNTs were

Table 25.1 The nanocomposite synthesis variations.

C-CNT	C-CNT-110	C-500	C-CNT-500
45 mL 5% oleum 30 g HMTA. The heat treatment for 2 h at a temperature of 180–200 °C Drying at 110 °C to constant weight	135 mL 5% oleum 90 g HMTA 12 g CNTs. The heat treatment for 2 h at a temperature of 180–200 °C Drying at 110 °C to constant weight	C-CNT + The heat treatment at 500 °C for 4 h in a tube furnace in a flow of argon	C-CNT-110 + The heat treatment at 500 °C for 4 h in a tube furnace in a flow of argon

added and mixed thoroughly until a homogeneous paste was formed. Then the beaker was placed into a thermostatic jacket and further processing was carried out, as described above for the synthesis without the addition of CNTs. After drying to constant weight at 110 °C, 64.17 g of product was obtained, in which the calculated content of CNTs was 18.7%. Further, this substance will be referred to as C-CNT-110.

The substances C-110 and C-CNT-110 in the native form turned out to be inconvenient for activation; as their heat treatment with potassium hydroxide led to strong foaming, the reaction mixture got out of the reactor. To eliminate this effect, preliminary heat treatment of the samples was carried out for 4 h at 500 °C in a tube furnace in a flow of argon. At the same time, volatile sulfur-containing compounds were released and the sulfur mass content in the products after pyrolysis was reduced to 0.3-0.5%. The mass of products after heat treatment at 500 °C was (from the initial) 68.1% for material C and 74.2% for material C-CNT. The calculated mass content of CNTs after heat treatment of the material at 500 °C is 25.2%.

For activation after heat treatment (at 500 °C), the substance was ground to a size of particles less than 0.5 mm. Then a portion of the substance (5 g) was placed in a steel beaker with an inner diameter of 68 mm, and a height of 88 mm, and potassium hydroxide pellets were added. The glass was purged with argon and closed with a cover containing an annular sluice through which, during the activation process, argon was passed to isolate the reaction zone from the atmosphere. This system was heated in a muffle furnace to 750 °C and held for 3 h. After cooling to room temperature, the reaction mixture was poured with water, and the precipitate was washed from alkali to neutral pH, then kept for 24 h with hydrochloric acid to dissolve the impurities of the metal compounds, washed again with water and dried at 110 °C to a constant weight. Further, the activated substances will be designated as C-500-A (x: 1), C-500-CNT-A (x: 1), where x: 1 is the mass ratio of pellets of 85% potassium hydroxide to the initial substance (C-500 or C-CNT-500). Table 25.1 is provided for a better understanding of the differences in materials synthesis technologies.

25.2.2 Material properties

The nanocomposite characteristics were determined by transmission electron microscopy (TEM) by a JEM-2010 instrument (JEOL Ltd., Tokyo, Japan), and scanning electron microscopy (SEM) by a MERLIN instrument (Carl Zeiss, Jena, Germany). The structural parameters were evaluated using an Autosorb-iQ instrument (Quantachrome Instruments). Mathematical models integrated into the instrument software were used to calculate the surface and porosity parameters. As the studied materials have both micro and mesoporosity, the density-functional theory (DFT) model was used as the most suitable in this case. The variant of the DFT model was chosen based on the best-fitting comparison with the experimental isotherm. For the samples studied, this was "N_2 on carbon at 77 K, slit/cylindrical pore, non-local density functional theory (NLDFT) equilibrium model."

25.2.3 Determination of adsorbent dosage

To determine the effective adsorbent weight, 30 mL of the initial concentration 1500 mg/L methyl orange (MO) solutions at pH 3.0, and 100 mg/L $Pb(NO_3)_2$ solutions at pH 6.0 (chemically pure grade, LavernaStroyEngineering Ltd., Moscow, Russia) in deionized water were used. The adsorbent dosage was as follows: 0.03, 0.05, 0.075, and 0.1 g. The mixtures were shaken for 10 min for MO and 40 min for Pb^{2+} at 100 rpm and room temperature on a programmable rotator Multi Bio RS-24 (Biosan, Riga, Latvia). After the adsorption procedure, the dye concentration was measured at 400 nm by a spectrophotometer PE 5400V instrument (Ekros, St. Petersburg, Russia). The Pb^{2+} concentration was measured by an atomic absorption spectrometer MGA-915MD instrument (Atompribor Ltd., Saint Petersburg, Russia). For the analysis, a hollow cathode lamp was used to determine the lead concentration with a characteristic wavelength of 380 nm.

25.2.4 pH investigation

For the experiments buffer systems with pH 3-9 were used. The pH was measured on a stationary pH meter HI 2210 (Hanna Instruments Deutschland GmbH, Vohrigen, Germany). To investigate the effect of the H^+ and OH^- concentration on the MO and Pb^{2+} adsorption, 30 mL of HCl and NaOH solutions of various concentrations were prepared. The chemical reagents were purchased from Laverna Lab Ltd. (Moscow, Russia) and ChemAgent Base No.1 CJSC (Staraya Kupavna, Moscow Region, Russia).

25.2.5 Determination of equilibrium time

For the kinetic investigation, experiments were carried out at room temperature with 0.03 g of the adsorbent and 30 mL of 1500 mg/L MO and 100 mg/L $Pb(NO_3)_2$ solutions. The mixtures were shaken at 100 rpm for 5, 7, 10, 20, 30, 40, and 60 min on a rotator

Multi Bio RS-24 (Biosan) and then filtered. After the adsorption procedure, the dye concentration was measured at 400 nm by a spectrophotometer PE 5400V instrument (Ekros, St. Petersburg, Russia). The Pb^{2+} concentration was measured by an atomic absorption spectrometer MGA-915MD instrument (Atompribor Ltd., Saint Petersburg, Russia).

25.2.6 Thermodynamic study

Adsorption experiments were carried out with 0.03 g of the adsorbent and 30 mL of 150-1500 mg/L MO and 50-500 mg/L Pb^{2+} (three replicates). The contact time was 10 min for the MO solutions and 40 min for the $Pb(NO_3)_2$ solutions. The temperature of the solutions was 25, 35, and 50 °C. The samples were shaken at 100 rpm on a programmable rotator Multi Bio RS-24 (Biosan). After the adsorption procedure, the dye concentration was measured at 400 nm by a spectrophotometer PE 5400V instrument (Ekros, St. Petersburg, Russia). The Pb^{2+} concentration was measured by an atomic absorption spectrometer MGA-915MD instrument (Atompribor Ltd., Saint Petersburg, Russia). The adsorption capacity of material for the dye and Pb^{2+} ions was calculated as

$$Q_e = \frac{(C_i - C_e)}{m} V \tag{25.1}$$

where Q_e is the dye or Pb^{2+} concentration on the nanocomposite at equilibrium (mg/g); C_i is the initial dye or Pb^{2+} concentration in the liquid phase (mg/L); C_e is the equilibrium dye or Pb^{2+} concentration in the liquid phase (mg/L) ; V is the solution volume (mL) ; m is the material mass (g).

25.3 Results and discussion

25.3.1 Materials investigation

The SEM image of the surface of the nanocomposite is presented in Fig. 25.1. It can be seen that the composite has intertwined the CNTs with an average diameter of about 10-20 nm, while the CNTs are homogeneously distributed throughout the volume of the material.

According to the TEM image (Fig. 25.2), the fragments of the CNTs with a diameter of about 25 nm and inner channel of 15 nm can be seen. The CNTs save their structural integrity after KOH high-temperature activation. This method was investigated by Raymundo-Piñero et al. (2005), wherein this fact was confirmed. But, short graphene plate fragments, not referred as CNTs, can be detected.

As a rule, the carbon materials with a developed surface are produced by physical or chemical activation of carbon-containing precursor materials. Physical activation is usually carried out with water vapor, carbon dioxide, or their mixture at a

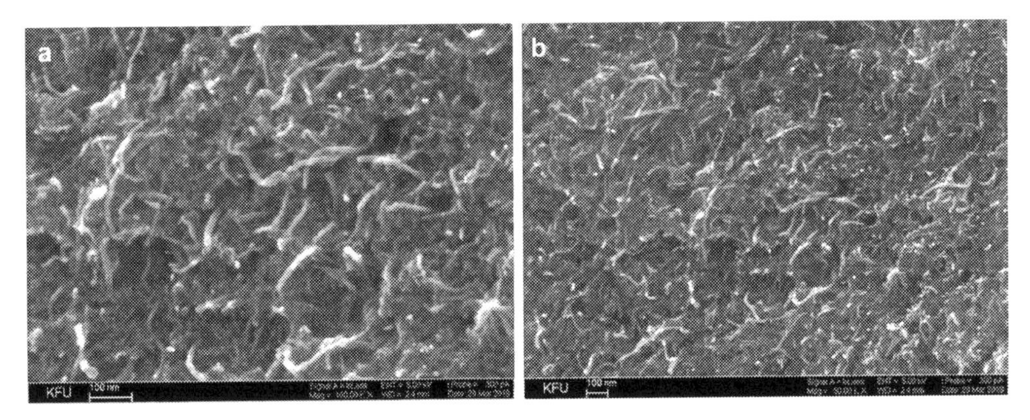

Figure 25.1 The morphology of the sorption material: SEM-image with magnification: 100.00× (A) and 50.00× (B).

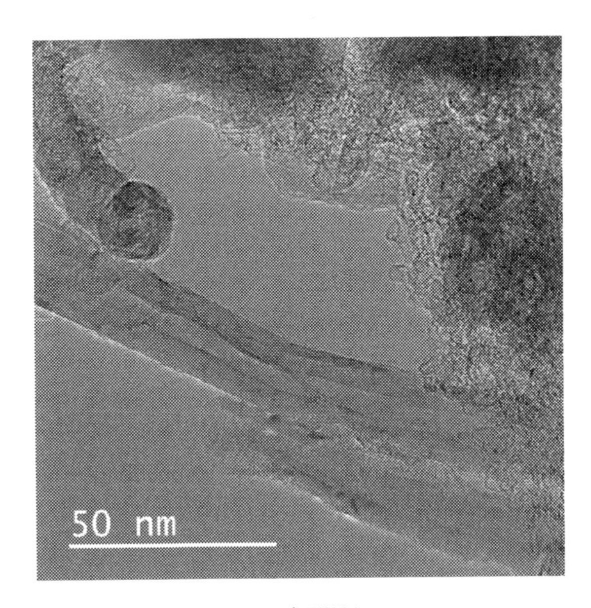

Figure 25.2 The structure of the sorption material: TEM image.

temperature of about 800–900 °C. For chemical activation, many reagents have been proposed—phosphoric acid, zinc chloride, potassium sulfide, sodium hydroxide, potassium hydroxide, potassium carbonate, and others.

Materials with the highest specific surface area (up to 3000–4000 m^2/g according to Brunauer–Emmett–Teller (BET) were obtained by activation with potassium hydroxide at a temperature of 700–900 °C. There are many research works in this area (Karri et al., 2020; Lingamdinne et al., 2020; Liu et al., 2020). The BET specific surface area values of carbon materials exceeding the specific surface area of a graphene monolayer (about 2630 m^2/g) (Marchesini et al., 2020) have been still considered as not physically realistic, to be rather effective values corresponding to the volume filling of micropores or ultramicropores. However, it was shown (Baburin et al., 2015) that, from perforated graphene layers, the structures for which a geometric surface area reaches 5100 m^2/g can be made. Microporous organometallic framework structures formed by metal ions and organic ligands, for which the effective BET specific surface area can exceed 4000 m^2/g, are also known (Shen et al., 2015).

Theoretically, the framework structures with a very high specific surface could be constructed from a pure carbon skeleton; however, such substances are not known to date. At the same time, it can be assumed, that during the chemical activation of carbon materials, the structures made of perforated graphene layers can be obtained. This assumption is based on the fact that the activation of carbon, in particular graphene materials with potassium hydroxide, begins primarily at structural defects, and, therefore, the formation of perforated and possibly skeletal structures is quite likely (Tadjenant et al., 2020). So, it is known that defective graphene, obtained by thermal or microwave decomposition of graphene oxide, can be effectively activated with potassium hydroxide (Murali et al., 2012; Zhu et al., 2011). At the same time, graphite and graphene nanoplatelets with a highly ordered structure, obtained by exfoliating graphite, as shown by our investigation, do not lend themselves to high-temperature activation with KOH, and the surface development and mass loss almost never occur. The same effect is known for CNTs—the more defective their structure, the better they can be activated with potassium hydroxide (Tkachev et al., 2017).

Earlier, we assumed that the carbon-containing substance formed during the polycondensation of HMTA in anhydrous sulfuric acid or dilute oleum (Melezhik et al., 2017; Tkachev et al., 2018) contains fragments of a polycumulene structure. The activation of such substances with potassium hydroxide has not been previously studied. It is possible that as a result of chemical activation, a framework structure, including polycumulene fragments, is formed with an irregular high specific surface.

Fig. 25.3 shows the pore size distribution for activated samples C-500-A (2: 1) and C-CNT-500-A (2: 1) synthesized as described above in the experimental part.

According to Fig. 25.3, the pore distribution of the synthesized materials is predominantly in the range up to 5 nm. At the same time, the total pore volume of the studied materials is practically identical. However, due to the decrease of the average pore size for the C-500-A material, its specific surface area is almost in two times higher (4491 m^2/g vs. 2308 m^2/g for the C-CNT-500-A).

Table 25.2 shows the surface and porosity parameters of the samples, as well as a yield of the activated products in wt% from the initial substances after 500 °C.

The table shows the surface and porosity parameters of the samples, as well as a yield of the activated products in wt% from the initial substances after 500 °C. Table 25.2 shows

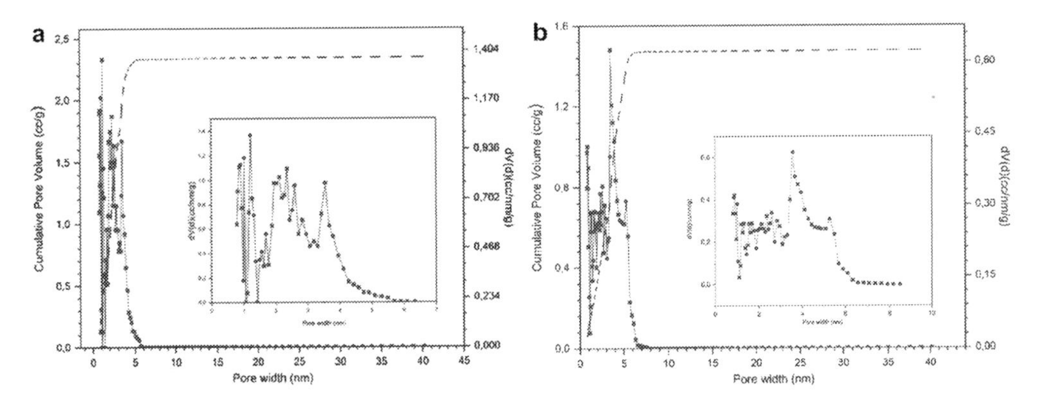

Figure 25.3 Integral and differential distribution of pore diameters for sample C-500-A (A) and C-CNT-500-A (B).

Table 25.2 The structural parameters of the activated materials.

Sample	Yield, wt.%	S_{BET}, m²/g	S_{DFT}, m²/g	Pore volume DFT, cm³/g	Pore volume DFT, <2 nm, cm³/g	Total pore volume, cm³/g
C-500-A(2:1)	25.4	4491	3699	2.345	0.851	2.441
C-500-A(3:1)	21.8	5346	4272	3.004	0.886	3.121
C-500-A(4:1)	17.2	4288	3603	3.428	0.630	3.540
C-500-CNT-A(2:1)	38.8	2308	1894	1.477	0.336	2.310
C-500-CNT-A(3:1)	34.0	2024	1840	2.256	0.191	2.310
C-500-CNT-A(4:1)	29.3	1517	1433	1.980	0.285	2.087

that the BET specific surface area of the samples C-500-A is very large and reaches a too high value of 5346 m^2/g for the sample C-500-A (3: 1). This specific surface area is much higher than the surface of the graphene layer (2630 m^2/g). It can be assumed that a framework carbon structure is formed based on perforated graphene layers or polycumulene fragments. The figures display the increase of the pore width, which is naturally associated with the burning of the pore walls when increasing the amount of activating reagent (KOH) for all materials. Adding CNTs to the composition of the activated material dramatically changes the pore size distribution in the direction of wider mesopores. The mass content of CNTs cannot be determined in the activated material because the CNTs in the activation conditions can also be activated with the burning of the mass part.

Interestingly, our attempts to produce wide-porous carbon materials by introducing CNTs into the original system dextrin-phenol-formaldehyde resin (Qiu et al., 2009), as

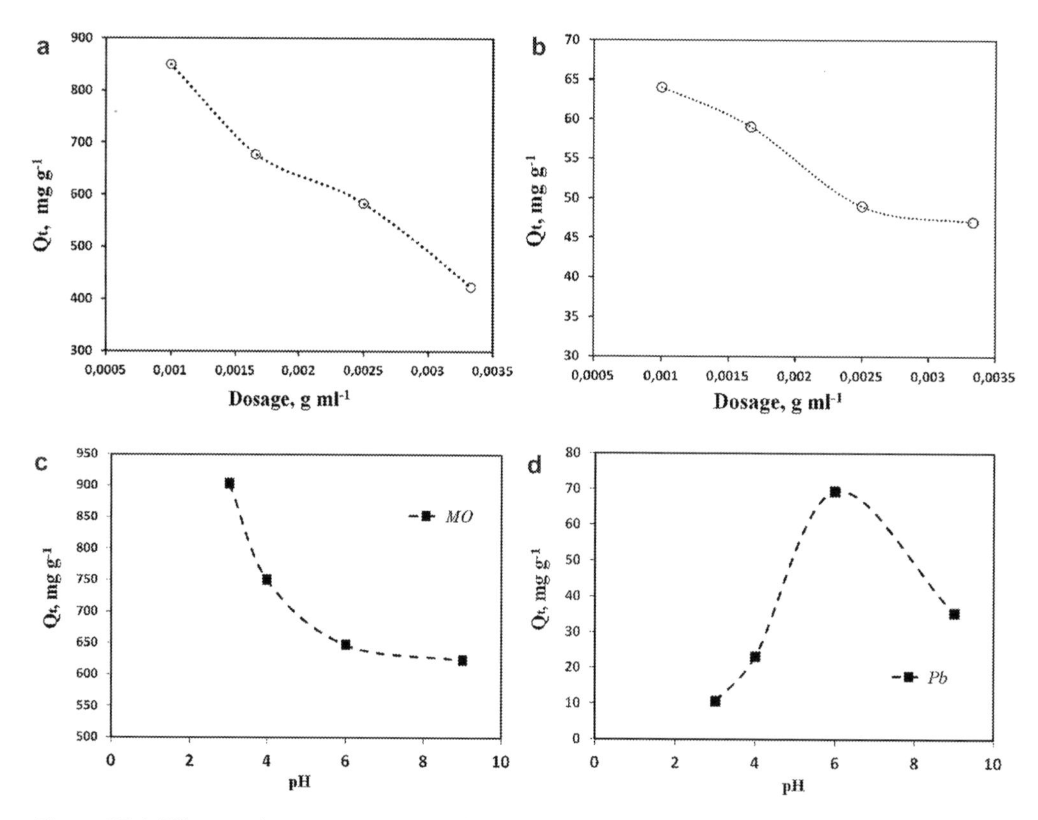

Figure 25.4 Effect on the nanocomposite adsorption capacity: adsorbent weight (MO (A) and Pb (B)) and solution pH (MO (C) and Pb (D)).

well as into the system dextrin-phenol-formaldehyde resin-graphene oxide, did not give significant results. It can be assumed that for the effective control of the pore size, CNTs should be well distributed in the matrix of carbon-containing material. Then the CNTs create a rigid framework, which, when a portion of the activated material is burned out, produces pores of increased size. If the nanotubes are present in the form of aggregates, then a significant part of the volume of the activated material is out of contact with the nanotubes, and the effect of CNTs on the pore size is negligible. Apparently, the solution of HMTA and its polycondensation products in anhydrous sulfuric acid well wets the surface of CNTs, which leads to their disaggregation.

25.3.2 Adsorption study

25.3.2.1 Adsorbent dosage and solution pH effect

The influence of the solution pH and adsorbent dosage on the MO and Pb^{2+} adsorption was elucidated herein. The results of this investigation are given in Fig. 25.4.

The nanocomposite adsorption capacity decreases with increasing adsorbent dosage (Figs. 25.4A,B). High material weight may affect the solid-liquid suspension physical characteristics, for example, by growing viscosity of the solution. Thereby, the nanocomposite weight of 0.03 g was chosen for this investigation.

The most efficient adsorption of dye molecules occurs at pH 3 (Figs. 25.4C, D). It should be noted that the MO solution has a red shade in a strongly acidic region (at pH$\leq$3). In this case, the largest number of molecules is in the protonated form, which is why the sorption activity increases. The equilibrium of the reaction will shift toward the formation of dissociated forms with pH increasing. When the number of disintegrated molecules exceeds the number of nondisintegrated ones, the solution will begin to change color. This process is initiated at pH $\sim$ 3.3–4.4 for MO. Under the chosen experimental conditions, the concentration of original solute molecules in the solution is much higher than the molecules that have dissociated. Thus, the further pH medium increasing leads to a decrease in the sorption activity of the synthesized material.

The analysis of the resulting dependence (Figs. 25.4C, D) allows affirming that the optimum pH value of the initial medium for the Pb^{2+} ions adsorption on the material under study is equal to 6, at which subsequent kinetic and thermodynamic studies were performed.

25.3.2.2 Kinetic study

The effect of the contact time (t, min) between the adsorbent and pollutants solution on the nanocomposite adsorption capacity is illustrated.

As seen in Fig. 25.5A, B, the studied material exhibited a high sorption capacity for the MO much faster (within 10 min) in comparison with the conventional materials (e.g., for activated carbons – within 30-50 min). The maximum adsorption capacity of 865 mg/g was achieved under contact equilibrium obtained within 10 min. For the Pb^{2+} ions, the adsorption capacity is gradually saturated and equilibrium (69 mg/g) was achieved within about 40 min. The results of the kinetics data description using the diffusion and chemical reaction model equations (Table 25.3) (Fares & R., 2021; Largitte & Pasquier, 2016) are presented in Fig. 25.5 and Tables 25.4 and 25.5.

It can be observed that the dependence (Fig. 25.5C) for the pseudo-first-order model represents at the initial period a straight line. This proposes that the nanocomposite contains functional groups that generate chemical bonds with the pollutants being removed. The pseudo-first-order equation provides a perfect fit only for Pb^{2+} ions (the maximum adsorption capacity overestimation is 8%), whereas for MO molecules, it is not satisfactory ($R^2 = 0.6804$ and 0.9411 for MO and Pb^{2+}, respectively).

As seen, based on the determination coefficient (R^2) values, the pseudo-second-order model (Fig. 25.5D) appears to give an excellent fit to the kinetic data for both samples

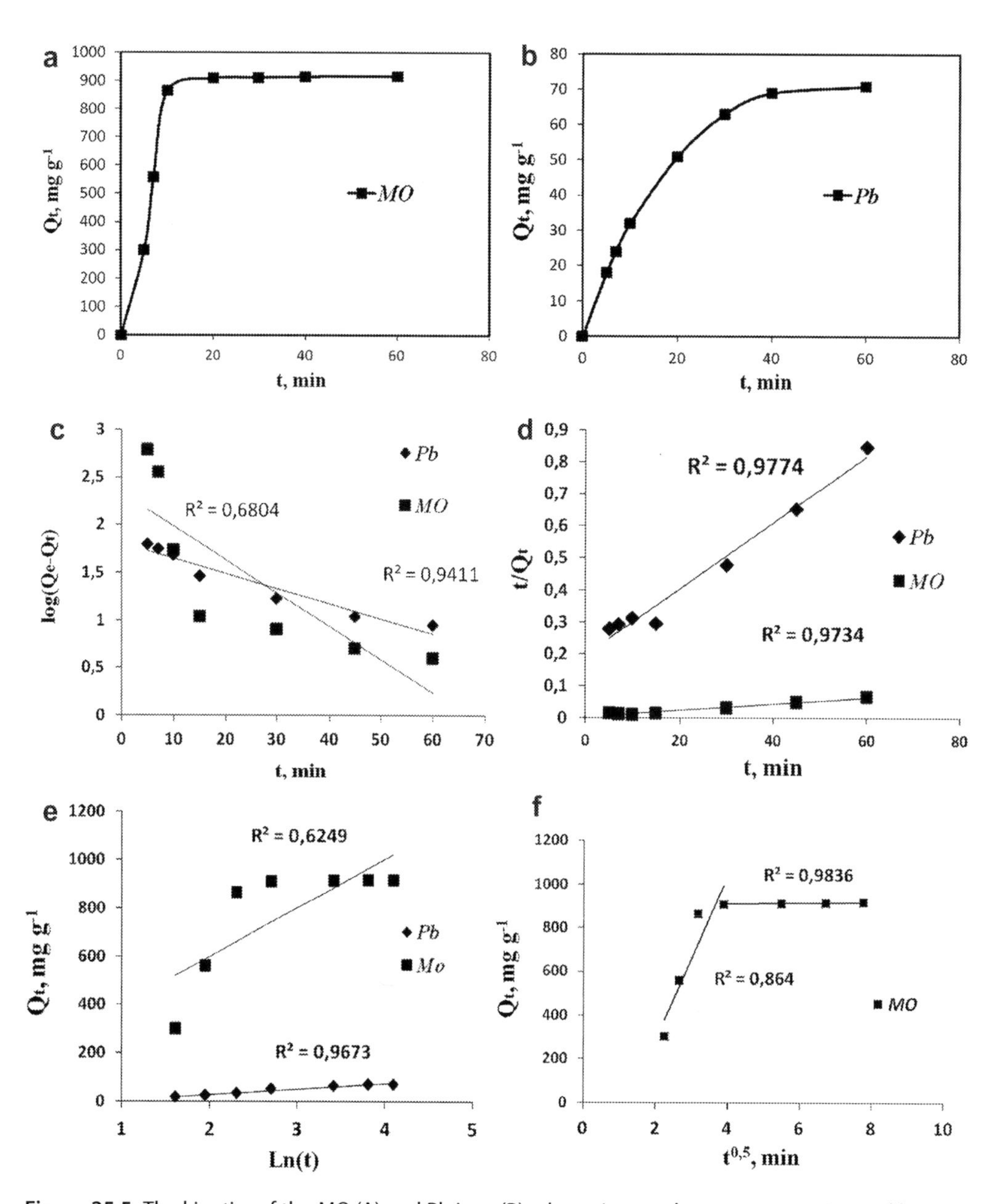

Figure 25.5 The kinetics of the MO (A) and Pb ions (B) adsorption on the nanocomposite, and kinetic models for describing the MO and Pb adsorption: pseudo-first-order (C), pseudo-second-order (D), intraparticle diffusion: MO (E) and Pb (F), external diffusion: MO (G) and Pb (H).

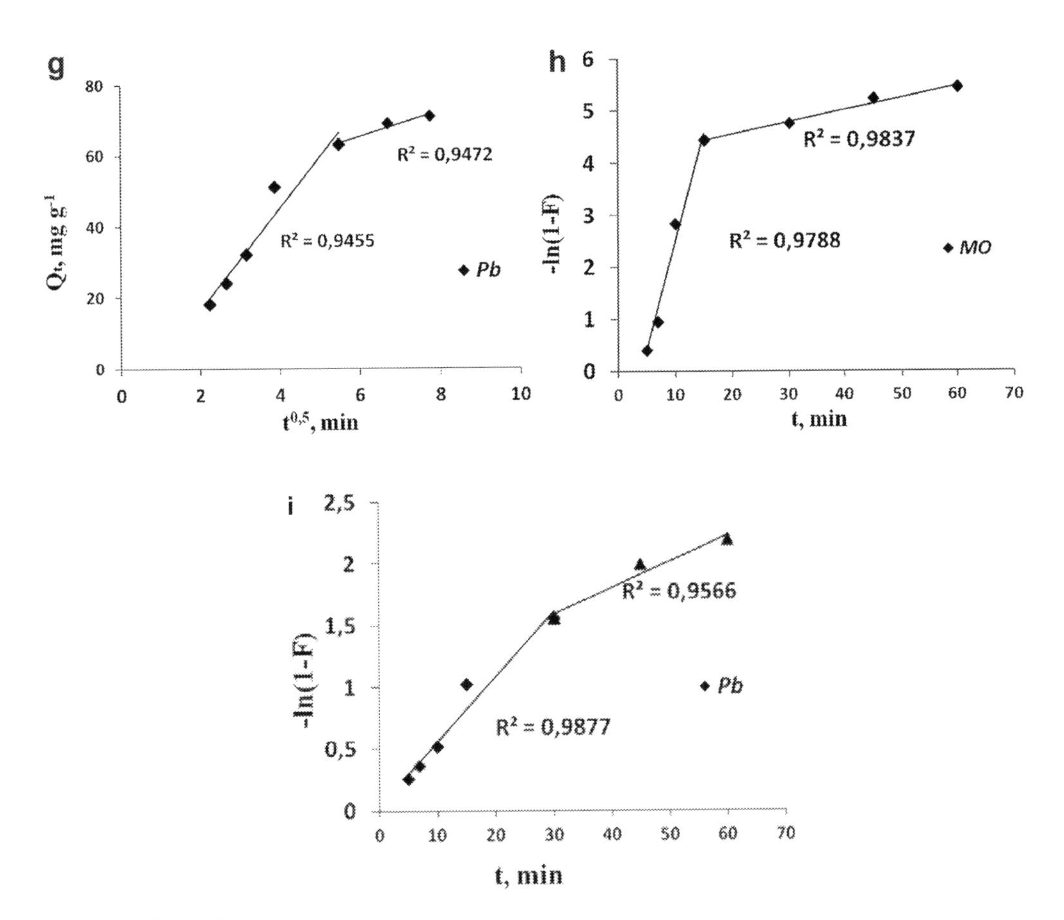

Figure 25.5, cont'd.

Table 25.3 Kinetic models describe the adsorption process.

Model	Equation
Pseudo-first-order	$\log(Q_e - Q_t) = \log(Q_e) - \frac{k_1 t}{2.303}$
Pseudo-second-order	$\frac{t}{Q_t} = \frac{1}{k_2 Q_e^2} + \frac{1}{Q_e} t$
Elovich	$Q_t = \frac{1}{\beta}\ln(\alpha\beta) + \frac{1}{\beta}\ln(t)$
Intraparticle diffusion	$Q_t = k_{id} t^{0.5} + C$

*Q_e is adsorption at equilibrium (mg/g); t is adsorption time (min); Q_t is adsorption at time t (mg/g); k_1 is pseudo-first-order adsorption rate constant, (min^{-1}); k_2 is pseudo-second-order adsorption rate constant (g/mg s); α is initial flow rate, (1/(min mg/g); β is desorption constant (degree of surface coverage and activation energy of chemisorption, (g/mg); k_{id} is intraparticle diffusion coefficient ($min^{-1/2}$); C is layer thickness constant (mg/g).

Table 25.4 Parameters of the chemical kinetics models.

	Experimental	Pseudo-first-order			Pseudo-second-order			Elovich		
Pollutants	Q_e, mg/g	Q_e	k_1	R^2	Qe	k_2	R^2	α	β	R^2
MO	865	216	0.08	0.6804	1000	0.0002	0.9734	523	0.005	0.6249
Pb^{2+}	69	63	0.04	0.9411	97	0.0005	0.9774	49	0.044	0.9673

Table 25.5 The diffusion models parameters.

Pollutants	k_{id1}	k_{id2}	C_1	C_2	R_1^2	R_2^2
MO	14.4	3.57	12.54	43.94	0.864	0.9836
Pb^{2+}	374	1.88	455.3	901.8	0.9455	0.9472

($R^2 = 0.9734$ and 0.9774 for MO molecules and Pb^{2+} ions, respectively). It means that the reaction between the functional groups and the molecules or ions occurs strictly stoichiometrically.

Regarding the other models, the Elovich model gives an excellent fit ($R^2 = 0.9673$) for the Pb^{2+} ions, whereas for the MO molecules, these equations are not so precise in explaining the process ($R^2 = 0.6249$). Relatively high agreement of the Elovich equation with the experimental data indicates that the adsorbent surface has heterogeneous adsorption sites (Largitte & Pasquier, 2016).

The dependence describing the intraparticle diffusion (Figs. 25.5E, F) performs a multilinear curve for both pollutants. But the approximation line does not pass through the origin; the diffusion through the boundary layer effect takes place. Thus, for both pollutants, the diffusion is not a limiting stage, and the superficial adsorption stage is laid on it. The diffusion rate for the Pb^{2+} is a few times higher than that for the MO (k_{id1}=374.03 and 14.412 mg/gmin$^{0.5}$, respectively). Fig. 25.5G, H presents the influence of the external diffusion on the standard rate of the adsorption process. It can be seen that the kinetics consisting of two line dependences related to the adsorption steps. The mass transfer is the limiting stage of the beginning of sorption if the first linear section does not pass through the origin. It can be supposed that the MO and Pb^{2+} adsorption on the nanocomposite is limited by external diffusion in the liquid film covering the nanocomposite surface (Qiu et al., 2009; Weber & Morris, 1963; Yao et al., 2011).

25.3.2.3 Thermodynamic study

To evaluate the effect of temperature on the adsorption of MO and Pb^{2+}, the free energy change (G^0), enthalpy change (H^0), and entropy change (S^0) were determined. In order for that, adsorption isotherms were plotted for each pollutant at different temperatures (Fig. 25.6).

The presented isothermal dependences let us conclude that the general nature of the process remains unchanged when the temperature increase. So, the presented isothermal curves reach an equilibrium state at the given concentration area. Under these

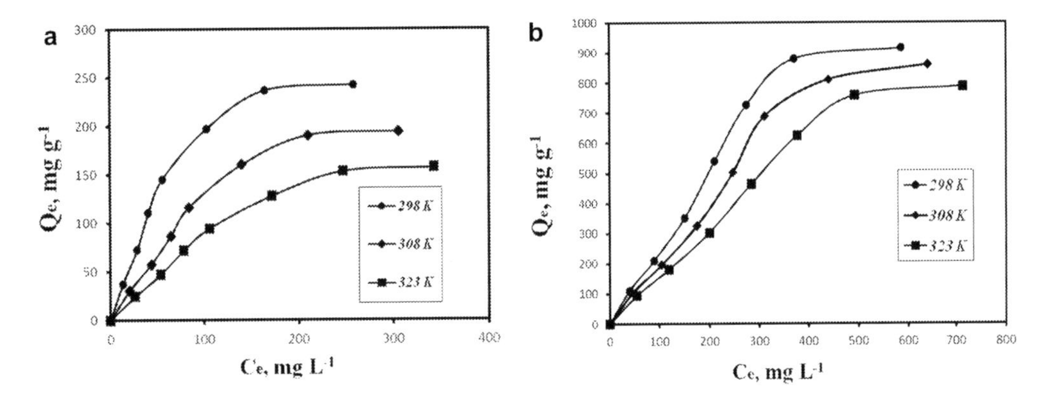

Figure 25.6 Isotherms of the Pb (A) and MO (B) adsorption on the nanocomposite at different temperatures.

circumstances the sorption capacity gradually decreases with increasing temperature. Decreasing the composite sorption activity under study, when the temperature increases, indicates the exothermic nature of the reaction. In this case, the changing of enthalpy should remain constantly at the investigated temperature range. The constant of the sorption process (K) should be calculated for calculating the main thermodynamic parameters of the sorption process. The Langmuir isotherm equation in a linear form is used for the calculation (Yao et al., 2011):

$$\frac{1}{Q_e} = \frac{1}{Q_{max}} + \frac{1}{Q_{max}K}\frac{1}{C}$$

where Q_{max} is maximum adsorption capacity (mg/g); C is equilibrium concentration (mg/L); K is the Langmuir equilibrium constant (L/mol).

For estimation of the maximum sorption capacity according to the Langmuir model and determining the process equilibrium constant, the obtained coefficients of the linear equations shown in the figure are used. So, calculating the partial change in the enthalpy of the sorption process is possible by using the obtained experimental data on the Van't Hoff isobar equation for a chemical reaction. Integrating this equation under the assumption that H° is constant and does not depend on temperature, which is valid in a narrow range that is considered in this experimental study, leads to (Yao et al., 2011):

$$\ln (K) = -\frac{H^0}{RT} + C$$

where R is gas constant (8.314 J $(\text{mol K})^{-1}$); T is temperature (K); C is integration constant.

Dependencies Ln(K) vs. 1/T are to be considered as linear for both pollutants. This confirms the available data about the independence of the enthalpy for the narrow temperature range concerned in this research. The resulted dependence makes it possible

Table 25.6 Basic thermodynamic parameters of the nanocomposite sorption process.

T, K	Ln(K)	H°, kJ/mol	S°, J/(mol K)	G°, kJ mol
Pb^{2+}				
298	5.89		−78.24	−14.6
308	5.28	−37.9	−79.21	−13.5
323	4.69		−78.33	−12.6
MO				
298	−7.08		−49.6	−1.5
308	−7.01	−16.3	−47.3	−1.73
323	−7.6		−49.4	−0.4

to calculate the enthalpy and entropy of the process, which are used to determine the Gibbs free energy. Experimental data are presented in Table 25.6.

The calculated data based on the experiment make it possible to assert that the adsorption of MO and Pb^{2+} is exothermic. The sorption activity decreases with increasing temperature. The entropy changing is also negative, which is linked with complex formation processes on the sorbent's surface. The Gibbs free energy of the system also decreases, which indicates the fundamental possibility of a spontaneous process. The obtained value of the enthalpy change confirms that the sorption process is exothermic and the sorption capacity decreases with an increase in temperature.

Concerning Pb^{2+} ions, it should be noted that the ions of the buffer solution are predominantly in the monovalent acetate complexes form. The obtained and calculated values of the enthalpy make it possible to assert that the mechanism of sorption extraction is based on chemical interaction with active centers of the surface. Most likely, complexation and ion exchange are parallel processes. The heat formation of these processes is relatively small that correlates entirely with the obtained thermodynamic parameters. In this case, the Gibbs free energy of the system also decreases that indicates the fundamental possibility of a spontaneous process to occur.

25.4 Comparison with other materials

The results of review studies in the application of different materials to the Pb^{2+} adsorption are shown in Table 25.7. The table shows the conditions for conducting experimental studies on the extraction process and the adsorption capacity values for various adsorbents (Babkin et al., 2019; Lee et al., 2015; L.P.Lingamdinne et al., 2017; Neskoromnaya et al., 2020; Wang et al., 2007).

Thus, it can be concluded that the nanocomposite obtained by the authors is more efficient for the adsorption, for example, heavy metal ions Pb^{2+} in comparison with the other carbon materials.

Table 25.7 Parameters of the lead ions adsorption on different carbon materials.

Materials	pH	T, K	C_i, mg/L	Adsorbent weight, g	t, min	Q_e, mg/g	Reference
Oxidized multiwalled carbon nanotubes	5	294	40	1	150	50	Wang et al., 2007
Graphene oxide	5	294	-	0.02	90	44.56	Lingamdinne et al., 2019
Graphene oxide–based ferrite nanocomposite	5	294	-	0.02	30	25.78	Lee et al., 2015
Mn_3O_4 precipitated on activated carbon	5	303	50	1 g/L	70	49.82	Neskoromnaya et al., 2020
Polyhydroquinone/ graphene nanocomposite	6	294	100	0.03	60	63.3	Babkin et al., 2018
C-500-A	6	298	100	0.03	45	69	Present work

25.5 Conclusion

A novel CNTs-based nanocomposite material was described above. The activated carbon materials with an anomalous high specific surface area were obtained, making it possible to assume the formation of a mesoporous carbon framework. The addition of CNTs to the initial system leads to a dramatic widening of the pores which is probably caused by forming a framework of CNTs. It should be noted that using of HMTA and its polycondensation products effectively moisten the surface of the applied CNTs. This allows increasing their disaggregation and the uniform distribution in the porous structure of the composite as a result. In its turn, the uniformity of distribution has an important influence on the mechanical characteristics of the carbon structure, as well as the broadening of pores.

As a result of the research, the obtained nanocomposite was investigated on adsorption properties toward MO molecules and Pb^{2+} ions from an aqueous solution. The adsorption capacity of the nanocomposite and equilibrium contact time were determined: MO—865 mg/g within 10 min, and Pb^{2+} – 69 mg/g within 40 min. The kinetics and thermodynamics parameters were evaluated for these processes. The Pb^{2+} ions adsorption on the nanocomposite was successfully fitted to the pseudo-first- and pseudo-second-order kinetic models and satisfactorily adjusted to the intraparticle and external diffusion and the Elovich models. For the MO molecules, the adsorption was highly adjusted to those models, except the pseudo-first-order equation. These findings confirm that the adsorption process is of chemical nature and pore diffusion is a rate-limiting stage for both pollutants. The chemical nature of the adsorption stands for the availability of more active adsorption sites able to adsorb more metal ions.

In this regard, the developed nanocomposite can be used as an adsorbent in water and wastewater treatment processes for effectively removing organic and inorganic compounds from aqueous solutions.

References

Ali, I., Alharbi, O.M.L., Tkachev, A., Galunin, E., Burakov, A., Grachev, V.A., 2018. Water treatment by new-generation graphene materials: hope for bright future. Environ. Sci. Pollut. Res. 25 (8), 7315–7329. https://doi.org/10.1007/s11356-018-1315-9.

Ali, I., Basheer, A.A., Mbianda, X.Y., Burakov, A., Galunin, E., Burakova, I., Mkrtchyan, E., Tkachev, A., Grachev, V., 2019. Graphene based adsorbents for remediation of noxious pollutants from wastewater. Environ. Int. 127, 160–180. https://doi.org/10.1016/j.envint.2019.03.029.

Awad, A.M., Jalab, R., Benamor, A., Nasser, M.S., Ba-Abbad, M.M., El-Naas, M., Mohammad, A.W., 2020. Adsorption of organic pollutants by nanomaterial-based adsorbents: an overview. J. Mol. Liq. 301, 112335. https://doi.org/10.1016/j.molliq.2019.112335.

Babkin, A.V., Burakova, I.V., Burakov, A.E., Kurnosov, D.A., Galunin, E.V., Tkachev, A.G., Ali, I., 2019. Removal of Cu 2+, Zn 2+ and Pb 2+ ions using a graphene-containing nanocomposite: a kinetic study, 693 https://doi.org/10.1088/1757-899x/693/1/012029.

Babkin, A.V., Melezhik, A.V., Kurnosov, D.A., Mkrtchyan, E.S., Burakova, I.V., Burakov, A.E., Galunin, E.V., 2018. Kinetics of the adsorption of synthetic dyes on a polyhydroquinone/graphene carbon nanocomposite. J. Phys. Conf. Ser. 1124 (8), 081030. https://doi.org/10.1088/1742-6596/1124/8/081030.

Baburin, I.A., Klechikov, A., Mercier, G., Talyzin, A., Seifert, G., 2015. Hydrogen adsorption by perforated graphene. Int. J. Hydrogen Energy 40 (20), 6594–6599. https://doi.org/10.1016/j.ijhydene.2015.03.139.

Burakov, A., Neskoromnaya, E., Babkin, A., 2019. Removal of the alizarin red s anionic dye using graphene nanocomposites: a study on kinetics under dynamic conditions. Mater. Today Proc., 11, pp. 392–397.

Burakova, I.V., Burakov, A.E., Tkachev, A.G., Troshkina, I.D., Veselova, O.A., Babkin, A.V., Aung, W.M., Ali, I., 2018. Kinetics of the adsorption of scandium and cerium ions in sulfuric acid solutions on a nanomodified activated carbon. J. Mol. Liq. 253, 277–283. https://doi.org/10.1016/j.molliq.2018.01.063.

Chauhan, M., Saini, V.K., Suthar, S., 2020. Enhancement in selective adsorption and removal efficiency of natural clay by intercalation of Zr-pillars into its layered nanostructure. J. Clean. Prod. 258, 120686. https://doi.org/10.1016/j.jclepro.2020.120686.

Dehghani, M.H., Karri, R.R., Yeganeh, Z.T., Mahvi, A.H., Nourmoradi, H., Salari, M., Zarei, A., Sillanpää, M., 2020. Statistical modelling of endocrine disrupting compounds adsorption onto activated carbon prepared from wood using CCD-RSM and DE hybrid evolutionary optimization framework: comparison of linear vs non-linear isotherm and kinetic parameters. J. Mol. Liq 302, 112526. https://doi.org/10.1016/j.molliq.2020.112526.

Fares, A., R., B.R, 2021. Bio-sorption of toxic metals from industrial wastewater by algae strains *Spirulina platensis* and *Chlorella vulgaris*: application of isotherm, kinetic models and process optimization. Sci. Total Environ. 755, 142654. https://doi.org/10.1016/j.scitotenv.2020.142654.

Garba, Z.N., Lawan, I., Zhou, W., Zhang, M., Wang, L., Yuan, Z., 2020. Microcrystalline cellulose (MCC) based materials as emerging adsorbents for the removal of dyes and heavy metals – a review. Sci. Total Environ. 717, 135070. https://doi.org/10.1016/j.scitotenv.2019.135070.

Hong, M., Yu, L., Wang, Y., Zhang, J., Chen, Z., Dong, L., Zan, Q., Li, R., 2019. Heavy metal adsorption with zeolites: the role of hierarchical pore architecture. Chem. Eng. J. 359, 363–372. https://doi.org/10.1016/j.cej.2018.11.087.

Huang, Q., Chai, K., Zhou, L., Ji, H., 2020. A phenyl-rich β-cyclodextrin porous crosslinked polymer for efficient removal of aromatic pollutants: Insight into adsorption performance and mechanism. Chem. Eng. J. 387, 124020. https://doi.org/10.1016/j.cej.2020.124020.

Huber, M., Welker, A., Helmreich, B., 2016. Critical review of heavy metal pollution of traffic area runoff: Occurrence, influencing factors, and partitioning. Sci. Total Environ. 541, 895–919. https://doi.org/10.1016/j.scitotenv.2015.09.033.

Kang, W., Cui, Y., Qin, L., Yang, Y., Zhao, Z., Wang, X., Liu, X., 2020. A novel robust adsorbent for efficient oil/water separation: magnetic carbon nanospheres/graphene composite aerogel. J. Hazard. Mater. 392,, 122499. https://doi.org/10.1016/j.jhazmat.2020.122499.

Karri, R.R., Sahu, J.N., 2018. Process optimization and adsorption modeling using activated carbon derived from palm oil kernel shell for Zn (II) disposal from the aqueous environment using differential evolution embedded neural network. J. Mol. Liq. 265, 592–602. https://doi.org/10.1016/j.molliq.2018.06.040.

Karri, R.R., Sahu, J.N., Meikap, B.C., 2020. Improving efficacy of Cr (VI) adsorption process on sustainable adsorbent derived from waste biomass (sugarcane bagasse) with help of ant colony optimization. Ind. Crops Prod. 143, 111927. https://doi.org/10.1016/j.indcrop.2019.111927.

Kashif, H., Saiqa, M., Jochen, B., Javed, C.H., 2017. Microbial biotechnology as an emerging industrial wastewater treatment process for arsenic mitigation: A critical review. J. Clean. Prod. 151, 427–438. https://doi.org/10.1016/j.jclepro.2017.03.084.

Kaur, M., Mubarak, N.M., Chin, Lai Fui, B., Khalid, M., Rao Karri, R., Walvekar, R., Abdullah, E., C., Amri Tanjung, F., 2020. Extraction of reinforced epoxy nanocomposite using agricultural waste biomass. IOP Conf. Ser. Mater. Sci. Eng. 943, 012021. https://doi.org/10.1088/1757-899x/943/1/012021.

Khan, F.S.A., Mubarak, N.M., Khalid, M., Walvekar, R., Abdullah, E.C., Mazari, S.A., Nizamuddin, S., Karri, R.R., 2020. Magnetic nanoadsorbents' potential route for heavy metals removal—a review. Environ. Sci. Pollut. Res. 27 (19), 24342–24356. https://doi.org/10.1007/s11356-020-08711-6.

Khan, F.S.A., Mubarak, N.M., Tan, Y.H., Karri, R.R., Khalid, M., Walvekar, R., Abdullah, E.C., Mazari, S.A., Nizamuddin, S., 2020. Magnetic nanoparticles incorporation into different substrates for dyes and heavy metals removal—A Review. Environ. Sci. Pollut. Res. 27 (35), 43526–43541. https://doi.org/10.1007/s11356-020-10482-z.

Largitte, L., Pasquier, R., 2016. A review of the kinetics adsorption models and their application to the adsorption of lead by an activated carbon. Chem. Eng. Res. Design. 109, 495–504. https://doi.org/10.1016/j.cherd.2016.02.006.

Lee, M.E., Park, J.H., Chung, J.W., Lee, C.Y., Kang, S., 2015. Removal of Pb and Cu ions from aqueous solution by Mn_3O_4-coated activated carbon. J. Ind. Eng. Chem. 21, 470–475. https://doi.org/10.1016/j.jiec.2014.03.006.

Li, H., Watson, J., Zhang, Y., Lu, H., Liu, Z., 2020. Environment-enhancing process for algal wastewater treatment, heavy metal control and hydrothermal biofuel production: a critical review. Bioresour. Technol. 298, 122421. https://doi.org/10.1016/j.biortech.2019.122421.

Liang, W., Wang, B., Cheng, J., Xiao, D., Xie, Z., Zhao, J., 2021. 3D, eco-friendly metal-organic frameworks@carbon nanotube aerogels composite materials for removal of pesticides in water. J. Hazard. Mater. 401, 123718. https://doi.org/10.1016/j.jhazmat.2020.123718.

Lingamdinne, L.P., Kim, I.S., Ha, J.H., Chang, Y.Y., Koduru, J.R., Yang, J.K., 2017. Enhanced adsorption removal of Pb(II) and Cr(III) by using nickel ferrite-reduced graphene oxide nanocomposite. Metals 7 (6), 225. https://doi.org/10.3390/met7060225.

Lingamdinne, L.P., Koduru, J.R., Karri, R.R., 2019. A comprehensive review of applications of magnetic graphene oxide based nanocomposites for sustainable water purification. J. Environ. Manage. 231, 622–634. https://doi.org/10.1016/j.jenvman.2018.10.063.

Lingamdinne, L.P., Vemula, K.R., Chang, Y.-Y., Yang, J.-K., Karri, R.R., Koduru, J.R, 2020. Process optimization and modeling of lead removal using iron oxide nanocomposites generated from bio-waste mass. Chemosphere 243, 125257. https://doi.org/10.1016/j.chemosphere.2019.125257.

Liu, Q., Li, Y., Chen, H., Lu, J., Yu, G., Möslang, M., Zhou, Y., 2020. Superior adsorption capacity of functionalised straw adsorbent for dyes and heavy-metal ions. J. Hazard. Mater. 382. https://doi.org/10.1016/j.jhazmat.2019.121040.

Lu, J., Zhou, Y., Lei, J., Ao, Z., Zhou, Y., 2020. Fe_3O_4/graphene aerogels: a stable and efficient persulfate activator for the rapid degradation of malachite green. Chemosphere 251, 126402. https://doi.org/10.1016/j.chemosphere.2020.126402.

Marchesini, S., Turner, P., Paton Keith, R., Reed Benjamen, P., Brennan, B., K. K., & Pollard, A. J. (2020). Gas physisorption measurements as a quality control tool for the properties of graphene/graphite powders. Carbon, 167, 585–595. https://doi.org/10.1016/j.carbon.2020.05.083

Melezhik, A.V., Alekhina, O.V., Gerasimova, A.V., Tkachev, A.G., 2017. The study of polycondensation of hexamethylenetetramine and the properties of the resulting products. Trans. TSTU 23 (3), 461–470.

Murali, S., Potts, J.R., Stoller, S., Park, J., Stoller, M.D., Zhang, L.L., Zhu, Y., Ruoff, R.S., 2012. Preparation of activated graphene and effect of activation parameters on electrochemical capacitance. Carbon 50 (10), 3482–3485. https://doi.org/10.1016/j.carbon.2012.03.014.

Nasrollahzadeh, M., Sajjadi, M., Iravani, S., Varma Rajender, S., 2021. Carbon-based sustainable nanomaterials for water treatment: state-of-art and future perspectives. Chemosphere, 263 https://doi.org/10.1016/j.chemosphere.2020.128005.

Neskoromnaya, E.A., Burakov, A.E., Melezhik, A.V., Babkin, A.V., Burakova, I.V., Kurnosov, D.A., Tkachev, A.G., 2020. Synthesis and evaluation of adsorption properties of reduced graphene oxide hydro- and aerogels modified by iron oxide nanoparticles. Inorgan. Mater. Appl. Res. 11 (2), 467–475. https://doi.org/10.1134/S2075113320020264.

Qiu, H., Lv, L., Pan, B.C., Zhang, Q.J., Zhang, W.M., Zhang, Q.X., 2009. Critical review in adsorption kinetic models. J. Zhejiang Univ. Sci. A 10 (5), 716–724. https://doi.org/10.1631/jzus.A0820524.

Raymundo-Piñero, E., Azaïs, P., Cacciaguerra, T., Cazorla-Amorós, D., Linares-Solano, A., Béguin, F., 2005. KOH and NaOH activation mechanisms of multiwalled carbon nanotubes with different structural organisation. Carbon 43, 786–795. https://doi.org/10.1016/j.carbon.2004.11.005.

Roop, K., Diane, P., Dattatraya, S.G., Ganesh, S.R., Romanholo, F.L.F., Muhammad, B., Ram, C., Naresh, B.R., 2021. Ecotoxicological and health concerns of persistent coloring pollutants of textile industry wastewater and treatment approaches for environmental safety. J. Environ. Chem. Eng. 9 (2), 105012. https://doi.org/10.1016/j.jece.2020.105012.

Ruthiraan, M., Mubarak, N.M., Abdullah, E.C., Khalid, M., Nizamuddin, S., Walvekar, R., Karri, R.R.Abd-Elsalam, K.A., Mohamed, M.A., Prasad, R. (Eds.), 2019. An overview of magnetic material: preparation and adsorption removal of heavy metals from wastewater.. Magnetic Nanostructures: Environmental and Agricultural Applications 131–159. https://doi.org/10.1007/978-3-030-16439-3_8.

Shen, J., Sulkowski, J., Beckner, M., Dailly, A., 2015. Effects of textural and surface characteristics of metal-organic frameworks on the methane adsorption for natural gas vehicular application. Micropor. Mesopor. Mater. 212, 80–90. https://doi.org/10.1016/j.micromeso.2015.03.032.

Sivaranjanee, R., Kumar, S.P., 2021. A review on cleaner approach for effective separation of toxic pollutants from wastewater using carbon sphere's as adsorbent: preparation, activation and applications. J. Clean. Prod. 291, 125911. https://doi.org/10.1016/j.jclepro.2021.125911.

Tadjenant, Ya., Dokhan, N., Barras, A., Addad, A., Jijie, R., Szunerits, S., Boukherroub, R., 2020. Graphene oxide chemically reduced and functionalized with KOH-PEI for efficient Cr(VI) adsorption and reduction in acidic medium. Chemosphere 258, 127316. https://doi.org/10.1016/j.chemosphere.2020.127316.

Tang, C.Y., Yu, P., Tang, L.S., Wang, Q.Y., Bao, R.Y., Liu, Z.Y., Yang, M.B., Yang, W., 2018. Tannic acid functionalized graphene hydrogel for organic dye adsorption. Ecotoxicol. Environ. Saf. 165, 299–306. https://doi.org/10.1016/j.ecoenv.2018.09.009.

Tkachev, A.G., Melezhik, A.V., Alekhina, O.V., 2018. Cumulene Substance, Method for its Production and Use. Patent RU 2 (661), 876.

Tkachev, A.G., Melezhik, A.V., Solomakho, G.V., 2017. The Method of Obtaining Mesoporous Carbon. Patent RU 2 (620), 404.

Wang, H.J., Zhou, A.L., Peng, F., Yu, H., Chen, L.F., 2007. Adsorption characteristic of acidified carbon nanotubes for heavy metal Pb(II) in aqueous solution. Mater. Sci. Eng. A 466 (1–2), 201–206. https://doi.org/10.1016/j.msea.2007.02.097.

Weber, J., Morris, C., 1963. Kinetics of adsorption on carbon from solution. J. Sanit. Eng. Div. 89, 31–60.

Worsley, M.A., Kucheyev, S.O., Mason, H.E., Merrill, M.D., Mayer, B.P., Lewicki, J., Valdez, C.A., Suss, M.E., Stadermann, M., Pauzauskie, P.J., Satcher, J.H., Biener, J., Baumann, T.F., 2012. Mechanically robust 3D graphene macroassembly with high surface area. Chem. Commun. 48 (67), 8428–8430. https://doi.org/10.1039/c2cc33979j.

Yang, Q., Wang, B., Chen, Y., Xie, Y., Li, J., 2019. An anionic In(III)-based metal-organic framework with Lewis basic sites for the selective adsorption and separation of organic cationic dyes. Chinese Chem. Lett. 30 (1), 234–238. https://doi.org/10.1016/j.cclet.2018.03.023.

Yao, Y., He, B., Xu, F., Chen, X., 2011. Equilibrium and kinetic studies of methyl orange adsorption on multiwalled carbon nanotubes. Chem. Eng. J. 170, 82–89. https://doi.org/10.1016/j.cej.2011.03.031.

Yazidi, A., Atrous, M., Edi Soetaredjo, F., Sellaoui, L., Ismadji, S., Erto, A., Bonilla-Petriciolet, A., Luiz Dotto, G., Ben Lamine, A., 2020. Adsorption of amoxicillin and tetracycline on activated carbon prepared from durian shell in single and binary systems: experimental study and modeling analysis. Chem. Eng. J. 379,, 122320. https://doi.org/10.1016/j.cej.2019.122320.

Yi, X., Tran, N.H., Yin, T., He, Y., Gin, K.Y.H., 2017. Removal of selected PPCPs, EDCs, and antibiotic resistance genes in landfill leachate by a full-scale constructed wetlands system. Water Res. 121, 46–60. https://doi.org/10.1016/j.watres.2017.05.008.

Zhou, Y., Gu, X., Zhang, R., Lu, J., 2014. Removal of aniline from aqueous solution using pine sawdust modified with citric acid and β-cyclodextrin. Ind. Eng. Chem. Res. 53 (2), 887–894. https://doi.org/10.1021/ie403829s.

Zhou, Y., Hu, Y., Huang, W., Cheng, G., Cui, C., Lu, J., 2018. A novel amphoteric B-cyclodextrin-based adsorbent for simultaneous removal of cationic/anionic dyes and bisphenol A. Chem. Eng. J. 341, 47–57. https://doi.org/10.1016/j.cej.2018.01.155.

Zhou, Y., Lu, J., Liu, Q., Chen, H., Liu, Y., Zhou, Y., 2020. A novel hollow-sphere cyclodextrin nanoreactor for the enhanced removal of bisphenol A under visible irradiation. J. Hazard. Mater. 384. https://doi.org/10.1016/j.jhazmat.2019.121267.

Zhu, Y., Murali, S., Stoller, M.D., Ganesh, K.J., Cai, W., Ferreira, P.J., Pirkle, A., Wallace, R.M., Cychosz, K.A., Thommes, M., Su, D., Stach, E.A., Ruoff, R.S, 2011. Carbon-based supercapacitors produced by activation of graphene. Science 332 (6037), 1537–1541. https://doi.org/10.1126/science.1200770.

CHAPTER 26

Magnetic nanoparticles and its composites toward the remediation of electromagnetic interference pollution

Rambabu Kuchi[a,b], Dongsoo Kim[a,b] and Jong-Ryul Jeong[c]
[a]Convergence Research Center for Development of Mineral Resources, Korea Institute of Geoscience and Mineral Resources, Daejeon, South Korea
[b]Powder & Ceramics Division, Korea Institute of Materials Science, Changwon, Gyeongnam, South Korea
[c]Department of Materials Science and Engineering, Graduate School of Energy Science and Technology, Chungnam National University, Daejeon, South Korea

26.1 Introduction

Nowadays, the electromagnetic (EM) wave radiation in the gigahertz (GHz) range has been subjected as a serious issue for advanced electronic technologies such as biological systems, industrial and military safety equipment, etc. As these EM radiations interact with the device (works in GHz range), input signals result in a noise that is known to be EM interference (EMI) pollution. For the modern technologies, the EMI pollution could be the most unwanted outcome and it could become dreadful to the human health, triggering many diseases including the headache, trepidation, and sleeping disorders (Higashiyama et al., 2012; Kawamura et al., 2012; Nojima & Tarusawa, 2002; Rao et al., 1999). Despite the hazardous problem to the public health, EMI could also deteriorate the durability, efficiency, and specific functionality of the electronic tools such as communication devices (e.g., cellphones, computers, laptops, and Bluetooth devices), commercial tools (e.g., microwave ovens, microwave circuits), and integrated electric circuits in automobile industries (Palisek & Suchy, 2011; Singh et al., 2018). Thus, this EMI pollution has come to be a severe global problematic issue so its control and remediation could be attained through the usage of EMI shielding materials (Jian et al., 2016). EMI shielding is achieved through reflection or absorption of EM wave radiation by a material that works as an obstacle against the interaction of EM wave radiation via passing through it. By this activity, the EMI shielding materials suppress or attenuate the EM radiation signals, and consequently those EM wave radiations cannot affect the features and performanence of the electronic devices.

Electrically, metal conductor materials showed good EMI shielding capabilities owing to their high reflectivity but these metallic conductors suffer a high density, high

Sustainable Nanotechnology for Environmental Remediation
DOI: https://doi.org/10.1016/B978-0-12-824547-7.00025-4

costs, and a lack of flexibility. Additionally, metals are corroding easily in the ambient environment, which leads to deterioration in the structure and loss of its shielding performance. Alternatively, the magnetic materials could permeate the EM signals due to their magnetic permeability. On the other hand, ferromagnetic materials have certain limitations and could be useful for a lower frequency region (below the GHz) so that impede their use for broad gigahertz range (Shukla, 2019). The practical application purposes are that the EMI shielding material should offer strong absorption or minimal reflection across a broad absorption band and have low density, low cost, and thin layer form (Zhao et al., 2015). From this viewpoint, the development of effective EM wave shielding materials has paid huge consideration. Therefore, need to explore the shieling material with good features disclosed earlier is still a great challenge task to apply for various applications (Agarwal et al., 2016; Kumar et al., 2015). Many of the recent studies on EMI shielding materials state that the magnetic particles and their magnetic composites have been used extensively as effective materials with improved shielding properties (Chen et al., 2017; Kuchi, Latif, et al., 2019; Kuchi, Sharma, et al., 2019; Kuchi et al., 2017, 2018). This chapter summarizes many of these materials including magnetic nanoparticles and their composites that have been designed and developed for the remediation of EMI pollution.

26.2 Objective

The demand for magnetic nanoparticles and their composites has been increased because of their extensive utilization in various applications. It is noticed that these materials are innovatively used in the remediation of EMI pollution owing to their excellent shielding behavior. Further development of high-performance EMI shielding materials comprising the magnetic components includes particles or composites that can be achieved from the study of shielding mechanism and design, synthesis approaches of magnetic particles and composites. In this chapter, we include all these and it could be particularly useful to the researchers who are about research in this area.

26.3 EMI shielding mechanism

EM radiation has two waves (sinusoidal waves) that contain electrical and magnetic components and they are vibrating perpendicular to each other. Like other sinusoidal waves, the EM wave has certain frequencies and corresponding energy so their spectrum could be characterized into several groups. It means that the EM waves at the lower end of the EM spectrum had lower frequencies and lower energy and these could increase gradually at the higher end of the EM spectrum. The incident EM wave with a certain frequency is passed through a medium (material) then its power could decrease after transmission due to the shielding efficiency (SE) of the material. The SE of the material

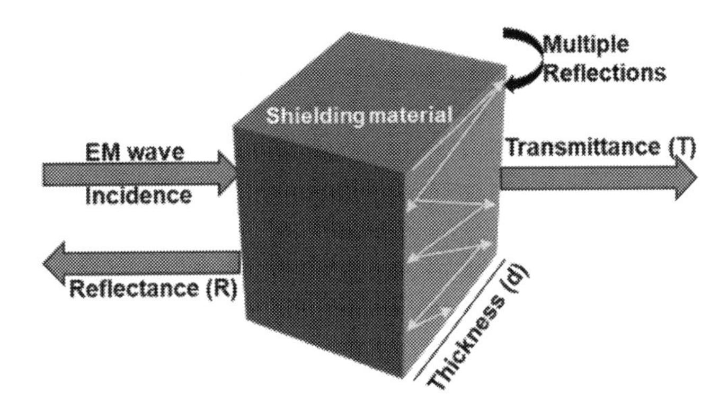

Figure 26.1 Representation of the EMI shielding mechanism process.

can be represented as the ratio between the incoming and outgoing power of EM wave expressed in terms of decibel (dB) loss, which measures the opacity of the material by interaction with EM waves at a certain frequency (Sanskaran et al., 2017). SE of the material is expressed as

$$SE = 10 \log_{10} (P_i/P_o) \tag{26.1}$$

where P_i and P_o refer to the power of input and output waves, respectively.

When the EM waves interact with the material, they undergo three processes (Fig. 26.1) namely reflectance (R), absorption (A), and transmission (T) (Clark et al., 2000; Das et al., 2008; Thostenson & Chou, 1999). The R consists of two components: one is from the front (incident) surface reflection and the second one is reflections from inside the material, which is internal or multiple reflections. The higher A results in the dissipation of EM wave energy in the form of heat energy, which results in raising the temperature at the shield (Kuchi, Sharma, et al., 2019; Zhao et al., 2016). The SE total (SE_T) for the finite-dimensional media is from the absorption (SE_A), reflection (SE_R), and multiple reflections (SE_M), but SE_M (from secondary reflections) was ignored for thicker media used for the practical case. Then SE_T can be changed in the following form:

$$SE_T = SE_R + SE_A \tag{26.2}$$

From Eq. (26.2), it is clear that the SE via the reflection and absorption by the sample under testing. The main mechanisms of EMI shielding are reflection and absorption. Reflection loss (RL) results from a mismatch of relative impedance between the EM waves and surface of shielding material and is given by

$$SE_R = 20 \log(Z_0/Z_{in}) = 39.5 + 10 \log \left(\sigma/2f\pi\mu\right) \tag{26.3}$$

where Z_0 and Z_{in} are impedance of vacuum and input by the material, μ is the permeability, f is frequency, is the material conductivity. Equation (26.3) showed that the SE_R α

means that SE_R decreases for a constant and values. Thus, the mobile carriers (electrons or holes) of the material can be responsible for the reflection of EM waves. Another mechanism is absorption loss (SE_A) and is figured out as the exponential decrease in the EM wave amplitude within the material when wave passes through it. Thereby obviously, the temperature increased via induced current in the medium. SE_A is stated as (Shukla, 2019)

$$SE_A = 20\ \log e^{d/\sigma} = 8.7d(f\pi\sigma\mu)^{1/2} \tag{26.4}$$

where d is the thickness of the absorbing medium or absorbance and the product of σ, μ is equal to the attenuation constant (α). So that the SE_A depends on the sample thickness and α, which means that magnetic conduction materials majorly undergo absorption rather than reflection.

It is obvious from Eqs. (26.3) and (26.4) that the shielding in nonmagnetic material is governed by reflection, whereas the magnetic material is dominated via absorption because the composite materials have heterogeneous structures that have various EM properties. In this case, permittivity and permeability denoted as ε_r and μ_r are the relative complex permittivity and permeability, respectively. These are expressed as $\varepsilon_r = \varepsilon' - j\varepsilon''$ and $\mu_r = \mu' - j\mu''$ (Kuchi, Latif, et al., 2019). ε' and μ are the real parts of permittivity and permeability which refer to the electric and magnetic energy storage. While the ε'', and μ'' are the imaginary parts of permittivity and permeability that relate to the loss ability of EM wave. Hence, these EM parameters have a complex dependency on the structure, conductivity, volume fraction, and size of each constituent in the composite. By adjusting permeability and permittivity to certain values, the reflection is lessened. It is essential for many applications so the effective EMI shielding materials have very low reflection and enhanced absorption. This is obtained when we avoid or minimize the impedance mismatch between the shielding medium and free space. From the transmission line theory, the intrinsic surface impedance of medium (shielding material) is related to its permeability and permittivity given by Eq. (26.5), then the EM absorption of the shielding material, given by Eq. (26.6) (Kuchi et al., 2018),

$$Z_{in} = Z_0\sqrt{\mu_r/\varepsilon_r}\ \tanh\left\{j\,(2\pi f d/c)\sqrt{\mu_r\varepsilon_r}\right\} \tag{26.5}$$

$$RL\ (dB) = 20\ \log|(Z_{in} - Z_0)/(Z_{in} + Z_0)| \tag{26.6}$$

can be higher if the impedance of free space and material is matched at the ideal condition, which means $Z_{in} = Z_o\ 377\ \Omega$. This can be possible at a specific matching thickness (t_m) and frequency (f_m) when effective broadband is achieved which could be tuned by 1/4 wavelength equation (Lv et al., 2015).

$$t_m = nc/4f_m\varepsilon u^{1/2} \tag{26.7}$$

where c is the velocity of light and n is the refractive index of the light. Based on the above relations it is concluded that the EMI shielding performance can be enhanced through controlling the relative permittivity and permeability to minimize impedance mismatch. Moreover, advanced technologies require high-performance EMI shielding materials with minimized RL, a wide bandwidth of frequency range, lower weight, good flexibility, and economically cheaper to process out.

26.3.1 Characterization of EMI shielding performance

It is measured by the evolution of the power and phase of the transmitted wave via the shielding material. The most commonly used methods are the shielded room method, shielded box method, open-field method, and coaxial transmission line method (Geetha et al., 2009). Among these methods, the coaxial transmission line method is the most generally used method for a wide range of frequencies importantly from 2 to 18 GHz as this frequency range could cover the many advanced technology systems. The coaxial transmission line method has been widely used among them for calculating the EMI SE of the materials. In this method, metal sample holder, transmitting and receiving coaxial cables, and a vector network analyzer (VNA) are used. This method could be used for long-range frequencies as the VNA can record EM wave intensities effectively in a wide frequency range by utilizing the coaxial cables that have smaller losses compared to antennas (Geetha et al., 2009, 2009). The results obtained using this method can be resolved into reflection, absorption, and transmitted components so that it gives important insight to understand the EMI shielding mechanism of the given material.

26.3.2 Important factors influencing the EMI shielding performance

26.3.2.1 EM properties

The relative permeability (μ_r) and permittivity (ε_r) are referred to as the EM properties. These are very crucial and play a significant role in the EMI shielding performance of the given material. This is observed in the shielding mechanism that these two parameters affect the reflection and absorption mechanism. The ε'' and μ'' confer the dielectric loss and magnetic loss in the material: the dielectric loss comes from conductivity and polarization losses and are two important parameters, whereas the magnetic loss originates from the exchange resonance, natural ferromagnetic resonance (FMR), and eddy current loss.

26.3.2.1.1 Dielectric loss

According to the free-electron theory the higher conductivity could be beneficial to have a higher dielectric loss. It was also proved in many previous studies on carbonaceous materials, which have high conductive networks (Cao et al., 2018; Song et al., 2012; Wen et al., 2014). Polarization loss is another contributor to dielectric loss and has ionic, electronic, dipole, and interfacial polarization. Ionic and electronic polarizations are ignored for microwave range as these work at a very a higher frequency from 1000 GHz (Shukla et al., 2019). Dipole polarization is caused by the defects and presence of residual groups in the material. These are mainly formed by the heating temperature and keeping time for the material synthesis process (Liu et al., 2015; Meng et al., 2018). The interfacial polarizations are generally observed in the composites or materials with heterogeneous atoms because of the presence of trapped space charges. They are also a respective relaxation process along with interfacial polarization, which is verified by a Cole–Cole semicircle plot (plot between ε' and ε''). In the plot, many semicircles represent the relaxation loss according to the Debye dipolar process (Xia et al., 2017). This phenomenon is observed in the multiinterface composites and hierarchical structures.

26.3.2.1.2 Magnetic loss

This mostly appears in the magnetic components through exchange resonance, FMR, and eddy current loss. The natural resonance frequency (f_r) depends on the anisotropy field (H_a) according to the equation: $f_r = \gamma H_a/2\pi$, $H_a = 2K/\mu_0 M_S$, where $\gamma/2\pi$ is the gyromagnetic ratio, μ_0 is vacuum permeability, M_S is the saturation magnetization and K is the anisotropy constant. These relations reveal that the higher M_S or lower H_a can enhance the natural FMR position toward higher f_r (redshift). Also, with a lower anisotropy constant (K) there is an improvement in the absorption bandwidth. It can also be observed that there is cut-off frequency f_r, which sharply decreases the permeability by Snoek's limit, $f_r\ (\mu - 1)\ \alpha\ M_S$ (Nakamura, 2000; O. & S., 2008). Magnetic metals and their alloys have the higher M_S, even though they tend to resonate at lower f_r and because of the highly conductive nature they could undergo the eddy current loss, which decreases the magnetic permeability. However, for the ferrites' semiconducting nature they have higher f_r (at a lower GHz range) compared to magnetic metals and alloys. However, these situations can be overcome by avoiding the eddy current loss which can be plausible if the materials have micro- or nanosized particles.

26.3.2.2 Surface morphology and size

The material surface morphology is very crucial to achieve the superior EMI shielding performance because it mainly affects the EM properties (ε_r and μ_r), which could contribute to EMI shielding performance. There is much literature regarding this aspect. C. Wang et al. reported the morphology-dependent EM properties of hierarchical cobalt assemblies such as spheres, flowers with dendritic petals, and flowers with sharp petals (Wang et al., 2010). Among these Co flowers with dendritic petals showed strong absorption of EM waves because it has higher coercivity (H_C) through its obvious shape anisotropy than flowers with sharp petals. Q. Liu et al., (2015) design and fabricate different surface morphologies of CoNi hierarchical structures including flower, urchin, ball, chain-like structures. The urchin-like structure exhibits a higher absorption bandwidth of 5.5 GHz along with minimal RL of about −33.5 dB RL owing to its high density of magnetic flux lines from electron holography (J. Liu et al., 2015; Meng et al., 2018). Bora et al. (2018) report the polyvinyl butyral (PVB)-MnO_2 nanocomposites with controlled MnO_2 morphology. PVB-MnO_2 nanorod was found to be more efficient for EMI shielding properties than PVB-MnO_2 nanosphere (NS) due to a higher degree of EM impedance matching and more effective permittivity of nanorod-loaded composite. Therefore, the shielded material surface morphology (geometry) has a decisive role in EM absorption properties as the size of the particle could directly affect the magnetic properties, which can control the EMI shielding performance. J. Liu et al. (2013) prepared the yolk-shell $Fe_3O_4/CuSio_3$ spheres with the controlled size of Fe_3O_4 spheres (core) and shell ($CuSio_3$) thickness. They found that the Fe_3O_4 spheres with a size of 450 nm had higher magnetic permeability than other Fe_3O_4 spheres with sizes 150 and 330 nm, respectively, and these spheres showed better absorption properties due to their higher magnetic permeability. Furtherly, the yolk-shell spheres consist of Fe_3O_4 spheres of 450 nm, and $CuSiO_3$ shell thickness of 63 nm showed enhanced EM absorption than that of Fe_3O_4 spheres. It is owing to the synergetic effect of core and shell, higher porosity, and larger surface area. Moreover, the complex structures (yolk-shell, composites, hierarchical structures, etc.) have multiinterfaces with bound charges causing the Maxwell–Wagner effect and then the interfacial polarization and relaxation process takes place. These processes could contribute to enhancing further the EMI shielding performance.

26.3.2.2 Reaction temperature and time

As the material synthesis was strongly dependent on the reaction temperature and time, thus these can control the EMI shielding performance. For the case of the ferrite, heat treatment could results in the formation of materials with defects as vacancies and dangling bonds, which are advantageous to minimize RL with higher attenuation of EM waves (Y. Liu et al., 2014). By varying the temperature and time, there could be the fabrication of composites or core-shell materials with improvement in structural properties. It is observed that the Fe_3O_4/single-walled carbon nanotube (SWCNT) composites annealed at 500 °C showed the better integration of the SWCNT with Fe_3O_4 particles followed by better microwave absorption performance than other samples annealed at high temperature [12]. The annealing time can also bring the structural changes similar to annealing temperature for FeCo/ZnO composites: 12 h reaction time composites displayed higher microwave absorption properties than 15 and 20 h (Lv et al., 2014).

26.3.2.3 Thickness

The quarter wavelength equation: the shielding material thickness (t_m) is nearly equal to a quarter of the propagating EM wavelength (λ_m) multiplied by an odd number, leading to the minimum RL. It indicated by the cancellation of the incident and reflected waves from the shielding material surface by the matching condition.

$$t_m = n\lambda_m/4 \text{ and } \lambda_m = \lambda_0 / \sqrt{\varepsilon\mu} \quad (26.8)$$

where n = (1, 3, 5....), n=1 represents the first dip. Therefore, the sample thickness increases the reflection peaks shift toward the lower frequencies because the number of dips increases as thickness increases.

26.3.2.4 Mass ratio

It follows the electrical properties of the sample, which has been to check for microwave absorption performance. Electrical properties are in according to the percolation threshold value of conductivity for the materials. For example, the low volume fraction of filler is sufficient in the case of elastomers composites if we take higher volume then the properties will decrease. So that the critical volume fraction for every material either simple particles or composites is variable, has been checked, and depends on many parameters According to many kinds of research, the mass ratio of the filler or absorber to the host matrix (paraffin wax or epoxy resin or other polymers) is related to many factors such as the morphology, conductivity, aspect ratio (composites) of the absorber. Besides, it also depends on the compatibility and distribution of the filler with the matrix (Saini & Arora, 2012).

26.4 EMI shielding materials

Semiconductor materials, carbon materials, magnetic materials, and composites have been used as EMI shielding materials over the past decade. However, the magnetic materials and their composites have attracted significant interest owing to their strong magnetic and mechanical properties, various morphological structures, and facile preparation methods.

26.4.1 Magnetic particles

Magnetic particles including the ferrites (Li et al., 2002), metals and their alloys (Kuchi et al., 2017; Zhao et al., 2015), and compound (Deng et al., 2018) have been drawn a specific attraction to be used as EMI shielding materials because of their successful preparation at various designable structures, higher magnetization value, large curie temperature, and displayed good EM wave absorption properties. Nevertheless, there are some limitations such as narrow absorption bandwidth, larger filler content, low yields, and complex formation process in certain cases (Q. Liu et al., 2015). To overcome these issues, several approaches were made in recent years including the design and synthesis of magnetic particles with various morphologies. It means that the absorber unit has a well-defined shape with controlled granularity making it possible to control the magnetic permeability, which affects EM absorption properties.

26.4.1.1 Magnetic metal and alloy particles

From the EM wave theory, tuning the surface morphology of the magnetic metallic particles could efficiently improve the EM wave absorption properties through the magnetic loss and surface scattering effects. It has been proved by many researchers; C. Wang et al. reported the Cobalt (Co) structures with controlled morphology include the spheres, flower with sharp petals, and flowers with dendritic petals. Among the various structures, the flower with dendritic petals was found to have the strongest absorption. They also prepared hierarchical nickel (Ni) with different structures and particularly the urchin-like Ni chains showed the best absorption properties over the smooth chains, rings, and hexagonal plate structures (Wang et al., 2010). Q. Liu et al. (2015) fabricated the CoNi hierarchical structures including flower, urchin, ball, chain-like structures. The urchin-like structure showed strong EM wave absorption performance.

The good EM wave absorbers based on magnetic particles are the metallic materials and their alloys due to their strong magnetic behavior and magnetic loss. Most of these materials exhibit the RL peaks at higher microwave frequency (12–18 GHz) thus it is a challenge to develop the simple synthesis approach to prepare the strong EM wave absorption material that has the RL peaks at low (2–6 GHz) and middle (6–12 GHz) frequency regions. For this purpose, the metallic cobalt (Co) has been studied as a good candidate over the other magnetic metals with a higher magnetic loss to enhance EM wave absorption properties. Despite the strong magnetic loss, Co has been facially prepared with diverse structures and was disperse uniformly in a host matrix (Kuchi et al., 2017). Furtherly, C. Wang et al. attempted the preparation of hierarchical Co assemblies by way of spheres, flowers with dendritic petals, and sharp petals. The Co flowers with dendritic petals showed RL of −13.6 dB, at 9 GHz, a flower with sharp petals shows the RL of −11.6 dB at 5.4 GHz, and spheres have the RL of −7 dB at 5 GHz. The magnetic properties of Co structures are related to their EM absorption; the flower with dendritic petals has shape anisotropy obvious to have higher coercivity (H_C) value than other structures and saturation magnetization values (Ms) values are different which are according to their pinned surface magnetic moments. These Co structure has covered the lower and middle microwave frequencies but the RL values are not strong enough.

To address this issue many researchers make an effort to prepare different Co morphologies: hollow porous spheres, microflower structures, hierarchical 3-D superstructures. All these structures are obtained by using expensive surfactants such as polyvinyl-pyrrolidine, sodium dodecylbenzene sulfonate, etc., and these processes need high temperatures and long reaction times. In this regard to overcome the mentioned limitations, R. Kuchi et al. (2017) reported the facile synthesis of Co superstructures with controlled morphology at room temperature in an aqueous medium. The morphology of Co has been tailored simply via varying the amounts of reducing agent, hydrazine hydrate. Varying the reducing agent amounts as 2, 4, 6, 8 mL results in the various structures shown in Fig. 26.2; the important and stable morphologies are microfoliage (4 mL) and isolated microfoliage (8 mL.

The formation mechanism for these structures could be explained based on diffusion-limited aggregation (DLA) and nucleation-limited aggregation processes. As the DLA mechanism initiated the violent stirring with a high concentration of hydrazine hydrate so the microfoliage structures are observed for 4 mL. While further increasing the hydrazine hydrate amount results in the formation of isolated microfoliage via cleavage of the branches in the microfoliage structure. The microfoliage and isolated foliage structures have different crystallite sizes because their formation kinetics is different. It found that the isolate foliage structure has a lower crystallite size and showed higher H_C value by its shape anisotropy. By this observation, it can be expected that the magnetic loss would be higher for the isolated foliage structures. This could be evaluated from the EM parameters where the real parts of permittivity and permeability stand for storage capacity, and imaginary parts refer to the loss abilities of propagated EM waves. From the results shown in Fig. 26.3, it is evident that the magnetic loss was higher for isolated foliage. Magnetic loss contributed by the natural (peaks at a lower frequency) and exchange (peaks at a higher frequency) resonances.

EM absorption performance was studied in terms of RL curves. From the RL calculation, using the EM parameters indicated that the isolated foliage Co structures exhibited a strong RL (–31.08 dB at 9.24 GHz) than the microfoliage (–17.76 at 11.14 GHz) structure shown in Fig. 26.3. The effective bandwidth (< –10 dB, important for practical use) is about 3.8 GHz and 3.5 GHz for the isolated microfoliage and aggregated foliage Co structures. The strong RL with wide effective absorption bandwidth to the isolated foliage structures ascribed to its better microwave attenuation, higher magnetic loss, and acted as quasi-antenna receivers. So that EM waves penetrate easily with the absorber. The magnetic metallic materials to be used as EMI absorbers are limited from their high density, poor flexibility, higher cost, and difficult to use at high temperatures due to their corrosion behavior through oxidation.

26.4.1.2 Magnetic metal oxide particles

To avoid the above restrictions, the magnetic metal oxides such as iron (Fe) oxide in the magnetite form are useful due to its wide nanostructure formations, higher abundance, and uses in many applications (Kuchi et al., 2018) because the magnetite is prepared in various morphologies such as flower-like, dendrite, urchin-like, nanotube, NS, nanowire, and hollow, porous NS. However, the porous structures have become important candidates due to their excellent properties such as low density, a large specific surface area, high magnetic condensation, and good mechanical and thermal stability. In general, templates have been used to prepare the porous structures but it is not impressive because of structural damage during template removal. Template-free methods such as Ostwald ripening, self-attachment, and Kirkendall effect mechanisms have been used

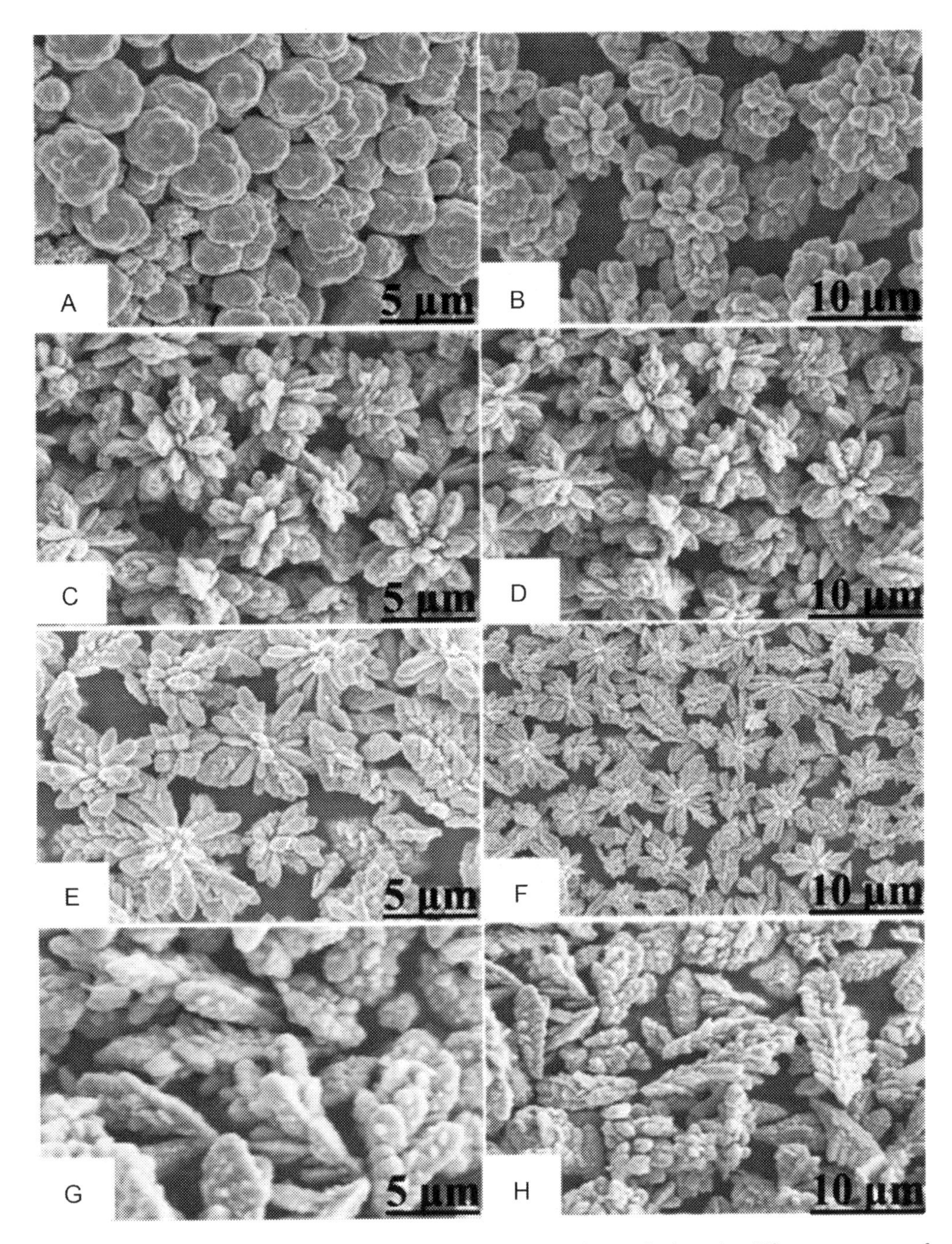

Figure 26.2 SEM images for the various Co superstructures obtained when the different amounts of hydrazine hydrate are added (A) Big sphere composed of small spheres, 2 mL; (B) nearly aggregated foliage, 3 mL; (C, D) aggregated foliage like 4 mL; (E, F) enlarged aggregated foliage, 6 mL; (G, H) isolated microfoliage, 8 mL (with permission, R. Kuchi., J.R. Jeong are the authors).

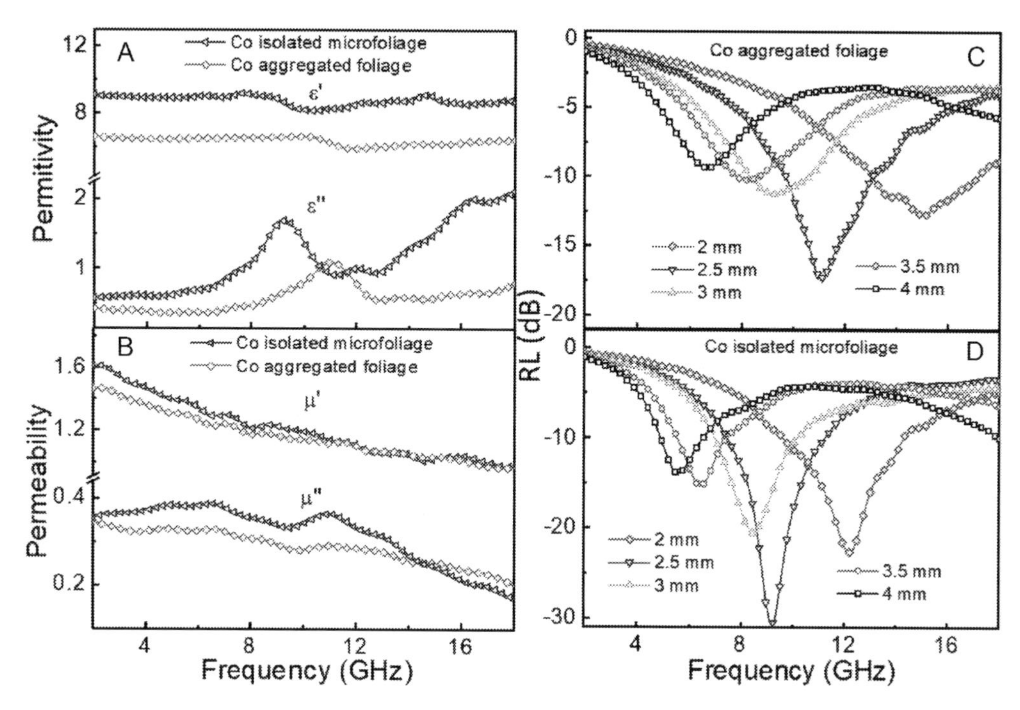

Figure 26.3 SEM properties of the microfoliage and isolated foliage Co structures (a) real part, the imaginary part of complex permittivity (b) real part, the imaginary part of complex permeability (c,d) RL plots (with permission, R. Kuchi., J.R. Jeong are the authors).

to obtain the desired porous spheres. Recently, porous Fe_3O_4 NS (p-Fe_3O_4 NS) with controlled porosity via a tune the size of primary nanoparticles, which took part in growth steps to attain the final structure. As the control of the primary nanoparticles, size is an efficient approach to have better EMI shielding performance. Hence, the p-Fe_3O_4 NS formation with controlled primary particles is resilient for EMI absorption applications. A gas bubble–assisted Ostwald ripening mechanism has been proposed to form porous structures. Urea has acted as a porogen (porous reagent) to create the bubbles in the Ostwald process. The p-Fe_3O_4 NS with different primary nanoparticles along with their morphological behavior is shown in Fig. 26.4. There is evidence of the control of primary nanoparticles as a function of the concentration of urea. It is also proved with magnetic hysteresis curves and FMR derivative curves. The M_S values and line widths of FMR curves increase as the size of primary particles increases.

It is verified that p-Fe_3O_4 NS obtained using 1 g urea has a higher surface area and porosity from N_2 adsorption isotherms. This would be favorable to absorb more EM wave radiation than other samples. The EMI absorption properties for various p-Fe_3O_4 NS have been evaluated and the results are displayed in Fig. 26.5. An excellent EM wave absorption of p-Fe_3O_4 NS is ascribed to dielectric loss through strong interfacial

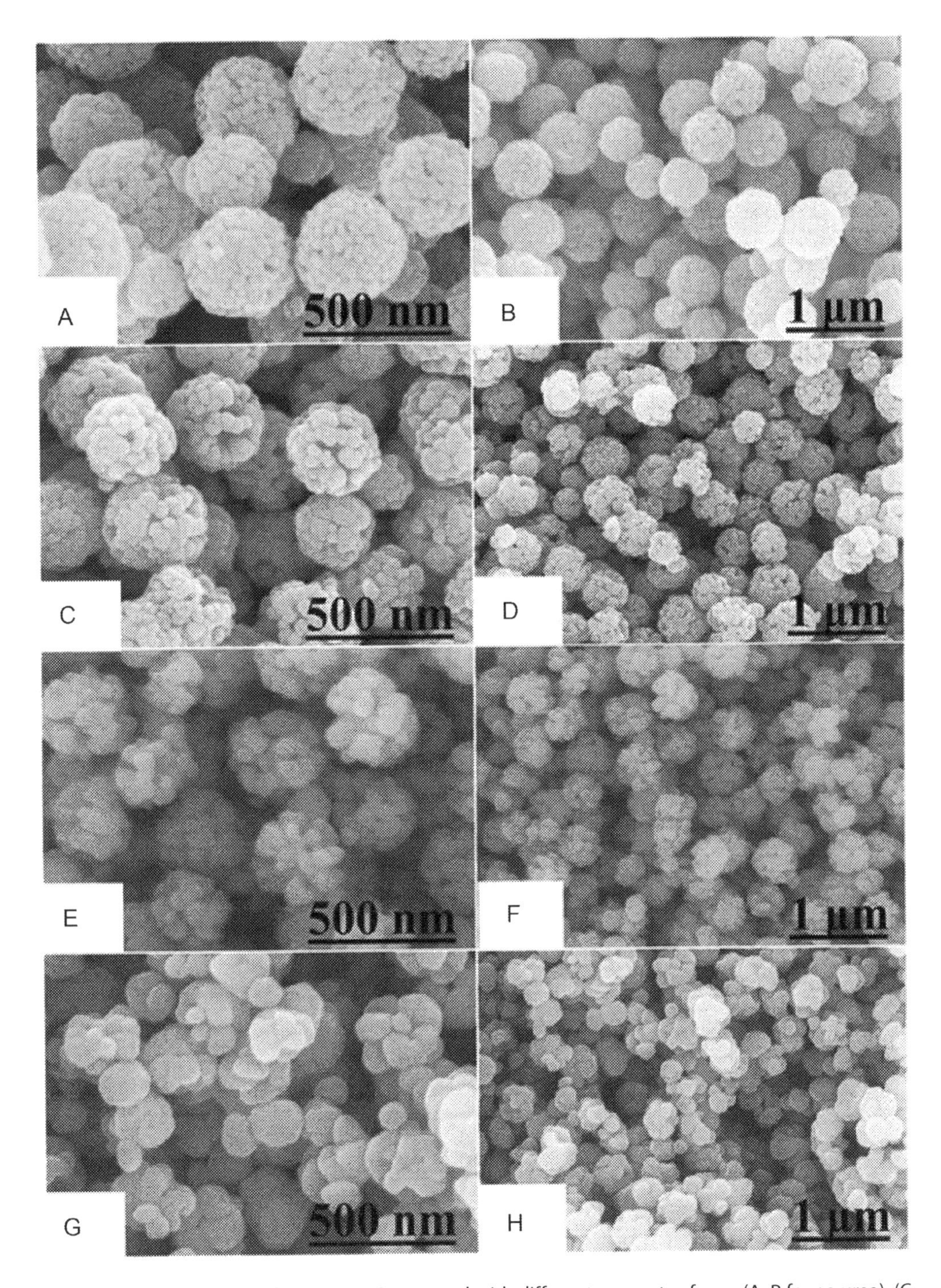

Figure 26.4 SEM images of p- Fe_3O_4 NS prepared with different amounts of urea (A, B for no urea), (C, D for 1 g), (E, F for 2 g), and (G, H for 4 g) (with permission, R. Kuchi., J.R. Jeong are the authors).

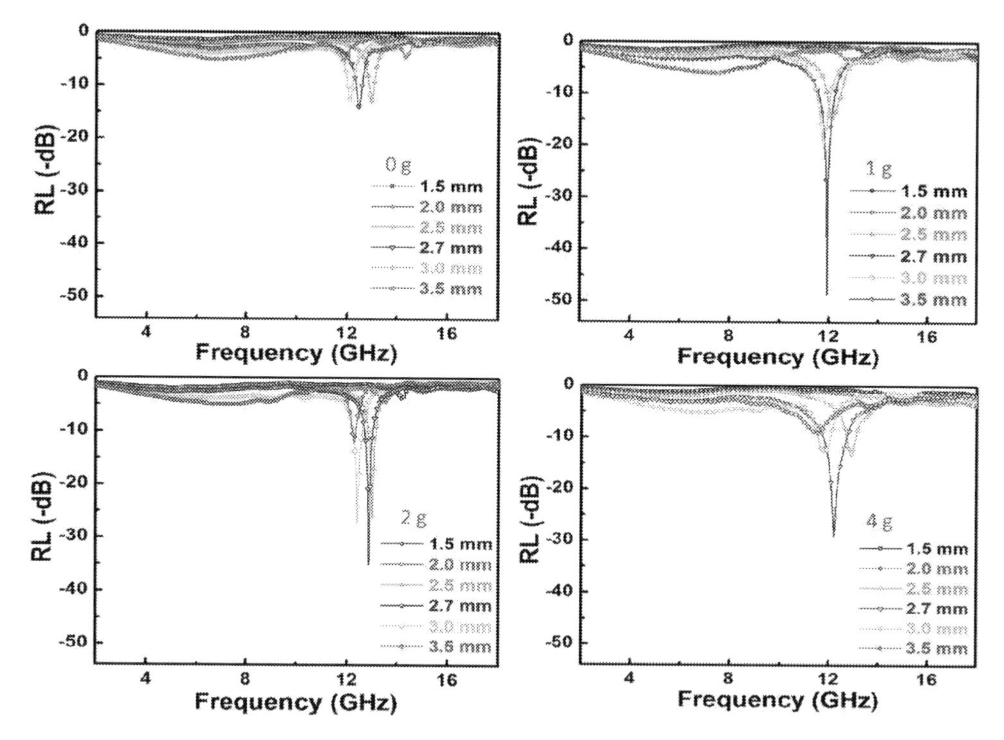

Figure 26.5 The RL for p- Fe_3O_4 NS prepared with different amounts of urea (with permission, R. Kuchi., J.R. Jeong are the authors).

Table 26.1 RL comparison of p-Fe.

Morphology	RL max (dB)	*f* max (GHz)	d (mm)	Reference
Fe_3O_4 spheres	−30.3	13.5	5.5	Jia et al., 2010
Fe_3O_4 flowers	−46	3.4	5.0	Guo et al., 2015
Fe_3O_4 urchins	−29.9	5.0	4.0	Tong et al., 2011
Fe_3O_4 porous structure	−28.3	13.2	2.0	Li et al., 2011
Fe_3O_4 hollow spheres	−43.5	4.0	5.0	Zhang and Zhang, 2014
Fe_3O_4/SnO_2 nanorods	−27.3	16.7	4.0	Chen et al., 2009
Porous Fe_3O_4 nanospheres	−49.2	11.9	2.7	Kuchi et al., 2018

polarization relaxation loss owing to their porous structure. p-Fe_3O_4 NS with 1 g urea reached the highest RL of −49.2 dB, at 11.9 GHz with a matching thickness of 2.7 mm, owing to the porosity, strong dielectric loss, and size effect of the primary nanoparticles. Besides, it has improved EM wave absorption performance compared to the other related reports on the Fe_3O_4 particles as shown in Table 26.1. This research emphasizes the effect of porosity could be very useful to enhance EM wave absorption.

However, the reported Fe_3O_4 structures and p-Fe_3O_4 NS do not have wide effective absorption bandwidth, and matching thickness is high. However, the magnetic metallic particles have other limitations as mentioned above. Therefore, magnetic particles composed of EMI shielding materials solely could not meet the practical requirements (strong RL along with wide absorption bandwidth, flexible, lower density, cost-effectiveness) to apply for EMI pollution remediation.

26.4.2 Magnetic composites

Development of the effective EM absorption material would be carried out through the design of the magnetic composites with a proper combination of ferromagnetic material and dielectric materials including carbon material, semiconducting metal oxides, and conducting polymers.

26.4.2.1 Magnetic composites with carbon

Ferromagnetic composites with carbon material proved as an impressive EM wave absorbing material owing to their control of the EM wave absorption properties via the tuning of composite structures as well as excellent chemical stability. The incorporation of Co with the carbon in the composites resulted in the improved EM wave absorption properties owing to high Snoek's limit and high Curie temperature and higher M_S of Co, which possibly will attenuate EM waves by the magnetic loss (Ding et al., 2017). As the microstructure of the composite material affects the EM wave absorption through the interfacial polarization and synergistic effects, so many researchers tried to design carbon-based composite with Co metallic material. Ding et al. (2017) prepared a core-shell type Co@C microspheres through *in-situ* polymerization followed by reduction at high temperature. It showed an RL of −40 dB at a matching frequency of 12 GHz. Lately, K. Wang et al. (2018) synthesized the Co@C core-shell composites through carbonization of a metal-organic framework at high temperatures and it has the RL of −62.12 dB at 11.8 GHz. These composites are prepared at high temperatures and have multiple steps. They also have flexibility and density problem to be used as advanced applications.

Recently, R. Kuchi et al., (2019) proposed a simple design by developing Co@C core-shell NS on carbon supports using a facile synthetic method that involved a one-pot thermal decomposition and low-temperature heat treatment. The carbon layers as supports and the nano size of Co@C NS favor making lightweight absorbers with a feature of flexibility. Using this approach could control the NS size by adjusting the initial Co precursor. This is observed from Fig. 26.6 that the NS were well dispersed on carbon supports and encapsulated in graphitic carbon layers. The crystallite sizes were calculated from the X-ray diffraction (XRD) patterns (Fig. 26.6) of three samples, which were 10.2, 20.4, and 29.5 nm, respectively. These values are almost similar to the sizes observed in transmission electron microscopy (TEM). The graphitic and distorted carbon natures are evident from the Raman spectra (Fig. 26.6).

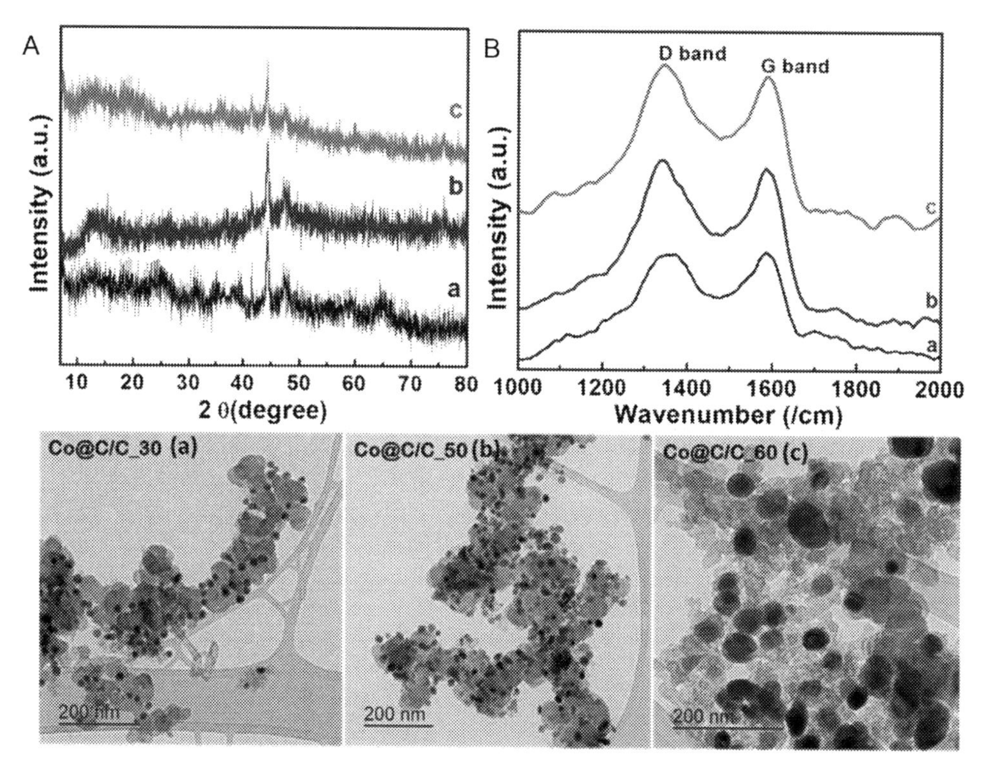

Figure 26.6 XRD pattern, Raman spectra, and TEM images for three different samples (with permission, R. Kuchi., J.R. Jeong are the authors).

Table 26.2 Magnetic properties of Co@C/C composites.

Samples	M_R (emu/g)	M_S (emu/g)	H_c (T)	Isotropic M_R/Ms
10_Co@C/C	6.25	28.5	0.02	0.21
20_Co@C/C	22.0	57.5	0.07	0.40
30_Co@C/C	22.5	78.0	0.06	0.30

The Co@C/C composite's magnetic properties are presented in Table 26.2. It shows the higher M_S value for the larger average size Co NSs. Nevertheless, 20_Co@C/C composite [particle size of Co ~ 20 nm] exhibited higher isotropic over the other samples, which will endow the improved EM wave absorption properties. In contrast, 10_Co@C/C composites [particle size of Co ~10 nm] had very low M_S and H_C values because of the high content of the carbon supports and it will decrease magnetic properties by its nonmagnetic properties. The magnetic and the structural properties could influence the EM properties from which the EM absorption properties can be evaluated. As shown in Fig. 26.7 the 20_Co@C/C composites showed the maximum

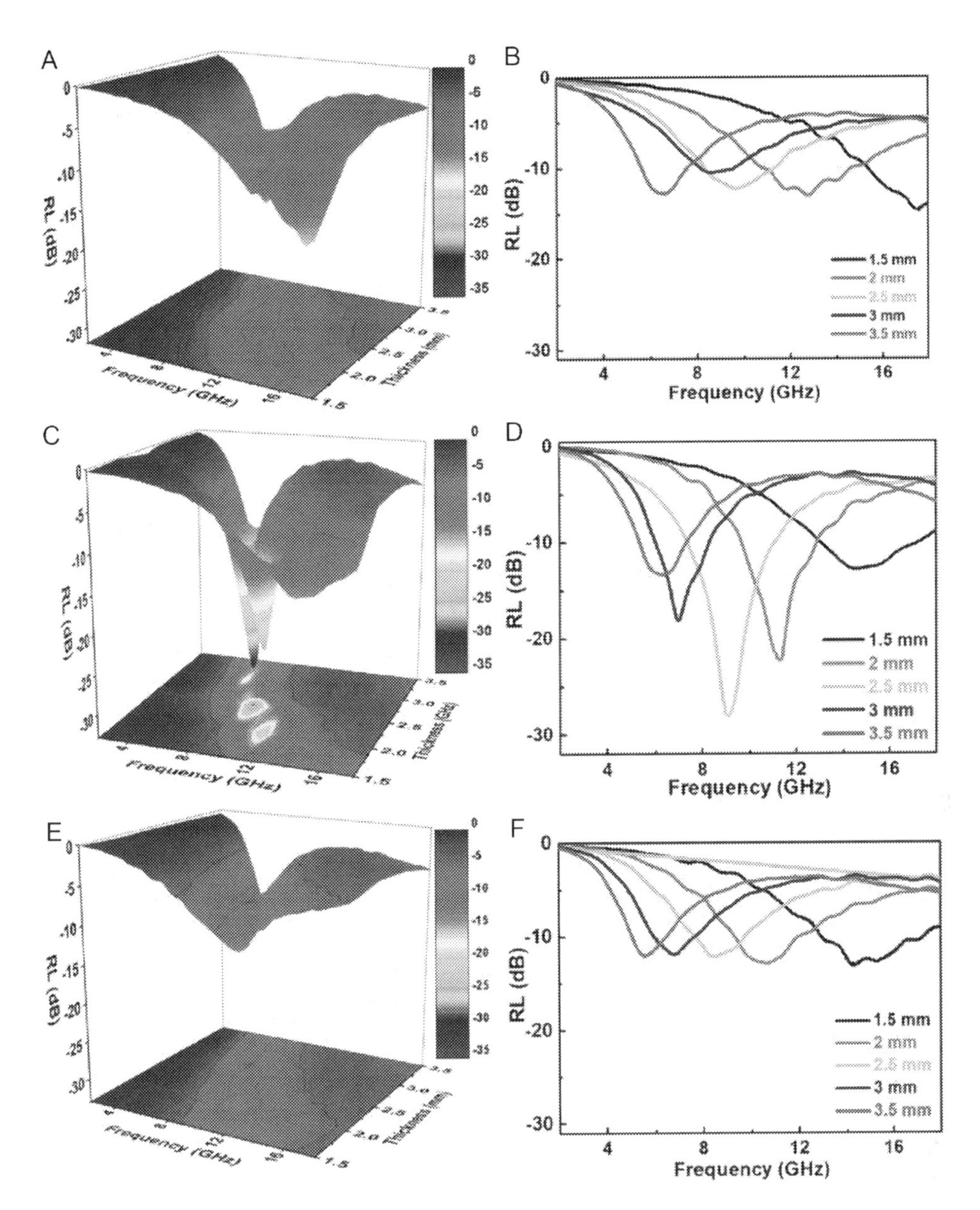

Figure 26.7 RL curves for (A,B) 10_Co@C/C, (C,D) 20_Co@C/C, and (E,F) 30_Co@C/C. (with permission, R. Kuchi., J.R. Jeong are the authors).

RL of −28.2 dB at 9.1 GHz (2.5 mm) along with the wider effective bandwidth (≤−10 dB) of 3.9 GHz (range: 7.28–11.18 GHz). This is better than other reported carbon-based composites. The enhanced EM wave absorption performance is ascribed mainly to the higher isotropic ratio of Co@C/C composites through tuning the size of Co

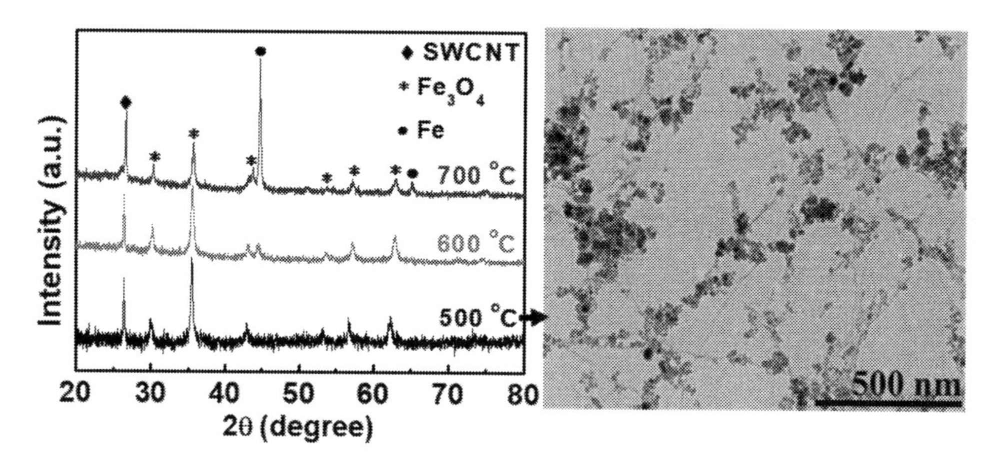

Figure 26.8 XRD and TEM image (500 °C) for the Fe_3O_4/SWCNT nanocomposites (with permission, R. Kuchi., J.R. Jeong are the authors).

particle, thereby the good impedance matching is achieved via the cooperation between the EM losses including dielectric and magnetic losses. This design and study of EM properties gave a new insight into the development of outstanding EM wave absorbers. However, this design needs a lighter absorber with enhanced EM wave absorption properties.

From impressive mechanical and electrical properties to the lightweight feature of carbon nanotubes (CNT), included magnetic composites attained particular interest among the carbonaceous magnetic composites. Notably, a combination of metals or metal oxides with CNT has shown to improve EM wave absorption performance by their combination of magnetic and electric properties. The big advantage is that the high permittivity of CNT can be easily adjustable to get good impedance matching. For example, CNT/Co, CNT/Fe, and CNT/Ni composites showed good EM wave absorption properties from their combination of permittivity and permeability (Wen et al., 2011). The properties can be further maximized by replacing the metals with metal oxide (Fe_3O_4) owing to high spin polarization at room temperature and high compatibility between CNT and Fe_3O_4 (Chen et al., 2017; Kuchi et al., 2017; Kuchi, Dongquoc, et al., 2018; Kuchi, Latif, et al., 2019; Kuchi, Sharma, et al., 2019). As stated earlier that the Fe_3O_4 was an environmentally friendly metal oxide that also possesses good dielectric and magnetic properties.

The Fe_3O_4/CNT composites have been studied by many researchers in recent years and used as an expensive commercially available CNT, the process was complex. An *in-situ* co-arc discharge method was reported to account for these issues. They fabricated the Fe_3O_4/SWCNT nanocomposites cost-effectively with good crystalline products. The different annealing temperatures to obtain the highly integrated composites are shown

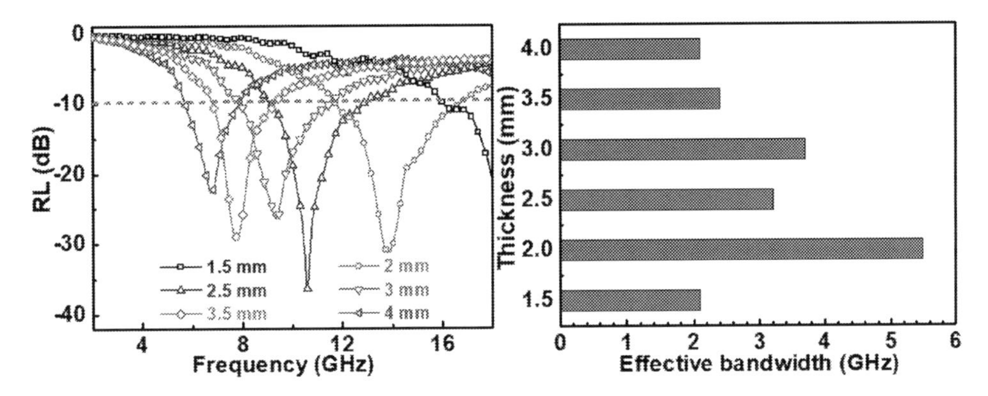

Figure 26.9 EM wave absorption analysis for the optimized Fe_3O_4/SWCNT nanocomposites (with permission, R. Kuchi., J.R. Jeong are the authors).

in Fig. 26.8. From XRD, morphological images show the low-temperature process results in highly integrated Fe_3O_4/SWCNT nanocomposites.

EM wave absorption performance for the optimized Fe_3O_4/SWCNT nanocomposites was carried out [Fig. 26.9]. Even with 30 wt% composites, the host matrix (paraffin wax) shows a strong RL of −36.9 dB at 10.5 GHz with a wide effective bandwidth of 5.5 GHz (11.5–17 GHz) at a 2.5 mm thickness. This is accredited to the superior integration of Fe_3O_4 nanoparticles on SWCNT walls, the cooperation among dielectric and magnetic parameters. Thus, this approach shows great promise for the synthesis and design of superior EM wave absorbers for practical applications. It will also extend to prepare another combination of material with SWCNT.

26.4.2.2 Magnetic composites with semiconductors

The core-shell composite microstructures is another strategy to increase EM wave absorption properties. In this type, we consider the Ni as the core material with different dielectric materials such as polymers, C, SnO_2, TiO_2, SiO_2, ZnO, ZnS, CuO as shell material. Zhao et al. designed Ni spheres wrapped with ZnS, these composites exhibited an RL of −42.4 dB with an effective bandwidth of 4.2 GHz (Zhao, Zhao, et al., 2015). They also reported core-shell composites as Ni@TiO_2 and Ni@SiO_2 and found the better EM wave absorption properties for the SiO_2 shell (Zhao, Shao, Fan, Zhao, Chen, et al., 2015). Deng et al. (2017) reported the composites consisting of Ni spheres coated with ZnO nanorods showed RL of −30.2 dB (). These reports revealed that the shell material plays an important role to achieve high EM wave absorption properties via cooperation between the EM parameters, synergistic effects of core and shell, impedance matching, and interfacial polarization. Enhanced interfacial polarization could be possible because of complex shell structure consisting of many atoms. Additionally, the shell thickness of the control could balance impedance matching. For this purpose, copper

Table 26.3 Ni-based core-shell composites EM wave absorption performance.

Sample	RL max (dB)	Thickness (mm)	Filler weight (percent)	Effective bandwidth (GHz)	Reference
Ni/SnO_2	−18.6	7	70	1.5	Zhao et al., 2014
Ni/polyaniline	−35	5	50	0.3	Dong et al., 2008
Ni/carbon	−13	2	20	2.6	Wu et al., 2016
Ni/ZnO	−30.2	2.2	40	2.5	Deng et al., 2017
Ni/ZnS	−42.4	2.2	40	4.2	Zhao et al., 2015
Ni/SiO_2	−40	1.5	70	3.5	Zhao et al., 2015
Ni/TiO_2	−35.4	4	70	1.0	Zhao et al., 2015
Ni/$CuSiO_3$	−39.5	2.5	30	4.8	Kuchi et al., 2020

silicate ($CuSiO_3$) is the better choice for the shell material to the magnetic core materials. It can be easily prepared and has good dielectric and antioxidation properties, and low cost. The Ni@$CuSiO_3$ composites with honeycomb-like core-shell structures were reported by R. Kuchi et al. (2020). They control the shell thickness by controlling the intermediate sacrificed SiO_2 layer thickness. A simple hydrothermal method combined with the sol–gel process is used to obtain the final composite and the formed different composites are shown in Fig. 26.10.

Controlling the $CuSiO_3$ shell thicknesses as 50, 30, and 20 nm was attained from the relative SiO_2 layer and composites were symbolized as NCS-1, NCS-2, and NCS-3, respectively. These are well crystalline and their EM properties are related to the shell thickness. Further, EM wave absorption analysis indicated the lower reflection for final composites than Ni and Ni@SiO_2 core-shell NSs, because the $CuSiO_3$ shell is effective than the simple Ni@SiO_2. Among the prepared composites, the NCS-2 (shell thickness ~30 nm), exhibited strong RL of −39.5 dB at 9.6 GHz (thickness is 2.5 mm) and widest effective absorbing bandwidth (RL below than 10 dB) of about 4.8 GHz [Fig. 26.11]. They also observed that these composites can be simply controlled EM absorption properties by adjusting the shell thickness.

Ni@$CuSiO_3$ composite NSs have good EM wave absorption performance (higher RL and wider effective bandwidth) than that of other Ni-based composites shown in Table 26.3. This superior performance suggests that the core-shell structure with honeycomb-like surface morphology contributed significantly to enhance EM wave absorption. $CuSiO_3$ shell material is composed of many atoms that create plenty of interfaces and thereby increase the interfacial polarization and efficient multiple scattering/reflection. The novel structure with a new shell material ($CuSiO_3$) could effectively dissipate EM wave energy owing to its high dielectric loss, high magnetic loss, as well as improved interfacial polarization and multiple reflections. This new design of the structure and control of the shell thickness is anticipated to contribute to the progress of high-performance EM wave absorber materials.

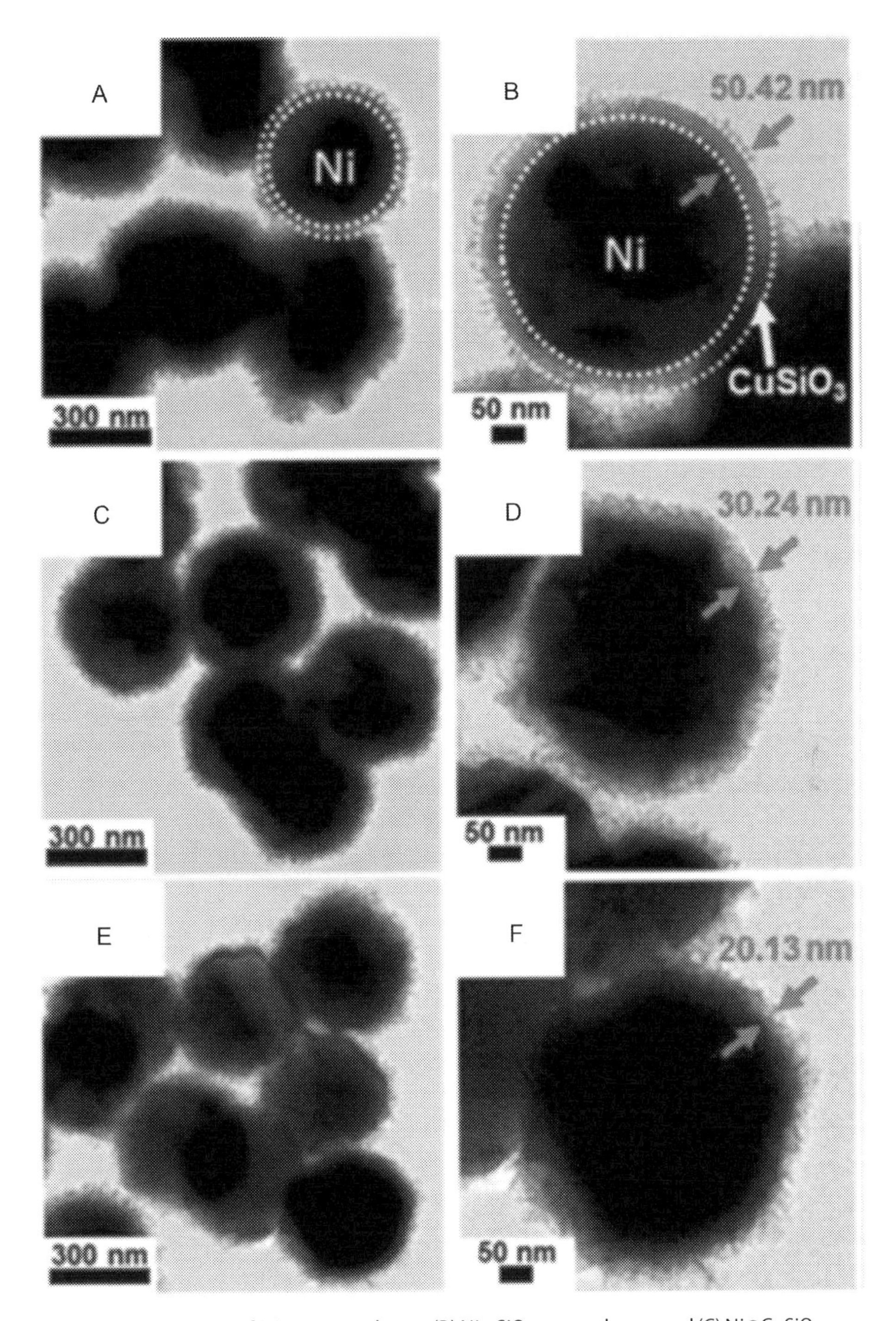

Figure 26.10 TEM images of (A) Ni nanospheres, (B) Ni@SiO_2 nanospheres, and (C) Ni@$CuSiO_3$ composite nanospheres (with permission, R. Kuchi., J.R. Jeong are the authors).

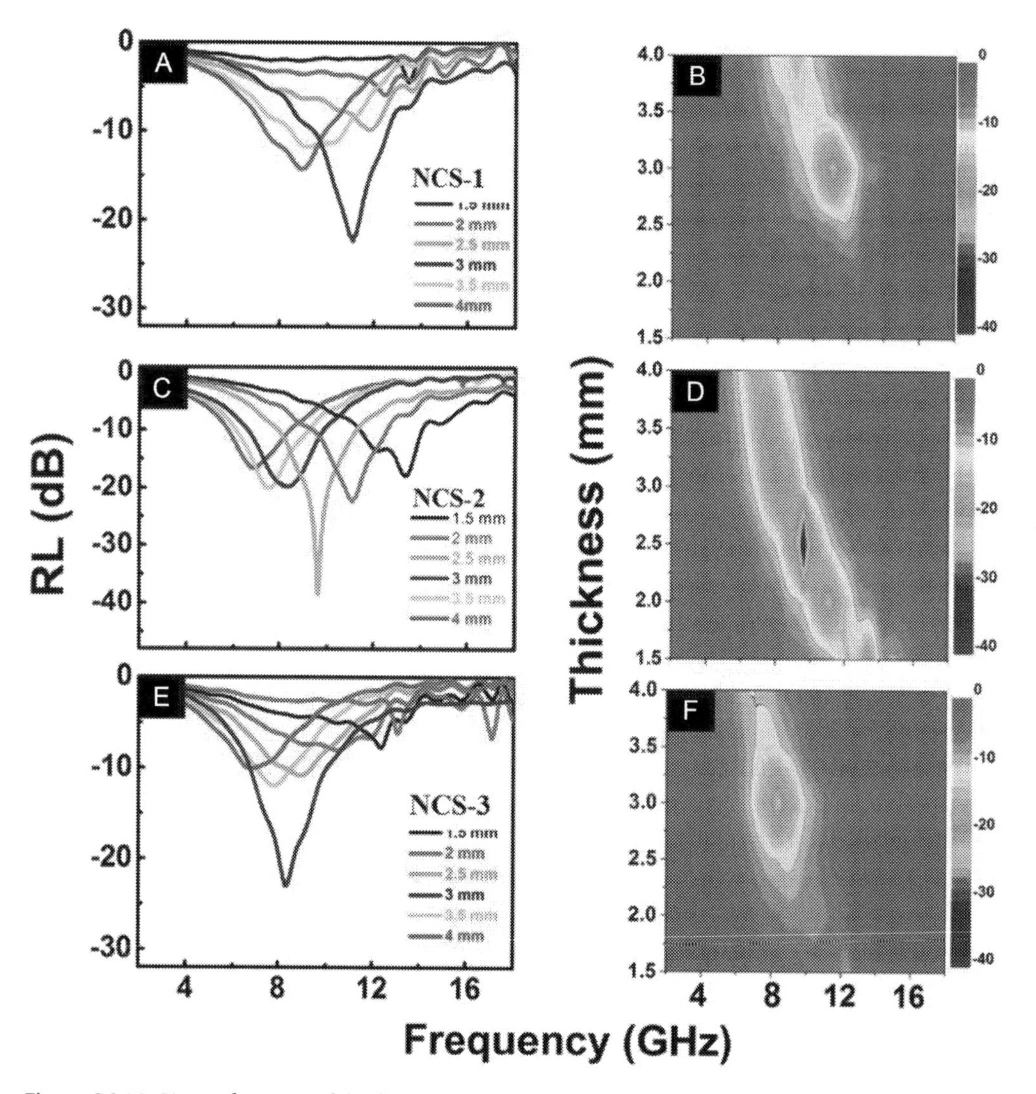

Figure 26.11 RL as a function of the frequency of composites with different shell thicknesses.

26.5 Conclusions

Magnetic particles and magnetic composites are very important to use for the remediation of EMI pollution by their superior EM wave absorption properties. This could be strongly affected by the EM properties and structural properties of the materials. Solely the magnetic particles and magnetic oxides with special structures showed higher EM wave absorption properties but they had certain shortcomings. However, the magnetic composites with the proper combinations consisting of the magnetic particles and different dielectric materials are exploited as potential candidates. In this regard, we demonstrated the various magnetic particles and magnetic composites as EM wave absorbers and their strategies to enhance the EM wave absorption properties. The well-designed composites with controlled morphology could optimize the EM properties then thereby increase EM wave attenuation and good impedance matching. With these features, the magnetic particles and magnetic composites exhibited higher EMI SE. There is still scope to design the new magnetic composites with various kinds of combinations by considering the above-stated conditions. Those magnetic composites would be lightweight with stronger RL and wide absorbing bandwidths to cover the major bands in the microwave range. Hence, this kind of research and development for magnetic composites will be needed for better usage of magnetic composites for environmental remediation purposes. The detailed description of the EMI shielding mechanisms and crucial parameters, which affected loss mechanisms and the various magnetic materials and magnetic composite material design routes, would be helpful to the researchers working in the EM wave absorbers to remediate the EMI pollution.

Acknowledgments

Authors Rambabu Kuchi and Dongsoo Kim would like to acknowledge and thanks National Research Council of Science & Technology (NST) grant by the Korea government (MSIT) (No.CRC-15-06-KIGAM), and Jong-Ryul Jeong would like to acknowledge and thanks the National Research Foundation of Korea (NRF) grant funded by the Korea government (No. 2020R1A2C100613611).

References

Agarwal, P.R., Kumar, R., Kumari, S., Dhakate, S.R., 2016. Three-dimensional and highly ordered porous carbon-MnO_2 composite foam for excellent electromagnetic interference shielding efficiency. RSC Adv. 6 (103), 100713–100722. https://doi.org/10.1039/c6ra23127f.

Bora, P.J., Azeem, I., Vinoy, K.J., Ramamurthy, P.C., Madras, G., 2018. Morphology controllable microwave absorption property of polyvinylbutyral (PVB)-MnO_2 nanocomposites. Compos. B: Eng. 132, 188–196. https://doi.org/10.1016/j.compositesb.2017.09.014.

Cao, M., Wang, X., Cao, W., Fang, X., Wen, B., Yuan, J., 2018. Thermally driven transport and relaxation switching self-powered electromagnetic energy conversion. Small 14 (29), 1800987. https://doi.org/10.1002/smll.201800987.

Chen, Y., Gao, P., Wang, R., Zhu, C., Wang, L., Cao, M., Jin, H., 2009. Porous Fe_3O_4/SnO_2 core/shell nanorods: synthesis and electromagnetic properties. J. Phys. Chem. C 113, 10061–10064.

Chen, N., Jiang, J.T., Xu, C.Y., Yuan, Y., Gong, Y.X., Zhen, L., 2017. Co_7Fe_3 and $Co_7Fe_3@SiO_2$ nanospheres with tunable diameters for high-performance electromagnetic wave absorption. ACS Appl. Mater. Interf. 9 (26), 21933–21941. https://doi.org/10.1021/acsami.7b03907.

Clark, D.E., Folz, D.C., West, J.K., 2000. Processing materials with microwave energy. Mater. Sci. Eng. A 287 (2), 153–158. https://doi.org/10.1016/s0921-5093(00)00768-1.

Das, S., Mukhopadhyay, A.K., Datta, S., Basu, D., 2008. Prospects of microwave processing: an overview. Bull. Mater. Sci. 31 (7), 943–956. https://doi.org/10.1007/s12034-008-0150-x.

Deng, J., Li, S., Zhou, Y., Liang, L., Zhao, B., Zhang, X., Zhang, R., 2018. Enhancing the microwave absorption properties of amorphous CoO nanosheet-coated Co (hexagonal and cubic phases) through interfacial polarizations. J. Colloid Interf. Sci. 509, 406–413. https://doi.org/10.1016/j.jcis.2017.09.029.

Deng, J., Wang, Q., Zhou, Y., Zhao, B., Zhang, R., 2017. Facile design of a ZnO nanorod-Ni core-shell composite with dual peaks to tune its microwave absorption properties. RSC Adv., 7 (15), 9294–9302. https://doi.org/10.1039/c6ra28835a.

Ding, D., Wang, Y., Li, X., Qiang, R., Xu, P., Chu, W., Han, X., Du, Y., 2017. Rational design of core-shell Co@C microspheres for high-performance microwave absorption. Carbon 111, 722–732. https://doi.org/10.1016/j.carbon.2016.10.059.

Dong, X.L., Zhang, X.F., Huang, H., Zuo, F., 2008. Enhanced microwave absorption in Ni/polyaniline nanocomposites by dual dielectric relaxations. Appl. Phys. Lett. 92, 013127.

Geetha, S., Kumar, K.K.S., Rao, C.R.K., Vijayan, M., Trivedi, D.C., 2009. EMI shielding: methods and materials—a review. J. Appl. Polym. Sci. 112 (4), 2073–2086. https://doi.org/10.1002/app.29812.

Guo, C., Xia, F., Wang, Z., Zhang, L., Xi, L., Zuo, Y., 2015. Flowerlike iron oxide nanostructures and their application in microwave absorption. J. Alloys Compd. 631, 183–191.

Higashiyama, J., Tarusawa, Y., Hikage, T., Nojima, T., 2012. EMI risk assessment of electromagnetic field from mobile phone in elevator cabin for implantable pacemaker. In: Proc. IEEE International Symposium on Electromagnetic Compatibility.

Jia, K., Zhao, R., Zhong, J., Liu, X., 2010. Preparation and microwave absorption properties of loose nanoscale Fe_3O_4 spheres. J. Magn. Magn. Mater. 322, 2167–2171.

Jian, X., Wu, B., Wei, Y., Dou, S.X., Wang, X., He, W., Mahmood, N., 2016. Facile synthesis of Fe_3O_4/GCs composites and their enhanced microwave absorption properties. ACS Appl. Mater. Interf. 8 (9), 6101–6109. https://doi.org/10.1021/acsami.6b00388.

Kawamura, Y., Hikage, T., Nojima, T., Fukui, K., Fujimoto, H., Toyoshima, T., 2012. Experimental estimation of EMI from electronic article surveillance on implantable cardiac pacemakers and implantable cardioverter defibrillators: interference distance and clinical estimation. Trans. Jpn. Soc. Med. Biol. Eng 50, 289–298.

Kuchi, R., Dongquoc, V., Kim, D., Yoon, S.G., Park, S.Y., Jeong, J.R., 2017. Large-scale room-temperature aqueous synthesis of Co superstructures with controlled morphology, and their application to electromagnetic wave absorption. Metal. Mater. Int. 23 (2), 405–411. https://doi.org/10.1007/s12540-017-6456-8.

Kuchi, R., Dongquoc, V., Surabhi, S., Kim, D., Yoon, S.G., Park, S.Y., Choi, J., Jeong, J.R., 2018. Porous Fe_3O_4 nanospheres with controlled porosity for enhanced electromagnetic wave absorption. Phys. Status Solidi A 215 (20), 1701032. https://doi.org/10.1002/pssa.201701032.

Kuchi, R., Latif, T., Lee, S.W., Dongquoc, V., Van, P.C., Kim, D., Jeong, J.R., 2020. Controlling the electric permittivity of honeycomblike core shell Ni/$CuSiO_3$ composite nanospheres to enhance microwave absorption properties. RSC Adv. 10, 1172–1180.

Kuchi, R., Nguyen, H.M., Dongquoc, V., Van, P.C., Surabhi, S., Yoon, S.G., Kim, D., Jeong, J.R., 2018. In-situ co-arc discharge synthesis of Fe_3O_4/SWCNT composites for highly effective microwave absorption. Phys. Status Solidi A 215 (20), 1700989. https://doi.org/10.1002/pssa.201700989.

Kuchi, R., Sharma, M., Lee, S.W., Kim, D., Jung, N., Jeong, J.R., 2019. Rational design of carbon shell-encapsulated cobalt nanospheres to enhance microwave absorption performance. Prog. Nat. Sci. Mater. Int. 29 (1), 88–93. https://doi.org/10.1016/j.pnsc.2019.03.013.

Kumar, A., Singh, A.P., Kumari, S., Srivastava, A.K., Bathula, S., Dhawan, S.K., Dutta, P.K., Dhar, A., 2015. EM shielding effectiveness of Pd-CNT-Cu nanocomposite buckypaper. J. Mater. Chem. A 3 (26), 13986–13993. https://doi.org/10.1039/c4ta05749j.

Li, Z.W., Chen, L., Ong, C.K., 2002. Studies of static and high-frequency magnetic properties for M-type ferrite $BaFe_{12-2x}Co_xZr_xO_{19}$. J. Appl. Phys. 92 (7), 3902–3907. https://doi.org/10.1063/1.1506387.

Li, X., Zhang, B., Ju, C., Han, X., Du, Y., Xu, P., 2011. Morphology-controlled synthesis and electromagnetic properties of porous Fe_3O_4 nanostructures from iron alkoxide precursors. J. Phys. Chem. C 115, 12350–12357.

Liu, J., Cao, W.Q., Jin, H.B., Yuan, J., Zhang, D.Q., Cao, M.S., 2015. Enhanced permittivity and multi-region microwave absorption of nanoneedle-like ZnO in the X-band at elevated temperature. J. Mater. Chem. C 3 (18), 4670–4677. https://doi.org/10.1039/c5tc00426h.

Liu, J., Cheng, J., Che, R., Xu, J., Liu, M., Liu, Z., 2013. Synthesis and microwave absorption properties of yolk-shell microspheres with magnetic iron oxide cores and hierarchical copper silicate shells. ACS Appl. Mater. Interf. 5 (7), 2503–2509. https://doi.org/10.1021/am3030432.

Liu, Y., Wei, S., Xu, B., Wang, Y., Tian, H., Tong, H., 2014. Effect of heat treatment on microwave absorption properties of Ni-Zn-Mg-La ferrite nanoparticles. J. Magnet. Magnet. Mater. 349, 57–62. https://doi.org/10.1016/j.jmmm.2013.08.054.

Lv, H., Ji, G., Liang, X.H., Zhang, H., Du, Y., 2015. A novel rod-like MnO_2@Fe loading on graphene giving excellent electromagnetic absorption properties. J. Mater. Chem. C 3 (19), 5056–5064. https://doi.org/10.1039/c5tc00525f.

Lv, H., Ji, G., Wang, M., Shang, C., Zhang, H., Du, Y., 2014. FeCo/ZnO composites with enhancing microwave absorbing properties: effect of hydrothermal temperature and time. RSC Adv. 4 (101), 57529–57533. https://doi.org/10.1039/c4ra09862e.

Meng, F., Wang, H., Huang, F., Guo, Y., Wang, Z., Hui, D., Zhou, Z., 2018. Graphene-based microwave absorbing composites: a review and prospective. Compos. B Eng. 137, 260–277. https://doi.org/10.1016/j.compositesb.2017.11.023.

Nakamura, T., 2000. Snoek's limit in high-frequency permeability of polycrystalline Ni-Zn, Mg-Zn, and Ni-Zn-Cu spinel ferrites. J. Appl. Phys. 88 (1), 348–353. https://doi.org/10.1063/1.373666.

Nojima, T., Tarusawa, Y., 2002. A new EMI test method for electronic medical devices exposed to mobile radio wave. Electron. Commun. Jpn 85 (4), 1–9. https://doi.org/10.1002/ecja.1085.

O., A., S., D, 2008. Generalization of Snoek's law to ferromagnetic films and composites. Phys. Rev. B 77, 104440. https://doi.org/10.1103/physrevb.77.104440.

Palisek, L., Suchy, L., 2011. High power microwave effects on computer networks. In: Proc. EMC Europe 2011 York - 10th International Symposium on Electromagnetic Compatibility, pp. 18–21.

Rao, S., Sathyanarayanan, A., Nandwani, U.K., 1999. EMI problems for medical devices. In: Proc. International Conference on Electromagnetic Interference and Compatibility. IEEE, pp. 21–24.

Saini, P., Arora, M., 2012. Microwave absorption and EMI shielding behavior of nanocomposites based on intrinsically conducting polymers, graphene and carbon nanotubes. In: Gomes, A.D. (Ed.), New Polymers for Special Applications. Intechopen, London, UK, pp. 71–112.

Sanskaran, S., Ravishankar, B.N., Sekhar, K.R., Dasgupta, S., Kumar, M.N., 2017. Syntactic foams for multifunctional applications. In: Kar, K.K. (Ed.), Composite Materials: Processing, Applications, Characterizations. Springer, Berlin, Heidelberg, pp. 281–314.

Shukla, V., 2019. Review of electromagnetic interference shielding materials fabricated by iron ingredients. Nanoscale Adv. 1 (5), 1640–1671. https://doi.org/10.1039/c9na00108e.

Singh, A.K., Shishkin, A., Koppel, T., Gupta, N., 2018. A review of porous lightweight composite materials for electromagnetic interference shielding. Compos. B Eng. 149, 188–197. https://doi.org/10.1016/j.compositesb.2018.05.027.

Song, W.L., Cao, M.S., Wen, B., Hou, Z.L., Cheng, J., Yuan, J., 2012. Synthesis of zinc oxide particles coated multiwalled carbon nanotubes: dielectric properties, electromagnetic interference shielding and microwave absorption. Mater. Res. Bull. 47 (7), 1747–1754. https://doi.org/10.1016/j.materresbull.2012.03.045.

Thostenson, E.T., Chou, T.W., 1999. Microwave processing: fundamentals and applications. Compos. A Appl. Sci. Manufact. 30 (9), 1055–1071. https://doi.org/10.1016/S1359-835X(99)00020-2.

Tong, G., Wu, W., Guan, J., Qian, H., Yuan, J., Li, W., 2011. Synthesis and characterization of nanosized urchin-like α-Fe_2O_3 and Fe_3O_4: Microwave electromagnetic and absorbing properties. J. Alloys Compd. 509, 4320–4326.

Wang, C., Han, X., Xu, P., Wang, J., Du, Y., Wang, X., Qin, W., Zhang, T., 2010. Controlled synthesis of hierarchical nickel and morphology-dependent electromagnetic properties. J. Phys. Chem. C 114 (7), 3196–3203. https://doi.org/10.1021/jp908839r.

Wang, K., Chen, Y., Tian, R., Li, H., Zhou, Y., Duan, H., Liu, H., 2018. Porous Co-C core-shell nanocomposites derived from Co-MOF-74 with enhanced electromagnetic wave absorption performance. ACS Appl. Mater. Interf. 10 (13), 11333–11342. https://doi.org/10.1021/acsami.8b00965.

Wen, B., Cao, M., Lu, M., Cao, W., Shi, H., Liu, J., Wang, X., Jin, H., Fang, X., Wang, W., Yuan, J., 2014. Reduced graphene oxides: light-weight and high-efficiency electromagnetic interference shielding at elevated temperatures. Adv. Mater. 26 (21), 3484–3489. https://doi.org/10.1002/adma.201400108.

Wu, N., Liu, X., Zhao, C., Cui, C., Xia, A., 2016. Effects of particle size on the magnetic and microwave absorption properties of carbon-coated nickel nanocapsules. J. Alloys Compd. 656, 628–634.

Xia, X., Mazzeo, A.D., Zhong, Z., Weng, G.J., 2017. An X-band theory of electromagnetic interference shielding for graphene-polymer nanocomposites. J. Appl. Phys. 122 (2), 025104. https://doi.org/10.1063/1.4992074.

Zhang, Y., Zhang, Z., 2014. Structure and electromagnetic properties of single-crystalline Fe_3O_4 hollow nanospheres. J. Nanosci. Nanotechnol. 14, 4664–4669.

Zhao, B., Guo, X., Zhao, W., Deng, J., Shao, G., Fan, B., Bai, Z., Zhang, R., 2016. Yolk-shell Ni@SnO_2 composites with a designable interspace to improve the electromagnetic wave absorption properties. ACS Appl. Mater. Interf. 8 (42), 28917–28925. https://doi.org/10.1021/acsami.6b10886.

Zhao, B., Shao, G., Fan, B., Li, W., Pian, X., Zhang, R., 2014. Enhanced electromagnetic wave absorption properties of Ni–SnO_2 core–shell composites synthesized by a simple hydrothermal method. Mater. Lett. 121, 118–121.

Zhao, B., Shao, G., Fan, B., Zhao, W., Chen, Y., Zhang, R., 2015. Facile synthesis of crumpled ZnS net-wrapped Ni walnut spheres with enhanced microwave absorption properties. RSC Adv. 5 (13), 9806–9814. https://doi.org/10.1039/c4ra15411h.

CHAPTER 27

Role of nanotechnology in enhancing crop production and produce quality

Muhammad Ashar Ayub[a,b], Asif Naeem[c], Muhammad Zia ur Rehman[a], Zia Ur Rahman Farooqi[a], Wajid Umar[d], Hina Fatima[e], Muhammad Nadeem[a] and Muhammad Shabaan[a]

[a]Institute of Soil and Environmental Sciences, Faculty of Agriculture, University of Agriculture Faisalabad, Pakistan
[b]Indian River Research and Education Center, Institute of Food and Agricultural Science, University of Florida, USA
[c]Institute of Plant Nutrition and Soil Science, Christian-Albrechts-Universität zu Kiel, Kiel, Germany
[d]School of Environmental Science, Hungarian University of Agriculture and Life Sciences, Gödöllő, Hungary
[e]School of Applied Biosciences, Kyungpook National University, Daegu, South Korea

27.1 Introduction

To fulfill the ever-increasing demand for food which is expected to increase by 70% until 2050 (Dimkpa & Bindraban, 2017), it is imperative to incorporate nanotechnology in agriculture. As nanotechnology as obvious from its name works at the scale of the one-billionth meter (Chhipa, 2017), it offers tremendous potential to revolutionize agriculture production. The concept of nanotechnology was first proposed by famous Physicist Richard Feynman (Feynman, 1960). However, the word "nanotechnology" was the first time used by K. Eric Drexler in his book *Engines of Creation: The Coming Era of Nanotechnology* (Eric, 1986). For the last four decades, scientists are eagerly working to explore new areas of nanoscience. Nanotechnology is extensively being used in medicine (Cheon et al., 2019), cancer treatment (Chitkara et al., 2018) engineering (Pacheco-Torgal, et al., 2018), and in the agriculture sector (Mitter and Karen, 2019). However, some consider it an emerging social challenge, because of potential environmental toxicity and human health hazards by nanoparticles (NPs).

Despite being a major source of food for both humans and domestic animals, agriculture is facing challenges such as low nutrient use efficiency (NUE), drought, crop pests, diseases, and overall low productivity. Like other fields, nanotechnology can revolutionize the agriculture sector as well. In agriculture, nanomaterials such as nanofertilizers and nanopesticides have done wonders (Adisa et al., 2019). First, there is a need to understand why nanomaterials perform better than commercial fertilizers. Nanomaterials are promising owing to their relatively lesser size, the greater surface to volume ratio, high performance, smart, and targeted delivery system (Shojaei et al., 2019).

Sustainable Nanotechnology for Environmental Remediation
DOI: https://doi.org/10.1016/B978-0-12-824547-7.00014-X

The especially small size of nanomaterials plays a game-changing role as this small size imparts a large surface area to the nanomaterials eventually becoming a reason to react with many other compounds. Moreover, they can easily penetrate through the leaves and roots on account of their extremely small size (Liscano et al., 2000). Hence, they improve the uptake, penetration rate, and NUE of minerals. Besides, nanomaterials are capable to release nutrients slowly thus increasing nutrient availability duration for crop plants and preventing it from loss. Also, this results in increased efficiency of nutrients and reduces the risk of environmental contamination (Singh, 2017). This chapter discusses the role of nanomaterials/NPs in crop growth management and agriculture sustainability. NPs of metal and metal oxides are extensively being used for soil, water, and even air remediation (Dhasmana et al., 2019). Few authors comprehensively summarized the role of NPs in solely water remediation (Saikia et al., 2019; Mohammad et al., 2019; Zhang et al., 2019) while others emphasize the importance of green synthesis of NPs and their use for remediation of both water and soil (Wang et al., 2019; Song et al., 2019). Also, the use of NPs can improve food quality and food safety, shelf-life, and may enhance the freshness of preserved foods (Dobrucka & Ankiel, 2019; Hoseinnejad et al., 2018). For example, the combination of silver and copper NPs (Ag-Cu) had fungicidal properties that reduces the negative effects on plants caused by *Listeria monocytogenes* and *Salmonella enterica sv Typhimurium* (Arfat et al, 2017).

Furthermore, metal oxides–based NPs (cerium oxide NPs) have an exclusive role in carbon sequestration, which results in lowering greenhouse gas emission especially carbon dioxide (CO_2) eventually preventing climate change. A bunch of studies (Ghosh & Ramaprabhu, 2019; Zhang et al., 2018; Upadhyay et al., 2019) has been carried out to reveal the potential of metal NPs for carbon sequestration. However, recently metal-organic framework NPs have emerged as a novel material to capture CO_2 and methane (CH_4) (Guo et al., 2018). Thus, these novel techniques have the potential to improve air quality and reduce air pollution. The purpose of this chapter is to provide a comprehensive review of nanomaterials (nanofertilizers, nanopesticides, nanofungicides and metal and metal oxides NPs) and their use in agriculture productivity, sustainability, food security, and improving soil, water, and air quality. This chapter also briefly describes the pros and cons of nanomaterials and the effect of their long-term use on the environment and human health.

27.2 Application of NPs in agriculture

27.2.1 Application of NPs for nutrition management

A robust review of published literature reveals a preponderance of micronutrient nanofertilizers (Cu, Zn, Ag, Fe, TiO_2) over macronutrient nanofertilizers (N, P, K, Ca, Mg, S) despite their importance in the crop productivity (Dimkpa & Bindraban, 2017).

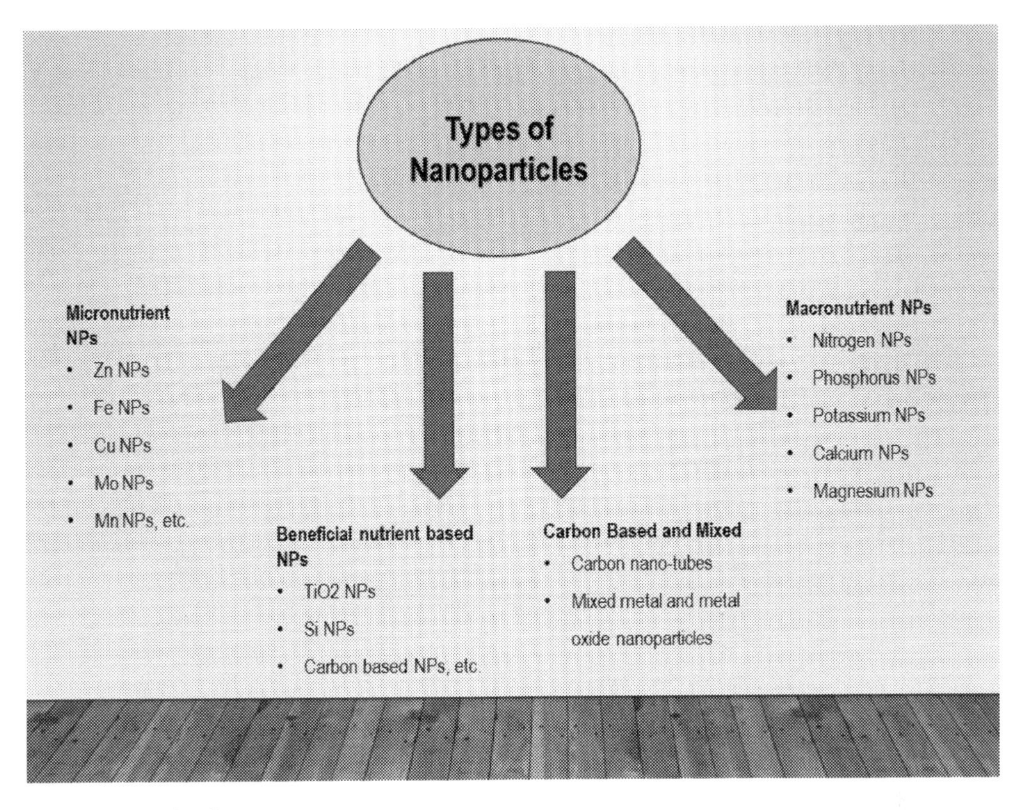

Figure 27.1 Role of NPs in crops.

However, for the past five years, researchers have started working on macronutrient nanofertilizers especially those encapsulating nitrogen and phosphorous. For example, the use of nanourea reduced nitrogen (N) losses up to 74% in rice crops when compared to conventional urea (Kottegoda et al., 2017). The utilization of nanophosphorus (P) would be eco-friendly as it is less soluble and less mobile thus reduces the chances of eutrophication eventually reducing the harm to life in water bodies. However, comparatively less is known about macronutrients encapsulating nanofertilizers and there is a need to explore more (Kottegoda et al., 2017). In contrast, micronutrients are required by plants in amounts less than 10 mg/kg soil but extensive research has been carried out on micronutrients containing nanofertilizers (Dimkpa & Bindraban, 2017) especially Zn, Cu, CuO, TiO_2, and CeO_2. Overall, the use of nano-enabled micronutrient fertilizers caused an improvement in plant metabolism and NUE thus resulting in improved plant growth and development (Adisa et al., 2019). Fig. 27.1 describes the schematic illustration of the processes through which NPs/nanofertilizers increase NUE and crop growth.

27.2.1.1 Macronutrient nanofertilizers

Macronutrient nanofertilizers are the nano-based materials of the primary plant nutrients (nitrogen, phosphorus, and potassium) used to increase NUE plant growth, and production. Uptake efficiencies of conventional fertilizers are generally low, and they are required in larger amounts. Low plant uptake efficiencies of nitrogenous and phosphatic fertilizers have two main reasons: (1) A quick change into the chemical forms that plant cannot uptake and (2) leaching, runoff, and environmental losses (Raliya et al., 2018). This results in the generation of potent greenhouse gases such as oxides of nitrogen, and eutrophication with negative impacts on the environment. Therefore, it is a dire need of the time to develop smart materials as fertilizers that are readily available and taken up by plants (Table 27.1).

27.2.1.2 Nitrogen fertilizer

Nitrogen is the primary macronutrient considered as one of the key nutrients for plant growth. Most of the agricultural soils are nitrogen deficient. Although N concentration in the environment is 78%, plants cannot use environmental nitrogen (N_2) (Raliya et al., 2018). Plants only uptake nitrogen in specific chemical forms such as ammonium and nitrates. Commercially all the N fertilizers are formed by the Haber and Bosch process in which hydrogen and environmental N are used as input materials (Esteves et al., 2015). Nitrogenous fertilizers play an important role in maintaining food supply by maintaining or enhancing plant growth and production (Jat et al., 2015). Urea is one of the common nitrogenous fertilizers that has very low use efficiency as 75% of urea fertilizer is lost as leaching and ammonium volatilization soon after the application. Current nitrogenous fertilizers have very low use efficiency ($<20\%$) which results in environmental pollution by eutrophication and greenhouse gas emission (Kahrl et al., 2010). Recently, a group of scientists formulated a nanotechnology-based nitrogenous fertilizer by coating hydroxyapatite NPs with urea for slow release and targeted delivery of N in the Sri Lankan Institute of nanotechnology (Kottegoda et al., 2017). The nanohybrids of urea-hydroxyapatite were formulated by 40% nitrogen weight, which released N 12 times slower than the conventional fertilizers. Field trials conducted by using above mentioned nanofertilizer showed that rice crop showed better yield at 50% lower urea concentration.

27.2.1.3 Phosphorus nanofertilizers

Along with N, P is also an important macronutrient essential for animals and plants. A major part of mined P ($> 90\%$) all over the world is used for food production (Dawson and Hilton, 2011). Rock phosphate is the base material used for most of the commercial fertilizer production. Worldwide rock phosphate reserves are restricted and according to geological survey reports, these reserves will only last for the coming 100 years. However, the increase in the global population increased the consumption of phosphatic fertilizers.

Table 27.1 Effect of macro, micro, and beneficial nutrient nano-fertilizers in plants.

Experiment	Nanofertilizer	Crop	Results	References
Macronutrient Nanofertilizer				
Greenhouse experiment	Nanozeourea	*Zeya mays*	28% more N contents in grains than urea fertilized plants	Manikandan and Subramanian, 2016
Greenhouse experiment	Chelates of nano nitrogen (NNC) and sulfur-coated NNC (SNNC)	*Solanum tuberosum*	Reduction in nitrate leaching by 35.72% and 41.56% against NNC and SNNC treatment	Zareabyaneh and Bayatvarkeshi, 2015
Growth room experiment	Clinoptilolite-NH_4	*Lolium multiflorum*	Increased use in nitrogen use efficiency as well as yield	Millán et al., 2008
Field experiment	Nanourea modified with hydroxyapatite	*Prunus dulcis*	Increased germination measurements, stem lengths and diameters, and other growth parameters due to the use of N-NPs @ 50% concentration. Besides, strong seedlings were produced due to the slow and continuous release of fertilizer	Badran and Savin, 2018
Field experiment	NPK nanofertilizers	*Solanum tuberosum*	High economic yield, starch rates, and NPK nutrient use efficiency were observed by NPK nanofertilizers @50% and 25% of recommended levels	Abd El-Azeim et al., 2020

(continued on next page)

Table 27.1 Effect of macro, micro, and beneficial nutrient nano-fertilizers in plants—cont'd

Experiment	Nanofertilizer	Crop	Results	References
Field experiment	Nano-nitrogen (nN)	*Punica granatum*	nN level of 1.8 kg N/ha increased fruit length, aril juice, total soluble solids, and titratable acidity. Also, fruit yield and fruit quality were improved too as compared to urea fertilization	Davarpanah et al., 2017
Glasshouse study	Hydroxyapatite nanoparticles	*Triticum aestivum*	Only 5% of phosphorous leaching losses	Montalvo et al., 2015
Greenhouse experiment	Synthetic apatite nanoparticles	*Glycine max*	The plant exhibited increased growth and yield in comparison with regular P fertilizer. Moreover, increment in biomass was also observed.	Liu and Lal, 2015
Glasshouse pot experiment	Hydroxyapatite nanoparticles (HA-NPs)	*Helianthus annuus*	Increase in shoot fresh biomass by16.5 folds and fresh root biomass by eightfold	Xiong et al., 2018
Pot experiment	Calcium phosphate nanoparticles (Ca-PNPs)	*Zeya mays*	Improvement in chlorophyll A contents and performance index. Also, growth promotion with simultaneous increment in root proliferation	Rane et al., 2015

(continued on next page)

Field experiment	NPK nanofertilizers	*Solanum tuberosum*	Observance of high economic yield (23.50 ton/ha), starch rates (79.62%), primary macronutrient use efficiency (67.74, 278.92, 118.54 kg potatoes/kg nutrient) were observed by NPK nanofertilizers @50% and 25% of recommended levels	Abd El-Azeim et al., 2020
Greenhouse experiment	Nanoparticles of potassium sulfate	*Medicago sativa*	Application 1/8 level of K_2SO_4-NPs caused the highest relative yield, root length, and dry weights of roots and shoots. Also, the Na/K ratio was significantly affected by nanoparticle application. In addition, reduction in electrolyte leakage, increment in catalase, and proline contents were observed with 1/8 level of K_2SO_4-NPs	El-Sharkawy et al., 2017

(continued on next page)

Table 27.1 Effect of macro, micro, and beneficial nutrient nano-fertilizers in plants—cont'd

Experiment	Nanofertilizer	Crop	Results	References
Pot experiment	Potassium-coated zeolite nanoparticle	*Ipomoea aquatic*	Better plant uptake of K in comparison with conventional fertilizers. In addition, improvement in soil physicochemical properties such as cation exchange capacity, moisture, pH, and available P and K	Rajonee et al., 2017
Field experiment	NPK nanofertilizers	*Solanum tuberosum*	High economic yield, starch rates, and NPK nutrient use efficiency were observed by NPK nanofertilizers @50% and 25% of recommended levels	Abd El-Azeim et al., 2020
Hydroponics experiment	Nanoscale calcium oxide particles (n-CaO)	*Hydroponics experiment*	Significant increase in calcium (Ca) contents in roots, shoots, and leaves	Deepa et al., 2015
Growth room experiment	Calcium phosphate NPs	*Oryza Sativa* L	Increase in root length by 3.4% and 6% @ concentration of 10 and 20 mg/L followed by a decline in the root length by 1% @ concentration of 50 mg/L. A similar trend was observed for shoot length, fresh and dry masses of roots and shoots. Similarly, reactive oxygen species production was decreased @ 10 and 20 mg/L in comparison with 50 mg/L.	Upadhyaya et al., 2017

(continued on next page)

Pot experiment	Calcium phosphate nanoparticles (Ca-PNPs)	*Zeya mays*	Improvement in chlorophyll A contents and performance index. In addition, growth promotion with simultaneous increment in root proliferation	Rane et al., 2015
Laboratory experiment	Calcium hydroxide ($Ca(OH)_2$) nanoparticles	-	In the context of CO_2 adsorption, NPs showed eight times higher adsorption capacity than regular Ca $(OH)_2$ particles. In addition, ($Ca(OH)_2$) nanoparticles showed the ability to treat acidic gases such as SO_x, HCl, and NO_x	Zhang et al., 2018
Field experiment	Nano-calcium	*Malus Domestica* L	Overall fruit quality was better in the treatments, where nano-calcium was applied as compared $CaCl_2$ treated plants	Ranjbar et al., 2018
Lab experiment	Mg-NPs	*Vigna unguiculata*	Foliar application of Mg-NPs caused improvements in almost all the growth traits as well as physiological traits such as plasma membrane stability, chlorophyll contents. In addition, stem Mg contents were also increased	Delfani et al., 2014

(*continued on next page*)

Table 27.1 Effect of macro, micro, and beneficial nutrient nano-fertilizers in plants—cont'd

Experiment	Nanofertilizer	Crop	Results	References
Pot experiment	Nano S modified zeolite	*Arachis hypogaea*	Yield attributes such as pod and haulm yield were significantly increased @ 30 kg S/ha by 12.4% and 26.8%, respectively. In addition, oil and crude protein contents were increased by 49.87% and 29.38%. Moreover, a significant increase in total free amino acids, methionine, and cystine was observed too.	Thirunavukkarasu and Subramanian, 2014
Greenhouse experiment	Sulfur nanoparticles (SNPs)	*Solanum Lycopersicum*	High-quality tomatoes were produced by the foliar application of SNPs @ concentrations of 200 ppm	Salem et al., 2016
Micronutrient Nanofertilizer				
Greenhouse experiment	ZnO nanoparticles	*Ocimum basilicum*	Enhancement in all the growth traits of plant and Zn contents	El-kereti et al., 2013
Field experiment	Nano-Fe, nano-Zn and nano-TiO_2	*Hordeum vulgare*	The application of all three types of nanoparticles caused an increment in the days to anthesis and maturity. In addition, an enormous improvement in the spike length, grain mass and grains number per spike, etc.	Janmohammadi et al., 2016

(continued on next page)

Growth room study	ZnO-NPs	*Cicer arietinum*	Overall improvement in the biomass accumulation in NPs treated seedlings. Low malondialdehyde contents	Burman et al., 2013
Solution-culture experiment	Nano-Zn oxide	*Zeya mays*	Improvement in almost all the growth parameters. Also, the enzyme activities of the maize plant were enhanced	Adhikari et al., 2016
Field experiment	Zinc complexed-chitosan nanoparticles (Zn-CNPs)	*Triticum aestivum*	Increase in grain Zn contents with a simultaneous positive effect on growth parameters	Dapkekar et al., 2018
Laboratory experiment	Nano-scale ZnO	*Arachis hypogaea*	Seedling vigor and seed germination were improved at a concentration of 1000 ppm, early flowering and higher chlorophyll contents	Prasad et al., 2012
Field experiment	ZnONPs	*Zeya mays*	ZnONPs @ 1500 ppm exhibited and highest germination percentage and (80%) and seedling vigor index (1923.20). In addition, 42% more yield than control	Subbaiah et al., 2016

(*continued on next page*)

Table 27.1 Effect of macro, micro, and beneficial nutrient nano-fertilizers in plants—cont'd

Experiment	Nanofertilizer	Crop	Results	References
Sand culture	ZnONPs	*Solanum lycopersicum*	Positive response of ZnONPs on seedling vigor, pigment, protein and sugar contents at lower concentrations	Singh et al., 2016
Pot experiment	ZnONPs	*Brassica oleracea* var. capitata, *Brassica oleracea* var. botrytis and *Solanum lycopersicum*	Increased antioxidant enzyme activities and improvement in plant biomass parameters in all the three crops	Singh et al., 2013
Field experiment	Nanochelate molybdenum (Mo)	*Arachis hypogaea*	MoNPs caused a considerable impact on growth parameters of plant with maximum pods and seed yield of 2320 and 3715 kg/ha	Manjili et al., 2014
Lab experiment	MoNPs	*Cicer arietinum*	Four times increment in nodulation in the treated seeds was observed in comparison with the control	Taran et al., 2014
Growth experiment	MoNPs	*Cicer arietinum*	Significant enhancement in the growth parameters of chickpea such as root length, root area, root diameter, no. of tips, biomass, and overall yield was observed with MoNPs @ 4 mg/kg	Thomas et al., 2017

(continued on next page)

Field experiment	Nanomolybdenum	*Cicer arietinum*	Increment in superoxide dismutase (SOD) activity by 15%, same trend was found for catalase enzyme	Taran et al., 2016
Greenhouse experiment	Mo nanoparticles	*Cicer arietinum*	Maximum number of microflora was observed in NPs treated plants. In addition, increase in the activities of peroxidase with a simultaneous decrease in the malondialdehyde activities was observed	Shcherbakova et al., 2017
Field experiment	Nano-Fe, nano-Zn and nano-TiO_2	*Hordeum vulgare*	Application of all the three types of nanoparticles caused an increment in the days to anthesis and maturity. In addition, enormous improvement in the spike length, grain mass, grains number per spike, etc.	Janmohammadi et al., 2016
Field experiment	Nano Zn-Fe oxide	*Triticum aestivum*	Increment in the SOD, POD, and polyphenol oxidase activities under nanooxide treatment. In addition, 17.40% improvement in grain yield was observed with nanooxide treatment then control	Babaei et al., 2017

(continued on next page)

Table 27.1 Effect of macro, micro, and beneficial nutrient nano-fertilizers in plants—cont'd

Experiment	Nanofertilizer	Crop	Results	References
Laboratory experiment	Iron nanoparticles	*Solanum tuberosum*	Increase in the sprout's length by 55.1%, root lengths by 34.4%, chlorophyll A content by 57% @ concentration of 0.0125 M	Mushinskiy et al., 2019
Germination experiment	Zero-valent iron nanoparticles (nZVI)	*Hordeum vulgare*	Germination was moderately affected @ concentration less than 250 mg/L and was completely inhibited at 1000–2000 mg/L of nZVI.	El-Temsah et al., 2016
Laboratory experiment	MnNPs	*Vigna radiata*	Better phosphorylation and oxygen evolution were observed by treated plants as compared to control.	Pradhan et al., 2013
Germination experiment	Manganese oxide nanoparticles (MO_x NPs)	*Lactuca sativa*	MO_xNPs @ concentration of <50 ppm stimulated growth of lettuce by 12%–54%	Liu et al., 2016
Solution culture experiment	Cu-NPs	*Zeya mays*	Increment in growth by 51% as compared to control. In addition, activity glucose-6-phosphate dehydrogenase	Adhikari et al., 2016
Petri dishes-laboratory experiment	Cu-nanoparticles	*Glycine max* L. and *Cicer arietinum* L.	CuO_2NPs having size <50 nm showed positive impact on germination up to 2000 ppm of applied Cu through CuO_2NPs	Adhikari et al., 2012

(continued on next page)

Laboratory experiment	CuNPs from *Citrus medica* fruit extract	*Allium cepa*	Enhancement in the mitotic index up to the concentration of 20 mg/L. In addition, with increasing the concentration, a slow reduction in abnormality and mitotic index was reported	Nagaonkar et al., 2015
Pot experiment	CuONPs	*Triticum aestivum*	Significant increment in yield and growth of wheat at 10–50 ppm with optimum effect being visible at 30 ppm of CuO-NPs in all the growth and physiological parameters	Hafeez et al., 2015
Field experiment	CuNPs	*Camellia sinensis*	Maximum leaf yield as well as soil macronutrients improvement was observed in nanocopper as compared to the bulk application. In addition, treated plants also showed significant improvement in resistance against red rot disease caused by *Poria hypolateritia*	Ponmurugan et al., 2016

(*continued on next page*)

Table 27.1 Effect of macro, micro, and beneficial nutrient nano-fertilizers in plants—cont'd

Experiment	Nanofertilizer	Crop	Results	References
Beneficial nutrients nanofertilizers				
Greenhouse experiment	Nano-TiO_2	*Brassica napus*	Promotion in seed germination and seedling vigor, large radicle and plumule growth	Mahmoodzadeh et al., 2013
Field experiment	TiO_2 NPs	*Triticum aestivum*	An increase in shoot and root length, biomass up to a certain concentration	Rafique et al., 2015
Laboratory experiment	TiO_2 NPs	*Triticum aestivum*	Enhancement in abscisic acid and jasmonic acid production and increase in the titanium (Ti) contents of wheat	Jiang et al., 2017
Pot experiment	TiO_2 NPs	*Vigna radiata*	Improvement in the root/shoot length, chlorophyll contents, microbial population and enzyme activities was observed	Raliya et al., 2015
Laboratory experiment	Nanoanatase-TiO_2	*Spinacia oleracea*	Enhancement in the activities of different enzymes related to metabolism of N such as glutamic-pyruvic transaminase, glutamine synthase, glutamate dehydrogenase, and nitrate reductase	Yang et al., 2006

(*continued on next page*)

Field experiment	Nano-Fe, nano-Zn, and nano-TiO_2	*Hordeum vulgare*	Application of all the three types of nanoparticles caused an increment in the days to anthesis and maturity. In addition, enormous improvement in the spike length, grain mass and grains number per spike, etc.	Janmohammadi et al., 2016
Field experiment	Nano-TiO_2	*Zeya mays*	An increment in the chlorophyll contents (A, B), anthocyanins, and carotenoids with a maximum pigments production at the reproductive stage	Morteza et al., 2013
Field study	TiO_2 nanoparticles	*Triticum aestivum*	Enhanced Ti concentration in wheat grains with proposed mobilization of TiO_2	Klingenfuss, 2014
Laboratory experiment	Anatase-TiO_2	*Arabidopsis thaliana*	Entrance of nanoconjugates in plant cells followed by their accumulation in subcellular locations	Kurepa et al., 2010

(continued on next page)

Table 27.1 Effect of macro, micro, and beneficial nutrient nano-fertilizers in plants—cont'd

Experiment	Nanofertilizer	Crop	Results	References
Pot experiment	TiO_2 NPs	*Triticum aestivum*	Enhanced shoot and root lengths also the phosphorous acquisition between 20 and 60 mg/kg in comparison with control and a higher chlorophyll contents then by 32.3% @ 60 mg/kg	Rafique et al., 2018
Petri dish laboratory experiment	TiO_2 NPs	*Vicia faba*	Internalization of NPs in plant root cells. Lower H_2O_2 contents in treated plants and highest POD contents in treated plants	Castiglione et al., 2016
Growth room experiment	Multiwalled carbon nanotubes (MWCNTs)	*Nicotiana tabacum*	Increase in the growth of 55%%–64% @concentration 5–500 μg/L	Khodakovskaya et al., 2012
Pot experiment	Multiwalled carbon nanotubes (CNTs)	*Solanum lycopersicum*	Flowers and fruits were increased significantly in NPs-treated plant than control plants and improvement in microbial biota was also observed	Khodakovskaya et al., 2013

(continued on next page)

Germination test	Multiwalled carbon nanotubes (MWCNTs)	*Triticum aestivum, Zeya mays, Arachis hypogaea, Allium sativum*	Improvement in plant growth and biomass, three to four times faster sprouting of seeds in the treated plants than control.	Anita and Rao, 2014
Hydroponic culture	Multiwalled carbon nanotubes (MWCNTs)	*Amaranthus dubius, Lactuca sativa, Oryza sativa, Cucumis sativus, Capsicum annuum*	Root/shoot lengths and overall biomass was enhanced @ concentrations less than 1000 mg/L	Begum et al., 2014
Germination test	Single-walled carbon nanohorns	*Hordeum vulgare, Zeya mays, Oryza sativa, Glycine max, Solanum lycopersicum*	Improvement in seed germination of plants was observed. In addition, growth of tobacco plant was increased significantly (78%) as compared to other plants. Moreover, expression of a number of stress responsive genes was increased in the treated plants	Lahiani et al., 2015
Greenhouse experiment	2-D graphite CNPs	*Lactuca sativa*	Reduction in nitrate leaching by 57% through increase in the soil hydraulic conductivities and increased nitrogen uptake	Pandorf et al., 2020

(continued on next page)

Table 27.1 Effect of macro, micro, and beneficial nutrient nano-fertilizers in plants—cont'd

Experiment	Nanofertilizer	Crop	Results	References
Hydroponic and pot experiment	Nano-SiO_2	*Zea mays*	18.2% silica accumulation with better nutrient alleviation in seeds, 6.5% increase in dry weight and 95.5% increase in germination percentage	Suriyaprabha et al., 2012
Field experiment	Silica nanoparticles (SNPs)	*Zeya mays*	Enhanced expression of different organic compound for example, proteins, chlorophyll, and phenols. More silica accumulation in leaves	Suriyaprabha et al., 2012
Laboratory conditions	SiNPs	*Lycopersicum esculentum*	Enhancement in germination of seed and time of germination, vigor index of seed, seed germination index, dry and fresh weights of seeds at concentration of 8 g/L	Siddiqui and Al-Waahaibi, 2014
Laboratory experiment	SiNPs	*Lupinus angustifolius, Triticum aestivum*	Facilitation of photosynthetic activity and plant growth. In addition, accumulation of NPs in different plant parts	Sun et al., 2016

(continued on next page)

Pot experiment	SiNPs	*Oryza sativa*	Improvement in the efficiency of genes that play role in the uptake of Si	Abdel-Haliem et al., 2017
—	SiNPs	*Oryza sativa*	Regulation in the functions of the genes responsible for the transport of cadmium to the vacuole and the uptake of silicon	Cui et al., 2017
Hydroponic experiment	SiNPs	*Trigonella foenum*	Increase in the uptake, translocation and accumulation, cell wall lignification and formation of stress enzymes than control	Nazaralian et al., 2017
Hydroponic experiment	SiNPs	*Triticum aestivum*	Alleviation of harmful impacts of UV-B radiation	Tripathi et al., 2017

Major problems related to P fertilizers are poor uptake efficiency and the conversion of plant-available forms into plant unavailable forms (Shen et al., 2011). Also, eutrophication is another problem for the environment (Liu and Lal, 2015). Previously different approaches such as vesicular-arbuscular mycorrhizae, phosphorus solubilizing bacteria, organic acid-producing fungi, and phosphatase and phytase-producing microorganisms have been investigated to enhance P uptake (Taktek et al., 2015). However, some factors make the use of microorganisms limited such as low organic carbon which makes the energy supply to microorganisms limited and higher water evaporation from the surface of the soil (Goebel et al., 2011). Nanoscience researchers are working to increase uptake efficiency of P, controlled release of fertilizers, reduce fixation, increase the plant-available forms. Phosphorus NPs were produced (fungal mediated) by using $Ca_3(PO_4)_2$ as precursor salt and it was confirmed by electron dispersive spectroscopy that 62% of the P was of 28 nm size (Tarafdar et al., 2012). Liu and Lal (2015) also synthesized hydroxyapatite NPs stabilized by carboxymethyl cellulose. The NPs were of 16 nm size and their effectiveness on soybean was investigated. Greenhouse experiments were conducted, and the NPs were applied into the soil. The results of the experiment showed that phenological development and yield were improved by 33% and 18%, respectively, relative to the plants to which conventional P fertilizer was applied. Hydroxyapatite NPs have weak relation with the surface of soil particles compared to the conventional charged fertilizer molecules; thus, enhances plant P uptake (Iqbal and Umar, 2019). According to recent reports, a phosphatic nanofertilizer known as water-phosphorite was prepared from natural raw phosphorite by ultrasonic dispersion (Sharonova et al., 2015). The size of the particles ranged from 60 to 120 nm. Seed treatment with the fertilizer gave significant results. Morphometric indices were improved for corn plants by P nanofertilizer in field and greenhouse tests. An increase in the biological yield was recorded up to 2.2-fold while an increase in grain yield was recorded up to 24.1%. It is stated that the total concentration of P in soil is adequate but is not available to plants (da Fonseca et al., 2005). To address the issue ZnONPs were used to solubilize or to convert immobile P to plant-available forms because the enzymes such as phytase and phosphatase require Zn for their functioning. It was recorded that ZnONPs foliar application enhances plant P uptake up to 11% in cereals and legumes without application of external P fertilizers (Raliya and Tarafdar, 2013; Tarafdar et al., 2014). Magnetite (Fe_2O_3) NPs were also capable of mobilizing native P to plant-available forms (Raliya et al., 2016). In summary, efforts to enhance P uptake by PNPs application or the mobilization of native P by micronutrient NPs (ZnO and Fe_2O_3) not only improve plant development and growth as well as reduce the environmental problems related to P fertilization (limited P availability, eutrophication).

27.2.1.4 Potassium nanofertilizers

Potassium (K) is also an important macronutrient necessary for enhanced plant growth and grain production. Potassium is not involved in the formation of organic compounds and is taken up by plants as a K^+. K is involved in stomatal conductance, water storage in the plant body, and the photosynthetic process as well (Abbasi et al., 2014). Modified K fertilizers are reported to be developed to enhance K use efficiency among which nanosized and coated fertilizers are important (Rameshaiah et al.,2015). Subbarao et al. (2013) conducted experiments on wheat and corn using K nanofertilizer and results showed that the yield of wheat and corn increased significantly compared to conventional fertilizers and effective nitrogen recovery was also noted. El-Sharkawy et al. (2017) conducted a greenhouse experiment on Alfa alfa using potassium sulfate (K_2SO_4) NPs for investigating the effects of K_2SO_4NPs on its growth and physiological attributes under salt stress. Results revealed that Alfa alfa showed better growth with K_2SO_4NPs compared to conventional K fertilizer. The ratio of Na/K was significantly affected by K_2SO_4NPs. The physiological response of plants was also enhanced under salt stress by reducing leakage of electrolyte, increasing proline, and catalase contents and it also enhanced the activity of antioxidant enzymes.

27.2.1.5 Secondary macronutrient nanofertilizers

Secondary nutrients are required in large amounts by the plants which include sulfur, calcium, and magnesium denoted as S, Ca, and Mg, respectively. Mg and S are required in equal amounts as phosphorus (P) (Jaffer et al., 2002). While the Ca requirement is greater than P, it mostly depends on plant species (Preetha and Balakrishnan, 2017). Reactions of S and N are dominated by microbial population and organic fraction in soil because both these nutrients behave similarly in the soil. In contrast, Ca and Mg come from clay fractions and both of these behave similarly to K. Supapron et al. (Supapron et al., 2002) revealed that zeolite acts as a slow-release fertilizer for Mg and Ca and also claimed that zeolite increased the Ca and Mg concentration in soil. It was also revealed that zeolites easily exchange Mg and Ca in soil (Fansuri et al., 2008). Li and Zhang (2010) conducted experiments to evaluate the possibility of the use of zeolite modified with a surfactant for the controlled release of sulfate. Results of column and batch experiments showed that it could be a good carrier of sulfate that can reduce the leaching of sulfate. Secondary nutrients are not studied much as their deficiencies are not much evident.

27.2.2 Micronutrient nanofertilizers

The essential plant nutrients required in very small quantities are called micronutrients. These include Zn, Fe, Mn, Cu, Mo, Ni, B, and Cl. Mostly these micronutrients are added to the crops with macronutrients at 5 ppm concentration (Jones et al., 2012).

27.2.2.1 Zn nanoparticles

Zinc (Zn) acts as a cofactor in different enzymes. It is an important part of carbonic anhydrase, superoxide dismutase, RNA polymerase, and alcohol dehydrogenase and it also plays an important role in plants in chlorophyll biosynthesis. Mahajan et al. (2011) experimented to check the effect of ZnONPs on *Vigna radiata* and *Cicer ariatium* (Chickpea). Results revealed that 20 ppm concentration of ZnONPs increased the root and biomass growth up to 42% and 41%, respectively. While compared to control, shoot length was increased up to 98%. In the case of chickpea, 1 ppm concentration of ZnONPs increased the length of shoot and root by 6% and 53%, respectively. Zhao et al. (2013a) also reported that 400–800 mg/kg concentration of ZnONPs in soil significantly increased the growth of *Cucumis sativus* (cucumber). Root enhancement was reported at 2 ppm ZnONPs concentration in *Brassica napus* (Rape) and *Raphanus sativus* (Radish) (Lin and Xing, 2007). A significant improvement in cluster bean growth was recorded at 20 ppm of biosynthesized NPs (Raliya and Tarafdar, 2013). They reported that 31.5% increase in the length of shoot, 27.1% in biomass, 73.5% root area, 66.3% root length, 73.5% phosphatase activity, and 27.1% insoluble proteins. Toxic effects of ZnONPs were also reported at higher concentrations (400–2000 ppm) (Lee et al., 2010; Zhao et al., 2013b).

27.2.2.2 Mo nanoparticles

Molybdenum (Mo) is an important micronutrient for plant photosynthetic systems. Many researchers are making efforts for the development of MoNPs. Taran et al. (2014) conducted an experiment in which a combination of nitrogen-fixing bacteria and MoNPs was used and their findings exhibited a significant increment in the growth compared to the sole application of water, Mo, and nitrogen-fixing bacteria, and it was determined that this combined application would be the better treatment for good plant growth. Thomas et al. (2017) also experimented on chickpea using MoNPs at 4 ppm concentration. It was concluded that the use of MoNPs significantly increased microbial activity in the rhizosphere, growth attributes, and grain yield of chickpea.

27.2.2.3 Fe nanoparticles

Iron (Fe) falls in the category of the most vital metal elements in terms of plant growth as it acts as a base for the development of chlorophyll and also carries different essential elements in the bloodstream during the blood circulation of human. Scientists are trying to make NPs of such important elements and their adaptability in the agriculture sector. Ghafariyan et al. (2013) conducted a greenhouse experiment using FeNPs on *Glycine max*. Results indicated that a noteworthy enhancement in chlorophyll contents. They concluded that FeNPs can act as Fe supplement for plants to reduce chlorotic diseases. Delfani et al. (2014) also experimented on black-eyed pea using FeNPs. Foliar application of FeNPs (500 ppm) was carried out. Results showed that the number of pods was increased on black-eyed pea plants and the Fe and chlorophyll contents were increased up to 34% and 10%, respectively, by FeNPs compared to control. Rui et al. (2016) carried out a study to estimate the efficiency of iron oxide NPs on peanut (*Arachis hypogea*). Results showed a considerable uplift in the growth attributes (plant height, root length, biomass) as well as the SPAD (Soil Plant Analysis Development) value after the application of FeONPs. FeONPs regulate the phytohormones and antioxidant enzyme activities, which in turn increased the plant growth. Iron contents were also recorded higher with FeONPs compared to control. The results of the experiment conducted by Yuan et al. (2018) showed that lower concentrations of FeNPs showed positive effects on plant growth. Moreover, results also revealed that FeNPs increased plant growth by increasing grana staking and chloroplast development, changing leaf anatomy, and regulating vascular bundle development. Boutchuen et al. (2019) reported that a 230%–830% increase in plant growth was recorded by hematite NPs drop. They also revealed that it increased the production of fruits per plant by twofold.

27.2.2.4 Mn nanoparticles

Manganese (Mn) is an important metal element required for plant growth and is a core component of the superoxide dismutase enzyme and plays an important role in photosynthesis at the water-splitting stage. Pradhan et al. (2013) conducted an experiment on *V. radiata* (mung bean) using MnNPs. Results showed an increase in plant growth and photosynthesis. They recorded that improvement in shoot length, root length, and dry biomass of up to 52%, 38%, and 100%, respectively, at a very low concentration of 0.5 ppm and concluded that MnNPs could be a good alternative material for Mn nutrition to plants compared to bulk $MnSO_4$. Liu et al. (2016) reported that a 12%–54% increase in the growth of seedlings of lettuce (*Lactuca sativa*) plants was recorded with MnNPs and also concluded that MnNPs can increase agronomic production and can be used as nanofertilizer.

27.2.2.5 Cu nanoparticles

Copper (Cu) is another essential plant nutrient and a vital component of different enzymes. Nekrasova et al. (2011) experimented using a very low concentration (0.25 ppm) of CuNPs in waterweed (*Elodea densa Planch*) and reported a 35% increase in photosynthesis activity. Similarly, an increase in the growth of lettuce plants of up to 40% was recorded when CuNPs were applied in the soil at 130 mg/kg(Shah and Belozerova, 2009). Lopez- Vargas et al. (2018) did a foliar application of CuNPs on tomato plants and the results showed that compared to control 2,2'-Azino-Bis-3-Ethylbenzothiazoline-6-Sulfonic Acid (biochemical reagent) antioxidant capacity and vitamin C lycopene were increased and the fruit was produced with great firmness and also the increase in superoxide dismutase and catalase enzyme was recorded. On the other side, toxic effects of CuNPs at 200–1000 ppm concentration were recorded on wheat (*Triticum aestivum*), mung bean, and on yellow squash (*Cucurbita pepo*) (Lee et al., 2008; Stampoulis et al., 2009; Musante and White, 2012). Figs. 27.1 and 27.2 describe a brief review of types of NPs being used in agro-ecosystem.

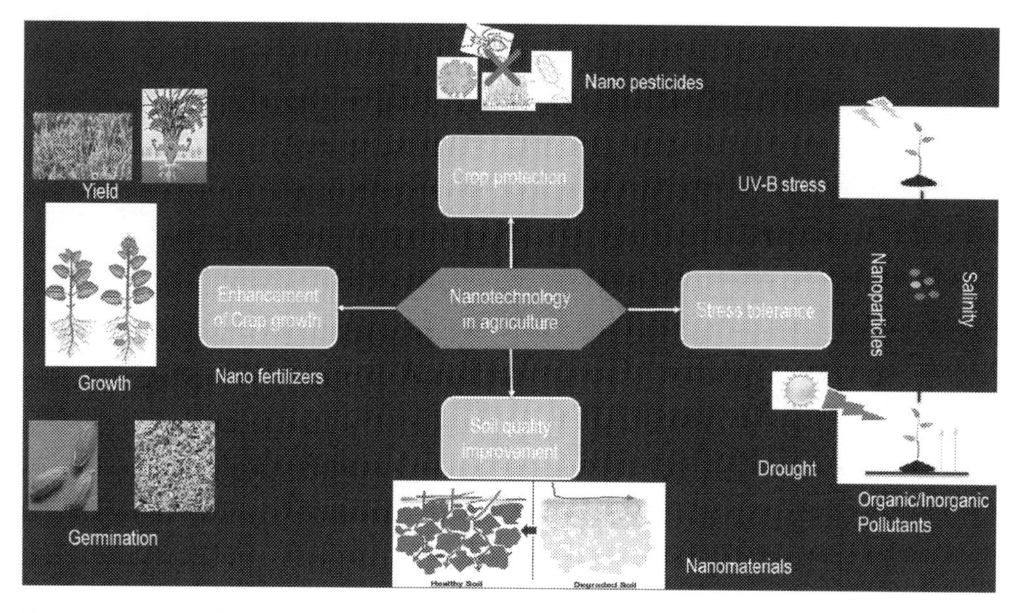

Figure 27.2 Types of NPs being used in agro-ecosystem.

27.2.3 Beneficial nutrients nanofertilizers

27.2.3.1 TiO_2 nanoparticles

As the commercial production of TiO_2 starts, it has been mostly used as a pigment in sunscreens and paints because of its photocatalytic ability to split water under UV light (Chen and Mao, 2007). Titanium can convert light energy into chemical or electrical under sunlight because of its photocatalytic activity. Due to this activity of TiO_2 researchers see an opportunity to enhance the photosynthetic activity of plants by using TiO_2NPs. Titanium NPs are the most widely studied NPs to evaluate the plant responses toward pest and disease management, pesticide degradation, seed germination, plant growth, etc. (Wang et al., 2016). Many experiments were conducted using TiO_2 on various crops such as lettuce, tomato, spinach, watermelon, millets, *Lemna minor*, wheat, and beans (Zahra et al., 2015; Raliya and Biswas, 2015; Raliya et al., 2015; Wang et al., 2013; Tarafdar et al., 2013; Feizi et al., 2012). The results of these studies concluded that TiO_2 increased chlorophyll content, yield, nutrient contents, photosynthetic activity, and germination rate. Plant responses were based on the species and applied concentration of TiO_2NPs. Scientists thought that TiO_2 can enhance the activity of ribulose 1,5 bisphosphate carboxylase (RUBISCO) and nitrogen metabolism, which is the main reason for increased physiological activities in plants (Linglan et al., 2008; Gao et al., 2008). The above-mentioned studies also demonstrated that photosynthetic activities increased when plants were exposed to TiO_2NPs which in turn increased the plant growth (Wang et al., 2016; Raliya et al., 2015).

27.2.3.2 Carbon nanotubes

Carbon nanotubes (CNTs) efficiently used as growth promoters in different plant systems (corn, carrot, grape, soybean, ryegrass, tomato, cabbage) and the positive or negative effect of CNTs depends on concentration and plant species. Lin and Xing (2007) conducted an experiment using CNTs (multiwalled) at 200 ppm concentration in corn, rape, and ryegrass. Results showed an increment in the root length of the germinating seeds. Elongation in root length was also reported in *Allium cepa* and cucumber when they are exposed to CNTs (single-walled) (Canas et al., 2008). Water channel protein expression was affected by CNTs in tomato plants and increased crop production by increasing water uptake. Khodakovsakaya et al. (2012) claimed that 50 ppm application of CNTs on tomato increased the production by twofold. Srinivasan and Saraswathi (2010) reported an increase in plant growth and fruit production when CNTs were applied in tomato crops. This is because CNTs enter into the germinating seeds and increase the water uptake. Mondal et al. (2011) studied the effect of CNTs at a concentration of 2.3–46 mg/L on mustard plants, and an increase in the root, shoot growth, and germination was recorded. The highest increase was recorded at the lowest concentration. It was also reported that as the concentration increased growth was reduced. Zhang et al. (2015) experimented to estimate the impact of CNT's application on the germination of tomato seed and reported a 26.6, 43.4, and 13.5% increase in germination on 2, 4, and 6 days of exposure.

27.2.3.3 Silicon nanoparticles

Silicon NPs (SiNPs) have been widely reported to impart beneficial impacts on photosynthesis as well as plant growth under normal as well as stressed environments (Mousavi et al., 2018; Cui et al., 2017; Hussain et al., 2019). Evidence of its beneficial role in terms of improvement in the germination of the seed, seedling growth, and overall plant growth was provided by various researchers (Sabaghnia and Janmohammadi 2014; Siddiqui and Al-Whaibi, 2014; Fitriani and Haryanti, 2016). Khalaki et al. (2016) stated that the application of nanosilver caused noteworthy progress in the germination rates and percentage, lengths, dry and fresh masses of roots and shoots of *Thymus kotschyanus*. Similarly, Janmohammadi and Sabaghnia (2015) stated that soaking and seed priming in SiNPs increased seedling shoot and root lengths, the biomass of seedling as well as vigor index in *Helianthus annuus*. It is suggested that Si accumulation in leaves keeps them upright, and stretched leaf surfaces attract more light and thereby, favor photosynthesis. Enhancement in the photosynthetic process will directly increase photosynthate production which can be accumulated in vegetative plant parts for example, roots, shoots, and leaves (Putri et al., 2017). In this regard, Sharifi-Rad et al. (2018) provided evidence that the application of SiNPs at all concentrations caused a significant increment in the chl "A," "B." and carotenoid contents in comparison with the control treatment. SiNPs improved fresh and dry masses of roots and shoots in *T. aestivum* (Karimi and Mohsenzadeh, 2016). Similarly, Ivani et al. (2018) observed an increment of 66% and 55%, respectively, in the shoot length and seedling length by application of SiNPs at 50 mg/L in fenugreek. Moreover, Yassen et al. (2017) stated that foliar application of SiNPs at different concentrations uplifted growth, yield, and chemical composition of cucumber. Table 27.1 summarizes role of nutrient NPs in plants.

27.3 Role of nanotechnology in disease and pest management

Nanoformulations are not only confined to nanofertilizers, but nanopesticides have also been used and are still being used against insect pests, fungi, and disease management (Anuar et al., 2017). Conventional pesticides have issues like polluting land and the aquatic system, as well as their supply of active ingredient, is usually dependent on the method of application. However, the Nanopesticides have the capability of slow-release and targeted delivery of active ingredient with reduced loss, improved bioavailability, and stability of active ingredient (Kah et al., 2018). NPs of metals and metals oxides drew significant attention due to their promising use against harmful pests (Pandey et al., 2018). Especially silver NPs have shown remarkable potential in comparison to the fungicides available commercially (Ismail et al., 2016; Nejad et al., 2016). Silver NPs have been successfully used to control serious diseases like *anthracnose* in mango (Nagaraju et al., 2020). Besides this CuNPs, ZnNPs and silica NPs have the capability to effectively control diseases like *curvularia leaf spot* (Choudhary et al., 2017), *cercospora leaf spot* disease of sugar beet (Farahat, 2018), and pest like *Callosobruchus maculatus* respectively (Silva-Castro et al., 2018). However, some researchers have expressed serious concerns about their use at a large level (Sun et al., 2019; Zhang et al., 2019). As nanopesticides can pose harmful effects on non-targeted organisms. For example, the use of nano-TiO_2 can be harmful to soil beneficial bacteria when killing harmful bacterial populations (Al-Mubaddel et al., 2017).

27.3.1 Zn nanoparticles

Zinc oxide NPs have been studied mostly as nanofertilizer for plant growth improvement rather than as a fungicidal agent. He et al. (2011) conducted a study to investigate the effect of ZnONPs as a fungicide on fruit mold and reported a great reduction in pathogen (*Botrytis cinerea* and *Penicillium expansum*) growth. Zn-encapsulated chitosan NPs displayed significant antifungal activity and growth promotion in maize crops. Moreover, it was found that foliar application, as well as seed treatment of ZnNPs, controlled CLS disease in maize (Choudhary et al., 2019). ZnNPs were also found useful in controlling Cercospora leaf spot disease in sugar beet (Farahat, 2018). Jayaseelan et al. (2012) also reported a growth reduction of *Aspergillus niger* and *Aspergillus flavus.* Wagner et al. (2016) concluded that ZnONPs can be used potentially as a nonpersistent, low dose, and economic antimicrobial agent against oomycete Peronospora *tabacina*. ZnONPs showed less toxicity to plants compared to AgNPs and have the potential to be used as a pesticide.

Trials for green synthesis of ZnNPs have also been practiced by various researchers. Ogunyemi et al. (2019) successfully synthesized ZnNPs using three different plant extracts to identify their antimicrobial activity against *Xanthomonas oryzae pv. Oryzae*. It was observed that the bacterial cell membrane of *X. oryzae pv. Oryzae* was significantly affected at NPs concentration of 16.0 mg/mL. Besides, antibacterial activity ZnONPs exhibited a significant efficiency to control fungi individually (Wagner et al., 2016; Savi et al., 2015) or in combination with MgONPs and TiO_2NPs (El-Argawy et al., 2017). An effort was made to compare Cu and Zn based on agrochemical efficacy as a biocide. It was reported that Zn-based agrochemicals had comparable higher antimicrobial efficiency with significantly lower phytotoxicity when compared to Cu-based agrochemicals (Rajasekaran *et al.* 2016).

27.3.2 Ag nanoparticles

Silver (Ag) is considered as an important metal concerning the antimicrobial properties. It reported that Ag can kill unicellular microbes by manipulating their enzyme system (Kim et al., 2017). It was concluded from the results of various studies that AgNPs can control different plant pathogens such as *Rhizoctonia solani, Bipolaris sorokiniana, Magnaporthe grisea, B. cinerea, Fusarium culmorum, Scalrotinia sclerotiorum, Colletotrichum gloeosporioides,* and *Pythium ultimum* (Gopal et al., 2011; Park et al., 2006). Jain and Kothari (2014) conducted an experiment and reported that foliar spray of AgNPs on bean leaves eliminate the sun-hemp rosette virus. Elbeshehy et al. (2015) also concluded that postinfection application of AgNPs in the case of bean yellow mosaic virus in faba bean showed remarkably good results compared to preinfection or right after the infection application. AgNPs have great potential in the agriculture sector to protect plants from pests and diseases, but some problems are associated with their use such as production, interaction with soil, and toxicity (Kah and Hofmann, 2014; Mishra and Singh, 2015). Carbendazim-conjugated AgNPs were used against *C. gloeosporioides*. Results declared that even a low concentration (0.1%) of 19–24 nm 24-nm-sized carbendazim coated AgNPs were effective for controlling anthracnose disease in mango (Nagaraju et al., 2020). AgNPs were also utilized to develop alternatives for commercial fungicide Kocide being used extensively for controlling early blight disease on the potato. Results showed that 12.7 nm $\times$7 nm sized AgNPs completely inhibited *Alternaria solani* (a causative agent of early blight disease) at AgNPs concentration 25 μg/mL. In contrast to AgNPs ,Kocide gave maximum inhibition at 600 μg/mL (Ismail et al., 2016; Nejad et al., 2016).

Additionally, green synthesized AgNPs have also been used as an effective biopesticide against insects (Murugan et al., 2017). For instance, AgNPs synthesized from *Moringa oleifera* leaf extract, *Bacillus amyloliquefaciens,* and *Bacillus subtilis* exhibited larvicidal and pupicidal toxicity against the house fly (Abdel-Gawad, 2018) and *Culex pipiens pallens,* respectively (Fouad et al., 2017).

27.3.3 Cu nanoparticles

Copper NPs (CuNPs) also have some antimicrobial properties and are reported as a good agent to control the diseases spread by *Xanthomonas* sp. of bacteria such as the bacterial blight of rice and leaf spot of mung (Chhipa and Joshi, 2016). Broad-spectrum antimicrobial properties of CuNPs were reported against bacteria (Gram-positive and Gram-negative) and fungi (Esteban-Tejeda et al., 2009), and CuNPs can also be used as a fungicide at lower concentrations (Giannousi et al., 2014). Similarly, Cioffi et al. (2005) also reported antifungal properties of CuNPs. CuNPs proved effective against some fungal species such as *Fusarium oxysporum* and *A. solani* which cause diseases in tomato (Saharan et al., 2015). For instance, in 2014, Kanhed and coworkers coated CuNPs with cetyltrimethylammonium bromide (CTAB) and used them against selected plant pathogenic fungi such as *Phoma destructiva, Curvularia lunata, Alternaria alternata,* and *F. oxysporum* to identify their antifungal efficiency (Kanhed et al., 2014). They found that CuNPs coated with CTAB exhibited significant antifungal activity as well as can be used as a potent disinfectant. Similarly, Viet *et al.* (2016) utilized CuNPs with support of CTAB using the chemical reduction method. It was reported that this nanomaterial at 450 ppm concentration reduced fungal growth up to 93.98%.

Moreover, eco-friendly approaches for the synthesis of CuNPs were also developed by various researchers. For instance, Shende *et al.* (2015) synthesized CuNPs from Citron juice (*Citrus medica* Linn.). This biogenic nanomaterial displayed significant inhibitory activity against bacterial species (*Pseudomonas aeruginosa, Escherichia coli,* and *Salmonella typhi*) also the pathogenic fungi (*F. culmorum, F. oxysporum,* and *Fusarium graminearum*). Likewise, biosynthesis of CuNPs utilizing endophytic actinomycetes isolates Ca of medicinal plant (*Convolvulus arvensis L.*) was also conducted. The synthesized nanomaterial was effective for phytopathogenic fungi, bacterial strains as well as some infectious insects. The antimicrobial effect was concentration dependent and maximum inhibition was observed at a maximum concentration (Hassan et al., 2018). The insecticidal aspect of CuNPs was also observed by various researchers. In 2016 Van and coworkers for the first time evaluated the effect of CuNPs on the insecticidal activity of transgenic cotton. They reported that CuNPs at low concentrations increased the expression of Bt toxin protein thus improving the pest resistance of transgenic cotton (Van et al., 2016).

27.3.4 Si nanoparticles

The effectiveness of silicon NPs (SiNPs) has been reported in various studies (Magda and Hussein 2016; Ziaee and Ganji, 2016; El-Bendary and El-Helaly, 2013). SiNPs can be used in two ways to control plant pathogens: (1) can be applied directly to plants to kill pests and insects and (2) can be used as a carrier to carry out different pesticides to increase their efficacy (Rastogi et al., 2019). Rouhani et al. (2013) reported the lethal properties of SiNPs against *C. maculatus.* It is reported that the effectiveness of SiNPs was more on mature insects compared to larvae because of the desiccating properties of Si which can result in digestive tract damage of insects or the enlargement of the integument (Rastogi et al., 2019). For instance, recently Derbalah and coworkers used fabricated mesoporous silica NPs and reported that these NPs were not only effective for increasing tomato growth but also could be used as an alternative for commercial fungicides being used to control early blight of tomato (Derbalah et al., 2018). In the following year, nanosilica with oligochitosan was reported as an effective nanomaterial to control brown spot disease in dragon fruit (Verma and Bhattacharya, 2018). SiNPs are not only being used as an alternative fungicide but their use as effective nanopesticides is also very common. For example, SiNPs have been used to control pests such as *Rhyzopertha dominica, Tribolium confusum* (Ziaee and Ganji, 2016) *Spodoptera littoralis* larvae (El-Helaly et al., 2016), and *C. maculatus* (Rouhani et al., 2013). Moreover, their antimicrobial activity against bacterial strains such as *Pseudomonas fluorescens, Pseudomonas syringae, Pectobacterium carotovorum* (Cadena et al., 2018) *Klebsiella pneumonia, Staphylococcus aureus, Proteus mirabilis, P. aeruginosa* (Mohammed et al., 2017), *E. coli* (Mohammed et al., 2017) *S. aureus, Aspergillus fumigatus* (Song et al., 2018) *B. subtilis, S aureus,* and *P. aeruginosa* (Tahmasbi et al., 2018) is well known. A brief review regarding published work on biotic stress management by NPs is presented in Table 27.2.

27.3.5 Role of NPs in the mitigation of abiotic stresses

Owing to the immobile behavior of the plant, they are exposed to various types of environmental stresses both abiotic (drought, heavy metal toxicity, nutrient deficiency, scorching heat, chilling injuries, salinity, and anaerobic condition) and biotic stresses (pathogens and pest). However, natural conditions (field) are often different from experimental conditions (lab) as in-field plants have to face more than one stress that is, drought, salinity, and heat (Suzuki et al., 2014). Therefore, maximum effort should be invested to mimic field conditions while performing lab experiments. Moreover, there is an urgent need to develop advanced tactics to deal with stress combinations for the food security of the incessant world population. Drought and heat are two major abiotic stresses threatening world food security as solely drought and heat stress were responsible for more than one billion dollar loss to the US agriculture sector from the year 1980 to 2017. NPs are effectively being used in alleviating different types of plant stresses as indicated in Table 27.3.

Table 27.2 Role of nanoparticles in management of biotic stress in plants.

Experiment	Nanofertilizer	Crop	Results	References
Field experiment	Calcium carbonate NPs	*Citrus reticulate*	Increased in Ca contents of plant and in addition, nanoparticles treated plants competed well against attack of *Aonidiella aurantia* and *Bactrocera dorsalis*	Deng et al., 2015
Pot experiment	Magnesium oxide nanoparticles (MgONPs)	*Solanum lycopersicum*	Enhanced resistance in the treated plant against *Ralstonia solanacearum* (responsible for bacterial wilt) in tomato	Imada et al., 2016
Laboratory experiment	Sulfur nanoparticles	*Zingiber officinale*	Maximum inhibition of *Fusarium oxysporum* activities causing soft rot of ginger (*Zingiber officinale*)	Athawale et al., 2018
Field experiment	CuNPs	*Camellia sinensis*	Maximum leaf yield as well as soil macronutrients improvement was observed in nanocopper compared to the bulk application. In addition, treated plants also showed significant improvement in resistance against red rot disease caused by *Poria hypolateritia*	Ponmurugan et al., 2016

(*continued on next page*)

Laboratory experiment	AgNPs and SeNPs	*Solanum tuberosum*	AgNPs @ concentration of 25 μg/mL completely inhibited *Alternaria solani*, responsible for early blight disease in potato, whereas Se-NPs caused complete inhibition of *A. solani* @ 800 μg/mL	Ismail et al., 2016
Glasshouse experiment	Ag-NPs	*Oryza sativa*	AgNPs @ concentration of 50 ppm significantly affected the activity of *Rhizoctonia solani*, responsible for causing sheath blight disease in rice. Moreover, at same concentration, AgNPs also positively affected fresh and dry weights of rice plant with a concomitant pronounced suppressive effect on the lesion formation in leaves	Nejad et al., 2016
In-vitro experiment	Carbendazim-conjugated silver nanoparticles	*Mangifera indica*	AgNPs @ concentration of 0.1% caused a significant decline in the activity of *C. gloeosporioides*, (causes mango anthracnose)	Nagaraju et al., 2020

(continued on next page)

Table 27.2 Role of nanoparticles in management of biotic stress in plants—cont'd

Experiment	Nanofertilizer	Crop	Results	References
Field experiment	Zn-chitosan nanoparticles (NPs)	*Zeya mays*	In field experiments, ZnNPs significantly controlled Cercospora leaf spot (CLS) disease with a simultaneous increase in the grain yield from 20.5% to 39.8% and improved grain Zn content to 62.21 μg/g dry weight.	Choudhary et al., 2019
Field experiment	ZnNPs	*Beta vulgaris*	Inhibition of Cercospora leaf spot (CLS) disease severity percentage of sugar beet plants.	Farahat, 2018
Bioassays	SiO_2NPs	*Triticum aestivum*	Effective control of *Tribolium confusum and Rhyzopertha dominica*	Ziaee and Ganji, 2016
Field experiment	hydrophobic nano-silica	*Solanum lycopersicum*	Nano-silica @ concentrations of 300 and 350 ppm induced more resistance in tomato plant against *Spodoptera littoralis*	El-Bendary and El-Helaly,

(continued on next page)

Laboratory experiment	SiO_2NPs and AgNPs	*Vigna unguiculata*	Both the NPs showed their effectiveness on the adults as well as larvae of *Callosobruchus maculatus* by 100% and 83%, respectively. Results also indicated that SiO_2NPs could be an effective tool in pest management program of *C. maculatus.*	Rouhani et al., 2013
Laboratory experiment	Mesoporous silica nanoparticles (MSN)	*Solanum lycopersicum*	High antifungal efficiency of MSN against *Alternaria solani*, responsible for early blight in tomato, with a simultaneous increase in the tomato growth parameters than control	Derbalah et al., 2018
Laboratory experiment	SiO_2NPs	*Zeya mays*	Higher resistance in maize against *Aspergillus niger* @ concentrations of 10 and 15 kg/ha. Moreover, nanosilica-treated plants exhibited increased hydrophobic potential and silica accumulation percentage than control.	Suriyaprabha et al., 2014

(continued on next page)

Table 27.2 Role of nanoparticles in management of biotic stress in plants—cont'd

Experiment	Nanofertilizer	Crop	Results	References
Greenhouse experiment	Nanosilica	-	Nanosilica @ concentration of 500 ppm suppressed *Spodoptera littoralis* pest	El-Helaly et al., 2016
Laboratory experiment	Mesoporous silica nanoparticles	*Pisum sativum*	Enhancement in the antimicrobial properties of peas against phytopathogen *Pseudomonas syringae*, the causal agent of pea bacterial blight	Cadena et al., 2018

Table 27.3 Abiotic stress alleviation by nanoparticles.

NPs	Crop	Stress type/control	Effect	Reference
Si	*Hordeum vulgare*	Drought	Augmentation in chlorophyll, carotenoids contents, and biomass of shoot	Ghorbanpour et al., 2020
Nanoanatase-TiO_2	*Dracocephalum moldavica*	Drought	Treated plants had high malondialdehyde (MDA) contents and H_2O_2 contents and more proline production under water deficit conditions	Mohammadi et al., 2016
ZnO	*Sorghum bicolor*	Drought	Enhanced nitrogen, potassium and zinc contents	Dimkpa et al., 2019
Si-NPs	*Crataegus aronia*	Drought	Alleviation in drought stress was observed because of more production of photosynthetic pigments and decrease in the MDA contents	Ashkavand et al., 2015
CNTs (single walled)	*Hyoscyamus niger*	Drought	Alleviation of drought stress induced germination and growth attributes. In addition, increased activities of various antioxidants such as superoxide dismutase (SOD), peroxidase dismutase (POD), catalase (CAT), ascorbate peroxidase (APX) under treated plants	Hatami et al., 2017
Iron oxides	*Triticum aestivum*	Drought	Increased photosynthesis rate and reduced oxidative stress	Adrees et al., 2020

(continued on next page)

Table 27.3 Abiotic stress alleviation by nanoparticles—cont'd

NPs	Crop	Stress type/control	Effect	Reference
Metal composite (Fe, Cu, Co, ZnO)	*Glycine max L.*	Drought	Enhanced tolerance against drought stress	Linh et al., 2020
Fullerenol	*Beta vulgaris L.*	Drought	Alleviation in the oxidative stress	Boríšev et al., 2016
Ag	Wheat (*Triticum aestivum*)	Heat stress	Augmented fresh and dry weight of wheat	Iqbal et al., 2019
Se	*Sorghum bicolor (L.) Moench*	Heat stress	Improved antioxidant activity	Djanaguiraman et al., 2018
Si	*Triticum aestivum*	Heat stress	Improved antioxidant defense system	Tripathi et al., 2017
Magnetite and zinc oxide	*Triticum aestivum*	Heat stress	Enhanced antioxidant enzymes activities and reduced lipid peroxidation rates	Hassan et al., 2018
Si	*Triticum aestivum*	Heat stress	Increased photosynthetic capacity and pigments also reduction in MDA	Younis et al., 2020
nano – Calcium (LITHOVIT)	*Solanum lycopersicum*	Salinity	A decline in the salinity induced reduction in stem diameter, chlorophyll contents and leaf area were observed at different salinity levels with maximum plant yield improvement (76%) was obtained with (LITHOVIT) @ 8 mS/cm	Sajyan et al., 2018

(continued on next page)

Manganese (III) oxide nanoparticles (MnNPs)	*Capsicum annuum L.*	Salinity	Root growth was improved in both the salt treated and nonsalt treated plants by MnNPs @ concentration of 0.1–1 mg/L	Ye et al., 2020
SiO_2 NPs	*Ocimum basilicum*	Salinity	Significant increase in physiological parameters and more proline production as an indication to induce salinity tolerance in plant	Kalteh et al., 2018
SiO_2 nanofertilizer	*Cucumis sativus*	Salinity	Foliar spray of SiO_2 nanoparticles @60 mg/L caused a significant increase for most growth parameters under salinity stress with a concomitant increase in N and P contents and a decline in Na uptake	Yassen et al., 2017
Nano-SiO_2	*Cucurbita pepo*	Salinity	An improvement in the seed germination under salinity stress and a decline in the MDA and hydrogen peroxide levels as well as electrolyte leakage. Increase in the stomatal conductance, net photosynthetic rate	Siddiqui et al., 2014

(continued on next page)

Table 27.3 Abiotic stress alleviation by nanoparticles—cont'd

NPs	Crop	Stress type/control	Effect	Reference
Si-NPs	*Lens culinaris*	Salinity	Improvement in growth and development of plant under salt stress	Sabaghnia and Janmohammadi, 2015
Iron phosphate nanoparticles	*Brassica oleracea*	Heavy metal stress	Cadmium (Cd) immobilization efficiency up to 81.3% with concomitant inhibition of Cd uptake by 44.8% and 70.2% in the below ground and above ground plant parts, respectively.	Qiao et al., 2017
Biochar-supported iron phosphate nanoparticle ($BC@Fe_3(PO_4)_2$)	*Brassica oleracea*	Heavy metal stress	Immobilization efficiency of Cd was 60.2%. Moreover, Cd uptake was reduced by 86.8%.	Xu et al., 2016
Fe-NPs	*Triticum aestivum*	Heavy metal stress	Improvement in the yield, chlorophyll contents, iron contents was observed with a concomitant decrease in the Cd contents. In addition, oxidative stress in the leaves was also alleviated.	Adrees et al., 2020

(continued on next page)

Nano-SiO_2	*Oryza sativa*	Heavy metal stress	Spray of nano-Si on foliage increased the growth, Fe, Zn, and Mg nutrition and chlorophyll contents under cadmium stress and caused a decline in the cadmium accumulation and its translocation to shoots	Wang et al., 2015
Si-NPs	*Pisum sativum*	Heavy metal stress	Protection of pea seedling against chromium phytotoxicity and upregulation of antioxidant defense system in plant	Tripathi et al., 2015

27.4 Application of NPs for soil health management and restoration

27.4.1 Drought stress

Drought occurs when soil and atmospheric temperature are high and humidity is low with an imbalance in plant's evapotranspiration flux and water intake (Lipiec et al., 2013). Various reports have reported the substantial potential of metal and metal oxide NPs on alleviating drought stress in plants. For instance, SiNPs in combination with silicate increased shoot biomass, total chlorophyll contents, and carotenoid contents of barley (*Hordeum vulgare*) by alteration of plant's osmolytes, metabolites, and oxidant enzymes (Ghorbanpour et al., 2020; Rastogi et al., 2019). Application of ZnONPs to drought-stressed sorghum (*Sorghum bicolor*) enhanced its growth, development, and yield. Also, ZnONPs were helpful to improve the N, K, and Zn contents of sorghum growing under moisture stress (Dimkpa et al., 2019; Dimkpa et al., 2020). Moreover, FeONPs were applied to wheat crops to analyzing the impact of FeONPs on crop growth, yield, and nutrient contents. It was observed that FeONPs alleviated both cadmium and drought stress simultaneously by enhancing photosynthesis rate and reducing oxidative stress. Further FeONPs reduced Cd concentration and augmented Fe contents in wheat grains (Adrees et al., 2020). Besides this, metal-based composite (Fe, Cu, Co, and ZnONPs) triggered drought-responsive genes in shoots and roots of soybean (*G. max (L.*) in turn improving drought tolerance of soya bean seedlings (Linh et al., 2020).

27.4.1.1 Heat stress

To survive in the constantly changing environment plants have to adopt several morphological, physiological, and biochemical responses (Huber & Bauerle, 2016). Physiological responses to heat stress include increased length of roots, leaf thickness, and reduced stomatal number and conductance to avoid loss of water by evapotranspiration (Goufo et al., 2017) and cuticular wax biosynthesis are the most common responses to heat and drought stress (Lee & Suh, 2013). However, among the most prevalent biochemical responses, activation of signaling cascade after the stress perception and generation of aquaporin proteins with alterations in the cytoplasmic pH, calcium, and phosphorylation level are most common (Johansson et al., 1998; Alleva et al., 2006). Besides, several studies have observed change in the signaling of root and shoot hormones after stress perception. Especially the hormones such as abscisic acid which have been reported as the most responsive hormone under drought, heat, and salinity stress (Lata et al., 2011). Various studies have been reported to emphasize the application of NPs and their alleviation effects on heat-stressed crops. For instance, a study conducted by Iqbal et al. (2019) revealed that the application of silver NPs (AgNPs) to the wheat seedlings enhanced plant fresh and dry weight with a significant increase in plant's leaf area and leaf number. Another study conducted on sorghum (*S. bicolor* (L.) Moench) disclosed the importance of SeNPs in heat stress alleviation. This study concluded that foliar application of SeNPs to sorghum not only increased crop yield but also enhanced the antioxidants activity in turn decreasing oxidants concentration and providing heat stress alleviation to sorghum plants (Djanaguiraman et al., 2018). Besides, SiNPs were applied on wheat (*T. aestivum*) seedlings to identify their counter effect on UV-B stress mitigations. It was observed that the application of SiNPs enhanced the antioxidant defense system by mediating nitric oxide signaling, and SiNPs were more effective compared to Si for UV-B stress alleviation (Tripathi et al., 2017).

27.4.1.2 Salinity stress

Salinity stress is one of the major abiotic stresses that restrain plant growth as well as their distribution even in their natural habitat (Tang et al., 2015). It affects approximately 800 mha of the arable lands and causes severe problems related to plant growth and productivity in arid and semi-arid regions of the world (Munns and Tester, 2008; Gohari et al., 2020). The application of NPs for mitigating the salinity stress in crops has gained attention these days and many researchers have provided sufficient evidence in this regard (Alsaeedi et al., 2019; El-Gazzar et al., 2020; Baz et al., 2020; Mahdy et al., 2020). Askary et al. (2017) evaluated the effect of different doses of FeONPs on the physiological parameters of peppermint (*Mentha piperita*) under salt stress. They noticed a substantial increment in the dry and fresh weight of leaves, K, P, Fe, and Zn concentration with a simultaneous decline in the uptake of sodium (Na^+). FeONPs were found more effective at a concentration of 30 μM. Similarly, Latef et al. (2017) investigated the impact of ZnONPs on lupine growth under 150 mM NaCl induced salinity stress. ZnONPs were also applied in different concentrations (20, 40, and 60 mg/L). Their findings depicted that salinity stress caused a negative impact on all the growth as well as physiological attributes of plants at all the concentrations. However, priming of the seeds with ZnONPs enhanced the development of the plant under stress conditions with a concomitant increase in the ascorbic acid, organic solutes, photosynthetic pigments, and Zn contents. In a similar experiment, Kalteh et al. (2018) performed a pot experiment for investigating the effect of SiNPs on the vegetative features of basil under salinity stress. Silicon was applied in the form of fertilizer as well as an NP. Results indicated a considerable decline in the growth and development under salinity stress; however, the application of SiONPs increased the leaf dry and fresh weights, chlorophyll, and proline contents, and this increment was attributed to the induction of SiONPs-induced tolerance in plants to salinity stress.

27.4.1.3 Heavy metal stress

Heavy metal remediation from the environment is very difficult on account of their nonbiodegradability and persistence for a longer period (Manikandan et al., 2015). Currently, extensive agricultural lands have been deteriorated due to high concentrations of heavy metals. Different strategies have been employed worldwide to tackle this issue, and the use of nanotechnology to combat the phytotoxicity of heavy metals is also gaining importance in this regard. Nanotechnology has the potential to uplift agricultural productivity in terms of nanofertilizers as well as other nano-based formulations (Mahakham et al., 2016; Rizwan et al., 2017). The use of NPs for alleviating the harms of heavy metals in crops has been reported by different researchers (Konate et al., 2017; Rahmatizadeh et al., 2019; Rizwan et al., 2019). Venkatachalam et al. (2017) investigated the effect of the application of ZnONPs on the oxidative stress induced by heavy metals (Cd and Pb) in *Leucaena leucocephala*. Their findings revealed that the application of AgONPs considerably enhanced seedling growth under heavy metal stress. Moreover, a significant improvement in the photosynthetic pigments and total soluble protein contents was observed with a concomitant decline in the malondialdehyde contents in the leaves. Similarly, Rizwan et al. (2019) evaluated the effect of seed priming with FeONPs and ZnONPs on the growth as well as Cd accumulation in wheat. Priming of wheat seeds was done with different concentrations of FeONPs and ZnONPs followed by their sowing in the soil that was subjected to Cd contamination as a result of long-term application of sewage water. Their findings indicated that the application of NPs particularly at higher rates caused a significant uplift in the spike length, plant height, dry weights of the root, shoot, spikes, and grains. Moreover, photosynthesis was also improved in wheat compared to the control treatment. In addition, the concentration of Cd in roots, shoots, and grains was decreased significantly. Hussain et al. (2019) assessed the seed priming effect with SiNPs under Cd stress in wheat. The parameters measured were Si and Cd accumulation in wheat, oxidative stress, yield, growth, and photosynthesis. The results revealed that SiONPs diminished the Cd-induced oxidative stress and caused a positive impact on wheat growth and chlorophyll contents in comparison with the control. Moreover, Cd concentrations in the grains were also declined with a simultaneous increase in the Si uptake by the plant. Fig. 27.2 is a pictorial presentation of the role of NPs in agriculture.

27.4.1.4 Application of NPs for soil health management and restoration

Agricultural productivity is hindered by various types of soil factors that affect soil health and its capacity to offer favorable conditions for healthy plant growth. One of the major factors that causes soil health deterioration is low-quality water (Wang *et al.*, 2016; Feng *et al.*, 2017; Díaz *et al.*, 2018). Soil irrigated with poor quality water limits the nutrients supply to the plants. Apart from that, different pH range, high amount of salts, and soil properties limit the nutrients supply to plants Egamberdieva *et al.*, (2017); Chiconato *et al.*, (2019); Guan *et al.*, (2019); Hayes *et al.*, (2019). Soil salinity is caused by various factors that is, weathering of parent material Daliakopoulos *et al.*, (2016), use of brackish water for irrigation Chhabra, (2017); Chen *et al.*, (2018), and excessive use of chemical fertilizers (Ritzema, 2016). In addition to salinity, heavy metals also play a role in the deterioration of soil health (He *et al.*, 2019; Tang *et al.*, 2019). Heavy metals affect soil productivity by reducing seed germination (Bae *et al.*, 2016; Seneviratne *et al.*, 2017) and plant growth by inducing oxidative stress, cellular damage, and altering cellular homeostasis (Dubey *et al.*, 2018) in turn diminishing overall crop productivity (O'Connor *et al.*, 2018; Hou *et al.*, 2019).

27.5 Role of NPs in soil remediation/clean up

Agriculture, food, and natural resources are among the top priority challenges that require sustainability. The induction of NPs in agriculture can decrease the amount of chemical use, lessen the losses of nutrients from fertilizers, and overall enhanced the production by nutrient and pest management. It can assist the food and agriculture industry by employing various nanotools such as enhancing the NUE through bionanofertilizers, nanofertilizers, and reducing the disease spread by nanopesticides. These NPs can also be used as sensors for monitoring agricultural soil quality (Prasad et al., 2017; Sangeetha et al., 2017). NPs can play a potential role in the improvement of soil health and overall rehabilitation of the degraded soils as NPs stabilize the heavy metals in soils and favor the construction of a stronger soil skeleton matrix in association with the cementitious materials (Thomé et al., 2015; Correia and Rasteiro, 2016). Besides, NPs help reduce soil heavy metals concentration (Cai et al., 2018) by coprecipitation (Wang, T. et al., 2019) and contaminant stabilization and capping (Wang, et al., 2019; Makselon et al., 2018). Various studies have been reported to reveal the remediation potential of metal and metal oxides NPs. For instance, NPs derived from selenium are being used as soil remediating agents by mercury immobilization (Wang, X. et al., 2019). Also, iron oxide NPs despite having a positive charge revealed the significant potential to adsorb both cations and anions. Among inorganic cations, bivalents (Cu, Pb, Ni, etc.) and trivalent (Al, Fe) are mostly adsorbed by iron oxide NPs through forming an inner-sphere complex ultimately accumulating in the soil. Not only inorganic cations, iron oxide NPs revealed the potential to sorb soil organic matter (humic substance) followed by degradation and oxidation reaction (Claudio et al., 2017). Moreover, recently Cui et al., (2017) conducted a study to explore the therapeutic effect of silicon NPs on Cd-stressed rice crops. In this regard, SiNPs of three (19 nm, 48 nm, and 202 nm) sizes were employed to investigate the size-dependent effect of SiNPs. It was found that SiNPs can augment the proportion of live cells up to 95.4% in the case of 19 nm sized SiNPs. The reduction in the Cd toxicity was correlated with the size of SiNPs that is, with the decrease in the size of NPs an increase in the proportion of the live cells was observed. Further, rice plants that were grown without treatment with SiNPs displayed severe damage in the organelle integrity of rice plants (Cui et al., 2017).

27.5.1 Role of NPs in soil rehabilitation

Nanomaterials are promising owing to their small size, large surface-to -volume ratio, high performance and smart, and targeted delivery system (Shojaei et al., 2019). The especially small size of nanomaterials plays a game-changing role as small size imparts the large surface area to the nanomaterial's eventually becoming a reason to react with a large number of other compounds. Moreover, due to their extremely small size, they can easily penetrate through the root and leaves (Liscano et al., 2000); hence, improving the uptake, penetration rate, and NUE of minerals. Besides, nanomaterials are capable to release nutrients slowly thus increasing nutrient availability duration for crop plants and preventing it from loss. Also, this results in increased nutrient efficiency and reduces the risk of environmental pollution (Singh, 2017).

Metals and metal oxides–derived NPs are being used as macro and micronutrient nanofertilizers, nutrient-loaded nanofertilizers, and plant-growth-enhancers, thus attempting to cope up with the future needs of the increasing population (Liu and Lal, 2015). The conventional belief of excessive use of fertilizers can lead to higher crop production has severely damaged the soil quality by polluting the soil environment. Instead, precision agriculture to employ optimum concentration of pesticides and fertilizers resulting in higher yields should be accomplished as excessive and unmonitored use of fertilizers can lead to problems such as eutrophication, which is hazardous for life in water bodies. Therefore, the use of nanofertilizers/nanoproducts could address the problem of excessive nutrient loss from the farming field with on-demand and targeted release of the nutrient in turn preventing the premature loss of the nutrients (Kim et al., 2018). A robust review of published literature reveals a preponderance of micronutrient nanofertilizers (Cu, Zn, Ag, Fe, TiO_2) over macronutrient nanofertilizers (N, P, K, Ca, Mg, S) despite their importance in the crop productivity (Dimkpa & Bindraban, 2017). Further, various studies claimed the benefits of using NPs of micronutrients in combination with basic essential macronutrients that is, nitrogen, phosphorus, and potassium. This strategy is not only helpful in increasing per hectare yield but also reduces the amounts of macronutrients required for fertilization. For instance, a study conducted by Kale and Gawade (2016) revealed that the use of ZnONPs in combinations with nitrogen, potassium, and phosphorus augmented brinjal (*Solanum melongena* L) yield by 91% and biomass by 45.3% compared to the recommended dose of fertilizers. Besides this, SiNPs in combination with zeolites have shown a substantial potential to hold water and release it slowly. Hence, SiNPs in combination with zeolites is efficient nanotools for increasing the water-holding capacity of soils suffering from issues such as droughts and high infiltration. Moreover, silica-based nanosensors have recently gained huge popularity owing to their better sensitivity, stability, and detection accuracy compared to other NPs (Rastogi et al., 2019). SiNPs alone or in combination with other NPs can be an effective sensing tool for agriculture use for monitoring soil and water contamination (Ion et al., 2010).

27.6 Prospects

Sustainable agriculture must be regarded as an ecosystem approach under which abiotic–biotic living beings coexist in equilibrium with food chains and their corresponding energy balances. To optimize agricultural productivity, emerging technology, modernization, expanded usage of nanochemicals, specialization, and government policies must be introduced. To get out of this crisis, new developments in the agriculture and food sectors must be implemented. As a result, nanotechnology, which has very unique properties in the food supply chain (from the field to the table: crop production, use of agro-chemicals such as nanofertilizers, nanopesticides, nanoherbicides, etc., precision farming techniques, intelligent feed, enhancement of food texture and quality, bioavailability/nutrient values, packaging and labeling, etc.), is the new and future technology. Many particular fields of agricultural nanotechnology or nanofoods research may need further focus in the immediate future:

- New environmental and safety distribution systems for specific plant nutrients, and other substances are being created.
- Bionanosensors and associated nanotechnology play an important function in insect pest management and agricultural food products. Consumers will now get real-time details about the status of a specific food commodity thanks to smart food packaging with nanosensors.
- Scale, dosage, exposure duration, surface chemistry, structures, immune reaction, aggregation, retention time, and other consequences of nanomaterials should all be closely examined. To identify, verify, and access the effects of each nanomaterial/nanofood in whole environments, new analytical methods must be created. Nanomaterials/nanofoods can be subjected to a life-cycle examination. For manipulation of this information, a broad databank as well as international cooperation for strategy, concept, and regulation are needed. The authorities can also have specific recommendations and roadmaps for reducing the hazards associated with the usage of nanotechnological devices.
- To address the impacts of this technology on human life, economy, and science, new contact networks and debates should be opened with involvement from various parties such as customers, academics, officials, manufacturing sectors, and others.

27.7 Conclusions

Nanotechnology is a leading discipline having multidisciplinary mode of action for the management of agriculture sustainability as it is a novel approach to manage insect pests, diseases, weeds, and improving the yields of crops by the application of nanopesticides, nanofungicides, and other nanomaterials such as nanofertilizers. Though a wide range of positive literature regarding the role of NPs in agriculture is reported, a huge debate on persistence and the long-term role of NPs in the agro-ecosystem persists. It is need of the hour to comprehensively understand the chemistry, behavior, and dynamics of NPs in agroecosystem with a special focus on their health risk for human.

References

A El-Kereti, M., A. El-feky, S., S. Khater, M., A. Osman, Y., A. El-sherbini, E.S., 2013. ZnO nanofertilizer and He Ne laser irradiation for promoting growth and yield of sweet basil plant. Recent Pat. Food, Nutr. Agric. 5 (3), 169–181.

Abbasi Khalaki, M., Ghorbani, A., Moameri, M, 2016. Effects of silica and silver nanoparticles on seed germination traits of *Thymus kotschyanus* in laboratory conditions. J. Rangeland Sci. 6 (3), 221–231.

Abbasi, G.H., Akhtar, J., Anwar-ul-Haq, M., Ali, S., Chen, Z., Malik, W., 2014. Exogenous potassium differentially mitigates salt stress in tolerant and sensitive maize hybrids. Pak. J. Bot 46 (1), 135–146.

Abd El-Azeim, M.M., Sherif, M.A., Hussien, M.S., Haddad, S.A., 2020. Temporal impacts of different fertilization systems on soil health under arid conditions of potato monocropping. J. Soil Sci. Plant Nutr. 20 (2), 322–334.

Abdel-Gawad, R.M., 2018. Insecticidal activity of *Moringa oleifera* synthesized silver and zinc nanoparticles against the house fly, *Musca domestica* L. Egypt. Acad. J. Biol. Sci. A, Entomol. 11 (4), 19–30.

Abdel-Haliem, M.E., Hegazy, H.S., Hassan, N.S., Naguib, D.M., 2017. Effect of silica ions and nano silica on rice plants under salinity stress. Ecol. Eng. 99, 282–289.

Adhikari, T., Kundu, S., Rao, A.S., 2016. Zinc delivery to plants through seed coating with nano-zinc oxide particles. J. Plant Nutr. 39 (1), 136–146.

Adhikari, T., Kundu, S., Biswas, A.K., Tarafdar, J.C., Rao, A.S., 2012. Effect of copper oxide nano particle on seed germination of selected crops. J. Agric. Sci. Technol. A 2 (6A), 815.

Adisa, I.O., Pullagurala, V.L.R., Peralta-Videa, J.R., Dimkpa, C.O., Elmer, W.H., Gardea-Torresdey, J.L., White, J.C., 2019. Recent advances in nano-enabled fertilizers and pesticides: a critical review of mechanisms of action. Environ. Sci. 6 (7), 2002–2030.

Adrees, M., Khan, Z.S., Ali, S., Hafeez, M., Khalid, S., Ur Rehman, M.Z., Hussain, A., Hussain, K., Chatha, S.A.S., Rizwan, M., 2020. Simultaneous mitigation of cadmium and drought stress in wheat by soil application of iron nanoparticles. Chemosphere 238, 124681.

Alleva, K., Niemietz, C.M., Sutka, M., Maurel, C., Parisi, M., Tyerman, S.D., Amodeo, G., 2006. Plasma membrane of *Beta vulgaris* storage root shows high water channel activity regulated by cytoplasmic pH and a dual range of calcium concentrations. J. Exp. Bot. 57 (3), 609–621.

Al-Mubaddel, F.S., Haider, S., Al-Masry, W.A., Al-Zeghayer, Y., Imran, M., Haider, A., Ullah, Z., 2017. Engineered nanostructures: a review of their synthesis, characterization and toxic hazard considerations. Arab. J. Chem. 10, S376–S388.

Alsaeedi, A., El-Ramady, H., Alshaal, T., El-Garawany, M., Elhawat, N., Al-Otaibi, A., 2019. Silica nanoparticles boost growth and productivity of cucumber under water deficit and salinity stresses by balancing nutrients uptake. Plant Physiol. BioChem. 139, 1–10.

Anita, S., Rao, D.P., 2014. Enhancement of seed germination and plant growth of wheat, maize, peanut and garlic using multiwalled carbon nanotubes. Eur. Chem. Bull. 3 (5), 502–504.

Anuar, M., E Hodson, M., Boxall, A., 2017. Differences in fate, behavior and uptake of conventional-and nano-pesticides. In: Proc. 19th EGU General Assembly, EGU2017, Vienna, Austria, p. 4859 23-28 April, 2017.

Arfat, Y.A., Ejaz, M., Jacob, H., Ahmed, J., 2017. Deciphering the potential of guar gum/Ag-Cu nanocomposite films as an active food packaging material. Carbohydr. Polym. 157, 65–71.

Ashkavand, P., Tabari, M., Zarafshar, M., Tomášková, I., Struve, D., 2015. Effect of SiO_2 nanoparticles on drought resistance in hawthorn seedlings. Forest Res. Papers 76 (4), 350–359.

Askary, M., Talebi, S.M., Amini, F., Bangan, A.D.B., 2017. Effects of iron nanoparticles on *Mentha piperita* L. under salinity stress. Biologija 63 (1), 65–75.

Athawale, V., Paralikar, P., Ingle, A.P., Rai, M., 2018. Biogenically engineered nanoparticles inhibit *Fusarium oxysporum* causing soft-rot of ginger. IET Nanobiotechnol. 12 (8), 1084–1089.

Babaei, K., Seyed Sharifi, R., Pirzad, A., Khalilzadeh, R., 2017. Effects of bio fertilizer and nano Zn-Fe oxide on physiological traits, antioxidant enzymes activity and yield of wheat (*Triticum aestivum* L.) under salinity stress. J. Plant Interact. 12 (1), 381–389.

Badran, A., Savin, I., 2018. Effect of nano-fertilizer on seed germination and first stages of bitter almond seedlings' growth under saline conditions. BioNanoScience 8 (3), 742–751.

Bae, J., Benoit, D.L., Watson, A.K., 2016. Effect of heavy metals on seed germination and seedling growth of common ragweed and roadside ground cover legumes. Environ. Pollut. 213, 112–118.

Baz, H., Creech, M., Chen, J., Gong, H., Bradford, K., Huo, H., 2020. Water-soluble carbon nanoparticles improve seed germination and post-germination growth of lettuce under salinity stress. Agronomy 10 (8), 1192.

Begum, P., Ikhtiari, R., Fugetsu, B., 2014. Potential impact of multi-walled carbon nanotubes exposure to the seedling stage of selected plant species. Nanomaterials 4 (2), 203–221.

Borišev, M., Borišev, I., Župunski, M., Arsenov, D., Pajević, S., Ćurčić, Ž., Vasin, J., Djordjevic, A., 2016. Drought impact is alleviated in sugar beets (*Beta vulgaris* L.) by foliar application of fullerenol nanoparticles. PLoS One 11 (11), e0166248.

Boutchuen, A., Zimmerman, D., Aich, N., Masud, A.M., Arabshahi, A., Palchoudhury, S., 2019. Increased plant growth with hematite nanoparticle fertilizer drop and determining nanoparticle uptake in plants using multimodal approach. J. Nanomater. 2019, 6890572.

Burman, U., Saini, M., Kumar, P., 2013. Effect of zinc oxide nanoparticles on growth and antioxidant system of chickpea seedlings. Toxicol. Environ. Chem. 95 (4), 605–612.

Cadena, M.B., Preston, G.M., Van der Hoorn, R.A., Townley, H.E., Thompson, I.P., 2018. Species-specific antimicrobial activity of essential oils and enhancement by encapsulation in mesoporous silica nanoparticles. Ind. Crops Prod. 122, 582–590.

Cai, L., Chen, J., Liu, Z., Wang, H., Yang, H., Ding, W., 2018. Magnesium oxide nanoparticles: effective agricultural antibacterial agent against *Ralstonia solanacearum*. Front. Microbiol. 9, 790.

Cañas, J.E., Long, M., Nations, S., Vadan, R., Dai, L., Luo, M., Ambikapathi, R., Lee, E.H., Olszyk, D., 2008. Effects of functionalized and nonfunctionalized single-walled carbon nanotubes on root elongation of select crop species. Environ. Toxicol. Chem. 27 (9), 1922–1931.

Castiglione, M.R., Giorgetti, L., Becarelli, S., Siracusa, G., Lorenzi, R., Di Gregorio, S., 2016. Polycyclic aromatic hydrocarbon-contaminated soils: bioaugmentation of autochthonous bacteria and toxicological assessment of the bioremediation process by means of *Vicia faba* L. Environ. Sci. Pollut. Res. 23 (8), 7930–7941.

Subbarao, C.V., Kartheek, G., Sirisha, D., 2013. Slow release of potash fertilizer through polymer coating. Int. J. Appl. Sci. Eng. 11 (1), 25–30.

Chen, W., Jin, M., Ferré, T.P., Liu, Y., Xian, Y., Shan, T., Ping, X., 2018. Spatial distribution of soil moisture, soil salinity, and root density beneath a cotton field under mulched drip irrigation with brackish and fresh water. Field Crops Res. 215, 207–221.

Chen, X., Mao, S.S., 2007. Titanium dioxide nanomaterials: synthesis, properties, modifications, and applications. Chem. Rev. 107 (7), 2891–2959.

Cheon, J., Chan, W. and Zuhorn, I., 2019. The future of nanotechnology: cross-disciplined progress to improve health and medicine. Acc. Chem. Res. 52(9), 2405

Chhabra, R., 2017. Soil Salinity and Water Quality, First. Routledge, London, pp. 1–300.

Chhipa, H., 2017. Nanofertilizers and nanopesticides for agriculture. Environ. Chem. Lett. 15 (1), 15–22.

Chhipa, H., Joshi, P., 2016. Nanofertilizers, nanopesticides and nanosensors in agriculture. In: Ranjan, S., Dasgupta, N., Lichtfouse, E. (Eds.), Nanoscience in Food and Agriculture 1. Springer, Cham, pp. 247–282.

Chiconato, D.A., Junior, G.D.S.S., dos Santos, D.M.M., Munns, R., 2019. Adaptation of sugarcane plants to saline soil. Environ. Exp. Bot. 162, 201–211.

Chitkara, D., Mittal, A., Mahato, R.I., 2018. Molecular Medicines for Cancer: Concepts and Applications of Nanotechnology. CRC Press, Boca Raton, FL.

Choudhary, R.C., Kumaraswamy, R.V., Kumari, S., Sharma, S.S., Pal, A., Raliya, R., Biswas, P., Saharan, V., 2019. Zinc encapsulated chitosan nanoparticle to promote maize crop yield. Int. J. Biol. Macromol. 127, 126–135.

Choudhary, R.C., Kumaraswamy, R.V., Kumari, S., Sharma, S.S., Pal, A., Raliya, R., Biswas, P., Saharan, V., 2017. Cu-chitosan nanoparticle boost defense responses and plant growth in maize (*Zea mays* L.). Sci. Rep. 7 (1), 1–11.

Cioffi, N., Torsi, L., Ditaranto, N., Tantillo, G., Ghibelli, L., Sabbatini, L., Bleve-Zacheo, T., D'Alessio, M., Zambonin, P.G., Traversa, E., 2005. Copper nanoparticle/polymer composites with antifungal and bacteriostatic properties. Chem. Mater. 17 (21), 5255–5262.

Claudio, C., Iorio, E.D., Liu, Q., Jiang, Z., Barrón, V., 2017. Iron oxide nanoparticles in soils: environmental and agronomic importance. J. NanoSci. Nanotechnol. 17 (7), 4449–4460.

Correia, A.A.S., Rasteiro, M.G., 2016. Nanotechnology applied to chemical soil stabilization. Procedia Eng. 143, 1252–1259.

Cui, J., Liu, T., Li, F., Yi, J., Liu, C., Yu, H., 2017. Silica nanoparticles alleviate cadmium toxicity in rice cells: mechanisms and size effects. Environ. Pollut. 228, 363–369.

Da Fonseca, A.F., Melfi, A.J., Montes, C.R., 2005. Maize growth and changes in soil fertility after irrigation with treated sewage effluent. Plant dry matter yield and soil nitrogen and phosphorus availability. Commun. Soil Sci. Plant Anal. 36 (13-14), 1965–1981.

Daliakopoulos, I., Tsanis, I., Koutroulis, A., Kourgialas, N., Varouchakis, A., Karatzas, G., Ritsema, C., 2016. The threat of soil salinity: a European scale review. Sci. Total Environ 573, 727–739.

Dapkekar, A., Deshpande, P., Oak, M.D., Paknikar, K.M., Rajwade, J.M., 2018. Zinc use efficiency is enhanced in wheat through nano-fertilization. Sci. Rep. 8 (1), 1–7.

Davarpanah, S., Tehranifar, A., Davarynejad, G., Aran, M., Abadía, J., Khorassani, R., 2017. Effects of foliar nano-nitrogen and urea fertilizers on the physical and chemical properties of pomegranate (*Punica granatum* cv. Ardestani) fruits. HortScience 52 (2), 288–294.

Dawson, C.J., Hilton, J., 2011. Fertiliser availability in a resource-limited world: production and recycling of nitrogen and phosphorus. Food Policy 36, S14–S22.

Deepa, M., Sudhakar, P., Nagamadhuri, K.V., Reddy, K.B., Krishna, T.G., Prasad, T.N.V.K.V., 2015. First evidence on phloem transport of nanoscale calcium oxide in groundnut using solution culture technique. Appl. NanoSci. 5 (5), 545–551.

Delfani, M., Baradarn Firouzabadi, M., Farrokhi, N., Makarian, H., 2014. Some physiological responses of black-eyed pea to iron and magnesium nanofertilizers. Commun. Soil Sci. Plant Anal. 45 (4), 530–540.

Deng, H., Wang, S., Wang, X., Du, C., Shen, X., Wang, Y., Cui, F., 2015. Two competitive nucleation mechanisms of calcium carbonate biomineralization in response to surface functionality in low calcium ion concentration solution. Regenerat. Biomater. 2 (3), 187–195.

Derbalah, A., Shenashen, M., Hamza, A., Mohamed, A., El Safty, S., 2018. Antifungal activity of fabricated mesoporous silica nanoparticles against early blight of tomato. Egypt. J. Basic Appl. Sci. 5 (2), 145–150.

Dhasmana, A., Uniyal, S., Kumar, V., Gupta, S., Kesari, K.K., Haque, S., Lohani, M., Pandey, J., 2019. Scope of nanoparticles in environmental toxicant remediation. In: Sobti, R.C., Arora, N.K., Kothari, R. (Eds.), Environmental Biotechnology: For Sustainable Future. Springer, Singapore, pp. 31–44.

Díaz, F.J., Grattan, S.R., Reyes, J.A., de la Roza-Delgado, B., Benes, S.E., Jiménez, C., Dorta, M., Tejedor, M., 2018. Using saline soil and marginal quality water to produce alfalfa in arid climates. Agric. Water Manage. 199, 11–21.

Dimkpa, C.O., Singh, U., Bindraban, P.S., Elmer, W.H., Gardea-Torresdey, J.L., White, J.C., 2019. Zinc oxide nanoparticles alleviate drought-induced alterations in sorghum performance, nutrient acquisition, and grain fortification. Sci. Total Environ. 688, 926–934.

Dimkpa, C., Andrews, J., Fugice, J., Singh, U., Bindraban, P.S., Elmer, W.H., White, J.C., 2020. Facile coating of urea with low-dose ZnO nanoparticles promotes wheat performance and enhances Zn uptake under drought stress. Front. Plant Sci. 11, 168.

Dimkpa, C.O., Bindraban, P.S., 2017. Nanofertilizers: new products for the industry? J. Agric. Food Chem. 66, 6462–6473.

Djanaguiraman, M., Belliraj, N., Bossmann, S.H., Prasad, P.V., 2018. High-temperature stress alleviation by selenium nanoparticle treatment in grain sorghum. ACS Omega 3 (3), 2479–2491.

Dobrucka, R., Ankiel, M., 2019. Possible applications of metal nanoparticles in antimicrobial food packaging. J. Food Safety 39 (2), e12617.

Dubey, S., Shri, M., Gupta, A., Rani, V., Chakrabarty, D., 2018. Toxicity and detoxification of heavy metals during plant growth and metabolism. Environ. Chem. Lett. 16 (4), 1169–1192.

Egamberdieva, D., Davranov, K., Wirth, S., Hashem, A., Abd_Allah, E.F., 2017. Impact of soil salinity on the plant-growth – promoting and biological control abilities of root associated bacteria. Saudi J. Biol. Sci. 24, 1601–1608.

El-Argawy, E.M.M.H., Rahhal, El-Korany, A., Elshabrawy, E.M., Eltahan, R.M., 2017. Efficacy of some nanoparticles to control damping-off and root rot of sugar beet in El-Behiera Governorate. Asian J. Plant Pathol. 11, 35–47.

El-Bendary, H.M., El-Helaly, A.A., 2013. First record nanotechnology in agricultural: silica nano-particles a potential new insecticide for pest control. App. Sci. Rep. 4 (3), 241–246.

Elbeshehy, E.K., Elazzazy, A.M., Aggelis, G., 2015. Silver nanoparticles synthesis mediated by new isolates of *Bacillus* spp., nanoparticle characterization and their activity against bean yellow mosaic virus and human pathogens. Front. Microbiol. 6, 453.

El-Gazzar, N., Almaary, K., Ismail, A., Polizzi, G., 2020. Influence of *Funneliformis mosseae* enhanced with titanium dioxide nanoparticles (TiO_2NPs) on *Phaseolus vulgaris* L. under salinity stress. PloS one 15 (8), e0235355.

El-Helaly, A.A., El-Bendary, H.M., Abdel-Wahab, A.S., El-Sheikh, M.A.K., Elnagar, S., 2016. The silica-nano particles treatment of squash foliage and survival and development of *Spodoptera littoralis* (Bosid.) larvae. J. Entomol. Zool. Stud. 4, 175–180.

El-Sharkawy, M.S., El-Beshsbeshy, T.R., Mahmoud, E.K., Abdelkader, N.I., Al-Shal, R.M., Missaoui, A.M., 2017. Response of alfalfa under salt stress to the application of potassium sulfate nanoparticles. Am. J. Plant Sci. 8 (08), 1751.

El-Temsah, Y.S., Sevcu, A., Bobcikova, K., Cernik, M., Joner, E.J., 2016. DDT degradation efficiency and ecotoxicological effects of two types of nano-sized zero-valent iron (nZVI) in water and soil. Chemosphere 144, 2221–2228.

Eric, D.K., 1986. Engines of Creation: The Coming Era of Nanotechnology. Anchor Book, IL.

Esteban-Tejeda, L., Malpartida, F., Esteban-Cubillo, A., Pecharromán, C., Moya, J.S., 2009. Antibacterial and antifungal activity of a soda-lime glass containing copper nanoparticles. Nanotechnology 20 (50), 505701.

Esteves, N.B., Sigal, A., Leiva, E.P.M., Rodríguez, C.R., Cavalcante, F.S.A., De Lima, L.C., 2015. Wind and solar hydrogen for the potential production of ammonia in the state of Ceará–Brazil. Int. J. Hydrogen Energy 40 (32), 9917–9923.

Fansuri, H, Prichard, D, Dong-Ke-Zhang, 2008. Manufacture of zeolites from flyash for fertilizer applications. Ph.D (Agri.) Thesis. Centre for fuels and energy. Curtain University of Technology, Australia.

Farahat, A.G., 2018. Biosynthesis of nano zinc and using of some nanoparticles in reducing Cercospora Leaf Spot disease of sugar beet in the field. Environ. Biodiv. Soil Security. 2, 103–117.

Feizi, H., Moghaddam, P.R., Shahtahmassebi, N., Fotovat, A., 2012. Impact of bulk and nanosized titanium dioxide (TiO_2) on wheat seed germination and seedling growth. Biol. Trace Element Res. 146 (1), 101–106.

Feng, G., Zhang, Z., Wan, C., Lu, P., Bakour, A., 2017. Effects of saline water irrigation on soil salinity and yield of summer maize (*Zea mays* L.) in subsurface drainage system. Agric. Water Manage. 193, 205–213.

Feynman, R.P., 1960. There's Plenty of Room at the Bottom. Engi. Sci. Magazine 23 (5), 22–36.

Fitriani, H.P., Haryanti, 2016. Effect of the use of nanosilica fertilizer on the growth of tomato plant (*Solanum lycopersicum*). Bul Anat. dan Fisiol. 24 (1), 34–41.

Fouad, H., Hongjie, L., Yanmei, D., Baoting, Y., El-Shakh, A., Abbas, G., Jianchu, M., 2017. Synthesis and characterization of silver nanoparticles using *Bacillus amyloliquefaciens* and *Bacillus subtilis* to control filarial vector *Culex pipiens* pallens and its antimicrobial activity. Artif. Cell. Nanomed. B. 45, 1369–1378.

Gao, F., Liu, C., Qu, C., Zheng, L., Yang, F., Su, M., Hong, F., 2008. Was improvement of spinach growth by nano-TiO_2 treatment related to the changes of Rubisco activase? Biometals 21 (2), 211–217.

Ghafariyan, M.H., Malakouti, M.J., Dadpour, M.R., Stroeve, P., Mahmoudi, M., 2013. Effects of magnetite nanoparticles on soybean chlorophyll. Environ. Sci. Technol. 47 (18), 10645–10652.

Ghorbanpour, M., Mohammadi, H., Kariman, K., 2020. Nanosilicon-based recovery of barley (*Hordeum vulgare*) plants subjected to drought stress. Environ. Sci. 7 (2), 443–461.

Ghosh, S., Ramaprabhu, S., 2019. Green synthesis of transition metal nanocrystals encapsulated into nitrogen-doped carbon nanotubes for efficient carbon dioxide capture. Carbon 141, 692–703.

Giannousi, K., Sarafidis, G., Mourdikoudis, S., Pantazaki, A., Dendrinou-Samara, C., 2014. Selective synthesis of Cu_2O and Cu/Cu_2O NPs: antifungal activity to yeast *Saccharomyces cerevisiae* and DNA interaction. Inorgan. Chem. 53 (18), 9657–9666.

GOEBEL, M.O., Bachmann, J., Reichstein, M., Janssens, I.A., Guggenberger, G., 2011. Soil water repellency and its implications for organic matter decomposition–is there a link to extreme climatic events? Global Change Biol. 17 (8), 2640–2656.

Gohari, G., Mohammadi, A., Akbari, A., Panahirad, S., Dadpour, M.R., Fotopoulos, V., Kimura, S., 2020. Titanium dioxide nanoparticles (TiO_2-NPs) promote growth and ameliorate salinity stress effects on essential oil profile and biochemical attributes of *Dracocephalum moldavica*. Sci. Rep. 10 (1), 912.

Gopal, M, Gogoi, R, Srivastava, C., Kumar, R., Singh, P.K., Nair, K.K., Yadav, S., Goswami, A., 2011. Nanotechnology and its application in plant protection. Plant Pathology in India: Vision 2030, Academia, pp. 224–232 Available at: https://www.academia.edu/28166636/Nanotechnology_and_its_application_in_plant_protection.

Goufo, P., Moutinho-Pereira, J.M., Jorge, T.F., Correia, C.M., Oliveira, M.R., Rosa, E.A., António, C., Trindade, H., 2017. Cowpea (*Vigna unguiculata* L. Walp.) metabolomics: osmoprotection as a physiological strategy for drought stress resistance and improved yield. Front. Plant Sci. 8, 586.

Guan, B., Xie, B., Yang, S., Hou, A., Chen, M., Han, G., 2019. Effects of five years' nitrogen deposition on soil properties and plant growth in a salinized reed wetland of the yellow river delta. Ecol. Eng. 136, 160–166.

Guo, A., Ban, Y., Yang, K., Yang, W., 2018. Metal-organic framework-based mixed matrix membranes: synergetic effect of adsorption and diffusion for CO_2/CH_4 separation. J. Membr. Sci. 562, 76–84.

Hafeez, A., Razzaq, A., Mahmood, T., Jhanzab, H.M., 2015. Potential of copper nanoparticles to increase growth and yield of wheat. J. Nanosci. Adv. Technol. 1 (1), 6–11.

Hassan, E.S., Salem. A. Fouda, S.S., Awad, M.A., El-Gamal, M.S., Abdu, A.M., 2018. New approach for antimicrobial activity and bio-control of various pathogens by biosynthesized copper nanoparticles using endophytic actinomycetes. J. Radiat. Res. App. Sci. 2018, 262–270.

Hatami, M., Hadian, J., Ghorbanpour, M., 2017. Mechanisms underlying toxicity and stimulatory role of single-walled carbon nanotubes in *Hyoscyamus niger* during drought stress simulated by polyethylene glycol. J. Hazard. Mater. 324, 306–320.

Hayes, S., Pantazopoulou, C.K., van Gelderen, K., Reinen, E., Tween, A.L., Sharma, A., de Vries, M., Prat, S., Schuurink, R.C., Testerink, C., Pierik., R., 2019. Soil salinity limits plant shade avoidance. Current Biol. 29, 1669–1676 e1664.

He, L., Liu, Y., Mustapha, A., Lin, M., 2011. Antifungal activity of zinc oxide nanoparticles against *Botrytis cinerea* and *Penicillium expansum*. Microbiol. Res. 166 (3), 207–215.

He, J., Yang, Y., Christakos, G., Liu, Y., Yang, X., 2019. Assessment of soil heavy metal pollution using stochastic site indicators. Geoderma 337, 359–367.

Hoseinnejad, M., Jafari, S.M., Katouzian, I., 2018. Inorganic and metal nanoparticles and their antimicrobial activity in food packaging applications. Crit. Rev. Microbiol. 44 (2), 161–181.

Hou, L., Tong, T., Tian, B., Xue, D., 2019. Crop yield and quality under cadmium stress. In: Hasanuzzaman, M., Prasad, M., Nahar, K. (Eds.), Cadmium tolerance in plants. Elsevier, Amsterdam, pp. 1–18.

Huber, A.E., Bauerle, T.L., 2016. Long-distance plant signaling pathways in response to multiple stressors: the gap in knowledge. J. Exp. Bot. 67 (7), 2063–2079.

Hussain, A., Rizwan, M., Ali, Q., Ali, S., 2019. Seed priming with silicon nanoparticles improved the biomass and yield while reduced the oxidative stress and cadmium concentration in wheat grains. Environ. Sci. Pollut. Res. 26 (8), 7579–7588.

Imada, K., Sakai, S., Kajihara, H., Tanaka, S., Ito, S., 2016. Magnesium oxide nanoparticles induce systemic resistance in tomato against bacterial wilt disease. Plant Pathol. 65 (4), 551–560.

Ion, A.C., Ion, I., Culetu, A., Gherase, D., 2010. Carbon-based Nanomaterials. Environmental Applications, Romania.

Iqbal, M., Umar, S., 2019. Nano-fertilization to enhance nutrient use efficiency and productivity of crop plants. In: Husen, A., Iqbal, M. (Eds.), Nanomaterials and Plant Potential. Springer, Cham, pp. 473–505.

Iqbal, M., Raja, N.I., Hussain, M., Ejaz, M., Yasmeen, F., 2019. Effect of silver nanoparticles on growth of wheat under heat stress. Iranian J. of Sci. and Technol., Trans. A 43 (2), 387–395.

Ismail, A.W.A., Sidkey, N.M., Arafa, R.A., Fathyand, R.M., El-Batal, A.I., 2016. Evaluation of in vitro antifungal activity of silver and selenium nanoparticles against *Alternaria solani* caused early blight disease on potato. Brit. Biotechnol. J. 12, 1–11.

Ivani, R., Sanaei Nejad, S.H., Ghahraman, B., Astaraei, A.R., Feizi, H., 2018. Role of bulk and Nanosized SiO_2 to overcome salt stress during Fenugreek germination (*Trigonella foenum-graceum* L.). Plant Signal. Behav. 13 (7), e1044190.

Jaffer, Y., Clark, T.A., Pearce, P., Parsons, S.A., 2002. Potential phosphorus recovery by struvite formation. Water Res. 36 (7), 1834–1842.

Jain, D., Kothari, S.L., 2014. Green synthesis of silver nanoparticles and their application in plant virus inhibition. J. Mycol. Plant Pathol. 44 (1), 21–24.

Janmohammadi, M., Sabaghnia, N., 2015. Effect of pre-sowing seed treatments with silicon nanoparticles on germinability of sunflower (*Helianthus annuus*). Botanica 21 (1), 13–21.

Janmohammadi, M., Amanzadeh, T., Sabaghnia, N., Dashti, S., 2016. Impact of foliar application of nano micronutrient fertilizers and titanium dioxide nanoparticles on the growth and yield components of barley under supplemental irrigation. Acta Agric. Slov. 107 (2), 265–276.

Jat, L.K., Singh, Y.V., Meena, S.K., Meena, S.K., Parihar, M., Jatav, H.S., Meena, R.K., Meena, V.S., 2015. Does integrated nutrient management enhance agricultural productivity. J. Pure Appl. Microbiol. 9 (2), 1211–1221.

Jayaseelan, C., Rahuman, A.A., Kirthi, A.V., Marimuthu, S., Santhoshkumar, T., Bagavan, A., Gaurav, K., Karthik, L., Rao, K.B., 2012. Novel microbial route to synthesize ZnO nanoparticles using *Aeromonas hydrophila* and their activity against pathogenic bacteria and fungi. SpectroChim. Acta A 90, 78–84.

Jiang, F., Shen, Y., Ma, C., Zhang, X., Cao, W., Rui, Y., 2017. Effects of TiO_2 nanoparticles on wheat (*Triticum aestivum* L.) seedlings cultivated under super-elevated and normal CO_2 conditions. PloS One *12* (5), e0178088.

Johansson, I., Karlsson, M., Shukla, V.K., Chrispeels, M.J., Larsson, C., Kjellbom, P., 1998. Water transport activity of the plasma membrane aquaporin PM28A is regulated by phosphorylation. Plant Cell 10 (3), 451–459.

Jones, R., Ougham, H., Thomas, H., Waaland, S., 2012. Molecular Life of Plants. Wiley-Blackwell, Hoboken.

Kah, M., Kookana, R.S., Gogos, A., Bucheli, T.D., 2018. A critical evaluation of nanopesticides and nanofertilizers against their conventional analogues. Nat. Nanotechnol. 13, 677.

Kah, M., Hofmann, T., 2014. Nanopesticide research: current trends and future priorities. Environ. Int. 63, 224–235.

Kahrl, F., Li, Y., Su, Y., Tennigkeit, T., Wilkes, A., Xu, J., 2010. Greenhouse gas emissions from nitrogen fertilizer use in China. Environ. Sci. Policy 13 (8), 688–694.

Kale, A.P., Gawade, S.N., 2016. Studies on nanoparticle induced nutrient use efficiency of fertilizer and crop productivity. Green Chem. Tech. Lett. 2, 88–92.

Kalteh, M., Alipour, Z.T., Ashraf, S., Marashi Aliabadi, M., Falah Nosratabadi, A., 2018. Effect of silica nanoparticles on basil (*Ocimum basilicum*) under salinity stress. J. Chem. Health Risks 4 (3), 49–55.

Kanhed, P., Birla, S., Gaikwad, S., Gade, A., Seabra, A.B., Rubilar, O., Duran, N., Rai, M., 2014. In vitro antifungal efficacy of copper against selected crop pathogenic fungi. Mater. Lett. 115, 13–17.

Karimi, J., Mohsenzadeh, S., 2016. Effects of silicon oxide nanoparticles on growth and physiology of wheat seedlings. Russ. J. Plant Physiol. 63 (1), 119–123.

Khodakovskaya, M.V., De Silva, K., Biris, A.S., Dervishi, E., Villagarcia, H., 2012. Carbon nanotubes induce growth enhancement of tobacco cells. ACS Nano 6 (3), 2128–2135.

Khodakovskaya, M.V., Kim, B.S., Kim, J.N., Alimohammadi, M., Dervishi, E., Mustafa, T., Cernigla, C.E., 2013. Carbon nanotubes as plant growth regulators: effects on tomato growth, reproductive system, and soil microbial community. Small 9 (1), 115–123.

Kim, D.Y., Kadam, A., Shinde, S., Saratale, R.G., Patra, J., Ghodake, G., 2018. Recent developments in nanotechnology transforming the agricultural sector: a transition replete with opportunities. J. Sci. Food Agric. 98 (3), 849–864.

Kim, K.-H., Kabir, E., Jahan, S.A., 2017. Exposure to pesticides and the associated human health effects. Sci. Total Environ 575, 525–535.

Klingenfuss, F., 2014. Testing of TiO_2 nanoparticles on wheat and microorganisms in a soil microcosm. (M.Sc. thesis). University of Gothenburg, Department of biology and environmental sciences, Göteborg.

Konate, A., He, X., Zhang, Z., Ma, Y., Zhang, P., Alugongo, G.M., Rui, Y., 2017. Magnetic (Fe_3O_4) nanoparticles reduce heavy metals uptake and mitigate their toxicity in wheat seedling. Sustainability 9 (5), 790.

Kottegoda, N., Sandaruwan, C., Priyadarshana, G., Siriwardhana, A., Rathnayake, U.A., Berugoda Arachchige, D.M., Kumarasinghe, A.R., Dahanayake, D., Karunaratne, V., Amaratunga, G.A., 2017. Urea-hydroxyapatite nanohybrids for slow release of nitrogen. ACS Nano 11 (2), 1214–1221.

Kurepa, J., Paunesku, T., Vogt, S., Arora, H., Rabatic, B.M., Lu, J., Wanzer, M.B., Woloschak, G.E., Smalle, J.A., 2010. Uptake and distribution of ultrasmall anatase TiO_2 Alizarin red S nanoconjugates in *Arabidopsis thaliana*. Nano Lett. 10 (7), 2296–2302.

Lahiani, M.H., Chen, J., Irin, F., Puretzky, A.A., Green, M.J., Khodakovskaya, M.V., 2015. Interaction of carbon nanohorns with plants: uptake and biological effects. Carbon 81, 607–619.

Lata, C., Bhutty, S., Bahadur, R.P., Majee, M., Prasad, M., 2011. Association of an SNP in a novel DREB2-like gene SiDREB2 with stress tolerance in foxtail millet [*Setaria italica* (L.)]. J. Exp. Bot. 62 (10), 3387–3401.

Latef, A.A.H.A., Alhmad, M.F.A., Abdelfattah, K.E., 2017. The possible roles of priming with ZnO nanoparticles in mitigation of salinity stress in lupine (*Lupinus termis)* plants. J. Plant Growth Regul. 36 (1), 60–70.

Lee, C.W., Mahendra, S., Zodrow, K., Li, D., Tsai, Y.C., Braam, J., Alvarez, P.J., 2010. Developmental phytotoxicity of metal oxide nanoparticles to *Arabidopsis thaliana*. Environ. Toxicol. Chem. 29 (3), 669–675.

Lee, W.M., An, Y.J., Yoon, H., Kweon, H.S., 2008. Toxicity and bioavailability of copper nanoparticles to the terrestrial plants mung bean (*Phaseolus radiatus)* and wheat (*Triticum aestivum*): plant agar test for water-insoluble nanoparticles. Environ. Toxicol. Chem. 27 (9), 1915–1921.

Lee, S.B., Suh, M.C., 2013. Recent advances in cuticular wax biosynthesis and its regulation in Arabidopsis. Mol. plant 6 (2), 246–249.

Li, Z., Zhang, Y., 2010. Use of surfactant-modified zeolite to carry and slowly release sulfate. Desalin. Water Treat. 21 (1-3), 73–78.

Lin, D., Xing, B., 2007. Phytotoxicity of nanoparticles: inhibition of seed germination and root growth. Environ. Pollut. 150 (2), 243–250.

Linglan, M., Chao, L., Chunxiang, Q., Sitao, Y., Jie, L., Fengqing, G., Fashui, H., 2008. Rubisco activase mRNA expression in spinach: modulation by nanoanatase treatment. Biol. Trace Element Res. 122 (2), 168–178.

Linh, T.M., Mai, N.C., Hoe, P.T., Lien, L.Q., Ban, N.K., Hien, L.T.T., Chau, N.H., Van, N.T., 2020. Metal-based nanoparticles enhance drought tolerance in soybean. J. Nanomater. 2020, 4056563.

Lipiec, J., Doussan, C., Nosalewicz, A., Kondracka, K., 2013. Effect of drought and heat stresses on plant growth and yield: a review. Int. AgroPhys. 27 (4), 463–477.

Liscano, J.F., Wilson, C.E., Norman-Jr, R.J., 2000. Zinc availability to rice from seven granular fertilizers, 963. Arkansas Agricultural Experiment Station.

Liu, R., Lal, R., 2015. Potentials of engineered nanoparticles as fertilizers for increasing agronomic productions. Sci. Total Environ 514, 131–139.

Liu, R., Lal, R., 2015. Synthetic apatite nanoparticles as a phosphorus fertilizer for soybean (*Glycine max*). Sci. Rep. 4, 5686.

Liu, R., Zhang, H., Lal, R., 2016. Effects of stabilized nanoparticles of copper, zinc, manganese, and iron oxides in low concentrations on lettuce (*Lactuca sativa*) seed germination: nanotoxicants or nanonutrients? Water Air Soil Pollut. 227 (1), 42.

López-Vargas, E., Ortega-Ortíz, H., Cadenas-Pliego, G., de Alba Romenus, K., Cabrera de la Fuente, M., Benavides-Mendoza, A., Juárez-Maldonado, A., 2018. Foliar application of copper nanoparticles increases the fruit quality and the content of bioactive compounds in tomatoes. Appl. Sci. 8 (7), 1020.

Magda, S., Hussein, M.M., 2016. Determinations of the effect of using silica gel and nano-silica gel against *Tutaabsoluta* (Lepidoptera: Gelechiidae) in tomato fields. J. Chem. Pharma. Res. 8 (4), 506–512.

Mahajan, P., Dhoke, S.K., Khanna, A.S., 2011. Effect of nano-ZnO particle suspension on growth of mung (*Vigna radiata*) and gram (*Cicer arietinum*) seedlings using plant agar method. J. Nanotechnol. 2011, 696535.

Mahakham, W., Theerakulpisut, P., Maensiri, S., Phumying, S., Sarmah, A.K., 2016. Environmentally benign synthesis of phytochemicals-capped gold nanoparticles as nanopriming agent for promoting maize seed germination. Sci. Total Environ. 573, 1089–1102.

Mahdy, A.M., Sherif, F.K., Elkhatib, E.A., Fathi, N.O., Ahmed, M.H., 2020. Seed priming in nanoparticles of water treatment residual can increase the germination and growth of cucumber seedling under salinity stress. J. Plant Nutr. 43 (12), 1862–1874.

Mahmoodzadeh, H., Nabavi, M. and Kashefi, H., 2013. Effect of nanoscale titanium dioxide particles on the germination and growth of canola (*Brassica napus).* J. Ornam. Plants, 3, pp. 25-32.

Makselon, J., Siebers, N., Meier, F., Vereecken, H., Klumpp, E., 2018. Role of rain intensity and soil colloids in the retention of surfactant-stabilized silver nanoparticles in soil. Environ. Pollut. 238, 1027–1034.

Manikandan, R., Sahi, S.V., Venkatachalam, P., 2015. Impact assessment of mercury accumulation and biochemical and molecular response of *Mentha arvensis*: a potential hyperaccumulator plant. Sci. World J. 2015, 715217.

Manikandan, A., Subramanian, K.S., 2016. Evaluation of zeolite based nitrogen nano-fertilizers on maize growth, yield and quality on inceptisols and alfisols. Int. J. Plant Soil Sci. 9 (4), 1–9.

Manjili, M.J., Bidarigh, S., Amiri, E., 2014. Study the effect of foliar application of nano chelate molybdenum fertilizer on the yield and yield components of peanut. Egypt. Acad. J. Biol. Sci. H. Bot. 5 (1), 67–71.

Millán, G., Agosto, F., Vázquez, M., 2008. Use of clinoptilolite as a carrier for nitrogen fertilizers in soils of the Pampean regions of Argentina. Int. J. Agric. Nat. Resour. 35 (3), 293–302.

Mishra, S., Singh, H.B., 2015. Biosynthesized silver nanoparticles as a nanoweapon against phytopathogens: exploring their scope and potential in agriculture. Appl. Microbiol. Biotechnol. 99 (3), 1097–1107.

Mitter, N., Karen, H., 2019. Moving policy and regulation forward for nanotechnology applications in agriculture. Nat. Nanotechnol. 14 (6), 508–510.

Mohammad, A., Ahmad, K., Rajak, R., Mobin, S.M.Martínez, L.M.T., Kharissova, O.V., Kharisov, B.I. (Eds.), 2019. Remediation of Water Contaminants. Handbook of Ecomaterials 373–391.

Mohammadi, H., Esmailpour, M., GHERANPAYE, A., 2016. Effects of TiO2 nanoparticles and water-deficit stress on morpho-physiological characteristics of dragonhead (*Dracocephalum moldavica* L.) plants. Acta Agric. Slov. 107 (2), 385–396.

Mohammed, J.G., Mohammed, M.Z., Ridha, D.M., 2017. The antimicrobial activity of silica oxide nanoparticles against some bacteria and fungi isolates. J. Glob. Pharm. Technol. 10, 498–502.

Mondal, A., Basu, R., Das, S., Nandy, P., 2011. Beneficial role of carbon nanotubes on mustard plant growth: an agricultural prospect. J. Nanoparticle Res. 13 (10), 4519.

Montalvo, D., Degryse, F., McLaughlin, M.J., 2015. Natural colloidal P and its contribution to plant P uptake. Environ. Sci. Technol. 49 (6), 3427–3434.

Morteza, E., Moaveni, P., Farahani, H.A., Kiyani, M., 2013. Study of photosynthetic pigments changes of maize (*Zea mays* L.) under nano TiO_2 spraying at various growth stages. SpringerPlus 2 (1), 247.

Mousavi, S.M., Motesharezadeh, B., Hosseini, H.M., Alikhani, H., Zolfaghari, A.A., 2018. Geochemical fractions and phytoavailability of zinc in a contaminated calcareous soil affected by biotic and abiotic amendments. Environ. Geochem. Health 40 (4), 1221–1235.

Munns, R., Tester, M., 2008. Mechanisms of salinity tolerance. Annu. Rev. Plant Biol. 59, 651–681.

Murugan, K., Anitha, J., Suresh, U., Rajaganesh, R., Panneerselvam, C., Tseng, L.C., Kalimuthu, K., Alsalhi, M.S., Devanesan, S., Nicoletti, M., Sarkar, S.K., 2017. Chitosan-fabricated Ag nanoparticles and larvivorous fishes: a novel route to control the coastal malaria vector *Anopheles sundaicus*? Hydrobiologia 797, 335–350.

Musante, C., White, J.C., 2012. Toxicity of silver and copper to *Cucurbita pepo*: differential effects of nano and bulk-size particles. Environ. Toxicol. 27 (9), 510–517.

Mushinskiy, A.A., Aminova, E.V., 2019. The effect of nanoparticles of iron, copper and molybdenum on the morphometric parameters of plants *Solanum tuberosum* L. Proc. IOP Conference Series: Earth and Environmental Science, 341. IOP Publishing.

Nagaonkar, D., Shende, S., Rai, M., 2015. Biosynthesis of copper nanoparticles and its effect on actively dividing cells of mitosis in *Allium cepa*. Biotechnol. Prog. *31* (2), 557–565.

Nagaraju, R.S., Sriram, R.H., Achur, R., 2020. Antifungal activity of Carbendazim-conjugated silver nanoparticles against anthracnose disease caused by *Colletotrichum gloeosporioides* in mango. J. Plant Pathol. 102, 39–46.

Nazaralian, S., Majd, A., Irian, S., Najafi, F., Ghahremaninejad, F., Landberg, T., Greger, M., 2017. Comparison of silicon nanoparticles and silicate treatments in fenugreek. Plant Physiol. BioChem. 115, 25–33.

Nejad, M.S., Bonjar, G.H.S., Khatami, M., Amini, A., Aghighi, S., 2016. In vitro and in vivo antifungal properties of silver nanoparticles against *Rhizoctonia solani*, a common agent of rice sheath blight disease. IET Nanobiotechnol 11, 236–240.

Nekrasova, G.F., Ushakova, O.S., Ermakov, A.E., Uimin, M.A., Byzov, I.V., 2011. Effects of copper (II) ions and copper oxide nanoparticles on Elodea densa Planch. Russ. J. Ecol. 42 (6), 458.

O'Connor, D., Peng, T., Zhang, J., Tsang, D.C., Alessi, D.S., Shen, Z., Bolan, N.S., Hou, D., 2018. Biochar application for the remediation of heavy metal polluted land: a review of in situ field trials. Sci. Total Environ 619, 815–826.

Ogunyemi, O.S., Abdallah, Y., Zhang, M., Fouad, H., Hong, X., Ibrahim, E., Masum, M.M.I., Hossain, A., Mo, J., Li, B., 2019. Green synthesis of Zinc oxide nanoparticles using different plant extracts and their antibacterial activity against *Xanthomonas oryzae* pv. oryzae. Artif. Cell Nanomed B. 7, 341–352.

Pacheco-Torgal, F., Diamanti, M.V., Nazari, A., Goran-Granqvist, C., Pruna, A., Amirkhanian eds, S., 2018. Nanotechnology in Eco-efficient Construction: Materials, Processes and Applications. Woodhead Publishing, UK.

Pandey, S., Giri, K., Kumar, R., Mishra, G., Rishi, R.R., 2018. Nanopesticides: opportunities in crop protection and associated environmental risks. Proc. Natl. Acad. Sci., India B Biol. Sci 88, 1287–1308.

Pandorf, M., Pourzahedi, L., Gilbertson, L., Lowry, G.V., Herckes, P., Westerhoff, P., 2020. Graphite nanoparticle addition to fertilizers reduces nitrate leaching in growth of lettuce (*Lactuca sativa*). Environ. Sci. 7 (1), 127–138.

Park, H.J., Kim, S.H., Kim, H.J., Choi, S.H., 2006. A new composition of nanosized silica-silver for control of various plant diseases. Plant Pathol. J. 22 (3), 295–302.

Ponmurugan, P., Manjukarunambika, K., Elango, V., Gnanamangai, B.M., 2016. Antifungal activity of biosynthesised copper nanoparticles evaluated against red root-rot disease in tea plants. J. Exp. NanoSci. 11 (13), 1019–1031.

Pradhan, S., Patra, P., Das, S., Chandra, S., Mitra, S., Dey, K.K., Akbar, S., Palit, P., Goswami, A., 2013. Photochemical modulation of biosafe manganese nanoparticles on *Vigna radiata*: a detailed molecular, biochemical, and biophysical study. Environ. Sci. Technol. 47 (22), 13122–13131.

Prasad, R., Bhattacharyya, A., Nguyen, Q.D., 2017. Nanotechnology in sustainable agriculture: recent developments, challenges, and perspectives. Front. MicroBiol. 8.

Prasad, T.N.V.K.V., Sudhakar, P., Sreenivasulu, Y., Latha, P., Munaswamy, V., Reddy, K.R., Sreeprasad, T.S., Sajanlal, P.R., Pradeep, T., 2012. Effect of nanoscale zinc oxide particles on the germination, growth and yield of peanut. J. Plant Nutr. 35 (6), 905–927.

Preetha, P.S., Balakrishnan, N., 2017. A review of nano fertilizers and their use and functions in soil. Int. J. Curr. Microbiol. App. Sci 6 (12), 3117–3133.

Putri, F.M., Suedy, S.W.A., Darmanti, S., 2017. The effect nanosilica fertilizer on numbers of stomata, chlorophyll content, and growth of black rice (*Oryza sativa* L. cv. Japonica). Bull. Anat. Physiol. 2, 72–79.

Qiao, Y., Wu, J., Xu, Y., Fang, Z., Zheng, L., Cheng, W., Tsang, E.P., Fang, J., Zhao, D., 2017. Remediation of cadmium in soil by biochar-supported iron phosphate nanoparticles. Ecol. Eng. 106, 515–522.

Rafique, R., Arshad, M., Khokhar, M.F., Qazi, I.A., Hamza, A., Virk, N., 2015. Growth response of wheat to titania nanoparticles application. NUST J. Eng. Sci. 7 (1), 42–46.

Rafique, R., Zahra, Z., Virk, N., Shahid, M., Pinelli, E., Park, T.J., Kallerhoff, J., Arshad, M., 2018. Dose-dependent physiological responses of *Triticum aestivum* L. to soil applied TiO_2 nanoparticles: alterations in chlorophyll content, H_2O_2 production, and genotoxicity. Agric. Ecosyst. Environ. 255, 95–101.

Rahmatizadeh, R., Arvin, S.M.J., Jamei, R., Mozaffari, H., Reza Nejhad, F., 2019. Response of tomato plants to interaction effects of magnetic (Fe_3O_4) nanoparticles and cadmium stress. J. Plant Interact. 14 (1), 474–481.

Rajasekaran, P., Kannan, H., Das, S., Young, M., Santra, S., 2016. Comparative analysis of Copper and Zinc based agrixhemical biocide products: materials characteristics, phytotoxicity and in vitro antimicrobial efficacy. AIMS Environ. Sci. 3, 439–455.

Rajonee, A.A., Zaman, S., Huq, S.M.I., 2017. Preparation, characterization and evaluation of efficacy of phosphorus and potassium incorporated nano fertilizer. Adv. Nanoparticles 6 (02), 62.

Raliya, R., Tarafdar, J.C., 2013. ZnO nanoparticle biosynthesis and its effect on phosphorous-mobilizing enzyme secretion and gum contents in cluster bean (*Cyamopsis tetragonoloba* L.). Agricultural Res. 2 (1), 48–57.

Raliya, R., Tarafdar, J.C., Pratim, B., 2016. Enhancing the mobilization of native phosphorus in the mung bean rhizosphere using ZnO nanoparticles synthesized by soil fungi. J. Agr. Food Chem. 64 (16), 3111–3118.

Raliya, R., Saharan, V., Dimkpa, C., Biswas, P., 2018. Nanofertilizer for precision and sustainable agriculture: current state and future perspectives. J. Agric. Food Chem. 66, 6487–6503.

Raliya, R., Biswas, P., 2015. Environmentally benign bio-inspired synthesis of Au nanoparticles, their self-assembly and agglomeration. RSC Adv. 5 (52), 42081–42087.

Raliya, R., Biswas, P., Tarafdar, J.C., 2015. TiO2 nanoparticle biosynthesis and its physiological effect on mung bean (*Vigna radiata* L.). Biotechnol. Rep. 5, 22–26.

Rameshaiah, G.N., Pallavi, J., Shabnam, S., 2015. Nano fertilizers and nano sensors – an attempt for developing smart agriculture. Int. J. Eng. Res. Gen. Sci. 3, 314–320 Volume.

Rane, M., Bawskar, M., Rathod, D., Nagaonkar, D., Rai, M., 2015. Influence of calcium phosphate nanoparticles, *Piriformospora indica* and *Glomus mosseae* on growth of *Zea mays*. Adv. Nat. Sci. 6 (4), 045014.

Ranjbar, S., Rahemi, M., Ramezanian, A., 2018. Comparison of nano-calcium and calcium chloride spray on postharvest quality and cell wall enzymes activity in apple cv. red delicious. Sci. Hortic. 240, 57–64.

Rastogi, A., Tripathi, D.K., Yadav, S., Chauhan, D.K., Živčák, M., Ghorbanpour, M., El-Sheery, N.I., Brestic, M., 2019. Application of silicon nanoparticles in agriculture. 3 Biotech 9 (3), 90.

Ritzema, H., 2016. Drain for gain: managing salinity in irrigated lands—a review. Agric. Water Manage 176, 18–28.

Rizwan, M., Ali, S., Ali, B., Adrees, M., Arshad, M., Hussain, A., ur Rehman, M.Z., Waris, A.A., 2019. Zinc and iron oxide nanoparticles improved the plant growth and reduced the oxidative stress and cadmium concentration in wheat. Chemosphere 214, 269–277.

Rizwan, M., Ali, S., Qayyum, M.F., Ok, Y.S., Adrees, M., Ibrahim, M., Zia-ur-Rehman, M., Farid, M., Abbas, F., 2017. Effect of metal and metal oxide nanoparticles on growth and physiology of globally important food crops: a critical review. J. Hazard. Mater. 322, 2–16.

Rizwan, M., Ali, S., ur Rehman, M.Z., Malik, S., Adrees, M., Qayyum, M.F., Alamri, S.A., Alyemeni, M.N., Ahmad, P., 2019. Effect of foliar applications of silicon and titanium dioxide nanoparticles on growth, oxidative stress, and cadmium accumulation by rice (*Oryza sativa*). Acta Physiol.Plant. 41 (3), 35.

Rouhani, M., Samih, M.A., Kalantari, S., 2013. Insecticidal effect of silica and silver nanoparticles on the cowpea seed beetle, *Callosobruchus maculatus* F. (Col.: Bruchidae). J. Entomol. Res. 4 (4), 297–305.

Rui, M., Ma, C., Hao, Y., Guo, J., Rui, Y., Tang, X., Zhao, Q., Fan, X., Zhang, Z., Hou, T., Zhu, S., 2016. Iron oxide nanoparticles as a potential iron fertilizer for peanut (*Arachis hypogaea*). Front. Plant Sci. 7, 815.

Sabaghnia, N., Janmohammadi, M., 2014. Graphic analysis of nano-silicon by salinity stress interaction on germination properties of lentil using the biplot method. Agric. Forestry/Poljoprivreda i Sumarstvo 60 (3), 29–40.

Sabaghnia, N., Janmohammadi, M., 2015. Effect of nano-silicon particles application on salinity tolerance in early growth of some lentil genotypes/Wpływ nanocząstek krzemionki na tolerancję zasolenia we wczesnym rozwoju niektórych genotypów soczewicy. Annales UMCS, Biologia 69 (2), 39–55.

Saharan, V., Sharma, G., Yadav, M., Choudhary, M.K., Sharma, S.S., Pal, A., Raliya, R., Biswas, P., 2015. Synthesis and in vitro antifungal efficacy of Cu–chitosan nanoparticles against pathogenic fungi of tomato. Int. J. Biol. Macromol. 75, 346–353.

Saikia, J., Gogoi, A., Baruah, S., 2019. Nanotechnology for water remediation. In: Dasgupta, N., Ranjan, S., Lichtfouse, E. (Eds.), Environmental Nanotechnology. Springer, Cham, pp. 195–211.

Sajyan, T.K., Shaban, N., Rizkallah, J., Sassine, Y.N., 2018. Effects of monopotassium-phosphate, nano-calcium fertilizer, acetyl salicylic acid and glycinebetaine application on growth and production of tomato (*Solanum lycopersicum*) crop under salt stress. Agron. Res. 16 (3), 872–883.

Salem, N.M., Albanna, L.S., Awwad, A.M., 2016. Green synthesis of sulfur nanoparticles using *Punica granatum* peels and the effects on the growth of tomato by foliar spray applications. Environ. Nanotechnol. Monit. Manage. 6, 83–87.

Sangeetha, J., Thangadurai, D., Hospet, R., Harish, E.R., Purushotham, P., Mujeeb, M.A., Shrinivas, J., David, M., Mundaragi, A.C., Thimmappa, S.C., Arakera, S.B., 2017. Nanoagrotechnology for soil quality, crop performance and environmental management. In: Prasad, R., Kumar, M., Kumar, V. (Eds.), Nanotechnology. Springer, Singapore, pp. 73–97.

Savi, D.G., Piacentini, K.C., Souza, S.R.D., Costa, M.E.B., Santos, C.M.R., Scussel, V.M., 2015. Efficacy of zinc compounds in controlling Fusarium head blight and deoxynivalenol formation in wheat (*Triticum aestivum* L.). Int. J. Food Microbiol. 98–104 2015.

Seneviratne, M., Rajakaruna, N., Rizwan, M., Madawala, H., Ok, Y.S., Vithanage, M., 2017. Heavy metal-induced oxidative stress on seed germination and seedling development: a critical review. Environ. Geochem. Health 41, 1813–1831.

Shah, V., Belozerova, I., 2009. Influence of metal nanoparticles on the soil microbial community and germination of lettuce seeds. Water Air Soil Pollut. 197 (1-4), 143–148.

Sharifi-Rad, J., Sharifi-Rad, M., Teixeira da Silva, J.A., 2018. Morphological, physiological and biochemical responses of crops (*Zea mays* L., *Phaseolus vulgaris* L.), medicinal plants (*Hyssopus officinalis* L., *Nigella sativa* L.), and weeds (*Amaranthus retroflexus* L., *Taraxacum officinale* FH Wigg) exposed to SiO_2 nanoparticles. J. Agr. Sci. Tech. 18, 1027–1040.

Sharonova, N.L., Yapparov, A.K., Khisamutdinov, N.S., Ezhkova, A.M., Yapparov, I.A., Ezhkov, V.O., Degtyareva, I.A., Babynin, E.V., 2015. Nanostructured water-phosphorite suspension is a new promising fertilizer. Nanotechnologies Russ. 10 (7-8), 651–661.

Shcherbakova, E.N., Shcherbakov, A.V., Andronov, E.E., Gonchar, L.N., Kalenskaya, S.M., Chebotar, V.K., 2017. Combined pre-seed treatment with microbial inoculants and Mo nanoparticles changes composition of root exudates and rhizosphere microbiome structure of chickpea (*Cicer arietinum* L.) plants. Symbiosis 73 (1), 57–69.

Shen, J., Yuan, L., Zhang, J., Li, H., Bai, Z., Chen, X., Zhang, W., Zhang, F., 2011. Phosphorus dynamics: from soil to plant. Plant Physiol. 156 (3), 997–1005.

Shende, S., Ingle, A.P., Gade, A., Rai, M., 2015. Green synthesis of copper nanoparticles by *Citrus medica* Linn (Idilimbu) juice and its antimicrobial activity. World J. Microbiol. Biotechnol. 31, 865–873.

Shojaei, T.R., Salleh, M.A.M., Tabatabaei, M., Mobli, H., Aghbashlo, M., Rashid, S.A., Tan, T., 2019. Applications of nanotechnology and carbon nanoparticles in agriculture. In: Rashid, S.A., Othman, R.N.I.R., Hussein, M.Z. (Eds.), Synthesis, Technology Applications. Carbon Nanomaterials. Elsevier, Amsterdam, pp. 247–277.

Siddiqui, M.H., Al-Whaibi, M.H., 2014. Role of nano-SiO_2 in germination of tomato (*Lycopersicum esculentum* seeds Mill.). Saudi J. Biol. Sci. 21 (1), 13–17.

Siddiqui, M.H., Al-Whaibi, M.H., Faisal, M., Al Sahli, A.A., 2014. Nano-silicon dioxide mitigates the adverse effects of salt stress on *Cucurbita pepo* L. Environ. Toxicol. Chem. 33 (11), 2429–2437.

Silva-Castro, I., Barreto, R.W., Rodriguez, M.C.H., Matei, P.M., Martín-Gil, J., 2018. Control of coffee leaf rust by chitosan oligomers and propolis. In: Proc. Agriculture for Life, Life for Agriculture Conference, 1, Sciendo, pp. 311–315.

Singh, A., Singh, N.B., Hussain, I., Singh, H., Yadav, V., Singh, S.C., 2016. Green synthesis of nano zinc oxide and evaluation of its impact on germination and metabolic activity of *Solanum lycopersicum*. J. Biotechnol. 233, 84–94.

Singh, M.D., 2017. Nano-fertilizers is a new way to increase nutrients use efficiency in crop production. Int. J. Agric. Sci. 9 (7), 0975–3710.

Singh, N.B., Amist, N., Yadav, K., Singh, D., Pandey, J.K., Singh, S.C., 2013. Zinc oxide nanoparticles as fertilizer for the germination, growth and metabolism of vegetable crops. J. NanoEng. NanoManufact. 3 (4), 353–364.

Song, Y., Jiang, H., Wang, B., Kong, Y., Chen, J., 2018. Silver-incorporated mussel-inspired polydopamine coatings on mesoporous silica as an efficient nanocatalyst and antimicrobial agent. ACS Appl. Mater. Interf. 10, 1792–1801.

Song, Y., Kirkwood, N., Maksimović, Č., Zhen, X., O'Connor, D., Jin, Y., Hou, D., 2019. Nature based solutions for contaminated land remediation and brownfield redevelopment in cities: a review. Sci. Total Environ. 663, 568–579.

Srinivasan, C., Saraswathi, R., 2010. Nano-agriculture–carbon nanotubes enhance tomato seed germination and plant growth. Curr. Sci. 99 (3), 274–275.

Stampoulis, D., Sinha, S.K., White, J.C., 2009. Assay-dependent phytotoxicity of nanoparticles to plants. Environ. Sci. Technol. 43 (24), 9473–9479.

Subbaiah, L.V., Prasad, T.N.V.K.V., Krishna, T.G., Sudhakar, P., Reddy, B.R., Pradeep, T., 2016. Novel effects of nanoparticulate delivery of zinc on growth, productivity, and zinc biofortification in maize (*Zea mays* L.). J. Agric. Food Chem. 64 (19), 3778–3788.

Sun, D., Hussain, H.I., Yi, Z., Rookes, J.E., Kong, L., Cahill, D.M., 2016. Mesoporous silica nanoparticles enhance seedling growth and photosynthesis in wheat and lupin. Chemosphere 152, 81–91.

Sun, Y., Liang, J., Tang, L., Li, H., Zhu, Y., Jiang, D., Song, B., Chen, M., Zeng, G., 2019. Nano-pesticides: a great challenge for biodiversity? Nano Today 28, 100757.

Supapron, J., Pitayakon, L., Kamalapa, W., Touchamon, Px., 2002. Effect of zeolite and chemical fertilizer on the change of physical and chemical properties on Lat Ya soil series for sugar cane. In: Proc. 17th WCSS Symposium, pp. 14–21 Aug.

Suriyaprabha, R., Karunakaran, G., Kavitha, K., Yuvakkumar, R., Rajendran, V., Kannan, N., 2014. Application of silica nanoparticles in maize to enhance fungal resistance. IET Nanobiotechnol 8, 133–137.

Suriyaprabha, R., Karunakaran, G., Yuvakkumar, R., Prabu, P., Rajendran, V., Kannan, N., 2012. Growth and physiological responses of maize (*Zea mays* L.) to porous silica nanoparticles in soil. J. Nanoparticle Res. 14 (12), 1294.

Suriyaprabha, R., Karunakaran, G., Yuvakkumar, R., Rajendran, V., Kannan, N., 2012. Silica nanoparticles for increased silica availability in maize (*Zea mays*. L) seeds under hydroponic conditions. Current NanoSci. 8 (6), 902–908.

Suzuki, N., Rivero, R.M., Shulaev, V., Blumwald, E., Mittler, R., 2014. Abiotic and biotic stress combinations. New Phytol. 203 (1), 32–43.

Tahmasbi, L., Sedaghat, T., Motamedi, H., Kooti, M., 2018. Mesoporous silica nanoparticles supported copper (II) and nickel (II) Schiff base complexes: synthesis, characterization, antibacterial activity and enzyme immobilization. J. Solid State Chem. 258, 517–525.

Taktek, S., Trépanier, M., Servin, P.M., St-Arnaud, M., Piché, Y., Fortin, J.A., Antoun, H., 2015. Trapping of phosphate solubilizing bacteria on hyphae of the arbuscular mycorrhizal fungus *Rhizophagus irregularis* DAOM 197198. Soil Biol. BioChem. 90, 1–9.

Tang, X., Mu, X., Shao, H., Wang, H., Brestic, M., 2015. Global plant-responding mechanisms to salt stress: physiological and molecular levels and implications in biotechnology. Crit. Rev. Biotechnol. 35 (4), 425–437.

Tang, J., Zhang, J., Ren, L., Zhou, Y., Gao, J., Luo, L., Yang, Y., Peng, Q., Huang, H., Chen, A., 2019. Diagnosis of soil contamination using microbiological indices: a review on heavy metal pollution. J. Environ. Manage 242, 121–130.

Tarafdar, A., Raliya, R., Wang, W.N., Biswas, P., Tarafdar, J.C., 2013. Green synthesis of TiO_2 nanoparticle using *Aspergillus tubingensis*. Adv. Sci., Eng. Med. 5 (9), 943–949.

Tarafdar, J.C., Raliya, R., Mahawar, H., Rathore, I., 2014. Development of zinc nanofertilizer to enhance crop production in pearl millet (*Pennisetum americanum*). Agric. Res. 3 (3), 257–262.

Tarafdar, J.C., Raliya, R., Rathore, I., 2012. Microbial synthesis of phosphorous nanoparticle from tri-calcium phosphate using *Aspergillus tubingensis* TFR-5. J. BionanoSci. 6 (2), 84–89.

Taran, N.Y., Gonchar, O.M., Lopatko, K.G., Batsmanova, L.M., Patyka, M.V., Volkogon, M.V., 2014. The effect of colloidal solution of molybdenum nanoparticles on the microbial composition in rhizosphere of *Cicer arietinum* L. Nanoscale Res. Lett. 9 (1), 289.

Taran, N., Batsmanova, L., Kosyk, O., Smirnov, O., Kovalenko, M., Honchar, L., Okanenko, A., 2016. Colloidal nanomolybdenum influence upon the antioxidative reaction of chickpea plants (*Cicer arietinum* L.). Nanoscale Res. Lett. 11 (1), 476.

Thirunavukkarasu, M., Subramanian, K.S., 2014. Surface modified nano-zeolite based sulphur fertilizer on growth and biochemical parameters of groundnut. Trends BioSci. 7 (7), 565–568.

Thomas, E., Rathore, I., Tarafdar, J.C., 2017. Bioinspired production of molybdenum nanoparticles and its effect on chickpea (*Cicer arietinum* L). J. BionanoSci. 11 (2), 153–159.

Thomé, A., Reddy, K.R., Reginatto, C., Cecchin, I., 2015. Review of nanotechnology for soil and groundwater remediation: Brazilian perspectives. Water Air Soil Pollut 226, 121.

Tripathi, D.K., Singh, S., Singh, V.P., Prasad, S.M., Dubey, N.K., Chauhan, D.K., 2017. Silicon nanoparticles more effectively alleviated UV-B stress than silicon in wheat (*Triticum aestivum*) seedlings. Plant Physiol. BioChem. 110, 70–81.

Tripathi, D.K., Singh, V.P., Prasad, S.M., Chauhan, D.K., Dubey, N.K., 2015. Silicon nanoparticles (SiNp) alleviate chromium (VI) phytotoxicity in *Pisum sativum* (L.) seedlings. Plant Physiol. BioChem. 96, 189–198.

Upadhyay, P.R., Gautam, P., Srivastava, V., 2019. Magnetic organic-silica hybrid supported Pt nanoparticles for carbon sequestration reaction. Chem. Papers 73, 2241–2253.

Upadhyaya, H., Begum, L., Dey, B., Nath, P.K., Panda, S.K., 2017. Impact of calcium phosphate nanoparticles on rice plant. J Plant Sci Phytopathol. 1, 1–10.

Van, L.N., Ma, C., Shang, J., Rui, Y., Liu, S., Xing, B., 2016. Effects of Cuo nanoparticles on insecticidal activity and phytotoxicity in conventional and transgenic cotton. Chemosphere 144, 661–670.

Venkatachalam, P., Jayaraj, M., Manikandan, R., Geetha, N., Rene, E.R., Sharma, N.C., Sahi, S.V., 2017. Zinc oxide nanoparticles (ZnONPs) alleviate heavy metal-induced toxicity in *Leucaena leucocephala* seedlings: a physiochemical analysis. Plant Physiol. BioChem. 110, 59–69.

Verma, J., Bhattacharya, A., 2018. Analysis on synthesis of silica nanoparticles and its effect on growth of *T. Harzianum* & *Rhizoctonia* species. Biomed. J. Sci. Tech. Res. 10, 7890–7897.

Viet, V.P., Nguyen, H.T., Cao, T.M., Hieu, L.V., 2016. *Fusarium* antifungal activities of copper nanoparticles synthesized by a chemical reduction method. J. Nanomater. 2016, 1957612.

Wagner, G., Korenkov, V., Judy, J.D., Bertsch, P.M., 2016. Nanoparticles composed of Zn and ZnO inhibit *Peronospora tabacina* spore germination in vitro and *P. tabacina* infectivity on Tobacco leaves. Nanomaterials 6, 50.

Wang, S., Wang, F., Gao, S., 2015. Foliar application with nano-silicon alleviates Cd toxicity in rice seedlings. Environ. Sci. Pollut. Res. 22 (4), 2837–2845.

Wang, H., Khezri, B., Pumera, M., 2016. Catalytic DNA-functionalized self-propelled micromachines for environmental remediation. Chem 1 (3), 473–481.

Wang, Q., Huo, Z., Zhang, L., Wang, J., Zhao, Y., 2016. Impact of saline water irrigation on water use efficiency and soil salt accumulation for spring maize in arid regions of china. Agric. Water Manage 163, 125–138.

Wang, S., Sun, H., Ang, H.M., Tadé, M.O., 2013. Adsorptive remediation of environmental pollutants using novel graphene-based nanomaterials. Chem. Eng. J. 226, 336–347.

Wang, T., Liu, Y., Wang, J., Wang, X., Liu, B., Wang, Y., 2019. In-situ remediation of hexavalent chromium contaminated groundwater and saturated soil using stabilized iron sulfide nanoparticles. J. Environ. Manage. 231, 679–686.

Wang, X., Pan, X., Gadd, G.M., 2019. Soil dissolved organic matter affects mercury immobilization by biogenic selenium nanoparticles. Sci. Total Environ. 658, 8–15.

Xiong, L., Wang, P., Hunter, M.N., Kopittke, P.M., 2018. Bioavailability and movement of hydroxyapatite nanoparticles (HA-NPs) applied as a phosphorus fertiliser in soils. Environ. Sci. 5 (12), 2888–2898.

Xu, Y., Fang, Z., Tsang, E.P., 2016. In situ immobilization of cadmium in soil by stabilized biochar-supported iron phosphate nanoparticles. Environ. Sci. Pollut. Res. 23 (19), 19164–19172.

Yang, F., Hong, F., You, W., Liu, C., Gao, F., Wu, C., Yang, P., 2006. Influence of nano-anatase TiO_2 on the nitrogen metabolism of growing spinach. Biol. Trace Element Res. 110 (2), 179–190.

Yassen, A., Abdallah, E., Gaballah, M., Zaghloul, S., 2017. Role of silicon dioxide nano fertilizer in mitigating salt stress on growth, yield and chemical composition of cucumber (*Cucumis sativus* L.). Int. J. Agric. Res 22, 130–135.

Ye, Y., Cota-Ruiz, K., Hernández-Viezcas, J.A., Valdés, C., Medina-Velo, I.A., Turley, R.S., Peralta-Videa, J.R., Gardea-Torresdey, J.L., 2020. Manganese nanoparticles control salinity-modulated molecular responses in *Capsicum annuum* L. through priming: a sustainable approach for agriculture. ACS Sustain. Chem. Eng. 8 (3), 1427–1436.

Younis, A.A., Khattab, H., Emam, M.M., 2020. Impacts of silicon and silicon nanoparticles on leaf ultrastructure and *TaPIP1* and *TaNIP2* gene expressions in heat stressed wheat seedlings. Biol. Plant. 64, 343–352.

Yuan, J., Chen, Y., Li, H., Lu, J., Zhao, H., Liu, M., Nechitaylo, G.S., Glushchenko, N.N., 2018. New insights into the cellular responses to iron nanoparticles in *Capsicum annuum*. Sci. Rep. 8 (1), 1–9.

Zahra, Z., Arshad, M., Rafique, R., Mahmood, A., Habib, A., Qazi, I.A., Khan, S.A., 2015. Metallic nanoparticle (TiO_2 and Fe_3O_4) application modifies rhizosphere phosphorus availability and uptake by *Lactuca sativa*. J. Agric. Food Chem. 63 (31), 6876–6882.

Zareabyaneh, H., Bayatvarkeshi, M., 2015. Effects of slow-release fertilizers on nitrate leaching, its distribution in soil profile, N-use efficiency, and yield in potato crop. Environ. Earth Sci. 74 (4), 3385–3393.

Zhang, H., Liu, R., Ning, T., Lal, R., 2018. Higher CO_2 absorption using a new class of calcium hydroxide ($Ca\ (OH)_2$) nanoparticles. Environ. Chem. Lett. 16, 1095–1100.

Zhang, P., Wang, H., Zhang, X., Xu, W., Li, Y., Li, Q., Wei, G., Su, Z., 2015. Graphene film doped with silver nanoparticles: self-assembly formation, structural characterizations, antibacterial ability, and biocompatibility. Biomater. Sci. 3 (6), 852–860.

Zhang, W., Zhang, D., Liang, Y., 2019. Nanotechnology in remediation of water contaminated by poly-and perfluoroalkyl substances: a review. Environ. Pollut. 247, 266–276.

Zhao, L., Hernandez-Viezcas, J.A., Peralta-Videa, J.R., Bandyopadhyay, S., Peng, B., Munoz, B., Keller, A.A., Gardea-Torresdey, J.L., 2013b. ZnO nanoparticle fate in soil and zinc bioaccumulation in corn plants (*Zea mays*) influenced by alginate. Environ. Sci. 15 (1), 260–266.

Zhao, L., Sun, Y., Hernandez-Viezcas, J.A., Servin, A.D., Hong, J., Niu, G., Peralta-Videa, J.R., Duarte-Gardea, M., Gardea-Torresdey, J.L., 2013a. Influence of CeO_2 and ZnO nanoparticles on cucumber physiological markers and bioaccumulation of Ce and Zn: a life cycle study. J. Agric. Food Chem. 61 (49), 11945–11951.

Ziaee, M., Ganji, Z., 2016. Insecticidal efficacy of silica nanoparticles against *Rhyzopertha dominica* F. and *Tribolium confusum* Jacquelin du Val. J. Plant Protect. Res. 56 (3), 250–256.

CHAPTER 28

Sustainable environmentally friendly approaches to the recycling of spent selective catalytic reduction (SCR) catalysts

Ana Belen Cueva-Sola[a,b], Pankaj Kumar Parhi[c], Jin-Young Lee[a,b] and Rajesh Kumar Jyothi[a,b]
[a]Convergence Research Center for Development of Mineral Resources (DMR), Korea Institute of Geoscience and Mineral Resources (KIGAM), Daejeon, Republic of Korea
[b]Resources Recycling major, Korea University of Science and Technology (UST), Daejeon, Republic of Korea
[c]Department of Chemistry, Fakir Mohan University, Balasore, Odisha, India

28.1 Introduction

During the past decades several types of wastes such as industrial scrap, spent catalyst, electronic waste, batteries have been generated; which contain a wide variety of valuable metals. The so-called urban mines contain a variety of wastes from where precious, heavy, and base metals can be recoverable due to their large concentrations alongside the depletion of primary grade ores (Hye-Rim et al., 2012; Ogi et al., 2016; Petter et al., 2014; Yanhua et al., 2012). Catalysts industrially used to comply with a strong regulations for de-sulphurization, de-ozonification, selective catalytic reduction (SCR) of nitrogen oxides are attractive as a secondary source; owing to the large concentration of strategic metals such as vanadium, tungsten, molybdenum, titanium, and others (Choi, 2018; Ma et al., 2019; Wu et al., 2018).

Due to its advantages and regeneration capabilities, the most used SCR catalyst for stationary applications (power plants and big industrial complexes) contains around 0.5–1.5% V_2O_5 as the catalytic agent, 7–10% WO_3 to improve durability, and 70–80% TiO_2 as the supporting oxide (Ferella, 2020; Wang et al., 2017). However, despite their long lifespan eventually spent catalyst cannot be regenerated anymore and need to be discarded, especially in landfills. Because of the high toxicity of both metals, the environmental burden of landfills, and the necessity of secondary sources of the title metals, spent SCR catalyst recycling and metal recovery is an urgent matter in the recycling field.

To recover the metals present in SCR catalyst, hydrometallurgical routes are usually employed. Up to now, the most common route consists of a leaching process followed by solvent extraction (Coca et al., 1990; Nakamura et al., 2009; Olazabal et al., 1992;

Sustainable Nanotechnology for Environmental Remediation
DOI: https://doi.org/10.1016/B978-0-12-824547-7.00008-4

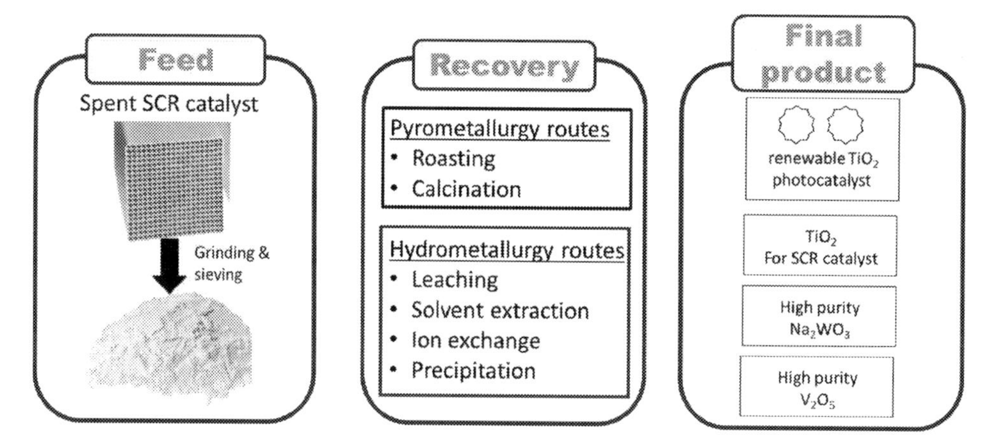

Figure 28.1 Metallurgical process of spent SCR catalyst.

Pandey et al., 2001; Zeng & Cheng, 2010). The investigation that gives origin to this chapter consists of the selective extraction of titanium from the spent SCR catalyst matrix followed by quantitative leaching of tungsten and vanadium carried out through soda roasting and alkali dissolution. The presence of vanadium and tungsten in the leach liquor leads to the main research issue during the recovery, the separation of both title elements due to their similar speciation behavior (Coca et al., 1990; Nakamura et al., 2009; Olazabal et al., 1992; Pandey et al., 2001; Zeng & Cheng, 2010).

Several researchers have been focused on the separation of both metals by selective solvent extraction, selective precipitation, and ion exchange among others while approaching a cost-effective and environmentally friendly process that will be described in the chapter (Fig. 28.1).

28.2 Nitrogen oxides (NO_x) problem and the current situation

In the last few years, environmental policies have become stricter toward emissions of solid, liquid, and gaseous effluents from industrial and urban sites. In the case of air pollutants, one of the most important noxious wastes is nitrogen oxides, mostly denominated NO_x (Cueva Sola et al., 2020; Jyothi et al., 2017; Radojevic, 1998). NO_x usually refers to nitrogen monoxide (NO), nitrous oxide (N_2O), and nitrogen dioxide (NO_2); however in the presence of air the majority of NO converts to NO_2 that is why regulations are usually specified in terms of NO_2 emissions (Mladenović et al., 2018). In the case of N_2O, the amounts produced are significantly lower than the other NO_x, therefore there is no current regulation in the emission of that compound even though it produces greenhouse effects and is known as very stable in the atmosphere.

Fig. 28.2 shows the different sources of NO_x emissions in the United States and Europe, it can be observed that the main contributors to the NO_x emissions are power

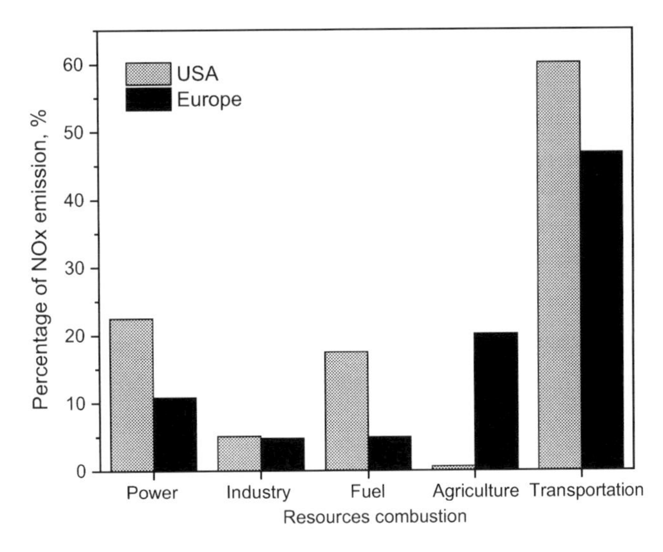

Figure 28.2 NO_x emission by the source in the US and Europe, Data from Roy, S., Hegde, M. S., & Madras, G. (2009). Catalysis for NO_x abatement. Appl. Energy, 86(11), 2283–2297. https://doi.org/10.1016/j.apenergy.2009.03.022.

generation and the transportation industries. Both industries heavily rely on the use of combustion as the main energy source, even though there have been incessantly efforts toward the development of alternative energy sources. Alongside many combustible materials, fossil fuels are the predominant material for energy generation contributing to currently 60% of the total energy generation by combustion and broadcasted as accounting for 80% of the energy supplies by 2035 (Energy and Dudley, 2016).

As discussed above, the predominant fuel for combustion is predicted to continue being fossil fuels processed using thermal combustion for energy generation (Foerter & Jozewicz, 2001; Zhang et al., 2018). The main issue regarding the process is not only the constant depletion of the primary sources of fossil fuels but the unavoidable generation of noxious air pollutants such as particulate matter, NO_x, sulfur oxides, etc. Among those compounds, NO_x is one of the most toxic agents due to being precursors of nitrates in particulate matter, which is associated with lung cancer and a series of respiratory diseases when exposed for a long period or high concentrations in air. Additionally, NO_x can react with volatile organic compounds to lead to the formation of ozone at ground level; ozone is responsible for a series of respiratory diseases and reduces lung capacity and is being considered as one of the most harmful compounds for human health. It has been proven by a series of researchers (Lu et al., 2016; Renzi et al., 2018; Strak et al., 2017) that exposure to high concentrations of NO_x for the long term can lead to cardiovascular and respiratory issues with significant statistical connection. Due to the health and environmental threat of NO_x, there has been a worldwide consensus toward

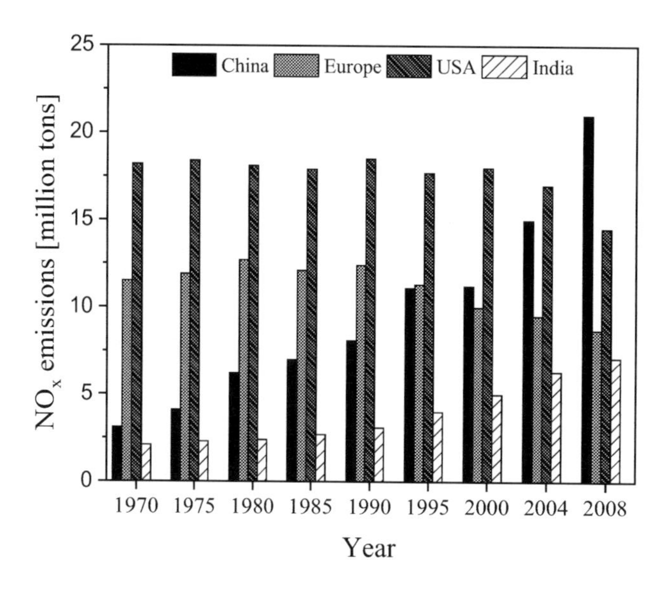

Figure 28.3 *NO_x* emission in time for different regions. Data from Communication from the Commission to the European Parliament, the Council, the European Economic and Social Committee and the Committee of the Regions on the 2017 list of Critical Raw Materials for the EU. (2017). Off J Eur Union COM.

the reduction of the current and the prevention of further emission of these compounds. As observed in Fig. 28.3 from 1970 until the late 90s the greatest contributors to the air pollution caused by NO_x were the United States and Europe (Janssens-Maenhout et al., 2017). However, an increasing trend of the noxious gases emission by China and India could be observed. The reasoning behind the continuous increase in the NO_x generation in the case of China and India is related to the development and industrialization of both countries alongside the lenient air regulation policies. On the other hand, since the late 90s it can be observed that Europe and the United States showed a constant decreasing trend in the emission of NO_x into the atmosphere, this is due to the strict environmental policies, which created a niche for the development of new technologies to reduce the emissions of NO_x in the industrial sector. For the year 2008, it could be observed in Fig. 28.3 that emissions from China have surpassed the emissions from the United States and became the biggest contributor to the environmental problem that NO_x presents in the world.

However, as shown in Fig. 28.4 even though the emissions in China peak for 2011 it can be observed that after 2011 the emissions start to slowly decrease every year following the trend of emissions in the United States (Bulletin of China's Environment State, 2016; Air Pollutant Emissions Trends Data, EPA, 2020). The main cause of the reducing trend is the latest annual standards set in China to participate in the Tier II category countries.

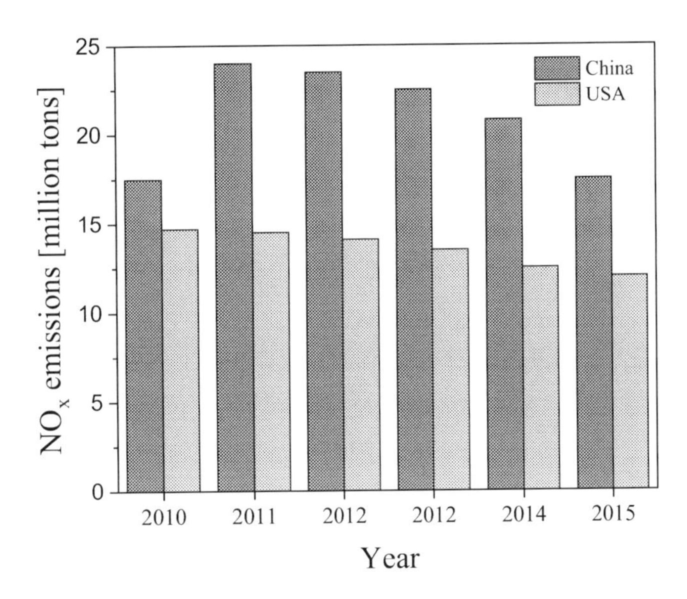

Figure 28.4 No_x emission in the 2000s. Data obtained from Bulletin of China's Environment State (2016) and Air Pollutant Emissions Trends Data, EPA (2020)

In that case, there has been a new set of regulations imposed by the Chinese government, those include the ultralow emission that establishes that thermal power boilers can emit 50 mg/m^3 of NO_x while incinerators can emit up to 150 mg/m^3 to the air. In addition, a 5-year plan (2016–2020) has been established to reduce the emissions of noxious gases to the environment (Gao et al., 2017). Japan is another country that has done incessant efforts toward the reduction of the emission of toxic gases to the environment; currently, the regulations establish that 15, 30, and 60 ppm can be emitted for large gas, oil, and coal-fired power plants, respectively, while gas turbines must emit less than 5 ppm to comply to the air quality standards set by the government (Forzatti, 2001).

28.3 Selective reduction catalyst process and its uses

Combustion technologies have been improving incessantly since they were first discovered. The main goal of the development is to improve efficiency and reduce the byproducts generated in the process, one of them being NO_x (Ferella, 2020; Radojevic, 1998). Before the regulations became very strict, combustion technologies were the main process used to reduce the amount of NO_x generated in combustion systems. However, currently using efficient and improved combustion technologies is not enough to fulfill the low emission requirements for nitrogen oxides (Forzatti, 2001; Weng et al., 2016).

The first parameter to consider when trying to reduce the NO_x emission is the type of fuel for combustion and the nitrogen content of it, being it fossil fuels, biomass, waste, among others, and the physical condition of the fuel: solid, liquid, and gaseous. Considering the first step one can start to control the emissions in the precombustion phase: by using a low nitrogen blend of fuels, using different additives and injection of O_2 or using oxygen-rich air to reduce the overall nitrogen contents to react (Ferella, 2020; Forzatti, 2001; Roy et al., 2009). During the combustion phase, the fluid dynamics of the combustion chamber could be improved and the design and operation parameters should be optimized. Finally, post combustion technologies are considered an add-on technology used to treat the effluents of combustion and in the case of NO_x to selectively convert it into N_2 to be emitted to the air. Among the postcombustion technologies, the most advanced, effective, and widely used for stationary applications is the SCR where usually a reductant such as ammonia (NH_3) or urea is introduced in the effluent stream to convert NO_x into N_2 with the aid of a catalytic agent. The reactions involved in the reduction of nitrogen oxides using ammonia and vanadium oxide as a catalyst are as follows:

$$4NO + 4NH_3 + O_2 \rightarrow 4N_2 + 6H_2O$$

$$8NH_3 + 6NO_2 \rightarrow 7N_2 + 12H_2O$$

$$2NH_3 + 3N2O \rightarrow 4N_2 + 3H_2O$$

where reaction (1) is the most predominant due to the rapid catalyst action at high temperatures (250–400 °C) when oxygen is introduced in excess in the chamber. The remaining two reactions only account for 5% of the NO_x emission in general, the great majority is NO.

SCR catalysts were first used in Japan in the late 1970s when regulations started to become more rigid against toxic emissions to the environment (Ferella, 2020; Forzatti, 2001; Kim et al., 2015). There are three types of catalysts developed for this technology and the classification depends on the temperature range of operation and the type of active catalytic agent, those are: (1) noble metals, (2) base metal oxides, and (3) zeolites.

The first type of catalyst developed was based on the noble metals in the early 1970s. They had the advantage of being very active in the selective reduction of NO_x; however, they could also oxidize effectively NH_3, which posed a big disadvantage for the process at high temperatures. Currently, these types of catalysts are used at low temperatures and for natural gas applications (Pereira et al., 1988), in addition, some of them are used for the combined reduction of NO_x and oxidation of CO at low temperatures (Heck et al., 1994).

Various base metal oxides have been investigated for their uses as a catalyst in the SCR; however, vanadium oxide (Kartte & Nonnenmaker, 1966) in a matrix of titanium oxide in the anatase phase (Spivey and Dooley, 2007) and enhanced with tungsten or molybdenum showed big advantages in the NO_x treatment and SO_2 oxidation. Vanadium is used as the active element in the reduction of NO but also has the undesirable capacity to oxidize

Table 28.1 Suppliers and types of industrial SCR catalysts (Radojevic, 1998).

Supplier	Types	Operation temp., °C
South Korea		
SK Innovation-Nano	Base metals/ceramic monolith	-
DAEYOUNG C&E	Base metals/ceramic monolith	-
Japan		
Babcock Hitachi	Base metals/ceramic monolith	250–416
Hitachi Zosen	Base metals/ceramic monolith	330–421
Ishikawajima-Harima	Base metals/ceramic monolith	204–400
Kawasaki	Base metals/ceramic monolith	300–400
Mitsubishi	Base metals/ceramic monolith	204–400
UBE	Base metals/ceramic monolith	250–400
US		
W. R. Grace	Noble metals/metallic substrate	225–275
Engelhard	Base metals/ceramic monolith	302–400
Johnson Matthey	Base metals/metallic substrate	343–427
Norton	Zeolite	221–521
Germany		
Badische Anilin und Soda Fabrik (BASF)	V_2O_5-WO_3/TiO_2/homogenous monolith	
Siemens	V_2O_5-WO_3/TiO_2/homogenous monolith	
Steuler	Zeolite	300–521

SO_2, these properties were first discovered around the 1960s (Kartte & Nonnenmaker, 1966), and because of that disadvantage vanadium content is usually kept low reaching even 1% when the presence of SO_2 is abundant. The role of titanium oxide in the anatase form is due to the weakly and reversible sulfation in the presence of SO_2 while the catalytic capacity of the V_2O_5 in TiO_2 is enhanced (Forzatti, 2001; Nakajima, 1978). Finally, the role of either tungsten (10%) or molybdenum (6%) is to increase the thermal stability and catalytic activity of the system while limiting the oxidation of SO_2 (Amiridis et al., 1999; Lietti et al., 1996, 1999).

Zeolite-based catalysts were introduced for high-temperature applications in gas-fired cogeneration plants, where the temperature could reach even 600 °C where other types of catalysts are unstable (Chen et al., 1995). However, the production cost of zeolite-based catalyst is very high, therefore the investigation for industrial purposes has been reduced significantly (Sorrels, J. 2019).

As observed in Table 28.1 the most commercially used SCR catalyst are the ones using base metals (especially vanadium) as the catalytic agent. This is due to their wide operation temperature, high NO_x conversion (90%–99%), the durability of poisonous agents, and regeneration capacity (Spivey and Dooley, 2007). In addition, the characteristics of the SCR catalyst depend on its applications. For stationary applications, the main parameter

Table 28.2 Comparison between the SCR catalysts based on its structure.

	Honeycomb type	Plate type	Corrugated type
Support	Overall extrusion	Stainless steel mesh	Glass fiber
Preparation	Uniform extrusion, calcined	Bilateral extrusion	Coating type
Catalytic activity	General	Low	High
Oxidation state	High	High	High
Pressure loss	High	General	Low
Corrosion resistance	General	High	Low
Antitoxic	Low	Low	High
Comprehensive cost	Low	Low	General
Features	Large surface area, high activity, long life, catalytic regeneration remains selective	Small surface area, simple production, high automation degree, conducive in smoke	Moderate surface area, light weight, easy installation
Scope	High and low dust	High and low dust	Low dust

to consider is the durability and regeneration capacity because of the high cost and labor involved in a big operation such as thermoelectric power plants and incinerators (Foerter & Jozewicz, 2001). On the other side, for mobile operations (vehicles), the most important parameter is the conversion rate of NO_x due to the strict regulations for air pollution. Regardless of the application of the SCR catalyst usually its removal efficiency is around 80%–90%, representing higher standards than other types of technologies (Ham & I-S, n.d.).

Commercial SCR catalysts structure is usually divided into three different types, each type offering different characteristics and advantages over the other, honeycomb monolith, plates, and corrugated (coated metal monoliths) types. The main advantage of using these types of structures over packed beds is the decrease in the pressure drop inside the column due to the larger open frontal area with different channels, higher surface area per unit volume for the catalytic reaction, and superior resistance and durability to poisons. Table 28.2 shows the characteristics and advantages of each type of SCR catalyst structure (Su et al., 2012).

28.4 Management and disposal of spent SCR catalyst: current approaches

Currently, the most common disposal method for spent SCR catalyst is in landfills, which are specially designated depending on the regulations of each country to consider if the waste is hazardous or not (Ferella, 2020; Forzatti, 2001). In the case of the United States, the institution in charge of the classification of hazardous or nonhazardous waste is the Federal Government; however, depending on the state and the local government there could be stricter and more regulations to comply (EPRI, 2008). The 40 CFR 261 article lists wastes that are considered potentially hazardous, nevertheless based on the local regulations and the characteristics and compositions of spent SCR catalysts, it could be considered as hazardous or not (CFR Part 261, 2020.). Spent SCR catalyst specifically, because of the toxicity of vanadium compounds and leachates it is most likely to be considered hazardous waste (EPRI, 2008).

In the case of the European Union, the European Waste Code (EWC) assigns spent SCR catalyst as part of the waste classification that includes "catalysts containing hazardous transition metals or compounds based on hazardous transition metals" (Waste Classification - Guidance on the Classification and Assessment of Waste (Edition 1.1) Technical Guidance WM3, 2018); nevertheless, the classification for the type of hazard these catalysts pose is determined by the concentration of the vanadium compounds (mutagenic, toxic for reproduction, toxic when aspirated, etc.) (Waste Classification - Guidance on the Classification and Assessment of Waste (Edition 1.1) Technical Guidance WM3, 2018). Usually, the concentration of vanadium pentoxide in SCR catalyst is around 1%–2% (Forzatti, 2001), being the highest 3% in some specific cases (Zheng et al., 2004); therefore the type of hazard and disposal method must be revised accurately case by case. The cost of disposal of catalyst for stationary applications usually rounds €350–500 per ton not including any extra cost for landfilling, transportation, and mechanical processes such as crushing, which can lead to a cost of €2000 per ton in some determined countries (Argyle & Bartholomew, 2015).

In the case of China, as being one of the countries largely dependent on coal combustion for power generation (70% of the total generation), the demand for spent SCR catalyst to fulfill the environmental regulations must be correspondingly high (Ferella, 2020; Report Linker, 2016). By January 2014, the annual production of SCR catalyst was around 400,000–500,000 m^3 and the total demand between 2010 and 2015 reached a peak of 850,000 m^3 (Report Linker, 2016). For the year 2018, it was estimated that around 38,000 tons per year of spent SCR catalysts were discarded. The composition of the waste was variable but usually contained 80%–90% titanium oxide, around 10% or less tungsten oxide, and 1%–3% vanadium oxide (Wu et al., 2018). Initially, the burden of collection, treatment, and disposal of the spent catalyst was for the catalyst producers; however, with the newly implemented strict regulations spent catalysts must be handled and disposed of by specific companies dedicated to the treatment of hazardous waste (Zhou et al., 2017). Nevertheless, based on the current attention that the spent catalyst is gaining as a secondary source of valuable metals, it is foreseeable that a well-designed and controlled facility would be built for the treatment and recycling of this waste.

South Korea is another country largely dependent on fossil fuel combustion as the primary source of energy (Korea Electric Power Corporation KEPCO, 2017) generation and based on the global trend of reduction of the toxic emissions to the environment the Korea Ministry of Environment launched a new regulation restricting NO_x emissions

Table 28.3 Generation and recycling rate for major catalysts in Korea 2014 (Korea Ministry of Environment, 2014).

Classification	Generation (ton)	Recycling (ton)	Landfill (ton)	Incineration (ton)
Desulfurization catalysts (domestics)	11,300	11,300	–	–
Desulfurization catalysts (imports)	11,000	11,000	–	–
FCC (Fluid catalytic cracking) catalysts	14,000	14,000	–	–
SCR catalysts	11,000	1,400	9,600	–
Automotive catalysts (domestic)	2,000	2,000	–	–
Automotive catalysts (imports),	3,000	3,000	–	–
etc.	10,772	9,056	1,679	37
Total	63,072	51,756	11,279	37

in various industries such as steel-making industry from 120–200 ppm to 100–170 ppm; cement industry from 330 ppm to 270 ppm; petroleum industry from 70–180 ppm to 50–130 ppm, among others (Jeong, 2017) . The regulations started to be applied in 2019 alongside the increasing demand for SCR catalysts to fulfill the regulations. As new regulations have to be implemented, the waste generated with the demand for SCR catalyst will indeed increase. Table 28.3 shows the amount of catalyst generated, recycled, and disposed of in the year 2014 in South Korea, as discussed in the case of other regions in the world recycling is not yet the main approach to the management of spent SCR catalyst waste; however, it can be observed that the trend has started to change with the growing interest in the use of these waste as a secondary source of vanadium, titanium, and tungsten (Korea Electric Power Corporation KEPCO, 2017).

28.5 Tungsten, vanadium, and titanium: demand, uses, and production methods

Vanadium, tungsten, and titanium possess great uses in a variety of high technology and modern industries. The great majority of the production of titanium oxide goes to the pigments and photocatalyst industries (Cueva Sola et al., 2020; Kim et al., 2015; Paulino et al., 2012). In addition, titanium is specially used in the alloy industry alongside different base metals such as aluminum, molybdenum, iron, and manganese. The use in this industry is due to its physical and mechanical properties such as high melting point, low weight, high mechanical strength, among others that make this metal fundamental for medical and aircraft applications. However, due to the severe reduction of the mechanical strength of the material when the temperature exceeds 426 °C it is not used for high-temperature applications (Zhang et al., 2018). The main sources of titanium and its oxide are ores such as ilmenite, rutile, leucoxene, and titanite, where rutile is the main source of titanium (Non-critical raw materials profiles, 2014).

The main uses for vanadium are in the alloy industries as ferrovanadium or the steel industries as an additive. In ferrovanadium alloys, usually, the concentration of vanadium varies from 35% to 85%, being Fe-V80 (80% vanadium) the most commonly produced alloy (Luo et al., 2003). Vanadium alloys with other base metals such as aluminum and titanium are used to produce jet engines and airframes, while steel alloys are implemented in other critical components in the engine such as crankshafts and axles. The nuclear industry is also using vanadium alloys due to their low neutron absorption capabilities and resistance to high temperatures. A great part of the demand for vanadium oxide goes to the catalyst industries, not only for SCR catalysts but also for desulphurization catalytic technologies (Ferella, 2020; Forzatti, 2001; J.W. Kim et al., 2015). However, the toxicity of vanadium compounds, especially V^{5+}, is a great concern for the environment and human health (Lenntech, n.d.). Several mineral deposits that contain vanadium are distributed around the world; however, the major producers according to the European Union between 2010 and 2014 were China, South Africa, and Russia with 53, 25, and 20% of the global production, respectively (List of Critical Raw Materials for the EU, 2017). The vanadium obtained from primary ores is usually a byproduct of bauxite, phosphate, and uranium extraction (Goonan, 2011). Nevertheless, the great majority of the production of V_2O_5 is obtained from the secondary sources, especially from slag in the magnetite iron ores as a coproduct in the steelmaking industry, reaching even 85% of the worldwide production (Polyak, 2018).

Tungsten is considered one of the most important refractory metals giving numerous applications in end-use products (Habashi, 1997; Lassner & Schubert, 1999) The first reported use for tungsten is in the filament inside light bulbs due to its high melting point, low vapor pressure, and break resistance at high temperatures. The lighting industry consumes around 4% of the global annual tungsten produced. Currently, the majority of tungsten is used for the production of superalloys, where tungsten is added to nickel, cobalt, or iron alloys combining different metals such as tantalum, rhenium, and molybdenum (Ferella, 2020; ITIA (International Tungsten Industry Association) (2019) Tungsten Applications, n.d.). The main uses of superalloys are in the aircraft engines industry, marine industry, and turbine blades for stationary power units. Additionally, tungsten is the only metal used for electron emitters due to its resistance to evaporation in the extreme conditions of an electric arc. The catalytic industry is another great consumer of tungsten for deNO_x catalysts, hydrocracking, hydro-dearomatization, among others. Finally, cemented carbides use the cementation process to mix tungsten monocarbide grains (an extremely hard material) with a nickel or cobalt alloy matrix to combine the resistance and strength of carbides while keeping the plastic and tough properties of alloys. Cemented carbides are used for cutting tools specially used for extremely hard materials such as diamond (ITIA (International Tungsten Industry Association) (2019) Tungsten Applications, n.d.). Tungsten can be mined from two primary sources scheelite ($CAWO_4$) and wolframite (MWO_4 where M=Fe or Mn) where hydrometallurgical processes are mainly applied for the purification of the metal due to its high boiling point

(Habashi, 1997). According to the European Union from 2010 to 2014, the main producer of tungsten was China with 84% of the worldwide production dispersedly followed by Russia which produced 4% of the demand (List of Critical Raw Materials for the EU, 2017).

Vanadium and tungsten are considered in the 27 critical raw metals issues updated in 2017 by the European Commission, thus their production for the primary and secondary sources is an important industrial and technological research topic.

28.6 Current recycling methods for spent SCR catalyst and future prospects

The first process to recover metals present in spent SCR catalyst after grinding and sieving is leaching, and through that process, some of the metals could be separated one from another or concentrated for further processing. There are three main ways to leach metals from spent SCR catalyst (Habashi, 1997, 2005; ITIA (International Tungsten Industry Association), 2019; Tungsten Applications, n.d.). The first being using strong mineral acids, the second using organic acids, and the third and most common and researched approach through the usage of alkalis such as sodium hydroxide and sodium carbonate.

In the case of leaching with mineral acids, such as hydrochloric, nitric, and sulfuric, the research is mostly directed toward the recovery of vanadium. It is important to consider that the vanadium aqueous chemistry is complex and a variety of compounds can be found depending on the alkaline or basic conditions of the solution and the concentration of vanadium and other ions present in the solution as shown in Fig. 28.5. Generally, in the case of the spent SCR catalyst vanadium is found as V_2O_5 supported in a titanium oxide matrix where the links between the matrix and the catalyst agent are relatively weak, therefore acid leaching could be effective due to the aforementioned conditions (Cueva Sola et al., 2020).

Li and his research group investigated the kinetics involved in the leaching of the spent SCR catalyst using sulfuric acid at atmospheric pressure. In their results, they concluded that vanadium recovery efficiency improves when the temperature increases alongside the concentration of acid, while the efficiency decreases when the solid/liquid ratio increases (Li et al., 2014). They also found that the kinetic mechanism is driven by diffusion and using a spent SCR catalyst of 74 μm grain size the effectiveness in leaching can reach values up to 40%–60%; however alongside the vanadium compound $VOSO_4$, $TiOSO_4$, and H_2WO_4 were also formed, decreasing the selectivity of the mechanism (Li et al., 2014).

The formation of a tungstic acid during the leaching and its deposition on the surface of the catalyst reduces the surface area for diffusion hindering the leaching process, based on this phenomenon it was concluded that acid leaching is not the most effective method to selectively recover valuable metals from the spent SCR catalyst (Ferella, 2020).

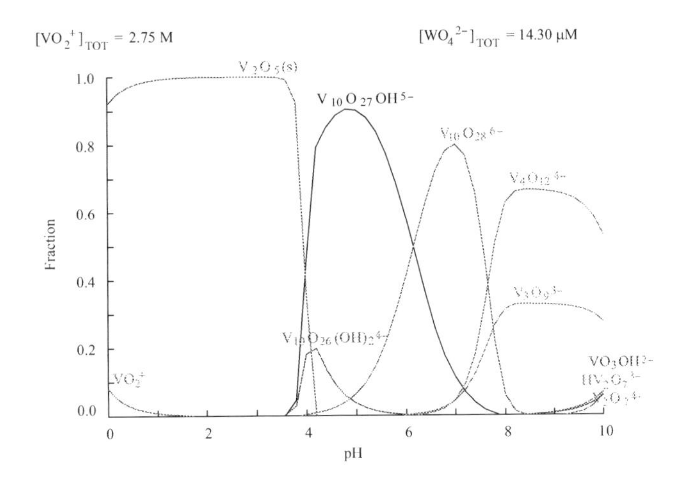

Figure 28.5 pH against fraction diagram plotted using Medusa software.

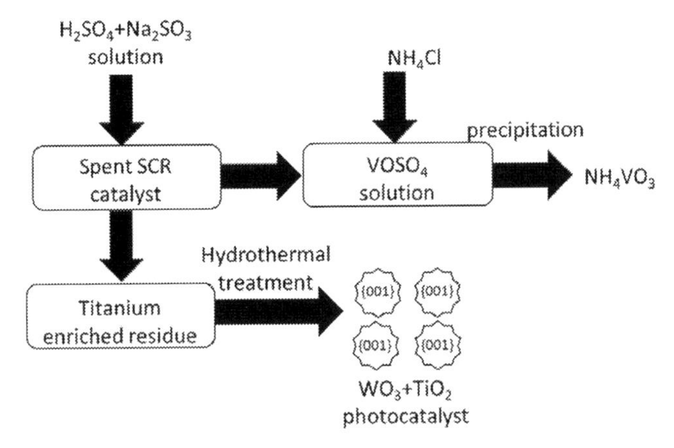

Figure 28.6 Recovery of vanadium and production of photocatalyst from spent SCR catalyst. Modified from Zhang, Q., Wu, Y., & Zuo, T. (2018). Green recovery of titanium and effective regeneration of TiO_2 photocatalysts from spent selective catalytic reduction catalysts. ACS Sustain. Chem. Eng., 6(3), 3091–3101. https://doi.org/10.1021/acssuschemeng.7b03038.

On the other hand, Zhang et al. based their research on the recovery of vanadium only from the spent catalyst while the titanium and tungsten carrier was reutilized (Foerter & Jozewicz, 2001; Zhang, Wu, & Zuo, 2018). The process is briefly shown in Fig. 28.6 where Na_2SO_3 is used because of their fast reaction with acid to produce SO_2, a potent reducing agent from V^{5+} to V^{4+}.

The equation describing the overall reaction is as follows

$$V_2O_5 + Na_2SO_3 + 2H_2SO_4 \rightarrow 2VOSO_4 + Na_2SO_4 + 2H_2O$$

By using the reduction ability of this method, they were able to achieve almost 100% removal of vanadium from the matrix, while the titanium-enriched residue went through a hydrothermal treatment to produce high-performance WO_3-TiO_2 photocatalysts with dominant {001} facets. The leaching parameters were optimized for the process and it was concluded that 5% acid solution with a 10% pulp density at 95 °C reacting for 2 h while 1 g of sodium sulfite is added every half hour was the optimum condition to reach nearly 100% leaching efficiency for the process (Zhang, Wu, Li, et al., 2018). The vanadium salt recovered after leaching was precipitated using ammonium chloride to obtain ammonium metavanadate as the final product.

Being titanium the major constituent of the spent SCR catalyst (~80%), its recovery is a very interesting subject for researchers, Ma et al. recovered around 99% of the titanium from the spent catalyst using roasting as a pretreatment to convert Ti and V into water-leachable compounds using Na_2CO_3 in a 1:3 ratio for 6 h at a temperature of 650 °C (Ma et al., 2019). The reactions to the roasting process are

$$TiO_2 + Na_2CO_3 \rightarrow Na_2TiO_3 + CO_2 \uparrow$$

$$V_2O_5 + Na_2CO_3 \rightarrow 2NaVO_3 + CO_2 \uparrow$$

After roasting the water-soluble compounds are leached with sulfuric acid at 60 °C for 6 h while Na_2TiO_3 remains in the solid phase to be converted into H_2TiO_3. The H_2TiO_3 is then decomposed with alkaline compounds in TiO_2 nanoparticles ready to be used as photocatalysts (Ma et al., 2019).

Another approach to the leaching process, however not very common, is the usage of organic acids to leach metals present in spent SCR catalyst. The main advantage of organic acids over inorganic acids is that they are produced by some bacteria making the whole process environmentally friendly in contrast to the highly pollutant inorganic acid production industry. However, they have the disadvantages of slow kinetics and poor dissolution capabilities (Ferella, 2020)

Wu et al. analyzed the leaching process of vanadium and iron from spent SCR catalyst using oxalic acid as the leaching agent (Wu et al., 2018). Temperature, acid concentration, time, particle size, and liquid to solid ratio were analyzed and under optimum conditions, 86.3% of vanadium and 100% of iron were leached. The optimum conditions for the process are an acid concentration of 1 mol/L at 90 °C during 3 h reaction time for a particle size smaller than 75 μm using an L/S ratio of 20 mL/g (Wu et al., 2018). The reaction mechanism of the leaching process was studied at a pH = 0.33 and the redox reactions between V^{5+}, V^{3+}, and Fe^{3+} with oxalic acid are described as follows:

$$V_2O_5 + V_2O_3 + 4H_2C_2O_4 \rightarrow 4VOC_2O_4 + 4H_2O$$

$$V_2O_5 + 3H_2C_2O_4 \rightarrow 2VOC_2O_4 + 3H_2O + 2CO_2$$

$$Fe_3O_4 + 7HC_2O_4^- + H_3O^+ \rightarrow 3Fe(C_2O_4)_2^{2-} + 5H_2O + 2CO_2$$

While in the case of titanium and tungsten dissolving complexation reactions took place, the amount of each metal leached was <1% and 18%, respectively. The solid residue after leaching was characterized and determined that contained TiO_2 in the anatase phase by X-ray Powder Diffraction (XRD) with a specific surface area of 67.38 m^2/g, which is suitable for the production of a new SCR catalyst matrix (Wu et al., 2018).

The most common procedure to separate metals in the spent SCR catalyst is using alkali reagents, sometimes for roasting and sometimes for direct leaching. The main advantage of this method is the enrichment of vanadium and tungsten due to their solubility in alkaline media while titanium is barely soluble (Choi et al., 2018, 2020; Cueva Sola et al., 2020). Zhang et al. for instance developed an environmentally friendly process to recover TiO_2 for usage in the photocatalyst industries starting from the premise that about 80% of the spent SCR catalyst is made of titanium matrix (Zhang, Wu, & Zuo, 2018). They used a molten salt process where they optimize several parameters such as temperature, a ratio of NaOH/spent catalyst, time for roasting, and water content in the process. Under the optimum conditions of 550 °C, NaOH/spent catalyst ratio=1.8/1, 10 minutes roasting using a 60%–80% m/m NaOH solution they determined that more than 98% of titanium could be recovered as α-Na_2TiO_3 (Zhang, Wu, & Zuo, 2018). The main reactions involved in the process are described as follows:

$$Ti - O - Ti + NaOH \rightarrow Ti - O - Na + Ti - OH$$

$$Ti - O - Na + H_2O \rightarrow Ti - OH + NaOH$$

$$Ti - OH + 2NaOH \rightarrow \alpha - Na_2TiO_3 + H_2O(g)$$

α-Na_2TiO_3 underwent a hydrothermal process where high purity (>99%) TiO_2 was produced and doped with either iron or P25 to analyze its photocatalytic activity, where iron-doped TiO_2 showed higher photocatalytic activity. The process shown in Fig. 28.7 is an environmentally friendly and cost-effective process to overcome one of the most important difficulties in the TiO_2 photocatalytic industry: the usage of very expensive chemicals such as $TiCl_4$, $TiOSO_4$, and titanium isopropoxide (Zhang, Wu, & Zuo, 2018).

As described in the previously mentioned study (Zhang, Wu, & Zuo, 2018), the main focus was on the recovery of titanium rather than other important metals such as vanadium and tungsten that were discarded as residues. Vanadium and tungsten being considered toxic for the environment must be recycled and the main process to do so has been NaOH roasting, where sodium vanadate and tungstate are produced as follows and these compounds are readily soluble in water. Thus, the main efforts are focused on the difficult separation of both metals due to their similar chemical and physical

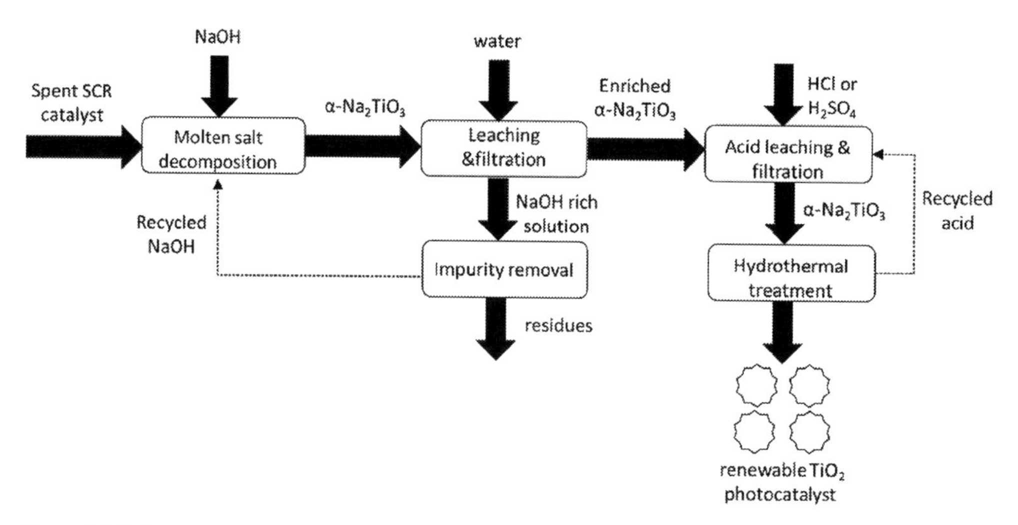

Figure 28.7 Regeneration process for TiO_2 from spent SCR catalyst. Modified from Zhang, Q., Wu, Y., Li, L., & Zuo, T. (2018). Sustainable approach for spent V_2O_5-WO_3/TiO_2 catalysts management: selective recovery of heavy metal vanadium and production of value-added WO_3-TiO_2 photocatalysts. ACS Sustain. Chem. Eng., 6(9), 12502–12510. https://doi.org/10.1021/acssuschemeng.8b03192.

properties.

$$WO_3(s) + NaOH(aq) \rightarrow Na_2WO_4(aq) + H_2O(l)$$

$$V_2O_5(s) + 2NaOH(aq) \rightarrow 2NaVO_3(aq) + H_2O(l)$$

In another case, Choi et al. established a method to completely treat spent SCR catalysts by recovering the three main metals: vanadium, tungsten, and titanium (Choi et al., 2019). Instead of using NaOH, they used sodium carbonate, which possesses the industrial benefit of being economically inexpensive. The optimum conditions for the process shown in Fig. 28.8 are the temperature of 950 °C, 20 min of the alkali fusion reaction, and 0.5 mole fraction ratio of MO_x/Na_2O+MO_x where M represents titanium, silicon, and tungsten. It was determined that decreasing the mole fraction results in decreasing the efficiency of leaching of vanadium and tungsten due to the formation of insoluble calcium vanadates and tungstates which cannot be redissolved during the alkali fusion process. Under the optimum conditions, it was obtained that the extraction efficiency of vanadium and tungsten reached more than 99% while titanium can be recovered as sodium titanate to be further processed to TiO_2 (Choi et al., 2019).

The same group of researchers (Choi et al., 2020) also proposed an industrially feasible process based on the lab-scale conclusions obtained previously; during that study, they were able to determine the feasibility, for the construction of an industrial commercial

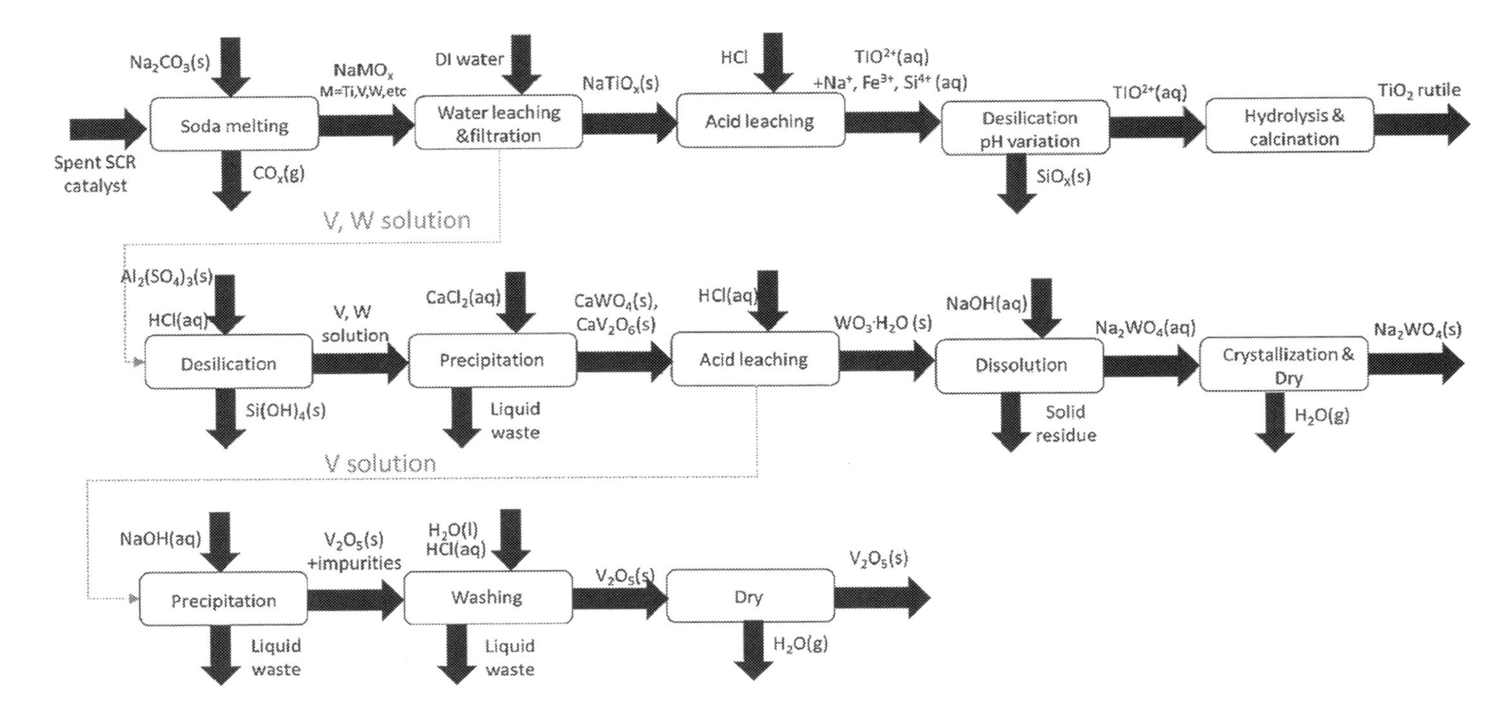

Figure 28.8 Complete treatment of spent SCR catalyst for V, W and Ti recovery. Modified from Choi, I. H., Cho, Y. C., Moon, G., Kang, H. N., Oh, Y. B., Lee, J. Y., & Kang, J. (2020). Recent developments in the recycling of spent selective catalytic reduction catalyst in South Korea. Catalysts, 10(2), 182. https://doi.org/10.3390/catal10020182.

plant to treat 3000 ton/year of spent SCR catalyst based on the semi-pilot and pilot-scale plants results. In the pilot-scale, they were able to produce titanium oxide (TiO_2) with a purity greater than 99%, V_2O_5 with a purity greater than 98%, and Na_2WO_4 with a purity of 99.3%. The pilot plant consisted of a furnace of 100 L for the soda melting process, 1 m^3 reactor for water leaching, desilication, and precipitation, a 100 L glass-lined reactor for H_2WO_4 acid, and a 300 L glass-lined reactor for titanium leaching and hydrolysis (Choi et al., 2020). Based on the pilot-scale experiments they were able to conclude an overall yield of 91.6% and 94.1% for vanadium and tungsten, respectively, while titanium had an overall recovery of 90% being it a circular economy project to recover spent SCR catalyst and obtain high purity commercial products.

Alkaline pressure leaching alongside roasting-water leaching is the two most commonly applied processes nowadays for the recovery of vanadium and tungsten present in the spent SCR catalyst, while the titanium matrix is either discharged as residue or treated for further applications. Choi et al. studied the optimum conditions for the pressure leaching of a spent SCR catalyst to obtain a leach liquor with a 91.5% recovery of vanadium and 87% recovery of tungsten (Choi et al., 2018). They discovered that this yield could be obtained under 38 bar pressure, 3 mol/L NaOH solution, 250 °C using a particle size smaller than 150 μm during 2 h of reaction with an S/L ratio of 40% w/v. However, the main issue with pressure leaching is still the high energetic burden while it is necessary to improve and deeply understand the separation techniques for vanadium and tungsten (Choi et al., 2018).

Various research efforts have been put toward the effective separation of vanadium and tungsten, for instance, Wu et al. used ion-exchange resins to separate tungsten from vanadium at an alkaline pH (Choi, 2018; Ma et al., 2019; W. Wu et al., 2018). After leaching optimization, they used the anion exchange resin Amberlite IRA900 at alkaline pH where tungsten is present as an anion ($WO_4{}^{2-}$) while vanadium was not adsorbed. To strip the adsorbed tungsten they used a sodium chloride solution obtaining a final leach liquor with 8.6 g/L of tungsten (W.C. Wu et al., 2016).

Solvent extraction has been one of the promising hydrometallurgical techniques to recover and separate vanadium from tungsten in different types of liquors including spent SCR catalyst leach liquor. For instance, Talla et al. used Cyanex 923 to separate molybdenum and tungsten from a leach liquor containing 5 ppm of tungsten, they used selective stripping with NaOH to separate both metals (Talla et al., 2010). However, one of the main drawbacks of this method is its lack of applicability for high concentration solutions such as the leach liquors from spent SCR catalyst (Talla et al., 2010). Various amines have been used for the separation of the title metals among them is N1923, a primary amine used to treat a liquor bearing around 1 g/L of tungsten and molybdenum at a pH of 7.07 and temperature of 15.5 °C. Even though their results showed a promising 329 separation factor, the low concentration of tungsten in the leach liquor has been proved as a disadvantage for highly concentrated solutions (Ning et al., 2009). One of the most successful studies for the separation of vanadium and tungsten was conducted by Nguyen and Lee who used LIX63 at pH 8 where vanadium is found in hydroxide form while tungsten is present as WO_4^{2-} anion. The concentrations used during the investigation were once again as follows: low concentrations of 0.1 g/L of vanadium and 1 g/L of tungsten (Bal et al., 2002; Nguyen & Lee, 2016). On the other hand, Cueva et al. proposed an environmentally friendly approach to the recovery and concentration of vanadium and tungsten from spent SCR catalyst leach liquor bearing concentrations around 0.7 g/L of vanadium and 7 g/L of tungsten. They used aliquot 336 a quaternary amine and elucidated the extraction mechanism and successful stripping of both metals reaching a sevenfold concentration from the original liquor (Cueva Sola et al., 2020); however, the separation was proposed by precipitation based on the studies performed by Choi et al. in the same research center (Choi et al., 2020) . Up to date, several studies have been performed for the separation of vanadium and tungsten from leach liquor bearing low concentrations of the metals, thus there is an extensive research area for extraction, separation, and concentration of liquors from those metals to prepare vanadium and tungsten compounds with commercial capabilities.

Acknowledgments

This study was supported by the R&D Center for Valuable Recycling (Global-Top R &D Program) of the Ministry of Environment (Project Number: 2019002230001), Korea.

References

Air Pollutant Emissions Trends Data, EPA, 2020. https://www.epa.gov/air-emissions-inventories/air-pollutant-emissions-trends-data.

Amiridis, M., Duevel, R.V., Wachs, I., 1999. The effect of metal oxide additives on the activity of V_2O_5/TiO_2 catalysts for the selective catalytic reduction of nitric oxide by ammonia. Appl. Catal. B Environ. 20, 101–105. https://doi.org/10.1016/S0926-3373(98.

Argyle, M.D., Bartholomew, C.H., 2015. Heterogeneous catalyst deactivation and regeneration: a review. Catalysts 5 (1), 145–269. https://doi.org/10.3390/catal5010145.

Bal, Y., Bal, K.E., Cote, G., 2002. Kinetics of the alkaline stripping of vanadium (V) previously extracted by Aliquat® 336. Miner. Eng. 15 (5), 377–379. https://doi.org/10.1016/S0892-6875(02)00044-4.

Bulletin of China's Environment State, 2016. http://english.mee.gov.cn/Resources/Reports/soe/ReportSOE/201709/P020170929573904364594.pdf.

CFR Part 261 - Identification and listing of hazardous waste. 2020. (https://www.law.cornell.edu/cfr/text/40/part-261).

Chen, J.P., Hausladen, M.C., Yang, R.T., 1995. Delaminated Fe_2O_3-Pillared clay: its preparation, characterization, and activities for selective catalytic reduction of NO by NH_3. J. Catal. 151 (1), 135–146. https://doi.org/10.1006/jcat.1995.1016.

Choi, I.H., 2018. Study on the recovery of vanadium and tungsten from spent V2O5-WO3/TiO2 catalyst, PhD Thesis, Dissertation was submitted to Resources Recycling Department. University of Science and Technology (UST), Korea.

Choi, I.H., Cho, Y.C., Moon, G., Kang, H.N., Oh, Y.B., Lee, J.Y., Kang, J., 2020. Recent developments in the recycling of spent selective catalytic reduction catalyst in South Korea. Catalysts 10 (2), 182. https://doi.org/10.3390/catal10020182.

Choi, I.H., Moon, G., Lee, J.Y., Jyothi, R.K., 2018. Extraction of tungsten and vanadium from spent selective catalytic reduction catalyst for stationary application by pressure leaching process. J. of Clean. Prod. 197, 163–169. https://doi.org/10.1016/j.jclepro.2018.06.196.

Choi, I.H., Moon, G., Lee, J.Y., Jyothi, R.K., 2019. Alkali fusion using sodium carbonate for extraction of vanadium and tungsten for the preparation of synthetic sodium titanate from spent SCR catalyst. Sci. Rep. 9 (1), 12316. https://doi.org/10.1038/s41598-019-48767-0.

Coca, J., Díez, F.V., Morís, M.A., 1990. Solvent extraction of molybdenum and tungsten by alamine 336 and DEHPA. Hydrometallurgy 25 (2), 125–135. https://doi.org/10.1016/0304-386X(90)90034-Y.

Communication from the Commission to the European Parliament, the Council, the European Economic and Social Committee and the Committee of the Regions on the 2017 list of Critical Raw Materials for the EU, 2017. Off J Eur Union COM. https://eur-lex.europa.eu/legal-content/EN/TXT/PDF/?uri=CELEX:52017DC0490&from=EN.

Cueva Sola, A.B., Parhi, P.K., Lee, J.Y., Kang, H.N., Jyothi, R.K, 2020. Environmentally friendly approach to recover vanadium and tungsten from spent SCR catalyst leach liquors using Aliquat 336. RSC Adv., 10 (34), 19736–19746. https://doi.org/10.1039/d0ra02229b.

Energy, B., Dudley, B., 2016. Energy Outlook 2035 launch.

EPRI, 2008. SCR Catalyst Disposal, Recycle, and On-Site Washing/Rejuvenation Options. EPRI. Palo Alto, CA, 1016397.

Ferella, F., 2020. A review on management and recycling of spent selective catalytic reduction catalysts. J. Clean. Prod 246, 118990.

Foerter, D., Jozewicz, W., 2001. Cost of Selective Catalytic Reduction (SCR) Application for NO_x Control on Coal-Fired Boilers. US Environ Protection Agency, p. 17.

Forzatti, P., 2001. Present status and perspectives in de-NOx SCR catalysis. Appl. Catal. A 222 (1–2), 221–236. https://doi.org/10.1016/S0926-860X(01)00832-8.

Forzatti, P., 2001. Present status and perspectives in de-NOx SCR catalysis. Applied Catalysis A: General 222.

Gao, F., Tang, X., Yi, H., Zhao, S., 2017. A review on selective catalytic reduction of NO_x by NH_3 over Mn–based catalysts at low temperatures: catalysts, mechanisms, kinetics and DFT calculations. Catalysts, 7 (7), 199. https://doi.org/10.3390/catal7070199.

Goonan, T.G., 2011. Vanadium recycling in the United States in 2004 (ver. 1.1, October 6, 2011), chap. S of In: Sibley, S.F. (Ed.), Flow studies for recycling metal commodities in the United States: U.S. Geological Survey Circular 1196, pp. S1–S17.

Habashi, F. (1997). Handbook of Extractive Metallurgy, Wiley-VCH, Weinheim ; New York.

Habashi, F. (2005). A short history of hydrometallurgy. Hydrometallurgy 79(1–2), 15–22. https://doi.org/10.1016/j.hydromet.2004.01.008.

Ham, S.-W., Nam, I-S., 2002. Selective catalytic reduction of nitrogen oxides by ammonia. Catalysis. 236–271. https://pubs.rsc.org/en/content/chapter/cl9780854042241-00236/978-0-85404-224-1.

Heck, R.M., Chen, J.M., Speronello, B.K., 1994. Operating characteristics and commercial operating experience with high temperature SCR NOx catalyst. In: Environmental Progress, 13. Wiley Online Library.

Hye-Rim, K., Jin-Young, L., Joon-Soo, K, 2012. Leaching of vanadium and tungsten from spent SCR catalysts for De-NO_x by soda roasting and water leaching method. J. Korean Inst. Resour. Recycling 65–73. https://doi.org/10.7844/kirr.2012.21.6.65.

International Tungsten Industry Association (ITIA), 2019. Tungsten applications. https://www.itia.info.

Janssens-Maenhout, G, Crippa, M, Muntean, M, Guizzardi, D, Schaaf, E, 2017. Emissions Database for Global Atmospheric Research. European Commission doi:10.2904/JRC_DATASET_EDGAR.JRC107275.

Jyothi, R.K., Moon, G., Kim, H.-R., Lee, J.-Y., Choi, I.H., 2017. Spent V_2O_5-WO_3/TiO_2 catalyst processing for valuable metals by soda roasting-water leaching. Hydrometallurgy 175, 292–299.

Jeong, D., 2017. Planning Chinese NOx emission standards apply 15 ppm, the world's highest level, starting this year, http://www.gasnews.com/news/articleView.html?idxno=76583.

Kartte, K., Nonnenmaker, H., 1966. US Patent 3,279,884.

Kim, J.W., Lee, W.G., Hwang, I.S., Lee, J.Y., Han, C., 2015. Recovery of tungsten from spent selective catalytic reduction catalysts by pressure leaching. J. Ind. Eng. Chem. 28, 73–77. https://doi.org/10.1016/j.jiec.2015.02.001.

Korea Electric Power Corporation (KEPCO), 2017. Statistics of powering energy sources.

Lassner, E., Schubert, W.-D., 1999. Tungsten: Properties, Chemistry, Technology of the Element, Alloys, and Chemical Compounds. Springer, US https://doi.org/10.1007/978-1-4615-4907-9.

Lenntech. (n.d.). Vanadium - V. Retrieved August 10, 2020, from https://www.lenntech.com/periodic/elements/v.htm.

Li, Q., Liu, Z., Liu, Q., 2014. Kinetics of vanadium leaching from a spent industrial V_2O_5/TiO_2 catalyst by sulfuric acid. Ind. Eng. Chem. Res. 53 (8), 2956–2962. https://doi.org/10.1021/ie401552v.

Lietti, L., Forzatti, P., Bregani, F., 1996. Steady-state and transient reactivity study of TiO_2-supported V_2O_5-WO_3 De-NO_x catalysts: relevance of the vanadium-tungsten interaction on the catalytic activity. Ind. Eng. Chem. Res. 35 (11), 3884–3892. https://doi.org/10.1021/ie960158l.

Lietti, L., Nova, I., Ramis, G., Dall'Acqua, L., Busca, G., Giamello, E., Forzatti, P., Bregani, F., 1999. Characterization and reactivity of V_2O_5-MoO_3/TiO_2 De-NO_x SCR catalysts. J. Catal. 187 (2), 419–435. https://doi.org/10.1006/jcat.1999.2603.

Lu, X., Yao, T., Li, Y., Fung, J.C.H., Lau, A.K.H., 2016. Source apportionment and health effect of NO_x over the Pearl River Delta region in southern China. Environ. Pollut. 212, 135–146. https://doi.org/10.1016/j.envpol.2016.01.056.

Luo, L., Miyazaki, T., Shibayama, A., Yen, W., Fujita, T., 2003. A novel process for recovery of tungsten and vanadium from a leach solution of tungsten alloy scrap. Miner. Eng. 16 (7), 665–670. https://doi.org/10.1016/S0892-6875(03)00103-1.

Ma, B., Qiu, Z., Yang, J., Qin, C., Fan, J., Wei, A., & Li, Y., 2019. Recovery of nano-TiO2 from spent SCR catalyst by sulfuric acid dissolution and direct precipitation. Waste Biomass Valoriz., 10(10), 3037–3044. https://doi.org/10.1007/s12649-018-0303-0

Mladenović, M., Paprika, M., Marinković, A., 2018. Denitrification techniques for biomass combustion. Renew. Sustain. Energy Rev. 82, 3350–3364. https://doi.org/10.1016/j.rser.2017.10.054.

Nakajima, F., 1978. US Patent 4,085,193.

Nakamura, T., Nishihama, S., Yoshizuka, K., 2009. A novel extractant based on d-glucosamine for the extraction of molybdenum and tungsten. Solv. Extr. Res. Dev. 16, 47–56.

Nguyen, T.H., Lee, M.S., 2016. A review on the separation of molybdenum, tungsten, and vanadium from leach liquors of diverse resources by solvent extraction. Geosyst. Eng. 19 (5), 247–259. https://doi.org/10.1080/12269328.2016.1186577.

Ning, P., Cao, H., & Zhang, Y., 2009. Selective extraction and deep removal of tungsten from sodium molybdate solution by primary amine N1923. Sep. Purif. Technol., 70(1), 27–33. https://doi.org/10.1016/j.seppur.2009.08.006

Non-critical raw materials profiles, 2014. European Commission.

Ogi, T., Makino, T., Okuyama, K., Stark, W.J., Iskandar, F., 2016. Selective biosorption and recovery of tungsten from an urban mine and feasibility evaluation. Ind. Eng. Chem. Res. 55 (10), 2903–2910. https://doi.org/10.1021/acs.iecr.5b04843.

Olazabal, M.A., Orive, M.M., Fernandez, L.A., Madariaga, J.M., 1992. Selective extraction of vanadium (V) from solutions containing molybdenum (VI) by ammonium salts dissolved in toluene. Solvent Extr. Ion Exch. 10 (4), 623–635. https://doi.org/10.1080/07366299208918125.

Pandey, B.D., Kumar, V., Bagchi, D., Jana, R.K., Premchand, 2001. Processing of tungsten preconcentrate from low grade ore to recover metallic values. Miner. Process. Extr. Metall. Rev. 22 (1–3), 101–120. https://doi.org/10.1080/08827509808962491.

Paulino, J.F., Afonso, J.C., Mantovano, J.L., Vianna, C.A., Silva Dias Da Cunha, J.W., 2012. Recovery of tungsten by liquid-liquid extraction from a wolframite concentrate after fusion with sodium hydroxide. Hydrometallurgy 127–128, 121–124. https://doi.org/10.1016/j.hydromet.2012.07.018.

Petter, P.M.H., Veit, H.M., Bernardes, A.M., 2014. Evaluation of gold and silver leaching from printed circuit board of cellphones. Waste Manage. 34 (2), 475–482. https://doi.org/10.1016/j.wasman.2013.10.032.

Pereira, C.J., Plumlee, K.W., Evans, M., Serovy, G.K., Fransson, T.H., 1988. Combined Cycle Technologies and Cogeneration. In: Proc. 2nd International Symposium on Tubomachinery, Montreux, Switzerland.

Polyak, D.E., 2018. Vanadium. In: 2017 Minerals Yearbook. U.S. Department of the Interior, U.S. Geological Survey, p. 5, November.

Radojevic, M., 1998. Reduction of nitrogen oxides in flue gases. Environ. Pollut., 102, pp. 685–689.

Renzi, M., Cerza, F., Gariazzo, C., Agabiti, N., Cascini, S., Di Domenicantonio, R., Davoli, M., Forastiere, F., Cesaroni, G., 2018. Air pollution and occurrence of type 2 diabetes in a large cohort study. Environ. Int. 112, 68–76. https://doi.org/10.1016/j.envint.2017.12.007.

Report Linker, 2016. China SCR Denitration Catalyst. Industry Report. https://www.researchandmarkets.com/research/lrt7ck/china_scr.

Roy, S., Hegde, M.S., Madras, G., 2009. Catalysis for NOx abatement. Appl. Energy 86 (11), 2283–2297. https://doi.org/10.1016/j.apenergy.2009.03.022.

Sorrels, J., 2019. Selective Non-Catalytic Reduction. https://www.epa.gov/economic-and-cost-analysis-air-pollution-regulations/chapter-1-selective-non-catalytic-reduction.

Spivey, J.J., Dooley, K.M (Eds.), 2007. Catalysis, Vol. 19. Royal Society of Chemistry, Cambridge https://pubs.rsc.org/en/content/ebook/978-0-85404-239-5.

Strak, M., Janssen, N., Beelen, R., Schmitz, O., Vaartjes, I., Karssenberg, D., van den Brink, C., Bots, M.L., Dijst, M., Brunekreef, B., Hoek, G., 2017. Long-term exposure to particulate matter, NO_2 and the oxidative potential of particulates and diabetes prevalence in a large national health survey. Environ. Int. 108, 228–236. https://doi.org/10.1016/j.envint.2017.08.017.

Su, S., Feng, S., Zhao, Y., Lu, Q., Cheng, W., & Dong, C., 2012. Comparison of three types of NH3-SCR catalysts. Appl. Mech. Mater. 130–134, 418–421. https://doi.org/10.4028/www.scientific.net/AMM.130-134.418.

Talla, R.G., Gaikwad, S.U., Pawar, S.D., 2010. Solvent extraction and separation of Mo(VI) and W(VI) from hydrochloric acid solutions using cyanex-923 as extractant. Indian J. Chem. Technol. 17 (6), 436–440. http://nopr.niscair.res.in/bitstream/123456789/10715/1/IJCT%2017%286%29%20436-440.pdf.

Wang, J., Miao, J., Yu, W., Chen, Y., Chen, J., 2017. Study on the local difference of monolithic honeycomb V_2O_5-WO_3/TiO_2 denitration catalyst. Mater. Chem. Phys. 198, 193–199. https://doi.org/10.1016/j.matchemphys.2017.05.055.

Waste Classification - Guidance on the classification and assessment of waste (Edition 1.1) Technical Guidance WM3, Scottish Environment Protection Agency, Scotland, UK. (2018).

Weng, Z., Haque, N., Mudd, G.M., Jowitt, S.M., 2016. Assessing the energy requirements and global warming potential of the production of rare earth elements. J. Clean. Prod. 139, 1282–1297. https://doi.org/10.1016/j.jclepro.2016.08.132.

Wu, W.C., Tsai, T.Y., Shen, Y.H., 2016. Tungsten recovery from spent SCR catalyst using alkaline leaching and ion exchange. Minerals 6 (4), 107. https://doi.org/10.3390/min6040107.

Wu, W., Wang, C., Bao, W., Li, H., 2018. Selective reduction leaching of vanadium and iron by oxalic acid from spent V_2O_5-WO_3/TiO_2 catalyst. Hydrometallurgy 179, 52–59. https://doi.org/10.1016/j.hydromet.2018.05.021.

Yanhua, Z., Shili, L., Henghua, X., Xianlai, Z., Jinhui, L., 2012. Current status on leaching precious metals from waste printed circuit boards. Procedia Environ. Sci. 16, 560–568. https://doi.org/10.1016/j.proenv.2012.10.077.

Zeng, L., Cheng, C.Y., 2010. Recovery of molybdenum and vanadium from synthetic sulphuric acid leach solutions of spent hydrodesulphurisation catalysts using solvent extraction. Hydrometallurgy 101 (3–4), 141–147. https://doi.org/10.1016/j.hydromet.2009.12.008.

Zhang, Q., Wu, Y., Li, L., Zuo, T., 2018. Sustainable approach for spent V_2O_5-WO_3/TiO_2 catalysts management: selective recovery of heavy metal vanadium and production of value-added WO_3-TiO_2 photocatalysts. ACS Sustain. Chem. Eng. 6 (9), 12502–12510. https://doi.org/10.1021/acssuschemeng.8b03192.

Zhang, Q., Wu, Y., Zuo, T., 2018. Green recovery of titanium and effective regeneration of TiO_2 photocatalysts from spent selective catalytic reduction catalysts. ACS Sustain. Chem. Eng. 6 (3), 3091–3101. https://doi.org/10.1021/acssuschemeng.7b03038.

Zheng, Y., Jensen, A.D., Johnsson, J.E., 2004. Laboratory Investigation of selective catalytic reduction catalysts: deactivation by potassium compounds and catalyst regeneration. Ind. Eng. Chem. Res. 43 (4), 941–947. https://doi.org/10.1021/ie030404a.

Zhou, B., Sun, C., Yi, H., 2017. Solid waste disposal in Chinese cities: an evaluation of local performance. Sustainability 9 (12), 2234. https://doi.org/10.3390/su9122234.

Index

Page numbers followed by "*f*" and "*t*" indicate, figures and tables respectively.

X

Y

Z

Printed in the United States
by Baker & Taylor Publisher Services